|  |  |  |  |  |  |  | Helium |
|---|---|---|---|---|---|---|---|
|  |  |  |  |  |  |  | 4.00260 |

| | | | III A | IV A | V A | VI A | VII A | |
|---|---|---|---|---|---|---|---|---|
| | | | 5 | 6 | 7 | 8 | 9 | 10 |
| | | | **B** | **C** | **N** | **O** | **F** | **Ne** |
| | | | Boron | Carbon | Nitrogen | Oxygen | Fluorine | Neon |
| | | | 10.81 | 12.011 | 14.0067 | 15.9994 | 18.998403 | 20.179 |
| | | | 13 | 14 | 15 | 16 | 17 | 18 |
| | | | **Al** | **Si** | **P** | **S** | **Cl** | **Ar** |
| | I B | II B | Aluminum | Silicon | Phosphorus | Sulfur | Chlorine | Argon |
| | | | 26.98154 | 28.0855 | 30.97376 | 32.06 | 35.453 | 39.948 |
| 28 | 29 | 30 | 31 | 32 | 33 | 34 | 35 | 36 |
| **Ni** | **Cu** | **Zn** | **Ga** | **Ge** | **As** | **Se** | **Br** | **Kr** |
| Nickel | Copper | Zinc | Gallium | Germanium | Arsenic | Selenium | Bromine | Krypton |
| 58.70 | 63.546 | 65.38 | 69.72 | 72.59 | 74.9216 | 78.96 | 79.904 | 83.80 |
| 46 | 47 | 48 | 49 | 50 | 51 | 52 | 53 | 54 |
| **Pd** | **Ag** | **Cd** | **In** | **Sn** | **Sb** | **Te** | **I** | **Xe** |
| Palladium | Silver | Cadmium | Indium | Tin | Antimony | Tellurium | Iodine | Xenon |
| 106.4 | 107.868 | 112.41 | 114.82 | 118.69 | 121.75 | 127.60 | 126.9045 | 131.30 |
| 78 | 79 | 80 | 81 | 82 | 83 | 84 | 85 | 86 |
| **Pt** | **Au** | **Hg** | **Tl** | **Pb** | **Bi** | **Po** | **At** | **Rn** |
| Platinum | Gold | Mercury | Thallium | Lead | Bismuth | Polonium | Astatine | Radon |
| 195.09 | 196.9665 | 200.59 | 204.37 | 207.2 | 208.9804 | (209) | (210) | (222) |

| 63 | 64 | 65 | 66 | 67 | 68 | 69 | 70 | 71 |
|---|---|---|---|---|---|---|---|---|
| **Eu** | **Gd** | **Tb** | **Dy** | **Ho** | **Er** | **Tm** | **Yb** | **Lu** |
| Europium | Gadolinium | Terbium | Dysprosium | Holmium | Erbium | Thulium | Ytterbium | Lutetium |
| 151.96 | 157.25 | 158.9254 | 162.50 | 164.9304 | 167.26 | 168.9342 | 173.04 | 174.967 |
| 95 | 96 | 97 | 98 | 99 | 100 | 101 | 102 | 103 |
| **Am** | **Cm** | **Bk** | **Cf** | **Es** | **Fm** | **Md** | **No** | **Lr** |
| Americium | Curium | Berkelium | Californium | Einsteinium | Fermium | Mendelevium | Nobelium | Lawrencium |
| (243) | (247) | (247) | (251) | (252) | (257) | (258) | (259) | (260) |

# Chemical
# Principles

**THIRD EDITION**

# Chemical Principles

## Robert S. Boikess

*Rutgers, The State University of New Jersey*

## Edward Edelson

**HARPER & ROW, PUBLISHERS,** New York
Cambridge, Philadelphia, San Francisco,
London, Mexico City, São Paulo, Singapore, Sydney

**Photo credits for chapter opening pages**

Chapter 1, *Gerhard E. Gscheidle,* © *Peter Arnold, Inc.,* Chapter 2, © *Sanford Weinstein,* Chapter 3, *Runk/Schoenberger, Grant Heilman,* Chapter 4, © *1985 Peter Menzel, Stock, Boston,* Chapter 5, © *Manfred Kage, Peter Arnold, Inc.,* Chapter 6, © *Bohdan Hrynewych, Stock, Boston,* Chapter 7, © *Sanford Weinstein,* Chapter 8, © *NIH/Science Source, Photo Researchers, Inc.,* Chapter 9, © *Tom McHugh, Photo Researchers, Inc.,* Chapter 10, *Manfred Kage,* © *Peter Arnold, Inc.,* Chapter 11, *Jen and Des Bartlett, Photo Researchers, Inc.,* Chapter 14, *Pamela R. Schuyler, Stock, Boston,* Chapter 15, *Grant Heilman,* Chapter 16, *Werner H. Müller,* © *Peter Arnold, Inc.,* Chapter 17, © *George Holton, Photo Researchers, Inc.,* Chapter 18, © *1972, Pierre Berger, Photo Researchers, Inc.,* Chapter 19, © *David Rosenfeld, Photo Researchers, Inc.,* Chapter 20, *Hale Observatories, Dec. 9, 1929, Photo Researchers, Inc.,* Chapter 21, © *Stephen L. Feldman, Photo Researchers, Inc.*

Sponsoring Editor: **Heidi Udell**
Project Editor: **Cynthia L. Indriso**
Text Design: **T. R. Funderburk**
Cover Design: **Hudson River Studio**
Cover Photo: **Robert S. Boikess**

Text Art: **J & R Services Inc.**
Photo Research: **Tobi Zausner**
Production: **Debra Forrest Bochner**
Compositor: **Progressive Typographers**
Printer and Binder: **R. R. Donnelley & Sons Company**

The cover illustration is a computer-generated model of a complex consisting of the enzyme carboxyl proteinase and its substrate, the drug pepstatin. The image shows a molecule of pepstatin that is bound to the active site of the enzyme. This kind of high-resolution computer model is now playing an important role in the development of drugs by enabling the design of compounds that are tailored to fit the active sites of biological molecules. Computer models also have become indispensable tools for the study of complex biological molecules such as proteins, nucleic acids, and carbohydrates.

The complex of pepstatin and proteinase was described by R. Bott, E. Subramanian, and D. Davies; the coordinates were provided by Dr. Joel Sussman and Dr. David Davies. The computer graphics were produced with the assistance of Professor R. Levy, Professor W. Olson, and Dr. J. Keepers of the biophysical chemistry group at Rutgers; the photographs were obtained by Professor P. Orenstein of the Department of Visual Arts at Rutgers. We acknowledge their kind assistance.

Chemical Principles, Third Edition

Copyright © 1985 by Harper & Row, Publishers, Inc.

**Library of Congress Cataloging in Publication Data**

Boikess, Robert S.
  Chemical principles.

  Includes index.
    1. Chemistry.   I. Edelson, Edward, 1932–
II. Title.
QD31.2.B63   1985          540          84-10798
ISBN 0-06-040805-7

Harper International Edition
ISBN 0-06-350201-1

84 85 86 87 9 8 7 6 5 4 3 2 1

To the students who will use their knowledge of science in the service of world peace and human justice.

# Brief Contents

# Contents

xii     Contents

Contents

# Selected Tables

# Topics for Today

# Preface

*Chemical Principles, Third Edition,* is written in the belief that a chemistry textbook must meet the needs of both students and teachers. As it was in the first two editions, our goal has been to meet those needs by writing a book that is both comprehensible and comprehensive, combining clarity of exposition with completeness of presentation.

*Chemical Principles* is intended for a full-year general chemistry course for students majoring in any science from agriculture to zoology, including those whose career goals are in the health professions, in engineering, and in the environmental sciences. Our approach does not use calculus, but it does require some familiarity with simple algebra, including facility with exponents and logarithms. An appendix covering mathematical procedures is included.

Throughout the book, we show the practical applications of the principles presented. We believe that students will master chemical ideas and develop chemical skills much more readily when they can see the purpose and value of those ideas and skills. Accordingly, we have included many such discussions in the body of the text. There are also a number of boxed topics of current interest illustrating the importance and usefulness of basic chemical principles.

We have chosen an approach that allows the teacher flexibility in selecting the order in which material is presented. The table of contents offers one of several possible sequences. We have written chapters or small groups of chapters as self-contained units. For example, although the topic of thermodynamics appears in the text before the topic of equilibrium, the teacher may reverse that sequence.

However, the table of contents has been structured with the student's need for timely introduction of topics and their reinforcement in mind. Early presentation of the writing of chemical formulas and chemical equations, naming of compounds, and classification of compounds and reactions may be referred to throughout the course. Early discussions of stoichiometry and of solutions permit the coordination of meaningful laboratory work with the material presented in class. Many illustrative examples are incorporated in the body of the text, in order to develop problem-solving skills.

We have also included substantial amounts of descriptive material as an integral part of the presentation of chemical principles. For example, an early chapter on the nonmetals is used to reinforce discussions of molecular structure, thermochemistry, and oxidation states. Those concepts are, in turn, used as the basis for the presentation and correlation of much of the chemistry of the nonmetals.

Because of the growing acceptance of the International System of Units (SI), it is important that students be introduced to this system of measurements. In *Chemical Principles,* SI units are used to the full extent that is compatible with chemical clarity. Although non-SI units are not used extensively, appropriate exercises and discussions are included to enable the student to master the use of scientific information that is presented in these units. For example, we use the liter as a unit of volume and the atmosphere as a unit of pressure. However, the milliliter is not used; instead, we use the equivalent cubic centimeter. The joule, rather than the calorie, is utilized in energy measurements, and the mole is defined and used as the unit of amount. We generally use the kelvin as the unit of temperature and the nanometer as the unit for atomic distances.

## New to This Edition

The third edition retains the basic framework of the first two, because of their popularity with students and teachers. Nonetheless, new material has been introduced throughout, and the book has been redesigned to improve readability. The changes have been based primarily on experience in the classroom.

We are grateful for the insightful comments we have received from chemistry teachers who have shared with us their thoughts on ways to make a successful textbook even better. We are also grateful for the detailed advice we have received from the colleagues who reviewed the book during the process of revision. Throughout that often arduous process, we have been keenly aware of the support of friends.

Some of the more important content changes in this third edition deserve special mention. The section on units of measurement and significant figures in Chapter 1 has been redone and expanded to help students master these essential topics. In Chapter 2, the section on ionic substances in solution has been expanded to strengthen and clarify the discussion of this subject. Much of Chapter 3 has been rewritten to provide many more solved examples that illustrate mass relationships and the stoichiometry of solutions, and to give an even clearer presentation of the important concepts of stoichiometry. A number of solved examples have also been added to the discussion of quantum numbers and the electronic structure of atoms in Chapter 6. Chapters 7 and 8 have been greatly revised to present material in a new sequence designed to improve students' understanding of chemical bonding and molecular structure. The treatment of covalent bonding, molecular geometry, hybridization, and molecular orbital theory has been completely reorganized. The section on weak interactions in Chapter 8 has been expanded considerably. In Chapter 16, the sections on electrochemical potential and its relationship to chemical behavior have been rewritten extensively, both for improved clarity and to incorporate developments that have occurred since publication of the second edition.

Another major change is the almost complete revision of the more than 1300 end-of-chapter exercises. The vast majority of the exercises in the second edition have been replaced. Throughout, the guiding principle has been to sharpen the focus of the exercises so that students can get even more understanding from them. As in the second edition, answers to some selected exercises are given in an appendix. Together, the exercises provide a review guide that allows students to test their understanding of concepts and to gain facility in solving numerical problems. The extensive tables of data that assist the teacher in formulating additional exercises or examination questions have been retained.

In addition to the major revisions, there has been rewriting throughout the text to further improve the clarity of presentation. Chapter summaries have also been added to help students review the material while it is fresh in their minds.

The Topics for Today, which are designed to illustrate the importance of chemistry in the everyday world, have been carefully reviewed for timeliness. A number of new Topics for Today have been added to describe some of the newest developments in chemistry, both basic and applied. The aim has been to illustrate for the student the close relationship between classroom studies and practical implications. For example, one new Topic for Today describes metalic glasses, which are increasingly being used in a number of technologies. Another discusses the climatic effects of the increasing concentration of carbon dioxide in the atmosphere. Still others describe the latest efforts to synthesize transuranium elements; the progress of the United States toward metrication; the strange, sulfur-dependent organisms that have been found in the ocean depths; and the newest developments in the global problem of acid rain. Most of the Topics for Today that have been retained from the second edition have been revised and updated.

Another notable change in this edition has been made possible by the swift advance of computer technology and programming ability. *Chemical Principles* now includes a number of computer-generated illustrations of molecular structures, which quite literally give students and teachers a new view of chemistry. We have taken advantage of the ability to generate an image of a complex molecule and study it from any desired angle to incorporate these illustrations in the third edition.

## Supplementary Materials

Supporting materials for this book are available to assist both the student and the teacher. For the teacher we have prepared an *Instructor's Manual* in which we analyze in detail every section of each chapter to help in the development of alternate sequences of presentation. In addition, all the end-of-chapter exercises are analyzed in terms of the topics they cover and the level of difficulty of each exercise.

A *Study Guide* that provides valuable assistance in the use and understanding of the text, in the development of skills, and in the delineation of learning objectives has been written by Professors Daniel L. Reger and Edward E. Mercer of the Department of Chemistry, University of South Carolina and Professor Robert S. Boikess.

*Chemical Principles in the Laboratory,* by Professors Robert Bryan of the Department of Chemistry, University of Virginia, and Robert S. Boikess of the Department of Chemistry, Rutgers University, is a laboratory manual that accompanies the text. The manual reinforces many of the principles presented in the text, developing them in greater depth and describing a number of instructive and interesting experimental procedures. The order of experiments and of topics discussed in the manual coincide with those of this book.

We owe a great deal to the many persons who have been involved in the planning, writing, and production of this newest edition. As has been true since the first edition, their comments and suggestions have provided invaluable help at every stage of the work, from the initial preparation of the general outline and approach to the final revisions. We hope that their help has enabled us to achieve our goal of combining conceptual rigor, a high standard of chemistry, a balanced approach, and a readable style to give both students and teachers what they want and need for a general chemistry course.

ROBERT S. BOIKESS

EDWARD EDELSON

# Acknowledgments

In the preparation of this third edition, we have again had the good fortune to be helped by many skilled persons. Their contributions are reflected on almost every page of the book. We are grateful for their time, their patience, and their willingness to help.

We express our gratitude to the many users of the second edition whose pithy responses to questionnaires helped us to focus on areas that required revision. We also thank those many individuals who read substantial portions of the manuscript for the third edition and gave us valuable suggestions on points of fact and style. We especially thank Professors Elliott Blinn, Robert Bryan, Bruce Bursten, Geoffrey Davies, Klaus Dichmann, James Hogan, Jerome Keister, L. Kelsey, Ronald E. McClung, Cortlandt Pierpont, Daniel Reger, A. Rheingold, Den Rusnak, and Barbara Sawrey.

Many colleagues at Rutgers gave us helpful advice and guidance. We especially thank Professors William Adams, Ramesh Agarwal, George Bird, Ken Breslauer, Martha Cotter, Lionel Goodman, Gregory Herzog, Rolfe Herber, Stephan Isied, Paul Kimmel, John Krenos, Karsten Krogh-Jespersen, Richard Laity, Joseph Potenza, Sidney Toby, and Alex Yacynych. Professors Wilma Olson and Ron Levy, and Dr. A. R. Srinivasan, Janet Cicariello, Ross Barnes, Won-Kyoo Cho, Gilbert Smith and Glenn van Slooten, who produced the computer-generated molecular structures for this edition, made an especially important contribution. Many students at Rutgers who used the book provided useful ideas and suggestions.

We cannot minimize our gratitude to the staff at Harper & Row, who were uniformly and unfailingly supportive. We cheerfully acknowledge the guidance, good advice, friendly prodding, and sound judgment of restaurants of our sponsoring editors, Malvina Wasserman and Heidi Udell, who smoothed our way considerably. Cindy Indriso, our project editor, contributed a remarkable balance of patience and impatience that helped transform the manuscript into a book most efficiently.

Finally, we thank the many teachers and students who used the book. We have learned much that is valuable from them.

**1**

# The Science
# of Chemistry

Preview    To understand the world we live in, you must know something about chemistry. In this chapter, we lay the foundation for that understanding. We begin by defining chemistry as the study of matter, and we give classifications of matter. We then describe basic systems of measurement—those used in the past, as well as the modern system. The chapter concludes with a description of some essential mathematical operations that we shall need, including the use of significant figures to express measurements.

Science is an organized body of knowledge based on a unique method of looking at the world. Any explanation of what is seen is based on the results of experiments and observations that must be verifiable by anyone who has the time and means to repeat them.

Because scientific theories are subject to test, they must be put forth precisely enough to avoid ambiguity. Because new experimental results may be obtained at any time, theory is always open to revision. Because an experiment must be verifiable, it is preferable to express results so that direct comparison between experiments is possible. Whenever possible, measurements, rather than broad descriptions, are preferred.

The scientific way of examining phenomena is rare in human history. Most societies have preferred methods that do not require experimentation and measurement. Many societies view the world as a mysterious place, governed by forces beyond human understanding. Even societies that believe that a rational understanding of these forces is possible usually do not demand that an explanation be verified by experiments. In ancient Greece, for example, philosophers developed many theories that, in modified form, are accepted today. The Greeks, however, generally were content with theorizing. It is only in the modern era of Western history, which began in Europe around A.D. 1500, that theories must be supported by observation and experimentation. The relatively few persons who have mastered the scientific method have achieved an amazingly thorough transformation of our planet.

## The Scientific Method

The scientific method is not easily summarized. No single set of rules can describe the great variety of scientific activities, but we can describe an idealized approach. Often, the beginning point is the development of a **hypothesis** about a specific phenomenon. The hypothesis may be tested by observation, by experimentation, or by a combination of the two, depending on the phenomenon that is studied. Chemists, for example, can perform experiments by attempting to carry out given chemical reactions in the laboratory. Astronomers can only observe the distant stars. They cannot experiment by changing them.

A hypothesis that is supported by observation or experiment may lead to a unifying explanation for many different phenomena: often this explanation is called a **law of nature.** Any law is subject to revision as new observations and experiments reveal new facts.

Science ordinarily is regarded as being done outside the influence of personality, economics, or social structure. In such a framework, the first step in science will be an experiment or an observation; theorizing begins only after data are gathered. Studies of the way that science actually is done indicate that this picture is rarely accurate. The progress of science is influenced by personality, money, and social forces. Scientific research

## THE SCIENTIFIC METHOD

"When I think of formal scientific method an image sometimes comes to mind of an enormous juggernaut, a huge bulldozer—slow, tedious, lumbering, laborious, but invincible. It takes twice as long, five times as long, maybe a dozen times as long as informal mechanic's technique, but you know in the end you're going to *get* it. There's no fault-isolation problem in motorcycle maintenance can stand up to it. When you've hit a really tough one, tried everything, racked your brains and nothing works, and you know that this time Nature has really decided to be difficult, you say, 'Okay, Nature, that's the end of the *nice guy*,' and you crank up the formal scientific method. . . ."

". . . That part of the formal scientific method called experimentation is sometimes thought of by romantics as all of science itself because that's the only part with much visible surface. They see lots of test tubes and bizzare equipment and people running around making discoveries. They do not see the experiment as part of a larger intellectual process and so they often confuse experiments with demonstrations, which look the same. A man conducting a gee-whiz science show with fifty thousand dollars' worth of Frankenstein equipment is not doing anything scientific if he knows beforehand what the results of his effort are going to be. A motorcycle mechanic, on the other hand, who honks the horn to see if the battery works is informally conducting a true scientific experiment. He is testing a hypothesis by putting the question to nature. The TV scientist who mutters sadly, 'The experiment is a failure; we have failed to achieve what we hoped for,' is suffering mainly from a bad scriptwriter. An experiment is never a failure solely because it fails to achieve predicted results. An experiment is a failure only when it also fails adequately to test the hypothesis in question, when the data it produces don't prove anything one way or another."

Robert M. Pirsig, *Zen and the Art of Motorcycle Maintenance*

generally begins with a quest for a specific piece of new knowledge, chosen by a combination of social, economic, and personal influence. The role of intuition (for lack of a better word to describe a mental process that is poorly understood) in the formulation of theories is dramatically important. Hunches, guesses, and even dreams have played parts in some of the greatest scientific discoveries. There is also a strong aesthetic element in scientific discovery. One of the highest forms of praise for a scientific experiment or hypothesis is to call it elegant, which implies not only admiration for its effectiveness but also appreciation of its beauty.

In the early stages of scientific discovery, the major effort is to describe phenomena and to classify them by their characteristics. Later, measurement—quantification—replaces the qualitative descriptions. Still later, the quantitative data can be described by a few concise statements or mathematical equations that are called **laws.** In some cases, it may be possible to construct a **theory** that explains many different laws by a few general principles. In biology, the theory of evolution is such a unifying principle. The atomic and molecular theory of matter is a unifying principle in chemistry.

Theories and laws are always subject to refinement and modification as new observations are made. For example, Newton's theory of gravitation was modified by Einstein's theories, which are themselves constantly being studied with a view toward modification.

## 1.2  CHEMISTRY AS A PHYSICAL SCIENCE

For convenience, science is divided into a number of disciplines; but the division never is as clear cut as it appears to be. All the sciences are related and draw on each other.

*Chemistry is defined as the branch of science that deals with the composition, structure, properties, and transformations of* **matter.** Matter occupies space, can be perceived by our senses, and is the constituent material of physical objects or the universe as a whole. Chemistry tends to study matter from an *atomic* and *molecular* point of view. Therefore, chemistry is closely related to physics, which can be defined in the most general way as the study of matter and energy and the interaction between them. Chemistry is also related to biology, which has as one of its major concerns the chemistry of living organisms. The relationship of chemistry to other sciences, such as astronomy and geology, has become clearer in recent years with the rise of such disciplines as astrochemistry and geochemistry.

In a hierarchy of the sciences, chemistry often is placed between physics and biology. Physics is viewed as taking the most general approach. It deals in all ways with matter, energy, and the interactions between them. Chemistry uses the information gathered by physics to study the properties and interactions of substances from a somewhat different point of view. Chemistry depends heavily on an understanding of the atom, which it regards as a basic unit of matter, to explain many interactions between substances. Biology uses the findings of both physics and chemistry to study living organisms.

### How Matter Is Classified

The matter that chemists deal with can be classified in several different ways. For example, a distinction is drawn between **homogeneous** material, which has the same properties throughout, and **heterogeneous** material, which has parts with visibly different properties (Figure 1.1). A lump

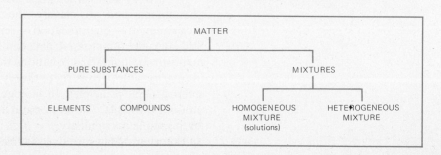

**Figure 1.1**
The classification of matter.

## OLD DOGMAS AND NEW TRICKS

Every scientific theory, however well established, is subject to revision as new data are gathered. Nothing so well illustrates the provisional nature of all "truths" in science as the history of molecular genetics in the 1960s and the 1970s, a period during which virtually all of the basic principles of the so-called central dogma of molecular genetics had to be modified because of new findings.

The central dogma was believed to supply a complete explanation of the way in which genetic information is passed on from generation to generation and is translated into instructions for the living cell. It was quickly established after the discovery of the structure of the deoxyribonucleic acid (DNA) molecule by Francis Crick and James Watson in 1953.

As the central dogma had it, genetic information is coded into the structure of the DNA molecule. Each section of a molecule is a gene which contains information for a specific enzyme or function. The information contained in the DNA molecule is used to make molecules of ribonucleic acid (RNA), which in turn are responsible for the production of the proteins that carry on the functions of the living cell. The one-way flow of information from DNA to RNA was said to occur in all living organisms.

The first major change in this picture came with the discovery that some viruses use RNA as their genetic material, and that the information contained in the RNA molecule is used to make DNA—just the reverse of the rule laid down in the central dogma.

Later it was discovered that in some living organisms, a single section of the DNA is not a single gene. Rather, one section can be read in two different ways, so two genes occupy the same section of the DNA molecule. Again, an important part of the central dogma of molecular genetics had to be modified.

Then it was found that the DNA molecules of higher organisms, including humans, contain large segments of repetitive DNA, in which the same sequence of subunits is repeated again and again. The function of these repetitive segments still has not been discovered.

Another modification in the central dogma was made with the discovery that many genes do not consist of unbroken stretches of DNA. Instead, such genes are groups of discontinuous segments. The active segments are separated from each other by inactive stretches of DNA.

Genes have also been found to be surprisingly mobile. Some genes rearrange themselves by shuffling segments of DNA, a phenomenon informally called "jumping genes." Recently, molecular biologists have found that genes that are closely associated with cancer can be moved from one organism to another by viruses. A number of laboratories are trying to learn how these genes contribute to the transformation of a normal cell to malignancy.

The rule that genetic information flows from DNA to RNA to protein still applies widely in molecular genetics. However, molecular biologists now must be careful to qualify all their statements by including the exceptions that have been found to exist. In all branches of science, theories are always subject to change in the light of new evidence from experience or observation.

---

of pure gold is a homogeneous material, while the rock in which the gold was found is a heterogeneous material.

A **pure substance** is a homogeneous material that has a reasonably definite chemical composition throughout. There are two kinds of substances. One kind can be changed into two or more different substances by simple decomposition; the other kind cannot. Ordinary table salt is a substance of the first kind. It can be decomposed into sodium and chlorine. Gold is a substance of the second kind. It cannot be decomposed and will not undergo chemical change without the intervention of a second substance. A substance that cannot be decomposed is called an **element.** Substances that can be decomposed by simple chemical changes are called **compounds** (Figure 1.2). There are many millions of compounds, but just over 100 elements.

Most of the materials around us are not single substances. Some are **mixtures,** which consist of a number of different substances. A bowl of vegetable soup is a mixture; so is a candy bar. Some are **solutions,** which are homogeneous combinations of different substances. The chief difference between a mixture and a solution is the homogeneous character of the solution. Any sample of a solution has the same composition, while the composition of a mixture is not the same throughout. Solutions can be gaseous, liquid, or solid. Carbon dioxide dissolved in water is a liquid solution; air is a gaseous solution; and sterling silver, an alloy of copper and silver, is a solid solution.

A distinction also must be drawn between a mixture and a compound. The elements making up a compound cannot be recovered without a chemical change. The substances making up a mixture or a solution can. The substances making up a mixture or a solution need not be elements. For example, we can prepare a solution by dissolving sugar, a compound, in water, another compound. In addition, the substances making up a mixture or a solution can be combined in varying proportions. The elements in a compound have fixed proportions.

### Properties of Matter

All substances have characteristics, or properties, by which we identify them. **Chemical properties** describe the way in which a substance can undergo change, either alone or in interactions with other substances, to form different materials. Such changes are called **chemical reactions.** We can list the chemical properties of any substance. Iron combines readily with oxygen to form the compound called rust. Gold is inert to nitric acid, while copper is not. Sulfur combines with silver to form a black compound that we call tarnish. And so on, almost endlessly.

**Physical properties** are characteristics by which we can describe a substance—color, density, hardness, melting point, and the like. Thus, we can describe gallium as a white metal, soft enough to be cut with a knife, that has a density 5.91 times greater than that of water, a melting point of 29.78°C, and a boiling point of 1983°C. We usually make a distinction between intensive properties and extensive properties (Figure 1.3). Intensive properties include density, melting point, boiling point, electrical conductivity, and temperature. If we have a sample of a substance, we can measure an intensive property anywhere in the system. The value for the entire system is the same as for any small part of the system. Extensive properties describe an entire sample of a substance. Mass, volume, length, and area are extensive properties. Their value for a

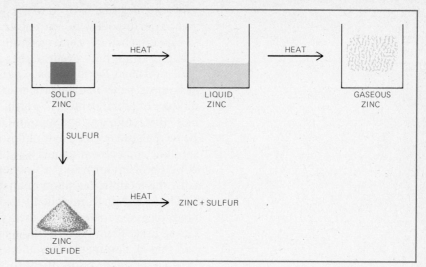

**Figure 1.2**
When an element such as zinc is heated, it may undergo a physical change from solid to liquid to gas, but it does not undergo a chemical change. A second substance, such as sulfur, is needed to bring about the chemical change of an element. The resulting compound, zinc sulfide, undergoes a chemical change when it is heated. It forms two substances, zinc and sulfur.

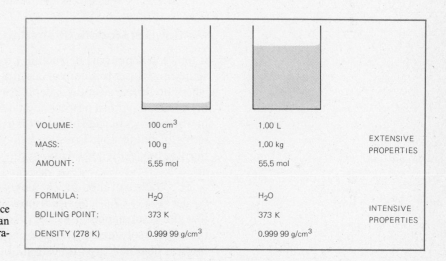

**Figure 1.3**
Intensive properties can be measured anyplace in the sample. Extensive properties describe an entire sample of the substance under consideration and therefore depend on quantity.

system is equal to the sum of the value for the individual parts of the system and therefore depends on quantity.

## 1.3 FROM ALCHEMY TO CHEMISTRY

Modern chemistry was born out of alchemy, a complex mixture of philosophy, astrology, mysticism, magic, real chemistry, and other ingredients. Alchemists were interested primarily in proving the truth of their philosophical system. Their patrons had more practical ends in mind—for example, the discovery of the philosophers' stone, which would transmute any base metal into gold. From the modern point of view, alchemy was valuable because it led to the development of laboratory procedures and to the discovery of new substances.

Modern chemistry emerged from alchemy in the seventeenth century. The most important event in the transformation was the publication in 1661 of Robert Boyle's book, *The Sceptical Chymist,* which rejected the alchemists' belief that the properties of all substances were derived from a mixture of four "elements" (earth, water, fire, and air) and four "principles" (hot, cold, dry, and moist). This philosophy went back to Aristotle. Boyle replaced it with a new philosophy, which described an element as a substance that always gained weight in a chemical change and that could not be broken down into a simpler substance. His description is essentially the modern definition of an element.

The transition from alchemy to chemistry took many years. One of the hindrances to progress was a theory put forward by a German physician, Georg Ernst Stahl (1660–1734), to explain combustion. Stahl's theory said that all flammable substances contain a material called phlogiston. When a substance burns, the theory said, phlogiston escapes into the air, leaving behind something called a calx.

### The Dawn of Modern Chemistry

It took the better part of a century to replace the phlogiston theory with something closer to a correct description of combustion. As late as 1774, when the British scientist Joseph Priestley (1733–1804) heated mercury to produce a reddish powder and then heated the powder to liberate a gas in which things burned brighter than in air, he described his experiment in terms of phlogiston. Priestley believed that the gas permitted brighter burning because it had no phlogiston, and so he called it "dephlogisticated air."

The "father of modern chemistry," the French scientist Antoine Lavoisier (1743–1794), finally disproved the phlogiston theory by observing that after a substance burns, the products of combustion weigh more than the original substance. Lavoisier said it was absurd to believe that phlogiston was liberated by combustion, since phlogiston would have to weigh less than nothing to explain these results. Lavoisier proposed that the burning substance combines with a portion of the air to form new compounds. Priestley had called that portion of the air "dephlogisticated air." Lavoisier called it oxygen, a name derived from the Greek words for "acid-former." Lavoisier's naming of the gas was not quite correct, since it was based on his belief that all acids contain oxygen. But the name, and Lavoisier's explanation of combustion, are still accepted.

Another major step toward modern chemistry was taken in the first decade of the nineteenth century when the English chemist John Dalton postulated that all elements are made up of atoms. Dalton used the word *element* in its modern sense. He pictured atoms as tiny, indestructible units that could combine with other atoms to form "compound atoms," or molecules. Dalton proposed that each element has its own kind of

An alchemist's laboratory, as seen in a seventeenth-century engraving. *(The Bettmann Archive)*

atom, that the atoms of different elements differ in essentially nothing but their masses, and that atoms combine in definite proportions. Dalton went on to determine the relative weights of atoms of several elements, including oxygen, hydrogen, and carbon. A new era had begun.

## 1.4 CHEMISTRY IN THE MODERN WORLD

We live in a world of synthetic chemicals. The clothes we wear, the paint on the wall, the furniture in our rooms probably were made by the chemical industry. In recent years, we have become aware that such chemicals may not always be beneficial.

The production of nitrogenous fertilizers affords a good example of the mixture of good and ill provided by modern chemistry. Millions of tons of fertilizer are produced annually by modern versions of the nitrogen fixing method developed before World War I by Fritz Haber and Karl Bosch, two German chemists. This fertilizer enables farmers to achieve the high yields that are needed to feed an increasing population. Without ample supplies of artificial fertilizers, the "green revolution," based on new, high-yielding strains of food grains, would be impossible.

Many scientists, however, have become concerned about the possible long-term ecological effects of fixing large amounts of nitrogen — that is, of converting the inert nitrogen in air to a biologically active form. When nitrogenous fertilizer is washed from a field into a lake or stream, it can serve as a nutrient for species of microscopic algae, which can multiply until these clear waters are clogged with a green scum. The resulting *eutrophication* can accelerate the natural aging process by which a lake is slowly filled in by organic matter.

Nuclear energy is another modern technology that has caused growing concern. Without chemical technology, nuclear energy could not have been developed. A chemical process for production of the compound uranium hexafluoride made possible the enrichment process that produced uranium first for the atomic bomb and later for nuclear reactor fuel rods. Nuclear energy today produces a significant proportion of the nation's electricity. Yet the future of nuclear energy has been placed in doubt by unanswered questions about disposal of radioactive wastes, the possible use of plutonium to make nuclear weapons, and reactor safety.

There are many examples of chemicals that are both useful and harmful. Phonograph records enable us to hear either Mozart or rock music at our convenience. But phonograph records are made of polyvinyl chloride, which is produced from vinyl chloride, a substance that can cause liver cancer and other diseases in industrial workers. Antibiotics, many of them synthesized in the laboratory, have virtually eliminated infectious diseases as a major cause of death in the developed nations. But overuse of antibiotics is threatening to create infectious agents that are resistant to the antibiotics on which we most rely.

The perception of the relative balance between benefits and problems caused by progress in chemistry changes with time. Modern insecticides increased the food supply and added to personal comfort, so they were used on a large scale. Today, we balance these benefits against the harm done by insecticides to birds, fishes, and useful insects. We have entered a new and complex era that requires a better understanding of the multitude of chemical reactions, both natural and artificial, that shape the world we live in.

## 1.5  MEASUREMENTS AND UNITS IN CHEMISTRY

We mentioned in Section 1.1 that science relies on quantitative data—numerical information. At the heart of any quantitative examination of our surroundings is the performance of operations called **measurement.**

Measurement is nothing more than counting. When we measure the value of a physical quantity, we either count that quantity or we count a ratio between two examples of the quantity. This counting usually is done with the help of some sort of measuring instrument.

The first measurements probably were those of length; they illustrate the counting aspect of measurement well. If you want to measure the length of a room, a simple way is to pace it off, counting the number of times you place one foot in front of the other as you go the length of the room. You can measure other rooms in the same way, comparing the lengths of different rooms by comparing the number of feet counted in pacing off each room. In each case, you are using your foot as the common standard of comparison, so it is easy to compare the lengths of different rooms.

But a problem arises when you try to learn the length of a room that you cannot measure yourself. Suppose a friend agrees to pace off his room and compare its length to that of your room. If your room's length is 16 of your feet and the length of your friend's room is 15 of his feet, you have only part of the answer. You cannot compare the lengths of your rooms unless you know the relative lengths of your feet.

In spite of such problems, most early units of measurement and many that are still used are based on some dimension of the human body. Table 1.1 lists some units of length. It also lists their equivalence in meters, an important unit of length in modern science.

## The Metric System

A table listing all the units of length that ever have been used around the world would go on for many pages. Such a table would be essential for any sort of international communication if we did not have an agreement to standardize measurements. There is such an agreement, in which virtually every major country has adopted the same units of length and many other basic physical quantities. The agreed-upon unit of length is the meter, and the other agreed-upon units are part of what originally was called the *metric system*.

The metric system was developed in France shortly after the French Revolution. It has two important features. One of them is an attempt to relate the units of measurement to natural phenomena. For example, the meter was originally defined as one ten-millionth of the length of a line running from the North Pole to the equator and passing through Paris.

The second important feature of the metric system concerns the relationship between different units for the same quantity. All these units are related decimally — that is, by powers of 10. This relationship usually is indicated by a prefix.

For example, the meter, the kilometer, and the centimeter are units of length in the metric system. One kilometer is 1000 meters and one centi-

TABLE 1.1   Units of Length

| Unit | Origin | Equivalent in Meters (m) |
|------|--------|--------------------------|
| inch | a thumb breadth or the length of three barleycorns, from the middle of the ear of barley | 0.0254 m |
| foot | 12 inches or a man's foot | 0.3048 m |
| cubit | the length from the tip of the middle finger to the elbow | 0.46 m |
| yard | 3 feet or the length from the nose to the thumb of King Henry I of England | 0.9144 m |
| furlong | 220 yards or the length of a furrow in a common field | 201.2 m |
| mile | 1760 yards; originally 1000 Roman double-step paces | 1609.3 m |

meter is one-hundredth of a meter. You can see that conversion between these units of length requires no more than a shift in the decimal point. The simplicity of this relationship is evident by comparison with the units listed in Table 1.1. To convert inches into feet, we must multiply by 12. To convert feet into yards, we multiply by 3. To convert yards into miles, we multiply by 1760. By comparison, we can convert centimeters into meters and meters into kilometers by using the appropriate powers of 10.

The metric system has been used universally by scientists for some time. As science developed and new phenomena were investigated, different units were introduced. It was not uncommon at one time for different scientists to be using different units derived from the metric system for the same quantity. Listings such as Table 1.1, giving the relationship between these units, became essential. One such table in a commonly used chemical handbook has more than 3000 entries and is far from complete.

## The SI

To avoid confusion, attempts to extend and improve the metric system have been made. The result is a system, first formulated in 1960, that has been adopted by many nations and international scientific bodies, including the one that represents chemists. This system is called the International System of Units, or SI (from the initials of its French name). The use of SI is designed to bring order out of the chaos that existed among scientific units of measurement. However, the adoption of the SI has been a slow process.

One reason is that old habits are hard to break, in science as in other fields. Another reason is that some units of the SI are not regarded as completely satisfactory by some scientists. Often SI units represent compromises between conflicting requirements of several disciplines. In chemistry, certain SI units seem very inconvenient, so the older non-SI units are frequently used. In Chapter 4, for example, we shall generally use the older unit for pressure, the atmosphere, which most chemists still use. The SI unit of pressure, the pascal, is not used extensively by chemists.

Finally, since a large body of chemical literature was written before the adoption of the SI, it is impossible to ignore the old units. Even today, much scientific literature includes non-SI units. We shall use the SI units as much as is consistent with current practice and convenience.

## The Seven SI Base Units

The SI starts with a set of **base units,** that is, a set of independent physical quantities. There are seven such base units in SI. The units for all other physical quantities are derived by multiplication and division of the base units, without the use of any numerical factors, not even powers of 10. Table 1.2 lists the names and symbols of the SI base units.

The first five units in Tables 1.2 are used in almost every branch of

# INCHING TOWARD METRIFICATION

The United States is the last major industrialized country that has not converted its system of weights and measures to metric units. Canada, for example, now uses metric units for almost all everyday purposes. The conversion is going on, however, although it is not immediately visible because of the way that metrification is being carried out.

In other countries, the conversion has been supervised by the government. In the United States, the Metric Conversion Act that was passed by Congress in 1975 provided specifically that metrification should be voluntary. A 17-member United States Metric Board, created by the act, did not begin work until 1978 and was eliminated in 1982 because of federal budget cuts.

Nevertheless, a substantial portion of American industry has begun the transition to metric units. The efforts are being coordinated by the American National Metric Council, which is privately financed. One of the council's major efforts is to help in the planning of industry-wide transitions to metric units.

For example, a committee made up of representatives of the chemical industry and its major customers concluded a 7-year effort in 1981 with a report that set 1984 as a "realistic date" for metrification. Under the voluntary plan, it was agreed that chemical products sold in the United States would be available in metric units by that year.

The automobile industry, which now routinely uses parts made in many countries for a single vehicle, is close to complete metrification. So is the electronics industry, whose influence is increasingly important.

Consumers can notice the transition to metrification in a number of areas. Alcoholic beverages now are sold in metric units, as are many soft drinks. Photographic equipment, such as lenses and film, use metric measures. The tire industry began using metric units early in the 1980s.

One factor that slowed the pace of the transition was the economic recession that gripped the United States at that time. Some companies were reluctant to pay the costs of metrification. In one of the most visible areas, gasoline sales, consumer psychology had an effect. The gasoline industry agreed in principle to convert from gallons to liters, at a time when prices were rising rapidly. When the time for conversion came, however, prices were falling slightly. Because U.S. consumers are not familiar with metric units, price changes are more evident when gasoline is pumped in gallons than in liters. The companies that preferred to ease the pain of price increases by metrification held back when gasoline prices held steady or dropped.

Metrification seems to be lagging because it is being done least in the most obvious areas: road signs, groceries, and the like. The growing use of metric units by industry, however, indicates that the United States eventually will complete the transition to the system being used by the rest of the world.

| TABLE 1.2  SI Base Units | | |
|---|---|---|
| **Physical Quantity** | **SI Unit** | **Symbol for Unit** |
| length | meter | m |
| mass | kilogram | kg |
| time | second | s |
| amount of substance | mole | mol |
| thermodynamic temperature | kelvin | K |
| electric current | ampere | A |
| luminous intensity | candela | cd |

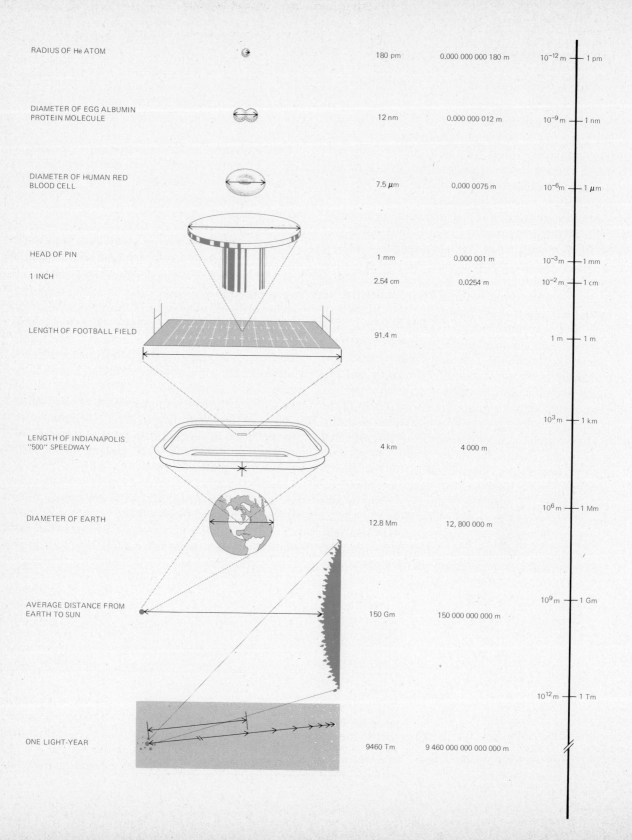

| | | | |
|---|---|---|---|
| RADIUS OF He ATOM | 180 pm | 0.000 000 000 180 m | $10^{-12}$ m — 1 pm |
| DIAMETER OF EGG ALBUMIN PROTEIN MOLECULE | 12 nm | 0.000 000 012 m | $10^{-9}$ m — 1 nm |
| DIAMETER OF HUMAN RED BLOOD CELL | 7.5 $\mu$m | 0.000 0075 m | $10^{-6}$ m — 1 $\mu$m |
| HEAD OF PIN | 1 mm | 0.000 001 m | $10^{-3}$ m — 1 mm |
| 1 INCH | 2.54 cm | 0.0254 m | $10^{-2}$ m — 1 cm |
| LENGTH OF FOOTBALL FIELD | 91.4 m | | 1 m — 1 m |
| LENGTH OF INDIANAPOLIS "500" SPEEDWAY | 4 km | 4 000 m | $10^{3}$ m — 1 km |
| DIAMETER OF EARTH | 12.8 Mm | 12,800 000 m | $10^{6}$ m — 1 Mm |
| AVERAGE DISTANCE FROM EARTH TO SUN | 150 Gm | 150 000 000 000 m | $10^{9}$ m — 1 Gm |
| | | | $10^{12}$ m — 1 Tm |
| ONE LIGHT-YEAR | 9460 Tm | 9 460 000 000 000 000 m | |

| TABLE 1.3 | SI Prefixes | | | | |
|---|---|---|---|---|---|
| Fraction | Prefix | Symbol | Multiple | Prefix | Symbol |
| $10^{-1}$ | deci | d | 10 | deka | da |
| $10^{-2}$ | centi | c | $10^2$ | hecto | h |
| $10^{-3}$ | milli | m | $10^3$ | kilo | k |
| $10^{-6}$ | micro | $\mu$ | $10^6$ | mega | M |
| $10^{-9}$ | nano | n | $10^9$ | giga | G |
| $10^{-12}$ | pico | p | $10^{12}$ | tera | T |
| $10^{-15}$ | femto | f | | | |

chemistry. The sixth, the unit for electric current, has more limited use. The seventh, the unit for luminous intensity, is rarely used in chemistry.

With the exception of the kilogram, the SI base unit for mass, all the base units are defined very precisely in terms of natural phenomena. The kilogram is defined as the mass of a specific sample of platinum-iridium alloy that is kept at the International Bureau of Weights and Measures at Sèvres in France. However, a way of defining the kilogram in terms of a natural phenomenon is being sought.

### The SI Prefixes

The SI also includes a set of prefixes that is used to form decimal multiples and decimal fractions of the SI units. Table 1.3 lists SI prefixes, their meanings, and their symbols. You will note that prefixes are not provided for every power of 10. Allowed fractions smaller than $10^{-3}$ and allowed multiples larger than $10^3$ all must have exponents that are divisible by 3 (Figure 1.4).

The symbol for an SI prefix is placed before the symbol for an SI base unit. The exception to this rule is the base unit of mass, the kilogram, which already has a prefix. For historical reasons, prefixes are added to the gram instead. Thus, one milligram is $\frac{1}{1000}$ of a gram and $\frac{1}{1\,000\,000}$ of a kilogram.

The SI base unit for length is the meter, whose symbol is m. A centimeter, whose symbol is cm, is one-hundredth ($10^{-2}$) of a meter. A micrometer, whose symbol is $\mu$m, is one-millionth ($10^{-6}$) of a meter. A kilometer, symbol km, is 1000 ($10^3$) meters, and a gigameter, symbol Gm, is one billion ($10^9$) meters.

### The SI Derived Units

The physical quantity area gives us an example of a derived unit. Area is (length) × (length). The unit for area is (meter) × (meter), or square meter. It is convenient to use an exponent for such a unit. Rather than writing "square meter," we write meter$^2$ or, using the symbol, m$^2$. An exponent that is used with a unit has the same meaning as an exponent

**Figure 1.4**
The use of prefixes allows us to avoid writing very large or very small numbers by selecting a unit of convenient size for a wide range of measurements.

that is used with a number or a symbol. Thus, m³ means (meter) × (meter) × (meter), or meter³, and m⁻¹ means 1/m, which can be read as "per meter."

It is important to select the proper unit for a given measurement. For example, if we want to express the area of a house or a farm, the square meter (m²) is a convenient unit. If we want to express the area of a country, however, the square meter is too small a unit. The square kilometer (km²), which means (kilometer) × (kilometer), is more suitable for such a measurement. For smaller areas, such as a page of this book, a more convenient unit is the square centimeter (cm²). You should note that *the base unit and the prefix are a single unit.* The exponent therefore operates on the entire symbol: cm² = cm × cm.

Units for all other physical quantities can be derived from combinations of one or more of the seven SI base units. Table 1.4 lists some physical quantities of interest to the chemist and the SI units for these quantities.

Each unit listed in Table 1.4 is named for a scientist who did research on the quantity measured by that unit. Each of these units has its own symbol. We can form decimal multiples and fractions of these units by using SI prefixes. For example, 1 kW, one kilowatt, is 1000 W, and 1 mJ, one millijoule, is 1/1000 J.

Many units used by chemists do not have special names or symbols. As we have seen, a unit that is derived directly from SI base units may not be convenient for a common measurement. In such a case, we can derive a suitable unit from units that have prefixes.

For example, the SI unit of volume derived from the base unit for length is the cubic meter, m³. This unit is not at all convenient for expressing the volumes encountered by most chemists. Instead chemists often use the cubic centimeter, cm³. A point to note is that the exponent applies to the cm, not to the m. The cm³ is *not* 1/100 of a m³. It is (cm) × (cm) × (cm), or 1/1 000 000 m³. Another unit of volume that is commonly used by chemists is symbolized as dm³. Formally, the dm³ is called the cubic decimeter. However, chemists have been using the dm³ for many years and have become accustomed to calling it the liter, whose symbol is

**TABLE 1.4   Some SI Derived Units**

| Physical Quantity | SI Unit | Symbol of Unit | Definition of SI Unit |
|---|---|---|---|
| energy | joule | J | $kg\ m^2\ s^{-2}$ |
| force | newton | N | $kg\ m\ s^{-2}$ (or $J\ m^{-1}$) |
| pressure | pascal | Pa | $kg\ m^{-1}\ s^{-2}$ (or $N\ m^{-2}$) |
| power | watt | W | $kg\ m^2\ s^{-3}$ (or $J\ s^{-1}$) |
| frequency | hertz | Hz (or $s^{-1}$) | $s^{-1}$ |
| electric charge | coulomb | C | $A\ s$ |
| electric potential difference | volt | V | $kg\ m^2\ s^{-3}\ A^{-1}$ (or $J\ s^{-1}\ A^{-1}$) |

L. The liter is accepted in SI and is allowed as an acceptable nickname for the $dm^3$. We shall use the liter extensively as a unit of volume in this book.

### Conversion of Units

Even using SI units, it is often necessary to convert a value for a physical quantity measured in one unit into a value measured in another unit. In addition, since many non-SI units are still in use, we must also be able to convert between such units. Conversion between SI units requires only multiplication by a power of 10. Conversion between non-SI units may require numerical factors that are not powers of 10. These numerical factors must be obtained from a table of conversion factors.

The use of conversion factors can be regarded as purely an arithmetic technique. To perform a conversion, we find the relationship between the two units. The prefix "c" in the symbol cm for centimeter tells us that 1 meter = 100 centimeters. This equation is the basis of all conversions between meters and centimeters. We can rewrite the equation in two ways:

$$\frac{1 \text{ meter}}{100 \text{ centimeters}} = 1 \quad \text{or} \quad \frac{100 \text{ centimeters}}{1 \text{ meter}} = 1$$

The fraction on the left side of each equation is a conversion factor that is used to convert between the two units, centimeters and meters. In a conversion from one unit to another, the given quantity is multiplied by the appropriate conversion factor. Note that this operation does not change the value of the quantity, because each conversion factor is equal to 1.

Now we must decide which conversion factor should be used for a given conversion. The numerator of the conversion factor must be the desired unit and the denominator must be the original unit, which becomes clear if the units are written in each step of the calculation. The first conversion factor is the correct one for converting from centimeters to meters, and the desired conversion is

$$183 \text{ cm} \times \frac{1 \text{ m}}{100 \text{ cm}} = 1.83 \text{ m}$$

You can see that the units of cm cancel, leaving only m, the desired unit.

---

**Example 1.1**  What is the mass in grams and in milligrams of a 50-kg student?

**Solution**  The relationship between grams (g) and kilograms (kg) is

$$1 \text{ kg} = 1000 \text{ g}$$

The appropriate conversion factor is 1000 g/1 kg, and

$$50 \text{ kg} \times \frac{1000 \text{ g}}{1 \text{ kg}} = 50\ 000 \text{ g}$$

The units of kilograms (kg) cancel, leaving only grams (g).
The relationship between grams and milligrams is

$$1 \text{ g} = 1000 \text{ mg}$$

and conversion from kilograms (kg) to milligrams (mg) can be carried out in one step by the use of two conversion factors:

$$50 \text{ kg} \times \frac{1000 \text{ g}}{1 \text{ kg}} \times \frac{1000 \text{ mg}}{1 \text{ g}} = 50\ 000\ 000 \text{ mg}$$

Again, all the units except milligrams (mg) cancel when the appropriate conversion factors are used. The use of conversion factors in the solution of chemistry problems will be developed more fully in Chapter 3.

To find relationships between derived SI units, we start with the relationship between the base unit and the unit with the prefix. The operation that converts the base unit to the derived unit is then carried out on both of these quantities. For example, suppose we want to find the relationship between cubic meters ($m^3$) and cubic decimeters ($dm^3$) or liters (L). The relationship between the meter and the decimeter is 1 m = 10 dm. We must raise both equivalent quantities in this equation to the third power.

$$(1 \text{ m})^3 = (10 \text{ dm})^3 \quad \text{or} \quad 1 \text{ m}^3 = 1000 \text{ dm}^3 = 1000 \text{ L}$$

**Example 1.2**    What is the volume in liters and in cubic decimeters of 568 $cm^3$ of water?

**Solution**    The relationship between the liter and the $cm^3$ is 1 L = 1000 $cm^3$. The factor for conversion from cubic centimeters to liters is 1 L/1000 $cm^3$ and

$$568 \text{ cm}^3 \times \frac{1 \text{ L}}{1000 \text{ cm}^3} = 0.568 \text{ L}$$

Since 1 m = 10 dm and 1 m = 100 cm, we can see that 10 dm = 100 cm or 1 dm = 10 cm. The relationship between the corresponding derived units for volume is $(1 \text{ dm})^3 = (10 \text{ cm})^3$ or 1 $dm^3$ = 1000 $cm^3$. Thus,

$$568 \text{ cm}^3 \times \frac{1 \text{ dm}^3}{1000 \text{ cm}^3} = 0.568 \text{ dm}^3$$

This result is expected, since the liter and the $dm^3$ are equal. Note that once again, all the units except those of the result cancel.

**TABLE 1.5**  Some Non-SI Units

| Physical Quantity | Name of Unit | Symbol of Unit | Definition[a] |
|---|---|---|---|
| length | angstrom | Å | $10^{-10}$ m |
| length | inch | in. | $2.54 \times 10^{-2}$ m |
| volume | quart (U.S.) | qt | 0.94 L |
| volume | quart (Brit.) | qt | 1.14 L |
| mass | pound | lb | 0.453 592 kg |
| force | dyne | dyn | $10^{-5}$ N |
| pressure | atmosphere | atm | 101 325 Pa |
| pressure | torr[b] | Torr | $\dfrac{101\ 325}{760}$ Pa |
| pressure | millimeters[b] of mercury | mmHg | $13.5951 \times 9.80665$ Pa |
| energy | erg | erg | $10^{-7}$ J |
| energy | calorie | cal | 4.184 J |

[a] We can change these definitions into equations by setting the unit equal to the definition. Thus, 1 lb = 0.453 592 kg.
[b] These two units have essentially the same value.

Table 1.5 lists some of the non-SI units that are still likely to be encountered in chemistry and their definitions in terms of SI units. You will find that SI prefixes are often used with these non-SI units. Thus, 1 kcal = 1000 cal. Although the calorie is not an SI unit, the prefix k has the SI meaning. Similarly, 1 liter (L) = 1000 milliliters (mL). Since 1 mL = 1 cm³, we shall prefer the latter SI unit.

To convert between the units listed in Table 1.5 and SI units, we use the definitions listed in the table and the method outlined in Example 1.1. We must also include any necessary nondecimal factors.

**Example 1.3**   The speed limit on American highways is 55 miles per hour. Express this speed in SI base units.

**Solution**   In Table 1.5, we find that 1 in. = $2.54 \times 10^{-2}$ m. This equation gives us the conversion factor we need to go from one system of units to the other. To make the conversion, we also must know the numerical factors in the non-SI system: 1 ft = 12 in. and 1 mi = 5280 ft. Since the SI base unit of time is the second, we also need the relationship between seconds and hours: 1 hr = 3600 s.

We can carry out the calculation by using these conversion factors and canceling units as we proceed:

$$55\ \frac{\text{mi}}{\text{hr}} \times \frac{5280\ \text{ft}}{1\ \text{mi}} \times \frac{12\ \text{in.}}{1\ \text{ft}} \times \frac{2.54 \times 10^{-2}\ \text{m}}{1\ \text{in.}} \times \frac{1\ \text{hr}}{3600\ \text{s}} = 25\ \frac{\text{m}}{\text{s}}$$

The need for several conversion factors to convert from miles to inches illustrates one major advantage of the SI system. To convert between SI units, we need to know only the meanings of the prefixes.

## Temperature Scales

One common conversion is that between different temperature scales. The unit of temperature in SI is the kelvin (K), which is used for the thermodynamic temperature scale. It is the only temperature scale scientists need use, but in practice the Celsius (°C) temperature scale is also used.

In the Celsius temperature scale, 0°C is the freezing point of water and 100°C is the boiling point of water. Thus, 1°C is ¹⁄₁₀₀ of the difference between these two points. The temperature interval of the Celsius scale is the same as that of the thermodynamic temperature scale: 1°C = 1 K. But the thermodynamic temperature $T$ is 273.15 K higher than the Celsius temperature $t$:

$$T = t + 273.15 \text{ K}$$

Thus, to convert from degrees Celsius (°C) to thermodynamic temperature in kelvins (K), add 273.15 to the Celsius value (Figure 1.5).

The Fahrenheit temperature scale is the one in everyday use in the United States. In the Fahrenheit scale, 32°F is the freezing point of water and 212°F is the boiling point of water. The temperature interval is $\frac{1}{180}$ of the difference between these two points. The degree Fahrenheit is smaller than the degree Celsius. The relationship between the two temperature scales is given by the equation:

$$F = 1.8C + 32$$

where $F$ is the Fahrenheit temperature and $C$ is the Celsius temperature.

## Significant Figures[1]

An expression of the magnitude of a physical quantity has two parts, a number and a unit. So far we have discussed the units. Let us now turn our attention to the numbers.

Some expressions of quantity are exact: for example, 12 eggs or $9.95. These numbers result from a count of each entity present. Other numbers are as exact as we want them to be. For example, we can write the values of $\pi$ and of the fraction $\frac{3}{7}$ with as many digits after the decimal point as we need. Most measurements of physical quantities are not exact. There is some error in the counting process. This error affects the numerical value of the measurement (Figure 1.6). The number used to express the magnitude of a physical quantity should also indicate the margin of error.

The simplest way to indicate the margin of error is by writing more or fewer digits in the number. The number of digits is called the number of

---

[1] A more detailed treatment of many of the arithmetic procedures used in general chemistry can be found in Appendix IV.

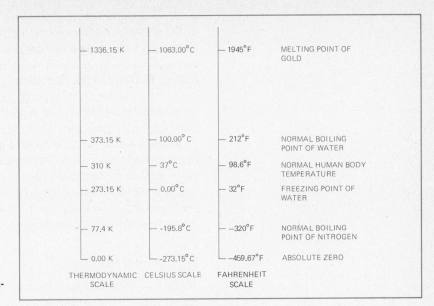

**Figure 1.5**
The relation between the thermodynamic, Celsius, and Fahrenheit temperature scales.

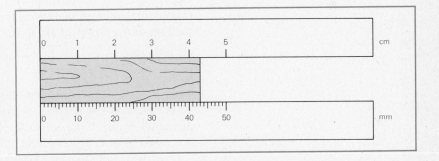

**Figure 1.6**
The precision of a measurement depends on the precision of the instrument used to make the measurement. The top scale is calibrated in centimeters (cm) and the bottom scale in millimeters (0.1 cm). The length of the piece of wood in the center is recorded as 4.3 cm if the top scale is used and 4.28 cm if the bottom scale is used. The measurement made with the more precise scale has more significant figures than that made with the less precise scale.

**significant figures.** The last digit written in the number is understood to be uncertain, usually by ± 1.

Thus, if we write that a sample has a mass of 12 g, we mean that its mass is between 11 g and 13 g. We can indicate this range directly by writing 12 ± 1 g. Either way, the uncertainty in the measurement is 1 g. If we report that the mass of the sample is 12.6 g, we indicate that there is less uncertainty in the measurement. We are saying that the mass is between 12.5 g and 12.7 g, so the uncertainty is 0.1 g. We can say that the second measurement is more *precise* than the first. If we report the mass of the sample as 12.60 g, the measurement is even more precise. We are saying that the mass is between 12.59 g and 12.61 g, and the uncertainty is 0.01 g.

You can see that the number of digits in the reported quantity describes the precision of the measurement. Generally, the number of digits used to report a measurement is the number of significant figures. We use significant figures to indicate the margin of error in a measurement.

Zeros sometimes are significant figures and sometimes are not. Zeros at the end of a number to the right of the decimal point are significant. Zeros that serve only to fix the decimal point in numbers less than 1 are not significant. Thus, 0.023, 0.0023, and 0.00023 all have two significant figures. But 0.0230 has three significant figures, because the last zero does not serve to fix the decimal point. It is there to indicate the precision of the measurement. However, zeros after the decimal point in numbers greater than 1 are all significant. There are four significant figures in 1.023.

When a whole number ends in a zero to the left of the decimal point, the number of significant figures can be ambiguous. For example, when we write that the distance between two cities is 2000 km, it is not possible to say whether the distance is between 1999 km and 2001 km, between 1990 km and 2010 km, or even between 1000 km and 3000 km. We need a way to eliminate this ambiguity.

## Exponential Notation

We can avoid ambiguity about significant figures by the use of **exponential notation.** Exponential notation is a method of writing numbers that is especially useful for dealing with very large or very small numbers. In exponential notation, a number is expressed as the product of an ordinary number and a power of 10, in the form

$$N \times 10^x$$

where $N$ is the ordinary number and $x$ is the exponent, or power to which 10 is raised. If the decimal point is moved one space to the left, $x$ is increased by one. If the decimal point is moved one space to the right, $x$ is decreased by one. For example, $25\,486 = 2.5486 \times 10^4$ and $0.000147 = 1.47 \times 10^{-4}$. Note that in numbers less than 1, the exponent, $x$, is negative, while in numbers greater than 1, $x$ is positive ($1 = 10^0$).

In chemistry, $N$ is written with only one integer to the left of the decimal point. The number of significant figures is one more than the number of integers to the right of the decimal point. Thus, the number 2000 can be written as $2 \times 10^3$ to show one significant figure, as $2.0 \times 10^3$ to show two significant figures, and as $2.00 \times 10^3$ to show three significant figures.

## Arithmetic Operations

When arithmetic operations are performed with numbers that represent measurements, the number of significant figures in the result can be determined by use of these rules:

**1.** In addition or subtraction, the number of significant figures that we keep to the right of the decimal point in the answer is the lowest number of digits to the right of the decimal point in any of the original numbers. The following examples illustrate this rule:

$$
\begin{array}{cccc}
4.983 & 12.2 & 4 \times 10^2 & 4.00 \times 10^2 \\
+1.02 & +\ 0.009 & -30 & -3 \\
\hline
6.00 & 12.2 & 4 \times 10^2 & 3.97 \times 10^2
\end{array}
$$

To understand the last two results, we must consider the number of significant figures in the exponential term: $4 \times 10^2$ is 400, with one significant figure. Subtraction of 30 gives 370, which is still 400 to one significant figure, or $4 \times 10^2$. On the other hand, $4.00 \times 10^2$ is 400, with three significant figures. Subtraction of 3 gives 397, with three significant figures.

2. In multiplication and division, the final product or quotient cannot have more significant figures than are in the original number with the least number of significant figures. The following examples illustrate this rule:

$$
4.3 \times 3.2 = 14 \qquad \frac{14.1}{13.8} = 1.02 \qquad 6.789 \times 0.000032 = 0.00022
$$

An extra significant figure usually is carried until calculations are completed. The answer is then rounded off. But in such a calculation as $3 \times 15.99876$, it is simpler to round off the original number before performing the calculation, since the answer will have only one significant figure. Always express the result of a calculation with the correct number of significant figures, because significant figures carry an implication about the uncertainty in the measurement. Remember, your calculator may give 10 figures in every answer, but usually not all these figures are meaningful. Do not simply copy all the numbers in the calculator's display.

Measurements are so central to chemistry that care must be taken to express both parts of any measurement correctly. The correct number of significant figures tells us about the error limits of the measurement. The units that are used make the numerical value of the measurement meaningful.

Knowledge about measurements is essential for the study of any science. With this knowledge at hand, we can now start to explore the science of chemistry in detail.

**Summary**

**C**hemistry is a scientific discipline whose subject is matter. We began by describing the **scientific method,** in which theories must be supported by observations and experiments. We then described how **matter** is classified into homogeneous and heterogeneous materials. We defined an **element** as a substance that cannot be decomposed by a simple chemical change and a **compound** as a material that can be decomposed by such a change. We described how modern chemistry grew out of alchemy and showed its importance in today's world. Since measurement is at the heart of modern chemistry, we described **SI units** that were adopted for scientific use from the metric system. We concluded by discussing some basic mathematical skills that are essential for chemistry, such as conversion from one unit to another, the meaning of **significant figures,** and the use of **exponential notation.**

## Exercises

**1.1** Distinguish between the following: (a) hypothesis and observation, (b) theory and law of nature, (c) observation and experimentation.

**1.2** Propose a hypothesis to account for the relatively large registration for your chemistry class. Suggest procedures to test your hypothesis.

**1.3**[2] Classify each of the following statements as a theory, a law, an observation, or a hypothesis:
(a) Combustion is a rapid reaction of a substance with oxygen.
(b) All flammable substances contain phlogiston.
(c) Matter is composed of very small particles that are in constant motion.
(d) There is a force of attraction between any two objects.

**1.4** Describe three everyday phenomena that you expect to understand better as a result of studying chemistry.

**1.5** Classify the following as elements, compounds, solutions, or mixtures: (a) peanut butter, (b) clean air, (c) nitrogen, (d) vitamin C, (e) a rock, (f) a glass of milk.

---

[2] The answers to exercises whose numbers are in color can be found in Appendix VII. The star indicates an exercise that is more challenging than average.

**1.6** Classify the following as intensive or extensive properties: (a) surface area, (b) density, (c) mass, (d) freezing point, (e) electrical resistance.

**1.7** Predict two important achievements of chemistry that may be made in the twenty-first century.

**1.8** Suggest a physical quantity for which each of the following units is appropriate: (a) $cm^2$, (b) $s^{-1}$, (c) ft/s, (d) cal, (e) mmHg.

**1.9** Suggest an SI unit that is convenient for measuring each of the following quantities: (a) the distance from the earth to the sun, (b) the capacity of a milk container, (c) the velocity of a jet aircraft, (d) the mass of a comma on this page, (e) the area of a comma.

**1.10** Here are three pairs of units. For each pair suggest two related physical quantities each of whose measurement is more convenient with each unit of the pair. (a) mm/s and Gm/s, (b) $mg/km^3$ and $g/cm^3$, (c) $nm^3$ and $m^3$.

**1.11** Starting with the SI base units, derive units for acceleration. Suggest the most convenient units to express the acceleration of a snail and the acceleration of a spacecraft leaving the earth.

**1.12** The newton (N), the SI unit of force, is derived from the SI base units by using the relationship $F = ma$, where $F$ is force, $m$ is mass, and $a$ is acceleration. The dyne is the unit of force in another system of units, in which mass is measured in grams and time is measured in seconds. Given that 1 dyne = $10^{-5}$ N, find the unit of length in this system of units.

**1.13** Momentum is defined as $mv$, where $m$ is mass and $v$ is velocity. Give the unit for momentum in SI base units. Find the relationship between this unit and the newton.

**1.14** The equation $PV = \frac{2}{3}E_k$ relates the pressure ($P$) and volume ($V$) of a gas to its kinetic energy ($E_k$). Show that this equation is consistent with the definitions of the SI derived units in Table 1.4.

**1.15** Find the volume of water in cubic meters ($m^3$) that would cover an area of 1 $km^2$ to a depth of 1 cm.

**1.16** Find the relationship between the cubic nanometer ($nm^3$) and the cubic kilometer ($km^3$).

**1.17** The density of water is 1 $g/cm^3$. Express the density of water in (a) kilograms per cubic meter ($kg/m^3$), (b) grams per liter (g/L), (c) nanograms per cubic nanometer (ng/nm³).

**1.18** The diameter of a hydrogen atom is 0.212 nm. Find the length in kilometers (km) of a row of $6.02 \times 10^{23}$ hydrogen atoms.

**1.19** Express your own mass in (a) kilograms, (b) grams, (c) tons, (d) ounces (1 lb = 16 oz).

**1.20** The area of Canada is 3 850 000 mi². Express the area in square kilometers (km²).

**1.21** There is 43 560 ft² in an acre. What is the area in square kilometers of a 375-acre tract of land?

**1.22** One quart contains 32 fluid ounces. Express the volume of an 8.0-oz glass in cubic centimeters.

**1.23** A good sprinter can run the 100-m dash in 10.2 s. Find the sprinter's average speed in miles per hour.

**1.24** If the current exchange rate gives 1.20 Canadian dollars to the American dollar and gasoline sells for 50 cents/L in Canada, find the price in U.S. dollars per gallon (U.S.) (1 gal = 4 qt).

**1.25★** A force of $4.62 \times 10^4$ N is applied to the face mask of a diving helmet whose area is 150 cm². Find the pressure (in atmospheres) on the face mask.

**1.26** The "calorie" used by dieters actually is equal to 1000 of the calories that are units of energy. One quarter-pound of hamburger has 421 nutritional calories. Find its energy content in joules.

**1.27** The velocity of sound in dry air at 20°C is 343 m/s. Find the velocity in miles per hour.

**1.28★** The density of water at 25°C is 0.997 g/cm³. Find the mass of 1 gal (U.S.) of water in pounds (1 gal = 4 qt).

**1.29** Astronomers have measured the average daytime temperature on the planet Saturn as 123 K. Give this temperature on the Celsius scale.

**1.30** Find the temperature at which the readings on the Celsius scale and the Fahrenheit scale are equal.

**1.31** Derive a relationship between temperature on the Kelvin scale and on the Fahrenheit scale.

**1.32** Find the temperature at which the readings on the Kelvin scale and the Fahrenheit scale are equal.

**1.33** What is normal human body temperature on the Kelvin scale?

**1.34** The boiling point of carbon is 4827°C. What is the boiling point on the Fahrenheit scale?

**1.35** How many significant figures are there in each of the following numbers? (a) 0.04560, (b) 1.04560, (c) 10 001, (d)10 000, (e) 10 000.0.

**1.36** Measure the length of this page with a meter stick that is marked off in millimeters. How many significant figures will there be in the result of your measurement?

**1.37** How many significant figures are there in each of the following expressions of quantity?
(a) That man must be one hundred years old.
(b) We will soon observe the two hundredth anniversary of the United States Constitution.
(c) Fifty million Frenchmen can't be wrong.
(d) You can buy that book for eleven dollars.
(e) A new car costs twelve thousand dollars.

**1.38** The relationship between the volume $V$ of a sphere and its radius $r$ is $V = \frac{4}{3}\pi r^3$. What determines the number of significant figures in a calculation of the volume of a sphere?

**1.39** Express the following numbers in exponential notation. Do not use a calculator.
(a) 12 345
(b) 0.000 000 000 500
(c) 1 000 000
(d) 45.54

**1.40** Using exponential notation, write expressions for the number 100 000 with one through six significant figures.

**1.41** Without using a calculator, carry out the following calculations:
(a) $(1 \times 10^5)/(1 \times 10^3)$
(b) $(1 \times 10^5)(1 \times 10^3)$
(c) $(1 \times 10^5)^3$
(d) $(1 \times 10^5)^{1/2}$

**1.42** Give the results of the following calculations with the correct number of significant figures:
(a) $1.0001 + 13.87 + 0.23680$
(b) $2.24 + 7.42 + 0.34$
(c) $325 - 26.0$
(d) $(555)(0.0002)$
(e) $(7777)(9999)/(1111)$

**1.43** Give the results of the following calculations with the correct number of significant figures:
(a) $3.0 \times 10^{-2} + 1.2 \times 10^{-3}$
(b) $(3.75 \times 10^4)(1.5 \times 10^{-6})$
(c) $3.4589 \times 10^4 - 1$
(d) $2.467 \times 10^{-3}/1.1 \times 10^{-1}$

**2**

# The Language of Chemistry

**Preview**

**N**ow you are ready to learn the language of chemistry. In this chapter, we begin by introducing the chemical elements of which all matter is made and the atoms of which all elements are made. We then describe the symbols that are used to represent the elements, and we go on to show how these symbols are used in chemical formulas, which describe the composition of substances. Next is a description of how names are assigned to ions, to substances made of ions, and to other substances, including acids and bases. We then outline the writing of balanced chemical equations that describe chemical changes concisely and accurately. The chapter concludes with an introduction to another topic of basic importance: energy, its transformations and conservation.

When you explore new territory, it helps if you know the language. You probably know a little of the language of chemistry. The aim of this chapter is to give you a wider and deeper knowledge of the symbols, numbers, and names that chemists use. English sentences are made of words. The "sentences" of chemistry are equations like those in algebra. They are made up of symbols and numbers. The language of chemistry is designed to convey a maximum amount of information in the most orderly and concise way.

## 2.1   ATOMS AND MOLECULES

The world is composed of a bewildering array of materials that the untrained mind cannot possibly classify. Historically, one major achievement of chemistry has been to simplify this complexity, to show that the tremendous variety of substances on earth can be explained as different combinations of a relatively small number of basic ingredients.

These basic ingredients are the **chemical elements.** There are not very many of them. To date, 108 elements have been identified; 90 of them were first found in nature; the rest are synthetic. Every object on earth can be described as a collection of substances made up of these elements.

A sample of a chemical element can be divided and subdivided only to a point. The division of such a sample can go on until you have a quantity far too small for the human senses to distinguish. But sooner or later, you arrive at the smallest quantity of an element that retains the identity of that element. **The basic unit of an element is an atom.** The chemical behavior of a single atom may not differ significantly from the behavior of a billion atoms of that element. But if you break an atom into pieces—a feat that became possible only a few years ago—you are left with entities of a completely different sort.

It is almost impossible to appreciate the minute size of the atom. A 1-cm cube of iron would fit comfortably on the tip of your thumb. This tiny cube contains $8.2 \times 10^{22}$ atoms of iron (82 000 000 000 000 000 000 000 atoms). That is more than all the grains of sand on all the beaches of earth. The thumb on which the cube rests contains a substantially larger number of atoms. The individual to which the thumb is attached contains even more atoms.

### The Variety of Molecules

Most substances do not exist as single atoms in nature: They are compounds of more than one element. We usually find that **atoms join together to form molecules** (Figure 2.1). Molecules usually form because they are more stable than the separate atoms. All the molecules of a pure substance have the same number of atoms attached in the same way. Each has the same number of atoms of each element in the molecule. The smallest molecules have only two atoms. The largest have many thousands.

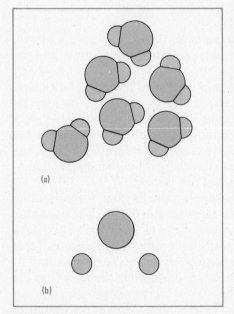

**Figure 2.1**
(a) Molecules of water. The large sphere represents an atom of oxygen, the small spheres atoms of hydrogen. The smallest drop of water contains billions of such molecules. (b) A molecule of water that has been split. It is no longer a molecule of water, but rather two separate hydrogen atoms and a single oxygen atom.

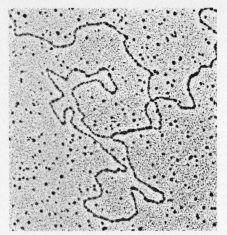

A molecule of nucleic acid, containing many tens of thousands of atoms. Nucleic acids and other molecules created by living organisms are the largest that are known to exist. *(Louise Tsi Chow, Coldspring Harbor Laboratories)*

Some elements normally exist as two-atom molecules. Among these are hydrogen, nitrogen, oxygen, fluorine, chlorine, bromine, and iodine. Some elements normally consist of molecules containing more than two atoms: Molecules of sulfur contain eight atoms, and molecules of phosphorus contain four atoms, for example.

Most of the largest molecules are produced by living organisms. Some of the largest of these molecules are in viruses, which are on the threshold of life. A virus typically consists of a nucleic acid with a protein coat. If the coat is stripped away, the long, stringy nucleic acid molecule can be seen under extremely high magnification in an electron microscope. This molecule contains perhaps half a million atoms, but it is insignificantly small by the standards of everyday life. You would need $10^{15}$ (one million billion) of these molecules to make a 1-cm cube.

The concepts of atoms and molecules allow us to describe any substance in terms of a few fundamental units of structure. If there were no such fundamental units, the study of chemistry would be much more difficult.

## The Structure of Atoms

Despite their small size, atoms are made of subunits. One of the great scientific advances of the twentieth century is the discovery and understanding of atomic structure.

Of the hundreds of subatomic particles that have been discovered, three are of primary interest to us: the **proton,** the **neutron,** and the **electron.**

Protons and neutrons account for almost all of the mass of the atom. The neutron has slightly more mass than the proton. The proton has an electrical change; the neutron has none. The nucleus of an atom of a given element has a fixed number of protons, which is called the **atomic number** of the element. Different nuclei of an element can have different numbers of neutrons. Atoms with the same number of protons but different numbers of neutrons are called **isotopes** of the element.

The electron is much smaller — so much smaller that its contribution to the mass of the atom is negligible. It has a negative electric charge that is exactly equal to the positive charge of the proton. An atom is electrically neutral because it has the same number of protons and electrons. Table 2.1 compares these three subatomic particles and an atom of iron.

In the atom, protons and neutrons are clustered in the nucleus at the center of the atom. The nucleus has almost all the mass of the atom but occupies only a very small fraction of its volume. Most of the volume of the atom is empty space around the nucleus. The electrons occupy this space.

Ordinary chemical changes result from changes in the electronic structure of atoms. One of the most basic chemical changes is the loss or gain of electrons by an atom or a group of atoms. Such a gain or loss produces an

**TABLE 2.1**   Comparison of Subatomic Particles

| Particle | Charge | Mass |
|---|---|---|
| proton | $+1$ | $1.673 \times 10^{-27}$ kg |
| neutron | $0$ | $1.675 \times 10^{-27}$ kg |
| electron | $-1$ | $9.110 \times 10^{-31}$ kg |
| iron atom | $0$ | $9.288 \times 10^{-26}$ kg |

**TABLE 2.2**   Elemental Symbols from Latin Names

| English Name | Symbol | Latin Name |
|---|---|---|
| antimony | Sb | stibium |
| copper | Cu | cuprum |
| gold | Au | aurum |
| iron | Fe | ferrum |
| lead | Pb | plumbum |
| mercury | Hg | hydrargentum |
| potassium | K | kalium |
| silver | Ag | argentum |
| sodium | Na | natrium |
| tin | Sn | stannum |

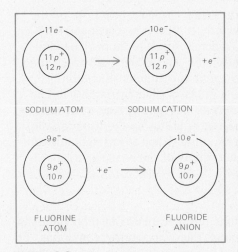

SODIUM ATOM     SODIUM CATION

FLUORINE ATOM     FLUORIDE ANION

**Figure 2.2**

The formation of ions from atoms. The sodium atom is neutral, with eleven protons ($p^+$) and eleven electrons ($e^-$). When it loses an electron it forms a cation of charge $+1$. Other atoms may lose two or more electrons to form cations of higher charge. When the fluorine atom gains one electron, it forms an anion of charge $-1$. Other atoms may accept more than one electron to form anions of greater negative charge.

**ion,** an atom or group of atoms with an electric charge. Electrons are negatively charged, so a gain of electrons produces a negatively charged ion called an **anion.** A loss of electrons results in a positively charged ion called a **cation** (Figure 2.2). We shall discuss the electronic structure of atoms in greater detail later.

## 2.2   CHEMICAL FORMULAS

### Elemental Symbols

To simplify work with atoms and molecules, a system of symbols is used. Each element is represented by a symbol of one or two letters. These symbols usually consist of the first letter of the English or Latin name[1] of the element; this letter is capitalized. Since the names of many elements start with the same letter, a second, uncapitalized, letter, is added to distinguish one element from another. Thus, the symbol for hydrogen is H; for helium He; for boron B; for beryllium Be. Examples of symbols derived from Latin names are listed in Table 2.2.

[1] The symbol W, for element 74, wolfram (called tungsten in English), is an exception. It comes from *wolfrumb,* the early German word for "wolf turnip," because the metal is found in turniplike lumps.

## ATOMS OBSERVED

The achievements of the twentieth-century microscopy have banished any remaining doubts about the existence of atoms by producing images of single atoms. In the 1950s, an instrument called the field ion microscope was used to create an image of the structure of the tip of a needle. The needle is placed in a tube, and its temperature is lowered to about 20 K. A gas such as helium is introduced into the tube, and a high positive voltage is applied to the tip of the needle. The combination of extreme cold and the high voltage causes the atoms of the gas to become electrically charged and to recoil from the atoms in the tip of the needle. The gas atoms strike a fluorescent screen, creating a pattern that is identical to the atomic structure of the tip of the needle. Magnifications of up to 750 000 are possible with the field ion microscope.

However, the field ion microscope is limited to substances that can be formed into filaments with very sharp points of about $1 \times 10^{-7}$ m in diameter. In 1976, physicists at the University of Chicago went one step further and produced motion pictures of the movement of single atoms. Using a scanning electron microscope of special design, the physicists were able to get images of single uranium atoms placed on a specimen of carbon 500 nm thick. The motion pictures show the uranium atoms as blurred bright spots against the dark background of the carbon, magnified about 5.5 million times. Images taken in sequence show the slow thermal motion of the uranium atoms.

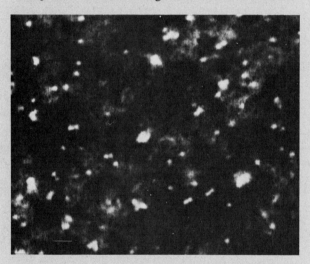

The high-resolution microscope used to take the picture can, in principle, visualize any atom on the carbon background. In the first black-and-white films, however, different kinds of atoms looked almost indistinguishable. In a later advance, the images were processed by computers that arbitrarily assigned colors to spots of different intensity. The processed images easily distinguished between different atoms. Researchers are using the technique to study the behavior of individual atoms in biological processes and in a variety of chemical changes.

A symbol can mean several things. It can be used as an abbreviation in a sentence. If we write, "Among the compounds containing Fe are . . . ," the symbol Fe stands for any quantity of iron. But the symbol Fe means not only the element iron but also a single atom of iron, the fundamental unit in any quantity of iron.

Symbols are used to indicate the elements in molecules and ions. Most molecules are made up of atoms of different elements. If we want to discuss the molecule made of one atom of oxygen and one atom of carbon, we can call it carbon monoxide. We can also call it CO, which is simpler. Similarly, it is easier to write "HCN" than to write "hydrogen cyanide," the name given to a molecule containing one atom of hydrogen, one atom of carbon, and one atom of nitrogen. The combination of symbols representing the atoms in a molecule of a compound is called a **formula.** For example, CO is the formula for a molecule of the compound carbon monoxide, HCN is the formula for a molecule of the compound hydrogen cyanide, and $HS^-$ is the formula for an anion of hydrogen and sulfur.

### Subscripts and Coefficients

Most molecules contain more than one atom of a given element. To specify the number of atoms of any element in a molecule we use **subscripts,** small numerals that are written below and after symbols in a formula. The subscript indicates the number of atoms of that element in the molecule.

For example, a molecule of the oxygen in the air we breathe consists of two joined atoms of oxygen. The formula for this molecule is $O_2$. We can also find small traces of atomic oxygen, whose symbol is O. A third form of oxygen that is important in atmospheric studies is ozone, a three-atom molecule whose formula is $O_3$. Again, the formula for hydrazine, an important rocket fuel, is $N_2H_4$. A single molecule of hydrazine contains two atoms of nitrogen and four atoms of hydrogen. You will not find the subscript 1 in a chemical formula. If a molecule has only one atom of an element, that symbol has no subscript, as in CO, the formula for carbon monoxide, or O, the formula for atomic oxygen.

Numerals called **coefficients** sometimes are written in front of chemical formulas. They specify the relative number of molecules taking part in a chemical reaction. It is easy to distinguish between coefficients and subscripts. In the formula $Cl_2$, the numeral 2 is a subscript that indicates that we have a two-atom molecule of chlorine. If we write 2Cl, we have two separate atoms of chlorine (Figure 2.3). Similarly, $CO_2$ is one molecule of carbon dioxide, while 2CO is two molecules of carbon monoxide. To repeat: **A coefficient** (a numeral written to the left of a chemical formula) **indicates the number of separate particles.** A subscript (a numeral written to the right of and below a symbol) **indicates the number of atoms of a given element in a single molecule or ion.**

**Figure 2.3**
The subscript 2 used with the symbol Cl indicates one molecule of $Cl_2$, left. The coefficient 2 used with the symbol Cl indicates two Cl atoms, right.

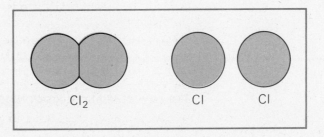

$Cl_2$        Cl        Cl

Claude-Louis Berthollet (1749–1822), born of poor parents and trained as a physician, was one of the first to accept Lavoisier's theories. He met Napoleon in 1798, taught him chemistry, and was made a senator and a count. Among his achievements were the determination of the composition of ammonia and the introduction of chlorine as a bleaching agent. *(Culver)*

## The Law of Constant Composition

The fact that substances are made of atoms and molecules tells us something significant about their composition. For example, since any sample of water consists of a large number of identical molecules, each containing two atoms of hydrogen and one atom of oxygen, then all samples of water have the same proportions of hydrogen and oxygen. The law of constant composition is a formal statement of this fact. It says that *all samples of a given pure substance contain the same elements in the same proportions.*

Before the existence of atoms was recognized, this fact was far from obvious. Indeed, it was the object of a major controversy that raged between Claude-Louis Berthollet (1749–1822) and Joseph-Louis Proust (1745–1826) in the late eighteenth century. Berthollet, one of the leading chemists of the day, believed that the composition of a pure substance could vary from sample to sample. Proust, a relative unknown, maintained the opposite. It took more than a decade of painstaking measurements before Proust could convince other chemists that he was right. Proust's work contributed to the formulation of the atomic theory.

The law of constant composition has a great practical advantage. Once we know the composition of a sample of a pure substance, we can easily calculate the composition of any other sample of that substance.

---

**Example 2.1**    Analysis of a sample of pure lithium chloride, whose mass is 6.61 g, shows that it contains 1.08 g of Li. Find the mass of lithium in a sample of 985 g of lithium chloride.

**Solution**    According to the law of constant composition, every sample of LiCl has the same proportion of Li. Using the results of the analysis of the 6.61-g sample, we can express the composition as a fraction:

$$\frac{1.08 \text{ g Li}}{6.61 \text{ g LiCl}}$$

We can then use this fraction as a conversion factor to calculate the mass of Li in any sample of LiCl:

$$\text{mass Li} = \frac{1.08 \text{ g Li}}{6.61 \text{ g LiCl}} \times 985 \text{ g LiCl} = 161 \text{ g Li}$$

We cancel the units of g LiCl in two expressions, and the units of the result are g Li.

---

## The Law of Multiple Proportions

Another law that seems obvious to us but was formulated only after many measurements early in the nineteenth century is the law of multiple proportions. It was first formulated by John Dalton (1766–1844). Dalton is remembered chiefly for his atomic theory, which we shall discuss in Chapter 5. The law of multiple proportions was an important step toward acceptance of the atomic theory.

Dalton was studying different compounds composed of the same two elements, such as carbon monoxide, CO, and carbon dioxide, $CO_2$. He measured the mass of the carbon and the oxygen in samples of each compound. From the twentieth-century vantage point, the relationship between the masses in the different compounds is evident. The formulas show that in $CO_2$, there are twice as many oxygen atoms per carbon atom as there are in CO. Therefore, there should be two times the mass of O for a given mass of C in $CO_2$ as in CO. We can list the data in a table (Table 2.3).

The last column in the table shows the mass of O that is combined with 1 g of C in each compound. You can see that these masses are in the ratio 2:1. This result does not surprise us, since we know that the formulas are CO and $CO_2$. In the nineteenth century, however, the formulas were unknown. Observations of the composition of such compounds were important bits of evidence for the existence of atoms and for simple formulas.

We can state the law of multiple proportions formally: If two elements combine to form more than one compound, the different masses of one element that combine with a fixed mass of the other element are in a ratio of small whole numbers.

The oxides of nitrogen provide a more extensive example of the law of multiple proportions. Table 2.4 lists a number of compounds, assigning

Joseph-Louis Proust (1754–1826) was the son of an apothecary and established himself in that business while he did research in chemistry. In addition to helping to establish the validity of the law of definite proportions, Proust also pioneered in the study of sugars, distinguishing several different kinds. He spent a great part of his life in Spain to avoid the upheavals that followed the French Revolution but returned to France to end his career. *(Edgar Fahs Smith Collection)*

**TABLE 2.3**   Composition of Compounds of Carbon and Oxygen

| | Mass of Sample (g) | Mass of O (g) | Mass of C (g) | Ratio of Mass O to Mass C |
|---|---|---|---|---|
| CO | 1.00 | 0.57 | 0.43 | 1.33 |
| $CO_2$ | 1.00 | 0.73 | 0.27 | 2.66 |

**TABLE 2.4**   The Law of Multiple Proportions

| Compound | Formula | Mass of Nitrogen | Mass of Oxygen | Relative Mass of Oxygen |
|---|---|---|---|---|
| nitrous oxide | $N_2O$ | 28 | 16 | 1 |
| nitric oxide | NO | 28 | 32 | 2 |
| dinitrogen trioxide | $N_2O_3$ | 28 | 48 | 3 |
| nitrogen dioxide | $NO_2$ | 28 | 64 | 4 |
| dinitrogen pentoxide | $N_2O_5$ | 28 | 80 | 5 |

John Dalton (1766–1844), the son of a weaver, was a practicing Quaker. His chemical studies began while he was teaching at a Quaker school. He first became interested in meteorology, which he studied with instruments he built himself. Weather observations led to studies of the composition of air and then of other gases. Dalton published his atomic theory in 1803. His Quaker principles made him refuse many proffered honors, but he did accept a pension from King William IV. *(Granger)*

each a fixed mass of nitrogen. You can see that the masses of O that combine with this fixed mass of N are in the ratio of $16:32:48:64:80$, or $1:2:3:4:5$.

The law of multiple proportions was particularly important when it was introduced because it promoted acceptance of Dalton's atomic theory. Acceptance did not always come readily. Some scientists steadfastly refused to believe that anything as small as an atom could exist. It took the experiments of twentieth-century physics to produce the evidence that finally showed that atoms are real things.

## 2.3  CHEMICAL NAMES AND FORMULAS

Most names used in chemistry are the result of a logical system based on chemical composition, but some originate from custom based on long usage. Many familiar substances have so-called common names. After all, could we call $H_2O$ anything but water? Who would use "sodium chloride" to refer to table salt? Baking soda, washing soda, aspirin — these are only a few of the compounds that are rarely called by their chemical names.

Common names have one major disadvantage. They convey a chemical meaning only to someone who already knows the meaning. You know what "water" is chemically, but do you know the chemical composition of *eau,* or *agua,* or *vann?* They are words for "water" in other languages.

Fortunately, most substances used by chemists are named systematically — that is, their names are derived from their composition according to fixed rules. By learning a relatively small number of rules, one can name a great many substances from their chemical formulas and obtain chemical formulas from names.

The study of the rules used to name substances is called **chemical nomenclature.** Learning chemical nomenclature is like learning to spell: You must learn both the rules and the exceptions. As you might expect, historical custom creates most of the exceptions. Usually, it is the more common compounds that have irregular chemical names. Chemists meet periodically to simplify the rules and eliminate as many irregularities as possible, but that work is far from complete.

The aim of chemical nomenclature is to give each compound a name that is unique and distinguishes it from all other compounds. The name must specify the composition of the substance and the number and types of atoms it contains. Since two substances may have the same type and number of atoms and still be different, because the atoms are arranged differently in the molecules, names sometimes must identify the exact arrangement of the atoms.

The rules of nomenclature that we shall study are primarily concerned with specifying the composition of substances. To arrive at the correct name of a compound, you must know more than its formula. You must also have a good understanding of how chemical elements behave, how they are classified, and how they combine to form compounds.

## NAMING THE ELEMENTS

In the past, elements have been named for individuals (curium), nations (francium), states (californium), and cities (berkelium). In the future, however, newly discovered elements will be given less colorful but more systematic names based on the Latin words for the atomic numbers, according to a decision made in 1979 by the International Union of Pure and Applied Chemistry (IUPAC).

The last element to be given a personal name was lawrencium, element 103, which was synthesized at the University of California and was named after the American physicist Ernest O. Lawrence. Element 104, IUPAC has decreed, will be named unnilquadium (un = 1, nil = 0, quad = 4). Element 105 will be named unnilpentium, element 106 will be named unnilhexium and so on. The system already has been put to use for element 107, unnilheptium, and element 109, unnilennium, a few atoms of which have been synthesized in accelerators.

A major advantage of the new system is that names can be assigned swiftly to any element. If element 999 is ever discovered or synthesized, its name is waiting: ennennennium. The disadvantage of the new system is the loss of the personal touch. This disadvantage, and the adoption of the new system, can be attributed to an international dispute about the credit for synthesizing element 104.

In 1964, Soviet scientists said they had synthesized element 104. They named it kurchatovium, after the leader of the Soviet atomic bomb program. Five years later, physicists at the University of California disputed the Soviet results and said that they had been the first to synthesize element 104. They named it rutherfordium, after the celebrated British physicist. Neither name was adopted formally, and IUPAC finally settled the dispute about element 104, and about the names of all heavier elements, by decreeing the new naming method.

### Classifying the Elements

All the elements are classified by an arrangement called the **periodic table** (see inside front cover). The periodic table is the most important tool ever devised for summarizing the behavior of the elements. We shall discuss it in detail in Chapter 6, but a brief introduction will help you understand the rules of chemical nomenclature.

Elements can be divided roughly into three types (Figure 2.4): the metals, the semimetals, and the nonmetals. This is not a rigid, precise

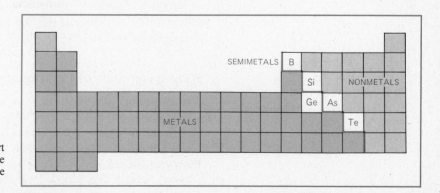

**Figure 2.4**
Distribution of types of elements in the short form of the periodic table. The metals are at the left, separated by a band of semimetals from the nonmetals at the right.

division; sometimes elements are not easily assigned to one group or another. In making the assignment, it is necessary to consider not only chemical behavior but also the physical structure and electrical properties of an element.

The electrical properties of an element are particularly important to this classification. Elements that are good conductors of electricity and whose conductivity decreases as the temperature rises are defined as **metals.** If an element is a poor conductor of electricity, and if its conductivity increases as its temperature rises, it is defined as a **semimetal.** (Semimetals, also called metalloids, include silicon and germanium. They are semiconductors, and modern electronics is based on their characteristics.) Any element that does not conduct electricity is a **nonmetal.**

As for physical state, all metals but one are solids at room temperature. The exception is mercury, which is a liquid at room temperature. The semimetals also are solids at room temperature, although their solid structure differs in important ways from that of the metals. Some nonmetals are gases at room temperature ($H_2$, $N_2$, $O_2$, $F_2$, $Cl_2$). Most of the others are solids. Iodine, for example, is a violet to black solid composed of $I_2$ molecules. Sulfur, which is commonly composed of $S_8$ molecules, is a yellow solid. The most common form of phosphorus is a white solid composed of $P_4$ molecules. The only nonmetal that is a liquid at room temperature is bromine, which is composed of $Br_2$ molecules; its color is an intense red-brown. The nonmetallic elements that occupy the column at the far right of the periodic table are all monatomic gases, called the *noble gases.*

Each of the three groups occupies a specific area of the periodic table. The nonmetals are found toward the top and right of the periodic table (hydrogen is an exception). The nonmetals include the halogens (fluorine, chlorine, bromine, and iodine), oxygen, sulfur, nitrogen, phosphorus, carbon, and the gases occupying the far right column, headed by helium.

The semimetals occupy a diagonal band starting at the upper left of the periodic table. They include silicon, germanium, arsenic, tellurium, and boron. A number of elements that are close to the semimetals in the periodic table have properties that make their classification as metals or nonmetals arbitrary.

When you have identified the nonmetals and semimetals, you will be left with the elements occupying the greater part of the periodic table. They are the metals.

## Writing Formulas of Compounds

The three-part classification of metal, semimetal, and nonmetal is the starting point for the rules that govern the naming of compounds and the way that formulas for compounds are written. You may not know these rules, but you probably have used them already. You would never write the formula for water as $OH_2$; if you happen to use the chemical name for table salt you almost certainly will call it "sodium chloride," not "chloride sodium." These rules work especially well for **binary compounds** (those containing two elements), and they are useful for all compounds. The rules are:

1. In a compound containing a metal and one or more nonmetals, the metal is written and named first. *Example:* NaCl, sodium chloride.
2. In binary compounds containing two nonmetals or semimetals, the nonmetals have the same order from left to right in the formula as they do in the periodic table. The one exception to this rule is oxygen, which always comes last in a formula except when combined with F. *Examples:* $NF_3$, $P_4S_3$, $SiCl_4$, $ClO_2$, $BrO_2$, $I_2O_5$, and $OF_2$.
3. When binary compounds contain nonmetals that are in the same column of the periodic table, the element closest to the bottom of the table is written first. *Examples:* $SO_2$, $SeS_2$, $ClF_3$, $BrCl_5$, SiC.
4. In a binary compound of hydrogen, the symbol for hydrogen is written *first* when hydrogen is combined with elements from columns VIA and VIIA of the periodic table and *last* when hydrogen is combined with elements from columns IA through VA. *Examples:* LiH, $B_2H_6$, $CaH_2$, $NH_3$, $H_2S$, HBr.

### Formulas and Names of Ions

Many substances that you will encounter in chemistry are classified as ionic compounds. An ionic compound can be described as a combination of anions and cations. The names and formulas of ionic compounds are based on the names and formulas of their component ions.

The most common cations are formed from single atoms; we say they are *monatomic.* These cations are formed when a metal atom loses one or more electrons. The cations take the names of the parent metal atoms.

The kind of cation formed by a metal is related to the position of the metal in the periodic table. Generally, the maximum charge of a cation is the same as the column number of its parent metal atom in the periodic table. Metals in column IA form monopositive cations, in which only one electron is lost. Examples are $Li^+$, the lithium ion; $Na^+$, the sodium ion; and $K^+$, the potassium ion. Metals in column IIA form dipositive cations, in which two electrons are lost. Examples are the magnesium ion, $Mg^{2+}$, and the calcium ion, $Ca^{2+}$.

Metals in the middle columns often form more than one cation and do not form cations with the maximum charge. For example, iron, in column VIIIB, forms the $Fe^{2+}$ and $Fe^{3+}$ ions. Tin, in column IVA, forms the $Sn^{2+}$ and $Sn^{4+}$ ions. To name these ions, we place the Roman numeral of the charge after the name of the metal. Thus we have the iron(II) and the iron(III) ions and the tin(II) and tin(IV) ions.

You may sometimes encounter an older method that uses suffixes to name these ions. For a metal that commonly forms only two ions, this

**TABLE 2.5   Stems Used in Naming Compounds of Nonmetals**

| Element | Stem | Element | Stem |
|---------|------|---------|------|
| C | carb | S | sulf |
| N | nitr | F | fluor |
| P | phosph | Cl | chlor |
| As | arsen | Br | brom |
| O | ox | I | iod |

method uses the suffix *-ous* for the ion with the lower charge and the suffix *-ic* for the ion with the higher charge. The suffixes sometimes are added to the name of the metal, as in *cobaltous* for the $Co^{2+}$ ion and *cobaltic* for the $Co^{3+}$ ion. Sometimes the name is derived from the Latin name of the metal, as in *stannous* for $Sn^{2+}$ and *stannic* for $Sn^{4+}$. This method is falling into disuse.

Many elements in columns VA, VIA, and VIIA of the periodic table form monatomic anions. Generally, such an anion has a maximum charge that we can calculate by subtracting 8 from the column number of the parent metal. Thus, nitrogen, in column VA, can form the $N^{3-}$ ion $(5 - 8)$. Oxygen, in column VIA, can form the $O^{2-}$ ion $(6 - 8)$ and fluorine, in column VIIA, the $F^-$ ion $(7 - 8)$.

Monatomic anions are named by addition of the suffix *-ide* to a stem derived from the name of the element: *nitride* for the $N^{3-}$ ion, *oxide* for the $O^{2-}$ ion, and so forth. The most common stems are given in Table 2.5.

Familiarity with the names and formulas of many monatomic cations and anions is essential in chemistry. Table 2.6 lists some monatomic ions.

## Polyatomic Ions

Many of the most common ions have more than one atom. They are called **polyatomic ions.** A polyatomic ion usually acts as a unit in chemical reactions.

There is no convenient system for naming all polyatomic ions, but there are some useful rules you should remember.

**TABLE 2.6   Important Monatomic Ions**

| Formula | Name | Formula | Name |
|---------|------|---------|------|
| $Li^+$ | lithium | $Cu^+$ | copper(I) |
| $Na^+$ | sodium | $Cu^{2+}$ | copper(II) |
| $K^+$ | potassium | $Ag^+$ | silver |
| $Mg^{2+}$ | magnesium | $Zn^{2+}$ | zinc |
| $Ca^{2+}$ | calcium | $Cd^{2+}$ | cadmium |
| $Ba^{2+}$ | barium | $Hg^{2+}$ | mercury(II) |
| $Al^{3+}$ | aluminum | $H^-$ | hydride |
| $Tl^{3+}$ | thallium(III) | $C^{4-}$ | carbide |
| $Sn^{2+}$ | tin(II) | $N^{3-}$ | nitride |
| $Sn^{4+}$ | tin(IV) | $P^{3-}$ | phosphide |
| $Pb^{2+}$ | lead(II) | $As^{3-}$ | arsenide |
| $Bi^{3+}$ | bismuth(III) | $O^{2-}$ | oxide |
| $Cr^{2+}$ | chromium(II) | $S^{2-}$ | sulfide |
| $Cr^{3+}$ | chromium(III) | $F^-$ | fluoride |
| $Mn^{2+}$ | manganese(II) | $Cl^-$ | chloride |
| $Fe^{2+}$ | iron(II) | $Br^-$ | bromide |
| $Fe^{3+}$ | iron(III) | $I^-$ | iodide |
| $Co^{2+}$ | cobalt(II) | | |
| $Co^{3+}$ | cobalt(III) | | |
| $Ni^{2+}$ | nickel(II) | | |

| TABLE 2.7 | Important Polyatomic Groups | | |
|-----------|---------------------|---------|------|
| Formula | Name | Formula | Name |
| $HCO_3^-$ | bicarbonate | $SO_4^{2-}$ | sulfate |
| $CO_3^{2-}$ | carbonate | $SO_3^{2-}$ | sulfite |
| $CN^-$ | cyanide | $ClO_4^-$ | perchlorate |
| $NO_3^-$ | nitrate | $ClO_3^-$ | chlorate |
| $NO_2^-$ | nitrite | $ClO_2^-$ | chlorite |
| $H_2PO_4^-$ | dihydrogen phosphate | $ClO^-$ | hypochlorite |
| $HPO_4^{2-}$ | hydrogen phosphate | $CrO_4^{2-}$ | chromate |
| $PO_4^{3-}$ | phosphate | $Cr_2O_7^{2-}$ | dichromate |
| $OH^-$ | hydroxide | $MnO_4^-$ | permanganate |
| $HSO_4^-$ | hydrogen sulfate | $NH_4^+$ | ammonium |
| | | $Hg_2^{2+}$ | mercury(I) |

As you can see from Table 2.7, most of the important polyatomic ions are anions that contain oxygen. These ions are called **oxyanions.** When an element forms two oxyanions, the ion with the larger number of oxygen atoms takes the suffix *-ate* and the ion with the smaller number of oxygen atoms takes the suffix *-ite.* For example, $NO_3^-$ is called *nitrate* and $NO_2^-$ is called *nitrite.* When an element forms more than two oxyanions, the prefix *per-* (meaning *more*) can be used with the suffix *-ate* and *hypo-* (meaning *less*) with the suffix *-ite,* to distinguish among anions, on the basis of the number of oxygen atoms each contains. The names of the four oxyanions of chlorine listed in Table 2.7 illustrate the use of these prefixes and suffixes.

Two or more anions may differ in the number of hydrogen atoms they contain. These anions can be distinguished by the use of either *hydrogen* or *dihydrogen* as a part of the name. For example, $PO_4^{3-}$ is phosphate; $HPO_4^{2-}$ is hydrogen phosphate; and $H_2PO_4^-$ is dihydrogen phosphate. In an older system, the prefix *bi-* was used to designate the presence of hydrogen, as in $HCO_3^-$, bicarbonate.

The formulas of compounds containing polyatomic ions are written to emphasize the fact that these ions are independent units. If a chemical formula contains more than one polyatomic ion, the formula for the ion is written in parentheses, with the subscript outside the parentheses. The subscript applies to the ion as a unit. Some examples are $Ba(HSO_4)_2$, $(NH_4)_2CO_3$, and $Ca_3(PO_4)_2$.

## Formulas and Names of Ionic Compounds

Many important compounds are classified as ionic. A **salt** is an ionic compound of a metal or polyatomic cation with either a nonmetal or polyatomic anion. Salts are solids at room temperature. An **oxide** is a binary compound of any element with oxygen. Most metal oxides are ionic solids. While many compounds of metals and nonmetals are not composed of ions, it is often useful to think of such compounds as ionic for the purposes of nomenclature.

We mentioned that an ionic compound can be described as a combination of cations and anions. **In an ionic compound, the sum of the charges of the cations must equal the sum of the charges of the anions.** For example, NaCl is composed of $Na^+$ cations and $Cl^-$ anions. Any amount of pure NaCl contains equal numbers of $Na^+$ and $Cl^-$ ions. If the numbers were not equal, pure sodium chloride in bulk would have a net electric charge. No such charge is detectable.

Another salt, $Ca(HSO_4)_2$, contains $Ca^{2+}$ cations and $HSO_4^-$ anions. Any quantity of $Ca(HSO_4)_2$ must contain half as many $Ca^{2+}$ as $HSO_4^-$ ions to be electrically neutral. Again, $NH_4Cl$ consists of equal numbers of $NH_4^+$ cations and $Cl^-$ anions, whereas $(NH_4)_3PO_4$ contains three times as many $NH_4^+$ cations as $PO_4^{3-}$ anions. A complex salt, $KAl(SO_4)_2$ consists of $K^+$, $Al^{3+}$, and $SO_4^{2-}$ in the ratio of $1:1:2$. The total negative charge of $-4$ is balanced by a total positive charge of $+4$.

An ionic compound is named by the use of the names of the component ions in the same order as in the formula. The names of the salts mentioned above are sodium chloride, NaCl; calcium hydrogen sulfate, $Ca(HSO_4)_2$; ammonium chloride, $NH_4Cl$; and potassium aluminum sulfate, commonly called alum, $KAl(SO_4)_2$.

---

**Example 2.2**

Write the formulas and names of (a) a binary compound of magnesium and bromine, (b) a binary compound of aluminum and sulfur, (c) a compound of barium with the $CO_3^{2-}$ ion, and (d) a compound of zinc with the $NO_3^-$ ion. Refer to Tables 2.6 and 2.7 as needed.

**Solution**

We write the formulas by using the ionic charges listed in Tables 2.6 and 2.7. The sum of the positive charges in each compound must equal the sum of the negative charges. We name the compounds by listing their component ions in the same order as in the formulas.

a. The $Mg^{2+}$ ion requires two $Br^-$ ions. The formula is $MgBr_2$ and the name is magnesium bromide.

b. The aluminum ion is $Al^{3+}$ and the sulfur ion is $S^{2-}$. The smallest number divisible by both 2 and 3 is 6. If we divide the numerical value of the charge on each ion into 6, we get its subscript in the formula. The formula is $Al_2S_3$. The name is aluminum sulfide. *Note:* The method works even though the compound is not ionic.

c. Since the charges on the $Ba^{2+}$ and $CO_3^{2-}$ ions have the same numerical value, the formula is $BaCO_3$. The compound is named barium carbonate.

d. The $Zn^{2+}$ ion requires two $NO_3^-$ ions. The formula is $Zn(NO_3)_2$ and the name is zinc nitrate. Note the use of parentheses to emphasize that the nitrate ion is an independent unit.

---

**Example 2.3**

Write the formulas and names of (a) the two binary compounds of copper with oxygen, (b) the two compounds of mercury with the $SO_4^{2-}$ ion.

Solution    a. The two ions of copper are $Cu^+$ and $Cu^{2+}$. The oxygen ion is $O^{2-}$. The two compounds are $Cu_2O$ and $CuO$. The name of each compound must designate the ion of copper it contains. The compounds are named copper(I) oxide and copper(II) oxide, respectively.

b. The two ions of mercury are $Hg_2^{2+}$ and $Hg^{2+}$. The two compounds have the formulas $Hg_2SO_4$ and $HgSO_4$. The first compound is called mercury(I) sulfate, since it has a charge of $+1$ per mercury atom. The second compound is called mercury(II) sulfate, since the mercury ion has a charge of $+2$.

Because solid salts exist in ionic form, there is a possible ambiguity in the interpretation of the formula of a salt. The formula NaCl, for example, might be regarded as representing a single "molecule" of sodium chloride. But solid sodium chloride does not consist of NaCl molecules. It consists of $Na^+$ and $Cl^-$ ions in a lattice that has each $Na^+$ ion surrounded by six $Cl^-$ ions and each $Cl^-$ ion surrounded by six $Na^+$ ions, as shown in Figure 2.5. The formula of a salt does not describe a molecule that is a discrete unit of structure. It defines the smallest whole-number ratio of ions that make up the salt. This does not affect the way we write the formulas of salts, but it does affect our interpretation of those formulas.

**Formulas and Names of Acids**

Classically, an acid is defined as a compound that is a source of $H^+$ ions. This is a useful introduction, although the definition is rather narrow and not quite correct, as you will see in Chapter 15. An acid usually is a compound of hydrogen and a nonmetal or a polyatomic anion.

We can write the formulas of acids by using the same principle used to write the formulas of salts and oxides. The sum of the charges of the anions in the formula must equal the sum of the charges of the cations. In writing the formula, we treat the hydrogens of the acid as $H^+$ ions. Most simply, an acid contains one anion and enough hydrogen ions to cancel the charge on the anion. Some typical acids are HF, $H_2S$, and $H_3PO_4$.

The names of acids are based on the anions they contain. A relatively small group of acids contain monatomic anions. When these acids are found in water solution, they are named by adding the prefix *hydro-* and the suffix *-ic* to the stem of the name of the anion. Examples are HF, hydrofluoric acid, and $H_2S$, hydrosulfuric acid. Most of these acids are gases when pure and are then named as if they were ionic compounds. Thus, HCl in the gaseous state is called hydrogen chloride and $H_2S$ is called hydrogen sulfide.

We name acids containing polyatomic anions whose names end in *-ate* by changing the suffix to *-ic*. Thus, $H_3PO_4$, an acid formed from the phosphate ion, is called phosphoric acid; $H_2CrO_4$, formed from the chromate ion, is called chromic acid; and $HClO_4$, formed from the perchlorate ion, is called perchloric acid. We name acids containing polyatomic anions whose names end in *-ite* by changing the suffix to *-ous*. Thus,

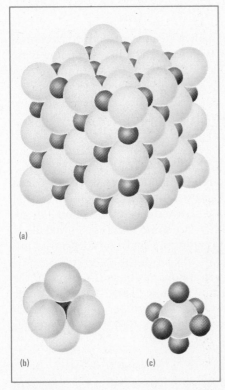

**Figure 2.5**
(a) A sodium chloride crystal consists of a lattice of sodium ions and chloride ions. (b) Each sodium ion is surrounded by six chloride ions. (c) Each chloride ion is surrounded by six sodium ions.

$HNO_2$, formed from the nitrite ion, is called nitrous acid; and $H_2SO_3$, formed from the sulfite ion, is called sulfurous acid.

**Example 2.4**     Name the acids that correspond to the four polyatomic ions of chlorine and oxygen listed in Table 2.7.

**Solution**     Two of the ions have names that end with the suffix *-ate*. The corresponding acids therefore have names that end in *-ic*. One is perchloric acid, $HClO_4$. The other is chloric acid, $HClO_3$. The other two polyatomic ions listed in Table 2.7 have names with the suffix *-ite*. The corresponding acids have names that end in *-ous*. One is chlorous acid, $HClO_2$. The other is hypochlorous acid, $HClO$.

## Formulas and Names of Bases

A useful, if limited, definition describes a base as a compound that is a source of $OH^-$ ions in water solution. By this definition, a compound of a cation and the $OH^-$ (hydroxide) anion is a base. Some bases are NaOH (sodium hydroxide), KOH (potassium hydroxide), and $Ba(OH)_2$ (barium hydroxide).

Compounds that do not contain hydroxide ions are defined as bases if they produce $OH^-$ ions by reaction with water. The best known of these compounds is ammonia, $NH_3$, which reacts with water to produce hydroxide ions by the reaction

$$NH_3 + H_2O \rightleftharpoons NH_4^+ + OH^-$$

## Compounds of Nonmetals

The rules that apply to ionic compounds cannot be used when we deal with many compounds that contain only nonmetals. We cannot predict the formulas of these compounds unless we know a great deal about them. But once the formulas are known, there are some simple rules that allow us to name these compounds. At this point, we shall consider only compounds of two nonmetals.

Binary compounds containing two nonmetals have the name of the first element in the formula followed by the stem (Table 2.5) of the name of the second element with the suffix *-ide*. The number of atoms of each element in the compound is indicated by a prefix. Some of these prefixes are

$$\text{mono} = 1 \qquad \text{tri} = 3 \qquad \text{penta} = 5$$
$$\text{di} = 2 \qquad \text{tetra} = 4 \qquad \text{hexa} = 6$$

(The prefix *mono* is often omitted; no prefix means that the compound contains only one atom of the element.) *Examples:* ICl, iodine chloride; $NO_2$, nitrogen dioxide; $Cl_2O$, dichlorine monoxide; $PCl_3$, phosphorus trichloride; $CF_4$, carbon tetrafluoride; $N_2O_5$, dinitrogen pentoxide; $P_4O_6$, tetraphosphorus hexoxide.

Some compounds of two nonmetals have nonsystematic common names that must be memorized. Important examples are nitrous oxide, $N_2O$, and nitric oxide, NO.

---

**Example 2.5**   Name these compounds of nonmetals: (a) $CCl_4$, (b) $N_2O_3$, (c) $PCl_5$, (d) $XeO_4$.

**Solution**   We can name these compounds by applying the rules listed above.

a. The name is carbon tetrachloride; the prefix *mono* is omitted.

b. The compound is called dinitrogen trioxide, an unambiguous name describing the presence of two nitrogen atoms and three oxygen atoms.

c. The name is phosphorus pentachloride. Again the prefix *mono* is omitted.

d. The name is xenon tetroxide. The final *a* of the prefix *tetra* is dropped when it precedes a stem that begins with a vowel.

---

## 2.4   BALANCED CHEMICAL EQUATIONS

Giving compounds proper formulas and names is only one part of chemical language. We also must describe how substances are converted into other substances by chemical reactions. We do so by writing balanced chemical equations.

A balanced chemical equation is a precise, nonverbal statement of how one or more substances, called **starting materials** or **reactants,** are converted into one or more new substances, called **products.**

Chemical equations are basically similar to the algebraic equations used in mathematics, with chemical formulas replacing the $x$ and $y$ of algebra. The most noticeable difference is that chemical equations use arrows instead of equal signs. The arrow indicates the direction of chemical change. By tradition, the formulas of the reactants appear to the left of the arrow and the formulas of the products to the right.

Ordinary chemical processes obey the **law of conservation of mass:** The total mass of the reactants must equal the total mass of the products. Furthermore, **transmutation of elements, the conversion of one element to another, does not occur in an ordinary chemical process.** The same number and type of atoms must appear on both sides of a chemical equation. All the atoms of all the elements in the reactants must be in the products of the chemical reaction.

### Balancing an Equation

As an example, consider an important reaction that occurs in the pollution control devices of automobiles: the conversion of carbon monoxide to carbon dioxide by reaction with oxygen. This change can be written

$$CO + O_2 \longrightarrow CO_2$$

The formula for oxygen is written as $O_2$ because oxygen occurs as a diatomic molecule in air. As it is written, the expression has three atoms of oxygen on the left and only two on the right. This expression is not balanced. It can be read to say that one molecule of CO reacts with one molecule of $O_2$ to produce one molecule of $CO_2$. This violates the law of conservation of mass. We cannot add new products, since we know the only product of the reaction is $CO_2$. We must balance the number of oxygen and carbon atoms on each side of the expression without changing either the reactants or the product.

We obtain a balanced equation by using **coefficients,** numerals that precede the formulas in the equation. We use the smallest whole-number coefficients that will give the same numbers of each type of atom on each side of the equation. Many equations are balanced by inspection. Others require careful calculation. There is no simple rule covering all equations.

In the unbalanced expression

$$CO + O_2 \longrightarrow CO_2$$

we could start by assuming that the right side is deficient in oxygen atoms, and that perhaps two molecules of $CO_2$ are formed. The expression would then read

$$CO + O_2 \longrightarrow 2CO_2$$

This expression is not balanced. It has four oxygen atoms and two carbon atoms on the right side, while there are three oxygen atoms and one carbon atom on the left. It can be balanced if we assume that two CO molecules take part in the reaction. The equation

$$2CO + O_2 \longrightarrow 2CO_2$$

then reads that two molecules of CO can combine with one molecule of $O_2$ to form two molecules of $CO_2$. This equation is consistent with conservation of mass, since it has the same numbers of each type of atom on each side of the equation. As you will see in Chapter 3, this equation also gives a great deal of information about the mass relationships of the substances in this reaction.

## Balancing in Several Steps

Sometimes several steps are needed to write correct equations, as in the case of the chemical equation for the combustion of propane. Combustion is the rapid reaction of a substance with oxygen. Propane is $C_3H_8$. It combines with $O_2$ to form $CO_2$ and $H_2O$. The reactants and products are

$$C_3H_8 + O_2 \longrightarrow CO_2 + H_2O$$

This expression is not balanced. Coefficients are needed for some or all of the formulas. A useful first step is to select the substance that has the largest number of atoms, in this case $C_3H_8$. Add coefficients to balance the numbers of atoms in this compound on both sides of the expression. We can see from the subscripts in the formulas $C_3H_8$, $CO_2$, and $H_2O$ that $C_3H_8 \rightarrow 3CO_2$ and that $C_3H_8 \rightarrow 4H_2O$. This gives

$$C_3H_8 + O_2 \longrightarrow 3CO_2 + 4H_2O$$

Another step is needed because the number of oxygen atoms is unbalanced; there are 2 oxygen atoms on the left and 10 on the right. The oxygen atoms can be balanced if we multiply the $O_2$ on the left by 5, giving a balanced equation:

$$C_3H_8 + 5O_2 \longrightarrow 3CO_2 + 4H_2O$$

Now consider the equation that describes the thermite reaction, by which aluminum reacts with metal oxides at high temperatures. The thermite reaction is ignited by a magnesium fuse and generates an immense amount of heat. It is used for welding and, in wartime, for incendiary bombs. The products of the reaction are aluminum oxide and the free metal from the metallic oxide, so the thermite reaction can be used industrially to prepare pure metals from oxides. The reaction for the production of manganese by this method is

$$Al + MnO_2 \longrightarrow Al_2O_3 + Mn$$

It is most convenient to start by considering the formula for aluminum oxide. There are two atoms of Al on the right and only one on the left, so we place a coefficient of 2 before the Al on the left. Balancing the oxygen atoms is not as easy. The oxygen-containing molecule on the left, $MnO_2$, has two oxygen atoms per molecule, while the oxygen-containing molecule on the right, $Al_2O_3$, has three oxygen atoms. A balance can be achieved if we use a fractional coefficient, $\frac{3}{2}$, on the left, giving $\frac{3}{2}MnO_2$, to balance the oxygen atoms, and the same fractional coefficient on the right to balance the manganese atoms:

$$2Al + \tfrac{3}{2}MnO_2 \longrightarrow Al_2O_3 + \tfrac{3}{2}Mn$$

Fractional coefficients are used occasionally. It is usually more convenient to multiply both sides of such equations by the denominator of the fraction (in this case, 2), to give whole-number coefficients:

$$4Al + 3MnO_2 \longrightarrow 2Al_2O_3 + 3Mn$$

## Net Chemical Equations

Every chemical equation must be balanced. A chemical equation should also be *net*—that is, it should contain only substances that undergo change. Writing the formulas for substances that do not undergo change clutters the equation without giving information about the reaction. In this respect, the chemical equation is analogous to an algebraic equation, which is also reduced to its simplest form. An equation written $2x + y = 6 + y$ would automatically be changed to $2x = 6$. Similarly, if hydrogen mixed with helium burns to form water, it is not proper to write

$$2H_2 + O_2 + He \longrightarrow 2H_2O + He$$

The correct equation is

$$2H_2 + O_2 \longrightarrow 2H_2O$$

**Example 2.6**    Convert the following expressions of chemical change into chemical equations:

a. $NH_4NO_3 \rightarrow N_2O + H_2O$,
b. $BF_3 + H_2O \rightarrow B(OH)_3 + HF$,
c. $P_4O_{10} + H_2O \rightarrow H_3PO_4$,
d. $NH_3 + O_2 \rightarrow N_2 + H_2O$.

**Solution**    a. First balance the H atoms:

$$NH_4NO_3 \longrightarrow N_2O + 2H_2O$$

This step is sufficient since the N atoms and the O atoms are also balanced
b. First balance the F atoms:

$$BF_3 + H_2O \longrightarrow B(OH)_3 + 3HF$$

Next balance the O atoms:

$$BF_3 + 3H_2O \longrightarrow B(OH)_3 + 3HF$$

Since the H and B atoms now are also balanced, this is the balanced equation.
c. First balance the P atoms:

$$P_4O_{10} + H_2O \longrightarrow 4H_3PO_4$$

Next balance the O atoms:

$$P_4O_{10} + 6H_2O \longrightarrow 4H_3PO_4$$

Since the H atoms are also balanced, this is the desired equation.

d. First balance the N atoms:

$$2NH_3 + O_2 \longrightarrow N_2 + H_2O$$

Next balance the H atoms

$$2NH_3 + O_2 \longrightarrow N_2 + 3H_2O$$

Finally balance the O atoms. Since there are two O atoms on the left and three on the right, we can use $\frac{3}{2}$ as the coefficient on the left and then multiply all the formulas by 2:

$$2NH_3 + \tfrac{3}{2}O_2 \longrightarrow N_2 + 3H_2O$$
$$4NH_3 + 3O_2 \longrightarrow 2N_2 + 6H_2O$$

You will find it easier to work with equations that have whole-number coefficients.

## 2.5   IMPORTANT TYPES OF CHEMICAL REACTIONS

The apparently bewildering variety of chemical reactions can be simplified if we classify reactions into specific categories. Many reactions can be placed into one of the following classes:

**1.** Addition or combination reactions: Two substances combine to form one:

$$2Na + Cl_2 \longrightarrow 2NaCl$$
$$P_4 + 3O_2 \longrightarrow P_4O_6$$
$$H_2O + C_2H_4 \longrightarrow C_2H_5OH$$
$$NO_2 + NO_2 \longrightarrow N_2O_4$$

**2.** Decomposition reactions: One compound breaks into two or more compounds or elements. These reactions are the reverse of addition or combination reactions.

$$CaCO_3 \longrightarrow CaO + CO_2$$
$$N_2O_4 \longrightarrow NO_2 + NO_2$$
$$C_2H_5OH \longrightarrow H_2O + C_2H_4$$

**3.** Displacement reactions: Substances exchange parts. There are many types of displacement reactions. One of the most important, often called **metathesis,** is the exchange of ions by two ionic compounds, with the anion of one compound joining the cation of the other compound and vice versa. A generalized formula for this exchange is:

$$AB + CD \longrightarrow AD + CB$$

Displacement reactions are especially important for compounds in aqueous solution. We shall discuss them in more detail in later chapters. One such reaction can be written as:

$$AgNO_3 + NaI \longrightarrow AgI + NaNO_3$$

Many other reactions resemble metathesis, although they do not necessarily involve ionic substances. Some of them are:

**a.** Hydrolysis, the reaction of a substance or ion with water. If we write the formula for water as HOH, to emphasize that water is a source of $H^+$ and $OH^-$ ions, we can predict the products of a hydrolysis reaction. The $OH^-$ anion combines with the positive portion of the compound that is hydrolyzed. This positive portion may be a cation or the atom that is written first in the formula of the compound. The $H^+$ cation combines with the negative portion of the compound, which may be an anion or the atom that is written last in the formula.

$$AlCl_3 + 3HOH \longrightarrow Al(OH)_3 + 3HCl$$
$$LiH + HOH \longrightarrow LiOH + H_2$$
$$PBr_3 + 3HOH \longrightarrow P(OH)_3 + 3HBr$$

**b.** Acid-base reactions. Generations of chemists have learned the catchword: Acid plus base gives water plus salt. This is true of bases that contain hydroxide. The reaction can be regarded as a simple switch of partners between compounds, as in these examples:

$$HCN + NaOH \longrightarrow NaCN + H_2O$$
$$2HF + Ca(OH)_2 \longrightarrow CaF_2 + 2H_2O$$

In Chapter 15, where more general definitions of acids and bases are given, you will see that acid-base reactions are not necessarily metatheses.

## 2.6 STATES OF SUBSTANCES

Until now, we have written chemical equations that convey information only about the relative quantities of reactants and products. We can write formulas that convey more information, especially about the physical states of the reactants and products.

Information about the state of a reactant is given by a letter in parentheses after the formula. The symbol (l) indicates that a substance is a liquid, as in $Br_2(l)$, $H_2SO_4(l)$, $C_2H_5OH(l)$. The symbol (s) indicates that a substance is a solid, as in NaCl(s), Ba(OH)$_2$(s), AgI(s). The symbol (g) indicates that a substance is a gas, as in H$_2$(g), NH$_3$(g), HCl(g).

For some reactions, we need information about the state of reactants and products at given conditions to write state symbols. For example, at 500 K, carbon monoxide and water form carbon dioxide and hydrogen:

$$CO(g) + H_2O(g) \longrightarrow CO_2(g) + H_2(g)$$

In this expression, H$_2$O is followed by (g) because water boils at 373 K, which means that it is a gas at 500 K. If the reaction occurred at 350 K, the equation would have H$_2$O(l) instead.

Reactions often occur at room temperature, which is assumed to be 298 K or 25°C. If the conditions are not specified, we can assume that the reaction occurs at or near room temperature. We must know the state of a substance at room temperature to assign the correct state symbol for such a reaction.

Most substances are solids at room temperature. All metals except mercury, all semimetals, and some nonmetals such as iodine, phosphorus, sulfur, and carbon are solids at room temperature. So are all salts and most other compounds that contain metals, most organic compounds, and many compounds of semimetals and nonmetals. A relatively small number of common substances are gases at room temperature. Some are listed in Table 2.8.

Some organic compounds are liquids at room temperature. Among them are carbon tetrachloride, CCl$_4$; chloroform, CHCl$_3$; and benzene, C$_6$H$_6$. Bromine, Br$_2$, is also a liquid at room temperature.

By using state symbols, we can even write chemical equations for changes in state. One change in state is the change from liquid to gas that occurs when water boils:

$$H_2O(l) \longrightarrow H_2O(g)$$

**TABLE 2.8**  Common Gases at Room Temperature

| Name | Formula | Name | Formula |
|---|---|---|---|
| hydrogen | H$_2$ | carbon monoxide | CO |
| nitrogen | N$_2$ | carbon dioxide | CO$_2$ |
| oxygen | O$_2$ | sulfur dioxide | SO$_2$ |
| fluorine | F$_2$ | nitrous oxide | N$_2$O |
| chlorine | Cl$_2$ | nitric oxide | NO |
| hydrogen chloride | HCl | nitrogen dioxide | NO$_2$ |
| hydrogen bromide | HBr | methane | CH$_4$ |
| hydrogen iodide | HI | ethane | C$_2$H$_6$ |
| hydrogen sulfide | H$_2$S | propane | C$_3$H$_8$ |
| ammonia | NH$_3$ | butane | C$_4$H$_{10}$ |
| elements from column VIIIA | He, Ne, Ar, Kr, and Xe | | |

## Substances in Aqueous Solution

State symbols can be used to indicate whether a substance is in aqueous solution. Any compound or ion whose formula is followed by the state symbol (aq) is dissolved in water. The equation

$$C_2H_5OH(l) \longrightarrow C_2H_5OH(aq)$$

describes alcohol dissolving in water. In the same way, the equation

$$C_6H_{12}O_6(aq) \longrightarrow C_6H_{12}O_6(s)$$

describes the crystallization of glucose from water.

When ionic substances dissolve in water, they almost always separate into their component ions. We call this change *dissociation*. Substances that undergo dissociation when they dissolve in water are called electrolytes. Equations for reactions in aqueous solution must indicate that such substances dissociate into ions. For example, NaCl dissociates almost completely in aqueous solution. Therefore, we write it as $Na^+(aq) +$ $Cl^-(aq)$, not as NaCl(aq), even though an extremely small amount of undissociated NaCl is part of the solution. By convention, we write the formula of the predominant form.

Most salts dissociate almost completely into ions in aqueous solution. Therefore, we write the formulas for the separate anions and cations, each followed by the state symbol (aq), for a salt in aqueous solution. Thus the formation of a solution of silver nitrate, a salt, is described by

$$AgNO_3(s) \longrightarrow Ag^+(aq) + NO_3^-(aq)$$

The precipitation of calcium fluoride, another salt, from solution is described by

$$Ca^{2+}(aq) + 2F^-(aq) \longrightarrow CaF_2(s)$$

A number of other types of substances also dissociate into ions in aqueous solution. You may recall our working definition of an acid as a substance that is a source of $H^+$ ions. When an acid dissolves in water, $H^+$ ions form in solution. Most acids produce only a relatively small amount of $H^+$ in aqueous solution. The formulas for aqueous solutions of these *weak acids* do not reflect this limited formation of ions. An example is hydrocyanic acid, a weak acid. The formula for hydrocyanic acid in aqueous solution is written HCN(aq). Out of solution at room temperature the formula is written HCN(l).

A relatively small group of acids dissociate virtually completely when they are dissolved in water. They are called *strong acids.* Table 2.9 lists the common strong acids. At this point, you should assume that any acid not listed in Table 2.9 is a weak acid.

| TABLE 2.9 Common Strong Acids | | |
|---|---|---|
| **Formula** | **Name** | **Ions in Water** |
| $HNO_3$ | nitric acid | $H^+(aq) + NO_3^-(aq)$ |
| $H_2SO_4$ | sulfuric acid | $H^+(aq) + HSO_4^-(aq)$ |
| $HClO_4$ | perchloric acid | $H^+(aq) + ClO_4^-(aq)$ |
| HCl | hydrochloric acid | $H^+(aq) + Cl^-(aq)$ |
| HBr | hydrobromic acid | $H^+(aq) + Br^-(aq)$ |
| HI | hydriodic acid | $H^+(aq) + I^-(aq)$ |

You can readily see the difference by comparing equations for the formation of solutions of strong and weak acids:

$$HCl(g) \longrightarrow H^+(aq) + Cl^-(aq) \qquad \text{strong acid}$$
$$HF(g) \longrightarrow HF(aq) \qquad \text{weak acid}$$

Another group of substances that undergo dissociation in aqueous solution is the *strong bases*. You may recall our working definition of a base as a compound that produces $OH^-$ ions in solution. Compounds of the hydroxide ion and the group IA and IIA metals (with the exception of Be) are strong bases. We write their formulas to show dissociation into ions in aqueous solution. Thus, the formation of a solution of barium hydroxide is written

$$Ba(OH)_2(s) \longrightarrow Ba^{2+}(aq) + 2OH^-(aq)$$

*Weak bases,* such as $NH_3$, produce only a relatively limited amount of $OH^-$ in solution. We do not show dissociation into ions in aqueous solution when we write their formulas. For example, the equation for the formation of an aqueous solution of $NH_3$ is written

$$NH_3(g) \longrightarrow NH_3(aq)$$

## 2.7 NET IONIC EQUATIONS

When writing a net ionic equation, you must write the formula for each substance to indicate:

1. Whether the substance is in solution. We use the state symbols (l), (s), or (g) if the substance is a liquid, solid, or gas, and the state symbol (aq) if it is in solution.
2. If the substance exists as ions in solution. We write the formula for the substance in solution either as an intact molecule or as ions.

One other detail is important. A correct net ionic equation must be *net* — that is, *it must include only substances that take part in the reaction.* The need to write net equations was mentioned earlier (Section 2.4), but it is not always easy to write a net equation.

Consider the metathesis reaction in which a solution of lead(II) nitrate is mixed with a solution of lithium iodide to yield a precipitate of lead(II) iodide and a solution of lithium nitrate. To write the net ionic equation for this process, we start by writing the formulas for the reactants and products in their proper states. A solution of lead(II) nitrate contains $Pb^{2+}$ and $NO_3^-$ ions. A solution of lithium iodide contains $Li^+$ and $I^-$ ions. A solution of lithium nitrate contains $Li^+$ and $NO_3^-$ ions. A precipitate of lead(II) iodide is $PbI_2(s)$. We can write

$$Pb^{2+}(aq) + 2NO_3^-(aq) + Li^+(aq) + I^-(aq) \longrightarrow$$
$$PbI_2(s) + Li^+(aq) + 2NO_3^-(aq)$$

This expression is easily balanced:

$$Pb^{2+}(aq) + 2NO_3^-(aq) + 2Li^+(aq) + 2I^-(aq) \longrightarrow$$
$$PbI_2(s) + 2Li^+(aq) + 2NO_3^-(aq)$$

This equation is not *net*. As you can see, $Li^+(aq)$ and $NO_3^-(aq)$ appear on both sides of the equation. Nothing has happened to them in the reaction. Therefore, they should be canceled from both sides of the equation. The correct net equation is

$$Pb^{2+}(aq) + 2I^-(aq) \longrightarrow PbI_2(s)$$

You might protest that this net equation omits a central feature of the reaction, the presence of lithium iodide. It is true that the final equation makes it impossible to tell whether sodium iodide, potassium iodide, or yet another iodide took part in the reaction. We can answer this objection by saying that, for the purposes we have in mind, most iodides are created equal. At the level of approximation we shall use, it makes no difference which iodide was the source of the $I^-$ ions — or, indeed, which lead(II) salt was the source of the $Pb^{2+}$ ions. The feature of interest of this reaction, the formation of a precipitate of lead iodide, is the same for all these solutions of salts, at least until one is working on a more sophisticated level.

This commonsense approach simplifies the discussion of many chemical reactions. Take the reaction in which hydrochloric acid is mixed with a solution of sodium hydroxide. If all the substances are included, the equation is

$$H^+(aq) + Cl^-(aq) + Na^+(aq) + OH^-(aq) \longrightarrow$$
$$H_2O + Na^+(aq) + Cl^-(aq)$$

but the net equation is

$$H^+(aq) + OH^-(aq) \longrightarrow H_2O$$

Or take the case in which a solution of nitric acid is mixed with a solution of barium hydroxide, forming water and barium nitrate as the products. If all the substances are included, the expression reads

$$2H^+(aq) + NO_3^-(aq) + Ba^{2+}(aq) + 2OH^-(aq) \longrightarrow$$
$$2H_2O + Ba^{2+}(aq) + NO_3^-(aq)$$

but the net equation is

$$H^+(aq) + OH^-(aq) \longrightarrow H_2O$$

Indeed, this is the net equation for the reaction of any strong acid with any strong base.

The method of writing equations described in this chapter makes it possible to convey a great deal of information in a small amount of space. Consider these four net ionic reactions involving the same two substances:

1. $\qquad NH_3(g) + HCN(l) \longrightarrow NH_4CN(s)$
2. $\qquad NH_3(g) + HCN(aq) \longrightarrow NH_4^+(aq) + CN^-(aq)$
3. $\qquad NH_3(aq) + HCN(l) \longrightarrow NH_4^+(aq) + CN^-(aq)$
4. $\quad NH_3(aq) + HCN(aq) \longrightarrow NH_4^+(aq) + CN^-(aq)$

Each equation describes a different form of the reaction of ammonia with hydrogen cyanide to form the salt ammonium cyanide. In Equation 1, the reaction occurs in the absence of water, and each substance has a state symbol that indicates its physical state. The following three equations describe different reactions in aqueous solutions. In Equation 2, ammonia gas is added to a solution of HCN, a weak acid, and the salt forms as dissociated ions in solution. In Equation 3, liquid HCN is added to a solution of ammonia, a weak base; and in Equation 4, both reactants are in solution. Many words are needed to convey differences that are described by small changes in the notation.

## 2.8   ENERGY CHANGES IN CHEMICAL PROCESSES

We do not know what energy really is, but we can define the energy of a body. It is the body's stored-up capacity for doing work. We also can state, as a fundamental law of nature, the principle of **conservation of energy: Energy can be changed from one form to another, but it can never be lost.**

There are many forms of energy. Among them are kinetic energy, gravitational energy, heat energy, chemical energy, potential energy, electrical energy, radiant energy, and nuclear energy. We can describe some of their characteristics.

**Potential energy** is the energy that a body has because of its position. A boulder at the top of a hill has **gravitational potential energy** because of its position in the earth's gravitational field. Charged bodies such as electrons have **electrical potential energy** because of their location with respect to other charged bodies or in electrical fields.

**Kinetic energy** (KE) is energy of motion. It is the work that a moving body can perform as it is brought to rest. The formula for calculating kinetic energy is

$$KE = \tfrac{1}{2}mv^2$$

where $m$ is the mass of the moving body and $v$ is its velocity.

**Chemical energy** is the energy associated with the arrangement of atoms into molecules or ions. Chemical energy includes the kinetic energy of the electrons moving within atoms. It also includes the electrical potential energy that results from the attraction between electrons and protons in atoms and from the attraction of atoms or ions for each other.

Another kind of energy is emitted from the surfaces of all bodies. This process is called radiation and the energy emitted is **radiant energy.**

Radiant energy exists as electromagnetic waves that travel with the velocity of light. Electromagnetic waves can be described as oscillating electric and magnetic fields. The radiant energy may be absorbed when the waves hit the surface of another body. Light is the most familiar form of radiant energy. There are many other forms, including X rays, infrared rays, and radio waves.

## Thermal Energy

**Thermal energy** or heat energy is simply the internal kinetic energy of a body. When we say that the thermal energy of a body increases, we mean that there is increased motion of its units of structure, its atoms or molecules or ions. Most forms of energy tend to change into thermal energy. In fact, any other kind of energy can be converted completely into thermal energy.

**Figure 2.6**
Energy is needed to move the boulder to a position of higher gravitational potential at the top of the hill. The potential energy is converted to kinetic energy (and some heat energy from friction) as the boulder rolls down the hill.

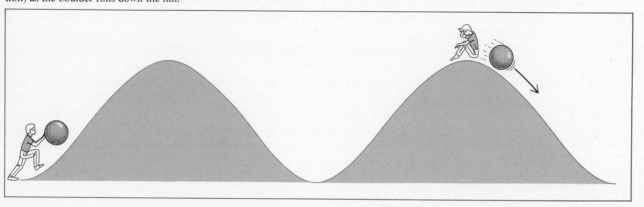

The original units of energy were units of heat. One such unit is the calorie (cal), which is equivalent to the quantity of heat needed to raise the temperature of 1 g of water from 14.5°C to 15.5°C. The calorie is now defined in terms of its relationship to the SI unit for energy, the joule (J), which is also the unit for both heat and work.

$$1 \text{ cal} = 4.184 \text{ J}$$

The joule is defined as 1 kg m$^2$/s$^2$ in SI base units. We shall learn more about energy relationships in Chapter 13.

### Energy Changes

Although the principle of conservation of energy states that energy is never created or destroyed, there are energy changes in chemical and physical processes. By "energy changes" we mean changes in either the form or the location of energy.

Energy often changes form (Figure 2.6). As something falls from a height, its gravitational potential energy changes to kinetic energy. When a sample of natural gas burns, some of the chemical energy of the gas and of the oxygen in the air changes to heat energy. When an electric current passes through a resistance such as wire, electrical energy changes to heat energy as the wire gets hotter.

When we talk about changes in the location of energy, we mean an energy transfer between substances undergoing physical or chemical change and their surroundings. There are two kinds of changes, those that increase the potential energy of the substances undergoing change and those that decrease their potential energy. *A physical or chemical change at room temperature that decreases the potential energy of substances undergoing the change usually is favorable.*[2] In general, the lower the potential energy, the more stable the substance.

For example, a weight that is released from a height falls to the surface by itself, with a decrease in its gravitational potential energy, but the weight will not fly back up by itself. To increase its gravitational potential energy, someone or something must lift the weight.

Energy usually is transferred from the changing substances to the surroundings mostly or wholly in the form of heat. There are many everyday examples. When gasoline burns, there is a spectacular energy transfer called **combustion.** Heat and light are liberated because the potential energy of the burning substances, gasoline and oxygen, is much greater than that of the products, $CO_2$ and $H_2O$.

On the atomic level, molecules of $H_2$ are formed when hydrogen atoms come together because the potential energy of the molecules is lower than that of the separate atoms. The energy lost when $2H \rightarrow H_2$ is transferred

[2] Another factor that determines whether a change is favorable is discussed in Chapter 13.

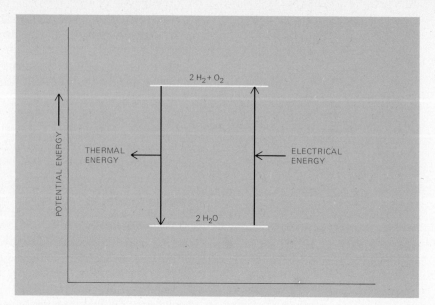

**Figure 2.7**

When $H_2$ and $O_2$ combine to form $H_2O$, at room temperature, the potential energy of the system decreases. The change occurs spontaneously and releases thermal energy to the surroundings. The reverse change, from $H_2O$ to $H_2$ and $O_2$, increases the potential energy of the system. It can take place only if energy is transferred to the system from the surroundings. One way to bring about this change is by a process called electrolysis in which electrical energy in the form of an applied electric current is provided to the $H_2O$.

to the surroundings in the form of heat. In general the formation of chemical bonds lowers the chemical energy of the system. Molecules are more stable than separated atoms.

Some chemical or physical changes cause an increase in the potential energy of the changing substances. These changes usually require the transfer of energy from the surroundings. Figure 2.7 shows a chemical change that can take place in either direction, depending on conditions.

The natural tendency of substances to proceed toward minimum potential energy is important in chemistry. As you will see in later chapters, an understanding of potential energy helps us to develop ideas about the structures of atoms and molecules and about chemical behavior.

**Summary**

In this chapter, we first defined the **chemical elements** as basic ingredients of matter. Then we defined the **atom** as the basic unit of an element. We described how atoms combine to form **molecules,** and discussed the structure of atoms briefly. Then we discussed how **chemical symbols** are used to represent atoms and molecules, how **subscripts** are used to specify the number of atoms in a molecule, and how **coefficients** are used to specify the number of molecules taking part in a chemical reaction. We introduced the **law of constant composition,** which states that all samples of a given substance contain the same elements in the same proportions, and the **law of multiple proportions,** which states that when two elements can form more than one compound, the different masses of one element that combine with a fixed mass of the other are in a ratio of small whole numbers. We described how elements are classified as **metals, semimetals,** or **nonmetals.** Then we outlined the rules for writing the **formulas** and **names** of compounds and ions. We then showed how to write and **balance** the **chemical equations** that describe reactions. We said that equations should include descriptions of the physical states of all the substances in the reaction. Then we showed how to write **net ionic equations.** Finally, we discussed the **energy changes** that occur in chemical reactions, describing the different forms of energy and their characteristics.

## Exercises

**2.1** Use the periodic table to find the number of protons and electrons in an atom of (a) tin, (b) argon, (c) potassium, (d) copper.

**2.2** Use the periodic table to find the number of protons and electrons in each of the following species: (a) Sb, (b) $Pb^{2+}$, (c) $O^{2-}$, (d) $Al^{3+}$.

**2.3** Give the number of independent particles that make up each of the following: (a) $NO_2 + N_2O_4$, (b) $2N + N_2$, (c) $C_6H_{12}O_6 + H_2$.

**2.4** The elemental symbols of (a) potassium, (b) tin, (c) copper are based on their Latin names. Propose two-letter elemental symbols based on their English names that do not duplicate existing symbols.

**2.5**[3] A sample of HCN has a mass of 3.92 g and is found to contain 0.145 g

---

[3] The answers to exercises whose numbers are in color can be found in Appendix VII.

of H and 2.03 g of N. Find the mass of C in the sample.

**2.6** A sample of 9.76 g of $H_2O$ is found to contain 1.09 g of H. Find the mass of H in a 4.88-g sample of $H_2O$.

**2.7** A sample of 9.76 g of $H_2O$ is found to contain 1.09 g of H and 8.67 g of O. Find the composition of a 100-g sample of $H_2O$.

**2.8** A sample of 5.89 g of $NO_2$ is found to contain 4.10 g of O. Find the mass of O in a 5.89-g sample of $N_2O_4$.

**2.9** A sample of 378 mg of NaF is found to contain 171 mg of F. Another sample of NaF is found to contain 286 mg of F. Find the mass of this sample.

**2.10** A sample of 5.55 g of MgO is found to contain 3.35 g of Mg. Another sample is found to contain 1.27 g of Mg. Find the mass of O in this sample.

**2.11** Analysis of a 1.34-g sample of

FeS, an ore of iron, indicates that it contains 0.85 g of iron. Find the mass of iron in 2000 kg of FeS.

**2.12** A sample of 4.22 g of $KNO_3$ is found to contain 1.63 g of K and 0.59 g of N. Find the mass of O in a 5.49-g sample of $KNO_3$.

**2.13** Carbon and hydrogen form many binary compounds. The compounds $C_2H_4$, $C_4H_8$, and $C_5H_{10}$ do not demonstrate the law of multiple proportions. Explain why not.

**2.14** The compounds $N_2O_4$ and $N_2O_6$ are two oxides of nitrogen that were not included in Table 2.4. Find the relative mass of oxygen that should be entered in the last column of the table for each of these two compounds.

**2.15** We wish to include the $NO_2^-$ ion and the $NO_3^-$ ion in Table 2.4. If the mass of nitrogen in the first column is 28, what is the mass of oxygen that should be entered in the second column for each ion?

**2.16**   In the compound $SO_2$, the ratio of the mass of S to the mass of O is $1.00:1.00$. In another binary compound of S and O, the ratio by mass of S to O is found to be $4.00:1.00$. Find the simplest formula of this compound.

**2.17**   The ratio of O to C by mass in CO is $1.33:1.00$. Find the ratio for the compound $C_2O_3$.

**2.18**   Classify the following elements as metals, nonmetals, or semimetals: (a) hydrogen, (b) thallium, (c) arsenic, (d) bismuth, (e) radon.

**2.19**   Arrange the following elements in order of increasing electrical conductivity: (a) antimony, (b) potassium, (c) selenium, (d) neon, (f) germanium.

**2.20**   Write the formula of a binary compound that each of the following elements forms with fluorine: (a) rubidium, (b) strontium, (c) boron, (d) silicon, (e) nitrogen, (f) oxygen, (g) chlorine.

**2.21**   Write the formula of the binary compound that each of the following elements forms with magnesium: (a) iodine, (b) selenium, (c) phosphorus, (d) carbon.

**2.22**   Which of the following formulas are written incorrectly? (a) $ClO_2$, (b) $S_2Te$, (c) ClBr, (d) $N_4P_4$, (e) $IF_7$.

**2.23**   Write the formula of the binary compound of each element with hydrogen: (a) arsenic, (b) silicon, (c) bromine, (d) aluminum, (e) beryllium, (f) lithium.

**2.24**   Write the formulas of two binary compounds of each element with oxygen: (a) tin, (b) iron, (c) copper.

**2.25**   Name the following ions: (a) $Al^{3+}$, (b) $Pb^{4+}$, (c) $P^{3-}$, (d) $Mn^{7+}$, (e) $As^{3-}$.

**2.26**   Predict the maximum negative charge found on anions of the follow-

ing elements: (a) phosphorus, (b) silicon, (c) iodine, (d) tellurium.

**2.27**   Predict the maximum positive charge found on cations of the following elements: (a) calcium, (b) gallium, (c) lead, (d) polonium.

**2.28**   Write the formulas of the polyatomic ions that are formed by the successive loss of $H^+$ from $H_4SiO_4$.

**2.29**   Which of the following formulas are incorrect: (a) $CaI_2$, (b) LiO, (c) $Al_2S_3$, (d) $MgN_3$, (e) $Rb_2Cl$.

**2.30**   Which of the following formulas are incorrect? (a) $K_2HSO_4$, (b) $CaCr_2O_7$, (c) $Hg_2Cl$, (d) $(HN_4)_2(PO_4)_3$, (e) $K_3Al(SO_4)_3$.

**2.31**   Find the charge on the polyatomic anions in the following compounds: (a) $Ca(H_2PO_3)_2$, (b) $Al_2(O_2)_3$, (c) $Na_2S_2O_3$.

**2.32**   Find the charge on the cation in each of the following compounds, by assuming (incorrectly) that they are ionic: (a) $OsO_4$, (b) $Mn_2O_7$, (c) $Cd_3N_2$, (d) $Al_2Cl_6$.

**2.33**   Write the formulas of the following compounds: (a) sodium sulfide, (b) calcium phosphide, (c) aluminum fluoride, (d) strontium oxide.

**2.34**   Write the formulas of the following compounds: (a) copper(I) oxide, (b) manganese(IV) oxide, (c) vanadium(V) oxide, (d) chromium(VI) oxide.

**2.35**   Write the formulas of the following compounds: (a) potassium hydrogen sulfate, (b) barium phosphate, (c) lithium sulfite, (d) ammonium dichromate.

**2.36**   Write the formulas of the following compounds: (a) iron(III) sulfate, (b) copper(II) phosphate, (c) titanium(IV) perchlorate, (d) gold(III) hydrogen phosphate.

**2.37**   Name   the   following   com-

pounds: (a) RbI, (b) SrS, (c) $Li_2O$, (d) AlAs, (e) $BaF_2$.

**2.38**   Name the following compounds: (a) $KMnO_4$, (b) $Li_2CrO_4$, (c) $(NH_4)_2S$, (d) $Ca(H_2PO_4)_2$, (e) $Mg(HCO_3)_2$.

**2.39**   Name the following compounds. (*Note:* Each metal can form more than one ion.) (a) $TiCl_4$, (b) $PbO_2$, (c) $Tl_2S$, (d) $HgBr_2$, (e) $WF_6$.

**2.40**   Name the following compounds. (*Note:* Each metal can form more than one ion.) (a) $Pt(CN)_4$, (b) $Hg_2SO_4$, (c) $Sn(NO_2)_2$, (d) $Cu(ClO_3)_2$.

**2.41**   Name the following acids in aqueous solution: (a) HI, (b) $H_2Se$, (c) $H_2O$, (d) HCN.

**2.42**   Name the following acids, which are derived from polyatomic anions: (a) $HMnO_4$, (b) $H_3PO_4$, (c) $H_2CrO_4$, (d) $H_2SO_3$.

**2.43**   Name the following acids, which are derived from polyatomic anions of iodine: (a) HIO, (b) $HIO_2$, (c) $HIO_3$, (d) $HIO_4$.

**2.44**   Write the formulas of the following acids: (a) nitrous acid, (b) hydrotelluric acid, (c) phosphorous acid, (d) hypophosphorous acid.

**2.45**   Name the following bases: (a) $Cr(OH)_3$, (b) CsOH, (c) $Co(OH)_2$, (d) $Fe(OH)_3$.

**2.46**   Write the formulas of the following bases: (a) ruthenium(III) hydroxide, (b) lithium hydroxide, (c) nickel(II) hydroxide, (d) aluminum hydroxide.

**2.47**   Name the following compounds of nonmetals (a) $N_2O_5$, (b) $P_4O_6$, (c) $ClF_3$, (d) $S_4N_4$.

**2.48**   Write the formulas of the following compounds of nonmetals: (a) phosphorus trichloride, (b) carbon tetrabromide, (c) dioxygen difluoride,

(d) tetraphosphorus hexoxide.

**2.49**  Iodine and fluorine form binary compounds in which one iodine atom combines with one, three, five, or seven fluoride atoms. Name these four compounds.

**2.50**  Balance the following expressions of chemical change:
(a) $C + CO_2 \rightarrow CO$
(b) $N_2 + Cl_2 \rightarrow NCl_3$
(c) $H_2 + N_2 \rightarrow NH_3$
(c) $Al + O_2 \rightarrow Al_2O_3$

**2.51**  Balance the following expressions of chemical change:
(a) $P_4 + Cl_2 \rightarrow PCl_5$
(b) $S_8 + O_2 \rightarrow SO_3$
(c) $NH_3 + O_2 \rightarrow NO + H_2O$
(d) $IF_5 + H_2O \rightarrow HF + HIO_3$

**2.52**  Balance the following expressions for decomposition:
(a) $N_2O_5 \rightarrow N_2O_4 + O_2$
(b) $AsCl_5 \rightarrow As_4 + Cl_2$
(c) $NI_3 \rightarrow N_2 + I_2$
(d) $HBrO_3 \rightarrow Br_2O_5 + H_2O$

**2.53**  Write balanced chemical equations for the combustion (the rapid reaction with $O_2$) of each of the following substances, assuming that $H_2O$ and $CO_2$ are the only products: (a) $CH_4$, (b) $C_3H_8$, (c) $C_2H_2$, (d) $C_8H_{18}$.

**2.54**  Write balanced chemical equations for the following processes:
(a) Ammonium nitrate decomposes to nitrogen, oxygen, and water.
(b) The incomplete combustion of $CH_4$ forms water and carbon monoxide.
(c) Aluminum and manganese(IV) oxide form manganese and aluminum oxide.
(d) Ammonia and oxygen form nitric oxide and water.

**2.55**  Fill in the missing substances to complete the balanced equations:
(a) $MgCO_3 \rightarrow CO_2 +$ _____
(b) $NH_4NO_3 \rightarrow 2H_2O +$ _____
(c) $P_4O_{10} + 6H_2O \rightarrow 4$_____

(d) $4KClO_3 \rightarrow KCl + 3$_____

**2.56**  Fill in the missing substances to complete the balanced equations:
(a) $2NaClO_3 + SO_2 + H_2SO_4 \rightarrow 2ClO_2 + 2$_____
(b) $5CO +$ _____ $\rightarrow I_2 + 5CO_2$
(c) $5$_____ $\rightarrow 4ClO_2 + HCl + 2H_2O$
(d) $4H_3PO_3 \rightarrow 3H_3PO_4 +$ _____

**2.57**  Classify each of the following reactions as an addition reaction, a decomposition reaction, or a displacement reaction:
(a) $BF_3 + 3H_2O \rightarrow H_3BO_3 + 3HF$
(b) $Cl_2 + C_2H_4 \rightarrow C_2H_4Cl_2$
(c) $C_6H_6 + Br_2 \rightarrow C_6H_5Br + HBr$
(d) $2KClO_3 \rightarrow 2KCl + 3O_2$

**2.58**  Classify each of the following reactions involving water as an addition, decomposition, or displacement reaction:
(a) $C_2H_4 + H_2O \rightarrow C_2H_5OH$
(b) $NH_3 + H_2O \rightarrow NH_4^+ + OH^-$
(c) $2H_2O \rightarrow 2H_2 + O_2$
(d) $Na_2O + 2H_2O \rightarrow 2NaOH + H_2$

**2.59**  Write the formulas of the following substances, including the appropriate state symbols: (a) water at 250 K, (b) methane at 298 K, (c) uranium hexafluoride at 298 K, (d) osmium (VIII) oxide at 298 K, (e) ozone at 298 K, (f) xenon at 298 K.

**2.60**  Write the formulas of the following substances when they are dissolved in water: (a) potassium chloride, (b) calcium nitrate, (c) sodium dihydrogen phosphate, (d) copper(II) bromide, (e) silver chloride, (f) magnesium perchlorate.

**2.61**  Write the formulas of the following substances when they are dissolved in water: (a) nitric acid, (b) phosphoric acid, (c) hydrocyanic acid, (d) ammonia, (e) cesium hydroxide, (f) hydrosulfuric acid, (g) sulfuric acid.

**2.62**  Write net ionic equations for the formation of aqueous solutions of the following substances: (a) aluminum ni-

trate, (b) nickel(II) perchlorate, (c) barium chloride, (d) ammonium dichromate, (e) chloroform.

**2.63**  Write net ionic equations for the formation of aqueous solutions of the following substances: (a) sodium hydroxide, (b) ammonia, (c) hydrogen iodide, (d) hydrogen sulfide, (e) bromine.

**2.64**  Write net ionic equations for the following reactions:
(a) Water at room temperature decomposes to hydrogen and oxygen.
(b) Calcium carbonate at 500 K decomposes to carbon dioxide and calcium oxide.
(c) Carbon monoxide and oxygen react in the atmosphere to form carbon dioxide.
(d) Hydrogen peroxide in water solution forms water and oxygen, which comes out of solution.

**2.65**  Write net ionic equations for the precipitations that take place when the following solutions are mixed:
(a) Strontium nitrate and sodium carbonate solutions are mixed to precipitate strontium carbonate.
(b) Sodium hydroxide and chromium(III) chloride solutions are mixed to precipitate chromium(III) hydroxide.
(c) Mercury(I) chlorate and potassium chlorate solutions are mixed to precipitate mercury(I) chloride.
(d) Sodium phosphate and magnesium iodide solutions are mixed to precipitate magnesium phosphate.

**2.66**  Write net ionic equations for the acid-base reactions that occur when:
(a) A solution of nitric acid is mixed with a solution of potassium hydroxide.
(b) A solution of hydrochloric acid is mixed with a solution of barium hydroxide.
(c) A solution of hydrochloric acid is mixed with a solution of ammonia.

(d) A solution of ammonia is mixed with a solution of hydrocyanic acid.

**2.67**   Write net ionic equations for the reactions that take place when:
(a) Pure ammonia is added to a solution of hydrobromic acid.
(b) Pure hydrogen bromide is added to a solution of ammonia.
(c) Hydrogen bromide and ammonia are mixed.

**2.68**   Write net ionic equations for the following processes:
(a) A solution of sodium carbonate is mixed with a solution of nitric acid, and bubbles of carbon dioxide form.
(b) Carbon dioxide is added to a solution of barium nitrate to form a precipitate of barium carbonate.

(c) Gaseous sulfur trioxide is added to a solution of lead(II) nitrate to precipitate lead(II) sulfate.

**2.69**   The joule is the SI unit of energy. Use the definition of kinetic energy to find the relationship between the joule and the SI base unit from which it is derived.

**2.70**   Find the kinetic energy in joules of an $O_2$ molecule of mass $5.31 \times 10^{-23}$ g moving with a velocity of 490 m/s.

**2.71**   Find the kinetic energy in joules of an automobile of mass 1500 kg moving with a velocity of 55 mi/hr.

**2.72**   The gravitational potential energy of an object of mass $m$ can be given as *mgh,* where $g$ is the acceleration of gravity. Show that this expression has the units of joules. (Acceleration is the rate of change of the velocity.)

**2.73**   Use both the old and the new definitions of the calorie to find the heat energy in joules required to raise the temperature of 180 g of water from 287.6 K to 288.6 K.

**2.74**   Give an example of each of the following energy transformations: (a) chemical energy to electrical energy, (b) nuclear energy to kinetic energy, (c) electrical energy to radiant energy, (d) radiant energy to kinetic energy.

**3**

# Stoichiometry

**Preview**

The simple act of counting is as important in chemistry as it is in finance. In this chapter, we describe some of the essential counting methods used in chemistry, starting with the basic unit of amount, the mole, and its relationship to Avogadro's number. We then discuss the meaning and use of atomic and molecular weights and go on to describe how the quantities of substances that take part in chemical reactions are designated. As part of that discussion, we show how a chemical formula and a balanced chemical equation give useful information about the masses and amounts of substances in chemical reactions. We then introduce some expressions of concentration that are helpful in laboratory work with solutions and conclude by describing methods for calculating the composition of mixtures.

Stoichiometry is the study of the quantities of reactants and products in chemical reactions. One important step in chemistry was the precise measurement of the relative proportions by mass of elements and compounds that participate in chemical reactions. Modern chemistry would be impossible without such measurements. Answers to problems as diverse as the minimum human daily requirement of a vitamin, the potential yield of metal from a mine, the life of an electric battery, or the explosive power of a stick of dynamite are all, in a broad sense, obtained from stoichiometry.

In this chapter we shall explore the methods used for specifying quantities of materials and the techniques by which some important types of stoichiometric calculations are carried out.

## 3.1   THE MOLE AND AVOGADRO'S NUMBER

A common way to specify a quantity of material is to state its **mass**. The mass of an object is a measure of its resistance to acceleration, that is, to having its speed or its position changed by the application of a force. Mass is defined by Newton's second law, the familiar expression:

$$F = ma$$

where $m$ is mass, $F$ is force, and $a$ is acceleration. In everyday usage, we often say "weight" when we really mean "mass." Mass and weight are not the same, although the two are closely related. Weight is the gravitational force exerted on a body by the earth. The relationship between weight and mass is

$$W = mg$$

where $W$ is weight and $g$ is the acceleration of gravity at the point on earth where the body is located.

We measure mass by using a balance to compare the weights of objects. Two objects that have the same weight have the same mass. Therefore, if the mass of one object is known, the mass of the other is also known. The SI unit for mass is the kilogram, whose symbol is kg, which is defined as the mass of a specific piece of platinum-iridium alloy that is stored in Sèvres, France.

We can use a second way to specify the quantity of a material that consists of discrete units or entities. We can specify either the *number* of entities or a number that is proportional to the number of entities (Figure 3.1). When we specify quantity in this way we are giving the **amount** of the material. We shall use the term amount in this limited way rather than in the loose way in which it is commonly used. *Amount means the number of entities.*

We use both methods of specifying quantity almost every day. If you

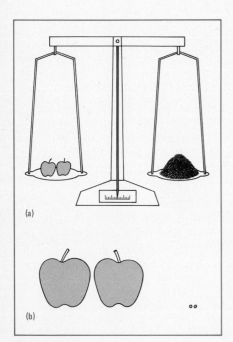

**Figure 3.1**

Mass and amount. (a) Equal masses of apples and peas. (b) Equal amounts of apples and peas.

buy some apples, you can indicate either the mass or the amount of the apples: 2.3 kg of apples or 24 apples, for instance. By custom, one method is more widely used than the other in specific situations. We always buy eggs by amount, one dozen eggs rather than 1 kg of eggs, because eggs tend to break when weighed. We buy grapes by mass, because it is impractical to count the many grapes in a bunch.

## Mass and Amount in Chemistry

When we deal with chemical substances whose composition is well defined, we can designate quantities in either units of mass or units of amount. Chemical substances consist of identifiable basic units, or entities—usually molecules, atoms, or ions. In the chemical laboratory, we most often deal with quantities whose mass is conveniently expressed in grams, g, rather than in kilograms. There are 1000 grams in one kilogram.

We can express the quantity of a chemical substance as an amount, by specifying the number of basic entities—for example, the number of molecules in a given quantity of substance. Counting molecules is much more difficult than counting the number of grapes in a bunch. There is an enormous number of molecules in any sample that is big enough to be visible. But there are compelling reasons for expressing the quantity of a chemical substance as an amount: A chemical equation describes the relative numbers of entities of the substances that participate in a chemical reaction.

The equations that express chemical changes include chemical formulas. A chemical formula is a representation of the basic unit of structure of a substance. For example, the equation

$$2H_2 + O_2 \longrightarrow 2H_2O$$

can be read as the formation of two molecules of water from two molecules of hydrogen and one molecule of oxygen. We tend to interpret all chemical changes in terms of individual molecules, atoms, or ions. However, an overwhelmingly large number of entities take part in most reactions. If we are to express the quantity of chemical substances in terms of amount, we need a way to handle such large numbers. When we deal with eggs, we talk about dozens. We usually say that we have bought one dozen eggs, rather than 12 eggs. When we deal with pencils, we may use the gross as the unit of amount—one gross of pencils, rather than 144 pencils.

## The Mole Defined

The mole (abbreviated mol) is the SI unit for amount. When we use the term amount in chemistry it will almost always be for a quantity whose units are moles. The mole is defined differently than the more common units of amount, because of the very large number of units that it describes. We can say that a dozen is 12 units and a gross is 144 units. There are so many units in a mole that it is not possible to specify the exact number. We define the mole by giving the quantity of a substance that contains that number of units. The formal definition is: The mole is the amount of substance of a system which contains as many elementary entities as there are atoms in exactly 12 g of $^{12}C$. (The symbol $^{12}C$ refers to the most common isotope of carbon. An isotope is one of two or more atoms of the same chemical element that have different atomic masses (Section 2.1). We shall discuss isotopes shortly.) This definition of the mole has a long, involved history that we need not explore. It is enough to say that in 1961, to settle a long-standing dispute between chemists and physicists, $^{12}C$ was adopted as the standard for defining the masses of all the elements. The definition of the mole is also based on this isotope.

The mole contains a very large number of entities indeed. The best measurements now available show that one mole of a substance contains $6.0220978 \times 10^{23}$ basic entities of that substance. This number, which is conveniently remembered to four significant figures as $6.022 \times 10^{23}$, is represented by the symbol $N_A$. It is called **Avogadro's number** in honor of a nineteenth-century chemist whose pioneering work on gases, largely ignored in his time, later proved to be valuable in the determination of accurate atomic weights.

We said that one mole of a substance contains Avogadro's number of entities of the substance. This means that one mole of Fe contains $6.022 \times 10^{23}$ atoms of Fe, while one mole of $O_2$ contains $6.022 \times 10^{23}$ molecules of $O_2$ and three moles of HCl contain $(3)(6.022 \times 10^{23})$ molecules of HCl. The principle is the same as when we talk about a dozen. Both the dozen and the mole are units of amount. The only difference between them is the number of basic entities in each. A dozen has 12, a mole has $6.022 \times 10^{23}$. Otherwise they are the same.

## The Relationship Between Mass and Amount

There are two ways to determine the quantity of everyday things such as eggs or baseballs or grapes. We can measure the mass of a sample with a balance or we can measure the amount by counting the number of eggs or baseballs or grapes in a sample. When we deal with samples of pure substances, whose basic entities are atoms or molecules, we cannot measure the amount by counting, because the number of basic entities is too large. We must measure mass. Still, we often want to know the number of entities in a sample. The relationship between mass and amount is then of particular interest (see Figure 3.2).

Everyday experience tells us that the relationship depends on the mass of a basic entity of the sample. If we know the mass of one baseball, we can easily find the mass of a dozen baseballs. One baseball has a mass of 100 g, so one dozen baseballs has a mass of 1200 g. It's the same with Ping-Pong balls. If the mass of one Ping-Pong ball is 5.0 g, the mass of three dozen Ping-Pong balls is

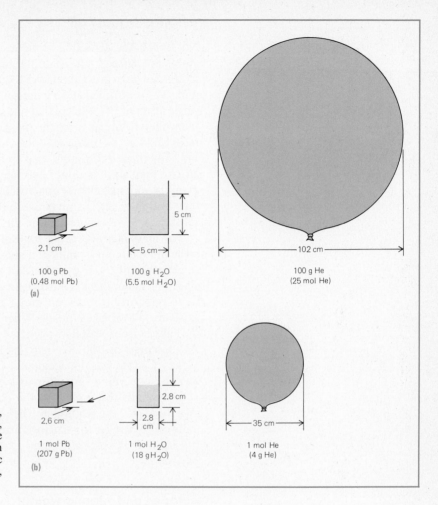

**Figure 3.2**

Mass and amount. (a) Equal masses of lead, water, and helium. (b) Equal amounts of lead, water, and helium. Unlike the equal masses, the cube of lead, the glass of water, and the balloon of helium all contain the same number of basic units of structure: $6.022 \times 10^{23}$ atoms of Pb, molecules of $H_2O$, and atoms of He.

$$3 \text{ dozen} \times \frac{12 \text{ balls}}{1 \text{ dozen}} \times \frac{5.0 \text{ g}}{1 \text{ ball}} = 180 \text{ g}$$

We use the same method with atoms and molecules, but the sheer number of entities creates a difficulty. It is easy to picture a dozen baseballs or three dozen Ping-Pong balls. It is very difficult to picture $10^{23}$ atoms or molecules.

When we deal with substances whose basic entities are atoms or molecules, the number of basic entities in a sample is very large and the mass of the basic entity is very small. Therefore, it usually is not convenient to use the same unit for the mass of the basic entity and for the mass of the sample. We need a unit for the basic entity that makes the conversion from mass to amount as simple as possible. Such a unit has been defined. It is the unified atomic mass unit.

## The Atomic Mass Unit

Not only is the definition of the mole based on $^{12}C$, but the masses of all atoms are defined *relative* to the mass of an atom of $^{12}C$. The masses of atoms can be expressed relative to $^{12}C$ in terms of the unified atomic mass unit (u). One unified atomic mass unit is exactly one-twelfth the mass of a single atom of $^{12}C$. We say that one atom of $^{12}C$ has a mass of exactly 12 u. If another atom has twice the mass of a $^{12}C$ atom, we can say that it has a mass of 24 u. An atom with one-third the mass of $^{12}C$ has a mass of 4 u.

Using these definitions, we can fix a relationship between the unified atomic mass unit, a unit of mass that is suitable for single atoms, and the gram, a unit suitable for macroscopic quantities of substances. We can then find the relationship between the mass and the amount of any given substance, if we can determine the mass of its basic unit.

**The mass in grams of one mole of $^{12}C$ atoms is numerically equal to the mass in atomic mass units of a single atom of $^{12}C$.** That is, the mass of a single atom of $^{12}C$ is exactly 12 u, and the molar mass of $^{12}C$ is exactly 12 g. The **molar mass** is the mass in grams of one mole of a substance. The same relationship between molar mass and the mass of a single atom exists for all other elements. When we find that the molar mass of $^1H$, the lightest isotope of hydrogen, is 1.007825 g, we know that one atom of $^1H$ has a mass of 1.007825 u. (Clearly, the assignment of 12 u as the mass of the $^{12}C$ atom was not accidental. The relative mass of the lightest atom known has a value very close to 1 because of this assignment.)

Since both the mole and the unified atomic mass unit are defined relative to $^{12}C$, the relationship between the mass of an atom of $^{12}C$ and a mole of $^{12}C$ applies for all substances. *The mass in grams of one mole of a substance is numerically equal to the mass of its basic entity in atomic mass units.*

## Isotopes and Atomic Weight

There is a complication. A sample of any given element usually contains different kinds of atoms of that element. These atoms have different masses. They are called isotopes of the element (Section 5.4).

For example, a mole of hydrogen atoms does not consist solely of $^1H$ atoms. It also contains some atoms of $^2H$, the heavier isotope of hydrogen

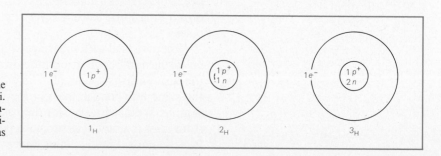

**Figure 3.3**
The isotopes of hydrogen differ only in the number of neutrons they have in their nuclei. The most common isotope, $^1H$, has no neutrons. Deuterium, $^2H$, has one neutron. Tritium, $^3H$, an unstable isotope of hydrogen, has two neutrons.

called deuterium (Figure 3.3). The fraction of $^2$H atoms (or of any isotope in a sample of any element) is called the **natural abundance** of the isotope.

A typical sample of hydrogen, H, of natural origin contains about 0.015% $^2$H atoms and 99.985% $^1$H atoms. A single $^1$H atom has a mass of 1.007825 u, and therefore the molar mass of $^1$H atoms is 1.007825 g. Similarly, a single $^2$H atom has a mass of 2.014102 u and the molar mass of $^2$H atoms is 2.014102 g. The molar mass of naturally occurring hydrogen is found by calculation of the weighted average of the molar masses of each isotope in a typical sample.

molar mass H = (fraction $^1$H)(molar mass $^1$H)

$$+ \text{(fraction } ^2\text{H)(molar mass } ^2\text{H)}$$

$$= \frac{99.985}{100}\,(1.0078 \text{ g}) + \frac{0.015}{100}\,(2.0141 \text{ g})$$

$$= 1.0079 \text{ g}$$

Because there is a small amount of $^2$H in naturally occurring hydrogen, the molar mass is slightly higher than the molar mass of pure $^1$H. This means that the mass of one mole of naturally occurring hydrogen, expressed in grams, is not the same as either the mass of one atom of $^1$H or the mass of one atom of $^2$H in unified atomic mass units. Rather, it corresponds to a weighted average that reflects the natural abundances and atomic mass of the two isotopes. Traditionally, this mass is called the **atomic weight** of the element.

Chemists recognize that the word *weight* is used incorrectly in this phrase. But since the term *atomic weight* is embalmed in tradition, it remains in use with the understanding that it refers to masses of elements relative to $^{12}$C, taking into account the existence of isotopes and their natural abundances. Table 3.1 lists the atomic weights of all the known elements. By convention atomic weights are given without units. But an entry in this table can also be read as the mass of an "average" atom in unified atomic mass units or as the molar mass (in grams) of the element. A table of atomic weights, such as Table 3.1, is of basic importance in chemistry. Since chemists usually work with naturally occurring elements, *the information in a table of atomic weights is assumed to be part of the data given in any problem dealing with quantities of material.*

The use of the mole as a unit of amount is not limited to atoms. We can have a mole of any entity — a mole of molecules, a mole of ions, a mole of electrons, and so on.

## How Chemists Use the Mole

The concept of a mole helps us understand many things in chemistry. Start with the simple example of the formula for water, $H_2O$. We know that one mole of $H_2O$ is composed of $6.022 \times 10^{23}$ molecules. Since each molecule has two hydrogen atoms and one oxygen atom, one mole of $H_2O$ consists of two moles, $(2)(6.022 \times 10^{23})$, of H atoms and one mole,

$6.022 \times 10^{23}$ of O atoms, combined in $H_2O$ molecules.

An analogous everyday situation might be this one: We have one dozen tricycles. Each one has two small rear wheels and one large front wheel. Therefore, there are one dozen, $1 \times 12$, large wheels and two dozen, $2 \times 12 = 24$, small wheels in the dozen tricycles.

The concept of a mole adds a new meaning to chemical formulas. A formula indicates not only the number of atoms in a molecule but also the number of moles of each atom in a mole of that substance. For example, we saw that one mole of $H_2O$ consists of two moles of hydrogen and one mole of oxygen. Similarly, one mole of $CF_4$ consists of one mole of carbon atoms and four moles of fluorine atoms. And while NaCl does not consist of molecules, one mole of NaCl does consist of one mole of $Na^+$ ions and one mole of $Cl^-$ ions.

For substances whose basic entities are molecules, we speak of **molecular weight,** the sum of the atomic weights of all the atoms in the molecule. For substances such as sodium chloride, which are not made up of mole-

**TABLE 3.1**   Table of Atomic Weights[a]

| Name | Symbol | Atomic Number | Atomic Weight | Name | Symbol | Atomic Number | Atomic Weight |
|---|---|---|---|---|---|---|---|
| actinium | Ac | 89 | 227.0278[b] | dysprosium | Dy | 66 | 162.50 |
| aluminum | Al | 13 | 26.98154 | einsteinium | Es | 99 | (252) |
| americium | Am | 95 | (243)[c] | erbium | Er | 68 | 167.26 |
| antimony | Sb | 51 | 121.75 | europium | Eu | 63 | 151.96 |
| argon | Ar | 18 | 39.948 | fermium | Fm | 100 | (257) |
| arsenic | As | 33 | 74.9216 | fluorine | F | 9 | 18.998403 |
| astatine | At | 85 | (210) | francium | Fr | 87 | (223) |
| barium | Ba | 56 | 137.33 | gadolinium | Gd | 64 | 157.25 |
| berkelium | Bk | 97 | (247) | gallium | Ga | 31 | 69.72 |
| beryllium | Be | 4 | 9.01218 | germanium | Ge | 32 | 72.59 |
| bismuth | Bi | 83 | 208.9804 | gold | Au | 79 | 196.9665 |
| boron | B | 5 | 10.81 | hafnium | Hf | 72 | 178.49 |
| bromine | Br | 35 | 79.904 | helium | He | 2 | 4.00260 |
| cadmium | Cd | 48 | 112.41 | holmium | Ho | 67 | 164.9304 |
| calcium | Ca | 20 | 40.08 | hydrogen | H | 1 | 1.00794 |
| californium | Cf | 98 | (251) | indium | In | 49 | 114.82 |
| carbon | C | 6 | 12.011 | iodine | I | 53 | 126.9045 |
| cerium | Ce | 58 | 140.12 | iridium | Ir | 77 | 192.22 |
| cesium | Cs | 55 | 132.9054 | iron | Fe | 26 | 55.847 |
| chlorine | Cl | 17 | 35.453 | krypton | Kr | 36 | 83.80 |
| chromium | Cr | 24 | 51.996 | lanthanum | La | 57 | 138.9055 |
| cobalt | Co | 27 | 58.9332 | lawrencium | Lr | 103 | (260) |
| copper | Cu | 29 | 63.546 | lead | Pb | 82 | 207.2 |
| curium | Cm | 96 | (247) | lithium | Li | 3 | 6.941 |

[a] The atomic weights of many elements are not invariant. They depend on the origin and treatment of the material. The values given here are from a 1981 report of the Commission on Atomic Weights of IUPAC. These values apply to elements as they exist on earth and to certain artificial elements. Values in parentheses are used for radioactive elements whose atomic weights cannot be quoted precisely without knowledge of origin.

[b] Relative atomic mass of the most commonly available long-lived isotope.

[c] A value in parentheses is the mass number of the least unstable isotope of a radioactive element whose atomic weight cannot be given more precisely.

cules, we speak of **formula weight,** the sum of the atomic weights of all the atoms in the formula. *The molecular weight or the formula weight is numerically equal to the molar mass, the mass of one mole of the substance in grams.*

**Example 3.1**  Find the molecular weight or the formula weight of (a) $C_6H_{12}O_6$, (b) $Ca(NO_3)_2$.

**Solution**  To find molecular weight or formula weight, we first find the atomic weights of the constituent atoms by using a table of atomic weights. We then multiply the atomic weight of each atom by the subscript of the atom in the formula. Finally, we add the individual sums.

a.
$$
\begin{aligned}
6 \times C &= 6 \times 12.011 = 72.066 \\
12 \times H &= 12 \times 1.0079 = 12.095 \\
6 \times O &= 6 \times 15.999 = \underline{95.994} \\
&\phantom{= 6 \times 15.999 = } 180.155
\end{aligned}
$$

**TABLE 3.1**  *(Continued)*

| Name | Symbol | Atomic Number | Atomic Weight | Name | Symbol | Atomic Number | Atomic Weight |
|------|--------|---------------|---------------|------|--------|---------------|---------------|
| lutetium | Lu | 71 | 174.967 | samarium | Sm | 62 | 150.4 |
| magnesium | Mg | 12 | 24.305 | scandium | Sc | 21 | 44.9559 |
| manganese | Mn | 25 | 54.9380 | selenium | Se | 34 | 78.96 |
| mendelevium | Md | 101 | (258) | silicon | Si | 14 | 28.0855 |
| mercury | Hg | 80 | 200.59 | silver | Ag | 47 | 107.8682 |
| molybdenum | Mo | 42 | 95.94 | sodium | Na | 11 | 22.98977 |
| neodymium | Nd | 60 | 144.24 | strontium | Sr | 38 | 87.62 |
| neon | Ne | 10 | 20.179 | sulfur | S | 16 | 32.06 |
| neptunium | Np | 93 | $237.0482^b$ | tantalum | Ta | 73 | 180.9479 |
| nickel | Ni | 28 | 58.70 | technetium | Tc | 43 | (98) |
| niobium | Nb | 41 | 92.9064 | tellurium | Te | 52 | 127.60 |
| nitrogen | N | 7 | 14.0067 | terbium | Tb | 65 | 158.9254 |
| nobelium | No | 102 | (259) | thallium | Tl | 81 | 204.37 |
| osmium | Os | 76 | 190.2 | thorium | Th | 90 | $232.038^b$ |
| oxygen | O | 8 | 15.9994 | thulium | Tm | 69 | 168.9342 |
| palladium | Pd | 46 | 106.4 | tin | Sn | 50 | 118.69 |
| phosphorus | P | 15 | 30.97376 | titanium | Ti | 22 | 47.90 |
| platinum | Pt | 78 | 195.09 | tungsten | W | 74 | 183.85 |
| plutonium | Pu | 94 | (244) | unnilennium$^d$ | Une | 109 | (266) |
| polonium | Po | 84 | (209) | unnilhexium$^d$ | Unh | 106 | (263) |
| potassium | K | 19 | 39.0983 | unnilpentium$^d$ | Unp | 105 | (262) |
| praseodymium | Pr | 59 | 140.9077 | unnilquadium$^d$ | Unq | 104 | (261) |
| promethium | Pm | 61 | (145) | unnilseptium$^d$ | Uns | 107 | (262) |
| protactinium | Pa | 91 | $231.0359^b$ | uranium | U | 92 | 238.029 |
| radium | Ra | 88 | $226.0254^b$ | vanadium | V | 23 | 50.9415 |
| radon | Rn | 86 | (222) | xenon | Xe | 54 | 131.30 |
| rhenium | Re | 75 | 186.207 | ytterbium | Yb | 70 | 173.04 |
| rhodium | Rh | 45 | 102.9055 | yttrium | Y | 39 | 88.9059 |
| rubidium | Rb | 37 | 85.4678 | zinc | Zn | 30 | 65.38 |
| ruthenium | Ru | 44 | 101.07 | zirconium | Zr | 40 | 91.22 |

$^d$ Starting with element 104, all new elements will be given systematic names based on their atomic numbers.

b.  $1 \times Ca = 1 \times 40.08 \quad = \quad 40.08$
$2 \times N \ = 2 \times 14.007 \quad = \quad 28.014$
$6 \times O \ = 6 \times 15.999 \quad = \quad \underline{95.994}$
$164.09$

These results also give us the mass of one mole of each substance. The mass of one mole of $C_6H_{12}O_6$ is 180.155 g and the mass of one mole of $Ca(NO_3)_2$ is 164.09 g.

## Calculations with the Mole

These relationships among moles, formulas, masses, and Avogadro's number are at the heart of many calculations in chemistry. In doing these calculations, you must remember that **a mole of any substance consists of $6.022 \times 10^{23}$ elementary entities of that substance and has a mass in grams that is numerically equal to the sum of the atomic weights of the atoms in its formula.**

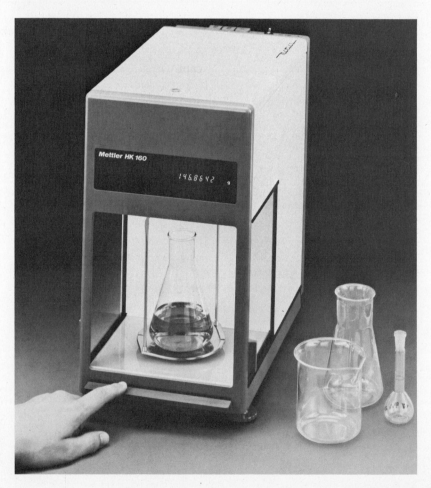

In the modern chemistry laboratory, mass is measured with an automatic balance such as this one. The mass of an object can be determined to seven significant figures in a matter of seconds. (*Mettler Instrument Corp.*)

## THE REVISION OF ATOMIC WEIGHTS

The International Union of Pure and Applied Chemistry (IUPAC) makes changes in the table of atomic weights from time to time. Revisions are required because of improved techniques of analysis, changes in the isotopic abundances of elements caused by human activities, and the laboratory synthesis of heavy elements.

In 1969, for example, the IUPAC Commission on Atomic Weights quoted values for radium, neptunium, and protactinium for the first time. While integral values for the atomic mass of the most stable isotope of these three elements had been printed in many tables before then, the commission had not previously accepted any of these values as the atomic weight because of the rarity of the elements.

At the same time, IUPAC refined the previously listed values of 22 elements. For 20 elements, the values were extended by one significant figure. But IUPAC also reduced the number of significant figures in the atomic weights of boron, carbon, hydrogen, lead, samarium, and sulfur, in most cases because of the variability of the isotopic content of natural sources.

The commission, which now meets biennially, made two changes in 1977. The atomic mass of vanadium was listed as 50.9415 instead of 50.9414, and the atomic mass of lutetium was listed as 174.967 instead of 174.97.

The commission on atomic weights also took note of a new factor: the effect of human activities on the listed values of atomic weights. It was noted that the atomic weights of uranium, boron, lithium, and several other elements are susceptible to wide variations because nuclear industry processes natural samples to separate out certain pure isotopes. The atomic weight values of some samples might be thrown into uncertainty if either the desired isotopes or spent material from which certain isotopes have been separated escape into the environment. The United States government, which conducts isotope separation activities on a large scale, suggested that chemists who had doubts about the isotopic composition of any compound should get a certificate of origin stating whether the material is from a natural source. Because of such human activities, the commission changed its definition of atomic weight. It is now defined as "the ratio of the mass per mole of atoms of that element to $\frac{1}{12}$ of the mass of one mole of $^{12}C$." The old definition had included a phrase referring to the "natural nuclidic composition" of an element. With the elimination of that phrase, the commission said, the definition could be used for elements whose isotopic composition had been made uncertain by human activity. It could also be used for a mixture that included isotopes synthesized in the laboratory.

The IUPAC commission continues to work toward more accurate atomic weights, since accurate determinations are lacking for some elements. For example, values for osmium, palladium, and samarium are known to only one decimal place, and the values for many other elements are known to only two decimal places.

For these calculations, we shall use the same approach as we did for unit conversions (Section 1.5). To convert between mass and amount, we get conversion factors by using the table of atomic weights. To convert between amounts of related substances, we get conversion factors by using formulas. We can even convert between amount in moles and number of basic entities by using the value of Avogadro's number in appropriate conversion factors. The following examples illustrate some of these conversions.

**Example 3.2**    Find the mass of 2.32 mol of sodium bicarbonate, $NaHCO_3$, which is commonly called baking soda.

**Solution**    We are given an amount, 2.32 mol, and are asked for a mass. To convert from amount to mass, we find the mass of one mole of the substance. We use this value to find the conversion factor.

To find the mass of one mole of a substance, we find the formula weight:

$$
\begin{array}{llll}
1 \times Na & = 1 \times 22.99 & = 22.99 \\
1 \times H & = 1 \times \phantom{0}1.008 & = \phantom{0}1.008 \\
1 \times C & = 1 \times 12.01 & = 12.01 \\
3 \times O & = 3 \times 16.00 & = \underline{48.00} \\
& & \phantom{=}\ 84.01
\end{array}
$$

The mass of one mole of the substance in grams is numerically equal to the formula weight. We can write this relationship as

$$1 \text{ mol } NaHCO_3 = 84.01 \text{ g } NaHCO_3$$

We can use this relationship to derive two conversion factors that allow us to convert between mass and amount:

$$\frac{1 \text{ mol } NaHCO_3}{84.01 \text{ g } NaHCO_3} \quad \text{or} \quad \frac{84.01 \text{ g } NaHCO_3}{1 \text{ mol } NaHCO_3}$$

Since we are given an amount and are asked for a mass, we use the conversion factor that gives us the result in grams, the unit of mass:

$$2.32 \text{ mol } NaHCO_3 \times \frac{84.01 \text{ g } NaHCO_3}{1 \text{ mol } NaHCO_3} = 195 \text{ g } NaHCO_3$$

It is very important to remember that these conversion factors are valid only when they include the formula of the specific substance. The reason 1 mol = 84.01 g is that we are referring to $NaHCO_3$.

In the previous example, we converted from amount to mass. Now we shall convert from mass to amount.

**Example 3.3**    Sulfuric acid, $H_2SO_4$, is the most widely used industrial chemical. Find the amount (in moles) of $H_2SO_4$ in 375 g of the pure acid.

**Solution**    Again, we first find the molecular weight of the substance by summing the atomic weights of its constituent atoms. For $H_2SO_4$, this sum is 98.07. Therefore

$$1 \text{ mol } H_2SO_4 = 98.07 \text{ g } H_2SO_4$$

The two conversion factors are

$$\frac{98.07 \text{ g } H_2SO_4}{1 \text{ mol } H_2SO_4} \quad \text{or} \quad \frac{1 \text{ mol } H_2SO_4}{98.07 \text{ g } H_2SO_4}$$

Since mass is given and we are asked for amount, we use the second factor

$$375 \text{ g } H_2SO_4 \times \frac{1 \text{ mol } H_2SO_4}{98.07 \text{ g } H_2SO_4} = 3.82 \text{ mol } H_2SO_4$$

The units of grams appear in both the numerator and denominator so they cancel; only the units of moles remain. When the proper conversion factor is chosen for a problem such as this, the units in which the solution is expressed — in this case, moles — will be in the numerator only. If the wrong conversion factor is used, a check will reveal the mistake immediately. If we had used

$$375 \text{ g } H_2SO_4 \times \frac{98.07 \text{ g } H_2SO_4}{1 \text{ mol } H_2SO_4} = 3680 \frac{\text{g}^2 \ H_2SO_4}{\text{mol } H_2SO_4}$$

the units would make no sense, and we would know that the result must be wrong.

The preceding two examples illustrate what are probably the two most common calculations in chemistry. One reason the table of atomic weights is so important is that it gives the data needed for these calculations. The key to doing these calculations is to use the molar mass of a substance, which is numerically equal to its molecular weight,[1] to get the needed conversion factor.

The molecular weight of a substance gives us the relationship between mass and amount for that substance. Once we have the molecular weight, we can also find relationships between mass and number of molecules by using Avogadro's number. If $N_A$ is Avogadro's number, then

$$1 \text{ mol} = N_A \text{ molecules}$$

We can use this equality to get conversion factors, as the next two examples show.

**Example 3.4**   Find the number of molecules in 0.0837 g of formaldehyde, $CH_2O$, which is used as a preservative.

**Solution**   When a conversion between mass and amount is part of a problem, it is usually best to start by finding the molecular weight of the substance. The molecular weight of $CH_2O$ is 30.03. We use the conversion factor that converts mass to amount:

[1] From this point on, we shall use the term *molecular weight* for all substances, even for those whose basic entities are not normally molecules.

$$0.0837 \ \text{g CH}_2\text{O} \times \frac{1 \ \text{mol CH}_2\text{O}}{30.03 \ \text{g CH}_2\text{O}} = 0.002787 \ \text{mol CH}_2\text{O}$$

To convert from moles to molecules, we use a conversion factor derived from the relationship between the mole and Avogadro's number:

$$0.002787 \ \text{mol CH}_2\text{O} \times \frac{6.022 \times 10^{23} \ \text{molecules}}{1 \ \text{mol}} = 1.68 \times 10^{21} \ \text{molecules CH}_2\text{O}$$

**Example 3.5**

Find the mass of $1.00 \times 10^{24}$ molecules of ozone, $O_3$, a gas that is a major air pollutant.

**Solution**

We are given an amount of ozone and are asked about its mass. We start by using the table of atomic weights to find the molecular weight of ozone, which is 48.00. We then use this number to get the factor needed to convert from amount in moles to mass. But since we are given the amount in molecules, we must first convert from molecules to moles, using a factor that includes Avogadro's number.

When we use more than one conversion factor in a calculation, we do not have to calculate intermediate results. We can list all the conversion factors in sequence, cancelling units as we go, and then do the arithmetic:

$$1.00 \times 10^{24} \ \text{molecules} \ O_3 \times \frac{1 \ \text{mol}}{6.022 \times 10^{23} \ \text{molecules}} \times \frac{48.00 \ \text{g O}_3}{1 \ \text{mol} \ O_3} = 79.7 \ \text{g O}_3$$

A chemical formula is a concise statement about the amounts of the elements that make up a compound. For example, we can say that the formula $C_6H_{12}O_6$ for glucose tells us that one molecule of glucose contains 6 atoms of carbon, 12 atoms of hydrogen, and 6 atoms of oxygen. This statement specifies the number of atoms in a single molecule. Almost any statement that is correct for atoms and molecules will still be correct if the phrase *moles of* is placed before the words *atom* and *molecule*. For example, we can say correctly that 1 mole of glucose molecules contains 6 moles of carbon atoms, 12 moles of hydrogen atoms, and 6 moles of oxygen atoms. If this idea seems unusual, think of a more familiar amount, the dozen, and a more familiar object, the automobile. One automobile has 4 wheels, 6 cylinders, and 2 headlights. One dozen automobiles has 4 dozen wheels, 6 dozen cylinders, and 2 dozen headlights.

We can use the amount relationships in a formula to write equalities which can then be used to form conversion factors. For example, we can write

$$1 \ \text{mol} \ C_6H_{12}O_6 = 6 \ \text{mol C}$$

This equation does not mean that 1 mol of glucose is the same as 6 mol of carbon. It means that these are equivalent amounts, that when we have 1 mol of glucose we have 6 mol of carbon. Given an amount of one substance, we can use this relationship as a conversion factor to find the amount of the other. The next group of examples illustrates the use of conversion factors that relate two amounts.

---

**Example 3.6**   Find the amount of oxygen in 0.891 mol of $Na_2Cr_2O_7$.

**Solution**   This problem requires what we can call an amount-to-amount conversion: We are given the amount of a substance and are asked to find the amount of one of its constituents. The subscripts of the formula give us the relationship we need to make the conversion. They tell us that

$$1 \text{ mol } Na_2Cr_2O_7 = 7 \text{ mol O}$$

From this equality, we get the conversion factor that allows us to convert from moles of $Na_2Cr_2O_7$ to moles of O:

$$0.891 \text{ mol } Na_2Cr_2O_7 \times \frac{7 \text{ mol O}}{1 \text{ mol } Na_2Cr_2O_7} = 6.24 \text{ mol O}$$

---

These amount-to-amount conversion factors can be worked into other types of mass-amount calculations.

---

**Example 3.7**   Find the number of atoms of B in a 35.8-g sample of $B_2O_3$, an oxide of boron that is used to make boron glasses.

**Solution**   Since we must convert from mass to amount, we start by using the table of atomic weights to find the molar mass of $B_2O_3$:

$$1 \text{ mol } B_2O_3 = 69.62 \text{ g } B_2O_3$$

We can then convert to amount of B by using the information in the formula

$$1 \text{ mol } B_2O_3 = 2 \text{ mol B}$$

When we have the amount of B in moles, we can find the number of B atoms by using Avogadro's number. In one step:

$$35.8 \text{ g } B_2O_3 \times \frac{1 \text{ mol } B_2O_3}{69.62 \text{ g } B_2O_3} \times \frac{2 \text{ mol B}}{1 \text{ mol } B_2O_3} \times \frac{6.022 \times 10^{23} \text{ atoms}}{1 \text{ mol}}$$
$$= 6.19 \times 10^{23} \text{ atoms B}$$

## 3.2  MASS RELATIONSHIPS IN CHEMICAL FORMULAS

A table of atomic weights is assumed to be part of the information given in any stoichiometry problem. Therefore, a chemical formula conveys information not only on the number of atoms in a molecule of a substance but also on the relative masses of the components of the molecule. We can also turn this around and say that if we know the relative masses of the components of an unknown substance, we can find out something about the chemical formula of the substance.

When we deal with a pure substance, we often want to know the mass of one component in a given mass of a sample. The next example shows one approach to this question.

**Example 3.8**

The main source of industrial fluorides and the chemical used to fluoridate water supplies is calcium fluoride, $CaF_2$, which is also called fluorspar. Find the mass of fluorine in a 453-g sample of $CaF_2$.

**Solution**

We are given a mass of $CaF_2$ and are asked about the mass of F, so we need a relationship between these two substances. We get a relationship from the formula. It is

$$1 \text{ mol } CaF_2 = 2 \text{ mol } F$$

But this relationship is one between amounts, not masses. We therefore start by converting the given mass, 453 g of $CaF_2$, to an amount. As usual, we do the conversion by using the table of atomic weights to find the molar mass of $CaF_2$:

$$1 \text{ mol } CaF_2 = 78.08 \text{ g } CaF_2$$

Once we know the amount of $CaF_2$ in the given mass, we can find the amount of F in the sample. When we know the amount of F, we can find the mass of F from its atomic weight.

$$1 \text{ mol } F = 19.00 \text{ g } F$$

In one step the calculation is

$$453 \text{ g } CaF_2 \times \frac{1 \text{ mol } CaF_2}{78.08 \text{ g } CaF_2} \times \frac{2 \text{ mol } F}{1 \text{ mol } CaF_2} \times \frac{19.00 \text{ g } F}{1 \text{ mol } F} = 220 \text{ g } F$$

We used a three-step approach in the previous example. First we converted a given mass to an amount, using a molar mass that we found from data in the table of atomic weights. Then we converted from the amount of the substance to the amount of one of its components, using the amount relationships given by the chemical formula. Finally, we converted from the amount of the component to its mass, using the table of atomic weights again. We shall discuss this type of calculation in more detail in Section 3.3

There is a more direct approach to this type of calculation. We can find the relative composition by mass of a substance by using its formula and the table of atomic weights. We can then use this information to find the composition of any sample of the substance. We illustrate this method in the next example, in which the composition of a substance is expressed as the mass fraction of each of its constituent elements.

**Example 3.9**   Calculate the fraction by mass of each element in the substance cryolite, $Na_3AlF_6$, which is used in glassmaking and in many metallurgical processes.

**Solution**   We start by calculating the molecular weight of the substance. We can simplify the calculation by setting out the data in an orderly manner:

$$
\begin{aligned}
3 \times Na &= 3 \times 22.99 = \phantom{0}68.97 \\
1 \times Al &= 1 \times 26.98 = \phantom{0}26.98 \\
6 \times F &= 6 \times 19.00 = \underline{114.00} \\
\text{molecular weight } Na_3AlF_6 &= 209.95
\end{aligned}
$$

This calculation gives us not only the molecular weight of the substance but also the mass of each element in a given mass of the substance. Since atomic and molecular masses are numerically equal to the molar mass in grams, the calculation can be read as if it dealt with masses. We can say that the mass of 1 mol of $Na_3AlF_6$, the molar mass, is 209.95 g and that it contains 68.97 g of Na, 26.98 g of Al, and 114.00 g of F. We can determine relative composition by dividing the mass of each component by the mass of the total.

$$
\text{fraction Na} = \frac{68.97 \text{ g Na}}{209.95 \text{ g Na}_3\text{AlF}_6} = \frac{0.3285 \text{ g Na}}{1 \text{ g Na}_3\text{AlF}_6}
$$

$$
\text{fraction Al} = \frac{26.98 \text{ g Al}}{209.95 \text{ g Na}_3\text{AlF}_6} = \frac{0.1285 \text{ g Al}}{1 \text{ g Na}_3\text{AlF}_6}
$$

$$
\text{fraction F} = \frac{114.00 \text{ g F}}{209.95 \text{ g Na}_3\text{AlF}_6} = \frac{0.5430 \text{ g F}}{1 \text{ g Na}_3\text{AlF}_6}
$$

A check shows that these three fractions sum to 1.00. We could also express these fractions as mass percents by multiplying each by 100%.

Once we have the relative composition, we can readily find the mass of any of the components in a given mass of the substance. For example, to find the mass of Al in 765 kg of cryolite, we use the mass fraction of Al:

$$
765 \text{ kg Na}_3\text{AlF}_6 \times \frac{0.1285 \text{ g Al}}{1 \text{ g Na}_3\text{AlF}_6} = 98.3 \text{ kg Al}
$$

We can use this fraction to find the mass of Al in any sample of $Na_3AlF_6$.

We can find a formula for an unknown compound by doing the same sort of calculation as in Example 3.9, but in reverse — that is, by starting with an unknown substance and determining the mass of each element in a sample of the substance. When we know the relative mass of each element, we can convert mass to amount, using the table of atomic weights to find the relative number of atoms of each element in the substance. This procedure gives a formula that indicates only the *relative* number of different atoms in a molecule of the substance. Such a formula is called an **empirical formula.**

**Example 3.10**    One area of current interest in chemistry is the study of the carboranes, a class of compounds composed of carbon, boron, and hydrogen. One carborane is found to contain 32.77% C and 59.00% B by mass. Calculate its empirical formula.

**Solution**    Since the empirical formula (or any other chemical formula) gives the relative number of moles of each constituent atom of a compound, the first step is to convert the data on mass into data on amount.

Since percentage is a fraction of 100, we can say that if a substance contains 32.77% C by mass, then 100 g of the substance contains 32.77 g of carbon.

$$32.77 \text{ g C} \times \frac{1 \text{ mol C}}{12.01 \text{ g C}} = 2.729 \text{ mol C}$$

This 100-g sample also contains 59.00 g of B:

$$59.00 \text{ g B} \times \frac{1 \text{ mol B}}{10.81 \text{ g B}} = 5.458 \text{ mol B}$$

The mass of H in the 100-g sample is obtained by subtraction:

$$\begin{aligned} \text{mass H} &= 100 \text{ g} - (\text{mass C}) - (\text{mass B}) \\ &= 100 \text{ g} - 32.77 \text{ g} - 59.00 \text{ g} \\ &= 8.23 \text{ g H} \end{aligned}$$

Thus there is

$$8.23 \text{ g H} \times \frac{1 \text{ mol H}}{1.008 \text{ g H}} = 8.165 \text{ mol H}$$

The ratio of moles of C to moles of B to moles of H is C:B:H = 2.729:5.458:8.165. We obtain the subscripts in the empirical formula for the compound by converting this ratio to an equivalent integer ratio. This can be done in two steps:

i. Divide all the terms of the ratio by the smallest number in the ratio:

$$\frac{2.729}{2.729} : \frac{5.458}{2.729} : \frac{8.165}{2.729} = 1:2:3$$

ii. In this case, division by the smallest number in the ratio gives a ratio containing only integers. If division by the smallest number produces a ratio that still includes decimals, multiply by the smallest factor that converts all the terms to integers. For example, the ratio 2.00 : 3.25 : 1.50 can be multiplied by 4 to give 8 : 13 : 6.

A check shows that the ratio 1 : 2 : 3 for C : B : H is reasonable for an empirical formula, since the numbers do not have any factors in common and the ratio is thus in lowest terms. The subscripts of the empirical formula are 1, 2, and 3, and the empirical formula is written

$$CB_2H_3$$

### Empirical Formula and Molecular Formula

An empirical formula may or may not be the same as a **molecular formula,** which gives the exact number of each kind of atom in a molecule. In the case of water, $H_2O$, the empirical formula and the molecular formula are identical. The empirical formula indicates that the ratio of hydrogen atoms to oxygen atoms is 2 : 1. The molecular formula shows that there are exactly two atoms of hydrogen and one atom of oxygen in a molecule of water. On the other hand, hydrogen peroxide has the empirical formula HO, which indicates that the ratio of hydrogen atoms to oxygen atoms in the substance is 1 : 1. But the molecular formula of hydrogen peroxide is $H_2O_2$. Each molecule actually has two atoms of hydrogen and two atoms of oxygen. There are many cases in which a number of substances, especially organic ones, have the same empirical formula but different molecular formulas (Figure 3.4).

The subscripts in a molecular formula are always integral multiples of the subscripts in the empirical formula. An empirical formula can be converted to a molecular formula if one more piece of information is available — the molecular weight of the substance. The molecular weight can be measured experimentally.

**Figure 3.4**

Types of formulas. Compounds with the same empirical formula may have different molecular formulas. The structural formula also indicates how the atoms in the molecule are bonded together.

| Empirical Formula | Molecular Formula | Structural Formula | Molecular Weight |
|---|---|---|---|
| CH | $C_2H_2$ | H—C≡C—H | 26.04 |
| CH | $C_6H_6$ | 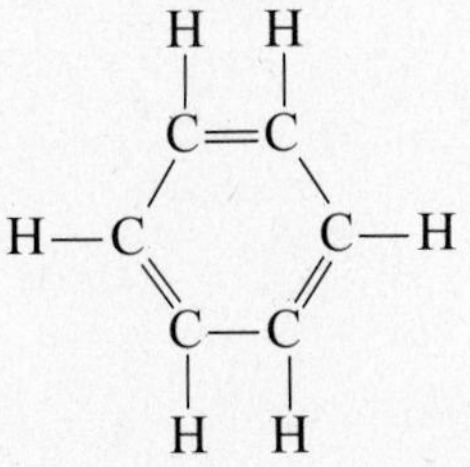 | 78.11 |

**Example 3.11**    The molecular weight of the carborane in Example 3.10 is found to be 72.5. Calculate the molecular formula of the compound.

**Solution**    We obtain the subscripts in a molecular formula by multiplying the subscripts in an empirical formula by an integer, usually a small one. The problem is to find the correct integer.

First, calculate the "empirical formula weight" of the empirical formula as if it were an ordinary chemical formula:

$$1 \times C = 1 \times 12.0 = 12.0$$
$$2 \times B = 2 \times 10.8 = 21.6$$
$$3 \times H = 3 \times 1.0 \ \ = \underline{3.0}$$
$$36.6$$

Divide this value into the experimentally determined molecular weight:

$$\frac{72.5}{36.6} = 1.98$$

and round off the result to the nearest integer, in this case 2. Then multiply each subscript in the empirical formula by this integer to obtain the molecular formula, in this case:

$$C_2B_4H_6$$

One technique used to find the composition of unknown substances in organic chemistry is combustion analysis (Figure 3.5). Compounds containing carbon and hydrogen burn to form carbon dioxide and water. We often can calculate the empirical formula of the substance that was burned by collecting and weighing the $CO_2$ and $H_2O$.

**Example 3.12**    A sample of a substance containing only carbon and hydrogen is subjected to combustion analysis. It produces 3.215 g of $CO_2$ and 1.097 g of $H_2O$ on complete combustion. Find the empirical formula of the substance.

**Solution**    To find the empirical formula we must find the amounts of carbon and hydrogen in the sample. We find the amount of carbon from the mass of $CO_2$. We first use the molar mass of $CO_2$, obtained with the aid of the table of atomic weights, to convert the given mass to amount of $CO_2$. Then we use the relationship between amount of $CO_2$ and amount of C, given by the formula $CO_2$, to find the amount of C:

$$3.125 \text{ g } CO_2 \times \frac{1 \text{ mol } CO_2}{44.010 \text{ g } CO_2} \times \frac{1 \text{ mol C}}{1 \text{ mol } CO_2} = 0.07305 \text{ mol C}$$

We find the amount of hydrogen from the given mass in the same way:

$$1.097 \text{ g H}_2\text{O} \times \frac{1 \text{ mol H}_2\text{O}}{18.15 \text{ g H}_2\text{O}} \times \frac{2 \text{ mol H}}{1 \text{ mol H}_2\text{O}} = 0.1218 \text{ mol H}$$

The molar ratio of C to H is $0.07305 : 0.1218$. We convert this decimal ratio to one with integers, as we did in Example 3.10.

i. Divide by the smaller amount:

$$\frac{0.07305}{0.07305} : \frac{0.1218}{0.07305} = 1 : 1.667$$

ii. Multiply by the smallest factor needed to convert $1.667$ ($\frac{5}{3}$) to an integer. This factor is 3, so the ratio is $3 : 5$ and the empirical formula is $C_3H_5$.

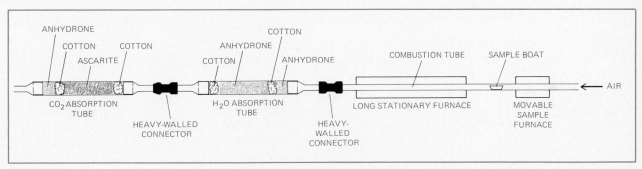

**Figure 3.5**
A combustion train for carbon-hydrogen analysis of organic compounds.

Combustion analysis can be carried out successfully on compounds containing other elements in addition to carbon and hydrogen. Many organic compounds contain oxygen. Even when we add oxygen to insure complete combustion, we can find the quantity of oxygen in the compound and its empirical formula by combustion analysis. We find the quantity of oxygen indirectly, not directly, from the amounts of $CO_2$ and $H_2O$ produced by combustion.

To do this, we use the basic principle that the whole is equal to the sum of its parts. If we know the mass of a sample and the mass of all but one of its components, we can find the mass of the unknown component by subtraction. We subtract all the known masses from the total mass. The remainder is the unknown mass. For a compound containing C, H, and O, we can find the mass of C from the mass of $CO_2$ and the mass of H from the mass of $H_2O$, as we did in Example 3.12. Then we subtract to find the mass of O. We can then obtain the empirical formula of the substance by converting masses of C, H, and O to amounts. The next example illustrates this method.

**Example 3.13**   A sample of a substance containing only C, H, and O has a mass of 1.037 g. On complete combustion in a stream of excess $O_2(g)$, it yields 1.900 g of $CO_2$ and 0.521 g of $H_2O$. Find the empirical formula of the substance.

Solution     We must find the mass of O in the sample by using the relationship

$$\text{total mass} = \text{mass C} + \text{mass H} + \text{mass O}$$

To find the mass of C, we use the mass fraction of C in $CO_2$, which we find by using the table of atomic weights. The molecular weight of $CO_2$ is 44.01 and the atomic weight of C is 12.01, so there is 12.01 g C in 44.01 g $CO_2$:

$$1.900 \text{ g } CO_2 \times \frac{12.01 \text{ g C}}{44.01 \text{ g } CO_2} = 0.5185 \text{ g C}$$

To find the mass of H, we use the mass fraction of H in $H_2O$, which we find by using the table of atomic weights. There is 2.016 g H in 18.02 g $H_2O$, the molar mass of water:

$$0.521 \text{ g } H_2O \times \frac{2.016 \text{ g H}}{18.02 \text{ g } H_2O} = 0.0583 \text{ g H}$$

We now find the mass of O by subtraction:

$$1.037 \text{ g} = 0.5185 \text{ g C} + 0.0583 \text{ g H} + \text{mass O}$$
$$\text{mass O} = 0.460 \text{ g O}$$

Now we convert the three masses to amounts and find the empirical formula.

$$0.5185 \text{ g C} \times \frac{1 \text{ mol C}}{12.01 \text{ g C}} = 0.0432 \text{ mol C}$$

$$0.0583 \text{ g H} \times \frac{1 \text{ mol H}}{1.008 \text{ g H}} = 0.0578 \text{ mol H}$$

$$0.460 \text{ g O} \times \frac{1 \text{ mol O}}{16.00 \text{ g O}} = 0.0288 \text{ mol O}$$

The molar ratio of the three elements is $C:H:O = 0.0432:0.0578:0.0288$. If we divide all the terms of the ratio by 0.0288, the smallest number in the ratio, we get $1.50:2.02:1.00$. We can convert this ratio to integers by multiplying by 2 and rounding off. This gives $C:H:O = 3:4:2$. The empirical formula thus is

$$C_3H_4O_2$$

## 3.3  MASS RELATIONSHIPS IN CHEMICAL REACTIONS

Anyone who practices chemistry must understand the mass relationships among substances involved in chemical changes. We cannot approach these relationships directly because of the way that we express information about chemical change. We use balanced chemical equations, which are statements about the *amounts* of substances in chemical changes. We do not measure amounts directly, however. Instead, we usually measure quantity by measuring *mass*. We use the table of atomic weights to find relationships between the mass and the amount of each substance in a chemical change. We use the balanced chemical equation to get amount relationships among the various substances in the reaction.

Mass relationships in a chemical process are not determined entirely by the amount relationships given in the chemical equation. They also depend on the quantities of reactants that are used. In the general process

$$2A + B \longrightarrow C$$

the equation can be taken to mean that two molecules of A combine with one molecule of B to form one molecule of C or that two moles of A combine with one mole of B to form one mole of C.

If we carry out this reaction by bringing together quantities of A and B whose molar ratio is exactly $2:1$, as it is in the equation, we say that we have **stoichiometric amounts** of A and B, and both of them theoretically can be completely consumed. *Stoichiometric amounts of reactants are those amounts that can combine completely, leaving no excess of any reactant.*

In practice, we rarely work with stoichiometric amounts. One reactant usually is present in excess. The quantity of the *other* reactant determines how much of the product is produced. For example, if two moles of A and two moles of B were mixed, B would be present in excess. The quantity of A would determine the amount of product produced. We would call A the **limiting quantity** or **limiting reagent.**

There is often a practical reason why an excess of one reagent is needed. The equation above says that one mole of C is obtained when two moles of A and one mole of B are mixed. The reaction is said to **go to completion:** All of A and B react to form C. In practice, many processes do not go to completion. Less than the "theoretical" amount of C is obtained because not all of the starting materials react. Impurities in the reactants and unwanted processes called side reactions may reduce the amount of product. These problems will be ignored at this point. Unless otherwise specified, the chemical equations associated with stoichiometric problems will be treated as going to completion.

### Amount-to-Amount Conversions

One basic skill needed in chemistry is the use of a chemical equation to find amount relationships among substances in a chemical change. The next example illustrates some typical amount-to-amount conversions.

**Example 3.14** Under ordinary conditions, ammonia reacts with the $O_2$ in the air to form $N_2$ and water. Find the amount of $N_2$ and water produced from the reaction of 0.69 mol of ammonia. Find the amount of $O_2$ consumed.

## THE MASS SPECTROMETER

The mass of an atom or a molecule can be measured with great accuracy with an instrument called a mass spectrometer. The atom or molecule is first ionized to a cation, usually by bombardment with electrons. The cation is accelerated by a high-voltage electric field, and then passes through a slit into a magnetic field. The cation is deflected by the magnetic field. For a magnetic field of given strength, the amount of deflection is determined by the mass and the charge of the cation. If two cations have the same electric charge, the heavier one will be deflected less. By adjustment of the strength of the magnetic field, any cation can be made to pass through a second slit and arrive at a collector. Since the strengths of the electric and magnetic fields are known, the mass of the cation can be determined with great accuracy.

It is possible to link a mass spectrometer with other instruments that enable fast analysis of complex samples. For example, a mass spectrometer often is linked to an instrument called a gas chromatograph. The gas chromatograph separates the substances of interest out of a sample containing many different compounds. These substances are then passed through the mass spectrometer to determine their masses. In one of the great feats of long-distance chemistry, such instruments were used by the two Viking spacecraft that landed on Mars in 1976 to

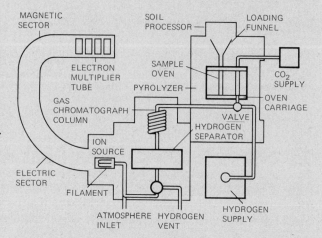

analyze samples of Martian soil. The absence of organic compounds in those samples was a major factor in convincing most members of the Viking scientific team that life does not exist on Mars.

One drawback of the mass spectrometer until recent years was a limit to the size of the molecules it could handle. Very large molecules were broken into smaller fragments by the electron bombardment that ionized them. Now a new method has been developed that uses a gentler bombardment of xenon atoms to ionize the samples. This technique is being used to determine the masses of large, complex biological molecules.

Solution    The essential first step in almost every calculation dealing with chemical change is to write a balanced chemical equation. The chemical change here is

$$NH_3(g) + O_2(g) \longrightarrow N_2(g) + H_2O(g)$$

The balanced equation is

$$4NH_3(g) + 3O_2(g) \longrightarrow 2N_2(g) + 6H_2O(g)$$

We can derive all the equalities between the substances of the reaction from this equation. For example, in this process

$$4 \text{ mol } NH_3 = 2 \text{ mol } N_2 \quad \text{and} \quad 4 \text{ mol } NH_3 = 6 \text{ mol } H_2O$$
$$4 \text{ mol } NH_3 = 3 \text{ mol } O_2 \quad \text{and} \quad 3 \text{ mol } O_2 = 2 \text{ mol } N_2$$

From equalities like these, we can derive the conversion factors that we need to make amount-to-amount conversions. The general rule is that these conversion factors include the substances of interest and their *coefficients* in the balanced equation.

In this example, we are asked to convert a given amount of $NH_3$ to amounts of other substances. To do this, we use conversion factors that have $NH_3$ in the denominator and the other substances in the numerator.

$$0.69 \text{ mol } NH_3 \times \frac{2 \text{ mol } N_2}{4 \text{ mol } NH_3} = 0.35 \text{ mol } N_2$$

$$0.69 \text{ mol } NH_3 \times \frac{6 \text{ mol } H_2O}{4 \text{ mol } NH_3} = 1.0 \text{ mol } H_2O$$

$$0.69 \text{ mol } NH_3 \times \frac{3 \text{ mol } O_2}{4 \text{ mol } NH_3} = 0.52 \text{ mol } O_2$$

## Mass-Amount Conversions

Sometimes we are given the mass of one substance in a reaction and are asked to find the amount of another. We can do such a calculation in two steps. We first convert the given mass to an amount, using the table of atomic weights. We then use the chemical equation to find the amount of the substance, as we did in Example 3.14. The next example illustrates this kind of mass-to-amount conversion.

**Example 3.15**    A pure metal usually is quite shiny, but the metals that we observe tend to be dull and grayish. The loss of sheen often is due to the formation of an oxide in a reaction between the metal surface and oxygen in the atmosphere. Lithium, the lightest metal, reacts rapidly with oxygen in the atmosphere to form lithium oxide. Find the amount of $O_2$ required to react with 0.493 g of Li metal.

**Solution**    The essential first step is to write the balanced chemical equation for the process:

$$4Li(s) + O_2(g) \longrightarrow 2 \text{ } Li_2O(s)$$

The conversion from mass of Li to amount of $O_2$ is done in two steps. We convert mass of Li to amount of Li, using its atomic weight, then convert amount of Li to amount of $O_2$, using the amount relationships in the equation:

$$0.493 \text{ g Li} \times \frac{1 \text{ mol Li}}{6.941 \text{ g Li}} \times \frac{1 \text{ mol } O_2}{4 \text{ mol Li}} = 0.0178 \text{ mol } O_2$$

We may also be given the amount of one substance in a reaction and asked to find the mass of another. Again, we do a two-step calculation. First we use the chemical equation to convert from the amount of the given substance to the amount of the other substance. We then find the mass of the second substance by using the table of atomic weights. The next example illustrates this kind of amount-to-mass conversion.

Example 3.16

In metallurgy, tin oxide, $SnO_2$, is treated with carbon to form tin metal and carbon monoxide. Find the mass of $SnO_2$ that produces 27.8 mol CO in this reaction.

Solution

We first write the balanced equation:

$$SnO_2(s) + 2C(s) \longrightarrow Sn(s) + 2CO(g)$$

We convert from amount of CO, which is given, to amount of $SnO_2$ by using the chemical equation. Then we convert from amount of $SnO_2$ to mass of $SnO_2$ by using the table of atomic weights:

$$27.8 \text{ mol CO} \times \frac{1 \text{ mol } SnO_2}{2 \text{ mol CO}} \times \frac{150.7 \text{ g } SnO_2}{1 \text{ mol } SnO_2} = 2090 \text{ g } SnO_2$$

## Mass-to-Mass Conversions

Because mass is easily measured, by far the most common calculation of mass relationships we perform in chemistry is to start with the mass of one substance in a reaction and to find the mass of one or more other substances. These calculations are done in three steps. The mass of the given substance is converted to an amount, using the table of atomic weights. The amount of the given substance then is converted to an amount of the other substance, using the coefficients in the balanced chemical equation. Finally, the amount of the other substance is converted to a mass, using the table of atomic weights. We can diagram this procedure:

$$\begin{array}{ll} \text{mass A} \longrightarrow \text{amount A} & \text{(table of atomic weights)} \\ \text{amount A} \longrightarrow \text{amount B} & \text{(chemical equation)} \\ \text{amount B} \longrightarrow \text{mass B} & \text{(table of atomic weights)} \end{array}$$

The next example illustrates this method.

Example 3.17

One important step in the metallurgy of copper is the reaction of copper (I) oxide with copper (I) sulfide to form copper metal and sulfur dioxide. Find the mass of copper that is produced from the reaction of 592 g of the oxide.

Solution

We start with the chemical equation for the process:

$$2 \, Cu_2O(s) + Cu_2S(s) \longrightarrow 6Cu(s) + SO_2(g)$$

To carry out the three-step procedure for this mass-to-mass conversion, we need three conversion factors which come from three equalities. First we need a factor to convert from mass of $Cu_2O$ to amount of $Cu_2O$. We use the table of atomic weights to find the molar mass of $Cu_2O$:

$$1 \text{ mol } Cu_2O = 143.1 \text{ g } Cu_2O$$

Next we need a factor to convert from amount of $Cu_2O$ to amount of Cu. From the equation we see that

$$2 \text{ mol } Cu_2O = 6 \text{ mol } Cu$$

Finally, the factor for converting amount of Cu to mass comes from the table of atomic weights, which gives us the molar mass of Cu:

$$1 \text{ mol } Cu = 63.55 \text{ g } Cu$$

We use these three conversion factors to carry out the three-step procedure:

$$592 \text{ g } Cu_2O \times \frac{1 \text{ mol } Cu_2O}{143.1 \text{ g } Cu_2O} \times \frac{6 \text{ mol } Cu}{2 \text{ mol } Cu_2O} \times \frac{63.55 \text{ g } Cu}{1 \text{ mol } Cu} = 789 \text{ g } Cu$$

$$\text{mass} \longrightarrow \text{amount} \longrightarrow \text{amount} \longrightarrow \text{mass}$$

We can use the same procedure to find the mass of the other substances in the reaction. For example, suppose we are asked to find the mass of $Cu_2S$ consumed in the reaction of the 592 g $Cu_2O$. We use the table of atomic weights to find the molar mass of $Cu_2S$, which is 159.2 g $Cu_2S$. We use the chemical equation to find the amount-to-amount relationship between $Cu_2S$ and $Cu_2O$. The three-step calculation is

$$592 \text{ g } Cu_2O \times \frac{1 \text{ mol } Cu_2O}{143.1 \text{ g } Cu_2O} \times \frac{1 \text{ mol } Cu_2S}{2 \text{ mol } Cu_2O} \times \frac{159.2 \text{ g } Cu_2S}{1 \text{ mol } Cu_2S} = 329 \text{ g } Cu_2S$$

We can use the general method outlined in Example 3.17 in many situations. In a number of cases, we can do conversions by a series of steps. If we write the equations for the steps in sequence, we can get amount-to-amount relationships successively from each equation. These relationships give us conversion factors that we can use to get mass-to-mass relationships between substances in different equations, as the next example shows.

**Example 3.18**   There is a sequence of reactions in which we start with $P_4O_{10}$ and eventually produce AgCl. The equations for this sequence are

$$P_4O_{10}(s) + 6H_2O \longrightarrow 4H_3PO_4(aq)$$
$$H_3PO_4(aq) + 3Ag^+(aq) \longrightarrow Ag_3PO_4(s) + 3H^+(aq)$$
$$Ag_3PO_4(s) + 3H^+(aq) + 3Cl^-(aq) \longrightarrow 3AgCl(s) + H_3PO_4(aq)$$

A total of 1.52 g of AgCl is isolated from this procedure. Find the starting mass of $P_4O_{10}$.

**Solution**   We must convert from mass of AgCl to mass of $P_4O_{10}$. As usual, we start with a mass-to-amount conversion of AgCl, using the table of atomic weights to find its molar mass:

$$1.52 \text{ g } AgCl \times \frac{1 \text{ mol } AgCl}{143.3 \text{ g } AgCl} = 0.01061 \text{ mol } AgCl$$

In the second step, we convert from amount of $AgCl$ to amount of $P_4O_{10}$. To carry out this conversion, we work our way from one equation to the next, getting an appropriate conversion factor from each:

$$0.01061 \text{ mol AgCl} \times \frac{1 \text{ mol Ag}_3\text{PO}_4}{3 \text{ mol AgCl}} \times \frac{1 \text{ mol H}_3\text{PO}_4}{1 \text{ mol Ag}_3\text{PO}_4} \times \frac{1 \text{ mol P}_4\text{O}_{10}}{4 \text{ mol H}_3\text{PO}_4}$$
$$= 8.842 \times 10^{-4} \text{ mol P}_4\text{O}_{10}$$

We then convert this amount of $P_4O_{10}$ to a mass of $P_4O_{10}$ in the usual way, finding its molar mass from the table of atomic weights:

$$8.842 \times 10^{-4} \text{ mol P}_4\text{O}_{10} \times \frac{283.9 \text{ g P}_4\text{O}_{10}}{1 \text{ mol P}_4\text{O}_{10}} = 0.251 \text{ g P}_4\text{O}_{10}$$

## The Limiting Reagent

For the common situation in which reactants are not present in exactly the correct molar ratios called the stoichiometric amounts, the reaction continues until one of the reagents is completely consumed. The reagent that is exhausted first is the *limiting reagent.*

The idea of a limiting reagent is familiar, although not by that name. We can encounter it while setting a dinner table. Each place requires two forks, one knife, and two spoons. If you have 10 forks, 10 knives, and 20 spoons, you can see that you will run out of forks after setting five places. The number of forks is the limiting factor. We use the same kind of reasoning when we work with limiting reagents in chemical equations.

When reactants are not present in stoichiometric amounts, an extra step is needed to determine the limiting reagent. The reactant or reagent that is not present in excess will be consumed before any of the others. When this happens, the reaction stops. Therefore, the one reactant that is not present in excess is the limiting reagent.

**Example 3.19**    Potassium nitrate, $KNO_3$, sometimes called saltpeter, is an important fertilizer. One process for its manufacture includes this reaction, carried out at elevated temperatures:

$$3KCl(s) + 4HNO_3(l) \longrightarrow Cl_2(g) + NOCl(g) + 2H_2O(g) + 3KNO_3(s)$$

What mass of $KNO_3$ will be produced when 101-g quantities of each reactant are combined?

**Solution**    Since the equation is in terms of amounts, we must convert the given masses of the reagents to amounts, using the appropriate molar masses:

$$\text{amount KCl} = 101 \text{ g KCl} \times \frac{1 \text{ mol KCl}}{74.55 \text{ g KCl}} = 1.355 \text{ mol KCl}$$

$$\text{amount HNO}_3 = 101 \text{ g HNO}_3 \times \frac{1 \text{ mol HNO}_3}{63.01 \text{ g HNO}_3} = 1.603 \text{ mol HNO}_3$$

Start with either of the reactants, say KCl, and calculate the amount of the other reactant needed to combine completely with the first reactant. If KCl is chosen:

$$\text{amount HNO}_3 \text{ needed} = 1.355 \text{ mol KCl} \times \frac{4 \text{ mol HNO}_3}{3 \text{ mol KCl}}$$

$$= 1.806 \text{ mol HNO}_3 \text{ needed}$$

Compare the calculated amount of the reactant with the amount that is actually present. If the amount present is less than the amount needed, this reactant is the limiting reagent. It will be consumed first.

In this case, 1.603 mol of $HNO_3$ is present, but 1.806 mol of $HNO_3$ is needed for conversion of all the KCl. There is insufficient $HNO_3$, so $HNO_3$ is the limiting reagent. If the amount of $HNO_3$ present had been greater than the calculated amount, then the other reactant, KCl, would have been limiting.

**The remainder of this calculation is based on the quantity of the limiting reagent.** The method of calculation is the same three-step procedure as that of Example 3.17.

$$101 \text{ g HNO}_3 \times \frac{1 \text{ mol HNO}_3}{63.01 \text{ g HNO}_3} \times \frac{3 \text{ mol KNO}_3}{4 \text{ mol HNO}_3} \times \frac{101.1 \text{ g KNO}_3}{1 \text{ mol KNO}_3} = 122 \text{ g KNO}_3$$

Note that 122 g of $KNO_3$ is produced from 101 g of $HNO_3$, no matter whether 1000 g or 1 000 000 g of KCl is present. The amount of the limiting reagent determines the quantities of products formed.

**Example 3.20**   What mass of KCl remains after the above reaction is completed?

**Solution**   First we calculate the mass of KCl consumed, using the procedure of Example 3.17:

$$101 \text{ g HNO}_3 \times \frac{1 \text{ mol HNO}_3}{63.01 \text{ g HNO}_3} \times \frac{3 \text{ mol KCl}}{4 \text{ mol HNO}_3} \times \frac{74.55 \text{ g KCl}}{1 \text{ mol KCl}} = 88.7 \text{ g KCl}$$

The quantity of KCl remaining after the reaction goes to completion is the original quantity less the quantity consumed: $101 \text{ g} - 88.7 \text{ g} = 12 \text{ g KCl}$ remaining. (Since KCl is present in excess, some must remain when the reaction is completed.)

We mentioned earlier that a reaction carried out in a laboratory will rarely produce the quantity calculated on the basis of 100% reaction, the *theoretical yield*. Almost invariably, the quantity of product actually obtained is less than the quantity expected theoretically. The reaction does not go to completion because of side reactions or for other reasons, such as poor experimental techniques on the part of the chemist. The quantity of product that is actually obtained is expressed as a **percent yield,** which is defined by the expression:

$$\text{percent yield} = \frac{\text{quantity obtained}}{\text{theoretical yield}} \times 100\% \qquad (3.1)$$

---

**Example 3.21**    Chromium metal can be produced from chromium(III) oxide by a high-temperature reaction with aluminum. Aluminum oxide is the other product. A sample of $Cr_2O_3$ of mass 19.7 g is treated with excess aluminum, and a total of 11.6 g of Cr is isolated. Find the percent yield of Cr.

**Solution**    We begin with the balanced equation for the process:

$$Cr_2O_3(s) + 2Al(s) \longrightarrow 2Cr(s) + Al_2O_3(s)$$

To find the percent yield of Cr, we must first find the theoretical yield. We can find it by converting from the given mass of $Cr_2O_3$, which is the limiting reagent, to the expected mass of Cr:

$$19.7 \text{ g } Cr_2O_3 \times \frac{1 \text{ mol } Cr_2O_3}{152.0 \text{ g } Cr_2O_3} \times \frac{2 \text{ mol Cr}}{1 \text{ mol } Cr_2O_3} \times \frac{52.00 \text{ g Cr}}{1 \text{ mol Cr}} = 13.5 \text{ g Cr}$$

This mass of Cr is the theoretical yield. To find the percent yield, we divide the theoretical yield into the mass of Cr actually obtained and multiply by 100%, as shown in Equation 3.1:

$$\frac{11.6 \text{ g Cr}}{13.5 \text{ g Cr}} \times 100\% = 85.9\%$$

---

The relationships implied by chemical formulas and chemical equations can be used in many different ways. In the nineteenth century, the atomic weights of newly discovered elements were determined by measurement of the quantities used in chemical conversions.

---

**Example 3.22**    Imagine that it is the middle of the nineteenth century and that a new element has been discovered whose chemical properties seem similar to those of sodium and potassium. Because of this similarity, we assume that the element forms ions of charge $+1$ in its salts. Because the element emits deep red light when heated in a flame, it has been named rubidium. Its symbol is Rb. A sample of rubidium carbonate (which presumably has the formula $Rb_2CO_3$, by analogy with $Na_2CO_3$ and $K_2CO_3$) is treated with a strong acid. The following reaction occurs:

$$Rb_2CO_3(s) + 2H^+(aq) \longrightarrow CO_2(g) + H_2O + 2Rb^+(aq)$$

We find that a 0.1475-g sample of $Rb_2CO_3$ produces 0.0281 g of $CO_2$. Assuming that the reaction goes to completion, that the atomic weight of C is known to be 12.01, and that the atomic weight of O is known to be 16.00, calculate the atomic weight of Rb.

**Solution**   In previous examples, we used the molar mass to find mass from amount or amount from mass. We can also use amount and mass to find molar mass. In this example, we are given the mass of $Rb_2CO_3$, a chemical equation, and the mass and molar mass of $CO_2$ that is formed. We can use this information to find the amount of $Rb_2CO_3$ with which we started. We convert from mass of $CO_2$ to amount of $Rb_2CO_3$:

$$0.0281 \text{ g } CO_2 \times \frac{1 \text{ mol } CO_2}{44.01 \text{ g } CO_2} \times \frac{1 \text{ mol } Rb_2CO_3}{1 \text{ mol } CO_2} = 0.000638 \text{ mol } Rb_2CO_3$$

Now that we have the mass and amount of a sample of $Rb_2CO_3$, we can do a mass-to-amount conversion and find its molar mass:

$$0.00638 \text{ mol } Rb_2CO_3 \times \text{molar mass } Rb_2CO_3 = 0.1475 \text{ g } Rb_2CO_3$$

$$\text{molar mass } Rb_2CO_3 = 231.0 \frac{\text{g } Rb_2CO_3}{1 \text{ mol } Rb_2CO_3}$$

We know that the molar mass is numerically equal to the molecular weight, which is equal to the sum of the atomic weights. If we let $x$ be the atomic weight of Rb, then:

$$2x + 12.01 + (3)(16.00) = 231.0$$
$$x = 85.5 = \text{atomic weight of Rb}$$

## 3.4   THE STOICHIOMETRY OF SOLUTIONS

Many chemical reactions occur in solutions, so we need convenient ways to indicate the composition of solutions. We shall not discuss solutions in detail until Chapter 12, but we introduce them here because many experiments you will do in chemistry laboratory are reactions in aqueous solution.

Most aqueous solutions are formed of relatively small quantities of substances called **solutes** that are dissolved in water, which is called the **solvent.** A solution is homogeneous, having the same composition throughout. It is defined by the relative quantities of solutes and solvent, the **concentration** of the solution.

The most direct way to describe concentration is to describe the **mass percent** of the solution, defined as:

$$\text{mass percent} = \frac{\text{mass of solute}}{\text{mass of solution}} \times 100\%$$

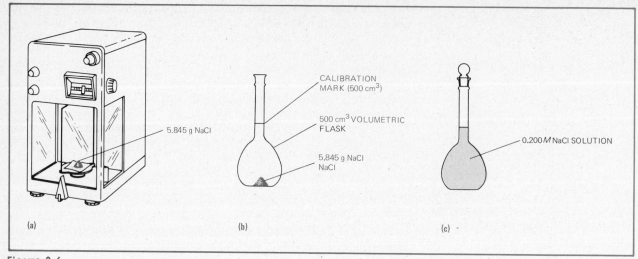

**Figure 3.6**
Preparation of a 0.200*M* solution of NaCl. (a) Using a balance, we measure 5.845 g of NaCl, which is 0.1000 mol of NaCl. (b) The NaCl is transferred to a volumetric flask that has a calibration mark on its neck to indicate a volume of 500 cm³. Water is added to the flask to dissolve the NaCl until the volume of the solution reaches the calibration mark. The concentration of the solution is 0.100 mol/0.500 L = 0.200*M*.

Thus, the expression "a 10% solution of sodium chloride" refers to a solution consisting of 10% NaCl, the solute, and 90% $H_2O$, the solvent. If we had 100 g of such a solution, it would contain 10 g of NaCl and 90 g of $H_2O$.

While nonchemists usually describe solutions in terms of mass percent, this expression is inconvenient in the chemical laboratory. It is usually easier to measure the volume of solution than to measure its mass. The expression that chemists most often use to indicate the quantity of solute in a given quantity of solution is **molarity**. Molarity, symbolized by **M**, is an expression of concentration defined as:

$$M = \frac{\text{amount of solute}}{\text{volume of solution}} \quad \text{or} \quad M = \frac{n}{V} \tag{3.2}$$

where the unit of amount is the mole and the unit of volume is the cubic decimeter (dm³), more commonly called the liter (L). That is, **molarity is the number of moles of solute in one liter of solution.**

Note that the denominator of the expression in Equation 3.2 refers to the entire solution, not just to the solvent. Thus, the expression "a 2*M* solution of NaCl" means that one liter of this solution contains two moles of dissolved NaCl. As we know (Section 2.6), the NaCl dissociates into $Na^+$ and $Cl^-$ ions in aqueous solution, so 1 L of this solution contains 2 mol of $Na^+$ ions and 2 mol of $Cl^-$ ions (Figure 3.6).

The relationship given as Equation 3.2 is often used in work with solutions. The equation has three terms: molarity, amount in moles, and volume in liters. Given any two terms, we can determine the third readily.

**Example 3.23**    What mass of $CaCl_2$ is needed to prepare 0.32 L of a 0.75*M* solution?

**Solution**   Let $n$ = amount of $CaCl_2$ (in moles) required. Substitution into Equation 3.2 gives:

$$0.75M = \frac{n}{0.32 \text{ L solution}}$$

$$n = 0.24 \text{ mol } CaCl_2$$

$$\text{mass } CaCl_2 = 0.24 \text{ mol } CaCl_2 \times \frac{111 \text{ g } CaCl_2}{1 \text{ mol } CaCl_2}$$

$$= 27 \text{ g } CaCl_2$$

---

**Example 3.24**   What volume of the above solution contains 38 g of $CaCl_2$?

**Solution**   Convert 38 g of $CaCl_2$ to amount (in moles):

$$38 \text{ g } CaCl_2 \times \frac{1 \text{ mol } CaCl_2}{111 \text{ g } CaCl_2} = 0.342 \text{ mol } CaCl_2$$

Let $V$ = the volume of solution required (in liters). If

$$M = \frac{n}{V} \quad \text{then} \quad V = \frac{n}{M}$$

and

$$V = \frac{0.342 \text{ mol } CaCl_2}{0.75M}$$

$$V = 0.46 \text{ L required}$$

---

In most cases, we have information about the volume and molarity of a solution, so we can find the amount of solute by application of Equation 3.2. Once the amount of solute is known, the mass of the solute can be found.

---

**Example 3.25**   When 0.100 L of a 1.30$M$ solution of $AgNO_3$ is mixed with an excess of an NaCl solution, an AgCl precipitate forms according to the reaction:

$$Ag^+(aq) + Cl^-(aq) \longrightarrow AgCl(s)$$

What mass of AgCl(s) is formed if essentially all the $AgNO_3$ is converted to AgCl(s)?

**Solution**   Since there is one $Ag^+$ ion in each unit of $AgNO_3$, the amount (in moles) of $Ag^+$ ion in solution is equal to the amount (in moles) of $AgNO_3$ dissolved. The equation shows that the amount of AgCl formed is equal to the amount of $Ag^+$ and therefore equal to the amount of $AgNO_3$. The amount of $AgNO_3$ in the solution can be calculated. Let $n$ = the amount of $AgNO_3$ (in moles). Then from Equation 3.2:

$$1.30M = \frac{n}{0.100 \text{ L solution}}$$

$$n = 0.130 \text{ mol AgNO}_3$$

Therefore, 0.130 mol of AgCl is formed.

$$\text{mass AgCl} = 0.130 \text{ mol AgCl} \times \frac{143.4 \text{ g AgCl}}{1 \text{ mol AgCl}}$$

$$= 18.6 \text{ g AgCl}$$

**Example 3.26**    What volume of $0.326M$ NaCl solution contains just the quantity of $Cl^-$ necessary for the conversion in Example 3.25?

**Solution**    From the chemical equation, we see that 0.130 mol of NaCl is required by 0.130 mol of $AgNO_3$. Let $V$ = volume of solution (in liters). Then, from Equation 3.2:

$$V = \frac{0.130 \text{ mol NaCl}}{0.326M}$$

$$V = 0.399 \text{ L solution}$$

We can formulate many variations of the previous examples. Solution problems of this type can be solved by use of the definition of molarity and its relationship to the mole. As soon as you know the number of moles of the reactants, the problem is one of ordinary stoichiometric relationships.

**Example 3.27**    One way to remove phosphates from sewage is by precipitation with calcium oxide, CaO. What mass of calcium oxide is needed to precipitate all the phosphate from $2.0 \times 10^3$ L of sewage that has a concentration of phosphate equal to $0.0022M$, assuming that the precipitation is complete? The relevant reactions are first a hydrolysis of the oxide:

$$CaO(s) + H_2O \longrightarrow Ca^{2+}(aq) + 2OH^-(aq)$$

then a precipitation of the salt:

$$3Ca^{2+}(aq) + 2PO_4{}^{3-}(aq) \longrightarrow Ca_3(PO_4)_2(s)$$

**Solution**    Let $n$ equal the amount of phosphate (in moles) in the quantity of solution:

$$0.0022M = \frac{n}{2.0 \times 10^3 \text{ L sewage}}$$

$$n = 4.4 \text{ mol PO}_4{}^{3-}$$

From the equations, we can write the conversion factors:

$$\frac{1 \text{ mol CaO}}{1 \text{ mol Ca}^{2+}} \quad \text{and} \quad \frac{3 \text{ mol Ca}^{2+}}{2 \text{ mol PO}_4{}^{3-}}$$

Therefore:

$$\text{amount CaO(s)} = 4.4 \text{ mol PO}_4{}^{3-} \times \frac{3 \text{ mol Ca}^{2+}}{2 \text{ mol PO}_4{}^{3-}} \times \frac{1 \text{ mol CaO}}{1 \text{ mol Ca}^{2+}}$$

$$= 6.6 \text{ mol CaO}$$

$$6.6 \text{ mol CaO} \times \frac{56.1 \text{ g CaO}}{1 \text{ mol CaO}} = 370 \text{ g CaO}$$

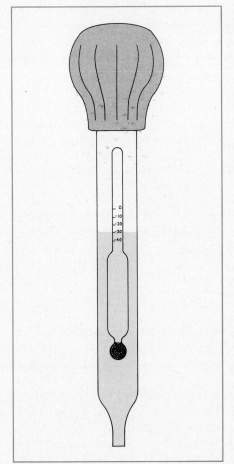

Figure 3.7
A radiator hydrometer. A hydrometer is used to check the antifreeze protection provided by the coolant in an automobile's radiator. By means of the rubber bulb, some of the coolant is drawn into the large tube containing the hydrometer. The depth at which the hydrometer floats indicates the density and thus the freezing point of the coolant solution.

## Density

Relationships between the mass and the volume of a solution often are important in calculations of the stoichiometry of solutions. We have already defined two units of concentration, mass percent and molarity, without defining the relationship between them. To find this relationship we must know the relationship between the mass and the volume of a given solution.

The relationship between mass and volume usually is expressed by the **density** ($d$) of the solution. Density can be understood to be the mass of a given volume of solution or of any homogeneous material:

$$d = \frac{\text{mass}}{\text{volume}}$$

For liquids and solids, mass is usually measured in grams and volume in cubic centimeters ($cm^3$). Density has the units grams per cubic centimeter. The density of liquid water, for example, has a value close to 1 $g/cm^3$. Table 3.2 lists the densities of some common substances.

**TABLE 3.2   Densities of Substances at Room Temperature**

| Liquids | | Solids | |
|---|---|---|---|
| Substance | Density (g/cm³) | Substance | Density (g/cm³) |
| butane, $C_4H_{10}$ | 0.58 | lithium, Li | 0.53 |
| octane, $C_8H_{18}$ | 0.70 | magnesium, Mg | 1.74 |
| ethanol, $C_2H_5OH$ | 0.79 | sodium chloride, NaCl | 2.16 |
| water, $H_2O$ | 1.00 | graphite, C | 2.27 |
| bromine, $Br_2$ | 3.12 | aluminum, Al | 2.70 |
| mercury, Hg | 13.6 | limestone, $CaCO_3$ | 2.71 |
| | | lime, CaO | 3.35 |
| | | diamond, C | 3.52 |
| | | corundum, $Al_2O_3$ | 3.99 |
| | | copper, Cu | 8.93 |
| | | lead, Pb | 11.3 |
| | | platinum, Pt | 21.5 |
| | | osmium, Os | 22.5 |

**Example 3.28**  A solution is prepared by dissolving 13.8 g of sucrose, $C_{12}H_{22}O_{11}$, in 38.4 g of $H_2O$. The solution has a volume of 0.0470 L. Find the density of the solution.

**Solution**  To express the volume of solution in cubic centimeters, we use the conversion factor 1000 cm³/1 L:

$$0.0470 \text{ L solution} \times \frac{1000 \text{ cm}^3}{1 \text{ L}} = 47.0 \text{ cm}^3 \text{ solution}$$

$$d = \frac{38.4 \text{ g } H_2O + 13.8 \text{ g sucrose}}{47.0 \text{ cm}^3 \text{ solution}}$$

$$= 1.11 \text{ g/cm}^3$$

**Example 3.29**  A water solution of $H_2SO_4$ is 28% $H_2SO_4$ by mass and has a density of 1.20 g/cm³. Calculate the molarity of the solution.

**Solution**  We can use density as a conversion factor to get the mass of one liter of solution. In problems of this type it is usually most convenient to consider one liter of solution.

$$1 \text{ L solution} \times \frac{1000 \text{ cm}^3 \text{ solution}}{1 \text{ L solution}} \times \frac{1.20 \text{ g solution}}{1 \text{ cm}^3 \text{ solution}} = 1200 \text{ g solution}$$

We are told that the solution is 28.0% $H_2SO_4$. Now that we know the mass of one liter of the solution, we can calculate that 1200 g of the solution contains:

$$1200 \text{ g solution} \times \frac{28.0 \text{ g } H_2SO_4}{100 \text{ g solution}} = 336 \text{ g } H_2SO_4$$

$$336 \text{ g } H_2SO_4 \times \frac{1 \text{ mol } H_2SO_4}{98.07 \text{ g } H_2SO_4} = 3.43 \text{ mol } H_2SO_4$$

Since 1200 g of solution has a volume of 1 L and there is 3.43 mol of $H_2SO_4$ in this solution, the concentration of $H_2SO_4$ is 3.43$M$.

The density of a solution can be measured with an instrument called a **hydrometer,** which is a long-necked sealed tube with a weighted bulb at one end. The upper part of the tube is calibrated in units of density (generally g/cm³). The hydrometer is placed in a solution, and density is measured by the depth to which it sinks.

One kind of hydrometer (Figure 3.7) is used to check the coolant in automobile radiators in the winter. When antifreeze is added, the density of the coolant in the radiator goes up and its freezing point goes down. By measuring the density of the coolant, a mechanic indirectly measures its freezing point. To make the task simpler, the hydrometer used in service stations is calibrated in degrees of temperature rather than in units of density, so that the freezing point of the coolant can be read directly.

## 3.5  THE COMPOSITION OF MIXTURES

Stoichiometric reasoning can be used to calculate the composition of mixtures. We shall present a general method. Once the general method for such calculations is mastered, it can be simplified in many cases.

To calculate the composition of a two-component mixture, we need two equations in two unknowns. One equation can be called the mass equation, since it usually depends on or expresses a relationship between the masses of the components and the total mass of the mixture. The second equation can be called the amount or molar relationship equation. This equation usually depends on or expresses a relationship between the number of moles of the components of the mixture and the number of moles of some product formed from them. To set up the amount equation, we usually need a chemical formula and a table of atomic weights. Information about a specific chemical reaction may also be needed.

If the identities of the components of a mixture are known, the composition of the mixture can be analyzed by conversion of the components to a substance whose mass can then be measured.

---

**Example 3.30**

A 2.00-g sample of a mixture of NaCl and NaBr is found to contain 0.75 g of Na. What is the fraction of NaCl in the mixture?

**Solution**

Let $x$ = mass of NaCl in grams and $y$ = mass of NaBr in grams. The mass equation follows directly:

$$x + y = 2.00$$

We can set up the molar relationship equation by following this line of reasoning: The mass of Na in the mixture is given. Therefore, we can find the amount of Na in the mixture by using the atomic weight of Na. The relationship between the amount of Na and the amounts of the other components is defined by the chemical formulas of the components of the mixture. There is one mole of Na in one mole of NaCl, and one mole of Na in one mole of NaBr. This relationship can be presented as:

$$\text{moles Na in NaCl} + \text{moles Na in NaBr} = \text{total moles Na}$$

or

$$\text{moles NaCl} + \text{moles NaBr} = \text{total moles Na}$$

$$\text{moles NaCl} = \frac{\text{mass NaCl}}{\mathcal{M}\ \text{NaCl}} = \frac{x}{58.4}$$

$$\text{moles NaBr} = \frac{\text{mass NaBr}}{\mathcal{M}\ \text{NaBr}} = \frac{y}{102.9}$$

$$\text{moles Na} = \frac{\text{mass Na}}{\mathcal{M}\ \text{Na}} = \frac{0.75}{23.0}$$

where $\mathcal{M}$ is the molar mass (mass of one mole in grams per mole).

Substituting these molar expressions into the above word equation gives the molar relationship equation:

$$\frac{x}{58.4} + \frac{y}{102.9} = \frac{0.75}{23.0}$$

The mass equation can be written as:

$$y = 2.00 - x$$

Substituting this expression for $y$ into the molar relationship equation gives:

$$x = 1.78 \text{ g NaCl} = \text{mass NaCl}$$
$$y = 0.22 \text{ g NaBr} = \text{mass NaBr}$$

$$\text{fraction NaCl} = \frac{\text{mass NaCl}}{\text{total mass}} = \frac{1.78 \text{ g NaCl}}{2.00 \text{ g mixture}} = 0.89$$

**Example 3.31**

A 3.20-g sample of a mixture of NaCl and $CaCl_2$ is dissolved in water and treated with excess silver nitrate solution. All the Cl in the mixture is converted to a precipitate of silver chloride, AgCl, according to the reaction:

$$Ag^+(aq) + Cl^-(aq) \longrightarrow AgCl(s)$$

The precipitate is found to have a mass of 7.94 g. Calculate the mass of sodium chloride in the mixture.

**Solution**

Let $x$ = mass of NaCl in grams and $y$ = mass of $CaCl_2$ in grams. The mass equation follows directly:

$$x + y = 3.20$$

We can set up the molar relationship by following this line of reasoning: The mass of the AgCl precipitate is given. Therefore, we can find the amount of AgCl by using the table of atomic weights:

$$7.94 \text{ g AgCl} \times \frac{1 \text{ mol AgCl}}{143.3 \text{ g AgCl}} = 0.0554 \text{ mol AgCl}$$

Each mole of NaCl forms one mole of AgCl, and each mole of $CaCl_2$ forms two moles of AgCl. Therefore, the amount of AgCl equals the amount of NaCl plus twice the amount of $CaCl_2$. That is:

$$\text{moles NaCl} = \frac{\text{mass NaCl}}{\mathcal{M} \text{ NaCl}} = \frac{x}{58.44}$$

$$\text{moles CaCl}_2 = \frac{\text{mass CaCl}_2}{\mathcal{M} \text{ CaCl}_2} = \frac{y}{111.0}$$

$$\text{moles NaCl} + (2)\text{moles CaCl}_2 = \text{moles AgCl}$$

or

$$\frac{x}{58.44} + \frac{2y}{111.0} = 0.0554$$

Solution of these two equations in $x$ and $y$ gives:

$$x = 2.49 \text{ g NaCl}$$

The calculation of molar relationships and mass relationships in chemical processes is basic to chemistry. The beginning student of chemistry should take pains to master these calculations. One observation may be helpful: If there is a common approach in these calculations, it is the *conversion from mass (grams) to amount (moles), and from amount to mass, with the use of the data given in the table of atomic weights.*

**Summary**

This chapter dealt with the precise measurement of masses and amounts that is essential to modern chemistry. We first distinguished between **mass,** which is based on resistance to acceleration, and **amount,** the number of entities in a sample. We then defined the **mole,** the SI unit of amount, as the amount of a substance equal to the number of atoms in exactly 12 g of carbon-12. We indicated that there is Avogadro's number, $6.022 \times 10^{23}$, elementary entities in one mole. We also defined **atomic weight** as the weighted average of the masses of the isotopes of an element as found in nature. We noted that the **molecular weight** of a substance is the sum of the atomic weights of all the atoms in the molecule. In discussing mass relationships in chemical formulas, we described the difference between an **empirical formula,** which gives the relative numbers of atoms in a molecule, and the **molecular formula,** the actual numbers of atoms in the molecule. We then introduced the concept of **stoichiometric amount,** the exact amounts of reactants that combine completely, and noted that when stoichiometric amounts are not present, the reagent in short supply is called the **limiting reagent,** which determines the amount of product produced. We went on to solutions, which are composed of substances called **solutes** that are dissolved in liquids called **solvents.** We said that the concentration of a solution could be described in terms of **mass percent,** the mass of solute dissolved in a mass of solution, or as **molarity,** the amount of solute dissolved in one liter of solution. We mentioned that **density** is the mass of a substance divided by its volume. Finally, we described a method for calculating the composition of mixtures.

## Exercises

**3.1**  The mass of 45 apples is 5025 g. Find the mass of an average apple.

**3.2**  The mass of an average orange is 65 g. Find the amount of oranges that has a mass of 5.4 kg.

**3.3**  The mass of an average grapefruit is 120 g. Find the mass of 3.5 dozen grapefruit.

**3.4**  The mass of 5.0 gross of pencils is 2.5 kg. Find the mass of an average pencil.

**3.5**  The mass of an average grain of sand is 4.3 mg. The mass of 1.0 $m^3$ of sand is $1.4 \times 10^2$ kg. There is $1.0 \times 10^6$ $m^3$ of sand on a certain beach. Find the number of grains of sand on the beach.

**3.6**[2]  The mass of an average grape is 4.1 g. Find the mass of Avogadro's number of grapes.

**3.7**  The mass of 1 mol of blueberries is $5.0 \times 10^{23}$ g. Find the mass of an average blueberry.

**3.8**  The mass of a truck is $4.6 \times 10^4$ kg. Find the number of trucks that has the same mass as 1.0 mol of blueberries, $5.0 \times 10^{23}$ g.

**3.9**  The diameter of a human red blood cell is 7.5 $\mu$m. Find the length of a row made up of 0.50 mol of red cells.

**3.10**  The mass of an average grape is 4.1 g. Find the mass of the grape in atomic mass units.

**3.11**  The mass of a molecule of human hemoglobin, the oxygen-carrying protein in blood, is 68 300 u. Find the mass of this molecule in grams.

**3.12**  The ratio of the mass of a mo-

[2] The answers to exercises whose numbers are in color can be found in Appendix VII. The star indicates an exercise that is more challenging than average.

lybdenum atom to the mass of an atom of $^{12}C$ is 8:1. Find the mass of 1.0 mol of the molybdenum atoms.

**3.13**  The ratio of the mass of one molecule of NO to the mass of one atom of $^{12}C$ is 5:2. The mass of an atom of O is 16 u. Find the mass of the N atom (in atomic mass units and grams).

**3.14**  The ratio by mass of S to O in $SO_3$ is 2:3. The ratio of O to $^{12}C$ by mass is 4:3. Find the mass of 1 mol of $SO_3$.

**3.15**  What is the disadvantage of using a bar of pure gold as the standard for defining an atomic weight scale?

**3.16**  Suppose that atomic weights were assigned relative to $^1H$, which is assigned a mass of exactly 1 u. Find the atomic weight of He on this scale. ($^1H$ has a mass of 1.007825 u on the $^{12}C$ scale.)

**3.17**  Before 1961, chemists used $^{16}O$ as the standard for atomic weights, assigning it a mass of exactly 16 u. Find the relative atomic mass of $^{12}C$ on that scale. (The mass of $^{16}O$ on the $^{12}C$ scale is 15.9949149.)

**3.18**  Suppose we redefine the mole to be the amount of substance with the same number of entities as there are in exactly 1 g of $^1H$ (relative atomic mass 1.007825 u). Find the new value of Avogadro's number.

**3.19**  A sample of a binary compound with the formula XY is found to contain X and Y in the ratio of 1.70:1.00 by mass. In a separate measurement, 1.00 mol of element X is found to have 11.7 times the mass of 1.00 mol of $^{12}C$. Find the atomic weight of element Y.

**3.20**  Naturally occurring lithium consists of two isotopes, $^6Li$ and $^7Li$, whose relative atomic masses are

6.0151 and 7.0160 and whose natural abundances are 7.42% and 92.58%, respectively. Use these data to calculate the atomic weight of lithium.

**3.21**  Naturally occurring magnesium consists of three isotopes: $^{24}Mg$, $^{25}Mg$, and $^{26}Mg$, whose relative atomic masses are 23.985, 24.986, and 25.983, and whose natural abundances are 78.70%, 10.13%, and 11.17%, respectively. Use these data to calculate the atomic weight of magnesium.

**3.22**  The atomic weight of boron is 10.81. It consists of two isotopes, $^{10}B$ and $^{11}B$. The relative atomic mass of $^{10}B$ is 10.013 and its natural abundance is 19.78%. Find the relative atomic mass of $^{11}B$.

**3.23**  Naturally occurring copper consists of two isotopes, $^{63}Cu$ and $^{65}Cu$, whose relative atomic masses are 62.930 and 64.928. The atomic weight of copper is 63.546. Find the natural abundances of the two isotopes.

**3.24**  Naturally occurring cobalt consists only of $^{59}Co$, whose atomic weight is 58.9332. A common synthetic isotope of cobalt that is used for radiation therapy for cancer is $^{60}Co$, which has a relative atomic mass of 59.9338. A sample of naturally occurring cobalt of mass 1.4982 g is accidentally contaminated with 0.0904 g of $^{60}Co$. Find the apparent atomic weight of the contaminated sample.

**3.25**  Find the molecular weight of each of these substances using the data in Table 3.1: (a) $N_2O_4$, (b) $POCl_3$, (c) nitrogen tribromide, (d) sulfurous acid.

**3.26**  Find the molar mass of each of these substances: (a) $Ba(NO_3)_2$, (b) $Mg_3(PO_4)_2$, (c) sodium bicarbonate, (d) potassium sulfate.

**3.27**  The general formula for the

class of hydrocarbons called the alkanes is $C_nH_{2n+2}$, where $n$ is the number of carbon atoms in a given member of the class. Derive a general formula that relates the molecular weight of an alkane to the number of carbon atoms it contains.

**3.28** Find the mass of 2.97 mol of each of these substances: (a) nitric acid, (b) $Al_2(SO_4)_3$, (c) phosphorus pentachloride.

**3.29** Find the mass of 0.639 mol of each of these substances: (a) glucose, $C_6H_{12}O_6$, (b) meperidine, $C_{15}H_{21}NO_2$, a widely used narcotic painkiller.

**3.30** Find the amount (in moles) of each of the following substances in a sample of mass 104 g: (a) $Ca_3(PO_4)_2$, (b) sodium fluoride, (c) $K_2HPO_4$.

**3.31** Find the amount (in moles) of each of the following substances in a sample of mass 0.0742 g: (a) barbituric acid, $C_4H_4N_2O_3$, (b) progesterone, $C_{21}H_{30}O_2$, an important steroid hormone.

**3.32** Under ordinary conditions, phosphorus exists primarily as $P_4$ molecules. It forms $P_2$ if it is heated to 800°C and it forms P at still higher temperatures. Find the amount (in moles) of each of these three forms of phosphorus that has a mass of 0.835 g.

**3.33** The mass of 0.0658 mol of a substance is 2.43 g. Find the molar mass of the substance.

**3.34** The mass of 0.00227 mol of the compound $XOF_3$ is 0.236 g. Find the atomic weight of element X.

**3.35** Two substances have molecular weights in a ratio of 2 : 1. Find the ratio of the masses of equal amounts of these substances. Find the ratio of the amounts of equal masses of these substances.

**3.36** Find the number of molecules in 11.6 mol of a substance.

**3.37** Find the number of molecules in a sample of $H_2O_2$ of mass 7.11 g.

**3.38** Find the number of molecules in a sample of sucrose, $C_{12}H_{22}O_{11}$, that has a mass of $5.64 \times 10^{-3}$ g.

**3.39** Express the amount of $8.17 \times 10^{25}$ molecules of $H_2$ in moles.

**3.40** Find the mass of $1.0 \times 10^{20}$ molecules of chloroform, $CHCl_3$.

**3.41** Find the mass of $1.0 \times 10^9$ molecules of a protein of molecular weight $1.0 \times 10^6$.

**3.42** Find the amount (in moles) of hydrogen in 1.67 mol of octane, $C_8H_{18}$.

**3.43** Find the amount (in moles) of O in 1.09 mol of each of the following: (a) $HClO_4$, (b) $Ca_3(PO_4)_2$, (c) $Na_2Cr_2O_7$.

**3.44** Find the number of H atoms in 0.0396 mol of the following: (a) $H_2O$, (b) $(NH_4)_2HPO_4$, (c) $C_6H_6O$.

**3.45** Find the total number of atoms in 0.00713 mol of morphine, $C_{17}H_{19}NO_3$.

**3.46** Find the number of H atoms in 1.03 g of lithium aluminum hydride, $LiAlH_4$, a reagent used to add hydrogen to many compounds.

**3.47** Find the total number of atoms in 17.8 mg of cocaine hydrochloride, $C_{17}H_{22}ClNO_4$, which is used as a topical anesthetic.

**3.48** The compound $NH_4NO_3$ is widely used in agriculture to add nitrogen to soil. Its price is $2.15 per kilogram. Find the cost per mole of nitrogen of this fertilizer.

**3.49** A sample of $Ag_2SO_4$ of mass 5.56 g is found to contain 3.85 g Ag, 0.572 g S, and 1.14 g O. Use these data to calculate the mass percent of each element in this compound.

**3.50** Calculate the fraction by mass of each element in $KNO_3$, saltpeter, which has many applications including

use in the manufacture of gunpowder and the pickling of meats.

**3.51** Calculate the fraction by mass of each element in the antibiotic penicillin, $C_{16}H_{18}N_2O_4S$.

**3.52** Find the percentage by mass of magnesium in these minerals: (a) magnesite, $MgCO_3$, (b) asbestos, $H_4Mg_3Si_2O_9$, (c) talc, $Mg_3Si_4O_{10}(OH)_2$.

**3.53** A 3.38-g sample of alum, $KAl(SO_4)_2$, a common astringent, is found to contain 0.839 g S and 1.68 g O. The masses of K and Al are in the ratio of 13.0 : 9.0. Use only these data to calculate the mass percent of each element in the compound.

**3.54** A sample of a compound of boron and chlorine is 9.23% B and 90.77% Cl by mass. Find its empirical formula.

**3.55** A compound is found to have the following composition by mass: 0.0369 H, 0.378 P, and 0.585 O. Find its empirical formula.

**3.56** A compound of nitrogen, chlorine, and fluorine is 16.02% nitrogen and 40.53% chlorine by mass. Find its empirical formula.

**3.57** A 2.09-g sample of pure Ru metal is heated in oxygen until its mass no longer increases. The resulting oxide has a mass of 2.75 g. Find its empirical formula.

**3.58** A 3.58-g sample of a compound of sodium, silicon, and oxygen is found to contain 1.35 g Na and 0.824 g Si. Find its empirical formula.

**3.59** The atomic weights of elements X and Y are in the ratio of 2.0 : 1.0. A compound of X and Y has masses of X and Y in the ratio of 1.5 : 1.0. Find the empirical formula of the compound.

**3.60★** The elements Q and Z form three different binary compounds. The first compound has the empirical formula $Q_2Z_3$ and forms from 10.00 g of Q and 17.14 g of Z. The second compound forms from 10.00 g of Q and 22.86 g of Z, and the third compound forms from 10.00 g of Q and 28.57 g of Z. Find the empirical formulas of the second and third compounds. Find the ratio of the atomic weights of Q and Z.

**3.61** A compound of a metal M and chlorine has the formula $MCl_3$. It is 67.2% Cl by mass. Find the atomic weight of the metal.

**3.62** A compound of a metal M and oxygen has the empirical formula $M_3O_4$ and is 72.36% M by mass. Find the atomic weight of M.

**3.63** Indicate which of the following are empirical formulas: (a) $C_5H_{12}$, (b) $C_6H_6N_2O$, (c) $C_9H_{12}Cl_3O_3$, (d) $C_{44}H_{56}N_8O_4$.

**3.64** The empirical formula of the compound putrescine is $C_2H_6N$. Its molecular weight is 88.2. Find its molecular formula.

**3.65** Find the number of carbon atoms in the compound whose empirical formula is $CH_2$ and whose molecular weight is 1022.

**3.66** A compound containing only C, H, Br, and O has a molecular weight of 150. The mass of C in the compound is 8.0 times the mass of H. Find its molecular formula.

**3.67** Upon analysis by combustion, a compound containing only C and H produces 2.402 g of $CO_2$ and 0.737 g of $H_2O$. Find its empirical formula.

**3.68** A compound containing only C and H produces $CO_2$ and $H_2O$ in the ratio of 4.0 : 1.0 by mass on complete combustion. Find its empirical formula.

**3.69** A 1.40-g sample of a compound containing only C, H, and O produces 2.56 g of $CO_2$ and 0.700 g of $H_2O$ on complete combustion. Find its empirical formula.

**3.70** A 1.66-g sample of a compound containing only C, H, and N produces 4.63 g of $CO_2$ and 0.928 g of $H_2O$ on complete combustion. Find its empirical formula.

**3.71** Cytosine, an important constituent of nucleic acids, contains C, H, N, and O. A 3.666-g sample is subjected to combustion analysis and produces 5.808 g of $CO_2$ and 1.484 g of $H_2O$. In a separate experiment, a 4.555-g sample of cytosine is found to contain 1.732 g of N. Find the empirical formula of cytosine.

**3.72** The compound $BrF_5$ decomposes to $Br_2$ and $F_2$ on heating. Find the amount of $F_2$ formed from the decomposition of (a) 1.0 mol of $BrF_5$, (b) 0.0821 mol of $BrF_5$.

**3.73** The reaction of $P_4$ with $Cl_2$ produces $PCl_3$. Find the amounts of $P_4$ and $Cl_2$ required to form (a) 1.0 mol of $PCl_3$, (b) 17.9 mol of $PCl_3$.

**3.74** When metallic tin dissolves in nitric acid ($HNO_3$), $SnO_2$, $NO_2$, and $H_2O$ are the only products. Find the amount of $NO_2$ that forms when 22.1 g of Sn dissolves.

**3.75** An important process used to remove phosphorus impurities during the manufacture of steel is the reaction between MgO and $P_4O_{10}$ to form $Mg_3(PO_4)_2$. Find the amount of MgO required to react with 73.8 kg of $P_4O_{10}$.

**3.76** The reactions of photosynthesis, the process by which green plants produce sugar, can be summarized in one reaction in which carbon dioxide and water form glucose, $C_6H_{12}O_6$, and $O_2$. Find the mass of $CO_2$ required to form 0.100 mol of glucose.

**3.77** The direct high-temperature reaction of $N_2$ and $H_2$ is used to manufacture $NH_3$. Find the mass of $H_2$ required to prepare $5.00 \times 10^3$ mol of $NH_3$.

**3.78** Limestone, $CaCO_3$, decomposes to CaO and $CO_2$ on heating. Find the mass of CaO that can be formed from the decomposition of 6.22 g of limestone.

**3.79** When $FeCr_2O_4$, an important ore of chromium, is heated with carbon, it forms iron, chromium, and carbon monoxide. Find the mass of C required to form $1.50 \times 10^3$ kg of Cr.

**3.80** Designers of nuclear weapons have made an intensive study of the fluorine compounds of uranium. One of them, $UF_5$, readily forms $UF_6$ and $U_2F_9$ if left standing. Find the mass of $UF_6$ that can form from the complete reaction of 2.32 g of $UF_5$.

**3.81** Para-dichlorobenzene, $C_6H_4Cl_2$, is an insecticidal fumigant often used against moths. It can be prepared by the reaction of benzene, $C_6H_6$, with $Cl_2$ to form $C_6H_4Cl_2$ and HCl. Find the mass of HCl produced during the manufacture of $6.85 \times 10^3$ kg of $C_6H_4Cl_2$.

**3.82** A sample of butane, $C_4H_{10}$, is subjected to complete combustion. It forms 0.843 g of $CO_2$. Find the mass of the sample of butane.

**3.83** Decomposition of 0.0887 mol of HI produces $H_2$ and $I_2$. The $I_2$ is then used to form $PI_3$ from $P_4$. Find the amount of $PI_3$ that forms.

**3.84** A sample of propane, $C_3H_8$, is subjected to complete combustion. The $CO_2$ that forms is collected and added to a solution of barium nitrate. A precipitate of $BaCO_3$ forms, is collected, and is heated to form 0.0535 mol of BaO. Find the amount of propane that was burned.

**3.85** Compounds of silicon and hydrogen are called silanes. One silane, $Si_2H_6$, reacts with water to form $SiO_2$ and $H_2$. When the $H_2$ from this reaction is used to convert $Cu_2O$ to Cu and $H_2O$, the mass of Cu obtained is 5.57 g. Find the amount of $Si_2H_6$.

**3.86** A 75.8-g sample of $Fe_2O_3$ is

treated with carbon monoxide to form iron metal and $CO_2$. The iron metal is treated with dilute HCl to form $FeCl_2$, which is treated with an excess of KCN to form potassium ferrocyanide, $K_4Fe(CN)_6$. Treatment of this substance with $Cl_2$ forms $Cl^-$ ions and potassium ferricyanide, $K_3Fe(CN)_6$, a compound that is used for making blueprints. Find the mass of potassium ferricyanide that forms.

**3.87★** The reaction of $NH_3$ and $O_2$ forms NO and water. The NO can be used to convert $P_4$ to $P_4O_6$, forming $N_2$ in the process. The $P_4O_6$ can be treated with water to form $H_3PO_3$, which forms $PH_3$ and $H_3PO_4$ when heated. Find the mass of $PH_3$ that forms from the reaction of 1.00 g of $NH_3$.

**3.88** Oxides of cobalt react with $H_2$ to form cobalt metal and water. A sample of 0.555 mol of an oxide of cobalt of unknown formula is found to react with 4.50 g of $H_2$ to form 98.7 g of cobalt. Find the formula of the oxide.

**3.89** To assemble a tricycle, a manufacturer requires one front wheel, two back wheels, and two handlebars. Find the number of tricycles that can be made from 40 front wheels, 50 back wheels, and 60 handlebars.

**3.90** Find the limiting reagent when equal masses of the following substances are mixed: (a) $H_2$ and $O_2$ to make water, (b) $Cl_2$ and $F_2$ to make $ClF_3$, (c) $S_8$ and $O_2$ to make $SO_2$.

**3.91** Find the mass of HI that forms when 1.00 g of $I_2$ is mixed with 0.0100 g of $H_2$.

**3.92** A 0.0875-g sample of Pt metal is treated with $Br_2$ to form 0.1105 g of $PtBr_4$. Find the limiting reagent.

**3.93** Silicon tetrachloride, $SiCl_4$, is prepared from $SiO_2$, $Cl_2$, and C in a reaction whose other product is CO. Find the mass of unreacted starting material after the complete reaction of 1.00 kg of each of the three reactants.

**3.94★** Triple superphosphate of lime is a fertilizer whose formula is $Ca(H_2PO_4)_2$. It is prepared from $Ca_3(PO_4)_2$, which costs 89.7 cents per kilogram, and $H_3PO_4$, which costs 75.3 cents per kilogram. To ensure that the reaction proceeds to completion, manufacturers use an excess of 10% by mass of one of the reagents. Which reagent should be used in excess to keep costs to a minimum?

**3.95** The reaction of CO and $H_2$ at high temperature and pressure to form methanol, $CH_3OH$, is important in industry. Find the yield of methanol when $23.9 \times 10^3$ kg of CO reacts to form $25.7 \times 10^3$ kg of $CH_3OH$.

**3.96** Arsine, $AsH_3$, is prepared from the reaction of $As_4O_6$ with zinc in strong acid. The balanced equation is $As_4O_6(s) + 12Zn(s) + 24H^+(aq) \rightarrow 4AsH_3(g) + 12Zn^{2+}(aq) + 6H_2O$. If a total of 56.7 mg of $AsH_3$ is isolated from the reaction of 87.1 mg of $As_4O_6$ with an excess of zinc, find the yield of $AsH_3$.

**3.97** Large quantities of formaldehyde, $CH_2O$, can be prepared in 87.1% yield from the controlled oxidation of methanol, $CH_3OH$, by $O_2$. Find the mass of methanol needed to prepare 675 kg of formaldehyde.

**3.98** Manganese can be isolated from $MnO_2$ by the thermite reaction, in which the oxide reacts with aluminum metal, forming Mn and $Al_2O_3$ and liberating a large amount of heat. A mixture of 227 g of $MnO_2$ and 73.4 g of Al reacts to form 92.0 g of Mn. Find the yield of Mn.

**3.99** A compound of a metal M and chloride is believed to have the formula $MCl_3$ and is found to be 67.2% Cl by mass. Find the atomic weight of M.

**3.100** The metal Q forms the oxide $Q_2O_5$ on complete reaction with $O_2$. A 27.9-g sample of Q requires 12.0 g of $O_2$ for complete reaction. Find the atomic weight of Q.

**3.101** A compound of a metal M and an oxyanion of a nonmetal X has the formula $M(XO_2)_2$. When 16.7 g of this compound is heated, it decomposes completely to $MX_2$ and 4.80 g of $O_2$. The $MX_2$ is dissolved in water and treated with NaOH, forming 9.12 g of $M(OH)_2$. Find the atomic weight of X.

**3.102** Find the mass of NaCl and the mass of water needed to prepare 750 g of a solution that is 11.0% NaCl by mass.

**3.103** Find the amount of $NH_4Cl$ that must be dissolved in 18.5 g of $H_2O$ to prepare a solution that is 33.3% $NH_4Cl$ by mass.

**3.104** Find the volume of a $1.45M$ solution of formic acid that contains 0.0685 mol of formic acid.

**3.105** Find the mass of KCl required to prepare 0.500 L of a $0.547M$ solution of KCl.

**3.106** Find the molarity of a solution that contains 1.94 g of $Na_2HPO_4$ in 75.0 cm³ of solution.

**3.107** Find the molarity of $Br^-$ in a solution prepared by mixing 53.3 cm³ of $0.43M$ KBr solution with 26.8 cm³ of $0.21M$ $CaBr_2$ solution.

**3.108** Find the amount of AgBr that precipitates when 0.151 L of $0.441M$ $AgNO_3$ solution is treated with an excess of NaBr solution.

**3.109** Find the mass of $SrF_2$ that precipitates when 1.05 L of $0.281M$ NaF solution is treated with an excess of $Sr(NO_3)_2$ solution.

**3.110** When 0.148 L of a $0.241M$ solution of $Mg(NO_3)_2$ is mixed with 0.162 L of a $0.287M$ solution of $Na_3PO_4$, a precipitate of $Mg_3(PO_4)_2$ forms. Find the limiting reagent.

**3.111** Find the volume of a $0.385M$ solution of HCl needed to react completely with 1.89 g of $Ba(OH)_2$. The reaction is $H^+(aq) + OH^-(aq) \rightarrow H_2O$.

**3.112** Find the volume of a $3.95M$ $H_2SO_4$ solution needed to react completely with 50.6 cm$^3$ of $0.87M$ $Ba(OH)_2$.

**3.113** A 23.7-cm$^3$ volume of $AgNO_3$ solution requires 47.2 cm$^3$ of a $0.438M$ NaCl solution to precipitate all the silver as AgCl. Find the molarity of the $AgNO_3$ solution.

**3.114** Use the data in Table 3.2 to find the mass of 25.0 cm$^3$ of (a) butane, (b) mercury, (c) diamond, (d) platinum.

**3.115** Use the data in Table 3.2 to find the volume of 25.0 g of (a) water, (b) ethanol, (c) lead, (d) osmium.

**3.116** When 15.7 g of $NH_4Cl$ is dissolved in 36.9 g of water, the volume of the resulting solution is 0.0412 L. Find the density of the solution.

**3.117** An aqueous solution is 16.2% glucose by mass. Its density is 1.23 g/cm$^3$. Find the mass of water in 1.00 L of the solution.

**3.118** An aqueous solution of NaOH is 11.2% NaOH by mass. Its density is 1.14 g/cm$^3$. Find the molarity of NaOH in the solution.

**3.119** A solution of ethanol in water has a density of 0.963 g/cm$^3$. The concentration of ethanol is $5.61M$. Find the mass percent of ethanol in the solution.

**3.120** Suggest an explanation for the observation that the molarity of a solution changes with temperature but the mass percent does not.

**3.121** A mixture of carbon and sulfur has a mass of 14.9 g. Complete combustion gives a mixture of $CO_2$ and $SO_2$ with a mass of 35.8 g. Find the mass of carbon in the original mixture.

**3.122** A mixture of CaO and $Al_2O_3$ has a mass of 388 g and is 41.6% O by mass. Find the mass of CaO in the mixture.

**3.123★** A mixture of 20.6 g of P and 79.4 g of $Cl_2$ reacts completely to form $PCl_3$ and $PCl_5$ as the only products. Find the mass of $PCl_3$ formed.

**3.124★** A solution contains $Ag^+$ and $Hg^{2+}$ ions. The addition of 0.100 L of $1.22M$ NaI solution is just enough to precipitate all the ions as AgI and $HgI_2$. The total mass of the precipitate is 28.1 g. Find the mass of AgI in the precipitate.

**3.125★** A mixture of $C_3H_8$ and $C_2H_2$ has a mass of 2.0 g. Complete combustion with excess $O_2$ gives a mixture of $CO_2$ and $H_2O$ that contains 1.5 times as many moles of $CO_2$ as of $H_2O$. Find the mass of $C_3H_8$ in the original mixture.

# 4

# States of Matter

**Preview**  **W**hy does matter on earth exist in three different phases, solid, gas, and liquid? We begin this chapter by answering that question. We then discuss how and why a vapor forms above a liquid or a solid, and we proceed to consider vapor pressure and the equilibrium state. We continue our study of the gaseous phase of matter by presenting the concept of an ideal gas and by describing the laws that relate temperature, pressure, and volume in the ideal gas. We then introduce an extremely useful model, the kinetic theory of gases, and show how it can be used to derive Boyle's law and to define relationships between temperature and kinetic energy of gases. The chapter defines and discusses some important aspects of the gaseous phase, including effusion and diffusion, the differences between the ideal gas and real gases, and critical temperature and critical pressure of gases.

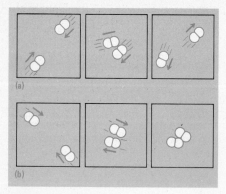

**Figure 4.1**
Interaction of two Br$_2$ molecules at different temperatures. At a relatively high temperature (a), the two molecules are attracted to one another, but their thermal motion is great enough to prevent their condensation. At a lower temperature (b), the thermal motion of the molecules is too small to overcome the attractive forces.

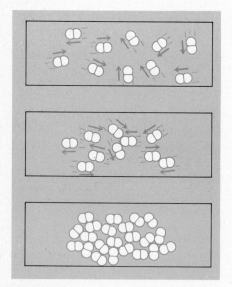

**Figure 4.2**
Formation of the condensed phase of a substance. At high temperatures, thermal motion keeps the molecules in the gas phase, moving randomly. As the temperature is lowered, the attractive forces between molecules become relatively more important, and some of them "stick" together. When the temperature is low enough, the molecules condense; the substance is in the liquid phase.

## 4.1  STATES OF MATTER

Matter on earth exists in three different states: solid, liquid, and gas. Although the vast majority of the substances on earth are solids, three-quarters of the earth's surface is covered by a single liquid; and the earth has an atmosphere composed of nitrogen, oxygen, argon, and other gases.

Temperature and pressure determine whether a substance is a solid, a liquid, or a gas. We can understand why by examining the microscopic structure of matter. By "microscopic structure" we mean the arrangement of the atoms, molecules, or ions of which a substance is composed.

### Why Three States of Matter?

The state of any substance is determined by a balance between two opposing tendencies. One acts to keep atoms, molecules, or ions apart. The other acts to bring them together.

Molecules[1] tend to separate because they are in constant motion. Thermal energy (Section 2.8), which increases as temperature rises, is responsible for this motion. The tendency to separate is counteracted by electrostatic forces of attraction, which become stronger as molecules come closer to each other.

In Figure 4.1(a), two Br$_2$ molecules at a relatively high temperature approach each other. Their thermal motion is great enough to overcome the attractive force between them, so the molecules remain apart.

In Figure 4.1(b), the temperature is lower. The thermal motion of the two Br$_2$ molecules is not great enough to overcome their mutual attraction, and they "stick" together.

### Characteristics of the States of Matter

The most significant characteristic of a **gas** is that its molecules move freely in space. They are in a completely random arrangement. A substance is in the gas phase when the temperature is high enough for thermal motion to overcome the attractive forces between its molecules. Because its molecules are widely separated, a gas is mostly empty space. Because its molecules are in constant random motion, a gas has no definite shape or volume. It assumes the shape and volume of its container. A gas is **expansible:** A small amount expands to fill a large container. A gas is **compressible:** A large amount can be squeezed into a small container.

When thermal motion is not great enough to overcome the attractive forces, molecules stick together in a **condensed phase** (Figure 4.2). One condensed phase is the **liquid state.** Thermal motion is still substantial in the liquid state, but the attractive forces prevent a completely random arrangement of the molecules.

[1] For convenience, in the following discussions we shall use the word *molecule* to describe all species, including atoms or ions, that are the basic units of structure of substances.

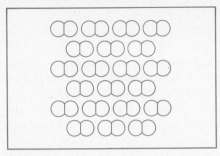

**Figure 4.3**
The solid phase. Molecules are in an ordered, nonrandom arrangement, rather than the relatively disordered arrangement of the gas phase. The arrangement of the molecules in the solid phase can influence the shape of the solid.

A liquid does not have a definite shape. Usually, it takes its shape from the portion of the container it occupies. A liquid does have a definite volume. While 1 g of gaseous $Br_2$ has the volume of the container it happens to occupy, 1 g of liquid $Br_2$ occupies a volume of 0.32 cm³ at 298 K and atmospheric pressure.

If the temperature of a liquid is lowered, a point is reached at which thermal motion becomes relatively unimportant. The molecules arrange themselves in a way that maximizes the attractive forces between them (Figure 4.3). The partial randomness of the liquid phase is replaced by a more ordered arrangement. In this condensed phase, called the **solid state,** the molecules generally are closer together than in the liquid state. The solid phase of a substance generally is denser than the liquid.

The major difference in the microscopic structure of gases, liquids, and solids involves randomness and order. A gas has essentially no order. A liquid has *short-range order.* The molecules in a small region of the liquid may have an orderly arrangement, but this arrangement is not repeated throughout the liquid. Most solids have *long-range order.* The arrangement of the molecules in a small region is repeated throughout the solid. The molecules in a solid are ordered in a regular array.

Solids usually are crystalline. Their elementary units—molecules, atoms, or ions—are in an ordered arrangement that creates a regularly repeating three-dimensional pattern. On the macroscopic level, such an ordered arrangement appears as a crystal that is a polyhedron, a solid whose faces are planes. Some crystals are very large, even several meters across. Some are very small and can be seen only with the help of an electron microscope. The three-dimensional pattern of a substance is always the same throughout the crystal, no matter how large or how small the crystal is.

The study of the shape of crystals and the arrangement of matter in them is called *crystallography.* We shall discuss it in detail in Chapter 10.

Not every substance can be classified neatly as a solid, liquid, or gas. Under appropriate conditions, some substances seem to have intermediate properties. A glass is a material that has characteristics of both a liquid and a solid. While a glass has the mechanical properties of a solid, it does not have a crystalline structure. Rather than melting at a given temperature to a liquid, it softens gradually when heated. Liquid crystals, a recent arrival on the scene, have both liquid and solid characteristics. As we shall see in Chapter 10, glasses and liquid crystals play important roles in technology and biology.

## 4.2 VAPOR PRESSURE

Temperature and pressure determine whether the elementary units of a substance form a condensed phase. Most chemistry is performed at normal atmospheric pressure. Therefore, we can discuss the effect of temperature changes with the assumption that the pressure of the system being considered is close to atmospheric pressure.

## Condensed Phases and Vapor

We said that when temperatures are high, thermal motion is great enough to overcome attractive forces, so a substance exists as a gas. As temperature is lowered, a condensed phase—usually but not always a liquid—forms from the gas. The system then includes not only a condensed phase, but also some gas.

For example, $Br_2$ is a dark reddish-brown liquid at room temperature. A red vapor, gaseous $Br_2$, can be seen rising from its surface. Indeed, a sample of liquid $Br_2$ that is left in an open container soon evaporates completely. If the container is closed and is not too large, the red vapor fills the container, although most of the $Br_2$ remains liquid.

A similar observations can be made with $I_2$, which is a solid at room temperature but gives off a purple vapor. Crystals of solid $I_2$ will disappear if left in an open container.

Observations such as these led, long ago, to the conclusion that some vapor or gas is always associated with a condensed phase. Thermal motion causes the vapor to form. Although we have implied that the thermal motion of any given molecule is proportional to temperature, in fact it is the *average* thermal motion of a large number of molecules that is proportional to temperature. At any given temperature, there is a distribution of the molecular energies associated with thermal motion (Section 4.10). Some molecules have less thermal motion than the average, and some have more. Every now and then, the thermal motion of a molecule on the surface of a liquid or a solid is great enough to overcome the attractive forces, and the molecule escapes into the gas phase (Figure 4.4). The process by which entities escape from the liquid phase to the gas phase is called **evaporation,** for example, $Br_2(l) \rightleftharpoons Br_2(g)$. The same process for a solid is called **sublimation,** for example, $I_2(s) \rightleftharpoons I_2(g)$.

Knowledge of these processes raises many questions. Why don't all solids sublime? Why don't all liquids (including the oceans) evaporate? In small, closed systems, why do we see such substances as $Br_2$ and $I_2$ existing partly in the condensed phase and partly in the gas phase?

Most solids don't sublime because the attractive forces between their units of structure are so great that only a negligibly small fraction of entities leave the surface at room temperature. The oceans do evaporate, constantly. But the water vapor forms clouds and falls as rain, which eventually finds its way back to the oceans. In a closed container, the molecules of $Br_2$ or $I_2$ that are in the gas phase cannot escape. Every now and then, a molecule strikes the surface of the condensed phase and is held on that surface by attractive forces. This process is called **condensation** (Figure 4.5). Examples are $Br_2(g) \rightleftharpoons Br_2(l)$ and $I_2(g) \rightleftharpoons I_2(s)$.

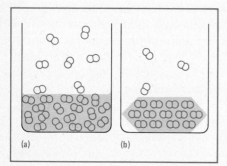

**Figure 4.4**
(a) Evaporation. Some molecules on the surface of a liquid have sufficient thermal motion to break free and enter the gas phase. (b) Sublimation. Some molecules on the surface of a solid have sufficient thermal motion to break free and enter the gas phase.

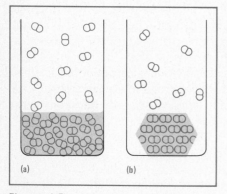

**Figure 4.5**
Condensation. Molecules in the gas phase come close to the surface of a liquid (a) or solid (b). Some of the molecules "stick" to the surface and become part of the condensed phase.

## The Equilibrium State

When a substance in a condensed state is placed in a closed container, evaporation or sublimation occurs. The rate at which vapor forms is determined primarily by a balance between attractive forces and thermal motion. The rate of condensation is determined by these two factors and by the rate at which molecules in the vapor collide with the condensed state.

When evaporation or sublimation begins, there are few molecules in the vapor, so the rate of condensation is small. As evaporation or sublimation continues, the number of molecules in the vapor increases, and so does the rate of condensation. Eventually, the rate of condensation equals the rate of evaporation or sublimation; the number of molecules going into the gas phase equals the number of molecules leaving the gas phase. To our eyes, the system does not seem to be changing in any way. The quantities of material in the vapor phase and the condensed phase remain the same.

When a system exists in a state that does not undergo a detectable change unless it is disturbed, the system is said to be in **equilibrium.** This system is in the equilibrium state.

The equilibrium system appears *static,* but only because we cannot follow individual molecules. The system is actually *dynamic.* On the molecular level, it is changing rapidly and continuously. The net result of all the changes on the molecular level, however, is no overall change on the macroscopic level, so the dynamic system appears static to us.

Any quantity of a condensed phase of a substance that is placed in a closed container that is not too large will reach such a dynamic equilibrium eventually. The actual quantity of the substance that goes into the vapor phase depends on several factors. One is the volume of the container. At a given temperature, the amount of a substance in the vapor phase increases with the volume of the container.

It is not always necessary to consider the actual volume of the container. Instead of trying to measure the *total* amount of the substance in the vapor phase, we can measure the amount in the vapor phase per unit of volume. If we measure by units of volume — for example, the number of gas molecules in 1 $cm^3$ of volume, or the number of moles of gas in 1 L — we can ignore the size of the container.

## Pressure

A direct measurement of the number of gas molecules in 1 $cm^3$ or the number of moles of gas per liter is not easy. In theory, we could first weigh the empty container and then get the mass of the gas by weighing the full container. This approach is impractical because the mass of a gas usually is very small compared to that of the container. Instead, a more easily measured quantity called **pressure** is customarily used to describe the amount of substance in the gas phase.

*Pressure is the force exerted by a gas on a given unit of area of surface with which the gas is in contact.* In a small, closed container, pressure is exerted on the walls of the container. On earth, the gases of the atmosphere exert pressure on the surface of the earth and everything exposed to the atmosphere.

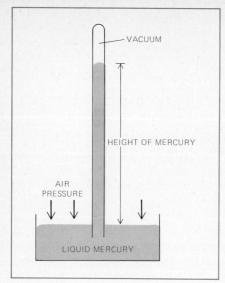

**Figure 4.6**
A barometer. The column of mercury is supported by the pressure of air on the open pool of mercury. As air pressure increases or decreases, the height of the column of mercury increases or decreases. Thus, air pressure can be measured by measurement of the height of the column of mercury.

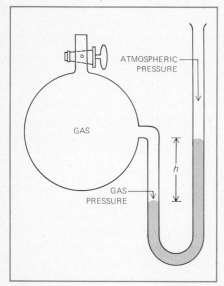

**Figure 4.7**
A manometer. Similar in principle to a barometer, the manometer allows measurement of gas pressure. Since the atmospheric pressure is known, the gas pressure can be determined by measurement of $h$, the difference in levels of mercury in the two arms of the tube. If $h = 0$, then the gas pressure is equal to atmospheric pressure.

The phenomenon of gas pressure is familiar to all of us. A child sees what pressure can do when a balloon explodes because it cannot withstand the force exerted by the gas within. Sucking on a straw that is dipped into a liquid creates a partial vacuum within the straw; atmospheric pressure then pushes the liquid up the straw. Evangelista Torricelli (1608–1647), a student of Galileo, first recognized that the pressure exerted by the atmosphere could be measured from such phenomena. Torricelli used the height of a column of liquid in the simple device called the **barometer** (Figure 4.6) to measure atmospheric pressure. A decrease in atmospheric pressure causes a drop in the height of the column, while an increase in atmospheric pressure increases its height.

The height of a column of liquid supported by atmospheric pressure depends in part on the density of the liquid. At sea level, the atmosphere can support a column of water more than 10 m high, or a column of mercury 0.76 m high. For convenience, mercury is used in most liquid-level barometers.

Gas pressure can also be measured by a U-tube manometer (Figure 4.7), which works on the same principle as a barometer. In a manometer, the pressure of a gas is related to the difference between the heights of two columns of liquid. The gas pressure in Figure 4.7 is the pressure of the atmosphere plus the pressure exerted by the height ($h$) of the mercury.

Chemists have commonly measured pressure in units of atmospheres (atm), millimeters of mercury (mmHg), or torr (named for Torricelli). Torr and mmHg are virtually identical. The unit atmosphere (atm) has a simple origin: Atmospheric pressure at sea level averages about 1 atm. The other units are related to the atmosphere: 1 atm = 760 mmHg = 760 Torr. Because so much chemistry is done at pressures close to 1 atm, we find it generally convenient for our purposes to use the atmosphere as the unit for pressures.

It would be more precise if we described units of pressure as so much force per area, since pressure is force per area. The SI uses such a unit. In the SI, the fundamental unit of force is called the newton (N); 1 N = 1 kg m s$^{-2}$. A newton is approximately equal to the force required to suspend a mass of 100 g above the ground against the pull of gravity. The unit of area is the square meter (m$^2$). The unit of pressure (force per area) is therefore the newton per square meter (N m$^{-2}$), which is called the pascal (Pa). A pressure of 1 atm is defined as exactly 101 325 Pa, while a pressure of 1 mmHg or 1 Torr is about 133 Pa.

## Vapor Pressure

The pressure of the vapor of a substance in equilibrium with its condensed phase is called the **vapor pressure** of the substance. Innumerable experiments have shown that *the vapor pressure of any substance increases as its temperature increases.* Vapor pressure also depends on the attractive forces between the units of structure in the condensed phase. *The weaker the attractive forces, the higher is the vapor pressure of a substance at a*

## WATER VAPOR IN THE AIR

We could describe the moisture content of the atmosphere in terms of vapor pressure, but this description is seldom used. Instead, we usually hear the moisture content of air stated in terms of humidity.

Humidity is simply another name for the water vapor content of air. Humidity varies greatly from place to place on earth. Water may constitute as little as $1/1000$ by mass of the cold winter air of Canada and as much as $18/1000$ by mass of the hot, moist air of the Brazilian rain forest. One measure of moisture content is the absolute humidity of air, the mass of water vapor in a given volume of air. Absolute humidity usually is expressed as grams of water vapor per cubic meter of air. At a temperature of 310 K, air can hold 47 g of water vapor per cubic meter. Air at this temperature is rarely this moist, however, because this relatively high temperature is most often found in dry areas.

Absolute humidity is not usually given in weather reports because the absolute humidity varies as changes in pressure cause changes in air density. As a given mass of air becomes less dense, its absolute humidity decreases, and vice versa. Weather reports usually describe the relative humidity, the ratio of the actual amount of water vapor present to the maximum quantity that a volume of air can hold, the saturation point. A relative humidity of 0 means that air is completely dry, and a relative humidity of 100 means that air is saturated.

The relative humidity of a given sample of air changes with temperature. If temperature goes up and moisture is neither added nor lost, relative humidity goes down. For example, air that has a relative humidity of 50% at 288 K will have a relative humidity of 20% if it is heated to 305 K without gaining or losing moisture.

In cold climates the heated air in homes can be exceptionally dry because of this effect. As cold air enters a house and is heated, its relative humidity goes down, sometimes to less than 1%. On the other hand, if air is cooled, its relative humidity may reach 100%, the saturation point, and the water may condense as fog.

*given temperature.* At room temperature, a crystal of $I_2$ molecules, which attract each other weakly, has a relatively high vapor pressure. A crystal of NaCl, which has very strong attractive forces between its $Na^+$ and $Cl^-$ ions, has a very low vapor pressure.

Since pressure is a measure of the amount of gas per unit of volume (Section 4.3), **the value of the vapor pressure depends only on the nature of the substance in the system and the temperature of the system.**

### Boiling Point

If a liquid in an open system is heated, eventually it reaches a temperature at which its vapor pressure equals atmospheric pressure. This temperature is called the *boiling point.* The boiling point depends on the atmospheric pressure at the place where the liquid is being heated. At sea level, where atmospheric pressure is 1 atm, water boils at 100°C (373 K). On a mountaintop, where atmospheric pressure is lower than at sea level, the boiling point is lower. At 3000 m, water boils at 363 K. Because of the lower boiling point at high altitudes, hard boiling an egg on a mountain can be a problem.

*The boiling point of a liquid is the highest temperature, at a given pressure, at which there can be a liquid component in a system at equilibrium.* At the boiling point, the liquid vaporizes, and bubbles form below its surface. Below the boiling point, the external pressure prevents such gas bubbles from forming. At temperatures above the boiling point, all of the substance is in the gas phase. Therefore, the boiling point can be regarded as the temperature at which the system changes from liquid to gas or from gas to liquid. The temperature at which the vapor pressure is 1 atm is called the *normal boiling point.*

## 4.3 THE GASEOUS STATE

Gases have been the subject of a great deal of chemical study, both theoretical and experimental. There is a theory of gases that is reasonably complete and highly useful, based on the random and essentially independent behavior of the basic units (atoms or molecules) of a gas.

### The Ideal Gas

Regardless of their chemical composition, all gases have some properties in common. To some degree, all gases follow certain simple, "ideal" laws of behavior. It is useful to imagine an **ideal gas** or **perfect gas** whose behavior can be explained completely by these ideal laws.

The behavior of any real gas only approximates that of an ideal gas. The closeness of the approximation depends on the chemical composition of a real gas and the conditions under which the gas is studied. The ideal laws are a good approximation of real-gas behavior at low pressures and reasonably high temperatures — that is, at atmospheric pressure or lower and at temperatures somewhat higher than that at which a gas condenses. As pressure increases, or as temperature decreases, real gases tend to deviate more and more from ideal behavior. Early observations of gases, made at ordinary conditions, led to laws that were believed to describe the behavior of real gases completely. In fact, these were ideal gas laws. The methods of measurement then in use could not detect the slight differences in behavior between real gases.

### Boyle's Law

Shortly after the barometer was invented, Robert Boyle (1627–1691) devised a simple apparatus to study the relationship between the pressure and volume of a gas (Figure 4.8). He used a U-shaped tube, closed at one end. The pressure on the gas in the closed end of the tube could be increased by addition of a liquid, such as mercury, to the open end. He obtained the change in pressure by noting the height of the column of mercury and the change in volume from a scale on the closed end of the tube.

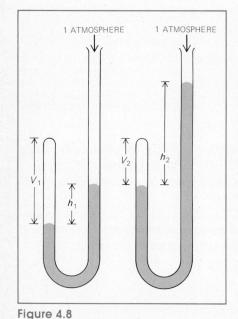

**Figure 4.8**

Boyle's device for studying the relationship between gas pressure and volume. Pressure on the gas can be increased by addition of mercury to the open end of the tube. The effect of the increased pressure can be determined by measurement of the change in the volume of the gas in the closed end of the tube. Here $P_1 = h_1 + P_{atm}$, $P_2 = h_2 + P_{atm}$, and $P_1V_1 = P_2V_2$, where $h_1$ and $h_2$ are the pressures exerted by the indicated heights of liquid.

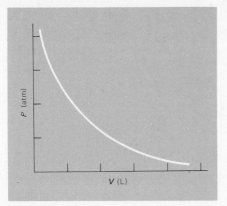

**Figure 4.9**
The relationship between gas pressure ($P$) and gas volume ($V$), at constant $T$ shown graphically. As pressure increases, the volume of a gas decreases; as pressure decreases, the volume of a gas increases. This is Boyle's law.

In 1662, Boyle published his results, which indicated that an increase in pressure is accompanied by a regular and predictable decrease in the volume of the gas. This observation is called Boyle's law. The modern statement of Boyle's law takes into account the fact that all of Boyle's observations were made at room temperature. The law is stated: **At constant temperature, the volume of a sample of gas is inversely proportional to its pressure.** This statement can be written as:

$$PV = k \qquad (4.1)$$

where $P$ is pressure, $V$ is volume, and $k$ is a proportionality constant. When $V$ increases, $P$ decreases; when $V$ decreases, $P$ increases. Figure 4.9 shows a graphical representation of this relationship.

Real gases obey this ideal relationship fairly well when $P$ is not too great. Equation 4.1 defines the relationship: For a given amount of gas at a given temperature, the product of pressure and volume is a constant. This relationship can be written in a form that is more suitable for calculations:

$$P_1 V_1 = P_2 V_2 \qquad (4.2)$$

where $P_1$ and $V_1$ represent one set of pressure and volume conditions and $P_2$ and $V_2$ represent another set of conditions for the same amount of gas at the same temperature.

---

**Example 4.1**   A cylinder whose volume is 2.5 L is filled with helium at a pressure of 5.3 atm. The helium is used to fill a balloon to a pressure of 1.0 atm at the same temperature. Find the volume of the balloon.

**Solution**   When there is a change in the pressure and volume of a sample of gas at a given temperature, we use Boyle's law. Since we know the pressure and volume of the helium under one set of conditions ($P_1$, $V_1$) and the other pressure ($P_2$), we can find the unknown, $V_2$, by using Equation 4.2:

$$P_1 = 5.3 \text{ atm} \qquad V_1 = 2.5 \text{ L}$$
$$P_2 = 1.0 \text{ atm} \qquad V_2 = ?$$

Substitution gives

$$(5.3 \text{ atm})(2.5 \text{ L}) = (1.0 \text{ atm})V_2$$
$$V_2 = 13 \text{ L}$$

---

**Example 4.2**   A sample of gas at room temperature is placed in an evacuated bulb of volume 0.51 L and is found to exert a pressure of 24 kPa. This bulb is connected to another evacuated bulb whose volume is 0.63 L, and the gas is allowed to fill both bulbs, as shown in Figure 4.10. What is the new pressure of the gas at room temperature?

Solution    List the information given:

$$V_1 = 0.51 \text{ L} \qquad\qquad P_1 = 24 \text{ kPa}$$
$$V_2 = 0.51 \text{ L} + 0.63 \text{ L} \qquad P_2 = ?$$

Substitute into Equation 4.2:

$$(24 \text{ kPa})(0.51 \text{ L}) = (1.14 \text{ L})P_2$$
$$P_2 = 11 \text{ kPa}$$

Note that $P_1$ and $P_2$ must have the same units, as must $V_1$ and $V_2$.

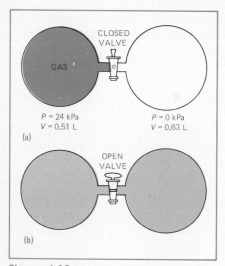

P = 24 kPa
V = 0.51 L
(a)

CLOSED VALVE
GAS

P = 0 kPa
V = 0.63 L

OPEN VALVE
(b)

**Figure 4.10**

Example 4.2 illustrated. At the beginning of the experiment, the bulb at the left contains a sample of gas, while the bulb at the right is evacuated. When the valve between the two bulbs is opened, some gas goes from the left bulb to the right bulb.

## The Law of Charles and Gay-Lussac

More than a century after Boyle's experiments, two French physicists, Jacques Alexandre Charles (1746–1823) and Joseph Louis Gay-Lussac (1778–1850) became interested in hot-air balloons. Their balloon work led them to study the relationship between the volume and the temperature of gases.

They knew from experience that a gas expands when heated and contracts when cooled. Charles measured the changes in volume. He found that gases expand by the same fractional amount of their original volume for equal increases in temperature. Later, Gay-Lussac showed that for every 1 °C change in temperature, any gas changes by ¹⁄₂₇₃ of its volume at 0 °C.

Suppose we start with a sample of gas whose volume is 273 cm³ at 0 °C. The data we get by changing the temperature of this sample and keeping the pressure constant are

| Temperature (°C) | 1 | 2 | 3 | −1 | −2 |
|---|---|---|---|---|---|
| Volume Change (cm³) | +1 | +2 | +3 | −1 | −2 |
| New Volume (cm³) | 274 | 275 | 276 | 272 | 271 |

If we plot these data, we obtain a straight line (Figure 4.11).

These observations, based on the Celsius temperature scale, can be expressed as:

$$V = V_0 + V_0\left(\frac{t}{273}\right) = V_0\left(1 + \frac{t}{273}\right) \tag{4.3}$$

where $V$ is the volume of the gas at a temperature of $t$ °C, $V_0$ is the volume of the gas at 0 °C, and $t$, the temperature of the gas on the Celsius scale, is also the number of degrees away from 0 °C.

Because the relationship is based on the Celsius temperature scale, the expression is more complicated than it need be. Volume and temperature appear to be proportional, since volume increases by the same amount for

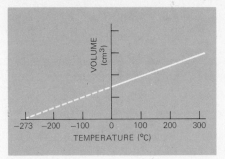

**Figure 4.11**
The relationship between temperature and gas volume. As temperature increases by 1°C, the volume of a gas increases by approximately ½₂₇₃ of its volume at 0°C. This is the law of Charles and Gay-Lussac.

equal increases in temperature. However, Equation 4.3 is not a simple expression of a direct proportionality.

A simpler relationship can be formulated with a different temperature scale, in which the interval of temperature — the degree — is the same Celsius degree but the zero point is 273°C below the Celsius zero point. If this temperature is called $T$, then:

$$T = t + 273 \tag{4.4}$$

where $t$ is the Celsius temperature.

Equation 4.3 can be written as:

$$V = V_0 \left( \frac{273 + t}{273} \right)$$

Combining this expression with Equation 4.4 gives:

$$V = V_0 \left( \frac{T}{273} \right) \tag{4.5}$$

Since $V_0$ is a constant, we can let $k = V_0/273$:

$$V = kT \tag{4.6}$$

which is a simple direct proportionality and a more convenient expression of the law of Charles and Gay-Lussac.

### The Kelvin Temperature Scale

The temperature scale used above is called **the absolute or kelvin temperature scale. It should be used in all calculations involving gases.** The interval of temperature on this scale is called the kelvin (K); 1 K = 1°C. The only difference between the kelvin scale and the Celsius scale is their zero points. The zero point on the kelvin temperature scale can be found from Equation 4.3. It is the temperature at which the volume ($V$) of an ideal gas is zero. Since the initial volume ($V_0$) of an ideal gas at 0°C is reduced by $V_0/273$ for every reduction of 1°C, it can be seen that 0 K = −273°C. Figure 4.12 shows this relationship graphically. Very accurate measurements give 0°C = 273.15 K.

The volume of a gas has never been reduced to zero. Any gas condenses to a liquid or a solid well before 0 K, the absolute zero, is reached. Theoreticians have proven that a temperature of 0 K can never be reached.

Equation 4.6 states that **the volume of a given amount of an ideal gas is directly proportional to its absolute temperature at constant pressure.** When $V$ increases, $T$ increases, and when $V$ decreases, $T$ decreases. Equation 4.7 expresses the same relationship in a form more suitable for calculations:

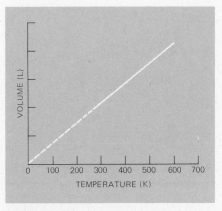

**Figure 4.12**
The relationship between temperature and gas volume at constant pressure extrapolated to zero. Theoretically, gas volume should be zero at 0 K (−273°C). Practically, zero gas volume and 0 K can never be achieved.

$$\frac{V_1}{T_1} = \frac{V_2}{T_2} \qquad\qquad (4.7)$$

where $V_1$ and $T_1$ represent one set of volume and absolute temperature conditions for a quantity of gas and $V_2$ and $T_2$ another set for the same quantity of gas at the same pressure.

**Example 4.3**   To what temperature must the gas inside a balloon at a temperature of 20°C be heated to triple the volume of the balloon? The pressure remains constant at 1 atm.

**Solution**   First convert the temperature to the absolute scale:

$$T = t + 273 = 20 + 273 = 293 \text{ K}$$

If the initial volume is $V_1$ and the final volume is $V_2$, the problem states that $V_2$ is triple $V_1$; that is, $V_2 = 3V_1$. Substituting into Equation 4.7 gives:

$$\frac{V_1}{293 \text{ K}} = \frac{3V_1}{T_2}$$

$$T_2 = (3)(293 \text{ K}) = 879 \text{ K}$$

This problem can be solved with less formal calculation once it is understood that gas volume and absolute temperature are directly proportional. If the gas volume is tripled, the absolute temperature also triples, since the pressure is constant.

**Example 4.4**   The air in a 51-L balloon is cooled from 450 K to 310 K at a constant pressure of 1 atm. Calculate the new volume of the balloon.

**Solution**   Equation 4.7 has four terms. Given the value of any three, the fourth can be found. In this case:

$$V_1 = 51 \text{ L} \qquad V_2 = ?$$
$$T_1 = 450 \text{ K} \qquad T_2 = 310 \text{ K}$$

$$\frac{51 \text{ L}}{450 \text{ K}} = \frac{V_2}{310 \text{ K}}$$

$$V_2 = 35 \text{ L}$$

The two laws for the behavior of ideal gases as expressed by Equations 4.2 and 4.7 can be combined into a single relationship. Suppose we have a gas with volume $V_1$, pressure $P_1$, and temperature $T_1$, and that these are changed to $V_2$, $P_2$, and $T_2$. Is there a simple relationship between the two sets of variables?

If the temperature is held at $T_1$ while the pressure is changed to $P_2$, Boyle's law applies. If the volume of the gas after this intermediate step is called $V_i$, then:

$$P_1V_1 = P_2V_i \quad \text{or} \quad V_i = \frac{P_1V_1}{P_2}$$

Now the pressure is held constant at $P_2$ while the temperature is changed from $T_1$ to $T_2$. The volume also changes from $V_i$ to $V_2$, and Charles's law applies:

$$\frac{V_i}{T_1} = \frac{V_2}{T_2}$$

We get the desired single relationship by substituting the value of $V_i$ found from Boyle's law into the expression above:

$$\frac{P_1V_1}{T_1} = \frac{P_2V_2}{T_2} \tag{4.8}$$

Equation 4.8 connects two sets of pressure, volume, and temperature conditions for any given quantity of an ideal gas. It can also be read as: The quantity $PV/T$ is a constant for a given amount of gas.

**Example 4.5**   An automobile tire is inflated to a total pressure of 190 kPa with air at 20°C. After five hours of driving, the volume of the tire has increased by 5.0% and the total pressure is 210 kPa. Calculate the temperature of the air inside the tire, assuming ideal gas behavior.

**Solution**   In such a problem, it is advisable to begin by listing all the data in suitable units:

$$P_1 = 190 \text{ kPa} \qquad P_2 = 210 \text{ kPa}$$
$$T_1 = 20 + 273 \text{ K} \qquad T_2 = ?$$

$$V_1 = V_1 \qquad\qquad V_2 = V_1 + \frac{5.0}{100}\, V_1$$

Using Equation 4.8, we get:

$$\frac{190 \text{ kPa}(V_1)}{293 \text{ K}} = \frac{(210 \text{ kPa})(1.05\ V_1)}{T_2}$$

The $V_1$ can be canceled, giving:

$$T_2 = \frac{(210 \text{ kPa})(293 \text{ K})(1.05)}{(190 \text{ kPa})}$$

$$= 340 \text{ K}$$

The increase in temperature is consistent with the increase in pressure, since the pressure of an ideal gas is also directly proportional to its absolute temperature.

**Example 4.6**    A balloon whose volume is 91 L contains helium at a temperature of 295 K and a pressure of 1.0 atm. When the balloon reaches an altitude of 20 km, the pressure is 0.083 atm and the temperature is 225 K. Calculate the volume of the balloon at that altitude.

**Solution**    List the data given:

$$P_1 = 1.0 \text{ atm} \qquad P_2 = 0.083 \text{ atm}$$
$$V_1 = 91 \text{ L} \qquad V_2 = ?$$
$$T_1 = 295 \text{ K} \qquad T_2 = 225 \text{ K}$$

Substitution into Equation 4.8 gives:

$$\frac{(1.0 \text{ atm})(91 \text{ L})}{295 \text{ K}} = \frac{(0.083 \text{ atm})V_2}{225 \text{ K}}$$
$$V_2 = 840 \text{ L}$$

The volume of the helium has increased by almost a factor of ten. This substantial increase in volume must be considered in designing high-altitude balloons.

**Example 4.7**    A steel cylinder contains $H_2$ at a pressure of 120 atm and a temperature of 298 K. The cylinder can withstand pressures up to 350 atm before exploding. What is the highest temperature at which the cylinder can be stored without exploding? Assume that the volume of the cylinder is constant.

**Solution**    List the data given:

$$P_1 = 120 \text{ atm} \qquad P_2 = 350 \text{ atm}$$
$$T_1 = 298 \text{ K} \qquad T_2 = ?$$

Since the volume is constant ($V_1 = V_2$), Equation 4.8 becomes

$$\frac{P_1}{T_1} = \frac{P_2}{T_2}$$

Substitution into this equation gives:

$$\frac{(120 \text{ atm})}{298 \text{ K}} = \frac{350 \text{ atm}}{T_2}$$
$$T_2 = 870 \text{ K}$$

For convenience, a standard set of conditions is defined for an ideal gas: a pressure of 1 atm and a temperature of 0°C (273 K). These conditions are referred to as standard temperature and pressure, **STP.**

---

**Example 4.8**   A sample of NO(g) occupies a volume of 3.2 L at a temperature of 298 K and a pressure of 0.50 atm. What is the volume of the sample at STP?

**Solution**   List the data given:

$$P_1 = 0.50 \text{ atm} \qquad P_2 = 1.0 \text{ atm}$$
$$V_1 = 3.2 \text{ L} \qquad V_2 = ?$$
$$T_1 = 298 \text{ K} \qquad T_2 = 273 \text{ K}$$

Substitution into Equation 4.8 gives:

$$\frac{(0.50 \text{ atm})(3.2 \text{ L})}{(298 \text{ K})} = \frac{(1.0 \text{ atm})V_2}{(273 \text{ K})}$$

$$V_2 = 1.5 \text{ L}$$

---

### Avogadro's Law

We have discussed the relationships between the pressure, volume, and temperature of an ideal gas without considering a fourth factor: the amount of gas. Yet $P$, $V$, and $T$ are related to the amount of the ideal gas in any sample.

In 1811, Amadeo Avogadro proposed a remarkable relationship. Avogadro said that at a given pressure and temperature, the number of molecules of gas in any sample depends only on the volume of the sample, not on the nature of the gas. This relationship can be stated as: **Equal volumes of different gases under the same conditions of temperature and pressure contain equal numbers of molecules.**

Amadeo Avogadro (1776–1856) *(left)* proposed two important hypotheses to explain gas behavior. He supported a proposal first made by Gay-Lussac, that equal volumes of gas contain equal numbers of molecules. He also said that molecules of gas could split into "half-molecules" when they combined chemically with other gases. Avogadro's hypotheses were rejected or ignored by chemists of the time. *(The Bettmann Archive)*

Stanislao Cannizzaro (1826–1910) *(right)* cleared up the confusion that existed among nineteenth-century scientists about the molecular weights of gases. The key to his achievement was his acceptance of Avogadro's proposal that equal volumes of gas contain equal numbers of molecules. Cannizzaro's work led to acceptance of Avogadro's theories, which had been rejected until then. *(American Chemical Society)*

At constant $T$ and $P$, the volume $V$ of a gas is directly proportional to the number of molecules $N$ of the gas:

$$V = kN \qquad (4.9)$$

When we increase the number of molecules in a sample of gas at constant temperature and pressure, the volume increases in direct proportion. For example, if the number of molecules is doubled, the volume occupied by the sample at the same total pressure doubles. The constant $k$ in Equation 4.9 is the same for all gases.

Since the amount in moles $n$ is directly proportional to the number of molecules $N$, it is more convenient to write:

$$V = kn \qquad (4.10)$$

or, in a form more suitable for calculations:

$$\frac{V_1}{n_1} = \frac{V_2}{n_2} \qquad (4.11)$$

**Example 4.9**     A balloon whose volume is 5.0 L contains 6.4 g of $N_2$. What mass of $H_2$ must be added to the balloon to expand its volume to 11 L at the same temperature and pressure?

**Solution**     List the data given:

$$V_1 = 5.0 \text{ L} \qquad\qquad\qquad\qquad V_2 = 11 \text{ L}$$

$$n_1 = \frac{6.4 \text{ g N}_2}{28.0 \text{ g N}_2/1 \text{ mol N}_2} = 0.23 \text{ mol N}_2 \qquad n_2 = \text{?}$$

Using Equation 4.11, we get:

$$\frac{5.0 \text{ L}}{0.23 \text{ mol N}_2} = \frac{11 \text{ L}}{n_2}$$

and $n_2 = 0.51$ mol, the total amount of gas required for the volume to be 11 L. Therefore

$$0.51 \text{ mol} - 0.23 \text{ mol N}_2 = 0.28 \text{ mol H}_2$$

which must be added, or

$$0.28 \text{ mol} \times \frac{2.02 \text{ g H}_2}{1 \text{ mol H}_2} = 0.57 \text{ g H}_2$$

must be added.

## 4.4   THE IDEAL GAS EQUATION

We have seen that $PV/T$ is a constant for a given quantity of ideal gas. The numerical value of the constant is proportional to the quantity of gas. Therefore, we can write:

$$\frac{PV}{T} = Rn \tag{4.12}$$

where $n$ is the amount in moles of an ideal gas and $R$ is the conventional symbol for the proportionality constant, which is called the *universal gas constant.*

The numerical value of the universal gas constant can be determined. This value depends on the units of measurement chosen for $V$ and $P$. While $T$ is always expressed in kelvins and $n$ is always expressed in moles, there is a choice of units for $V$ and $P$. If we measure $V$ in liters and $P$ in atmospheres, the value of $R$ is 0.082057 L atm mol$^{-1}$ K$^{-1}$. If we measure $V$ in liters and $P$ in kilopascals, $R$ is 8.3144 L kPa mol$^{-1}$ K$^{-1}$.

Equation 4.12 usually is written in the form:

$$PV = nRT \tag{4.13}$$

and is called the **ideal gas equation.** The ideal gas equation combines in one relationship the laws of Boyle, Charles and Gay-Lussac, and Avogadro. It provides essentially a complete description of the state of an ideal gas and is thus one of the most useful relationships in chemistry. There are four unknowns in the ideal gas equation. When the values of any three are specified, the value of the fourth is fixed.

**Example 4.10**   Calculate the volume of one mole of an ideal gas at STP.

**Solution**   List the data given:

$$P = 1.00 \text{ atm} \qquad n = 1.00 \text{ mol}$$
$$V = ? \qquad\qquad T = 273 \text{ K}$$

Do the calculation, using Equation 4.13. Be sure that all units are consistent with the units of $R$:

$$(1.00 \text{ atm})V = (1.00 \text{ mol})\left(0.0821 \frac{\text{L atm}}{\text{mol K}}\right)(273 \text{ K})$$

$$V = 22.4 \text{ L}$$

The volume 22.4 L is called the *standard molar volume* of an ideal gas.

We can use the ideal gas equation to solve a large number of problems, if the problems specify or imply the values of any three of the four variables in the equation.

**Example 4.11**     A mouse is placed in a gas bulb that is filled with $O_2$ and contains some KOH to absorb $CO_2$ and $H_2O$. The volume of the bulb, after we allow for the mouse and the KOH, is 1.02 L. The pressure of the $O_2$ is 0.953 atm, and the temperature of the gas is 37°C. After two hours, the pressure in the bulb drops to 0.774 atm. What mass of $O_2$ was consumed by the mouse?

**Solution**     Assume that the consumption of $O_2(g)$ is the only reason for the drop in pressure. Then the pressure of $O_2(g)$ consumed is 0.953 atm − 0.774 atm = 0.179 atm. The other data given are $V = 1.02$ L and $T = 37 + 273$ K. Substituting these data into the ideal gas equation gives:

$$(0.179 \text{ atm})(1.02 \text{ L}) = n \left( 0.0821 \frac{\text{L atm}}{\text{mol K}} \right)(310 \text{ K})$$

$$n = 0.00717 \text{ mol } O_2$$

$$\text{mass } O_2 = 0.00717 \text{ mol } O_2 \times \frac{32.0 \text{ g } O_2}{1 \text{ mol } O_2}$$

$$= 0.230 \text{ g } O_2(g) \text{ consumed}$$

We can get more information about ideal gases by using the ideal gas equation to derive different but equivalent relationships. To obtain these new relationships, we combine the ideal gas equation with expressions that define variables such as $n$. For example, we know that the amount in moles of a sample of a substance is related to the mass of the sample and the molar mass of the substance. The relationship can be expressed as:

$$n = \frac{m}{\mathcal{M}}$$

where $m$ is the mass in grams and $\mathcal{M}$ is the molar mass in grams per mole. If we substitute this expression for $n$ into the ideal gas equation, we get:

$$PV = \frac{mRT}{\mathcal{M}} \tag{4.14}$$

If we study a real gas under conditions where it displays ideal gas behavior, we can find the molecular weight of the gas (which is numerically equal to the molar mass) by making the appropriate measurements and using Equation 4.14.

**Example 4.12**     A sample of an unknown liquid whose mass is 1.35 g is introduced into an evacuated bulb of volume 502 cm³. After all of the liquid is vaporized, the gas pressure is 0.477 atm at 333 K. Assuming ideal gas behavior, calculate the molecular weight of the liquid.

**Solution**   List the given data in appropriate units:

$$V = 502 \text{ cm}^3 \times \frac{1 \text{ L}}{1000 \text{ cm}^3} = 0.502 \text{ L} \qquad T = 333 \text{ K}$$

$$P = 0.477 \text{ atm} \qquad\qquad m = 1.35 \text{ g}$$

Using Equation 4.14, we can calculate:

$$(0.477 \text{ atm})(0.502 \text{ L}) = \frac{1.35 \text{ g}}{\mathcal{M}}\left(0.0821 \frac{\text{L atm}}{\text{mol K}}\right)(333 \text{ K})$$

$$\mathcal{M} = 154 \text{ g/mol}$$

The molecular weight is 154 since it is numerically equal to the molar mass.

We can derive another version of the ideal gas equation by recalling that the density $d$ of a gas is expressed as the mass in grams per liter of the gas:

$$d = \frac{m}{V}$$

where $m$ is the mass in grams and $V$ is the volume in liters. Equation 4.14 can then be rewritten as:

$$\frac{m}{V} = \frac{P\mathcal{M}}{RT}$$

or
$$d = \frac{P\mathcal{M}}{RT} \tag{4.15}$$

which expresses the relationship between the molar mass and the density of a gas at a given temperature and pressure.

**Example 4.13**   The density of an unknown gas at STP is 9.91 g/L. Calculate the molar mass of the gas.

**Solution**   List the data given:

$$d = 9.91 \text{ g/L} \qquad P = 1.00 \text{ atm} \qquad T = 273 \text{ K}$$

Using Equation 4.15, we get:

$$9.91 \text{ g/L} = \frac{(1.00 \text{ atm})\mathcal{M}}{(0.0821 \text{ L atm/mol K})(273 \text{ K})}$$

$$\mathcal{M} = 222 \text{ g/mol}$$

The problem can be approached in another way. At STP, one mole of ideal gas occupies a volume of 22.4 L. Therefore, the mass of 22.4 L of the gas at STP is the molecular weight of the gas. We are given the density, which is the mass of 1 L of the gas. Therefore, the molar mass of the gas can be obtained by multiplication of the density of the gas by 22.4 L/mol; that is, $\mathcal{M} = (22.4 \text{ L/mol})(d)$ at STP.

$$\mathcal{M} = 22.4 \text{ L/mol} \times 9.91 \text{ g/L} = 222 \text{ g/mol}$$

The ideal gas equation is often used to determine the amount of gas consumed or produced in chemical processes. If we assume ideal gas behavior, we can determine the quantity of the gas by measuring other variables, such as pressure, volume, and temperature.

**Example 4.14**    The thermal decomposition of potassium chlorate once was a standard (although potentially hazardous) laboratory procedure for the production of oxygen by the reaction:

$$2KClO_3(s) \longrightarrow 2KCl(s) + 3O_2(g)$$

A quantity of $KClO_3$ is heated and partially reacts. The oxygen that forms is collected in a balloon at a pressure of 1.0 atm and a temperature of 300 K. The volume of the balloon is found to be 0.79 L. What mass of $KClO_3$ has reacted?

**Solution**    The amount of $O_2$ liberated can be calculated from the given data. The chemical equation will then give us the relationship between the amount of $O_2$ and the amount of $KClO_3$.
List the data:

$$P = 1.0 \text{ atm} \qquad T = 300 \text{ K} \qquad V = 0.79 \text{ L}$$

The ideal gas equation gives the relationship between pressure, volume, temperature, and amount of gas:

$$(1.0 \text{ atm})(0.79 \text{ L}) = n \left( 0.0821 \frac{\text{L atm}}{\text{mol K}} \right) (300 \text{ K})$$
$$n = 0.032 \text{ mol } O_2$$

Then, to find the mass of $KClO_3$ that reacted:

$$\text{mass } KClO_3 = 0.032 \text{ mol } O_2 \times \frac{2 \text{ mol } KClO_3}{3 \text{ mol } O_2} \times \frac{122.6 \text{ g } KClO_3}{1 \text{ mol } KClO_3}$$
$$= 2.6 \text{ g } KClO_3 \text{ reacted}$$

**Example 4.15**    A 0.52-L bulb is filled with $CO_2(g)$ at a pressure of 0.86 atm and a temperature of 298 K. A quantity of $CaO(s)$ of negligible volume is introduced into the bulb. After some time, the pressure of $CO_2(g)$ is found to fall to 0.39 atm because it is consumed by the process:

$$CaO(s) + CO_2(g) \longrightarrow CaCO_3(s)$$

What is the mass of the calcium carbonate that is formed?

**Solution** First use the ideal gas equation to calculate the amount (in moles) of $CO_2(g)$ consumed. The drop in pressure, which is $0.86 \text{ atm} - 0.39 \text{ atm} = 0.47 \text{ atm}$, corresponds to the consumption of $CO_2(g)$. Since we are given the pressure, temperature, and volume, we can calculate the amount of gas:

$$(0.47 \text{ atm})(0.52 \text{ L}) = n \left( 0.0821 \frac{\text{L atm}}{\text{mol K}} \right) (298 \text{ K})$$

$$n = 0.010 \text{ mol } CO_2 \text{ consumed}$$

We use the chemical equation to find the mass of $CaCO_3$ from the amount of $CO_2$ consumed.

$$\text{mass } CaCO_3(s) = 0.010 \text{ mol } CO_2 \times \frac{1 \text{ mol } CaCO_3}{1 \text{ mol } CO_2} \times \frac{100.1 \text{ g } CaCO_3}{1 \text{ mol } CaCO_3}$$

$$= 1.0 \text{ g } CaCO_3(s) \text{ formed}$$

## 4.5 DALTON'S LAW OF PARTIAL PRESSURE

Early in the nineteenth century, John Dalton's experiments led him to recognize that **the pressure exerted by a gas in a mixture of a number of different gases is the same as if the gas were present alone.** We now know that this is ideal gas behavior. In a mixture of two or more gases, all displaying ideal behavior, each gas can be regarded as acting independently of all the others.

When a mixture of gases is present in a container, it is convenient to define a **partial pressure** of each gas—the pressure exerted by that gas alone. The total pressure of all the gases in the mixture is the sum of the partial pressures. For example, if a container holds a quantity of He with partial pressure $P_{He}$, a quantity of Ne with partial pressure $P_{Ne}$, and a quantity of Ar with partial pressure $P_{Ar}$, the total pressure $P_T$ is given by:

$$P_T = P_{He} + P_{Ne} + P_{Ar}$$

Usually, only $P_T$ can be measured directly.

Dalton's law of partial pressures, which states that **the total pressure of a mixture of gases is the sum of the partial pressures of the gases,** has a broad range of applications. It is useful in the study of systems where substances in the gas phase are in contact with condensed phases of other volatile substances. Figure 4.13 shows such a system, which is frequently encountered: the collection of a gas by displacement of water. The gas is produced by a chemical reaction and is directed into a water-filled tube. As the gas collects in the container, it mixes with the water vapor. The

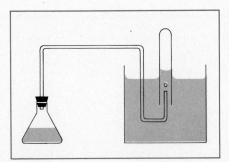

**Figure 4.13**

Dalton's law of partial pressures can be used to calculate the pressure of a gas collected by displacement of water. Some of the water vaporizes in the closed end of the tube. We can determine the pressure exerted by the gas that is being collected by subtracting the vapor pressure of water from the total pressure in the tube.

**TABLE 4.1    Vapor Pressure of Water**

| $T$ (K) | $P$ (atm) | $P$ (mmHg) | $T$ (K) | $P$ (atm) | $P$ (mmHg) |
|---|---|---|---|---|---|
| 273 | 0.0061 | 4.6 | 298 | 0.0313 | 23.8 |
| 274 | 0.0065 | 4.9 | 299 | 0.0332 | 25.2 |
| 275 | 0.0070 | 5.3 | 300 | 0.0351 | 26.7 |
| 276 | 0.0075 | 5.7 | 301 | 0.0372 | 28.3 |
| 277 | 0.0080 | 6.1 | 302 | 0.0395 | 30.0 |
| 278 | 0.0086 | 6.5 | 303 | 0.0418 | 31.8 |
| 279 | 0.0092 | 7.0 | 304 | 0.0443 | 33.7 |
| 280 | 0.0099 | 7.5 | 305 | 0.0470 | 35.7 |
| 281 | 0.0105 | 8.0 | 306 | 0.0496 | 37.7 |
| 282 | 0.0113 | 8.6 | 307 | 0.0525 | 39.9 |
| 283 | 0.0121 | 9.2 | 308 | 0.0555 | 42.2 |
| 284 | 0.0129 | 9.8 | 313 | 0.0728 | 55.3 |
| 285 | 0.0138 | 10.5 | 318 | 0.0946 | 71.9 |
| 286 | 0.0147 | 11.2 | 323 | 0.1217 | 92.5 |
| 287 | 0.0158 | 12.0 | 328 | 0.1553 | 118.0 |
| 288 | 0.0168 | 12.8 | 333 | 0.1966 | 149.4 |
| 289 | 0.0179 | 13.6 | 338 | 0.2467 | 187.5 |
| 290 | 0.0191 | 14.5 | 343 | 0.3075 | 233.7 |
| 291 | 0.0204 | 15.5 | 348 | 0.3804 | 289.1 |
| 292 | 0.0217 | 16.5 | 353 | 0.4672 | 355.1 |
| 293 | 0.0230 | 17.5 | 358 | 0.5705 | 433.6 |
| 294 | 0.0246 | 18.7 | 363 | 0.6918 | 525.8 |
| 295 | 0.0261 | 19.8 | 368 | 0.8341 | 633.9 |
| 296 | 0.0278 | 21.1 | 373 | 1.0000 | 760.0 |
| 297 | 0.0295 | 22.4 | 378 | 1.1922 | 906.1 |

total gas pressure is the sum of the partial pressure of the gas that is being collected and the partial pressure of water vapor (more exactly, of the vapor pressure of water at the temperature of the experiment).

Table 4.1 lists the vapor pressure of water at a number of temperatures. To find the actual pressure of the gas being collected, we subtract the vapor pressure of water from the total measured pressure.

**Example 4.16**    One method of preparing $H_2(g)$ is by the reaction of an active metal with acid, for example:

$$Mg(s) + 2H^+(aq) \longrightarrow Mg^{2+}(aq) + H_2(g)$$

The hydrogen gas can be collected over water, as shown in Figure 4.13. The volume of the gases in the collecting tube is found to be 0.350 L when the water levels inside and outside the bottle are equal. The total pressure of gases in the bottle is equal to the pressure in the room (Figure 4.7), which is found by barometer measurement to be 1.041 atm. The temperature is 299 K. What mass of $H_2(g)$ is collected?

**Solution**    Since the volume and temperature of the $H_2(g)$ are given and the mass is to be found, the ideal gas equation can be used if $P_{H_2}$, the pressure of $H_2$, is known. We can find the $P_{H_2}$ by using Dalton's law and the data in Table 4.1.

At 299 K, the vapor pressure of $H_2O$ is 0.0332 atm.

$$P_T = P_{H_2} + P_{H_2O}$$

Substituting:

$$1.041 \text{ atm} = P_{H_2} + 0.033 \text{ atm}$$
$$P_{H_2} = 1.008 \text{ atm}$$

Since we now have three of the variables in the ideal gas equation, we can find the fourth unknown.

$$(1.008 \text{ atm})(0.350 \text{ L}) = \frac{m}{2.016 \text{ g } H_2/1 \text{ mol } H_2}\left(0.0821 \frac{\text{L atm}}{\text{mol K}}\right)(299 \text{ K})$$

$$\text{mass } H_2 = 0.0290 \text{ g } H_2 \text{ is collected}$$

We see from the ideal gas laws that the pressure of an ideal gas at constant volume and temperature is directly proportional to the amount in moles of the gas:

$$P = kn$$

Two or more gases in the same container have the same volume and temperature. The law of partial pressures states that the pressure exerted by each gas is not affected by the presence of the other gases. Therefore, the partial pressure of each gas is directly proportional to the amount of that gas in the container. When we have such a mixture, we can determine the amount of a gas by finding its partial pressure. Since it is relatively easy to measure the total pressure of a mixture of ideal gases, we can often use partial pressures for stoichiometric calculations.

**Example 4.17**   A sample of air collected over Los Angeles during a smog alert occupies a volume of 1.0 L and contains 0.00010 g of CO(g). If the temperature is 308 K, what is the partial pressure (in pascals) of CO in the sample?

**Solution**   The partial pressure of CO in the mixture of gases is the same as if CO were present alone. Since we know the values of four of the variables in Equation 4.14, we can find the value of the fifth:

$$P(1.0 \text{ L}) = \frac{0.00010 \text{ g}}{28.0 \text{ g CO}/1 \text{ mol CO}}\left(8.31 \frac{\text{L kPa}}{\text{mol K}}\right)(308 \text{ K})$$

$$P_{CO} = 9.1 \times 10^{-3} \text{ kPa}$$

$$= 9.1 \times 10^{-3} \text{ kPa} \times 1.0 \times 10^3 \frac{\text{Pa}}{\text{kPa}}$$

$$= 9.1 \text{ Pa}$$

**Example 4.18**    A sample of $NCl_3(l)$ is introduced into a container of volume 3.00 L that contains $Ar(g)$ at a pressure of 1.00 atm. The temperature is 29°C. When the $NCl_3(l)$ is exposed to light, it decomposes completely to $N_2(g)$ and $Cl_2(g)$. After the temperature returns to 29°C, the total pressure in the container is found to be 1.88 atm. Find the amount of $NCl_3(l)$ that was in the container.

**Solution**    The equation for the reaction is

$$2NCl_3(l) \longrightarrow N_2(g) + 3Cl_2(g)$$

If we knew the amount of either gaseous product, we could find the amount of $NCl_3(l)$ that decomposed. The amount of $N_2(g)$ or $Cl_2(g)$ can be found if the partial pressure of either gas is found, since their volume and temperature are given. We can solve the problem by using the total pressure (which is given) and the stoichiometric relationships to calculate $P_{N_2}$.

After the reaction is over:

$$P_T = P_{Ar} + P_{N_2} + P_{Cl_2} = 1.88 \text{ atm}$$

From the stoichiometry, 2 mol of $NCl_3$ produce 1 mol of $N_2$ and 3 mol of $Cl_2$. Therefore $n_{Cl_2} = 3n_{N_2}$; and since partial pressure is proportional to amount, $P_{Cl_2} = 3P_{N_2}$. Thus:

$$P_T = P_{Ar} + P_{N_2} + 3P_{N_2}$$

and since $P_{Ar}$ is 1.00 atm,

$$1.88 \text{ atm} = 1.00 \text{ atm} + 4P_{N_2}$$
$$P_{N_2} = 0.22 \text{ atm}$$
$$(0.22 \text{ atm})(3.00 \text{ L}) = n_{N_2} \left( 0.0821 \frac{\text{L atm}}{\text{mol K}} \right)(29 + 273 \text{ K})$$
$$n_{N_2} = 0.027 \text{ mol } N_2$$
$$0.027 \text{ mol } N_2 \times \frac{2 \text{ mol } NCl_3}{1 \text{ mol } N_2} = 0.54 \text{ mol } NCl_3$$

**Example 4.19**    One process related to the formation of smog is

$$2NO(g) + O_2(g) \longrightarrow 2NO_2(g)$$

A convenient way to study this reaction in the laboratory is to monitor the total pressure of a mixture of the gases. Since the reaction converts every 3 mol of reactant gases to 2 mol of gaseous products, the total pressure of a mixture of NO and $O_2$ will drop as the reaction proceeds.

A bulb of volume 0.510 L is filled with a mixture of $NO(g)$ and $O_2(g)$. The initial total pressure is 1.22 atm at 298 K. After six hours, the total pressure is 0.82 atm. What mass of $NO_2(g)$ has been formed?

Solution  The chemical equation defines the relationship between the amounts of the gases. Therefore, it also defines the relationship between the changes in their partial pressures. In a problem of this kind, it is often convenient to tabulate the relationships under the formulas in the chemical equation.

$$2NO(g) \quad + O_2(g) \quad \longrightarrow 2NO_2(g)$$

before reaction         $P_{NO}$          $P_{O_2}$                    0

If we let $x$ = the pressure (in atm) of NO that undergoes reaction:

after reaction          $(P_{NO} - x)$          $\left(P_{O_2} - \dfrac{x}{2}\right)$          $x$

Here is how you can reason to get the terms after reaction.

The equation states that two moles of NO reacts with one mole of $O_2$ to form two moles of $NO_2$. From Dalton's law, the same statement can be made with units of pressure substituted for moles. If we begin with a quantity of NO represented as $P_{NO}$ (in atmospheres), and a quantity $x$ (in atmospheres) of NO undergoes reaction, then $(P_{NO} - x)$ remains. Similarly, if we begin with $P_{O_2}$ and a quantity $x/2$ undergoes reaction, then $(P_{O_2} - x/2)$ remains. Further, when $x$ of NO undergoes reaction, $x$ of $NO_2$ is formed.

The total pressure of gases in the system after reaction is the sum of the partial pressures of NO, $O_2$, and $NO_2$. An expression for each of these partial pressures is tabulated under the formula of the gas in the chemical reaction.

$$P_T = (P_{NO} - x) + \left(P_{O_2} - \frac{x}{2}\right) + x = 0.82 \text{ atm}$$

or

$$P_T = P_{NO} + P_{O_2} - \frac{x}{2} = 0.82 \text{ atm}$$

From the initial conditions:

$$P_{NO} + P_{O_2} = 1.22 \text{ atm}$$

Therefore:

$$1.22 \text{ atm} - \frac{x}{2} = 0.82 \text{ atm}$$

$$x = 0.80 \text{ atm}$$

Since we now have the values of four of the variables in Equation 4.14, we can calculate the value of the fifth, the mass of $NO_2$:

$$(0.80 \text{ atm})(0.510 \text{ L}) = \frac{m}{46.0 \text{ g } NO_2/1 \text{ mol } NO_2}\left(0.0821 \frac{\text{L atm}}{\text{mol K}}\right)(298 \text{ K})$$

$$\text{mass } NO_2 = 0.77 \text{ g}$$

Sometimes we can solve a problem of this type by reasoning directly from the chemical equation. In the formation of $NO_2$ from $O_2$ and $N_2$, 3 atm of gas is required to form 2 atm of gas. Therefore, when the total pressure drops by 1 atm, 2 atm of the product has been produced.

Once this relationship is known, we can calculate the amount of product that is formed if we know the change in pressure. The change in pressure is 1.22 atm − 0.82 atm = 0.40 atm. Therefore, (2)(0.40 atm) = 0.80 atm of the product, $NO_2$, has been formed.

In any case where a simple relationship between change in total pressure and the pressure of product formed can be recognized, the relationship can be used for calculations.

## 4.6   THE KINETIC THEORY OF GASES

We can predict the behavior of ideal gases by starting with a simple picture of their structure on the molecular level, called a *model*. This model can predict all aspects of ideal gas behavior. It also helps us understand why real gases do not behave like ideal gases.

A model for an ideal gas should be consistent with such observations as Boyle's law, Charles's law, and Avogadro's law. It should show why gases are compressible and expansible. The important features of a model, called the **kinetic model,** that can account for these observations are as follows:

1. An ideal gas consists of molecules (or atoms) in constant, random motion.
2. These molecules do not exert forces on one another except when they collide.
3. The volume of the molecules is negligible compared to the space between them. An ideal gas consists primarily of empty space. Its molecules can be regarded as point masses that have no volume.
4. The molecules of an ideal gas collide with the walls of their container and with each other. All of these collisions are elastic — that is, there is no loss in total kinetic energy because of the collisions.

Many features of this model cannot be correct for real gases. This is an *ideal* model. The deviations of real gas behavior from ideal gas behavior can be understood by a comparison of the ideal model with a real gas.

Some qualitative features of an ideal gas follow directly from the model. A gas is compressible because it is primarily empty space. A gas is expansible because intermolecular forces are negligible. Ideal gases behave independently of each other in mixtures (Dalton's law of partial pressures) because of the negligible intermolecular forces.

The pressure of a gas results from the collisions of its molecules with the walls of the container. As the volume of a sample of gas is decreased, the frequency of collisions with the walls increases and the pressure increases. As the temperature of a gas is increased, the velocity of the molecules increases, the force of the collisions with the walls increases, and the pressure increases. Quantitative aspects of ideal gas behavior also can be derived from the model.

## 4.7  DERIVATION OF BOYLE'S LAW

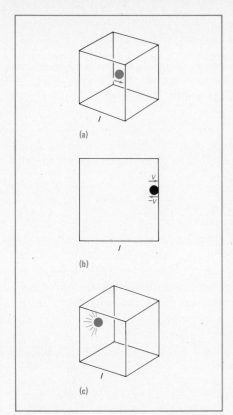

**Figure 4.14**
The pressure exerted by the gas results from the collisions by gas molecules with walls of the container. Here we see one of the molecules bouncing back and forth between the left and right walls of the container.

By making some helpful simplifying approximations about the random motion of gas molecules, we can derive Boyle's law from the ideal model. Consider an ideal gas in a cubical container of side $l$ (Figure 4.14). The cube contains a large number, $N$, of molecules moving randomly. We divide the molecules into three equal sets, each containing $N/3$ molecules. To simplify the mathematics, we make the assumption that one set of molecules bounces back and forth perpendicular to the left and right walls of the cube, another set bounces back and forth perpendicular to the top and bottom walls, and the third set bounces perpendicular to the front and back walls.

We can find the pressure of the gas on the walls of the container by finding the pressure on any one wall. We know from observation that a gas exerts pressure equally on all the walls of its container. Since pressure is defined as force/area, let us begin by finding the force imparted to the right-hand wall by one molecule. According to Newton's second law:

$$f = ma \qquad (4.16)$$

where $f$ is the force imparted by one molecule, $m$ is the mass of a molecule, and $a$ is its acceleration. Since acceleration is the change in velocity in an elapsed time:

$$f = m\frac{\Delta v}{\Delta t} \qquad (4.17)$$

where $\Delta v$ is the change in velocity and $\Delta t$ is the elapsed time.

To obtain a useful expression for the force imparted to the wall by collisions with a molecule of mass $m$ and velocity $v$, we must answer two questions: How much does the velocity of the molecule change as a result of the collision? How often does a collision occur?

The molecule hits the wall perpendicularly, and the collision is elastic. Therefore, the molecule changes direction by $180°$, and its velocity changes from $v$ to $-v$ (Figure 4.14(b)). Thus:

$$\Delta v = v - (-v) = 2v \qquad (4.18)$$

The time between collisions depends on the velocity of the molecule and the distance it travels before colliding with the same wall again. After a collision with the wall, the molecule moves a distance $l$ from right to left. It collides with the left-hand wall (Figure 4.14(c)), bounces back, travels a distance $l$ again, and collides with the right-hand wall. The distance it travels between right-hand-wall collisions is $2l$.

The time between collisions is the distance traveled divided by the velocity. Thus:

$$\Delta t = \frac{2l}{v} \tag{4.19}$$

Substitution of Equations 4.18 and 4.19 into Equation 4.17 gives:

$$f = \frac{m(2v)}{2l/v} = \frac{mv^2}{l} \tag{4.20}$$

as the force imparted to the wall by one molecule. Since $N/3$ molecules are pictured as hitting the wall and the velocities of the molecules are not all equal, the total force imparted is

$$F = \left(\frac{N}{3}\right)\left(\frac{m\overline{v^2}}{l}\right) \tag{4.21}$$

where $\overline{v^2}$ is the average value of $v^2$ of the $N/3$ molecules and $F$ is the force imparted by many molecules.

We get the pressure on one wall, and therefore the pressure of the gas, by dividing the force on the wall by the area of the wall, which is $l^2$. Thus:

$$P = \frac{(N/3)(m\overline{v^2}/l)}{l^2} = \left(\frac{N}{3}\right)\left(\frac{m\overline{v^2}}{l^3}\right) \tag{4.22}$$

Since the volume ($V$) of the cube is $l^3$, Equation 4.22 can be written as:

$$P = \frac{N}{3}\frac{m\overline{v^2}}{V} \tag{4.23}$$

where $P$ is the pressure.[2] This equation can be rewritten as:

$$PV = \frac{1}{3}Nm\overline{v^2} \tag{4.24}$$

In Section 2.8, we saw that kinetic energy, the energy of motion, is $\frac{1}{2}mv^2$. We can rewrite Equation 4.24 to relate it to kinetic energy by introducing the factor $\frac{2}{3}$:

$$PV = \frac{2}{3}N\left(\frac{1}{2}m\overline{v^2}\right) \tag{4.25}$$

---

[2] Those with the necessary background will recognize that the velocity of each molecule has three components: $v_x$, $v_y$, and $v_z$ and that $v^2 = v_x^2 + v_y^2 + v_z^2$. If the motion of a large number of bodies is random, the averages of the squares of the three components will be equal to each other and to one-third the average of the square of the total velocity:

$$\overline{v_x^2} = \overline{v_y^2} = \overline{v_z^2} = \frac{1}{3}\overline{v^2}$$

This equation leads to the same result we obtained using our approximation.

Since $m$ is the mass of a molecule of ideal gas and $\overline{v^2}$ is the average of the squares of the velocities of all the molecules, the term $\frac{1}{2}m\overline{v^2}$ is the average kinetic energy per molecule of gas. The total kinetic energy of all the molecules, $E_k$, is the number of molecules $N$ multiplied by the average kinetic energy per molecule, so Equation 4.25 gives us a simple relationship between $PV$ and the total kinetic energy:

$$PV = \frac{2}{3} E_k \qquad (4.26)$$

At a given temperature, the total kinetic energy of a sample of ideal gas is a constant. Equation 4.26 is therefore equivalent to Boyle's law, which states that $PV$ is a constant for a sample of gas at a given temperature.

## 4.8 KINETIC ENERGY AND TEMPERATURE

It is instructive to examine some aspects of the kinetic energy of an ideal gas. We can show that the kinetic energy of a sample of ideal gas is a property *only* of temperature. Therefore, it can be used to define temperature. For convenience, we can define a temperature scale in which the kinetic energy of a sample of ideal gas is directly proportional to the temperature. This temperature scale, as you may have guessed, is the absolute or kelvin scale, which we introduced in connection with Charles's law (Section 4.3).

You can see that the temperature scales are the same by the following argument. We used the kinetic model to derive a relationship between $PV$ and the kinetic energy, given in Equation 4.26: $PV = \frac{2}{3}E_k$. The ideal gas equation gives us the relationship between $PV$ and the kelvin temperature: $PV = nRT$. Therefore

$$\frac{2}{3} E_k = nRT \qquad (4.27)$$

or rearranging:

$$E_k = \frac{3}{2} nRT \qquad (4.28)$$

You can think of Equation 4.28 as a definition of absolute temperature based on the kinetic model; or you can think of it as the bringing together of the kinetic model, the theory, with the ideal gas equation, which sums up the experimental observations.

### Absolute Temperature

You might wonder why kinetic energy and volume of a given quantity of an ideal gas are both directly proportional to the absolute temperature.

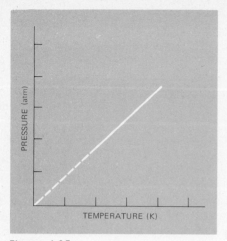

**Figure 4.15**
The relationship between pressure of a gas and temperature. As temperature decreases, the pressure of the gas decreases. According to the ideal gas law, gas pressure should become zero at 0 K.

The explanation is as follows:

Charles's law indicates that when a given quantity of gas is held at constant pressure and the temperature is lowered, the volume should become zero at 0 K. Suppose we perform an imaginary experiment in which the volume of a gas is instead held constant, the temperature is lowered, and the pressure is allowed to vary. As Figure 4.15 shows, $P$ would decrease as $T$ decreased. However, Boyle's law states that $PV$ is a constant at a fixed temperature, so $PV$ must be equal to zero at 0 K. We know that $V$, which is being held constant, is not zero. But $PV$ equals zero. Therefore, $P$ must equal zero.

Thus, for a given quantity of gas in a container of fixed volume, the pressure will be zero at 0 K. If the pressure is zero, no molecules are colliding with the walls of the container. Therefore, the molecules are no longer moving in the container; that is, $v = 0$ and so $\frac{1}{2}mv^2 = 0$. Since kinetic energy is zero at 0 K, it increases in direct proportion to increases in absolute temperature.

All the other aspects of ideal gas behavior that we discussed in Sections 4.3 through 4.5 can be derived from the kinetic model through relatively simple reasoning and algebra.

## 4.9  EFFUSION AND DIFFUSION

The first postulate of the kinetic model is that molecules are in constant random motion. We can use the kinetic model to derive an expression that permits us to calculate the average speed of the molecules, using easily measured quantities.

Equation 4.28 shows that the total kinetic energy of one mole of gas molecules ($n = 1$) is $\frac{3}{2}RT$. The total kinetic energy is the average kinetic energy per molecule multiplied by the number of molecules. Since there are Avogadro's number, $N_A$, of molecules in a mole, the total kinetic energy of one mole of molecules is $N_A\frac{1}{2}m\overline{v^2}$. If we call $\mathcal{M}$ the molar mass, then $\mathcal{M} = N_A m$. Thus we can substitute into Equation 4.28 to get

$$\frac{1}{2}\mathcal{M}\overline{v^2} = \frac{3}{2}RT \tag{4.29}$$

Taking the square root of both sides of this equation and solving for $\overline{v}$ gives:

$$\overline{v} = \sqrt{\frac{3RT}{\mathcal{M}}} \tag{4.30}$$

The quantity $\overline{v}$ is called the root-mean-square speed. It is the square root of $\overline{v^2}$, which is the average value of $v^2$. The root-mean-square speed is almost the same as the average speed of the gas molecules, and we shall use the terms *average speed* and *root-mean-square speed* interchangeably.

## VAPOR PRESSURE AND THE TRANSPLUTONIUM ELEMENTS

An ingenious method that uses vapor pressure and diffusion has been developed to determine the chemical and physical properties of the elements that are heavier than plutonium. Such information about the so-called transplutonium elements is difficult or impossible to obtain by conventional methods because these elements usually are available only in very small quantities. The transplutonium elements decay so rapidly that they are not found naturally on earth. Yet information about their properties is desirable because these elements may be produced in future nuclear generating plants.

The measurements are made in a diffusion system that is designed for extremely small samples. The metal is placed in a cell and is heated until it melts. Some of the sample evaporates, and an equilibrium between the metal and its vapor phase is established. A tiny amount of the vapor diffuses through a hole that is small enough so that the vapor loss is negligi-ble. The metal vapor that escapes through the hole can be deposited on a target plate or directed into a mass spectrometer for analysis. The amount of metal that escapes can be used to determine the vapor pressure in the cell. The value of the vapor pressure in turn can be used to determine the metal's heat of vaporization. If measurements are made at different temperatures, the properties of the liquid, solid, and gas forms of the metal can be found.

This system has been used to investigate americium, curium, berkelium, and californium. Perhaps the most impressive achievement was the determination of the properties of einsteinium, element 99, in early 1983 by investigators at Los Alamos National Laboratory. They found that the metal is divalent and has the lowest heat of vaporization of any divalent element, 128 kJ/mol. The measurement was made on what was then the entire supply of einsteinium in the non-Communist world — 100 $\mu$g.

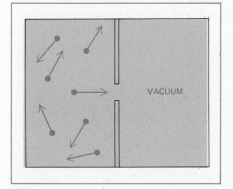

**Figure 4.16**
Effusion. If a small hole is made in a wall that separates a gas and a vacuum, the gas will pass through the hole, by a process called effusion. The rate of effusion is proportional to the speed of the gas molecules.

From Equation 4.30, we can show that if two different ideal gases are at the same temperature, the average speed of their molecules is inversely proportional to the square root of their molar masses or their molecular weights. For example, when $H_2$ ($MW = 2$) and $O_2$ ($MW = 32$) are at the same temperature:

$$\frac{(\bar{v}H_2)}{(\bar{v}O_2)} = \frac{\sqrt{3RT/2}}{\sqrt{3RT/32}} = \sqrt{\frac{32}{2}} = 4$$

That is, the $H_2$ molecules, on the average, move four times faster than the $O_2$ molecules at the same temperature.

Some aspects of gas behavior are related to the speed of the gas molecules. One of the first of these properties to be studied is illustrated in Figure 4.16. If a small hole is made in the wall of the container holding a gas, and if the hole leads to a vacuum, the gas in the container will pass through the hole. This phenomenon is called **effusion**. It is reasonable to assume that for a given gas pressure, the rate of effusion will be directly proportional to the speed of the gas molecules. Therefore, the rate of effusion is inversely proportional to the square root of the molar mass or molecular weight of the gas. The relative rates of effusion of two different gases, A and B, can be expressed as:

$$\frac{\bar{v}_A}{\bar{v}_B} = \frac{r_A}{r_B} = \sqrt{\frac{MW_B}{MW_A}} \qquad (4.31)$$

where $r_A$ and $r_B$ are the rates of effusion of gas A and gas B, $MW_A$ and $MW_B$ are the molecular weights of the gases, and $v_A$ and $v_B$ are their root-mean-square speeds.

**Example 4.20**    A sample of $N_2$ is placed in an effusion apparatus of the type shown in Figure 4.16. After 15 min, the pressure in the evacuated portion of the apparatus has risen to 0.21 atm. The $N_2$ is removed from the apparatus and the experiment is repeated with a quantity of $Br_2(g)$ at the same temperature. What is the pressure of $Br_2$ in the evacuated portion of the apparatus after 15 min?

**Solution**    The rate of increase in pressure depends on the rate of effusion. The rate of effusion depends on the molecular weights of the gases, as Equation 4.31 indicates. The molecular weight of $Br_2$ is 160 and of $N_2$ is 28. Equation 4.31 gives the relationship between molecular weight and relative rate of effusion:

$$\frac{r_{N_2}}{r_{Br_2}} = \sqrt{\frac{160}{28}} = 2.4$$

The $N_2$ effuses 2.4 times faster than the $Br_2$. If the pressure of $N_2$ after 15 min is 0.21 atm, then the pressure of $Br_2$ after the same period is (0.21 atm)(1/2.4) = 0.088 atm.

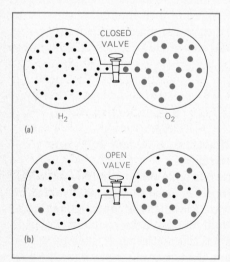

**Figure 4.17**
Diffusion. If we introduce a gas into a container that already holds a different gas by opening a stopcock, the introduced gas will begin to spread — diffuse — throughout the container.

The relationship between the rate of effusion and the chemical nature of a gas was first observed by Thomas Graham (1805–1869) before the kinetic theory of gases was developed. He found that the relative rates of effusion of two gases under the same conditions of temperature and pressure are inversely proportional to the square root of their densities:

$$\frac{r_A}{r_B} = \sqrt{\frac{d_B}{d_A}} \qquad (4.32)$$

This relationship is called **Graham's law.** It is consistent with Equation 4.31, since the densities of gases are proportional to their molecular weights.

A closely related aspect of gas behavior is the mixing of gases. If we introduce a gas into a container that already holds another gas — for example, by opening a stopcock as shown in Figure 4.17 — it eventually spreads throughout the container. The process by which a gas spreads throughout a space is called **diffusion**. Again, it is reasonable to suppose that the rate of diffusion of a gas is proportional to the speed of its molecules. Therefore, Equations 4.31 and 4.32 also apply to the relative rates of diffusion of two gases.

The type of apparatus shown in Figure 4.18 is often used to study what is commonly called diffusion but is really relative effusion. A mixture of gases is allowed to pass through a porous wall. The relative rates of diffusion determine the relative amounts of the gases that pass through the barrier in a given time.[3] An experiment of this kind can be used to find the molecular weight of a gas.

**Example 4.21**

A mixture of carbon dioxide and an unknown gas, each with the same partial pressure, is placed in the left side of the apparatus shown in Figure 4.18. The right side of the apparatus is evacuated. After a short period, the pressure in the right side of the apparatus is found to be 0.041 atm. The $CO_2$ is then removed by absorption with a solid, and the pressure on the right side drops to 0.016 atm. Calculate the molecular weight of the unknown gas.

**Solution**

In this experiment, pressure measurements are used to determine relative amounts of the gases. The pressure of $CO_2$ is indicated by the drop in the total pressure when the $CO_2$ is removed:

$$P_{CO_2} = 0.041 \text{ atm} - 0.016 \text{ atm} = 0.025 \text{ atm}$$

The pressure of the unknown gas is 0.016 atm, the remainder. Therefore, the relative rates of diffusion of $CO_2$ and the unknown gas are, from Equation 4.31:

$$\frac{r_{CO_2}}{r_{unknown}} = \frac{0.025}{0.016} = \sqrt{\frac{MW_{unknown}}{MW_{CO_2}}}$$

The $MW_{CO_2}$ is known to be 44. Therefore, we have the values of three of the four unknowns in the equation. Solving for the fourth gives

$$MW_{unknown} = 110$$

You can see that when gases diffuse through a barrier, the relative amount of the lighter gas — in this case $CO_2$ — is higher on the other side of the barrier than it was in the original mixture. The lighter gas passes through the barrier more readily than the heavier gas. We can use an apparatus of this kind to separate gases of different molecular weights. The most important application of this method occurred during World War II.

Relatively pure $^{235}_{92}U$ was needed to make an atomic bomb. Naturally occurring uranium consists almost entirely of $^{238}_{92}U$, which is not suitable bomb material; $^{235}_{92}U$ is a minor isotope. The two uranium isotopes could not be separated chemically because isotopes of an element have virtually the same chemical behavior. The problem was solved by the use of uranium hexafluoride, $UF_6$, a compound that has a high vapor pressure at room temperature.

[3] As a simplification, we shall assume that once gas has diffused through the barrier, it does not diffuse back.

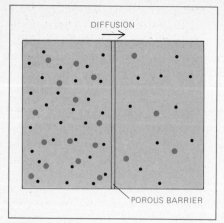

DIFFUSION

POROUS BARRIER

**Figure 4.18**

An apparatus for studying the diffusion of gases. A mixture of two or more gases is placed in the left side of the container and is allowed to pass through the porous barrier into the right side of the container. The relative amount of each gas that passes through the barrier in a given period of time indicates the rate of diffusion of the gas.

Vapors of $UF_6$ consist of $^{235}UF_6$, $MW = 235 + 6(19) = 349$, and $^{238}UF_6$, $MW = 238 + 6(19) = 352$. When these vapors diffuse through a porous barrier, the lighter gas comes through slightly faster:

$$\frac{r_{(235UF_6)}}{r_{(238UF_6)}} = \sqrt{\frac{352}{349}} = 1.004$$

Therefore, the fraction of $^{235}UF_6$ is very slightly higher in the vapor that passes through the porous barrier. If this diffusion process is repeated thousands of times, $^{235}U$ of the desired purity can be obtained.

## 4.10    MOLECULAR SPEEDS

One interesting aspect of kinetic theory is that it allows us to calculate the average velocity of gas molecules of known molecular weight at any temperature. We simply substitute the appropriate data into Equation 4.30. To make the calculation, we must express $R$ in appropriate units.

We saw in Equation 4.28 that $R$ is a proportionality constant between kinetic energy on the molar scale and kelvin temperature. In the SI, the units of energy are joules, where $1 \text{ J} = 1 \text{ kg m}^2 \text{ s}^{-2}$. We can see from Equation 4.28 that $R$ has units of $\text{J mol}^{-1} \text{ K}^{-1}$ or of $\text{kg m}^2 \text{ s}^{-2} \text{ mol}^{-1} \text{ K}^{-1}$. The numerical value of $R$, expressed in these units, is 8.3144. Since this value of $R$ is expressed in terms of length (m) and time (s), it can be used to calculate speed ($\text{m s}^{-1}$).

**Example 4.22**    Calculate the average speed of $N_2$ molecules in the atmosphere at a temperature of 300 K.

**Solution**    Speed is expressed in units of $\text{m s}^{-1}$, so we must take care that all the units cancel except $\text{m s}^{-1}$. Since $R$ has units of $\text{kg m}^2 \text{ s}^{-2} \text{ mol}^{-1} \text{ K}^{-1}$, the molar mass of $N_2$ must be expressed in $\text{kg mol}^{-1}$. Equation 4.30 gives the expression for the average speed:

$$\bar{v} = \sqrt{\frac{3RT}{\mathcal{M}}}$$

$$\bar{v} = \sqrt{\frac{(3)(300 \text{ K})(8.31 \text{ kg m}^2 \text{ s}^{-2} \text{ mol}^{-1} \text{ K}^{-1})}{0.028 \text{ kg mol}^{-1}}}$$

$$= \sqrt{267\ 000 \text{ m}^2 \text{ s}^{-2}}$$

$$\bar{v} = 517 \text{ m s}^{-1}$$

The average speed of $N_2$ molecules calculated in Example 4.22 is quite large. For comparison, the speed of sound in air is 330 $\text{m s}^{-1}$.

# THE USES OF DIFFUSION

Diffusion, the process by which molecules or other particles mix because of their random thermal motion, occurs not only in gases but also in liquids, solutions, and even solids. Diffusion is seen in many chemical processes, in both natural and synthetic systems. The rates of chemical transformations depend on the concentration of the reactants. As the initial concentration is depleted, it is nearly always replenished by some diffusion process. The rate of diffusion thus can determine the rate of the chemical transformation.

One commercial system that puts diffusion to practical use is called the iodine lamp. The hot tungsten filament of the lamp produces a relatively high vapor pressure of tungsten, whose atoms diffuse to the glass wall of the lamp. As tungsten accumulates on the glass, it lessens light output and shortens lamp life. If a trace of iodine vapor is added to the bulb, tungsten iodide tends to form in the relatively cool region near the glass. The tungsten iodide diffuses toward the filament, where it decomposes to release the tungsten. Thus, the rate at which tungsten accumulates on the glass is lessened.

Diffusion also plays an important role in microbes. The cell of a microbe can be regarded as a unit in which a number of different chemical reactions occur — the reactions by which a microbe maintains itself. For microbes, the diffusion of oxygen through a liquid medium into the cell is of major importance. It has been shown that the rate of oxygen diffusion into yeast cells can be the controlling factor in the activity of the cells under some conditions. These studies have been used to help develop the most efficient methods for a new food technology of great promise. Certain strains of yeast can ferment petroleum and produce edible proteins. The rate at which oxygen is supplied is an important part of the technology of such systems.

Diffusion also plays an interesting role in the process by which a spermatozoon, or sperm cell, manages to fertilize an egg. A spermatozoon "swims" toward the egg by wiggling its tail. The energy for the wiggling is derived from a molecule called adenosine triphosphate (ATP). However, ATP is produced in the body of the sperm cell, at the front of the tail. The spermatozoon is able to wiggle forward successfully because ATP diffuses all through the tail. One study of diffusion in the spermatozoon resulted in the calculation that diffusion could supply an adequate amount of ATP only over a length of $5 \times 10^{-3}$ cm to $9 \times 10^{-3}$ cm. In fact, the length of the tail of a sperm cell is about $5 \times 10^{-3}$ cm.

## The Maxwell-Boltzmann Equation

The average speed, or root-mean-square speed, is a convenient way to describe molecular motion in gases. But any macroscopic sample of a gas contains an enormous number of molecules moving with many different speeds. Because of the large number of molecules, it is impossible to specify their individual speeds. In the nineteenth century, however, James Clerk Maxwell (1831–1879) and Ludwig Boltzmann (1844–1906) found a statistical relationship between the temperature of a gas and the distribution of molecular speeds.

Maxwell and Boltzmann derived a complex equation that can be used to calculate the fraction of molecules that have a given range of speeds at a given temperature. A detailed discussion of the Maxwell-Boltzmann equation is beyond our scope. We shall say only that the results of calcula-

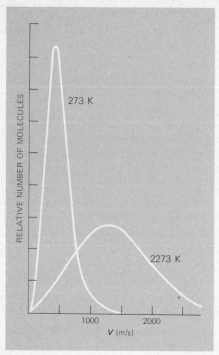

**Figure 4.19**
The distribution of molecular speeds at different temperatures, calculated from the Maxwell-Boltzmann equation. The maximum of each curve gives the most probable speed of a gas molecule at a given temperature, while the curve as a whole shows the distribution of molecular speeds for a given temperature.

tion with this equation can be plotted. Figure 4.19 shows some of these curves, which can be regarded as probability curves. The maximum of the curve gives the most probable speed for a molecule. The theoretical predictions of the distribution of molecular speeds made by the Maxwell-Boltzmann equation have been verified experimentally.

## 4.11 REAL GASES

The ideal gas model proposed in Section 4.6 leads to an equation of state that describes the properties of an ideal gas. It has been observed that a real gas behaves most like an ideal gas when the temperature is relatively high in relation to the boiling point of the gas and when the pressure is relatively low — not much higher than atmospheric pressure.

One way to evaluate the extent to which a real gas deviates from ideal gas behavior is to measure four variables: pressure, volume, temperature, and amount. When these measured values are substituted into the expression

$$\frac{PV}{nT}$$

the value obtained for an ideal gas is 0.082 L atm mol$^{-1}$ K$^{-1}$. A different value is obtained for a real gas that deviates from ideal gas behavior. If the value is greater than 0.082, the real gas is said to display a *positive deviation*. If the value is less than 0.082, the real gas is said to display a *negative deviation*. The behavior of some real gases is shown in Figure 4.20, in which experimentally observed values of $PV/nT$ are plotted for some real gases at various pressures. Each curve represents the indicated gas at a given temperature.

**Figure 4.20**
The behavior of some real gases, shown by substituting measured values into the expression $PV/nT$. For an ideal gas, the value of this ratio is 0.082 L atm mol$^{-1}$ K$^{-1}$. Each curve represents the indicated real gas at a given temperature. Hydrogen always shows a positive deviation. Other gases show a negative deviation at low pressure but a positive deviation at high pressure.

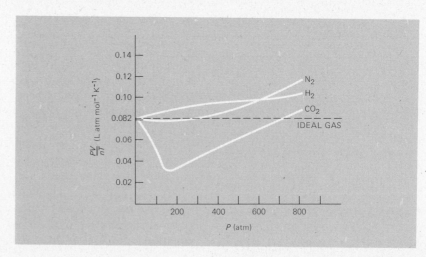

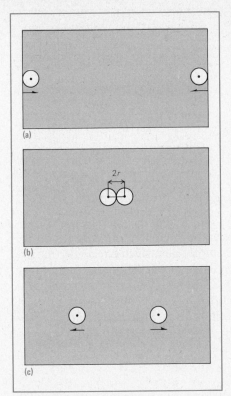

**Figure 4.21**

Why real gases can have higher pressures than are calculated for an ideal gas. The ideal gas approximation assumes that molecules of the gas do not occupy any volume. In a real gas, when molecules collide with one another, the distance that the molecules must travel between collisions with the container wall is less by $2r$, where $r$ is the radius of a gas molecule. This results in an increased frequency of collisions with the walls and therefore a higher pressure.

## Approximations in the Ideal Gas Model

We can understand how and why real gases deviate from ideal behavior by studying the approximations in the ideal gas model. The molecules of the ideal gas are assumed to have no volume, so the calculated volume of an ideal gas is only the volume of the empty space. The molecules of a real gas have volume, so the measured volume of a real gas is larger. At high pressures, the molecules of the real gas are closer to each other than at low pressures, so a larger fraction of the gas volume is occupied by the molecules. When the pressure is high a real gas shows a positive deviation, as can be seen in Figure 4.20. The measured volume is greater than the ideal gas volume at the same pressure, so $PV/nT > 0.082$.

The volume of the molecules of a real gas also affects the gas pressure. The measured pressure of a real gas is greater than the pressure of the same volume of ideal gas. Pressure results from collisions by molecules on a container wall; more frequent collisions mean higher pressure. For an ideal gas, the frequency of collisions is calculated on the assumption that the entire volume is empty space. The molecules of a real gas have volume. We must exclude their volume when we calculate the distance that the molecules travel between collisions with the walls. A molecule of a real gas travels less distance than a molecule of an ideal gas before colliding with the wall. The actual pressure of a real gas is therefore higher than the calculated pressure of the same volume of an ideal gas.

Figure 4.21 illustrates the effect of the volume of gas molecules on gas pressure. A molecule leaves the right-hand wall of the container at the same time that another molecule leaves the left-hand wall. The two molecules collide in the center of the container and rebound toward the walls they left originally. Because the molecules have volume, the distance each one travels before colliding with a wall is less by $2r$, $r$ being the radius of one molecule. Therefore, the frequency of collisions is higher. The volume of the gas molecules becomes the dominant factor at high pressures, where real gases display positive deviations.

The ideal gas model also assumes that gas molecules do not attract each other. In fact, there are rather substantial attractive forces between molecules in the gas phase. To a great degree, the attractive force is related to the size of the gas molecules; larger molecules have greater attractive forces.

The major effect of these attractive forces is to reduce pressure. Figure 4.22 shows two molecules that pass near one another on the way to opposite walls of a container. Their mutual attraction deflects them from their straight paths, so they travel a longer distance before they hit the walls. The frequency of collisions with the walls is reduced, so gas pressure is reduced. Thus, attractive forces cause negative deviations of real gases from ideal gas behavior.

Most gases show negative deviations when pressure is not high enough to make molecular volume an important factor, as Figure 4.20 shows. The

**TABLE 4.2    Van der Waals Constants**

| Substance | $a$ (atm L$^2$ mol$^{-2}$) | $b$ (L mol$^{-1}$) | Substance | $a$ (atm L$^2$ mol$^{-2}$) | $b$ (L mol$^{-1}$) |
|---|---|---|---|---|---|
| He | 0.034 | 0.0237 | $CO_2$ | 3.592 | 0.0427 |
| Ne | 0.211 | 0.0171 | HCl | 3.667 | 0.0408 |
| $H_2$ | 0.244 | 0.0266 | $NH_3$ | 4.170 | 0.0371 |
| NO | 1.340 | 0.0279 | $C_2H_2$ | 4.390 | 0.0514 |
| Ar | 1.345 | 0.0322 | $H_2S$ | 4.431 | 0.0429 |
| $O_2$ | 1.360 | 0.0318 | $H_2O$ | 5.464 | 0.0305 |
| $N_2$ | 1.390 | 0.0391 | $Cl_2$ | 6.493 | 0.0562 |
| CO | 1.485 | 0.0399 | Hg | 8.093 | 0.0170 |
| $CH_4$ | 2.253 | 0.0428 | $C_2H_5OH$ (ethanol) | 12.02 | 0.0841 |
| Kr | 2.318 | 0.0398 | | | |

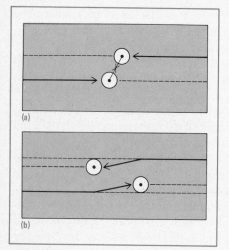

(a)

(b)

**Figure 4.22**

Molecular attractions in real gases. Two molecules attract one another, increasing the distance they must travel before hitting the walls of the container and thus lowering the measured pressure. The ideal gas law assumes that no such attractions take place.

extent of the negative deviation at a given temperature depends on the magnitude of the attraction between molecules, which is related to the size of the molecules. In gases that have large molecules and are not at high pressures, the intermolecular forces are more important than the volume of the molecules. Only gases made up of very small molecules, such as $H_2$ and He, always show positive deviations. In such gases, the force of intermolecular attraction is quite small.

## The van der Waals Equation

Attempts to make the ideal gas equation provide a better description for real gases use an empirical approach. Many measurements are made of real gases, and the results are used to give a corrected version of the ideal gas equation. One of the first and most successful attempts was made by Johannes van der Waals (1837–1923). He corrected the measured pressure of a gas for attractive forces and the measured volume of the gas for the volume of the molecules. The changes can be expressed as a constant for each gas. Table 4.2 lists the constants obtained in this manner. The constant $a$ is related to molecular attractions and corrects $P$. The constant $b$ is related to the volume of the molecules and corrects $V$. The new equation of state, called the *van der Waals equation,* is written as:

$$\left(P + \frac{n^2a}{V^2}\right)(V - nb) = nRT \tag{4.33}$$

**Example 4.23**    A heavy-walled evacuated container with a volume of 1.98 L is filled with 215 g of dry ice, $CO_2$(s). After a time, all the dry ice sublimes to $CO_2$(g), and the temperature of the gas is found to be 299.2 K. Using the data of Table 4.2, calculate the pressure of the $CO_2$(g) and compare it to the ideal gas pressure.

**Solution**    The van der Waals equation, like the ideal gas equation, has four variables, $P, V, n$, and $T$. The values of three of the variables are given in the problem. The van der

Waals equation has three constants, $R$, $a$, and $b$, whose values are known. Therefore, the value of the unknown variable, $P$, can be calculated. First list the data:

$$V = 1.98 \text{ L} \qquad\qquad R = 0.0821 \text{ L atm mol}^{-1} \text{ K}^{-1}$$
$$T = 299.2 \text{ K} \qquad\qquad a = 3.59 \text{ L}^2 \text{ atm mol}^{-2}$$
$$n = \frac{215 \text{ g CO}_2}{44.01 \text{ g/mol CO}_2} = 4.89 \text{ mol} \qquad b = 0.0427 \text{ L mol}^{-1}$$

Then carry out the calculations:

$$\left[ P + \frac{(4.89 \text{ mol})^2 (3.59 \text{ L}^2 \text{ atm mol}^{-2})}{(1.98 \text{ L})^2} \right]$$
$$[1.98 \text{ L} - (4.89 \text{ mol})(0.0427 \text{ L/mol})]$$
$$= (4.89 \text{ mol}) \left( 0.0821 \frac{\text{L atm}}{\text{mol K}} \right) (299.2 \text{ K})$$
$$P = 45.9 \text{ atm}$$

If we use the same values of $n$, $T$, and $V$ in the ideal gas equation, we obtain a different value:

$$P(1.98 \text{ L}) = (4.89 \text{ mol}) \left( 0.0821 \frac{\text{L atm}}{\text{mol K}} \right) (299.2 \text{ K})$$
$$P = 60.7 \text{ atm}$$

The measured pressure of $CO_2(g)$ under these conditions is found to be 44.8 atm; the van der Waals equation gives an answer closer to the measured pressure than does the ideal gas equation. The measured pressure is substantially less than the ideal pressure; $CO_2(g)$ shows a negative deviation from ideality. The negative deviation is consistent with the relatively strong intermolecular attraction between $CO_2$ molecules, which are relatively large.

## 4.12  CRITICAL PHENOMENA

Suppose we want to liquefy a sample of a real gas. We can lower the temperature until it is at or below the condensation point, so that the gas condenses spontaneously. Theoretically, we can liquefy the gas by raising the external pressure without lowering the temperature. As the pressure increases, the molecules come closer together, until the attractive forces between them are strong enough to cause condensation.

Some gases will not liquefy at room temperature, no matter how high the pressure. This is true of most of the common gases, such as $N_2$, $O_2$, $H_2$, and $CH_4$. We can liquefy these gases by raising the external pressure, but only if the temperature is lowered. Experiments show just how low the temperature must be for each gas to liquefy. While $N_2$ can be liquefied at 154 K if the pressure is increased sufficiently, $H_2$ cannot be liquefied with high pressure until the temperature is decreased to 33 K.

TABLE 4.3 Critical Temperatures and Pressures

| Substance | Critical $T$ (K) | Critical $P$ (atm) |
|---|---|---|
| He | 5.3 | 2.26 |
| $H_2$ | 33.3 | 12.8 |
| Ne | 44.5 | 26.9 |
| $N_2$ | 126 | 33.5 |
| CO | 133 | 34.5 |
| $F_2$ | 144 | 55 |
| Ar | 151 | 48 |
| $O_2$ | 155 | 50.1 |
| NO | 180 | 64 |
| $CH_4$ | 191 | 45.8 |
| Kr | 209 | 54.3 |
| $CO_2$ | 304 | 72.9 |
| $C_2H_6$ | 305 | 48.2 |
| $C_2H_2$ | 309 | 61.6 |
| HCl | 325 | 82.1 |
| $H_2S$ | 374 | 88.9 |
| $NH_3$ | 406 | 112.5 |
| $Cl_2$ | 417 | 76.1 |
| $C_2H_5OH$ | 515 | 63 |
| $H_2O$ | 647 | 218.3 |
| Hg | 1823 | 200 |

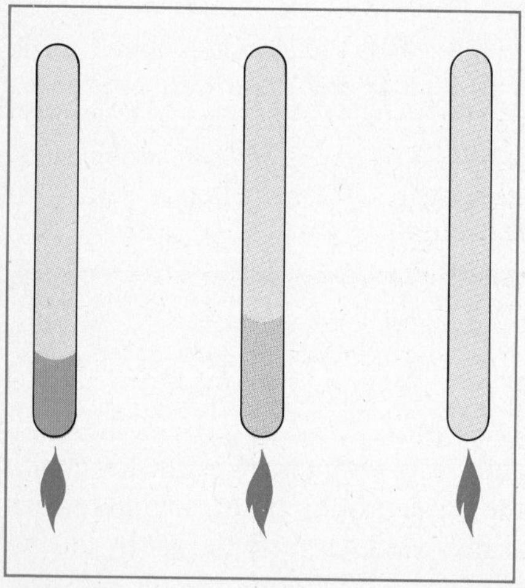

**Figure 4.23**
Critical temperature. As the temperature of a liquid in a closed container is raised, the density of the liquid decreases and the density of the gas increases. When the temperature is high enough, the density of the liquid and the density of the gas phase are equal and the two phases become one. This occurs at a temperature just above the critical temperature of the substance. *Note:* This demonstration should *not* be attempted because the increased vapor pressure can shatter the glass.

Every gas has a characteristic maximum temperature, called the **critical temperature,** above which liquefaction is impossible no matter how high the pressure. The minimum pressure required to liquefy a gas at the critical temperature is called the **critical pressure.** Table 4.3 lists the critical temperatures and pressures of some common substances.

The existence of critical temperatures and pressures can be explained in several ways. Above the critical temperature, no amount of pressure can push the molecules close enough so that the attractive forces overcome thermal motion. Table 4.3 shows that gases with small molecules (and hence weak attractive forces), such as $H_2$ and He, have very low critical temperatures. Gases with larger molecules (and stronger attractive forces), such as $CO_2$, $H_2O$, or $NH_3$, have higher critical temperatures. The critical temperature is an indicator of the relative attractive forces for any gas. It can be shown to be related to the van der Waals constant $a$.

The existence of a critical temperature can also be observed and explained macroscopically. Suppose a liquid is placed in a sealed container, as shown in Figure 4.23. The liquid stays on the bottom of the container, while the vapor phase of the substance fills the rest of the container. There is a division between the two phases, gas and liquid, called a *meniscus.*

If the temperature of the sealed system is raised, the vapor pressure increases. The quantity of the substance in the gas phase increases, while its volume does not change appreciably. The density of the gas, which is mass divided by volume, thus increases as the temperature rises. The liquid, meanwhile, expands as the temperature rises, so its density decreases as the temperature increases.

When the temperature is high enough, the density of the gas equals the density of the liquid. The meniscus disappears, and the liquid and the gas are indistinguishable. The temperature at which this happens is just above the critical temperature. At this or any higher temperature, there is no difference between the liquid state and the gas state, and we can say that it is impossible to liquefy a gas.

As you can see from our discussion of gases, the ideal gas law and the model of gas behavior on which it is based provide the framework for an understanding of the behavior of real gases. However, the deviations of real gases from ideal gas behavior can be explained only by detailed experimental study of the nature of real gases and of the interactions between the molecules and atoms in the gaseous state.

**Summary**

We began this chapter by distinguishing between a **gas,** whose molecules move freely in space; a **liquid,** whose entities have some short-range **order;** and a **crystalline solid,** whose entities have an ordered arrangement. We then described how **thermal motion** can cause **evaporation,** the movement of molecules from the liquid to the gas phase, or **sublimation,** the movement of molecules from the solid to the gas phase. We noted that a system at **equilibrium** appears static on the macroscopic scale but actually undergoes constant change on the microscopic scale. We defined **pressure** as the force exerted by a gas on a given unit area of surface. We introduced the concept of an **ideal gas,** which follows ideal gas laws such as **Boyle's law** — that the volume of a gas at constant temperature is inversely proportional to its pressure — and the **law of Charles and Gay-Lussac** — that the volume of a gas at constant pressure is directly proportional to its **absolute temperature.** We mentioned **Avogadro's law,** which says that equal volumes of different gases at the same temperature and pressure contain equal numbers of molecules, and then described the **ideal gas equation,** which relates the pressure, volume, temperature, and amount of a sample of an ideal gas. We then introduced **Dalton's law of partial pressures,** which states that the total pressure of a mixture of gases is the sum of the partial pressures of the gases. We noted the difference between **effusion,** the movement of gas molecules through a hole in a container, and **diffusion,** the spread of gas molecules through a space. We showed why **real gases** differ from an ideal gas. Finally, we described the **critical temperature,** above which a gas cannot be liquefied at any pressure, and the **critical pressure,** the pressure needed to liquefy a gas at its critical temperature.

## Exercises

**4.1**  Suggest an example and an exception for these generalizations, made in Section 4.1:
(a) A solid melts when it is heated.
(b) A liquid does not have a definite shape.
(c) A solid is denser than a liquid.
(d) A gas liquefies when it is cooled.

**4.2**  Illustrate the differences between the three states of matter on the microscopic level by drawing three different arrangements of 15 to 20 spheres.

**4.3**[4]  A liquid is in a sealed, spherical container in equilibrium with its vapor at 300 K. Predict whether the amount of vapor increases, decreases, or remains the same for each of these changes:
(a) The temperature goes up to 310 K.
(b) More liquid is added to the container.
(c) The container is opened.
(d) The container is changed to a cube, with no change in volume.

**4.4**  Use the data in Table 3.2 to find the height of a column of ethanol that is supported by 1 atm of pressure.

**4.5**  Some pumps used for wells operate by creating a vacuum. Find the maximum depth from which water can be raised by such a pump operating at sea level.

**4.6**  The tops of two flasks are connected. One of them is filled with a liquid, and the other is cooled. After some time, most of the liquid is found in the colder flask. Suggest a mechanism on the molecular level that explains this occurrence.

**4.7**  A cylinder whose volume is 4.3 L contains compressed helium at a pressure of 75 atm. The helium is used to

fill balloons to a pressure of 1.0 atm. Find the total volume of the balloons filled from this cylinder.

**4.8**  A cylinder with a movable piston is filled with a gas at a pressure of 0.97 atm. Weights are placed on the piston, and it moves from a height of 73 cm to 14 cm above the floor of the cylinder. Find the pressure of the gas.

**4.9**  The compression ratio of an automobile engine is the ratio of the maximum pressure of gasoline and air in the cylinder to the pressure of gasoline and air as the cylinder fills. In an automobile engine with a compression ratio of 6:1, the volume of the cylinder at maximum pressure is 91 cm³. Find the volume of the cylinder when it fills with gasoline and air.

**4.10**  A balloon contains gas at a temperature of 298 K. You are asked to expand the volume of the balloon from 1.3 L to 6.1 L. To what temperature must the gas be heated?

**4.11**  On a winter day when the temperature is −7°C, a storage tank is filled with natural gas so that the tank is expanded to 87% of its maximum volume. The gas is stored until a summer day when the temperature rises to 35°C. Is the tank safe?

**4.12**  If $V_F$ is the volume of an ideal gas at 0°F, find its volume at 1°F at the same pressure.

**4.13**  If a sealed container of gas is cooled it may sometimes collapse. Explain why.

**4.14**  State whether each of the following statements about a given amount of gas is true or false. Justify your answer.
(a) If the pressure of a gas is doubled and the volume is doubled, the temperature of the gas is unchanged.
(b) If the pressure of a gas is doubled and the temperature goes from 100 K to 200 K, the volume of the gas is unchanged.

(c) If the pressure of a gas is halved and the temperature goes from 100°C to 200°C, the volume of the gas is decreased.

**4.15**  Which of the following changes cause an increase in the pressure of a given amount of an ideal gas?
(a) The volume increases from 1 L to 2 L and the temperature increases from 300°C to 800°C.
(b) The volume increases from 2 L to 4 L and the temperature decreases from 400 K to 200 K.
(c) The volume decreases from 2 L to 1 L and the temperature decreases from 600 K to 300 K.

**4.16**  A gas in a 750-cm³ container at a pressure of 765 mmHg and a temperature of 399 K is transferred to a container of volume 50.0 cm³ at a temperature of 110 K. Find the new pressure of the gas.

**4.17**  After a tire is inflated, the pressure gauge reading increases from 30 psi to 50 psi and the volume of the tire is found to have increased by 11%. The original temperature of the air in the tire was 283 K. Find the new temperature.

**4.18**  A gas occupies a volume of 1.05 L at a pressure of 0.447 atm and a temperature of 298 K. Find its volume at STP.

**4.19**  The pressure of a gas at 22°C is increased from 173 kPa to 197 kPa while the volume remains constant. Find the new temperature of the gas.

**4.20**  As a result of a chemical reaction, the amount (moles) of gas in a balloon of volume 652 cm³ increases by 50%. The gas then returns to the temperature and pressure it had before the reaction. Find the volume of the balloon.

**4.21**  Derive simple relationships between (a) the amount of an ideal gas

[4] The answers to exercises whose numbers are in color can be found in Appendix VII. The star indicates an exercise that is more challenging than average.

and its pressure at constant volume and temperature, (b) the amount of an ideal gas and its temperature at constant volume and pressure.

**4.22** A 1.1-L bulb contains $N_2$ and a 3.3-L bulb contains $NO_2$. Both gases are at the same temperature and pressure. Find the ratio of $N_2$ to $NO_2$ by (a) amount, (b) mass.

**4.23\*** When the amount of gas in a system is tripled, the volume of the system doubles but the temperature does not change. The original pressure of gas was 1.0 atm. Find the new pressure.

**4.24** Find the value of $R$ in units of mmHg cm$^3$/molecule K.

**4.25** What size cylinder is needed to store 10.0 mol of $O_2$ at a pressure of 121 atm and a temperature of 299 K?

**4.26** Find the amount of $CO_2$ collected in a balloon of volume 279 cm$^3$ at a temperature of 25°C and a pressure of 758 mmHg.

**4.27** A 9.76-mg sample of $CO_2$ is placed in a 25-cm$^3$ container. The pressure is found to be 155 mmHg. Find the temperature of the $CO_2$.

**4.28** A neon sign is made from glass tubing that is 2.17 m long and has a diameter of 1.31 cm. The pressure of neon inside the tube is 1.78 mmHg at 297 K. Find the mass of neon in the tube.

**4.29** A 0.551-g sample of volatile liquid is vaporized completely in a bulb of volume 125 cm$^3$ at a pressure of 1.00 atm and a temperature of 20°C. Find the molecular weight of the liquid.

**4.30** A 7.07-mg sample of a gaseous form of phosphorus is contained in a bulb of volume 50.1 cm$^3$ at a pressure of 30.0 mmHg and a temperature of 422 K. Find the formula of the gas.

**4.31** The density of a gas is 6.71 g/L at a temperature of 35°C and a pressure of 1.00 atm. Find the molar mass of the gas.

**4.32** Find the density of $CH_4$ at STP.

**4.33** The density of $Cl_2$ at a certain temperature and pressure is 2.88 g/L. Find the density of HCl at the same temperature and twice the pressure.

**4.34\*** A quantity of CO gas occupies a volume of 0.48 L at 1.0 atm and 275 K. The pressure of the gas is lowered and its temperature is raised until its volume is 1.3 L. Find the density of the CO under the new conditions.

**4.35** The decomposition of $CaCO_3$ produces $CO_2(g)$ and $CaO(s)$. A sample of calcium carbonate decomposes at 700 K to form enough $CO_2$ to fill a 1.00-L container to a pressure of 2.47 atm. Find the mass of $CaCO_3$ that decomposed.

**4.36** A 50.2-cm$^3$ container is filled with $F_2$ at a pressure of 0.209 atm and a temperature of 188 K. When a small sample of $P_4$ is introduced into the container, an immediate reaction takes place that consumes all the $P_4$ and forms $PF_3$. When the container cools to 188 K after the reaction, the pressure inside the bulb is found to be 0.162 atm. Find the mass of the $P_4$. (Neglect the vapor pressure of $PF_3$).

**4.37** Find the volume of $O_2$, measured at STP, needed for the complete combustion of 175 g of octane, $C_8H_{18}$, to $CO_2$ and $H_2O$.

**4.38** A mixture of $2.3 \times 10^{23}$ molecules of $N_2$ and $4.1 \times 10^{23}$ molecules of NO has a total pressure of 1.8 atm under certain conditions. Find the partial pressure of each gas in the mixture.

**4.39** A mixture of 17.1 g of $CO_2$ and 8.65 g of Ne exerts a pressure of 549 mmHg. Find the partial pressure of each gas in the mixture.

**4.40** A 1.0-L bulb that contains Ar at a pressure of 0.63 atm is connected by a stopcock to a 1.2-L bulb that contains Ne at a pressure of 0.78 atm. The stopcock is opened and the gases mix. Find the partial pressure of each gas in the

mixture, assuming no temperature change.

**4.41\*** When 0.583 g of neon is added to an 800-cm$^3$ bulb containing a sample of argon, the pressure of the gases is found to be 1.17 atm at a temperature of 22°C. Find the mass of argon in the bulb.

**4.42** Oxygen gas prepared by the decomposition of $KClO_3$ is collected over water, as shown in Figure 4.13. Find the amount of $O_2$ that is collected in a volume of 0.488 L at a pressure of 758 mmHg and a temperature of 296 K.

**4.43** When $CH_3Na$ reacts with water, $CH_4(g)$ is liberated. When a sample of $CH_3Na$ reacts with a large excess of water, a volume of 143 cm$^3$ of $CH_4(g)$ is collected above the water at a pressure of 0.988 atm and a temperature of 308 K. Find the mass of the sample of $CH_3Na$.

**4.44** Carbon and chlorine react at high temperatures to form carbon tetrachloride. The reaction is $C(s) + 2Cl_2(g) \rightarrow CCl_4(g)$. A sample of carbon reacts with 0.552 atm of $Cl_2$ at 600 K. After all the carbon reacts, some $Cl_2$ remains and the pressure is 0.416 atm. Find the pressure of $CCl_4$.

**4.45** In the gas phase, $N_2O_5$ decomposes by the reaction $2N_2O_5(g) \rightarrow 4NO_2(g) + O_2(g)$. A sample of $N_2O_5$ is placed in an evacuated 0.996-L container and is heated to 500 K. After complete decomposition, the pressure is found to be 0.352 atm. Find the mass of $N_2O_5$ that decomposed.

**4.46** A sample of $H_2$ gas is placed in a container at 300 K. The pressure is found to be 1.00 atm. The $H_2$ is heated to 2700 K, and the pressure is found to be 9.36 atm. Find the partial pressure of H, which forms according to the reaction $H_2(g) \rightarrow 2H(g)$ at 2700 K.

**4.47** A mixture of $C_2H_2$ and $C_2H_4$ gases has a total pressure of 0.219 atm. Just enough $O_2$ is added to the mixture for complete combustion of the gases to $CO_2(g)$ and $H_2O(g)$. After combustion, the total pressure is 0.743 atm. Assuming constant temperature and volume, find the partial pressure of $C_2H_2$ in the original mixture.

**4.48** At high temperatures, both $NO_2(g)$ and $N_2O_3(g)$ decompose to $N_2$ and $O_2$. A gaseous mixture of $NO_2$ and $N_2O_3$ has a total pressure of 1.00 atm at 298 K. When the mixture is heated to 1100 K, the two gases decompose completely. The total pressure of the products $N_2$ and $O_2$ is 7.85 atm. Find the partial pressure of $N_2O_3$ in the original mixture.

**4.49** The statement that the molecules of an ideal gas are in constant, random motion is based on many observations of gases. Give some of these observed properties of gases.

**4.50** Use the derivation of Boyle's law in Section 4.7 to predict and explain how each of the following changes affects the pressure of a gas: (a) an increase in temperature, (b) an increase in the size of the container, (c) an increase in the amount of gas.

**4.51** Use the conversion factor 1 L atm = 101 J to find the kinetic energy in joules of 1.0 mol of an ideal gas at 300 K.

**4.52** Find the total kinetic energy of a sample of gas in a 1.0-L container at a pressure of $1.0 \times 10^5$ Pa.

**4.53** Find the ratio of the average speed of He atoms to that of Xe atoms at the same temperature.

**4.54** Find the ratio of the average speed of $H_2$ molecules at 300 K to their average speed at 3000 K.

**4.55** Find the ratio of the average speed of $O_2$ molecules at 300 K to that of $CO_2$ molecules at 500 K.

**4.56** A sample of He is placed in the left side of the apparatus shown in Figure 4.18 and is found to have a pressure of 0.155 atm. After 10.0 min, the pressure of He on the right is 0.112 atm. When a sample of an unknown gas at the same temperature and pressure is introduced into the left side of the same apparatus, the pressure on the right after 10.0 min is found to be 0.0264 atm. Find the molecular weight of the unknown gas.

**4.57** A sample of Ne on one side of the apparatus shown in Figure 4.18 has a pressure of 0.284 atm. After 175 s, its pressure falls to 0.164 atm. Find the time required for the pressure of a sample of $Cl_2$ with the same pressure at the same temperature in the same apparatus to fall to 0.164 atm.

**4.58** A 1.0-L balloon filled with an unknown gas is punctured. The balloon deflates in 2.3 s. Another 1.0-L balloon filled with 1.3 g of $N_2$ at the same temperature and pressure deflates in 3.8 s when it is punctured. Assuming the same size hole in both punctures, find the density of the unknown gas under these conditions.

**4.59★** A sample of Ne with a pressure of 0.443 atm and a temperature of 297 K is introduced into one side of the apparatus shown in Figure 4.18. After 220 s, the pressure of Ne on the other side of the apparatus is 0.223 atm. When 0.443 atm of an unknown gas is introduced into one side of this apparatus at 297 K, the pressure on the other side after 220 s is found to be 0.111 atm. Find the density of the unknown gas.

**4.60** A mixture of equal amounts of CO and $CO_2$ is placed in one side of the apparatus shown in Figure 4.18. The total pressure is 0.838 atm. Find the pressure on the other side after half of the CO has diffused.

**4.61** The rate of diffusion of a hydrocarbon gas is found to be 97.6% of the rate of diffusion of argon under the same conditions. The hydrocarbon gas is 85.9% carbon by mass. Find its formula.

**4.62** Find the average speed of $O_2$ molecules at 298 K.

**4.63** Find the temperature of an H atom that has a velocity just below the speed of light ($2.998 \times 10^8$ m/s).

**4.64** One mole of gas in a container whose volume is 0.0686 L exerts a pressure of 378 atm at a temperature of 273 K. Find whether this gas displays a positive or a negative deviation from ideality.

**4.65** Eight moles of each of the following gases are placed in a 1.0-L container at 300 K. Predict which gas in each pair will have the higher measured pressure: (a) $H_2$ and $CO_2$, (b) He and CO, (c) $NH_3$ and $Cl_2$.

**4.66** Predict which gas in each pair behaves most like an ideal gas: (a) $CO_2$ at 300 K and 1 atm or $H_2$ at 200 K and 1 atm, (b) $NH_3$ at STP or Kr at STP.

**4.67★** A Kr atom is a sphere whose radius is $1.1 \times 10^{-10}$ m. Find the volume of the Kr atoms in a 1.0-L container at STP and in a 1.0-L container at 273 K and 200 atm. (Assume ideal gas behavior.)

**4.68** Use the data in Table 4.2 to find the pressure when 1.33 mol of $CH_4(g)$ is placed in a 10.2-L container at 299 K. Find the pressure using the ideal gas equation. Account for the difference between the two pressures.

**4.69** Use the data in Table 4.2 to find the pressure of $H_2O$ when 2.66 mol of steam is in a 1.02-L container at 400 K. Find the pressure using the ideal gas equation. Account for the difference between the two pressures.

**4.70** The critical volume is defined as the volume of one mole of gas at the critical temperature and pressure (Table 4.3). Find the critical volume of $O_2$, using the ideal gas equation. Compare your result to the experimental value, which is 0.075 L/mol, and account for the difference.

# 5

# The Structure
# of Atoms

**Preview** **N**ow we step into the world of the atom, the basic unit of all elements. We first tell how the concept of the atom evolved and then describe the structure of the atom, introducing the subatomic particles: the protons and neutrons in the atomic nucleus and the electrons that orbit the nucleus. We next describe the genesis of one of the great unifying principles of modern physics, the quantum theory, and explain how the theory has helped to define the modern picture of atomic structure. Finally, we show how two apparently conflicting views of subatomic phenomena, the wave picture and the particle picture, were reconciled in a quantum mechanical description of the atom, and how that description was used to develop a model of the atom that is essential to modern chemistry.

" " **I**f in some cataclysm, all of scientific knowledge were to be destroyed, and only one sentence passed on to the next generation of creatures, what statement would contain the most information in the fewest words? I believe it is the atomic hypothesis (or the atomic fact, or whatever you wish to call it) that *all things are made of atoms — little particles that move around in perpetual motion, attracting each other when they are a little distance apart, but repelling upon being squeezed into one another.* In that one sentence you will see there is an enormous amount of information about the world, if just a little imagination and thinking are applied."

> Richard P. Feynman, Robert B. Leighton, and Matthew Sands
> *The Feynman Lectures on Physics*

## 5.1   THE DEVELOPMENT OF THE ATOMIC THEORY

We can describe chemistry as the study of how atoms combine to form molecules and how molecules are transformed into other molecules. To study how atoms combine, we need to know what atoms are and what the structure of an atom is.

In Western culture, the concept of the atom — the idea that matter is composed of extremely small, ultimately indivisible units — goes back to 500 B.C. The Greek natural philosopher Leucippus and his student Democritus speculated that the universe consisted of the void and atoms, that atoms of different substances were fundamentally the same, and that atoms were in constant motion.

These ideas were debated by other classical philosophers. Lucretius accepted them; Aristotle rejected them. But the debate was purely philosophical. None of the philosophers thought of testing the theory by trial in the real world. The ancient Greeks were not experimental scientists. Democritus had made a lucky guess with the theory of atoms. The failure to test the theory by experiment made it just one of many competing theories about the nature of matter. Some twenty centuries had to pass before the atomic theory was supported by experimental data.

### Dalton's Atomic Theory

It was John Dalton, in the early years of the nineteenth century, who first enunciated the atomic theory in what could be called modern terms. Dalton's picture of the atom is not the one we have today, but it did sum up some valid basic principles about the composition and behavior of chemical substances. These principles were consistent with the data available to Dalton and other scientists of the time. The major assumptions of Dalton's atomic theory were:

**1.** Elements are composed of small indivisible particles called atoms.
**2.** All the atoms of a given element are identical; they have the same mass,

size, and properties. Atoms of one element have properties different from those of all other elements.

**3.** Atoms of two or more elements can combine to form new substances in which the atoms are held firmly together. The relative numbers of atoms in a given pure compound are constant.

These assumptions opened fruitful lines of investigation. For example, the assumption that all atoms of a given element have the same mass made it meaningful to calculate the mass of atoms—but only relative masses, since atoms are so small that an individual atom could not be weighed.

Using the assumption that the relative numbers of atoms in any pure compound are constant, Dalton set out to assign relative weights to atoms. He gave the hydrogen atom a relative weight of 1 and measured all other atoms by that standard. For example, experiments showed that 1 g of hydrogen would combine with 8 g of oxygen to form water. Using Occam's razor—which says essentially that the most economical explanation should be adopted whenever possible—Dalton proposed that each molecule of water had one atom of oxygen and one atom of hydrogen. He therefore assigned the oxygen atom a weight of 8 relative to the hydrogen atom. He also noted that 3 g of carbon combined with 1 g of hydrogen to form methane. Assuming methane to be CH (not $CH_4$ as we now know it to be), he assigned carbon the relative weight of 3, meaning that one carbon atom was three times heavier than one hydrogen atom.

## 5.2 THE ELECTRON

Dalton thought of atoms as tiny, featureless spheres, rather like infinitesimally small, sticky billiard balls, which combined in some mysterious way to form molecules. Today we know that the combination of atoms to form molecules is governed by a number of factors. The most important is the distribution of electrons in atoms. Much of the work of physicists in the nineteenth and twentieth centuries has been devoted to the structure of the atom and the nature of the forces in and between atoms.

### The Nature of Electricity

The most primitive human knows that an object thrown into the air falls to the ground. Newtonian physics say that the fall is caused by the gravitational attraction between the object and the earth. Physicists have measured and defined gravitational force. It is always positive; objects are attracted but never repelled. It is directly proportional to the masses of the two objects and inversely proportional to the square of the distance between them, and it is relatively small. Gravitational attraction is important for relatively large objects, such as apples and planets. It is inconsequentially small for very small objects, such as atoms and molecules.

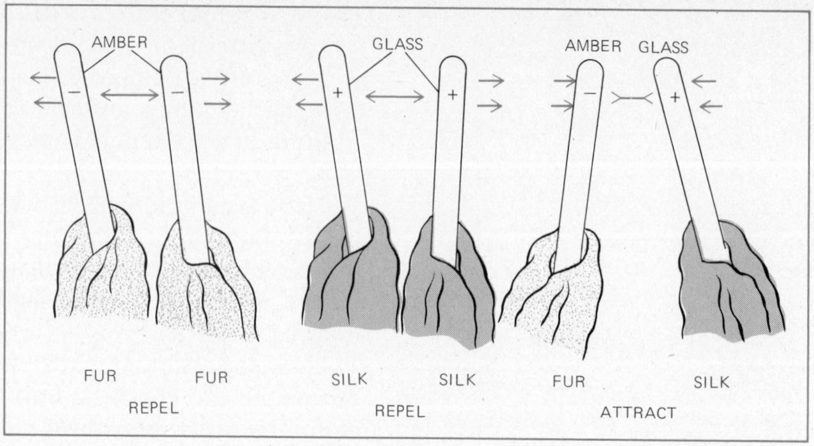

**Figure 5.1**
Like electric charges repel each other and unlike charges attract. Two pieces of amber rubbed with fur develop negative charges and repel each other. Two rods of glass rubbed with silk develop positive charges and repel each other. A charged piece of amber and a charged rod of glass attract each other.

Gravity is not the only force that acts on objects. If a child rubs a balloon vigorously against a wall, the balloon will cling to the wall without falling. There is an attractive force between the balloon and the wall that overcomes the gravitational attraction that would otherwise make the balloon fall.

The ancient Greeks demonstrated the existence of this second attractive force by rubbing a piece of amber (the petrified sap of a tree) with wool or fur (Figure 5.1). Once rubbed, the amber attracts light objects such as feathers. If a second piece of amber is rubbed and is held near the first, the two bits of amber repel each other. If a piece of glass is rubbed with silk and held near the amber, the glass and the amber attract each other.

We can explain this behavior by saying that there are *electrical forces* between the pieces of amber and glass, and that the rubbing causes the amber and the glass to be electrically charged. As there is a gravitational force that always attracts, there is an electrical force that is more powerful than gravitational force and that may either attract or repel.

By convention, electricity was described as positive (+) or negative (−). Rubbed amber was described as having a negative electric charge, while rubbed glass had a positive charge. Like electrical charges repel each other, so anything that repelled rubbed glass had a positive charge and anything that repelled amber had a negative charge. The magnitude of the charge is measured by determination of the force of attraction or repulsion between two charged objects.

## Coulomb's Law

The SI unit of charge is the coulomb, symbol C, named after the French scientist Charles de Coulomb (1736 – 1806). It is defined as the quantity of charge that flows through a cross section of wire in 1 s if there is a current of 1 ampere (A). Experiments by eighteenth-century scientists, notably Coulomb, described how the force between two charges varies with the magnitude of the charges and the distance between them. The relationship, known as **Coulomb's law,** is

$$\text{force} = \frac{kq_1q_2}{r^2} \qquad (5.1)$$

In Equation 5.1, the magnitude of one charge is represented by $q_1$ and the magnitude of the other charge by $q_2$. The distance between the charges is $r$. (The charges are considered to be concentrated at points in space.) The symbol $k$ is a constant whose value depends on the choice of units for $q$ and $r$ and on the nature of the medium between the two charges. Coulomb's law is often stated: *The force of attraction between two opposite charges is inversely proportional to the square of the distance between them and directly proportional to their magnitudes.*

The resemblance between Coulomb's law and the law of gravity should be noted. Gravitational attraction is directly proportional to the masses of two objects and inversely proportional to the square of the distance between them.

The participants in the 1927 Solvay conference, held in Brussels, pose for a group portrait. The conferences were sponsored by the Solvay Institute, established by Ernest Solvay, who became wealthy by developing a process for manufacturing sodium carbonate. Only the greatest chemists and physicists of the era were invited to the conferences, and the scientists in the picture represent an almost unmatchable gathering of genius. Many of these names will be mentioned in this and later chapters. *(Institute International De Physique Solvay)*

A. PICCARD    E. HENRIOT   P. EHRENFEST   Ed. HERZEN   Th. DE DONDER   E. SCHRÖDINGER   E. VERSCHAFFELT   W. PAULI   W. HEISENBERG   R.H. FOWLER   L. BRILLOUIN

P. DEBYE    M. KNUDSEN    W.L. BRAGG    H.A. KRAMERS    P.A.M. DIRAC    A.H. COMPTON    L. de BROGLIE    M. BORN    N. BOHR

I. LANGMUIR    M. PLANCK    Mme CURIE    H.A. LORENTZ    A. EINSTEIN    P. LANGEVIN    Ch.E. GUYE    C.T.R. WILSON   O.W. RICHARDSON

Absents : Sir W.H. BRAGG, H. DESLANDRES et E. VAN AUBEL.

## The Discovery of the Electron

Nineteenth-century experiments on electricity showed that the atom is not the indivisible particle pictured by Dalton. It was found that the atom contains subunits—particles smaller than atoms.

The British physicist Michael Faraday (1791–1867) played a major role in this work. In his experiments, Faraday studied the relative weights assigned by Dalton and later scientists to atoms of different elements. He found a relationship between these relative weights and the quantity of electricity needed to free a given amount of an element from a compound. We shall discuss electrochemical reactions in detail in Chapter 16.

On the basis of Faraday's experiments, G. J. Stoney (1826–1911) suggested in 1874 that electricity exists in units associated with atoms, and that atoms are neutral because they contain not only such units of electricity but also units of opposite charge. Stoney proposed in 1891 that the unit of electricity be called an **electron.** The name was adopted, but not without controversy. Many scientists of the time doubted not only the existence of units within the atom but also the existence of the atom itself.

The existence of atoms and electrons was established beyond doubt by a long series of experiments in many laboratories. Much of this work concerned the discharge of electricity through gas-filled tubes.

When an electric potential is applied across two electrodes at either end of a tube filled with gas at low pressure, the gas emits light. At very low pressure the gas does not emit light, but a glow can be seen from the part of the tube that is directly opposite to the negative electrode, called the cathode. The first phenomenon is used in neon lights. The second is used in television tubes.

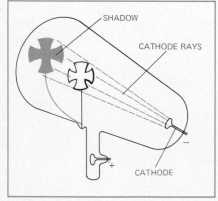

**Figure 5.2**
A Crookes tube. The rays emitted by the cathode do not strike the glass behind the metal cross. This result shows that cathode rays travel in straight lines.

## The Crookes Tube

The forerunner of the television tube was invented by the English physicist William Crookes (1832–1919). In 1879, Crookes showed that the glow of the glass opposite the cathode is caused by rays emitted by the cathode. In the experiment shown in Figure 5.2, Crookes demonstrated that cathode rays travel in straight lines. He placed a piece of metal in the shape of a Maltese cross between the cathode and the wall of the tube. When the electric current was turned on, the glass around the cross glowed from the rays emitted by the cathode, while a shadow was cast on the glass behind the metal cross.

One aspect of the Crookes tube experiment could be interpreted in two different ways. The cathode rays could be either waves or particles. A later series of experiments showed that cathode rays could be deflected by magnetic and electric fields, as particles would be. In one experiment, a fluorescent screen was placed in the tube, so that the path of the cathode rays could be seen (Figure 5.3). When a magnet was placed near the tube, the beam was bent in a way consistent with the behavior of negatively charged particles.

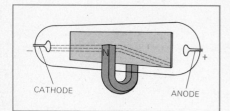

**Figure 5.3**
A cathode ray experiment. The rays are deflected by the magnet, in a way that can be shown to indicate that cathode rays are negatively charged.

## THE EVOLUTION OF TELEVISION

The modern television tube is a remarkably sophisticated version of the devices that William Crookes and J. J. Thomson used in the discovery of the electron. The essential elements of the cathode ray tube used in a television set consist of a source of electrons, a focusing coil that causes the electron beam to converge at a spot on a screen, other coils that deflect the electron beam horizontally and vertically, and a fluorescent screen that emits light when it is struck by the electrons.

The image on the screen is formed one line at a time. One deflecting coil sends the electron beam from right to left, while the other shifts it down one line at a time. Variations in the intensity of the electron beam produce darker or lighter parts of the image. In the United States, South America, and Japan, television images are produced at a rate of 30 complete frames per second, with 525 lines for each frame. In Europe and most of the rest of the world, the standards are 25 frames per second and 625 lines per frame.

In color television transmission, the light entering the camera is separated into red, green, and blue components. Electrical information about the brightness of each of the three colors is superimposed on the brightness signal. The tube in the color television set has three electron beams, one for each color, projected through a mask that has some 200 000 precisely positioned holes. Each electron beam is directed at a different set of phosphors, one for each color. The eye blends the three images to perceive a full-color picture.

The flat, portable television sets, small enough to fit in a shirt pocket, that made their appearance in the early 1980s were made possible by a television tube of radical new design. In conventional television tubes, the electron source is directly behind the fluorescent screen. In pocket-sized television sets, the tube has its electron source underneath the screen and parallel to it. The beam of electrons is deflected at a right angle to its source. The result is a two-inch television screen that is barely more than an inch thick. While the technology is highly advanced, the basic principles of the cathode ray tube have not changed.

The experiment that did the most to confirm the existence of such negatively charged particles — electrons — was performed in 1897 by J. J. Thomson (1856 – 1940). Thomson is given the credit for discovering the electron, because he was the first to measure one of its properties, the ratio of the charge of the electron to its mass, represented as $e/m$.

### The Electron Charge to Mass Ratio

Thomson built an apparatus that could determine the angle of deflection and the velocity of a beam of cathode rays as it passed through an electric or magnetic field of known strength. Figure 5.4 shows the apparatus.

Thomson found that cathode rays move at about one-fifth the speed of light. Given this speed and the measured angle of deflection, he was able to calculate a value of $e/m$. He then compared the value of $e/m$ obtained for the electron with the values of $e/m$ for several ions. The largest value of $e/m$ that had been measured until then was for hydrogen. It was known to be $e/m = 96\ 485\ C/1.0079\ g\ H$. The value of $e/m$ for the electron was found to be almost 2000 times larger than for hydrogen — to be exact,

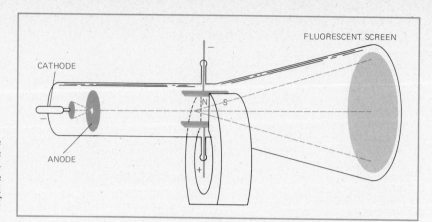

**Figure 5.4**

Thomson's experiment. A beam of cathode rays is deflected simultaneously by an electric field and a magnetic field and strikes the fluorescent screen. By altering the strengths of the magnetic and electric fields, Thomson could calculate the ratio of the charge to the mass of the electrons comprising the cathode ray beam.

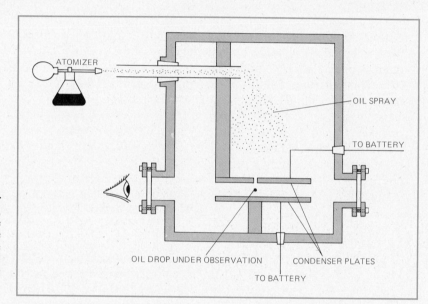

**Figure 5.5**

Millikan's oil drop experiment. By increase or decrease of the charge on the condenser plates, the rate of fall of charged particles of oil can be changed, so that it is possible to determine the charge on the drops. Millikan found that the charge was always $1.6 \times 10^{-19}$ C or an integral multiple of that value, which he assumed to be the charge of an electron.

1837 times larger. The magnitude of the charge $e$ is the same for the hydrogen nucleus, which is one proton, and for the electron. Therefore, the mass of the electron is 1837 times smaller than that of hydrogen.

Thomson could determine only the relative mass of the electron. To measure the actual mass, it was necessary to get an exact measurement for either $e$ or $m$; if one is known, the value of the other can be calculated. There were a number of attempts to make such measurements. Robert A. Millikan (1868–1953) of the University of Chicago succeeded with the apparatus shown in Figure 5.5.

## The Millikan Oil Drop Experiment

Millikan's experiment started with a spray of small drops of oil into a chamber. Some of the drops acquired an electrical charge by collision with ions in the air. The normal rate at which the drops fell was measured. Then the drops were allowed to fall between two oppositely charged electrical plates. Uncharged drops will all fall at the same rate. If a drop acquires a charge, it will be attracted to the oppositely charged plate, and its velocity will change. By turning the charge on the plates off and on, Millikan could make the drops rise and fall while their velocities were measured. From the relative velocities, and from the known charge on the plates, Millikan could calculate the total charge on the drops.

He found that the smallest value of the charge was $1.6 \times 10^{-19}$ C, and that the charge on any drop was either this value or a simple multiple of it. Some drops had a charge of $3.2 \times 10^{-19}$ C, some had a charge of $4.8 \times 10^{-19}$ C, and so on. But there was no drop whose charge was a fractional value of $1.6 \times 10^{-19}$ C. Millikan proposed that the value of $e$ is this smallest value, $1.6 \times 10^{-19}$ C. Considering that the work was done in 1909 with relatively unsophisticated equipment, Millikan's result was remarkably accurate. The latest figure for the charge on the electron, made with the most modern equipment, is $e = 1.602\ 1892 \times 10^{-19}$ C.

With $e$ known, the mass of the electron could be calculated. The mass is $m = 9.109\ 534 \times 10^{-31}$ kg, which is $1/1837$ the mass of a hydrogen atom. Since the hydrogen atom is the smallest known atom, the electron has to be a subunit, or a subatomic particle.

## 5.3   X RAYS AND RADIOACTIVITY

While cathode rays were being studied, scientists were discovering and investigating a number of other rays that proved to be powerful tools for examining the details of atomic structure.

In 1895, Wilhelm Roentgen (1845–1923) discovered that a fluorescent screen placed near a cathode ray tube glows when the tube is in operation. The glow is caused by rays emitted by the walls of the cathode ray tube. Since the nature of the rays was unknown, Roentgen called them *X rays,* a name that has stuck.

Roentgen found that X rays had great penetrating power. They could cause a photographic plate to darken even when relatively thick objects were placed between the source of the X rays and the plate — a discovery that was quickly put to medical use. X rays were found to travel in straight lines. Unlike cathode rays, they are not deflected by electric or magnetic fields. Eventually, X rays were shown to be radiation of the same sort as ordinary light, but of much higher energy. In addition to their medical value, X rays are also extremely useful in the determination of the structure of molecules (Chapter 10).

Soon after Roentgen discovered X rays, Henri Becquerel (1852–1908) observed that a sample of a uranium compound, left in a drawer with a photographic plate, produced rays that darkened the plate. These rays were studied by a number of investigators, including Marie Sklodowska Curie (1867–1934) and her husband Pierre Curie (1859–1906). It was

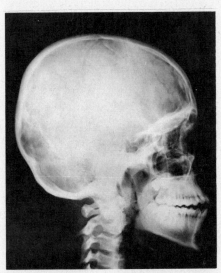

X rays were being used in medicine within months of their discovery by Wilhelm Roentgen in 1895. Modern medical practice would be impossible without X-ray images such as this one. *(Omikron, Photo Researchers, Inc.)*

found that other compounds and pure elements give off similar rays, a property that was named *radioactivity*. Eventually, it was shown that radioactive elements can emit three different kinds of rays, which were called alpha, beta, and gamma rays.

These three rays are distinguished by their behavior in a magnetic field, as shown in Figure 5.6. Gamma rays, which are the same sort of radiation as X rays but with even higher energies, do not change direction in the field. Beta rays, which are high-speed electrons, are deflected in a way consistent with their negative charge. Alpha rays are deflected in a way indicating that they are positively charged. Alpha rays are completely ionized helium atoms: that is, helium atoms with no electrons, which are called alpha particles.

## 5.4 THE NUCLEAR ATOM

By the beginning of the twentieth century, scientists accepted the notion that atoms exist. One subatomic particle, the electron, was known. Therefore, it was evident that the atom has a structure. The effort to determine that structure began.

Atoms were known to be electrically neutral. They also were known to contain electrons, which are negatively charged. Atoms thus had to contain subatomic particles with positive charges that cancel out the electrons' negative charges. But how were these negative and positive charges distributed within the atom?

Several models were proposed. One had the atom resembling the planet Saturn, with a large, positively charged center sphere surrounded by a "halo" of electrons. Another model, proposed by Thomson, compared the atom to a raisin pudding, with the electrons scattered through a positively charged mass like raisins embedded in a pudding.

**Figure 5.6**
The behavior of rays emitted by radioactive substances in a magnetic field. The alpha rays, which are ionized helium atoms, are positively charged. The beta rays, high-speed electrons, are negatively charged. Gamma rays, high-energy photons, have no charge and are not deflected by the field.

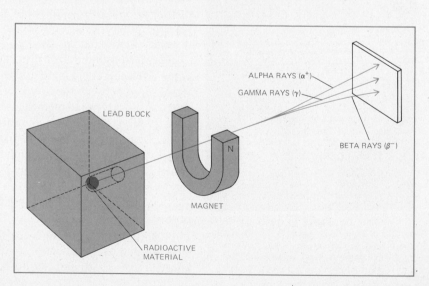

ALPHA RAYS ($\alpha^+$)
GAMMA RAYS ($\gamma$)
LEAD BLOCK
N
BETA RAYS ($\beta^-$)
MAGNET
RADIOACTIVE MATERIAL

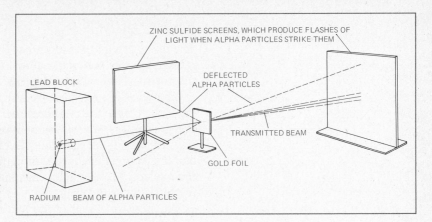

**Figure 5.7**
Rutherford's experiment. Most alpha particles pass through the thin sheet of gold foil with little or no change of path. A few alpha particles are deflected at large angles. On the basis of this experiment, Rutherford proposed that the atoms of gold in the foil have almost all of their masses concentrated in extremely small, dense nuclei.

## The Rutherford Model

Thomson's "raisin pudding" atom seemed to fit the known facts, but experiments were needed to test the theory. The great physicist Ernest Rutherford (1871–1937) designed such an experiment, probably the most important ever done in the study of atomic structure.

The experiment was simple in principle. The results were a staggering surprise. A beam of alpha particles was directed at a thin sheet of gold foil, with a fluorescent screen behind the foil (Figure 5.7). Most of the alpha particles went straight through the foil, with little or no deflection. A few of the alpha particles were deflected sharply, some of them ricocheting off at angles of 90° or more. As Rutherford described the results: ". . . as if you fired a fifteen-inch shell at a piece of tissue paper and it came back and hit you."

Only one conclusion was possible. The atoms in the foil have almost all their mass concentrated in a very small positively charged region, which Rutherford called the **nucleus.** The tiny nucleus is surrounded by negatively charged electrons, whose distance from the nucleus is great on the subatomic scale. To explain why the negatively charged electrons are not pulled into the positively charged nucleus, Rutherford proposed that the electrons orbit the nucleus much as planets orbit the sun.

Rutherford's picture explained why most of the alpha particles passed through the foil undeflected, while a few were deflected at large angles. Most of the atom is empty space, through which the alpha particles passed unaffected. A few particles collided with the atomic nuclei and bounced off at large angles (Figure 5.8). By counting the percentage of alpha particles that were deflected, Rutherford was able to calculate the relative diameter of the atomic nucleus. He found it to be only $1/10\,000$ of the atom's total diameter.

The Rutherford model provided the framework for the modern description of the atom. However, it later underwent some major modifications.

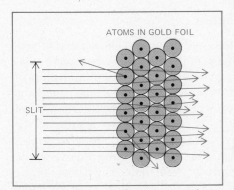

**Figure 5.8**
The Rutherford experiment on the atomic level. The majority of alpha particles pass through the layers of gold atoms without hitting the nucleus of an atom and are deflected only slightly or not at all. An occasional alpha particle comes close to an atomic nucleus and is deflected sharply.

## The Atomic Nucleus

The discovery that the atom has a nucleus that carries a positive charge raised a number of new questions. To begin with, what is the magnitude of the positive charge of the nucleus?

That question was answered by H. G. J. Moseley (1887–1915), working at the University of Manchester, England. Moseley studied the X rays that are emitted when an element is bombarded with high-energy cathode rays. He found that each element emits a characteristic pattern of X rays, and that this pattern is related to the magnitude of the positive charge of the nucleus. Moseley then calculated the value of the positive charge of the nuclei of many elements.

The positive charge of the nucleus is attributed to positively charged particles called **protons.** The number of protons in the nucleus is equal to the number of electrons in the neutral atom. The number of protons in the nucleus of a given element is called the **atomic number** of the element.

This information is not enough to determine the mass of the atomic nucleus. From detailed studies of the effect of alpha particles on various elements, Rutherford concluded that the nucleus of a hydrogen atom is nothing more than a single proton. He also concluded that the nucleus of any atom heavier than hydrogen consists of a number of protons held together in some unknown manner. However, the total mass of the nucleus relative to hydrogen is substantially greater than the atomic number.

In 1932, when James Chadwick (1891–1974) discovered a new particle in the nucleus, the difference between the atomic number and the atomic mass of the nucleus was accounted for. The new particle, called the **neutron,** has no electric charge and has a mass that is slightly larger than that of a proton.

## Atomic Number and Mass Number

The nucleus consists of both protons and neturons. **The number of protons, and therefore the positive charge of the nucleus, is the atomic number of the element.** If an atom gains or loses protons, it becomes an atom of a different element. The number of protons in the nucleus equals the number of electrons in the electrically neutral atom. **The total number of protons and neutrons in the nucleus is called the mass number of the nucleus.** The relative atomic mass of the atom is very close to the mass number. (The electron, whose mass is only $1/1837$ that of a proton, can be ignored in this reckoning.) Thus, since an atom of fluorine has an atomic number of 9 and a relative atomic mass that is very close to 19, we know that the nucleus of a fluorine atom has nine protons and $19 - 9 = 10$ neutrons.

The details of nuclear structure still are not understood completely, but for most purposes, we can picture the nucleus as consisting of particles called **nucleons,** held together by extremely strong forces (Section 20.1).

Some of the nucleons are protons, which have positive electric charge. The others are neutrons, with no charge. Figure 5.9 shows this picture for several different nuclei. The number of nucleons in each nucleus approximates the relative atomic mass of the atom.

## Isotopes

In Section 3.2, we discussed the relationship between the atomic weight of an element and the existence of two or more isotopes of the element. You will recall that the atomic weight of an element is a weighted average that reflects the natural abundances and atomic masses of the isotopes of the element.

Every naturally occurring element has at least two isotopes. Some elements have only one stable isotope; the other isotopes decay by radioactive processes. Among these elements are Be, F, Al, P, and As. Some elements have no known stable isotopes. These include technetium (atomic number 43), promethium (atomic number 61), and all elements of atomic number greater than 83.

All isotopes of a given element have very similar chemical behavior. The number and energy of the electrons and the nuclear charge of an element are mainly responsible for its chemical behavior. Since all isotopes of a given element have the same number of protons and the same number of electrons in the neutral atom, they have virtually the same chemical behavior.

There is a system of notation that specifies the atomic number and the mass number of any atom. We start with the symbol for the element. Two numbers precede the symbol. A superscript, the uppermost number, indicates the mass number of the isotope. A subscript indicates the atomic number of the isotope. This notation can be represented as:

$$^{A}_{Z}X$$

where X is the symbol for the element, $A$ is the symbol for mass number, and $Z$ is the symbol for atomic number (or nuclear charge, or the number of protons in the nucleus). Thus, the two isotopes of boron are designated $^{10}_{5}B$ and $^{11}_{5}B$. To calculate the number of neutrons in each of these isotopes, subtract the atomic number from the mass number, since:

$$A = Z + N \qquad (5.2)$$

where $N$ is the number of neutrons.

Elements in nature usually are mixtures of isotopes. The combination of isotopes in the mixture may vary slightly depending on the source of the sample. Thus there is some difficulty in establishing a very precise scale of relative atomic weights. Furthermore, the slight differences in mass between a proton and a neutron, and the small mass of the electrons, are

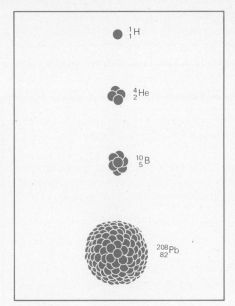

**Figure 5.9**
The hydrogen nucleus consists of a single proton. The helium nucleus contains two protons and two neutrons. The boron nucleus has five protons and five neutrons. The lead nucleus contains 82 protons and 126 neutrons.

noticeable when atomic masses of pure isotopes are measured to more than three decimal places. Fractional atomic weights are the rule. An exception is the isotope $^{12}_{6}C$ which, as was mentioned earlier, has been assigned an atomic mass of exactly 12. All other atomic weights are calculated relative to this mass.

**Example 5.1**     The two naturally occurring isotopes of iridium have mass number of 191 and 193. Find the number of neutrons in the nucleus of each isotope.

**Solution**     The number of neutrons in a nucleus is the mass number less the number of protons in the nucleus. The number of protons is the same as the atomic number. The periodic table shows that the atomic number of Ir is 77. Therefore

$$N = A - Z = 191 - 77 = 114$$

for one isotope and

$$N = 193 - 77 = 116$$

for the other.

## 5.5   THE DEVELOPMENT OF THE QUANTUM THEORY

The atomic model in which most of the atom's mass was concentrated in a small, positively charged nucleus surrounded by orbiting electrons fit the results of Rutherford's experiments on alpha particle scattering; but it was inconsistent with some other observed properties of substances, and with the well-established laws of electricity and magnetism.

At the beginning of the twentieth century, scientists were faced with many experimental results that seemed to violate inviolable laws or that did not agree with other equally valid experimental results. They eventually solved these dilemmas, but only by arriving at a new understanding of the nature of matter and energy.

### Electromagnetic Radiation and Waves

The modern view is that the universe is composed of matter and radiant energy, and that there is an equivalence between the two. Radiant energy travels with the speed of light, while matter moves more slowly.

The work that led to this view began in the seventeenth century with the first serious investigations of light, one form of radiant energy. Out of these investigations came two contradictory theories of the nature of light. One school of thought, led by Sir Isaac Newton (1642–1727), held that light consists of small particles, or corpuscles. The second theory, held by

Christian Huygens (1629 – 1695), was that light consists of waves. By the middle of the nineteenth century, the wave theory of light was generally accepted. Later, when it was found that other forms of radiant energy, such as X rays, gamma rays, and radio waves, are essentially similar to visible light, they were also assumed to consist of waves.

## The Nature of Waves

A wave can be described as a disturbance that travels through a medium in a given direction. The medium itself is not carried along. For example, consider a swimmer floating in the ocean. The swimmer bobs up and down with each passing wave, but is not carried toward the shore. If molecules of water in the ocean could be followed, it would be seen that they do not travel toward shore, although the peaks of the waves do. (Objects are washed up on a beach by currents or turbulence, not by waves.)

Similarly, sound waves do not produce any substantial net movement of the gas molecules that make up air. This fact can be shown if we float a balloon in front of a rock band at a concert. The balloon will not move toward the back of the hall, no matter how loud the sound.

Because a wave is a moving disturbance, its location cannot be specified exactly. The disturbance is a series of peaks and valleys, crests and troughs, whose position changes constantly as the wave travels. Because a wave is spread out over a region of space, we can describe it only by specifying a number of its features.

Figure 5.10 shows an idealized wave. One feature of a wave is the distance between two adjacent peaks, the **wavelength.** It is commonly represented by $\lambda$, the Greek letter lambda. Another characteristic of a wave is the distance from a horizontal midline to either the peak or the trough. This distance is called the **wave amplitude.**

Wavelength and wave amplitude are not enough to describe a wave. Information is also needed on the velocity of the wave, which can be described as the rate of motion of the peak (or any other point) in the direction of propagation. The value of the velocity, $c$, for light waves and all other electromagnetic radiation is approximately $3.00 \times 10^8$ m/s.

The **frequency** of the wave, which is represented by $\nu$, the Greek letter nu, is directly related to wavelength and velocity. Picture an observer looking at a fixed point as a wave goes by (Figure 5.11). If the observer counts the number of peaks that pass the fixed point in a given time period, the frequency of the wave can be specified as so many units per second. The units of frequency are $s^{-1}$ or hertz (Hz). If there is a long distance between peaks — that is, if the wavelength is long — fewer peaks will pass in a given time period. In other words, *the frequency of a wave is inversely proportional to the wavelength.* The number of peaks that pass by in a given time period also depends on the velocity of the wave. Therefore, frequency is a function of two characteristics: the velocity $c$ of

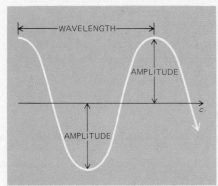

**Figure 5.10**
An idealized wave. The distance from peak to peak is the wavelength. The distance from the horizontal midline to the peak or the trough is the wave amplitude. The wave is traveling from left to right across the page.

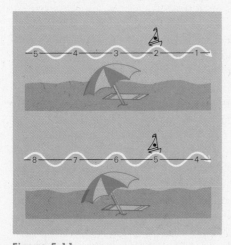

**Figure 5.11**
Wave frequency. An observer sitting under the beach umbrella and looking at the boat riding at anchor counts the number of times the boat bobs up and down in a given time period to calculate the wave frequency.

the wave and the wavelength $\lambda$. The relationship is

$$v = c/\lambda \quad \text{or} \quad c = \lambda v \tag{5.3}$$

Since the velocity $c$ of light is known, the wavelength of light can be calculated if the frequency is known, and vice versa.

---

**Example 5.2**    A man on the deck of a ship anchored in the ocean observes that the crests of passing waves are 11 m apart and that a crest hits the bow of the ship every 3.0 s. Calculate the velocity of the waves.

**Solution**    Since the values of $\lambda$ (wavelength) and $v$ (frequency) of the waves are known, we can calculate the velocity $v$:

$$v = \lambda v = (11 \text{ m})\left(\frac{1}{3.0 \text{ s}}\right) = 3.7 \text{ m/s}$$

---

**Example 5.3**    Calculate the frequency of light whose wavelength is $5.0 \times 10^{-7}$ m.

**Solution**    The value of $c$ is known to be $3.00 \times 10^8$ m/s. The value of $\lambda$ is given. Equation 5.3 gives the relationship between speed, wavelength, and frequency:

$$c = \lambda v$$
$$3.00 \times 10^8 \text{ m/s} = (5.0 \times 10^{-7} \text{ m})v$$
$$v = 6.0 \times 10^{14} \text{ s}^{-1}$$

The three parameters, $c$, $\lambda$, and $v$, are related. If we are given any two for a wave, the value of the third can be calculated. For light waves, or any other waves traveling at fixed velocity, wavelength and frequency are inversely proportional. As wavelength increases, frequency decreases, and vice versa.

---

## Electromagnetic Waves

If light is a wave, an important question about the physical universe is raised. Since a wave is a disturbance in a medium, what is the medium through which light travels? A great deal of effort was devoted to that question in the late nineteenth century and early years of the twentieth century. The answer proved surprising.

The work of James Clerk Maxwell (1831–1897) and Heinrich Hertz (1857–1894) showed that visible light is one of many different waves, all of which are electromagnetic disturbances. One crucial experiment that supported this view showed that even an oscillating electric charge produces electromagnetic waves. We now know that all radiant energy is electromagnetic waves, which travel at a velocity of $c = 3.00 \times 10^8$ m/s.

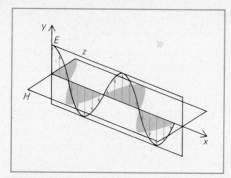

**Figure 5.12**
Electromagnetic radiation has two components, an electric field and a magnetic field, oscillating at right angles to each other and to the direction in which the radiation is traveling.

An electromagnetic wave has two components, an electric field and a magnetic field. The two fields oscillate at right angles to each other and at a right angle to the direction of travel (Figure 5.12). For several decades, it was believed that these waves were a disturbance in something called the "ether," which filled all of space but was undetectable. One of Albert Einstein's major achievements was to show that the ether does not exist. Ocean waves cannot exist without water, sound waves cannot exist without air, but no medium is necessary for electromagnetic waves.

Electromagnetic radiation is classified according to wavelength or frequency in an orderly arrangement called the *electromagnetic spectrum.* For convenience, the different wavelength (or frequency) regions of the electromagnetic spectrum often are given names, such as visible light, X rays, gamma rays, ultraviolet rays, infrared rays. Figure 5.13 shows a part of the electromagnetic spectrum. Among other things, Figure 5.13 shows that visible light, the only part of the spectrum that we humans can detect with our eyes, is a very small sliver of the entire range of electromagnetic radiation.

We must understand radiant energy to understand the modern model of the atom, since much of the information on which this model is based comes from studies of the interaction between matter and radiant energy.

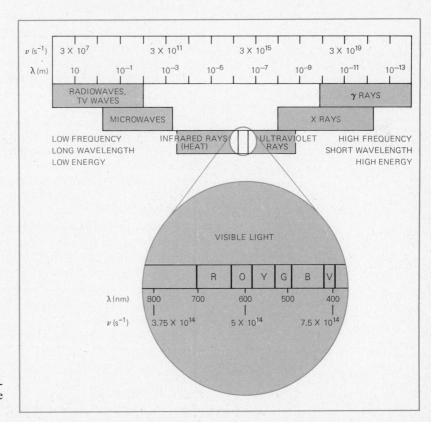

**Figure 5.13**
The electromagnetic spectrum. Visible light occupies a small section toward the center of the spectrum.

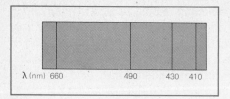

$\lambda$ (nm) 660          490      430   410

**Figure 5.14**

The visible portion of the atomic spectrum of hydrogen. Each element has a unique atomic spectrum and can be identified by the pattern of emission lines.

## Atomic Spectra

Atoms of a pure element in the gas phase can emit light when heated. When studies of this phenomenon began in the mid-nineteenth century, it was expected that the results would be similar to those obtained in an experiment done two centuries earlier by Isaac Newton.

Newton passed ordinary sunlight through a prism. He found that a prism breaks the sunlight into a continuous spectrum that contains the colors of visible light. But the same sort of continuous spectrum is not obtained when light is emitted by gaseous atoms of a pure element. Instead, the emitted light consists of a relatively small number of narrow bands (or lines) of color, with large dark spaces between them (see the color plate facing page 200).

These lines are called *spectral lines.* Each element has a characteristic pattern of spectral lines. The pattern is called an **atomic spectrum** and serves as a "fingerprint" to identify the element (Figure 5.14). Spectral lines are emitted not only in the visible region of the electromagnetic spectrum, but also in the ultraviolet region, at wavelengths shorter than those of visible light, and in the infrared region, at wavelengths longer than those of visible light.

## The Balmer Formula

Nineteenth-century physics could not explain why different elements have different spectral patterns. The first step toward an understanding of atomic spectra was made by an obscure Swiss schoolteacher, J. J. Balmer (1825–1898). In 1885, Balmer developed a formula for the wavelengths of the visible lines in the atomic spectrum of hydrogen:

$$\frac{1}{\lambda} = R_\text{H}\left(\frac{1}{2^2} - \frac{1}{n^2}\right) \tag{5.4}$$

where $\lambda$ is the wavelength of a given spectral line, $n$ is any small integer greater than 2, and $R_\text{H}$ is a proportionality constant, now called the *Rydberg constant,* whose value is $1.097\ 373\ 177 \times 10^7$ m$^{-1}$, one of the most accurately measured quantities known.

By substituting different integers for $n$ in this formula, one could calculate wavelengths that correspond to those in the observed spectrum of hydrogen. The formula even had a predictive value. It gave not only the wavelengths of known spectral lines in the visible region but also the wavelengths of several ultraviolet lines of hydrogen, which had not been detected at the time.

Later, Balmer's original formula was generalized to account for new sets of lines of hydrogen that were discovered in the ultraviolet and infrared regions. The more generalized formula reads:

$$\frac{1}{\lambda} = R_\text{H}\left(\frac{1}{n_f^2} - \frac{1}{n_i^2}\right) \tag{5.5}$$

in which the original term $1/2^2$ has been replaced by $1/n_f^2$ and $1/n^2$ has been replaced by the term $1/n_i^2$. In this formula, $n_f$ and $n_i$ may be any integers, as long as $n_i$ is greater than $n_f$. If integer values are set for $n_f$, a series of spectral lines can be found for the allowed values of $n_i$. To find the allowed values, we let $n_i = n_f + 1$; $n_i = n_f + 2$; $n_i = n_f + 3$; and so on.

Balmer's formula had one major weakness. It had no theoretical basis at all. The formula worked, but no one had any idea why. Classical physics could not explain why atomic spectra consisted of narrow spectral lines, rather than a continuous spectrum.

---

**Example 5.4**     Calculate the wavelengths of the two spectral lines with the longest wavelengths (called the first two lines) in the visible region of the atomic spectrum of hydrogen.

**Solution**     Equation 5.4, the Balmer formula, can be used to calculate the wavelengths of visible lines in the atomic spectrum of hydrogen. To calculate the wavelengths of the first two lines, we use the two smallest allowed integers, $n = 3$ and $n = 4$, in the Balmer formula:

$$\frac{1}{\lambda} = R_H \left( \frac{1}{2^2} - \frac{1}{n^2} \right) = 1.1 \times 10^7 \text{ m}^{-1} \left( \frac{1}{4} - \frac{1}{9} \right)$$

$$= 1.1 \times 10^7 \text{ m}^{-1}(0.14)$$

$$\lambda = \frac{1}{1.1 \times 10^7 \text{ m}^{-1}(0.14)}$$

$$\lambda = 6.5 \times 10^{-7} \text{ m}$$

$$\frac{1}{\lambda} = R_H \left( \frac{1}{2^2} - \frac{1}{n^2} \right) = 1.1 \times 10^7 \text{ m}^{-1} \left( \frac{1}{4} - \frac{1}{16} \right)$$

$$\lambda = 4.8 \times 10^{-7} \text{ m}$$

---

**Example 5.5**     In the original Balmer formula, the wavelengths of lines in the visible region of the atomic spectrum of hydrogen were calculated with $n_f = 2$. The ultraviolet region of the atomic spectrum of hydrogen has lines of shorter wavelengths than the lines in the visible region. Find the value of $n_f$ that defines a series of lines in the ultraviolet region, and then assign values to $n_i$ to predict the positions of the two lines of longest wavelength in the ultraviolet region.

**Solution**     Since the left side of Equation 5.5 is a fraction that has $\lambda$ in the denominator, the value of $\lambda$, the wavelength, decreases as the value of the term on the right side of the equation increases. Since the ultraviolet region is of shorter wavelength than the visible region, the solution requires values for $n_f$ and $n_i$ that increase the value of the term

$$R_H \left( \frac{1}{n_f^2} - \frac{1}{n_i^2} \right)$$

Since $n_f$ is in the denominator of a fraction, a smaller value for $n_f$ means a larger value for the entire term. The smallest value that we can given to $n_f$ is 1.

To find the longest wavelengths corresponding to $n_f = 1$, we use the smallest allowed values of $n_i$ which are $n_i = 2$ and $n_i = 3$. If we use these values to calculate values for $\lambda$, the results are

$$\frac{1}{\lambda} = R_H \left( \frac{1}{1^2} - \frac{1}{2^2} \right)$$

$$\lambda = 1.2 \times 10^{-7} \text{ m}$$

$$\frac{1}{\lambda} = R_H \left( \frac{1}{1^2} - \frac{1}{3^2} \right)$$

$$\lambda = 1.0 \times 10^{-7} \text{ m}$$

## Blackbody Radiation

If a bar of iron is heated, it first glows dull red, then yellowish, and eventually white-hot. The radiation emitted by the bar of iron, or any other solid, differs from the radiation emitted by gaseous atoms. The radiation emitted by a solid does not have individual spectral lines separated by dark spaces. Instead it is emitted in a continuous spectrum through the ultraviolet, visible, and infrared regions.

There is a practical problem to overcome when studying radiation from solids. Most solids reflect a considerable proportion of the light that falls on them. When we study the radiation from a hot solid, we may thus be studying a mixture of emitted light and reflected light. To get around this difficulty, we need something that reflects none of the light falling on it. Because black objects reflect the least light, this imaginary object is called a *blackbody*. An ideal blackbody does not exist; but we can create a close approximation of the hypothetical blackbody by using a heated cavity inside a block of metal, with a tiny hole that allows radiation to escape.

Early studies of blackbody radiation concentrated on the way in which emitted energy is distributed over the range of wavelengths. Curves such as those in Figure 5.15 were drawn to show how the wavelength distribution changed with temperature. The curves show that as temperature rises, the region of maximum emission of energy moves from the infrared region to the visible. These curves are consistent with the everyday observation that the light emitted by, say, a bar of metal changes from red to yellow — from longer to shorter wavelenghts — as the bar is heated.

Classical theories could not explain the wavelength distribution of blackbody radiation. One theory explained the short wavelength distribution but failed for longer wavelengths. Another theory successfully explained the longer wavelengths but failed spectacularly with short wavelengths. The second theory, which was consistent with all the known laws of physics, predicted that a blackbody would emit an infinite amount of energy in the ultraviolet region, which is a physical impossibility. This prediction was so disturbing to physicists that it was called the "ultraviolet catastrophe."

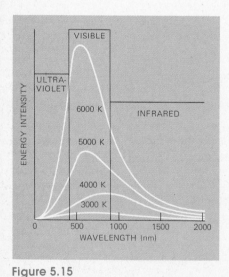

**Figure 5.15**
Curves showing the change in the emission of radiation as a blackbody is heated. As the temperature goes up, the total amount of radiation emitted by the blackbody increases. Both the amount and fraction of the total radiation that is in the visible part of the spectrum increase as temperature rises.

## SEPARATING ISOTOPES WITH LASER LIGHT

Isotopes of the same element have identical chemical properties; they cannot be separated chemically. Isotope separation has been a complicated and expensive undertaking, most notably in the World War II program to make an atomic bomb by separating fissionable uranium-235 from nonfissionable uranium-238 (Section 4.9).

But, the different isotopes of an element have very slightly different patterns of spectral lines, which means that they emit and absorb slightly different wavelengths of light. The possibility of exploiting the slight differences in absorption spectra to separate the two isotopes of uranium was examined and rejected, because the available light sources were unsuitable. In recent years, however, the development of lasers has provided a source of light whose wavelength may be controlled precisely enough so that one isotope will absorb the energy while another isotope will not. In addition, laser light is intense enough to be suitable for efficient isotope separation.

Laser light differs from ordinary light because it is "coherent"; that is, its waves all have the same wavelength, frequency, and orientation. Laser light is produced when a large number of molecules are induced to emit radiation of the same wavelength simultaneously. If the light is of the appropriate wavelength, it can be used to add energy to only one isotope of an element but not other isotopes. The more energetic isotopes could then be separated from the others.

Potentially, laser isotope separation of uranium is three to five times less expensive than gaseous diffusion separation. Laser separation not only uses less energy but also can be done with less expensive equipment. Work toward building a practical laser separation system is proceeding.

In 1982, after considering a number of alternative proposals, the Department of Energy announced that its first choice was a method that uses three different lasers, all emitting visible light. In the method, uranium is first vaporized. Three photons, one from each of the lasers, combine to ionize the fissionable uranium atoms but not the atoms of uranium-238. The ions can then be collected by sending the uranium vapor through a magnetic field. The uncharged uranium-238 atoms are not deflected by the field, but the uranium ions are.

An alternative method that is also being explored uses a vapor of uranium hexafluoride, $UF_6$, and two different lasers. An infrared laser first adds energy to the $UF_6$ molecules containing uranium-235. An ultraviolet laser causes the excited molecules to break up, so they can be separated from the other molecules chemically.

The Department of Energy plans to have a small-scale plant to demonstrate the feasibility of laser separation in operation by 1987. A full-scale plant could be in operation by the end of the century.

### The Quantum Theory

The dilemma was resolved by Max Planck (1858–1947), who introduced a principle that required a basic revision of classical theory. In classical theory, it was assumed that the radiant energy emitted by a blackbody could have any value within a continuous range. Planck proposed instead that radiant energy can be emitted only in certain fixed quantities. The word that Planck used to describe the smallest such quantity was **quantum** (plural *quanta*), from the Latin word for "how much." The energy emitted by a blackbody is always an integral multiple of the quantum and is never less than a quantum.

Planck said that the quantum of energy is given by:

$$E = hv \qquad (5.6)$$

where $E$ is the energy of the quantum, $v$ is the frequency of the emitted radiation, and $h$ is a constant with the value $6.6262 \times 10^{-34}$ J s.

By substitution from Equation 5.3, Equation 5.6 can be written as:

$$E = \frac{hc}{\lambda} \qquad (5.7)$$

Equations 5.6 and 5.7 tell us that **the magnitude of a quantum of energy is directly proportional to the frequency of the radiant energy,** and that **the magnitude of the quantum is inversely proportional to the wavelength of the radiant energy.** Planck's hypothesis explained the distribution of frequencies in blackbody radiation, which had been inexplicable by classical theory.

Planck also proposed that the total radiant energy of any given frequency $v$ had to be $E = nhv$, where $n$ could have only integer values: 1, 2, 3, 4, . . . . A blackbody could emit $2hv$, $3hv$, $4hv$, or any other integral multiple of $hv$, but it could not emit $3.27hv$, $5.9hv$, or any other fractional value of $hv$.

The concept of the quantum—the idea that energy is not infinitely divisible but rather is emitted in discrete units—is fundamental to modern science. On a different scale, quantum phenomena are common in everyday life. The production of eggs by chickens is a quantized phenomenon. A flock of chickens must lay any integral number of eggs in a day; a chicken cannot lay a fraction of an egg.

The idea that radiant energy is not infinitely divisible was intellectually disturbing to many scientists of the day. Planck himself worked unsuccessfully for years to find an alternative explanation. Today, the idea of a quantum is a cornerstone of science. In particular, the idea of the quantum is behind today's model of atomic structure.

## 5.6  THE PHOTOELECTRIC EFFECT

In 1905, Albert Einstein (1879–1955), a poorly paid clerk in the Swiss patent office, published three scientific papers that revolutionized physics. One paper outlined the special theory of relativity. The second explained Brownian motion, the random movement of small objects in liquid. The third was a major step forward in the understanding of the nature of light. The publication of these three papers was a spectacular achievement; historians speak of 1905 as Einstein's "wonder year." While the theory of relativity had the greatest repercussions, it was the paper on the nature of light that was cited when Einstein was awarded the Nobel Prize.

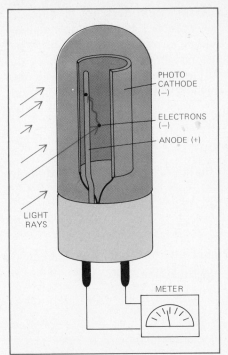

**Figure 5.16**
The photoelectric effect. If the frequency of radiation striking the cathode is low, no electrons are emitted. If the frequency of the radiation increases above a threshold, electrons are emitted. This cell can be used to measure the number of electrons that are emitted and their kinetic energy.

In this paper Einstein applied the quantum theory to a phenomenon called the *photoelectric effect,* which had been unexplained until then. The photoelectric effect is the emission of electrons from the surface of a metal that is struck by light or other electromagnetic radiation. An instrument such as the one shown in Figure 5.16 is used to study it.

Experiments show that the emission of electrons depends simply on the frequency—that is, the energy—of the radiation. If the electromagnetic radiation that strikes the surface of the metal is of low frequency, no electrons are emitted by the metal. If energy of progressively higher frequency is used one suddenly arrives at a "threshold"—a frequency at which electrons are first emitted by the metal. No electrons are emitted at frequencies lower than the threshold frequency, no matter how much radiation of that frequency strikes the metal. That is, below the threshold frequency, increasing the intensity of the radiation has no effect.

Once the threshold frequency is reached, the number of electrons given off by the metal does not change if the frequency is increased still higher. What does change is the energy of the electrons that are emitted. As the *frequency* of the incident radiation increases, only the *energy* of the emitted electrons increases. As the *intensity* of the incident radiation above the threshold frequency is increased, only the *number* of electrons emitted increases.

## Light Consists of Photons

These results were a puzzle in 1905 because they could not be reconciled with the wave theory of light. Einstein solved the puzzle by proposing that radiant energy exists as quanta called **photons.** In doing so, Einstein made the first practical application of Planck's quantum theory. Einstein not only used the same term, quantum, to describe a "bundle" of light, but he also used Planck's constant, $h$, in the formula for the energy of a quantum of light. The energy of a quantum of light, Einstein said, is $E = h\nu$; that is, the energy is the frequency times Planck's constant.

The idea that light is quantized was revolutionary. The quantum description of light conveys the idea that light consists of particles or corpuscles, an idea that had long been discarded in favor of the wave theory. Nevertheless, the wave theory could not explain the photoelectric effect, but the quantum theory could.

To say that light consists of small bundles—quanta—of energy is the same as saying that light consists of particles, the photons. One can visualize electrons being knocked loose from a metal by a beam of photons. We can picture an electron as being held by the nucleus of an atom with a certain energy, called the *binding energy,* $\epsilon$. Only a photon with enough energy to overcome the binding energy can knock the electron loose. In this visualization, if the photon that hits the electron does not have enough energy, it will bounce away and the electron will remain in the atom.

The fact that the energy of a photon is proportional to its frequency

($E = h\nu$) explains why low-frequency radiation does not produce the photoelectric effect. Photons of low-frequency radiation do not have enough energy to overcome the binding energy of the electron. At the threshold frequency, the photon has just enough energy to knock the electron loose. If the photon has more energy than is required to overcome the binding energy, the excess energy gives the departing electron more energy of motion.

## Binding Energy and Kinetic Energy

If the threshold frequency is known, the binding energy of the electron can be found. It is

$$h\nu_0 = \epsilon \qquad (5.8)$$

where $\nu_0$ is the threshold frequency and $\epsilon$ is the binding energy. The energy of motion imparted to the electron, called kinetic energy (Section 4.6), is

$$\text{K.E.} = \tfrac{1}{2}mv^2 \qquad (5.9)$$

where $m$ is the mass and $v$ is the velocity of the moving object.

If the radiation hitting the metal has a frequency greater than the threshold frequency, the energy $E$ of a photon goes in part to overcome the binding energy $\epsilon$ and in part to give extra kinetic energy to the emitted electron:

$$E = \epsilon + \tfrac{1}{2}mv^2 \qquad (5.10)$$

or, substituting from Equations 5.6 and 5.8:

$$h\nu = h\nu_0 + \tfrac{1}{2}mv^2 \qquad (5.11)$$

where $\nu$ is the frequency of the incident radiation and $\nu_0$ is the threshold frequency.

A large-scale model of the photoelectric effect is shown in Figure 5.17. A carnival game requires the player to knock a tenpin off a shelf by shooting at it. The tenpin is glued to the shelf. If corks are used for ammunition, they do not have enough energy to overcome the binding energy of the glue. The corks are like photons whose frequency is below the threshold frequency. If light metal shot is used for ammunition, a hit just knocks the tenpin off the shelf. The light shot has just enough energy to overcome the binding energy of the glue, as do photons at the threshold frequency. If heavy shot is used as ammunition, a hit not only knocks the tenpin off the shelf but sends it flying, just as a photon above the threshold frequency sends the electron off with kinetic energy.

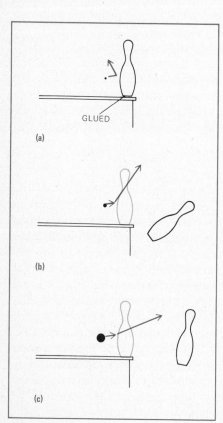

GLUED

(a)

(b)

(c)

**Figure 5.17**
The carnival game of knock-the-tenpin-off-the-shelf can be used as a model of the photoelectric effect. (a) Low-energy corks cannot overcome the binding energy of the glue holding the tenpin to the shelf. (b) Light shot has just enough energy to overcome the binding of the glue and knock the tenpin off the shelf. (c) Ammunition of higher energy not only knocks the tenpin loose but sends it flying off with kinetic energy.

**Example 5.6**   If the threshold frequency, $v_0$, of a metal is $6.7 \times 10^{14}$ s$^{-1}$, calculate the kinetic energy of a single electron that is emitted when radiation of frequency $v = 1.0 \times 10^{15}$ s$^{-1}$ strikes the metal.

**Solution**   The relationship between $v_0$, $v$, and kinetic energy is given in Equation 5.11. Solving the equation for kinetic energy gives:

$$\begin{aligned}
\text{K.E.} = \tfrac{1}{2}mv^2 &= h(v - v_0) \\
&= (6.6 \times 10^{-34} \text{ J s})(1.0 \times 10^{15} \text{ s}^{-1} - 6.7 \times 10^{14} \text{ s}^{-1}) \\
&= (6.6 \times 10^{-34} \text{ J s})(3.3 \times 10^{14} \text{ s}^{-1}) \\
&= 2.2 \times 10^{-19} \text{ J}
\end{aligned}$$

By using Planck's quantum theory to explain the photoelectric effect, Einstein posed as many questions as he answered. How can all the experiments that supported the wave theory of light be explained away? Can light be both a wave and a particle? Some answers came from experiments and theories that developed a new model of the atom.

## 5.7   THE BOHR MODEL OF THE HYDROGEN ATOM

A new model of the atom was needed because the Rutherford model was far from satisfactory. In fact, given the classical laws of physics, such an atom could not exist. The model described negatively charged electrons moving in a closed orbit around a positively charged nucleus. A body moving in a closed orbit must have acceleration to keep it going. According to classical theory, a negatively charged body such as the electron must radiate energy under such conditions of acceleration. As the electron lost energy, it would spiral down toward the nucleus. The atom would radiate a continuous spectrum of radiant energy and it would collapse rapidly. Observation of atoms gives a different picture. Atoms emit a noncontinuous spectrum of radiant energy, they do so only at high temperatures, and they stubbornly continue to exist. The great Danish physicist Niels Bohr (1885–1962) developed a successful new model of the hydrogen atom. Bohr made the crucial assumption that not all of the classical laws of electrodynamics apply to phenomena on the atomic scale.

### Bohr's Postulates

Bohr's model is based on several postulates that seemed to have no justification.

**1.** The energy of a hydrogen atom is not a continuous function that can have any value within a range. Rather, the hydrogen atom **can exist only in a limited number of energy states.** In other words, **the energy of**

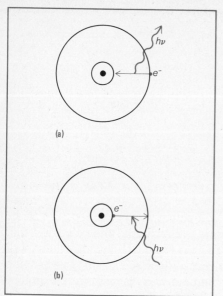

**Figure 5.18**
The simple picture of the hydrogen atom from Bohr's three postulates. (a) Radiation is emitted when the electron goes from a high orbit to a low orbit. (b) Radiation is absorbed when the electron goes from a low orbit to a high orbit. In this figure $\Delta E$ is the same for emission and absorption, so the frequency of the emitted radiation is the same as the frequency of the absorbed radiation.

**the hydrogen atom is quantized,** a concept that is an extension of the Planck-Einstein quantum hypothesis. Each allowed energy state of the hydrogen atom is called a **stationary state.** There is a stationary state of minimum energy, called the **ground state.** All higher states are called **excited states.**

2. The atom does not radiate or absorb energy as long as it remains in a stationary state. The atom goes from a higher energy state to a lower energy state by emitting radiation. It goes from a lower energy state to a higher energy state by absorbing radiation. The different energy states correspond to different orbits of the electron (only one electron, since this is a model of the hydrogen atom). The orbits of the electron are circular. A change in energy state corresponds to a jump of the electron from one circular orbit to another. When energy is absorbed, the electron jumps to a higher orbit. When energy is emitted, the electron jumps to a lower orbit. The discontinuous jumping of the electron cannot be visualized or explained by the classical laws of physics. It is a quantum phenomenon that occurs only on the atomic level.

3. Because the atom emits and absorbs energy only when the electron moves from one orbit to another, the energy emitted and absorbed by the atom is quantized. The energy of a quantum of such radiation is given by the formula $E = h\nu$. This quantum of energy is the difference between the lower and higher energy states of the atom. This relationship, called the *Bohr frequency rule,* is:

$$E_{\text{high}} - E_{\text{low}} = \Delta E = h\nu \qquad (5.12)$$

where $\Delta E$ is the difference in energy between the high and low states, $h$ is Planck's constant, and $\nu$ is the frequency of a quantum of emitted or absorbed energy. The picture that follows from these postulates is shown in Figure 5.18.

On the basis of these three postulates, Bohr dealt with much of the observed behavior of the hydrogen atom. He could even calculate the differences in energy between some stationary states.

**Example 5.7**    The wavelength of one line in the visible region of the atomic spectrum of hydrogen is $6.5 \times 10^{-7}$ m. This radiation is emitted when a hydrogen atom goes from a high energy state to a lower energy state. Calculate the difference in energy between the two states.

**Solution**    The difference in energy is related to the frequency of the radiation. The frequency is related to the wavelength.

$$\Delta E = h\nu \quad \text{and} \quad \nu = c/\lambda$$

Thus

$$\Delta E = \frac{hc}{\lambda} = \frac{(6.6 \times 10^{-34}\ \text{J s})(3.0 \times 10^8\ \text{m s}^{-1})}{(6.5 \times 10^{-7}\ \text{m})}$$
$$= 3.0 \times 10^{-19}\ \text{J}$$

However, Bohr was able to go much further than this relatively simple description of the absorption and emission of energy by the hydrogen atom. His model had one more postulate.

**4.** The allowed orbits are defined most simply in terms of the *angular momentum* of an electron in the orbit. (Momentum is a measure of the tendency of a moving body to keep moving; angular momentum is a measure of the tendency of a body in rotational motion to keep rotating.) Since only certain orbits are allowed, only certain values of the angular momentum of the electron are allowed. Bohr postulated that the allowed values of the angular momentum are integral multiples of $h/2\pi$. There was no deep theoretical reason for the use of the value $h/2\pi$. Bohr chose it because it gave the right answers. In the ground state, the state of lowest energy, the angular momentum of the electron is $h/2\pi$. The angular momentum of an electron in any higher orbit is an integral multiple of $h/2\pi$. The angular momentum of the electron is $nh/2\pi$, where $n$ is an integer (1, 2, 3, 4, . . .) corresponding to a given orbit. For the ground state, $n = 1$. For the next higher state, $n = 2$, and the angular momentum of an electron in this orbit is $2h/2\pi$. For the next higher state, $n = 3$, and so on. There can be no fractional value of $h/2\pi$. Thus, the integer $n$ can be used to designate an orbit and a corresponding energy state; $n$ is called the atom's *principal quantum number.*

## Calculating Electron Orbits

Using these postulates and some classical laws of physics, Bohr could calculate the radius of each orbit of the hydrogen atom, the energy of the state associated with each orbit, and the wavelength of the radiation emitted in transitions between orbits. The wavelengths calculated by this method agreed with those in the atomic spectrum of hydrogen.

Bohr started with the knowledge that the coulombic attraction between the electron and the nucleus must be balanced exactly by the force of acceleration that keeps the electron in orbit.

From Coulomb's law, the coulombic attraction is $e^2/r^2$, where $e$ is the magnitude of the negative charge on the electron and the positive charge on the hydrogen nucleus and $r$ is the radius of the circular orbit.[1] The electron, like any body moving in a closed orbit, has an acceleration. The

---

[1] We shall not show the proportionality constants needed to reconcile units in the following derivation.

acceleration is $v^2/r$, where $v$ is the velocity of the electron and $r$ is the radius of the orbit. From $F = ma$, the force is $mv^2/r$, where $m$ is the mass of the electron. The two forces are equal:

$$\frac{e^2}{r^2} = \frac{mv^2}{r} \tag{5.13}$$

or, multiplying both sides by $r$:

$$\frac{e^2}{r} = mv^2 \tag{5.14}$$

Both $e$, the charge on the electron, and $m$, the mass of the electron, had been determined by experiment. If $v$, the velocity of the electron, were known, then $r$, the radius of the orbit, could be calculated.

The velocity of the electron is known to be related to the angular momentum. The angular momentum of a body moving in a circular orbit is $mvr$, according to Newton's laws of motion. From one of Bohr's postulates, the angular momentum of the electron is $nh/2\pi$. Setting these as equal gives:

$$mvr = \frac{nh}{2\pi} \tag{5.15}$$

or, after algebraic manipulation:

$$v = \frac{nh}{2\pi mr}$$

Substituting this expression for $v$ in Equation 5.14 gives:

$$\frac{e^2}{r} = m\left(\frac{nh}{2\pi mr}\right)^2 \tag{5.16}$$

and solving for $r$ gives:

$$r = \frac{n^2 h^2}{4\pi^2 m e^2} \tag{5.17}$$

Since values for all the quantities on the right-hand side of Equation 5.17 are known, values for $r$, the radii of the allowed orbits of the electron, can be calculated. Putting in the known values of $h$, $m$, and $e$ and using the necessary unit conversion factors gives a simple relationship between the size of the radius and the principal quantum number of the orbit:

$$r = 5.29 \times 10^{-11}\, n^2 \quad \text{(where } r \text{ is in meters)} \tag{5.18}$$

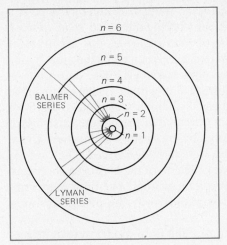

**Figure 5.19**
The hydrogen spectrum diagrammed. The Balmer lines are caused by the emission of specific amounts of energy as the electron drops from higher orbits to the orbit $n = 2$. The Lyman series of lines is generated by the energy emitted as the electron drops from higher energy levels to the orbit $n = 1$. Each individual transition produces one photon with a wavelength and frequency characteristic of the energy change of the atom.

Substituting $n = 1$ gives $5.29 \times 10^{-11}$ m as the raduis of the hydrogen orbit in the ground state, a value that is consistent with other information on the size of atoms. Substituting $n = 2$ gives $(2^2)(5.29 \times 10^{-11})$ m as the size of the orbit in the first excited state. We can diagram the allowed orbits of the hydrogen atom as a series of concentric circles of known size (Figure 5.19), with the nucleus at the center and the radii increasing by factors of $2^2$, $3^2$, $4^2$, and so on.

### The Energy of the Electron

It can be shown from classical physics that the total energy of the electron in an orbit of the Bohr atom is

$$E = \frac{-e^2}{2r} \tag{5.19}$$

The negative sign in Equation 5.19 means that the energy of the electron in the atom is lower than the energy of a free electron and is negative. As $r$ increases, the numerical value of $-e^2/2r$ decreases; the energy becomes less negative, which means that the energy increases. In other words, the energy of the electron increases as its distance from the nucleus increases.

Substituting the expression for $r$ from Equation 5.17 into Equation 5.19 gives the energy as:

$$E = \frac{-2\pi^2 m e^4}{n^2 h^2} \tag{5.20}$$

Equation 5.20 allows us to calculate the energy for each orbit, since the values of $m$, $e$, and $h$ and the necessary unit conversion factors are known, while the appropriate integer can be assigned for $n$, the principal quantum number.

### Bohr and Balmer

The most impressive result of the Bohr model came when the expression for $E$ given in Equation 5.20 was substituted into Equation 5.12, the Bohr frequency rule, and $c/\lambda$ was substituted for $v$. If $n_f$ is the principal quantum number of the final, or lower energy state, and $n_i$ is the principal quantum number of the initial, or higher energy state of emission, the resulting equation is:

$$\frac{1}{\lambda} = \frac{2\pi^2 m e^4}{c h^3} \left( \frac{1}{n_f^2} - \frac{1}{n_i^2} \right) \tag{5.21}$$

an expression that appears to be the same as Equation 5.5, the Balmer expression for the atomic spectrum of hydrogen, if the term $R_H$ in that expression equals $2\pi^2 m e^4/c h^3$. Substituting experimentally determined values of $m$, $e$, $c$, and $h$ and the necessary unit conversion factors gives a number that is in excellent agreement with the measured value of $R_H$.

In the Bohr model, each series of lines in the atomic spectrum of hydrogen is the radiant energy emitted in transitions of the atom's lone electron from different high-energy orbits to a single low-energy orbit. As Figure 5.19 shows, the Balmer series, those lines in the visible region of the atomic spectrum of hydrogen predicted by Balmer's formula, results from transitions to the orbit $n = 2$. Bohr was able to predict the existence of another series of lines, those produced by the electron's transitions to the orbit defined by $n = 1$. Those lines were later discovered. They are the Lyman series of lines in the ultraviolet region of the hydrogen spectrum. Balmer's formula could and did predict new lines. The Bohr theory not only predicted new lines but also explained why the predictions were correct.

While the quantitative predictions of the Bohr model were quite successful for the hydrogen atom, they were unsuccessful for every other atom. The Bohr model would work only for an atom with one electron. Relationships such as Equation 5.17 could be extended to a monatomic cation with a nuclear charge greater than $+1$, but only if the cation had just one electron. Thus the radius of a one-electron cation of nuclear charge $Z$ is

$$r = \frac{n^2 h^2}{4\pi^2 m e^2 Z} \tag{5.22}$$

The energy of the electron in a one-electron species of nuclear charge $Z$ is obtained by extension from Equation 5.20:

$$E = -\frac{2\pi^2 m e^4 Z^2}{n^2 h^2} \tag{5.23}$$

The reason the Bohr model would not work for anything with more than one electron was found with the answer to the question posed in Section 5.6: How can something appear to be both a particle and a wave?

## 5.8  WAVES AND PARTICLES

The suggestion that a particle and a wave can be confused with one another seems absurd. At the beginning of the twentieth century, the definition of each was fixed precisely, with no danger of overlap. A wave was seen as a disturbance that moves through space. A particle was seen as something that occupies space and has definite boundaries.

The fact that a wave is in motion causes some difficulty in describing it precisely, but this is not an insuperable difficulty. We can measure the velocity of a particle by following it in motion. We measure the velocity of a wave by picking a point on the wave — a particular peak, for instance —

and following that point for a given time.

These rules work well enough for particles as big as baseballs and waves of the ocean. But the rules do not work well for particles as small as atoms and molecules. When one tries to describe objects on the atomic scale, the rules of the game change. The fact that the principles that apply to macroscopic objects may not apply to objects on the atomic scale is the key to an understanding of the atom. Once we stop trying to understand the atom by analogy with baseballs and ocean waves, progress can be made.

First we must deal with an apparent paradox. We seem to say that atoms are fundamentally different from baseballs and ocean waves are completely different from light waves. Yet baseballs and ocean waves consist of atoms. Logically, there should be a fundamental similarity, not a fundamental difference.

In fact, a basic similarity between subatomic phenomena and everyday objects has been found. Baseballs, atoms, ocean waves, and light waves all have aspects in common. If the world is viewed properly, there are no waves and no particles. Instead, the world is made of wave particles, one phenomenon with two complementary aspects. On the everyday scale, one aspect or the other is dominant. The wave aspect of baseballs is undetectable and the particle aspect of ocean waves is also undetectable. But on the atomic scale, both aspects are detectable and important. Electrons, which were first thought of as particles, have an equally important wave aspect. Electromagnetic radiation, which was first thought of as waves, has an equally important particle aspect. The conception of a given object as either wave or particle depends on the point of view of the observer.

## The de Broglie Hypothesis

By the 1920s, experiments had established the dual wave-particle nature of electromagnetic radiation. Even though momentum (the product of mass and velocity, $mc$) classically was associated only with particles and not with waves, it was possible to calculate the momentum of a photon, a particle of light with wave characteristics. Combining Einstein's famous $E = mc^2$ with the relationship $E = h\nu$ gives:

$$h\nu = mc^2 \qquad (5.24)$$

which can be rewritten as:

$$mc = \frac{h\nu}{c}$$

and, since $\nu/c = 1/\lambda$, we can write:

$$mc = \frac{h}{\lambda} \quad \text{or} \quad \lambda = \frac{h}{mc} \qquad (5.25)$$

In 1924, Louis de Broglie (1892–      ) made a suggestion that was simple but startling: If light waves have particle characteristics, then particles should have wave characteristics. De Broglie noted that if a value for $\lambda$ is substituted in Equation 5.25, it is possible to calculate a value for $m$, the mass of a photon of this wavelength $\lambda$. De Broglie suggested a different calculation: Replace $m$ with the known mass of an electron, and replace $c$, the velocity of light, with $v$, the velocity of an electron. A value for $\lambda$ can then be calculated. This value is the wavelength of an electron, an entity that had been regarded solely as a particle.

The calculation suggested by de Broglie was performed. Experiments produced proof that a moving electron has an associated wavelength — a wavelength whose value was found to be the value calculated from Equation 5.2 by de Broglie's method.

## The Wavelengths of Particles

The de Broglie relationship can be made more general. A wavelength can be calculated for any particle moving with a velocity $v$. The formula is:

$$\lambda = \frac{h}{mv} \tag{5.26}$$

Equation 5.26 means that any moving particle, even a baseball, has a wavelength associated with it. The formula also shows why the wavelengths associated with baseballs and other large-scale objects usually are undetectable.

The value of $h$ is small: $6.6 \times 10^{-34}$ J s, or $6.6 \times 10^{-34}$ kg m$^2$ s$^{-1}$. The wavelength of a moving particle will also be small unless the $mv$ term in Equation 5.26 is of the same order of magnitude as $h$. The mass $m$ of any large object, such as a baseball, is always large compared to the mass of a particle like the electron — so large that the momentum, $mv$, is enormous compared to $h$.

For example, a baseball thrown by a good pitcher has a momentum of about 5 kg m s$^{-1}$ and a wavelength, $\lambda = (6.6 \times 10^{-34}$ kg m$^2$ s$^{-1})/(5$ kg m s$^{-1}) = 1.3 \times 10^{-34}$ m. This means that the peaks of the waves associated with the baseball are separated from each other by a distance that is $10^{24}$ times smaller than the size of an atom. The waves associated with a heavy object are undetectable. Only on the atomic scale, where values of $m$ are very small, does $\lambda$ become significant.

The wave aspect of the electron explains why the angular momentum of an electron is quantized. If a wave goes around in a circular path, the circumference of the circle must be an integral multiple of the wavelength:

$$2\pi r = \lambda n \tag{5.27}$$

A circle whose circumference is a fractional part of the wavelength is not an allowed orbit. The peaks and troughs of the wave would overlap and

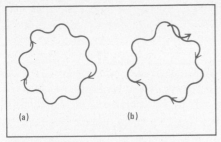

**Figure 5.20**
The electron as a wave. Only orbits whose circumference is an integral multiple of the electron's wavelength are possible. In other orbits the overlapping waves would interfere and cancel each other.

the wave would cancel itself out, as Figure 5.20 shows. If we substitute into Equation 5.27 the value of the electron's wavelength $\lambda$ given in Equation 5.26, we get:

$$2\pi r = \frac{nh}{mv} \tag{5.28}$$

or, rearranging the terms:

$$mvr = \frac{nh}{2\pi}$$

which is identical with Equation 5.15, part of the Bohr derivation.

Looking back at the long history that led to the discovery of the wave-particle nature of energy and matter, it seems rather a historical accident that light was first regarded as waves and electrons were first regarded as particles. Though the road to the present view of energy and matter has been long, it has reached the point where light, electrons, and all other "waves" and "particles" can be viewed as having the same essential nature: They are wave-particles moving with different velocities.

### The Heisenberg Uncertainty Principle

Given this *wave-particle duality,* a new way was needed to describe the structure of the atom. The description must include not only electron particles but also electron waves.

The wave pictured in Figure 5.10 is a pure harmonic wave with a regular pattern extending to infinity. Such pure waves probably do not exist in the real world. A real wave starts from a point and ends at another point. Real waves tend to pile up on themselves to produce what are called *wave packets.* Figure 5.21 shows such a wave packet, whose wavelength and amplitude cannot be defined precisely. Because the packet extends through a region of space, its position also cannot be defined precisely.

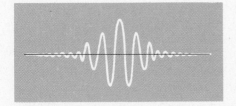

**Figure 5.21**
A wave packet, with varying wavelengths and amplitudes. Compare with the idealized wave in Figure 5.10. Such wave packets moving through space are used to describe the motion of photons, whose velocity is $c$, $3 \times 10^8$ m/s, and of matter, whose velocity is always less than $c$.

The wave associated with a particle is such a wave packet, since a moving particle starts at one point and finishes at another point. The wavelength associated with a moving particle cannot be defined precisely. There is a narrow range of wavelengths associated with the particle. We can say that there is an *uncertainty* of wavelength within that range. That is not the only uncertainty. From Equation 5.26, it can be seen that the momentum of a particle is $h$, a constant, divided by $\lambda$, the wavelength. If there is uncertainty in the wavelength, there must also be uncertainty in the momentum. Since a particle such as an electron is actually moving as a wave packet, there is also uncertainty about its position.

This thought—that there is a fundamental uncertainty about the wavelength, momentum, and position of wave-particles—is one of the more unsettling discoveries of twentieth-century science. Uncertainty now has been built into the accepted picture of natural phenomena.

## EINSTEIN AND UNCERTAINTY

"God does not play dice with the universe," Albert Einstein said. It was his brief way of rejecting the view of the universe that accompanied acceptance of the Heisenberg uncertainty principle. The implication of the uncertainty principle is that the universe is indeterminate on the most basic level. Since there is uncertainty in the most elementary events, no exact cause-and-effect relationship can be established. Instead, in the words of de Broglie, quantum physics appears to be "governed by statistical laws, and not by any causal mechanisms, hidden or otherwise."

Einstein could never accept this implication of quantum theory, a point of view that put him increasingly at odds with other physicists as the years went by. In 1944, in a letter to Max Born, Einstein expressed his beliefs directly:

"You believe in the God who plays dice, and I in complete law and order in a world which objectively exists, and which I, in a wildly speculative way, am trying to capture. I firmly *believe,* but I hope that someone will discover a more realistic way, or rather a more tangible basis than it has been my lot to do.

Even the great initial success of the quantum theory does not make me believe in the fundamental dice game, although I am well aware that our younger colleagues interpret this as a consequence of senility."

Einstein made a number of attempts to describe "thought experiments" that would show that cause-and-effect exists on the subatomic level. A thought experiment is an experiment that can never be performed, only imagined, but that can nevertheless test a theory. But in every case, other physicists were able to show flaws in his thought experiments, and thus to uphold the uncertainty principle.

Einstein's viewpoint was summed up by another statement he made more or less casually when he heard of an experimental result that purported to disprove the theory of relativity. "God is subtle, but he is not malicious," Einstein said.

However, quantum mechanics and the uncertainty principle remain cornerstones of modern physics. If Einstein's world view is correct, it is on a deeper level than we have reached as yet.

Werner Heisenberg (1901–1976), in his uncertainty principle, set forth the minimum amount of uncertainty to be expected in any description of the universe. The uncertainty principle can be stated in several ways. One statement is that the uncertainty in the momentum of any particle multiplied by the uncertainty of position can never be less than $h$, Planck's constant; that is:

$$\Delta(mv_x) \times \Delta x \ge h \qquad (5.29)$$

where $mv_x$ is the momentum in the $x$ direction of the particle, $x$ is the position, and $\Delta$ is the uncertainty or error in each.

### The Consequences of Uncertainty

The uncertainty principle has many consequences. Equation 5.29 can be interpreted to mean that it is impossible to know exactly both the position and the momentum of a particle at any moment. Any attempt to make an exact measurement of position and momentum must fail, because there is no way of making the measurement without changing one of the characteristics that is being measured.

We can illustrate this concept by inventing a party game, "Find the Balloon." In this game, a person is in a dark room and must find a floating balloon without turning on the lights. One method is to throw a dart, listen for a pop, and thus learn where the balloon was — except that the balloon is no longer there (and no longer a balloon). Another possible method is to line up some friends in the room, throw peanuts, and calculate the location by the interaction between peanuts, balloon, and friends. But if you hit the balloon, both the balloon and the peanut are deflected, and the location is still uncertain.

If the player cheats by turning on the light for a quick peek, the results are better. The balloon can be seen and located instantly. It is true that the location of the balloon is changed because photons bounce off it, but the change is negligible for the purpose of the game.

If the game is "Find the Electron," turning on the light — using photons in the search — causes problems. If a photon with the same momentum as the electron is used, the collision deflects the electron measurably. If a photon of long wavelength is used, accuracy is decreased because the position of the electron can be determined only within the limit of the wavelength. Therefore, one wants a photon with low momentum and short wavelength — which is an impossibility. According to the de Broglie relationship, momentum and wavelength are inversely proportional; one increases as the other decreases. The nature of the wave-particles in the game sets an unconquerable limit on the accuracy of the measurement.

## 5.9 THE QUANTUM MECHANICAL DESCRIPTION OF THE HYDROGEN ATOM

This new information showed why the Bohr model of the atom was only partially successful. First, the model did not take into account the wave aspect of the electron. Second, it violated the uncertainty principle by defining the motion and position of the electron too precisely.

The impossibility of defining the electron's position precisely is now recognized. The best that can be done is to define a *probability* for the electron being in a given location. It is not possible to say that at a given moment, the electron is at point A rather than at point B. It is possible to say that at some moment the electron is ten times as likely to be at point A as at point B.

Max Born (1882–1970) proposed that these probabilities could be obtained by a study of the wave packet associated with the moving electron. In the quantum mechanical description of the atom that is used today, the idea of an electron particle is replaced by the concept of the electron wave. A mathematical function called the *wave function,* represented by $\psi$, the Greek letter psi, is used to describe the wave packet. The value of the wave function $\psi$ at any point in space is related to the amplitude of the wave, its intensity of vibration, at that point. The probability that an electron is at a given point in space is proportional to the *square* of the wave function, $|\psi|^2$, at that point.

## The Schrödinger Wave Equation

The simplest equation that can be used to calculate the electron wave function $\psi$ was developed in 1926 by Erwin Schrödinger (1887–1961). The Schrödinger wave equation for the motion of a particle in any one direction is:

$$-\left(\frac{h^2}{8\pi^2 m}\right)\frac{d^2\psi}{dx^2} + V\psi = E\psi \qquad (5.30)$$

Equation 5.30 seems complex, but analysis shows it to be a single equation in two unknowns, $\psi$ and $E$. All the other quantities, such as $m$, the mass of the particle, and $V$, its potential energy, are known. Thus, the Schrödinger equation gives the relationship between the total energy $E$ of the electron and its wave function, $\psi$.

When the Schrödinger equation is applied to a real system such as a hydrogen atom, solutions are obtained only at certain values of the total energy $E$. These values of $E$ are related to each other by small integers — the quantum numbers. Thus, quantum numbers and quantized energy of the electron are automatic consequences of the mathematics of the Schrödinger equation and of the fact that electron waves in atoms have restrictions imposed on their motion.

The solutions of the Schrödinger equation give a set of allowed energies of the electron, with one or more wave functions $\psi$ corresponding to each energy. How can we picture such an electron?

## The Orbital

The simplest and most widely used picture replaces an orbit, which is a path, with an **orbital**. An orbital can be described as a picture of the probability of finding the electron at every point in space.

We can develop a graphical representation of an orbital in a number of different ways. One simple way is to make a three-dimensional plot of all the points in space where the probability of finding the electron has an arbitrarily assigned constant value. Such a plot is a continuous closed surface, called a *surface of constant probability,* which encloses a region of space.

The efficient course of action is to concentrate on the region where the electron is likely to be found most of the time. The typical orbital figure is made to enclose the region of space where the electron is found at least 90% of the time. The ground state orbital of hydrogen is shown in Figure 5.22.

Other graphical representations of an orbital can be used, but the

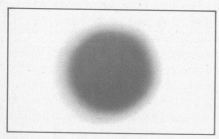

**Figure 5.22**

A graphical representation of the ground state orbital of hydrogen. The surface of constant probability is a sphere. The sphere is drawn large enough to enclose the region of space where the electron in this orbital is found at least 90% of the time. Orbitals corresponding to higher energy states have different shapes or sizes.

surface of constant probability is the most common one. When we use the term *orbital*, we shall also take it to mean a graphical representation of the orbital.

A "charge cloud" picture of the electron in an atom can also be visualized. The electron is pictured as a cloud that takes its shape from the orbital. The cloud is most dense where the probability of finding the electron is highest. The cloud thins out where the probability is low.

The quantum mechanical description of the atom has extremely important implications for chemistry. A detailed picture of the electronic structure of atoms has been developed, and it has been used to explain not only the observed characteristics of atoms but also the way in which atoms join to form molecules. The use of knowledge about the electronic structure of atoms to explain chemical behavior will be examined in Chapters 6 and 7.

**Summary**

This chapter introduced us to the **atom,** the basic unit of an element. We began by showing how John Dalton's **atomic theory** was a key step toward acceptance of the idea that atoms really exist. We then described how research into the nature of electricity led to the discovery of the **electron,** a negatively charged subunit of the atom, and to the measurement of the **charge** and **mass** of the **electron.** We told how **radioactivity** was discovered at the turn of the century, and how work by Ernest Rutherford led to an understanding of the structure of an atom, which consists of a small but massive **nucleus,** containing **protons** and **neutrons,** around which electrons orbit. The **atomic number** is the number of protons in a nucleus, while the **mass number** is the total number of protons and neutrons. We mentioned that all elements have **isotopes,** whose nuclei have the same number of protons but different numbers of neutrons. We then described the evolution of **quantum theory,** which says that energy is emitted only in certain fixed quantities. We outlined Albert Einstein's utilization of quantum theory to explain that visible light and other electromagnetic energy consists of quanta called **photons.** Finally, we described Neils **Bohr's model** of the hydrogen atom, which proposed that electrons could have only a limited number of energy states. We described later work that led to the realization that all entities have both **particle and wave** characteristics, and we showed how Werner **Heisenberg** used that work to demonstrate the impossibility of making an exact measurement of both the momentum and the location of a subatomic particle. Finally, we noted how the **Schrödinger wave equation** leads to the idea of an **orbital,** which gives the probability of finding an electron at any point in space.

## Exercises

**5.1** Give an evaluation of the three major assumptions of Dalton's atomic theory, based on your general knowledge. State whether any of Dalton's assumptions are incorrect.

**5.2** Suppose Dalton had studied the compound ethane, which has a ratio of carbon to hydrogen by mass of 4:1. What formula would he have proposed for ethane on the basis of his ideas about atomic weights?

**5.3** The value of $k$ in Coulomb's law is $9.0 \times 10^9$ N m$^2$ C$^{-2}$. The charge on an electron is $1.6 \times 10^{-19}$ C. Find the force of attraction between a proton and an electron at a distance of $5.3 \times 10^{-11}$ m.

**5.4** The charge on the electron is $1.602 \times 10^{-19}$ C, and 96 485 C of charge is needed to produce 1.008 g of $H_2$ from water. Find the number of electrons in 96 485 C and the number of H atoms in 1.008 g of $H_2$.

**5.5** The relative atomic mass of a deuterium atom is 2.014. Find $e/m$ for the deuterium nucleus.

**5.6²** The monopositive cation X$^+$ has $e/m = 1\ 909.2$ C/g. Find the relative atomic mass of X. (*Note:* The relative atomic mass of a hydrogen is 1.0079.)

**5.7** An alpha particle, $^4$He$^{2+}$, has a relative atomic mass of 4.00151. Find the value of $e/m$ in coulombs per gram of an alpha particle.

**5.8** Find the number of neutrons and electrons in the following atoms and ions: (a) $^{14}_6$C, (b) $^{232}_{90}$Th, (c) $^{37}_{17}$Cl$^-$, (d) $^{39}_{19}$K$^+$, (e) $^{56}_{26}$Fe$^{3+}$.

**5.9** Nuclei with the same mass num-

---

bers but different numbers of protons are called *isobars*. Isobars of mass number 56 are known for all the elements from Cr to Ni. Using the system of notation given in Section 5.4, write symbols for all of these isobars.

**5.10** Nuclei with the same number of neutrons but different numbers of protons are called *isotones*. Using the notation of Section 5.4, write the symbols for the four isotones of $^{235}$U.

**5.11** Identify the elements that have the following nuclear compositions: (a) $A = 40, N = 20$, (b) $A = 226, N = 138$, (c) $A = 169, N = 100$.

**5.12** Indium, whose atomic weight is 114.82, exists as two different isotopes, $^{113}$In and $^{115}$In, with relative atomic masses of 112.904 and 114.904. Find the natural abundances of the two isotopes.

**5.13** Find the wavelength of X rays whose frequency is $3.1 \times 10^{19}$ s$^{-1}$.

**5.14** Find the frequency of the television waves whose wavelength is 9.7 m.

**5.15** Find the velocity of a sound wave with a frequency of $1.5 \times 10^2$ s$^{-1}$ and a wavelength of 2.3 m.

**5.16** The frequency range on the FM dial is from 88.0 to 108 MHz. Find the range of wavelengths of FM waves. (*Note:* 1 MHz $= 10^6$ Hz.)

**5.17** A light bulb that is 275 m from an observer emits light of frequency $5.1 \times 10^{14}$ s$^{-1}$. Find the number of crests in the light wave between the bulb and the observer.

**5.18** In Example 5.4, we calculated the wavelengths of the first two lines in the visible spectrum of hydrogen. Calculate the wavelengths of the next two lines.

**5.19** Find the frequency of the short-

est wavelength radiation that can be emitted by hydrogen.

**5.20** The first series of lines in the infrared region of the atomic spectrum of hydrogen is called the Paschen series. Find the two longest wavelengths in this series.

**5.21** Helium was discovered in the sun before it was found on earth. How was this possible?

**5.22** Find the energy of a quantum of X rays whose frequency is $4.65 \times 10^{18}$ s$^{-1}$.

**5.23** A quantum of radiation has an energy of $4.14 \times 10^{-19}$ J. Find the wavelength of the radiation.

**5.24** The energy required to melt 1.0 mol of ice is 6.0 kJ. Find the number of quanta of infrared radiation of frequency $5.25 \times 10^{12}$ s$^{-1}$ required to melt the ice.

**5.25** A common source of ultraviolet radiation in the chemistry laboratory is a mercury lamp which produces radiation of wavelength 253.7 nm. Find the energy of 1.00 mol of photons of this wavelength.

**5.26** A metal has a threshold frequency of $5.8 \times 10^{14}$ s$^{-1}$ for the photoelectric effect. Predict the effect of each of the following changes on the number of electrons emitted and their kinetic energy:
(a) The frequency of radiation striking the metal increases from $5.2 \times 10^{14}$ s$^{-1}$ to $5.6 \times 10^{14}$ s$^{-1}$.
(b) The frequency of radiation striking the metal is lowered from $6.2 \times 10^{14}$ s$^{-1}$ to $5.9 \times 10^{14}$ s$^{-1}$.
(c) The metal is moved closer to a source of radiation of frequency $6.1 \times 10^{14}$ s$^{-1}$.

**5.27** The threshold frequency of a metal is $1.23 \times 10^{15}$ s$^{-1}$. Find the binding energy of an electron in the metal.

**5.28** When radiation of frequency $9.2 \times 10^{14}\,s^{-1}$ strikes a metal, the metal emits electrons with kinetic energy $3.1 \times 10^{-19}$ J. Find the threshold frequency of the metal.

**5.29** Find the threshold frequency of a metal whose binding energy is 259 kJ/mol.

**5.30**★ Using the value of the mass of the electron given in Appendix I, find the velocity of an electron emitted when a metal with threshold frequency $3.71 \times 10^{14}\ s^{-1}$ is exposed to light of wavelength 422 nm.

**5.31** A neon atom emits light of wavelength 640 nm. Find the difference in energy between the two energy states involved in the emission.

**5.32** Find the angular momentum of an electron in the orbit corresponding to the third excited state ($n = 4$) of hydrogen. Find the radius of this orbit.

**5.33** Use the mass of the electron to find the velocity of an electron in the ground state orbit of the Bohr hydrogen atom.

**5.34** When we use Equation 5.20 to calculate the energy in joules corresponding to an orbit in the Bohr atom, the numerator must include $k^2$, where $k$ is the constant in Coulomb's law and has the value $9.0 \times 10^9$ N m$^2$ C$^{-2}$. Obtain the necessary data from Appendix I and calculate the energy of the ground state orbit of the Bohr hydrogen atom.

**5.35** Using the Bohr model, give the relationship between nuclear charge and the radius of the ground state orbit in one-electron monoatomic cations. Use Coulomb's law to explain this relationship.

**5.36** Find the wavelength associated with an automobile of mass 1500 kg moving at a velocity of 24 m/s.

**5.37** The velocity of an electron in the ground state orbit of the Bohr hydrogen atom is $2.18 \times 10^6$ m/s. Find the wavelength of this electron, using the mass of the electron given in Appendix I.

**5.38** Using Equation 5.27, find the longest allowed wavelength of the electron in the first excited state orbit of the Bohr hydrogen atom.

**5.39** Find the longest allowed wavelength of a wave traveling in a circular orbit of radius 17.9 m.

**5.40** How can you use the de Broglie relationship to help you measure the mass of a neutral subatomic particle when the mass cannot be found directly?

**5.41**★ Heisenberg's uncertainty principle can be formulated in terms of different pairs of variables, such as momentum and position or energy and time. In the Bohr model, the energy of each orbit is defined very accurately. What problem is created by this accuracy?

**6**

# Atoms, Electrons, the Periodic Table

**Preview**

This chapter picks up where the last one left off, describing the quantum numbers that define the allowed orbitals of electrons in atoms and showing how a rule called the Aufbau principle can be used to work out the electronic configuration of atoms. We go on to show how the properties of atoms are related to their electronic configuration—especially to the outermost, or valence, electrons. This discussion helps introduce the periodic table of the elements, the single most important unifying concept in modern chemistry. Finally, we give several examples of periodic trends: in the properties of a group of elements such as the alkali metals and in the properties of classes of compounds such as the oxides, halides, and hydrides.

**C**hapter 5 described a quantum mechanical picture that defines regions in space around an atom where the probability of finding the electron is high. By enclosing such regions with a surface, we can better visualize the "shapes" of orbitals. We must, however remember that an orbital is really a total picture of the probability of finding the electron at every point in space.

Now we shall see why and how the electronic structure of atoms determines their chemical behavior. We shall also see how a knowledge of the electronic structure of atoms is used to explain an important unifying principle in chemistry, a principle that is embodied in the periodic table of the elements.

## 6.1   THE QUANTUM MECHANICAL DESCRIPTION OF THE HYDROGEN ATOM

The Schrödinger wave equation has been solved completely for all the allowed energy states of the hydrogen atom. These solutions can be used as a basis for describing more complex atoms.

### Quantum Numbers

The Schrödinger equation gives quantum numbers corresponding to allowed energy states for the electron in a hydrogen atom. These quantum numbers can be used to define the allowed orbitals and to describe the behavior of an electron in an orbital. Each orbital corresponds to an allowed energy state. We describe an electron of a hydrogen atom by assigning values to four interrelated quantum numbers. The quantum numbers are

1. *The Principal Quantum Number, n.* The principal quantum number can have any integral value greater than zero: $n = 1, 2, 3, 4, \ldots$ . We can specify the energy of the electron in the hydrogen atom by assigning it a value of $n$. The principal quantum number is the same as Bohr's quantum number. It is also part of the solution of the Schrödinger wave equation. If we solve the Schrödinger equation for the hydrogen atom, we get an expression for the total energy of the electron,

$$E = \frac{-2\pi^2 m e^4}{n^2 h^2} \tag{6.1}$$

in terms of $n$, the principal quantum number. The expression is identical with Equation 5.20, which was the result obtained by Bohr. Once a value is assigned to $n$, the energy can be calculated.

Again, note that the total energy of the electron in an atom is negative and is lower than the energy of a separated electron and proton. A decrease in the numerical value of $E$ means that the energy has become

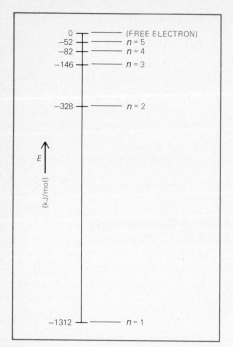

**Figure 6.1**

The energy of the electron of the hydrogen atom is less than that of the free electron and is negative. As *n* increases, the energy increases, since it becomes less negative.

less negative—that is, the energy has increased. As *n*, the principal quantum number, increases, the energy *E* of the electron increases, as shown in Figure 6.1

**2.** *The Angular Momentum Quantum Number, $\ell$.* The values of $\ell$, which is also called the azimuthal quantum number, depend on the value of the principal quantum number *n*. The quantum number $\ell$ has integral values from 0 to $n - 1$. If $n = 1$, then there is only one possible value, $\ell = n - 1 = 0$. If $n = 2$, there are two possible values, $\ell = 0$ or 1. If $n = 3$, there are three possible values, $\ell = 0$, 1, or 2. And so on.

| $n$ | $\ell$ |
|---|---|
| 1 | 0 |
| 2 | 0, 1 |
| 3 | 0, 1, 2 |
| 4 | 0, 1, 2, 3 |
| ⋮ | ⋮ |

The value of $\ell$ determines the shape of an orbital and the angular momentum (Section 5.7) of an electron occupying that orbital. In a hydrogen atom, the value of $\ell$ does not affect the total energy, which is determined completely by the value of *n*. In atoms with more than one electron, the energy of an electron in an orbital, while described primarily by *n*, also depends to some extent on the value of $\ell$. In multi-electron atoms, the energy of electrons increases as the value of $\ell$ increases. When electrons have the same quantum number *n*, the energy of the electron with the quantum number $\ell = 2$ is higher than that of the electron with $\ell = 1$; the electron with $\ell = 3$ has a higher energy level than the electron with $\ell = 2$.

**3.** *The Magnetic Quantum Number, $m_\ell$.* The magnetic quantum number is allowed all integral values, including 0, from $-\ell$ to $+\ell$. If $\ell = 0$, then $m_\ell = 0$. If $\ell = 1$, then $m_\ell$ has three possible values, $-1$, 0, and $+1$. If $\ell = 2$, then $m_\ell$ has five possible values, $-2$, $-1$, 0, $+1$, and $+2$.

| $\ell$ | $m_\ell$ |
|---|---|
| 0 | 0 |
| 1 | $-1$, 0, $+1$, |
| 2 | $-2$, $-1$, 0, $+1$, $+2$ |
| ⋮ | ⋮ |

The value of the quantum number $m_\ell$ does not affect the energy of an electron. Rather, $m_\ell$ is most simply described as a directional quantum number. It distinguishes between orbitals that have the same shape and energy but point in different directions. The numerical value of $m_\ell$ is not important. What is important is the number of different values of $m_\ell$ that correspond to a given value of $\ell$. When $\ell$ is 0, $m_\ell$ has only one value. When $\ell$ is 1, $m_\ell$ has three allowed values. When $\ell$ is 2, $m_\ell$ has five allowed values.

**Electron Spin**

The three quantum numbers $n$, $\ell$, and $m_\ell$ are sufficient to describe the orbitals in the hydrogen atom. However, another aspect of the behavior of the electron must be considered. The electron behaves as if it were spinning. This property of the electron, called its **spin,** was discovered in 1925.

Spin is best visualized as a manifestation of the particle aspect of the electron. While it is difficult to visualize a spin associated with a wave packet, we can easily picture a particle spinning like a top. The electron can spin in one of two ways, clockwise or counterclockwise. The spin is specified by the *spin quantum number, $m_s$*. Since the electron has only two allowed directions of spin, $m_s$ has only two values, which by convention are $+\frac{1}{2}$ and $-\frac{1}{2}$. There is no relationship between the spin quantum number and the values of $n$, $\ell$, and $m_\ell$.

We should not make the mistake of associating a particular value of $m_s$ with a particular direction of spin. As we shall see, $m_s$ describes relative, not absolute, direction of spin.

*Quantum Numbers*

$$n = 1, 2, 3, \ldots$$
$$\ell = 0, 1, 2, \ldots (n-1)$$
$$m_\ell = 0, \pm 1, \ldots (\pm \ell)$$
$$m_s = +\tfrac{1}{2} \text{ or } -\tfrac{1}{2}$$

**Example 6.1**    An electron in a multielectron atom has the principal quantum number 3. How many different sets of quantum numbers can this electron have?

**Solution**    The principal quantum number limits the values of $\ell$. Since $n$ is 3, the value of $\ell$ may be 0, 1, or 2. Let us take each case separately.

If $\ell = 0$, then the value of $m_\ell$ must be 0 and there are two allowed sets of quantum numbers: 3, 0, 0, $+\frac{1}{2}$ and 3, 0, 0, $-\frac{1}{2}$.

If $\ell = 1$, then the value of $m_\ell$ can be $-1$, 0, or $+1$. Therefore, there are six allowed sets of quantum numbers. In three sets, $m_s = -\frac{1}{2}$ and the sets are 3, 1, $-1$, $-\frac{1}{2}$; 3, 1, 0, $-\frac{1}{2}$; and 3, 1, $+1$, $-\frac{1}{2}$. The other three sets, in which $m_s = +\frac{1}{2}$, are 3, 1, $-1$, $+\frac{1}{2}$; 3, 1, 0, $+\frac{1}{2}$; and 3, 1, $+1$, $+\frac{1}{2}$.

If $\ell = 2$, then $m_\ell$ can be $-2, -1, 0, +1$, or $+2$. Five values of $m_\ell$ lead to ten sets of quantum numbers for the electron, five with $m_s = +\frac{1}{2}$ and five with $m_s = -\frac{1}{2}$.

Totaling, we find that there are 18 allowed sets of quantum numbers for an electron with $n = 3$.

**The Set of Allowed Orbitals**

The customary way of identifying orbitals is first to write the principal quantum number $n$, and then to use a letter to identify the value of $\ell$. The

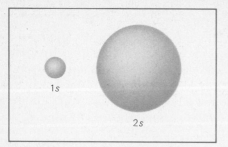

**Figure 6.2**
Two s orbitals, with different principal quantum number n. The orbital on the left has n = 1, the orbital on the right has n = 2. All s orbitals are represented as spheres whose radii are determined by the value of n. The 2s orbital is larger than the 1s orbital. A 3s orbital would be even larger than the 2s orbital.

letters $s$, $p$, $d$, and $f$ are used for historical reasons. They once were used to describe features of atomic spectra, and have never been abandoned. When $\ell = 0$, the letter $s$ is used. When $\ell = 1$, the letter $p$ is used. When $\ell = 2$, the letter $d$ is used. When $\ell = 3$, the letter $f$ is used. For values of $\ell$ greater than 3, the lettering proceeds alphabetically from $f$. In tabular form:

| $\ell$ | 0 | 1 | 2 | 3 | 4 | 5 | ... |
|--------|---|---|---|---|---|---|-----|
| type | $s$ | $p$ | $d$ | $f$ | $g$ | $h$ | ... |

Table 6.1 shows how the value of $n$ and the letter corresponding to the value of $\ell$ are used to name orbitals. The three $p$ orbitals, which possess the same values of $n$ and $\ell$, can be distinguished by the use of the subscripts $x$, $y$, and $z$ to designate the three values of $m_\ell$. Since these three orbitals differ only in their spatial orientation, the three subscripts can be considered to describe the relative orientation of the orbitals along three axes at right angles in space. For $d$ orbitals, more complicated subscripts are used, as Table 6.1 indicates. Note that there is no correspondence between a given subscript and a given value of $m_\ell$. Only the number of values of $m_\ell$ is important.

### Shape and Location of the Orbitals

A given set of quantum numbers completely characterizes the shape, size, and orientation of an orbital.

A change in the value of $n$, the principal quantum number, has a direct effect on the orbital description. As the value of $n$ increases, we must go further from the nucleus to get a surface to enclose a region of space that includes more than 90% of the electron density. In other words, a higher value of $n$ means that the region of highest probability for the electron is further from the nucleus. Our picture of a 3s orbital is larger than a 2s orbital, which is larger than a 1s orbital.

The orbital shapes defined by low values of $\ell$ are easily described. For an $s$ orbital ($\ell = 0$), the surface that represents the orbital is a sphere

**TABLE 6.1    A Portion of the Set of Allowed Orbitals**

| $n$ | $\ell$ | $m_\ell$ | Orbital Designation |
|-----|--------|----------|---------------------|
| 1 | 0 | 0 | $1s$ |
| 2 | 0 | 0 | $2s$ |
|   | 1 | $-1, 0, +1$ | $2p_x, 2p_y, 2p_z$ |
| 3 | 0 | 0 | $3s$ |
|   | 1 | $-1, 0, +1$ | $3p_x, 3p_y, 3p_z$ |
|   | 2 | $-2, -1, 0, +1, +2$ | $3d_{z^2}, 3d_{x^2-y^2}, 3d_{xy}, 3d_{xz}, 3d_{yz}$ |

whose radius depends on the value of $n$ (Figure 6.2). When $\ell = 0$, $m_\ell$ has only one value, since spheres are not directional.

For $p$ orbitals ($\ell = 1$), the geometry is more complicated. We can think of these orbitals as having two spherical lobes, as shown in Figure 6.3. The three $p$ orbitals are identical in shape and size but differ in orientation. The three possible orientations of $p$ orbitals are usually called $p_x$, $p_y$, and $p_z$. The surfaces drawn to represent the three $p$ orbitals enclose different regions of space. If these three regions are superimposed, the resulting space can be enclosed by a spherical surface.

There are five $d$ orbitals. Figure 6.4 shows a convenient way to represent their shapes. Figure 6.4 appears to show two different shapes for a $d$ orbital. One shape is four-lobed, the other two-lobed with an associated torus (the doughnut-shaped figure). But all $d$ orbitals have the same shape. They only seem to be different because of the methods used to draw orbitals. Figure 6.5 shows that the two "different" $d$ orbital shapes are not really different. If two four-lobed $d$ orbitals are superimposed, the result is the alternate shape, a two-lobed orbital with an associated torus. If all the $d$ orbitals are superimposed, they enclose a spherical region of space, just as the $p$ orbitals do.

The $d$ orbitals get their names from the orientation of their maximum electron density. Three of the orbitals $d_{xy}$, $d_{xz}$, and $d_{yz}$, have maximum electron density in the plane indicated by their subscripts. These regions of maximum electron density are centered between the coordinate axes.

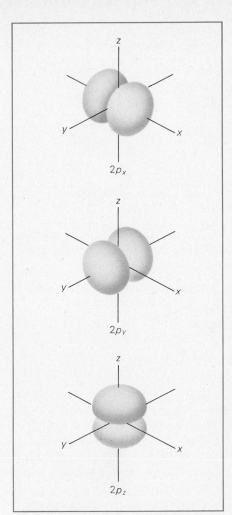

**Figure 6.3**
The three $2p$ orbitals of the hydrogen atom, each represented as two spheroidal lobes, but with a different orientation for each orbital.

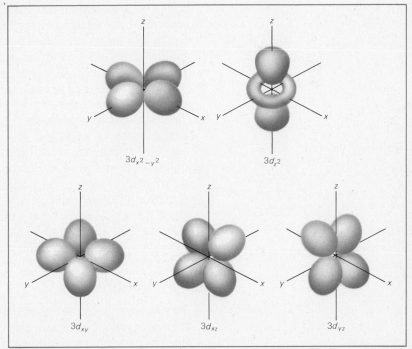

**Figure 6.4**
The five $3d$ orbitals of the hydrogen atom. If all five orbitals are superimposed, the enclosed space is spherical.

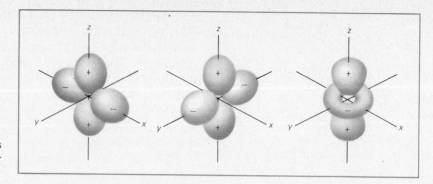

**Figure 6.5**
The superposition of two four-lobed $d$ orbitals creates a $d$ orbital with two lobes and an associated torus.

The regions of maximum electron density of the other two $d$ orbitals, $d_{x^2-y^2}$ and $d_{z^2}$, are centered on the coordinate axes, the $d_{x^2-y^2}$ orbital on the $x$ and $y$ axes, and the $d_{z^2}$ orbital on the $z$ axis.

## 6.2 ELECTRONIC CONFIGURATION OF MULTIELECTRON ATOMS

The description of the single electron in the hydrogen atom is relatively simple, since we must consider only the interaction between the electron and the nucleus. An exact quantum mechanical description of any atom with more than one electron is a formidable mathematical challenge. In atoms with more than one electron, interactions between the electrons must also be considered. The effect of these interactions on the electronic structure of atoms cannot yet be completely included in the description of multielectron atoms. But the description is close enough to serve most purposes, as experimental results show.

The set of allowed orbitals in a multielectron atom can be approximated by the use of the set of allowed orbitals in the hydrogen atom. This "hydrogenlike approximation" is used to build a picture of the electronic configuration of any atom. But unlike the situation in hydrogen, the energy of an orbital in a multielectron atom depends on $\ell$ as well as $n$. We use the approximation to determine the sets of quantum numbers that describe, as completely as possible, the orbitals occupied by an atom's electrons. We shall see that a description of an atom's electronic configuration helps us to understand the chemical properties of the atom. Generally, we shall be interested in the electronic configuration that corresponds to the lowest electronic energy, or *ground state,* of the atom.

### The Aufbau Principle

To work out the electronic configuration of atoms, we use what is called the **Aufbau principle,** from the German word for "building up." The Aufbau principle can be stated simply: The electronic configuration of lowest energy is the one in which each electron is placed in the lowest-energy hydrogenlike orbital that is available.

Let us start with the hydrogen atom. The lone electron has a principal quantum number $n = 1$, so the quantum number $\ell$ must have a value of 0, and the quantum number $m_\ell = 0$. The electron can have a spin quantum number $m_s$ of either $+\frac{1}{2}$ or $-\frac{1}{2}$. With quantum numbers $n = 1$ and $\ell = 0$, the electron is in the $1s$ orbital.

Next consider the simplest multielectron atom, helium, which has two electrons. The lowest-energy orbital is the $1s$ orbital, and one electron goes there. Does the second electron go into the $1s$ orbital, or does it go into a higher-energy orbital with principal quantum number $n = 2$?

One possible view is that, since the region of space that corresponds to the principal quantum number 1 already has an electron, the second electron must have a principal quantum number of 2. But experimental evidence indicates that neither electron in helium has a principal quantum number of 2. We can approximate the electronic configuration of helium in the ground state as one with both electrons in the $1s$ orbital. The electronic configuration of helium is written $1s^2$, the superscript 2 indicating the number of electrons in the $1s$ orbital. (The superscript is pronounced "two," not "squared," even though the notation is the same as for an exponent in mathematics. This rule applies to all superscripts describing the number of electrons in a set of orbitals.)

When we say that the two electrons of helium are in the same orbital, we are specifying three quantum numbers for each election: $n = 1$, $\ell = 0$, and $m_\ell = 0$. We must also specify the spin quantum number for each electron. The most important point about $m_s$ is that it describes *relative* spins. If we picture two electrons as spinning in opposite directions, as in Figure 6.6(a), we can say that one has $m_s = +\frac{1}{2}$ and the other has $m_s = -\frac{1}{2}$. In such a case, we say that the spins of the electrons are *paired*. If we picture the two electrons as spinning in the same direction, as in Figure 6.6(b), we can say that they have the same value of $m_s$. Whether that value is $-\frac{1}{2}$ or $+\frac{1}{2}$ does not matter, as long as both electrons have the same value of $m_s$. In such a case, we say that the spins of the electrons are *parallel,* or *unpaired.* The concept of paired and unpaired electronic spins is important in the description of many atoms.

### The Pauli Exclusion Principle

One of the basic rules of atomic structure, formulated by Wolfgang Pauli (1900–1958), states that **in a given atom, no two electrons can have the same four quantum numbers.** This is the **Pauli exclusion principle.** The principle means that **when two electrons reside in the same orbital, their spins must be paired.** The principle also means that an orbital, which is described by three quantum numbers, $n$, $\ell$, and $m_\ell$, can hold a maximum of two electrons that must have different values of $m_s$.

There is room in the $1s$ orbital for a second electron because there are two allowed values of $m_s$. Both electrons in the ground state of helium

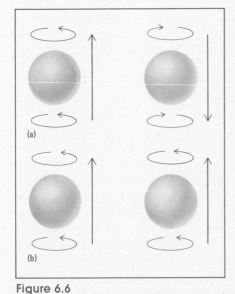

**Figure 6.6**
Two electrons may occupy the same orbital if they have different spins. In (a), the electron on the right has spin $+\frac{1}{2}$ and the electron on the left has spin $-\frac{1}{2}$. In (b), the electron on the right has spin $-\frac{1}{2}$ and the electron on the left also has spin $-\frac{1}{2}$. These two electrons cannot be in the same orbital. The signs are arbitrary and do not indicate actual directions of spin.

have the quantum numbers $n = 1$, $\ell = 0$, $m_\ell = 0$. They differ in the fourth quantum number. One electron has $m_s = +\frac{1}{2}$, the other electron has $m_s = -\frac{1}{2}$.

Now consider the lithium atom, atomic number 3. In accordance with the Pauli exclusion principle, the third electron in a lithium atom cannot also be in the $1s$ orbital. It is in a higher-energy orbital, where $n = 2$. We ask: Which is the lowest-energy orbital available for this electron?

In the hydrogen atom, the $2s(n = 2, \ell = 0)$ orbital and the $2p(n = 2, \ell = 1)$ orbital have the same energy, since they have the same principal quantum number $n$. But in multielectron atoms the relative energy of an electron in a given orbital depends on the value of two quantum numbers, $\ell$ and $n$. For a given value of $n$, a lower value of $\ell$ means lower energy. The $2s$ orbital ($n = 2$, $\ell = 0$) is of lower energy than a $2p$ orbital ($n = 2$, $\ell = 1$), so the third electron of lithium goes into the $2s$ orbital. The electronic configuration of lithium is $1s^2 2s^1$.

In beryllium, atomic number 4, the fourth electron is also in the $2s$ orbital. In accordance with the Pauli exclusion principle, the two $2s$ electrons of beryllium have paired spins. With the addition of the fourth electron, the $2s$ orbital is filled. The next orbital to fill is a $2p$ orbital. In boron, atomic number 5, one electron is in the $2p$ orbital. We can now summarize the electronic configurations of the ground states of the first five elements.

$$\begin{aligned}
&\text{H} : && 1s^1 \\
&\text{He:} && 1s^2 \\
&\text{Li} : && 1s^2 2s^1 \\
&\text{Be} : && 1s^2 2s^2 \\
&\text{B} : && 1s^2 2s^2 2p^1
\end{aligned}$$

Note that in the electronic configuration of boron, it is not necessary to specify whether the $2p$ electron is in the $2p_x$, $2p_y$, or $2p_z$ orbital. All three orbitals correspond to the same energy and are indistinguishable.

By following Table 6.2, you can draw up your own list showing how orbitals are filled. As Table 6.2 shows, the maximum number of electrons of any given quantum number $n$ is $2n^2$. For example, there can be as many as $2(3)^2 = 18$ electrons of $n = 3$. We expect that there can be as many as $2(4)^2 = 32$ electrons of principal quantum number $n = 4$ in a given atom.

The order in which orbitals fill can be predicted accurately for elements of relatively low atomic number. For the lower-energy orbitals, $1s$, $2s$, $2p$, $3s$, and $3p$, the relative energies of orbitals (and therefore the order in which they fill) parallel those predicted from the hydrogen atom, except that for any given $n$, a higher value of $\ell$ leads to a slightly higher energy. It should be noted that the difference in energy between a $1s$ orbital and a $2s$ orbital, which have different principal quantum numbers, is much greater than the difference between a $2s$ and a $2p$ orbital, which have the same principal quantum number, as shown in Figure 6.7.

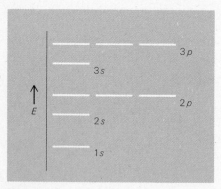

**Figure 6.7**

Qualitative energy differences between orbitals for a multielectron atom. There is a greater gap between the energy level of a $1s$ orbital and a $2s$ orbital than between the $2s$ and $2p$ orbitals or the $3s$ and $3p$ orbitals. In this group, energy differences are greatest between orbitals with different principal quantum numbers. This diagram is not drawn to scale, since the energy differences between orbitals vary among different elements.

**Example 6.2**     Write the ground state electronic configuration of fluorine.

**Solution**     We can apply the Aufbau method in a straightforward way for fluorine, atomic number 9.

   Two electrons fill the $1s$ orbital. Two electrons fill the $2s$ orbital. The remaining five electrons are in the three $2p$ orbitals. We can write the electronic configuration as:

$$1s^2 2s^2 2p_x^2 2p_y^2 2p_z^1$$

or for convenience we can group the three $2p$ orbitals and write

$$1s^2 2s^2 2p^5$$

   There are complications once we get to element 19, potassium, the first of the elements with more electrons than can be accommodated in the $1s$ through $3p$ orbitals. We cannot predict the electronic configuration of the heavier elements on the basis of simple theory alone. Instead, electronic configurations must be assigned in a way that is consistent with experimental data, including atomic spectra and X-ray spectra, as well as the magnetic and chemical properties of the element.

**TABLE 6.2**   The Electronic Distribution in Allowed Orbitals from 1s to 3d

| Principal Quantum Number $n$ | $\ell$ | $m_\ell$ | $m_s$ | Orbital Designations | Total Electrons |
|---|---|---|---|---|---|
| 1 | 0 | 0 | $+\frac{1}{2}, -\frac{1}{2}$ | $1s$ | 2 |
| 2 | 0 | 0 | $+\frac{1}{2}, -\frac{1}{2}$ | $2s$ | 8 |
|  | 1 | $-1$ | $+\frac{1}{2}, -\frac{1}{2}$ | $2p_x$ |  |
|  |  | $0$ | $+\frac{1}{2}, -\frac{1}{2}$ | $2p_y$ |  |
|  |  | $+1$ | $+\frac{1}{2}, -\frac{1}{2}$ | $2p_z$ |  |
| 3 | 0 | 0 | $+\frac{1}{2}, -\frac{1}{2}$ | $3s$ | 18 |
|  | 1 | $-1$ | $+\frac{1}{2}, -\frac{1}{2}$ | $3p_x$ |  |
|  |  | $0$ | $+\frac{1}{2}, -\frac{1}{2}$ | $3p_y$ |  |
|  |  | $+1$ | $+\frac{1}{2}, -\frac{1}{2}$ | $3p_z$ |  |
|  | 2 | $-2$ | $+\frac{1}{2}, -\frac{1}{2}$ | $3d_{xy}$ |  |
|  |  | $-1$ | $+\frac{1}{2}, -\frac{1}{2}$ | $3d_{xz}$ |  |
|  |  | $0$ | $+\frac{1}{2}, -\frac{1}{2}$ | $3d_{yz}$ |  |
|  |  | $+1$ | $+\frac{1}{2}, -\frac{1}{2}$ | $3d_{z^2}$ |  |
|  |  | $+2$ | $+\frac{1}{2}, -\frac{1}{2}$ | $3d_{x^2-y^2}$ |  |

**TABLE 6.3** The Physicist's Shells

| Shell | Subshells |
|-------|-----------|
| K | $1s$ |
| L | $2s, 2p$ |
| M | $3s, 3p, 3d$ |
| N | $4s, 4p, 4d, 4f$ |
| O | $5s, 5p, 5d, 5f, 5g$ |
| P | $6s, 6p, 6d, 6f, 6g, 6h$ |

## Shells and Subshells

To help work out electronic configurations, it is convenient to group orbitals into *shells* and *subshells.* There are two ways to describe orbital shells. One is favored by physicists. The other is useful to chemists. The physicist uses the letter designation of shells shown in Table 6.3. Each value of the principal quantum number $n$ is associated with a shell designation. For $n = 1$, the shell is designated K. For $n = 2$, the shell is designated L. For $n = 3$, the shell is designated M, and so on. The K shell contains only the $1s$ orbital. Table 6.3 shows that each shell after K contains one more type of orbital.

We used the idea that each shell consists of subshells in Table 6.2. Each subshell includes all the orbitals that have the same value of $n$ and $\ell$. The L shell has two subshells: $2s$, which can hold 2 electrons, and $2p$, which can hold 6 electrons. The M shell has the $3d$ subshell, which can hold up to 10 electrons. The N shell has the $4f$ subshell, which can hold up to 14 electrons, and so on.

The chemist's description of electronic configuration, shown in Table 6.4, is more closely related to the chemical properties of the elements. Each shell in this description is named for a noble gas, and each shell contains the orbitals that are filled in the designated noble gas atom. The helium shell of the chemist corresponds to the K shell of the physicist; each contains the $1s$ orbital. The neon shell of the chemist corresponds to the L shell of the physicist; each contains the $2s$ and $2p$ orbitals that are filled in neon ($1s^2 2s^2 2p^6$). However, the argon shell of the chemist is not the same as the M shell of the physicist. The argon shell includes only the $3s$ and the three $3p$ orbitals, which are filled in the argon atom ($1s^2 2s^2 2p^6 3s^2 3p^6$). The $3d$ orbitals that are in the M shell of the physicist are found in the krypton shell of the chemist.

The chemist's description is related more closely to the arrangement of elements in the periodic table. It is also closer to the order in which orbitals are actually filled with increasing atomic number. One noble gas shell is filled before electrons start filling the next noble gas shell. Also, in general one subshell is filled before electrons begin filling the orbitals in the next subshell.

**TABLE 6.4** Noble Gas Shells

| Name | Subshells | Maximum Number of Electrons in Shell |
|------|-----------|--------------------------------------|
| helium | $1s$ | 2 |
| neon | $2s, 2p$ | 8 |
| argon | $3s, 3p$ | 8 |
| krypton | $3d, 4s, 4p$ | 18 |
| xenon | $4d, 5s, 5p$ | 18 |
| radon | $4f, 5d, 6s, 6p$ | 32 |
| eka-radon | $5f, 6d, 7s, 7p$ | 32 |

## Noble Gas Electronic Cores

As we mentioned, we can use the hydrogen atom as a model to predict the order in which subshells are filled until the $3p$ subshell is completed. Thus, we begin filling the $2p$ subshell at boron, after the $2s$ subshell is completed. Each element from boron, atomic number 5, to neon, atomic number 10, adds another electron to the $2p$ subshell, which is filled at neon, whose configuration is $1s^2 2s^2 2p^6$. The $3s$ orbital begins filling with sodium, atomic number 11, whose electronic configuration is $1s^2 2s^2 2p^6 3s^1$.

There is an easier way to symbolize the electronic configuration of sodium. It is simply the electronic configuration of neon with one additional electron, which is in the next noble gas shell. We say that the electronic configuration of sodium consists of a neon core, represented as [Ne], with one $3s$ electron, so we can write it as $[Ne]3s^1$. Similarly, the electronic configuration of magnesium, atomic number 12, is $[Ne]3s^2$.

When the $3s$ orbital is filled, the $3p$ subshell begins filling at aluminum, atomic number 13, whose electronic configuration is $[Ne]3s^2 3p^1$. The $3p$ subshell continues filling until the noble gas shell is completed at argon, element 18, which has the noble gas configuration $[Ne]3s^2 3p^6$. This electronic configuration is called the argon core and is represented as [Ar].

**Example 6.3**   Write the ground state electronic configurations of sulfur and chlorine, using the appropriate noble gas cores.

**Solution**   We start by finding the atomic number of each element, which gives us the number of electrons that must be accommodated. The atomic number of sulfur is 16. The appropriate noble gas shell is that of neon, which has 10 electrons. For sulfur, six electrons must be placed in orbitals outside the neon shell. Two are in the $3s$ orbital and the remaining four are in $3p$ orbitals. We can write the configuration as:

$$[Ne]3s^2 3p^4$$

The atomic number of chlorine is 17. The seven electrons outside the completed neon shell are accommodated in the same orbitals as in sulfur. The ground state electronic configuration of chlorine is:

$$[Ne]3s^2 3p^5$$

When the argon shell is completed, the krypton shell begins to fill. In potassium, atomic number 19, the ground state electronic configuration is $[Ar]4s^1$. For calcium, atomic number 20, the ground state electronic configuration is $[Ar]4s^2$. In these elements, the $4s$ subshell fills before the $3d$ subshell, something that is not predicted by the hydrogenlike approximation.

**Facing Page**

As shown in the top portion of this figure, a beam of white light that passes through a prism is spread out as a spectrum. The spectrum is continuous because white light is made up of all visible wavelengths, from red to blue. The bottom portion of the figure shows: (1) A spectrum from an incandescent solid is also continuous. (2) The sun's spectrum shows several dark lines due to absorption by the indicated elements. The discontinuous emission spectra of sodium, hydrogen, calcium, mercury, and neon are also shown. *[Figs. 4-5, 4-6 "color spectrum" from* General College Chemistry *6th ed. (1980) by Charles W. Keenan, Donald C. Kleinfelter, and Jesse H. Wood. Copyright © 1980 by Charles W. Keenan, Donald C. Kleinfelter, Jesse H. Wood (Harper & Row, Publishers, Inc.) By permission of the authors.]*

**TABLE 6.5**    Filling Order of Subshells[a]

| | | Sequence | | | | | | | |
|---|---|---|---|---|---|---|---|---|---|
| | | 1 | 2 | 3 | 4 | 5 | 6 | 7 | 8 |
| K | $1s$ | | | | | | | | |
| L | $2s$ | $2p$ | | | | | | | |
| M | $3s$ | $3p$ | $3d$ | | | | | | |
| N | $4s$ | $4p$ | $4d$ | $4f$ | | | | | |
| O | $5s$ | $5p$ | $5d$ | $5f$ | | | | | |
| P | $6s$ | $6p$ | $6d$ | | | | | | |
| Q | $7s$ | $7p$ | | | | | | | |

[a] The numerals at the top indicate the sequence that is followed in filling subshells. Subshells on diagonal 2 are filled before those on diagonal 3, which are filled before those on diagonal 4, and so on.

Table 6.5 relates the physicist's shells to the chemist's shells. The order of filling of subshells is indicated by the diagonal lines. But the rules are not always followed. Experimental data are needed to be sure of the exact electronic configuration of the heavier elements.

**Example 6.4**    Using Table 6.5, find the set of quantum numbers for the electron in an incomplete subshell in (a) Y, (b) Ga.

**Solution**    a. The atomic number of Y is 39. There are thus three electrons outside the complete krypton shell. In krypton, the $4p$ subshell is the last to fill. Table 6.5 shows that the $5s$ subshell fills after the $4p$. Two electrons fill the $5s$ subshell. The remaining electron in Y goes into the $4d$ subshell, which starts filling after the $5s$. The principal quantum number of this electron is $n = 4$. The $\ell$ quantum number for a $d$ orbital is 2. We can choose any of the allowed values of $m_\ell$, such as $-2$. The value of $m_s$ may be either $-\frac{1}{2}$ or $+\frac{1}{2}$. One allowed set of quantum numbers is thus 4, 2, $-2$, $+\frac{1}{2}$. There are others.

b. The atomic number of Ga is 31. There are thus 13 electrons outside the completed argon shell, in which the $3p$ subshell is the last to fill. The next two electrons fill the $4s$ subshell, and the subsequent 10 electrons fill the $3d$ subshell. The last remaining electron goes into the $4p$ subshell, which begins filling after the $3d$. One allowed set of quantum numbers for an electron in a $4p$ orbital is 4, 1, 0, $-\frac{1}{2}$.

## Hund's Rule

A problem arises when we attempt to write the correct electronic configuration of carbon, atomic number 6. The first five electrons of the carbon

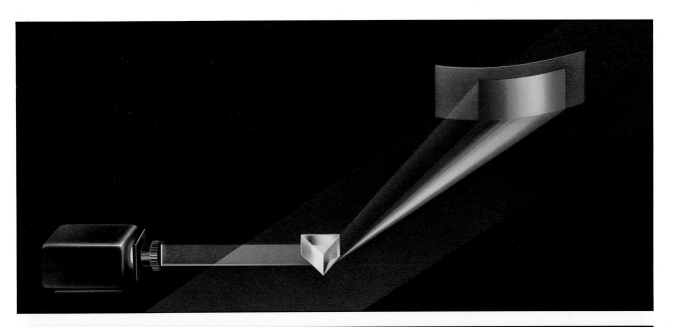

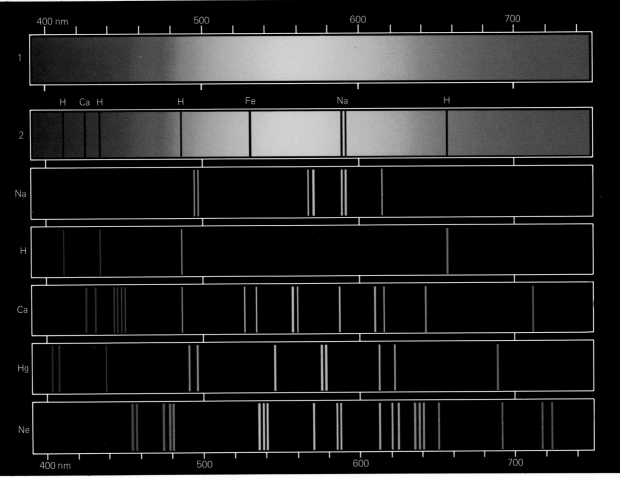

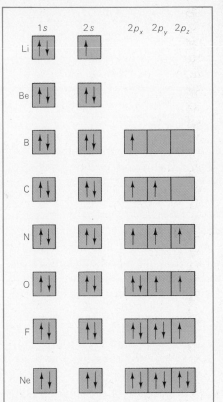

**Figure 6.8**
Ground state electronic configuration of the elements from lithium to neon showing electronic spins.

atom are readily assigned to the $1s$, $2s$, and one of the $2p$ orbitals, say the $2p_x$ orbital. We know that the sixth electron must go into a $2p$ orbital. Should it go into the $2p_x$ orbital, pairing with the electron that is already there? Or should it go into an unoccupied $2p$ orbital, say $2p_y$, giving two half-filled $2p$ orbitals?

From experiment and theory, we find that the latter choice is correct. A guideline called **Hund's rule,** tells us that electrons go into unoccupied orbitals of a subshell whenever possible, rather than pairing with electrons in partially occupied orbitals. In more amplified form, Hund's rule says:

1. When electrons are filling a subshell, the lowest energy state is obtained if the largest number of different orbitals in the subshell are used. When electrons are placed in different orbitals, unfavorable interactions between electrons are minimized. In the ground state of carbon, there are two half-filled $2p$ orbitals, rather than one completely filled $2p$ orbital.
2. When there are two or more electrons in half-filled orbitals, the state of lowest energy is the one in which the spins of all the electrons are parallel.

Relative spins of electrons can be designated by arrows. Arrows pointing in the same direction indicate parallel spins. Arrows pointing in opposite directions indicate paired spins. The ground state electronic configuration of carbon is represented as $1s^22s^22p_x\uparrow 2p_y\uparrow$. This notation helps distinguish the ground state of carbon from a higher electronic energy state, in which the electron spins are paired, such as $1s^22s^22p_x\uparrow 2p_y\downarrow$.

We can write the electronic configuration of the ground state of nitrogen, atomic number 7, to indicate the spins of $2p$ electrons: $1s^22s^22p_x\uparrow 2p_y\uparrow 2p_z\uparrow$. Since Hund's rule is understood to apply, this usually is simplified: $1s^22s^22p^3$. Similarly, the ground state configuration of oxygen, atomic number 8, is $1s^22s^22p_x^2 2p_y\uparrow 2p_z\uparrow$, or $1s^22s^22p^4$.

Figure 6.8 diagrams the ground state electronic configuration of these elements, showing electronic spins.

---

**Example 6.5**   Show the ground state electron configuration, including electron spins, of the electrons in the unfilled subshells of (a) P, (b) Fe.

**Solution**   a. To find the electrons in the unfilled subshell, we start with the ground state electronic configuration of the atom. Phosphorus, element 15, has the configuration $[\text{Ne}]3s^23p^3$. The unfilled subshell is the $3p$, which has three orbitals. Each of the three electrons goes into a different orbital with parallel spins, according to Hund's rule:

$$3p_x\uparrow\ 3p_y\uparrow\ 3p_z\uparrow$$

b. The ground state electron configuration of iron, element 26, is $[\text{Ar}]4s^23d^6$. There are five $3d$ orbitals and six electrons, so one of the orbitals has two electrons and is filled. Each of the other four $3d$ orbitals has one electron. These electrons have parallel spins, according to Hund's rule:

$$3d_{xy} \uparrow \downarrow 3d_{xz} \uparrow 3d_{yz} \uparrow 3d_{z^2} \uparrow 3d_{x^2-y^2} \uparrow$$

## Magnetic Properties and Electronic Spin

Much of what we know about electronic spin comes from the study of magnetic properties. A substance that is slightly repelled by a magnetic field is said to be **diamagnetic.** The vast majority of substances are diamagnetic. All the electrons in a diamagnetic substance are paired. A relatively small number of substances are strongly attracted by a magnetic field. They are classified as **paramagnetic.** A paramagnetic substance has one or more unpaired electrons. An experiment of the kind shown in Figure 6.9 can demonstrate whether a substance is paramagnetic or diamagnetic and can measure the number of unpaired electrons. Most elements are paramagnetic. Of the elements we have discussed so far, He, Be, and Ne are diamagnetic.

## Stable Electronic Configurations

Certain electronic configurations tend to be especially stable. One stable configuration is a half-filled subshell. Each orbital of the subshell holds one electron, and all the electrons in the subshell have parallel spins. Another especially stable electronic configuration is one in which a subshell is completely filled.

Table 6.6 lists the experimentally determined ground state electronic configurations of all the elements. Generally, the electronic configurations are those predicted by simple theory. There are exceptions, some of which reflect the stability associated with half-filled and filled subshells.

For example, the electronic configuration of vanadium, element 23, is as expected, $1s^2 2s^2 2p^6 3s^2 3p^6 4s^2 3d^3$ or [Ar]$4s^2 3d^3$, with the $4s$ subshell filling before the $3d$ subshell. But in chromium, element 24, the configuration is [Ar]$4s^1 3d^5$. The $4s$ and the $3d$ subshells are both half-filled, showing the extra stability associated with a half-filled subshell. The ground state configuration of copper, element 29, is [Ar]$4s^1 3d^{10}$, another illustration of the same effect. Stability is associated with one filled subshell and a half-filled subshell.

In other words, the set of orbitals derived from the model of the hydrogen atom should not be considered a set of pigeonholes that are filled in precise sequence according to simple rules. The order in which orbitals fill is altered by the magnitude of the nuclear charge and by the presence of the other electrons in the atom. Therefore, the assignment of electrons to orbitals in some atoms is far from simple.

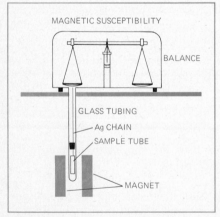

**Figure 6.9**
A device for determining the magnetic properties of substances. The substance is suspended in the sample tube by a slender copper wire attached to a silver chain and is placed in a magnetic field. A substance that is repelled by the magnetic field is diamagnetic; one that is attracted by a magnetic field is paramagnetic.

## CLEANING UP WITH PARAMAGNETISM

We can readily separate a mixture of iron filings and charcoal powder by passing a magnet over the mixture. The iron is attracted to the magnet; the charcoal is not. Until recently, this technique of magnetic separation was possible on a large scale only for mixtures containing three elements: iron, nickel, or cobalt, the three ferromagnetic elements. These elements and some of their compounds are strongly magnetized by a weak magnetic field. Now, a new technique of magnetic separation has been developed that is potentially usable with the many elements and compounds that are paramagnetic. The technique could greatly increase the earth's iron resources, help purify water supplies, and even be used to remove sulfur from coal.

The successful magnetic attraction of paramagnetic substances requires not only a strong magnetic field but also one with a very steep gradient — that is, a rapid increase in the intensity of the field over a very short distance. The magnet must also have a very large surface for collecting the material. The development of strong magnets with steep gradients and large surfaces in recent years has made possible the separation of mixtures containing paramagnetic substances.

One practical application of this method is the separation of impurities out of kaolin, a white clay that is used as a coating for papers. These impurities, the most important of which is titanium dioxide, are extremely small paramagnetic particles. A single large magnetic separator now can treat thousands of kilograms of kaolin in an hour.

The use of the technique to separate iron from low-grade taconite ore is being explored. Most of the iron ore that has been mined until now contains magnetite, a compound of iron that is ferromagnetic. Magnetite is separated from iron ore with conventional magnetic separation techniques. But supplies of magnetite-containing iron ore are running low. Taconite ore contains hematite, a paramagnetic compound of iron. If hematite can be obtained magnetically from the ore by the new technique, enormous savings in energy and a large increase in usable iron reserves will result.

The same sort of process could be used for water purification. Many water pollutants are paramagnetic. Others can be made paramagnetic if the water is "seeded" with appropriate paramagnetic particles. Coliform bacteria, which are leading causes of waterborne diseases, are known to adhere to the surface of paramagnetic iron oxide particles. In experiments, large quantities of water containing coliform bacteria, viruses, and asbestos particles have been purified by magnetic separation. However, widespread use of this technique will require the development of magnets that use much less energy than those now in common use.

The difficulty of assigning electrons to orbitals is also shown in the study of ions, electrically charged species in which an atom has lost or gained one or more electrons. Table 6.6 shows that the $4s$ subshell begins filling before the $3d$ subshell. In the *first transition series,* the elements from scandium (atomic number 21) to copper (atomic number 29), the $3d$ subshell is being filled. There is considerable evidence that the energy of the $3d$ subshell relative to the $4s$ subshell becomes progressively lower with increasing atomic number in this series of elements. This phenomenon, which is caused by the increased charge on the nucleus of each succeeding element, is in part responsible for the observation that a $4s$

**TABLE 6.6**  Electronic Configurations of the Elements[a]

| Atomic No. | Element | 1s | 2s | 2p | 3s | 3p | 3d | 4s | 4p | 4d | 4f | 5s | 5p | 5d | 5f | 6s | 6p | 6d | 6f | 7s | 7p | 7d | 7f |
|---|---|---|---|---|---|---|---|---|---|---|---|---|---|---|---|---|---|---|---|---|---|---|---|
| 1 | H | 1 | | | | | | | | | | | | | | | | | | | | | |
| 2 | He | 2 | | | | | | | | | | | | | | | | | | | | | |
| 3 | Li | 2 | 1 | | | | | | | | | | | | | | | | | | | | |
| 4 | Be | 2 | 2 | | | | | | | | | | | | | | | | | | | | |
| 5 | B | 2 | 2 | 1 | | | | | | | | | | | | | | | | | | | |
| 6 | C | 2 | 2 | 2 | | | | | | | | | | | | | | | | | | | |
| 7 | N | 2 | 2 | 3 | | | | | | | | | | | | | | | | | | | |
| 8 | O | 2 | 2 | 4 | | | | | | | | | | | | | | | | | | | |
| 9 | F | 2 | 2 | 5 | | | | | | | | | | | | | | | | | | | |
| 10 | Ne | 2 | 2 | 6 | | | | | | | | | | | | | | | | | | | |
| 11 | Na | 2 | 2 | 6 | 1 | | | | | | | | | | | | | | | | | | |
| 12 | Mg | 2 | 2 | 6 | 2 | | | | | | | | | | | | | | | | | | |
| 13 | Al | 2 | 2 | 6 | 2 | 1 | | | | | | | | | | | | | | | | | |
| 14 | Si | 2 | 2 | 6 | 2 | 2 | | | | | | | | | | | | | | | | | |
| 15 | P | 2 | 2 | 6 | 2 | 3 | | | | | | | | | | | | | | | | | |
| 16 | S | 2 | 2 | 6 | 2 | 4 | | | | | | | | | | | | | | | | | |
| 17 | Cl | 2 | 2 | 6 | 2 | 5 | | | | | | | | | | | | | | | | | |
| 18 | Ar | 2 | 2 | 6 | 2 | 6 | | | | | | | | | | | | | | | | | |
| 19 | K | 2 | 2 | 6 | 2 | 6 | | 1 | | | | | | | | | | | | | | | |
| 20 | Ca | 2 | 2 | 6 | 2 | 6 | | 2 | | | | | | | | | | | | | | | |
| 21 | Sc | 2 | 2 | 6 | 2 | 6 | 1 | 2 | | | | | | | | | | | | | | | |
| 22 | Ti | 2 | 2 | 6 | 2 | 6 | 2 | 2 | | | | | | | | | | | | | | | |
| 23 | V | 2 | 2 | 6 | 2 | 6 | 3 | 2 | | | | | | | | | | | | | | | |
| 24 | Cr[b] | 2 | 2 | 6 | 2 | 6 | 5 | 1 | | | | | | | | | | | | | | | |
| 25 | Mn | 2 | 2 | 6 | 2 | 6 | 5 | 2 | | | | | | | | | | | | | | | |
| 26 | Fe | 2 | 2 | 6 | 2 | 6 | 6 | 2 | | | | | | | | | | | | | | | |
| 27 | Co | 2 | 2 | 6 | 2 | 6 | 7 | 2 | | | | | | | | | | | | | | | |
| 28 | Ni | 2 | 2 | 6 | 2 | 6 | 8 | 2 | | | | | | | | | | | | | | | |
| 29 | Cu[b] | 2 | 2 | 6 | 2 | 6 | 10 | 1 | | | | | | | | | | | | | | | |
| 30 | Zn | 2 | 2 | 6 | 2 | 6 | 10 | 2 | | | | | | | | | | | | | | | |
| 31 | Ga | 2 | 2 | 6 | 2 | 6 | 10 | 2 | 1 | | | | | | | | | | | | | | |
| 32 | Ge | 2 | 2 | 6 | 2 | 6 | 10 | 2 | 2 | | | | | | | | | | | | | | |
| 33 | As | 2 | 2 | 6 | 2 | 6 | 10 | 2 | 3 | | | | | | | | | | | | | | |
| 34 | Se | 2 | 2 | 6 | 2 | 6 | 10 | 2 | 4 | | | | | | | | | | | | | | |
| 35 | Br | 2 | 2 | 6 | 2 | 6 | 10 | 2 | 5 | | | | | | | | | | | | | | |
| 36 | Kr | 2 | 2 | 6 | 2 | 6 | 10 | 2 | 6 | | | | | | | | | | | | | | |
| 37 | Rb | 2 | 2 | 6 | 2 | 6 | 10 | 2 | 6 | | | 1 | | | | | | | | | | | |
| 38 | Sr | 2 | 2 | 6 | 2 | 6 | 10 | 2 | 6 | | | 2 | | | | | | | | | | | |
| 39 | Y | 2 | 2 | 6 | 2 | 6 | 10 | 2 | 6 | 1 | | 2 | | | | | | | | | | | |
| 40 | Zr | 2 | 2 | 6 | 2 | 6 | 10 | 2 | 6 | 2 | | 2 | | | | | | | | | | | |
| 41 | Nb[b] | 2 | 2 | 6 | 2 | 6 | 10 | 2 | 6 | 4 | | 1 | | | | | | | | | | | |
| 42 | Mo[b] | 2 | 2 | 6 | 2 | 6 | 10 | 2 | 6 | 5 | | 1 | | | | | | | | | | | |
| 43 | Tc[b] | 2 | 2 | 6 | 2 | 6 | 10 | 2 | 6 | 6 | | 1 | | | | | | | | | | | |
| 44 | Ru[b] | 2 | 2 | 6 | 2 | 6 | 10 | 2 | 6 | 7 | | 1 | | | | | | | | | | | |
| 45 | Rh[b] | 2 | 2 | 6 | 2 | 6 | 10 | 2 | 6 | 8 | | 1 | | | | | | | | | | | |
| 46 | Pd[b] | 2 | 2 | 6 | 2 | 6 | 10 | 2 | 6 | 10 | | | | | | | | | | | | | |
| 47 | Ag[b] | 2 | 2 | 6 | 2 | 6 | 10 | 2 | 6 | 10 | | 1 | | | | | | | | | | | |
| 48 | Cd | 2 | 2 | 6 | 2 | 6 | 10 | 2 | 6 | 10 | | 2 | | | | | | | | | | | |
| 49 | In | 2 | 2 | 6 | 2 | 6 | 10 | 2 | 6 | 10 | | 2 | 1 | | | | | | | | | | |
| 50 | Sn | 2 | 2 | 6 | 2 | 6 | 10 | 2 | 6 | 10 | | 2 | 2 | | | | | | | | | | |
| 51 | Sb | 2 | 2 | 6 | 2 | 6 | 10 | 2 | 6 | 10 | | 2 | 3 | | | | | | | | | | |

[a] Electrons outside a completed noble gas shell are shown in color.
[b] Does not follow the order predicted by Table 6.5.

**TABLE 6.6** (Continued)

| Atomic No. | Element | 1s | 2s | 2p | 3s | 3p | 3d | 4s | 4p | 4d | 4f | 5s | 5p | 5d | 5f | 6s | 6p | 6d | 6f | 7s | 7p | 7d | 7f |
|---|---|---|---|---|---|---|---|---|---|---|---|---|---|---|---|---|---|---|---|---|---|---|---|
| 52 | Te | 2 | 2 | 6 | 2 | 6 | 10 | 2 | 6 | 10 |  | 2 | 4 |  |  |  |  |  |  |  |  |  |  |
| 53 | I | 2 | 2 | 6 | 2 | 6 | 10 | 2 | 6 | 10 |  | 2 | 5 |  |  |  |  |  |  |  |  |  |  |
| 54 | Xe | 2 | 2 | 6 | 2 | 6 | 10 | 2 | 6 | 10 |  | 2 | 6 |  |  |  |  |  |  |  |  |  |  |
| 55 | Cs | 2 | 2 | 6 | 2 | 6 | 10 | 2 | 6 | 10 |  | 2 | 6 |  |  | 1 |  |  |  |  |  |  |  |
| 56 | Ba | 2 | 2 | 6 | 2 | 6 | 10 | 2 | 6 | 10 |  | 2 | 6 |  |  | 2 |  |  |  |  |  |  |  |
| 57 | La[b] | 2 | 2 | 6 | 2 | 6 | 10 | 2 | 6 | 10 |  | 2 | 6 | 1 |  | 2 |  |  |  |  |  |  |  |
| 58 | Ce | 2 | 2 | 6 | 2 | 6 | 10 | 2 | 6 | 10 | 2 | 2 | 6 |  |  | 2 |  |  |  |  |  |  |  |
| 59 | Pr | 2 | 2 | 6 | 2 | 6 | 10 | 2 | 6 | 10 | 3 | 2 | 6 |  |  | 2 |  |  |  |  |  |  |  |
| 60 | Nd | 2 | 2 | 6 | 2 | 6 | 10 | 2 | 6 | 10 | 4 | 2 | 6 |  |  | 2 |  |  |  |  |  |  |  |
| 61 | Pm | 2 | 2 | 6 | 2 | 6 | 10 | 2 | 6 | 10 | 5 | 2 | 6 |  |  | 2 |  |  |  |  |  |  |  |
| 62 | Sm | 2 | 2 | 6 | 2 | 6 | 10 | 2 | 6 | 10 | 6 | 2 | 6 |  |  | 2 |  |  |  |  |  |  |  |
| 63 | Eu | 2 | 2 | 6 | 2 | 6 | 10 | 2 | 6 | 10 | 7 | 2 | 6 |  |  | 2 |  |  |  |  |  |  |  |
| 64 | Gd[b] | 2 | 2 | 6 | 2 | 6 | 10 | 2 | 6 | 10 | 7 | 2 | 6 | 1 |  | 2 |  |  |  |  |  |  |  |
| 65 | Tb | 2 | 2 | 6 | 2 | 6 | 10 | 2 | 6 | 10 | 9 | 2 | 6 |  |  | 2 |  |  |  |  |  |  |  |
| 66 | Dy | 2 | 2 | 6 | 2 | 6 | 10 | 2 | 6 | 10 | 10 | 2 | 6 |  |  | 2 |  |  |  |  |  |  |  |
| 67 | Ho | 2 | 2 | 6 | 2 | 6 | 10 | 2 | 6 | 10 | 11 | 2 | 6 |  |  | 2 |  |  |  |  |  |  |  |
| 68 | Er | 2 | 2 | 6 | 2 | 6 | 10 | 2 | 6 | 10 | 12 | 2 | 6 |  |  | 2 |  |  |  |  |  |  |  |
| 69 | Tm | 2 | 2 | 6 | 2 | 6 | 10 | 2 | 6 | 10 | 13 | 2 | 6 |  |  | 2 |  |  |  |  |  |  |  |
| 70 | Yb | 2 | 2 | 6 | 2 | 6 | 10 | 2 | 6 | 10 | 14 | 2 | 6 |  |  | 2 |  |  |  |  |  |  |  |
| 71 | Lu | 2 | 2 | 6 | 2 | 6 | 10 | 2 | 6 | 10 | 14 | 2 | 6 | 1 |  | 2 |  |  |  |  |  |  |  |
| 72 | Hf | 2 | 2 | 6 | 2 | 6 | 10 | 2 | 6 | 10 | 14 | 2 | 6 | 2 |  | 2 |  |  |  |  |  |  |  |
| 73 | Ta | 2 | 2 | 6 | 2 | 6 | 10 | 2 | 6 | 10 | 14 | 2 | 6 | 3 |  | 2 |  |  |  |  |  |  |  |
| 74 | W | 2 | 2 | 6 | 2 | 6 | 10 | 2 | 6 | 10 | 14 | 2 | 6 | 4 |  | 2 |  |  |  |  |  |  |  |
| 75 | Re | 2 | 2 | 6 | 2 | 6 | 10 | 2 | 6 | 10 | 14 | 2 | 6 | 5 |  | 2 |  |  |  |  |  |  |  |
| 76 | Os | 2 | 2 | 6 | 2 | 6 | 10 | 2 | 6 | 10 | 14 | 2 | 6 | 6 |  | 2 |  |  |  |  |  |  |  |
| 77 | Ir | 2 | 2 | 6 | 2 | 6 | 10 | 2 | 6 | 10 | 14 | 2 | 6 | 7 |  | 2 |  |  |  |  |  |  |  |
| 78 | Pt[b] | 2 | 2 | 6 | 2 | 6 | 10 | 2 | 6 | 10 | 14 | 2 | 6 | 9 |  | 1 |  |  |  |  |  |  |  |
| 79 | Au[b] | 2 | 2 | 6 | 2 | 6 | 10 | 2 | 6 | 10 | 14 | 2 | 6 | 10 |  | 1 |  |  |  |  |  |  |  |
| 80 | Hg | 2 | 2 | 6 | 2 | 6 | 10 | 2 | 6 | 10 | 14 | 2 | 6 | 10 |  | 2 |  |  |  |  |  |  |  |
| 81 | Tl | 2 | 2 | 6 | 2 | 6 | 10 | 2 | 6 | 10 | 14 | 2 | 6 | 10 |  | 2 | 1 |  |  |  |  |  |  |
| 82 | Pb | 2 | 2 | 6 | 2 | 6 | 10 | 2 | 6 | 10 | 14 | 2 | 6 | 10 |  | 2 | 2 |  |  |  |  |  |  |
| 83 | Bi | 2 | 2 | 6 | 2 | 6 | 10 | 2 | 6 | 10 | 14 | 2 | 6 | 10 |  | 2 | 3 |  |  |  |  |  |  |
| 84 | Po | 2 | 2 | 6 | 2 | 6 | 10 | 2 | 6 | 10 | 14 | 2 | 6 | 10 |  | 2 | 4 |  |  |  |  |  |  |
| 85 | At | 2 | 2 | 6 | 2 | 6 | 10 | 2 | 6 | 10 | 14 | 2 | 6 | 10 |  | 2 | 5 |  |  |  |  |  |  |
| 86 | Rn | 2 | 2 | 6 | 2 | 6 | 10 | 2 | 6 | 10 | 14 | 2 | 6 | 10 |  | 2 | 6 |  |  |  |  |  |  |
| 87 | Fr | 2 | 2 | 6 | 2 | 6 | 10 | 2 | 6 | 10 | 14 | 2 | 6 | 10 |  | 2 | 6 |  |  | 1 |  |  |  |
| 88 | Ra | 2 | 2 | 6 | 2 | 6 | 10 | 2 | 6 | 10 | 14 | 2 | 6 | 10 |  | 2 | 6 |  |  | 2 |  |  |  |
| 89 | Ac[b] | 2 | 2 | 6 | 2 | 6 | 10 | 2 | 6 | 10 | 14 | 2 | 6 | 10 |  | 2 | 6 | 1 |  | 2 |  |  |  |
| 90 | Th[b] | 2 | 2 | 6 | 2 | 6 | 10 | 2 | 6 | 10 | 14 | 2 | 6 | 10 |  | 2 | 6 | 2 |  | 2 |  |  |  |
| 91 | Pa[b] | 2 | 2 | 6 | 2 | 6 | 10 | 2 | 6 | 10 | 14 | 2 | 6 | 10 | 2 | 2 | 6 | 1 |  | 2 |  |  |  |
| 92 | U[b] | 2 | 2 | 6 | 2 | 6 | 10 | 2 | 6 | 10 | 14 | 2 | 6 | 10 | 3 | 2 | 6 | 1 |  | 2 |  |  |  |
| 93 | Np[b] | 2 | 2 | 6 | 2 | 6 | 10 | 2 | 6 | 10 | 14 | 2 | 6 | 10 | 4 | 2 | 6 | 1 |  | 2 |  |  |  |
| 94 | Pu | 2 | 2 | 6 | 2 | 6 | 10 | 2 | 6 | 10 | 14 | 2 | 6 | 10 | 6 | 2 | 6 |  |  | 2 |  |  |  |
| 95 | Am | 2 | 2 | 6 | 2 | 6 | 10 | 2 | 6 | 10 | 14 | 2 | 6 | 10 | 7 | 2 | 6 |  |  | 2 |  |  |  |
| 96 | Cm[b] | 2 | 2 | 6 | 2 | 6 | 10 | 2 | 6 | 10 | 14 | 2 | 6 | 10 | 7 | 2 | 6 | 1 |  | 2 |  |  |  |
| 97 | Bk | 2 | 2 | 6 | 2 | 6 | 10 | 2 | 6 | 10 | 14 | 2 | 6 | 10 | 9 | 2 | 6 |  |  | 2 |  |  |  |
| 98 | Cf | 2 | 2 | 6 | 2 | 6 | 10 | 2 | 6 | 10 | 14 | 2 | 6 | 10 | 10 | 2 | 6 |  |  | 2 |  |  |  |
| 99 | Es | 2 | 2 | 6 | 2 | 6 | 10 | 2 | 6 | 10 | 14 | 2 | 6 | 10 | 11 | 2 | 6 |  |  | 2 |  |  |  |
| 100 | Fm | 2 | 2 | 6 | 2 | 6 | 10 | 2 | 6 | 10 | 14 | 2 | 6 | 10 | 12 | 2 | 6 |  |  | 2 |  |  |  |
| 101 | Md | 2 | 2 | 6 | 2 | 6 | 10 | 2 | 6 | 10 | 14 | 2 | 6 | 10 | 13 | 2 | 6 |  |  | 2 |  |  |  |
| 102 | No | 2 | 2 | 6 | 2 | 6 | 10 | 2 | 6 | 10 | 14 | 2 | 6 | 10 | 14 | 2 | 6 |  |  | 2 |  |  |  |
| 103 | Lr | 2 | 2 | 6 | 2 | 6 | 10 | 2 | 6 | 10 | 14 | 2 | 6 | 10 | 14 | 2 | 6 | 1 |  | 2 |  |  |  |
| 104 | Rf | 2 | 2 | 6 | 2 | 6 | 10 | 2 | 6 | 10 | 14 | 2 | 6 | 10 | 14 | 2 | 6 | 2 |  | 2 |  |  |  |

electron is removed more easily from these atoms than a $3d$ electron. In these elements, cations have ground state electronic configurations in which the $3d$ subshell fills before the $4s$ subshell. In general, when an ion is formed, electrons with higher principal quantum number $n$ are removed first.

For example, if vanadium, $[Ar]4s^2 3d^3$, loses two electrons, the dipositive cation $V^{2+}$ has the configuration $[Ar]3d^3$. The electrons have been lost from the $4s$ subshell. In the zinc atom, $[Ar]4s^2 3d^{10}$, further along in the series, the $3d$ subshell is much lower in energy than the $4s$ subshell. The cation $Zn^{2+}$, which has a completed $3d$ subshell and the configuration $[Ar]3d^{10}$, forms rather easily, but it is exceedingly difficult to remove more than two electrons from the Zn atom.

## 6.3   ELECTRONS IN MULTIELECTRON ATOMS

Before we try to relate electronic configurations to the properties of atoms, it will be helpful to discuss how electrons in hydrogenlike orbitals are influenced by the nuclear charge and by the presence of other electrons in multielectron atoms.

So far, we have used a graphical representation of an orbital, in which a surface of constant probability is plotted. Arbitrarily, the surface is drawn to enclose a region of space in which the electron is found at least 90% of the time.

If we are interested in the probability distribution of the electron in an orbital, we can use a different kind of representation. The probability of finding an electron at a given distance from the nucleus can be calculated. We can express the results of such a calculation by a plot of what is called the *radial probability function* against the distance from the nucleus. Each point on the plot gives the probability of finding the electron in a very thin spherical shell at a distance $r$ from the nucleus.

Figure 6.10 shows the curves for the radial probability functions of the $1s$, $2s$, $2p$, $3s$, $3p$, and $3d$ orbitals. These curves reveal two important aspects of probability distributions within orbitals.

First, the region of maximum probability moves away from the nucleus as the principal quantum number increases. In other words, the distance of the electron from the nucleus increases as the principal quantum number increases. We have already noted (Section 5.7) that the energy of the electron increases with its increasing distance from the nucleus and with an increase in principal quantum number.

Second, Figure 6.10 shows that there is a significant probability of finding an $s$ electron in a region close to the nucleus, even if the $s$ electron has a high principal quantum number. There is a lesser (but definite) probability of finding a $p$ electron close to the nucleus. It is much less probable that electrons in orbitals with higher values of $\ell$, such as those in $d$ orbitals ($\ell = 2$) or $f$ orbitals ($\ell = 3$), are close to the nucleus. Thus, $s$ orbitals and, to a lesser extent, $p$ orbitals can be designated as *penetrating*

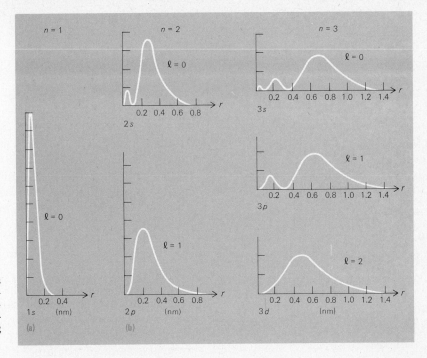

**Figure 6.10**
Radial probability functions for different values of the principal quantum number $n$ of the hydrogen atom. The region of maximum probability moves away from the nucleus as $n$ increases, but there is some probability of finding an $s$ electron close to the nucleus.

*orbitals,* while $d$ and $f$ orbitals can be designated as *nonpenetrating orbitals.* This distinction is important when we account for some properties of multielectron atoms.

Therefore, two important characteristics of orbitals are revealed by the radial probability functions. A higher value of $n$ means a greater average distance of the electron from the nucleus. A higher value of $\ell$ means a lower probability of the electron even being close to the nucleus. These two characteristics can help us understand how nuclear charge and the presence of other electrons affect relative electronic energies in multielectron atoms.

### Nuclear Charge and Electronic Energy

The total energy of the electron in a hydrogen atom is lower (more negative) than the energy of a free electron because of the coulombic attraction between the negatively charged electron and the positively charged nucleus. To study the effect of an increased nuclear charge on electronic energy, we can consider a monatomic cation that has only one electron and a nuclear charge (more simply, an atomic number) designated as $Z$. Examples of such cations are $He^+$ ($Z = 2$), $Li^{2+}$ ($Z = 3$), and $Be^{3+}$ ($Z = 4$). At any given distance from the nucleus, the total energy of the single electron will be more negative than the energy of the electron in a hydrogen atom. An exact relationship for the energy of an electron in such an atomic species is found by a suitable modification of Equation 6.1:

$$E = \frac{-2\pi^2 me^4 Z^2}{n^2 h^2} \qquad (6.2)$$

where $Z$ is the nuclear charge (or the atomic number) of the one-electron cation.

It is logical that the value of $E$ depends on the value of $Z$. A larger nuclear charge $Z$ means a greater coulombic attraction between the nucleus and the electron, and thus a lower value of $E$. The most important feature of Equation 6.2 is simply that as $Z$ increases, the energy becomes more negative, or lower, and as $n$ increases, the energy becomes less negative, or higher. The numerical value of the energy is directly proportional to the square of the nuclear charge and inversely proportional to the square of the principal quantum number:

$$E \propto \frac{Z^2}{n^2} \qquad (6.3)$$

Suppose we want to extend this relationship to a multielectron atom. Generally, the atomic number $Z$ is much higher than the highest principal quantum number $n$ of any electron in the ground state of the atom. In cesium, for example, $Z = 55$ and $n = 6$. It would seem that every electron in the cesium atom would be of lower energy than the electron in the hydrogen atom. But this is not so. To understand why, we must consider the effect of the other electrons on the interaction between any given electron and the nucleus in a multielectron atom.

## The Screening Effect

Figure 6.10 indicates that electrons with higher principal quantum numbers tend to be further from the nucleus than those with lower principal quantum numbers. In a three-electron atom, the $1s$ electrons tend to lie between the nucleus and the $2s$ electron. Therefore, the negatively charged $1s$ electrons are said to screen, or shield, the $2s$ electrons from the coulombic attraction of the nucleus.

Figure 6.11 shows this screening effect for a lithium atom. The nucleus, with a charge of $+3$, is surrounded by a sphere of two electrons, with a total charge of $-2$. This charge of $-2$ essentially reduces the nuclear charge felt outside the sphere of the two electrons from $+3$ to $+1$. The $2s$ electron is outside this sphere and thus feels the attraction of nuclear charge, which we call the *effective* nuclear charge, just slightly larger than $+1$. Therefore, the total energy of the $2s$ electron is not nearly as negative as is expected from the value of $Z^2/n^2 = 3^2/2^2$.

You might expect that the effective nuclear charge felt by the $2s$ electron would be exactly $+1$, the $+3$ nuclear charge reduced by the $-2$ charge of the $1s$ electrons. The radial probability distributions of Figure 6.10 show why the effective nuclear charge is somewhat greater than $+1$. The $2s$ obital is a penetrating orbital. For some part of the time, the

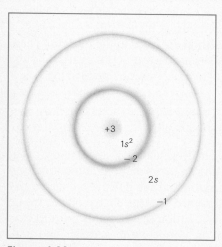

**Figure 6.11**

Electron shielding shown schematically for the lithium atom. The outermost $2s$ electron is shielded from the attractive force of the nucleus by the two $1s$ electrons, making the effective nuclear charge only slightly larger than $+1$.

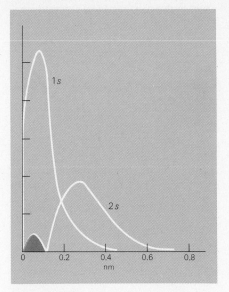

**Figure 6.12**
The effect of penetration on the effective nuclear charge of the lithium atom. Because the 2s is a penetrating orbital, as shown by the radial probability distribution curve, the outermost electron is not fully shielded by the two 1s electrons, making the effective nuclear charge larger than +1.

electron in the 2s orbital will be close to the nucleus, penetrating the shield set up by the 1s electrons (Figure 6.12). The electron in the 2s orbital is not fully screened by the 1s electrons. It feels an effective nuclear charge somewhat larger than +1.

The phenomenon of screening is more noticeable for electrons in p orbitals, which are less penetrating than s orbitals. The penetration of d orbitals and f orbitals is negligible, so the screening of them by electrons in orbitals nearer the nucleus is almost completely effective. It has also been found that electrons in penetrating orbitals are better at screening the nucleus than electrons in nonpenetrating orbitals. For a given value of n, electrons in s orbitals screen best. Electrons in p orbitals are less effective and electrons in d orbitals are still less effective.

Differences in penetration explain why the energy of electrons depends on both the value of ℓ and the value of n in multielectron atoms. We said that the energy of the electron in the hydrogen atom depends only on the value of n. The hydrogen atom has only one electron, so there is no screening. In other atoms, screening and penetration must be considered, so electronic energy depends on both n and ℓ.

*The chemical behavior of the elements depends primarily on the behavior of the outermost, or* **valence electrons.** *The valence electrons generally are those which lie outside the highest completed noble gas shell in the ground state electronic configuration of the atom.* Any description of the behavior of valence electrons must take into account not only the actual nuclear charge but also the effective nuclear charge that the valence electrons feel through the screen of the inner electrons.

## 6.4   IONIZATION ENERGY

One of the most fundamental properties of an atom is its **ionization energy,** sometimes called ionization potential. *Ionization energy is the energy required to remove the electron with the highest total energy from the ground state of an atom in the gas phase.* The ionization energy is a positive quantity, and it provides a direct measure of the energy of this electron. Higher ionization energy means greater coulombic attraction between the electron and the nucleus and lower (more negative) energy of the electron.

For convenience, ionization energies can be expressed in megajoules per mole (MJ/mol), which can be converted to kilojoules per mole (kJ/mol): $1 \text{ MJ} = 10^3 \text{ kJ}$. Another unit that is sometimes used for ionization energy is the electron volt (eV), which is convenient for atomic phenomena: $1 \text{ eV} = 1.60 \times 10^{-19} \text{ J}$.

Table 6.7 shows ionization energies for some elements. Note that data are available on the energy required to remove a second electron from an atom that has lost one electron, a third electron from an atom that has lost two electrons, and so on. These energies are called the second ionization energy, the third ionization energy, and so on. The first ionization energy

**TABLE 6.7**    Ionization Energies (*MJ/mol*)

| Z | Element | I | II | III | IV | V | VI | VII | VIII | IX | X |
|---|---------|------|------|------|------|------|------|------|------|------|------|
| 1 | H | 1.3120 | | | | | | | | | |
| 2 | He | 2.3723 | 5.2504 | | | | | | | | |
| 3 | Li | 0.5203 | 7.2981 | 11.8149 | | | | | | | |
| 4 | Be | 0.8995 | 1.7571 | 14.8487 | 21.0065 | | | | | | |
| 5 | B | 0.8006 | 2.4270 | 3.6598 | 25.0257 | 32.8266 | | | | | |
| 6 | C | 1.0864 | 2.3526 | 4.6205 | 6.2226 | 37.8304 | 47.2769 | | | | |
| 7 | N | 1.4023 | 2.8561 | 4.5781 | 7.4751 | 9.4449 | 53.2664 | 64.3598 | | | |
| 8 | O | 1.3140 | 3.3882 | 5.3004 | 7.4693 | 10.9895 | 13.3264 | 71.3345 | 84.0777 | | |
| 9 | F | 1.6810 | 3.3742 | 6.0504 | 8.4077 | 11.0227 | 15.1640 | 17.8677 | 92.0378 | 106.4340 | |
| 10 | Ne | 2.0807 | 3.9523 | 6.122 | 9.370 | 12.178 | 15.238 | 19.999 | 23.069 | 115.3791 | 131.4314 |
| 11 | Na | 0.4958 | 4.5624 | 6.912 | 9.544 | | | | | | |
| 12 | Mg | 0.7377 | 1.4507 | 7.7328 | 10.540 | | | | | | |
| 13 | Al | 0.5776 | 1.8167 | 2.7448 | 11.578 | | | | | | |
| 14 | Si | 0.7865 | 1.5771 | 3.2316 | 4.3555 | | | | | | |
| 15 | P | 1.0118 | 1.9032 | 2.912 | 4.957 | | | | | | |
| 16 | S | 0.9996 | 2.251 | 3.361 | 4.564 | | | | | | |
| 17 | Cl | 1.2511 | 2.297 | 3.822 | 5.158 | | | | | | |
| 18 | Ar | 1.5205 | 2.6658 | 3.931 | 5.771 | | | | | | |
| 19 | K | 0.4189 | 3.0514 | 4.411 | 5.877 | | | | | | |
| 20 | Ca | 0.5898 | 1.1454 | 4.9120 | 6.474 | | | | | | |
| 21 | Sc | 0.631 | 1.235 | 2.389 | 7.089 | | | | | | |
| 22 | Ti | 0.658 | 1.310 | 2.6525 | 4.1746 | | | | | | |
| 23 | V | 0.650 | 1.414 | 2.8280 | 4.5066 | | | | | | |
| 24 | Cr | 0.6528 | 1.496 | 2.987 | 4.74 | | | | | | |
| 25 | Mn | 0.7174 | 1.5091 | 2.2484 | 4.94 | | | | | | |
| 26 | Fe | 0.7594 | 1.561 | 2.9574 | 5.29 | | | | | | |
| 27 | Co | 0.758 | 1.646 | 3.232 | 4.95 | | | | | | |
| 28 | Ni | 0.7367 | 1.7530 | 3.393 | 5.30 | | | | | | |
| 29 | Cu | 0.7455 | 1.9579 | 3.554 | 5.33 | | | | | | |
| 30 | Zn | 0.9064 | 1.7333 | 3.8327 | 5.73 | | | | | | |
| 31 | Ga | 0.5788 | 1.979 | 2.963 | 6.2 | | | | | | |
| 32 | Ge | 0.7622 | 1.5372 | 3.302 | 4.410 | | | | | | |
| 33 | As | 0.944 | 1.7978 | 2.7355 | 4.837 | | | | | | |
| 34 | Se | 0.9409 | 2.045 | 2.9737 | 4.1435 | | | | | | |
| 35 | Br | 1.1399 | 2.10 | 3.5 | 4.56 | | | | | | |
| 36 | Kr | 1.3507 | 2.3503 | 3.565 | 5.07 | | | | | | |
| 37 | Rb | 0.4030 | 2.633 | 3.9 | 5.08 | | | | | | |
| 38 | Sr | 0.5495 | 1.0643 | 4.21 | 5.5 | | | | | | |
| 39 | Y | 0.616 | 1.181 | 1.980 | 5.96 | | | | | | |
| 40 | Zn | 0.660 | 1.267 | 2.218 | 3.313 | | | | | | |
| 41 | Nb | 0.664 | 1.382 | 2.416 | 3.69 | | | | | | |
| 42 | Mo | 0.6850 | 1.558 | 2.621 | 4.477 | | | | | | |
| 43 | Tc | 0.702 | 1.472 | 2.850 | | | | | | | |
| 44 | Ru | 0.711 | 1.617 | 2.747 | | | | | | | |
| 45 | Rh | 0.720 | 1.744 | 2.997 | | | | | | | |
| 46 | Pd | 0.805 | 1.875 | 3.177 | | | | | | | |
| 47 | Ag | 0.7310 | 2.074 | 3.361 | | | | | | | |
| 48 | Cd | 0.8677 | 1.6314 | 3.616 | | | | | | | |
| 49 | In | 0.5583 | 1.8206 | 2.705 | 5.2 | | | | | | |
| 50 | Sn | 0.7086 | 1.4118 | 2.9431 | 3.9303 | | | | | | |

*Source:* J. E. Huheey, *Inorganic Chemistry,* 3rd Ed., New York: Harper & Row, 1983.

**TABLE 6.7** *(Continued)*

| Z | Element | I | II | III | IV | V | VI | VII | VIII | IX | X |
|---|---------|------|------|------|------|---|----|-----|------|----|---|
| 51 | Sb | 0.8316 | 1.595 | 2.44 | 4.26 | | | | | | |
| 52 | Te | 0.8693 | 1.79 | 2.698 | 3.610 | | | | | | |
| 53 | I | 1.0084 | 1.8459 | 3.2 | | | | | | | |
| 54 | Xe | 1.1704 | 2.046 | 3.10 | | | | | | | |
| 55 | Cs | 0.3757 | 2.23 | | | | | | | | |
| 56 | Ba | 0.5029 | 0.96526 | | | | | | | | |
| 57 | La | 0.5381 | 1.067 | 1.8503 | 4.820 | | | | | | |
| 58 | Ce | 0.528 | 1.047 | 1.949 | 3.543 | | | | | | |
| 59 | Pr | 0.523 | 1.018 | 2.086 | 3.761 | | | | | | |
| 60 | Nd | 0.530 | 1.034 | 2.13 | 3.900 | | | | | | |
| 61 | Pm | 0.536 | 1.052 | 2.15 | 3.97 | | | | | | |
| 62 | Sm | 0.543 | 1.068 | 2.26 | 4.00 | | | | | | |
| 63 | Eu | 0.547 | 1.085 | 2.40 | 4.11 | | | | | | |
| 64 | Gd | 0.592 | 1.17 | 1.99 | 4.24 | | | | | | |
| 65 | Tb | 0.564 | 1.112 | 2.11 | 3.84 | | | | | | |
| 66 | Dy | 0.572 | 1.126 | 2.20 | 4.00 | | | | | | |
| 67 | Ho | 0.581 | 1.139 | 2.20 | 4.10 | | | | | | |
| 68 | Er | 0.589 | 1.151 | 2.19 | 4.11 | | | | | | |
| 69 | Tm | 0.5967 | 1.163 | 2.284 | 4.12 | | | | | | |
| 70 | Yb | 0.6034 | 1.175 | 2.415 | 4.22 | | | | | | |
| 71 | Lu | 0.5235 | 1.34 | 2.022 | 4.36 | | | | | | |
| 72 | Hf | 0.654 | 1.44 | 2.25 | 3.21 | | | | | | |
| 73 | Ta | 0.761 | | | | | | | | | |
| 74 | W | 0.770 | | | | | | | | | |
| 75 | Re | 0.760 | | | | | | | | | |
| 76 | Os | 0.84 | | | | | | | | | |
| 77 | Ir | 0.88 | | | | | | | | | |
| 78 | Pt | 0.87 | 1.7911 | | | | | | | | |
| 79 | Au | 0.8901 | 1.98 | | | | | | | | |
| 80 | Hg | 1.0070 | 1.8097 | 3.30 | | | | | | | |
| 81 | Tl | 0.5893 | 1.9710 | 2.878 | | | | | | | |
| 82 | Pb | 0.7155 | 1.4504 | 2.0815 | 4.083 | | | | | | |
| 83 | Bi | 0.7033 | 1.610 | 2.466 | 4.37 | | | | | | |
| 84 | Po | 0.812 | | | | | | | | | |
| 85 | At | | | | | | | | | | |
| 86 | Rn | 1.0370 | | | | | | | | | |
| 87 | Fr | | | | | | | | | | |
| 88 | Ra | 0.5094 | 0.97906 | | | | | | | | |
| 89 | Ac | 0.49 | 1.17 | | | | | | | | |
| 90 | Th | 0.59 | 1.11 | 1.93 | 2.78 | | | | | | |
| 91 | Pa | 0.57 | | | | | | | | | |
| 92 | U | 0.59 | | | | | | | | | |
| 93 | Np | 0.60 | | | | | | | | | |
| 94 | Pu | 0.585 | | | | | | | | | |
| 95 | Am | 0.578 | | | | | | | | | |
| 96 | Cm | 0.581 | | | | | | | | | |
| 97 | Bk | 0.601 | | | | | | | | | |
| 98 | Cf | 0.608 | | | | | | | | | |
| 99 | Es | 0.619 | | | | | | | | | |
| 100 | Fm | 0.627 | | | | | | | | | |
| 101 | Md | 0.635 | | | | | | | | | |
| 102 | No | 0.642 | | | | | | | | | |

always refers to the electron that is most easily removed from the intact atom. The second ionization energy refers to the electron most easily removed from the monopositive cation. The third ionization energy refers to the electron most easily removed from the dipositive cation.

$$A(g) + energy \longrightarrow A^+(g) + e^- \quad \text{(first ionization energy)}$$
$$A^+(g) + energy \longrightarrow A^{2+}(g) + e^- \quad \text{(second ionization energy)}$$
$$A^{2+}(g) + energy \longrightarrow A^{3+}(g) + e^- \quad \text{(third ionization energy)}$$

The measurement of ionization energy is a valuable tool for determining the electronic structure of atoms. An electron can be removed from an atom by bombardment with a beam of photons or free electrons (Section 5.6, The Photoelectric Effect). The energy of these photons or free electrons can be measured with accuracy. A common technique for measurement of ionization energies is to find the minimum energy of a photon or free electron required to change a neutral atom into a cation. This minimum energy is the ionization energy. Ionization energies are always measured on single atoms in the gas phase.

### The Periodicity of Ionization Energies

Table 6.7 shows that the first ionization energy varies in a predictable way as we move down the list of elements by atomic number. There is a general rise, then a sudden drop followed by another general rise and a sudden drop, again and again. Ionization energy is just one of the many periodic properties of the elements. The way in which chemical properties vary periodically with atomic number is made evident by the arrangement of the elements into the periodic table.

Figure 6.13 shows the periodicity of ionization energies graphically. If first ionization energies are plotted as a function of atomic number, we get a pattern of fairly regular rises and falls. The first ionization energy is lowest for the group IA metals, which begin each horizontal row of the periodic table. The energy generally increases from left to right across a row, reaching a maximum for the noble gases at the far right of the row. As we go from the noble gas at the right of one row to the group IA metal at the left of the next row down, there is a noticeable drop in first ionization energy.

There is also a tendency for first ionization energy to decrease with increasing atomic number in any given column of the periodic table, although there are exceptions to this trend.

### Ionization Energy and Electronic Configuration

We can interpret ionization energies better when we see how they are related to electronic configuration. The ionization energy of hydrogen is 1.31 MJ/mol. Recalling that $E \propto Z^2/n^2$, we would predict that the first

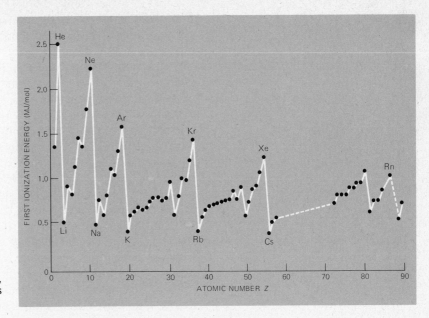

**Figure 6.13**
The first ionization energy of the elements, showing the periodicity of ionization energy as a function of atomic number.

ionization energy of the helium atom ($Z = 2$) will be $2^2 = 4$ times greater than that of the hydrogen atom. But the first ionization energy of helium is 2.31 MJ/mol, only about twice as large as that of hydrogen. It is easier to remove an electron from helium than simple theory predicts because the two electrons of the helium atom repel each other. The second ionization energy of helium is 5.25 MJ/mol, four times larger than the ionization energy of hydrogen, as the theory predicts. When one of the electrons of the helium atom is removed, the repulsion no longer exists and the ionization energy increases considerably.

The first ionization energy of lithium is lower than that of hydrogen because of the screening effect of the $1s$ electrons. The effective nuclear charge felt by the $2s$ electrons is not $+3$ but is closer to $+1$, and the $2s$ electron is removed relatively easily because it is further from the nucleus than is the electron in hydrogen. The second and third ionization energies of lithium are very large. In both cases, a $1s$ electron must be removed from an atom with a nuclear charge of $+3$, so we expect a large ionization energy.

The first ionization energy of beryllium is greater than that of lithium, consistent with the greater nuclear charge of Be. Electron-electron repulsions in the Be atom keep the first ionization energy from being even larger. The second ionization energy of Be is much less than the second ionization energy of Li, because in Be the $2s$ electron is lost. In Li, it is the $1s$ electron that is lost.

For boron, the first ionization energy is lower than that of Be because a $2p$ electron is removed from the B atom, while a $2s$ electron is removed from the Be atom. The $2p$ orbital is less penetrating, and therefore more effectively screened by the $1s$ electrons, than the $2s$ orbital, so the $2p$ electron is removed more readily.

The first ionization energies of the next two elements, carbon ($Z = 6$) and nitrogen ($Z = 7$) increase slightly, but the first ionization energy of oxygen ($Z = 8$) is less than that of nitrogen. In nitrogen, the three $2p$ electrons are in separate orbitals, so electron-electron repulsions are minimized. There are four $2p$ electrons in the oxygen atom, and two of them must share an orbital. The increase in electron-electron repulsion resulting from the presence of two electrons in the same orbital offsets the effect of the larger nuclear charge of the oxygen atom.

We have noted the increased stability associated with half-filled electronic subshells. In nitrogen, the $2p$ subshell is exactly half full, and the first ionization energy of nitrogen is relatively high. The second ionization energy of oxygen is also relatively high. A singly ionized oxygen atom, with seven electrons remaining, has the same half-filled subshell as a nitrogen atom with seven electrons. We can say that N and $O^+$ are *isoelectronic,* which means that the two species have the same electronic configuration. Although the ionization energies of isoelectronic species are qualitatively similar, they are not equal. The difference results from the difference in nuclear charge.

The increase in first ionization energy continues with fluorine and reaches a maximum with neon. The large first ionization energy of neon reflects the stability associated with a completely filled electron shell.

## Ionization Energies and Chemical Properties

The chemical properties of any atom are related to the configuration of its outermost electrons. This configuration is reflected in the atom's ionization energies. In the second row of the periodic table, from Li on the left to F on the right, there is a trend toward higher first ionization energies — that is, toward increasing difficulty in removing an electron. This trend is reflected in the chemical properties of the elements. Cations formed by electron removal play an important role in the chemistry of the elements in columns IA and IIA of the periodic table. But cations of oxygen (column VIA) and fluorine (column VIIA) are almost never found under ordinary conditions because of the high first ionization energies of these elements. The relationship between electronic configuration and the formation of chemical bonds will be examined in detail in Chapter 7.

The trends in ionization energy in the third row of the periodic table parallel those of the second row. The actual ionization energies are lower for each third-row element. The atoms of the third-row elements are larger than those of the second-row elements, so the outermost electrons are further away from the nuclei. Therefore, relatively less energy is required to remove these electrons. The stability of half-filled shells and subshells is demonstrated by two interruptions in the steady increase of ionization energies in the third row: at Al, where the $3p$ subshell starts filling, and at S, where the second half of the $3p$ subshell starts filling.

## The Transition Metals

A substantial block of elements from Sc ($Z = 21$) to Cu ($Z = 29$) are very close in first ionization energy. These elements are metals and have quite similar chemical properties. They are called *transition metals*. The small, steady increase in ionization energies in the group is consistent with increases in nuclear charge, but only if the electron is being removed from the same subshell in each element. By using ionization energy data and information from spectroscopic studies, we can form a picture of the electronic configuration of these elements and of the cations derived from them.

We mentioned in Section 6.2 that the $4s$ orbital begins filling before the $3d$ orbital does. But in the series of elements from Sc through Cu, each first ionization energy refers to the removal of a $4s$ electron. This simply reflects the fact that the electron in the $4s$ orbital is always the one of highest energy even though in K and Ca the *total* energy of the atom is lower when the $4s$ rather than the $3d$ orbital is occupied.

The explanation can be based on the differences in radial probability functions (Figure 6.10) between the $4s$ and $3d$ orbitals and the effect of an increase in $Z$ on the relative energy of electrons in these orbitals. The electron in the $4s$ orbital is at a greater average distance from the nucleus than the electron in the $3d$ orbital. But the $4s$ electron penetrates closer to the nucleus, and thus is not screened from the attractive force of the nuclear charge as much as the $3d$ electron. In elements 19 and 20, screening by the electrons in the filled Ar shell makes the effective nuclear charge relatively small for the outermost electrons. The penetration of the $4s$ orbital thus is important; and $[Ar]4s^1$ is more stable than $[Ar]3d^1$. In element 21 and succeeding elements, the effective nuclear charge outside the screen of the Ar core increases substantially. Penetration of the $4s$ orbital is then not as important a factor in determining electronic energy. The electrons in the $3d$ orbitals, whose average distance is closer to the nucleus, are of lower energy than the $4s$ electrons. Thus, the $4s$ electrons are easier to remove than $3d$ electrons.

There is a noticeable increase in ionization energy at Zn ($Z = 30$), reflecting the stability of completed subshells. The electronic configuration of Zn is $[Ar]4s^23d^{10}$. With Ga ($Z = 31$), a new electronic subshell, the $4p$, begins to fill, and there is the predictable drop in its first ionization energy.

There is another apparent anomaly in ionization energy trends in the transition elements in the sixth row of the periodic table. The ionization energies of these metals are unusually high. For example, the first electron to be removed from both Ba ($Z = 56$) and W ($Z = 74$) is a $6s$ electron. Yet the ionization energy of W is almost 60% higher than that of Ba.

The high ionization energies of W and other transition elements result from their relatively high effective nuclear charges. As we go down a column of the periodic table, we find a difference of 18 in the atomic number, and therefore the nuclear charge, between the fourth and fifth rows, but a difference of 32 between the fifth and sixth rows, beginning with element 72, hafnium. This larger difference in atomic number is caused by the presence of the lanthanoid series of elements, atomic numbers 57 to 71, in which the 4f electronic subshell is being filled. The 14 electrons in the 4f subshell are not very effective at screening the outermost electrons from the nucleus. Thus, the effective nuclear charge beginning at element 72 is relatively high, so the ionization energy is relatively high.

## 6.5  ELECTRON AFFINITY

**Electron affinity** is most simply defined as the energy required to remove an electron that has been added to the atom in the gas phase. Electron affinity is thus the energy associated with the process:

$$A^-(g) \longrightarrow A(g) + e^-$$

A positive value of the energy means a positive electron affinity and vice versa. Electron affinity is important because it can be used to predict and explain the chemical behavior of elements.

Electron affinities are positive for many atoms, meaning that $A^-(g)$ is more stable than $A(g) + e^-$; the acceptance of an electron by the neutral atom is a favorable process. Unlike ionization energies, however, electron affinities can also have negative values. A negative value means that the anion is less stable than the neutral atom and the free electron. In such a case, the repulsion between the electrons of the neutral atom and the extra electron is greater than the attraction between the nucleus and the extra electron.

There are periodic trends for electron affinity, just as there are for ionization. Table 6.8, which gives some electron affinities, shows this periodicity.

---

**TABLE 6.8**  Electron Affinities (kJ/mol)

| | | | | | | | | |
|---|---|---|---|---|---|---|---|---|
| H  72.9 | | | | | | | | He <0 |
| Li 59.8 | Be <0 | B 23 | C  122 | N 0 | O 141 | F  322 | | Ne <0 |
| Na 52.9 | Mg <0 | Al 44 | Si 120 | P 74 | S 200 | Cl 349 | | Ar <0 |
| K  48.4 | | | Ge 116 | As 77 | Se 195 | Br 325 | | Kr <0 |
| Rb 46.9 | | | Sn 121 | Sb 101 | Te 190 | I 295 | | Xe <0 |
| Cs 45.5 | | | Pb 100 | Bi 100 | | | | |

---

*Source:* James E. Huheey, *Inorganic Chemistry,* 2nd Ed., New York: Harper & Row, 1978.

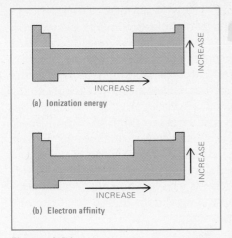

**Figure 6.14**

Trends in (a) ionization energy and (b) electron affinity with position in the periodic table.

In a general way, electron affinity values increase from left to right across any row of the periodic table, as does ionization energy (Figure 6.14). The maximum value of electron affinity is not found in column VIIIA, the furthest column on the right, but in column VIIA. This characteristic is to be expected. All the elements in column VIIIA have the stability that accompanies completely filled noble gas electron shells. Each element in column VIIA is just one electron short of having such a completely filled shell. Adding an electron to a column VIIA element makes that element isoelectronic with the corresponding column VIIIA element. For example, the anion $F^-$ has the electronic configuration $[He]2s^2 2p^6$, identical with that of Ne.

As with ionization energies, the values of electron affinity usually decrease in going down a group of the periodic table. One important exception to this general trend is the low electron affinity of some second-row elements. This effect is most evident in N, O, and F, which all have substantially lower electron affinities than the corresponding third-row elements. The small size of the second-row elements causes serious electron-electron repulsions in the anion that forms when the atom accepts an electron. These repulsions lead to the lower electron affinities. The effect of electron-electron repulsion on the value of electron affinity is especially important in dianions. For example, the electron affinity of $O^-$ has a negative value. The process

$$O^{2-}(g) \longrightarrow O^-(g) + e^-$$

is favorable, which means that the $O^{2-}$ anion is unstable relative to the $O^-$ anion and an electron and cannot exist by itself in the gas phase. The same is true for the $S^{2-}$ and $N^{3-}$ anions. Such ions can exist only if they are stabilized by the surroundings, which can occur in solution or in a crystalline lattice.

## 6.6   THE PERIODIC TABLE OF THE ELEMENTS

So far, we have discussed the properties of atoms entirely from the point of view of twentieth-century chemistry. Discoveries made in this century make it possible to state that the chemical properties of elements are determined primarily by the behavior of valence electrons, the electrons in the outermost shell of an atom. As orbitals are filled according to the Aufbau principle, the electronic configuration of valence shells tends to repeat. The chemical and physical properties of elements with the same valence electronic configurations tend to be similar. For example, each of the elements

$$
\begin{array}{ll}
\text{Li:} & [He]2s^1 \\
\text{Na:} & [Ne]3s^1 \\
\text{K:} & [Ar]4s^1 \\
\text{Rb:} & [Kr]5s^1 \\
\text{Cs:} & [Xe]6s^1
\end{array}
$$

has one *s* electron in the valence shell. Therefore, all these elements should have similar properties. The periodicity of ionization energy and electron affinity is also seen in a number of other important chemical properties. When the elements are listed in order of increasing atomic number, elements with similar properties recur at predictable intervals. The recurrence of elements with similar properties reflects the recurrence of similar valence electronic configurations.

There is a tendency for today's students to think that the idea of periodicity derives from knowledge about electrons and orbitals. In fact, the descriptions of orbitals and electrons are relatively recent concepts that provide a theoretical framework for all of the observations and correlations made by generations of chemists who had no knowledge of electronic configurations. Indeed, these observations helped to create our current picture of electronic configuration.

## Early Periodic Tables

Early in the nineteenth century, chemists began to recognize similarities in the chemical behavior of the elements, and to suggest systematic groupings of certain elements based on their properties. The earliest arrangement was made by J. W. Döbereiner (1780–1849). In 1829, Döbereiner set forth a scheme of triads, groups of three elements of similar properties. The atomic weight of the middle element in each triad was the average of the atomic weights of the first and third elements. Among the triads proposed by Döbereiner were

| I | II | III |
|---|---|---|
| Lithium | Calcium | Chlorine |
| Sodium | Strontium | Bromine |
| Potassium | Barium | Iodine |

Most attempts to find numerical relationships based on nineteenth-century atomic weights were not successful, partly because there were errors and uncertainties in the values of atomic weights and partly because many elements were then undiscovered.

It was a Russian chemist, Dmitri I. Mendeleev (1834–1907), who in 1869 developed the essential principles of the periodic table we use today. Although the German chemist Lothar Meyer (1830–1895) made the same discovery at almost the same time, Mendeleev is recognized today as the discoverer of the periodic table of the elements.

In essence, Mendeleev found that when the elements are arranged in order of increasing atomic weight, there is a repetition of properties at approximately regular intervals. This periodicity is most apparent when the elements are listed in a table. Mendeleev did more than just list elements by column and row. His genius enabled him to recognize underlying principles that make the periodic table a powerful tool of modern chemistry.

Dmitri Ivanovich Mendeleev (1834–1907) was born in Siberia, went to college in France and Germany, and returned to Russia to become a professor of chemistry in St. Petersburg. In addition to developing the periodic table of the elements, Mendeleev wrote an outstanding chemistry textbook and did much to improve chemistry education in Russia. In 1906, he failed by one vote to receive the Nobel Prize in chemistry. *(Culver)*

## Mendeleev's Periodic Table

Table 6.9 shows a portion of Mendeleev's periodic table of 1871, with the atomic weights he assigned to the elements. To start with, Mendeleev separated hydrogen from the rest of the elements. He also left blank spaces where he believed that elements remained to be discovered. Mendeleev could not list the elements in order of increasing atomic number, as today's periodic table does, because atomic numbers were not known at that time. He did recognize that a listing of elements by atomic weight was not satisfactory. Mendeleev assumed that the *observed chemical properties* of the elements were more important than the atomic weights assigned to the elements—a good assumption, considering the suspect nature of many atomic weights of the time.

For example, the atomic weight of the element indium was put at 76.6, which would place indium between As (75) and Se (78). But on the basis of their chemical properties, As had to be in group V and Se had to be in group VI, leaving no room between them for indium. Mendeleev suggested that the reported atomic weight of indium was incorrect, because it was calculated on the basis of a valence (combining power) of 2. If the valence of In was in fact 3, its atomic weight would be 115, putting In in a space between cadmium and tin. Mendeleev showed that compounds of indium resembled compounds of aluminum and thallium, which are both in group III of the periodic table. This observation was further evidence that In belonged in the seventh row of group III. He made similarly accurate placements for other elements, including uranium (92) and beryllium (4), which had been misplaced in other efforts to develop periodic listings of the elements.

In other cases, Mendeleev stated that the accepted values of atomic weights for some elements were incorrect. He placed gold, whose atomic weight was believed to be 196.2, after osmium, iridium, and platinum,

**TABLE 6.9**   A Portion of Mendeleev's Periodic Table

| Row | Group I | Group II | Group III | Group IV | Group V | Group VI | Group VII | Group VIII |
|---|---|---|---|---|---|---|---|---|
| 1 | H = 1 | | | | | | | |
| 2 | Li = 7 | Be = 9.4 | B = 11 | C = 12 | N = 14 | O = 16 | F = 19 | |
| 3 | Na = 23 | Mg = 24 | Al = 27.3 | Si = 28 | P = 31 | S = 32 | Cl = 35.5 | |
| 4 | K = 39 | Ca = 40 | — = 44 | Ti = 48 | V = 51 | Cr = 52 | Mn = 55 | Fe = 59, Co = 59, Ni = 59 |
| 5 | Cu = 63 | Zn = 65 | — = 68 | — = 72 | As = 75 | Se = 78 | Br = 80 | |
| 6 | Rb = 85 | Sr = 87 | Yt = 88 | Zr = 90 | Nb = 94 | Mo = 96 | — = 100 | Ru = 104, Rh = 104, Pd = 106 |
| 7 | Ag = 108 | Cd = 112 | In = 115 | Sn = 118 | Sb = 122 | Te = 125 | I = 127 | |
| 8 | Cs = 133 | Ba = 137 | Di = 138? | Ce = 140? | — | — | — | ——— |
| 9 | — | — | — | — | — | — | — | |
| 10 | — | — | Er = 178? | La = 180? | Ta = 182 | W = 184 | — | Os = 195, Ir = 197, Pt = 198 |
| 11 | Au = 199 | Hg = 200 | Tl = 204 | Pb = 207 | Bi = 208 | | | |

whose atomic weights were believed to be 198.6, 196.7, and 196.7 respectively, contending that these values were wrong. Mendeleev's prediction, based on the chemical properties of those elements, was right. The modern values for the atomic weights of these elements are Os, 190.2; Ir, 192.2; Pt, 195.1; Au, 197.0. Mendeleev was sometimes wrong in assuming that the atomic weights of elements were incorrect, but most of his judgments based on chemical properties were correct.

## Mendeleev's Predictions

Mendeleev's correction of atomic weights on the basis of his observations of chemical properties was impressive. His predictions of the existence, and even the properties, of undiscovered elements were spectacular. In at least three instances, Mendeleev said that a new element would be found and described its properties so accurately that he could even suggest where and how it could be found.

In Table 6.9, space is left for an element in group III intermediate in properties between Al and In. Mendeleev called this unknown element *eka aluminum* (*eka* means "first" in Sanskrit), and he formulated an elaborate description of it. The element eventually was discovered. It was named gallium, because the discovery was made in France and Gallic patriotism demanded a name that honored French science. Every prediction made by Mendeleev was found to be correct as is seen in Table 6.10.

Mendeleev made equally accurate predictions about the properties of the unknown elements he named *eka-boron,* with a predicted atomic weight of 44, and *eka-silicon,* atomic weight 72. Eka-boron was discovered in 1879 in Scandinavia, and was named scandium. Eka-silicon was discovered in 1886 in Germany, and was named germanium.

## The Modern Periodic Table

As more elements were discovered and better experimental data became available, the periodic table was improved and expanded. The modern periodic table systematizes and rationalizes so much chemical knowledge that a familiarity with its important features is essential. Figure 6.15 shows

**TABLE 6.10**  Properties of Gallium

|  | **Predicted Properties** | **Observed Properties** |
|---|---|---|
| atomic weight | 68 | 69.72 |
| density (relative to water) | 5.9 | 5.91 |
| melting point | low | 29.78°C |
| boiling point | high | 1983°C |
| product with oxygen | $X_2O_3$ | $Ga_2O_3$ |
| hydroxide | $X(OH)_3$ | $Ga(OH)_3$ |
| product with chlorine | $XCl_3$ | $GaCl_3$ |

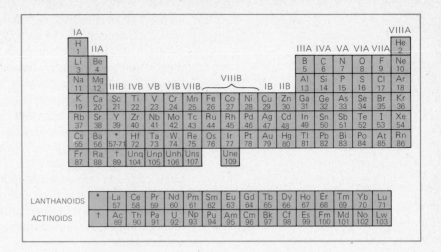

**Figure 6.15**
The periodic table of the elements.

a frequently used version of the periodic table, designed to emphasize the grouping of similar elements. Some important features of this table are as follows:

1. The arrangement of elements in order of increasing atomic number from left to right and from top to bottom of the table is interrupted to remove two horizontal rows, or periods, of elements (57 through 71 and 89 through 103) from the main body of the table. The elements from 57 through 71 are called the *lanthanoids,* or *rare-earth metals.* The elements from 89 through 103 are called the *actinoids.* One reason for listing these elements separately is practical: If this were not done, the table would be too wide to fit on most pages. The separate listing also emphasizes the great similarity in the properties of elements in each of the two groups. This similarity in chemical behavior can be explained by similarities in electronic configuration.

   The lanthanoids have a filled $6s$ subshell, a filled $5p$ subshell, and $5d$ subshells that either are empty or hold only one electron. It is the $4f$ subshell that is filling in this series. Since the $4f$ subshell, with its relatively low principal quantum number, is essentially buried in the large electron clouds of these elements, the $4f$ electrons have little effect on chemical behavior.

2. Each column of the periodic table is designated by a number, usually with a letter attached. Eight columns begin with elements in the first or second row, or period, of the table. The number of the column gives the number of valence electrons of each element in the column, as can be seen by a comparison of Table 6.6 and Figure 6.15. The letter A is added to the designation of eight of these columns to indicate that each column starts from the top of the table.

   The elements in each vertical column can be thought of as a family, or group. The elements in a column have similar chemical properties, with regular variations in those properties as the atomic number increases. Elements in the same family are called *congeners*. The elements of family IA are called the *alkali metals*. Those of family IIA are the *alkaline-earth metals*. Those of VA are the *pnictogens;* those of VIA the *chalcogens*. Those of family VIIA are the *halogens;* those of family VIIIA are the *noble gases*.

3. Eight more families, or groups, begin with the fourth row of the periodic table. These families are designated by numerals, followed by the letter B to distinguish them from the A groups. The elements in column IB and columns IIIB through VIIIB are called the *transition metals*. Each of these elements either has an incomplete *d* subshell or can readily give rise to cations that have an incomplete *d* subshell. The elements of column IIB have completed *d* subshells and do not readily form cations with an incomplete *d* subshell. They are called *post-transition metals*.

Column VIIIB is unusual because it is three elements wide. All the members of group VIIIB have similar properties, so they are grouped together.

4. All the elements in a given family resemble one another. For example, group IB, the coinage metals, includes copper, silver, and gold, which have similar properties. There is also some resemblance between elements in columns with the same numeral but different letter designations. For example, strontium, in group IIA, and cadmium, in group IIB, have a number of similarities, including the tendency to form dipositive cations. Such similarities do not always exist. Cesium, in group IA, and gold, in group IB, are soft metals that form monopositive cations. The resemblance stops there. Cesium is one of the most reactive metals, reacting explosively with water and nonmetals. Gold is one of the least reactive metals. It can be dissolved only by aqua regia, a potent mixture of concentrated nitric acid and hydrochloric acid.

As we observed in Chapter 2, elements to the right and near the top of the periodic table tend to be nonmetals, while most elements, those to the left and near the bottom of the table, are metals. A diagonal band of semimetals starts with boron and runs to tellurium.

Figure 6.16 shows another way of dividing the elements into broad blocks — on the basis of the filling of electronic subshells. The elements in the *s* block to the left or the *p* block to the right are called the *representative* or *main group elements*. The *d* block contains transition elements, while the *f* block contains inner transition elements. Elements in a given block often have some chemical characteristics in common.

The main emphasis in the periodic table is on vertical relationships — the similarities in members of the same group. The tendency of the properties of elements in a group to vary in a regular way is called a *group trend*.

There are regular changes in properties going from left to right across the table. These changes are called *horizontal trends*. There are also diagonal trends among some elements. Two elements often show similarities in

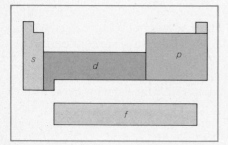

**Figure 6.16**

The periodic table divided into blocks of elements on the basis of the filling of electron subshells. Elements in the *s*, *d*, and *f* blocks often have chemical characteristics in common.

## TOWARD THE "ISLAND OF STABILITY"

The heavier an atomic nucleus is, the more likely it is to be unstable. No element of greater atomic number than uranium, element 92, is known to occur naturally on earth at this time. However, nuclear theory indicates that there should be an "island of stability" —or at least relative stability—in which there are some superheavy elements of atomic numbers around 110, 116, 126, and beyond, up to element 184.

Hopes for the production of superheavy elements were increased when a West German research team announced in September 1982, that element 109 had been created at a particle accelerator. The announcement caused other research centers to use the technique that had produced element 109 to attempt to create even heavier elements.

One of the most unusual aspects of the West German achievement was that instruments had been able to detect a single atom of element 109, the entire yield of the experiment. That atom was produced by firing a beam of iron-58 nuclei, which have 26 protons and 32 neutrons, at a foil made of bismuth-209 (83 protons and 126 neutrons). The intent was to fuse two nuclei to form a compound nucleus with 109 protons and 158 neutrons. The iron nuclei had to be given an energy of exactly 5.15 million electron volts. Below that energy, electrostatic repulsion would prevent the nuclei from fusing. Above that energy, the nuclei would shatter.

As the beam of iron nuclei was directed at the bismuth foil, the products of the collisions were aimed at a device, 12 m long, that used electric and magnetic fields to separate nuclei of different masses and charges. On the very last day of the experiment, instruments indicated that a single nucleus with the predicted properties of element 109 had entered the detector. The nucleus had decayed in exactly the way predicted for element 109. Nevertheless, it took a month of analysis before the researchers were certain enough to announce their accomplishment. Soon afterward, an effort began in the United States to produce element 114, the first of the postulated stable superheavy elements, by beaming calcium-48 nuclei at a target of curium-248. Several detection methods were being used in the effort. One relied on chemical analysis to detect any atoms of element 114, whose properties are predicted to resemble those of lead.

There have been some false starts in the search for the superheavy elements. In 1976, a group of physicists working at Florida State University reported evidence for the existence of elements 116, 124, 126, and other superheavies in samples of the mineral mica found in Madagascar. However, a number of other researchers who attempted to reproduce their findings got negative results. The current belief is that if the superheavy elements are found, it will be in the laboratory, not in nature.

behavior when the heavier element is one column to the right and one row below the lighter element. Lithium and magnesium have such a diagonal relationship, as do beryllium and aluminum.

## 6.7  PERIODIC TRENDS IN ATOMIC SIZES

One of the first periodic variations in the properties of the elements to be recognized was that of atomic size. As early as 1870, Lothar Meyer published a graph plotting a quantity called "atomic volume" against atomic weight. Atomic volume is only an approximate indicator of the relative size of atoms; nevertheless, Meyer's graph was a striking indication of the periodic variation in atomic size.

The calculation or measurement of actual atomic size is not easy, because atoms do not have sharply defined boundaries. To determine atomic size, we must decide arbitrarily where the atom ends. This judgment is based primarily on the way the atom interacts with what is around it. Therefore, there are several ways to express the size of atoms. These methods may produce different values, but they all show periodic trends that are consistent with what we know about quantum numbers, electronic shells, and nuclear charge.

For example, we have seen (Section 6.4) that the effective nuclear charge is less important than the principal quantum number $n$ when we compare elements in a family. Ionization energies tend to decrease from the top to the bottom of a column of the periodic table. The greater importance of the principal quantum number $n$ is also responsible for the trend in atomic sizes: The size of atoms in a family tends to *increase* as the atomic numbers increase. One measure of atomic size is the atomic radius. In groups IIA, the atomic radii in nanometers (nm) are: Be, 0.105; Mg, 0.15; Ca, 0.18; Sr, 0.20; and Ba, 0.22. The regular increase reflects the steady increase in the value of $n$ for the valence electrons from Be, whose electronic configuration is $[He]2s^2$, to Ba, whose configuration is $[Xe]6s^2$.

In general, as we go across a row of the periodic table, atomic size *decreases* as atomic number increases. The valence electrons of the elements in one row of the table have the same principal quantum number. The increase in effective nuclear charge in successive elements tends to pull the electron cloud in a little tighter. This phenomenon is reminiscent of the increase in ionization energy from left to right in a row of the periodic table. The tendency for atomic size to decrease is seen in the data for atomic radii, in nanometers (nm), of the second-row elements: Li, 0.15; Be, 0.11; B, 0.085; C, 0.070; N, 0.065; O, 0.060; F, 0.050. Figure 6.17 shows the periodic variation of atomic size.

The effect of an increase in nuclear charge on size is even more evident in ions. The radius of an ion is of special interest because it is often related to the chemical properties of ionic compounds. We can measure the radii of a series of ions that are isoelectronic — that is, all the ions have the same electronic configuration but different nuclear charges.

The ions $O^{2-}$, $F^-$, $Na^+$, and $Mg^{2+}$ are isoelectronic. They have the electronic configuration $1s^2 2s^2 2p^6$, the filled neon shell. The ionic radii of these species are: $O^{2-}$ ($Z = 8$), 0.140 nm; $F^-$ ($Z = 9$), 0.136 nm; $Na^+$ ($Z = 11$), 0.095 nm; and $Mg^{2+}$ ($Z = 12$), 0.065 nm. The steady decrease in ionic radius reflects a steady increase in nuclear charge that is not offset by electronic screening.

The data for the ions of a group show another interesting trend. For example, the radii of the halogens in nanometers (nm) are: $F^-$, 0.136; $Cl^-$, 0.181; $Br^-$, 0.195; $I^-$, 0.216. The increase in radius is greater from $F^-$ to $Cl^-$ than from $Cl^-$ to $Br^-$ or from $Br^-$ to $I^-$. The same is true in a number of other families: The increase in radius is much greater between elements in

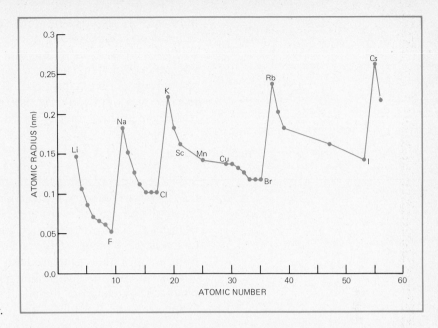

**Figure 6.17**

A plot of atomic radius against atomic number.

the second and third row than between elements in the third and fourth row. This trend reflects the fact that nuclear charge increases by 8 between elements in the second and third rows, but by 18 between elements in the third and fourth rows. The extra nuclear charge tends to contract the larger ions.

## 6.8   A GROUP TREND EXEMPLIFIED: THE ALKALI METALS

The physical and chemical properties of the group IA elements, the alkali metals, exemplify the vertical group trends of the periodic table more than those of any other family of elements.

The alkali metals are lithium, sodium, potassium, rubidium, and cesium. They are all soft metals, of great chemical reactivity, with low boiling points. They are silvery white, are good conductors of electricity, and tarnish rapidly when exposed to air.

However, we cannot predict all the physical and chemical properties of a family of elements, even the alkali metals, from the behavior of isolated atoms in the gas phase. In looking at isolated atoms, we primarily see the effect of atomic size and principal quantum number on the properties of the elements. We must also look at solids or liquids in which atoms interact with each other, or at substances in which different atoms are combined. These interactions have important effects on physical and chemical properties.

The great chemical reactivity of the alkali metals reflects their relatively low first ionization energies. Predictably, their chemistry involves primarily the monopositive cations $Li^+$, $Na^+$, $K^+$, $Rb^+$, and $Cs^+$, formed by removal of one electron from the neutral atom. We mentioned that ionization energy decreases going down the column from Li to Cs. Therefore, reactivity increases as we go from Li to Cs. All the alkali metals react violently with water. The reaction of lithium and water is vigorous. The reaction of sodium and water is noisy and is accompanied by fire. The reaction of cesium and water is explosive. The general equation for the reaction of an alkali metal (M) with water is:

$$2M(s) + 2H_2O \longrightarrow 2M^+(aq) + H_2(g) + 2OH^-(aq)$$

Such a reaction is accompanied by the release of a great deal of heat. With the exception of lithium, the reaction of an alkali metal with water releases enough heat to ignite the liberated $H_2(g)$, which reacts with the $O_2(g)$ in air.

Because alkali metals are so highly reactive, they are never found uncombined in nature. To obtain a pure alkali metal, we must liberate it from a salt. Lithium and sodium usually are prepared by passing an electric current through a molten salt, most commonly LiCl or NaCl. The monopositive cation of the alkali metal gains an electron to form the metal:

$$Na^+ + e^- \longrightarrow Na$$

while the anion of the salt (in this case chloride) gives up an electron:

$$Cl^- \longrightarrow \tfrac{1}{2}Cl_2 + e^-$$

Chemical transformations brought about by an electric current are called **electrolysis.** We shall discuss electrolysis in detail in Chapter 16.

The heavier alkali metals can also be prepared by electrolysis. However, the process usually is impractical on an industrial scale, because the greater reactivity of the heavier alkali metals makes it more difficult to collect them. Methods in which the metal can be removed as a vapor are used. For example, cesium is prepared by a displacement reaction at temperatures of 1000 K to 1200 K:

$$Ca(s) + 2CsCl(l) \longrightarrow CaCl_2(s) + 2Cs(g)$$

At these temperatures, cesium is a gas that is easily separated from the other components of the reaction, which are solids or liquids. A high conversion is achieved by use of excess Ca and removal of Cs as it forms.

### Properties of the Alkali Metals

We have seen that there is a fairly regular trend in such properties as ionization energy, electron affinity, atomic radius, and ionic radius in the family of alkali metals as atomic number increases. Table 6.11 lists other physical properties of the pure alkali metals. The data in the table show

**TABLE 6.11**   Properties of the Alkali Metals

| Element | Valence Shell Electronic Configuration | M.P. (K) | B.P. (K) | Atomic Volume (cm³/mol) |
|---------|--------------|----------|----------|----------------|
| Li | $2s^1$ | 453.7 | 1638 | 13.1 |
| Na | $3s^1$ | 371.0 | 1156 | 23.7 |
| K  | $4s^1$ | 336.4 | 1030 | 45.3 |
| Rb | $5s^1$ | 312.2 | 961 | 55.9 |
| Cs | $6s^1$ | 301.7 | 954 | 70.0 |

the effect of increases in atomic size on physical properties. Often, the attractive forces between neighboring atoms are greatest when the atoms are smallest. Since both the melting point and the boiling point of an alkali metal are determined by the attractive forces between atoms, melting points and boiling points decrease as the size of the atoms increases.

For example, the melting point of Rb is lower than that of K. The Rb atom is larger than the K atom, so the forces holding the Rb atoms together in the solid are not as great as the forces holding the K atoms together. Less thermal energy is needed to separate the Rb atoms, so they separate — melt — at a lower temperature.

We mentioned earlier that the change in a property is greater between second-row and third-row elements in a family than between elements in other rows. Table 6.11 shows that the alkali metals demonstrate this trend. The decrease in melting point from Li (second row) to Na (third row) is more than 80 K. The total decrease from Na (third row) to Cs (sixth row) is about 70 K. This difference reflects the greater difference in relative size between second-row and third-row elements of a family than between elements in other rows.

## Uses of the Alkali Metals

Although the high reactivity of the alkali metals makes them useless as structural metals, they are important in technology. Their uses depend on the properties of each alkali metal and the cost of producing it.

Sodium is the least expensive alkali metal. In fact, its low cost of $0.40/kg and low density make sodium the cheapest of all metals by volume. Sodium chloride is abundant, and producing sodium is much less costly than producing the heavier alkali metals. Thus, whenever an alkali metal is needed in industry, sodium is the first choice. Most uses of metallic sodium are based on its low first ionization energy, which makes sodium valuable as a supplier of electrons in chemical processes.

Sodium is used to produce titanium metal by the high-temperature reaction:

$$4Na(l) + TiCl_4(g) \longrightarrow Ti(s) + 4NaCl(s)$$

Sodium metal is also used to manufacture some sodium compounds that cannot be obtained readily by other methods. One process burns sodium in air

$$2Na(s) + O_2(g) \longrightarrow Na_2O_2(s)$$

to form the bleaching agent sodium peroxide.

The relatively high cost of the other alkali metals limits their use. Lithium is a starting materal in many chemical processes. Compounds of lithium are used in the manufacture of high-strength glass and glass-ceramics for cookware. Lithium is also being used in newer high-power-density batteries for digital watches and hand-held calculators. Some alloys of lithium with aluminum or magnesium are lightweight and maintain strength at high temperatures, which makes them valuable in aerospace applications. One such alloy, LA 141, contains 14% Li, 1% Al, and 85% Mg.

Potassium is used primarily in processes where sodium is unsatisfactory. Sodium-potassium alloys that contain 40% to 90% potassium by weight have an application with important implications for future energy supply. While both sodium and potassium are solids at room temperature, these NaK (called "nack" in the trade) alloys are liquid at room temperature. Alkali metals are among the best available conductors of heat, and NaK alloys are used as heat exchange liquids in fast-breeder nuclear reactors.

One combustion product of potassium has a major practical application. Unlike sodium, which burns to give the peroxide, potassium burns to give the superoxide:

$$K(s) + O_2(g) \longrightarrow KO_2(s)$$

The most useful property of potassium superoxide is summed up by the equation:

$$4KO_2(s) + 2H_2O(l) + 4CO_2(g) \longrightarrow 4KHCO_3(s) + 3O_2(g)$$

This process consumes water and carbon dioxide, the products of human respiration, and produces oxygen, which makes this compound invaluable for breathing masks. Potassium superoxide is a store of an emergency supply of oxygen, which is released as the person wearing the mask needs it.

The uses of rubidium and cesium are based primarily on the ease with which they are ionized. Both are used in small amounts in radio tubes and photocells. Because of their low ionization energies, they have a low threshold frequency for the photoelectric effect. Therefore, they emit electrons even when exposed to low-energy visible light. Both cesium and rubidium hold promise for use in magnetohydrodynamic generators and thermionic generators. Magnetohydrodynamic generators produce electricity by passing ions through a magnetic field, and thermionic generators produce an electric current by "boiling" electrons off suitable substances. Cesium is also being investigated as a possible fuel for ion propulsion engines, which could propel spacecraft in outer space. The current high prices for Cs and Rb restrict their use.

Generally, the compounds of alkali metals, not the pure elements, are of greatest importance. Compounds of both sodium and potassium are essential human nutrients. Potassium compounds are essential constituents of fertilizers. The industrial uses of sodium and potassium compounds are legion. Lithium compounds also have a wide range of uses. One of the most striking, developed recently, is the medical use of lithium salts to treat manic depression. The uses of rubidium and cesium salts are more limited.

## 6.9   PERIODIC TRENDS IN THE OXIDES, HYDRIDES, AND HALIDES OF ELEMENTS

The properties of the elements affect the properties of their compounds. We find a number of correlations between the properties of compounds and the positions of their constituent elements in the periodic table. Such correlations are especially clear in the oxides, binary compounds of oxygen; and in the hydrides, binary compounds of hydrogen. Correlations are also seen in the halides, binary compounds of the halogens.

### Oxides

With the exception of some of the lighter noble gases, every element forms at least one binary compound with oxygen. Some elements form a number of different oxides. Vanadium, a transition metal, forms $VO$, $V_2O_3$, $VO_2$, and $V_2O_5$. Chlorine, a nonmetal, forms $Cl_2O$, $Cl_2O_3$, $ClO_2$, $Cl_2O_6$, and $Cl_2O_7$. The nature of the oxide or oxides of an element is determined primarily by the position of that element in the periodic table.

A number of periodic trends can be discerned in the behavior of oxides. Oxides of elements on the left side of the periodic table, such as those of the alkali metals and the heavier members of the alkaline-earth metals of group IIA, generally are ionic solids. Oxides of the metallic elements toward the middle of the table and of the semimetals also are solids, but often are not ionic. Oxides of nonmetals are discrete, separate molecules, generally existing as liquids or gases at room temperature.

The oxides of the metals on the left side of the periodic table generally are easily soluble in water and give alkaline solutions because of the formation of $OH^-(aq)$ ions. For example, lithium oxide dissolves readily in water:

$$Li_2O(s) + H_2O \longrightarrow 2Li^+(aq) + 2OH^-(aq)$$

## MAGNETOHYDRODYNAMICS: ION POWER

In today's conventional generators, electric current is produced by movement of a conductor (a wire loop called an armature) through a magnetic field. An alternative method proposed for future generating plants would use a hot ionized gas, "seeded" with a metal such as cesium, as the conductor. Such a generator would have no moving parts and would generate 50% more electricity for a given amount of fuel than today's generating plants do, proponents say.

However, this magnetohydrodynamic or MHD generator, as it is called, is difficult to bring into creation because it places unusual demands on parts and materials. To generate electricity the MHD way, a gas would be heated to about 2700 K, compared to the 750 K operating temperature of conventional plants. Cesium or another metal that is easily ionized would be added to increase the conductivity of the gas, which would be completely ionized at these high temperatures. The seeded gas would be expanded through a nozzle and would be sent at the speed of sound down a channel lined with electrodes. The combination of great heat, erosion caused by the speed of the gas, and corrosion caused by the metal particles means that the channel would have to be built of ceramics that are far more durable than the best available today.

The total operating experience with MHD generators amounts to only a few hundred hours, but the rapid increase in oil prices that took place in the 1970s has stimulated efforts to put MHD to large-scale use. The Soviet Union has had an MHD generator tied into a working electric power grid for several years. This generator, located near Moscow, has run for only brief periods because of problems with the durability of its channel. In the 1970s, the United States and the Soviet Union began a cooperative program of MHD research. Although foreign policy considerations put the future of the cooperative effort in doubt, the future of MHD in a fuel-conscious world seems bright.

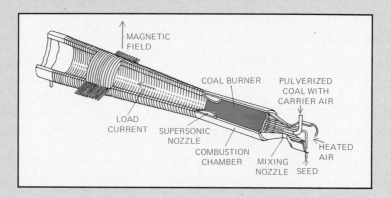

A diagram of a magnetohydrodynamic generator.

The oxides of the nonmetals on the right of the periodic table generally dissolve in water to form acidic solutions, as in the case of dichlorine heptoxide, which dissolves in water

$$Cl_2O_7(l) + H_2O \longrightarrow 2H^+(aq) + 2ClO_4^-(aq)$$

to form a solution of perchloric acid. Sulfur trioxide dissolves in water

$$SO_3(g) + H_2O \longrightarrow H^+(aq) + HSO_4^-(aq)$$

to form a sulfuric acid solution.

The behavior of the oxides of the main group elements in columns IIA to VA of the periodic table is less easily summarized, but there is a trend of increasing acidity and decreasing basicity from left to right across the table. Some oxides formed by elements close to the left side of the periodic table are insoluble in water but dissolve in acidic solutions. They are classed as basic oxides. As an example, magnesium oxide (which in its hydrated form is the familiar milk of magnesia) dissolves only slightly in water but completely in acid:

$$MgO(s) + 2H^+(aq) \longrightarrow Mg^{2+}(aq) + H_2O$$

Some elements further to the right but still in the middle of the periodic table form oxides that do not dissolve in water to any appreciable extent but do dissolve in both acidic and basic solutions. Such oxides are said to be *amphoteric,* meaning that they behave both as acids and bases. Aluminum oxide is amphoteric:

$$Al_2O_3(s) + 6H^+(aq) \longrightarrow 2Al^{3+}(aq) + 3H_2O$$
$$\text{and} \quad Al_2O_3(s) + 2OH^-(aq) + 3H_2O \longrightarrow 2Al(OH)_4^-(aq)$$

It dissolves in acid to produce the equivalent of the $Al^{3+}$ ion in solution and it dissolves in alkaline solution to produce the complex ion $Al(OH)_4^-$ in solution.

Further to the right in the periodic table are elements whose oxides do not dissolve in water but are considered to be acidic because they dissolve in alkaline solution. Silica, $SiO_2$, dissolves slowly in a strong alkaline solution:

$$SiO_2(s) + 2OH^-(aq) \longrightarrow SiO_3^{2-}(aq) + H_2O$$

Since glass is made primarily of $SiO_2$, strongly alkaline solutions are stored in polyethylene containers rather than in glass bottles.

The trend of increasing acidity and decreasing basicity of oxides as we move from left to right across the periodic table is clear. There is a secondary trend of decreasing acidity and increasing basicity in moving down a given column of the periodic table. In group IIA, the oxide of magnesium, MgO, dissolves in acid solution but not to any appreciable extent in water. The oxide of barium, BaO, three rows lower in the same column, dissolves readily in water to produce an alkaline solution:

$$BaO(s) + H_2O \longrightarrow Ba^{2+}(aq) + 2OH^-(aq)$$

On the other side of the periodic table, such an increase in basicity can be seen in the oxides formed when the elements of group VA combine with the maximum number of oxygen atoms. The oxide of nitrogen,

$N_2O_5$, dissolves in water to produce a strongly acidic solution:

$$N_2O_5(s) + H_2O \longrightarrow 2H^+(aq) + 2NO_3^-(aq)$$

The oxide of phosphorus, $P_4O_{10}$, dissolves in water to form phosphoric acid, an acidic solution of moderate strength:

$$P_4O_{10}(s) + 6H_2O \longrightarrow 4H_3PO_4(aq)$$

The corresponding oxide of arsenic, $As_2O_5$, behaves much the same as $P_4O_{10}$. But the oxides of antimony and bismuth, $Sb_2O_5$ and $Bi_2O_5$, are rather unstable and do not dissolve readily in water. They are neither markedly acidic nor markedly basic, although there is some evidence suggesting that the oxide of bismuth is basic.

By classifying an oxide as acidic, basic, or amphoteric on the basis of its position in the periodic table, we can not only predict its behavior with water, but we can also get a better understanding of other aspects of its chemical behavior. For example, acidic oxides and basic oxides react to form complex oxides that are actually salts:

$$Na_2O(s) + SO_3(g) \longrightarrow Na_2SO_4(s)$$
and
$$CaO(s) + SiO_2(s) \longrightarrow CaSiO_3(s)$$

The periodic correlation of oxides is complicated by the tendency of some elements, notably some transition metals and nonmetals, to form a number of different oxides. For nonmetals, the correlation of the properties of an oxide is best when the nonmetal is combined with the maximum number of oxygen atoms. The properties of oxides such as $CO_2$, $N_2O_5$, $SO_3$, and $Cl_2O_7$ are closest to the periodic trends. Furthermore, there is a trend in a series of oxides of the same element: As the number of oxygen atoms combined with the element increases, the acidity of the oxide increases. Thus, while both oxides of sulfur, a nonmetal, are acidic, $SO_3$ is much more acidic than $SO_2$.

This type of relationship is especially important among the transition elements, which tend to form many different oxides. The oxides of vanadium are a good example:

| Oxide | Oxide Atoms per V atom | Properties |
|---|---|---|
| VO | 1 | basic; dissolves only in strongly acidic solutions |
| $V_2O_3$ | $1\frac{1}{2}$ | basic, dissolves only in strongly acidic solutions |
| $VO_2$ | 2 | amphoteric; dissolves in both acidic and basic solutions but not in water |
| $V_2O_5$ | $2\frac{1}{2}$ | acidic; dissolves slightly in water to produce an acidic solution; but is much more soluble in acid solution than in water, so it resembles an amphoteric oxide |

## Hydrides

Hydrides are binary compounds of hydrogen. The correspondence between the position of an element in the periodic table and the properties of its hydrides is not as strong as it is for oxides.

The hydrides of the alkali metals and the alkaline-earth metals are saltlike compounds of metal cations and hydride ($H^-$) anions. They react with water to liberate hydrogen gas and form hydroxide ion, producing alkaline solutions. Sodium hydride reacts with water:

$$NaH(s) + H_2O \longrightarrow H_2(g) + OH^-(aq) + Na^+(aq)$$

The reaction of lithium hydride with water

$$LiH(s) + H_2O \longrightarrow H_2(g) + OH^-(aq) + Li^+(aq)$$

was put to life-saving use in World War II, when it provided the hydrogen gas for inflating balloons that were used as antennas for emergency radios.

The hydrides of the main group elements toward the middle of the periodic table have complicated molecular structures, are often difficult to prepare, and are thermally unstable; that is, they decompose on being heated. The hydrides of the transition metals usually are hard compounds with high melting points. They tend to have indefinite compositions, with varying numbers of hydrogen atoms fitted into spaces between the metal atoms in the solid. Hydrides of the semimetals usually are gases or liquids with low boiling points, usually not very stable thermally. Hydrides of the nonmetals also tend to be gases or liquids with low boiling points. They begin displaying some acid properties at group VA and are clearly acids at groups VIA and VIIA.

In addition to the general trend of increasing acidity of hydrides as we move from left to right across the periodic table, there are rough periodic trends in the stability of nonionic hydrides of the main group elements. If we exclude the hydrides of the alkaline metals and the heavy alkaline-earth metals, which are ionic, we find a tendency for increasing stability as we move from left to right across the periodic table. There is an offsetting tendency toward decreasing stability as we move down a given column of the table. The hydrides of the heavier elements in groups IIIA through VIA are not stable and often are virtually impossible to prepare.

The hydrides that are most important to chemists include many of the hydrides of the nonmetals and semimetals. Aqueous solutions of the hydrogen halides, HF, HCl, HBr, and HI, are important acids. Other familiar hydrides are methane, $CH_4$, the major component of natural gas; ammonia, $NH_3$, which has many practical uses; hydrogen sulfide, $H_2S$, whose rotten-egg smell is unmistakable; diborane, $B_2H_6$, an important rocket fuel; and $H_2O$, the most important hydride on earth.

## Halides

The halides are binary compounds of an element and a halogen, one of the members of group VIIA. The common halides are the fluorides, the chlorides, the bromides, and the iodides. With the exception of the light noble gases, all the elements form compounds with halogens.

Going from left to right across the periodic table there is a general trend in the properties of halogen compounds. Elements on the left side of the table, such as the alkali metals and the heavy alkaline-earth metals, tend to form stable ionic halide salts with high melting points. Elements on the right side of the table tend to form unstable nonionic halides that are gases or low-boiling liquids.

Halide salts do not react with water, but they usually are soluble in water. However, the nonionic halides of the transition elements and of many main group elements in the middle of the periodic table usually react with water, undergoing vigorous hydrolysis reactions:

$$TiCl_4(l) + 2H_2O(l) \longrightarrow TiO_2(s) + 4HCl(g)$$
$$AlCl_3(s) + 3H_2O(l) \longrightarrow Al(OH)_3(s) + 3HCl(g)$$
$$PCl_5(s) + 4H_2O(l) \longrightarrow H_3PO_4(l) + 5HCl(g)$$

On the other hand, $CCl_4$ is a common solvent that does not undergo hydrolysis, and some other halides, such as $NF_3$ and $SF_6$, undergo hydrolysis very slowly. Often, it is difficult to discern clear periodic trends in the hydrolysis or general reactivity of the halides because of the interplay of a number of factors.

Of the halides formed by elements in the second row of the periodic table, ionic lithium chloride does not undergo a hydrolysis reaction; both $BeCl_2$ and $BCl_3$ hydrolyze very readily; $CCl_4$ is almost completely unreactive toward water; $NCl_3$ and $ClO_2$ hydrolyze as readily as $BCl_3$; and $ClF$ does not hydrolyze readily. On the basis of the position of these elements in the periodic table, the only pattern is one of unpredictability.

The same irregularity is sometimes seen in the behavior of halides formed by elements in a given column of the periodic table. Of the halides formed by elements in group IVA, $CCl_4$, as we have seen, is unreactive toward water; $SiCl_4$ is very reactive; $GeCl_4$ undergoes only a mild hydrolysis reaction; and $SnCl_4$ is not hydrolyzed by pure water but is hydrolyzed by alkaline solution.

The complex pattern of behavior of the halides illustrates the danger of looking for simple relationships between the properties of compounds and the positions of the constituent elements in the periodic table. Nevertheless, the periodic table is by far the best single tool we have for correlating patterns of chemical behavior.

**Summary**

In this chapter, we showed how and why the **electronic structure of atoms** determines their chemical behavior. We started by describing the **quantum numbers** that define the **orbitals** of electrons in atoms: The **principal quantum number,** the **angular momentum quantum number,** and the **magnetic quantum number.** We noted that a set of three quantum numbers completely defines the shape, size, and orientation of an orbital, and we listed the letters used to designate orbitals. We also mentioned electron spin and the spin quantum number. We described the **Aufbau principle,** which says that in working out the electronic configuration of an atom, we put each electron into the lowest-energy orbital that is available. We mentioned the **Pauli exclusion principle,** which states that no two electrons in the same atom can have the same four quantum numbers. We noted that electrons in atoms are arranged in **shells** and **subshells,** and described **Hund's rule,** which says that electrons go into unoccupied orbitals of a subshell whenever possible. We discussed how electrons in an atom are influenced by the presence of other electrons, and noted the vital point that chemical behavior depends primarily on the outermost or **valence electrons** of atoms. We described **ionization energy,** the energy needed to remove an electron from an atom, and noted that it varies in a predictable way as we move along the list of elements. We then described the **periodic table** of the elements, one of the most powerful tools of modern chemistry, and told how it was developed by Dmitri Mendeleev. We noted how the periodic table shows **trends** in various properties of the elements. We concluded by describing such trends in the group of elements called the **alkali metals** and in the oxides, hydrides, and halides of elements.

## Exercises

**6.1** Find the number that correctly completes each sentence:
(a) When $n = 3$, $\ell$ may have _____ different values.
(b) When $\ell = 3$, $m_\ell$ may have _____ different values.
(c) When $n = 6$, $m_s$ may have _____ different values.

**6.2** Find the number that correctly completes each sentence:
(a) The number of orbitals with $n = 2$, $\ell = 1$ is _____.
(b) The number of orbitals with $n = 4$ is _____. _____

[1] The answers to exercises whose numbers are in color can be found in Appendix VII. The star indicates an exercise that is more challenging than average.

(c) The number of orbitals with $n = 5$, $\ell = 5$ is _____.
(d) The number of orbitals with $n = 3$, $\ell = 1$, $m_\ell = 0$ is _____.

**6.3**[1] Find the number of electrons of an atom that can have the quantum numbers $n = 4$, $\ell = 2$.

**6.4** Indicate which of these orbital designations is incorrect: (a) $4s_x$, (b) $3p_y$, (c) $2d_{xy}$, (d) $8s$, (e) $5g$.

**6.5** Name all of the orbitals that can correspond to the following sets of quantum numbers: (a) 1, 0, 0; (b) 2, 1, 1; (c) 3, 2, 2; (d) 4, 1, 0.

**6.6** Arrange the following orbitals of a multielectron atom in order of in-

creasing energy. Group orbitals of the same energy together. $4d_{z^2}$, $3d_{yz}$, $3d_{x^2-y^2}$, $4p_y$, $3p_x$, $3p_z$, $2p_y$, $3s$, $2s$, $1s$.

**6.7** Sketch a $2p_x$, a $2p_z$, and a $3p_y$ orbital, using the same scale for each.

**6.8** We wish to define a set of shells analogous to the noble gas shells that are completed when an $s$ subshell fills rather than when a $p$ subshell fills. Using Table 6.4 as a model, list the subshells in each completed shell.

**6.9** Without consulting a periodic table, write the ground state electronic configuration of the elements in the third row of the periodic table.

**6.10** Without consulting a periodic

table, write the atomic numbers of each element in column IA, the alkali metals.

**6.11** Write the ground state electronic configurations of the elements with the following atomic numbers: (a) 20, (b) 34, (c) 13, (d) 53.

**6.12** Write the ground state electronic configurations of the elements with the following atomic numbers: (a) 28, (b) 29, (c) 30, (d) 76, (e) 77.

**6.13** Name the element that corresponds to each description: (a) the first halogen with $d$ electrons, (b) the first element with a $4f$ electron, (c) the element with a half-filled $4p$ subshell, (d) the heaviest alkali metal without $d$ electrons.

**6.14★** The last noble gas listed in Table 6.4 is eka-radon. Find the subshells and the number of electrons in the next larger noble gas shell. Find the atomic number of this noble gas.

**6.15** Write the electronic configurations of the elements with the following atomic numbers, using arrows to show spins of electrons and directional subscripts to identify orbitals in incomplete subshells: (a) 15, (b) 34, (c) 25, (d) 40.

**6.16** Find the number of unpaired electrons in each element from atomic numbers 39 to 48.

**6.17** Identify the element with the most unpaired electrons in each of the four first rows of the periodic table.

**6.18** Predict which of the A columns of the periodic table contain elements that are paramagnetic as single atoms.

**6.19** Name the element or elements corresponding to the following electronic configurations: (a) three unpaired $5p$ electrons, (b) two unpaired $4p$ electrons, (c) five unpaired $4d$ electrons, (d) one $6s$ electron.

**6.20** Palladium does not have the

electronic configuration predicted by Table 6.5. What is its most likely configuration?

**6.21** Write the quantum numbers of each electron in the N atom.

**6.22** Write the electronic configurations of the following ions: (a) $Br^-$, (b) $Br^+$, (c) $Br^{5+}$, (d) $N^{3-}$, (e) $N^{5+}$.

**6.23** Write the electronic configurations of the cations that form from Cl by the loss of one to seven electrons. Indicate which ions are paramagnetic.

**6.24** Name an element that fits each description:
(a) Its monopositive cation has the same electronic configuration as the trinegative anion of N.
(b) Its dianion has the same electronic configuration as the tripositive cation of Sc.
(c) Its monoanion and its monocation have two unpaired electrons each.

**6.25** Predict the number of unpaired electrons in each of the following ions: (a) $Cr^+$, (b) $Cr^{3+}$, (c) $Cr^{6+}$, (d) $Pd^{4+}$.

**6.26** The heaviest element listed in column IIA of the periodic table is radium, element 88. Predict the atomic number of the next element in the column.

**6.27** Without consulting a periodic table, predict the electronic configuration of the elements just below elements (a) 6, (b) 16, (c) 29 in the periodic table.

**6.28** Predict the chemical properties of element 119.

**6.29** Without consulting a periodic table, predict (a) which element between 50 and 53 is most similar in properties to element 34, (b) which element between 13 and 17 is most similar to element 51.

**6.30** Without consulting a periodic table, predict which elements between 28 and 33 are (a) most similar, (b) least

similar to element 42. Justify your predictions.

**6.31** It has been suggested that the lanthanoid series should begin with La, element 57, and end with Yb, element 70. In such a case, Lu, element 71, would replace La in column IIIB. Using the data in Table 6.6, decide whether this change is consistent with electronic configurations. Suggest what should be done with elements 89 to 103 if the proposal is adopted.

**6.32** Find the relationships between the number of valence electrons of an atom and the group number of the atom in the periodic table.

**6.33★** We cannot be certain of the electronic configuration of element 106, which has not been obtained in sufficient quantity for study. What are the two most likely configurations?

**6.34★** Imagine a universe in which the electron can have three spin values, corresponding to the values of $+\frac{1}{2}, -\frac{1}{2}$, and 0 for $m_s$. The Pauli exclusion principle and all other aspects of the Aufbau method are valid in this universe, and the elements have the same names and atomic numbers; only their electronic configurations have changed. Name the elements that meet these descriptions: (a) the second noble gas, (b) the first element to have a $d$ electron, (c) the first element in which the $3p$ subshell is filled, (d) the element below carbon in the new periodic table.

**6.35** Find the ratio of the energy of the electron in the ground state of hydrogen to the energy of (a) the electron in the ground state of $He^+$, (b) the electron in the ground state of $Li^{2+}$, (c) the electron in the first excited state of $Be^{3+}$.

**6.36** Assuming complete screening of the outermost electrons by the electrons in completed subshells, find the ratio of the energy of the electron in the ground state of hydrogen to the energy of (a) the $3s$ electron in Na, (b) the $3p$ electron in Al. Given that the screening is actually incomplete, predict which of these two ratios is closer to the value found by experiment. Justify the prediction.

**6.37** Assuming complete screening of the outermost electrons by electrons in completed subshells, find the ratio of the energy of the electron in the ground state of hydrogen to the energy of (a) the $2p$ electron in B, (b) the $2p$ electron in F. Which ratio do you think is closer to the experimental value? Why?

**6.38** Account for the observation that the first ionization energies of Mn and Zn are relatively high compared to those of the preceding elements, Cr and Cu.

**6.39** Predict the value of the eleventh ionization energy of sodium.

**6.40**⋆ The smooth decrease in ionization energies of the group IIA metals is interrupted because radium has a higher ionization energy than barium. Explain this observation.

**6.41** The electronic configurations of elements 89 to 93 do not follow the order predicted by Table 6.5. It is not until element 94 that we find the predicted number of electrons in the $5f$ shell and none in the $6d$ shell, which should fill after the $5f$. Reasoning by analogy with the electronic configurations of K and Ca, explain this observation.

**6.42** Account for the observation that the electron affinity of F is less than the ionization energy of Ne.

**6.43** Account for the observation that the electron affinity of lithium is greater than that of sodium, while the electron affinity of fluorine is less than that of chlorine.

**6.44** Arrange the following isoelectronic species in order of increasing (a) ionization energy, (b) electron affinity, (c) ionic radius: $N^{3-}$, $O^{2-}$, $F^-$, Ne, $Na^+$, $Mg^{2+}$.

**6.45** Arrange the elements S, Cl, and Br in order of increasing (a) ionization energy, (b) electron affinity, (c) atomic radius.

**6.46** Arrange the elements Li, Na, and Mg in order of increasing (a) ionization energy, (b) electron affinity, (c) atomic radius, (d) reactivity with water.

**6.47** Without consulting the periodic table, write the electronic configurations of (a) the fourth alkali metal, (b) the third halogen, (c) the second element in group IIB, (d) the third element in group IB.

**6.48** Without consulting the periodic table, write the electronic configuration of the representative elements in the fifth row of the table.

**6.49** Predict as much as you can about the physical and chemical properties of element 117, which has not yet been discovered.

**6.50** Table 6.11 does not include the element francium (Fr), the heaviest alkali metal, because all of its isotopes are unstable. No one has ever prepared enough francium to weigh, but some of its properties have been measured by special techniques. Predict the values of the entries for Fr in Table 6.11. Predict the nature and products of the reaction of Fr with (a) water, (b) chlorine, (c) oxygen.

**6.51** Arrange these oxides in order of increasing acidity in aqueous solution: $N_2O_3$, $P_4O_6$, $As_4O_6$, $Sb_4O_6$, $Bi_2O_3$.

**6.52** Arrange these oxides in order of increasing acidity in aqueous solution: $Cl_2O_7$, $Cl_2O_6$, $ClO_2$, $Cl_2O_3$, $Cl_2O$.

**6.53** Arrange these oxides in order of decreasing basicity: lithium oxide, sodium oxide, magnesium oxide, silicon dioxide, thallium(I) oxide, thallium(III) oxide, tetraarsenic decoxide, selenium trioxide, dibromine heptoxide, diiodine heptoxide.

**6.54** Write equations for the reactions of each oxide with water: (a) potassium oxide, (b) barium oxide, (c) carbon dioxide, (d) sulfur trioxide.

**6.55** Write equations for the reaction of the hydride of each of the following elements with an excess of water: (a) Rb, (b) Al, (c) C, (d) N, (e) F. If no reaction takes place, write "no equation."

**6.56** Write equations for the reactions of the chlorides of the following elements with water: (a) boron, (b) arsenic, (c) zirconium, (d) silicon.

**6.57** You believe that you have cracked a code that uses elemental symbols to spell words. The code uses numbers to designate the elemental symbols. Each number is the sum of the atomic number and the principal quantum number of the highest occupied orbital of the element whose symbol is part of the code. You have determined that messages can be written normally or backward. Using Table 6.6, decipher these messages:
(a) 10, 12, 58, 11, 7, 44, 63, 66
(b) 9, 99, 30, 95, 19, 47, 79
(c) 30, 21, 109, 57, 96, 2, 74, 19, 80, 95, 97, 58, 11, 44, 10, 99, 36, 79, 9, 19, 80, 74, 19
(d) 40, 10, 80, 81, 27, 10, 99, 10, 44, 11, 58, 18, 99, 48, 37, 47, 79

**7**

# The Chemical Bond

Preview    **N**ow we take up one of the most important topics in chemistry: the way that atoms join to form molecules. We begin by describing the two idealized forms of chemical bonding: the ionic bond, in which electrons are transferred from one atom to another; and the covalent bond, in which electrons are shared. We then show how a property called electronegativity is used to classify bonds into three groups: ionic, polar, or nonpolar. Next we discuss Lewis structures, diagrams that illustrate covalent bonding. We then describe resonance, which exists when we can write two or more Lewis structures for the same compound and those structures differ only in the distribution of multiple bonds. The chapter ends with a discussion of oxidation numbers and how we can use them to learn more about the way elements combine.

Chemistry only begins with isolated atoms. Most of the time, chemistry studies atoms that are held together by attractive forces and the nature of these forces.

In simplest terms, the attractive force between two atoms can be regarded as the coulombic attraction between electric charges of opposite sign. The potential energy of a system is *lowered* (made more negative) as the charges come closer to each other. Lower potential energy is synonymous with greater stability (Figure 6.1). *The lower the potential energy of a system, the more stable it is.* The attractive forces between atoms can be classified by the extent to which the potential energy of the system is lower than the potential energy of the isolated atoms. When the potential energy is lowered by 40 kJ/mol or more, we say that the atoms are held together by a chemical bond.

Two atoms form a chemical bond when the net attractive forces make it more favorable for them to be close to each other than to be apart. While the theory of chemical bonding can be quite complex, there is a useful simplification based on two idealized types of bonding, which are opposite extremes. They are

1. ionic bonds, in which there is a complete transfer of one or more electrons between the atoms that form the bond;
2. covalent bonds, in which two atoms share the electrons of the bond equally.

Most chemical bonds are somewhere between the extremes of ionic and covalent bonding, closer to one than the other but having characteristics of each. In practice, a bond usually is described as either ionic or covalent, depending on which extreme the bond comes closest to matching.

## 7.1  IONIC BONDS

In an ionic bond, the entities that are bonded together can be described as ions of opposite charge. An ionic bond is the result of the electrostatic attraction between positively charged cations and negatively charged anions. According to Coulomb's law, the force of this attraction is

$$F \propto \frac{e^- e^+}{r^2}$$

and the potential energy of the system of charges is

$$E \propto \frac{e^- e^+}{r}$$

where $e^-$ and $e^+$ are the magnitudes of the anionic and cationic charge and $r$ is the distance between the centers of the ions. The energy $E$ has a negative value, since $e^-$ and $e^+$ are of opposite sign, and the stability of the ionic bond increases as $E$ becomes more negative.

For two atoms to form an ionic bond, they must transfer electrons. One atom loses one or more electrons to become a cation; the other gains one or more electrons to become an anion. Consider sodium chloride, which is best described as an ionic compound consisting of $Na^+$ cations and $Cl^-$ anions. An observation of the reaction that forms sodium chloride tells us that it is a stable compound. When sodium metal and chlorine gas are mixed, there is a rapid, even violent reaction. Much heat is liberated and sodium chloride is formed:

$$Na(s) + \tfrac{1}{2}Cl_2(g) \longrightarrow NaCl(s) + energy$$

Sodium chloride can be converted back to elemental sodium and elemental chlorine, but a good deal of energy, usually in the form of electricity, is needed to decompose the compound. The fact that a great deal of energy is liberated when sodium chloride is formed and that a great deal of energy is needed to decompose the compound tells us that sodium chloride is stable.

## Formation of Ions and Ion Pairs

To understand the stability of a salt such as sodium chloride, we can analyze its formation as the result of several different processes. Some of these processes require the input of energy and are unfavorable for the formation of sodium chloride. Others liberate energy and are favorable for its formation. If we study the overall process in the way that an accountant goes over a balance sheet, totaling the energy "debits" and "credits," we find that the balance sheet favors sodium chloride.

Let us consider the transformations that $Na(s)$ and $Cl_2(g)$ might undergo to become sodium chloride. We can picture three major transformations: the formation of $Na^+(g)$ ions from $Na(g)$; the formation of $Cl^-(g)$ ions from $Cl(g)$; and the formation of the ionic bond between the two ions. (For the moment, we shall neglect the formation of $Na(g)$ from $Na(s)$ and of $Cl(g)$ from $Cl_2(g)$, which must also occur.)

The energy associated with the formation of $Na^+$ ions from Na in the gas phase is simply the ionization energy of Na (Section 6.4). We can write

$$496 \text{ kJ/mol} + Na(g) \longrightarrow Na^+(g) + e^-$$

because an input of energy is needed to remove an electron from a sodium atom. By convention, energy is given a positive sign when it is absorbed. The ionization energy of Na is $+496$ kJ/mol.

From the electron affinities in Table 6.8, we can see that adding an electron to a Cl atom actually releases energy

$$Cl(g) + e^- \longrightarrow Cl^-(g) + 349 \text{ kJ/mol}$$

By convention, energy is given a negative sign when it is liberated. The energy associated with the formation of $Cl^-$ ions is $-349$ kJ/mol.

Sodium and chlorine form ions by separate processes. The sodium atom loses an electron; the chlorine atom gains an electron. Yet we can picture these two processes as occurring simultaneously: The electron is removed from the sodium atom and is transferred to the chlorine atom.

It is often said that sodium chloride is a stable compound because both the sodium ion and the chlorine ion in NaCl have the stable electronic configuration of a noble gas; $Na^+$ has the neon configuration, $[He]2s^22p^6$ and $Cl^-$ has the argon configuration, $[Ne]2s^22p^6$. However, the energy data cited above show that the completion of noble gas shells cannot account for the stability of sodium chloride. Adding the two reaction equations and the associated energies, we find that energy must be added to bring about the electron transfer:

| Reaction | Energy Required |
|---|---|
| $Na(g) \longrightarrow Na^+(g) + e^-$ | $+496$ kJ/mol |
| $Cl(g) + e^- \longrightarrow Cl^-(g)$ | $-349$ kJ/mol |
| $Na(g) + Cl(g) \longrightarrow Na^+(g) + Cl^-(g)$ | $+147$ kJ/mol |

The process in which an electron is transferred from a gaseous Na atom to a gaseous Cl atom requires an input of 147 kJ/mol. The energy gained by the Cl atom is less than the energy cost of removing an electron from an Na atom. Such an unfavorable energy balance is not unique to NaCl. In fact, there is always an unfavorable energy balance in the formation of a pair of ions.

Tables 6.7 and 6.8 show that the lowest ionization energy (376 kJ/mol for Cs) is greater than the highest electron affinity (349 kJ/mol for Cl). Thus, an input of energy is always needed to form an anion and a cation from a pair of elements. But we can see in the laboratory that the formation of most ionic compounds is accompanied by the release of energy.

We can eliminate this paradox if we include in our calculations the favorable effect of the electrostatic attractions between the anions and the cations. The process in which the two separated ions come together

$$Na^+(g) + Cl^-(g) \longrightarrow NaCl(g)$$

releases 585 kJ/mol. This energy release more than offsets the 147 kJ/mol required to form the ions. The release of energy results from the coulombic attraction of the oppositely charged ions when the distance between their nuclei is reduced to that observed in NaCl(g), about 0.24 nm. The $Na^+$ cation and the $Cl^-$ anion do not approach any closer because of coulombic repulsions between the inner-shell electrons.

**Ion Crystal Structures**

So far, we have assumed that ionic compounds exist as ion pairs in the gas phase. In fact, they do not. Under normal conditions, ionic compounds

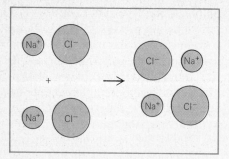

**Figure 7.1**
Formation of part of a sodium chloride crystal structure from ion pairs. In this structure, the distance between like ions generally is greater than the distance between unlike ions, maximizing the favorable coulombic forces.

are solids. The ions lie at the points of a regular, repeating three-dimensional arrangement called a *lattice.*

The stability of ionic compounds results primarily from favorable coulombic attractions. These attractions are much greater when ions are in a crystalline solid than when they exist in the gas phase. In the solid phase there are no discrete ion pairs; each ion interacts with many ions of opposite charge. These extended interactions are responsible for the stability of ionic compounds.

Consider the simplest case, two sodium chloride ion pairs that come together as shown in Figure 7.1. Each $Na^+$ ion attracts two $Cl^-$ ions, while each $Cl^-$ ion attracts two $Na^+$ ions. Instead of the two favorable coulombic attractions that exist for two separate ion pairs in the gas phase, there are a total of four attractions. These attractions are offset somewhat by the repulsions that result when two $Na^+$ ions are brought close together and two $Cl^-$ ions are brought close together. But the gain of stability caused by the attractive forces is much greater than the loss caused by repulsive forces.

If we regard the four ions as being at the corners of a square, the distance between the centers of like ions ($Na^+$ and $Na^+$ or $Cl^-$ and $Cl^-$) is greater than the distance between the centers of the unlike ions. The energy resulting from coulombic attraction or repulsion is proportional to $1/r$. Therefore, the loss in stability from repulsion is not as great as the gain from the attraction.

Ion pairs can be assembled into a three-dimensional array. The energy advantage increases every time an ion pair is added to the crystal structure, but only if the arrangement of the ions results in greater attraction than repulsion. The geometry of the ions in the lattice plays an important role in determining whether an arrangement of ions is more stable than isolated ion pairs. A crystal structure forms only if it is more stable than the same number of ion pairs.

Figure 7.2 shows two common types of crystal structures. In the sodium chloride structure, each $Na^+$ ion is surrounded by six $Cl^-$ ions, which are its nearest neighbors, and each $Cl^-$ ion is surrounded by six $Na^+$ ions. We say that each ion has a *coordination number* of 6. In the cesium chloride structure, each ion has a coordination number of 8, because each $Cs^+$ ion is surrounded by eight $Cl^-$ ions and each $Cl^-$ ion is surrounded by eight $Cs^+$ ions.

The formation of crystal structures, three-dimensional arrays of ions arranged at the points of a lattice for the maximum coulombic attraction between ions, is the major reason for the stability of ionic compounds.

## Lattice Energy

We mentioned that when an ionic lattice forms from isolated ions in the gas phase, energy is released.

$$Na^+(g) + Cl^-(g) \longrightarrow NaCl(s) + 770 \text{ kJ/mol}$$

This process releases 770 kJ/mol of energy. The energy released by such a process is called the **lattice energy;** we say that the lattice energy of sodium chloride is 770 kJ/mol.

Lattice energy can be taken as a measure of the strength of ionic bonds. The greater the lattice energy, the stronger the bonds. As expected, there is a greater gain in stability from the formation of the solid lattice of sodium chloride (770 kJ/mol) than from the formation of ion pairs of sodium chloride in the gas phase (585 kJ/mol).

We can get the overall process by which an ionic solid is formed from gaseous atoms by combining two processes. The first is the formation of ions in the gas phase from atoms of their elements. The second is the formation of the crystal structure from ions in the gas phase. The associated energies can also be combined, taking care to give energy released a negative sign.

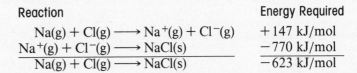

| Reaction | Energy Required |
|---|---|
| $Na(g) + Cl(g) \longrightarrow Na^+(g) + Cl^-(g)$ | $+147$ kJ/mol |
| $Na^+(g) + Cl^-(g) \longrightarrow NaCl(s)$ | $-770$ kJ/mol |
| $Na(g) + Cl(g) \longrightarrow NaCl(s)$ | $-623$ kJ/mol |

There is a difference of about 200 kJ/mol between the energy calculated by this two-step process and the energy actually liberated by the reaction between sodium and chlorine. The difference is the energy needed to form $Na(g)$ from $Na(s)$ and $Cl(g)$ from $Cl_2(g)$. We shall take this point up in more detail in Section 8.7.

### Geometry and Lattice Energy

The geometry of the crystal structure is an important influence on the magnitude of the lattice energy. We can define a geometric factor that is independent of the ionic charge and the distance between ions. This factor relates the geometry of the structure to the difference in energy between ions in the crystal and ion pairs in the gas phase.

In sodium chloride crystal, for example, we can calculate the attractions between an ion and its six nearest neighbors, then add the repulsions between the same ion and its twelve next nearest neighbors, then include the interactions with more distant ions. Eventually, we can calculate that the coulombic attraction per ion pair is 1.75 times greater in the crystal than for a single NaCl ion pair. Any ionic crystal with the same lattice geometry as sodium chloride has this factor of 1.75. Other crystal structures have different factors.

When we talk about the sodium chloride ionic bond, we are talking not about the attraction in an $Na^+Cl^-$ pair but about the attractions among many $Na^+$ and $Cl^-$ ions in the crystal. There really is no single anion-cation association. Rather, the sodium chloride crystal can be visualized as a giant unit that includes all the $Na^+$ and $Cl^-$ ions. We cannot conveniently write the formula for such a giant collection of ions. Instead, we write the

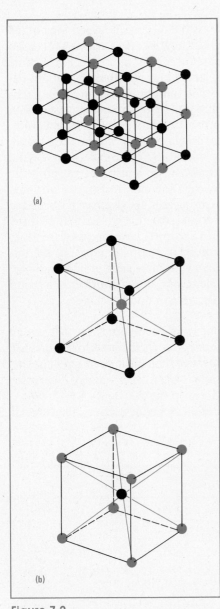

(a)

(b)

**Figure 7.2**

Two common types of ionic crystal structures. In the sodium chloride structure (a), each ion is surrounded by six ions of opposite charge. In the cesium chloride structure (b), each ion is in the center of a cube, surrounded by eight ions of opposite charge at the corners of the cube.

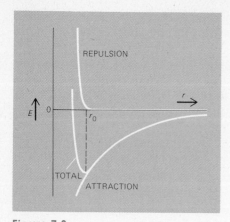

**Figure 7.3**
Variation of the potential energy of oppositely charged ions with distance. The point at which any further increase in the coulombic attraction of the ions with a decrease in distance is offset by the repulsion of the ions' inner electrons is the most favorable distance of separation for the ions in a lattice, and is the minimum of the curve.

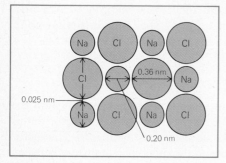

**Figure 7.4**
A section of an NaCl crystal showing ionic radii and distance between ions. In this and most other ionic crystal structures, the distance between the centers of two adjacent ions is roughly the sum of their ionic radii.

formula NaCl to indicate that the ratio of $Na^+$ and $Cl^-$ ions in the ionic crystal is $1:1$.

## Ionic Radius

A second influence on lattice energy is the distance between ions in the crystal lattice. From Coulomb's law, we know that the attraction increases and the energy of the system decreases as ions of opposite charge come closer together. We also know that as the ions get closer, repulsion between their inner electrons becomes more important (Figure 7.3). Eventually, the repulsion between the inner electrons increases faster with decreasing distance than the coulombic attraction does. There is a point, $r_0$ (in Figure 7.3), which is the most favorable distance of separation for the ions. The distance that represents the minimum on the energy curve is the most likely distance between the centers of two ions in the gas phase.

If certain assumptions are made, each ion can be assigned an ionic radius that is a measure of its size. In most cases, the distance between the centers of any two adjacent ions in a crystal structure is close to the sum of their ionic radii, as Figure 7.4 shows. In Section 6.7, we discussed trends in ionic radius as a function of position in the periodic table. In a given column, ionic radius usually increases with increasing atomic number. Ionic radius tends to decrease with increasing atomic number in a group of isoelectronic ions because of the increase in nuclear charge. Table 7.1 shows some of these trends.

Figure 7.5 shows the relative sizes of a number of ions. Note that anions generally are much larger than cations. In a crystal structure, all other things being equal, lattice energy is highest when the ions are smallest and therefore closest together. Since $E \propto e^+ e^-/r$, a smaller value of $r$ increases the lattice energy.

Table 7.2 gives the lattice energy of alkali halides, that is, the value of the energy released in the reaction

**TABLE 7.1   Ionic Radii (nm)[a]**

| | | | | | |
|---|---|---|---|---|---|
| $Li^+$ | $Be^{2+}$ | $B^{3+}$ | $N^{3-}$ | $O^{2-}$ | $F^-$ |
| 0.060 | 0.031 | 0.020 | 0.171 | 0.140 | 0.136 |
| $Na^+$ | $Mg^{2+}$ | $Al^{3+}$ | $P^{3-}$ | $S^{2-}$ | $Cl^-$ |
| 0.095 | 0.065 | 0.050 | 0.212 | 0.184 | 0.181 |
| $K^+$ | $Ca^{2+}$ | | | $Se^{2-}$ | $Br^-$ |
| 0.133 | 0.099 | | | 0.198 | 0.195 |
| $Rb^+$ | $Sr^{2+}$ | | | $Te^{2-}$ | $I^-$ |
| 0.148 | 0.113 | | | 0.221 | 0.216 |
| $Cs^+$ | $Ba^{2+}$ | | | | |
| 0.169 | 0.135 | | | | |

[a] Atomic scale measurements are often reported in angstroms (Å), which is a non-SI unit; 1 nm = 10 Å.

$$M^+(g) + X^-(g) \longrightarrow MX(s)$$

where M is an alkali metal and X is a halogen. The data show that for a given negative ion, lattice energy decreases as the atomic number of the alkali metal increases. For a given positive ion, lattice energy decreases as the atomic number of the halogen increases. The decrease in lattice energy is a result of the larger ionic radii of elements of higher atomic number.

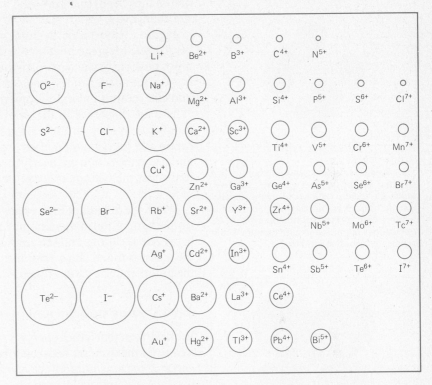

**Figure 7.5**
The relative sizes of common ions. Anions tend to be larger than cations.

**TABLE 7.2   Lattice Energies of Alkali Halides (kJ/mol)[a]**

| | | | |
|---|---|---|---|
| LiF | LiCl | LiBr | LiI |
| 1030 | 840 | 781 | 718 |
| NaF | NaCl | NaBr | NaI |
| 914 | 770 | 728 | 681 |
| KF | KCl | KBr | KI |
| 812 | 701 | 671 | 632 |
| RbF | RbCl | RbBr | RbI |
| 780 | 682 | 654 | 617 |
| CsF | CsCl | CsBr | CsI |
| 744 | 630 | 613 | 585 |

[a] Expressed as positive numbers. Energy actually is released when the crystal forms from gas phase ions.

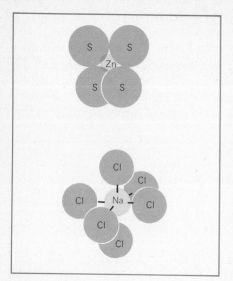

**Figure 7.6**
The influence of ionic radius on coordination number. In the sodium chloride structure, where the sodium cations are slightly smaller than the chloride anions, there is a coordination number of 6. In the ZnS structure, where the zinc cations are much smaller than the sulfur anions, the coordination number is 4.

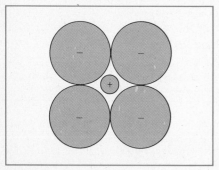

**Figure 7.7**
The effect of differences in size on ionic lattice formation. If the positive ion is much smaller than the negative ion, the anions will touch and the radius of the anions will determine the internuclear distances in the structure.

Some of these crystal structures have the sodium chloride geometry. Others have the cesium chloride geometry. There is no noticeable difference in lattice energy associated with this difference in geometry.

A good rule to remember is this one: The stability due to coulombic interactions between ions in a crystal is greater for small ions. The relative size of the ions in a crystal also affects crystal structure. For example, the radius of an ion and of its oppositely charged neighbor affects the coordination number of that ion — the number of its nearest neighbors of opposite charge.

Figure 7.6 illustrates the influence of ionic radii on coordination number. If there is a great difference in the size of two ions of opposite charge, it is not possible to place many of the larger ions around the smaller ion. In most of the alkali halides, where the anions are larger than the cations, we find the sodium chloride geometry, with a coordination number of 6. Where the difference in the size of the ions is greater, the coordination number is smaller. In ZnS, the great difference in size between the $Zn^{2+}$ cation and the $S^{2-}$ anion leads to a coordination number of 4. When the anion and the cation are nearly the same size, as in CsCl, CsBr, or CsI, the ions form the cesium chloride structure, with its coordination number of 8.

When the difference in size between ions is even more pronounced, the simple sum of the ionic radii may not be enough to indicate the distance between the cations and the anions in a structure. As Figure 7.7 shows, if the anion is much larger than the cation, the anions will "touch" before the anion and the cation touch.

## Ionic Charge and the Ionic Bond

We have seen that lattice energy is affected by the geometry of the lattice as well as by the distance between ions. But *the major determinant of lattice energy is the magnitude of the charge on the ions in the structure.* It follows that lattice energy will be greater in crystals of ions with multiple charges.

The only ionic solid we have examined closely is NaCl, which is made up of cations of charge 1+ and anions of charge 1−. There are ionic compounds in which the ions have greater charges, such as the halides of the alkaline-earth metals ($MgCl_2$, $BaF_2$, $CaBr_2$, etc.). The halide ions each have a charge of 1−; but the alkaline-earth metal cations each have a charge of 2+. The general formula for such compounds is $MX_2$, where M is any alkaline-earth metal and X is any halide. In the crystal structure, there are two halide anions for each alkaline-earth metal cation.

The increased charge of the cations increases the lattice energy. We can see the magnitude of the effect by comparing the lattice energies of the alkaline-earth metal halides, in which the cations have a charge of 2+, and the alkali metal halides, whose cations have a charge of 1+. The lattice energy of each alkaline-earth metal halide is substantially larger than the lattice energy of the corresponding alkali metal halide.

The lattice energy of NaCl is 770 kJ/mol. The lattice energy of $MgCl_2$, the corresponding alkaline-earth metal chloride, is 2500 kJ/mol. The lattice energies of other alkaline-earth metal chlorides are $CaCl_2$, 2260 kJ/mol; $SrCl_2$, 2130 kJ/mol; $BaCl_2$, 2050 kJ/mol. The lattice energies of the corresponding alkali chlorides are lower in every instance. An even more striking example of the effect of ionic charge on lattice energy can be seen in calcium oxide, CaO. Here the anion has a charge of $2-$ and the cation has a charge of $2+$. The lattice energy of CaO is 3600 kJ/mol, much larger than the lattice energies of the halides.

## Noble Gas Electronic Configurations

We have shown that the stability of ionic compounds is not a result of the noble gas electronic configurations of the ions making up such compounds. Yet the ionic compounds of main group elements almost always consist of ions with noble gas electronic configurations.

The best way to explain this observation is to note why some compounds are *not* formed. For example, there are no salts in which an alkali metal forms a cation of charge $2+$, such as $NaCl_2$, although we would expect the dipositive $Na^{2+}$ cation to produce a much larger lattice energy than is found in NaCl.

This predicted increase in lattice energy for $NaCl_2$ is not great enough to offset the very substantial amount of energy needed to form the $Na^{2+}$ ion. Table 6.7 shows that the second ionization energy of sodium is enormous, 4562 kJ/mol. This second ionization energy is much greater than the anticipated increase in lattice energy of $NaCl_2$.

Table 6.7 shows that an extremely large quantity of energy is needed to remove an electron from any species with a noble gas electron configuration. The table lists very large values for the second ionization energies of the group IA metals, the third ionization energies of the group IIA metals, the fourth ionization energies of the group IIIA metals, and so on.

These high ionization energies can be explained by the relatively low principal quantum number of the electron that must be removed from a completed electron shell. For example, the first ionization of sodium removes the $3s$ electron, while the second ionization removes a $2p$ electron. The second ionization of Ba removes a $6s$ electron, while the third ionization removes a $5p$ electron. The ionization energies are so high that species such as $Ba^{3+}$ are rarely formed. Therefore, we can expect that salts of metals will be composed of metal cations that have the noble gas electronic configuration. We can also expect that the maximum positive charge on the metal cation will be the same as the group number of that metal in the periodic table.

The same reasoning can be applied to anions. An excessive amount of energy is needed to added electrons to an anion with a noble gas electronic configuration. Therefore, compounds such as $Na_2Cl$, which include a $Cl^{2-}$ anion, do not exist. The energy needed to add an electron to a species such as $Cl^-$ is very large because the added electron would be in a $4s$ orbital. The relatively small coulombic attraction between this electron and the Cl nucleus is not nearly as great as the electron-electron repulsions introduced by the addition of the electron to the $Cl^-$ anion.

The same repulsions work against the addition of electrons to other anions with noble gas electronic configurations. Therefore, anions found in salts tend to have the number of electrons in a noble gas electron shell. We can find the maximum negative charge on an anion in an ionic compound by subtracting 8 from the group number of the element. Group VIIA anions have a charge of 1 −, group VIA anions have a charge of 2 −, and so on.

Ionization energies alone do not explain why there are no ionic solids made up of ions on the "other side" of noble gas electronic configurations — for example, an $Mg^+$ cation, which has one more electron than a noble gas configuration. A compound such as MgCl(s), which contains the $Mg^+$ cation, should be quite stable. The first ionization energy of Mg is not very large. It is readily offset by the electron affinity of Cl and the lattice energy of MgCl(s). Yet MgCl(s) is not encountered, because the compound quickly reacts with itself to produce $MgCl_2(s)$, which has substantially greater lattice energy:

$$MgCl(s) + MgCl(s) \longrightarrow Mg(s) + MgCl_2(s) + energy$$

In the solid phase, MgCl(s) is so reactive that it cannot be prepared. But in the gas phase, the chance of two MgCl molecules reacting with each other is much lower, and small amounts of MgCl(g) have been observed.

Compounds in which one ion is on the "wrong side" of a noble gas electronic configuration do not exist because there is a substantial gain in lattice energy when they are transformed into compounds whose ions have the noble gas electronic configuration. They may be *stable* — lower in energy than their constituent elements — but they are excessively *reactive* because a still more stable possibility exists.

## 7.2    THE COVALENT BOND

The ionic bond is one extreme model of bonding. The two species that form the bond are ions of opposite charge, held together by coulombic forces of attraction, often in an ionic crystal in which each ion is associated with many neighboring ions of opposite charge.

The covalent bond is at the opposite extreme. The covalent bond model emphasizes the *sharing of electrons* between two neutral atoms, rather than the transfer of electrons to form ions. We can say that the pure covalent bond is characterized by a symmetrical distribution of electrical charge. In the ionic bond, we study a large crystal structure of many atoms. In the covalent bond, we focus our attention only on the two atoms that are bonded to each other.

## HARDNESS AND THE CHEMICAL BOND

Hardness is not easily defined. The traditional method of judging the hardness of materials is based on the Mohs scale, which was introduced about 1812 by a German mineralogist. Ten minerals are arranged in order of increasing hardness, and each is assigned a hardness value from 1 to 10:

| Mineral | Formula | Mohs Value |
|---|---|---|
| talc | $Mg_3Si_4O_{10}(OH)_2$ | 1 |
| rock salt | $NaCl$ | 2 |
| calcite | $CaCO_3$ | 3 |
| fluorite | $CaF_2$ | 4 |
| apatite | $Ca_5(PO_4)_3F$ | 5 |
| feldspar | $KAlSi_3O_8$ | 6 |
| quartz | $SiO_2$ | 7 |
| topaz | $Al_2SiO_4$ | 8 |
| corundum | $Al_2O_3$ | 9 |
| diamond | C | 10 |

Other materials are assigned values by comparison with the ten standard minerals. For example, on the Mohs scale, the hardness of a steel file is about 6 or 7. Window glass is about 5 Mohs, a knife blade is about 5 Mohs, a copper penny is about 3 Mohs, and a human fingernail is about 2 Mohs.

However, the Mohs scale is unsatisfactory because the differences in hardness between the ten standard minerals are not equal. A new scale has been developed in which hardness is interpreted in terms of binding energy per unit volume. Substances that bond ionically, such as sodium chloride, are relatively soft. All very hard substances, such as diamond and corundum, consist of crystals whose atoms are bonded covalently, In addition, if a substance is very hard, it cannot have weak bonds between any of its units of structure.

The second-hardest substance known is a synthetic material named cubic boron nitride. The atoms in a crystal of cubic boron nitride are covalently bonded in virtually the same way as the atoms in diamond. When we deal with materials as hard as cubic boron nitride or diamond, the Mohs scale does not give an adequate indication of relative hardness. On a more modern version of the Mohs scale, corundum retains its old value of 9 and harder substances are given proportionally larger values. On this scale, cubic boron nitride has a hardness value of 19 and diamond has a hardness value of 42+. These values can be correlated with binding energy per unit volume.

In theory, it is possible to manufacture a harder material than diamond by finding a substance whose binding energy per unit volume is greater than the binding energy of diamond. However, no such substance is known to exist. Conceivably, diamond is the hardest material that can exist, because of the nature and strength of its chemical bonds.

A completely ionic bond, characterized by a complete transfer of one or more electrons, does not exist. But we can describe the bonds in such diatomic molecules as $H_2$, $Cl_2$, and even $Na_2$ as completely covalent in the sense that neither atom of the molecule is relatively positive or relatively negative with respect to the other. The two identical atoms share electrons equally.

The sharing of electrons by two atoms in a covalent bond is energetically favorable. It is advantageous for these two atoms to be close to one another rather than to be separated. Energy is liberated when two isolated atoms form a covalent bond:

$$2H(g) \longrightarrow H_2(g) + energy$$

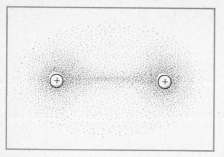

**Figure 7.8**

The $H_2$ molecule. The darker regions represent high probability for the electrons. The electrons tend to lie between the two nuclei. The two electrons are attracted by both nuclei and offset the repulsion between the nuclei.

Unlike the ionic bond, the covalent bond cannot be understood fully on the basis of simple electrostatic theory. A full explanation of covalent bonding requires a quantum mechanical treatment that is beyond our scope. However, we can still examine some characteristics of the covalent bond.

**1.** The electrons tend to lie between the two bonded nuclei as shown in Figure 7.8. In this position, the two electrons are attracted by both nuclei. The repulsion of the nuclei is offset by this attraction.
**2.** We know that two electrons occupying the same orbital in an atom must have opposite spin, and that no more than two electrons can occupy the same oribital. By analogy, when the two electrons are in a covalent bond, the electrons will have opposite spin.
**3.** A covalent bond usually has two electrons. There are a few compounds with one-electron bonds, but almost all covalent bonds have electron pairs — two electrons for each covalent bond.
**4.** Two atoms that form a covalent bond keep approaching each other until the energy of the system reaches a minimum. Figure 7.9 plots energy against internuclear distance in a covalent bond. Note that it resembles Figure 7.3, a similar plot for ions of opposite charge. If the atoms get too close, the forces of repulsion between the two nuclei and between the inner electron shells (for all atoms but H) become significant compared to the force of attraction, and the energy of the system begins to increase. The point at which energy is at a minimum is the point of most favorable internuclear separation.

## 7.3 ELECTRONEGATIVITY AND BONDING

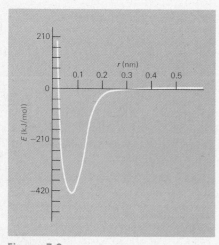

**Figure 7.9**

Variation of energy with internuclear distance in a covalent bond. Note the resemblance to Figure 7.3.

Both the ionic bond and the covalent bond are simplified models. No substance has a completely ionic bond, and only homonuclear diatomic molecules — those consisting of two identical atoms — can even be regarded as having completely covalent bonds. Most real bonds lie somewhere between these two extremes.

An idealized ionic bond includes the complete transfer of one or more electrons from an atom that forms a cation to an atom that forms an anion. An electron pair thus is associated completely with the anion. The ideal covalent bond has equal sharing of the bonding pair between the two atoms.

In a covalent bond between two different atoms, the bond is partially ionic in character and the bonding electron pair is more closely associated with one of the atoms. As a result, each of the two atoms that are bonded has a partial electrical charge. We call this pair of charges a **dipole.** The atom with the greater share of the bonding electron pair has a relative excess of negative charge, while the atom with the lesser share has a relative excess of positive charge. There are two equal but opposite partial charges, and an electrical polarity is associated with the bond. Such a bond is called a **polar bond.**

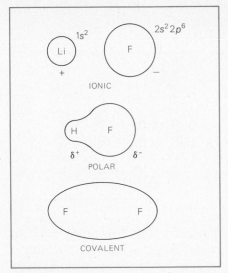

**Figure 7.10**
Three types of bonds of the fluorine atom. The LiF bond is largely ionic, but with some electron sharing. The HF bond is polar. The $F_2$ bond is completely covalent.

Figure 7.10 shows three types of bonds of fluorine. The $F_2$ bond is completely covalent. The HF bond is a polar bond. The LiF(g) bond is described as an ionic bond, but measurements show that there is some electron sharing by the atoms.

We can classify a bond as ionic, polar, or nonpolar, by the extent to which the bonding electron pair is associated with one of the two atoms. If the electron pair is closely associated with one atom, the bond is ionic; the closer the association, the more ionic the bond. If the two atoms share the electron pair more or less equally, the bond is nonpolar; the more equal the sharing, the more nonpolar is the bond. The sharing in polar bonds is somewhere between these extremes.

Ionization energy and electron affinity are measures of the facility with which an isolated atom in the gas phase loses or gains electrons. It would be desirable to have a similar measurement for the more complicated situation in which an atom that is combined with other atoms gains or loses electrons. Unfortunately, no such simple measurement can be made. To fill the need, a property of atoms called **electronegativity** was devised.

**Trends in Electronegativity Values**

*The electronegativity of an atom is a measure of the relative ability of that atom to attract shared electrons to itself when combined with other atoms.* An atom with a high electronegativity value has a relatively great attraction for electrons when it is combined with other atoms. A number of measured properties can be used to assign electronegativity values to atoms, and the exact value of the electronegativity depends on which properties are used. One method uses ionization energy and electron affinity values. Trends in electronegativity parallel the trends in ionization energy and electron affinity. Electronegativity increases from left to right across the periodic table. It also increases from the bottom to the top of the periodic table. In general, the increase is more pronounced from left to right than from bottom to top, as shown in Figure 7.11.

**Electronegativity Values and Bond Type**

Electronegativity gives us a measure of the bond type. If there is a great difference between the electronegativity values of two atoms, those atoms are likely to form ionic bonds. If there is a very small difference, the bonds are relatively nonpolar.

Table 7.3 lists the electronegativity values calculated for most of the elements by Linus Pauling (1901–     ). There is an arbitrary rule of thumb to help us distinguish between bond types. When the two bonded atoms have an electronegativity difference of 0.5 or less, we shall call the bond nonpolar. In polar bonds, the electronegativity difference is between 0.5 and 1.7. In ionic bonds, the difference is greater than 1.7.[1]

**Figure 7.11**
Trends in electronegativity with position in the periodic table. In general, the increase from left to right is more pronounced than the increase from bottom to top. Elements that are widely separated horizontally in the periodic table tend to form binary ionic compounds.

---

[1] This rule does not work too well for F, which has a very large electronegativity. Many of its polar compounds are predicted to be ionic by this rule.

**TABLE 7.3**    Electronegativities of the Elements

| H 2.2 | | | | | | | | | | | | | | | | |
|---|---|---|---|---|---|---|---|---|---|---|---|---|---|---|---|---|
| Li 1.0 | Be 1.6 | | | | | | | | | | | B 2.0 | C 2.6 | N 3.0 | O 3.4 | F 4.0 |
| Na 0.9 | Mg 1.3 | | | | | | | | | | | Al 1.6 | Si 1.9 | P 2.2 | S 2.6 | Cl 3.2 |
| K 0.8 | Ca 1.0 | Sc 1.4 | Ti 1.5 | V 1.6 | Cr 1.7 | Mn 1.6 | Fe 1.8 | Co 1.9 | Ni 1.9 | Cu 1.9 | Zn 1.7 | Ga 1.8 | Ge 2.0 | As 2.2 | Se 2.6 | Br 3.0 |
| Rb 0.8 | Sr 1.0 | Y 1.2 | Zr 1.3 | Nb 1.6 | Mo 2.2 | Tc 1.9 | Ru 2.2 | Rh 2.3 | Pd 2.2 | Ag 1.9 | Cd 1.7 | In 1.8 | Sn 1.8 | Sb 2.1 | Te 2.1 | I 2.7 |
| Cs 0.7 | Ba 0.9 | La 1.0 | Hf 1.3 | Ta 1.5 | W 2.4 | Re 1.9 | Os 2.2 | Ir 2.2 | Pt 2.3 | Au 2.5 | Hg 2.0 | Tl 1.6 | Pb 1.9 | Bi 2.0 | | |

It is convenient to classify substances by bond type, using standard criteria. At one extreme, a substance is classified as ionic if it contains at least one ionic bond, which will be between two atoms that differ by more than 1.7 in electronegativity. At the other extreme is a substance that is nonpolar and contains only bonds between identical atoms or atoms that are relatively close in electronegativity.

When we say that a binary compound has ionic bonding, we do not mean that there is a complete transfer of one or more electrons from one atom to another, but just that the electron transfer is sufficient to make the ionic model the best description of the bond. No compound is completely ionic, not even CsF, the binary compound of the least and the most electronegative elements. In the same way, a nonpolar bond between two unlike atoms has some electron transfer, but so little that the nonpolar description is better.

Using the values in Table 7.3, we can see that the ionic description is best for binary compounds in which elements with the lowest electronegativity values—the alkali metals or the heavier alkaline-earth metals—combine with the elements that have the highest electronegativity values, the halogens and oxygen. Other binary compounds between metals and nonmetals have ionic characteristics, but the ionic description is not always the best one for such compounds. In particular, many binary compounds of the transition metals with the nonmetals can be described as nonionic.

## Polar and Nonpolar Molecules

Molecules that have polar bonds tend to have an overall polarity. Such molecules are called *polar molecules.* However, the polarity of the individual bonds sometimes cancels, and the molecule as a whole is nonpolar. Among such molecules are $CO_2$, $BF_3$, and $CCl_4$. We shall see why they are nonpolar when we discuss their geometry in Chapter 8.

Nonpolar compounds include homonuclear diatomic molecules,

some binary compounds of nonmetals, and most compounds of hydrogen and carbon. Polar compounds include covalent compounds of metals and nonmetals, compounds of hydrogen with nonmetals, compounds of oxygen with nonmetals, and some compounds of halogens with nonmetals.

We can describe the properties of ionic, polar, and nonpolar compounds in general terms. Ionic compounds, such as $NaCl$ and $CaF_2$, are hard, brittle solids with high melting points and high boiling points. They conduct electricity when they become liquid. They dissolve in polar solvents such as $H_2O$ or liquid $NH_3$ but are insoluble in nonpolar solvents.

Compounds with polar bonds, such as $H_2O$, $CHCl_3$, $CH_3OH$, and $POCl_3$, usually are solid or liquid. If solid, they usually are soft at room temperature. Their boiling points and melting points are much lower than those of ionic compounds. They may conduct electricity to a limited extent. They dissolve in polar solvents and, to a lesser degree, in nonpolar solvents.

Nonpolar compounds can be gases, liquids, or solids at room temperature. The solids have low melting points, generally are soft and waxy, and do not conduct electricity. They dissolve in nonpolar organic solvents, such as gasoline or kerosene.

The solubility of substances is described by a simple rule: Like dissolves like. Nonpolar substances dissolve best in nonpolar solvents; ionic or polar substances dissolve best in polar solvents. When we make salad dressing out of oil (nonpolar) and vinegar (polar), we see the effect of this rule. Oil and vinegar don't mix.

## 7.4  LEWIS STRUCTURES

Until 1916, chemists drew diagrams of molecules in which the bonds between atoms were indicated by lines, called valence bonds, whose actual significance was not understood. In these diagrams, $H_2$ was written H—H; $H_2O$ was written H—O—H. There was no feeling that the lines represented physical reality.

In 1916, G. N. Lewis (1875–1946), an American chemist, suggested that the dashes represented pairs of shared electrons. The suggestion led to a new explanation of chemical bonding.

Lewis emphasized the importance of the valence shell, the atom's highest-energy electron shell, in covalent species. He developed a method of diagraming the details of electronic structure in compounds with covalent bonds, using dots to represent electrons. These diagrams, only slightly modified, still are used to show the structure of covalent molecules. They are called **Lewis structures,** or **electron dot structures.**

A mastery of the technique of constructing Lewis structures is important for the study of chemistry. *A correctly drawn Lewis structure indicates how all the atoms of a molecule are attached to one another and accounts for all the valence electrons of all the atoms in the molecule.* It does not, however, give information about the shape of a molecule. Lewis structures can be constructed by these rules:

1. Only valence (outer-shell) electrons are shown. For the main group elements in columns IA through VIIIA of the periodic table, the valence electrons are the electrons of highest principal quantum number. *The number of valence electrons of a main group element is the same as the column number of the element in the periodic table.* For example, as we go across the second row of the periodic table, the number of valence electrons of each element is

| | | |
|---|---|---|
| Li | 1 | Li· |
| Be | 2 | ·Be· |
| B | 3 | ·B· |
| C | 4 | ·C· |
| N | 5 | ·N· |
| O | 6 | ·O· |
| F | 7 | :F· |
| Ne | 8(or 0) | :Ne: |

where each dot represents a valence electron.

The post-transition metals in group IIB generally have two valence electrons. The number of valence electrons in a transition metal is not as easily defined.

2. A shared electron pair can be represented either by two dots or by a line between two atoms. Although Lewis used only dots, modern structures generally use lines to represent shared electron pairs. We shall call such structures Lewis structures, even though they are actually modified Lewis structures. When there is one line between two atoms, the atoms are said to be joined by a single bond. There are many compounds in which atoms share two or even three pairs of electrons; this arrangement is called **multiple bonding.** A double bond, in which two pairs of electrons are shared, is shown by two lines between the atoms. A triple bond is represented by three lines between the atoms.

3. The valence electrons that are not included in covalent bonds are called *nonbonding electrons.* In a Lewis structure, nonbonding electrons are assigned to specific atoms and are represented by dots drawn next to the symbols for these atoms. Nonbonding electrons, like bonding electrons, almost always come in pairs.

Figure 7.12 shows some Lewis structures. You will note that nonbonding electrons are grouped into pairs, with each pair placed on one side of the symbol for the element.

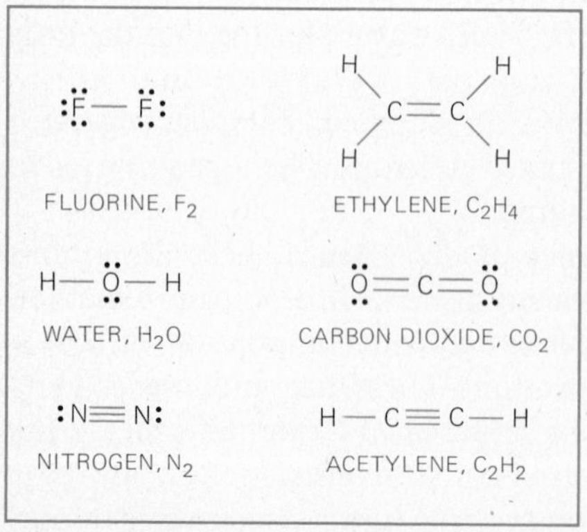

**Figure 7.12**
Lewis structures of some simple molecules. The lines represent bonds, and the dots represent nonbonding valence electrons.

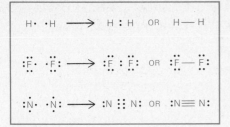

**Figure 7.13**
Electron sharing allows each atom to achieve a noble gas electronic configuration.

## The Octet Rule

Most Lewis structures satisfy the **octet rule,** meaning that *each atom in the molecule achieves a noble gas electronic configuration by covalent bonding.* The octet rule is so named because the noble gases Ne, Ar, Kr, and Xe all have the valence shell electron configuration $ns^2np^6$, an octet, or group of eight outer-shell electrons. (Since helium is $1s^2$, hydrogen achieves the noble gas configuration when it is surrounded by only two electrons.)

The octet rule is quite helpful in the writing of Lewis structures — so much so that the stability of the covalent bond often is incorrectly ascribed to the stability of the noble gas configuration. Figures 7.12 and 7.13 show how noble gas electronic configurations can be achieved by covalent bonding.

When two hydrogen atoms come together, each with one electron, each nucleus becomes associated with two electrons and can be regarded as having the electronic configuration of He, $1s^2$. When two fluorine atoms come together, each with seven valence electrons ($2s^22p^5$), each fluorine atom gives one of its valence electrons to the bond. Each fluorine atom is now associated with eight valence electrons: two bonding electrons, which are shared, and six nonbonding electrons, which are unshared. Each fluorine atom in such a bond can be regarded as having the electronic configuration of neon, $[He]2s^22p^6$. The octet rule is satisfied.

The same reasoning can be extended to multiple covalent bonds. Each isolated nitrogen atom ($[He]2s^22p_x^12p_y^12p_z^1$) has five valence electrons, two paired and three unpaired. When two nitrogen atoms come together, each atom can be associated with eight electrons if their unpaired electrons are shared. Thus, each atom in the $N_2$ molecule is associated with two nonbonding and six bonding electrons, and the octet rule is satisfied.

One helpful way to explain covalent bonding is to say that the separate atoms do not have enough electrons to form complete octets. The insufficiency is overcome by the sharing of electrons in covalent bonds, so that the atoms complete their octets.

We can say that each atom of a single bond has an unpaired electron to contribute to the formation of the bond. In a double bond between two atoms, each atom contributes two electrons. In a triple bond, each atom contributes three electrons. This concept is helpful but not essential in describing covalent bonds.

The octet rule can even help to predict the composition of molecules. Figure 7.14 shows the Lewis structures of the hydrides of some elements in the second row of the periodic table. We see that HF has one bond and six nonbonding electrons grouped in three pairs, that $H_2O$ has two bonds and four nonbonding electrons grouped in two pairs, that $NH_3$ has three bonds and one pair of nonbonding electrons, and that $CH_4$ has four bonds. The element forming each hydride has achieved a noble gas structure by sharing electrons in one or more covalent bonds with one or more hydrogen atoms. The number of H atoms in each hydride is equal to the number of electrons that the other atom needs to complete its octet.

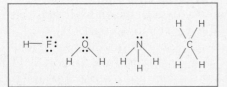

**Figure 7.14**
Hydrides of some elements in the second row of the periodic table. The octet rule is obeyed.

## Drawing a Lewis Structure

As an illustration, let us draw the correct Lewis structure for nitrous acid, $HNO_2$.

1. The Lewis structure must account for all valence electrons. The first step is to find the number of valence electrons in the molecule by totaling the valence electrons of the atoms. In $HNO_2$, H has one valence electron; N (group VA) has five valence electrons; each O (group VIA) has six valence electrons. The total for the molecule is $1 + 5 + (2 \times 6) = 18$ valence electrons.

2. Now construct the skeleton of the molecule by connecting the symbols for the atoms by single lines, indicating single bonds. This initial step may require a combination of experimental data, an understanding of chemical bonding, and even chemical intuition. With some experience, we can construct the skeleton for most molecules without too much trouble. There are rules that help in this construction:

   a. A hydrogen atom can accommodate only one electron pair, so it is always bonded by a single bond to one of the other atoms in the compound.

   b. Second-row elements never bond to more than four atoms. (They may bond to fewer than four atoms.)

   c. When a nonmetal bonds with an element of lower electronegativity than itself in a neutral compound, the number of bonds *usually* can be calculated by subtraction of the group number of the nonmetal from 8. Thus, the halogens, from column VIIA, tend to form one bond when bonded to less electronegative elements. Oxygen, sulfur, and other elements from column VIA tend to form two bonds. Nitrogen, phosphorus, and other elements from column VA tend to form three bonds.

   d. In compounds with more than one oxygen atom, there usually are no bonds between the oxygen atoms. If there is an oxygen-oxygen bond, the name of the compound usually includes the word *peroxide* or *superoxide*. As a rule, we do not connect oxygen atoms to one another in a Lewis structure unless there is specific information that such bonds exist.

   e. Rings of atoms are not written unless there is specific information that they exist in a molecule.

   f. Most oxyacids, acids containing oxygen and hydrogen, have the hydrogen atoms bonded to the oxygen atoms.

   g. A halogen atom forms only one single bond, unless it is bonded to an oxygen atom or to a halogen of lower atomic number.

   Using these rules, we can assign a skeleton structure to nitrous acid. The molecule will have an O—H bond (rule f), no O—O bonds (rule d), and no rings of atoms (rule e). Only one structure is possible:

$$O—N—O—H$$

**3.** As the next step, count the electrons that have been used to construct the skeleton structure. Each dash is a single bond, representing two electrons. Subtract this number from the total number of valence electrons. In the $HNO_2$ skeleton structure, there are three dashes, representing six electrons. Since 18 valence electrons were available, $18 - 6 = 12$ electrons remain to be accounted for.

**4.** Assume that all the atoms obey the octet rule and determine the number of electrons needed to complete all the octets in the molecule.

The hydrogen atom in a skeleton structure can never accommodate more than two electrons. The nitrogen atom has two bonds, representing four electrons, so four more electrons will complete its octet. The oxygen atom that is bound to the H atom and the N atom has two bonds; four more electrons will complete its octet. The other oxygen atom has one bond; it requires six electrons to complete its octet. Adding, we find that $4 + 4 + 6 = 14$ electrons are needed to complete all the octets in $HNO_2$.

**5.** Compare the number of valence electrons that are available (step 3) with the number needed to complete the octets (step 4). If the number from step 3 is the same as the number from step 4, complete the Lewis structure by drawing in enough dots to give each atom a complete octet. If the number of electrons found in step 3 is less than the number found in step 4, one additional bond must be drawn in the Lewis structure for each difference of two electrons.

For $HNO_2$, 14 electrons are needed to complete all the octets but only 12 valence electrons remain after the single-bond structure is drawn. For all the octets in $HNO_2$ to be complete, some of the remaining valence electrons must be shared. (As Figure 7.13 shows, sharing allows all the atoms in a molecule to have complete octets even though there are fewer than eight electrons for each atom.)

In $HNO_2$, there is a two-electron difference, so one more bond is needed. This bond *cannot* be placed between the H atom and the O atom because H never has more than one bond. It is best *not* placed between the N atom and the O atom that is bonded to the H atom to give:

$$O\text{---}N\text{==}O\text{---}H$$

because O usually has only two bonds (rule c), rather than the three it would have in this structure. In general, we avoid drawing structures with three bonds to O in neutral molecules. By elimination, the double bond is placed to form the structure:

$$O\text{==}N\text{---}O\text{---}H$$

With the placement of the double bond established; there are $12 - 2 = 10$ valence electrons in the $HNO_2$ molecule remaining to be accounted for. The number of electrons needed to complete all octets is

$14 - 4 = 10$. It remains only to place the dots representing these electrons around the appropriate symbols in the Lewis structure, giving:

$$\ddot{\text{O}}=\ddot{\text{N}}-\ddot{\text{O}}-\text{H}$$

as the final Lewis structure of $HNO_2$. To see whether you have written a correct structure, you can count the total number of valence electrons and also note whether the octet rule is obeyed for each atom.

## Lewis Structures and the Octet Rule

The task of drawing Lewis structures is simplified by the fact that there are many compounds in which the atoms obey the octet rule. Some observations related to the octet rule help us to formulate many Lewis structures:

1. The atoms of elements of the second row of the periodic table (Li through Ne) are never associated with more than eight valence electrons. An atom that is associated with more than eight electrons in a covalent compound is said to have an *expanded octet. Atoms of elements in the second row never have expanded octets.* In other words, atoms of these elements never have more than the eight valence electrons needed to complete the neon shell, $2s^2 2p^6$.
2. The atoms of elements of group IIIA often disobey the octet rule by being associated with only six electrons in covalent compounds. They are said to have *incomplete octets*.
3. A compound with an odd number of valence electrons must have at least one atom that does not satisfy the octet rule. Such a compound also has at least one unpaired electron.
4. Atoms of elements whose atomic number is 14 or greater often may be surrounded by more than eight electrons in covalent compounds. We say that these atoms have expanded octets.

---

**Example 7.1**    Draw a Lewis structure for nitrogen trichloride, $NCl_3$.

**Solution**    **Step 1.** First calculate the number of valence electrons. The N atom has five and each Cl atom has seven, giving $5 + (3 \times 7) = 26$ valence electrons.

**Step 2.** Determine the skeleton structure. Since Cl forms only one bond unless it is bonded to O or F, the skeleton structure must be

$$\begin{array}{c} \text{Cl}-\text{N}-\text{Cl} \\ | \\ \text{Cl} \end{array}$$

**Step 3.** The number of electrons used to form the three bonds in the skeleton structure is 6, leaving $26 - 6 = 20$ electrons to complete the octets in the molecule.

**Step 4.** The number of electrons needed to complete all octets is 2 (for the N atom) $+ (3 \times 6)$ (for the Cl atoms) $= 20$ electrons.

**Step 5.** Since the number of electrons needed to complete all octets (Step 4) is the same as the number of available electrons (Step 3), the Lewis structure is finished by completing the octets, writing in the nonbonding valence electrons as pairs of dots:

$$:\ddot{C}l - \underset{\underset{:\ddot{C}l:}{|}}{\ddot{N}} - \ddot{C}l:$$

**Step 6.** As a final step, check to be sure that the Lewis structure has the correct number of electrons. The structure we have written for $NCl_3$ has 26 electrons, 6 bonding and 20 nonbonding, which is the number of valence electrons.

**Example 7.2**  Draw the Lewis structure of nitric acid, $HNO_3$.

**Solution**  **Step 1.** There are $1 + 5 + (3 \times 6) = 24$ valence electrons.
**Step 2.** The most likely skeleton structure is

$$\begin{array}{ccc} O & & O \\ \diagdown & & \diagup \\ & N & \quad H \\ & | & \\ & O & \end{array}$$

which avoids O—O bonds and includes the O—H bond expected for an oxyacid.
**Step 3.** The number of electrons remaining to complete octets is $24 - 8 = 16$.
**Step 4.** The number of electrons needed to complete octets is 18, 6 for each of the two O atoms with one bond, 2 for the N atom, 4 for the O atom bonded to H, 0 for the H atom.
**Step 5.** Since the number of remaining electrons is two less than the number of electrons needed to complete all octets, one more bond is required. Placing the bond in a way that avoids a ring structure or three bonds to an O atom gives:

$$\begin{array}{ccc} O & & O \\ \diagdown & & \diagup \\ & N & \quad H \\ & \| & \\ & O & \end{array}$$

This drawing places the double bond between the N atom and the lower O atom. The double bond could just as well be placed between the N atom and the upper O atom, since these two O atoms are equivalent. Completing the octets gives the Lewis structure:

$$\begin{array}{ccc} :\ddot{O} & & \ddot{O}-H \\ \diagdown & & \diagup \\ & N & \\ & \| & \\ & :\ddot{O}: & \end{array}$$

**Step 6.** The number of electrons in the Lewis structure, 24, equals the number of valence electrons.

**Example 7.3**    Draw the Lewis structure for formic acid, $CH_2O_2$.

**Solution**    **Step 1.** The number of valence electrons is $4 + 2 + (2 \times 6) = 18$.

**Step 2.** There are several possible skeletons. But since we know that $CH_2O_2$ is an acid, at least one O—H bond is necessary. Therefore, the only two possibilities are:

$$
\begin{array}{ccc}
& O\text{—}H & & O \\
& \diagup & & \diagup \\
C & & \text{or} \quad H\text{—}C & \\
& \diagdown & & \diagdown \\
& O\text{—}H & & O\text{—}H
\end{array}
$$

Only 10 electrons remain, while 12 are needed to complete all the octets. An additional bond is needed so that all the octets can be completed. Adding a bond in the first structure will give an O atom that has three bonds. The second structure therefore seems best. This choice is verified by experimental evidence. The structure with the required double bond and all the completed octet is

$$
\begin{array}{c}
:\overset{\displaystyle ..}{O}: \\
\parallel \\
H\text{—}C\text{—}\overset{\displaystyle ..}{\underset{\displaystyle ..}{O}}\text{—}H
\end{array}
$$

The formula of formic acid is often written HCOOH to reflect the bonding in this structure.

The same procedure can be used to write Lewis structures for polyatomic ions. For such species, the initial count of valence electrons is modified to include the charge on the ion. For a mononegative ion such as $HCO_3^-$, the count should include one more than the sum of the valence electrons of the neutral atoms. For $HCO_3^-$, we calculate four valence electrons for the C atom, one for the H atom, $3 \times 6 = 18$ for the three O atoms, and one for the negative charge on the ion, giving $4 + 1 + (3 \times 6) + 1 = 24$ valence electrons in all. A dinegative anion such as $SO_4^{2-}$ has two more valence electrons than are present in the neutral atoms. The count is $6 + (4 \times 6) + 2 = 32$ valence electrons. For cations, the number of valence electrons is less than the number present in the atoms. Again, the difference is the charge on the ion. Thus, $NH_4^+$ has $5 + (1 \times 4) - 1 = 8$ valence electrons, while $CH_3^+$, the carbonium ion, has $4 + (1 \times 3) - 1 = 6$ valence electrons.

**Example 7.4**    Draw a Lewis structure for the bicarbonate ion, $HCO_3^-$.

**Solution**    **Step 1.** The number of valence electrons of the atoms is 23, plus one for the negative charge, giving a total of 24.

**Step 2.** The best skeleton structure is

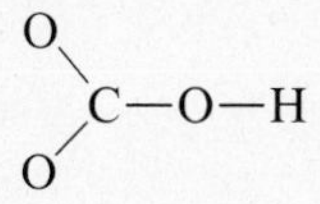

$$
\begin{matrix}
O \\
\quad\diagdown \\
\qquad C\text{—}O\text{—}H \\
\quad\diagup \\
O
\end{matrix}
$$

All three O atoms must be attached to the C atom to avoid O—O bonds. The H atom must also be attached to an O atom. If the H atom were attached to the C atom, no further bonds would be possible, since a C atom can have no more than four bonds.

   **Step 3.** The number of electrons remaining is $24 - 8 = 16$.
   **Step 4.** The number of electrons needed to complete all octets is $(2 \times 6) + 2 + 4 = 18$.
   **Step 5.** One additional bond is needed. It is best placed between the C atom and either of the O atoms not bonded to H:

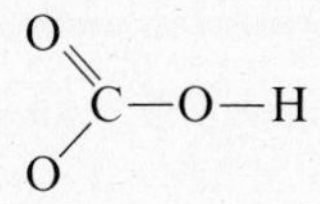

$$
\begin{matrix}
O \\
\quad\diagdown\!\!\diagdown \\
\qquad C\text{—}O\text{—}H \\
\quad\diagup \\
O
\end{matrix}
$$

**Step 6.** Completing the octets gives the structure:

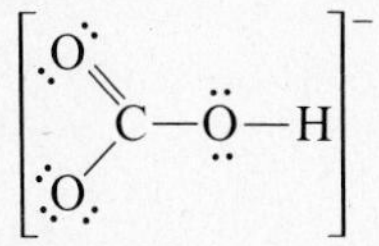

$$
\left[
\begin{matrix}
.\ddot{O}. \\
\quad\diagdown\!\!\diagdown \\
\qquad C\text{—}\ddot{O}\text{—}H \\
\quad\diagup \\
:\ddot{O}.
\end{matrix}
\right]^{-}
$$

which accounts for the required 24 electrons and indicates the negative charge of the species.

## Incomplete Octets

In writing some Lewis structures, we may find that there are not enough valence electrons to complete octets for all the atoms in the skeleton structure, but that none of the atoms can accommodate multiple bonds. In such cases, some octets must be left incomplete. Generally, incomplete octets are associated with atoms that have fewer than four valence electrons.

   In the Lewis structure for $BF_3$, the skeleton structure is

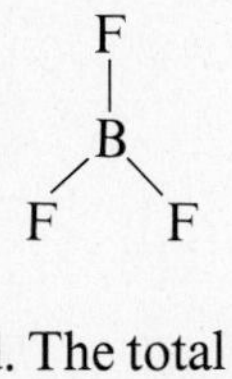

$$
\begin{matrix}
F \\
| \\
B \\
\diagup \quad \diagdown \\
F \qquad F
\end{matrix}
$$

since F can have only one bond. The total number of valence electrons is $3 + (3 \times 7) = 24$, of which six are accounted for in the skeleton structure.

That leaves 18 valence electrons to complete all the octets. However, a count shows that $2 + (3 \times 6) = 20$ valence electrons are needed to complete all octets. The further sharing of electrons is ruled out, since we shall assume that F cannot form double bonds. The only possible conclusion is that one of the atoms in the $BF_3$ molecule will be surrounded by only six electrons (three pairs), rather than by a complete octet. The B atom is the atom with the incomplete octet, since it has lower electronegativity than the F atoms. The correct structure is

---

**Example 7.5**    Draw the Lewis structure of $HgCl_2$.

**Solution**    **Step 1.** The number of valence electrons is 16; $(2 \times 7) = 14$ from Cl and 2 from Hg (which is in group IIB).

**Step 2.** Since Cl can form no more than one covalent bond with Hg, the only possible structure is Cl—Hg—Cl.

**Step 3.** The skeleton structure leaves $16 - 4 = 12$ electrons to be accounted for.

**Step 4.** A total of $6 + 4 + 6 = 16$ electrons is needed to complete all octets. Only 12 electrons are available. Ordinarily, two multiple bonds would be added because of the shortage of four electrons. Since Cl can form only one bond, this solution is not possible. The $HgCl_2$ molecule must therefore have an incomplete octet. The Hg atom, which is less electronegative than Cl, has the incomplete octet, giving the Lewis structure:

$$:\overset{..}{\underset{..}{Cl}}—Hg—\overset{..}{\underset{..}{Cl}}:$$

---

## Octet Expansion

We mentioned earlier that many compounds containing atoms of elements whose atomic number is 14 or greater can have expanded octets — that is, more than eight valence electrons (four pairs) associated with a single atom.

We can explain why expanded octets exist in these elements but not in elements of atomic number 13 or less. The $2s$ and $2p$ orbitals of the valence shell of a second-row element are filled by an octet of electrons. To accommodate more than eight electrons, such an element must use $3s$ and $3p$ orbitals. There is a large energy difference in these elements between the $2s$ and $2p$ orbitals and the $3s$ and $3p$ orbitals, which have different principal quantum numbers, and the use of the higher orbitals is energetically disadvantageous.

In the third-row elements, whose valence shell includes the $3s$ and $3p$ orbitals, the octet can be expanded if the electrons go into the next available orbitals. These are the $3d$ orbitals, which have the same principal quantum number as the other orbitals in the valence shell. The energy required to place electrons in these orbitals is often relatively small. The formation of extra bonds as a result of octet expansion releases much more energy. Since many heavy elements have empty orbitals whose energy is close to that of the orbitals occupied by the valence electrons, expanded octets are common in these elements.

The number of electrons in an expanded octet is limited by electron-electron repulsion. An expanded octet usually includes no more than 12 electrons; there usually will be no more than six bonds around any atom.

The presence of expanded octets in a compound can be recognized either from the rules of valence or, after a skeleton structure is written, by the presence of more valence electrons than are needed to complete octets for all the atoms in a molecule.

In the $PCl_5$ molecule, for example, the P atom must have an expanded octet. All five Cl atoms must be attached to the P atom because they cannot be attached to each other. Any single atom that is covalently bonded to more than four other atoms must have an expanded octet. Therefore, the structure of $PCl_5$ is

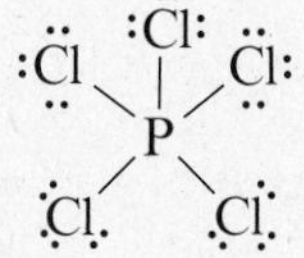

with the P atom accommodating 10 valence electrons (five pairs).

In formulating a Lewis structure for $BrF_3$, the skeleton structure:

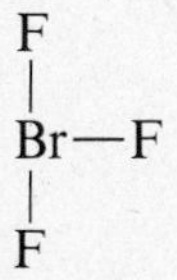

requires $(3 \times 6) + 2 = 20$ valence electrons to complete all octets. Since there are $28 - 6 = 22$ valence electrons remaining, one of the atoms must have an expanded octet. Usually, it is the atom that already has the most bonds, Br in $BrF_3$. The complete structure is

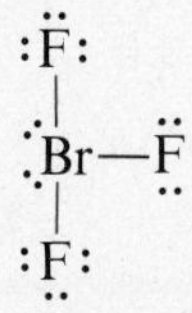

**Example 7.6**    Draw the Lewis structure for $SF_6$.

**Solution**    Since F can have only one bond, the skeleton must be:

$$
\begin{array}{ccc}
 & F & \\
F & | & F \\
\diagdown & S & \diagup \\
\diagup & | & \diagdown \\
F & F & F
\end{array}
$$

Octet expansion is necessary. The number of valence electrons is $6 + (6 \times 7) = 48$. The number remaining to complete octets is $48 - 12 = 36$, which is also the number required to complete the octets of all the F atoms. The complete structure will include three pairs of electrons around each F atom:

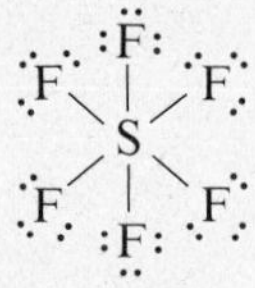

$$
\begin{array}{ccc}
 & :\ddot{F}: & \\
:\ddot{F} & | & \ddot{F}: \\
\diagdown & S & \diagup \\
\diagup & | & \diagdown \\
:\ddot{F} & :\ddot{F}: & \ddot{F}:
\end{array}
$$

The S atom has six pairs of valence electrons.

---

**Example 7.7**    Draw the Lewis structure of $IF_5$.

**Solution**    **Step 1.** The number of valence electrons is $7 + (5 \times 7) = 42$.

**Step 2.** Since only I can form more than one bond (it is bonded to another halogen of lower atomic number), the skeleton must be

$$
\begin{array}{ccc}
 & F & \\
 & | & \\
F & \!\!-\!I\!-\!\! & F \\
\diagup & & \diagdown \\
F & & F
\end{array}
$$

**Step 3.** The number of electrons remaining to complete octets is $42 - 10 = 32$.

**Step 4.** The number of electrons needed to complete the octets of the F atoms is $5 \times 6 = 30$.

**Step 5.** Since there are two more electrons than are necessary to complete octets, there will be even more octet expansion than is found in the skeleton structure. The I atom is large enough to accommodate the two extra electrons even though it is already surrounded by 10 electrons. The structure is:

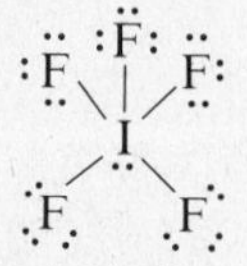

$$
\begin{array}{ccc}
 & :\ddot{F}: & \\
:\ddot{F} & | & \ddot{F}: \\
\diagdown & \ddot{I} & \diagup \\
\diagup & & \diagdown \\
:\ddot{F} & & \ddot{F}:
\end{array}
$$

The I atom has six pairs of valence electrons.

## Formal Charge and the Electroneutrality Principle

We noted earlier that in covalent molecules that include nonmetals with complete octets, atoms of a given element tend to form a specific number of bonds. Usually, the number of bonds formed by an atom that is bonded to less electronegative atoms will be equal to 8 minus its group number in the periodic table.

The number of bonds is often the same as the number of unpaired valence electrons of each atom that are available for bonding. For example, the ground state electronic configuration of nitrogen, which forms three bonds, is $[He]2s^2 2p_x^1 2p_y^1 2p_z^1$; that of oxygen, which forms two bonds, is $[He]2s^2 2p_x^2 2p_y^1 2p_z^1$.

Nonmetals can form compounds in which they do not have the expected number of bonds. Nitric acid, $HNO_3$, is such a compound. In the Lewis structure of $HNO_3$ (Example 7.2), the N atom has four bonds and one of the O atoms has one bond. The N atom has five valence electrons. Three of these electrons are used to form three normal bonds, in which the atom bonded to N contributes an electron to the shared pair, as does the N atom. Two electrons are left for the fourth bond formed by the N atom (Figure 7.15).

By this method of electron bookkeeping, the fourth bond formed by the N atom seems to differ from the other three. It uses two electrons from the N atom and none from the other atom of the bond, the O.

A bond in which one atom seems to contribute both electrons is called a *dative* or *coordinate-covalent bond.* Such a bond does not differ in physical properties from a normal bond. We create this difference by the way we do our electron bookkeeping. The existence of a dative bond in a compound is noted by the use of **formal charges,** pluses or minuses written next to the symbols. "Formal" is meant literally. *The formal charges do not indicate actual charge distribution* within the molecule. They are devices that help us in electron bookkeeping.

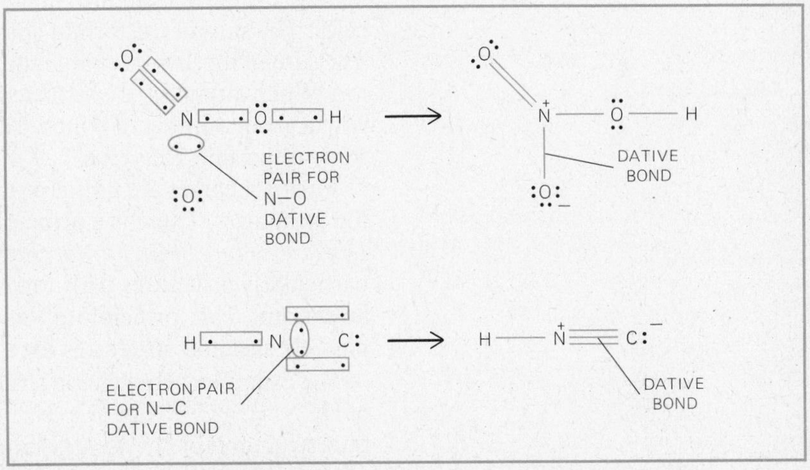

**Figure 7.15**

Electron bookkeeping and dative bonds. Usually a bond is formed from one electron on each atom as shown for all but one bond in each molecule. In $HNO_3$, we call the N—O single bond a dative bond because we can account for its formation by using two electrons from the N and none from the O. In HNC, we can call one of the three NC bonds a dative bond because both of the electrons of the bond are from the N.

We can calculate the formal charge on an atom by dividing the electrons in a bond equally between the atoms and finding the charge that results:

formal charge = (valence electrons) − (nonbonding electrons)
$$-\tfrac{1}{2} \text{ (bonding electrons)}$$

For example, we can calculate the formal charge of the N atom of $HNO_3$ by using this relationship:

$$\text{formal charge} = 5 - 0 - \tfrac{1}{2}(8) = +1$$

The formal charge of the O atom of $HNO_3$ can be calculated as

$$\text{formal charge} = 6 - 6 - \tfrac{1}{2}(2) = -1$$

The structure of $HNO_3$ thus can be written:

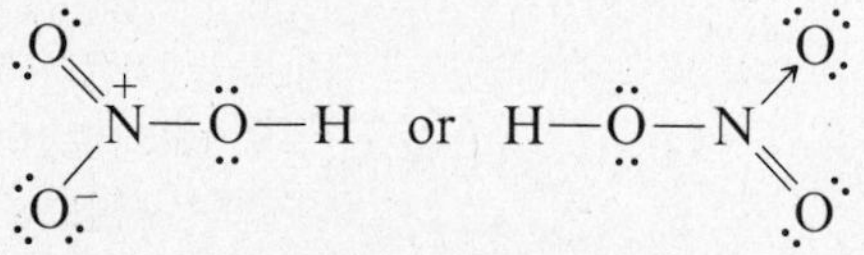

The arrow between the O and the N atoms represents a covalent bond. We can take it to mean that both electrons in the N—O bond have been contributed by the N atom.

Similarly, the structure of ozone, $O_3$, can be written:

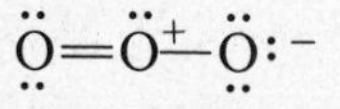

$$\ddot{O}{=}\overset{+}{\ddot{O}}{-}\ddot{\underset{\cdot\cdot}{O}}{:}^{-}$$

The central O atom has the formal charge of $6 - 2 - \tfrac{1}{2}(6) = +1$. The O atom with one bond has the formal charge of $6 - 6 - \tfrac{1}{2}(2) = -1$.

If we examine structures that have formal charges, we note these useful rules: The sum of the formal charges in a neutral molecule must be zero. The sum of the formal charges in an ion must be equal to the charge of the ion. When atoms have completed octets, we can associate a formal charge with a given number of bonds. Table 7.4 summarizes this relationship for three important elements.

Formal charges are useful primarily as a guide in writing correct skeleton structures. There is a principle of **electroneutrality,** which states that *Lewis structures with formal charges are to be avoided whenever possible,* particularly structures with formal charges greater than +1 and −1 on any atom. This principle often helps us make a choice between two possible skeleton structures for a molecule.

For example, two skeleton structures are possible in the Lewis structure of HCN: H—C—N or H—N—C. These two skeleton structures lead to final structures of H—C≡N: and H—N≡C:. The second structure has a formal charge of −1 on the C atom and a formal charge of +1 on the N

**TABLE 7.4**   Number of Bonds and Formal Charge of Atoms with Completed Octets

| Element | Number of Bonds | Formal Charge |
|---|---|---|
| C | 3 | $-1$ |
| N | 4 | $+1$ |
|   | 2 | $-1$ |
| O | 3 | $+1$ |
|   | 1 | $-1$ |

atom. The principle of electroneutrality says that the first structure, which has no formal charges, should be chosen. This choice is borne out by experimental evidence, which shows that the first structure is the more stable one.

In other cases, both possible skeleton structures are known to exist. The electroneutrality principle can help us predict which one is more stable.

For $H_2SO_4$, the Lewis structure could be:

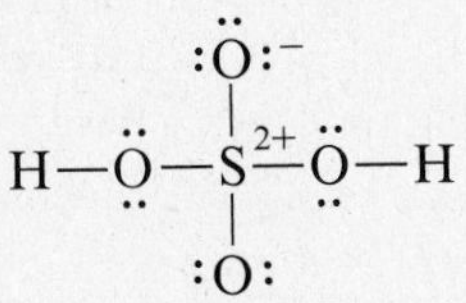

$$
\begin{array}{c}
:\ddot{O}:^{-} \\
| \\
H-\ddot{O}-\overset{2+}{S}-\ddot{O}-H \\
| \\
:\underset{..}{O}:^{-}
\end{array}
$$

Because this structure gives each atom a complete octet, it requires formal charge of $-1$ for two O atoms, which have only one bond each. The S atom has a formal charge of $6 - 0 - \frac{1}{2}(8) = +2$. One possible way to avoid these formal charges is to expand the octet of the S atom as in the structure:

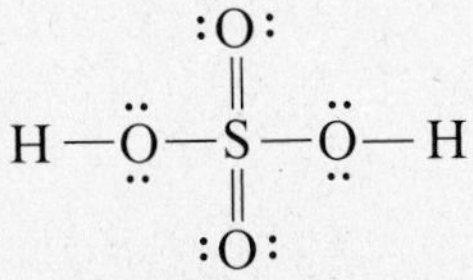

$$
\begin{array}{c}
:O: \\
\| \\
H-\ddot{O}-S-\ddot{O}-H \\
\| \\
:O:
\end{array}
$$

Here the choice is not between different skeleton structures but between different representations. You should note that both representations have exactly the same number of electrons; only the location of two pairs of electrons is different. Either structure can be regarded as correct.

**Example 7.8**   Draw the most likely Lewis structure for formaldehyde, $CH_2O$.

**Solution**   There are $(2 \times 1) + 4 + 6 = 12$ valence electrons. We must decide between two possible skeletons:

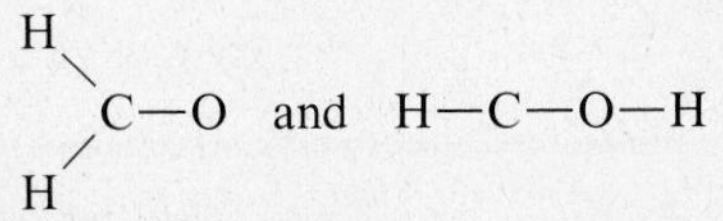

$$
\begin{array}{c}
H \\
\diagdown \\
\phantom{x}C-O \quad \text{and} \quad H-C-O-H \\
\diagup \\
H
\end{array}
$$

The number of electrons remaining to complete octets is $12 - 6 = 6$. The number of electrons needed to complete octets is 8 in either structure. One more bond is needed. When the bond is added and the octets are completed, the two structures are

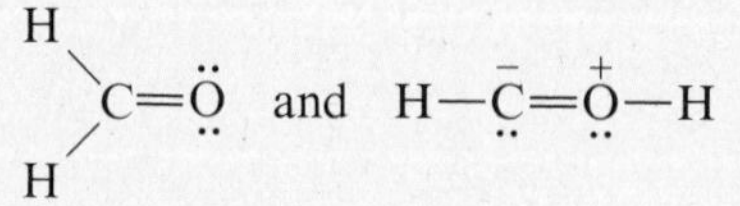

$$
\begin{array}{c}
\text{H} \\
\phantom{}\diagdown \\
\phantom{xx}\text{C}=\ddot{\text{O}} \quad \text{and} \quad \text{H}-\overset{-}{\underset{\phantom{x}}{\text{C}}}=\overset{+}{\ddot{\text{O}}}-\text{H} \\
\phantom{}\diagup \\
\text{H}
\end{array}
$$

The first structure is preferred because it does not have formal charges. This choice is verified by experiment.

---

**Example 7.9**    Draw the Lewis structure of chloric acid, $HClO_3$.

**Solution**    There are $1 + 7 + (3 \times 6) = 26$ valence electrons. Since this molecule is an acid, the skeleton will include an O—H bond. It will also include more than one bond to Cl, which is permissible when Cl is bonded to O. The best structure is

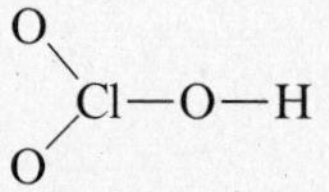

$$
\begin{array}{c}
\text{O} \\
\phantom{}\diagdown \\
\phantom{xx}\text{Cl}-\text{O}-\text{H} \\
\phantom{}\diagup \\
\text{O}
\end{array}
$$

The number of electrons remaining to complete octets is $26 - 8 = 18$. The number of electrons needed to complete octets is $(2 \times 6) + 2 + 4 = 18$. We can therefore write the structure:

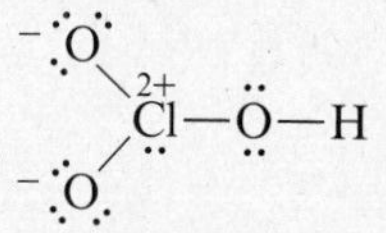

$$
\begin{array}{c}
{}^{-}\!\ddot{\text{O}}\!\!: \\
\phantom{}\diagdown{}^{2+} \\
\phantom{xx}\ddot{\text{Cl}}-\ddot{\text{O}}-\text{H} \\
\phantom{}\diagup \\
{}^{-}\!:\!\ddot{\text{O}}\!\!:
\end{array}
$$

Each of the singly bonded O atoms in this structure has a formal charge of $-1$, while the Cl atom has a formal charge of $7 - 2 - \frac{1}{2}(6) = +2$. Since expansion of the octet of Cl is permissible, we may write Lewis structures that reduce the formal charge.

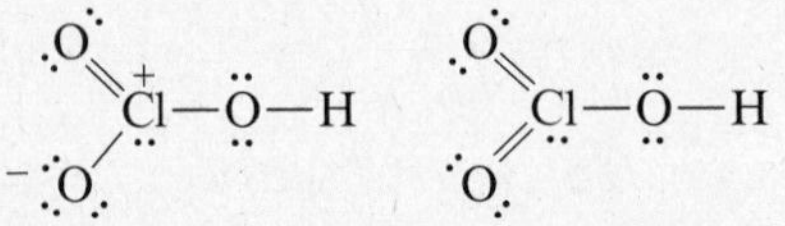

$$
\begin{array}{cc}
\begin{array}{c}
:\ddot{\text{O}} \\
\phantom{}\diagdown{}^{+} \\
\phantom{xx}\ddot{\text{Cl}}-\ddot{\text{O}}-\text{H} \\
\phantom{}\diagup \\
{}^{-}:\ddot{\text{O}}:
\end{array}
&
\begin{array}{c}
:\ddot{\text{O}} \\
\phantom{}\diagdown \\
\phantom{xx}\text{Cl}-\ddot{\text{O}}-\text{H} \\
\phantom{}\diagup \\
\ddot{\text{O}}:
\end{array}
\end{array}
$$

All three of these structures can be regarded as correct.

can write two structures:

$$O{=}N{-}O \longleftrightarrow O{-}N{=}O$$

In distributing the nonbonding electrons, we can write structures in which the N atom has an incomplete octet and has the unpaired or odd electron:

$$\ddot{O}{=}\dot{N}{-}\ddot{O}{:} \longleftrightarrow {:}\ddot{O}{-}\dot{N}{=}\ddot{O}$$

or structures in which the O has the incomplete octet, such as:

$$\ddot{O}{=}\ddot{N}{-}\ddot{O}{\cdot}$$

The first two structures, in which the N atom has an incomplete octet, make the most important contribution. Structures such as the last one, in which the O atom has the incomplete octet, are less important, since the more electronegative element usually has the completed octet.

The important point to remember is that resonance hybrids, when they are examined experimentally, are found to have bonds that are truly *intermediate* between those of the contributing structures, even though the contributing structures are imaginary constructs drawn for the sake of convenience. In $HNO_3$, both N—O bonds are found by experiment to be midway in characteristics between a single and double bond, and both O atoms in these bonds are found to be identical. In $NO_3^-$, the nitrate ion, a resonance hybrid with three contributing structures:

all three O atoms are found to be identical, as are all three N—O bonds.

## 7.6 OXIDATION NUMBERS

Every atom in every molecule can be assigned an *oxidation number,* a positive or negative integer that is related to the electronic structure of the molecule. We first imagine that the electrons of a bond are transferred completely to the more electronegative element of the bond. Then we calculate the resulting charge on each atom. That charge is the oxidation number. In any diatomic molecule, the more electronegative atom will have a negative oxidation number and the less electronegative atom will have a positive oxidation number.

For example, in the formation of LiF an electron is transferred almost completely from Li to F, so we think in terms of $Li^+F^-$. Thus, Li has an oxidation number of $+1$ and F has an oxidation number of $-1$. In HF, an electron is partially transferred from the H atom to the F atom, whose electronegativity is higher; H has an oxidation number of $+1$ and F has an oxidation number of $-1$. In the N$=$O molecule, the oxidation numbers are $+2$ for N and $-2$ for O. In $F_2$, each F atom has an oxidation number of 0.

**Example 7.12**    Calculate the oxidation numbers of the atoms in (a) $PCl_5$, (b) $BrF_3$, (c) $NO_2$.

**Solution**    a. The Lewis structure of $PCl_5$ is

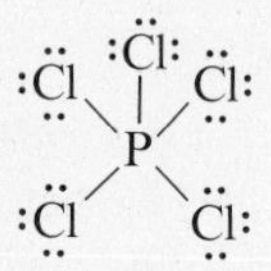

The Cl atom is more electronegative than the P atom. Let us assume that the pair of electrons in each P—Cl bond is transferred totally to the Cl atom. Its configuration will then be $:\ddot{\text{C}}\text{l}:$. A neutral chlorine atom has seven electrons, while $:\ddot{\text{C}}\text{l}:$ has eight. Therefore, $:\ddot{\text{C}}\text{l}:$ has a charge of $-1$ and the oxidation number of each Cl atom in $PCl_5$ is $-1$. If the transfer is total, the P atom will be left with no valence electrons. Since the atom originally had five valence electrons, its charge, and thus its oxidation number, is now $+5$.

b. The Lewis structure of $BrF_3$ is:

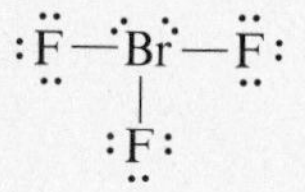

The F atom is more electronegative than the Br atom. If the bonding electron pairs are transferred completely to the F atom, its configuration is $:\ddot{\text{F}}:$, which corresponds to an oxidation number of $-1$. The Br atom loses three of its seven electrons. Therefore, its oxidation number is $+3$.

c. An important contributing structure for $NO_2$ is

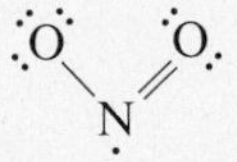

A complete transfer of the bonding electrons to the more electronegative O atoms gives each O atom eight electrons. Since the neutral O atom has only six electrons, the oxidation number of O is $-2$. The N atom is left with one electron. Since the neutral N atom has five electrons, the oxidation number of N in $NO_2$ is $+4$.

Oxidation numbers are purely arbitrary numbers calculated by certain rules. They are useful because they give information about both the combining power of elements and the bonding in molecules. They also provide a convenient framework for systematizing a great deal of data about chemical behavior.

A positive oxidation number means that the atom, on balance, loses electrons in the molecule. A negative oxidation number means that the atom gains electrons. The value of the oxidation number gives the number of electrons that the atom gains or loses, either partially or wholly. An oxidation number of $+2$ means that the atom contributes two electrons, or more than an equal share of two electrons, to bonds. A $-2$ oxidation number means that the atom gains two electrons, or more than an equal share of two electrons, from bonds.

## Assigning Oxidation Numbers

It is not always easy to calculate oxidation numbers for all atoms in a polyatomic molecule. Sometimes it is tedious to calculate the hypothetical charge on every atom. We can avoid this procedure by using some simple rules to assign oxidation numbers:

1. The sum of all the oxidation numbers of the atoms in an electrically neutral chemical substance must be zero. In $H_2S$, for example, each H atom has an oxidation number of $+1$, and the oxidation number of the S atom is $-2$. The sum is $2(+1) + (-2) = 0$.
2. In polyatomic ions, the sum of oxidation numbers must equal the charge on the ion. For example, in $NH_4^+$, each H atom has an oxidation number of $+1$, and the oxidation number of the N atom is $-3$, making the total $4(+1) + (-3) = +1$, which is the charge on the ion. In $SO_4^{2-}$, each O atom has an oxidation number of $-2$ (see rule 5). The sum of oxidation numbers for the ion must be $-2$, which is the charge on the ion. Since the sum of the oxidation numbers of the four O atoms is $-8$, the oxidation number of the S atom must be $+6$.
3. The oxidation number of an atom in a monatomic ion is its charge. In $Na^+$, sodium has an oxidation number of $+1$. In $S^{2-}$, sulfur has an oxidation number of $-2$.
4. The oxidation number of an atom in a single-element neutral substance is 0. Thus, the oxidation number of every sulfur atom in $S_2$, $S_6$, and $S_8$ is 0; the oxidation number of chlorine in $Cl_2$ is 0; the oxidation number of oxygen in $O_2$ or $O_3$ is 0. Every element has at least one form in which its oxidation state is 0.
5. Some elements have the same oxidation number in all or nearly all of their compounds. When F combines with other elements, its oxidation number is always $-1$. The halogens Cl, Br, and I have the oxidation number $-1$ except when they combine with oxygen or a halogen of lower atomic number. Oxygen usually has the oxidation number $-2$, except when it combines with F or with itself in such compounds as the peroxides or superoxides. A metal in group IA of the periodic table always has an oxidation number of $+1$. A metal in group IIA always has an oxidation number of $+2$.
6. Hydrogen always has the oxidation number $+1$ when combined with nonmetals and $-1$ when combined with metals.
7. Metals almost always have positive oxidation numbers.
8. A bond between identical atoms in a molecule makes no contribution

274 CHAPTER 7 The Chemical Bond

to the oxidation number of that atom because the electron pair of the bond is divided equally. In hydrogen peroxide, $H_2O_2$, for example, the two O atoms are bonded to one another. We can calculate the oxidation number of O by determining the contribution of the two H atoms, each of which has an oxidation number of $+1$. Since the sum of the oxidation numbers of the H atoms is $+2$ and the molecule is neutral, the sum of the oxidation numbers of the two O atoms is $-2$, giving each an oxidation number of $-1$.

**Example 7.13**

Calculate the oxidation number of all the atoms in (a) $KMnO_4$, (b) $H_5IO_6$, (c) $HPO_4^{2-}$.

**Solution**

a. Since $KMnO_4$ is neutral, the sum of the oxidation numbers is zero. The oxidation number of oxygen is $-2$, making its contribution $4(-2) = -8$. The oxidation number of K (group IA) is $+1$. If we let $x =$ the oxidation number of Mn, then $(-8 + 1) + x = 0$; $x = +7 =$ oxidation number of Mn.

b. The sum of the oxidation numbers in the neutral molecule $H_5IO_6$ is zero. The oxidation number of H is $+1$. The oxidation number of O is $-2$. We can obtain the oxidation number of I by letting $x =$ the oxidation number of I and solving $(5)(+1) + (6)(-2) + x = 0$; $x = +7$.

c. The sum of the oxidation numbers for $HPO_4^{2-}$ is $-2$, the charge on the ion. If $x =$ the oxidation number of P, solving for $(+1) + (4)(-2) + x = -2$ gives $x = +5$.

**Example 7.14**

Calculate the oxidation numbers of all the atoms in (a) $Cr_2O_7^{2-}$, dichromate ion, and (b) $Fe_3O_4$, magnetite.

**Solution**

a. Applying the method of Example 7.13, $2Cr + 7(-2) = -2$; $2Cr = +12$, and each Cr atom has an oxidation number of $+6$.

b. The method of Example 7.13 does not give a simple answer for $Fe_3O_4$, because Fe seems to have a fractional oxidation number of $+8/3$. A better approach is to assume that there are two types of Fe atoms in this molecule, an assumption that is consistent with the experimental evidence. If $Fe_3O_4$ is considered to be $FeO + Fe_2O_3$, then one Fe atom has the oxidation number $+2$ and two Fe atoms have the oxidation number $+3$. We shall encounter other examples where either a fractional oxidation number or two different oxidation numbers are found for an element. Often, experimental evidence is needed to make correct assignments of oxidation numbers in such situations.

Oxidation numbers define **oxidation states.** The two terms often are used interchangeably. Oxidation states help us to understand chemical processes in which there is oxidation and reduction of elements.

The terms *oxidation* and *reduction* are defined as changes in oxidation state. If a chemical reaction changes the oxidation states of some atoms in the reactants, the process is called an **oxidation-reduction** or **redox** reac-

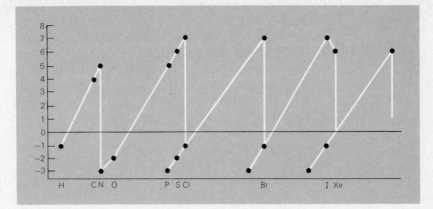

tion. The substance in which an atom has an *increase* in oxidation state is said to be **oxidized.** The substance in which an atom has a *decrease* in oxidation state is said to be **reduced.** Oxidation and reduction always occur together. There can be no oxidation without an accompanying reduction.

In Section 6.8, we discussed a redox process that is brought about by an electric current, the electrolysis of molten sodium chloride:

$$NaCl \longrightarrow Na + \tfrac{1}{2}Cl_2$$

In this reaction, the oxidation number of Na goes down from $+1$ to $0$ and the oxidation number of Cl goes up from $-1$ to $0$. Sodium is reduced; chlorine is oxidized. We shall discuss redox processes in detail in Chapter 16.

As Figure 7.16 shows, the characteristic oxidation states of the main group elements are related directly to the position of the elements in the periodic table. This relationship is to be expected, since the oxidation number of any atom is related to its electronic configuration. The oxidation number of a main group element often is the same as the number of electrons the element must gain or lose to achieve a noble gas electronic configuration. However, some atoms can lose *p* electrons but retain a complete *s* subshell. For example, the heavier elements of groups IIIA and IVA can have oxidation numbers of $+1$ and $+2$, respectively, in addition to the expected values of $+3$ and $+4$. Thus, Tl can form either TlCl or $TlCl_3$, while Pb can form either PbO or $PbO_2$.

It is very rare for any element to have a positive oxidation number that is higher than the number of its group in the periodic table. The highest oxidation number of P is $+5$; the highest oxidation number of S is $+6$; and the highest oxidation number of Cl is $+7$. In general, negative oxidation numbers are improbable in elements with low electronegativity values. Likewise, the most negative oxidation number that a nonmetal may have is its group number minus 8. The lowest oxidation number of P is $5 - 8 = -3$, the lowest oxidation number of C is $4 - 8 = -4$, and the lowest oxidation number of S is $6 - 8 = -2$.

**Summary**

We began this chapter by describing two idealized types of chemical bonding: **ionic bonds,** in which one atom loses electrons to become a cation and another gains electrons to become an anion; and **covalent bonds,** in which two atoms share the electrons of the bond completely. We noted that while we usually describe a bond as either covalent or ionic, most bonds have some characteristics of each type. In a discussion of ionic compounds, we said that they normally are solids in which the ions are in a regular, repeating arrangement called a **crystal.** We noted that the **lattice energy,** the energy released when the crystal is formed, is a measure of the strength of an ionic bond. We described the factors such as **ionic radius** and **ionic charge** that influence lattice energy. We then differentiated between **polar bonds,** in which two atoms that are bonded covalently have partial electrical charge, and **nonpolar bonds,** in which the atoms have no such partial charges. We introduced **electronegativity,** a measure of the ability of an atom to attract electrons to itself when combined with other atoms, and showed how it gives us a measure of bond type. We then described **Lewis structures,** diagrams of covalently bonded species that use dots to represent electrons. We stressed that most Lewis structures satisfy the **octet rule,** meaning that each atom in the structure achieves a noble gas electronic configuration, and we outlined the rules for writing Lewis structures. We mentioned that **formal charges** are sometimes useful in writing Lewis structures, but noted that formal charges are to be avoided whenever possible — a rule called the **electroneutrality principle.** We then defined **resonance,** a phenomenon that occurs when we can write two or more Lewis structures, called contributing structures, for the same compound, that differ only in the location of multiple bonds. We added that contributing structures do not actually exist, and that **resonance hybrids** — species that display resonance — actually are blends of all their contributing structures. We concluded with a discussion of **oxidation numbers,** positive and negative integers that are related to the electronic structures of molecules and are very useful for correlating a great deal of chemical behavior.

## Exercises

**7.1** Write the formulas of the following ionic solids: (a) calcium chloride, (b) sodium phosphate, (c) ammonium carbonate, (d) cesium nitride.

**7.2** What is the effect on the potential energy of a system of many opposite charges of an increase in (a) cationic charge, (b) anionic charge, (c) distance

[^2]: ² The answers to exercises whose numbers are in color can be found in Appendix VII. The star indicates an exercise that is more challenging than average.

between charges. Explain your answers.

**7.3** Rank the following cations by the ease with which they form from atoms of their elements in the gas phase, starting with the most easily formed: $Li^+$, $Na^+$, $K^+$, $Ca^+$, $Ca^{2+}$, $Mg^{2+}$, $Al^{3+}$, $Li^{2+}$. (*Hint:* Consult Table 6.7.)

**7.4** Rank the following anions by the ease with which they form from atoms of their elements in the gas phase, starting with the most easily formed: $F^-$,

$Cl^-$, $I^-$, $Cl^{2-}$, $O^{2-}$, $N^{3-}$, $S^{2-}$. (*Hint:* Consult Table 6.8.)

**7.5**² From each set of ions, select the separated pair that requires the least energy to form from atoms of the elements in the gas phase: (a) $Na^+$ and $Cl^-$, $K^+$ and $Cl^-$, $Rb^+$ and $Cl^-$; (b) $Na^+$ and $F^-$, $Na^+$ and $Cl^-$, $Na^+$ and $Br^-$; (c) $Ca^{2+}$ and $O^{2-}$, $Ca^{2+}$ and $O^-$, $Ca^{2+}$ and $N^{3-}$.

**7.6** Using the data in Tables 6.7 and 6.8, calculate the energy required to form the following separated pairs of ions from the corresponding atoms in the gas phase: (a) $Cs^+$ and $F^-$, (b) $Mg^{2+}$ and $I^-$, (c) $Al^{3+}$ and $Cl^-$.

**7.7** List the following pairs of gas phase ions in the order of the quantity of energy released when they are formed from completely separated gas phase ions: $K^+Cl^-$, $Ca^{2+}Cl^-$, $Al^{3+}O^{2-}$, $Ca^{2+}O^{2-}$.

**7.8** List the following gas phase ion pairs in order of the quantity of energy required to separate them into single ions, starting with the most easily separated: $Li^+F^-$, $Na^+F^-$, $Na^+Br^-$, $K^+Br^-$, $Rb^+Br^-$.

**7.9** The potential energy of a system of charges is $E = kq_1q_2/r$, where $k$ is a proportionality constant equal to $9.0 \times 10^9$ Nm$^2$/C$^2$. The charge on the electron is $1.6 \times 10^{-19}$ C. Find the energy released when 1.0 mol of $Li^+F^-$ ion pairs forms. The distance between the ions is 0.20 nm.

**7.10*** The force of attraction between the two members of an ion pair with charges of $\pm 1$ is found to be $5.8 \times 10^{-9}$ N. Find the difference between the force of attraction of two isolated ion pairs and two ion pairs arranged with the ions at the corners of a square, as shown in Figure 7.1

**7.11** Assuming that the following solids have the same geometry, arrange them in order of increasing lattice energy: $MgF_2$, $MgCl_2$, $CaCl_2$, $CaBr_2$, $MgO$.

**7.12** The energy released by the formation of one mole of the ion pair $Na^+Br^-$ in the gas phase is 478 kJ/mol. The geometry of the sodium-bromide crystal is the same as that of the NaCl crystal. Calculate the lattice energy of NaBr from these data.

**7.13** The energy released when an $Na^+Cl^-$ ion pair forms in the gas phase is 585 kJ/mol. The geometric factor for the NaCl structure is 1.75. The lattice energy of the solid is observed to be less than $1.75 \times 585$ kJ/mol. Explain why.

**7.14** Suggest an explanation for the observation that anions generally have larger ionic radii than cations.

**7.15** The three major influences on lattice energy are (i) ionic charge, (ii) ionic radius, (iii) geometry. Which influence is most responsible for each of these observations:
(a) The lattice energy of MgS is less than that of CaO.
(b) The lattice energy of MgS is greater than that of NaCl.
(c) The lattice energy of MgS is greater than that of BeS.

**7.16** The lattice energy of NaCl is 770 kJ/mol and the lattice energy of $MgCl_2$ is 2500 kJ/mol, yet the charge on $Mg^{2+}$ is only twice that of $Na^+$. Explain why the lattice energy of $MgCl_2$ is so much greater.

**7.17** The lattice energy of CaBr is calculated to be greater than that of KBr, yet solid CaBr cannot be prepared. Account for this observation.

**7.18*** Ionic solids of the $O^-$ and $O^{3-}$ anions do not exist, while ionic solids of the $O^{2-}$ anion are common. Explain.

**7.19** List the oxides of these elements in order of increasing ionic bond character: potassium, calcium, gallium, zinc, manganese.

**7.20** List the following compounds in order of increasing ionic bond character: NaF, RbCl, $SrCl_2$, $AlBr_3$, $PBr_3$.

**7.21** List the following substances in order of decreasing ionic bond character: HBr, $Br_2$, LiBr, $NBr_3$, $PBr_3$, $CBr_4$.

**7.22** Classify the following substances as ionic, polar, or nonpolar: (a) NaF, (b) $NF_3$, (c) $NBr_3$, (d) ClBr, (e) HBr.

**7.23** Arrange the following substances in order of increasing melting point: KBr, $F_2$, $SO_3$, $C_3H_8$.

**7.24** Elements 3 through 9 each tend to form a characteristic number or numbers of bonds. Find the relationship between the group number of an element and the number of bonds it forms.

**7.25** Find the relationship between the electronic configuration of the $p$ subshell in a nonmetal and the number of bonds that the nonmetal usually forms.

**7.26** Draw Lewis structures for the following chlorine compounds: (a) $CCl_4$, (b) $NCl_3$, (c) $Cl_2O$, (d) ClF.

**7.27** Draw Lewis structures for the following hydrides: (a) $SiH_4$, (b) $PH_3$, (c) $H_2S$, (d) HCl.

**7.28** Draw Lewis structures for the following compounds: (a) HClO, (b) $H_2S_2$, (c) $H_2SO_3$, (d) HNO.

**7.29** Draw Lewis structures for the following compounds: (a) $C_2H_2$, (b) $CO_2$, (c) $As_2O_3$ (which contains an As—As bond), (d) $S_8$ (a cyclic compound).

**7.30** Draw Lewis structures for the following oxyacids: (a) $H_2CO_3$, (b) $H_3PO_4$, (c) $HClO_2$, (d) $H_4SiO_4$.

**7.31** Draw Lewis structures for the following ions: (a) $O_2^{2-}$, (b) $CN^-$, (c) $HPO_4^{2-}$, (d) $HCO_2^-$ (formate).

**7.32** Draw Lewis structures for the following cations: (a) $NH_4^+$, (b) $CH_3^+$, (c) $PCl_4^+$, (d) $C_2H_4Br^+$.

**7.33** There are two compounds with the formula $C_2H_6O$. Their skeleton structures are different. Neither has multiple bonds. Draw the Lewis structures of these two compounds.

**7.34** A molecule with the formula $C_3H_6$ has no multiple bonds but satisfies the octet rule. Write its Lewis structure.

**7.35*** Three compounds have the formula $C_2H_4O$. Two have multiple bonds; one does not. Draw the Lewis structures of these three compounds.

**7.36** Draw Lewis structures for the following molecules: (a) $BI_3$, (b) $AlCl_3$, (c) $ZnBr_2$, (d) $CuI$.

**7.37** Draw Lewis structures for the following ions: (a) $NO^+$, (b) $NO_2^+$, (c) $Hg_2^{2+}$, (d) $AlCl_4^-$.

**7.38** Which of the following molecules has no expanded octets around any of its atoms: (a) $BrF$, (b) $ClF_3$, (c) $N_2H_2$, (d) $HBrO_3$.

**7.39** Draw Lewis structures for the following molecules: (a) $PCl_5$, (b) $TeF_6$, (c) $IF_7$.

**7.40** Draw Lewis structures for the following molecules: (a) $SF_4$, (b) $BrCl_3$, (c) $H_5IO_6$.

**7.41** Draw Lewis structures for the following ions: (a) $SiF_6^{2-}$, (b) $BrO_3^-$, (c) $I_3^-$.

**7.42** Which of the following cannot completely satisfy the octet rule? (a) $ClF$, (b) $ClO_2$, (c) $NO_2$, (d) $N_2O_5$.

**7.43** Draw Lewis structures for the following species: (a) $NO$, (b) $ClO$, (c) $O_2^-$.

**7.44** Draw Lewis structures for the following molecules, indicating any formal charges in each structure: (a) $N_2O_3$ (N—N bond), (b) $AlCl_3NH_3$ (Al—N bond), (c) $CH_3NC$ (all H atoms attached to the same C atom), (d) $CO$.

**7.45** The compound $SO_3$ has three O atoms on a central sulfur atom. Write the structure of this compound in two ways: (a) with octet expansion and no formal charges, (b) with formal charges and no octet expansion.

**7.46** Draw two possible Lewis structures for the compound with the formula $CNBr$. Predict which structure is more likely to be correct.

**7.47** Phosgene, a compound with the structure $COCl_2$, was used as a poison gas in World War I. Draw its Lewis structure, which contains no formal charges.

**7.48** Draw two structures corresponding to the compound $NH_3O$. Indicate whether each structure can exist.

**7.49** Draw the important contributing structures for the following molecules: (a) $N_2O_4$ (N—N bond), (b) $CH_3NO_2$ (3 C—H bonds), (c) $O_3$.

**7.50** Draw the important contributing structures for the following ions: (a) $NO_2^-$, (b) $HPO_3^{2-}$ (P—H bond), (c) $CO_3^{2-}$.

**7.51★** Diazomethane, $CH_2N_2$, is poisonous and explosive. Both H atoms are attached to the C atom. The compound has two major contributing structures, both with formal charges. Draw them.

**7.52** Find the oxidation number of

the metal in (a) $TiCl_4$, (b) $HgO$, (c) $Mn_2O_7$, (d) $Hg_2Cl_2$, (e) $Nb_2O_5$, (f) $Na_2O_2$.

**7.53** Find the oxidation number of the metal in each of the following ions: (a) $CoCl_6^{3-}$, (b) $Cr_2O_7^{2-}$, (c) $CrO_4^{2-}$, (d) $Mo_3Br_4^{2+}$.

**7.54** Find the oxidation numbers of all the atoms in (a) $SO_2Cl$, (b) $SO_2Cl_2$, (c) $NH_4HSO_4$, (d) $Pt(NH_3)_2Br_2$, (d) $KAuF_4$.

**7.55** Chlorine forms binary monoanions with oxygen corresponding to the oxidation numbers $+1$, $+3$, $+5$, and $+7$ for Cl. Draw the Lewis structures of these ions.

**7.56** Find the oxidation numbers of each atom in (a) $H_2S_3$, (b) $Fe_3O_4$ (the O atoms all have the oxidation number, $-2$).

**7.57** Indicate whether each conversion is an oxidation, a reduction, or neither: (a) aluminum metal to aluminum cation, (b) bromine to bromide, (c) $H_2O$ to $H_2O_2$, (d) $H_2SO_4$ to $H_2SO_5$, (e) $H_2SO_4$ to $H_2S_2O_7$, (f) $NO_2$ to $N_2O_5$.

**7.58★** The thiosulfate ion $S_2O_3^{2-}$ has a number of possible structures. The two sulfur atoms act as if they have two oxidation numbers, $+6$ and $-2$. Draw a structure consistent with this information.

# 8

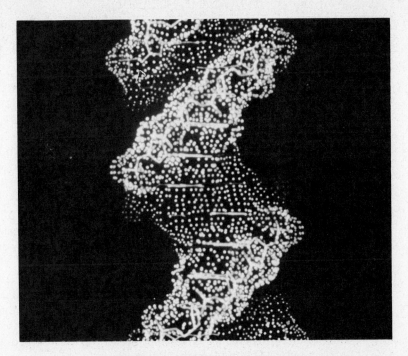

# Molecular Structure and Stability

**Preview**

**O**ur discussion of the chemical bond was a prelude to the study of the structure of the molecules that are formed by chemical bonding. We first show how molecular structure is described in terms of bond lengths and bond angles and note that spectroscopy and other experimental methods are used to measure them. We then describe the valence shell electron pair repulsion (VSEPR) model for molecular geometry. We show how covalent bonds can be pictured as forming from the overlap of atomic orbitals. We then describe the hybridization model, an alternative to the VSEPR model. We outline the relationship between the structure of molecules and such properties as dipole moment and bond length. Then we show how molecules can be described in terms of molecular orbitals that are formed by the overlap of atomic orbitals. We discuss how thermochemistry, the study of the heat changes that occur during chemical reactions, is related to molecular structure and stability. Finally, we describe weak interactions, which are attractive forces between molecules.

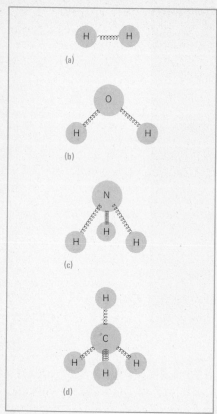

**Figure 8.1**
Ball-and-spring models of molecules. The balls are atoms, the springs are the bonds between atoms. (a) $H_2$, (b) $H_2O$, (c) $NH_3$, (d) $CH_4$.

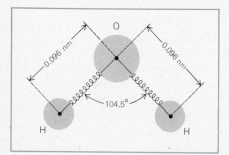

**Figure 8.2**
Bond angle and bond length illustrated for the $H_2O$ molecule. Bond length is the distance between the centers of the nuclei of the O and the H atoms, 0.096 nm in this molecule. Bond angle is the angle formed by imaginary lines connecting the O atom with the two H atoms, 104.5°.

The development of methods for representing molecular structures led to the growth of a major branch of chemistry called structural chemistry. It deals with the shapes of molecules and the nature of the bonds that hold them together.

In this chapter we shall discuss some details of the structure of molecules and some properties of bonds that give us information about molecular structure. We shall also present some of the theory that explains the experimental data on molecular structure.

Molecules are fairly rigid. One common way to picture a molecule is as a collection of balls, the atoms, connected by springs, the bonds (Figure 8.1). The "springs" are not very flexible and they allow only limited movement of the atoms relative to one another.

Molecular structure is usually discussed in terms of **bond length** and **bond angle.** *Bond length is the mean distance between the centers of the nuclei of two bonded atoms,* as shown in Figure 8.2. A definition of bond angle requires a molecule with at least three atoms. *Bond angle is the mean angle formed by the two imaginary lines that connect the central atom with the two atoms on either side of it.* In the water molecule H—O—H, for example, the bond angle is the angle formed by the lines representing the two O—H bonds. We can describe the geometry of the water molecule by saying that the H—O—H bond angle is 104.5° and the H—O bond length is 0.096 nm (0.96 Å).

Less precisely, we can just say that $H_2O$ is a bent molecule. A molecule such as $CO_2$, whose O—C—O bond angle is 180°, is characterized as linear, rather than bent. Both $CO_2$ and $H_2O$ are planar, meaning that the centers of the three atoms lie in a plane. All three-atom molecules are planar.

In molecules with more than three atoms, it is often necessary to indicate the relative orientation of atoms that are not bonded directly to one another. Many three-dimensional representations are used to describe such molecules. The branch of chemistry that is concerned with the three-dimensional structure, bond angles, and bond lengths of molecules is called **stereochemistry.**

The relative stability of a molecule can often give us valuable information about its structure. The energy change that occurs in the formation of a molecule is a measure of its stability. When bonds form, energy is liberated, usually in the form of thermal energy (heat). To break bonds, energy must be supplied, usually in the form of thermal energy. The branch of chemistry that measures the heat effects associated with chemical processes is called **thermochemistry.**

## 8.1   EXPERIMENTAL METHODS

Most of the experimental methods that help us to understand the structure of molecules are based on the interaction of matter and radiant energy. These methods usually require complicated instruments that have become available only in the past few decades. The enormous increase in chemical knowledge in recent years has resulted in large part from the development of sophisticated chemical instrumentation, especially in combination with the computer. Today, a team of chemists working on a large research project usually includes several members whose specialty is instrumentation.

One powerful group of methods for structure determination uses radiation or subatomic particles. When radiation such as X rays or particles such as electrons or neutrons are directed at a crystalline substance, the interaction can create a diffraction pattern. By studying this pattern, experts can determine the molecular structure responsible for it. Generally, diffraction experiments give the location of atomic nuclei relative to one another in a molecule quite accurately. One of the more memorable uses of diffraction patterns in recent decades was the determination of the structure of the DNA (deoxyribonucleic acid) molecule, the core material of the gene, an intellectual adventure described vividly in James Watson's book, *The Double Helix.*

Diffraction studies are the closest available approach to the direct observation of molecular structure. Other highly specialized methods are available for the measurement of dipole moments and thermochemical quantities.

## Spectroscopy

Spectroscopy, which studies the interaction of matter with electromagnetic radiation, is the technique most often used to explore molecular structure. In Chapter 5, electromagnetic radiation was described as an oscillating electric-magnetic field. Ordinary spectroscopy studies the interaction of the atom's electrical charges with the electric component of electromagnetic radiation. The branch of spectroscopy called magnetic resonance studies the interaction of the magnetic component of electromagnetic radiation with the magnetic dipoles within molecules.

Usually, a relatively narrow wavelength range of electromagnetic radiation is used for a specific spectroscopic study. We speak of infrared spectroscopy or ultraviolet-visible spectroscopy or microwave spectroscopy, depending on the wavelengths of the radiation used for the study.

In a typical experiment, a beam of electromagnetic radiation is directed at a sample of a substance. The wavelength of the radiation is varied continuously within the limits of the instrument, and the extent to which radiation of a given wavelength is absorbed by the substance is measured. The results can be recorded as a graph of the extent of absorption as a function of wavelength. This graph is called an *absorption spectrum* (Figure 8.3).

The determination of a spectrum once was a tedious job. Today, spectra are determined by instruments called *spectrometers,* which even draw the final graph. An ultraviolet spectrum of the type in Figure 8.3 can be obtained in 10 minutes on a spectrometer that can be operated by a

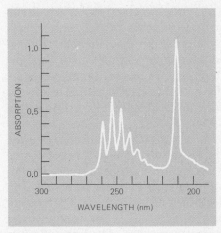

**Figure 8.3**

An absorption spectrum of a molecule. The peaks show the wavelengths of radiation absorbed by the molecule. Analysis of such spectra gives valuable information about the structure of molecules.

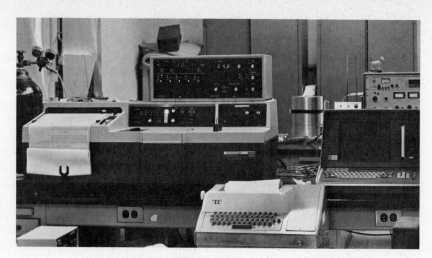

A typical example of the sophisticated equipment used in today's chemistry laboratory to investigate the structure and properties of molecules. *(Robert Boikess)*

competent chemistry student after 30 minutes of training.

Spectroscopy is a valuable method of examining molecules because any given molecule absorbs only selected wavelengths of electromagnetic radiation. Parts of the absorption spectrum are unique for each compound, as fingerprints are unique for each human. Analysis of the absorption spectrum pattern can yield a great deal of information about the molecule.

## 8.2 THE VALENCE SHELL ELECTRON PAIR REPULSION (VSEPR) MODEL

In Chapter 7 we presented Lewis structures, which provide a very useful method for representing the structures of covalent molecules. Lewis structures, however, do not give any information about the actual geometry of the molecule. To understand the geometry of molecules, we must look at them in different ways. These ways of looking at molecules are called models. We shall present the two simplest models for molecular geometry: the VSEPR model and the directed valence or hybridization model.

The *valence shell electron pair repulsion (VSEPR) model* explains the observed geometry of molecules and polyatomic ions by an electrostatic picture based primarily on the Lewis structure, without reference to orbitals. The model assumes that each electron pair in the valence shell of an atom in a molecule occupies a well-defined region of space. The concept of a central atom is important in the VSEPR model. A central atom is any atom that is bonded to two or more other atoms. The first and most important rule of the VSEPR model is that *the bond angles about a central atom are those that minimize the total repulsion between the electron pairs in the valence shell of the atom.* We shall use the word *geometry* to mean the arrangement of the atoms about the central atom.

## CONSTRUCTING MOLECULAR MODELS

Most college bookstores sell relatively inexpensive collections of balls and sticks, reminiscent of the Tinker Toys of our younger days, called molecular model kits. With a molecular model kit you can build a three-dimensional model of a molecule that can help you visualize spatial relationships between atoms not directly bonded to each other. In these "ball-and-stick" models, the balls represent atoms and the sticks represent bonds. The holes in the balls hold the sticks at angles similar to the actual bond angles in the molecule.

Ball-and-stick models give us an approximate picture of molecules. We can construct very precise scale models of molecules by using other types of molecular model kits, which are usually quite expensive. One kit consists of stainless steel rods that snap into stainless steel sleeves to represent bonds. The bond lengths and bond angles are made accurately to scale in these kinds of models, but the atoms are not shown at all. They are understood to be at the intersections of the bonds. There are also space-filling models, which take the opposite approach. In space-filling models, the atoms are represented by truncated balls that are held together by snap fasteners and the bonds are not shown. The atomic radii are made accurately to scale in these models. Each type

of model emphasizes different aspects of molecular geometry, as you can see in the three models of ethane, $C_2H_6$, shown below.

Although these models are extremely useful, they have drawbacks. A model of a large, complicated molecule is tedious and expensive to construct and is fragile. In recent years, chemists have been able to obtain better models of molecules by using increasingly sophisticated computer graphics programs.

A typical computer program includes a memory store called a data base that contains the basic bonding characteristics of many elements, the characteristic bond angles of the atoms, and similar data. To construct a molecular model, the user specifies the chemical composition of the molecule. The computer program generates a three-dimensional display in which bonds are shown as lines and atoms are indicated as the points at which the lines intersect, much like a framework model. Many programs allow the user to rotate the model, so that it can be viewed from different angles. More advanced programs can give stereo models, so that the three-dimensional structure of exceptionally complicated molecules can be studied in detail, or allow the user to see how the substitution of different atoms affects the structure of the molecule.

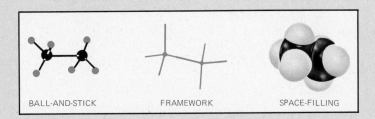

BALL-AND-STICK          FRAMEWORK          SPACE-FILLING

There is a simple procedure for predicting the geometry of many molecules:

1. Draw the Lewis structure of the molecule.
2. Identify the central atom or atoms. (For simplicity, we shall start with molecules that have only one central atom.)
3. Count the number of atoms and the number of lone pairs (nonbonding

| TABLE 8.1 | Electron Pairs and Ideal Geometries |
|---|---|
| **Number of Pairs** | **Geometry** |
| 2 | linear |
| 3 | trigonal planar |
| 4 | tetrahedral |
| 5 | trigonal bipyramidal |
| 6 | octahedral |

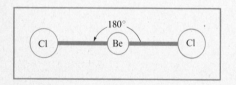

**Figure 8.4**
The $BeCl_2$ molecule has linear geometry. This arrangement minimizes the repulsions between the two valence shell electron pairs.

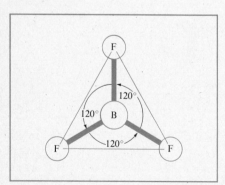

**Figure 8.5**
The $BF_3$ molecule has trigonal planar geometry. This arrangement minimizes the repulsions among the three valence shell electron pairs.

valence shell electron pairs) about the central atom. This total is considered to be the number of valence shell electron pairs. An atom bonded by a double or triple bond is counted only once.

4. The arrangement of the atoms around the central atom corresponds to a simple geometric construction called an *ideal geometry*. Using Table 8.1, identify the ideal geometry associated with the number of *valence shell electron pairs*. This ideal geometry usually is close to the actual geometry of the molecule. A few molecules have exactly the ideal geometry.

5. Correct the ideal geometry in molecules that have (a) multiple bonds, (b) lone pairs (nonbonding valence electron pairs) around the central atom, and (c) more than one type of atom bonded to the central atom.

**Molecules with Ideal Geometry**

Several kinds of molecules have the ideal geometries described by the VSEPR model. They include:

1. Compounds with two valence shell electron pairs around a central atom. The halides of group IIA and group IIB metals are examples of such compounds. For example, the Lewis structure of $BeCl_2$ is

$$: \ddot{C}l - Be - \ddot{C}l :$$

The central Be atom has two atoms bonded to it and no lone pairs. The VSEPR model correctly predicts that such a molecule will have linear geometry (Table 8.1). The three atoms of this molecule lie on a straight line. As Figure 8.4 shows, the Cl—Be—Cl angle is 180°.

2. Compounds with three valence shell electron pairs around a central atom. Compounds of boron and some compounds of aluminum are examples. The Lewis structure of $BF_3$ is

The central B atom has three atoms bonded to it and no lone pairs. The VSEPR model correctly predicts trigonal planar geometry for such a molecule (Table 8.1). The three F atoms lie at the corners of an equilateral triangle, with the B atom at the center of the triangle. The F—B—F bond angles are 120° and all the atoms are in the same plane, as Figure 8.5 shows.

3. Compounds with four valence shell electron pairs around a central atom. Compounds of carbon with no multiple bonds are examples. So are some compounds of the other group IVA elements. Unlike the compounds of Be and B described above, compounds with four va-

lence shell electron pairs around a central atom have completed octets. A typical example is $CCl_4$, which has the Lewis structure

$$:\ddot{C}l:$$
$$:\ddot{C}l\!-\!\underset{\displaystyle |}{C}\!-\!\ddot{C}l:$$
$$:\ddot{C}l:$$

The central C atom has four atoms bonded to it and no lone pairs. The VSEPR model correctly predicts tetrahedral geometry (Table 8.1). The four Cl atoms lie at the corners of a regular tetrahedron, as Figure 8.6 shows. The Cl—C—Cl bond angles are all 109°28′.

4. Compounds with five valence shell electron pairs around a central atom. Examples include compounds of phosphorus in the +5 oxidation state with no multiple bonds and compounds of the heavier group VA elements. For example, the Lewis structure of $PCl_5$ is

$$:\ddot{C}l:$$
$$:\ddot{C}l\!-\!P\!-\!\ddot{C}l:$$
$$:\ddot{C}l:\quad:\ddot{C}l:$$

The VSEPR model correctly predicts trigonal bipyramidal geometry (Table 8.1). The five Cl atoms of $PCl_5$ lie at the corners of a geometric figure called a trigonal bipyramid, shown in Figure 8.7. Unlike those found in the tetrahedral, planar, and linear geometries, the five positions — and hence the five bonds — of a trigonal bipyramid are not all equivalent. Two of the positions are called *axial*. In the $PCl_5$ molecule they form a linear system, Cl—P—Cl. The other three positions are called *equatorial*. They lie in a plane at the corners of an equilateral triangle,

$$\underset{Cl\quad\quad Cl}{\overset{\displaystyle Cl}{\underset{\displaystyle |}{P}}}$$

similar to that in $BF_3$.

The bond angle between the equatorial Cl atoms is 120°, the bond angle between the two axial Cl atoms is 180°, and the bond angle between an axial Cl atom and an equatorial Cl atom is 90°. The axial bonds are somewhat longer than the equatorial bonds in such molecules. The two axial Cl atoms and the three equatorial Cl atoms in $PCl_5$ can be distinguished experimentally because of the different bond angles and bond lengths.

5. Compounds with six valence shell electron pairs around a central atom.

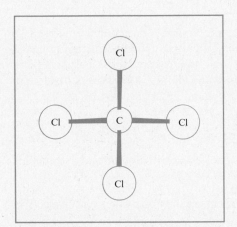

**Figure 8.6**
The geometry of the $CCl_4$ molecule. The four Cl atoms lie at the corners of a tetrahedron, with the C atom at its center. This configuration minimizes the repulsive forces among the valence-shell electron pairs of the C atom.

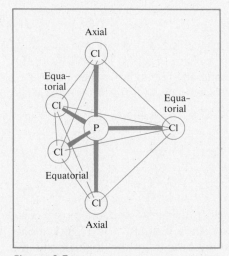

**Figure 8.7**
In the $PCl_5$ molecule, the five Cl atoms lie at the five corners of the trigonal bipyramid. The repulsion of the five valence shell electron pairs is minimized in this arrangement.

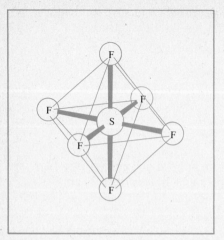

**Figure 8.8**

In the $SF_6$ molecule, the six F atoms lie at the six corners of the octahedron. The repulsion of the six valence shell electron pairs is minimized in this arrangement.

Halides of the group VIA elements in the $+6$ oxidation state are examples. The Lewis structure of $SF_6$ is

$$\begin{array}{ccc} :\ddot{F}: & & \ddot{F}: \\ & \diagdown \ \diagup & \\ :\ddot{F}- & \hspace{-0.3em}S\hspace{-0.3em} & -\ddot{F}: \\ & \diagup \ \diagdown & \\ :\ddot{F} & & \ddot{F}: \end{array}$$

The VSEPR model correctly predicts octahedral geometry (Table 8.1). The six F atoms lie at the corners of a regular octahedron, as Figure 8.8 shows. An octahedron is a regular solid figure whose corners are all identical. Therefore, all the bonds and all the F atoms are equivalent in the $SF_6$ molecule. The F—S—F bond angles are 90° or 180°.

## Molecules with Multiple Bonds

The VSEPR model works well for molecules with double and triple bonds if we introduce two rules:

*Rule 1.* A double or triple bond counts as only one valence shell electron pair.

*Rule 2.* Actual bond angles that include multiple bonds generally are larger than those of the ideal geometry.

The geometry of the HCN molecule illustrates Rule 1. The Lewis structure is

$$H-C\equiv N:$$

Because the triple bond counts as only one electron pair, we count two valence shell electron pairs around the central C atom. The VSEPR model correctly predicts linear geometry and an HCN bond angle of 180°.

The geometry of the phosgene molecule, $Cl_2CO$, illustrates Rule 2. The Lewis structure is

$$\begin{array}{c} :\ddot{Cl} \\ \hspace{1em}\diagdown \\ \hspace{2em}C=\ddot{O} \\ \hspace{1em}\diagup \\ :\ddot{Cl} \end{array}$$

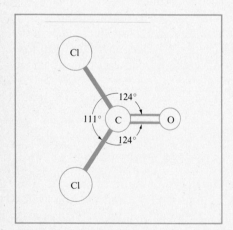

**Figure 8.9**

The experimentally determined bond angles of phosgene are close to the value in ideal trigonal planar geometry. The VSEPR model correctly predicts that the ClCO angle is slightly larger than 120° and the ClCCl angle is smaller.

The double bond is counted as one electron pair, so we count three pairs of valence shell electrons around the central C atom. The VSEPR model predicts trigonal planar geometry. But because of the double bond, the model also predicts that the Cl—C—O bond angles will be slightly larger than the ideal value of 120°, and that the Cl—C—Cl bond angles will be somewhat smaller than 120°. Experimental evidence verifies these predictions, as Figure 8.9 shows.

**Example 8.1**   Use the VSEPR model to predict the geometry of the following molecules or ions: (a) $AsF_5$, (b) $CH_3^+$, (c) $CO_2$, (d) $C_2H_4$.

**Solution**   a.  **Step 1.** Draw the Lewis structure.

$$
\begin{array}{c}
\ddot{\text{:}}\ddot{\text{F}}\text{:} \\
| \\
\text{:}\ddot{\text{F}}\!\!-\!\!\text{As}\!\!-\!\!\ddot{\text{F}}\text{:} \\
\diagup \quad \diagdown \\
\text{:}\ddot{\text{F}} \qquad \ddot{\text{F}}\text{:}
\end{array}
$$

**Step 2.** Identify the central atom or atoms. In this molecule, the As atom is the only central atom.

**Step 3.** Count the number of valence shell electron pairs around the central atom. There are five atoms about the As atom, and it has no lone pairs. Therefore, it has five valence shell electron pairs.

**Step 4.** Identify the ideal geometry. We predict trigonal bipyramidal geometry for a central atom with five valence shell electron pairs.

**Step 5.** Correct the ideal geometry. No correction is necessary in this case. The As atom has no lone pairs and no multiple bonds, and all the atoms around the As atom are identical. Thus, each F atom lies at the corner of a trigonal bipyramid.

b.  **Step 1.**

$$
\begin{array}{c}
\text{H} \\
| \\
\text{C}^+ \\
\diagup \quad \diagdown \\
\text{H} \qquad \text{H}
\end{array}
$$

**Step 2.** The central atom is the C atom.

**Step 3.** There are three atoms around the C atom, which has no lone pairs. Therefore, we count three valence shell electron pairs for the C atom.

**Step 4.** The ideal geometry is trigonal planar.

**Step 5.** No correction is necessary. The H—C—H bond angles are all 120° and all four atoms lie in the same plane.

c.  **Step 1.**

$$
\ddot{\text{O}}\!\!=\!\!\text{C}\!\!=\!\!\ddot{\text{O}}
$$

**Step 2.** The C atom is the central atom.

**Step 3.** There are two atoms around the C atom, which has no lone pairs. Each double bond counts as only one electron pair (Rule 1), so we count two valence shell electron pairs for the C atom.

**Step 4.** The ideal geometry is linear.

**Step 5.** No correction is needed for two valence shell electron pairs. The O=C=O bond angle is 180°.

d.  **Step 1.**

$$
\begin{array}{c}
\text{H} \qquad\qquad \text{H} \\
\diagdown \qquad\quad \diagup \\
\text{C}\!\!=\!\!\text{C} \\
\diagup \qquad\quad \diagdown \\
\text{H} \qquad\qquad \text{H}
\end{array}
$$

**Step 2.** The two C atoms are central atoms, since each is bonded to more than one atom. Each C atom is bonded to another C atom and to two H atoms, so the geometry for each central atom is the same.

**Step 3.** We count three valence shell electron pairs for each C atom, since the double bond counts as only one pair (Rule 1).

**Step 4.** The ideal geometry around each C atom is trigonal planar. Each C atom and its three bonded atoms lie in the same plane.

**Step 5.** The ideal geometry must be corrected because of the double bond (Rule 2). The H—C—C angle should be larger than the value of 120° in the ideal geometry, and the H—C—H angle should be smaller. Experimentally, we find that the H—C—C angle is 121° and the H—C—H angle is 118°

## Nonbonding Valence Shell Electron Pairs: Lone Pairs

Perhaps the greatest success of the VSEPR model is in predictions of the geometry of molecules with central atoms that have lone pairs; that is, nonbonding valence shell electron pairs. An additional rule is needed to make these predictions:

*Rule 3.* A lone pair occupies more space than does a bonding electron pair.

This rule reflects the fact that electrons in a bond tend to be concentrated in a small region of space between two nuclei. Alternately, we can say that repulsions between a nonbonding electron pair and a bonding pair are greater than repulsions between bonding pairs.

When we describe the geometry of molecules whose central atoms have one or more lone pairs, we may specify either the bond angles or the positions of the atoms. For simplicity, we do not give the positions of the nonbonding pairs, although they help to determine the geometry of the molecule.

The $NH_3$ molecule illustrates these guidelines. The Lewis structure is

$$H-\ddot{N}-H$$
$$|$$
$$H$$

There are four valence shell electron pairs around the central N atom. For four electron pairs, Table 8.1 predicts tetrahedral geometry with bond angles of 109°28′. We must make two modifications in this picture.

Only the positions of the atoms are needed to describe the molecular geometry. Therefore, the ammonia molecule is described as pyramidal. The N atom lies at the apex and the three H atoms lie at the corners of a pyramid with a triangular base. Also, repulsions between the lone pair and the bonding electron pairs are greater than repulsions between bonding pairs (Rule 3). The VSEPR model thus predicts that the H—N—H bond angles will be less than 109°28′. Experimentally, we find that the

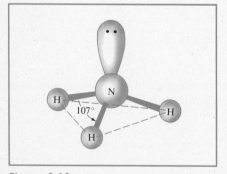

**Figure 8.10**

The four atoms of $NH_3$ lie at the corners of a pyramid. The HNH bond angles are less than the value in ideal tetrahedral geometry because of the larger size of the nonbonding valence electron pair.

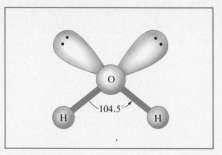

**Figure 8.11**
The three atoms of $H_2O$ are in the same plane, but not on the same line. The HOH bond angle is less than the HNH bond angle in $NH_3$ because of the two nonbonding valence shell electron pairs.

H—N—H bond angles are 107°. Figure 8.10 shows the structure of the ammonia molecule.

We can also apply Rule 3 to the $H_2O$ molecule. The Lewis structure is

$$H—\ddot{\underset{..}{O}}—H$$

There are four electron pairs around the central O atom. For four electron pairs, Table 8.1 predicts tetrahedral geometry, with bond angles of 109°28′. We must make two modifications in this picture.

The water molecule is planar and bent; the three atoms lie in the same plane but not on the same line. Since there are two lone pairs in the $H_2O$ molecule, compared to one in the $NH_3$ molecule, the VSEPR model predicts that the H—O—H bond angle should be even smaller than the H—N—H angle in $NH_3$. Experimentally, we find that the H—O—H bond angle in the water molecule is 104.5°, as shown in Figure 8.11.

These two examples suggest another way of stating Rule 3. We see that the repulsion between two nonbonding pairs (NBP) is greater than the repulsion between a nonbonding pair and a bonding pair (BP), which in turn is greater than the repulsion between two bonding pairs, or:

$$NBP–NBP > NBP–BP > BP–BP$$

Therefore, an alternate way to state Rule 3 is as follows: When nonbonding pairs are present, the bond angles generally are smaller than the values in the ideal geometry.

**Example 8.2**   Predict the geometry of the following molecules or ions: (a) $GeF_2$, (b) NOCl, (c) $CCl_3^-$.

**Solution**   a. **Step 1.**

$$:\ddot{\underset{..}{F}}—\ddot{Ge}—\ddot{\underset{..}{F}}:$$

We cannot avoid the incomplete octet of Ge, since F does not form a double bond.
**Step 2.** The central atom is the Ge atom.
**Step 3.** There are two atoms and one lone pair on the Ge atom, so we count three valence shell electron pairs.
**Step 4.** Table 8.1 predicts trigonal planar geometry for three electron pairs.
**Step 5.** The Ge atom and the two F atoms lie in the same plane but not on the same line. Because of the lone pair, the F—Ge—F angle is smaller than the ideal value of 120°.
b. **Step 1.**

$$:\ddot{\underset{..}{Cl}}—\ddot{N}=\ddot{\underset{..}{O}}$$

**Step 2.** The central atom is the N atom.
**Step 3.** There are two atoms and one lone pair on the N atom, so we count three valence shell electron pairs.

**Step 4.** Table 8.1 predicts trigonal planar geometry for three electron pairs.

**Step 5.** The atoms lie in the same plane, but not on the same line. A lone pair generally affects the structure more than does a double bond. In this case, the measured Cl—N—O bond angle is 116°.

c. **Step 1.**

$$\left[\begin{array}{c} :\overset{..}{\underset{|}{C}l}: \\ :\overset{..}{\underset{..}{C}l}\!\!-\!\!\overset{|}{\underset{|}{C}}: \\ :\overset{..}{\underset{..}{C}l}: \end{array}\right]^{-}$$

**Step 2.** The central atom is the C atom.

**Step 3.** There are three atoms and one lone pair on the C atom, so we count four valence shell electron pairs.

**Step 4.** Table 8.1 predicts tetrahedral geometry for four electron pairs.

**Step 5.** The atoms of this ion lie at the corners of a pyramid. Because of the lone pair, the bond angle is smaller than the ideal value of 109°28′.

Table 8.2 summarizes the geometry of molecules with four, three, and two valence shell electron pairs.

When the VSEPR model predicts trigonal bipyramidal geometry and one or more of the five valence shell electron pairs are nonbonding, we must ask another question: Do the nonbonding electron pairs occupy the equatorial or the axial positions of the trigonal bipyramid? Both experiment and theory give the same answer. The nonbonding pairs occupy the equatorial positions and the atoms occupy the axial positions.

The $I_3^-$ ion provides a striking example of this arrangement. The Lewis structure is

$$[\,:\!\overset{..}{\underset{..}{I}}\!\!-\!\!\overset{..}{\underset{..}{I}}\!\!-\!\!\overset{..}{\underset{..}{I}}\!:\,]^-$$

The central I atom has two atoms and three lone pairs about it, so we count five valence shell electron pairs. The two end I atoms lie in the axial positions of a trigonal bipyramid and the three lone pairs lie in the equatorial positions. As Figure 8.12 shows, the three I atoms lie on the same line; the I—I—I bond angle is 180°.

We shall discuss the geometry of other molecules with expanded octets and nonbonding valence electron pairs in Chapter 9.

## Limitations of the VSEPR Model

While the VSEPR model makes accurate predictions about many species, it is not always successful. For example, the ion $[PtCl_4]^{2-}$ has four atoms attached to the central Pt atom. Its structure is square planar, as shown in Figure 8.13, rather than tetrahedral as predicted by the VSEPR model. The VSEPR model does not work well for the transition metals.

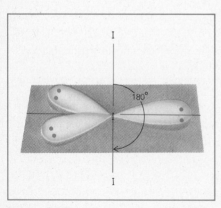

**Figure 8.12**
The three atoms of the $I_3^-$ ion lie on a straight line.

TABLE 8.2  Geometry Around Atoms with Four, Three, and Two Valence
           Electron Pairs

| Total Number of Valence Electron Pairs[a] | Number of Lone Pairs | Bond Angle | Atomic Geometry | Example[b] |
|---|---|---|---|---|
| 4 | 0 | 109°28′ | tetrahedral | $\underline{C}H_4$, $\underline{Si}F_4$, $\underline{N}H_4^+$ |
|  | 1 | <109° | pyramidal | $\underline{N}H_3$, $\underline{P}Cl_3$, $H_3\underline{O}^+$ |
|  | 2 | <109° | bent | $H_2\underline{O}$, $\underline{O}F_2$, $\underline{N}H_2^-$ |
| 3 | 0 | 120° | trigonal | $\underline{B}F_3$, $\underline{Al}Cl_3(g)$, $\underline{C}H_3^+$ |
|  | 1 | <120° | bent | $\underline{Ge}F_2$, $\underline{Sn}Cl_2$ |
| (double bond) | 0 | ~120° | planar | $\underline{C}_2H_4$, $\underline{C}H_2O$, $H\underline{N}O_3$ |
| (double bond) | 1 | <120° | bent | $\underline{N}OCl$, $\underline{N}O_2^-$ |
| 2 | 0 | 180° | linear | $\underline{Be}Cl_2$, $\underline{C}O_2$, $\underline{C}_2H_2$, $\underline{N}O_2^+$ |

[a] Electrons in double and triple bonds are counted as one pair.
[b] The underlined atom is the central atom of the bond angle.

Many compounds whose central atom is in the third row and lower in the periodic table do not follow the VSEPR rules. The VSEPR model is especially unsuccessful when the atoms about the central atom have low electronegativities. For example, the VSEPR model does not work well for hydrides of elements such as P, S, and As. The model predicts that compounds such as $PH_3$ and $H_2S$, whose central atoms have four valence shell electron pairs, should have the same geometry as $NH_3$ and $H_2O$. Instead of the predicted bond angles of approximately 109°28′, we find that these compounds have bond angles close to 90°.

One puzzle is the geometry of some halides of the heavier group IIA metals. These compounds form ionic lattices in the solid phase but are discrete molecules in the gas phase. The VSEPR model predicts that a molecule such as $BaF_2$, which has two valence shell electron pairs around the central Ba atom, should be linear. Experimentally, the molecule is found to be bent.

## 8.3  OVERLAP OF ATOMIC ORBITALS

Before we can understand the hybridization model for molecular structure, we must develop a description of covalent bonding that gives us a better picture of how electrons form covalent bonds. The modern description of electrons in atoms is based on orbitals. Covalent bonds can be understood better if the description of the electrons of these bonds is also based on the concept of orbitals.

We describe a covalent bond as the sharing by two atoms of two electrons. This description suggests that we can use a picture in which the

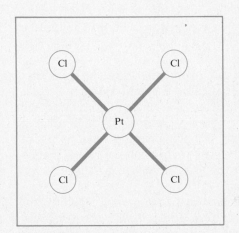

Figure 8.13
The ion $PtCl_4^{2-}$ has all five atoms lying in a plane, with the four Cl atoms at the corners of a square. This square planar geometry is found in molecules containing certain transition metals.

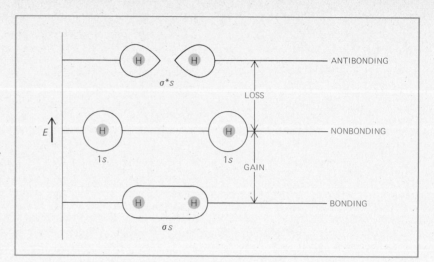

**Figure 8.14**

A covalent bond pictured as the combination of orbitals from two atoms. Each H atom contributes a $1s$ electron to the bond, and the resultant orbital is formed by the overlap of the two $1s$ orbitals. The new orbital is lower in energy than either of the original isolated nonbonding orbitals. Along with the bonding orbital, an antibonding orbital of higher energy than the atomic orbitals also forms.

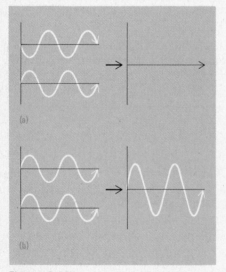

**Figure 8.15**

Schematic representation of the two basic ways in which electron waves can combine. Their peaks and crests can cancel by interference (a) or reinforce one another by superimposition (b).

electrons are found in an orbital formed by the combination of two orbitals, one from each atom. In the $H_2$ molecule, for example, each H atom contributes a $1s$ electron to the bond. Figure 8.14 shows that the resultant orbital can be pictured most simply as being formed by the overlap or combination of the two $1s$ orbitals.

This picture is consistent with the known features of covalent bonds. The new orbital that forms in a covalent bond is located primarily between the two nuclei. It can hold a maximum of two electrons of opposite spin (obeying the Pauli exclusion principle) and corresponds to a lower energy state than either of the isolated $1s$ orbitals.

Such an orbital is called a **bonding molecular orbital.** The bonding orbital is spread out in space, but there is a high probability of finding the two electrons between the two nuclei. We can say that there is an **overlap** of the two atomic orbitals.

Two atomic orbitals can combine in two ways, so there is another allowed orbital for the electron. It is called the **antibonding molecular orbital.** The bonding molecular orbital is of lower energy than either of the atomic orbitals. The antibonding orbital is of higher energy than either of the atomic orbitals. The probability of finding electrons that are in an antibonding orbital between the nuclei is low. In a homonuclear diatomic molecule, the energy decrease in the bonding orbital is equal to the energy increase in the antibonding orbital.

When two or more atomic orbitals overlap to form molecular orbitals, *the number of molecular orbitals created always equals the number of atomic orbitals that have combined.* A highly simplified explanation for orbital overlap can be based on the wave nature of electrons. As Figure 8.15 shows, there are two basic ways in which waves can overlap, or combine. The waves can be superimposed in a way that heightens the crests and troughs or in a way that cancels the crests and troughs by interference. Since electrons in orbitals have wave characteristics, the orbitals can be combined in the same two ways. The bonding orbital corresponds to the superimposition that heightens the wave crests. The antibonding orbital corresponds to the superimposition that cancels the wave crests.

This concept of overlapping orbitals gives an oversimplified picture of molecular structure. The picture of bond formation by overlap between $s$, $p$, and $d$ orbitals must be modified to be consistent with experimental evidence about the structure of real molecules.

The need for modifying the simplified picture is evident even in the study of such simple molecules as methane, $CH_4$. The Lewis structure:

$$
\begin{array}{c}
\text{H} \\
| \\
\text{H} - \text{C} - \text{H} \\
| \\
\text{H}
\end{array}
$$

and what we know about electronic configuration can be interpreted as showing that the carbon atom forms four bonds by overlap of one of its valence orbitals with the $1s$ orbital of each hydrogen atom. The carbon atom has one $2s$ and three $2p$ valence orbitals, so one might suggest that there should be two types of C—H bonds. In three of the bonds in the $CH_4$ molecule, the three $2p$ orbitals of the carbon atom should overlap with the $1s$ orbitals of three H atoms. The fourth bond should be formed by overlap of the $2s$ orbital of the C atom and the $1s$ orbital of the fourth H atom. Since the three $2p$ orbitals of the C atom are perpendicular to each other, we expect these three C—H bonds to form H—C—H angles of 90° relative to one another.

Every experimental result obtained for the structure of $CH_4$ is at variance with this hypothetical structure. All four C—H bonds are found to be identical, all the H—C—H bond angles are found to be 190°28′, and all the C—H bonds are found to have the same length, 0.109 nm (Figure 8.16).

The simple orbital picture is therefore wrong. The hybridization or directed valence model modifies the simple picture to explain the experimental results.

## 8.4 THE HYBRIDIZATION OR DIRECTED VALENCE MODEL

The *hybridization* or *directed valence model* accounts for the geometry of molecules and polyatomic ions by modifying the description of atomic orbitals when they overlap to form bonds. This modified description then is used to predict molecular geometry. The hybridization model uses the concept of a *hybrid orbital,* an intermediate atomic orbital formed by the mixing of simple atomic orbitals.

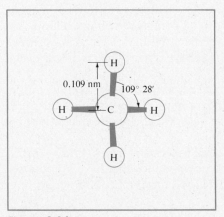

**Figure 8.16**

A $CH_4$ molecule. All four C—H bond lengths are 0.109 nm and all four C—H bond angles are 109°28′. All bond angles are identical and all bond lengths are identical.

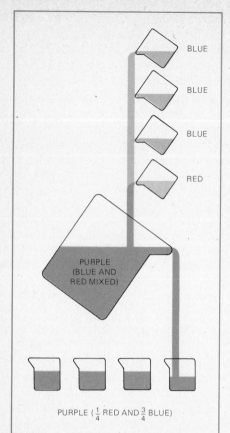

**Figure 8.17**
The different "pure" atomic orbitals mix to form identical hybrid orbitals just as the "pure" red and blue colors mix to form a uniform intermediate color. The number of orbitals is the same before and after hybridization.

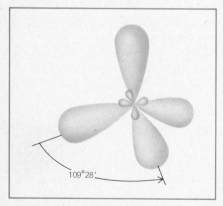

**Figure 8.18**
Hybrid orbitals point along the internuclear line. The $sp^3$ hybrid orbitals of $CH_4$ are directed so that the bond angles are 109°28′.

## Molecules with Ideal Geometry

A number of different hybrid orbitals can be described. Let us start by applying the hybridization model to the structure of $CH_4$, which has a central C atom bonded to four H atoms.

*$sp^3$ Hybridization.* In $CH_4$, we use what is called $sp^3$ hybridization. We can account for the four bonds that are formed in $CH_4$ by proposing that the four valence orbitals of the carbon atom "mix" to form four new, equivalent hybrid orbitals. The four hybrid orbitals about the C atom are formed by the mixing of one $2s$ orbital and three $2p$ orbitals. Therefore, each hybrid orbital is composed of three parts of $p$ orbital to one part of $s$ orbital. In notation, the contribution of each orbital is indicated by a superscript, with the superscript 1 omitted. The four hybrid orbitals of the $CH_4$ molecule are indicated by the notation $2sp^3$. The coefficient 2 is used to indicate the principal quantum number.

*When hybrid orbitals are formed from atomic orbitals, the number of orbitals does not change.* As Figure 8.17 indicates, hybridization may be considered to be simply the mixing of various "pure" orbitals to form the same number of hybrid orbitals.

The use of four equivalent $sp^3$ orbitals for the four bonds of $CH_4$ is consistent with the equivalency of the four bonds. Furthermore, it can be shown mathematically that the use of $sp^3$ orbitals for bonding results in bond angles of 109°28′ and tetrahedral geometry.

We use hybrid orbitals as a convenient concept to help us understand and visualize bonding. We construct hybrid orbitals in molecules because they allow the formation of stronger bonds than do unhybridized orbitals. We can say that these bonds are stronger because our hybridized orbitals are less symmetrical than unhybridized ones — they "point" in various directions, as shown in Figure 8.18. When such a hybrid orbital forms a bond, its region of high electron probability is directed toward the bonded atom. We can say that the hybrid orbital points at the bonded atom, so that there is a high probability that the electrons in the orbital lie between the two bonded atoms. The resulting picture is one of better orbital overlap, as shown in Figure 8.19.

A number of different hybrid orbital descriptions of bonding are possible in many molecules. The hybrid orbitals that we shall discuss are convenient because they are simple and give adequate explanations of observed properties of molecules.

The description of the $CH_4$ molecule in terms of four hybrid $sp^3$ orbitals also works well for other carbon compounds that have four atoms bonded to the carbon atom. Ethane, $C_2H_6$, has the Lewis structure:

$$
\begin{array}{ccc}
& H & H \\
& | & | \\
H- & C-C & -H \\
& | & | \\
& H & H
\end{array}
$$

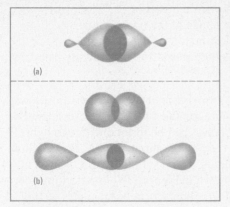

**Figure 8.19**
(a) Overlap of hybrid orbitals. The high electron density in one direction allows effective overlap. (b) Unhybridized orbitals have a symmetrical electron density. The extent of overlap is not as great.

For the C—H bonds, the $sp^3$ orbital of the carbon atom overlaps with the $1s$ orbital of hydrogen. In the C—C bonds, two $sp^3$ orbitals overlap. The bond angles in ethane are all close to $109°28'$.

In many cases the bonding of any four atoms to any single central atom can be described in terms of four $sp^3$ orbitals of the central atom. The logic of this description is clear: Four orbitals are needed to form four bonds. A nitrogen atom is the central atom in the ammonium ion, which is isoelectronic with $CH_4$ and has the Lewis structure:

$$\left[\begin{matrix} & H & \\ & | & \\ H- & N & -H \\ & | & \\ & H & \end{matrix}\right]^+$$

The bonding in the ammonium ion or in any other compound in which N is bonded to four other atoms is conveniently described by four $sp^3$ orbitals. Indeed, the $sp^3$ explanation usually is satisfactory for any atom of a nontransition element that is bonded to four other atoms and has no nonbonding electrons in its valence shell.

**Example 8.3**   Indicate which orbitals overlap to form each of the bonds in $CH_3\overset{+}{N}H_3$.

**Solution**   **Step 1.** Draw the Lewis structure:

$$\begin{matrix} & H & H & \\ & | & | & \\ H- & C & -\overset{+}{N} & -H \\ & | & | & \\ & H & H & \end{matrix}$$

**Step 2.** All the bonds of the H atoms are formed by overlap of the $1s$ orbitals of the H atoms and suitable orbitals of the other atoms. Note how many bonds each of the other atoms forms. Both the C atom and the N atom form four bonds. Both are $sp^3$ hybridized, and hence each bond to an H atom is formed by the overlap of an $sp^3$ orbital with a $1s$ orbital. The C—N bond is formed by the overlap of two $sp^3$ orbitals. All the bond angles in this ion are close to $109°28'$.

*$sp^2$ Hybridization.* We can extend the hybridization model to molecules or ions that have a central atom bonded to only three atoms. Consider the molecule $BF_3$. Three hybrid orbitals are needed to form the three B—F bonds. To form three hybrid orbitals, we need three pure atomic orbitals: the $2s$ orbital and two of the boron atom's $2p$ orbitals. Mixing these three orbitals creates three hybrid orbitals, each two parts $p$ and one part $s$; in notation, $sp^2$. It can be demonstrated mathematically that $sp^2$ orbitals

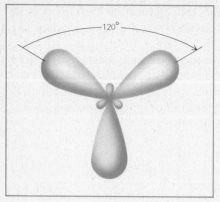

**Figure 8.20**

The $sp^2$ hybrid orbitals in a molecule such as $BF_3$ are directed so that the bond angles are 120°.

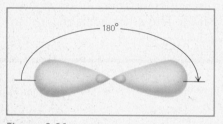

**Figure 8.21**

The $sp$ hybrid orbitals in a molecule such as $BeCl_2$ are directed so that the bond angles are 180°.

form bond angles of 120°, as shown in Figure 8.20. These three $sp^2$ orbitals are used to form the three bonds of the boron atom.

The structures of many molecules and ions in which an atom of a main group element is bonded to three other atoms and has no lone pairs are consistent with $sp^2$ hybridization.

*sp Hybridization.* When a central atom is bonded to only two atoms, we need two hybrid orbitals to account for the two bonds. Consider the molecule $BeCl_2$. The two Be—Cl bonds can be formed with two hybrid orbitals. Each hybrid orbital is formed by the mixing of the $2s$ orbital of the Be atom with one of its $2p$ orbitals. The resulting hybrid orbitals are each one part $s$ and one part $p$, as shown in Figure 8.21; in notation, $sp$. It can be shown mathematically that $sp$ orbitals form bond angles of 180°. The structures of many molecules and ions in which a main group element is bonded to two other atoms and has no nonbonding valence shell electrons are consistent with $sp$ hybridization.

## Sigma Orbitals and Pi Orbitals

The overlap between orbitals that we have discussed so far is one way in which two atomic orbitals can overlap. These bonding orbitals have cylindrical symmetry with respect to an imaginary line joining the nuclei of the two atoms of the bond. This kind of symmetrical overlap is called sigma ($\sigma$) overlap and the molecular orbital is called a $\sigma$ orbital. A $\sigma$ orbital is the most stable kind of bonding orbital. The antibonding orbital that corresponds to the $\sigma$ bonding orbital is called a "sigma star" ($\sigma^*$) orbital. If two orbitals are pointed at each other, $\sigma$ overlap takes place. Indeed, there are very few exceptions to the rule that any single bond between two atoms is a $\sigma$ bond. *Only one $\sigma$ bond is possible between two atoms.*

A second kind of orbital overlap can occur in covalent bonds formed by $p$ orbitals. When two $p$ orbitals are parallel to each other, there can be a kind of orbital overlap called pi ($\pi$) overlap. The bond formed in this way is called a $\pi$ bond, shown in Figure 8.22. The most probable region for the electrons of a $\pi$ bond is outside the region of the $\sigma$ bonding orbital. Two $p$ orbitals can overlap to form either a $\sigma$ or a $\pi$ bond, but only if the $p$ orbitals have the same directional subscript.

When a $\pi$ bonding orbital forms, so does a $\pi^*$ antibonding orbital. Since $\sigma$ overlap usually is more favorable than $\pi$ overlap, a $\pi$ bonding orbital usually is not as low in energy as the corresponding $\sigma$ bonding orbital, and a $\pi^*$ orbital is not as high in energy as the $\sigma^*$ orbital.

The $\sigma$-$\pi$ description can be used to explain the formation of double and triple bonds, as in the case of the : N≡N : molecule (Figure 8.23). Each N atom contributes three unpaired electrons, which are in the $2p_x$, $2p_y$, and $2p_z$ orbitals. A $\sigma$ bond can be formed by the overlap of the two $2p_x$ orbitals. No more $\sigma$ bonds are possible. The other two electron pairs in the triple bond exist in two $\pi$ bonding orbitals, one formed between the two $2p_y$ orbitals and the other formed between the two $2p_z$ orbitals. In both of

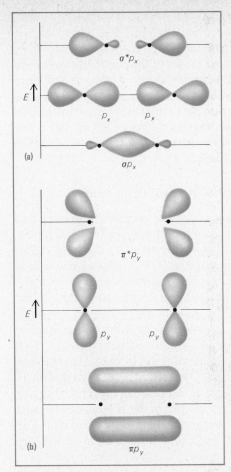

Figure 8.22

The two types of $p$ orbital overlap. The $p$ orbitals can form (a) a $\sigma$ bond or (b) a $\pi$ bond.

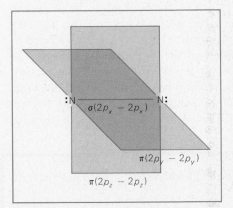

Figure 8.23

Bond formation in the $N_2$ molecule. The $\pi$ bonds are represented by the two shaded planes, at right angles to each other and to the $\sigma$ bond.

these $\pi$ bonding orbitals, the most probable region for the electrons is outside the region of the $\sigma$ bond.

The triple bond can thus be described as one $\sigma$ bond in one direction and two $\pi$ bonds at right angles to each other and to the direction of the $\sigma$ bond. The $\pi$ bonds are not as strong as the $\sigma$ bonds, because $\pi$ overlap is not as effective as $\sigma$ overlap. Therefore, a triple bond generally is not three times stronger than a single bond. In the same way, a double bond that consists of one $\sigma$ bond and one $\pi$ bond usually is not twice as strong as a single bond.

The hybridization model accommodates the $\sigma$-$\pi$ description of bonding very nicely, with one complication. The shape and directional character of the various hybrid orbitals we have described are very well suited for $\sigma$ overlap, but not for $\pi$ overlap. In $sp^2$ and $sp$ hybrids, however, there remain unhybridized $p$ orbitals that can be used for $\pi$ bonding.

For example, in ethylene, $C_2H_4$, which has a double bond, each carbon atom forms three $\sigma$ bonds. Therefore, only three $sp^2$ hybrid orbitals are required for each atom. Only three of the four valence orbitals of carbon are used to form the three hybrid $sp^2$ orbitals. The fourth valence orbital is an unhybridized $2p$ orbital that is perpendicular to the plane of the $sp^2$ hybrid orbitals, as shown in Figure 8.24. This unhybridized $2p$ orbital is used for the $\pi$ bond. There is $\pi$ overlap between the two parallel unhybridized $p$ orbitals in $C_2H_4$. The complete orbital picture of $C_2H_4$ is shown in Figure 8.25. However, this simple hybridization model does not account for the slight deviation of the bond angles in $C_2H_4$ from the ideal value of $120°$.

The $sp$ hybridization model works well when an atom forms two $\pi$ bonds, as it does in a triple bond or two double bonds. Thus the carbon atoms in $H$—$C$≡$C$—$H$ or $O$=$C$=$O$ are $sp$ hybridized. In $C_2H_2$ or $CO_2$ there are two unhybridized $2p$ orbitals on each carbon atom, perpendicular to each other and to the line of the $sp$ orbitals (Figure 8.26). The two $\pi$ bonds of the $C$≡$C$ are constructed from pairs of parallel, unhybridized $2p$ orbitals, one from each C atom (Figure 8.27).

## Molecules with Nonideal Geometry

Each of the three hybrid orbitals mentioned thus far is associated with an ideal bond angle: $sp^3$ hybrid orbitals with a bond angle of $109°28'$; $sp^2$ hybrid orbitals with a bond angle of $120°$; $sp$ hybrid orbitals with a bond angle of $180°$. The hybridization model can be extended to molecules that do not have this ideal geometry.

This rule helps us relate hybridization to Lewis structure: The hybridization of an atom *without expanded octets* and with *no lone pairs* generally is related to the number of $\sigma$ bonds that the atom forms. Four $\sigma$ bonds indicate $sp^3$ hybridization. Three $\sigma$ bonds indicate $sp^2$ hybridization. Two $\sigma$ bonds indicate $sp$ hybridization.

A hybridization model for the atoms in a molecule with nonideal geometry can also be based on measured bond angles. If the angles are

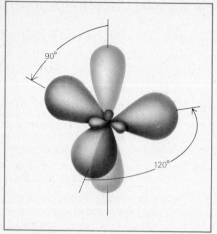

**Figure 8.24**

Each carbon in the $C_2H_4$ molecule has three hybrid $sp^2$ orbitals, with bond angles of 120°, and one unhybridized $2p$ orbital, perpendicular to the plane of the hybrid orbitals.

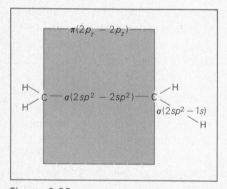

**Figure 8.25**

The complete orbital picture of $C_2H_4$, with three $\sigma$ bonds and one $\pi$ bond associated with each carbon atom. The designation of the directional subscripts of the $p$ orbitals is arbitrary.

closest to 109°28′, $sp^3$ hybridization of the central atom is indicated. If the angles are closest to 120°, $sp^2$ hybridization of the central atom is indicated. If the angles are closest to 180°, $sp$ hybridization of the central atom is indicated. In a molecule such as $CH_3Cl$, which we discussed earlier, the geometry is not ideal. Still, we say that the carbon atom is $sp^3$ hybridized because the bond angles are closer to 109° than to 120° or 180°. The deviation from the ideal value results from the different distribution of the electrons in the C—H and C—Cl bonds.

### Nonbonding Valence Shell Electron Pairs: Lone Pairs

The hybridization model can account for the observed geometry and bond angles of molecules that have atoms with lone pairs. The simplest approach is to count each lone pair as the equivalent of a pair of electrons in a $\sigma$ bond. Thus for the oxygen atom in H—Ö—H, we count four valence electron pairs, two in the $\sigma$ bonds and two lone pairs. For the nitrogen atom in :$NH_3$, we count four valence electron pairs, three in the $\sigma$ bonds and one lone pair.

The hybridization description for atoms with nonbonding valence electron pairs is essentially the same as for atoms without them. In $H_2O$, for example, we can say that the two $\sigma$ bonds result from overlap between two of the four $sp^3$ hybrid orbitals of the O atom and the $1s$ orbital of the H atoms, and that the two lone pairs of the O atom are in the remaining two $sp^3$ orbitals. In $NH_3$, three $sp^3$ hybrid orbitals of the N atom can be used for the three $\sigma$ bonds, and the remaining $sp^3$ orbital for the lone pair. The $sp^3$ hybridization of the O atom of $H_2O$ and of the N atom of $NH_3$ are consistent with the measured bond angles, which are close to 109°28′.

In general, we can predict the hybridization of nontransition elements that have both lone pairs and bonds by *totaling the $\sigma$ bonds and the lone pairs*. If the total is four, the hybridization is $sp^3$. If the total is three, the hybridization is $sp^2$. If the total is two, the hybridization is $sp$.

Thus, while the N atom in $NH_3$ is $sp^3$, the N atom in the molecule $CH_2NH$ (Figure 8.28) is better described as $sp^2$: It has only two $\sigma$ bonds and one nonbonding valence electron pair, a total of three (the $\pi$ bond is not counted). The $sp^2$ hybridization of the nitrogen atom suggests a C—N—H bond angle close to 120°. In the H—C≡N molecule, we can say that the N atom is $sp$ hybridized, since it has one $\sigma$ bond and one lone pair. We can assign a hybridization to an atom with one $\sigma$ bond, but we cannot check the assignment, since two $\sigma$ bonds are needed to define a bond angle.

---

**Example 8.4**    For the molecule $HNO_2$, indicate the hybrid orbitals that overlap to form each bond and estimate the bond angles.

**Solution**    **Step 1.** Draw the Lewis structure:

$$H—\ddot{O}—\ddot{N}=\ddot{O}$$

**Step 2.** Taking one atom at a time, total the number of $\sigma$ bonds and lone pairs for each. From left to right:

The first O atom has two $\sigma$ bonds and two lone pairs. The total of four suggests $sp^3$ hybridization of the O atom and an H—O—N bond angle of about $109°28'$. (Note again that the bonding of the central atom sets the value of the bond angle.) The O—H bond forms from overlap of the $1s$ orbital of the H atom and the $sp^3$ orbital of the O atom. The two lone pairs of the O atom are in $sp^3$ orbitals.

The N atom forms two $\sigma$ bonds (and one $\pi$ bond that is not counted) and has one lone pair. The total of three suggests $sp^2$ hybridization of the N atom and an O—N—O bond angle close to $120°$. The single O—N bond is formed by $sp^3$-$sp^2$ overlap.

The remaining O atom has only one $\sigma$ bond, so hybridization is not absolutely necessary for its description. However, for convenience we can say that the O atom is $sp^2$ hybridized, since it also has one $\sigma$ bond and two lone pairs, a total of three. The $\sigma$ bond between the N atom and this O atom is formed by $sp^2$-$sp^2$ overlap. The remaining lone pairs of the N atom and of this O atom are in hybrid $sp^2$ orbitals. The $\pi$ bond forms by overlap of two parallel unhybridized $2p$ orbitals, one on the N atom and one on the O atom. These $2p$ orbitals must have the same directional subscript, for example, $2p_z$-$2p_z$. The orbital diagram for this molecule is shown in Figure 8.29.

We know that the bond angles in molecules with lone pairs on the central atom are smaller than those of the ideal geometry. The hybridization model uses the same rule as the VSEPR model to accommodate this observation: A nonbonding valence shell electron pair (lone pair) occupies more space than a bonding pair.

Table 8.3 summarizes the hybridization in molecules or ions with no expanded octets.

## Molecules with Expanded Octets

Any atom that satisfies the octet rule can be assigned one of the three idealized hybridizations, $sp^3$, $sp^2$, or $sp$. The maximum number of hybrid orbitals required for such atoms is four, as in the $sp^3$ hybridization, since only $s$ and $p$ orbitals are involved. Elements of atomic number 14 and higher can expand their octets. In such cases, we may use hybrid orbitals that include $d$ orbitals to account for the bond angles.

Generally, either one or two $d$ orbitals can mix with $s$ and $p$ orbitals to form hybrid orbitals. If one $d$ orbital is mixed with one $s$ orbital and three $p$ orbitals, five hybrid orbitals are formed, each of which is called $sp^3d$. Such an orbital is 20% $d$, 20% $s$, and 60% $p$ in character. This hybridization can be used to describe the geometry about an atom in which the sum of $\sigma$ bonds and lone pairs is five. An example is the $PCl_5$ molecule, whose Lewis structure:

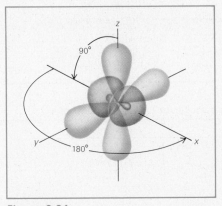

**Figure 8.26**

Each carbon of the $C_2H_2$ (acetylene) molecule has two $sp$ hybrid orbitals (shown on the $x$-axis), with $180°$ bond angles, and two $2p$ orbitals (shown in the $yz$ plane) at right angles to each other and to the hybrid orbitals.

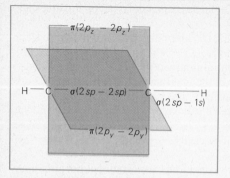

**Figure 8.27**

Orbital diagram of the $C_2H_2$ molecule, showing two $\sigma$ bonds and two $\pi$ bonds associated with each carbon atom. The $\pi$ bonds are the shaded planes perpendicular to each other and to the plane of the paper, which contains the $\sigma$ bonds.

with five $\sigma$ bonds to the P atom suggests $sp^3d$ hybridization for phosphorus.

The geometry associated with five $sp^3d$ hybrid orbitals is the same one that the VSEPR model describes as minimizing repulsions between five valence electron pairs, the trigonal bipyramid, shown in Figure 8.7.

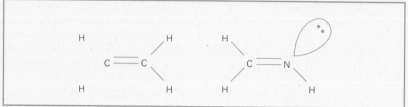

**Figure 8.28**

$CH_2NH$ is similar to $CH_2CH_2$. The lone pair on the nitrogen atom replaces one hydrogen atom of $CH_2CH_2$.

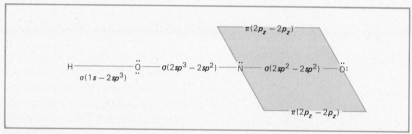

**Figure 8.29**

Orbital diagram of HONO. The $\pi$ bond is perpendicular to the plane of the paper.

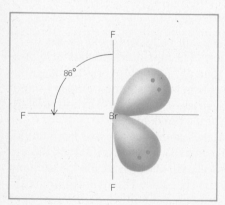

**Figure 8.30**

The structure of $BrF_3$ is based on the trigonal bipyramid associated with $sp^3d$ hybridization. The large size of the two lone pairs causes them to occupy the equatorial positions of the trigonal bipyramid and to make the F—Br—F bond angle smaller than the value of $90°$ in ideal trigonal bipyramidal geometry.

**TABLE 8.3**    Hybridization of Atoms

| Number of $\sigma$ Bonds | Number of $\pi$ Bonds | Number of Nonbonding Valence Electron Pairs | Hybridization | Example[a] |
|---|---|---|---|---|
| 4 | 0 | 0 | $sp^3$ | $CH_4$, $\underline{Si}F_4$, $\underline{N}H_4{}^+$, $\underline{B}F_4{}^-$ |
| 3 | 0 | 1 | $sp^3$ | $\underline{N}H_3$, $\underline{P}Cl_3$, $H_3\underline{O}^+$, $\underline{C}H_3{}^-$ |
| 2 | 0 | 2 | $sp^3$ | $H_2\underline{O}$, $O\underline{F}_2$, $\underline{S}Cl_2$, $\underline{N}H_2{}^-$ |
| 3 | 0 | 0 | $sp^2$ | $\underline{B}F_3$, $\underline{Al}Cl_3(g)$, $\underline{C}H_3{}^+$ |
| 3 | 1 | 0 | $sp^2$ | $\underline{C}_2H_4$, $\underline{C}H_2O$, $H\underline{N}O_3$ |
| 2 | 0 | 1 | $sp^2$ | $\underline{Ge}F_2$, $\underline{Sn}Cl_2{}^-$ |
| 2 | 1 | 1 | $sp^2$ | $\underline{N}OCl$, $\underline{N}O_2{}^-$ |
| 2 | 0 | 0 | $sp$ | $\underline{Be}Cl_2$, $\underline{Hg}Cl_2$ |
| 2 | 1 | 0 | $sp$ | $H\underline{C}^+{=}O$ |
| 2 | 2 | 0 | $sp$ | $\underline{C}O_2$, $\underline{C}_2H_2$, $\underline{N}O_2{}^+$ |

[a] Hybridization is given for the underlined atom.

**Example 8.5**   Predict the hybridization and geometry of $BrF_3$.

**Solution**   **Step 1.** Draw the Lewis structure:

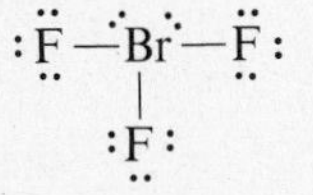

$$:\ddot{F}-\ddot{Br}-\ddot{F}:$$
$$|$$
$$:\ddot{F}:$$

**Step 2.** The sum of the $\sigma$ bonds and the lone pairs on the central Br atom is five. The hybridization of the atom is thus $sp^3d$. The geometry is a trigonal bipyramid, with the Br atom at its center.

**Step 3.** The three fluorine atoms attached to the central Br atom and the two lone pairs must be distributed among the five possible positions of the trigonal bipyramid. Experimental evidence indicates that it is more favorable for the lone pairs to be in the equatorial positions. Therefore, two of the three F atoms lie in the axial positions. The four atoms form a distorted "T" as shown in Figure 8.30.

If two $d$ orbitals are mixed with the $s$ and $p$ orbitals, the result is six hybrid orbitals, designated $sp^3d^2$. In the compound $SF_6$, the Lewis structure:

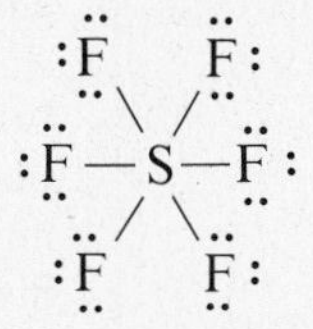

$$:\ddot{F}\quad\ddot{F}:$$
$$\search\nearrow$$
$$:\ddot{F}-S-\ddot{F}:$$
$$\nearrow\searcW$$
$$:\ddot{F}\quad\ddot{F}:$$

indicates that the S atom, with six $\sigma$ bonds, has the hybridization $sp^3d^2$. The geometry associated with this hybridization is the same octahedral geometry, shown in Figure 8.8, which the VSEPR model describes as minimizing repulsions between six valence electron pairs.

**Example 8.6**   Use the VSEPR model and the hybridization model to (a) predict the geometry and (b) describe the bonding in $HNO_3$.

**Solution**   a. **Step 1.** The molecule is a resonance hybrid:

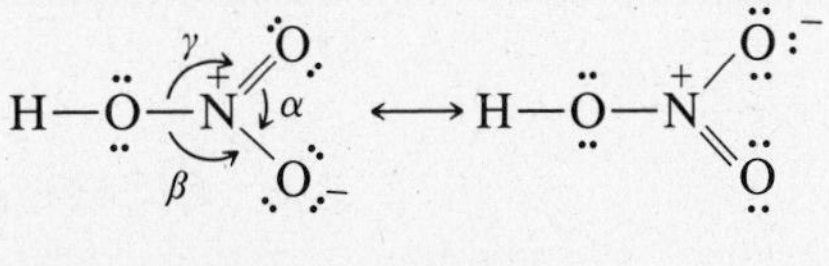

$$H-\ddot{O}-N\underset{\beta}{\overset{\gamma}{\rightleftharpoons}}\overset{\ddot{O}}{\underset{:\ddot{O}:^-}{}}\quad\longleftrightarrow\quad H-\ddot{O}-\overset{+}{N}\overset{\ddot{O}:^-}{\underset{\ddot{O}}{}}$$

We can work with either contributing structure.

**Step 2.** There are two central atoms, the N atom and the O atom of the OH group.

**Step 3.** In each contributing structure. the N atom is bonded to three atoms and has no lone pairs. The O atom is bonded to two atoms and has two lone pairs.

Thus, we count a total of three valence shell electron pairs for the N atom and four for the O atom.

**Step 4.** The ideal geometry about the N atom is trigonal planar. The ideal geometry about the O atom is tetrahedral.

**Step 5.** Since the real structure of $HNO_3$ is a blend of the two contributing structures, both the N—O bonds have substantial double bond character. Therefore, we expect some deviation from the ideal angles of 120° around the N atom. The VSEPR model tells us that bond angles that include multiple bonds generally are larger than those of the ideal geometry (Rule 2). We can therefore predict that the O—N—O bond angle ($\alpha$) will be larger than 120°. Since the three O atoms and the N atom are in the same plane, the sum of the angles around the N atom must be 360°. If the O—N—O bond angle ($\alpha$) is greater than 120°, the other O—N—O bond angles ($\beta$ and $\gamma$) must be smaller than 120°. Experimentally, we find that the O—N—O angle ($\alpha$) is 130° and the other two O—N—O angles are each about 115°.

Because the central O atom has two lone pairs, we expect the H—O—N bond angle to be smaller than the ideal value of 109°28′ (Rule 3). Experimentally, this angle is found to be 102°. The N atom is more electronegative than the H atom, so the electron pair in the O—N bond occupies less space than the electron pair in the O—H bond. This observation helps to explain why the H—O—N bond angle is smaller than the H—O—H bond angle in $H_2O$, whose value is 104.5°.

b. To formulate a hybridization picture of this molecule, we must total the $\sigma$ bonds and the lone pairs on each atom in one of the contributing structures. From left to right in the first structure:

The H atom has one $\sigma$ bond and uses its $1s$ orbital for bonding. The O atom has two $\sigma$ bonds and two lone pairs. The total of four is consistent with $sp^3$ hybridization and an H—O—N bond angle close to 109°28′. The N atom has three $\sigma$ bonds, which is consistent with $sp^2$ hybridization and bond angles close to 120°. Since each of the other O atoms forms only one $\sigma$ bond, they are not central atoms. We can picture them as unhybridized if we choose.

The H—O bond forms from $\sigma$ overlap of the $1s$ orbital of the H atom with the $2sp^3$ orbital of the O atom. The O—N bond forms from $\sigma$ overlap of the $2sp^3$ orbital of the O atom with the $2sp^2$ orbital of the N atom. The N—O single bond forms from $\sigma$ overlap of the $2sp^2$ orbital of the N atom with a $2p$ orbital of the unhybridized O atom. The N=O bond forms from $\sigma$ overlap of the $2sp^2$ orbital of the N atom with a $2p$ orbital of the O atom and from $\pi$ overlap of the $2p_z$ orbital of the N atom with the $2p_z$ orbital of the O atom.

We can modify this orbital picture to accommodate the resonance; the two electrons of the $\pi$ bond are placed in an orbital formed by overlap of the $2p_z$ orbital of the N atom with a $2p_z$ orbital of each of the O atoms. (Figure 8.41). Figure 8.31 summarizes the geometry and bonding in $HNO_3$.

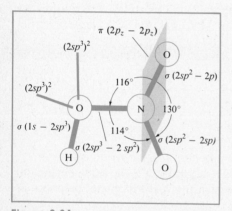

**Figure 8.31**

The geometry and bonding in $HNO_3$. The lone pairs on the unhybridized O atoms, which are not shown, fill unhybridized $2s$ and $2p$ orbitals. The nonbonding valence-shell electron pairs on the O of the OH are shown in $2sp^3$ orbitals. The $\pi$ bond is the shaded plane perpendicular to the plane of the N and O atoms.

## 8.5   MOLECULAR PROPERTIES AND GEOMETRY

The goal of the structural chemist is not just to make the predictions or measurements needed for an accurate picture of a molecule. It is to understand the relationship between the structure of a molecule and its properties. Two basic relationships are of particular interest to us: the relationship between bond angles and the polarity of molecules and between bond lengths and the stability of molecules.

### Polarity of Molecules

In Section 7.3, we said that the bonds between atoms of different electronegativities are polar. The atoms share the electrons of the bond unequally, and the result is a separation of electric charge that produces the polarity. Electrical polarity is expressed as a *dipole moment,* which is a measure of the magnitude of the separated charges and the distance between them. The dipole moment of a molecule is determined by several factors. We can describe a molecular dipole moment at least approximately as the sum of all the individual dipole moments of the bonds in the molecule. Such a sum is called the vector sum. The individual dipole moments are called *bond moments.* The vector sum must include not only the magnitude of the bond moments but also their direction.

Using this description, we can understand why some molecules that have atoms of different electronegativities do not have molecular dipole moments. These molecules have polar bonds. They are nonpolar molecules because the bond moments of the polar bonds are oriented in such a way that they can cancel each other.

One such molecule is $BeCl_2$. Each Be—Cl bond is polar, because the two atoms have very different electronegativities, but $BeCl_2$ is nonpolar overall. This nonpolarity is a result of the linear geometry of $BeCl_2$. If we represent the bond moments by arrows pointing at the negative end of the dipole, $BeCl_2$ can be shown as

$$:\!\ddot{C}l \longleftarrow Be \longrightarrow \ddot{C}l\!:$$

The two bond moments are of equal magnitude, but they point in exactly opposite directions and so cancel each other. By contrast, the structure of the $SCl_2$ molecule can be shown as

$$
\begin{array}{c}
\ddot{\ddot{S}} \\
{:}\!\ddot{C}l \diagup \quad \diagdown \ddot{C}l\!: 
\end{array}
$$

The central S atom has two lone pairs, and so the molecule is bent. Therefore, its bond moments do not cancel (Figure 8.32); $SCl_2$ is polar.

Another molecule that has polar bonds but is nonpolar is $BF_3$, whose geometry is trigonal planar. The three bond moments are of equal magnitude. They point to the corners of an equilateral triangle:

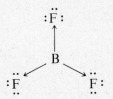

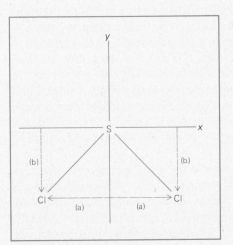

**Figure 8.32**

The $x$ components (a) of the bond dipoles are equal in magnitude but opposite in direction so they cancel. But the $y$ components (b) of the bond dipoles are in the same direction. They add, giving the molecule a net dipole moment.

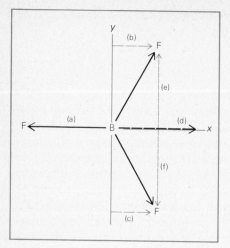

**Figure 8.33**

The $x$ components (b, c) of two of the bond dipoles add to give (d), which cancels the third bond dipole (a). The $y$ components of the two bond dipoles (e, f) are equal in magnitude and opposite in direction, so they cancel each other. The dipole moment of the molecule is zero.

Because of the symmetry of the molecule, the bond moments cancel (Figure 8.33).

Four identical bond moments will cancel in tetrahedral geometry. One molecule with tetrahedral geometry that has polar bond moments but is nonpolar overall is $CCl_4$. In $CCl_4$, four bond moments of equal magnitude are directed toward the corners of a regular tetrahedron:

The four bond moments cancel each other. If one of the atoms around the central C atom is changed, the new molecule has overall polarity. The geometry is no longer exactly tetrahedral and the bond moments no longer cancel each other.

## Bond Lengths

Bond lengths are important in the geometry of a molecule. They help us to understand the configuration of a molecule. They also tell us about the stability of a molecule, because bond lengths are related to bond strengths. The rule is: As bond length increases, bond strength decreases.

A number of factors influence the length of bonds in covalent molecules. One is the size of the two atoms of the bond. In general, *larger atoms form longer bonds.* We can see the effect of atomic size in the lengths of the bonds in the homonuclear diatomic halogen molecules listed in Table 8.4.

The length of the bond between two atoms of similar size is affected in an important way by the *bond order*. For most of the bonds we shall discuss, *bond order is the number of electron pairs shared between the two atoms of the bond.* A small bond order means a longer bond. A single bond is longer than a double bond, and a double bond is longer than a triple bond. For example, the C—C bond length in $C_2H_6$ is 0.154 nm, while the C=C bond length in $C_2H_4$ is 0.134 nm and the C≡C bond length in $C_2H_2$ is 0.120 nm.

The length of a bond of a given bond order also depends on the fraction of $s$ character in the hybrid orbital forming the bond. A larger fraction of $s$ character means a shorter bond. Single bonds of $sp^3$ orbitals, which are one-quarter $s$ in character, are longer than single bonds of $sp^2$ orbitals, which are one-third $s$ in character. These single $sp^2$ bonds, in turn, are longer than single bonds of $sp$ orbitals, which are one-half $s$ in character.

This hybridization effect can be seen in a series of C—H bonds. The C—H bonds in $C_2H_6$, where the C atom is $sp^3$, are 0.110 nm long. The C—H bonds in $C_2H_4$, where the C atom is $sp^2$, are 0.107 nm long. Those in $C_2H_2$, where the C atom is $sp$, are 0.106 nm long.

**TABLE 8.4**   Bond Lengths in the Halogens

| Bond | Length (nm) |
| --- | --- |
| F—F | 0.142 |
| Cl—Cl | 0.199 |
| Br—Br | 0.229 |
| I—I | 0.266 |

## 8.6 THE MOLECULAR ORBITAL METHOD

Everything said thus far about bonding has been based on the assumption that the electrons in atoms can be divided into two groups, bonding and nonbonding, and that the two groups can then be considered separately. For example, in the Lewis structure $:\ddot{F}-\ddot{F}:$ and even in a simple orbital description of the $F_2$ covalent bond, the two shared electrons are described as being in a bonding orbital formed from the overlap of the two $2p_x$ orbitals. The interactions of the six nonbonding electrons of each F atom with the bonding electrons and with each other are not considered.

This approach is called the **valence bond (VB) method**. Lewis structures are at the heart of the VB method. A second approach, called the **molecular orbital (MO) method,** also gives a good description of most molecules. We shall use the VB method in most cases, because it is easier. However, it is worth knowing how the MO method works and when it is more successful than the VB method.

The basic approach of the molecular orbital method resembles that of the Aufbau method, which is used to determine the electronic configuration of atoms. The MO approach assumes that there is a set of allowed orbitals for the molecule, rather than different sets of orbitals for each atom in the molecule. The allowed orbitals are called molecular orbitals, and they give the method its name. Molecular orbitals (MOs) are not restricted to one or two atoms; they are associated with the entire molecule. The electrons of the molecule are placed in the set of allowed orbitals by the Aufbau method — that is, the orbitals are filled in order of increasing energy. As the Aufbau method gives the ground state electronic configuration of an atom, the MO method gives the ground state electronic configuration of a molecule. Just as an atomic orbital gives a probability distribution for an electron in the atom, a molecular orbital gives a probability distribution for an electron in the molecule.

However, there is a certain lack of accuracy in the MO method. Just as an exact set of orbitals cannot be constructed for a multielectron atom, an exact set of orbitals cannot be constructed for a molecule by the MO method. Constructing even an approximate set of orbitals, and arranging the orbitals in order of relative energy, can be a complex task.

### The MO Description of Homonuclear Diatomics

Consider the MO description of the simplest homonuclear diatomics of the first two elements. The set of molecular orbitals associated with two hydrogen nuclei is shown in Figure 8.14. Since each atom has only one valence orbital, there are only two molecular orbitals that we need consider, the $\sigma 1s$ orbital of relatively low energy and the $\sigma^* 1s$ orbital of relatively high energy. The same set of two MOs is associated with any diatomic combination of H or He atoms.

Now consider the electronic configuration of diatomic species of H or He. The $H_2^+$ ion has one electron. As Figure 8.34 shows, the electron is in

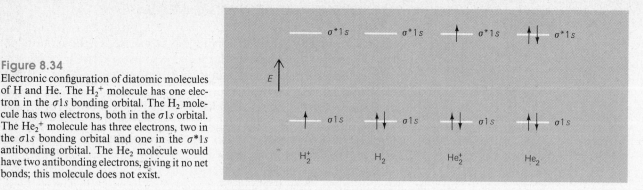

**Figure 8.34**
Electronic configuration of diatomic molecules of H and He. The $H_2^+$ molecule has one electron in the $\sigma 1s$ bonding orbital. The $H_2$ molecule has two electrons, both in the $\sigma 1s$ orbital. The $He_2^+$ molecule has three electrons, two in the $\sigma 1s$ bonding orbital and one in the $\sigma^* 1s$ antibonding orbital. The $He_2$ molecule would have two antibonding electrons, giving it no net bonds; this molecule does not exist.

the $\sigma 1s$ bonding orbital. Since a covalent bond consists of two electrons in a bonding orbital, $H_2^+$ is said to have half of one bond. The observation that $H_2^+$ exists is consistent with this description.

The $H_2$ molecule has two electrons, both of which can be placed in the $\sigma 1s$ orbital. Two electrons equal one full bond, and the observed stability of the $H_2$ molecule corresponds to the existence of one full bond.

The $He_2^+$ ion has three electrons. There is room for only two electrons in the $\sigma 1s$ bonding orbital. The third electron must go into the orbital of next lowest energy, the $\sigma^*$ orbital. Thus, the electronic configuration of $He_2^+$ is $(\sigma 1s)^2(\sigma^* 1s)^1$. But an antibonding orbital is destabilizing. Placing one electron in an antibonding orbital cancels the favorable effect of placing an electron in a corresponding bonding orbital. For bookkeeping purposes, placing an electron in an antibonding orbital cancels half of one bond. Thus, the number of bonds is given by:

$$\tfrac{1}{2}(\text{bonding electrons} - \text{antibonding electrons})$$

In $He_2^+$, there are two electrons in bonding orbitals and one in an antibonding orbital, giving $\tfrac{1}{2}(2-1) = \tfrac{1}{2}$ bond. This calculation is confirmed by observation. The stability of $He_2^+$ is found to be comparable to the stability of $H_2^+$, which also has half of one bond.

The power of the MO method is demonstrated by its correct prediction for $He_2$. This hypothetical species has four electrons and the electronic configuration $(\sigma 1s)^2(\sigma^* 1s)^2$, in which there are two bonding and two antibonding electrons. There are $\tfrac{1}{2}(2-2) = 0$ bonds in this electronic configuration. A molecule with no net stabilization through bond formation should not exist. In fact, no one has ever found evidence that the $He_2$ molecule exists.

## Second-Row Diatomic Molecules

The MO treatment of the homonuclear diatomic molecules of the second-row elements from Li to F is more complicated, even though the $1s$ orbitals and electrons can be omitted from our calculations on the grounds that only valence orbitals need be considered.

Each atom of an element in the second row has four valence orbitals: $2s$, $2p_x$, $2p_y$, and $2p_z$. Therefore, a total of eight MOs are considered in a diatomic molecule.

One bonding orbital and one antibonding orbital correspond to each of these four valence orbitals. The types of orbitals to be expected for the diatomic molecules of these atoms are $\sigma 2s$ and $\sigma^*2s$; $\sigma 2p$ and $\sigma^*2p$; and $\pi 2p$ and $\pi^*2p$. One pair of $2p$ orbitals overlaps to form one $\sigma 2p$ orbital and one $\sigma^*2p$ orbital. The remaining two pairs of $2p$ orbitals overlap to form two $\pi 2p$ orbitals of equal energy and two $\pi^*2p$ orbitals of equal energy. Orbitals of equal energy are said to be *degenerate*.

The eight MOs must be arranged in order of increasing energy. While the finer points of this arrangement are beyond the scope of this book, some main points can be described. The $s$ orbitals are considerably lower in energy than $p$ orbitals, so the MO of lowest energy is the $\sigma 2s$ orbital. The next lowest is the $\sigma^*2s$ orbital. The highest-energy orbital is the $\sigma^*2p$ MO. Next highest are the two degenerate $\pi^*2p$ MOs. The ranking by energy of the $\sigma 2p$ orbital and the two $\pi 2p$ orbitals presents some difficulty. Two arrangements of these orbitals are possible, as shown in Figures 8.35 and 8.36.

Figure 8.35 shows an energy level diagram that applies to diatomic molecules of Li, Be, B, C, and N. In these molecules, the $\sigma 2p$ MO is believed to be of higher energy than the two degenerate $\pi 2p$ orbitals.

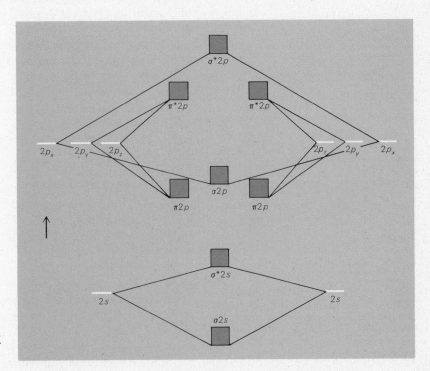

**Figure 8.35**

An energy level diagram for $Li_2$, $Be_2$, $B_2$, $C_2$, and $N_2$, in which the $\sigma 2p$ orbital is of higher energy than the two $\pi 2p$ orbitals.

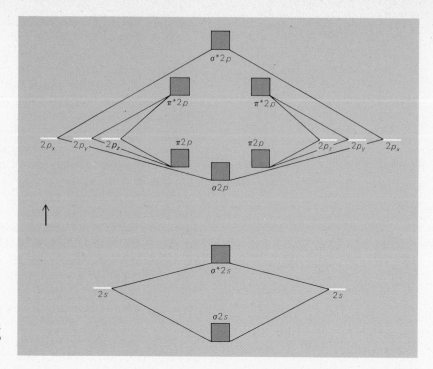

**Figure 8.36**

An energy level diagram for $O_2$ and $F_2$, in which the $\sigma 2p$ orbital is of lower energy than the two $\pi 2p$ orbitals.

Figure 8.36 is an energy level diagram for the MOs in diatomic molecules of O and F, in which the $\sigma 2p$ MO is of lower energy than the two degenerate $\pi 2p$ MOs.

The results of applying the Aufbau method to the set of allowed orbitals of the homonuclear diatomics of the second-row elements are shown in Table 8.5.

The $Li_2$ molecule (Figure 8.37) has two valence electrons, one $2s$ electron from each Li atom. These two electrons can be placed in the $\sigma 2s$ bonding orbital. Therefore, $Li_2$ has $\frac{1}{2}(2 - 0) = 1$ bond and should be stable. The $Li_2$ molecule has been found to exist in the gas phase and to be moderately stable.

The $Be_2$ molecule has four valence electrons, two $2s$ electrons from each Be atom. Two of these electrons can be placed in the $\sigma 2s$ bonding orbital. The other two are placed in the $\sigma^* 2s$ antibonding orbital (Figure 8.37). Therefore, $Be_2$ has $\frac{1}{2}(2 - 2) = 0$ bonds. A molecule with no bonds should not exist. No evidence has been found for the existence of the $Be_2$ molecule.

The MO method predicts an unusual feature for the $B_2$ molecule: two unpaired electrons, as shown in Figure 8.38. Each B atom has three valence electrons, so there are six electrons that must be placed in the molecular orbitals of the $B_2$ molecule. Two electrons fill the $\sigma 2s$ molecular orbital. Two more fill the $\sigma^* 2s$ MO. The remaining two electrons fill the two degenerate (equal-energy) $\pi 2p$ orbitals. Each electron half fills one of the orbitals, consistent with Hund's rule.

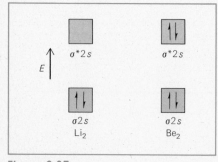

**Figure 8.37**

Molecular orbital diagrams for $Li_2$ and $Be_2$. The $Li_2$ has two electrons in the $\sigma 2s$ bonding orbital and is stable. The $Be_2$ has two electrons in the bonding orbital and two electrons in an antibonding orbital, and is not stable. The vertical spacing of the orbitals is not to scale.

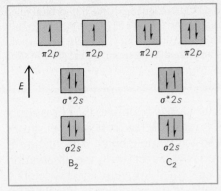

**Figure 8.38**
Molecular orbital diagrams for the $B_2$ and $C_2$ molecules. The $B_2$ has unpaired electrons, while the $C_2$ has only paired electrons. The vertical spacing of the orbitals is not to scale.

**TABLE 8.5**   Electronic Configuration of Second-Row Homonuclear Diatomics

| Molecule | Lewis Structure | Number of Electrons | | | | | Bond | Unpaired Electrons |
|---|---|---|---|---|---|---|---|---|
| | | $\sigma 2s$ | $\sigma^* 2s$ | $\pi 2p$ | $\pi 2p$ | $\sigma 2p$ | | |
| $Li_2{}^a$ | Li—Li | 2 | | | | | 1 | 0 |
| $Be_2{}^b$ | | 2 | 2 | | | | 0 | 0 |
| $B_2{}^a$ | $(:B—B:)^c$ | 2 | 2 | 1 | 1 | | 1 | 2 |
| $C_2{}^a$ | :C=C: | 2 | 2 | 2 | 2 | | 2 | 0 |
| $N_2$ | :N≡N: | 2 | 2 | 2 | 2 | 2 | 3 | 0 |

| | | Number of Electrons | | | | | | | |
|---|---|---|---|---|---|---|---|---|---|
| | | $\sigma 2s$ | $\sigma^* 2s$ | $\sigma 2p$ | $\pi 2p$ | $\pi 2p$ | $\pi^* 2p$ | $\pi^* 2p$ | $\sigma^* 2p$ |
| $O_2$ | $(:O=O:)^c$ | 2 | 2 | 2 | 2 | 2 | 1 | 1 | | 2 | 2 |
| $F_2$ | :F̈—F̈: | 2 | 2 | 2 | 2 | 2 | 2 | 2 | | 1 | 0 |
| $Ne_2{}^b$ | | 2 | 2 | 2 | 2 | 2 | 2 | 2 | 2 | 0 | 0 |

$^a$ Rarely encountered, very reactive.
$^b$ Nonexistent molecule.
$^c$ Incorrect structures.

There is a major difference between the $B_2$ electonic configuration of the MO method and that of the VB (Lewis structure) method. Several different Lewis structures can be written for the $B_2$ molecule. All have incomplete octets; none of them need have unpaired electrons. The MO method indicates that the $B_2$ molecule has two unpaired electrons. There are four electrons in bonding orbitals and only two electrons in the $\sigma^* 2s$ antibonding orbital, giving $\frac{1}{2}(4-2) = 1$ bond, enough to hold the molecule together.

We can determine which method gives the more accurate picture by examining the magnetic properties of the $B_2$ molecule. As we discussed in Section 6.2, a substance that is slightly repelled by a magnetic field is diamagnetic and has all its electrons paired. A relatively small number of substances are paramagnetic and are attracted by a magnetic field. Such substances have one or more unpaired electrons. Experimentally, we find that $B_2$ is paramagnetic. Therefore, it has unpaired electrons. The MO picture of the structure of $B_2$ is more consistent with this observation.

The $B_2$ molecule is rare. So is the $C_2$ molecule, whose electronic configuration, including eight electrons, is shown in Figure 8.38. All the electrons are paired, so $C_2$ should be diamagnetic. Experiments have confirmed that $C_2$ is diamagnetic.

**Example 8.7**   Show how the fact that $B_2$ is paramagnetic and $C_2$ is diamagnetic confirms that the two degenerate $\pi 2p$ MOs are of lower energy than the $\sigma 2p$ MO and fill first.

**Solution**   Figure 8.38 shows the electronic configurations that result when the $\pi 2p$ MOs have lower energy than the $\sigma 2p$ MO, as experimental results suggest. If the $\sigma 2p$ orbital filled first, $B_2$ would have the electronic configuration $(\sigma 2s)^2(\sigma^* 2s)^2(\sigma 2p)^2$, with no unpaired electrons, so $B_2$ would be diamagnetic. If the $\sigma 2p$ orbital filled first, $C_2$ would have the electronic configuration $(\sigma 2s)^2(\sigma^* 2s)^2(\sigma 2p)^2(\pi 2p)^1(\pi 2p)^1$, with two unpaired electrons, one in each of the degenerate $\pi 2p$ MOs, so $C_2$ would be paramagnetic. Experimental results indicate that these are not the ground state configurations of $B_2$ and $C_2$.

The MO picture of the $N_2$ molecule closely resembles the Lewis structure, as shown in Figure 8.39. The $N_2$ molecule has six bonding electrons in the $2p$ molecular orbitals, and so has a triple bond. This picture is consistent with the Lewis structure $:N\equiv N:$ and with the observation that $N_2$ is diamagnetic.

The MO picture of the $O_2$ molecule, however, differs markedly from the VB picture. Historically, one of the first major successes of MO theory was the correct prediction of the properties of the $O_2$ molecule, which VB theory could not predict in any simple way.

A simple Lewis structure with completed octets is written for $O_2$:

$$\ddot{O}=\ddot{O}$$

This structure, in which all the valence electrons are paired, is inconsistent with experimental evidence, which indicates that $O_2$ is paramagnetic and has two unpaired electrons. By contrast, the MO method predicts that two of the 12 valence electrons in the $O_2$ molecule will be unpaired, as shown in Figure 8.39.

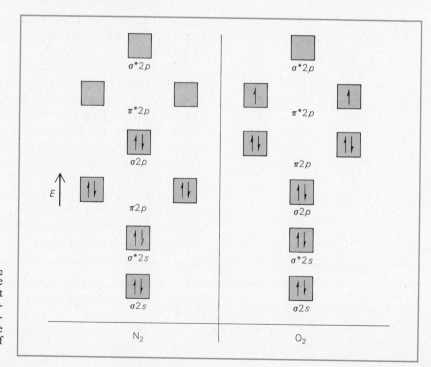

**Figure 8.39**
Molecular orbital diagrams of the $N_2$ and $O_2$ molecules. The MO picture of the $N_2$ molecule is consistent with the Lewis structure of $N_2$, but the MO configuration of the $O_2$ molecule, unlike the Lewis structure, shows unpaired electrons. Experiment shows the MO model to be more accurate for $O_2$. The vertical spacing of the orbitals is not to scale.

of multiple conjugation. Such molecules often absorb visible light and are intensely colored. Most organic dyes have this type of extended conjugation. Vitamin A is conjugated, as are the pigments that give carrots and tomatoes their distinctive colors.

We know (Sections 5.5 and 5.6) that the spectrum of atomic hydrogen consists of discrete lines that correspond exactly to changes in the energy levels of the electron. In the same way, the absorption of a quantum of light can promote an electron in a molecule from an occupied molecular orbital to an unoccupied, usually antibonding, molecular orbital. Consider the case of $F_2$. Absorption of a photon of the correct frequency can promote an electron from the filled $\sigma$ orbital to the empty $\sigma^*$ orbital. The net bonding is the algebraic sum of bonding and antibonding orbitals; in this case, $1 - 1 = 0$, or no bond. Thus, light can cause a molecule to

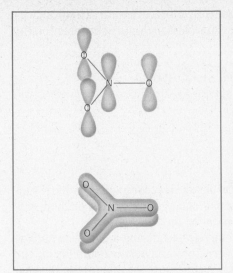

**Figure 8.42**
The molecular orbital method applied to the $NO_3^-$ anion. Molecular orbitals that encompass all four atoms can be constructed by the use of the $p$ orbitals that are shown. One molecular orbital is shown. It is perpendicular to the plane of the paper.

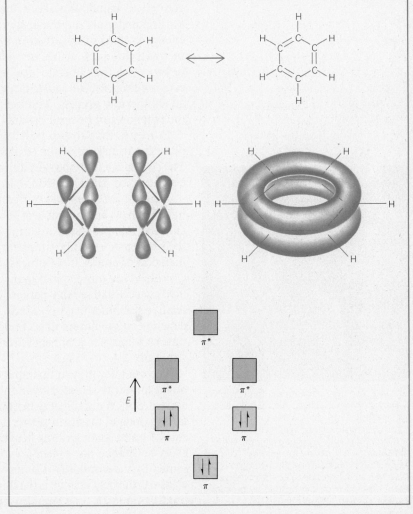

**Figure 8.43**
The molecular orbital method applied to the benzene molecule, $C_6H_6$. The molecule forms six molecular orbitals, three bonding and three antibonding. One bonding molecular orbital is shown.

fragment into atoms. Such photochemical processes are extraordinarily important in the atmosphere. One example is the sequence of reactions by which ozone, $O_3$, forms from $O_2$:

$$O_2 + light \longrightarrow O + O$$
$$O + O_2 \longrightarrow O_3$$

## 8.7   THERMOCHEMISTRY

Almost every chemical and physical change is accompanied by the evolution or absorption of heat, or thermal energy. The branch of chemistry that studies these thermal effects is called thermochemistry.

The quantity of heat absorbed or released during a chemical change can tell us something about the structure and relative stability of the starting materials and the products. We measure the quantity of thermal energy evolved or absorbed in a chemical or physical change by calorimetric techniques, which we shall discuss in Chapter 13.

Processes that liberate heat cause a rise in the temperature of the surroundings and are said to be **exothermic.** The combustion of gasoline is an exothermic process. The temperature rise in the region surrounding the reaction can be used productively, as it is in an automobile engine.

Processes that absorb heat are said to be **endothermic.** Endothermic processes usually result in a drop in the temperature of the surroundings. The coolness felt on the skin during the evaporation of a liquid with a low boiling point, such as alcohol, is the result of an endothermic process.

A useful *but by no means rigorous* generalization is that most exothermic processes are favorable at room temperature (about 25 °C) and most endothermic processes are unfavorable at room temperature. That is, if the process A → B liberates heat, compound B probably is more stable than compound A. If the process C → D absorbs heat, the reverse is true: C probably is more stable than D. This generalization does not always hold. As we shall see in Chapter 13, heat changes alone do not determine relative stabilities of compounds. The relative stability of any compound depends on the nature of its bonds and other details of molecular structure, so a knowledge of heat effects can give information about molecular structure.

The quantity of heat liberated or absorbed in a given process depends on the amounts of materials and the conditions under which the process is carried out. We shall deal primarily with processes that occur in containers open to the atmosphere. These processes occur at a *constant* pressure of 1 atm. In describing heat changes, it is convenient to imagine that every substance has a *heat content,* called its **enthalpy,** which is represented by the symbol $H$. For any process A → B carried out at constant pressure, the heat change is the difference between $H_B$, the enthalpy of the products, and $H_A$, the enthalpy of the reactants. The Greek capital letter

The test tube contains iron titanium hydride, which can store very large amounts of hydrogen gas. Hydrogen can be used as a fuel in place of oil or gas, but storing the hydrogen safely is one of the problems that must be overcome before a transition to a "hydrogen economy" can become feasible. Stored correctly, hydrogen is less explosive than gasoline. The hydrogen gas is easily released from the hydride by moderate heating. *(Department of Energy)*

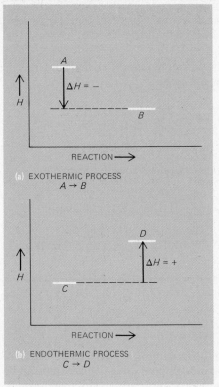

**Figure 8.44**
An exothermic process liberates heat and thus has a negative $\Delta H$ (a). An endothermic process absorbs heat and has a positive $\Delta H$ (b).

$\Delta$, *delta, is used to indicate a difference between a final and an initial state.* Thus:

$$\Delta H = H_B - H_A \qquad (8.1)$$

The quantity $\Delta H$, called the *enthalpy change,* is the heat absorbed or released by a process that occurs at constant pressure. By the definitions cited above, an exothermic process has a negative $\Delta H$ and an endothermic process has a positive $\Delta H$ (Figure 8.44).

Some features of $H$ and $\Delta H$ should be noted:

1. The actual value of the enthalpy, $H$, of a single substance cannot be measured. Only the difference in the enthalpy, $\Delta H$, between the reactants and products of a chemical reaction or between two states of a substance can be measured.

2. Since the value of $\Delta H$ depends on the quantity of material reacted or produced, $\Delta H$ usually is expressed in units that indicate the amount of material as well as the quantity of energy. The SI units for $\Delta H$ are joules per mole (J/mol) or kilojoules per mole (kJ/mol), depending on the magnitude of the heat change. The most common non-SI units for $\Delta H$ are calories per mole or kilocalories per mole: 1 cal = 4.184 J. The amounts of material sometimes are implied by the equation for a process, as in $2B \rightarrow D$, which indicates that two moles of B undergo reaction.

3. If the magnitude and sign of $\Delta H$ are known for a process, $\Delta H$ of the reverse process has the same magnitude but the opposite sign. In the evaporation of water:

$$H_2O(l) \longrightarrow H_2O(g)$$

$\Delta H$ is $+ 44$ kJ/mol at 298 K. The process is endothermic. Therefore, the reverse process, the condensation of steam:

$$H_2O(g) \longrightarrow H_2O(l)$$

is exothermic and $\Delta H$ is $- 44$ kJ/mol at 298 K, if all conditions are the same.

4. The value of $\Delta H$ for a process depends on the physical state of each component, as is evident from the two processes discussed above. Therefore, it is essential to indicate the state of every component in a process by an appropriate state symbol (Section 2.6).

When we do calculations about a chemical reaction, we can treat the quantity of heat that is released or absorbed in the same way that we treat the quantity of product or reactant. Only the units are different. We measure the quantity of heat in joules or calories, not in the grams or moles we use for reactants. The next example shows such a calculation.

Example 8.8    When 1.00 mol of $CH_4$ is burned completely to $CO_2(g)$, 890 kJ of heat is liberated. Find the heat liberated when 375 g of $CH_4$ is burned.

Solution    We can use the given relationship between the amount of $CH_4$ that burns and the amount of heat that is liberated to get conversion factors:

$$\frac{1 \text{ mol } CH_4}{890 \text{ kJ}} \quad \text{or} \quad \frac{890 \text{ kJ}}{1 \text{ mol } CH_4}$$

Using the appropriate conversion factor, we can handle this calculation in the same way that we handled the stoichiometry calculations in Chapter 3:

$$375 \text{ g } CH_4 \times \frac{1 \text{ mol } CH_4}{16.04 \text{ g } CH_4} \times \frac{890 \text{ kJ}}{1 \text{ mol } CH_4} = 20\,800 \text{ kJ liberated}$$

The answer can also be given as $-20\,800$ kJ. The minus sign indicates that heat is liberated.

## Heat of Formation

The heat content $H$ of single substances cannot be measured. But standard, well-defined processes can be used to give values of $\Delta H$ that can be used to tabulate the relative enthalpies of substances. The formation of one mole of a substance from its constituent elements provides values of $\Delta H$ that can help us to assess the relative stability of the substance and to calculate the values of $\Delta H$ for other processes in which the substance takes part.

The exact conditions of the process by which one mole of a substance forms from its constituent elements must be carefully specified to make comparisons meaningful. For convenience, a specific set of conditions called a **standard state** is defined for this process.

The standard state value for pressure is 1 atm (101.325 kPa), which is close to the average atmospheric pressure in most laboratories. The physical and chemical state of an element that is most stable at this pressure and a specified temperature is defined as the standard state of the element. We shall specify the temperature as 298.15 K (25°C) for the purposes of this discussion of thermochemistry. Some elements in their standard states are: $H_2(g)$, $O_2(g)$, $Br_2(l)$, $I_2(s)$, $Na(s)$, $Mg(s)$, $Fe(s)$, $Hg(l)$, $C(s)_{graphite}$, $Pt(s)$, and $S_8(s)$.

The value of $\Delta H$ for the reaction, carried out under standard conditions, in which one mole of a substance is formed from its constituent elements in their standard states is called the **standard heat of formation** or the standard enthalpy of formation of the substance. This value is written $\Delta H_f^\circ$, in which the superscript refers to the standard state and the subscript refers to *formation*. Appendix II lists values of $\Delta H_f^\circ$ for various substances.

The units of $\Delta H_f^\circ$ usually are kilojoules per mole (kJ/mol). For example, $\Delta H_f^\circ$ of water is the $\Delta H$ of the process:

$$H_2(g) + \tfrac{1}{2}O_2(g) \longrightarrow H_2O(l)$$

and has the value $-286$ kJ/mol at 298 K. (The fractional coefficient is used for $O_2$ so that only one mole of $H_2O(l)$ is formed.)

By definition, $\Delta H_f^\circ$, *the standard heat of formation of any element in its standard state, must be zero.* Chemical forms other than the standard state of the element have positive values of $\Delta H_f^\circ$.

A number of common substances also have positive values of $\Delta H_f^\circ$. Most of them are unstable relative to their elements. The gaseous binary compounds of oxygen and nitrogen are well-known examples. One oxide of nitrogen is formed in the process.

$$\tfrac{1}{2}N_2(g) + \tfrac{1}{2}O_2(g) \longrightarrow NO(g)$$

The $\Delta H_f^\circ$ of NO(g) is $+90$ kJ/mol. Nitric oxide, NO, is unstable relative to its elements, as the positive value of its $\Delta H_f^\circ$ suggests. Such substances should decompose to their elements. Many of them can exist because they decompose very slowly. It is important to note that thermally unstable substances can exist when the process by which they change is exceedingly slow. When we say that a substance is unstable, we mean only that it will decompose without help from the surroundings. We do not say anything about the time needed for the process to occur. We shall discuss these points in greater detail in Chapter 13.

Water is a more typical molecule than NO. Its standard heat of formation is negative; water is stable with respect to its elements and the process by which it forms from its elements is exothermic. We can use the $\Delta H_f^\circ$ of water listed in Appendix II as a guide to its stability. In general, the stability of a substance is high if its $\Delta H_f^\circ$ has a large negative value.

Liquid water, $H_2O(l)$, has a $\Delta H_f^\circ$ of $-286$ kJ/mol. The $\Delta H_f^\circ$ of hydrogen peroxide, $H_2O_2(l)$, is $-188$ kJ/mol, which is less negative than the $\Delta H_f^\circ$ of $H_2O(l)$. We can predict that $H_2O_2(l)$ probably will decompose to $H_2O(l)$ and $O_2(g)$ at 298 K, while the reverse process will not occur to any appreciable extent. The prediction is borne out by observation. Similarly, we note that for $H_2O(g)$, the $\Delta H_f^\circ$ is $-242$ kJ/mol. The difference between this value and the $\Delta H_f^\circ$ for $H_2O(l)$ is the quantity of heat required for one mole of $H_2O$ to evaporate at standard conditions.

We can use $\Delta H_f^\circ$ to calculate the standard heat of reaction, a quantity whose symbol is $\Delta H^\circ$. The standard heat of reaction is the difference in enthalpy between the products and the reactants in any process carried out at 1 atm. We shall specify the temperature as 298 K. If $\Delta H_f^\circ$ is known for all the components in a process, then:

$$\Delta H^\circ = \Sigma \, \Delta H_f^\circ(\text{products}) - \Sigma \, \Delta H_f^\circ(\text{reactants}) \qquad (8.2)$$

(The symbol $\Sigma$, the Greek capital letter sigma, means "the sum of.") Equation 8.2 is equivalent to Equation 8.1.

---

**Example 8.9**     Calculate $\Delta H°$ for the reaction:

$$2H_2O(l) \longrightarrow H_2O_2(l) + H_2(g)$$

**Solution**     **Step 1.** Calculate the $\Sigma \Delta H_f°$ of the products. For $H_2O_2(l)$, $\Delta H_f°$ is $-188$ kJ/mol. For $H_2(g)$, $\Delta H_f°$ is zero, since $H_2(g)$ is an element in its standard state. Thus, $\Sigma \Delta H_f°$ of the products is $-188$ kJ/mol.

**Step 2.** Calculate the $\Sigma \Delta H_f°$ of the reactants. The $\Delta H_f°$ of $H_2O(l)$ is $-286$ kJ/mol. Since there are two moles of $H_2O(l)$ as reactants, $\Sigma \Delta H_f°$ (reactants) $= (2$ mol $H_2O(l))(-286$ kJ/mol $H_2O(l)) = -572$ kJ. (In calculations of this type, every $\Delta H_f°$ is expressed in units of kilojoules per mole (kJ/mol) and must be multiplied by the coefficient of the substance, since the coefficient indicates the number of moles of the substance in the reaction.)

**Step 3.** From Steps 1 and 2, $\Delta H° = -188$ kJ $- (-572$ kJ$) = +384$ kJ. The positive value of $\Delta H°$ indicates that this process is endothermic and probably does not occur spontaneously at room temperature.

---

### Hess's Law

Enthalpy is one of a number of properties that are called *state functions.* Some other state functions are temperature, pressure, and volume. We will discuss state functions in more detail in Chapter 13. For the moment, we shall say that one important fact about a state function is that its value does not depend on past history. The value of a state function for a system in a given state is completely independent of the way in which the system got to that state.

We use the symbol $\Delta$ preceding a state function to designate the value of a change in the state function, in other words, the difference between the two values of the state function for two different states of the system. As we said above, the value of this difference does not depend on how the system went from one state to the other. The value of a quantity such as $\Delta H$ is independent of the path followed by the system. It depends only on the initial and final states.

This principle was first observed experimentally for chemical reactions in 1840 by G. H. Hess. Specifically, he found that $\Delta H$ for any reaction is independent of any intermediate reactions that may occur. His observations have been called Hess's law of constant heat summation, which is often stated: *The heat change ($\Delta H$) that accompanies a given chemical reaction is the same whether the reaction occurs in one step or in several steps.*

One important consequence of this law is that chemical reactions can be manipulated in the same way as algebraic equations. They can be added to or subtracted from one another or multiplied by a common

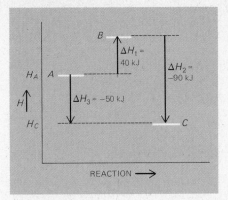

**Figure 8.45**
The heat change for any overall reaction, such as A → C, is always the same, no matter how many steps occur in the reaction.

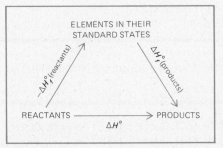

**Figure 8.46**
Equation 8.2 follows directly from Hess's law.

factor to obtain new equations. If the $\Delta H$ is known for each of a starting set of equations and the same arithmetic operations performed with these equations are performed with their $\Delta H$'s, then we obtain the $\Delta H$ for the new equation.

A simple example is a two-step process:

1. $\qquad\qquad\qquad\qquad$ A $\longrightarrow$ B $\qquad \Delta H_1 = +40$ kJ
2. $\qquad\qquad\qquad\qquad$ B $\longrightarrow$ C $\qquad \Delta H_2 = -90$ kJ

We can perform a simple addition of equations to obtain:

$$A + B \longrightarrow B + C$$

so,

3. $\qquad$ A $\longrightarrow$ C $\quad$ and $\quad \Delta H_3 = \Delta H_1 + \Delta H_2$
$$= +40 \text{ kJ} + (-90 \text{ kJ}) = -50 \text{ kJ}$$

In this example, there are two steps between A and C. No matter how many steps there are, $\Delta H$ for the overall reaction A → C is always the same and is always the sum of the enthalpy changes for all the individual steps. This relation is shown in Figure 8.45. The value of $\Delta H$, the difference in $H$ between A and C, does not depend on the path taken to get from A to C but only on the values of $H_A$ and $H_C$.

If you look at Equation 8.2, which relates the heat of a reaction to the heats of formation of the reactants and products, you can see that it is simply a special case of Hess's law. As Figure 8.46 shows, using this equation is the same as carrying out the reaction by a two-step pathway that goes from the reactants to the elements in their standard states to the products. The $\Delta H°$ of this pathway is the same as that of any other pathway from reactants to products. More complete manipulations of equations are shown in the following example.

**Example 8.10**

Given the data:

1. $\qquad\qquad\qquad$ N$_2$O$_4$(g) $\longrightarrow$ 2NO$_2$(g) $\qquad \Delta H_1 = 58$ kJ
2. $\qquad\quad$ NO(g) + $\frac{1}{2}$O$_2$(g) $\longrightarrow$ NO$_2$(g) $\qquad \Delta H_2 = -56$ kJ

calculate $\Delta H$ for the process:

3. $\qquad\qquad$ 2NO(g) + O$_2$(g) $\longrightarrow$ N$_2$O$_4$(g) $\qquad \Delta H_3 = ?$

**Solution**

All the substances in Equation 3 appear in Equations 1 and 2. By appropriate manipulations of Equations 1 and 2, Equation 3 can be obtained. The same manipulations can then be repeated for the $\Delta H$ value of each equation to obtain $\Delta H_3$.

**Step 1.** The first substance on the left side of Equation 3 is 2NO(g). This substance appears in Equation 2 as NO(g). Multiplying Equation 2 by two, we get:

$$2NO(g) + O_2(g) \longrightarrow 2NO_2(g) \qquad 2\Delta H_2 = 2(-56 \text{ kJ}) = -112 \text{ kJ}$$

**Step 2.** The next substance on the left side of Equation 3 is $O_2(g)$. This substance is already included in Step 1.

**Step 3.** The product on the right side of Equation 3 is $N_2O_4(g)$, which appears on the left side of Equation 1. We can reverse Equation 1, which means that we must reverse the sign of its $\Delta H$. This procedure gives us $-$ (Equation 1):

$$2NO_2(g) \longrightarrow N_2O_4(g) \qquad -\Delta H_1 = -58 \text{ kJ}$$

**Step 4.** Combining the equations in Steps 1 and 3 gives:

$$[2 \times (\text{Equation 2})] + [-(\text{Equation 1})]$$

or $\qquad 2NO(g) + O_2(g) + 2NO_2(g) \longrightarrow N_2O_4(g) + 2NO_2(g)$

or, simplifying:

3. $$2NO(g) + O_2(g) \longrightarrow N_2O_4(g)$$

Since Equation 3 = $[2 \times (\text{Equation 2})] + [-(\text{Equation 1})]$

$$\begin{aligned}\Delta H_3 &= [2 \times (\Delta H_2)] + (-\Delta H_1) \\ &= 2(-56 \text{ kJ}) - (58 \text{ kJ}) \\ &= -170 \text{ kJ}\end{aligned}$$

In the next example we see how the state of a substance is related to the $\Delta H_f^\circ$.

**Example 8.11**    Given that 44 kJ of heat is needed to evaporate one mole of water at 298 K, calculate $\Delta H_f^\circ$ of $H_2O(g)$, given that $\Delta H_f^\circ$ of $H_2O(l) = -286$ kJ/mol.

**Solution**    The $\Delta H_f^\circ$ of $H_2O(g)$ refers to the process:

$$H_2(g) + \tfrac{1}{2}O_2(g) \longrightarrow H_2O(g)$$

We are given that $\Delta H$ for the process

1. $$H_2(g) + \tfrac{1}{2}O_2(g) \longrightarrow H_2O(l)$$

is $-286$ kJ/mol, and that $\Delta H$ for the process

2. $$H_2O(l) \longrightarrow H_2O(g)$$

is 44 kJ/mol.

We can sum these two equations to get:

$$H_2(g) + \tfrac{1}{2}O_2(g) + H_2O(l) \longrightarrow H_2O(l) + H_2O(g)$$

or, simplifying:

3. $$H_2(g) + \tfrac{1}{2}O_2(g) \longrightarrow H_2O(g)$$

Summing $\Delta H$ for Equations 1 and 2 gives:

$$\Delta H_1 + \Delta H_2 = -286 \text{ kJ/mol} + 44 \text{ kJ/mol} = -242 \text{ kJ/mol} = \Delta H_3$$

as shown in Figure 8.47.

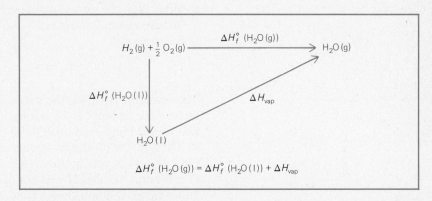

**Figure 8.47**
$\Delta H_f^\circ(H_2O(g)) = \Delta H_f^\circ(H_2O(l)) + \Delta H_{vap}$

In Example 8.11, Equation 2 represents a change in state or phase, rather than a chemical change. There are a number of such phase changes, and there is a specific name for the heat change associated with each phase change.

*The process in which a liquid becomes a gas is called vaporization.* The heat change associated with it is called the **heat of vaporization** ($\Delta H_{vap}$). *The heat of vaporization always has a positive value.*

*The reverse process, in which a gas becomes a liquid, is called condensation.* Under the same conditions, the **heat of condensation** is equal in magnitude to the heat of vaporization and has the opposite sign. As the heat of vaporization always has a positive value, *the heat of condensation always has a negative value.*

*The heat change associated with the change from solid to gas is the* **heat of sublimation** ($\Delta H_{sub}$). *The heat change associated with the change from solid to liquid is the* **heat of fusion** ($\Delta H_{fus}$). *Often a heat change accompanies the formation of a solution and is called the* **heat of solution** ($\Delta H_{sol}$). The heat liberated by the complete combustion of a substance is called the *heat of combustion.* The heat liberated by the reaction $CH_4(g) + 2O_2(g) \longrightarrow CO_2(g) + 2H_2O(l)$ is the heat of combustion of methane.

### Average Bond Energy

The way in which thermochemical observations give us information about molecular structure is illustrated by the reaction:

$$H_2(g) + Cl_2(g) \longrightarrow 2HCl(g)$$

This process is exothermic at 298 K, so the product is more stable than the reactants. The increased stability of two HCl molecules as compared to separate $H_2$ and $Cl_2$ molecules is related to the molecular structure of these species; more specifically, to the differences in bond strengths of the molecules.

Examining Lewis structures, we see that $H_2$ has a single H—H bond, $Cl_2$ has a single Cl—Cl bond, and HCl has a single H—Cl bond. Since the process by which 2HCl is formed is exothermic, two HCl bonds must be stronger than the sum of one H—H bond and one Cl—Cl bond.

We can use this kind of analysis on many reactions, such as the reaction of Example 8.9:

$$2H_2O(l) \longrightarrow H_2O_2(l) + H_2(g)$$

The reactant, $H_2O$, has the structure H—O—H, which has two O—H bonds. The product $H_2O_2$ has the structure H—O—O—H, which has two O—H bonds and one O—O bond, while $H_2$ has one H—H bond.

The conversion shown by this reaction can be described in this way: Four O—H bonds (from $2H_2O$) are broken. Two O—H bonds, one O—O bond, and one H—H bond form. The net result is that two O—H bonds break and one O—O bond and one H—H bond form. Since the process is endothermic, the bonds on the left side of the equation are stronger, in sum, than the bonds on the right side of the equation. This does not necessarily mean that the compounds on the right side of the equation are unstable. In this case, they are stable relative to their elements, but the compounds on the left side of the equation have stronger bonds and so are even more stable. It is an essential principle of chemistry that **the formation of chemical bonds between two atoms is always an exothermic process.** Otherwise no bonds would be formed.

We could calculate the heat of a reaction if we knew how much heat is needed to break each bond and how much is liberated by the formation of each bond in the reaction. The quantity of heat needed to break a bond between two atoms of a molecule in the gas phase is called the *bond dissociation energy*.[1] It is always a positive quantity.

The bond energy of the H—Cl bond is defined by $\Delta H$ for the process:

$$HCl(g) \longrightarrow H(g) + Cl(g) \qquad \Delta H = 432 \text{ kJ/mol}$$

while the bond energy of the I—I bond is defined by $\Delta H$ for the process:

$$I_2(g) \longrightarrow 2I(g) \qquad \Delta H = 151 \text{ kJ/mol}$$

Note that standard states are not necessarily involved in defining bond energies. All substances must be in the gas phase to define bond energies, and the gas phase is not the standard state of many substances.

---

[1] Strictly speaking, we should use the term *enthalpy.* "Bond dissociation energy" is a form of chemical slang.

## FUELS FOR THE FUTURE

As the earth's supply of petroleum dwindles, the search has begun for an alternative fuel, made from renewable sources, to replace gasoline. Major attention has focused on hydrogen and on alcohol — either methanol, $CH_3OH$, or ethanol, $C_2H_5OH$. While the studies done thus far indicate that either alcohol or hydrogen could replace gasoline and other petroleum fuels, they also show why gasoline was chosen as the fuel for most internal combustion engines: Its properties are not easy to match.

The heat of combustion of gasoline — the amount of energy released when it burns — compares favorably with that of the other fuels. The heat of combustion of typical unleaded gasoline is 43 megajoules per kilogram (MJ/kg). For methanol, the heat of combustion is 20 MJ/kg; for ethanol, it is about 39 MJ/kg; and for hydrogen, it is about 115 MJ/kg.

Hydrogen seems to offer the most advantages as an alternative to gasoline. In addition to its high heat of combustion, hydrogen is also a literally inexhaustible fuel. It can be extracted from water and it burns to produce water, from which it can again be extracted. However, it is the most distant possibility. The cost of obtaining hydrogen by the electrolysis of water is prohibitively high. Almost all $H_2$ is obtained from natural gas, itself a fuel in short supply. Efforts are being made to develop low-cost methods of decomposing water. The future of the proposed "hydrogen economy" based on the use of $H_2$ gas as fuel, depends on the success of these efforts.

Other apparent drawbacks, such as the low density of hydrogen and the danger of explosions, are more easily overcome. Studies have shown that the danger of hydrogen explosions is no greater than the danger of natural gas explosions. The fact that hydrogen cannot be liquefied above 33.3 K has created interest in special systems for storage. One such system is based on the ready absorption of hydrogen by such metals as iron and titanium. The hydrogen is easily released by a small amount of heat as needed.

Alcohol already is being used on a relatively small scale as an alternative fuel. Several major oil companies in the United States are marketing what was originally called "gasohol," a fuel consisting of 90% gasoline and 10% alcohol (methanol). Enthusiasm for gasohol was at its peak in the late 1970s, when world petroleum prices reached their highest levels. Gasohol lost some of its attractiveness as oil prices went down in the early 1980s, since alcohol costs substantially more than gasoline. Oil companies now advertise alcohol primarily as a fuel additive that increases the octane rating of gasoline, rather than as a competing fuel.

Several objections have been raised about the use of alcohol as an auto fuel. Some authorities have calculated that more energy is used to produce a gallon of alcohol than is extracted from the alcohol when it is used as fuel. Another objection arises from the use of ethanol made from wheat or corn as fuel. Critics say that a major increase in alcohol fuel use would drive food prices to unacceptably high levels. The possibility of using less valuable plant material, such as weeds and stems, is being explored.

It is relatively simple to find the bond energy for a diatomic molecule, because it has only one bond. It is more difficult to assign bond energies to bonds that do not occur in diatomic molecules. We do not try to measure the heat needed to break just one bond in a molecule with a number of bonds. Instead, we find the energy needed to break all of the bonds in the molecule and then divide by the number of bonds. This procedure gives what is called the *average bond energy*. The process:

$$H\text{—}O\text{—}H(g) \longrightarrow 2H(g) + O(g) \qquad \Delta H = 926 \text{ kJ/mol } H_2O$$

can define the average bond energy of the O—H bond. Since there are two O—H bonds, the average bond energy is $\Delta H/2 = 463$ kJ/mol.

The process:

$$CH_4(g) \longrightarrow C(g) + 4H(g) \qquad \Delta H = 1650 \text{ kJ/mol } CH_4$$

can define the average bond energy of the C—H bond. Since there are four C—H bonds, the average bond energy is $\Delta H/4 = 413$ kJ/mol. The average bond energy is not based on $\Delta H$ for the process $CH_4(g) \rightarrow CH_3(g) + H(g)$, because this process does not break all the bonds in the $CH_4$ molecule. *Only those processes that break all the bonds in a polyatomic molecule can be used to calculate average bond energies.* The $\Delta H$ for the process $CH_4(g) \rightarrow CH_3(g) + H(g)$, which is 423 kJ/mol, is the bond dissociation energy for one C—H bond in $CH_4$.

The usefulness of average bond energies is based on the idea that a given type of bond has roughly the same strength in many different compounds. We use the methane molecule to measure the average bond energy of the C—H bond, but this value applies wherever we encounter a C—H bond. In the same way, once we have an average bond energy for a C=C bond, we use this value for the C=C bond in any compound.

A table of average bond energies is found in Appendix III. Such a table is useful because it describes the relative strengths of different types of bonds in a straightforward way: The greater the bond energy, the stronger the bond. The fact that the bond energy of the N≡N bond in $N_2(g)$ is 946 kJ/mol, a high value, is consistent with the observation that $N_2$ in air is highly unreactive.

Average bond energies can be used to estimate approximate values of the $\Delta H$ of reactions in a way similar to that in which $\Delta H_f^\circ$ is used, although calculations using $\Delta H_f^\circ$ give more accurate results. The two methods often are complementary. If a process is analyzed in terms of bonds broken and bonds made, then:

$$\Delta H = \Sigma \text{ bond energy of bonds broken}$$
$$- \Sigma \text{ bond energy of bonds made} \qquad (8.3)$$

A simple way to remember this formula is to bear in mind that bond breaking makes a positive contribution and bond making makes a negative contribution to the $\Delta H$ of reaction.

**Example 8.12**    Using the average bond energy values in Appendix III, estimate $\Delta H$ for the reaction:

$$CH_4(g) + Cl_2(g) \longrightarrow CH_3Cl(g) + HCl(g)$$

**Solution**    If we analyze changes in bonds by using Lewis structures, we find that in this reaction, one C—H bond and one Cl—Cl bond are broken, and one H—Cl bond and one C—Cl bond are formed. Since all molecules in the reaction are in the gas phase, average bond energies can be used directly in the calculation. Equation 8.3 defines $\Delta H$ in terms of bonds broken and bonds made:

$$\Delta H = [(413 \text{ kJ/mol}) + (243 \text{ kJ/mol})] - [(328 \text{ kJ/mol}) + (432 \text{ kJ/mol})]$$
$$= -104 \text{ kJ}$$

This type of calculation can be combined with $\Delta H_f^\circ$ data to determine other average bond energies.

**Example 8.13**   Calculate the average bond energy of an N—F bond in $NF_3$ using only the following data: $\Delta H_f^\circ$ of $NF_3(g) = -114$ kJ/mol; bond energy of $N\equiv N = 946$ kJ/mol; bond energy of F—F = 158 kJ/mol.

**Solution**   The $\Delta H_f^\circ$ is the $\Delta H$ of the reaction:

$$\tfrac{1}{2}N_2(g) + \tfrac{3}{2}F_2(g) \longrightarrow NF_3(g)$$

The $\Delta H$ of this reaction can also be defined in terms of bonds made and bonds broken. In the reaction as written, $\tfrac{1}{2}$ mol of $N\equiv N$ bonds and $\tfrac{3}{2}$ mol of F—F bonds are broken, while 3 mol of N—F bonds is formed. (Each $NF_3$ has three N—F bonds.) The average bond energies of the $N\equiv N$ bonds and the F—F bonds are known, while the average bond energy of the N—F bond is not known. If we designate the N—F bond energy as $x$ and use Equation 8.3 to calculate $\Delta H$ for the reaction, we get:

$$\Delta H = [\tfrac{1}{2}(946 \text{ kJ/mol}) + \tfrac{3}{2}(158 \text{ kJ/mol})] - [3x] = -114 \text{ kJ/mol}$$
$$x = 275 \text{ kJ/mol}$$

Although average bond energies are defined for the gas phase, they can also be used in calculations involving other phases, provided that the $\Delta H$ of the relevant phase change is known.

**Example 8.14**   Calculate the average bond energy of the O—H bond using only the following data: $\Delta H_{vap}$ of $H_2O$ at 298 K = 44 kJ/mol; $\Delta H_f^\circ$ of $H_2O(l) = -286$ kJ/mol; bond energy of H—H bond = 436 kJ/mol; bond energy of O=O bond = 498 kJ/mol.

**Solution**   The average bond energy can be calculated from $\Delta H$ for the reaction:

$$H_2(g) + \tfrac{1}{2}O_2(g) \longrightarrow H_2O(g)$$

where two O—H bonds are made and one H—H and one half O=O bond are broken, all in the gas phase. In Example 8.11, the $\Delta H$ of this reaction was calculated from $\Delta H_f^\circ$ and $\Delta H_{vap}$ to be $-242$ kJ. If we designate the O—H bond energy as $x$ and use Equation 8.3 to calculate $\Delta H$, we get:

$$\Delta H = [(436 \text{ kJ/mol}) + \tfrac{1}{2}(498 \text{ kJ/mol})] - [2x] = -242 \text{ kJ/mol}$$
$$x = 463 \text{ kJ/mol}$$

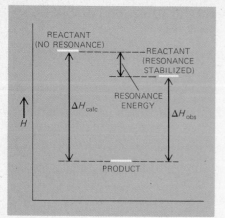

**Figure 8.48**

Calculation of resonance energy for a resonance hybrid starts with the calculation of the $\Delta H$ of a reaction based on an important contributing structure of the molecule. This calculated value is then compared with the measured value of $\Delta H$ of the reaction. The difference between the two values of $\Delta H$ is the resonance energy of the molecule.

## Resonance Energy

A number of factors other than bond strength affect the stability of molecules. For example, we mentioned in Section 7.5 that resonance hybrids have greater stability than would be predicted on the basis of bond energies alone. The extra stability, called the resonance energy, can be given a numerical value. By convention, resonance energy is given a positive value to simplify calculations with bond energies, which also have positive values.

To find the value of the resonance energy of a molecule, we must find the difference between the $\Delta H$ that is calculated from the bond energies of an important contributing structure of the resonance hybrid and the $\Delta H$ that is determined experimentally.

The determination of resonance energy starts with the calculation of $\Delta H$ for a given reaction of the molecule on the basis of bond energies alone. If a molecule has resonance energy, the calculation of $\Delta H$ on the basis of bond energies alone will be incorrect by an amount equal to the resonance energy. If the correct value of $\Delta H$ is determined experimentally, the difference between the observed $\Delta H (\Delta H_{obs})$ and the calculated value $(\Delta H_{calc})$ is the resonance energy. Figure 8.48 shows this process graphically.

---

**Example 8.15**

Calculate the resonance energy of $N_2O$ from the following data: $\Delta H_f^\circ$ of $N_2O = 82$ kJ/mol; bond energy of $N\equiv N = 946$ kJ/mol; bond energy of $O=O$ (in $O_2$) $= 498$ kJ/mol; bond energy of $N=O = 607$ kJ/mol; bond energy of $N=N = 418$ kJ/mol.

**Solution**

The measured value of $\Delta H_f^\circ$ refers to the reaction:

$$N_2(g) + \tfrac{1}{2}O_2(g) \longrightarrow N_2O(g)$$

The value of $\Delta H$ for this reaction can be calculated from the bond energies. Draw an important contributing structure for $N_2O$:

$$\ddot{N}=N=\ddot{O}$$

On the basis of this contributing structure, we can say that in this reaction, an $N=N$ bond and an $N=O$ bond are made and that an $N\equiv N$ bond and one-half of an $O=O$ bond are broken. We can use the values of the bond energies and Equation 8.3 to calculate $\Delta H$:

$$\Delta H_{calc} = [(946 \text{ kJ/mol}) + \tfrac{1}{2}(498 \text{ kJ/mol})] - [(607 \text{ kJ/mol}) + (418 \text{ kJ/mol})]$$
$$= 170 \text{ kJ/mol}$$
$$\Delta H_{obs} = 82 \text{ kJ/mol, the } \Delta H_f^\circ$$
$$\text{resonance energy} = \Delta H_{calc} - \Delta H_{obs}$$
$$= 170 \text{ kJ/mol} - 82 \text{ kJ/mol} = 88 \text{ kJ/mol}$$

Molecular stability can also be influenced by interactions between portions of a molecule that are not directly attached to one another. These interactions are important in large, complex organic molecules such as proteins and nucleic acids. The calculation of such nonbonded interactions is beyond the scope of this book.

### The Born-Haber Cycle

One important application of the principles of thermochemistry, and of Hess's law in particular, is the analysis of a chemical process as a series of thermochemical steps. One such analysis is based on what is called the Born-Haber cycle.

The Born-Haber cycle is often used to analyze the formation of an ionic solid. When $\Delta H$ can be measured for an overall process and for all but one of the steps in the process, the missing $\Delta H$ can then be calculated. The factors that affect the stability of a substance can be understood with the help of a Born-Haber cycle.

To show how a Born-Haber cycle can be used, let us construct one for sodium chloride. We can measure $\Delta H_f^\circ$ for NaCl:

$$\text{Na(s)} + \tfrac{1}{2}\text{Cl}_2\text{(g)} \longrightarrow \text{NaCl(s)} \qquad \Delta H_f^\circ = -411 \text{ kJ/mol} \qquad (8.4)$$

Clearly, NaCl(s) is a stable species.

The source of its stability was discussed semiquantitatively in Section 7.1. The Born-Haber cycle gives a more precise answer. We know that the stability of NaCl(s) results primarily from lattice energy, which is the heat liberated by the formation of the NaCl lattice from gaseous ions:

$$\text{Na}^+\text{(g)} + \text{Cl(g)} \longrightarrow \text{NaCl(s)} \qquad (8.5)$$

Lattice energy cannot be measured directly, but we can find $\Delta H$ for the processes that lead to formation of $\text{Na}^+\text{(g)}$ from Na(s) and of $\text{Cl}^-\text{(g)}$ from $\tfrac{1}{2}\text{Cl}_2\text{(g)}$. Using Hess's law and the measured $\Delta H_f^\circ$, we can then calculate the lattice energy for NaCl.

Two steps, each with measurable $\Delta H$, lead from Na(s) to $\text{Na}^+\text{(g)}$:

1. $\text{Na(s)} \longrightarrow \text{Na(g)}$        $\Delta H_{\text{sub}}$ (heat of sublimation) = 109 kJ
2. $\text{Na(g)} \longrightarrow \text{Na}^+\text{(g)} + e^-$        $\Delta H_{\text{ion}}$ (ionization energy) = 496 kJ

Two steps, each with measurable $\Delta H$, lead from $\tfrac{1}{2}\text{Cl}_2\text{(g)}$ to $\text{Cl}^-\text{(g)}$:

3.     $\tfrac{1}{2}\text{Cl}_2\text{(g)} \longrightarrow \text{Cl(g)}$     $\tfrac{1}{2}\Delta H_{\text{diss}}$ ($\tfrac{1}{2}$ Cl—Cl bond energy) = 122 kJ
4.    $\text{Cl(g)} + e^- \longrightarrow \text{Cl}^-\text{(g)}$      $-\Delta H_{\text{e.a.}}$ (electron affinity) = $-349$ kJ

According to Hess's law, $\Delta H$ for the process that is the sum of steps 1 through 4 is the sum of the $\Delta H$ values of these four steps:

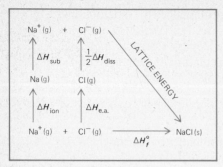

**Figure 8.49**
The Born-Haber cycle for NaCl.

$$Na(s) + \tfrac{1}{2}Cl_2(g) \longrightarrow Na^+(g) + Cl^-(g) \qquad \Delta H = 378 \text{ kJ} \qquad (8.6)$$

The sum of Equation 8.6 and Equation 8.5

$$Na^+(g) + Cl^-(g) \longrightarrow NaCl(s)$$

which is the process in which the NaCl lattice is formed from gaseous ions, is

$$Na(s) + \tfrac{1}{2}Cl_2(g) \longrightarrow NaCl(s)$$

which is identical with Equation 8.4. Since the sum of Equations 8.5 and 8.6 is Equation 8.4, by Hess's law:

$$\Delta H(8.5) + \Delta H(8.6) = \Delta H(8.4)$$
or
$$\Delta H(8.5) + 378 \text{ kJ} = -411 \text{ kJ}$$
$$\Delta H(8.5) = -789 \text{ kJ}$$

The lattice energy is 789 kJ/mol. Figure 8.49 shows the cyclical nature of this thermochemical construction.

The Born-Haber cycle analysis shows that the major reason for the stability of NaCl(s) is its lattice energy, which has a large negative value. All the other factors, except the electron affinity of Cl, have positive $\Delta H$ values, and therefore are actually destabilizing influences.

**Example 8.16**

Acetylene, $C_2H_2(g)$, is unstable. Give an explanation in terms of energy for this instability with the aid of a thermochemical cycle, using data in Appendix II and Appendix III.

**Solution**

The reported value of $\Delta H_f^\circ$ for $C_2H_2$ is $+227$ kJ/mol. The large positive value means that acetylene is unstable relative to its constituent elements, carbon and hydrogen, in their standard states. Acetylene exists only because its rate of decomposition at room temperature is very slow. The $\Delta H_f^\circ$ is defined by the reaction:

$$2C(s)_{graphite} + H_2(g) \longrightarrow C_2H_2(g)$$

We can identify the reason for the instability of $C_2H_2$ by breaking this reaction down into a series of steps and constructing a thermochemical cycle.

The atoms of the $C_2H_2$ molecule are held together by strong bonds. Indeed, the process in which $C_2H_2$ forms from gaseous atoms is exothermic:

1.      $2C(g) + 2H(g) \longrightarrow C_2H_2(g) \qquad \Delta H = -\Sigma$ bonds made

The Lewis structure of $C_2H_2$ is H—C≡C—H. In step 1, two C—H bonds and one C≡C bond are formed. Thus: $\Delta H = -[2(413 \text{ kJ}) + 812 \text{ kJ}] = -1638$ kJ. This is a highly favorable process. The reason for the relative instability of $C_2H_2$ must lie elsewhere.

The steps required to produce the gaseous atoms on the left side of the reaction in step 1 are

2. $\qquad H_2(g) \longrightarrow 2H \qquad \Delta H_{diss} = $ H—H bond energy $= 436$ kJ
3. $\qquad 2C(s)_{graphite} \longrightarrow 2C(g) \qquad \Delta H_{sub} = 2(718 \text{ kJ}) = 1436$ kJ

The large positive value of the $\Delta H$ of step 3 gives the major reason $C_2H_2$ is unstable relative to its elements. In graphite, the standard state of carbon, the C atoms are held together tightly. A large amount of energy is needed to separate them before compounds of carbon can be formed.

The sum of steps 1 through 3 defines $\Delta H_f^\circ$ for $C_2H_2$. The sum of the values of $\Delta H$ for each of these steps is 234 kJ/mol, which is close to the experimentally determined value, 227 kJ/mol. A similar line of reasoning was behind the statement that reactions of $N_2$ are difficult to bring about because of the strength of the nitrogen-nitrogen triple bond.

The stability of molecules can be described in an approximate way in terms of the quantity of heat that is absorbed or released as a result of chemical changes in which they participate. By measuring heat changes that accompany chemical processes, we can describe the stability of the substances that are formed or react as a result of these processes. We can also calculate the strength of bonds in specific molecules, and we can predict the relative stability of these molecules. A thorough understanding of heat changes will give us a deeper understanding of chemical processes and molecular stability.

## 8.8  WEAK INTERACTIONS

In addition to the strong forces that hold atoms together in chemical bonds, there are a number of weaker attractive forces between atoms that are not bonded to each other. These weak interactions can have a profound effect on the physical and chemical properties of substances. Indeed, the existence of condensed phases is due in large part to these attractive forces. Usually, weak interactions are intermolecular — that is, they are attractions between portions of different molecules. But there can also be weak intramolecular attractive forces between different portions of the same molecule.

All these weak interactions are electrostatic, the attraction of opposite charges. As we shall see, the charges may result from the distribution of electrons in a molecule or from changes in electronic distribution induced by external electric charges or electric fields. Weak interactions are substantially weaker than chemical bonds. For example, $Br_2$ is a liquid at room temperature because of attractions between $Br_2$ molecules. The heat of vaporization, the heat required to overcome these attractions, is 31 kJ/mol. The bond energy of the Br—Br bond is 193 kJ/mol.

The magnitude of these weak interactions is very sensitive to distance.

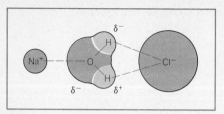

**Figure 8.50**
Ion-dipole interaction. The polar water molecule attracts both anions and cations in a solution of sodium chloride.

While the energy of the interaction between a cation and an anion is simply a function of $1/r$, where $r$ is the distance between their centers, the energy of some weak interactions varies as $1/r^6$, as we shall see. Therefore, weak interactions are consequential only when the interacting species are very close.

## Ion-Dipole Interactions

When an ion and a polar molecule with a dipole are near each other, there will be an attraction between the ionic charge and the end of the dipole of opposite charge. Such attractions are stronger and less sensitive to distance than the other weak interactions. The energy is a function of $1/r^2$, where $r$ is the distance between the center of the charge and the center of the dipole. These interactions are very important in solutions of ionic substances in polar solvents (Section 12.3). Figure 8.50 shows the ion-dipole interactions in a solution of sodium chloride in water. Ion-dipole interactions can also be used to explain some aspects of complex ions (Section 18.3).

## Dipole-Dipole Interactions

Polar molecules, which have permanent dipoles, can attract each other if the dipoles of the molecules are oriented correctly. Figure 8.51 shows how two HCl molecules can attract each other. The relatively high melting points and boiling points of many polar compounds can be attributed to dipole-dipole attractions. In condensed phases, polar molecules tend to orient themselves favorably. The energy of this interaction is a function of $1/r^3$. In the gas phase or in the liquid phase at relatively high temperatures, the orientation of polar molecules is random. The energy of this interaction then is a function of $1/r^6$ and is considerably less important.

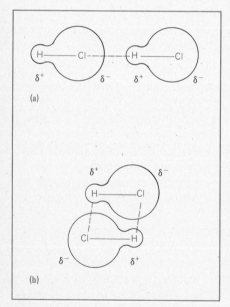

**Figure 8.51**
The simplest case of a weak attraction. Dipole-dipole attraction, the coulombic attraction between the electrical dipoles of two polar diatomic molecules.

## Induced Dipole Interactions

When an ion or a molecule with a dipole approaches an atom or molecule that has no dipole, a dipole can be induced. An interaction can then take place between the ion and the induced dipole or between the dipole and the induced dipole, as shown in Figure 8.52. These interactions tend to be quite weak and quite sensitive to distance. The energy of the ion-induced dipole interaction is a function of $1/r^4$. That of the dipole-induced dipole interaction is a function of $1/r^6$. These interactions are important in solutions of ionic or polar substances in nonpolar solvents and in some gas mixtures.

## London Forces

There can also be attractive forces even between completely nonpolar substances. The existence of such forces can be deduced from the obser-

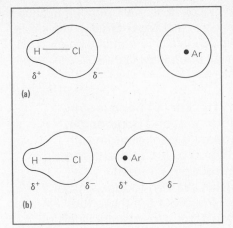

**Figure 8.52**

Dipole-induced-dipole attraction. The polar HCl molecule induces a dipole in the ordinarily nonpolar Ar atom. The induced dipole is oriented so that there is a weak attraction with the HCl molecule.

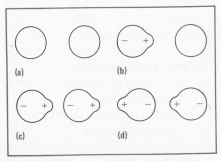

**Figure 8.53**

Instantaneous dipole-induced-dipole attraction. A momentary displacement of electron density creates a dipole in a nonpolar molecule. This "instantaneous dipole" creates a dipole in a second nonpolar molecule, causing an attraction between the molecules.

vation that nonpolar substances such as $I_2$ and $Br_2$ are liquid or solid at room temperature. There must be some relatively strong intermolecular forces that hold the molecules of these substances together.

These intermolecular forces are called *London forces.* They are believed to arise from momentarily unequal distributions of electron density in molecules, even nonpolar molecules. At any instant, there may be an excess of electron density in one region of the molecule and a corresponding deficiency of electron density elsewhere in the molecule. The uneven distribution of electron density can create an "instantaneous dipole."

Suppose there is another nonpolar molecule nearby. As Figure 8.53 shows, a dipole is induced in this second molecule by the instantaneous dipole of the first molecule. An attractive force results. This attraction between molecules can be maintained even though the instantaneous dipole disappears. A shift in the electron density of the first molecule can create another instantaneous dipole with a different orientation. If the induced dipole shifts in synchronization with the instantaneous dipole, the attraction between the molecules is maintained. This model is used to account for the attractive forces between nonpolar species.

The energy of the interaction between an instantaneous dipole and an induced dipole is a function of $1/r^6$. But these interactions can still be relatively substantial, especially in large molecules. The heat of vaporization of $I_2$, a fairly large nonpolar molecule, is 62 kJ/mol. Except some small polar molecules, London forces are the most important weak interactions.

The extent to which a dipole can be induced in a nonpolar molecule depends on two factors: the force that induces the dipole and the ease with which the electron distribution in the molecule can be distorted. The responsiveness of the electron distribution of a molecule to a distorting influence is called the *polarizability.* Several factors influence polarizability. An increase in the volume of a molecule or an increase in the number of electrons in the molecule results in an increase in the polarizability of the molecule. Since high polarizability indicates stronger attractive forces between molecules, molecules of the larger members of any family of elements will attract each other more strongly.

The physical properties of the halogens illustrate the increase in London forces in the larger members of a family of elements. At room temperature, $Cl_2$ is a gas, $Br_2$ is a liquid, and $I_2$ is a solid. The $Cl_2$ molecule is the smallest of the three and $I_2$ is the largest. The intermolecular forces between the $Cl_2$ molecules in the liquid or solid are weaker than the forces between the $I_2$ molecules in the liquid or solid.

All the weak interactions between permanent dipoles, instantaneous dipoles, and induced dipoles are known collectively as **van der Waals forces.** They are responsible for deviations of real gases from ideal gas behavior.

A special kind of weak interaction is the hydrogen bond. Hydrogen bonds occur primarily in compounds that contain O—H bonds or N—H bonds, although HF also has a hydrogen bond. Generally, compounds with hydrogen bonds display properties that are consistent with strong intermolecular attractions. Ion-dipole interactions and hydrogen bonding will be discussed at greater length in Chapter 11.

**Summary**

**W**e began our discussion of structural chemistry by defining **bond length,** the distance between the centers of two bonded atoms, and **bond angle,** the angle formed by two imaginary lines that connect a central atom with two atoms to which it is bonded. We mentioned some of the methods used to study the structure of molecules, such as spectroscopy. We then introduced the **valence shell electron pair repulsion (VSEPR)** model, whose important rule is that the bond angles about a central atom are those that minimize repulsions between valence shell electron pairs. We described the **ideal geometries** that the VSEPR describes for the arrangement of different numbers of atoms and electron pairs around a central atom. We noted that the VSEPR model has some limitations, and that an alternative method, the **hybridization** model, which is based on the mixing of atomic orbitals to form **hybrid orbitals** and the overlap of these hybrid orbitals to form bonds, is better for describing some molecules. We described the ideal geometries predicted by the hybridization model for a number of different species. We then showed how the **polarity** of molecules is related to their geometry and how bond length affects the stability and geometry of molecules. We then discussed the **molecular orbital method,** which can be used to describe chemical bonding by assuming that there is a set of allowed orbitals for a molecule and assigning electrons to orbitals on the basis of increasing energy. We moved on to **thermochemistry,** which deals with the heat absorbed or released by chemical reactions. A reaction that liberates heat is **exothermic** and one that absorbs heat is **endothermic.** We defined **enthalpy** as the heat content of a substance and the enthalpy change as the heat absorbed or released by a process that occurs at constant pressure. To help measure enthalpy change, a specific set of conditions called the **standard state** is defined. The **standard heat of formation** is the enthalpy change when one mole of a substance is formed from its elements in their standard states. We then mentioned **Hess's law,** which says that the heat change that accompanies a chemical reaction is the same whether the reaction occurs in one step or in a number of steps. We defined **average bond energy** as the total quantity of energy needed to break all the identical bonds in a molecule divided by the number of those bonds. We then described the **Born-Haber** cycle, which helps us to understand the factors that affect the stability of a substance. We concluded with a discussion of **weak interactions,** electrostatic attractions between different molecules.

# Exercises

**8.1** Draw the Lewis structure of the molecule $C_2H_3Cl$ and indicate the number of different bonds and bond angles in this molecule.

**8.2** Use the VSEPR model to predict which of these triatomic molecules are linear and which have some other shape: (a) HOBr, (b) $SF_2$, (c) $ZnF_2$, (d) $BeCl_2$.

**8.3** Use the VSEPR model to predict which of these tetraatomic molecules are linear and which have some other shape: (a) $H_2S_2$, (b) $Hg_2F_2$, (c) $C_2Cl_2$.

**8.4** Using the VSEPR model, predict which of the following molecules have ideal trigonal planar geometry, which have close to ideal trigonal planar geometry, and which have some other geometry: (a) $PF_3$, (b) $AlBr_3$, (c) $BFCl_2$, (d) $NH_2Cl$.

**8.5**[2] Indicate which ions have ideal trigonal planar geometry, which have close to ideal trigonal planar geometry, and which have some other geometry: (a) $CH_3^+$, (b) $CH_3^-$, (c) $NH_2^-$, (d) $CH_2Br^+$, (e) $BF_4^-$.

**8.6** Indicate which of the following molecules have ideal tetrahedral geometry, which have close to ideal tetrahedral geometry, and which have some other geometry: (a) $CHCl_3$, (b) $CBr_4$, (c) $SCl_4$, (d) $NH_2OH$, (e) $CH_3I$.

**8.7** Indicate which of the following ions have tetrahedral geometry, which have close to tetrahedral geometry, and which have some other geometry: (a) $CH_3^-$, (b) $NH_4^+$, (c) $BCl_4^-$, (d) $H_3O^+$.

**8.8** All of the following molecules have multiple bonds. Which are linear? (a) $N_2H_2$, (b) $CO_2$, (c) $C_2HCl$, (d) NOF.

[2] The answers to exercises whose numbers are in color can be found in Appendix VII. The star indicates an exercise that is more challenging than average.

**8.9** All of the following ions have multiple bonds. Which are linear? (a) $N_3^-$, (b) $NO_2^-$, (c) $NO_2^+$, (d) $C_2H^-$.

**8.10** Give the closest ideal geometry for these molecules: (a) $CH_2NH$, (b) $H_2SO_4$, (c) $MgF_2(g)$, (d) $COCl_2$.

**8.11** Give the closest ideal geometry for these ions: (a) $PO_4^{3-}$, (b) $HCOO^-$, (c) $CO_3^{2-}$, (d) $BF_3Cl^-$.

**8.12** All of the following molecules have expanded octets. Give the closest ideal geometry for each: (a) $TeF_6$, (b) $SF_3Cl_3$, (c) $PCl_3F_2$, (d) $SOCl_4$.

**8.13** Fluorine forms a compound or an ion with each element from 31 to 35. Each of these species has ideal octahedral geometry. Draw their structures.

**8.14★** There are two possible geometrical arrangements for $SF_4I_2$. Draw them and predict which is more likely to be stable. Explain your prediction.

**8.15** Predict the geometry of the following molecules: (a) $SOCl_2$, (b) $SO_2Cl_2$, (c) $S_2Cl_2$, (d) $SO_2FCl$.

**8.16** Predict the geometry of the following ions: (a) $NO_2^+$, (b) $HSO_4^-$, (c) $ClO_2^-$, (d) $PH_4^+$.

**8.17** Use the designations of geometry given in Table 8.2 to describe the geometry of the following molecules: (a) $GaF_3$, (b) $PbI_2$, (c) CHOCl (C—H bond), (d) ClCN.

**8.18** Use the designations of geometry in Table 8.2 to describe the geometry of the following ions: (a) $SO_3^-$, (b) $ClO_3^-$, (c) $NCO^-$.

**8.19★** The bond angles in $NF_3$ are $102.1°$. The bond angle in $OF_2$ is larger, $103.2°$. This is opposite to the trend in $NH_3$ and $H_2O$. Rationalize this observation.

**8.20** Using a model of simple overlap of unhybridized orbitals, predict the geometry of $NH_3$ and $H_2O$.

**8.21** Identify the atomic orbitals that may overlap to form the bonds in (a) HF, (b) NaH, (c) $Cl_2$, (d) $H_2S$.

**8.22** Identify the atomic orbitals that may overlap to form the bonds in (a) $OH^-$, (b) HBr, (c) $S_8$, (d) $ZnI_2$.

**8.23** Indicate the most likely hybridization for the carbon atom in each compound: (a) $CH_2I_2$, (b) $C_2H_2Cl_2$, (c) $CH_3^-$, (d) HCN.

**8.24** Indicate which hybrid and atomic orbitals overlap to form the bonds in each of the following compounds: (a) $CHCl_3$, (b) $GeH_4$, (c) $CF_4$.

**8.25** Indicate which orbitals overlap to form each of the bonds in $CH_3CH_2Cl$.

**8.26** Indicate which hybrid and atomic orbitals overlap to form the bonds in each of the following ions: (a) $AsH_4^+$, (b) $BF_4^-$, (c) $SO_4^{2-}$.

**8.27** Describe the bonding in the following molecules in terms of $\sigma$ or $\pi$ bonds. Identify the atomic orbitals that may overlap to form these bonds. (a) NO, (b) $N_2O$ (consider just one contributing structure), (c) $NO_2$.

**8.28** Draw a bonding diagram of the type shown in Figure 8.23 for CO, which has the same number of electrons as $N_2$.

**8.29** Draw a bonding diagram of the type shown in Figure 8.23 for the $CN^-$ anion.

**8.30** Indicate which orbitals overlap to form the $\sigma$ bonds in (a) $AlF_2Cl$, (b) $CH_2NH$, (c) $NO_2^-$.

**8.31** Indicate which orbitals overlap to form the $\sigma$ bonds in (a) $BeF_2$,

(b) HNCO, (c) $HgBr_2$.

**8.32** Draw an orbital picture of the type shown in Figures 8.25 and 8.27 for $CH_2O$.

**8.33** Draw an orbital picture of the type shown in Figures 8.25 and 8.27 for ClCN.

**8.34** Draw an orbital picture of the type shown in Figures 8.25 and 8.27 for $CO_2$.

**8.35** Draw an orbital picture of the type shown in Figures 8.25 and 8.27 for formic acid, HCOOH.

**8.36** Indicate the most likely hybridization for the starred atoms in each of the following molecules: (a) $N^*Cl_3$, (b) $HBrO^*$, (c) $N_2^*H_4$, (d) $N_2^*O_5$.

**8.37** Indicate the most likely hybridization for the starred atoms in each of the following ions: (a) $N^*H_2^-$, (b) $N^*O_3^-$, (c) $H_3O^{*+}$.

**8.38** Allene is a hydrocarbon whose formula is $C_3H_4$. The three carbon atoms lie on a straight line, but allene has no triple bonds. Draw its Lewis structure.

**8.39** Carbon suboxide, $C_3O_2$, is an unusual oxide of carbon that has linear geometry. Write a reasonable Lewis structure for carbon suboxide that is consistent with this geometry.

**8.40★** Diazomethane, $CH_2N_2$, is an explosive, highly poisonous gas that is nonetheless commonly used in the synthesis of organic compounds. The CNN bond angle is 180° and the HCN bond angle is close to 120°. Write a Lewis structure for $CH_2N_2$ that is consistent with these observations. (*Hint:* Formal charges cannot be avoided.)

**8.41** Propose a reasonable hybridization for the central atom in (a) $PCl_3$, (b) $PCl_5$, (c) $SF_4$, (d) $SeCl_6$.

**8.42** Propose a reasonable hybridization for the central atom in (a) $SiF_4$, (b) $SiF_6^{2-}$, (c) $ClF_3$, (d) $ClF_5$.

**8.43** Predict the geometry of each of the following phosphorus-containing species: (a) $PCl_3$, (b) $POCl_3$, (c) $PCl_5$, (d) $PCl_6^-$.

**8.44** All of the following substances have five valence shell electron pairs around the central atom, but each has a different geometry. Predict the geometry of each and account for the differences. (a) $SbF_5$, (b) $TeF_4$, (c) $IF_3$, (d) $XeF_2$.

**8.45** All of the following substances have six valence shell electron pairs around the central atom, but each has a different geometry. Predict the geometry of each and account for the differences. (a) $SF_6$, (b) $BrF_5$, (c) $XeF_4$.

**8.46** The ion $CH_5^+$ can be produced in a mass spectrometer under special conditions. Propose a hybridization for the carbon atom and predict the geometry of the ion.

**8.47** Use both the VSEPR model and the hybridization model to predict the geometry of and to describe the bonding in $HSO_4^-$.

**8.48** Use both the VSEPR model and the hybridization model to predict the geometry of and to describe the bonding in $ICl_3$.

**8.49★** Neither the VSEPR model nor the hybridization model is able to account for the experimental observation that $BaF_2(g)$ is nonlinear. Suggest a possible explanation for this observation.

**8.50** Arrange the following molecules in order of increasing dipole moment: $CHBr_3$, $CHCl_3$, $CHF_3$, $CHI_3$.

**8.51** Indicate which of the following species should have no dipole moment: (a) $CH_2Br_2$, (b) $CH_3F$, (c) $CH_3^+$, (d) $NH_3$.

**8.52** Either $ClF_2^-$ or $O_3$ has a dipole moment of zero. Select the one and justify your selection.

**8.53** Arrange the following species in order of increasing length of the bonds between the carbon atom and the halogen atoms: $CHBr_3$, $CHCl_3$, $CHF_3$, $CHI_3$.

**8.54** Predict the relative lengths of the bonds between the nitrogen and oxygen atoms in $HNO_3$.

**8.55** Arrange the following molecules in order of increasing length of their C—H bonds and C—N bonds: HCN, $CH_2NH$, $CH_3NH_2$.

**8.56** Use the molecular orbital method to find the number of bonds in each of these ions: (a) $Li_2^{2-}$, (b) $Be_2^{2+}$, (c) $Be_2^{2-}$.

**8.57** Use the molecular orbital method to find the number of bonds and the number of unpaired electrons in (a) $O_2^+$, (b) $O_2^{2+}$, (c) $O_2^-$, (d) $O_2^{2-}$.

**8.58** Predict the electronic configurations (Table 8.5) and magnetic properties of (a) $N_2^{2+}$, (b) $F_2^{2+}$, (c) $F_2^{2-}$, (d) $N_2^{2-}$.

**8.59** The molecular orbitals used for homonuclear diatomic species can be used for heteronuclear species whose atoms are close in atomic number. Write electronic configurations of the type shown in Table 8.5 for the following heteronuclear diatomic molecules: (a) CO, (b) NO, (c) NF, (d) CN.

**8.60** Account for the relative stability of the $CO_3^{2-}$ ion (a) in valence bond terms, (b) in molecular orbital terms.

**8.61** At 298 K, 44.0 kJ of heat is needed to evaporate 1.00 mol of $H_2O$. Find the heat needed to evaporate 1.00 kg of water.

**8.62** The $\Delta H_f^\circ$ of BaO(s) is $-558$ kJ/ mol. Find the mass of BaO that forms when a sample of solid Ba reacts completely with $O_2(g)$, with the liberation of 1870 kJ of heat.

**8.63** When 1.00 L of gasoline undergoes complete combustion, $3.1 \times 10^4$ kJ of heat are liberated. When 1.0 kg of uranium undergoes complete fission,

$1.4 \times 10^{12}$ kJ of heat is liberated. Find the volume of gasoline necessary to liberate the heat obtained by the fission of 1.0 g of uranium.

**8.64** Find[3] the standard heat of reaction for $HCl(g) + NH_3(g) \rightarrow NH_4Cl(s)$.

**8.65** Find[3] the standard heat of reaction for the decomposition of 2.00 mol of $CaCO_3(s)$ to $CO_2(g)$ and $CaO(s)$.

**8.66** Find[3] $\Delta H^\circ$ for the reactions:
(a) $CH_4(g) + Cl_2(g) \rightarrow CH_3Cl(g) + HCl(g)$
(b) $CH_4(g) + 2Cl_2(g) \rightarrow CH_2Cl_2(l) + 2HCl(g)$
(c) $CH_4(g) + 3Cl_2(g) \rightarrow CHCl_3(l) + 3HCl(g)$
(d) $CH_4(g) + 4Cl_2(g) \rightarrow CCl_4(l) + 4HCl(g)$
The $\Delta H_f^\circ$ of $CH_3Cl(g)$ is $-82.0$ kJ/mol and of $CH_2Cl_2(l)$ is $-117$ kJ/mol.

**8.67** Find the $\Delta H^\circ$ of the reaction:
$4NH_3(g) + 7O_2(g) \rightarrow 4NO_2(g) + 6H_2O(g)$.

**8.68** The $\Delta H_f^\circ$ for $SO_2(g)$ is $-297$ kJ/mol and the $\Delta H^\circ$ of the reaction $2SO_2(g) + O_2(g) \rightarrow 2SO_3(g)$ is $-196$ kJ. Find $\Delta H_f^\circ$ for $SO_3(g)$.

**8.69** Find[3] the heat of combustion when $CH_2O(g)$ burns to form $H_2O(g)$ and $CO_2(g)$.

**8.70** The combustion of 1.00 g of propane, $C_3H_8(g)$, to $CO_2(g)$ and $H_2O(g)$ releases 45.9 kJ of heat. Find[3] the $\Delta H_f^\circ$ of propane.

**8.71** The heat of fusion of ice is 6.0 kJ/mol. The heat of evaporation of water is 44 kJ/mol. Find the heat of sublimation of ice.

**8.72** The reaction $N_2O_4(g) \rightarrow 2NO_2(g)$ absorbs 24.3 kJ of heat. The reaction $NO(g) + NO_2(g) \rightarrow N_2O_3(g)$ liberates 40.2 kJ of heat. Find $\Delta H^\circ$ of the reaction $2N_2O_3(g) \rightarrow N_2O_4(g) + 2NO(g)$.

**8.73** The $\Delta H_f^\circ$ of HBr(g) is $-36$ kJ/mol. The $\Delta H^\circ$ of vaporization of $Br_2(l)$ is 31 kJ/mol. Find the $\Delta H^\circ$ of the reaction $H_2(g) + Br_2(g) \rightarrow 2HBr(g)$.

**8.74** The $\Delta H^\circ$ of the reaction $2HI(g) \rightarrow H_2(g) + I_2(g)$ is 10 kJ. Find[3] the heat of sublimation of $I_2$.

**8.75** Calculate $\Delta H^\circ$ for the reaction $CH_4 \rightarrow C(g) + 4H(g)$, using tabulated values of $\Delta H_f^\circ$. Repeat the calculation using the average C—H bond energy.

**8.76** The *heat of atomization* is the heat needed to convert a molecule in the gas phase into its constituent atoms in the gas phase. Find[4] the heat of atomization of $CH_2=C=O$.

**8.77** Calculate[4] $\Delta H$ for the following gas phase reactions:
(a) $C_2H_4 + HBr \rightarrow C_2H_5Br$
(b) $H_2O_2 \rightarrow H_2 + O_2$
(c) $3F_2 + N_2 \rightarrow 2NF_3$

**8.78** The BrBr and FF bond energies are 193 kJ/mol and 158 kJ/mol, respectively. The $\Delta H_f^\circ$ of $BrF_3(g)$ is $-256$ kJ/mol. The $\Delta H^\circ$ of evaporation of $Br_2$ is 31 kJ/mol. Calculate the average bond energy of the Br—F bond in $BrF_3$.

**8.79** Repeat the calculation of Exercise 8.78 for $BrF_5(g)$, whose $\Delta H_f^\circ$ is $-429$ kJ/mol. Account for the apparent difference in the BrF bond strengths in $BrF_3$ and $BrF_5$.

**8.80★** The standard state of sulfur is $S_8(s)$. Use the values given in Appendix II for $\Delta H_f^\circ$ of S(g), O(g), and $SO_2(g)$ to find the average bond energy of the SO bond in $SO_2$.

**8.81** Calculate[4] the resonance energy of $NO_2(\ddot{O}=\ddot{N}—\ddot{O}:)$, using the measured $\Delta H_f^\circ$ of 34 kJ/mol.

**8.82** Calculate[4] the $\Delta H_f^\circ$ of $N_2O_5(g)$, given that its resonance energy is 296 kJ/mol.

**8.83** The measured $\Delta H^\circ$ for the combustion of benzene, $C_6H_6(g)$, to $CO_2(g)$ and $H_2O(g)$ is $-3170$ kJ/mol. Calculate[4] the $\Delta H^\circ$ for this reaction using bond energies, and then calculate the resonance energy of benzene.

**8.84** The $\Delta H_f^\circ$ of KF(s) is $-563$ kJ/mol, the ionization energy of K(g) is 419 kJ/mol, and the heat of sublimation of potassium is 88 kJ/mol. The electron affinity of F(g) is 322 kJ/mol and the FF bond energy is 158 kJ/mol. Calculate the lattice energy of KF(s) using a Born-Haber cycle.

**8.85★** The $\Delta H_f^\circ$ of $CaBr_2$ is $-675$ kJ/mol. The first ionization energy of Ca(g) is 590 kJ/mol, and its second ionization energy is 1145 kJ/mol. The heat of sublimation of calcium is 178 kJ/mol. The bond energy of $Br_2$ is 193 kJ/mol, the $\Delta H_{vap}$ of $Br_2(l)$ is 31 kJ/mol, and the electron affinity of Br(g) is 325 kJ/mol. Calculate the lattice energy of $CaBr_2(s)$.

**8.86★** The $\Delta H_f^\circ$ of $PI_3(s)$ is $-24.7$ kJ/mol and the PI bond energy is 184 kJ/mol. Construct a thermochemical cycle of the type illustrated in Example 8.16 using the necessary data in Appendix II and Appendix III. Calculate the heat of sublimation of $PI_3(s)$ with the aid of the cycle.

**8.87** Explain the regular increase in boiling point with the increase in atomic number of the noble gases.

**8.88** Explain the regular decrease in melting point with the increase in atomic number of the alkali metals.

**8.89** Explain the observation that ClF is polar but has a lower boiling point than $Cl_2$, which is nonpolar.

---

[3] The necessary values of $\Delta H_f^\circ$ can be found in Appendix II.

[4] The necessary values of the bond energies can be found in Appendix III.

# 9

# The Chemistry of the Nonmetals

**Preview**  The nonmetals are as familiar to us as the air we breathe — which, in fact, is made up of nonmetals. In this chapter, we shall use the concepts you have learned to describe the chemistry of the nonmetals, introducing the properties of the most important nonmetals and describing some of the compounds they form. We begin with hydrogen, the most abundant element in the universe, and then discuss carbon, the element that is the basis for life on earth. We go on to discuss four nonmetals that are important both industrially and in living things: nitrogen, phosphorus, oxygen, and sulfur. We conclude by describing two important families of nonmetals, the halogens and the noble gases.

The nonmetals are a relatively small but extremely important group of elements found on the right side and toward the top of the periodic table (Figure 9.1). The nonmetals and the compounds they form play a major role in chemistry. Most of the atoms in the universe are nonmetals, as are most of the atoms in the human body and the bodies of all other living organisms.

In their elemental form, the nonmetals usually are relatively small molecules, although they can also form quite large molecules. The nonmetals can form polar or nonpolar compounds with each other. In either case, the bonding is predominantly covalent.

In this chapter, we shall describe the behavior of the nonmetals and the compounds they form with each other. We shall focus our attention primarily on the most common nonmetals: hydrogen; carbon, which is in group IVA; nitrogen and phosphorus, which are in group VA; oxygen and sulfur, in group VIA; the halogens (fluorine, chlorine, bromine, and iodine) in group VIIA; and the noble gases (helium, neon, argon, krypton, xenon, and radon) in group VIIIA.

## 9.1 HYDROGEN

### Periodic Classification

Alone among the elements, hydrogen cannot be classified as a member of any chemical family. Because the electronic configuration of hydrogen is $1s$, hydrogen is sometimes classified as a member of group IA, the alkali metals, which have the valence electronic configuration $ns^1$. But the alkali metals have many properties that are quite different from those of hydrogen. Because hydrogen is just one electron short of a noble gas configuration, some have classified it as a member of group VIIA, the halogens, whose electronic configuration, $ns^2np^5$, is also one electron short of a noble gas configuration. Again, the substantial differences between the properties of hydrogen and those of the halogens make this classification

**Figure 9.1**

The nonmetals are grouped at the right side and the top of the periodic table, in the colored boxes. The semimetallic elements and the metallic elements which are found in the same columns of the periodic table are in the gray boxes.

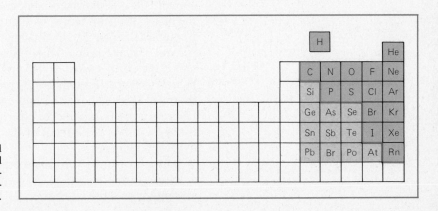

The extreme flammability of hydrogen was demonstrated by the fire that destroyed the German airship Hindenburg at Lakehurst, New Jersey, on May 6, 1937. *(Mary Evans Picture Library, Photo Researchers, Inc.)*

untenable. Finally, because hydrogen has a half-filled noble gas shell, some have tried to place it in group IVA, whose elements have a half-filled $ns^2np^6$ shell. Hydrogen does have virtually the same electronegativity as carbon, a member of group IVA. But again, the differences between hydrogen and carbon are greater than the similarities. Hydrogen cannot really be assigned to any chemical family. It is unique because it has only one electron.

### Discovery and Occurrence

While observations of "inflammable air" go back to the sixteenth century, hydrogen was first isolated in 1766 by Henry Cavendish (1731–1810). Hydrogen is by far the most abundant element in the universe. Many stars consist mainly of hydrogen. The fusion process by which hydrogen is converted to helium is the major energy source of stars and the first step in the series of fusion reactions that leads to the formation of the heavier elements (Chapter 20). Hydrogen is relatively abundant on earth. Fifteen percent of all the atoms in the atmosphere, the oceans, and the top kilometer of the earth's crust are hydrogen atoms. Because hydrogen is the lightest element, it is only 1% the mass of the oceans, the atmosphere, and the crust. Most of the earth's hydrogen is combined in water and minerals. Hydrogen gas, $H_2$, escapes easily from the earth's gravitational pull and so is only a minor constituent of the atmosphere, but the oceans are 10% hydrogen by mass.

In its pure form, hydrogen exists over a wide range of conditions as $H_2$, a colorless, odorless, tasteless gas. Its critical temperature is 33 K. No amount of pressure will liquefy $H_2$ when it is above this temperature. At atmospheric pressure, $H_2$ becomes a liquid at 20 K. It freezes at 14 K. The H—H bond has a large bond energy, 436 kJ/mol, so $H_2$ dissociates into H atoms only at extremely high temperatures. Even at 2000 K, less than 0.1% of $H_2$ is dissociated.

## Preparation

Hydrogen is widely used commercially. For industrial purposes, a major source of $H_2$ is methane, $CH_4$, from natural gas. Another important source of hydrogen is the water-gas reaction, the reaction of red-hot coke with steam:

$$H_2O(g) + C(s) \longrightarrow CO(g) + H_2(g)$$

For laboratory needs, hydrogen can be bought in small quantities, or it can be made by several methods. The most common is the reaction of a metal, such as zinc or iron, with a strong acid such as hydrochloric acid:

$$Zn(s) + 2H^+(aq) \longrightarrow Zn^{2+}(aq) + H_2(g)$$

Most hydrogen is produced as a by-product of petroleum refining. High-purity hydrogen is prepared by the electrolysis of brine solutions:

$$H_2O(l) \longrightarrow H_2(g) + \tfrac{1}{2}O_2(g)$$

Most industrial hydrogen is consumed in the synthesis of ammonia, which has many uses. Hydrogen is used to hydrogenate vegetable oils, changing liquid oils to solid fats for products such as margarine. Hydrogen is also used in the hydrogenation of crude oil to produce gasoline, in the manufacture of hydrogen chloride, and as a fuel in oxyhydrogen torches and in rocket engines. Hydrogen must be handled with care because mixtures of hydrogen and air can explode in the presence of a spark. The most famous of these explosions occurred at Lakehurst, N.J., on May 6, 1937, destroying the zeppelin *Hindenburg* and effectively ending the era of the lighter-than-air ship.

## Chemical Properties

Because the H—H bond is so strong, $H_2$ is relatively unreactive at room temperature. It does combine vigorously with fluorine, even at low temperatures, but does not react with other halogens, with oxygen, or with the other nonmetals under ordinary conditions. In general, $H_2$ must be activated to participate in chemical reactions. Hydrogen can be activated by heat, by being adsorbed on the surface of a solid (usually a metal), or by treatment with a reactive chemical species.

The reaction of $H_2$ and $Cl_2$ is typical. At room temperature, mixtures of these two gases are quite stable. When the temperature is raised to 650 K, a reaction takes place rapidly:

$$H_2(g) + Cl_2(g) \longrightarrow 2HCl(g) \qquad \Delta H^\circ = -184 \text{ kJ}$$

Mixtures of $H_2$ and $Cl_2$ react rapidly—indeed, explosively—at room temperature under powerful illumination. The energy from a bright light acts in much the same way as thermal energy. Both start the reaction by causing the dissociation of some $Cl_2$:

$$Cl_2(g) \rightleftharpoons 2Cl(g)$$

When the mixture of $H_2(g)$ and $Cl_2(g)$ is heated, the $Cl_2$ dissociates, rather than the $H_2$, because the Cl—Cl bond is weaker than the H—H bond. When the mixture is illuminated, the $Cl_2$ dissociates because it absorbs light and $H_2$ does not. We can tell that hydrogen does not absorb radiation in the visible part of the spectrum because $H_2$ is colorless. The greenish yellow color of $Cl_2$ shows that it absorbs some wavelengths of visible light. The Cl atoms that form from the dissociation of $Cl_2$ are very reactive, and the reaction:

$$Cl(g) + H_2(g) \longrightarrow HCl(g) + H(g) \qquad \Delta H^\circ = 4.6 \text{ kJ}$$

occurs rapidly. The H atoms formed in this step are also very reactive, and the reaction:

$$H(g) + Cl_2(g) \longrightarrow HCl(g) + Cl(g) \qquad \Delta H^\circ = -188 \text{ kJ}$$

takes place. These two steps, taken together, are called a *chain reaction,* because the Cl atoms formed in the second step immediately participate in the first reaction, which forms more H atoms that take part in the second reaction, and so on. The first step is very slightly endothermic because the H—Cl bond is slightly weaker than the H—H bond. The second step is exothermic because the H—Cl bond is stronger than the Cl—Cl bond.

Hydrogenation, the addition of hydrogen atoms to a molecule, is an important reaction. The simplest reaction of this sort is the hydrogenation of ethylene to ethane:

$$C_2H_4(g) + H_2(g) \longrightarrow C_2H_6(g) \qquad \Delta H^\circ = -137 \text{ kJ}$$

In the manufacture of margarine, hydrogenation converts vegetable oils, which are liquid at room temperature, to fats, which are solid at room temperature. Hydrogenation reactions can be carried out at relatively low temperatures and pressures in the presence of metals such as platinum or palladium. The $H_2$ molecule adheres to the surface of the metal, and the H—H bond is weakened. The molecule that is to be hydrogenated approaches the surface of the metal, and two or more hydrogen atoms are

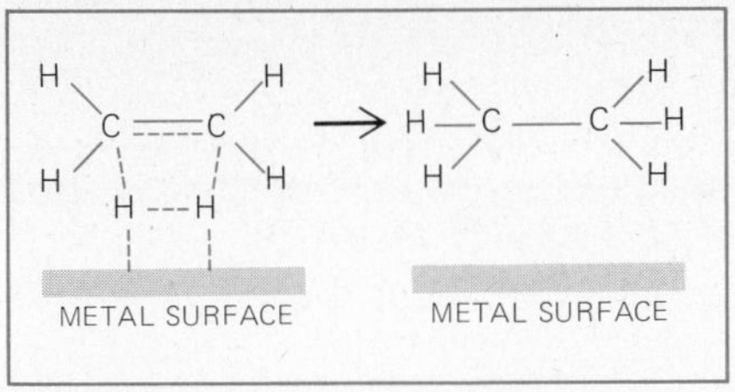

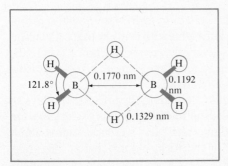

**Figure 9.2**
The hydrogenation of an unsaturated compound can occur under relatively mild conditions in the presence of a powdered metal, such as palladium or platinum.

added to the molecule as shown in Figure 9.2. In hydrogenation, the hydrogen atoms are added to molecules that have multiple bonds. Any compound with a multiple bond is said to be *unsaturated*. A compound with no multiple bonds is said to be *saturated*. Hydrogenation produces saturated compounds from unsaturated compounds.

## Hydrides

A binary compound of hydrogen and another element is called a hydride. In Section 6.9, we mentioned that almost all elements form hydrides, and we discussed some periodic trends in the behavior of hydrides. Some of the most important compounds in nature are hydrides. We shall devote an entire chapter to water, the most common and the most important of the hydrides (Chapter 11). Another chapter will discuss the hydrocarbons, hydrides of carbon and hydrogen (Chapter 21). Virtually every organic compound includes both carbon and hydrogen. In Chapter 15, we shall discuss the many hydrides that are acids. In this section, we shall discuss some other hydrides.

One interesting group of hydrides is the boranes, hydrides of boron. More than 20 simple boranes have been identified. Many of them have unusual structures and properties. The simplest borane that exists as a stable compound is diborane, $B_2H_6$. The compound $BH_3$ seems to have only a brief existence, even under extraordinary conditions. More complex boranes include $B_5H_9$, $B_{10}H_{14}$, and $B_{20}H_{16}$.

Formulating the structure of diborane, shown in Figure 9.3, is not easy. It cannot be done with only ordinary covalent bonds. By conventional rules, the structure of $B_2H_6$ requires a minimum of seven bonds (six B—H bonds and one B—B bond) and therefore a minimum of 14 valence electrons. But the compound has only 12 valence electrons, so an unusual kind of bond must exist in diborane. The ordinary covalent bond has two electrons shared by two atoms. Two of the bonds in diborane have two electrons that are shared by *three* atoms. These bonds are called *three-center bonds*. In each, an electron pair is shared by two boron atoms and one hydrogen atom, as shown in Figure 9.3. The hydrogen atom in this bond is called a bridge atom. The other hydrogen atoms in diborane are called terminal atoms. Some other boranes have three-center bonds in which one electron pair is shared by three boron atoms. Three-center bonds also are found in compounds of other elements.

The chemistry of the boranes has been studied extensively, and not only because of their unusual structure. They have some important practical applications. They are valuable reagents for the synthesis of many organic compounds, which are otherwise difficult to prepare. Boranes also undergo combustion reactions that are similar to those of hydrocarbons but release more heat. The heat of combustion is the quantity of heat liberated when one mole of a substance undergoes the combustion reaction with oxygen. For ethane, $C_2H_6$, the heat of combustion is 970 kJ/mol. For diborane, the heat of combustion is 2020 kJ/mol. The reaction is

**Figure 9.3**
The structure of diborane. There are two three-center bonds, in each of which two boron atoms and one hydrogen atom share an electron pair.

$$B_2H_6(g) + 3O_2(g) \longrightarrow B_2O_3(s) + 3H_2O(l)$$

Because they release so much heat, boranes have been considered for use as high-energy rocket fuels. Unfortunately, they cost too much to be used as rocket propellants.

## 9.2  CARBON

### Periodic Classification

Carbon is the lightest element in group IVA of the periodic table, and the only nonmetal in the group. Silicon and germanium, the next two elements in group IVA, are best classified as semimetals. Tin and lead, the two heaviest elements in the group, are metals.

The valence electronic configuration of carbon is $2s^2 2p^2$, which means that the carbon atom requires more electrons to achieve the noble gas electronic configuration than any other nonmetal. To achieve the electronic configuration of neon, a carbon atom has to gain four electrons. There are some rare compounds called carbides, such as $Be_2C$ and $Al_4C_3$, in which carbon does seem to exist as the $C^{4-}$ anion. In theory, a carbon atom could achieve the electronic configuration of helium by losing four electrons, but such a loss is never observed under ordinary conditions.

Carbon forms an unusually large number of compounds, in large measure because a carbon atom forms strong covalent bonds with other carbon atoms and with most of the common nonmetals. Carbon atoms easily form long chains, a process called *catenation*. Often, many hydrogen atoms are attached to the carbon atoms of such a chain. This ability to form stable long chains even when combined with other elements is unique to carbon. It helps to account for the importance of carbon in the chemistry of living things.

### Discovery and Occurrence

As charcoal or soot, carbon in relatively pure form was known in primitive times. But it was not until the late eighteenth century that carbon was recognized as an element. The discovery was made in France, where the word *carbone* was introduced as the name of the element, to distinguish it from *charbon*, the French word for charcoal.

Carbon is the third most abundant element in the universe. It plays an important part in the energy-producing fusion reactions of the stars. On earth, carbon is much less abundant. The earth's crust contains about 0.027% carbon by mass, mainly in the form of carbonates. The most common of these are $CaCO_3$ and $FeCO_3$. The former, calcium carbonate, occurs in many different forms, including limestone, chalk, marble, calcite, and seashells. A small fraction of the earth's atmosphere, about 0.013% by mass, consists of carbon, mainly in the form of carbon dioxide.

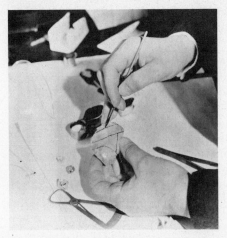

A large diamond is marked for cutting. After such a gem diamond is cleaved, diamond saws and disks coated with diamond dust will be used to create the facets that bring out its brilliance. *(American Museum of Natural History)*

This relatively minuscule percentage of carbon provides most of the raw material for life on earth. Carbon is constantly being recycled through living organisms, which use it for the complex chemicals of life. The oil, coal, and natural gas that provide almost all the energy used by our modern technological society consist of the fossilized remains of ancient living organisms. These fossil fuels contain only a small fraction of the carbon in the earth's crust, but it is an important part to civilization.

## Allotropes of Carbon

Different forms of the same element, especially in the same phase, are called **allotropes.** The two most prominent solid allotropes of carbon are graphite and diamond. Amorphous carbon, which includes such substances as coke, wood charcoal, and carbon black, consists of aggregates of microscopically small graphite particles and various surface impurities.

The differences between graphite and diamond could not be more striking. Graphite is black and so soft that it can be used as a lubricant. Pure diamonds are extremely hard, colorless crystals. Yet both consist only of carbon atoms.

Graphite gets its name from the Greek word *graphein,* meaning "to write." This Greek word indicates that the ancients used this soft, black, almost greasy substance in the same way we do today, for the "lead" in lead pencils. Graphite has other useful properties. It has great resistance to heat and thus is used in furnaces and in similar applications. It conducts electricity (unlike the other nonmetals) and thus is used in electrodes, dynamos, and microphones. Graphite is manufactured when an amorphous form of carbon, such as coke, is heated in an electric furnace to temperatures of 2800 K. Impurities are driven off and only pure graphite is left.

Diamonds were also known to the ancients (although the "diamonds" mentioned in the Old Testament probably were corundum, a hard, colorless form of aluminum oxide). The name is derived from the Greek word *adamas,* meaning "invincible." Most diamonds are found in South Africa, where the largest diamond known thus far, the Cullinan diamond, with a mass of 570 g — 2850 carats — was found in 1905. Aside from their value as gems, diamonds have important industrial uses.

A great deal of effort has gone into developing methods for making synthetic diamonds. These methods generally use temperatures above 1200 K and pressures greater than 70 000 atm. Even under these conditions, diamonds of gem quality cannot be synthesized. Indeed, the way in which diamonds are created naturally remains a mystery.

Figure 9.4 shows the structure of diamond. A diamond can be thought of as a single giant molecule made up of nothing but carbon atoms. All the carbon atoms are $sp^3$ hybridized. Each is bonded to four other carbon atoms that are at the corners of a regular tetrahedron. In this crystal structure, all the C—C bonds are identical. Both the strength and number of the C—C bonds give diamond its hardness. The strength of the C—C

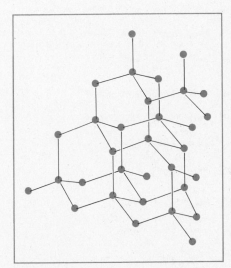

**Figure 9.4**
The structure of diamond, a giant molecule made up of carbon atoms in a regular tetrahedral array.

bonds in diamond is comparable to the strength of a C—C bond in an ordinary molecule such as ethane. But the C—C bonds in diamond extend throughout the crystal. The carbon atoms in the diamond crystal are held together by covalent bonds, which are much stronger forces of attraction than those in most other crystals. To deform a solid, it is necessary to overcome some of the forces holding its units of structure together. These forces are more difficult to overcome in diamond than in almost any other substance.

Despite the great strength of its crystal structure, diamond is the less stable allotrope of carbon under ordinary conditions. The standard state of carbon at 298 K and 1 atm is graphite. The standard heat of formation $(\Delta H_f^\circ)$ C(s)$_{graphite}$ → C(s)$_{diamond}$ of diamond is a postive quantity, 1.9 kJ/mol. It is only at pressures much greater than 1 atm that diamond is the more stable allotrope.

Diamonds form naturally at the high pressures found in the interior of the earth. Diamonds can be synthesized by application of high pressure to graphite, but there is a complicating factor. Diamonds cannot be synthesized at low temperature because the rate of conversion is too slow. As temperature increases, the stability of diamond relative to graphite decreases, even at high pressures. When temperatures are high, pressures must be extraordinarily high so that diamond will be the more stable allotrope.

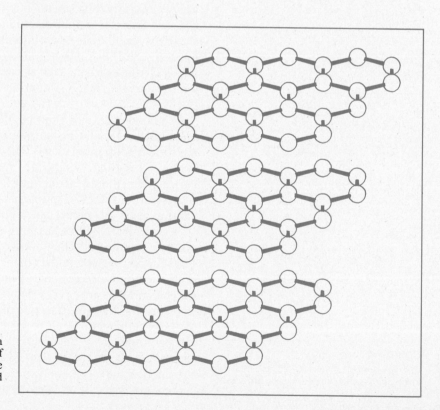

**Figure 9.5**
The structure of graphite, which is made up of a stack of flat molecules that are composed of carbon atoms. The graphite molecules slide readily over one another, so graphite is soft and slippery. No double bonds are shown.

## The Graphite Molecule

A diamond crystal represents a single large molecule. Graphite, by comparison, consists of a large number of flat molecules stacked rather loosely on one another. As Figure 9.5 shows, each carbon atom in a flat graphite molecule is $sp^2$ hybridized and is bonded to three other carbon atoms that lie at the corners of an equilateral triangle. All the bonds are equivalent, and the bonding between the carbon atoms in a molecule is quite strong. If graphite absorbs water, the adsorbed film allows the flat molecules to slide over each other, a property that gives graphite its softness and makes it useful as a lubricant. In the absence of water, graphite is abrasive. The structure of graphite is much more open than the structure of diamond, so graphite has a lower density than does diamond. High pressure favors the diamond structure because it is the more compact, denser form.

The greater stability of graphite under ordinary conditions can be explained by thermochemical reasoning. If we picture a single carbon atom of a graphite molecule, we see that it forms two single C—C bonds and one C=C double bond. We know the bond energies of these bonds, so we can estimate that 2(346 kJ/mol) + (615 kJ/mol) = 1307 kJ/mol is liberated when these bonds form. If we picture a single carbon atom of a diamond molecule, we see that it forms four single C—C bonds, which would liberate 4(346 kJ/mol) = 1384 kJ/mol of heat. This result is not what we expect. Arguing from bond energy alone, diamond appears to be more stable than graphite by $\frac{1}{2}$(1384 kJ/mol − 1307 kJ/mol) = 39 kJ/mol. (The difference in bond energy is divided by two because each bond in a complete molecule includes two atoms, rather than the single atom of our hypothetical case.)

The greater stability of graphite has another explanation. All the bonds in graphite are intermediate in length between a C—C single bond and a C=C double bond. This intermediate bond length suggests that graphite has delocalized bonding and is a resonance hybrid of contributing structures with alternating single and double bonds. Figure 9.6 shows two contributing structures for a portion of the graphite molecule. Graphite is more stable than diamond by 1.9 kJ/mol because graphite has resonance energy of about 42.4 kJ/mol.

While carbon forms strong covalent bonds with itself and with other nonmetals, many carbon compounds still have relatively small negative heats of formation. As Table 9.1 shows, some carbon compounds even have positive heats of formation. The fact that graphite has such strong bonds explains why carbon compounds often are not much more stable than the elements from which they are formed. Example 8.16 illustrated this effect for the formation of acetylene.

## Carbon Dioxide and the Carbon Cycle

Life on earth is based on carbon. The major source of carbon for the compounds that make life on earth possible is the carbon dioxide in the

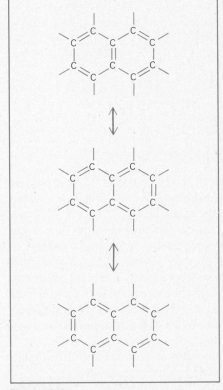

**Figure 9.6**
Three contributing structures of a portion of the graphite molecule.

**TABLE 9.1    Standard Heats of Formation of Carbon-Containing Substances at 298 K**

| Substance | Name | $\Delta H_f^\circ$ (kJ/mol) |
|---|---|---|
| $C(s)$ | graphite | 0 |
| $C(s)$ | diamond | 1.9 |
| $C(g)$ | | 718 |
| $C_2(g)$ | | 808 |
| $CO(g)$ | carbon monoxide | −110.5 |
| $CO_2(g)$ | carbon dioxide | −393.5 |
| $CH_4(g)$ | methane | −75 |
| $C_2H_2(g)$ | acetylene | 227 |
| $C_2H_4(g)$ | ethylene | 52 |
| $C_2H_6(g)$ | ethane | −85 |
| $CCl_4(g)$ | carbon tetrachloride | −107 |
| $CHCl_3(g)$ | chloroform | −100 |
| $CBr_4(g)$ | carbon tetrabromide | 50 |
| $CS_2(g)$ | carbon disulfide | 115 |
| $CH_2O(g)$ | formaldehyde | −116 |

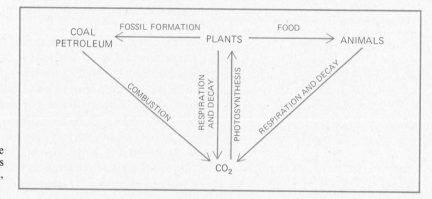

**Figure 9.7**

The carbon cycle. Plants obtain $CO_2$ from the atmosphere for photosynthesis. The carbon is returned to the atmosphere by respiration, decay, or combustion of organic material.

atmosphere. The group of reactions by which $CO_2$ is removed from and returned to the atmosphere is called the *carbon cycle*. Figure 9.7 shows some of the major features of the carbon cycle.

Photosynthesis in plants removes carbon dioxide from the atmosphere, in a process that converts solar energy to chemical energy. In photosynthesis, sunlight is used to produce glucose, a sugar, and other organic compounds from carbon dioxide and water. The plants that carry on photosynthesis and the animals that eat those plants use the chemical energy produced by photosynthesis, converting the organic substances back to carbon dioxide in the process. When organisms die, their carbon compounds are returned to the atmosphere in the form of $CO_2$ as part of the process called decay. Some dead plants are converted to fossil fuels. When we burn these fuels, $CO_2$ is added to the atmosphere.

The atmosphere contains a relatively small amount of carbon dioxide. Only 0.0325% of the atmosphere—325 parts per million (ppm) by volume—is $CO_2$. In recent decades, the concentration of $CO_2$ in the atmosphere has been going up by about 1 ppm per year. The increase appears to be caused primarily by the burning of fossil fuels. On a global scale, the rising $CO_2$ content of the atmosphere could have serious long-term consequences. If the century-long pattern of rising consumption of fossil fuels continues, the amount of $CO_2$ in the atmosphere could double by early in the twenty-first century. Climatologists calculate that such an increase could raise the global temperature by an average of between 0.5 K and 1 K. For the first time, human activity will be on a scale great enough to change the climate, with unknown consequences.

Chemically, carbon dioxide is quite stable, with $\Delta H_f^\circ = -393.5$ kJ/mol. It is formed by the oxidation of carbon or carbon-containing compounds in the presence of excess $O_2$:

$$C(s) + O_2(g) \longrightarrow CO_2(g) \qquad \Delta H^\circ = -393.5 \text{ kJ/mol}$$
$$CH_4(g) + 2O_2(g) \longrightarrow CO_2(g) + 2H_2O(l) \qquad \Delta H^\circ = -890.3 \text{ kJ/mol}$$

The formation of the very stable compound $CO_2$ explains why combustion reactions of organic compounds generally are strongly exothermic.

Carbon dioxide can also be formed by the heating of metal carbonates. A typical reaction is:

$$CaCO_3(s) \longrightarrow CO_2(g) + CaO(s)$$

Carbon dioxide dissolves in water to form a weakly acidic solution that contains carbonic acid (Section 15.6) and thus has a slightly sour taste. Solid carbon dioxide is familiar to us as dry ice, a common refrigerant.

The carbon dioxide molecule has a linear structure; the O=C=O bond angle is 180°. However, the C=O bond in $CO_2$ is both shorter and stronger than the C=O bond in other compounds. A possible explanation is that $CO_2$ has delocalized bonding and is really a resonance hybrid of the structure above and two other contributing structures, which have triple bonds.

$$^-\!:\!\ddot{O}\!-\!C\!\equiv\!O\!:\!^+ \longleftrightarrow \ ^+\!:\!O\!\equiv\!C\!-\!\ddot{O}\!:\!^-$$

## Carbon Monoxide

Carbon monoxide is a colorless, odorless, tasteless gas that forms when carbon or carbon-containing compounds burn with an insufficiency of oxygen:

$$C(s) + \tfrac{1}{2}O_2(g) \longrightarrow CO(g) \qquad \Delta H^\circ = -110.5 \text{ kJ/mol}$$

Carbon monoxide is stable relative to its elements, but it is unstable relative to carbon dioxide and is readily converted to $CO_2$ in an exothermic reaction with $O_2$:

$$CO(g) + \tfrac{1}{2}O_2(g) \longrightarrow CO_2(g) \qquad \Delta H^\circ = -283.0 \text{ kJ/mol}$$

It also can react with itself to form $CO_2$:

$$2CO(g) \longrightarrow CO_2(g) + C(s) \qquad \Delta H^\circ = -172.5 \text{ kJ/mol}$$

While this reaction is favorable, it occurs rapidly only at high temperatures. The reaction becomes important only above 750 K. Thus, even though carbon monoxide can convert to the very stable $CO_2$, it is itself a stable compound at room temperature.

Carbon monoxide has the same number of electrons as $N_2$, but the carbon-oxygen bond in CO is even stronger than the nitrogen-nitrogen triple bond in $N_2$. In fact, the CO bond is the strongest bond known, with a bond energy of 1070 kJ/mol. The great strength of this bond results in part from resonance between two important contributing structures: $:C=\ddot{O}: \longleftrightarrow \;^-:C\equiv O:^+$. This picture of CO is consistent with the experimental observation that it has a small dipole moment. There is very little net electric charge separation between the atoms of CO, with a slight negative charge on carbon. The O atom is much more electronegative than the C atom, and we might expect to find a relative negative charge on the O atom and a relative positive charge on the C atom. But the contribution of the structure with the formal charges more than cancels the electronegativity difference. The CO molecule thus is relatively nonpolar.

Carbon monoxide is an important industrial raw material, prepared in large quantities by the water-gas reaction, the reaction between coke and steam. Reactions of CO with such substances as hydrogen, water, or simple organic compounds are used to manufacture many more complex organic compounds. Because CO has a nonbonding valence electron pair on the carbon atom, it can react with transition metals to form a useful class of compounds called metal carbonyls.

The carbon monoxide content in the atmosphere under natural conditions is quite small, from 0.1 to 0.2 ppm. About 70% of this CO is produced by human activities, primarily in industrial areas. Most CO is produced in combustion processes where the ratio of fuel to oxygen is too high; the internal combustion engine is a major source of carbon monoxide. The CO content of the atmosphere in large cities with many automobiles often is as high as 10 to 20 ppm, concentrations that can be of medical concern.

Red blood cells transport oxygen in the human body. They contain a molecule called hemoglobin, which forms a complex with $O_2$ and carries it to the cells. Carbon monoxide interferes with this process, since its affinity for hemoglobin is 300 times greater than that of oxygen. A CO molecule holds so tightly to a hemoglobin molecule that it cannot perform its normal function. Exposure to high concentrations of CO can be fatal. Long-term exposure to lower levels of CO is believed to be damaging. For example, one reason cigarette smoking is believed to increase the risk of heart disease is the high concentration of carbon monoxide in cigarette smoke.

## 9.3 NITROGEN

### Periodic Classification

Nitrogen is the lightest member of group VA of the periodic table. Nitrogen and phosphorus are the only nonmetals in this group. Arsenic and antimony are semimetals, and bismuth, the heaviest member of the group, is a metal. The electronic configuration of nitrogen is $2s^2 2p^3$. A nitrogen atom thus needs three electrons to achieve the noble gas electronic configuration. Nitrogen, like carbon, forms primarily covalent bonds. An exception to this rule is a small group of compounds called nitrides, which seem to have the $N^{3-}$ ion, but these are rare. There are no long chains of nitrogen atoms similar to the long chains of carbon atoms, because the N—N single bonds that would exist in such chains are relatively weak. There are only a few compounds in which two or more nitrogen atoms are bonded to each other.

Most nitrogen compounds are organic compounds; their nitrogen atoms are attached to a framework of carbon atoms.

### Discovery and Occurrence

Although nitrogen compounds were known to the ancients, nitrogen was not recognized to be an element until 1772. The identification of nitrogen as an element usually is credited to a physician, Dr. Daniel Rutherford (1749–1819), an uncle of Sir Walter Scott, the famous novelist. Nitrogen, which exists as $N_2$ under ordinary conditions, is a colorless, odorless, tasteless gas. The earth's atmosphere is 78.09% nitrogen by volume, and the atmosphere also contains trace amounts of various nitrogen compounds. The earth's crust and the oceans contain compounds of nitrogen, mainly nitrate salts such as $KNO_3$ and $NaNO_3$. Nitrogen is a component of many compounds that are essential to life, including proteins and nucleic acids.

The normal boiling point of $N_2$ is 77.36 K and the critical temperature of $N_2$ is 126.6 K. The Lewis structure of $N_2$ is $:N \equiv N:$. The bond energy of the triple bond is 946 kJ/mol. This triple bond is so strong that $N_2$ does not dissociate into nitrogen atoms except under the most extreme chemical conditions. Single nitrogen atoms usually play no role in the chemistry of nitrogen.

### Nitrogen Fixation and the Nitrogen Cycle

Living organisms seeking nitrogen for proteins (which average 16% nitrogen), nucleic acids, and other organic compounds can be compared to a thirsty mariner adrift in a salty sea. There is an abundance of $N_2$ in the atmosphere, but to be biologically usable, $N_2$ must be converted into compounds of nitrogen. This conversion is called *nitrogen fixation*. It is a difficult conversion, because the $N \equiv N$ bond must be cleaved.

Some microorganisms can carry out the essential first step in nitrogen fixation by converting $N_2$ to $NH_3$ or an equivalent substance. Soil bacteria, yeasts, and blue-green algae can perform this transformation by themselves. Another important source of biologically useful nitrogen compounds is the cooperative effort of plants and bacteria, a process called symbiotic nitrogen fixation. The bacteria that take part in this process are found in the root nodules of the leguminous plants, such as beans, peas, alfalfa, and clover. Other plants use relatively simple inorganic compounds of nitrogen in the soil to make more complex nitrogen compounds. Without nitrogen fixation processes, however, the supply of nitrogen compounds in the soil would be depleted quickly and life would cease.

The natural replenishment of nitrogen in the soil is inadequate for modern agriculture. Farmers add simple compounds of nitrogen to the soil in the form of fertilizers. The first fertilizers were animal wastes. Later, inorganic nitrates such as Chile saltpeter, $NaNO_3$, were mined for use as fertilizer. Now farmers rely on industrial processes for nitrogen fixation.

The most important of these is the Haber process, developed originally to produce munitions in Germany just before World War I by Fritz Haber (1868–1934). In the Haber process, $N_2$ and $H_2$ react at 700 to 850 K and $10^2$ to $10^3$ atm to form ammonia, $NH_3$, which can be used directly for fertilizer or can be converted to other compounds of nitrogen. Since the major source of $H_2$ for this process is natural gas, which is increasingly costly, and since the process uses a great deal of energy, the development of methods for conversion of nitrogen to ammonia at room temperature and atmospheric pressure is of considerable interest.

Nitrogen fixation is just one step in the sequence of processes by which nitrogen is made available to living organisms and is eventually returned to the atmosphere in the form of $N_2$. This group of processes is called the

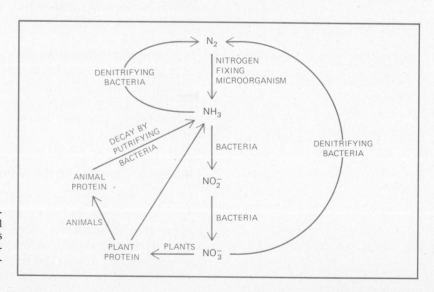

**Figure 9.8**
The nitrogen cycle. Nitrogen gas from the atmosphere is converted to biologically useful substances by nitrogen-fixing bacteria. Plants and animals make use of these nitrogen compounds. The nitrogen is returned to the atmosphere by the activity of denitrifying bacteria.

# NITROGEN FIXATION: IMPROVING ON NATURE

The revolutionary new technique of genetic manipulation that has resulted from recombinant DNA research has raised the possibility that food production could be increased enormously by giving crops such as wheat and corn the ability to do their own nitrogen fixation. The first steps toward this revolution have been taken, although great obstacles remain to be overcome.

The technique for producing recombinant DNA, developed in the early 1970s, makes it possible to transfer a gene for a specific trait from one organism to another. This kind of genetic engineering already has become the basis of a major industry, in which bacteria or yeasts are given the genes that enable them to produce hormones or other medically useful substances. Genetic engineers also are working to give plants valuable properties, such as nitrogen fixation.

Biological nitrogen fixation is still far more important than human fertilizer production. It is estimated that microorganisms fix some 160 billion kg of nitrogen each year, about four times more than the quantity of nitrogen fertilizers produced annually. Because the rising price of energy has caused a comparable increase in the price of nitrogen fertilizer, the search for more efficient methods of nitrogen fixation is being pursued actively.

One such method would be to take the *nif* genes that control nitrogen fixation out of microorganisms and put them into plants. Thus far, it has been possible to transfer *nif* genes out of nitrogen-fixing bacteria into other bacteria, yeasts, and algae. Molecular biologists have also been able to determine the detailed structures of several genes that are required for nitrogen fixation.

However, the great leap to transferring nitrogen fixation capability to the cells of higher plants is yet to be accomplished. One basic problem is that biologists still have not identified all the genes that are involved in the process of nitrogen fixation. In addition, it is not enough to transplant the genes. They must also be made to function in their new environment. A plant cell is far different from the microorganisms in which the *nif* genes ordinarily function. For one thing, the enzyme nitrogenase, which plays a central role in nitrogen fixation, is acutely sensitive to oxygen and must be protected from destruction. Just as important, simply transferring a gene accomplishes nothing unless the recipient plant cell is also given the biological components needed for the gene to function efficiently.

Another approach is to improve the nitrogen fixation efficiency of existing bacteria through genetic engineering. Biologists have identified an enzyme, found in some bacteria, that appears to help them fix nitrogen more efficiently. Efforts to transfer the gene for that enzyme to other strains of bacteria are being pursued.

Research is also being conducted on nonbiological methods of nitrogen fixation that would use less energy than the Haber process. Nitrogen fixation at, or near, room temperature would be a revolution almost as great as successful transfer of the *nif* genes. Some success has been reported with methods using complexes of transition metals and nitrogen. However, a practical low-energy system of nitrogen fixation seems to be years away.

*nitrogen cycle,* which is outlined in Figure 9.8. In the nitrogen cycle, ammonia usually is converted to nitrate, which moves more freely through the soil. Plants use the nitrate to produce protein. Animals eat the plants and are eaten by other animals in turn. Animal wastes are rich in nitrogen compounds, which are converted back into ammonia by soil bacteria. These bacteria also convert nitrogen compounds into ammonia during the decay of dead animals or plants. Other bacteria then convert ammonia or nitrate back to $N_2$, completing the cycle.

## Compounds of Nitrogen

There are simple inorganic compounds that correspond to every oxidation state of nitrogen between $-3$ and $+5$. Table 9.2 lists some of them and their standard heats of formation. Note that $\Delta H_f^\circ$ for many of these compounds is positive, suggesting that they probably are unstable relative to the standard states of their constituent elements. These standard heats of formation generally reflect the great strength of the NN triple bond in $N_2$.

Positive oxidation states for nitrogen are possible only when it combines with oxygen and fluorine, the only two elements clearly more electronegative than nitrogen.

A discussion of the inorganic compounds of nitrogen could—and does—fill many volumes. We shall survey some main features of a few of these compounds.

## Ammonia

Ammonia, $NH_3$, perhaps the most important simple compound of nitrogen, is a colorless gas at room temperature. It has a distinctive, irritating odor. With a boiling point of 239.7 K and a critical temperature of 405.6 K, ammonia can be condensed rather easily. Liquid ammonia is made in large quantities and has many uses. In the chemical laboratory, ammonia is used as a polar solvent for many reactions for which water is unsuitable.

**TABLE 9.2**  Standard Heats of Formation of Nitrogen Compounds at 298 K

| Oxidation State | Compound | Name | $\Delta H_f^\circ$ (kJ/mol) |
|---|---|---|---|
| $-3$ | $NH_3(g)^a$ | ammonia | $-46$ |
| $-2$ | $N_2H_4(l)$ | hydrazine | 50 |
| $-1$ | $NH_2OH(s)$ | hydroxylamine | $-107$ |
| 0 | $N_2(g)$ | nitrogen | 0 |
| $+1$ | $N_2O(g)$ | nitrous oxide | 82 |
| $+2$ | $NO(g)$ | nitric oxide | 90 |
|  | $N_2F_4(g)$ | tetrafluorohydrazine | $-7$ |
| $+3$ | $N_2O_3(g)$ | dinitrogen trioxide | 84 |
|  | $HNO_2(aq)^b$ | nitrous acid | $-119$ |
|  | $NF_3(g)$ | nitrogen trifluoride | $-114$ |
| $+4$ | $NO_2(g)$ | nitrogen dioxide | 34 |
|  | $N_2O_4(g)$ | dinitrogen tetroxide | 10 |
| $+5$ | $N_2O_5(s)$ | dinitrogen pentoxide | $-42$ |
|  | $HNO_3(aq)$ | nitric acid | $-207$ |

$^a$ State symbol gives the state at 298 K.
$^b$ The state symbol (aq) denotes a $1M$ aqueous solution.

One property that makes ammonia valuable as a solvent is its ability to dissolve substantial quantities of some metals, including the alkali and alkaline-earth metals. Solutions of these metals in liquid ammonia have a beautiful deep blue color. They also have the highest electrical conductivity of any known solutions, apparently because dissolution in ammonia separates the metal and its valence electrons. For example, it is believed that when sodium dissolves in liquid ammonia, the $3s$ electron leaves the Na atom, which becomes the $Na^+$ ion. The independent, or "solvated," electron apparently becomes loosely associated with a number of $NH_3$ molecules. Many interesting reactions can be brought about by these solvated electrons. If $O_2$ is bubbled into the solution, for instance, an electron joins the $O_2$ molecule to form superoxide ion, $O_2^-$. The reaction is $O_2(g) + e^- \rightarrow O_2^-$, where $e^-$ represents a solvated electron.

Ammonia has a large number of applications, ranging from the manufacture of synthetic fibers to its everyday use in water solution as a household cleaner. Ammonia is used as a fertilizer. It is also the starting material for the manufacture of many commercially important nitrogen compounds.

One of the most widely used of these compounds is nitric acid, $HNO_3$, which is produced by what is called the Ostwald process. The first and key step in this process is

$$4NH_3(g) + 5O_2(g) \longrightarrow 4NO(g) + 6H_2O(g) \qquad \Delta H° = -906 \text{ kJ}$$

In this reaction, the ammonia is oxidized. The oxidation state of its nitrogen atom changes from $-3$ to $+2$. The reaction is highly exothermic, with a $\Delta H°$ of $-906$ kJ for 4 mol of $NH_3$. The formation of NO as a product of this reaction is something of a surprise, since the data in Table 9.1 indicate that nitric oxide has the highest positive $\Delta H_f°$ of all the oxides of nitrogen. As the data in Table 9.1 show, there are several reactions with more favorable enthalpy changes: the oxidation of $NH_3$ to $N_2O$ in the $+1$ state, to $NO_2$ in the $+4$ state, or, most strikingly, to $N_2$ in the 0 oxidation state. For this last reaction:

$$4NH_3(g) + 3O_2(g) \longrightarrow 2N_2(g) + 6H_2O(g) \qquad \Delta H° = -1268 \text{ kJ}$$

The oxidation of ammonia takes the desired pathway to NO only under special conditions. In the Ostwald process, a preheated mixture of ammonia and air is passed quickly over a platinum-rhodium gauze at temperatures above 925 K. Some of the platinum is lost because the metal has a high vapor pressure at these temperatures, so the process is rather expensive. The nitric oxide formed in this step is converted to nitric acid in subsequent steps of the Ostwald process.

Because ammonia has a nonbonding valence electron pair, $NH_3$ can readily accept a proton to form the ammonium cation:

$$NH_3(aq) + H^+(aq) \rightleftharpoons NH_4^+(aq)$$

Fertilizer is applied to farmland. Heavy application of nitrogen-containing fertilizers helps farmers get high yield from their lands. *(Grant Heilman)*

One of the many ammonium salts is ammonium nitrate, $NH_4NO_3$, which ranks in the top ten products of the American chemical industry. It is used chiefly as a fertilizer, but it also is an explosive. High temperatures or the shock wave from a detonator can start a reaction in which ammonium nitrate is converted very rapidly to $N_2(g)$, with an enormous release of heat:

$$NH_4NO_3(s) \longrightarrow N_2(g) + \tfrac{1}{2}O_2(g) + 2H_2O(l)$$

It is the formation of the extremely strong triple bond in $N_2$ that causes the great release of heat and gives ammonium nitrate and other nitrogen compounds explosive properties. One of the worst disasters in American history occurred in 1947, when an explosion of $NH_4NO_3$ killed 576 people in Texas City, Texas.

The decomposition of ammonium nitrate can be controlled by careful heating. The product is nitrous oxide:

$$NH_4NO_3(s) \longrightarrow N_2O(g) + 2H_2O(g)$$

If ammonium nitrate is heated even more slowly, it can revert to ammonia and pure nitric acid:

$$NH_4NO_3(s) \longrightarrow NH_3(g) + HNO_3(g)$$

### Nitrous Oxide

Nitrous oxide, $N_2O$, is best known as laughing gas, from its ability to induce hysterical excitement and laughter when inhaled. Mixed with oxygen, nitrous oxide is used as an anesthetic in dentistry and obstetrics.

Because it is highly soluble in cream, nitrous oxide is also used as a propellant in canned whipped cream imitations. As was mentioned earlier, it is manufactured by the controlled decomposition of ammonium nitrate.

Nitrous oxide has delocalized bonding and is a resonance hybrid of two important contributing Lewis structures, each of which has formal charges:

$$:N\equiv N—\ddot{\underset{..}{O}}:^{-} \longleftrightarrow ^{-}:\ddot{N}=N=\ddot{\underset{..}{O}}$$

The molecule is linear. The length of the NO bond is intermediate between that of a single and a double bond, while the length of the NN bond is intermediate between that of a double and a triple bond. While nitrous oxide has a positive heat of formation and is relatively unstable, it is also relatively unreactive, and so is easily transported and is widely available.

### Nitric Oxide

Nitric oxide, NO, is the simplest molecule known that has an odd number of electrons and is relatively unreactive. The length of the NO bond is intermediate between the length of a double bond and that of a triple bond. As in CO, there is not much charge separation between the atoms in NO; the molecule is relatively nonpolar. This molecule, like $O_2$, is one for which the molecular orbital method (Section 8.6) is more successful than the valence bond description. It is possible to write contributing Lewis structures for NO, such as

$$\cdot\ddot{N}=\ddot{\underset{..}{O}} \longleftrightarrow :\ddot{N}=\ddot{O}\cdot$$

but this approach does not explain some of the properties of the NO very well.

Nitric oxide is a colorless gas at room temperature and a blue liquid or solid at low temperatures. It is manufactured by the oxidation of ammonia by the Ostwald process. Its most common reaction is further oxidation to $NO_2$:

$$2NO(g) + O_2(g) \longrightarrow 2NO_2(g) \qquad \Delta H° = -112 \text{ kJ}$$

Nitric oxide has one more electron than $N_2$ or CO, both of which are very stable molecules. It is not surprising, therefore, that NO readily gives up an electron to form the nitrosyl cation, $NO^+$. The nitrosyl cation forms bonds to metals similar to those formed by carbon monoxide. Using a molecular orbital description of nitric oxide, we would say that the odd electron is in an antibonding orbital (Section 8.3). Since $NO^+$ is formed by the loss of this electron, the theory predicts that the NO bond in the nitrosyl cation will be stronger than the NO bond in nitric oxide. This prediction is consistent with experimental observations.

At 298 K, NO does not form from $N_2$ and $O_2$. But at much higher temperatures, the reaction

$$N_2(g) + O_2(g) \longrightarrow 2NO(g) \qquad \Delta H° = 180 \text{ kJ}$$

proceeds rather easily. If the hot mixture of these three gases is cooled rapidly to room temperature, the nitric oxide remains. This process occurs in automobile engines: Nitric oxide forms from the hot air in the cylinders and the gases cool rapidly as they are exhausted. Nitric oxide from automobiles is a major component of photochemical smog.

## Nitrogen Dioxide

We cannot discuss nitrogen dioxide, $NO_2$, without mentioning dinitrogen tetroxide, $N_2O_4$. Whenever $NO_2$ is present, we find some $N_2O_4$. The relative amount of $N_2O_4$ decreases as the temperature rises. The relevant reaction is

$$2NO_2(g) \rightleftharpoons N_2O_4(g) \qquad \Delta H° = -58 \text{ kJ}$$

$NO_2$ is reddish brown and $N_2O_4$ is colorless, so the composition of a mixture of the two gases can be determined by measurements of color intensity. At low temperatures, both oxides exist together as liquids, but $N_2O_4$ alone is found in the solid.

At 298 K, $\Delta H°$ for the formation of $N_2O_4$ from $2NO_2$ is $-58$ kJ/mol. The release of energy results from the formation of the NN bond in $N_2O_4$. Like NO, $NO_2$ has an odd number of electrons, and the NN bond in $N_2O_4$ is formed by the unpaired electron of each $NO_2$ molecule:

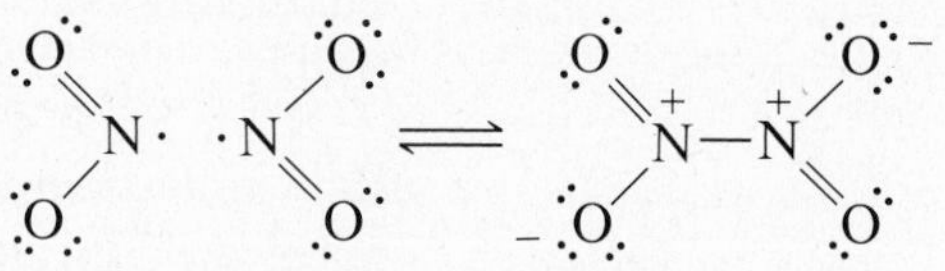

Nitrogen dioxide is toxic. Prolonged exposure to 5 ppm in air is harmful to humans, and 100 ppm in air can be fatal. The dirty brown color of the air over many large cities results in part from $NO_2$. Nitrogen dioxide is also an occupational health hazard, since $NO_2$ is a side product of high-temperature processes such as welding.

## Nitric Acid

Nitric acid, $HNO_3$, was known to the alchemists, who used it to refine precious metals. It remains one of the most widely used strong acids. Nitric acid is produced by the Ostwald process; the final product of the oxidation of ammonia in this process is nitric acid. Nitric oxide formed in the first step of the Ostwald process

$$4NH_3(g) + 5O_2(g) \longrightarrow 4NO(g) + 6H_2O(g)$$

is then oxidized to nitrogen dioxide:

$$NO(g) + \tfrac{1}{2}O_2(g) \longrightarrow NO_2(g)$$

The nitrogen dioxide then reacts with water in an absorption tower to form a solution of nitric acid and gaseous nitric oxide:

$$3NO_2(g) + H_2O(l) \longrightarrow 2H^+(aq) + 2NO_3^-(aq) + NO(g)$$

The nitric oxide is then recycled. This process produces what is called concentrated nitric acid, which is about 55% to 60% $HNO_3$ by mass in water, and which is subsequently concentrated to 70% $HNO_3$. Pure nitric acid can be obtained if potassium nitrate is heated with pure sulfuric acid:

$$2KNO_3(s) + H_2SO_4(l) \longrightarrow 2HNO_3(g) + K_2SO_4(s)$$

Cooling the $HNO_3$ vapor yields solid nitric acid, a white crystalline material with a low melting point.

If nitric acid is treated with dehydrating agents such as $P_2O_5$, the product is dinitrogen pentoxide, the *anhydride* of nitric acid:

$$2HNO_3(s) + P_2O_5(s) \longrightarrow N_2O_5(s) + 2HPO_3(s)$$

An anhydride of a compound (from *anhydrous,* meaning "without water") is formed from the compound by the removal of water.

If water is added to $N_2O_5$, which is a solid at room temperature, the product is nitric acid:

$$N_2O_5(s) + H_2O \longrightarrow 2H^+(aq) + 2NO_3^-(aq) \qquad \Delta H^\circ = -86 \text{ kJ}$$

Interestingly, the structure of $N_2O_5$ alters with its physical state. In the gas phase, $N_2O_5$ has a covalent structure that can be represented by Lewis structures such as

In the solid phase, $N_2O_5$ has a saltlike structure, composed of the $NO_2^+$ cation and the $NO_3^-$ anion.

## 9.4  PHOSPHORUS

### Periodic Classification

Phosphorus lies immediately below nitrogen in the periodic table. Its valence electron configuration is $3s^2 3p^3$. There are similarities between phosphorus and nitrogen. Phosphorus, like nitrogen, rarely forms the 3— ion, although it does so in some compounds called phosphides. The chemistry of both phosphorus and nitrogen is that of the covalent bond. The composition of some compounds of phosphorus resembles the composition of the corresponding compounds of nitrogen. For example, phosphine, $PH_3$, the hydride of phosphorus, corresponds to ammonia, $NH_3$. However, the differences between phosphorus and nitrogen are greater than the similarities.

The differences start with the types of covalent bonds formed by each. The chemistry of nitrogen abounds with compounds that have a strong double or triple bond between two nitrogen atoms, or between nitrogen and another atom. The triple bond in $N_2$ and the NO double bond in oxides and oxyacids are outstanding examples. Phosphorus does not form this kind of multiple bond.

The multiple bonds in nitrogen compounds result from $\pi$ overlap between $2p$ orbitals. In the NO double bond, for example, we can picture $\pi$ overlap occurring between a $2p$ orbital of the N atom and the corresponding $2p$ orbital of the O atom. The $3p$ orbitals of phosphorus rarely form bonds by $\pi$ overlap, so phosphorus almost never forms the kind of multiple bonds formed by nitrogen.

A drag line in operation at a phosphate rock strip mine in Florida. Large quantities of phosphorus-containing ores are mined for use in fertilizers that are essential to modern agriculture. *(Glinn, Magnum)*

**TABLE 9.3** Standard Heats of Formation of Phosphorus-Containing Substances at 298 K

| Substance | Name | $\Delta H_f^\circ$ (kJ/mol)[a] |
|---|---|---|
| P(g) | | 334 |
| $P_2$(g) | | 179 |
| $P_4$(s) | white phosphorus | 18 |
| $PH_3$(g)[b] | phosphine | 23 |
| $PCl_3$(l) | phosphorus trichloride | −300 |
| $PCl_5$(s) | phosphorus pentachloride | −400[c] |
| $POCl_3$(l) | phosphorus oxychloride | −578 |
| $PBr_3$(l) | phosphorus tribromide | −167 |
| $PBr_5$(s) | phosphorus pentabromide | −293 |
| $P_4O_6$(s) | phosphorus trioxide | −1593 |
| $P_4O_{10}$(s) | phosphorus pentoxide | −2940 |
| $P_4S_3$(s) | tetraphosphorus trisulfide | −155 |
| $HH_2PO_2$(aq) | hypophosphorous acid | −592 |
| $H_2HPO_3$(aq) | phosphorous acid | −955 |
| $H_3PO_4$(s) | phosphoric acid | −1260 |

[a] Based on red phosphorus as the standard state.
[b] State symbol gives the state at 298 K; the state symbol (aq) denotes a $1M$ aqueous solution.
[c] Estimated from $\Delta H_f^\circ$ of $PCl_5$(g).

The $3d$ orbitals of phosphorus, which are close in energy to the occupied valence orbitals in its compounds, are important in its chemistry. We often find strong multiple bonds in which there is $\pi$ overlap between a $3d$ orbital of P and a $p$ orbital of another element. In addition, the $3d$ orbitals of phosphorus allow octet expansion. There are many compounds with five or even six groups bonded to a single P atom. By contrast, we do not find octet expansion in nitrogen, which has no valence shell $d$ orbitals.

The similarities between phosphorus and the next two elements in group VA, arsenic and antimony, both of which are classified as semimetals, are closer than those between nitrogen and phosphorus. Table 9.3 lists some phosphorus-containing substances.

### Discovery and Occurrence

About 1669, a Dr. Brand, a physician and alchemist living in Hamburg, Germany, got the idea that human urine might contain a substance capable of converting silver to gold. He did not find such a substance, but he did succeed in isolating from a large quantity of urine a small quantity of a waxy white substance that glowed in the dark and burst into flame when exposed to air. The substance was later named *phosphorus* from the Greek words for "light I bear."

Phosphorus is estimated to be the twelfth most abundant element in the earth's crust. Large deposits of phosphorus, in the form of phosphates such as $Ca_3(PO_4)_2$ and $Ca_5(PO_4)_3F$, are found in many areas. Most of these deposits consist of the accumulated shells, bones, and tissues of marine organisms that collected on the bottom of ancient oceans.

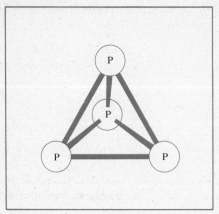

**Figure 9.9**

The structure of $P_4$. The four phosphorus atoms lie at the corners of a regular tetrahedron. All the P—P bonds are identical.

Phosphorus and its compounds have been obtained from these deposits for more than a century. Earlier, phosphorus-containing substances were obtained from biological materials, such as guano—deposits of bird droppings—and fish. Both animal and human bones were used as a source of phosphorus during the early nineteenth century, when bones were collected from the battlefields of Europe. Guano, bones, and similar biological materials contain phosphates that are necessary for plant growth.

Every method for producing phosphorus, from Dr. Brand's original process for isolating phosphorus from urine to the technique used today for manufacturing phosphorus from phosphate rock, is essentially the same. The phosphorus is reduced from the +5 to the 0 oxidation state by the use of coke at a temperature of 1500 K, in the presence of $SiO_2$, ordinary sand. The overall reaction can be represented as

$$Ca_3(PO_4)_2 + 3SiO_2 + 5C \longrightarrow 3CaSiO_3 + 5CO(g) + P_2(g)$$

Phosphorus is produced in large quantities for industrial use. In addition to the many compounds of phosphorus used by industry, the element itself is used in the striking surface for safety matches, as an ingredient for incendiary devices, and in the caps for children's cap pistols.

## Allotropes of Phosphorus

The $P_2(g)$ that is the product of the process for extracting phosphorus from phosphate rock exists only at very high temperatures. When $P_2$ is cooled, it is converted to $P_4$, a much more stable form of phosphorus. The $P_2$ molecule can be represented by the Lewis structure $:P\equiv P:$, which is analogous to the structure of $N_2$. But the energy of the PP bond is only 490 kJ/mol, which is much less than the bond energy of $N_2$. The relative weakness of the bond in $P_2$ results from poor $\pi$ overlap between $3p$ orbitals.

The structure of $P_4$ is shown in Figure 9.9. The four phosphorus atoms lie at the corners of a regular tetrahedron. The bonds are identical. The same $P_4$ structure is found in liquid phosphorus and in white phosphorus, the most common solid form of the element, which is formed when $P_4(g)$ is condensed under water. White phosphorus is the most reactive form of the element. It bursts into flame spontaneously when exposed to air, so it must be stored under water.

When white phosphorus is heated for some time, it is converted into another solid allotrope called red phosphorus, an inexact term used to describe a number of different forms that are more or less red in color and that are much less reactive than white phosphorus. The various kinds of red phosphorus seem to consist of three-dimensional networks of phosphorus atoms that are covalently bonded more or less randomly, to form giant molecules. It has been proposed that the change from white phosphorus to red phosphorus starts when one of the P—P bonds in the $P_4$ tetrahedron breaks and long-chain molecules of many $P_4$ units form.

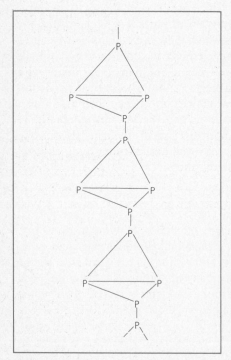

**Figure 9.10**

A section of a chain formed from many $P_4$ units. Red phosphorus can consist of such long-chain molecules.

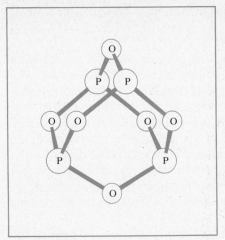

**Figure 9.11**
The structure of $P_4O_6$, phosphorus trioxide. We can visualize $P_4O_6$ as being formed by the insertion of an oxygen atom into each P—P bond of the $P_4$ molecule shown in Figure 9.9.

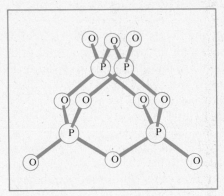

**Figure 9.12**
The structure of $P_4O_{10}$, phosphorus pentoxide. We can visualize $P_4O_{10}$ as being formed by the addition of an oxygen atom to each of the phosphorus atoms of the $P_4O_6$ molecule shown in Figure 9.11.

Figure 9.10 shows a section of such a chain.

The most stable form of phosphorus is a third allotrope, black phosphorus, which is quite stable in air and is not easily ignited by a flame. Black phosphorus is difficult to prepare from red or white phosphorus and is therefore the least commonly encountered allotrope. It consists of P atoms, each bonded to three other P atoms in a giant double-layered network. These double layers are stacked one on top of the other in the crystal. The structure of black phosphorus resembles the structure of graphite.

### Phosphorus Oxides

Figure 9.11 shows the structure of the oxide $P_4O_6$, tetraphosphorus hexoxide, which is prepared by the controlled reaction of phosphorus with oxygen. Its common name is phosphorus trioxide, from its empirical formula, $P_2O_3$. The structure of $P_4O_6$ is closely related to the structure of $P_4$. In $P_4O_6$, each of the four P atoms is at the corner of a regular tetrahedron, with an O atom between each pair of P atoms. Thus, we can visualize a structure for the $P_4O_6$ molecule by starting with a $P_4$ molecule and inserting an O atom into each PP bond. Since there are six PP bonds in $P_4$, six O atoms are added.

A more important oxide of phosphorus is $P_4O_{10}$, tetraphosphorus decoxide, commonly called phosphorus pentoxide because its empirical formula is $P_2O_5$. The structure of $P_4O_{10}$ can be derived from that of $P_4O_6$. Each of the four P atoms bears an additional O atom, giving the structure shown in Figure 9.12.

Phosphorus pentoxide is prepared by burning phosphorus in an excess of oxygen. It has a great affinity for water and is widely used in the chemistry laboratory as a drying and dehydrating agent. For example, it converts $HNO_3$ to $N_2O_5$ and $H_2SO_4$ to $SO_3$. Phosphorus pentoxide represents the $+5$ oxidation state of phosphorus. It is used as a starting material for the production of phosphates and phosphoric acids, which are compounds of the $+5$ oxidation state of phosphorus.

### Oxyacids of Phosphorus

Figure 9.13 shows the three most important oxyacids of phosphorus, which correspond respectively to the $+1$, $+3$, and $+5$ oxidation states of phosphorus: hypophosphorous acid ($+1$), phosphorous acid ($+3$), and phosphoric acid ($+5$). All these compounds have tetrahedral structures. In each molecule, there is a central phosphorus atom surrounded by four groups that lie roughly at the corners of a regular tetrahedron. The hydrogen atoms that are bonded to the P atom are not acidic. Only the hydrogen atoms of the OH groups may ionize in polar solvents. We can make this distinction by separating the ionizable hydrogens from the others when writing the formulas of the oxyacids. Thus, the formula $H_2[HPO_3]$ indicates that phosphorous acid has two ionizable hydrogens, and the formula

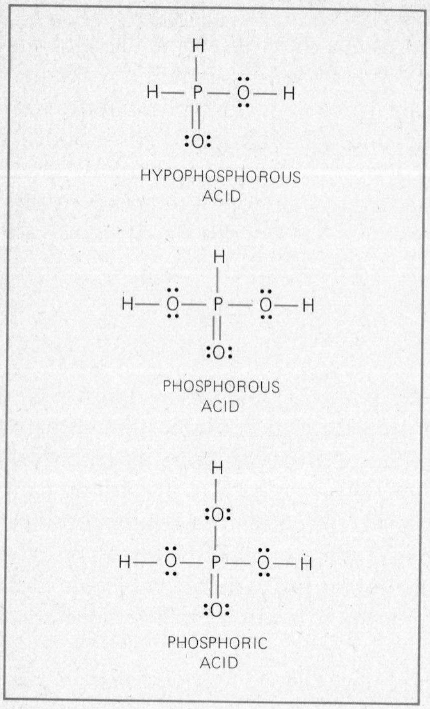

**Figure 9.13**

Three oxyacids of phosphorus. They represent respectively the $+1$ oxidation state of phosphorus (hypophosphorous acid), the $+3$ oxidation state (phosphorous acid), and the $+5$ oxidation state (phosphoric acid).

$H[H_2PO_2]$ indicates that hypophosphorous acid has one ionizable H atom.

All three oxyacids have a double bond between P and O. This $P=O$ bond, which is formed by $\pi$ overlap between a $2p$ orbital of oxygen and a $3d$ orbital of phosphorus, is much stronger than a P—O single bond. The $P=O$ bond is characteristic of many phosphorus-oxygen compounds.

Phosphoric acid, $H_3PO_4$, the $+5$ oxyacid, and substances related to it, are by far the most important compounds of phosphorus. If we compare phosphoric acid to the corresponding nitrogen compound, nitric acid, we find major differences. Nitric acid, $HNO_3$, is a good oxidizing agent and is easily reduced to lower oxidation states of nitrogen. Phosphoric acid is not an oxidizing agent under ordinary conditions, and the $+5$ state of phosphorus is clearly the most stable state of phosphorus in its compounds with oxygen.

Phosphoric acid is prepared by burning phosphorus and treating the resulting oxide with water:

$$P_4O_{10}(s) + 6H_2O(l) \longrightarrow 4H_3PO_4(l)$$

or by treating phosphate rock with $H_2SO_4$. Both processes yield a syrupy liquid that is 85% phosphoric acid. Pure $H_3PO_4$ is a solid at room temperature and melts at 315 K.

Phosphoric acid is used extensively in foods. It gives cola drinks and root beer their tart taste. It is mixed with molasses as an animal feed supplement. It is also used in acid baths to polish aluminum or stainless steel.

## The Phosphates and Phosphate Esters

The phosphates are substances in which one or more of the hydrogen atoms of phosphoric acid are replaced by other chemical species. Inorganic phosphates in which metal atoms or other cations replace the hydrogen atoms are common and are of great practical importance. In the human body, several calcium phosphates, the most notable of which is hydroxyl apatite, $Ca_5(PO_4)_3OH$, help form bones and teeth. In agriculture, calcium phosphates, especially $Ca(H_2PO_4)_2$, are used as fertilizers. Sodium phosphates have many applications. The acidic properties of $NaH_2PO_4$ make it useful in effervescent laxative tablets; $Na_2HPO_4$ is used to pickle hams and to manufacture American cheese; $Na_3PO_4$ is added to scouring powders because its high alkalinity helps remove kitchen grease.

Phosphate esters are compounds in which one or more of the hydrogen atoms of phosphoric acid are replaced by organic groups. We can picture their formation as the joining together of an OH group of the phosphoric acid molecule and an OH group of the organic substance, with an $H_2O$ molecule being formed in the process. Figure 9.14 shows the formation of methyl phosphate, a phosphate ester of methyl alcohol, in this way. As Figure 9.14 also shows, the process can be repeated to form dimethyl

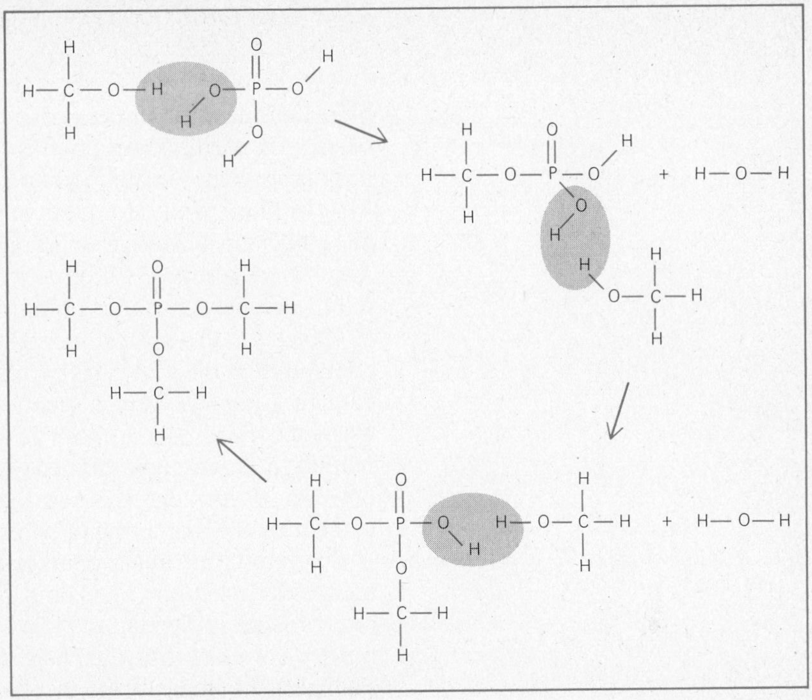

**Figure 9.14**
The formation of methyl phosphate. The OH group of phosphoric acid joins to the OH group of methyl alcohol. An $H_2O$ molecule is eliminated as the two molecules join. The process can be repeated twice more with the same phosphoric acid molecule, to form trimethyl phosphate.

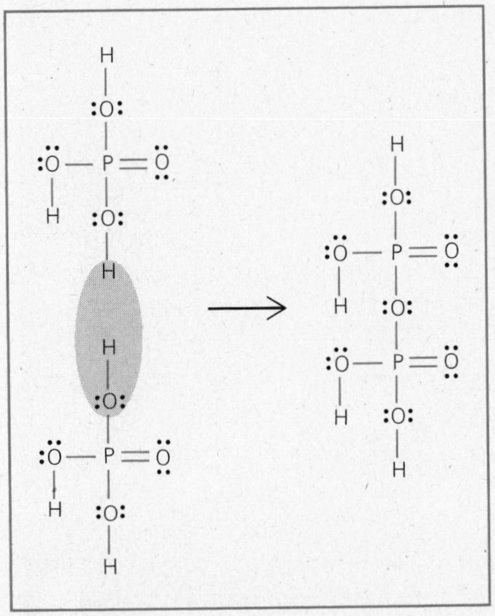

**Figure 9.15**
The formation of pyrophosphoric acid. The structure of the molecule is best understood if we visualize it as a combination of two phosphoric acid molecules, with the elimination of a molecule of $H_2O$.

phosphate, a diester, or trimethyl phosphate, a triester. Sugars have OH groups, and phosphate esters of sugars are of great biological importance. Two of these esters, DNA (deoxyribonucleic acid) and RNA (ribonucleic acid), the genetic materials of living organisms, are long-chain molecules whose backbones consist of phosphate esters of the sugars deoxyribose or ribose.

The elimination of $H_2O$ can also occur between OH groups of two phosphoric acid molecules. When phosphoric acid is heated, such a reaction occurs:

$$2H_3PO_4(l) \longrightarrow H_4P_2O_7(s) + H_2O(g)$$

The product is pyrophosphoric acid; the prefix *pyro* means "heat." If we visualize $H_4P_2O_7$ as two $H_3PO_4$ molecules that have been joined by the elimination of one $H_2O$, its structure is easily understood (Figure 9.15). Reactions in which more than two molecules of $H_3PO_4$ condense in this way are common. Esters of such condensed phosphoric acids get their names from the number of phosphoric acid units in the molecule. Esters of pyrophosphoric acid are called diphosphates, and esters in which three phosphoric acid molecules are joined are called triphosphates.

A substance called adenosine can form either a diphosphate ester or a triphosphate ester—adenosine diphosphate and adenosine triphosphate, usually abbreviated ADP and ATP. These compounds play a pivotal role in the basic energy processes in most living cells (Section 16.8). The reaction $ATP + H_2O \rightarrow ADP + H_2PO_4^-$ is crucial in biology.

## 9.5  OXYGEN

Oxygen is only a minor constituent of the universe as a whole, but it is the most abundant element on the earth's surface. The great abundance of oxygen on earth is presumed to be one factor that makes this planet a hospitable place for life. The crust of the earth is 46.5% oxygen by mass. Most of the oxygen is found in oxides of metals or, combined with metals and silicon, in silicates. The water of the oceans is 88.8% oxygen by mass. The atmosphere is 23.0% oxygen by mass. Most of the oxygen in air exists in the elemental form, as $O_2(g)$.

Oxygen is the lightest element of group VIA. Its valence electronic configuration is $2s^2 2p^4$; it is two electrons short of the noble gas electronic configuration. Oxygen is second in electronegativity only to fluorine. Oxygen occasionally displays positive oxidation states, primarily in compounds in which it is combined with fluorine. In the vast majority of cases, however, oxygen displays negative oxidation states, either when in covalent compounds or as the $O^{2-}$ ion, which is found in many metal oxides.

Oxygen forms binary compounds called oxides with all the other elements except helium, neon, and argon. We discussed trends in the behavior of oxides in Section 6.9, and we shall discuss many of these oxides further as we describe the chemistry of the elements. In this section, we shall discuss some forms of elemental oxygen and some compounds

Fed by oxygen in the air, a fire burns out of control. Fire fighters extinguish such blazes by using a material that prevents oxygen from reaching the combustible substance. They use water because it can cut off the supply of oxygen, not because it is wet. *(Wide World)*

TABLE 9.4  Unusual Oxidation States of Oxygen

| State | Species | Name |
|---|---|---|
| +2 | $OF_2$ | oxygen difluoride |
| +1 | $O_2F_2$ | dioxygen difluoride |
| $+\frac{1}{2}$ | $O_2^+$ | dioxygenyl cation |
| 0 | $O_3$ | ozone |
| $-\frac{1}{2}$ | $O_2^-$ | superoxide anion |
| -1 | $H_2O_2$ | hydrogen peroxide |

in which oxygen is not in the $-2$ oxidation state. Table 9.4 lists some of these compounds.

### Molecular Oxygen

The $O_2$ molecule, which is the stable form of oxygen at ordinary conditions, is usually, but imprecisely, called either oxygen or molecular oxygen. As work with other forms of oxygen has become more common, the systematic name dioxygen has been proposed for the $O_2$ molecule.

At room temperature, $O_2$ is a colorless, odorless gas. Its normal boiling point is 90.18 K and its critical temperature is 154.8 K. It condenses to a blue liquid. As gas, liquid, or solid, $O_2$ has the unusual property of being attracted fairly strongly by a magnetic field. It is one of the small group of substances of simple structure that are paramagnetic (Section 6.2). The presence of two unpaired electrons in the $O_2$ molecule is predicted by the molecular orbital description of $O_2$ (Section 8.6), which places the two highest-energy electrons of the molecule in two different molecular orbitals of the same energy. By Hund's rule, these electrons have parallel spins in the ground state of $O_2$. Any electronic state in which we find two unpaired electrons with parallel spins is called a *triplet* state; $O_2$ is quite unusual because it is a molecule whose ground state, or most stable electronic configuration, is a triplet state.

If one of the unpaired electrons in $O_2$ flips its spin, as represented in Figure 9.16, a higher energy state called a *singlet state* is formed, in which all the electron spins of the molecule are paired. The energy of visible light can bring about the formation of singlet $O_2$ from triplet $O_2$. Singlet $O_2$ is much more reactive than triplet $O_2$. It is believed to play a major role in oxidations with $O_2$ that are brought about by light, including some that occur in living organisms.

### Ozone

Ozone, $O_3$, is an allotrope of oxygen. It is a blue gas at room temperature. It liquefies at 161 K to a dark blue liquid. When it freezes, at 81 K, ozone is a dark purple solid. Ozone is highly explosive and quite poisonous, so it is rarely handled in pure form. Its distinctive sharp odor is often notice-

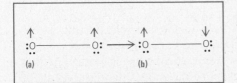

**Figure 9.16**
An $O_2$ molecule goes from (a) the triplet state, in which it has two unpaired electrons, to (b) the singlet state, in which these two electrons have paired spins, when one of the electrons changes its spin.

able after a thunderstorm, when lightning forms ozone from $O_2$ in the atmosphere.

Ozone has delocalized bonding and is a resonance hybrid of several contributing structures. The two major contributing structures are

$$\ddot{\ddot{O}}\overset{+}{} \quad \overset{+}{\ddot{O}}$$

The OOO bond angle has been measured; it is 116.6°. Ozone, $O_3$, is thermally unstable with respect to $O_2(g)$, the standard state of oxygen. The standard heat of formation of $O_3(g)$ is $\Delta H_f^\circ = 142$ kJ/mol.

For laboratory or industrial use, ozone usually is prepared in low yield by passage of an electric spark through a stream of $O_2$. In the atmosphere, most ozone is formed by the action of sunlight on $O_2$. Although ozone is a minor constituent of the lower atmosphere, it plays a significant role in the well-being of living organisms on earth. The concentration of ozone is highest in the upper atmosphere. It reaches a maximum at some 24 km above the surface of the earth, a region called the ozone layer. At this altitude, ozone is formed in a two-step process in which short-wavelength ultraviolet light ($\sim 200$ nm) acts on $O_2$:

$$O_2 + \text{light} \longrightarrow O + O$$
$$O + O_2 \longrightarrow O_3$$

The ozone layer effectively absorbs ultraviolet radiation in the wavelength region from 200 nm to 350 nm by the process

$$O_3 + \text{light} \longrightarrow O_2 + O + \text{heat}$$

which is doubly important. First, the ozone layer prevents this ultraviolet radiation, which is extremely damaging to living things, from reaching the earth's surface. Second, the heat released by the chemical change creates a warm layer in the upper atmosphere that helps to regulate the temperature of the lower atmosphere. In recent years, atmospheric chemists have warned about threats to the ozone layer from high-flying supersonic aircraft, the release of fluorocarbons into the atmosphere, nuclear explosions, and other human activities.

The industrial uses of ozone are based on its powerful oxidizing action. Ozone is used in increasing quantities in water purification, in bleaching, and in chemical manufacturing processes where a strong oxidizing agent is needed. This oxidizing ability makes ozone an unwelcome constituent of the lower atmosphere. Ozone reacts with oxides of nitrogen and hydrocarbons in the formation of photochemical smog. As an air pollutant, ozone causes rubber to crack and is a hazard to health. Some vestiges of the old belief that *ozone* is synonymous with *fresh air* still survive, but the truth is that ozone is a dangerous constituent of industrial smog.

# THE CHEMISTRY OF PHOTOCHEMICAL SMOG

The word smog was coined near the beginning of the twentieth century to describe a mixture of coal smoke and fog. Now it is used to refer to many kinds of atmospheric pullution. One kind of smog has reducing properties. It is called sulfurous smog because of its high concentrations of oxides of sulfur. Another kind of smog has oxidizing properties. It is called photochemical smog because sunlight plays a crucial role in its formation. Although many details of the chemical reactions in photochemical smog are uncertain, the major reactions are reasonably well known.

One of the major pollutants in photochemical smog is the oxidant ozone, $O_3$. Ozone is formed naturally in relatively large amounts in the stratosphere, at altitudes of 25 km to 40 km. Air currents bring some of this ozone down to the surface. While only relatively small traces of natural ozone are found in air, these small amounts seem to play an essential role in the chemistry of smog formation. A chain of reactions is begun when $O_3$ absorbs ultraviolet radiation in a process that leads to formation of an excited O atom (O*):

$$O_3 + h\nu \longrightarrow O^* + O_2$$

In a small percentage of cases, the excited oxygen atom reacts with water vapor to yield the highly reactive hydroxyl radical:

$$O^* + H_2O \longrightarrow 2OH$$

The hydroxyl radicals can then combine with hydrocarbons, which can be denoted as HRH, where R contains at least one carbon atom:

$$OH + HRH \longrightarrow H_2O + RH$$

A complex series of steps follows. The RH reacts with $O_2$ to form products that react with nitric oxide from automobiles. One product of these reactions is nitrogen dioxide, $NO_2$, which is decomposed by the action of sunlight:

$$NO_2 + h\nu \longrightarrow NO + O$$

The reaction of these O atoms with $O_2$ is responsible for the formation of the $O_3$ that is a significant pollutant in photochemical smog.

This sequence of reactions is not the only pathway by which ozone is formed. The OH radicals can also react with carbon monoxide to begin a similarly complex series of steps. The overall reaction in this sequence is given by:

$$CO + 2O_2 \longrightarrow CO_2 + O_3$$

Photochemical smog was first identified in Los Angeles in the late 1940s and early 1950s. Years of research were needed for us to learn that automobile exhaust fumes play a key role in the formation of this kind of smog. Today, the air over Los Angeles remains badly polluted despite the most stringent automobile emission controls in the world. During the worst times of the year, it is common for children to be kept out of playgrounds and for outdoor athletics to be canceled because of pollution alerts.

## Hydrogen Peroxide

Water is not the only possible product of the reaction between $H_2$ and $O_2$. Hydrogen peroxide, $H_2O_2$, is another stable compound of hydrogen and oxygen that can form under suitable conditions. The standard heat of formation of aqueous hydrogen peroxide is $\Delta H_f^\circ = -191$ kJ/mol. For $H_2O_2(l)$, $\Delta H_f^\circ = -188$ kJ/mol. But $H_2O_2$ is rarely encountered in pure form because it is unstable with respect to the formation of water and oxygen:

$$H_2O_2(l) \longrightarrow H_2O(l) + \tfrac{1}{2}O_2(g) \qquad \Delta H° = -98 \text{ kJ/mol}$$

This reaction is exothermic, but it proceeds very slowly. Pure hydrogen peroxide or very concentrated solutions of it can be prepared at room temperature. However, even small amounts of impurities such as dust, grease, or metals can cause $H_2O_2$ to decompose with explosive speed. While 3% solutions of $H_2O_2$ are used as antiseptics and 30% solutions are commonly used in the chemistry laboratory, solutions more concentrated than 50% $H_2O_2$ must be handled with extreme caution.

The oxygen in $H_2O_2$ can be regarded as being in the $-1$ oxidation state, intermediate between the $-2$ state of oxygen in oxides and the zero state in $O_2$. The Lewis structure of the hydrogen peroxide molecule is H—$\ddot{\text{O}}$—$\ddot{\text{O}}$—H. The HOO bond angle is 94.8°.

Hydrogen peroxide forms salts in which its hydrogen atoms are replaced by metal atoms. Two such salts are sodium peroxide, $Na_2O_2$, and barium peroxide, $BaO_2$. Hydrogen peroxide is used industrially as an oxidizing agent and as a bleach. It is produced commercially by the oxidation of water; peroxydisulfate, $S_2O_8^{2-}$, is used as a reagent in a process that yields an aqueous solution that is 30% to 35% $H_2O_2$. Aqueous solutions of $H_2O_2$ can also be manufactured by the reduction of $O_2$ by $H_2$ under special conditions.

It is produced as an intermediate in the reduction of $O_2$ to $H_2O$, a reaction that occurs as a part of many biological energy-supplying processes. Because $H_2O_2$ is toxic, living cells have developed pathways that remove it efficiently.

## Superoxide

The superoxide anion, $O_2^-$, is found combined with a number of cations. It represents an oxidation state of oxygen midway between that of $O_2$ and that of $H_2O_2$. Since oxygen is in the zero oxidation state in $O_2$ and the $-1$ oxidation state in $H_2O_2$, it might appear difficult to assign an oxidation state to the oxygen atoms in superoxide. We can solve the problem either by assigning one oxygen atom in $O_2^-$ the oxidation number $-1$ and the other the oxidation number 0, or by assigning each oxygen atom an oxidation number of $-\tfrac{1}{2}$. Since experimental evidence indicates that both oxygen atoms in the superoxide anion are equivalent, the use of fractional oxidation numbers is preferable. The existence of species in which fractional or mixed oxidation states occur is quite common. Other examples include $N_3^-$, the azide anion, and $O_2^+$, the dioxygenyl cation.

Lewis structures are readily drawn for the superoxide anion by the addition of one electron to the $O_2$ molecule:

$$\left[:\ddot{O}-\dot{O}: \longleftrightarrow :\dot{O}-\ddot{O}:\right]^-$$

The superoxide anion is paramagnetic, which is consistent with the fact that it has an odd number of electrons.

Superoxides react vigorously with water to form peroxides and oxygen:

$$2O_2^-(aq) + H_2O \longrightarrow O_2(g) + HO_2^-(aq) + OH^-(aq)$$

There is evidence that the superoxide anion also forms in living organisms as an intermediate in the biological reduction of $O_2$ to $H_2O$. The superoxide anion is believed to be highly toxic. Its lifetime in cells is extremely short since it is converted to $O_2$ and $H_2O_2$ almost as quickly as it forms.

## 9.6 SULFUR

Sulfur is the second element of group VIA. It is found in its uncombined state in many parts of the earth, and has been known since the earliest times. As "brimstone," sulfur is mentioned frequently in the Bible. Its fiery associations led the Hebrews to associate it with torment and suffering. In our industrial age, sulfur is mined in quantity for industrial use. The largest known deposits of elemental sulfur are in Texas and Louisiana. These are underground deposits, apparently formed by the reduction of sulfur compounds by bacteria. To obtain sulfur from these deposits, superheated steam is forced into the ground. The sulfur melts and is forced to the surface by air pressure. The sulfur obtained by this method, the Frasch process, can be 99.5% pure.

Sulfur atoms have a tendency to form rings or chains, not only in elemental sulfur but also in many of its compounds. Oxygen, the lightest member of group VIA, displays no such tendency.

The stable form of elemental sulfur at room temperature is cyclooctasulfur, a molecule whose formula is $S_8$ and whose eight sulfur atoms are arranged in a crown-shaped ring (Figure 9.17). There are also less stable molecular forms of sulfur consisting of rings containing from 6 to 12 sulfur atoms. Yet another form can be obtained if molten sulfur is poured into ice water. This "plastic sulfur" resembles rubber. It can be stretched into long, elastic fibers and consists of long-chain molecules of sulfur atoms. Left alone, it changes spontaneously but slowly to $S_8$.

Sulfur has a substantial vapor pressure, which reaches 1 atm at 717.8 K. The composition of the vapor depends on temperature and pressure. At very high temperatures ($>2500$ K) and very low pressures ($<1 \times 10^{-8}$ atm), sulfur vapor consists predominantly of S atoms. At lower temperatures, the predominant form is the $S_2$ molecule, which, like $O_2$, is paramagnetic and which gives sulfur vapors their blue color. At still lower temperatures, the vapor contains a mixture of different forms of

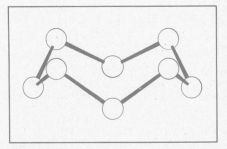

**Figure 9.17**
The structure of cyclo-octasulfur, $S_8$, the stable form of sulfur at room temperature. The eight sulfur atoms form a ring.

sulfur containing from two to eight atoms. Below 717 K, sulfur forms liquid sulfur, whose unusual behavior and structure will be discussed in Section 10.1. Sulfur is a solid below 392.4 K.

One difference between sulfur and oxygen has already been mentioned; sulfur has a tendency to catenation and oxygen does not. There are other differences. Sulfur has the valence electronic configuration $3s^2 3p^4$ and, like oxygen, often displays the $-2$ oxidation state. Since sulfur is less electronegative than oxygen, compounds of sulfur in the $-2$ oxidation state usually are less ionic than the corresponding oxides.

The $3d$ orbitals of sulfur may often be involved in its chemistry, whereas there are no available $d$ orbitals in oxygen that can be used for bonding. Because the $3d$ orbitals are available for bonding in sulfur, we find many compounds in which the coordination number of sulfur is as high as 6. By contrast, the coordination number of oxygen is usually only 2 or 3. There is evidence that the $3d$ orbitals of sulfur can form $\pi$ bonds, as do the $3d$ orbitals of phosphorus.

Sulfur, unlike oxygen, often displays positive oxidation states in many of its compounds. The $+4$ and $+6$ states are the most common. The combination of the lower electronegativity of sulfur and the accessibility of the $3d$ orbitals in sulfur is believed to be responsible for these positive oxidation states.

Table 9.5 lists standard heats of formation of compounds of sulfur. Before discussing some of these compounds in detail, we shall consider briefly the other elements of group VIA: selenium, tellurium, and polonium. Selenium conducts electricity when exposed to light and thus behaves somewhat as a semimetal does, but it is best classified as a nonmetal.

**TABLE 9.5  Standard Heats of Formation of Sulfur-Containing Substances at 298 K**

| Substance | Name | $\Delta H_f^\circ$ (kJ/mol)[a] |
|---|---|---|
| S(g) | | 277 |
| S$_2$(g) | | 129 |
| H$_2$S(g)[b] | hydrogen sulfide | $-20.4$ |
| S$_2$Cl$_2$(l) | disulfur dichloride | $-60$ |
| SF$_4$(g) | sulfur tetrafluoride | $-780$ |
| SF$_6$(g) | sulfur hexafluoride | $-1220$ |
| SOCl$_2$(g) | thionyl chloride | $-210$ |
| SO(g) | sulfur monoxide | 6.9 |
| SO$_2$(g) | sulfur dioxide | $-297$ |
| SO$_3$(g) | sulfur trioxide | $-396$ |
| H$_2$SO$_3$(aq) | sulfurous acid (SO$_2$ + H$_2$O) | $-633$ |
| H$_2$SO$_4$(l) | sulfuric acid | $-814$ |
| H$_2$SO$_4$(aq) | sulfuric acid | $-908$ |
| FSO$_3$H(l) | fluorosulfonic acid | $-800$ |

[a] Based on S$_8$(s) as the standard state.
[b] State symbol gives the state at 298 K; state symbol (aq) means that the substance is in $1M$ aqueous solution.

## SULFUR-BASED LIFE IN THE SEA

Oceanographers using research submersibles to explore the bottom of the sea have found some of the most unusual colonies of living things on earth: Animals flourishing in scalding hot, mineral-laden water that gushes from vents in the ocean floor. The animals feed on bacteria that get the energy they need for their life processes by oxidizing inorganic chemicals, chiefly hydrogen sulfide.

One of these colonies is located at a depth of 2500 m at 21° north latitude on the East Pacific Rise, where two of the giant plates that make up the earth's crust are spreading apart. Undersea explorers found mineral-rich streams of water pouring from vents at temperatures of at least 300°C. These remarkable undersea fountains, descriptively called "black smokers," are surrounded by beds of mussels, giant clams, sea worms several meters long, and other animals in surprising profusion. The primary food source of these abundant colonies is the sulfur-oxidizing bacteria.

It is believed that seawater is heated and picks up substantial concentrations of minerals as it circulates through molten rock that wells up from the interior of the earth. Analyses have found that the heated water contains varying concentrations of oxygen along with hydrogen sulfide and other minerals. Given the oxygen, the bacteria can oxidize the hydrogen sulfide to release energy. The process had been seen previously in shallow waters, but its occurrence in the extreme conditions of temperature and pressure found on the ocean floor had not been suspected.

Studies showed that some species of worms and clams found in the deep-sea colonies were able to accumulate hydrogen sulfide at concentrations that would be fatal to almost any other organism, apparently because their bodies contain an unusual sulfide-bonding enzyme. Hydrogen sulfide is poisonous because it inhibits cytochrome oxidase, an enzyme that uses oxygen produced by metabolism. In these undersea species, another enzyme apparently binds the sulfide, making it harmless.

One of the most provocative findings about the sulfur-oxidizing bacteria has expanded assumptions about the conditions under which life is possible. Until these bacteria were found, it was generally believed that no living organism could survive temperatures of several hundred degrees Celsius. Biologists working with bacteria from the East Pacific Rise colonies, however, found that they not only survived but flourished when they were placed in chambers where seawater was heated to 250°C at 265 atm, the ambient pressure at the depths where they live. At this pressure, water remains liquid even when heated far above its normal boiling point.

One possibility raised by the study is that the deep-sea vents may have been one site where life on earth originated. The bacteria found at the vents belong to a primitive family that is believed to have evolved early in the earth's history. An even more intriguing possibility is that life can exist in environments on other planets whose conditions until now have been thought to be too extreme for any living things.

Tellurium is a semimetal that exists only as a long-chain molecule. Polonium is a radioactive metal that is quite rare. As we look over group VIA, we see a gradual increase in the metallic nature of its elements, moving from the lightest to the heaviest.

### Hydrogen Sulfide

The rotten-egg smell of hydrogen sulfide is one of the most distinctive on earth. Hydrogen sulfide is extremely poisonous; more so than hydrogen

These two photographs of a statue, taken some 60 years apart, show the corrosive power of air pollutants such as sulfur oxides. When the top picture was taken in 1908, the statue at Harten Castle in Westphalia, Germany, had survived almost intact for more than two centuries. The bottom photograph was taken in 1969 and shows the damage done by an increasingly polluted atmosphere. (From E. M. Winkler, *Stone: Properties, Durability in Man's Environment,* Springer-Verlag, New York, 1973. Reproduced with permission.)

cyanide, for example, but its odor gives a useful warning signal well before $H_2S$ reaches lethal concentrations. In passing, we should note that many compounds of sulfur, selenium, and tellurium have unpleasant odors, especially compounds in which these elements are in the $-2$ oxidation state. Many of these odors are associated with decay and putrefaction. An organic derivative of sulfur, *n*-butyl mercaptan, gives a skunk its odor. Other mercaptans are used to give a warning smell to home heating gas.

In aqueous solution, $H_2S$ is a weak acid. Binary compounds called sulfides form between sulfur and the metals. With the exception of the sulfides of the group IA and group IIA metals, these sulfides are very insoluble in water. They play an important role in qualitative analysis of inorganic cations. Hydrogen sulfide tarnishes silver objects. Even small amounts of hydrogen sulfide in the air can cause tarnishing, which is the formation of a black compound, $Ag_2S$, on silver surfaces.

## Oxides of Sulfur

Sulfur dioxide, $SO_2$, and sulfur trioxide, $SO_3$, are both of great practical interest. Sulfur dioxide is a gas at room temperature and liquefies at 263 K. Its structure can be written as a resonance hybrid of two contributing structures with formal charges, and a third with an expanded octet of sulfur:

$$\overset{\cdot\cdot}{\underset{\ddots}{S}}{}^{+} \quad \longleftrightarrow \quad \overset{\cdot\cdot}{\underset{\ddots}{S}}{}^{+} \quad \longleftrightarrow \quad \overset{\cdot\cdot}{\underset{\ddots}{S}}$$

In this third structure, one of the $S{=}O$ $\pi$ bonds can be visualized as forming by $\pi$ overlap between a $3d$ orbital of the S atom and the corresponding $2p$ orbital of the O atom. Both electrons in this $\pi$ bond are donated by O, canceling the formal charges. The observed OSO bond angle of $119.5°$ is consistent with this picture.

Sulfur dioxide has many uses—as a bleach, a preservative, and a disinfectant, among others. It is prepared commercially by the combustion of sulfur:

$$\tfrac{1}{8}S_8(s) + O_2(g) \longrightarrow SO_2(g) \qquad \Delta H° = -297 \text{ kJ/mol}$$

Sulfur dioxide dissolves readily in water to form a solution that behaves as a weak acid. This solution usually is given the formula $H_2SO_3$ and the name sulfurous acid, although there is no evidence that a sulfurous acid molecule exists. Salts of sulfurous acid, such as sodium bisulfite, $NaHSO_3$, and sodium sulfite, $Na_2SO_3$, are well known.

The unintentional production of sulfur dioxide by the combustion of sulfur compounds in coal and oil is of major concern in most urban areas. As an air pollutant, sulfur dioxide is irritating in even small concentrations and is believed to be cumulatively toxic. As little as 1 ppm $SO_2$ can damage plants. Many cities mandate the use of low-sulfur fuels as a central part of their clean-air efforts.

Sulfur trioxide, $SO_3$, is more difficult to prepare than $SO_2$ under ordinary conditions, even though it is more stable, because the reactions by which it forms are slow. The conversion of $SO_2$ to $SO_3$ is a key step in the manufacture of sulfuric acid, one of the most widely used industrial chemicals. The reaction can be represented as:

$$SO_2(g) + \tfrac{1}{2}O_2(g) \longrightarrow SO_3(g) \qquad \Delta H° = -99 \text{ kJ/mol}$$

and is brought about at high temperatures in the presence of a substance such as platinum or $V_2O_5$. It is carried out commercially on a large scale.

Because sulfur trioxide has a low boiling point, it is often handled as a gas. In the gas phase, the $SO_3$ molecule is planar. All three sulfur-oxygen bonds are equivalent, and the OSO bond angles thus are $120°$. The structure of sulfur trioxide can be written as a resonance hybrid of three contributing structures that obey the octet rule but have formal charges and a fourth structure with no formal charges but with an expanded octet of sulfur:

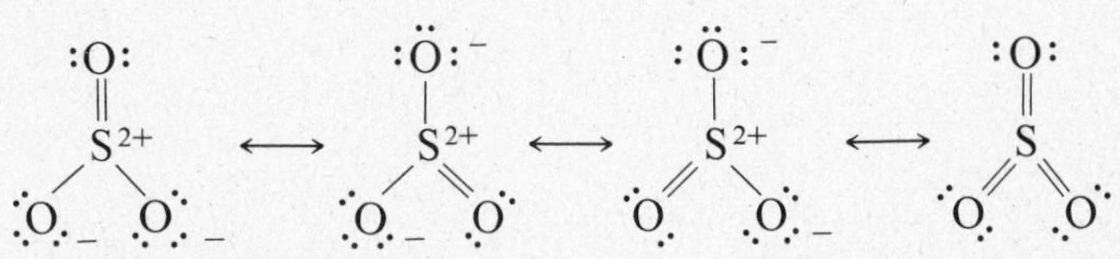

The extra $\pi$ bonds form from electron pairs on the O atoms. They involve $3d$ orbitals of sulfur.

Sulfur trioxide is an extremely reactive substance in the presence of reagents that can add to an $S{=}O$ double bond to form two single bonds. It even reacts with itself. There is a form of solid $SO_3$ that has a fibrous, asbestoslike structure. The fibers consist of long-chain molecules made up of $SO_3$ subunits, as shown in Figure 9.18.

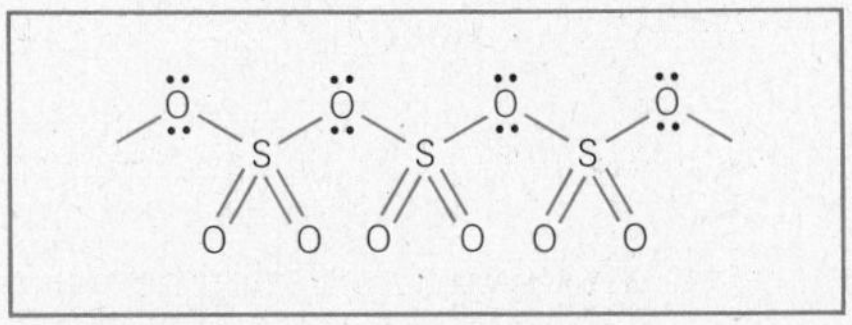

## Sulfuric Acid

Sulfuric acid is formed when sulfur trioxide reacts with water. Pure sulfuric acid, $H_2SO_4$, is a colorless, oily liquid that has long had many uses. A method for preparing $H_2SO_4$ was recorded as long ago as the thirteenth century, when the compound was called "oil of vitriol." Sulfuric acid today is not only the most prominent compound of sulfur but also the most widely used strong acid. More than 30 million tons of sulfuric acid are produced annually in the United States, and $H_2SO_4$ is so essential to industry that the level of sulfuric acid production can be used as an accurate economic barometer. Most sulfuric acid is manufactured by the contact process, in which sulfur is oxidized by air in two steps; first to $SO_2$ and then to $SO_3$. In the final step, the $SO_3$ is dissolved in sulfuric acid that contains some water:

**Figure 9.18**
A fibrous form of $SO_3$. The sulfur trioxide molecules can link together to form a long chain.

$$SO_3(g) + H_2O(l) \longrightarrow H_2SO_4(l)$$

Since the raw materials for this process are sulfur, water, and air, sulfuric acid can be produced cheaply.

The uses of sulfuric acid are legion. A full 40% of sulfuric acid production goes to produce phosphate fertilizers. The phosphoric acid needed to produce fertilizer is prepared by treatment of phosphate rock with $H_2SO_4$. Any industrial process that requires a strong acid is likely to use sulfuric acid. Its uses include the manufacture of drugs and dyes, petroleum refining, the cleaning of steel, and the manufacture of other strong acids. The high boiling point of sulfuric acid makes it especially suitable for the preparation of lower-boiling compounds, such as nitric acid, which can be separated from the reaction mixture by boiling them out.

Many applications of $H_2SO_4$ are based on its well-known affinity for water. Gases can be dried by bubbling through $H_2SO_4$. Sulfuric acid can remove water from organic compounds that contain hydrogen and oxygen, and it can be used to promote reactions in which water is produced, such as

$$C_3H_5(OH)_3 + 3HNO_3 \longrightarrow C_3H_5(NO_3)_3 + 3H_2O$$

in which the explosive nitroglycerine is produced from glycerine. The nitric acid used in this process has some sulfuric acid added to it, and the affinity of $H_2SO_4$ for water enhances the formation of the products.

Sulfuric acid must be handled carefully. A good deal of heat, about 880 kJ/mol, is evolved when sulfuric acid is mixed with water. This heat can readily vaporize the water, so the mixing must be done cautiously to avoid dangerous splashing. Sulfuric acid is always added to water with stirring to disperse the heat. For safety's sake, water should never be added to sulfuric acid.

## The $H_2SO_4$ Molecule

In the $H_2SO_4$ molecule, the four oxygen atoms are at the corners of a tetrahedron around a central sulfur atom. A Lewis structure with formal charges or one in which two of the SO bonds are represented as double bonds can be drawn.

The measured bond lengths are consistent with some $\pi$ overlap between the vacant $3d$ orbitals of the S atom and the $2p$ orbitals of the two O atoms.

A comparison with the Lewis structure of $SO_3$ suggests why $H_2SO_4$ is formed so readily by the reaction of $SO_3$ with water. One S=O double bond of $SO_3$ becomes two single S—O bonds in sulfuric acid. The bond energy of two S—O bonds is substantially greater than that of one S=O. The same thing happens when $SO_3$ adds to $H_2SO_4$ to form disulfuric acid, $H_2S_2O_7$, whose Lewis structure is

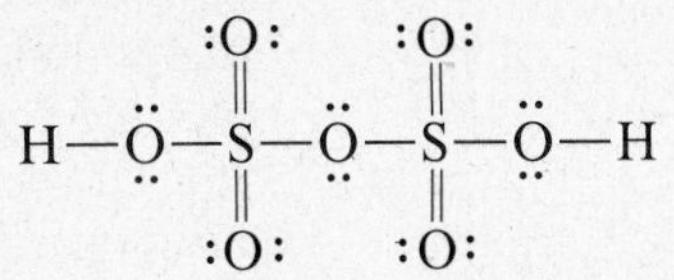

$$H-\overset{\cdot\cdot}{\underset{\cdot\cdot}{O}}-\overset{\overset{\displaystyle :O:}{\|}}{\underset{\underset{\displaystyle :O:}{\|}}{S}}-\overset{\cdot\cdot}{\underset{\cdot\cdot}{O}}-\overset{\overset{\displaystyle :O:}{\|}}{\underset{\underset{\displaystyle :O:}{\|}}{S}}-\overset{\cdot\cdot}{\underset{\cdot\cdot}{O}}-H$$

The process can be repeated to form trisulfuric acid, $H_2S_3O_{10}$, and higher sulfuric acids. Mixtures of sulfuric acid and sulfur trioxide are called oleum or fuming sulfuric acid, because they emit white fumes of $SO_3$.

## Sulfur-Sulfur Bonds

The strong tendency of sulfur atoms to catenate — form bonds with other sulfur atoms — also appears in compounds of sulfur. There are, for example, the sulfanes, each of which consists of a straight chain of sulfur atoms with a hydrogen atom at each end of the chain. The general formula for a sulfane is $H_2S_x$, where $x$ is an integer from 2 to 6. The Lewis structure of one sulfane, $H_2S_4$, is

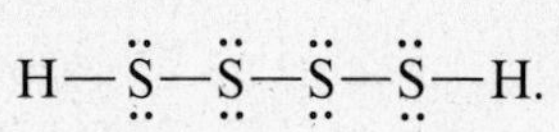

$$H-\overset{\cdot\cdot}{\underset{\cdot\cdot}{S}}-\overset{\cdot\cdot}{\underset{\cdot\cdot}{S}}-\overset{\cdot\cdot}{\underset{\cdot\cdot}{S}}-\overset{\cdot\cdot}{\underset{\cdot\cdot}{S}}-H.$$

The sulfanes are unstable, decomposing to $H_2S$ and elemental sulfur.

Another family of sulfur compounds, the polysulfide salts, can be prepared by a number of methods, including the addition of sulfur to boiling aqueous solutions of metal sulfides, or the direct reaction of sulfur with a group IA or IIA metal. A polysulfide ion has the general formula $S_x^{2-}$, where $x$ is an integer from 2 to 6. Polysulfide ions have the same straight-chain structure as the sulfanes.

The thiosulfate ion, $S_2O_3^{2-}$, can be prepared by addition of sulfur to a boiling solution of $SO_3^{2-}$ ion. The reaction can be represented as:

$$SO_3^{2-}(aq) + S(s) \longrightarrow S_2O_3^{2-}(aq)$$

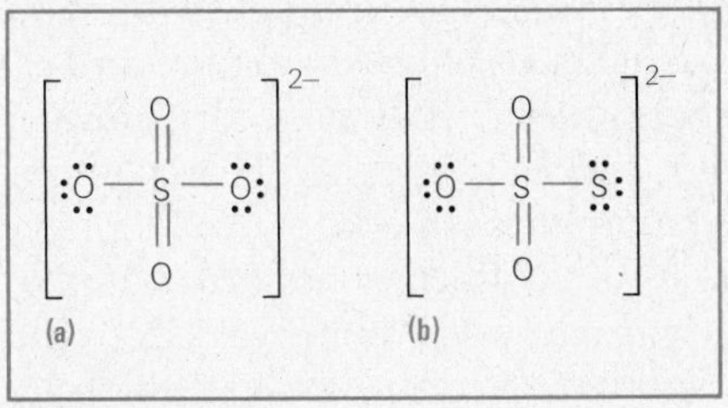

The acid $H_2S_2O_3$, corresponding to the thiosulfate ion, is unstable under ordinary conditions, but there are several stable salts of the ion. The best known of these salts is sodium thiosulfate, $Na_2S_2O_3$, often called hypo (for sodium hypo sulfite, an older name), which is used to develop photographic film (Chapter 19). Sodium thiosulfate helps dissolve the unreacted silver bromide from film after exposure. We can visualize the thiosulfate ion as being formed from sulfate ion, $SO_4^{2-}$, by formal replacement of one O atom by the S atom, as shown in Figure 9.19. The prefix *thio* is commonly used to indicate the replacement of one O atom of a compound by an S atom in a new compound. On this basis, the central S

**Figure 9.19**

The thiosulfate ion (b) can be regarded as being formed by replacement of one of the oxygen atoms in a sulfate ion (a) by a sulfur atom.

atom in thiosulfate has a +6 oxidation number, just as it has in sulfate. If this is so, the other S atom has the same −2 oxidation number as the O atom it replaces.

Sulfur is vital in the chemistry of life, and carbon-sulfur bonds are prominent in organic chemistry. We often find a disulfide group, an S—S bridge, linking two parts of a protein molecule. These sulfur bonds maintain the convoluted yet organized structure that enables protein molecules to perform their vital functions. There is a complex sequence of reactions, the sulfur cycle, in which sulfur passes through different oxidation states while being cycled from organic sulfur compounds in plants and animals through elemental sulfur and sulfate compounds and back to plants and animals. The sulfur cycle is carried on by bacteria in water and soil.

## 9.7  THE HALOGENS

### The Elements of Group VIIA

The halogens are the elements of group VIIA of the periodic table: fluorine, chlorine, bromine, iodine, and astatine. The halogens all are nonmetals. While many properties of these elements change predictably with increasing atomic size (Section 6.7), there is a discontinuity between the behavior of fluorine and that of the other halogens. This behavior is typical of a number of second-row members of other families of elements.

Astatine, the least common halogen, is a fast-decaying radioactive element. It has not been possible to prepare enough of this element or any of its compounds for detailed study, but it is believed that astatine has properties similar to those of the other halogens.

The halogens, with a valence electron configuration of $ns^2np^5$, are each one electron short of the noble gas electronic configuration. Both ionic and covalent compounds of the −1 oxidation state, in which one electron is added to a halogen, are common. Salts of the halide anions $X^-$ (where X symbolizes any halogen), as well as many compounds with other nonmetals, are well known. Fluorine is the most electronegative element and has the −1 oxidation state in all of its compounds. The other halogens can display positive oxidation states when they combine with more electronegative elements—oxygen or halogens of lower atomic number. The most common positive oxidation states encountered for chlorine, bromine, and iodine are +1, +3, +5, and +7. Note that the maximum oxidation state, +7, is equal to the group number.

At room temperature, the halogens are covalently bonded diatomic molecules, $X_2$, in which each X atom achieves the noble gas electronic configuration by electron sharing. Fluorine, $F_2$, is a yellow gas; chlorine, $Cl_2$, is a greenish yellow gas; bromine, $Br_2$, is a dark red-brown liquid; and iodine, $I_2$, is a shiny black solid at room temperature.

Table 9.6 lists some properties of the halogen atoms, X, and the halogen molecules, $X_2$. The regular decrease in ionization energy and in

**TABLE 9.6**  Properties of the Halogens

| Property | F | Cl | Br | I |
|---|---|---|---|---|
| **Atomic Properties** | | | | |
| atomic number | 9 | 17 | 35 | 53 |
| atomic weight | 19.0 | 35.5 | 79.9 | 126.9 |
| valence electronic configuration | $2s^22p^5$ | $3s^23p^5$ | $4s^24p^5$ | $5s^25p^5$ |
| ionization energy (kJ/mol) | 1681 | 1251 | 1140 | 1008 |
| electron affinity (kJ/mol) | 322 | 348 | 324 | 295 |
| electronegativity | 3.98 | 3.16 | 2.96 | 2.66 |
| ionic radius of $X^-$ (nm) | 0.133 | 0.182 | 0.198 | 0.220 |
| $\Delta H_f^\circ$ of $X(g)$ (kJ/mol) | 78.99 | 121.3 | 111.8 | 106.8 |
| $\Delta H^\circ$ of hydration of $X^-$ (kJ/mol) | $-510$ | $-372$ | $-339$ | $-301$ |
| **Properties of $X_2$** | | | | |
| melting point (K) | 53.5 | 172 | 266 | 387 |
| boiling point (K) | 85.0 | 239 | 332 | 458 |
| bond energy of $X-X$ (kJ/mol) | 158 | 243 | 193 | 151 |
| bond length of $X-X$ (nm) | 0.142 | 0.198 | 0.228 | 0.266 |
| $\Delta H^\circ$ of vaporization (kJ/mol) | | | 30.91 | 62.42 |

electronegativity as atomic number increases is predictable. So is the increase in melting points and boiling points, resulting from the stronger interactions between molecules of larger atoms. We also observe the expected increase in ionic and covalent size with increasing atomic numbers. Fluorine, as $F^-$ or $F_2$, is much smaller than the other halogens. One important group property of the halogens is not revealed by the table. Each halogen can oxidize the halides below it in the periodic table. For example, $Cl_2$ can oxidize bromides or iodides but not fluorides. A typical reaction is

$$Cl_2(aq) + 2Br^-(aq) \longrightarrow Br_2(aq) + 2Cl^-(aq)$$

## The Fluorine Atom

A number of consequences stem from the small size of the fluorine atom. Looking at the trend for the rest of the halogens, we expect fluorine to have a larger electron affinity than chlorine. In fact, the electron affinity of fluorine is smaller than that of chlorine. We find a similar anomaly in bond strength. Based on the trends in the other halogens, the bond energy of the $F-F$ bond should be greater than that of the $Cl-Cl$ bond. But the $F-F$ bond is weaker than the $Cl-Cl$ bond. Both of these effects result from the unusually small size of fluorine.

In such a small atom, the repulsions between an added electron and the valence electrons are great enough to offset, at least partially, the favorable interaction between the added electron and the nucleus in $F^-$. The interactions between the added shared electron and the nonbonding valence electrons weaken the $F-F$ bond. This weakening has important chemi-

## FLUORIDATION AND FALLOUT IN TEETH AND BONES

More than 100 million Americans drink water to which small amounts of fluoride compound have been added to reduce tooth decay. The addition of fluorides to water supplies resulted from observations that there is less tooth decay in communities whose water is naturally fluoridated. More recent studies have found that fluoridation achieves its effect by strengthening the crystal structure of the enamel, the tough outermost material of teeth.

An essential part of both bones and teeth is a hexagonal crystal called apatite, a complex phosphate of calcium. Several kinds of apatite are found in varying mixtures in bones and teeth. These include hydroxylapatite, $Ca_5(PO_4)_3OH$, which contains a hydroxyl group, and fluorapatite, $Ca_5(PO_4)_3F$, which contains fluorine. Tooth enamel is about 99% apatite, while dentine, the material of the interior of teeth, is about 78% apatite.

Most of the apatite in teeth is hydroxylapatite. The percentage of fluorapatite is small, but apparently it is of great importance in protecting against tooth decay. Cavities are caused by acid that is produced by bacteria in plaque, a sticky substance that coats the teeth, and fluorapatite seems to be more resistant to acid attack than is hydroxylapatite. Apparently, the fluorapatite concentration is fixed during the early years of dental development. Water that contains about one part per million of fluoride compounds has been found to produce a reduction of tooth decay of about 75% in children who drink it during these years, by increasing the proportion of fluorapatite in the enamel. The protection against tooth decay provided by fluoridation appears to be lifelong.

The fact that apatite is an essential part of bone helps explain some of the concern expressed about radioactive fallout from nuclear testing. One element in fallout is $^{90}Sr$, a radioactive isotope of strontium. Calcium and strontium are in the same group of the periodic table, and they form similar compounds. In fact, a little of the calcium in bone is always replaced by strontium, so there is some strontium apatite among the calcium apatite of bone. If radioactive $^{90}Sr$ is incorporated in bone instead of the stable isotope $^{88}Sr$, radiation is emitted into the bone marrow, where blood cells are formed, as the $^{90}Sr$ decays. Since the bones of children are growing rapidly, they will incorporate more $^{90}Sr$ than the bones of adults. The realization that fallout is most damaging to the children of the world is one of the major factors in the attempt to eliminate all nuclear bomb tests.

cal consequences. Fluorine, $F_2$, is the most reactive element, combining explosively with a large number of substances. It is also one of the strongest oxidizing agents known. The great reactivity of fluorine results at least in part from the weakness of the F—F bond.

### Interhalogen Compounds

Table 9.7 lists the standard heats of formation of some halogen compounds. The most common of these compounds are the halides, in which the halogens have the $-1$ oxidation state. Periodic trends in the behavior of the halides were discussed in Section 6.9.

There is a large group of binary compounds of one halogen with another, called interhalogen compounds. In such a compound, the halogen of lower electronegativity is in a positive oxidation state and the halogen of higher electronegativity is in the $-1$ state. Diatomic interhalogen compounds include ClF, BrCl, and IBr. The more complex

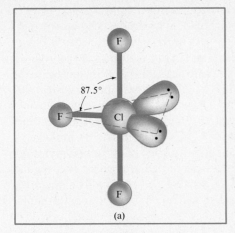

87.5°

(a)

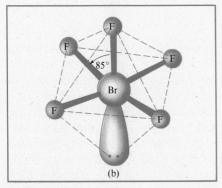

85°

(b)

**Figure 9.20**
The structure of (a) $ClF_3$ and of (b) $BrF_5$. In $ClF_3$, the three fluorine atoms are at three of the five corners of a trigonal bipyramid, with the chlorine atom at the center. In $BrF_5$, the five fluorine atoms are at five of the six corners of a regular octahedron, with the bromine atom at the center. In each molecule, the presence of nonbonding valence electrons affects the bond angles. The dotted lines outline the geometrical figures.

interhalogen compounds are fluorides, with the sole exception of $ICl_3$. They include $ClF_3$, $ClF_5$, $BrF_3$, $BrF_5$, $IF_3$, $IF_5$, and $IF_7$.

The geometries of the complex interhalogen compounds are worth studying. We can understand the structure of these compounds quite well by the VSEPR model, which assumes that their geometry minimizes the repulsion between the bonding electrons and the relatively larger lone pairs (Section 8.2). The Lewis structure of $ClF_3$, for example, shows a central chlorine atom bearing three substituent atoms and two lone pairs (nonbonding valence electron pairs). As we mentioned earlier, five valence shell electron pairs around a central atom lie at the corners of a trigonal bipyramid (Figure 8.20), and the geometry of $ClF_3$ can be explained on this basis.

Repulsions are minimized in trigonal bipyramid structures when the lone pairs of the central atom lie in the equatorial positions. The two lone pairs of the Cl atom in $ClF_3$ do indeed seem to occupy equatorial positions. The three F atoms must therefore lie at the remaining three corners of the trigonal bipyramid. As Figure 9.20(a) shows, the $ClF_3$ molecule is observed to be roughly T-shaped. The FClF bonds do not have the ideal 90° angles of a T, since the relatively large lone pairs push the atoms together.

**TABLE 9.7**   Standard Heats of Formation, $\Delta H_f^\circ$, of Halogen Compounds at 298 K (kJ/mol)

| Compound | Name[a] | X is: | F | Cl | Br | I |
|---|---|---|---|---|---|---|
| HX(g) | hydrogen chloride | | −273 | −92 | −36 | 26 |
| KX(s) | potassium chloride | | −529 | −437 | −394 | −328 |
| XF(g) | chlorine fluoride | | | −51 | −59[b] | −95[b] |
| XF$_3$(g) | chlorine trifluoride | | | −159 | −256 | −485[b] |
| XF$_5$(g) | chlorine pentafluoride | | | −240 | −429 | −840 |
| XF$_7$(g) | iodine heptafluoride | | | | | −961 |
| XO(g) | chlorine monoxide | | 109[b] | 101[b] | | |
| XO$_2$(g) | chlorine dioxide | | | 103 | c | |
| X$_2$O(g) | dichlorine monoxide | | 28 | 88 | c | |
| X$_2$O$_5$(s) | iodine pentoxide | | | | c | −158 |
| X$_2$O$_7$(l) | dichlorine heptoxide | | | 272 | | |
| HXO(aq) | hypochlorous acid | | | −121 | −113 | −138 |
| XO$^-$(aq) | hypochlorite ion | | | −107 | −94 | −108 |
| HXO$_2$(aq) | chlorous acid | | | −52 | | |
| XO$_2^-$(aq) | chlorite ion | | | −67 | c | c |
| HXO$_3$(aq) | chloric acid | | | −98 | −40 | −230 |
| XO$_3^-$(aq) | chlorate ion | | | −104 | −67 | −220 |
| XO$_4^-$(aq) | perchlorate ion | | | −128 | 13 | −145 |
| KXO$_4$(s) | potassium perchlorate | | | −432 | −287 | −461 |
| HXO$_4$(l) | perchloric acid | | | −41 | | |
| H$_5$XO$_6$(s) | periodic acid | | | | | −834 |

[a] The name given is that of the chlorine compound, if there is one. The name of the corresponding fluorine, bromine, or iodine compound is obtained by substitution of the appropriate stem for *chlor*.
[b] Very reactive and not isolable under normal conditions.
[c] Substance is known but $\Delta H_f^\circ$ has not been measured.

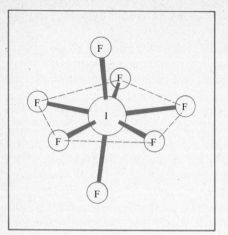

**Figure 9.21**

The structure of $IF_7$. The single iodine atom is in the center of an almost planar pentagon of fluorine atoms. The other two fluorine atoms lie directly above and below the iodine atom.

In $BrF_5$, we find a Lewis structure in which the central bromine atom bears five fluorine atoms and one lone pair. Six valence shell electron pairs around a central atom lie at the corners of a regular octahedron (Figure 8.8). The observed geometry of $BrF_5$ has the five F atoms at the corners of a slightly distorted pyramid with a square base. The Br atom is close to the center of the base. Again, the FBrF bonds have less than the ideal 90° angles because of the relatively large lone pair on the Br atom.

Another interhalogen compound, $IF_7$, is interesting because its geometry cannot be predicted by the simple VSEPR model. The $IF_7$ molecule is found to have the geometry of a slightly distorted pentagonal bipyramid. Five of the F atoms lie almost in a plane at the corners of a regular pentagon. The other two F atoms lie above and below this plane, on a perpendicular line through the I atom, which is at the center of the pentagon (Figure 9.21).

## Halogen-Oxygen Compounds

There are two classes of binary compounds between halogens and oxygens. One is the oxyfluorides, in which oxygen is in a positive oxidation state and fluorine is in the $-1$ oxidation state. The second is the halogen oxides, compounds between oxygen and chlorine, bromine, or iodine, in which oxygen generally is in the $-2$ oxidation state and the halogens are in positive oxidation states. About 20 halogen-oxygen compounds have been identified and characterized—a remarkably large number, since these compounds (with the exception of some oxides of iodine) have positive standard heats of formation and are thermally unstable. Most halogen-oxygen compounds are exceedingly reactive; many are explosive.

Many halogen oxides are of theoretical interest. One of them, the gas $ClO_2$, is of practical interest. This gas is widely used as a bleach and an oxidant; about 10 000 tons is consumed annually in the bleaching of paper pulp. The compound is quite explosive, and it must be handled with considerable care. Table 9.7 includes many halogen-oxygen compounds.

Chlorine, bromine, and iodine form a number of oxyacids in aqueous solution. There are four known oxyacids of chlorine: hypochlorous acid, $HClO$; chlorous acid, $HClO_2$; chloric acid, $HClO_3$; and perchloric acid, $HClO_4$. Perchloric acid has been isolated in pure form; the other three oxyacids are known only in solution. These four acids represent, respectively, the $+1$, $+3$, $+5$, and $+7$ oxidation states of chlorine. In theory, these oxidation states can be produced consecutively by a sequence of *disproportionations,* oxidation-reduction reactions in which a compound reacts with itself.

When $Cl_2$ is dissolved in water, an oxidation-reduction reaction occurs. The chlorine is oxidized to the $+1$ state and reduced to the $-1$ state:

$$Cl_2(aq) + H_2O \rightleftharpoons H^+(aq) + Cl^-(aq) + HClO(aq)$$

The reaction occurs much more readily in basic solution:

$$Cl_2(aq) + 2OH^-(aq) \rightleftharpoons Cl^-(aq) + ClO^-(aq) + H_2O$$

The $ClO^-$ can then react with itself, so the chlorine goes up to the $+3$ state and down to the $-1$ state:

$$2ClO^-(aq) \longrightarrow Cl^-(aq) + ClO_2^-(aq)$$

In practice, this reaction does not stop here. If chlorous acid or chlorites, salts of $ClO_2^-$, are desired, they are prepared from $ClO_2(g)$. The isolable products of the reaction in which hypochlorite ion, $ClO^-$, disproportionates are the chlorate ion, in which chlorine is in the $+5$ state, and chloride ion, in which chlorine is in the $-1$ state:

$$3ClO^-(aq) \longrightarrow 2Cl^-(aq) + ClO_3^-(aq)$$

The $+7$ state can then be formed by disproportionation of the $+5$ state:

$$4ClO_3^-(aq) \longrightarrow Cl^-(aq) + 3ClO_4^-(aq)$$

However, this reaction occurs slowly, so perchlorates, salts of $ClO_4^-$, are prepared by other methods.

Perchloric acid is used commercially as an oxidizing agent of organic material — with care, because it reacts so powerfully that it is potentially explosive. Many perchlorates must also be handled with care for the same reason. Perchlorate salts often are used in the laboratory to study the behavior of metal cations in solution, since $ClO_4^-$ and metal cations generally do not interact appreciably in aqueous solution. The structure of $ClO_4^-$ is tetrahedral. It can be represented by a number of different Lewis structures in which there may be varying numbers of formal charges and of double bonds arising from $\pi$ overlap between the $3d$ orbitals of the chlorine atom and $2p$ orbitals of oxygen. There is disagreement among chemists about the extent to which the $d$ orbitals of Cl can form $\pi$ bonds.

It is often more difficult to prepare the oxygen acids of bromine and iodine, and the salts of these acids, than the corresponding compounds of chlorine. One such compound, perbromic acid, $HBrO_4$, provides a striking illustration of the rule that *negative experimental evidence must be interpreted with great care*. For years, all attempts to prepare perbromic acid and perbromate salts, in which bromine displays the $+7$ oxidation state, ended in failure. Elaborate theories were devised to explain why these compounds were unstable and could not exist. In the late 1960s, however, perbromic acid and perbromate salts were prepared by several methods. These compounds proved to be quite stable. For example, $KBrO_4$ must be heated to 550 K before it decomposes.

Periodic acid, the $+7$ oxyacid of iodine, differs somewhat from per-

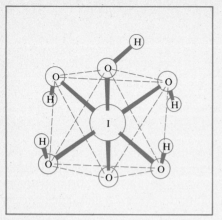

**Figure 9.22**
The structure of periodic acid. The formula $H_5IO_6$ best fits the molecule, which has six oxygen atoms at the corners of an octahedron, with the single iodine atom at the center. Hydrogen atoms are on five of the oxygen atoms.

chloric and perbromic acids. In aqueous solution, periodic acid is best represented as $H_5IO_6$ rather than by the formula $HIO_4$, which corresponds to the $HClO_4$ of perchloric acid and the $HBrO_4$ of perbromic acid. The iodine atom is large enough to accommodate six oxygen atoms, which lie at the corners of an octahedron (Figure 9.22). If $H_5IO_6$ loses two $H_2O$ molecules, then $HIO_4$ remains. While $HIO_4$ is unstable with respect to $H_5IO_6$, salts of the $IO_4^-$ anion are common.

## 9.8  THE NOBLE GASES

Until 1962, the elements of column VIIIA of the periodic table, the noble gases, were believed to be chemically inert. They were called "noble" in the belief that they would not join chemically with lesser elements. The lighter noble gases—helium, neon, and argon—remain chemically inert. But in one of the more notable achievements of recent chemical research, it has been found possible to prepare compounds of krypton, xenon, and radon.

All the elements of column VIIIA exist as monatomic gases that are present in the atmosphere and the earth's crust. Helium is a constituent of natural gas, from which it is separated in commercial quantities. Radon is one of the daughter products of the radioactive decay of radium. The other noble gases are produced commercially by separation from the atmosphere.

The existence of the noble gases was first suspected in the late eighteenth and early nineteenth centuries, when a small proportion of the atmosphere was found to be chemically inert. The first noble gas to be identified was helium, which was detected, not on earth, but in the sun. Hence its name, from *helios,* the Greek word for "sun." Helium was identified because of the presence of a strange spectral line in the sun. Later, the same spectral line was found in gases emitted from Mt. Vesuvius, confirming the presence of helium on earth. The other noble gases were discovered in the late nineteenth century as minor components of air.

Helium is second only to hydrogen in cosmic abundance. Between them, hydrogen (76%) and helium (23%) are believed to make up about 99% of the atoms of the universe. Helium is relatively rare on earth, since it is so light that it can drift out of the earth's gravitational pull. Argon is more abundant in the atmosphere. Air is 0.94% argon.

Table 9.8 gives some physical properties of the noble gases. The trends in these properties with increasing atomic number are quite regular.

The special characteristics of helium make it useful for a number of applications. Because of its low boiling point, helium is used as a coolant in low-temperature systems. Because of its low density and nonflammability, helium is used in balloons and other lighter-than-air craft. Because of its low solubility, helium is used instead of nitrogen in breathing mixtures for deep-sea divers to prevent the "bends," the formation of gas

**TABLE 9.8   Physical Properties of the Noble Gases**

| Property | Helium | Neon | Argon | Krypton | Xenon | Radon |
|---|---|---|---|---|---|---|
| atomic number | 2 | 10 | 18 | 36 | 54 | 86 |
| boiling point (K) | 4.23 | 27.1 | 87.3 | 119.8 | 165.0 | 211 |
| critical temperature (K) | 5.26 | 44.5 | 150.9 | 209.4 | 289.8 | 378 |
| melting point (K) | 0.96[a] | 24.5 | 84.0 | 116 | 161 | 196 |

[a] At 26 atm. Helium does not solidify at 1 atm.

bubbles in the blood of divers who surface too rapidly. The other noble gases have one application that is quite visible. They are used in advertising signs, generically known as neon signs. When an electric discharge is passed through a tube containing neon, the neon atoms emit a red light that has delighted advertising agencies. Argon is used to provide a chemically inert atmosphere for laboratory and industrial purposes. The noble gases can be used as anesthetics, but other agents have been found to be more suitable.

## Compounds of the Noble Gases

Before 1962, this section could not have existed; there were no compounds of noble gases. In that year, the compound $XePtF_6$, a yellow crystalline solid, was prepared. Immediately, a new field of chemical research came into existence. As early as 1963, when most reference books were still describing the noble gases as inert, a 400-page volume, *Noble Gas Chemistry,* was published on the known compounds of xenon, radon, and krypton.

These three noble gases all display positive oxidation states and expanded octets in their compounds. Powerful oxidizing agents are needed to make the noble gases react, so their compounds always include oxygen or fluorine or both. One of the simplest methods of preparing a noble gas compound is to allow xenon and fluorine to react at high temperatures, obtaining a mixture of three compounds in which xenon has the $+2, +4,$ and $+6$ oxidation states:

$$Xe(g) + F_2(g) \longrightarrow XeF_2(g) + XeF_4(g) + XeF_6(g)$$

The relative quantities of the three products can be altered by changes in the relative quantities of the reactants and in the conditions. All three of these fluorides are colorless solids at room temperature. The $\Delta H_f^\circ$ of each is negative. In all three compounds, the Xe—F bond has an energy of about 134 kJ/mol. The compound $XeF_8$ has not yet been prepared, but the $+8$ oxidation state of xenon is found in perxenate salts such as sodium perxenate, $Na_4XeO_6$, and in $XeO_4$. Such compounds are among the strongest oxidizing agents known.

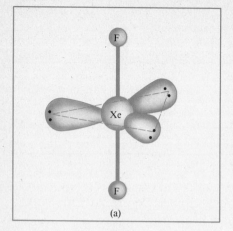

(a)

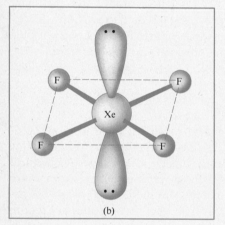

(b)

**Figure 9.23**
(a) The structure of $XeF_2$, with three nonbonding valence electron pairs at the corners of an equilateral triangle. The xenon atom is in the center, with the two fluorine atoms above and below the plane of the electron pairs. The geometry is linear. (b) The $XeF_4$ molecule. The Xe atom lies in the center of a square, with four fluorine atoms at the corners. The nonbonding electron pairs are above and below the plane of the square.

The structures of the noble gas compounds have been studied extensively. In most cases, we find structures consistent with minimum valence electron pair repulsion. The Lewis structure of $XeF_2$ is

$$:\!\ddot{F}\!-\!\ddot{X}\!\ddot{e}\!-\!\ddot{F}\!:$$

The two fluorine atoms and three lone pairs of Xe lie at the corners of a trigonal bipyramid, with the two F atoms at the axial positions. The observed geometry is linear, as shown in Figure 9.23(a). The structure of $XeF_2$ is quite similar to the linear structure of $I_3^-$ (Section 8.2), which has the same number of valence electrons. The Lewis structure of $XeF_4$ is:

The four fluorine atoms and the two lone pairs lie at the corners of an octahedron. The two electron pairs lie on the same axis, and the four fluorine atoms lie at the corners of a square, as shown in Figure 9.23(b). In $XeF_6$, there are six fluorine atoms and one lone pair around the Xe atom, resulting in a more complicated geometry. In $XeO_4$, in which Xe has no lone pairs, we find the expected tetrahedral arrangement of four groups around a central atom.

**Summary**

In this chapter, we discussed the **nonmetals,** the small but important group of elements on the right and toward the top of the periodic table. We began with **hydrogen,** the lightest and most common element in the universe. We mentioned the importance of hydrogenation, the addition of hydrogen to a molecule, and noted that binary compounds of hydrogen and other elements are called **hydrides.** We then discussed **carbon,** the most important element in the chemistry of living things. We noted that the two most common forms of carbon are diamond and graphite, whose properties differ dramatically, and then outlined the **carbon cycle** by which carbon dioxide moves from the atmosphere to living things and back again. We next discussed **nitrogen,** which makes up 80% of the earth's atmosphere. We described how some living organisms can convert the unreactive nitrogen in air to a reactive form that is incorporated in organic compounds that are essential to life. We mentioned some important compounds of nitrogen, including **ammonia,** some nitrogen oxides, and **nitric acid.** We then discussed **phosphorus,** which lies below nitrogen in the periodic table but differs from it in important ways. We described several phosphorus oxides, the oxyacids of phosphorus, and the **phosphates** and phosphate esters. We then discussed **oxygen,** the most abundant element on the earth's surface, which usually exists as $O_2$ but is also found as $O_3$, or **ozone.** We went on to discuss **sulfur** and the compounds it forms, including hydrogen sulfide, the oxides of sulfur, and **sulfuric acid,** an important industrial chemical. We then described the **halogens,** the elements of group VIIA of the periodic table, and concluded with a discussion of the **noble gases,** found in column VIIIA, which once were believed to be chemically inert but now are known to form compounds.

## Exercises

**9.1**   List the nonmetals (excluding the noble gases) in order of (a) increasing electron affinity, (b) increasing ionization potential.

**9.2**   The large difference in ionization potential between hydrogen and the alkali metals is one argument against placing hydrogen in column IA of the periodic table. Account for this difference.

**9.3**   The large difference in electron affinity between hydrogen and the halogens is one argument against placing hydrogen in column VIIA of the

periodic table. Account for this difference.

**9.4**[1]   Write equations for the following reactions, each of which produces hydrogen as a product: (a) methane and steam, (b) nitric acid and magnesium, (c) sodium metal and water, (d) lithium hydride and water.

**9.5**   The chain reaction shown in Section 9.1 for the formation of HCl from $H_2$ and $Cl_2$ begins with the formation of Cl atoms. Write the chain reaction that begins with the formation of H atoms. Calculate the $\Delta H$ for each step in the chain, using the bond energies in Appendix III.

**9.6**   Explain why substances such as platinum or palladium must be used to

bring about hydrogenation reactions at room temperature even though these reactions are quite exothermic.

**9.7**   When diborane reacts with water, one of the products is boric acid, $H_3BO_3$. Write a chemical equation for this reaction.

**9.8**★   The structure of $B_4H_{10}$, one of the simpler boranes, resembles that of diborane. It has four three-center bonds, in each of which two boron atoms and one hydrogen atom share two electrons. Unlike $B_2H_6$, however, $B_4H_{10}$ has one B—B bond. The two boron atoms of this bond are bonded to one hydrogen atom each by ordinary two-center B—H bonds. Each of the other two boron atoms is bonded to

---

[1] The answers to exercises whose numbers are in color can be found in Appendix VII. The star indicates an exercise that is more challenging than average.

two hydrogen atoms by two-center BH bonds. Draw a structure for $B_4H_{10}$.

**9.9** Write all the possible ways in which six carbon atoms may be bonded to each other to form a molecule without including ring structures or multiple bonds.

**9.10** Write the general formula for a chain of carbon atoms in which each carbon atom is connected to the others only by single bonds, and in which each carbon atom has sufficient hydrogen atoms attached so that the octet rule is obeyed. Let $n$ equal the number of carbon atoms.

**9.11** Using the value of 346 kJ/mol for the C—C bond energy, find the heat of sublimation of diamond.

**9.12** The heat of sublimation of graphite is 718 kJ/mol. Using 346 kJ/mol and 615 kJ/mol for the single and double CC bond energies respectively, calculate the resonance energy of graphite.

**9.13** Use the bond energies given in Appendix III to calculate the heat of combustion of $C_4H_{10}$ to give $CO_2(g)$ and $H_2O(g)$.

**9.14** Draw the important contributing structures of the carbonate ion. Indicate which orbitals overlap to form each bond in this ion. Predict the geometry of the ion.

**9.15** Find the mass of CO that is produced in the water-gas reaction from 643 kg of coke if the yield is 90%.

**9.16** Find the volume of $CO_2$ at STP that is generated by the complete decomposition of 385 g of calcium carbonate.

**9.17** Breathing air that contains 0.13% CO(g) by volume for 30 minutes will cause death. The volume of air inside an automobile is approximately 7.5 m³. Find the mass of CO necessary to produce a fatal volume of carbon monoxide in an automobile at 1.0 atm and 298 K.

**9.18** Use the average bond energies in Appendix III to find $\Delta H_f^\circ$ of the compound $N_3H_5$. Using this result, discuss the behavior of nitrogen with regard to catenation.

**9.19** In one step of the nitrogen cycle, $NH_3$ is oxidized to $NO_2^-$ by bacteria. Write the equation for this reaction when it is brought about by $O_2$ in the presence of $OH^-$.

**9.20** Draw the Lewis structures of $NH_2OH$ and $HN_3$. Predict which is likely to be more stable.

**9.21** Draw the Lewis structures of $N_2H_4$ and $N_2F_4$. Use the data in Appendix III to account for the observation that the latter compound is stable with respect to formation from its elements while the former is not.

**9.22** The formula of diimide is $N_2H_2$. Like many other nitrogen compounds, diimide has a positive heat of formation. Unlike some other nitrogen compounds, however, it is extremely reactive and can be prepared only at very low temperatures. Account for these observations.

**9.23** Two compounds with the molecular formula $H_2N_2O_2$ are known. One, hyponitrous acid, has two O—H bonds. The other, nitramide, has no O—H bonds. Draw Lewis structures of these two compounds. There is a third possible compound with this structure that has one O—H bond. Draw its Lewis structure.

**9.24** Calculate the mass of NO(g) that can be produced from the oxidation of 569 kg of $NH_3$ by the Ostwald process if the yield is 94.2%.

**9.25** Predict the product of the reaction that occurs when water is added to these oxides of nitrogen: (a) $N_2O_5$, (b) $N_2O_3$, (c) $N_2O$.

**9.26** Predict the ONO bond angles in each of the following species, and justify your predictions: (a) $NO_2^-$, (b) $NO_2$, (c) $NO_2^+$.

**9.27** Propose a hybridization picture and a likely geometry for $NO_2F$, which has an N—F bond.

**9.28** Find the mass of phosphorus that is produced from treatment of $2.00 \times 10^3$ kg of calcium phosphate with excess sand and coke. The yield of this process is 98.5%.

**9.29** Find the PPP bond angle in $P_4$. Predict the hybridization of each P atom in $P_4$, considering only the number of lone pairs and the number of atoms to which each P is bonded. Using the same considerations, account for the great reactivity of $P_4$.

**9.30** Draw an orbital picture of $POF_3$, indicating the orbitals that can overlap to form each bond. Predict the geometry of $POF_3$.

**9.31** Propose an explanation of the observation that the +5 oxyacid of nitrogen is $HNO_3$, rather than $H_3NO_4$ as it should be by analogy with the comparable oxyacid of phosphorus.

**9.32** Write chemical equations for: (a) the hydrolysis of phosphorus trioxide, (b) the combustion of white phosphorus to phosphorus trioxide, (c) the hydrolysis of phosphorus trichloride, (d) the hydrolysis of phosphorus pentabromide.

**9.33** Phosphorus forms compounds with sulfur whose structures are similar to those of the oxides of phosphorus and can be derived by the insertion of S atoms into the $P_4$ tetrahedron. One such substance is $P_4S_3$, which is used in safety matches. Draw its structure.

**9.34** The adenosine portion of the ATP molecule has the formula $C_{10}H_{12}N_5O_3$. The triphosphate portion has the formula $P_3O_{10}$. Find the mass percent of phosphorus in ATP.

**9.35** A 1.83-g sample of $PCl_5$ is placed in a sealed container of volume 1.00 L and is heated to 436 K, so that it vaporizes. Some of the $PCl_5$ decomposes to $PCl_3$ and $Cl_2$. After complete vaporization, the total pressure is found to be 0.391 atm. Find the fraction of $PCl_5$ that decomposed.

**9.36**  Suggest a sequence of reactions for preparing $POBr_3$ from its elements.

**9.37**  Using the bond energies given in Appendix III, calculate the $\Delta H_f^\circ$ of $O_3(g)$. Account for the difference between this value and the measured value listed in Appendix II.

**9.38**  Draw a bonding diagram showing the orbitals that overlap to form the bonds in $H_2O_2$.

**9.39**  Ozone is a powerful oxidizing agent. When it oxidizes a substance, ozone is reduced. Identify the product of the reduction of ozone.

**9.40**  Predict the species listed in Table 9.4 that are likely to be paramagnetic.

**9.41**  Write chemical equations for: (a) the reaction of excess sodium with hydrogen peroxide, (b) the reaction of excess hydrogen peroxide with potassium, (c) the reaction of potassium superoxide with carbon dioxide to form $O_2$ and potassium carbonate, (d) the reaction of potassium superoxide with water.

**9.42**  The ozonide anion has the formula $O_3^-$. Draw a Lewis structure for this ion and assign each O atom in the structure an oxidation number.

**9.43**  The products of the reaction of $CH_4$ with pure hydrogen peroxide are $CO_2(g)$ and $H_2O(l)$. Calculate $\Delta H^\circ$ of this reaction and compare it to the heat of combustion of methane.

**9.44**  Predict the atomic number, electronic configuration, and chemical properties of the as yet undiscovered element that follows polonium in column VIA of the periodic table.

**9.45**  A 1.00-g sample of sulfur is placed in a 10.0-L container and heated to 2000 K, forming a mixture of $S(g)$ and $S_2(g)$. The pressure is found to

be 0.358 atm. Find the amount (in moles) of $S_2$ in the vapor.

**9.46**  Explain why a flat arrangement of the S atoms in $S_8$ would be less stable than the crownlike arrangement shown in Figure 9.17.

**9.47**  Draw the Lewis structure of the hypothetical molecule $H_2SO_3$, which could form from the addition of $H_2O$ to $SO_2$. By analogy with this molecule, draw a structure of the compound that could form from the addition of HCl to $SO_2$.

**9.48**  Thionyl chloride, $SOCl_2$, is a reagent that is often used in organic chemistry to convert OH groups in various types of compounds to Cl groups. For example, when $CH_3OH$ is treated with thionyl chloride, it gives $CH_3Cl$. Draw the Lewis structure of thionyl chloride. Predict the other product of its reaction with $CH_3OH$.

**9.49**  The compound $S_2O$ is unstable and very reactive, but it has been isolated at 77 K. The central atom in this compound is an S atom. Draw three contributing structures for this oxide.

**9.50**  The formula of peroxysulfuric acid is $H_2SO_5$. Draw its Lewis structure and find the oxidation number of sulfur in this compound.

**9.51**  Write equations for: (a) the formation of phosphoric acid from sodium phosphate and sulfuric acid, (b) the formation of nitric acid from sodium nitrate and sulfuric acid. What property of sulfuric acid permits the use of such processes for the manufacture of these acids?

**9.52**  Use the data in Appendix II to find the heat liberated when 115 g of $H_2SO_4(l)$ is dissolved in enough water to make a $1.00M$ solution.

**9.53**  Predict the products of the hydrolysis of $SCl_6$ and $SCl_4$.

**9.54**  The compound $H_2S_2O_6$, dithionic acid, is moderately stable and has

an S—S bond and a symmetrical structure. Draw its Lewis structure and find the oxidation number of sulfur in this compound.

**9.55**  Predict the geometry of $H_2S_4$.

**9.56**  Account for the observation that the electron affinity of fluorine is less than that of chlorine while the electronegativity of fluorine is greater than that of chlorine.

**9.57**  Explain the steady decrease from $F^-$ to $I^-$ in the heat released when the halide ions are dissolved in water. Why is the decrease much greater from $F^-$ to $Cl^-$ than from $Cl^-$ to $I^-$.

**9.58**  The atomic properties of At are not listed in Table 9.6. Predict those properties.

**9.59**  Write equations for (a) the reaction of $F_2$ with $I_2$, (b) the reaction of $BrF_3$ with water, (c) the reaction of $IF_7$ with water, (d) the reaction of chlorine with aluminum.

**9.60**  What factors explain why KF has a much more negative $\Delta H_f^\circ$ than KCl?

**9.61**  Predict the geometry of the following ions: (a) $BrF_2^-$, (b) $BrF_4^-$, (c) $BrF_6^-$.

**9.62**  Write the formulas and the names of the anhydrides of each acid: (a) perbromic acid, (b) chloric acid, (c) iodous acid, (d) hypochlorous acid.

**9.63**  Write equations for the following disproportionations in basic solution: (a) $Br_2$ to the $+1$ and $-1$ states, (b) $ClO^-$ to the $-1$ and $+7$ states, (c) $ClO_2^-$ to the $-1$ and $+5$ states.

**9.64**  Draw the Lewis structures and predict the geometry of $XeO_3$ and $XeO_6^{4-}$ ion.

**9.65**  Draw the Lewis structures and predict the geometry of $XeOF_2$ and $XeOF_4$.

**9.66**  Explain the trend in the boiling points of the noble gases.

# 10

# States of Matter: Condensed States

**Preview**

**N**ow we discuss the liquid and solid states of matter in which most objects on earth are found. We begin by describing some basic characteristics of the liquid state, and then discuss how the solid, liquid, and gaseous states are related. We next show how the combinations of pressure and temperature that determine the state in which a substance can exist can be pictured in phase diagrams. Focusing next on the solid state, we describe the crystalline form of solids and show how crystals can be pictured as three-dimensional arrays of unit cells or, in some cases, as collections of closely packed spheres. We then explain how X-ray crystallography is used to gather information about crystal structure. Finally we describe the unusual substances called liquid crystals, and we discuss the defects found in crystal structures.

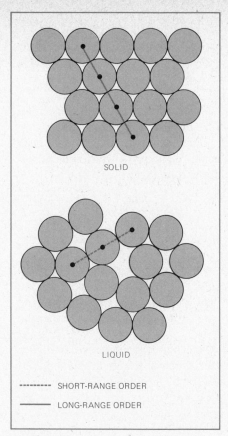

SOLID

LIQUID

- - - - - - - - SHORT-RANGE ORDER
——————— LONG-RANGE ORDER

**Figure 10.1**
Order and disorder in solids and liquids. While some molecules in one part of a liquid may have a regular, ordered arrangement (bottom), this order is not maintained throughout the liquid. Molecules in a crystalline solid (top) are in an ordered, regular arrangement.

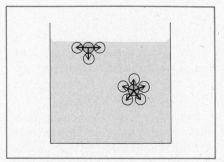

**Figure 10.2**
Surface tension. Molecules in the interior of a liquid are surrounded by other molecules and experience equal attractions from all directions. Molecules on the surface of the liquid are subjected to unequal forces, being attracted only by liquid molecules below the surface.

In Chapter 4, we discussed some of the similarities and differences among the three most important states of matter on earth: solid, liquid, and gas. We went on to discuss the gaseous state in detail. A model to account for the observed properties of gases can be developed relatively easily. Models of the solid and liquid states are more complex. In this chapter, we shall discuss the liquid and solid states in more detail, and we shall try to account for some characteristics of these states of matter. You should review Sections 4.1 and 4.2 before you begin this chapter.

## 10.1 THE LIQUID STATE

The structural units of a liquid can be atoms, as in $Hg(l)$. They can be molecules, as in $Br_2(l)$ or $H_2O(l)$. They can even be ions, as in NaCl, which exists as a liquid at temperatures above 1100 K. For convenience, we shall use "molecule" to refer to all these possibilities.

As temperature falls, the thermal motion of molecules in the gas phase decreases. When the temperature falls low enough, the thermal motion of the molecules cannot overcome the attractive forces between them, and the gas condenses to a liquid. However, thermal motion still prevents the molecules from assuming the ordered state of a crystalline solid.

Figure 10.1 shows that a small group of molecules in one part of a liquid may have a regular arrangement. But this regularity is not repeated throughout the liquid. The attractions between the molecules of a liquid are not great enough to create a completely ordered arrangement.

One fundamental difference between molecules in the gas phase and molecules in the liquid phase is their energy content. Thermal energy is liberated when molecules in the gas phase form a liquid. Thermal energy must be added to the liquid phase to form the gaseous state. The heat of vaporization is always positive.

### Surface Tension

In Section 8.8, we mentioned the various kinds of attractive forces between molecules. As Figure 10.2 suggests, molecules at the surface of a liquid have fewer attractive interactions with other molecules than molecules that are below the surface. Attractions between molecules lower the potential energy of condensed molecules compared to separated ones, in the same way that the formation of chemical bonds lowers the relative energy of atoms. Therefore, molecules at the surface of a liquid have higher energy than those in the interior, but lower energy than those in the gas phase. A liquid tends to assume a shape that minimizes the number of molecules at the surface. This shape maximizes the number of attractions between molecules and thus results in the lowest energy for the liquid. The lowest-energy shape for a liquid is a sphere. Drops of a liquid or bubbles of a gas tend to be spherical because a sphere has the smallest surface area for a given volume (the smallest surface-to-volume ratio) of all shapes.

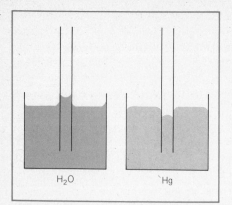

**Figure 10.3**
Capillarity. The forces between molecules of water and the walls of a glass tube (left) are relatively strong compared to the attractive forces between the water molecules. Water wets glass; the liquid surface becomes concave as the liquid rises in the tube. The forces between atoms of mercury and a glass tube are relatively weak compared to the forces between mercury atoms. Mercury does not wet glass; the surface of the liquid mercury becomes convex as the liquid is pulled down within the tube.

The surface of a liquid can be visualized as being under tension, much as if it were a stretched membrane made of an elastic material. **Surface tension** *is the force that pulls the surface inward and resists an increase in surface area.*

Surface tension is related to the attractive forces between molecules. Liquids with large attractive forces have relatively large surface tensions. The surface tension of a given liquid decreases as the temperature rises and the thermal motion of the molecules increases. Surface tension becomes zero at close to the critical temperature (Section 4.12).

## Capillary Action

When a liquid is in contact with a solid surface such as glass, there are forces of attraction between the molecules of the liquid and the surface of the solid. When these forces are relatively strong compared to the attractions in the liquid itself, the liquid is said to *wet* the solid. If these forces are not too strong, the liquid does not wet the solid. The curvature of the liquid surface, the *meniscus,* upward or downward, indicates whether the liquid wets the solid.

When water is in a glass tube, its surface is concave, as shown in Figure 10.3. The water wets the glass, and the shape of the meniscus tends to increase the contact between the liquid and the solid surface. Mercury does not wet the glass. Its meniscus is convex, which minimizes the contact between the liquid and the solid surface.

If a capillary tube, a glass tube of narrow bore, is placed in a vessel containing liquid, there is a readily apparent difference between the height of the liquid in the tube and in the vessel. A liquid that wets the glass rises in the tube. A liquid that does not wet the glass falls in the tube. Surface tension affects the difference between the level of the liquid in the tube and the liquid in the vessel. Measuring this difference in height is one way to measure surface tension.

## Viscosity

Liquids flow. Viscosity is the resistance of liquids to flow. Viscosity is a property with great practical consequences. For example, a major consideration in making motor oils for specific applications is to provide the desired viscosity in a given temperature range.

Flow can be regarded as the movement of layers of molecules in a regular way with respect to one another (Figure 10.4). A liquid that flows with ease is said to be *mobile,* one that does not is said to be *viscous.* To some extent, we can picture a mobile liquid as one whose molecules flow smoothly over each other, with few tangles between layers. In a viscous liquid, tangles between molecules in different layers interfere with the smooth flow.

Liquids whose molecules consist of long chains of atoms often are very viscous, because tangles occur easily between layers of such molecules.

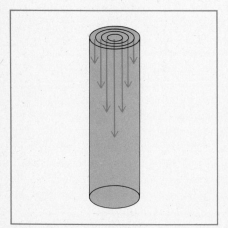

**Figure 10.4**
Flow. When layers of molecules of a liquid move in a regular way with respect to one another, the liquid is flowing in an ideal way.

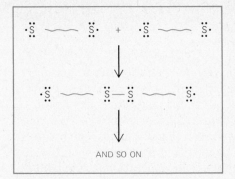

**Figure 10.5**
The viscosity of sulfur increases above 160°C, as short-chain molecules join to form long-chain molecules, decreasing the ability of the liquid to flow. Above 200°C, the long chains begin to break, and the viscosity of sulfur decreases.

For example, gasoline, lubricating oil, and tar all consist predominantly of chains of carbon atoms with hydrogen atoms attached. Gasoline flows easily, lubricating oil is less mobile, and tar is quite viscous. Gasoline has the shortest molecules, tar the longest, and lubricating oil molecules are of intermediate length.

The viscosity of a liquid usually decreases as the temperature rises. An interesting exception to this rule is sulfur. At room temperature, sulfur is a solid composed of $S_8$ molecules in which the eight S atoms are joined in a ring (Figure 9.17). When $S_8(s)$ is heated slowly, it melts to a mobile, straw-colored liquid at 119°C. If this liquid is heated further, it becomes more mobile at first. But at about 160°C, the viscosity of liquid sulfur starts to increase, reaching a maximum at about 200°C.

This behavior results from a change in molecular structure in $S_8$ just above 160°C. At this temperature, the $S_8$ rings break, and the eight S atoms in each ring form a chain. Each end of the chain has an S atom with only seven valence electrons:

$$\cdot \ddot{S}-\ddot{S}-\ddot{S}-\ddot{S}-\ddot{S}-\ddot{S}-\ddot{S}-\ddot{S}\cdot$$

These chains join to form long-chain molecules, as shown in Figure 10.5, causing the increase in viscosity. When the temperature goes above 200°C, the long chains begin breaking, and the liquid sulfur becomes less viscous again.

## 10.2   PHASE TRANSITIONS AND PHASE EQUILIBRIA

Chapter 4 had a brief discussion of the way in which changes in temperature determine the most stable phase of a given substance. We know from experience that if the pressure remains constant and the temperature goes down, most substances change from gas to liquid to solid. There are exceptions to the rule. At atmospheric pressure, for example, $CO_2$ changes directly from gas to solid. Even ordinary phase changes are interesting to study.

Figure 10.6 shows a *warming curve,* which shows how the temperature of a sample of water goes up as thermal energy is added at a constant rate. As expected, the temperature of solid water, ice, at first increases linearly with time as heat is added. But at 273 K (0°C), the melting point of ice, the temperature stops rising as heat is added. The temperature remains the same because the heat melts the ice. The quantity of heat needed to melt the ice is the heat of fusion. When all the ice has melted, the temperature of the water again rises as heat is added.

It is logical to assume that if we started with liquid water and removed heat at a constant rate, we would get a plot identical to the warming curve but running in the other direction. In fact, there are differences.

Figure 10.7 shows a plot, called a *cooling curve,* in which heat is removed at a constant rate from a sample of water. The unexpected dip in

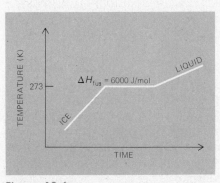

**Figure 10.6**
The warming curve of water. As heat is added at a constant rate, the temperature of solid water — ice — increases steadily until it reaches 273 K, then remains constant for a time as the heat melts the ice. When the ice is melted, the temperature of the liquid water increases as heat continues to be added. The length of the horizontal part of the warming curve depends on the size of the sample of water and the rate at which heat is added.

# THE LIQUID ROCK BENEATH US

While rock on the earth's surface appears to be strong and brittle, under extremely great pressures and high temperatures rock can flow slowly. Such conditions exist in the earth's interior. Near the surface, the temperature rises at a rate of about 1 K for each 30 m of depth. While the temperature gradient is not as steep deeper within the earth, it is estimated that the temperature is 1400 K 100 km beneath the surface and 2000 K at a depth of 1000 km. The pressure at a depth of 1000 km is estimated to be 400 000 atm. At the very center of the earth, the pressure is estimated to be 3.5 million atm.

By studying the transmission of earthquake shock waves through the earth, geologists have determined that the rigid outer layer of rock, which is called the crust, is only 70 km thick under the oceans and 150 km thick under the continents. Below the crust is a region called the athenosphere, in which increasing temperature and pressure cause the rock to become increasingly fluid with depth. In essence, the crust is floating on melted rock.

In the past decade, this picture of the earth's structure has been elaborated into the theory of plate tectonics, which is supported by a mass of evidence. The earth's surface is divided into a number of plates of rigid rock of unequal size. These plates sometimes rub against one another and sometimes collide head-on. Many of the phenomena studied in geology can be explained in terms of plate tectonics.

For example, the San Andreas fault, which runs through part of California, is a boundary where two plates are sliding past each other. The Pacific plate, the largest plate of the Pacific Ocean, is moving northward at a rate of about 6.5 cm a year. At the present rate of motion, Los Angeles will be at the same latitude as San Francisco in about ten million years. Along part of the fault, the slippage occurs smoothly. However, the two plates tend to stick together in many areas and then to slip suddenly to release the growing strain. An earthquake results.

The ridges that run down the center of some oceans, including the Atlantic, have been identified as regions where molten rock wells up from the interior, and the crustal plates are moving apart. There are also ocean trenches, where plates move together. In these trenches, the edge of one plate is forced under the other, creating a deep valley. The rock of the crustal plate melts deep in the interior. In many parts of earth, the transformation of rock between liquid and solid phases goes on continuously.

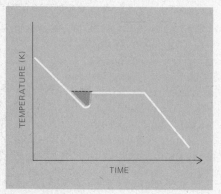

**Figure 10.7**

The cooling curve of water. The unexpected dip is caused by supercooling—failure of the liquid to freeze because its molecules cannot achieve the proper ordered arrangement.

the cooling curve occurs because a liquid does not necessarily solidify when its temperature reaches the freezing point. Water often remains liquid when its temperature is well below 273 K. This phenomenon is called *supercooling*. It occurs because molecules must be arranged in a regular manner to form a crystalline solid. When the freezing point is reached, some time may pass before the necessary pattern of molecules forms in the liquid. Until this pattern forms, the substance remains liquid even though the temperature goes down.

Eventually, however, a small crystal forms in the liquid. This crystal acts as a template, or seed, for the rapid crystallization of part of the liquid. The release of heat that accompanies the crystallization causes the temperature of the liquid to rise to the freezing point. The temperature then remains at the freezing point until the entire sample crystallizes. Only then does the temperature begin to drop again.

Several techniques are used to promote crystallization. A liquid can be stirred vigorously to create more motion and hasten the formation of the molecular arrangement needed for crystallization. Solid surfaces, such as dust particles or, ideally, small seed crystals, can provide sites to promote the crystalline pattern. If water is kept both still and dust-free, it can remain liquid at 235 K, nearly 40 K below is freezing point.

## Supercooling and Glasses

Normally, crystallization is induced easily in supercooled liquids. But there are some important exceptions to this rule. The forces of attraction between the molecules or atoms of some liquids are strong, even though long-range order does not exist. When such liquids are supercooled, it may be virtually impossible for the molecules to arrange themselves into the regular pattern needed for crystallization. Such a substance will remain liquid no matter how low the temperature goes.

We know that the viscosity of a liquid increases as the temperature goes down. The viscosity of a supercooled liquid may increase so much that the rate of flow is unobservable by any ordinary means. The supercooled liquid has the appearance of a solid and many of the properties of a solid. But it lacks long-range order characteristic of a crystalline solid, and its cooling curve and warming curve do not display the kind of flat regions characteristic of the phase change from liquid to solid. Such supercooled liquids are called *amorphous solids* or *glasses,* the latter name coming from the best known of these materials.

Ordinary glass is a supercooled liquid prepared from a mixture of molten silica, $SiO_2$, sodium carbonate, and calcium carbonate. There are strong forces of attraction between the units of structure of glass. But these units of structure are not arranged in the regular repeating pattern expected in a crystal. Rather, they become parts of very large arrays of atoms that are not in a regular pattern. Glass does flow, although the rate of flow is so small as to be unnoticeable in most cases.

## Superheating

The persistence of the liquid state that manifests itself as supercooling also can be found when we examine the behavior of liquids on warming. We expect the warming curve of a liquid to resemble that of a solid. There should be a flat portion where added heat does not raise the temperature but instead causes boiling, the change from liquid to gas. A substantial amount of heat, the heat of vaporization, may be needed to cause this phase change.

But boiling may not occur when the temperature of a liquid goes above the boiling point. This phenomenon is called *superheating.* When a liquid boils, bubbles of its vapor form below its surface. Bubbles tend to form on a site such as a particle of dust or a sharp edge. If no such site exists, bubbles do not form readily and the liquid may not boil. Very pure water has been heated 150 K *above* its normal boiling point without boiling. When a superheated liquid finally does boil, bubbles can form so violently that the liquid spatters out of its container. This potentially dangerous behavior is called bumping. In many laboratory operations that require boiling, inert solids with sharp edges, called boiling chips, are added to a liquid to prevent superheating and bumping.

## Equilibrium States

Supercooled liquid and superheated liquid represent conditions different from those we have discussed until now. They are nonequilibrium states, and this is reflected in their unusual behavior.

In Section 4.1, the equilibrium state of a substance was described as a function of temperature and pressure. For a given pressure, there is a temperature at which the liquid phase and the vapor phase of a substance are in equilibrium. The temperature at which the vapor pressure of a liquid is 1 atm is called the *normal boiling point* of the liquid. For water, the normal boiling point is 373.15 K (100°C). A liquid in a system where the pressure is 1 atm, such as an open container at sea level, boils when the temperature reaches the normal boiling point.

At the normal boiling point, the liquid and its vapor are in equilibrium at a pressure of 1 atm. Below this temperature, the liquid and its vapor are in equilibrium at a vapor pressure of less than 1 atm. Above this temperature, no liquid exists at equilibrium if the pressure is only 1 atm. Thus, superheated liquid is not at equilibrium.

Similarly, at a given pressure there is a temperature called the *freezing point* at which solid and liquid are at equilibrium. The freezing point of water is 273.15 K (0°C) at 1 atm. Above this temperature, no solid is present at equilibrium. Below this temperature, no liquid is present at equilibrium. If liquid exists below the freezing point, the system is not at equilibrium. While supercooled water can exist for some time in an isolated system, it cannot exist in contact with ice.

## Phase Diagrams

The equilibrium state of a system depends on temperature and pressure. The study of the variations of the equilibrium state of a system with changes in pressure and temperature is important in chemistry. Figure 10.8 shows an idealized apparatus for studying these variations. We can vary the temperature of the system by changing the temperature of the bath, and we can vary the pressure by changing the force on the piston, which is assumed to be weightless and frictionless. The system is airtight and contains only water.

Suppose the system in Figure 10.8 is at a temperature of 300 K and a pressure of 0.5 atm. Only liquid is present, because the vapor pressure of water at 300 K is only 0.035 atm (Table 4.1). The vapor pressure is not enough to push the piston off the surface of the liquid, so there is no vapor space. If the pressure on the piston is lowered to just below 0.035 atm, the vapor pushes the piston up, as shown in Figure 10.9. At a pressure of 0.035

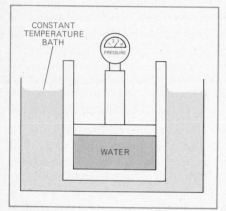

**Figure 10.8**

An idealized apparatus for studying the effect of variations in temperature and pressure on the equilibrium state of a system. We can control temperature by heating or cooling the water in the bath, and we can control pressure by changing the force on the piston.

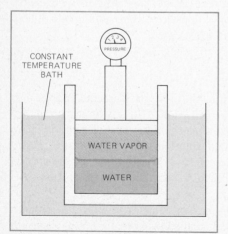

**Figure 10.9**

The measurement of the vapor pressure of water at a given temperature. The pressure on the piston is lowered to the point where it is equal to the vapor pressure and the piston therefore rises to create vapor space.

atm and a temperature of 300 K, liquid and vapor are in equilibrium. If the external pressure is kept at 0.5 atm and the temperature is raised, vapor appears at 355 K. At this temperature, the vapor pressure of water is 0.5 atm. Liquid and vapor are in equilibrium at $T = 355$ K and $P = 0.5$ atm.

The same sort of reasoning can be applied to the changes that occur as the temperature goes down. If supercooling does not occur, solid appears at the temperature at which solid and liquid are in equilibrium when $P = 0.5$ atm. For water, this temperature is very slightly above 273.15 K, the freezing point of water in air at a pressure of 1 atm.

Data from such experiments are most conveniently shown in graphical form. Figure 10.10 shows such a graph for water. All the points corresponding to values of $P$ and $T$ at which vapor and liquid are in equilibrium fall on a smooth curve, which is called the *vapor pressure curve* or *boiling curve* of liquid water. All the points corresponding to values of $P$ and $T$ at which vapor and solid are in equilibrium fall on another smooth curve, which is called the *vapor pressure curve* (Section 4.2) or *sublimation curve* of solid water, or ice. All the points corresponding to values of $P$ and $T$ at which solid and liquid are in equilibrium fall on a third smooth curve, which defines the melting behavior of ice.

### The Phase Diagram of Water

When all three of these curves are drawn using the same set of axes, we have what is called a **phase diagram.** Figure 10.10 is the phase diagram of water. The three curves intersect at only one point, the **triple point.** The triple point marks the temperature and pressure at which solid, liquid, and gas all are in equilibrium. For water, the triple point occurs at $T =$

**Figure 10.10**

The phase diagram of water. All the values of $P$ and $T$ at which liquid and vapor are in equilibrium fall on the vapor pressure curve of liquid water. All the points at which vapor and solid are in equilibrium fall on the vapor pressure curve of solid water. All points at which solid and liquid are in equilibrium fall on the solid-liquid curve. The triple point is at the intersection of the three curves. It is the temperature and pressure at which all three phases are in equilibrium.

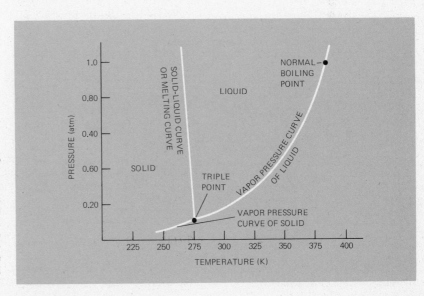

273.16 K and $P = 0.006025$ atm, or 610.5 Pa. For any system composed of one chemical substance, there is only one point at which three phases can be in equilibrium. Note that the triple point of water is the temperature at which the vapor pressure of solid and the vapor pressure of liquid are equal.

All $P$, $T$ points that fall on curves of a phase diagram correspond to conditions where two phases are in equilibrium. But most pressure-temperature combinations do not fall on these curves. The regions between curves in phase diagrams correspond to combinations of pressure and temperature at which just one phase of a substance can exist at equilibrium.

For example, the region to the left of the melting curve and above the solid vapor pressure curve represents temperatures lower than the melting point and pressures higher than the vapor pressure. This region corresponds to a state that has only solid. The region between the melting curve and the liquid vapor pressure curve — where $T$ is higher than the melting point and $P$ is higher than the vapor pressure — corresponds to an equilibrium state in which only liquid exists. The region to the right of and below the vapor pressure curve, representing higher temperature than the boiling point and lower pressure than vapor pressure, corresponds to an equilibrium state in which only vapor exists. These regions are labeled accordingly in the phase diagram.

### The Phase Diagram of $CO_2$

The ideas we use to discuss the phase diagram of water can be applied to other systems with only one component. Figure 10.11 shows the phase diagram for $CO_2$. The diagram has a general resemblance to the phase diagram for water, with three curves intersecting at a triple point. But

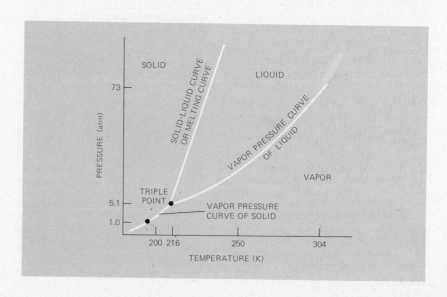

Figure 10.11
The phase diagram for carbon dioxide.

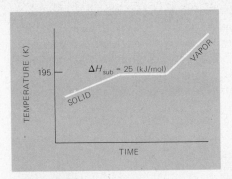

**Figure 10.12**
The warming curve of solid carbon dioxide at 1 atm. The phase change in this case is from solid to vapor, and the flat part of the curve represents the heat of sublimation.

there are significant differences between the two phase diagrams. These differences reflect the differences between the properties of water and carbon dioxide.

The phase diagram for $CO_2$ is shifted to lower temperatures. At room temperature and pressure, water is a liquid but $CO_2$ is a gas. The pressure at the triple point of $CO_2$ is 5.1 atm and the temperature is 216.6 K. At pressures lower than 5.1 atm, it is possible to have solid $CO_2$ or $CO_2$ vapor or an equilibrium between the two but it is not possible to have liquid $CO_2$. Liquid $CO_2$ is never observed at atmospheric pressure. "Dry ice," solid $CO_2$, does not melt as it warms at 1 atm. Rather, gas is formed directly, as the phase diagram shows.

It is interesting to consider the warming curve for solid $CO_2$ at 1 atm pressure. Figure 10.12 shows that the warming curve of $CO_2$ looks very much like the warming curve of ice. There is the same flat part of the curve where an input of thermal energy does not raise the temperature but changes the phase of the material. The phase change for $CO_2$ is from solid to vapor. The heat required for the change is the heat of sublimation, $\Delta H_{sub}$ at 1 atm. The phase change occurs at 195 K ($-78\,°C$), the temperature at which the vapor pressure of the solid is 1 atm.

The vapor pressure curve of liquid $CO_2$ is shown as ending at 304 K, the critical point (Section 4.12). There is no need to continue the line past the critical point, since there is no distinction between liquid and vapor above the critical point.

## 10.3 THE CRYSTALLINE STATE

In everyday language, we call a substance a solid if we cannot see the flow and change in shape that we associate with liquids. We can distinguish crystalline and amorphous solids by the presence or absence of long-range order on the molecular scale. Long-range order usually is visible to the unaided eye in crystalline solids, which occur in characteristic shapes called **crystals.** The existence of crystals is a consequence of long-range order, or repetition of basic units, on the atomic scale. Amorphous solids or glasses do not have the crystalline form and the accompanying long-range atomic order of crystalline solids.

### Classification of Crystals

The structural entities of a crystalline solid, such as atoms, molecules, or ions, are in essentially fixed positions that represent the most favorable balance between attractive and repulsive forces. We can classify solids by the nature of the attractive forces between these entities.

We have already discussed the coulombic forces in *ionic crystals,* where the repeating units contain ions of opposite charge. In *molecular crystals,* the repeating units are molecules, and the attractive forces may be any of the weak interactions. In one especially important type of molecular crystal, the forces of attraction between the units of structure are hydrogen bonds (Chapter 11). Ice is one such crystal.

There are other crystals, called *covalent network solids,* in which the forces of attraction are chemical bonds between atoms. Well-known examples of covalent network solids are diamond and quartz. Still another important type of crystal is one composed only of metal atoms. The forces

(a)

(b)

(c)

(d)

A variety of crystals: (a) tourmaline, (b) stibnite, (c) copper, (d) sulfur. *(American Museum of Natural History)*

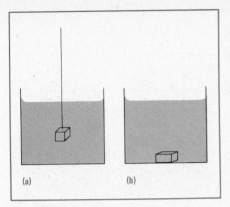

**Figure 10.13**
A crystal of sodium chloride grown by suspension in the center of a solution will have a symmetrical, cubic shape, as all faces of the crystal grow equally (a). If the crystal is grown on the bottom of a container, it will grow faster horizontally than vertically (b).

of attraction in a *metallic crystal* will be described in detail in Chapter 19.

Crystals range in size from those large enough for their faces or cleavage planes to be evident at a glance to those so small that they seem to be powders. But many powders are, in fact, heaps of tiny crystals. A magnifying device, perhaps one as powerful as an electron microscope, or X-ray diffraction techniques, may be used to detect their crystalline structure. The way in which a crystal grows affects its size. Size is in inverse proportion to the rate of growth. If a crystal grows slowly from the liquid state or from solution, there is greater opportunity for large crystals to form.

Crystals exist in a variety of shapes. The way in which a crystal of a given substance is formed may affect its shape. A sodium chloride crystal that is grown suspended in a solution of NaCl, as shown in Figure 10.13(a), will be cubical, because NaCl deposits at an equal rate on all sides of the crystal. But if the crystal grows on the bottom of the container (Figure 10.13(b)), material is not deposited on the face that touches the bottom. Therefore, the crystal grows faster horizontally than vertically. The shape of a crystal of sodium chloride can be changed even further by other variations in the conditions of growth.

## 10.4  UNIT CELLS AND LATTICES

We find the same angles between corresponding faces of all the crystals of a given substance because of the long-range order that is characteristic of the solid state. In fact, a crystal can be defined as an object whose basic units of structure are arranged in a regularly repeating three-dimensional order.

The idea of a regularly repeating pattern is familiar from everyday life. We can understand some characteristics of crystals by examining a tile floor (Figure 10.14), which represents a regularly repeating two-dimensional pattern. To lay the tile floor shown in Figure 10.14, we must repeat the basic unit of design, the tile that forms one complete pattern. We can describe the tile floor by describing the pattern and then stating that this pattern is repeated to cover the entire area of the floor.

The shape of the area covered by the tiles depends on how the tiles are laid. We can make a square floor by laying tiles equally on all four sides and a long, narrow floor by laying tiles to the two narrow ends of the floor. We can make an irregularly shaped floor by varying the way in which the tiles are laid (Figure 10.15).

Another feature of tile floors is worth noting. Only a limited number of basic shapes can cover an area completely, with no gaps. An area can be covered entirely by square tiles or by tiles that are triangles, parallelograms, or regular hexagons. An area cannot be covered entirely by tiles that are regular pentagons or heptagons, or tiles of a single irregular shape. In fact, only a relatively small number of polygons will cover an area completely.

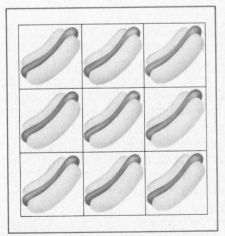

**Figure 10.14**
A tile floor consisting of a regularly repeating two-dimensional pattern, in this case a hot dog. No matter how large an area it covers, the pattern of this floor will always be a repetition of hot dogs.

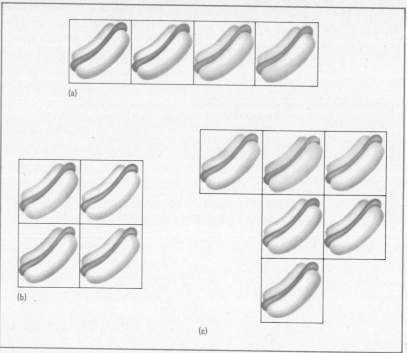

**Figure 10.15**

Variations on the theme of a regular two-dimensional pattern. This tile floor consists of tiles fitted together to depict hot dogs. No matter how the pattern is shaped—as a rectangle (a), a square (b), or any irregular shape (c)—the angles between corresponding sides of the repeating unit are still the same.

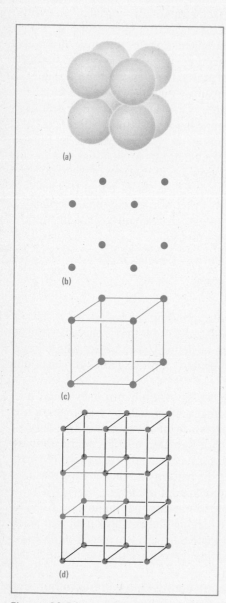

**Figure 10.16**

The unit cell of a crystal. The crystal is formed by repetition of a group of eight atoms (a). It can be represented by points at the centers of those atoms (b). The points can be connected by lines to form a cube (c), and that cubical pattern can be extended through space to form a lattice (d).

Tile floors are two-dimensional, but they have a number of features in common with three-dimensional crystals. *A crystal also has a characteristic pattern based on a relatively small number of basic structural units,* which can be atoms, ions, or molecules. This basic repeating three-dimensional pattern is called the **unit cell.** Just as only tiles of certain shapes can cover a floor area completely, only unit cells of certain shapes can fill three-dimensional space completely.

But the analogy between floor tiles and crystals breaks down if it is carried too far. A floor tile is a clearly defined object with real sides. The unit cell of a crystal is an idealized shape. It is generated by imaginary lines connecting some or all of the points that form the regularly repeating pattern called the *crystal lattice* (Section 7.1).

Suppose that a crystal is formed by repetition of the group of eight spherical atoms shown in Figure 10.16(a). The most convenient way to describe the repeating pattern of the crystal is to select a set of points corresponding to this group. For the group of atoms shown in Figure 10.16(a), we can conveniently place these points in the centers of the atoms, as shown in Figure 10.16(b). When these points are connected by

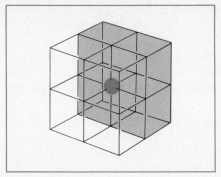

**Figure 10.17**
In a crystal made up of cubic unit cells, each point in the lattice is at a corner of eight adjoining cubes.

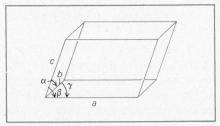

**Figure 10.18**
We can classify unit cells and the crystal lattices that result from them by specifying the relationship between three sides of a unit cell (*a*, *b*, and *c*) and three angles of the unit cell (α, β, and γ).

lines, a cube is formed, as shown in Figure 10.16(c). When this pattern of points is continued in three dimensions, it forms a lattice whose unit cell is a cube, as shown in Figure 10.16(d). Note that each point at the corner of a unit cell actually belongs to eight cubes in the lattice, as shown in Figure 10.17. Also note that the points in a representation of a unit cell do not convey information about the size of the atoms. In most crystals, there is very little empty space, and the units of structure are always touching.

It might seem that a great variety of unit cell shapes is possible. But the variety of shapes is restricted sharply by the requirement that the unit cells must fill space completely when repeated in three dimensions.

### Bravais Lattices and Unit Cells

*All unit cells are parallelepipeds,* solids that have six faces, all of which are parallelograms. In the nineteenth century, Auguste Bravais (1811–1863) devised a classification of unit cells and the crystal lattices that can result from them. The classification is based on the symmetry of the crystals. It is designed to clarify the arrangement of the points of the lattice. Bravais lattices and unit cells are most conveniently classified if we specify the relationship between three sides and three angles of the parallelepiped, as shown in Figure 10.18.

The seven *primitive* (or *simple*) unit cells are shown in Figure 10.19,

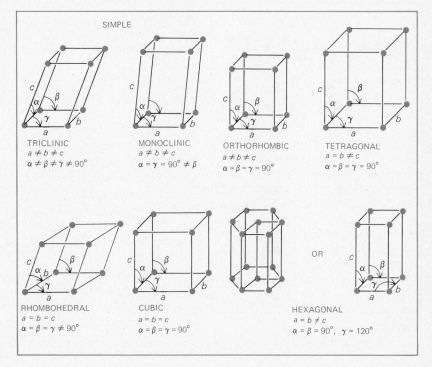

**Figure 10.19**
The seven primitive unit cells of a crystal lattice.

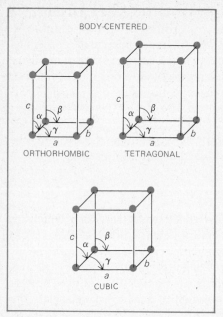

**Figure 10.20**

The three unit cells of a body-centered crystal lattice. The extra point is in the center of the solid.

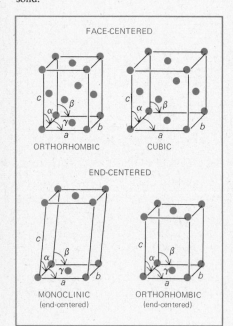

**Figure 10.21**

The four unit cells with lattice points in the faces.

which also classifies them by the relationship between the three sides and the three angles of each. At one extreme is the cube, the unit cell in which all three sides are equal ($a = b = c$) and all three angles are 90° ($\alpha = \beta = \gamma$). The least symmetrical unit cell, called triclinic, has three sides of unequal length ($a \neq b \neq c$) and three unequal angles ($\alpha \neq \beta \neq \gamma$). The unit cells between these two extremes have various combinations of equal and unequal sides and angles.

In the seven primitive unit cells, the lattice points lie only at the corners of the unit cells, and each lattice point belongs to eight unit cells. There is another type of unit cell, which has an additional lattice point in the center of the unit cell. There are three such **body-centered** unit cells, which are shown in Figure 10.20. Note that the point in the center of a body-centered unit cell belongs only to that cell.

Still other unit cells have one or more lattice points centered in the faces, in addition to the points at the corners. There are two **face-centered** unit cells and two **end-centered** cells, which are shown in Figure 10.21. A point on the face of a unit cell belongs to two unit cells, which share a face in the lattice (Figure 10.22).

The relationship between the unit cell and the number of lattice points that it contains is important. The seven primitive unit cells have one lattice point per unit cell (eight lattice points, each of which is shared by eight cells). The three body-centered unit cells each have an additional lattice point in the center, and therefore a total of two lattice points per unit cell. The two face-centered cells have six additional shared lattice points on the faces, giving a total of four lattice points per unit cell. There are two end-centered cells, which have two additional shared lattice points, one in each end, giving them a total of two lattice points per unit cell.

The crystal lattices observed in solids can be accommodated by fewer than 14 lattices. The Bravais lattice system was devised to help recognize the symmetry of crystals. Being aware of these 14 lattices can help us interpret crystal structures.

*All lattice points are equivalent.* Intuitively, it might seem that some points are more important than others. For example, the single point inside a body-centered unit cell might seem different from the points at the cell's corners, which belong to other cells as well. But if we ignore the underlying unit cell and look only at the three-dimensional lattice, there is no difference between any of its points. It is only when we draw imaginary lines to outline unit cells that some lattice points appear to be different from others.

Since macroscopic quantities of solid substances contain very large numbers of repeating unit cells, the properties of the unit cell are related to properties of the solid. We saw earlier that Avogadro's number often defines a relationship between a measurement on the atomic scale and a measurement on the macroscopic scale. Avogadro's number can be determined by measurements of unit cells in a crystalline solid.

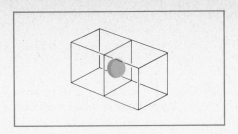

**Figure 10.22**
A point in the face of a unit cell belongs to two
cells in the lattice.

Example 10.1    The unit cell in a crystal of metallic calcium is found to be a face-centered cube
with a side of 0.556 nm and atoms located at the lattice points. The density of
calcium is 1.54 g/cm³ and its atomic weight is 40.08. Calculate Avogadro's number from these data.

Solution    The volume of a cube of side $a$ is $a^3$. Therefore the volume of this cubic unit cell is

$$(0.556)^3 \text{ nm}^3 = 0.172 \text{ nm}^3$$

The mass of this volume of calcium can be found from the density:

$$\text{mass unit cell} = 0.172 \text{ nm}^3 \times (10^{-7})^3 \frac{\text{cm}^3}{\text{nm}^3} \times 1.54 \frac{\text{g}}{\text{cm}^3}$$

$$= 2.649 \times 10^{-22} \text{ g}$$

Since this face-centered cubic unit cell has four atoms, the mass of a single atom is

$$\text{mass atom} = \frac{2.649 \times 10^{-22} \text{ g}}{4} = 6.622 \times 10^{-23} \text{ g}$$

The ratio of the mass of one mole of a substance to the mass of one atom of a
substance is Avogadro's number:

$$\frac{40.08 \text{ g/mol}}{6.622 \times 10^{-23} \text{ g/atom}} = 6.05 \times 10^{23} \text{ atoms/mol}$$

which is the experimental value of Avogadro's number found from these measurements on calcium.

The relationship between the macroscopic properties of a solid and the
properties of the unit cell is illustrated further in the following example.

Example 10.2    Tungsten has a density of 19.35 g/cm³ and an atomic weight of 183.85. The unit
cell in the most important crystalline form of tungsten is the body-centered cubic
unit cell with atoms located at the lattice points. Calculate the length of a side of
the unit cell.

Solution    Since there are two atoms per unit cell in this body-centered cubic lattice, the mass
of the unit cell is the mass of two atoms of tungsten:

$$\text{mass unit cell} = 2 \frac{\text{atoms}}{\text{cell}} \times 183.85 \frac{\text{g}}{\text{mol}} \times \frac{1}{6.0221 \times 10^{23}} \frac{\text{mol}}{\text{atom}}$$

$$= 6.106 \times 10^{-22} \text{ g/cell}$$

The volume of the unit cell can be found from the density:

$$6.106 \times 10^{-22} \text{ g/cell} \times \frac{1 \text{ cm}^3}{19.35 \text{ g}} = 3.156 \times 10^{-23} \text{ cm}^3/\text{cell}$$

We find the length of the side of a cube of known volume by taking the cube root of the volume:

$$(0.03156 \times 10^{-21} \text{ cm}^3)^{1/3} = 0.3160 \times 10^{-7} \text{ cm} = 0.3160 \text{ nm}$$

## 10.5  CLOSEST-PACKED STRUCTURES

The structure of any crystal, no matter how complex, can be described by lattice points, unit cells, and symmetry properties. The method works for a crystal of a complicated organic molecule, such as a protein or a nucleic acid, as well as it does for a crystal of a simple monatomic substance, such as a metal.

There is a simpler, alternative method of picturing the crystal structure of many substances. We can describe the crystal as an assembly of closely packed spheres. This alternative method works very well when the basic structural unit is an atom. Some molecular substances can also be described in this way, but only if the molecules are simple enough to be roughly spherical. Many minerals also lend themselves to a description of this sort.

The basic approach is to see how spheres of identical size can be packed to fill a volume as completely as possible. This problem is encountered in everyday life, as when we try to pack oranges or other spherical objects in a crate.

There are two ways to pack spheres most efficiently while preserving long-range order. (*Efficiently* means that the smallest possible fraction of the volume is empty space.)

Figure 10.23 shows one layer of spheres packed as closely as possible. There are small empty spaces, or *holes*, between spheres. These holes, sometimes called *interstices*, are almost as important as the spheres for understanding close-packed structures.

In Figure 10.23, each sphere is surrounded symmetrically by six other spheres, its nearest neighbors. In three dimensions, close-packed arrangements are made by layers of spheres stacked on one another. As Figure 10.24 shows, this stacking is done most efficiently when the spheres of the top layer are fitted into the holes of the bottom layer. Figure 10.24 also

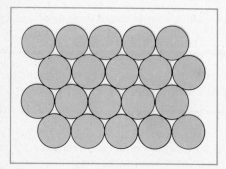

**Figure 10.23**

One layer of spheres packed as closely as possible. Most of the volume is occupied by the spheres, but an irreducible volume is represented by the holes, or interstices, between spheres.

shows that only half of the holes in the bottom layer are covered by spheres in the top layer. The other half of the holes in the bottom layer lie directly below holes in the top layer.

We can say that there are two "types" of holes in the bottom layer. There are holes that are covered by spheres in the second layer; these are labeled *t* in Figure 10.24. And there are holes that lie below holes in the second layer; these are labeled *o*.

When we place layers of spheres together, we create not only two types of holes in each layer but also different types of space between the layers. These spaces are called *sites*. In Figure 10.24, the spaces labeled *t* are called *tetrahedral sites*, because each is formed by four spheres whose centers lie at the corners of a tetrahedron (Figure 10.25(a)). Three of the spheres in a tetrahedron are in one layer and the fourth is in the other layer. The spaces labeled *o* are called *octahedral sites*, because they are formed by six spheres whose centers lie at the corners of an octahedron, as shown in Figure 10.25(b). Three of the spheres are in the bottom layer and the other three are in the top layer.

## Hexagonal Closest Packing and Cubic Closest Packing

If we want to place a third layer on top of the second, there are two ways to do it. One way is to have each sphere in the third layer directly above a

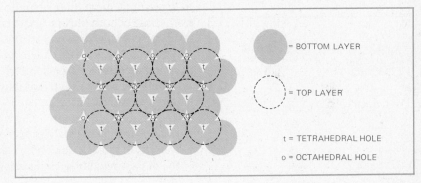

**Figure 10.24**
When one layer of close-packed spheres is placed atop another, two types of holes are found in the bottom layer: tetrahedral (*t*) holes, which are covered by spheres in the top layer, and octahedral (*o*) holes, which are not covered by spheres in the top layer.

= BOTTOM LAYER

= TOP LAYER

t = TETRAHEDRAL HOLE

o = OCTAHEDRAL HOLE

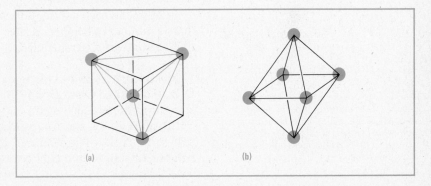

**Figure 10.25**
The tetrahedral (*t*) sites in layers of close-packed spheres (see Figure 10.24) are formed by four spheres whose centers lie at the corners of a tetrahedron (a). Octahedral sites are formed by six spheres whose centers lie at the corners of an octahedron (b).

(a)                    (b)

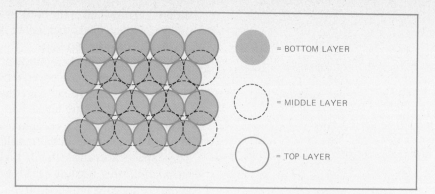

= BOTTOM LAYER

= MIDDLE LAYER

= TOP LAYER

**Figure 10.26**

An *ababab* . . . pattern in three layers of close-packed spheres. Each sphere in the top layer lies directly over a sphere in the bottom layer. The pattern can be extended through an indefinite number of layers.

sphere of the first layer, as shown in Figure 10.26. In this method, the layers of spheres have an *ababab* . . . pattern; the spheres of the first, third, fifth, seventh, . . . layers lie directly above each other, and the spheres of the second, fourth, sixth, eighth, . . . layers also lie directly above each other. Figure 10.27 shows another view that helps to explain why this arrangement is called *hexagonal closest packing*. This kind of packing corresponds to the hexagonal Bravais lattice, but not all the spheres are at lattice points.

Another way is to have the spheres of the third layer lie above the holes labeled *o*, the octahedral sites of the first layer, as shown in Figure 10.28. In this arrangement, the orientation of the spheres in the third layer differs from those of both the first and second layers. This is an *abcabc* . . . pattern, since it does not repeat until the fourth layer. Figure 10.29 shows why this arrangement is called *cubic closest packing*. Cubic closest packing corresponds to a face-centered cubic unit cell (see Figure 10.21).

A detailed study shows that these two arrangements have a number of features in common:

**1.** In both arrangements, just over 74.0% of a given volume is occupied by spheres and just under 26.0% is empty space. It can be shown that this is the closest regular packing arrangement possible for spheres of equal size.
**2.** Each sphere has 12 nearest neighbors, six in its own horizontal layer and three in each adjacent horizontal layer.
**3.** The number of octahedral sites is the same as the number of spheres. The number of tetrahedral sites is twice the number of spheres.

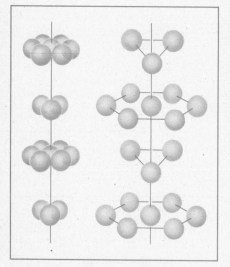

**Figure 10.27**

Another view of the layers of spheres shown in Figure 10.26, showing that the layers can be regarded as alternating units made up of hexagons whose corners are defined by spheres. This arrangement is sometimes called hexagonal closest packing.

Metals such as Be and Mg in group IIA, the metals in groups IIIB and IVB, and Zn and Cd in group IIB have hexagonal closest-packed crystal structures. Metals such as Ca and Sr in group IIA, Ni and Pt in group VIIIB, and Cu, Ag, and Au in group IB have cubic closest-packed crystal structures. At low temperatures, the noble gases (with the exception of helium) crystallize in a cubic closest-packed arrangement. Some small,

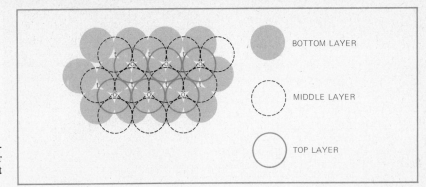

BOTTOM LAYER

MIDDLE LAYER

TOP LAYER

**Figure 10.28**
An *abcabc* . . . pattern in three layers of close-packed spheres. The spheres in the third layer lie above the octahedral (*o*) holes in the first layer.

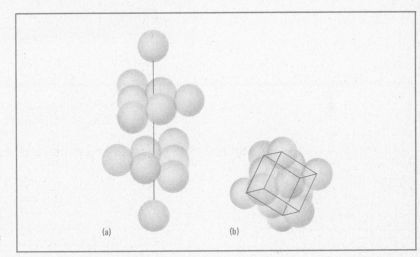

(a)                         (b)

**Figure 10.29**
Another view of the layers of spheres shown in Figure 10.28, showing the cubic symmetry. This arrangement is sometimes called cubic closest packing.

more or less spherical, molecules such as $H_2$ have closest-packed crystalline arrangements. The major advantage of a closest-packed arrangement is that the attractive forces are maximized, because each structural unit has a large number of nearest neighbors.

Some metals, such as those in groups IA, VB, and VIB, as well as Ti and Ba, crystallize with a structure that is not packed quite as closely as the two mentioned thus far. This structure has the body-centered cubic unit cells shown in Figure 10.20. Each sphere in the body-centered cubic structure has only eight nearest neighbors as compared to 12 in the two closest-packed arrangements. However, in the body-centered cubic arrangement there are six next-nearest neighbors, which lie relatively close to each sphere. In this arrangement, spheres occupy 68% of the available volume.

In the next example, we see once again how an understanding of closest-packed structures on the atomic level allows us to relate macroscopic properties and atomic properties.

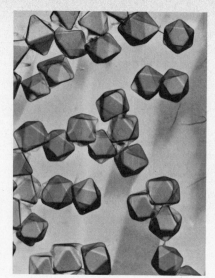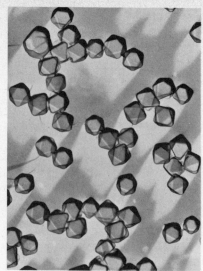

These two electron micrographs show silver halide crystals from two color films, magnified 10 000 times. The smaller grains on the right permit faster developing time for the film. Note that all the silver halide crystals have a symmetrical octahedral structure that is visible only at high magnification. *(Eastman Kodak Company)*

**Example 10.3**   Magnesium crystallizes in a hexagonal closest-packed structure with a density of 1.74 g/cm³. Calculate the volume and the radius of a magnesium atom.

**Solution**   The volume occupied by one mole of magnesium is:

$$V_{molar} = \frac{1\ cm^3}{1.74\ g\ Mg} \times \frac{24.3\ g\ Mg}{1\ mol\ Mg} = \frac{14.0\ cm^3}{mol\ Mg}$$

This volume of Mg consists of $6.02 \times 10^{23}$ (Avogadro's number) of spherical atoms of Mg and of empty space between the atoms. In a closest-packed arrangement, 26.0% is empty space and 74.0% is the actual volume of the Mg atoms:

$$\frac{14.0\ cm^3}{mol\ Mg} \times \frac{74.0\%}{100\%} = \frac{10.3\ cm^3}{mol\ Mg}$$

The volume of an individual atom is

$$\frac{10.3\ cm^3}{mol\ Mg} \times \frac{1\ mol}{6.02 \times 10^{23}\ atoms} = \frac{1.72 \times 10^{-23}\ cm^3}{atom\ Mg}$$

The individual atom is a sphere. To find the radius of a sphere from its volume, we employ the relationship:

$$V = \frac{4}{3}\pi r^3 \quad or \quad r = \left(\frac{3V}{4\pi}\right)^{1/3}$$

where $V$ is the volume and $r$ is the radius of the sphere. For this spherical atom, $r$ can be found to be $1.60 \times 10^{-8}$ cm = 0.160 nm.

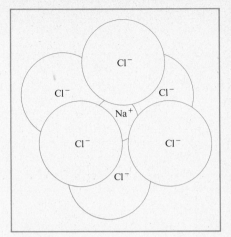

**Figure 10.30**
Detail of an NaCl crystal structure, showing that each Na$^+$ ion has six Cl$^-$ ions as nearest neighbors.

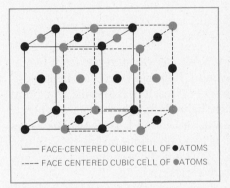

—— FACE-CENTERED CUBIC CELL OF ●ATOMS
---- FACE CENTERED CUBIC CELL OF ●ATOMS

**Figure 10.31**
Another view of an NaCl crystal structure, showing that it can be regarded as being made up of two different interlocked face-centered cubic unit cells. The points of one unit cell are connected by solid lines, the points of the other by dashed lines.

## The Structures of Ionic Solids

Picturing crystal structures as close-packed arrangements can also help us to understand the structures of many ionic solids.

Monatomic ions can be regarded as spheres. In Chapter 7, we mentioned that anions usually are considerably larger than cations. The structure of many salts made up of ions with a large difference in size can be pictured as a close-packed arrangement of the anions, with the much smaller cations placed in the holes.

Sodium chloride has such a structure. The relatively large Cl$^-$ anions are arranged in a cubic closest-packed structure. The relatively small Na$^+$ cations are located in a regular way throughout this close-packed lattice.

We know that there are an equal number of Na$^+$ cations and Cl$^-$ anions in NaCl. The number of octahedral sites is equal to the number of spheres in a closest-packed lattice. In an NaCl crystal, the Na$^+$ ions are found in the octahedral sites of the Cl$^-$ lattice. Since an octahedral site is formed by six spheres, each Na$^+$ ion has six Cl$^-$ ions as nearest neighbors, as shown in Figure 10.30. Each Cl$^-$ ion also has six Na$^+$ ions as nearest neighbors.

Cubic closest packing corresponds to a face-centered cubic cell. The Cl$^-$ anions in NaCl make up such a unit cell. The Na$^+$ cations in the octahedral sites also define a face-centered cubic unit cell. Figure 10.31 shows that sodium chloride can be regarded as two interlocking lattices made up of face-centered cubic unit cells. However, it is easier for most people to visualize the Na$^+$ ions occupying the octahedral sites of a cubic closest-packed Cl$^-$ lattice.

Many salts and oxides have the same crystal structure as NaCl. This type of structure appears in substances with equal numbers of cations and anions. A compound such as Na$_2$O cannot have the sodium chloride lattice because it has two cations for each anion, and there are not enough octahedral sites to hold all the cations. Many substances of this kind have a structure in which the anion, O$^{2-}$ in this case, forms a cubic closest-packed lattice, and the cations occupy the tetrahedral sites. This arrangement is possible because there are twice as many tetrahedral sites as spheres. Figure 10.32 shows that each Na$^+$ ion in Na$_2$O has only four nearest neighbors. Each O$^{2-}$ ion can be seen to have eight Na$^+$ nearest neighbors if the lattice is extended.

## 10.6   X-RAY CRYSTALLOGRAPHY

Information about the arrangement of units of structure in crystals is obtained primarily by a method called X-ray crystallography, which is the study of the interaction between X rays and crystals. We can get a general understanding of X-ray crystallography by recalling certain characteristics of electromagnetic radiation.

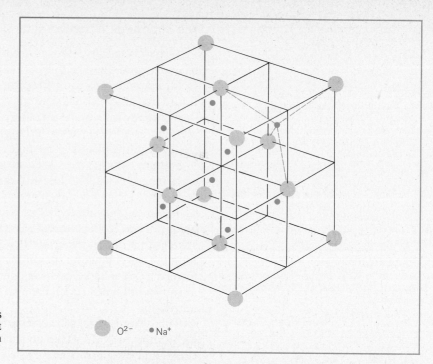

**Figure 10.32**

An $Na_2O$ crystal structure. The dashed lines connect the $Na^+$ ion to the four $O^{2-}$ ions that form a tetrahedral site in which the $Na+$ ion fits.

Suppose beams of light of a given wavelength start out from two or more points and arrive at the same point. If the light waves are in phase — that is, if the maxima and minima of each wave strike the point simultaneously — we will see a bright light. If the light waves are in phase, the waves reinforce each other, as was shown in Figure 7.17. The waves will reach a given point in phase if two conditions are met: (a) all the beams start at the same time and (b) the difference of the distances they travel is an integral multiple of the wavelength.

If the waves are out of phase when they strike the point, they will interfere, or cancel, to a greater or lesser extent. If the beams are completely out of phase, so that a maximum of one wave and a minimum of another reach the point simultaneously, there will be a complete cancellation of the light, as was shown in Figure 7.17. A diffraction pattern is produced by wave interference of visible light. The bright bands are caused by wave reinforcement, the dark bands by wave cancellation.

Experimentally, we can meet the two conditions described above by splitting one light source into several beams. Figure 10.33 shows a simple method that uses two slits as starting points for the light, creating an interference pattern. A number of slits would create a more complex diffraction pattern.

In 1912, a group of physicists at the University of Munich showed that a diffraction pattern could be produced if a beam of radiation was directed at a crystalline substance. Directing the beam of radiation at the regularly

## X-RAY CRYSTALLOGRAPHY AND THE HEMOGLOBIN MOLECULE

The first protein whose structure was determined by X-ray crystallography was myoglobin. While the structure of myoglobin appears complex, it is relatively simple compared to that of hemoglobin, which was an even greater challenge for X-ray crystallographers. Myoglobin is a single chain of subunits with one iron-containing heme group, while hemoglobin consists of four chains and four heme groups, containing a total of more than 10 000 atoms. The same set of techniques, developed in British laboratories in the 1950s, was used to determine the structure of myoglobin and hemoglobin.

Perhaps the most important part of the technique is the use of heavy atoms, such as those of mercury, to serve as reference points. The heavy atoms, attached at specific sites in the hemoglobin molecule, change the diffraction patterns. If we attach different atoms at different sites and compare literally thousands of diffraction maxima, we can build up a three-dimensional image of the molecule, layer by layer. Millions of arithmetic operations must be performed to obtain the image, and the task would be impossible

without computers to perform the calculations.

The structure of the myoglobin molecule was determined in 1957. The first accurate three-dimensional image of the hemoglobin molecule was obtained 2 years later. Somewhat to the surprise of the investigators, the basic structure of both myoglobin and the hemoglobin molecule proved to be essentially the same. Each of the four chains of the hemoglobin molecule has a shape closely resembling that of the myoglobin molecule. It has been found that this similarity in shape does not reflect a similarity of composition. The sequences of the amino acid subunits that make up the hemoglobin and myoglobin molecule are 90% different; that is, the same amino acid occurs in only 10% of the sites in the two chains. It is believed that the two molecules had a common ancestor, and that the differences result from the different evolutionary paths that were followed. Thus the determination of the structure of hemoglobin and myoglobin helps us to understand how they function on the molecular level and also tells us something about their evolution.

spaced units of structure of the crystal has the same effect as sending radiation through regularly spaced slits. In each case, a diffraction pattern is created.

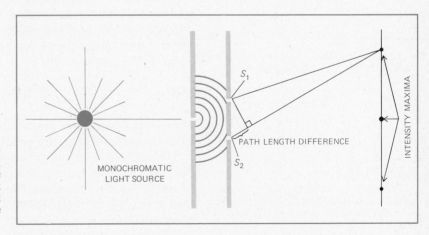

**Figure 10.33**
A simple apparatus for creating a diffraction pattern. Monochromatic light from a single source is passed through two slits, which break up the beam. An interference pattern is created if the distance between the two slits is about the same as the wavelength of the light.

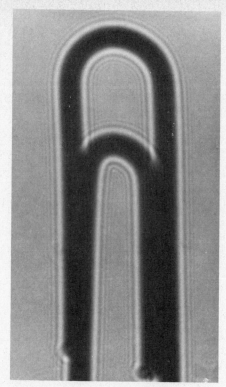

A diffraction pattern produced as light passes by the edges of a paper clip. The picture was taken without a lens in the camera, since the interference bands would be destroyed if the image were focused through a lens. *(Fundamental Photographs, Granger)*

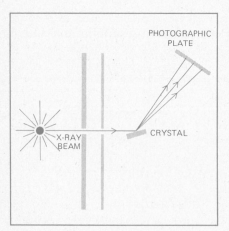

**Figure 10.34**
A diagram of the apparatus used by William and Lawrence Bragg to study crystal structure by X-ray diffraction. The X-ray beam is directed at the crystal, and the diffraction pattern is captured on the photographic plate.

The physicists did not use visible light in their experiment. To create a diffraction pattern, the "slits" through which the radiation travels must be about as far apart as the wavelength of the radiation. In the experiment, the "slits" were the spaces between units of structure of the crystal. The units are separated by about $10^{-1}$ nm, so the X rays that were used, which have wavelengths in this range, caused a diffraction pattern.

The characteristics of an X-ray diffraction pattern are related to the structure of the crystal, the wavelength of the X rays, and the angle at which the radiation strikes the crystal. Two English physicists, William Bragg (1862–1942) and his son Lawrence (1890–1971), devised experimental methods for studying X-ray diffraction patterns and developed the basic theory for interpreting the results. Figure 10.34 shows a schematic diagram of their apparatus.

**The Bragg Equation**

The theory developed by the Braggs is based on the idea that if two waves arrive at the same point in phase—that is, if they produce maximum intensity at this point—then the difference between the distances they have traveled is an integral multiple of their wavelength.

Figure 10.35 shows how one beam of X rays interacts with two units of structure in successive layers of a crystal. Using a picture of this type, the Braggs showed how to calculate the distance between these layers.

The distance is calculated by measurement of the angle between the crystal and a spot of intense radiation in the diffraction pattern. This angle is labeled $\theta$ in the diagram. The intense spot is produced when waves of X rays that are "reflected" (the process taking place in diffraction) by different units of structure in the crystal arrive at one point on the detector in phase. For this to happen, the difference between the distances traveled by the waves before and after they are reflected by the crystal must be an integral multiple of the wavelength, $n\lambda$.

The geometric construction shown in Figure 10.35 allows us to find the relationship between the angle $\theta$, the difference in path length, and the distance between layers of the crystal. If the distance between layers is called $d$, then the difference between the path lengths of the two waves is $2d \sin \theta$. We can thus write:

$$n\lambda = 2d \sin \theta \qquad (10.1)$$

where $n$ is any integer. This relationship is called the Bragg equation. Experimentally, $\theta$ and $\lambda$ are measured, and $d$ then can be calculated.

The study of X-ray diffraction patterns can give much more information than just the spacing between planes of the lattice points in the crystal. If the crystal consists of molecules, the diffraction pattern can be used to map the electron density in the unit cell from which the structure of the molecules may be derived. The structure of the molecule can be determined from the electron density.

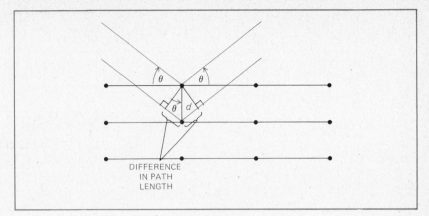

DIFFERENCE
IN PATH
LENGTH

**Figure 10.35**

The interaction of a beam of X rays with units of structure in two layers of a crystal. By analyzing the diffraction pattern created by such an interaction, we can determine the distance between layers of the crystal. If the distance between layers is $d$, the difference between the path lengths of the two waves is $2d \sin \theta$.

X-ray crystallography is one of the most powerful methods available for determining molecular structure. Computers can be used to analyze diffraction patterns and to obtain maps of the structure of large and complex molecules in living organisms. Among the major successes of X-ray crystallography are the determination of the structures of complex protein molecules such as myoglobin, the oxygen-transporting protein of muscle cells, and lysozyme, the antibacterial enzyme present in many body fluids. The double helix structure of DNA was proposed by Watson and Crick on the basis of X-ray diffraction data.

## 10.7   LIQUID CRYSTALS

Some substances do not undergo the simple transition from the solid phase to the liquid phase that has been described for water. There can be phases whose structures lie between the complete long-range order of a crystal and the almost complete long-range disorder of a liquid. These phases are most commonly observed in some substances made up of long-chain molecules.

Figure 10.36 shows how this occurs. In Figure 10.36(a), a crystal of long-chain molecules is ordered in three dimensions. In Figure 10.36(b), the molecules all have the same orientation and they are in equispaced planes, but the arrangement within the planes is irregular. The substance no longer is a true crystal, although it still has more order than a liquid. It is in what is called a smectic state. When the molecules no longer are in equispaced planes (Figure 10.36(c)), more disorder is introduced, although some order is maintained if the molecules all have the same orientation; this is called the nematic state. When even this order is lost, the substance is in the liquid state (Figure 10.36(d)).

Substances that have the partially ordered arrangements are said to be in the *paracrystalline state.* They are called liquid crystals. A liquid crystal is a distinct phase, like a solid or a liquid. Substances that display the

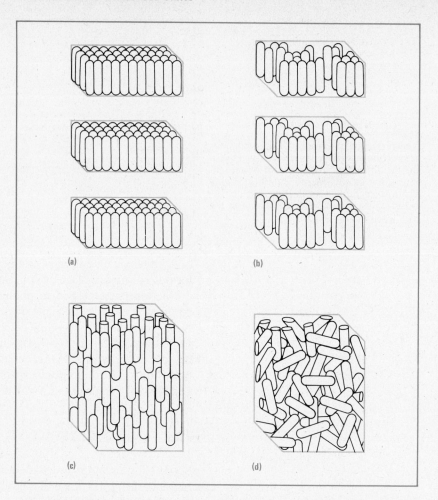

**Figure 10.36**
Four different states of the same substance. (a) Crystalline. (b) Smectic structure. (c) Nematic structure. (d) Liquid structure.

changes shown in Figure 10.36 undergo sharp transitions from crystal to liquid crystal, from liquid crystal to liquid, or even from the smectic to the nematic state. These transitions are comparable to the melting or freezing of solids and liquids.

Some substances have a number of distinct paracrystalline phases between solid and liquid. Changes from one phase to another can occur abruptly because of a slight change in temperature. Often, the different paracrystalline states have different colors. A growing number of practical applications have been found for liquid crystals whose color changes in response to very slight temperature changes.

Liquid crystals can be used in thin, lightweight display devices that require very little energy. In recent years, liquid crystal display devices have been used increasingly in digital watches, pocket calculators, and similar instruments. A typical display device has a film of liquid crystal no more than a thousandth of an inch thick between two pieces of glass, one of which is coated on one side with a conductive material. The liquid

## TEMPERATURE IN LIVING COLOR

A physician paints a black liquid onto a patient's skin. In moments, the skin is alive with a pattern of vivid reds and blues that give a startlingly good picture of the blood vessels in the body.

A nurse in an intensive care unit for newborn babies checks the color of a small circle of material on an infant's stomach. That quick glance tells the nurse whether the baby's body temperature is within normal limits.

A business executive who isn't feeling quite right holds a small square of plastic to his forehead. By looking in the mirror to see what appears on the plastic, the executive can tell whether he's running a fever.

All three of these devices use liquid crystals that change color in response to slight changes in temperature. The liquid painted on the patient's body shows warmer areas as blue, cooler areas as red, and intermediate temperatures as green or yellow. The dot on the baby's stomach is blue-green at normal temperature but turns bronze if the infant is too cold; if the baby is running a fever, the dot turns blue. The small piece of plastic used by the executive has been designed so that a warning letter will appear if the person using it is feverish.

The medical uses of temperature-sensitive liquid crystals are only a few of the applications of this new technology. In the Apollo missions to the moon, liquid crystal markers were placed on the scientific instruments that were left on the surface of the moon. The Apollo astronauts could tell at a glance whether the instruments were overheating. The "mood rings" that were a brief craze in 1976 consisted of liquid crystal patches. The changes in mood that the rings were supposed to mirror by color changes were actually temperature variations of the body. In a less emotional application, digital thermometers have been designed in which the temperature responses of liquid crystals are carefully orchestrated. As the temperature goes up or down, one number fades out and another fades in. The color of the number that is displayed gives a more precise reading of temperature: A blue 22 indicates that the temperature is closer to 23°C, while a white 22 indicates a reading closer to 21°C. Liquid crystal producers see another potential market among wine lovers, who monitor the temperature at which their wine is stored constantly to be sure that the wine is aging properly. A tiny liquid crystal dot on each bottle of wine could give assurance that the temperature is in the right range for proper aging.

Liquid crystal displays also have made a major impact on the field of consumer electronics, notably in the displays used for digital watches and hand-held calculators. When these instruments first became popular, they used light-emitting diodes to display time or calculations. These displays, however, used a relatively large amount of electricity, could not be seen in bright daylight, and make it necessary to push a button to see the time. Most digital watches and calculators now use liquid crystals, which can be seen in bright light and which use so little electricity that batteries last for thousands of hours.

The same advantages have made larger liquid crystal displays usable as the screens in television sets that are small enough to be slipped into a pocket or worn on the wrist, and to display data in the newest generation of portable computers. Auto manufacturers have also begun to replace conventional dashboard instruments with liquid crystal displays, insuring continued growth in their use.

crystal molecules are polar, so the application of a small electric current can align them to present a number or a letter. The image can be changed by alteration of the voltage.

The paracrystalline state has features of both liquids and solids. A substance in the paracrystalline state flows and mixes readily, but it also

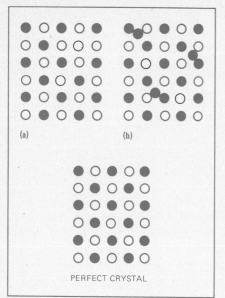

(a)                    (b)

PERFECT CRYSTAL

**Figure 10.37**
Point defects in a crystal. (a) A unit of structure can migrate to the surface of the crystal, leaving an empty space. In an ionic crystal, overall neutrality of electric charge is maintained despite such migrations. (b) A second type of point defect occurs when a unit of structure migrates to a hole, leaving a vacant site.

maintains an ordered arrangement of its units of structure. Such a combination of properties often is desirable for living systems. Some recent work indicates that portions of cell membranes and organelle membranes are liquid crystals. The lipid, or fat, molecules in the membtanes are in the long-chain shape that favors liquid crystal formation.

## 10.8  DEFECT CRYSTALS

So far, we have described what might be called ideal or perfect crystal structures. Real crystals have defects or imperfections. Many properties of the solid state can be understood only in terms of the defects in real crystals. Moreover, many important practical applications of the properties of the solid state, especially in photography and in semiconductors, result from crystal defects.

There are two main categories of crystal defects: structural imperfections, which are flaws in the crystal lattice; and chemical imperfections, the presence of impurities in the crystal.

There are two types of structural imperfections, point defects and linear defects. Figure 10.37 illustrates the two main kinds of point defects. In one type, a unit of structure migrates to the surface of the crystal, leaving a *vacant site* (Figure 10.37(a)). In an ionic crystal, vacant sites occur in such a way as to preserve electrical neutrality. Such defects account for some aspects of the conduction of electricity by ionic solids, and for diffusion in the solid state.

The second type of point defect is more complicated. An example is shown in Figure 10.37(b), in which a unit of structure migrates to an interstitial site, or hole, in the lattice, leaving behind a vacant site. Such defects occur in salts such as AgBr, in which an $Ag^+$ ion migrates to the interstitial site. This phenomenon is put to practical use in photographic film, which consists of AgBr in a gelatin coating. The changes that occur in a small crystal of AgBr when it is struck by light are used to create a photographic negative. These changes are believed to result from the migration of $Ag^+$ ions to interstitial sites and the accompanying creation of vacant sites.

The second major kind of defect, the linear defect, is also called a *dislocation.* The two important types of linear defects are edge dislocation and screw dislocation. Both result from the imperfect orientation of planes with respect to one another in the crystal.

Figure 10.38(a) shows an edge dislocation, which can be described as the insertion of an extra plane of structural units partway into the crystal. A screw dislocation (Figure 10.38(b)) results from stresses that can occur while the crystal is forming. A screw dislocation resembles a spiral staircase within the crystal.

Recent evidence indicates that the structural strength of solids, especially metals, is closely related to the number and type of dislocations in

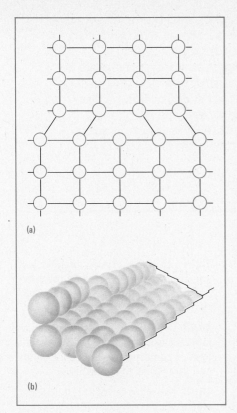

(a)

(b)

the solids. Any dislocation reduces structural strength considerably. Even chemical reactions that occur on the surface of a crystal tend to occur at the sites of dislocations. One approach to creating stronger structural materials is the effort to reduce or eliminate dislocations in metals such as steel.

Defects caused by impurities in crystals play a vital role in electronics. Impurities often are added deliberately to make semiconductors, substances whose electrical conductivity increases with increasing temperature. Two important semiconductors are silicon and germanium. When either of these substances is extremely pure, its electrical conductivity is quite low. The deliberate addition of tiny amounts of impurities, such as arsenic or boron, a process called "doping," causes a substantial increase in the conductivity of Si and Ge, which then become semiconductors. The electronics industry today is based on semiconductors such as doped silicon and doped germanium. We shall discuss how doping is effective in creating semiconductors in Chapter 19.

**Figure 10.38**
Linear defects, or dislocations, in a crystal. An edge dislocation can be pictured as the partial insertion of an extra plane of units in the crystal lattice (a). A screw dislocation resembles a spiral staircase within the crystal (b).

**Summary**

The two **condensed states** of matter were the subjects of this chapter. We began by describing some properties of the **liquid state: surface tension,** the force that pulls the surface of a liquid inward; **capillary action,** which is related to surface tension; and **viscosity,** the resistance of a liquid to flow. We then discussed **phase transitions** from solid to liquid to gas. We described **supercooling,** in which a liquid does not become a solid at its freezing point, and **superheating,** in which a liquid does not boil at its boiling point. We mentioned **phase equilibrium,** the given conditions of temperature and pressure at which a solid and a liquid, a solid and its vapor, or a liquid and its vapor are at equilibrium. We showed some **phase diagrams,** which give phase equilibria of substances as a set of curves on a graph of temperature and pressure. We defined the **triple point** as the temperature and pressure at which the solid, liquid, and gas phase of a substance are in equilibrium. Then we moved on to the **solid state,** describing **crystals,** the characteristic structures of crystalline solids. We said that a crystal consists of a regular, repeating, three-dimensional set of **unit cells,** and that the unit cells can contain ions, molecules, or atoms. All unit cells are parallelepipeds, solids with six faces, and we described the 14 unit cells in the **Bravais** classification system. We then discussed a simpler way to describe crystals, as an assembly of closely packed spheres. We described **hexagonal closest packing,** in which the layers of spheres have an *ababab* . . . pattern, and **cubic closest packing,** in which the layers are in an *abcabcabc* . . . pattern. We showed how the structure of a crystal can be determined by **X-ray crystallography,** in which the **diffraction** pattern produced when X rays pass through a crystal is analyzed. We mentioned **liquid crystals,** which have structures between those of solids and liquids. Finally, we discussed the structural and chemical imperfections that are found in many crystals.

## Exercises

**10.1**  Explain why heat is always absorbed when a solid melts or a liquid boils.

**10.2**  Some insects can walk on water even though their bodies are denser than water. Explain why.

**10.3**  What is the relationship between the viscosity of a liquid and the rate at which a solid falls through the liquid? Explain your answer.

**10.4**  Draw a warming curve for water from 373 K to 400 K that does not show superheating. (*Hint:* Figure 10.6.)

**10.5**  Draw a warming curve for water from 273 K to 400 K in which the water superheats to 385 K before it boils.

**10.6**[1]  Find the quantity of heat released when 100 g of water at 273 K freezes to ice at 273 K.

**10.7**  The heat of fusion of a solid is always smaller than the heat of vaporization of the corresponding liquid. Explain this observation.

**10.8**  Draw a cooling curve for carbon dioxide from 225 K to 175 K at 1 atm.

**10.9**  Describe the phase changes that take place when the temperature of

[1] The answers to exercises whose numbers are in color can be found in Appendix VII. The star indicates an exercise that is more challenging than average.

water is increased from 250 K to 350 K at a constant pressure of (a) $1 \times 10^{-4}$ atm, (b) 0.006025 atm, (c) 0.1 atm.

**10.10**  Describe the changes that take place when the pressure on water is increased from $1.0 \times 10^{-4}$ atm to 1.1 atm at a constant temperature of (a) 270 K, (b) 273.15 K, (c) 273.16 K, (d) 280 K.

**10.11**  Under what conditions does ice sublime?

**10.12**  Describe the phase changes that take place when the temperature of $CO_2$ is increased from 190 K to 310 K at a constant pressure of (a) 0.8 atm, (b) 1 atm, (c) 5.1 atm, (d) 15 atm, (e) 75 atm.

**10.13**  Describe the phase changes that take place when the pressure on $CO_2$ is increased from 0.5 atm to 75 atm at a constant temperature of (a) 194 K, (b) 216.6 K, (c) 217 K, (d) 304 K.

**10.14**  Draw a phase diagram for $O_2$ from the following data: normal melting point, 54.8 K; normal boiling point, 90.2 K; triple point, 54.3 K; 0.0020 atm; critical temperature, 155 K; critical pressure, 50.1 atm; vapor pressure at 74.3 K, 0.13 atm.

**10.15**  Classify crystals of each of the following substances by the attractions between their units of structure: (a) $I_2$, (b) NaCl, (c) Fe, (d) $NH_4NO_3$, (e)

$C_6H_{12}O_6$, (f) graphite.

**10.16**  Examine crystals of table salt and sugar, with the help of a magnifying glass if necessary. Describe their shapes.

**10.17**  Build models of the Bravais lattices, using miniature marshmallows or gumdrops for the lattice points and toothpicks to connect them.

**10.18**  The unit cell of a crystal of gold is a face-centered cube with atoms at all the lattice points. The cube has a side of 0.408 nm. The density of gold is 19.3 g/cm$^3$ and its atomic weight is 197. Calculate Avogadro's number from these data.

**10.19**  The unit cell in a crystal of barium is a body-centered cube with atoms at all the lattice points. The cube has a side of 0.506 nm. Find the molar volume of barium.

**10.20**  The unit cell of diamond belongs to a cubic system. The volume of the unit cell is 0.0454 nm$^3$ and the density of diamond is 3.52 g/cm$^3$. Find the number of carbon atoms in a unit cell of diamond.

**10.21**  The unit cell of osmium is a face-centered cube with atoms at all the lattice points. The density of osmium is 22.6 g/cm$^3$. Find the length of a side of the cubic unit cell.

**10.22**  The unit cell of iron is a face-centered cube with atoms at all the lattice points. It has a volume of 0.04714 $nm^3$. The iron atom centered in the face of the cube is a sphere that touches the four atoms at the corners of that face. Find the atomic radius of iron.

**10.23** A metal crystalizes in a body-centered cube with atoms at all the lattice points. The edge of the cube is 0.28840 nm and the density of the metal is 7.1987 $g/cm^3$. Find the atomic weight of the metal.

**10.24** The atomic radius of copper is 0.1285 nm. Its unit cell is a face-centered cube with atoms at all the lattice points. Find the density of copper.

**10.25★** The unit cell of platinum is a face-centered cube with atoms at all the lattice points. The density of platinum is 21.45 $g/cm^3$. Find the atomic radius of platinum.

**10.26** Obtain a quantity of marbles or other spheres and use them to produce the regular close-packed structures.

**10.27** Cubic closest packing corresponds to a face-centered cubic lattice. Calculate to five significant figures the fraction of empty space in a face-centered cube with spheres at all the lattice points.

**10.28** When spheres of radius $r$ are packed in a body-centered cubic arrangement, they occupy 68.0% of the available volume. Calculate the value of $a$, the length of the edge of the cube, in terms of $r$.

**10.29** Copper crystalizes with cubic closest packing. It has a density of 8.93 $g/cm^3$. Find the volume and radius of a copper atom.

**10.30** The atomic radius of titanium is 0.147 nm. It crystalizes with hexagonal closest packing. Calculate the density of titanium.

**10.31** The atomic radius of lead is 0.175 nm. Its density is 11.34 $g/cm^3$. Does lead have a close-packed structure?

**10.32** The atomic radius of vanadium is 0.136 nm. It crystalizes in a body-centered cubic structure. Find the molar volume of vanadium.

**10.33** The atomic radius of an unknown metal is 0.137 nm. Its density is 21.02 $g/cm^3$ and it crystalizes in a hexagonal closest-packed structure. Find the atomic weight of the metal.

**10.34** The atomic radius of the $Br^-$ ion is 0.196 nm. Consider a closest-packed structure in which $Br^-$ anions are just touching. Find the radius of the cation that just fits in the octahedral holes of this lattice of anions.

**10.35** A cation with a radius of 0.075 nm just fits in an octahedral hole of a close-packed structure of $Cl^-$ anions that are just touching. The radius of the cation that just fits in the tetrahedral holes is 0.041 nm. Ionic solids are more stable if the anions do not touch. Using this information, suggest a relationship between cation size and the structure of ionic solids.

**10.36**[2] The smallest angle of reflection observed in the X-ray diffraction pattern of strontium is 16.0° when the wavelength of the incident X rays is 0.167 nm. Calculate the distance between the reflecting planes of strontium atoms.

---

[2] We shall assume, incorrectly in some cases, that $n = 1$ in our calculations using the Bragg equations.

**10.37** The distance between planes of atoms that are parallel to the face of a face-centered cube is half the length of the edge of the cubic unit cell. The smallest angle of reflection observed in the X-ray diffraction pattern of calcium, which is in a face-centered cubic lattice, is 14.2° when the wavelength of the incident X rays is 0.141 nm. Find the atomic radius of calcium.

**10.38** The distance between planes of ions that are parallel to the face of the unit cell of potassium bromide is 0.330 nm. The smallest angle of reflection observed in the X-ray diffraction pattern is 4.65°. Find the wavelength of the incident X rays.

**10.39** Find the minimum angle of reflection in the X-ray diffraction patterns of a crystal in which the distance between planes of atoms is 0.222 nm when the wavelength of the incident X rays is 0.241 nm.

**10.40** The maximum value of the sine of an angle is 1. Use the Bragg equation to find the minimum distance between planes of atoms that gives a diffraction pattern with X rays of wavelength $\lambda$.

**10.41** Although long molecules seem best at forming liquid crystals, many of them show no tendency to liquid crystal formation. Suggest a reason why a long molecule such as $CH_3(CH_2)_nCH_3$, where $n$ is at least 20, does not form liquid crystals while a molecule whose chain is much shorter but has N—O groups substituted on it does.

**10.42** Predict how the two types of point defects shown in Figure 10.37 will affect the measured density of a crystalline solid.

# 11

# Water

**Preview**

**W**ater is so common a substance that its extraordinary features often go unnoticed. In this chapter we shall explain why water is one of the most unusual as well as one of the most abundant substances on earth. We begin by explaining why many properties of water, such as its melting point and its boiling point, are out of the ordinary, and we then show how the hydrogen bonding that occurs between water molecules accounts for these properties. We then discuss water as a solvent and go on to talk about the composition of fresh water and seawater in terms of dissolved substances. The chapter concludes with a topic of pressing importance: the water pollution caused by human activities and the techniques used to remove pollutants from water

"**B**lood is a strange juice," Goethe said. Water is nearly as strange. Because water is all around us, we tend to forget that it has very unusual properties indeed, including some that make it unique among liquids. These properties are not just minor curiosities. If water behaved the way other liquids do, the earth would be a much different planet. The earth's inhabitants would also be different — if there were any. As you will learn in this chapter, this is one familiar substance that is also quite extraordinary.

## 11.1   PROPERTIES OF WATER

To start with the obvious, water on earth usually exists as a liquid that appears transparent in small amounts and bluish green in large quantities. At a pressure of 1 atm, the boiling point of water is 100°C (373.15 K) and its melting point is 0°C (273.15 K).

We are not surprised that water and ice exist simultaneously on earth. We should be. Water is the only common inorganic substance that exists as both a liquid and a solid at temperatures that normally occur on the surface of the earth. This unique characteristic arises from a property of water that does not strike us as unusual only because it is so familiar. The boiling point and the melting point of water are much higher than those of similar compounds, notably other hydrides of nonmetals. Table 11.1 lists data for these compounds. It can be seen that methane, $CH_4$ (MW = 16), boils at a temperature 264 K lower than water (MW = 18) and freezes at a temperature 182 K lower than water. Again, $H_2S$, the hydride of the element just below oxygen in group VIA of the periodic table, boils 161 K lower and freezes 86 K lower than $H_2O$. None of the boiling points and melting points of the hydrides in Table 11.1 are close to those of water. The unusually high boiling point and melting point of $H_2O$ result from an unusually high degree of intermolecular attraction in water, which we shall discuss in Section 11.2.

**TABLE 11.1**   Boiling Points and Melting Points of Hydrides of Nonmetals

| Compound | Boiling Point (K) | Melting Point (K) |
|----------|-------------------|-------------------|
| $H_2O$   | 373 | 273 |
| $CH_4$   | 109 | 91 |
| $NH_3$   | 240 | 195 |
| HF       | 293 | 190 |
| $SiH_4$  | 161 | 88 |
| $PH_3$   | 185 | 140 |
| $H_2S$   | 212 | 187 |
| HCl      | 188 | 158 |
| $GeH_4$  | 185 | 108 |
| $SbH_3$  | 256 | 185 |
| $H_2Se$  | 232 | 213 |
| HBr      | 206 | 185 |

It is not only the phase transition temperatures of water that are unusual. The general thermal behavior of water is also distinctive.

We know that when heat is added to a substance, its temperature will increase unless the substance is undergoing a phase transition. The temperature rise that occurs when a given quantity of heat is added to a substance depends on the nature of the substance and the mass of the sample that is being heated. If an unusually large amount of heat is needed to cause a given increase in temperature, we say that a substance has a high heat capacity.

Water has a very high heat capacity. It absorbs a great deal of heat for a given temperature increase and releases a great deal of heat for a given temperature decrease. In addition, water has a very high heat of vaporization compared to similar substances; a great deal of heat must be added to evaporate a given quantity of water. Even the heat of fusion of water is relatively high, so a substantial quantity of heat is needed to melt a given mass of ice. To some degree, the same intermolecular attractions that explain the high boiling point and melting point of water also explain its unusual thermal properties.

## Water and Life on Earth

Life on earth is influenced profoundly by the unusual thermal properties of water. The high heat capacity and high heat of vaporization of water are of major importance in creating favorable conditions for life on our planet. We can start with temperature changes. If the earth had no oceans, it would have extreme temperature fluctuations daily. A great deal of the heat that the earth gets from the sun every day is absorbed by the ocean, whose temperature does not increase appreciably because of the high heat capacity of water. At night, the water that covers more than 70% of our planet releases much of this heat. As a result, temperature fluctuations are minimized.

The effect of water on temperatures on earth should not be underestimated. On the moon, where there is no water, the temperature can vary more than 250 K from day to night. On earth, a variation of 25 K is regarded as large. The large heat of vaporization of water also plays a role in moderating temperature fluctuations. A good deal of the sun's heat goes into the evaporation of water, one of the processes responsible for the conditions called weather.

Evaporation of water is also important as a temperature regulator for the higher animals. The properties of many compounds that are essential to life, such as proteins, are quite sensitive to temperature changes. Humans and other mammals can withstand substantial external temperature changes primarily because they have a system for regulating body temperature that is based on the thermal properties of water. The human body is cooled by the evaporation of water through pores in the skin. A high heat of vaporization means that the evaporation of a relatively small amount of water causes considerable cooling. Therefore, humans and

## HOW TREES GET WATER

Atmospheric pressure can support a column of water less than 10 m high. Yet plants can move water many time higher; the sequoia tree can pump water to its very top, more than 100 m above the ground. Until the end of the nineteenth century, the movement of water in trees and other tall plants was a mystery. Some botanists hypothesized that the living cells of plants acted as pumps. But many experiments demonstrated that the stems of plants in which all the cells are killed can still move water to appreciable heights. Other explanations for the movement of water in plants have been based on root pressure, a push on the water from the roots at the bottom of the plant. But root pressure is not nearly great enough to push water to the tops of tall trees. Furthermore, the conifers, which are among the tallest trees, have unusually low root pressures.

If water is not pumped to the tops of tall trees, and if it is not pushed to the tops of tall trees, then we may ask: How does it get there? According to the currently accepted cohesion-tension theory, water is pulled there. The pull on a rising column of water in a plant results from the evaporation of water at the top of the plant. As water is lost from the surface of the leaves, a negative pressure, or tension, is created. The evaporated water is replaced by water moving from inside the plant in unbroken columns that extend from the top of a plant to its roots. The same forces that create surface tension in any sample of water are responsible for the maintenance of these unbroken columns of water. When water is confined to tubes of very

small bore, the forces of cohesion (the attraction between water molecules) are so great that the strength of a column of water compares with the strength of a steel wire of the same diameter. This cohesive strength permits columns of water to be pulled to great heights without being broken.

The theory has been tested both in the laboratory and in the forest. In laboratory experiments, the pull exerted by evaporating water has been used to draw a fine column of mercury 226 cm high in a fine glass tube, the equivalent of maintaining a column of water about 30 m high. And in 1960, a team of plant physiologists in the jungles of northeastern Australia carried out several experiments on rattan vines, which can climb up to 50 m high in trees. When the base of a vine was severed and then immersed in a basin of water, the vine continued to draw up water at a steady rate. It was found that a rattan vine 2 cm in diameter will take up about 12 cm$^3$ of water per minute indefinitely. The pull exerted by the rattan vine, the sequoia, and the 100-m eucalyptus tree of Australia can be equivalent to a pressure of 100 atm.

The same sort of phenomenon is responsible for much of the water content of soil. The water table ordinarily is rather far below the surface. In gravels and coarse soils, the water moves only a few centimeters, but in fine silts it can move several meters. As water evaporates from the surface, more is drawn up from below. The attractive forces between water molecules can also maintain unbroken columns of water in the soil.

other mammals can maintain body temperature within narrow limits without drinking inordinate amounts of water.

Some organisms do not have such water-based regulatory systems. For example, many marine organisms can function only in a very narrow range of external temperatures. They survive because they live in large bodies of water whose temperature normally varies by only a few degrees. The addition of heated water from industrial processes to natural bodies of water, called thermal discharges or thermal pollution, can cause serious problems for some organisms.

## The Density Behavior of Water

Another unusual property of water is the way that its density changes with temperature. Water is one of the very few exceptions to the generalization that the solid phase of any substance is denser than the liquid phase.

In any mixture of the liquid and solid phases of most substances, the solid sinks to the bottom. Not so with water. When water freezes, it expands. Icebergs float on the surface of the ocean and ice cubes float on the surface of a drink. Ice — water in solid form — is *less* dense than liquid water.

The fact that ice is less dense than liquid water has important implications for living organisms. If the density behavior of water were like that of most liquids, ice would form on the bottoms of rivers, lakes, and ponds in the winter. A heavy coat of ice would interfere with the rich growth of organisms on the bottom and the diversity of life in these bodies of water would be reduced. Instead, the layer of ice that forms on the surface insulates the lower layers of water, helping fishes and other aquatic animals to survive cold weather.

The change in volume of $H_2O$ with the addition of heat is also unusual. Most substances expand when heated. As the temperature of a substance increases, its volume increases and its density decreases. Water follows this rule over much of the temperature range in which it is liquid. But the density of liquid water *increases* as its temperature increases from 273 K (0°C) to 277 K (4°C). In this temperature range, water contracts as its temperature goes up. Above 277 K, water follows the usual rule. Its density decreases as the temperature increases.

Other unusual properties of water include its surface tension and viscosity. All these anomalous properties result from the structure and composition of the water molecule.

## 11.2    THE HYDROGEN BOND

When a hydrogen atom is bonded covalently to an atom of an element with high electronegativity, such as nitrogen, fluorine, or oxygen, the bond is highly polar. There is a partial positive charge on the H atom and a partial negative charge on the atom of the more electronegative element. We can represent these partial charges as $\delta^+$ and $\delta^-$ to show this charge distribution, as for example

$$-\ddot{O}^{\delta-} - H^{\delta+}$$

In Section 8.8, we mentioned that whenever such polar bonds exist, there are dipole-dipole intermolecular attractions, particularly in condensed phases. For several reasons, the dipole-dipole attractions are unusually large for a hydrogen atom in a polar bond. The most important factor is the small size of the hydrogen atom. The dipole-dipole attraction is very large if the atom with the partial positive charge is very close to the atom with the partial negative charge. The closeness of the approach is limited by repulsions between the valence electrons and inner-shell electrons of the atoms.

A hydrogen atom in a polar bond has no inner-shell electrons. Therefore, a highly electronegative atom with a partial negative charge can get closer to a hydrogen atom than to any other atom. The strength of dipole-dipole interactions involving hydrogen atoms thus is appreciably larger than those of other atoms.

In addition, the electronegative atom has one or more nonbonding valence shell electron pairs. One of these electron pairs can be directed toward the positively charged hydrogen atom of another molecule, providing a strong localized negative charge that strengthens the dipole-dipole interaction.

The unusually strong coulombic attraction that can occur between a hydrogen atom bonded to an atom of great electronegativity and the lone pair on such an atom is called a *hydrogen bond*. A hydrogen bond is substantially weaker than a covalent or ionic bond, but it is substantially stronger than the weak interactions described in Section 8.8.

In discussing water, we shall be especially interested in hydrogen bonds between water molecules. A hydrogen atom in one molecule can attract a highly electronegative atom of another molecule. In water, the hydrogen atom of one $H_2O$ molecule attracts the oxygen atom in another $H_2O$ molecule.

This interaction makes for strong cohesive forces between molecules of the substance. Figure 11.1 shows three common examples of this kind of hydrogen bond. The ordinary covalent bond is represented by a solid line, the hydrogen bond by a dashed line. The three atoms of the hydrogen bond — the two electronegative atoms and the hydrogen — usually lie on a straight line.

Intermolecular hydrogen bonds are important for many substances. Water is the most common of these substances. It is also the substance in which the effect of hydrogen bonding is most pronounced. Other substances in which hydrogen bonds are important are ammonia, $NH_3$, and hydrogen fluoride, $HF$. We can predict that the phase transition temperatures of both substances will be relatively high, since energy is required to break the hydrogen bonds. The data in Table 11.1 confirm this prediction. However, the melting point and boiling point of $H_2O$ are still considerably higher than those of $NH_3$ and $HF$.

## Hydrogen Bonding in Water

Before considering hydrogen bonding in other substances, we should ask why hydrogen bonding is especially strong in water. The answer is that $H_2O$ molecules are uniquely suited for the formation of intermolecular

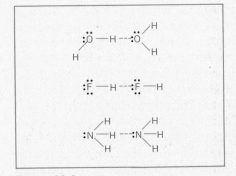

**Figure 11.1**

Three examples of hydrogen bonds between identical molecules. In each example, the hydrogen atom of one molecule forms a hydrogen bond (dashed line) with an electronegative atom of another molecule. All the hydrogen bonds are weaker than the covalent bonds (solid lines).

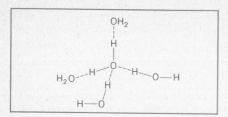

**Figure 11.2**
Why hydrogen bonding is important in water. The oxygen atom of the $H_2O$ molecule has two nonbonding valence electron pairs and can form two hydrogen bonds. Each hydrogen atom can form one bond, making a total of four hydrogen bonds for each $H_2O$ molecule.

hydrogen bonds with each other. The partial positive charge on the H atom in $H_2O$ is substantial because oxygen is highly electronegative; only fluorine is more electronegative. More important, each water molecule has two hydrogen atoms and two nonbonding valence electron pairs, so each water molecule can form four hydrogen bonds, as shown in Figure 11.2. The water molecule acts as both a donor and an acceptor. It supplies hydrogen atoms for two of the bonds and also accepts a hydrogen atom with each of its nonbonding electron pairs. The oxygen atom of a water molecule thus is surrounded by four hydrogen atoms—two of its own and two from two other $H_2O$ molecules. By contrast, ammonia and hydrogen fluoride form only two hydrogen bonds per molecule.

The four hydrogen atoms surrounding the oxygen atom of a hydrogen-bonded water molecule lie at the corners of a tetrahedron. In a large array of water molecules, each oxygen atom is in the center of a tetrahedron whose four corners are defined by four hydrogen atoms. Two of the hydrogen atoms are covalently bonded to the O atom and two are hydrogen-bonded to the O atom, as shown in Figure 11.3. The hexagonal pattern of this structure is reflected in the hexagonal patterns of snowflakes.

## The Crystalline Structure of Ice

The crystalline structure of ice is a hexagonal one that is relatively open compared to the tightly packed crystals of many other substances. The "open" arrangement of the ice crystal maximizes favorable coulombic attractions by permitting the formation of a maximum number of hydrogen bonds. The relatively low density of ice is a result of this open crystal structure.

This open structure is also responsible for the abnormal density behavior of water. When the temperature of ice reaches 273 K, the thermal motion of the $H_2O$ molecules is great enough to start collapsing the open crystal structure by breaking some hydrogen bonds. As the crystal structure collapses, the density increases. The density of ice is 0.9168 $g/cm^3$ and that of liquid water at 273 K is 0.9998 $g/cm^3$. But even when all the ice is melted, there still are some aggregates of $H_2O$ molecules that have open, tetrahedral-type structures. These structures collapse as the temperature rises from 273 K to 277 K. In this temperature range, the density of water increases as the temperature rises. Above 277 K, the normal thermal expansion of water is more important than the collapse of the remaining tetrahedral structures, and density decreases as temperature rises. Figure 11.4 shows how the density of water changes in this temperature range.

If the pressure is increased while the temperature remains below 273 K, the hydrogen bonds begin to bend. When the pressure is great enough, ordinary ice (sometimes called ice I) collapses. It is replaced by forms of ice whose tetrahedral structures are deformed because the hy-

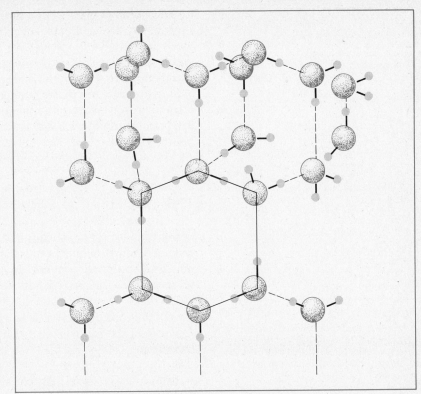

**Figure 11.3**
The hydrogen bonding of ice. Each oxygen atom, represented by a large sphere, is in the center of a tetrahedron whose corners are defined by hydrogen atoms, the small spheres. The overall pattern is hexagonal. Snowflakes are six-sided because their geometry is related to this structure.

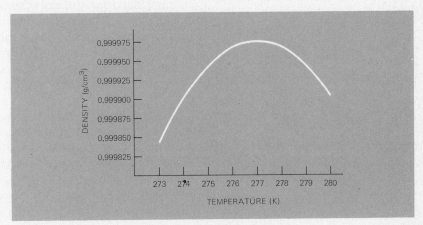

**Figure 11.4**
The density of liquid water increases as temperature rises from 273 K (0°C) to 277 K (4°C) because of continuing collapse of the open tetrahedral hydrogen-bonded structures. Above 277 K, water behaves as other liquids do and expands with increasing temperature.

drogen bonds are bent. The amount of bending depends on the amount of pressure. Seven other forms of ice have been identified as stable phases at pressures from about 2000 atm to 25 000 atm.

These high-pressure forms of ice can be stable at temperatures well above 273 K. Ice VII (density = 1.65 g/cm$^3$) is stable at 373 K, the normal

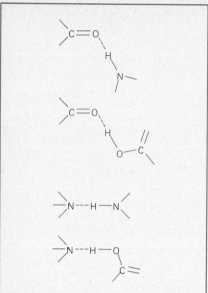

**Figure 11.5**

A variety of hydrogen bonds. Hydrogen atoms that are bonded to oxygen or nitrogen atoms can form bonds (dashed lines) with other O and N atoms, in other molecules or in the same molecule.

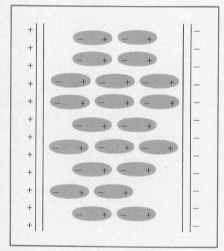

**Figure 11.6**

When a polar substance is placed in an electric field, its molecules are oriented so that their positive ends face the negative pole and their negative ends face the positive pole. This alignment partially neutralizes the field. The dielectric constant of a substance can be regarded as a measure of the extent of neutralization.

boiling point of water, but exists only at pressure above 24 000 atm. Fortunately, the imaginary substance called ice IX, which purports to be stable at a temperature well above 273 K at a pressure of 1 atm, exists only within the pages of Kurt Vonnegut's novel *Cat's Cradle,* in which the world comes to a warm but frozen end.

A hydrogen atom that is bonded to an oxygen or a nitrogen atom will form hydrogen bonds with other oxygen or nitrogen atoms. Such hydrogen bonding can occur between two different molecules. It can also occur between atoms that are in different parts of the same molecule. Figure 11.5 shows some of the possibilities.

## 11.3  WATER AS A SOLVENT

Water is an excellent solvent. It dissolves a wide range of materials that will not dissolve in other liquids. For example, ionic compounds (salts) do not dissolve in such common organic solvents as gasoline, kerosene, turpentine, and cleaning fluids. But many salts dissolve readily in water. So do many inorganic polar substances and even a variety of nonionic organic substances, such as sugars and alcohols of low molecular weight.

For a substance to dissolve in a solvent, some of the things that must happen are

1. The attractive forces between the units of structure of the substance must be overcome, so that it can separate into atomic or molecular units that disperse evenly through the solution.
2. Some of the attractive forces of the solvent must also be overcome, to allow dispersal of the dissolving substance.
3. The "cost" in energy of disrupting these attractive forces must be paid back at least in part by creation of new attractive forces between the solvent and the dissolved substance.

By using this general picture, we can see why water is a good solvent for ionic salts. In these salts, the forces of attraction are the ionic bonds that we discussed in Section 7.1. The ionic bond is the attractive force between charges of opposite sign in a lattice. This force is proportional to the product of the magnitudes of the charges divided by the square of the distance between them. Coulomb's law, which expresses this relationship, can be written to include a proportionality constant:

$$F = \frac{1}{K} \frac{e^+ e^-}{r^2} \tag{11.1}$$

### The Dielectric Constant

The constant $K$ is called the *dielectric constant.* The dielectric constant is a characteristic of the medium in which the charges are found. From Equa-

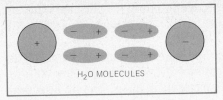

**Figure 11.7**
One reason water is a good solvent for ionic substances. The highly polar $H_2O$ molecules reduce the attractions between cations and anions.

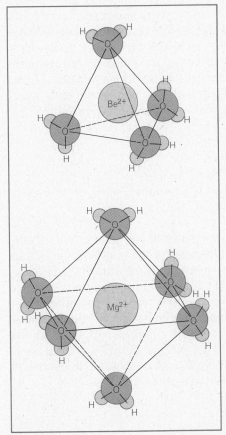

**Figure 11.8**
The interactions of two cations with water. The $Be^{2+}$ ion is surrounded by $H_2O$ molecules, whose oxygen atoms are directed at the cation. The larger $Mg^{2+}$ ion attracts six $H_2O$ molecules. In general, larger cations attract a larger number of $H_2O$ molecules. The lines between the water molecules simply outline geometry; they do not represent forces between the molecules.

tion 11.1, we can see that an increase in the dielectric constant causes a decrease in the force between charges.

The dielectric constant is a measure of the ability of a substance to dissipate an applied electric field. An electric field can be dissipated if the molecules of a substance are oriented so that their positively charged ends face the negative pole of the electric field (which means that their negatively charged ends face the positive pole). The degree to which the electric field is dissipated is related directly to the polarity of the substance (Figure 11.6). Polar substances thus have relatively high dielectric constants. The dielectric constant of a vacuum is 1, and the dielectric constant of air is close to 1. The dielectric constant of water has a high value—about 80.

Water has a high dielectric constant not only because it is a polar substance but also because of its hydrogen bonds. In general, liquids whose molecules form hydrogen bonds have dielectric constants that are higher than might otherwise be expected.

In practical terms, the fact that water has a dielectric constant of 80 means that the attractive forces between the anions and cations in a salt are reduced by a factor of 80 when water is the medium between them. As shown in Figure 11.7, water molecules come between the anions and the cations. The water molecules line up in the electric field, with their negatively charged ends directed at the cations and their positively charged ends directed at the anions. Interposing the water molecules reduces the coulombic attractions between the anions and the cations.

Once the ions are separated, they interact with water molecules in a number of ways. A cation is surrounded by water molecules whose oxygen atoms, which have a relative negative charge, are directed at the positive ion. Because of the favorable ion-dipole interactions in this arrangement, water molecules are held very tightly by cations, especially by bipositive and terpositive cations. These ions are said to be *hydrated*. Usually, a given cation holds a specific number of water molecules. The number is determined primarily by the size of the cation. A small ion, such as $Be^{2+}$, is surrounded by four water molecules. Most cations are larger and can hold more water molcules. For example, $Mg^{2+}$ can accommodate six water molecules, as Figure 11.8 shows.

The number of water molecules held by an ion is called the *ligancy* or coordination number or hydration number of the ion. Thus, the ligancy of $Be^{2+}$ is four and the ligancy of $Mg^{2+}$ is six.

In many cases, the water molecules accompany the cations when ionic solids crystallize from aqueous solution. This water is called the *water of crystallization*. Many crystalline salts, especially those of dipositive and terpositive cations, have stoichiometric quantities of water of crystallization. Their formulas are written to indicate the presence of water, with $H_2O$ separated from the formula of the salt by a dot. Some examples are $BeSO_4 \cdot 4H_2O$; $MgCl_2 \cdot 6H_2O$; and $Fe(NO_3)_2 \cdot 6H_2O$, in which each cation retains all its water of hydration; $FeSO_4 \cdot 7H_2O$, in which an additional water molecule is associated with the $SO_4^{2-}$ anion; and $KAl(SO_4)_2 \cdot 12H_2O$, or alum, in which each of the two cations accommodates six $H_2O$ molecules. Salt crystals can also form in which the water of crystallization has been removed partly or completely, as with $MgSO_4 \cdot H_2O$ and $MgSO_4$.

Attraction between negative ions and water molecules is not nearly as important as attraction between cations and water molecules. Anions generally are much larger than cations, so their charge is more dispersed and the benefit of association with the water molecule is reduced. Nevertheless, anions in solution are surrounded by water molecules, whose positively charged ends point toward the anions. The water molecules may even form hydrogen bonds with the anions.

Nonionic substances such as sugars, alcohols, or amines are soluble in water primarily because of the formation of hydrogen bonds. Molecules of these substances may contain one or more $-\ddot{O}-H$ or $-\ddot{N}-H$ groups whose hydrogen atoms can form hydrogen bonds with water molecules. In addition, the nonionic substances form hydrogen bonds with the hydrogen atoms in water molecules, as Figure 11.9 shows. Some other substances, such as formaldehyde, $H_2C=O$, are soluble in water because the water molecules can form hydrogen bonds with them, although the formaldehyde molecules cannot form hydrogen bonds with each other. The hydrogen bond between formaldehyde and water is represented by a dashed line:

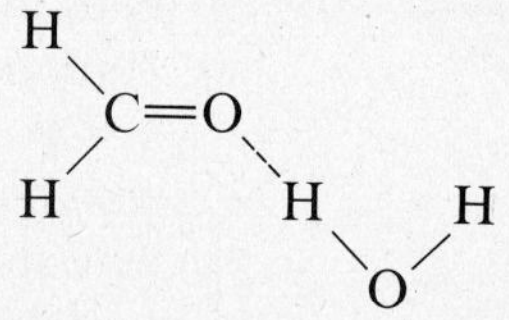

$$
\begin{matrix}
H \\
\quad \diagdown \\
\quad\quad C=O \cdots H \quad H \\
\quad \diagup \quad\quad\quad\quad \diagdown \diagup \\
H \quad\quad\quad\quad\quad O
\end{matrix}
$$

Many substances do not dissolve in water to any great extent. A substance dissolves in water primarily because of the attractive forces that arise when the substance goes into solution. These attractions must be greater than the attractive forces that are lost between the units of structure of the pure substance and between water molecules. A substance will not dissolve in water if the attraction between its units of structure and water molecules are not strong enough. Such is the case with some cations that form salts with substantial lattice energies but do not form good hydrates. The ions in these lattices do not separate in water, so these salts are not soluble in water. Many silver salts are in this class.

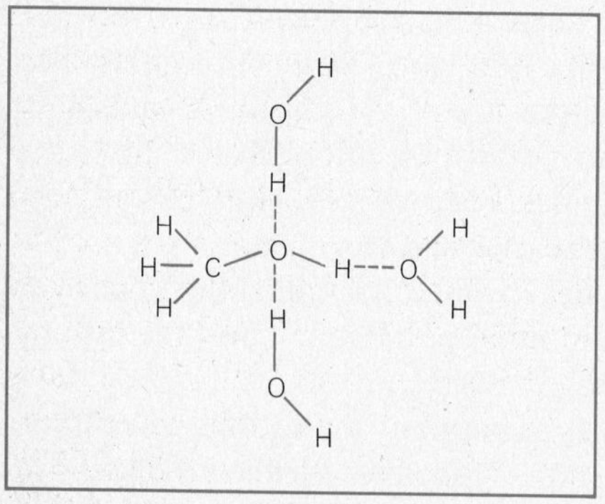

## Clathrates

Nonpolar substances usually do not dissolve in water, because there are no appreciable favorable interactions between the nonpolar molecules and the water molecules. Under special conditions, however, many nonpolar substances of simple structure can form crystalline hydrates with water. One such substance is $CH_4$, which does not dissolve appreciably in water but does form crystals of a hydrate whose composition is $CH_4 \cdot 5\frac{3}{4}H_2O$.

**Figure 11.9**

The solubility of nonionic substances can be explained by the formation of hydrogen bonds. A hydrogen atom of the $CH_3OH$ molecule forms a hydrogen bond with the oxygen atom of a water molecule, while the oxygen atom of the $CH_3OH$ molecule forms two hydrogen bonds with hydrogen atoms of water molecules. The strength of these attractions allows the substance to dissolve.

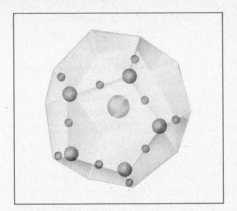

**Figure 11.10**

The structure of xenon hydrate, a clathrate crystal. The water molecules form a hydrogen-bonded three-dimensional network. The xenon atoms occupy the cavities in the structure.

Many other substances, such as the noble gases, simple hydrocarbons, chloroform, chlorine, and bromine, form crystalline hydrates with complex structures. All these hydrates have a hydrogen-bonded tetrahedral array of water molecules similar to that of ice but even more open. These crystal structures can be pictured as a three-dimensional array of cavities whose walls are made of water molecules. Each cavity holds an atom or a molecule of the substance that forms the hydrate. These crystals are called clathrate crystals — *clathrate* is from the Latin word for "cage"— because the molecule of the substance seems to be enclosed in a cage of molecules. Figure 11.10 shows the partial structure of one such hydrate. Clathrate crystals can also be formed by molecules other than water.

Aside from their purely theoretical interest, clathrate hydrates are of practical importance because they may play a role in anesthesia. Some gases, such as the noble gases, are highly effective anesthetics even though they are chemically inert under physiological conditions. It has been suggested that these gases depress the functioning of the nervous system by forming clathrate hydrate crystals in the brain.

## 11.4 THE COMPOSITION OF NATURAL WATERS

Water is almost never found in a pure state because it is such a good solvent. Water dissolves a great variety of substances. It is also a medium for the suspension of many kinds of undissolved particles. The type and amount of substances that are dissolved or suspended in water are a function of the immediate environment.

Any organisms that live in water have a vital interest in the composition of the water surrounding them. Humans do not live in water, but we do use a great deal of water for both personal and industrial purposes. It is important that the composition of water be suitable for these purposes. Failing that, the composition must be known so that the water can be treated to make it usable.

The substances found in water can be divided into two categories: those that are present as a result of natural biogeochemical cycles, such as the ions found in seawater; and those that have been introduced by human activity. Some species, such as phosphates, fit in both categories. The second category includes both substances put into water for useful purposes and those that get into water as a by-product of industrial or municipal activity. Many of the latter substances can harm living things. They are pollutants.

### Seawater

Most of the liquid water on earth is in the oceans. Seawater is a solution of many different substances, including salts, gases, and organic materials.

In discussing the composition of seawater, we should remember that when salts dissolve in water, they exist as dissociated ions. Thus, a solution prepared from equal amounts of NaCl(s) and KBr(s) will contain Na$^+$ and K$^+$ cations and Cl$^-$ and Br$^-$ anions. A solution prepared from equal amounts of KCl(s) and NaBr(s) will have the same composition. The composition of a solution of salts thus is discussed only in terms of the ions in solution.

Seawater contains 11 major ionic species and a large number of minor constituents whose mass is negligible compared to that of the major ions. The quantity of salts dissolved in seawater is measured by *salinity,* which is the total mass in grams of dissolved salts in 1 kg of seawater. Typically, the salinity of seawater is 35 g of salts per kilogram of seawater. Although the salinity of samples of seawater may vary, the *relative* quantities of the major dissolved ionic species in water from the open sea are remarkably invariant and appear to have been so for the past 100 million years. There can, however, be variations in some of the major ionic species, such as Ca$^{2+}$ and Mg$^{2+}$, near land. Table 11.2 lists the relative quantities of the major ionic species in a sample of seawater. The two species that are present in the largest quantities are Na$^+$ and Cl$^-$.

There is no similar constancy of composition for the minor ionic species, such as Ni$^{2+}$, Co$^{2+}$, I$^-$, and many others that are present in very small concentrations. The two reasons for these variations in different samples of seawater are the way in which the ions reach the sea and their role in biological processes.

The relative ionic composition of seawater is similar to that of the blood and other body fluids of many animals. Seawater and the body fluids of marine vertebrates and land animals have roughly the same relative proportions of the major dissolved ions, although seawater has a higher concentration of ions than the body fluids. The body fluids of marine invertebrates such as lobsters and clams resemble seawater closely, both in salinity and in the relative concentrations of dissolved ions. The constancy of ionic concentrations in seawater and in body fluids is one of several pieces of evidence that all life originated in the sea.

**TABLE 11.2    Major Ionic Species in Seawater**

| Ion | g/kg H$_2$O | Percent of Total Salt Content |
| --- | --- | --- |
| Cl$^-$ | 18.98 | 55.04 |
| Na$^+$ | 10.56 | 30.61 |
| SO$_4^{2-}$ | 2.65 | 7.68 |
| Mg$^{2+}$ | 1.27 | 3.70 |
| Ca$^{2+}$ | 0.40 | 1.16 |
| K$^+$ | 0.38 | 1.10 |
| HCO$_3^-$ | 0.14 | 0.41 |
| Br$^-$ | 0.065 | 0.19 |
| Sr$^{2+}$ | 0.013 | 0.04 |
| F$^-$ | 0.001 | 0.003 |

## Fresh Water

Freshwater bodies such as lakes, rivers, and ponds account for less than 0.03% of the earth's surface water. Lakes and rivers do contain dissolved salts. However, the salinity of fresh water is usually much lower than that of the sea, and the major dissolved salts in fresh water often are different from those in seawater. In addition, the composition of individual freshwater bodies is much more variable than the composition of seawater.

Water is regarded as being fresh if it contains less than 0.1% dissolved solids (1.0 g of salts per 1 kg of water), but the usual United States standard for drinking water requires no more than 0.05% dissolved solids. These salts are not necessarily those found in seawater. The major dissolved ions in river water are not $Na^+$ and $Cl^-$, but rather cations such as $Mg^{2+}$ and $Ca^{2+}$ and anions such as $HCO_3^-$ and $SO_4^{2-}$. In addition, silicon is found in fresh water in various forms. Most of these constituents are dissolved from soil in the drainage basins of rivers.

While $Ca^{2+}$ is a major ionic constituent of rivers, it is found in relatively low concentrations in the sea because of the activities of living organisms. The shells formed by marine invertebrates consist mainly of $CaCO_3$. These invertebrates collect $Ca^{2+}$ from seawater to form their shells, thus reducing the concentration of $Ca^{2+}$ in seawater.

The composition of the water in lakes is much more variable than that of either river water or seawater. Some lakes have negligibly small concentrations of dissolved salts. Others, such as the Great Salt Lake and the Dead Sea, are much more saline than the oceans. The composition of lake water depends not only on the type of soil surrounding the lake, which influences the amount of salts washed into it, but also on the age of the lake, the living organisms in the lake, and the rate of evaporation of lake water. Older lakes tend to have a higher concentration of dissolved substances.

## Gases in Water

Natural waters also contain varying amounts of dissolved gases, some of which are of major importance to living organisms. The three most important dissolved gases are $O_2$, $CO_2$, and $N_2$. The first two are essential to marine life. Many problems of water pollution center on the concentrations of dissolved $O_2$ and $CO_2$ in bodies of water.

The concentration of $O_2$ in water is much lower than the concentration of $O_2$ in the air. Air is about 20% $O_2$, while water has no more than 0.0008% to 0.001% of $O_2$ by mass, the equivalent of about 7.5 cm$^3$ of $O_2$ at STP per liter of water. Marine animals that get their oxygen from water have evolved mechanisms for extracting a sufficient supply even though the concentration of oxygen is low. The most widely used mechanism is the gill, in which large amounts of water pass over a relatively thin membrane through which gas exchange takes place.

# EL NIÑO, WEATHER, AND FOOD FROM THE SEA

The cold water of the Peru Current that flows from the Antarctic northward along the western coast of South American wells to the surface just off Peru. The water carries with it a rich supply of phosphates, nitrates, and other nutrients, the remnants of dead marine plants and animals that have accumulated on the ocean bottom. When these nutrients reach the surface, marine plants grow in great profusion. These plants support a food chain consisting of billions of animals, ranging in size from microscopic crustaceans to large fishes. The most numerous of these animals are anchovies, which abound in these waters. More than a fifth of the world's total fish catch is made off the coast of Peru in normal years.

But one year in every seven or eight is not a normal year. Cold water does not well up, and the nutrient supply is reduced drastically. The food chain is broken, and the number of anchovies decreases dramatically. This catastrophic event is called El Niño, the Spanish name for the Christ Child, because it often begins around Christmas time.

For decades, it was believed that El Niño was caused by the weakening of the strong winds that normally blow offshore. According to this theory, these winds normally pushed coastal water seaward, so that cold water welled up to fill the gap. More recently, oceanographers have been able to link El Niño to global patterns of wind and water, and to explain changes in the world's weather by these patterns.

The new theory says that El Niño occurs when the trade winds that blow from the southeast across the Pacific Ocean become weaker than usual. Normally, the winds push surface water away from the western coast of South America, and cold water wells up to replace it. As the winds weaken, the surface water begins to flow toward South America, in the form of a warmer current. This warmer water reaches the South American coast to cause El Niño.

It has been found that this pattern of events can have drastic effects on the earth's weather. For example, an unusually strong reversal of the trade wind pattern occurred in the winter of 1982–83. The reversal produced an exceptionally large body of warm water in the Pacific that led to weather abnormalities around the globe. California and neighboring parts of the far West were struck by a succession of storms that normally would have gone further north, to Oregon and Washington. The northeastern United States had a warmer-than-normal winter, while the cold air that would normally reach that region went to the center of the nation. Elsewhere, there was a severe drought in Australia, the death of millions of birds on Christmas Island in the Indian Ocean caused by an El Niño-like phenomenon, and unusually heavy monsoon rains in southeastern Asia.

The discovery of the large-scale pattern that causes El Niño has enabled oceanographers to give advance notice of the phenomenon. The Peruvian government has taken steps to preserve the anchovy fishery, after several years of reduced yields. But the fishery has not fully recovered, and it is not certain yet whether the anchovies will ever return in their old numbers.

The solubility of a gas in water depends on a number of factors, including the partial pressure of the gas and the temperature of the liquid. The solubility of any gas increases with an increase in pressure and decreases with an increase in temperature. The surface layer of any body of water generally is rich in oxygen, which dissolves from the atmosphere. Water near the surface of the ocean is also rich in oxygen because of phytoplankton, which carry out photosynthesis and release oxygen as a waste product. Typically, surface water contains 5 cm$^3$ of oxygen per liter, while the oxygen content at a depth of 1 km may be 1 cm$^3$/L.

This oxygen is consumed in part by living marine organisms and in part by the decay of dead organisms and of animal wastes. The microbes that decompose these substances consume oxygen. The dumping of raw sewage into rivers and lakes is undesirable because so much oxygen is consumed by the decay of these wastes that there is none left to support higher forms of life.

## 11.5   WATER POLLUTION

Humans produce many different wastes. One kind is domestic sewage, the waste material from our everyday activities. Domestic sewage consists of human excrement and its associated microorganisms, some of which cause disease; food wastes; soap and detergents; and a variety of miscellaneous materials that we wash down the sink, flush down the toilet, or dump down the drain.

Domestic sewage contains organic matter that is decomposed by microorganisms in the water. These microorganisms are of two main types: aerobes, which require free $O_2$ to live, and anaerobes, which do not re-

The activities of a beaver are part of the natural cycle by which bodies of water age. Human activities can speed up the aging process for many bodies of water by adding excess nutrients that cause eutrophication. *(Wide World)*

quire free $O_2$ and may even be harmed by it. Aerobes decompose organic matter into $CO_2$, $H_2O$, and various ions that usually are not harmful. Anaerobes produce substances that are unpleasant or toxic, such as $H_2S$ and $CH_4$.

It is clearly desirable for sewage to undergo aerobic decomposition. But when this happens, substantial amounts of dissolved $O_2$ are consumed. The rate at which the microbial population consumes $O_2$ is called the *biochemical oxygen demand* (BOD). When large amounts of organic wastes are dumped in a body of water, BOD becomes high and $O_2$ is consumed faster than it can be replenished. When oxygen content drops below about 3 $cm^3/L$, all fish die. If the $O_2$ supply is exhausted, only anaerobes remain and the water becomes septic and noxious.

Agricultural wastes are also a source of water pollution. Substances such as fertilizers, pesticides, and animal wastes are carried into bodies of water by surface runoff.

Industrial waste is much more variable in composition and unusual in content. Most domestic sewage and agricultural waste are potentially dangerous because of their sheer volume, which can overwhelm the natural cleansing processes of bodies of water. Industrial waste often consists of synthetic materials that can produce unexpected results when introduced into bodies of water. For example, the first detergents to be put on the market passed through sewage treatment plants untouched and created mountains of foam in rivers and streams. The microorganisms in sewage treatment plants can decompose the natural fats in soap, but they cannot decompose the molecules of the original synthetic detergents, which are derived from petroleum chemicals. The detergent industry had to convert to "biodegradable" products, which can be decomposed by the microorganisms in sewage treatment plants.

Some pesticides, such as the chlorinated hydrocarbons used as insecticides. are almost immune to decomposition by natural methods. The chlorinated hydrocarbons, including DDT, each have a chlorine-carbon bond that is rare in nature and that few microorganisms can break. Relatively small amounts of pesticides, introduced accidentally into rivers, have killed millions of fishes. The chlorinated hydrocarbons are relatively insoluble in water, but they are soluble in fats. Therefore, they tend to accumulate in fatty tissues of animals, where they can build to poisonous concentrations.

The effects of pollutants often are subtle and surprising. What seems to be small quantities of industrial waste may be extremely harmful. An example is mercury-containing waste.

At one time, it was believed that the release of mercury and inorganic mercury products into water was relatively harmless. The mercury was believed to sink harmlessly to the bottom of any body of water. It has now been found that bacteria convert inorganic mercury to organic mercury compounds, such as dimethyl mercury, $(CH_3)_2Hg$, which are extremely toxic. These mercury compounds can be concentrated in body tissues as large animals feed on smaller organisms, until a toxic level of mercury is reached for animals at the top of the food chain.

Other heavy-metal ions in industrial wastes, such as the ions of cadmium, lead, and nickel, can also be retained in body tissues. The concentration of these ions increases as they are passed up through the food chain. Smaller organisms, the primary consumers in the food chain, may not be damaged. But the large predators at the top of the food chain may be poisoned.

### Eutrophication

One effect of domestic sewage, agricultural fertilizers, and some industrial wastes is eutrophication, a serious problem for some lakes. Eutrophication is caused by an excess of phosphates, nitrates, and other nutrients. This excess leads to a number of objectionable developments, including an acceleration of the normal process by which a lake "ages."

When a lake is first formed by geological activity, it consists of relatively pure water that can support very little life. As nutrients from the soil find their way into the lake, more and more organisms can be sustained in the water. As these organisms die, their decay is carried out by aerobic microbes, with the consumption of $O_2$. The solid residue that is left sinks to the bottom to form a sediment. Over many centuries, the sediment formed by dead organic material builds up, slowly filling in the lake until it becomes first a marsh and then finally dry land.

If nutrients such as phosphorus and nitrogen are added to the lake by human activities, the lake is able to support more living organisms. Organic sediments build up faster and the lake grows "old" too quickly. The overgrowth of living organisms causes many other problems. The decomposition of dead organisms can increase the BOD level so much that the $O_2$ content of the water is decreased below the concentration needed to support aerobic organisms. Anaerobes take over, with all the unpleasant consequences they bring with them. The deeper levels of a lake may become completely depleted of oxygen, so that fish habitats are destroyed and desirable species are eliminated. Use of the lake for drinking water or human recreation may become impossible. The best solution seems to be a limit on the addition of phosphates from detergents, nitrates from agricultural fertilizers, and the varied nutrients from human sewage and other sources.

## 11.6   WATER TREATMENT AND PURIFICATION

Water intended for human use must be treated to remove any objectionable or dangerous constituents. Water that has been used for domestic or industrial purposes requires treatment to remove any added impurities. The techniques used in water purification and sewage treatment have many features in common, but there are differences related to the amount and quality of water that is being treated and the purpose of the treatment.

## Drinking Water

Water for human use must not only be safe, it must also be free of odors, color, particulates, and other unpleasant impurities. Excessive "hardness"—the presence of appreciable quantities of dissolved ions—is also undesirable. Water for domestic consumption goes through several treatment processes to meet modern standards.

The simplest way to remove particulate matter is by sedimentation. If water is allowed to stand, most solids settle to the bottom. Sedimentation occurs naturally in reservoirs or in settling tanks in water treatment plants. Solid particles that are too small to settle out can be removed by

(a) In a typical municipal water treatment plant, chemicals are added to soften the water, remove tastes and odors, and kill bacteria. (b) Large paddles are used to insure that the added chemicals mix completely. Impurities in the water combine with the chemicals to form a substance called floc. (c) The floc settles to the bottom. The basin must be drained periodically to remove the sediment. (d) More impurities are removed when the water is allowed to flow through a series of filtering materials. (e) The treated water is stored in a reservoir, where it will be chlorinated again before use. (Photographs taken by Philip R. Pryde. From Lucy T. Pryde, *Environmental Chemistry: An Introduction,* Copyright © 1973 by Cummings Publishing Company, Inc., Menlo Park, California.)

(a)

(b)

(c)

(d)

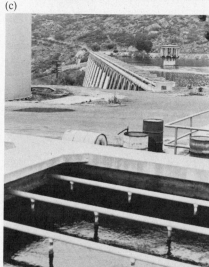

(e)

chemical coagulation. If aluminum sulfate, $Al_2(SO_4)_3$, is added to the water, $Al^{3+}$ ions are released. The ions react with water molecules to form a gelatinous precipitate of $Al(OH)_3(s)$, to which small particulates and other substances become attached. These substances are removed with the precipitate, which is called floc. Water that is treated by this method, called flocculation, may also be passed through filter beds of sand to remove even finer solid particles and other impurities.

There are a number of ways of removing unpleasant tastes and odors from water. Aeration, mixing air with water, is a method that eliminates volatile bad-smelling substances such as $H_2S$. Many organic substances can be removed if water is passed over powdered activated charcoal, which adsorbs the substances. Adsorption is the process by which a substance sticks to a solid surface. Activated charcoal consists of very fine particles of carbon that collectively have a very large surface area and so provide ample surface for adsorption.

Perhaps the best-known method of water treatment is chlorination. Chlorine or chlorine compounds kill microorganisms and oxidize some undesirable organic compounds. Relatively uncontaminated water can be disinfected by less than one part per million of chlorine. Several parts per million may be needed for contaminated water. Chlorine sometimes causes an objectionable smell or taste, which can be eliminated by activated charcoal treatment.

Although chlorination is credited with eliminating epidemics of waterborne diseases such as typhoid, its safety has been questioned recently. Sophisticated instruments have found that chlorination causes the formation of a number of apparently undesirable compounds, such as chloroform and carbon tetrachloride, in drinking water. The concentrations of these compounds usually are no greater than a few hundred parts per billion, and their effect on human health is unknown. However, some communities have turned to an alternative method of disinfection called ozonization, in which ozone, $O_3$, is used instead of chlorine to kill microorganisms.

## Water Softening

While fresh water is much less saline than seawater, it may still contain dissolved ionic material in concentrations that are too high for some uses. Water that contains appreciable quantities of such ions as $Ca^{2+}$, $Mg^{2+}$, $Mn^{2+}$, and $Fe^{2+}$ is called hard water. Hard water makes washing with soap difficult because these ions form an unpleasant, scummy precipitate with soap. Hard water is also an industrial problem. Salts of calcium, such as $CaCO_3$ and $CaSO_4$, tend to precipitate and form scale in boilers, hot water pipes, and water-holding tanks.

A common method of softening hard water uses lime and soda ash to precipitate the calcium and magnesium as carbonate and hydroxide, which can be removed by filtration. Another less common method uses substances called *zeolites,* which soften hard water by a process called *ion exchange.* A zeolite is a mineral that contains aluminum, silicon, oxygen, and a metal such as sodium or calcium. A typical empirical formula for a zeolite is $NaAlSi_2O_6$.

In general, a zeolite is a crystal whose aluminosilicate portions form a lattice structure with sizable vacancies. The metal cations, such as $Na^+$, are found in these corridorlike vacancies, and have considerable freedom of movement. When water flows over a collection of zeolite crystals, dipositive cations in the water displace the $Na^+$ cations from the zeolite crystal. Unlike the dipositive cations that are responsible for the "hardness" of water, the $Na^+$ cations do not form a precipitate with soap or form scale in boilers. Replacing the dipositive cations with $Na^+$ cations "softens" the water, in a process that can be represented by the reaction:

$$2Na^+Z^-(s) + Ca^{2+} \longrightarrow Ca^{2+}(Z^-)_2(s) + 2Na^+$$

where $Z^-$ represents the zeolite framework.

When most of the sodium in the zeolite has been displaced, the zeolite can be regenerated if it is placed in contact with concentrated sodium chloride solution, in the form of brine. When this is done, $Na^+$ ions reenter the zeolite, and the dipositive cations are washed away with the $Cl^-$ anions.

Until recently, water softening was regarded as harmless. But epidemiological studies in Great Britain during the 1960s found a relatively high death rate from heart disease in communities with water softened by this or other processes, compared to communities with hard water. Subsequent studies have produced conflicting results. The role of water softening in heart disease remains a matter of controversy.

Other methods of removing dissolved ionic material are available for processes that require water of exceptionally high purity. Since ionic impurities in water can interfere with or divert the course of chemical reactions, water that is pure enough to drink is sometimes further purified by distillation for laboratory use.

Another method uses an ion exchange resin that incorporates acid groups in its structure. All the cations in water can be exchanged with the $H^+$ ions of the acid. A general representation for one kind of organic acid used in these resins is RCOOH, where COOH represents the acid and R the organic portion. A typical reaction can be represented as:

$$RCOOH(s) + Na^+ \longrightarrow RCOO^-Na^+(s) + H^+$$

Water that is treated in this way can be passed over another resin that incorporates organic basic groups in its structure, to remove all anions and neutralize the $H^+$ ions. The process can be represented as:

$$(RNH_3)^+(OH)^-(s) + Cl^- + H^+ \longrightarrow (RNH_3)^+Cl^-(s) + H_2O$$

This ion removal process can be carried out with columns containing the two kinds of resin in series. The process is expensive and is used only when water of unusual purity is needed, as in pharmaceutical manufacturing.

## Sewage Treatment

Sewage treatment varies greatly from community to community in the United States. Many suburban and rural areas have no central sewage treatment systems and rely on septic tanks. Some communities dump raw sewage into rivers or lakes. A few have sewage treatment systems capable of producing water that is pure enough for swimming or drinking. Most communities are somewhere in between.

Sewage treatment can be classified as primary, secondary, or tertiary. Primary treatment is the removal of suspended and floating solids by screening and sedimentation. Primary treatment collects about 60% of the total solids in sewage, in a form called primary sludge. Usually, sludge is decomposed by anaerobic bacteria. This process produces methane, $CH_4$, which can be burned to supply part of the energy needs of the sewage plant. However, the anaerobic decomposition is not complete. The remaining sludge may be used as fertilizer or landfill, or it may simply be dumped into the ocean.

Secondary treatment uses microorganisms to remove many of the impurities that remain after primary treatment. The sewage can be exposed to the microorganisms in sand beds, in tanks where compressed air is diffused through the sewage, or in oxidation ponds. The microorganisms metabolize the organic matter to such substances as $CO_2$, $NH_4^+$, $NO_3^-$, $PO_4^{3-}$, and $SO_4^{2-}$.

Treated sewage may also be chlorinated to kill the coliform group of bacteria, which can cause disease. Some communities also add another step, tertiary treatment, in which a high percentage of the solids and organic matter remaining after primary and secondary treatment are removed. Tertiary treatment uses a variety of chemical processes, each designed to remove specific impurities. In a few cases, tertiary treatment of sewage produces water that is suitable for human use.

**Summary**

This chapter began by describing some of the **unusual properties** of **water.** It has an unusually high boiling point and freezing point and a very great heat capacity, and it expands when it freezes, so ice is less dense than liquid water. We then explained how these remarkable properties result from the structure and composition of the water molecule. Much of the explanation is based on the **hydrogen bond,** a polar bond that is weaker than a covalent or ionic bond but is nonetheless important. We noted that a water molecule forms unusually strong hydrogen bonds with other water molecules, and that these bonds give ice a relatively open crystal structure. The collapse of this structure causes liquid water to be denser than ice. We said that water is an excellent **solvent** for polar substances because it has a high **dielectric constant,** which is a measure of the ability of a substance to dissipate an applied electric field, and thus to separate ions. We mentioned that some nonpolar compounds can form crystalline hydrates, called **clathrates,** with water. We then described the major ionic species that are dissolved in **seawater** and noted that **fresh water** is defined as water containing less than 0.1% dissolved solids. We noted many gases dissolve in water, and that dissolved oxygen supports marine animals. Finally, we described the nature and source of **water pollutants** and the treatments such as filtration, chlorination, screening, sedimentation, and exposure to microorganisms that are used to purify water.

**Exercises**

---

**11.1** Use the data in Table 11.1 to estimate what the boiling points of $H_2O$, $NH_3$, and HF would be if there were no hydrogen bonding.

**11.2** The metric system originally defined the gram as the mass of 1 cm$^3$ of water at 4°C. Why was 4°C chosen?

**11.3** The electronegativity difference between fluorine and carbon is greater than that between oxygen and hydrogen, yet $CF_4$ is a gas at room temperature while $H_2O$ is a liquid. Explain.

**11.4** Bromine and nitrogen have almost identical electronegativity values. Bromine does not participate appreciably in hydrogen bonding and nitrogen does. Explain this observation.

**11.5[1]** Which of the following com-

pounds can form intermolecular hydrogen bonds? (a) $H_2CO_3$, (b) $H_3PO_2$, (c) $NH_4Br$, (d) $CH_2F_2$, (e) $NH_2OH$.

**11.6** Which of the following compounds can form hydrogen bonds with water? (a) $N_2O_3$, (b) $N_2H_4$, (c) $CH_3OBr$, (d) LiF, (e) BrI.

**11.7** Arrange the following three compounds in order of increasing boiling point and explain your ranking:

$$CH_3NCH_3, CH_3NHCH_2CH_3,$$
$$\underset{\displaystyle CH_3}{|} \qquad CH_3CH_2CH_2NH_2.$$

**11.8** A 0.033-g sample of HF is vaporized at 320 K in a container of volume 0.10 L. The pressure is found to be 0.16 atm, substantially less than the pressure of 0.43 atm calculated from the ideal gas equation. The major reason for the discrepancy is the formation in the vapor of hydrogen-bonded complexes with the formula $H_6F_6$. Cal-

culate the amounts of HF and $H_6F_6$ in the vapor.

**11.9** The compound $NH_5F_2$ is a salt of a monocation and a monoanion. Write the structures of these ions.

**11.10** The salt $NH_4F$ has a crystal structure resembling that of ice, while $NH_4Cl$, $NH_4Br$, and $NH_4I$ have crystal structures resembling the alkali halides. Explain why.

**11.11** Solid HF consists of long zigzag chains made up of many HF molecules. Draw the structure of part of such a chain.

**11.12** Which of the following pairs of compounds can form hydrogen bonds between each other? (a) $NH_3$ and NaF, (b) $CH_4$ and $CH_3F$, (c) $CH_3OH$ and $CH_2O$.

**11.13** The boiling point of HCN is 26°C while the boiling point of ClCN is 13°C. Explain the difference.

---

**11.14*** The formula of oxalic acid is HOOCCOOH. All the C and O atoms have completed octets and there is a C—C single bond. Oxalic acid can form intramolecular hydrogen bonds. Draw the structure of oxalic acid, indicating an intramolecular hydrogen bond.

**11.15** Draw a diagram of a solution of NaCl in water, showing the interaction between the water molecules and the dissolved species.

**11.16** Explain the observation that ammonia is very soluble in water while $NCl_3$ is almost insoluble in water.

**11.17** The oxides of many metals are much less soluble in water than their chlorides, even though oxygen forms much stronger hydrogen bonds than chlorine. Explain why.

**11.18** Methanol, $CH_3OH$, is soluble in water in all proportions, while octanol, $C_8H_{17}OH$, is insoluble in water. Explain why.

**11.19** Predict the solubility of KF in liquid HF. Justify your prediction.

**11.20*** Lithium fluoride is less soluble in water than lithium chloride. Cesium fluoride is more soluble in water

than cesium chloride. Explain why.

**11.21** Describe the geometries of hydrated $K^+$ ion and hydrated $Li^+$ ion.

**11.22** A 1.887-g sample of hydrated $Co_2(SO_4)_3$ is heated to drive off all the water. The mass of salt remaining is 1.665 g. Find the number of water molecules associated with each $Co_2(SO_4)_3$ molecule.

**11.23** What prevents the rapid decomposition of the clathrates of gases such as xenon and krypton? Which of the two would you predict to decompose more readily?

**11.24** The density of seawater is 1.0 $g/cm^3$. Find the volume of seawater that contains 10 g of $Ca^{2+}$.

**11.25** The volume of the Atlantic Ocean is $3.2 \times 10^8$ $km^3$. The concentration of radium in seawater is 0.09 mg per $10^6$ $m^3$ of water. Find the mass of radium in the Atlantic Ocean.

**11.26** Find the amount (in moles) of the seven major ionic species in 1.0 L of seawater whose density is 1.0 $g/cm^3$.

**11.27** Indicate how each of the following substances can harm a natural body of water: (a) sodium nitrate (agri-

cultural waste), (b) phosphates (from detergents), (c) chlorinated hydrocarbons (insecticides), (d) mercury (industrial waste), (e) domestic sewage.

**11.28** Write equations for the precipitation of calcium ion and magnesium ion as carbonates by the addition of a soda to hard water.

**11.29** Write the chemical equation for the regeneration of a zeolite that has been used to remove calcium ions from water.

**11.30*** One major task in the maintenance of a marine aquarium is to prevent the buildup of ammonia or nitrites that form from partial decomposition of metabolic waste products of fishes. When such an aquarium is first established, the ammonia concentration may reach 7 ppm and the nitrite concentration may reach 10 ppm. With time, bacteria flourish that oxidize ammonia and nitrite to nitrate, which is not so harmful to the inhabitants of the aquarium. Calculate the volume of $O_2$ at STP needed to oxidize the ammonia and nitrite in a tank containing 200 L of water when each of these substances reaches its maximum concentration.

# The Properties
# of Solutions

**Preview**     **T**he previous chapter on water appears to be a natural introduction to solutions—but you are reminded that a solution can be gaseous, liquid, or solid. We begin this chapter with the definitions needed to understand solutions, and we next define the expressions commonly used for concentrations of solutions. We then discuss the factors that affect the solubility of solids and gases in liquid solvents and examine the colligative properties of liquid solutions—vapor pressure, boiling point, freezing point, and osmotic pressure—before discussing the properties of solutions of electrolytes. We conclude with a description of the mixtures called colloids.

**P**ure substances are rare. Most materials that are familiar to us are mixtures of two or more substances. One type of mixture that is of special interest is the **solution.** The air, the oceans, and the fluids in our bodies all are more or less solutions.

Most reactions that chemists study and virtually all the reactions that biologists study take place in systems that are solutions or have many of the characteristics of solutions. In this chapter, we shall discuss the different sorts of solutions and their properties.

## 12.1 TERMINOLOGY

*A solution is a homogeneous mixture of two or more substances. Homogeneous* means that a sample taken from any part of a solution has the same relative composition as the entire solution, even if the sample is very small. You probably think of a solution as a liquid that contains dissolved substances. But a solution can be a solid, a liquid, or a gas. The air we breathe is a solution of gases (although air is not entirely homogeneous, since its composition can vary from place to place). Alloys such as brass, bronze, and sterling silver are solid solutions of metals.

In this chapter, we shall deal primarily with solutions that are liquids under the conditions of temperature and pressure at which they are studied. The liquid component, which is usually the major component of such a solution, is called the **solvent.** A substance that is dissolved in the solvent is called a **solute.** A single solution may contain a number of different solutes.

The terminology for a liquid solution is not always clear-cut, especially when all the components of a solution are liquids. For example, 100-proof vodka contains almost equal quantities of water and ethanol. We can ask whether vodka is a solution of water in alcohol or of alcohol in water. The best answer is to avoid the question by saying that vodka is a homogeneous system with two components.

We can use several terms to describe some aspects of the composition of a solution. The **solubility** of a substance is the maximum quantity of the substance that can dissolve in a given quantity of a solvent under a specified set of conditions. A solution that contains the maximum amount of solute under a specified set of conditions is called a *saturated solution.*

For example, the solubility of NaCl in water at 273 K is 35.7 g of NaCl in 100 $cm^3$ of water. If solid NaCl is added to a solution containing 35.7 g of dissolved NaCl in 100 $cm^3$ of water, which is already a saturated solution, no additional NaCl will dissolve. If a solution of NaCl in water has less than 37.5 g of NaCl to 100 $cm^3$ of water, it is said to be *unsaturated.* If solid NaCl is added to an unsaturated solution, more NaCl will dissolve until the solution is saturated.

The concepts of solubility and saturation are useful for solid solutes. They are not always necessary for liquid-liquid solutions, because some liquids are soluble, or miscible, in one another in all proportions. For example, water and ethanol are miscible in all proportions and can form solutions of any desired composition.

## 12.2   EXPRESSIONS OF CONCENTRATION

In Section 3.4, we introduced some methods of expressing concentration of solutions. We shall now discuss these methods in detail.

An expression of concentration is a ratio, or fraction, that describes the quantity of solute in a given quantity either of solvent or of solution. The quantity of solute is the numerator in the fraction. Either the quantity of solvent or the total quantity of solution is the denominator. We can express these quantities in several ways: in units of mass, such as grams; in units of amount, such as moles; or in units of volume, such as liters. Expressions of concentration can be classified by the type of units they employ.

### Mass-Mass Expressions

**Mass fraction** and **mass percent** are expressions that describe the mass of solute in a given mass of *solution:*

$$\text{mass fraction} = \frac{\text{mass of solute}}{\text{mass of solution}}$$

$$\text{mass percent} = \frac{\text{mass of solute}}{\text{mass of solution}} \times 100\% \tag{12.1}$$

These expressions can be used for any kind of solution. Usually, the masses of solute and of solution are expressed in the same units.

**Example 12.1**   An alloy of copper and aluminum is prepared from 65.6 g of Cu and 423.1 g of Al. Calculate the mass percent of each component in this solid solution.

**Solution**   The data tell us that the total mass of the solution is 65.6 g of Cu + 423.1 g of Al = 488.7 g of solution, and that there is 65.6 g of Cu in the solution. Therefore:

$$\text{mass percent Cu} = \frac{65.6 \text{ g Cu}}{488.7 \text{ g solution}} \times 100\% = 13.4\%$$

$$\text{mass percent Al} = 100\% - 13.4\% = 86.6\%$$

Note that the sum of the percents of the components must equal 100%.

For solutions with very low concentrations, units such as parts per million (ppm) or parts per billion (ppb) can be used. One part per million of a solute is $1 \times 10^{-6}$ g of solute in 1 g of solution. One part per billion is $1 \times 10^{-9}$ g of solute in 1 g of solution.

$$1 \text{ ppm} = \frac{\text{mass of solute}}{\text{mass of solution}} \times 10^6$$

$$1 \text{ ppb} = \frac{\text{mass of solute}}{\text{mass of solution}} \times 10^9$$

## Amount-Amount Expressions

**Mole fraction** and **mole percent** are exactly the same as mass fraction and mass percent, except that they use the mole, the unit of amount. Just as the mass of a solution is the sum of the masses of all the components, so the number of moles in a solution is the sum of the number of moles of each component in the solution.

$$\text{mole fraction} = \frac{\text{moles of solute}}{\text{total moles of solution}}$$

$$\text{mole percent} = \frac{\text{moles of solute}}{\text{total moles of solution}} \times 100\% \qquad (12.2)$$

For a two-component solution:

$$X_1 = \frac{n_1}{n_1 + n_2} \qquad X_2 = \frac{n_2}{n_1 + n_2} \qquad (12.3)$$

and $$X_1 + X_2 = 1$$

where $X_1$ and $X_2$ are the mole fractions of the two components, $n_1$ is the number of moles of component 1, and $n_2$ is the number of moles of component 2. These relationships can be extended for solutions with more than two components.

**Example 12.2** Calculate the mole fraction of sugar, $C_{12}H_{22}O_{11}$, in an aqueous solution that is 5.30% sugar by mass.

**Solution** To calculate the mole fraction, we must find the number of moles of sugar and of water in some fixed quantity of solution. From the given mass percent, we can say that there is 5.30 g of sugar in 100 g of solution. Since the solution consists of only sugar and water, 100 g of solution must contain 100 g − 5.30 g = 94.70 g of water. A table of atomic weights shows that the molecular weight of water is 18.01 and the molecular weight of sugar is 342.3. Therefore, in the 100-g sample of solution there is:

$$5.30 \text{ g sugar} \times \frac{1 \text{ mol sugar}}{342.3 \text{ g sugar}} = 0.0155 \text{ mol sugar}$$

and $$94.70 \text{ g H}_2\text{O} \times \frac{1 \text{ mol H}_2\text{O}}{18.02 \text{ g H}_2\text{O}} = 5.26 \text{ mol H}_2\text{O}$$

We can substitute these data into Equation 12.3, which gives the mole-fraction relationships for a two-component solution:

$$X_{\text{C}_{12}\text{H}_{22}\text{O}_{11}} = \frac{0.0155 \text{ mol sugar}}{0.0155 \text{ mol sugar} + 5.26 \text{ mol H}_2\text{O}} = 0.00294$$

$$X_{\text{H}_2\text{O}} = 1 - X_{\text{C}_{12}\text{H}_{22}\text{O}_{11}} = 1 - 0.00294 = 0.99706$$

The mole is just the unit of amount, or of numbers of particles. Therefore, any property of a solution that is determined by the relative number of molecules of solute often can be calculated by the use of mole fractions.

### Amount-Mass and Amount-Volume Expressions

**Molality** is an expression of concentration that differs from those we have discussed because its denominator is a quantity of *solvent,* rather than a quantity of solution. Molality is defined by:

$$m = \frac{n_{\text{solute}}}{kg_{\text{solvent}}} \tag{12.4}$$

where $m$ is the common abbreviation for molality and $kg$ is the mass in kilograms. Molality is the amount of solute (in moles) that is dissolved in 1 kg of solvent.

**Example 12.3**    Calculate the molality of sugar in the solution of Example 12.2.

**Solution**    In Example 12.2 we determined that there is 0.0155 mol of sugar in a quantity of solution that also contains 94.7 g of water. Using Equation 12.4:

$$m = \frac{0.0155 \text{ mol sugar}}{94.7 \text{ g H}_2\text{O}} \times 1000 \frac{\text{g}}{\text{kg}} = 0.164 \frac{\text{mol}}{\text{kg}}$$

The solution is said to be 0.164$m$ in sugar.

Molality is commonly used in the calculation of the boiling point and freezing point of solutions.

**Molarity** is the most commonly used expression of concentration in the research laboratory. It is defined by:

$$M = \frac{n_{\text{solute}}}{V_{\text{solution}}} \tag{12.5}$$

where $M$ expresses molarity and $V$ is the volume of the *solution* in liters.

The molarity of a solute is the amount (in moles) of that solute dissolved in 1 L of solution. Some calculations of molarity were carried out in Section 3.4.

**Example 12.4**    A solution is prepared from 22.2 g of glucose, $C_6H_{12}O_6$, and enough water for a volume of 251 cm³. Find the molarity of the glucose.

**Solution**    To find the molarity, we express the quantity of solute in moles:

$$22.2 \text{ g glucose} \times \frac{1 \text{ mol glucose}}{180.1 \text{ g glucose}} = 0.123 \text{ mol glucose}$$

and the volume of the solution in liters:

$$251 \text{ cm}^3 \times \frac{1 \text{ L}}{1000 \text{ cm}^3} = 0.251 \text{ L solution}$$

Using Equation 12.5,

$$M = \frac{0.123 \text{ mol glucose}}{0.251 \text{ L solution}} = 0.491 \text{ } M$$

To determine molarity, the volume of a liquid must be measured. One reason molarity is used so often is that it is easier to measure the volume of a liquid than to measure the mass of a liquid with reasonable accuracy. (However, it is easier to measure the mass of a liquid with extreme accuracy.) The molarity of a solution changes with temperature, since the volume of a liquid generally increases as the temperature rises and decreases as the temperature falls. This complication is not encountered when we use molality or other mass-mass expressions.

You may occasionally encounter the unit called formality ($F$). This unit exists because some chemists dislike talking about the molarity of a solution of a salt such as NaCl when there are no NaCl molecules in solid NaCl. Aside from the need to calculate the molecular weight of the formula (the formula weight) for formality, there is no difference between formality and molarity. When we say "a $1M$ NaCl solution," we mean that a mass of NaCl corresponding to the sum of the atomic weights of Na and Cl is dissolved in 1 L of solution. We shall not use the term formality.

Another unit that closely resembles molarity is normality ($N$), which was devised to simplify stoichiometric calculations of reactions in solutions. We shall not use this unit.

### Volume-Volume Units

Volume-volume units are convenient for liquid-liquid solutions.

**Volume percent** is similar to mass percent but refers to the volumes rather than the masses of the components. It is defined by:

$$\text{volume percent} = \frac{\text{volume of pure solute}}{\text{sum of the volumes of each component}} \times 100\% \quad (12.6)$$

For a two-component solution:

$$\text{volume percent}_i = \frac{V_1}{V_1 + V_2} \times 100\% \quad (12.7)$$

where $V_1$ is the volume of component 1 when it is pure and $V_2$ is the volume of component 2 when it is pure. In general, the volume of the solution is *not* the sum of the volumes of the individual components. For example, a mixture of 0.500 L water and 0.500 L ethanol occupies a volume of only 0.965 L.

**Example 12.5**    What volume of diethyl ether must be mixed with 0.125 L water to form a solution in which the volume percent of the diethyl ether is 3.2%?

**Solution**    Let $x$ equal the required volume (in liters) of diethyl ether. Then the volume percent of diethyl ether is defined by Equation 12.7 as:

$$\frac{x}{x + 0.125 \text{ L}} \times 100\% = 3.2\%$$

$$x = 0.0041 \text{ L diethyl ether}$$

One volume percent measurement that appears frequently in everyday life is *proof,* which is equal to twice the volume percent of ethanol in water. Thus, a wine that is 12% alcohol by volume is 24 proof. A rum that is 150 proof contains 75% alcohol by volume. A 3.2% beer is 6.4 proof.

## Conversion Between Concentration Units

You will often find it desirable to convert from one unit of concentration to another. For some of these calculations, we must know the relationship between a given volume of solution and its mass. This quantity, which we discussed in Chapter 3, is the density:

$$\text{density} = \frac{\text{mass}}{\text{volume}} \quad (12.8)$$

which is conveniently defined in units of grams per cubic centimeter ($g/cm^3$) for aqueous solutions.

Example 12.6    The concentration of ethanol in the solution called 86-proof vodka is 6.5$M$. The density of the solution is 0.95 g/cm³. Calculate the molality and the mole fraction of ethanol, $C_2H_6O$, in the vodka.

Solution    The mass of 1 L of the solution is found from the density:

$$\text{mass solution} = 0.95 \ \frac{g}{cm^3} \times 1000 \ cm^3 = 950 \ g$$

Since we are told that 1 L of solution contains 6.5 mol of ethanol, and we know that the mass of 1 L of solution is 950 g, we can see that 950 g of solution contains 6.5 mol of ethanol. The molecular weight of ethanol is 46.1, so the mass of 6.5 mol of ethanol is

$$6.5 \ \text{mol ethanol} \times \frac{46.1 \ \text{g ethanol}}{1 \ \text{mol ethanol}} = 299 \ \text{g ethanol}$$

The mass of the water in 950 g of solution can be determined by subtraction:

$$950 \ \text{g solution} - 299 \ \text{g ethanol} = 651 \ \text{g } H_2O$$

We now can calculate the molality, using Equation 12.4:

$$m = \frac{6.5 \ \text{mol ethanol}}{651 \ \text{g } H_2O} \times 1000 \ \frac{g}{kg} = 10 \ \frac{mol}{kg}$$

This solution is 10$m$ in ethanol.

   To calculate the mole fraction of ethanol, we must know the number of moles of ethanol and of water in a given quantity of solution. In 1 L of this solution there is 6.5 mol of ethanol and

$$651 \ \text{g } H_2O \times \frac{1 \ \text{mol } H_2O}{18.0 \ \text{g } H_2O} = 36 \ \text{mol } H_2O$$

Thus    $$X_{\text{ethanol}} = \frac{6.5 \ \text{mol ethanol}}{6.5 \ \text{mol ethanol} + 36 \ \text{mol } H_2O} = 0.15$$

## 12.3    SOLUBILITY

We mentioned in Chapter 11 that several things must happen when a solute, especially a solid, dissolves in a liquid solvent (Figure 12.1). The attractions between the structural units of the solute must be overcome so that the units may be dispersed throughout the solution. Some interactions between solvent molecules must also be overcome to make room for the solute molecules. The cost in energy of overcoming these attrac-

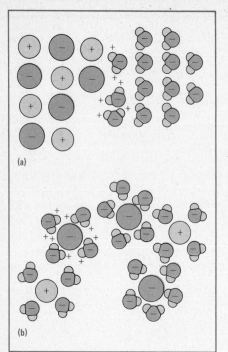

**Figure 12.1**
How an ionic solid dissolves in water. The attractions between the polar molecules and the ions making up the crystal lattice of the solid are greater than the attractive forces that hold the lattice together. The ions become dispersed among the water molecules. Note how the water molecules orient their positive ends toward the negative ions and their negative ends toward the positive ions.

tions is repaid by favorable interactions between solvent and solute molecules and by the advantages that accompany the mixing of substances (Chapter 13).

### Like Dissolves Like

Since at least these three interrelated factors determine solubility, we can make only qualitative predictions about the solubility of one substance in another. The best rule for such predictions is that *like dissolves like,* meaning that a solute tends to dissolve in solvents that are chemically similar to it. The most important characteristics in determining similarity are polarity and electrical properties. The formation of hydrogen bonds also has an important effect on solubility.

Let us study the solubility of substances in three liquids. The first is octane, $C_8H_{18}$, a nonpolar hydrocarbon found in gasoline. The second is water, a highly polar hydrogen-bonding solvent. The third is acetone, an organic solvent with the structure:

$$\begin{array}{ccccc} & \text{H} & :\!\overset{\displaystyle ..}{\text{O}}\!: & \text{H} & \\ & | & \| & | & \\ \text{H}\!-\!&\text{C}&\!-\!\text{C}\!-\!&\text{C}&\!-\!\text{H} \\ & | & & | & \\ & \text{H} & & \text{H} & \end{array}$$

Water and octane are essentially insoluble in one another. Octane is nonpolar and water is polar, and octane cannot form hydrogen bonds, so there is no good way for the molecules of these two substances to interact strongly with each other. There are no favorable solute-solvent interactions to compensate for the solute-solute attractions and solvent-solvent attractions that must be overcome if the solute is to disperse in the solvent. Octane and water are not miscible.

Water and acetone, on the other hand, are miscible. The acetone molecule has a relatively negative oxygen atom that can form a hydrogen bond with the hydrogen of a water molecule, as shown in Figure 12.2. Acetone and octane are also miscible. The interactions between molecules of octane and acetone are similar in magnitude to those between molecules of acetone alone or of octane alone, since acetone cannot form hydrogen bonds with itself. When acetone and octane are mixed, there is nothing to prevent the molecules of the two substances from mixing freely.

The rule that like dissolves like can be applied to predict the solubility of a salt in these three solvents. Lithium chloride, LiCl, is an ionic, highly polar salt. It is most soluble (about 80 g/100 g $H_2O$) in water, the polar solvent, for several reasons. The attractive forces between the anions and the cations of LiCl are reduced in water because of the high dielectric constant of water (Section 11.3). In addition, ion-dipole attractions occur between the $Li^+$ cations and the relatively negative oxygen atoms of the water molecules. Hydrogen bonding between the $Cl^-$ anions and the

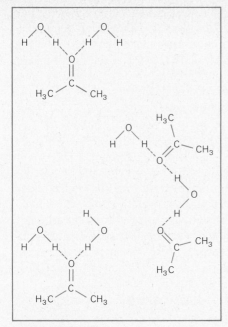

**Figure 12.2**

The miscibility of acetone and water. Hydrogen bonds form between the water molecules and the acetone molecules.

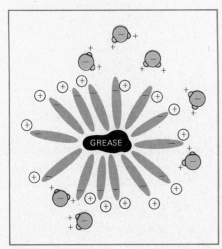

**Figure 12.3**

Soap and/or detergent molecules in action. Each molecule has a polar end and a nonpolar end. The polar end becomes associated with water molecules. The nonpolar end becomes associated with nonpolar substances, such as grease or other forms of dirt. The soap-water-dirt complex is washed away, leaving a clean fabric.

water molecules also occurs. These favorable solute-solvent interactions more than compensate for the loss of the attractions in pure water and pure LiCl.

Lithium chloride is less soluble in acetone (about 5 g/100 g of acetone). The acetone molecules do not interact as favorably with dissolved ions as do water molecules, and acetone has a lower dielectric constant than water. Finally, LiCl does not dissolve at all in octane, because there are no favorable interactions between the charged ions of LiCl and the nonpolar molecules of octane.

The behavior of a nonpolar hydrocarbon solute such as naphthalene, $C_{10}H_8$ (mothballs) is opposite to that of an ionic salt. Naphthalene, a solid, dissolves in octane. The interactions between molecules of naphthalene and molecules of octane are similar in magnitude to those between molecules of naphthalene alone or of octane alone. Naphthalene is less soluble in acetone. There are some polar attractions between molecules of acetone, and naphthalene does not interact well enough with acetone to compensate for the loss of these attractions. Naphthalene is virtually insoluble in water. There are no favorable attractions between water molecules and the nonpolar naphthalene molecules to compensate for the energy required to break the strong hydrogen bonds between the water molecules.

A polar organic solid such as glucose, $C_6H_{12}O_6$, has five OH groups per molecule. It dissolves in water because of the hydrogen bonds that form between the water molecules and these OH groups. Glucose does not dissolve in octane. There are no favorable interactions between glucose molecules and octane molecules to compensate for the energy required to break the hydrogen bonds between glucose molecules in the solid.

The like-dissolves-like rule should not be interpreted to mean that a solute will dissolve in any solvent that is chemically similar. Many ionic substances are extremely insoluble in polar solvents, even water. Many nonpolar substances are extremely insoluble in nonpolar solvents such as octane. The rule tells us only *relative* solubilities. For example, a salt that has low solubility in water generally will be even less soluble in nonpolar solvents.

Soaps and detergents provide an interesting application of the like-dissolves-like rule. A molecule of soap or detergent consists of a long nonpolar hydrocarbon chain with an ionic or hydrogen-bonding group at one end. The nonpolar portion of the molecule tends to dissolve in nonpolar substances, such as grease. The polar end of the molecule tends to dissolve in polar solvents, such as water. In a system containing both water and grease, the hydrocarbon portion of the molecule becomes associated with the grease, while the polar group is associated with the water molecules. The soap or detergent molecule is carried away with the water, carrying grease and other nonpolar dirt with it, as shown in Figure 12.3.

The rule that like dissolves like also helps explain the orientation of large organic molecules in aqueous media — for example, of proteins in a living cell or of fats in the membrane surrounding the cell. Proteins and

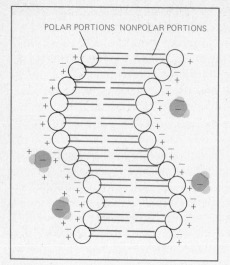

**Figure 12.4**

A large organic molecule such as a protein can have both polar and nonpolar portions. In aqueous solution, the conformation of the protein molecule is influenced by the surrounding polar water molecules. The protein molecule is folded so that its polar portions interact with the water molecules while its nonpolar portions are tucked away.

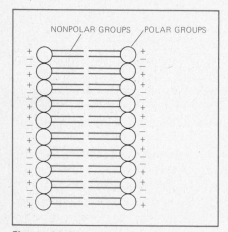

**Figure 12.5**

The schematic structure of a cell membrane. The structural units of the membrane are long-chain fatty acids. The polar ends of the fatty acid molecules are oriented toward the water molecules inside and outside the cell, while the nonpolar ends are in the middle of the membrane.

fats are long molecules with both nonpolar hydrocarbon segments and polar subunits. A protein often folds itself in a way that buries the nonpolar units within the molecule and presents the polar subunits to the water that surrounds the protein, as Figure 12.4 shows. The cell membrane is believed to consist of long-chain phospholipid molecules whose polar ends point toward the aqueous medium outside the membrane and whose nonpolar portions are within the membrane (Figure 12.5).

## Solubility of Ionic Compounds in Water

Much of the chemistry we carry out in the laboratory occurs in aqueous solution. Accordingly, it is useful to have a working knowledge of the solubilities of salts and related compounds in water.

For the sake of convenience, we shall divide solutes into three arbitrary categories: soluble, slightly soluble, and insoluble. We shall regard a substance as *soluble* if more than 1 g dissolves in 100 cm³ of water, as *insoluble* if no more than 0.1 g dissolves in 100 cm³ of water, and as *slightly soluble* if the mass that dissolves in 100 cm³ of water is between 0.1 g and 1 g. The dividing line between slightly soluble and the other categories is understood to be arbitrary.

Some general rules for the solubility of the more common salts and hydroxides at room temperature are listed below. The listing is made easier by the fact that almost all the salts of some anions and cations tend to have the same solubility behavior.

The common salts of the following anions are soluble in water (exceptions are indicated):

1. All nitrates ($NO_3^-$) are soluble.
2. All chlorates ($ClO_3^-$) are soluble.
3. All acetates ($CH_3COO^-$ or $OAc^-$) are soluble.
4. All sulfates ($SO_4^{2-}$) are soluble, except $BaSO_4$, $SrSO_4$, and $PbSO_4$, which are insoluble, and $CaSO_4$, $Ag_2SO_4$, and $Hg_2SO_4$, which are slightly soluble.
5. All chlorides ($Cl^-$), bromides ($Br^-$), and iodides ($I^-$) are soluble, except those of silver ($AgCl$, $AgBr$, $AgI$) and those of mercury(I) ($Hg_2Cl_2$, $Hg_2Br_2$, $Hg_2I_2$), which are insoluble, and those of lead ($PbCl_2$, $PbBr_2$, $PbI_2$), which are slightly soluble.

The salts of some other groups of anions that are usually insoluble are listed below, with the exceptions to the rule:

6. All hydroxides ($OH^-$) are insoluble, except those of group IA metals ($LiOH$, $NaOH$, $KOH$, etc.) and $Ba(OH)_2$, which are soluble, and $Sr(OH)_2$ and $Ca(OH)_2$, which are slightly soluble.
7. All carbonates ($CO_3^{2-}$) and phosphates ($PO_4^{3-}$) are insoluble, except those of group IA metals and $(NH_4)_2CO_3$ and $(NH_4)_3PO_4$, which are all soluble.

**8.** All sulfides ($S^{2-}$) are insoluble, except those of the group IA and IIA metals and $(NH_4)_2S$, all of which are soluble.

**9.** As the above rules show, almost all the common salts of the cations $NH_4^+$, $Na^+$, and $K^+$ are soluble, a point that is worth remembering.

Solubility data for a large number of compounds can be found in standard compilations such as *The Handbook of Chemistry and Physics*, which is published annually by the Chemical Rubber Company.

### Solubility of Gases

Most gases dissolve at least partially in liquids. One reason is that there are almost no intermolecular attractions to be overcome in a gas.

The solubility of a given gas in a given solvent is determined primarily by two factors: the partial pressure of the gas in contact with the liquid and the temperature of the system. In 1803, William Henry (1775–1836) found a simple relationship between the partial pressure of a gas above a liquid and the solubility of the gas in the liquid. There is a direct proportionality between these two measurements that can be expressed for a gas A as:

$$X_A = kP_A \tag{12.9}$$

where $P_A$ is the partial pressure of gas A, $X_A$ is the mole fraction of the gas in solution, and $k$ is a proportionality constant, called the Henry's law constant.

Table 12.1 gives some values of Henry's law constants for gases in water. The higher the value of $k$, the greater is the solubility of the gas. Henry's law is an idealized description that works well only if the partial pressure of the gas is not too high and the gas is not extremely soluble.

Some practical consequences of the increase in the solubility of a gas with an increase in pressure are well known. Carbonated soft drinks are bottled under a pressure of carbon dioxide slightly greater than 1 atm. As

**TABLE 12.1**  Henry's Law Constants for Gases in Water

| Gas | $T$ (K) | $k$ (atm$^{-1}$) |
|---|---|---|
| $O_2$ | 273 | $4.5 \times 10^{-5}$ |
| | 298 | $2.3 \times 10^{-5}$ |
| | 333 | $1.6 \times 10^{-5}$ |
| $N_2$ | 273 | $1.9 \times 10^{-5}$ |
| | 298 | $1.2 \times 10^{-5}$ |
| Ar | 298 | $2.7 \times 10^{-5}$ |
| He | 298 | $6.8 \times 10^{-6}$ |
| $CO_2$ | 273 | $1.4 \times 10^{-3}$ |
| | 298 | $6.1 \times 10^{-4}$ |
| | 333 | $2.9 \times 10^{-4}$ |

long as the cap is on, the pressure of $CO_2$ in the bottle remains greater than 1 atm. When the bottle is opened, the pressure of $CO_2$ above the soda suddenly is lowered to 0.0003 atm, the $CO_2$ pressure of the atmosphere. The solubility of $CO_2$ decreases, and bubbles appear as the gas comes out of solution.

A more dangerous manifestation of the same phenomenon is decompression sickness, commonly called the bends. Large quantities of gases, especially nitrogen, dissolve in the blood of divers when they are subjected to high pressures deep under water. When a diver comes to the surface, the lower pressure decreases the solubility of the gases. If the pressure is lowered too quickly, bubbles of gas can form in body fluids, just as they do in soda when a bottle is opened. The results can be painful or even fatal.

## Effect of Temperature

Temperature usually has an important effect on the solubility of any substance. The effect depends on the nature of the solute and the solvent and on the specific temperature range.

The solubility of most salts in water increases with increasing temperature (Figure 12.6). However, there are many exceptions to this rule. For example, the solubility of sodium sulfate decreases as the temperature goes up. Even different hydrates of the same salt may behave differently. The stable hydrate of sodium carbonate at room temperature is $Na_2CO_3 \cdot 10H_2O$. The solubility of this hydrate increases as the temperature goes up to about 305 K. At this temperature, $Na_2CO_3 \cdot 10H_2O$ is no longer stable. The new stable hydrate is $Na_2CO_3 \cdot H_2O$, whose solubility decreases as the temperature continues to go up.

The solubility of gases usually decreases with an increase in temperature. You can see the effect of temperature on solubility in your own

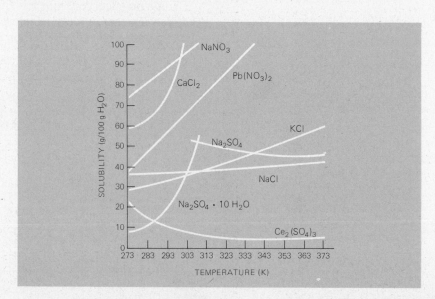

Figure 12.6
Solubility of ionic solids in water at different temperatures.

kitchen. Heat a pot of water on the stove. As the temperature of the water rises, bubbles form. Dissolved air is coming out of solution because the solubility of the air in water has been decreased by the increase in temperature. By the time the water boils, almost all the dissolved air has been expelled. If you turn off the heat, air will dissolve back into the water rather slowly after the water cools. You may notice a subtle sort of "flatness" if you drink some water that was boiled recently. If you put a fish into water that was boiled recently, it may suffocate.

## 12.4   VAPOR PRESSURE OF SOLUTIONS

In Chapter 4, we used the concept of an ideal gas as an approach to a discussion of gases. The behavior of a real gas is often close to that of an ideal gas. Even when real gases deviate from ideal gas behavior, the ideal gas provides a standard against which real gas behavior can be judged.

In our discussion of solutions, we shall use the concept of an **ideal solution** as a model for the behavior of real solutions. The ideal solution is a model in which all the interactions between all the components of the solution are the same. In an ideal solution of A and B, the interactions between A and A, between A and B, and between B and B are the same.

We can come close to this uniformity of interactions if the components of a real solution are quite similar chemically. For example, benzene, $C_6H_6$, and toluene, $C_7H_8$, are very similar chemically. A solution of benzene and toluene has properties that are close to the ideal. At the other extreme, a solution of an ionic solid in water deviates sharply from ideal solution behavior, because water and ionic solids are quite different chemically.

In general, solutions of solids in liquids do not follow the ideal solution model. But if such solutions are dilute, their behavior comes fairly close to that of an ideal solution. In a dilute solution, we deal primarily with solvent-solvent interactions—that is, with interactions between identical molecules. There are only a relatively small number of solvent-solute interactions, and they do not cause much deviation from ideal solution behavior.

One of the most important properties of any solution is the vapor pressure of its volatile components. To describe the relationship between the composition and the vapor pressure of an ideal solution, we start with the vapor pressure of a pure liquid (Section 4.2). The magnitude of the vapor pressure depends on the rate of evaporation and the rate of condensation. Several factors determine these two rates. One factor is the number of molecules on the surface of the liquid.

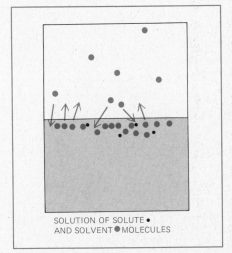

SOLUTION OF SOLUTE ●
AND SOLVENT ● MOLECULES

**Figure 12.7**
The vapor pressure of a solvent is related to the rate at which the solvent evaporates. The evaporation rate, in turn, depends on the number of solvent molecules on the surface of the solution. Solute molecules occupy part of the surface, reducing the evaporation rate of the solvent.

### Raoult's Law

In an ideal solution, part of the surface is occupied by solute particles, as shown in Figure 12.7. The evaporation rate of the solvent depends on the relative number of solvent molecules on the surface. Therefore, the evaporation rate decreases in relation to the fraction of the surface that is occupied by the solute. The portion of the surface that is occupied by solvent molecules is related to the proportion of solvent molecules in the entire solution. Since the relative number of solvent molecules is expressed by the mole fraction of solvent, we can write:

$$r_e = k_e X_A \tag{12.10}$$

where $r_e$ is the rate of evaporation, $X_A$ is the mole fraction of solvent A, and $k_e$ is a proportionality constant.

The rate of condensation of the solvent A is proportional to the partial pressure of A in the vapor above the solution. That is:

$$r_c = k_c P_A \tag{12.11}$$

where $r_c$ is the rate of condensation, $P_A$ is the partial pressure of A, and $k_c$ is another proportionality constant. At equilibrium, $P_A$ is the vapor pressure and the rate of evaporation is equal to the rate of condensation:

$$r_e = r_c$$

or, combining Equations 12.10 and 12.11:

$$k_e X_A = k_c P_A$$

or

$$P_A = \frac{k_e}{k_c} X_A \tag{12.12}$$

Equation 12.12 states that the vapor pressure of a solvent above a solution is directly proportional to the mole fraction of the solvent. We can evaluate the proportionality constant $k_e/k_c$ if we know the vapor pressure that corresponds to a given value of the mole fraction of A. When $X_A = 1$, then $P_A$ is the vapor pressure of the pure liquid, which is written as $P_A^0$. Substituting into Equation 12.12 for this case gives:

$$P_A^0 = \frac{k_e}{k_c} (1)$$

or

$$\frac{k_e}{k_c} = P_A^0 \tag{12.13}$$

Combining Equation 12.13 and Equation 12.12 gives:

$$P_A = P_A^0 X_A \tag{12.14}$$

which states that *at a given temperature, the vapor pressure of a volatile component of a solution is equal to its vapor pressure when pure at this temperature multiplied by its mole fraction in the solution.* This relationship was discovered in 1886 by François-Marie Raoult (1830–1901) and is called **Raoult's law**.

# DISTILLING COGNAC AND ARMAGNAC

The word brandy is said to derive from the Dutch term *brandewijn,* or burnt wine, which refers to the fire used to heat wine in the distillation process. Possibly the two greatest brandies in the world are Cognac and Armagnac, which come from regions of France that are only 130 km apart, and which are made by two different methods of distillation.

By law, Cognac must be distilled twice. The law requires use of a traditional and unique copper pot still. This still consists of a boiler in which the wine is placed, an onion-shaped head, and an elongated pipe called a *col de cygne,* or swan's neck, which leads to a curved copper coil that is surrounded by cold water.

In the first distillation, a fire is lit under the boiler. The volatile elements boil off first and are collected in the condenser, in the form of a milky liquid called the *broulis,* whose alcohol content is about 28%. The *broulis* is about one-third the quantity of the original wine. In the second distillation, the *broulis* is placed in the boiler. The fraction that distills first, called the head, is discarded. So is the fraction that distills last, the tails. Only the "heart" of the second distillation, a clear liquid of about 70% alcohol content, is retained. Several years of aging in oak barrels is required to tame the fiery liquid and to give it the distinctive color and aroma of fine Cognac, which comes from substances leached out of the barrel.

At one time, Armagnac was distilled in small, portable stills that were drawn from farmyard to farmyard by horse. Today, almost all Armagnac is distilled in what is called a continuous still, which was invented in about 1830 by an Irishman, Aeneas Coffey.

In a continuous still, wine passes from a storage tank into an adjoining unit, which consists of two or three boilers above a fire. The volatile fraction of the wine is vaporized by the heat of the fire and rises through several perforated plates, which stop some of the less volatile components. The vapor that condenses has an alcohol content of between 52% and 70%. Once again, several years of aging are required to give the distillate the smoothness and color that are typical of fine Armagnac.

The pot stills of the type used for Cognac once were the only stills available. Their antiquity is evidenced by their traditional name of *alembic,* which comes from the Arabic. Today, almost all brandy and other distilled spirits are prepared in continuous stills. These stills require much less labor than the pot stills, which must be watched round the clock. But the continuous still does not allow the distiller to separate out the best fraction of the distillate, as is possible with the pot still. The best malt whiskey also is prepared in pot stills, rather than in continuous stills.

This derivation of Raoult's law reveals an important characteristic of the vapor pressure of a solvent above a solution. We considered only the *number* of solvent molecules displaced from the surface by solute particles. Assuming ideal solution behavior, the vapor pressure is not influenced by the nature of the dissolved substance. The properties of solutions that depend on the *number* of solute particles, but not on their specific nature, are called **colligative properties.** Vapor pressure is a colligative property.

Raoult's law is exact only for an ideal solution. But it does produce results close to the data observed experimentally for dilute solutions of nonionic solids in liquids and for solutions of two or more chemically similar liquids. As we shall see in Section 12.7, Raoult's law also works well, with suitable modifications, for dilute aqueous solutions of ionic solids.

**Example 12.7**    The vapor pressure of water at 343 K is 0.308 atm. Calculate the vapor pressure of water above a solution prepared from 215 g of water and 25.0 g of glucose, $C_6H_{12}O_6$ (MW = 180.2) at 343 K.

**Solution**    Calculate the amount of each component in the solution, so that the mole fraction of water in the solution can be found:

$$n_{H_2O} = 215 \text{ g H}_2\text{O} \times \frac{1 \text{ mol H}_2\text{O}}{18.02 \text{ g H}_2\text{O}}$$

$$= 11.9 \text{ mol H}_2\text{O}$$

$$n_{glucose} = 25.0 \text{ g glucose} \times \frac{1 \text{ mol glucose}}{180.2 \text{ g glucose}}$$

$$= 0.139 \text{ mol glucose}$$

$$X_{H_2O} = \frac{n_{H_2O}}{n_{H_2O} + n_{glucose}}$$

$$= \frac{11.9 \text{ mol H}_2\text{O}}{11.9 \text{ mol H}_2\text{O} + 0.139 \text{ mol glucose}}$$

$$= 0.988$$

Since the vapor pressure of pure water is given and the mole fraction of water in the solution is known, Raoult's law can be used to find the vapor pressure of water above the solution:

$$P_{H_2O} = 0.308 \text{ atm} \times 0.988 = 0.304 \text{ atm}$$

---

**Example 12.8**    Assuming ideal behavior, what mass of naphthalene, $C_{10}H_8$, would have to be dissolved in 200 g of octane, $C_8H_{18}$, to lower the vapor pressure of pure octane by 20%?

**Solution**    If the vapor pressure of pure octane is $P_{oct}^0$, then the desired vapor pressure above the solution is 80% of $P_{oct}^0$, or 0.80 $P_{oct}^0$. According to Raoult's law:

$$0.80 \ P_{oct}^0 = P_{oct}^0 X_{oct}$$
$$X_{oct} = 0.80$$

$$= \frac{n_{oct}}{n_{oct} + n_{C_{10}H_8}}$$

The molecular weight of octane is 114, and the molecular weight of naphthalene is 128. Thus:

$$0.80 = \frac{\dfrac{200 \text{ g octane}}{114 \text{ g octane/1 mol octane}}}{\dfrac{200 \text{ g octane}}{114 \text{ g octane/1 mol octane}} + \dfrac{\text{mass } C_{10}H_8}{128 \text{ g } C_{10}H_8/1 \text{ mol } C_{10}H_8}}$$

mass $C_{10}H_8$ = 56 g

Solutions consisting of a number of volatile components are interesting for several reasons. Let us compare a solution of ethanol and water, such as a glass of Scotch and water, with a solution of sugar and water. We find only water vapor above the solution of sugar and water. But the vapor above the Scotch and water consists of two gases, ethanol and water (as well as some minor components that give Scotch its smell and that we shall ignore). Given the proper information, we can find the composition of this mixture of gases.

**Example 12.9**     Calculate the composition of the vapor above a solution containing 735 g of water and 245 g of ethanol, $C_2H_5OH$, at 323 K. At this temperature, the vapor pressure of pure ethanol is 0.292 atm and the vapor pressure of pure water is 0.122 atm. Ignore any other gases that may be present.

**Solution**     First calculate the mole fraction of each component:

$$X_{H_2O} = \frac{\dfrac{735 \text{ g } H_2O}{18.0 \text{ g } H_2O/1 \text{ mol } H_2O}}{\dfrac{735 \text{ g } H_2O}{18.0 \text{ g } H_2O/1 \text{ mol } H_2O} + \dfrac{245 \text{ g ethanol}}{46.0 \text{ g ethanol}/1 \text{ mol ethanol}}}$$

$$= 0.885$$
$$X_{ethanol} = 1 - X_{H_2O} = 1 - 0.885 = 0.115$$

According to Raoult's law:

$$P_{H_2O} = (0.122 \text{ atm})(0.885) = 0.108 \text{ atm}$$
$$P_{ethanol} = (0.292 \text{ atm})(0.115) = 0.0336 \text{ atm}$$

The vapor above the solution is a solution of ideal gases, so we can apply Dalton's law. The mole fraction of each component in the vapor can be calculated directly, since the partial pressure of a gas is directly proportional to the amount (moles) of the gas:

$$X_{H_2O(g)} = \frac{P_{H_2O(g)}}{P_{H_2O(g)} + P_{ethanol(g)}}$$

$$= \frac{0.108 \text{ atm}}{0.108 \text{ atm} + 0.0336 \text{ atm}} = 0.763$$

$$X_{ethanol(g)} = 1 - 0.763 = 0.237$$

Example 12.9 illustrates an important point: We find a higher concentration of the more volatile component in the vapor than in the solution. The mole fraction of ethanol in Example 12.9 increases from 0.115 in solution to 0.240 in the vapor. This phenomenon is the basis of an important set of methods of separation and purification called *distillation*.

If the vapor above the solution described in Example 12.9 is condensed, it gives us a new solution that has the same composition as the

462

CHAPTER 12 The Properties of Solutions

vapor; that is, $X_{H_2O} = 0.763$ and $X_{ethanol} = 0.273$. There will be vapor above this new solution, since part of the solution will evaporate. We can repeat the calculation of Example 12.9 to find the composition of this vapor. By Raoult's law (Equation 12.14):

$$P_{H_2O} = (0.122 \text{ atm})(0.763) = 0.0931 \text{ atm}$$
$$P_{ethanol} = (0.292 \text{ atm})(0.237) = 0.0692 \text{ atm}$$

The mole fraction of ethanol in this vapor is

$$X_{ethanol} = \frac{0.0692 \text{ atm}}{0.0692 \text{ atm} + 0.0931 \text{ atm}} = 0.426$$

and the $X_{H_2O} = 1 - 0.426 = 0.574$.

If this vapor is again removed and condensed, we have a new solution, further enriched in the more volatile component, ethanol. By carrying the solution through many of these evaporation-condensation-evaporation steps, we can remove the more volatile component in extremely pure form.

Distillation is used for many purposes. In the chemical laboratory a distillation apparatus can be used for the slow but highly efficient separation of small amounts of material. The use of a still to convert large quantities of wine (about 25 proof) to brandy, a more concentrated solution (about 85 proof) of ethanol in water, is also distillation.

## 12.5 BOILING POINT AND FREEZING POINT OF SOLUTIONS

The phase transition temperatures of a substance are closely related to its vapor pressure, so the boiling point and freezing point of liquid solutions are also colligative properties.

The boiling point is defined as the temperature at which the vapor pressure of the liquid is equal to the external pressure. The vapor pressure of water is 1 atm at 373 K, and this temperature is the normal boiling point of water. But an aqueous solution of a nonvolatile solute that is heated to 373 K at 1 atm does not boil. The vapor pressure of water above the solution is always lower than the pressure of pure water, as Figure 12.8 shows. At 373 K, the vapor pressure of water above a solution is less than 1 atm.

The temperature at which an aqueous solution boils depends on the vapor pressure of water above the solution. The vapor pressure of water is determined by the concentration of solute in the solution. As the concentration of solute goes up, the vapor pressure of water goes down and the boiling point of the solution goes up. In other words, an increase in the concentration of solute means a boiling point *elevation*.

In an ideal solution, the magnitude of the boiling point elevation for the solvent depends on two factors: the nature of the solvent and the

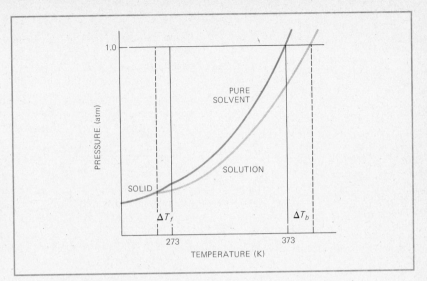

**Figure 12.8**
Pure water that is heated to 373 K at 1 atm will boil. An aqueous solution of a nonvolatile solute heated to the same temperature at the same pressure will not boil. The boiling point is the temperature at which the vapor pressure equals the external pressure, and the vapor pressure of a solution is lower than the vapor pressure of pure water at any given temperature. Pure water that is cooled to 273 K freezes, while an aqueous solution does not freeze until its temperature falls below 273 K.

number of solute particles (that is, the concentration of the solute). This relationship usually is written in the approximate form:

$$\Delta T = K_b m \qquad (12.15)$$

where $\Delta T$ is the difference between the boiling point of the pure solvent and the boiling point of the solvent in solution, $m$ is the molality of solute, and $K_b$ is a proportionality constant whose value is determined by the nature of the solvent. The proportionality constant $K_b$ is called the *molal boiling point constant*. It is a measure of the way in which the boiling point of a solvent changes in response to the addition of a solute. Values of $K_b$ for various solvents can be found in Table 12.2.

Equation 12.15 works best for dilute solutions of nonvolatile, nonionic solutes in liquids. With some modifications, it also works for dilute aqueous solutions of ionic solids.

**TABLE 12.2   Molal Boiling Point and Freezing Point Constants**

| Solvent | $K_b{}^a$ | B.P. (K) | $K_f{}^a$ | M.P. (K) |
|---|---|---|---|---|
| water | 0.512 | 373.15 | 1.86 | 273.15 |
| chloroform | 3.63 | 334.9 | | |
| benzene | 2.53 | 353.3 | 4.90 | 278.7 |
| carbon tetrachloride | 5.03 | 349.7 | | |
| acetic acid | 3.07 | 391.1 | 3.90 | 298.8 |
| ethanol | 1.22 | 351.7 | | |
| *n*-octane | 4.02 | 398.8 | | |
| naphthalene | | | 6.8 | 353.7 |
| camphor | | | 40.0 | 453.0 |

$^a$ The units of these constants are K kg$_{solvent}$/mol$_{solute}$.

**Example 12.10**    Using the data in Table 12.2, calculate the boiling point of a solution of 18.2 g of DDT ($C_{14}H_9Cl_5$), a nonvolatile nonionic substance, in 342 g of chloroform, $CHCl_3$.

**Solution**    First calculate the molality of DDT in the solution. The molecular weight of DDT is 354.5. The molality is given by:

$$\frac{18.2 \text{ g DDT}}{342 \text{ g CHCl}_3} \times \frac{1 \text{ mol DDT}}{354.5 \text{ g DDT}} \times 1000 \frac{\text{g}}{\text{kg}} = 0.150m$$

From the molality calculated here and the value of $K_b$ for chloroform in Table 12.2, we can use Equation 12.15 to calculate the boiling point elevation of chloroform that is caused by the addition of 18.2 g of DDT:

$$\Delta T = (3.63)(0.150) = 0.54 \text{ K}$$

The boiling point of pure chloroform, given as 334.9 K in Table 12.2, is raised by 0.54 K in this solution. The boiling point for the chloroform in the solution therefore is

$$334.9 \text{ K} + 0.54 \text{ K} = 335.4 \text{ K}$$

**Example 12.11**    Using the data in Table 12.2, calculate the mass of glucose, $C_6H_{12}O_6$, which must be dissolved in 100 g water to raise the boiling point of the water to 374.20 K.

**Solution**    One way to perform this calculation is to combine the definition of molality given in Equation 12.4 with the definition of boiling point elevation given in Equation 12.15:

$$\Delta T = K_b \frac{n_{C_6H_{12}O_6}}{kg_{H_2O}}$$

Substituting the given data into this expression:

$$374.20 \text{ K} - 373.15 \text{ K} = 0.512 \frac{n_{C_6H_{12}O_6}}{100 \text{ g H}_2\text{O}} \times 1000 \frac{\text{g}}{\text{kg}}$$

$$n_{C_6H_{12}O_6} = 0.205 \text{ mol } C_6H_{12}O_6$$

The molecular weight of glucose is 180.2, and the required mass of glucose therefore is:

$$\text{mass glucose} = 0.205 \text{ mol glucose} \times 180.2 \frac{\text{g glucose}}{\text{mol glucose}}$$

$$= 36.9 \text{ g glucose}$$

## The Freezing Behavior of Solutions

The freezing behavior of liquid solutions is more complicated than the boiling behavior. It is well known that the freezing point of a solution of a nonvolatile solute is lower than the freezing point of the pure solvent. The salt water of the sea does not freeze as readily as freshwater lakes and rivers do. Let us examine how a solution of sodium chloride in water behaves as the temperature drops.

A solution of sodium chloride in water does not freeze when the temperature reaches 273 K, the freezing point of pure water. The solution freezes at a lower temperature that depends on the concentration of NaCl in the solution. When the appropriate temperature is reached, ice starts to form. This ice consists of pure $H_2O$. The remaining solution becomes more concentrated, since it contains the same amount of NaCl in a smaller amount of liquid $H_2O$.

Pure water maintains a constant temperature as it freezes, but a solution behaves differently. The temperature of a solution drops steadily as ice freezes out. The freezing point of the solution decreases as the concentration of the solute increases. Eventually, the solution becomes saturated in NaCl, and its temperature can go no lower. The solution becomes saturated at 252 K, which is called the *eutectic temperature* of the solution.

If the solution is maintained at the eutectic temperature, a solid with the composition $NaCl \cdot 2H_2O$ crystallizes out of the solution. Eventually, the entire solution freezes, giving a solid called the eutectic solid. This solid consists of a mixture of ice and $NaCl \cdot 2H_2O$.

The lowered freezing point in a solution of salt and water is related to the lower vapor pressure of a solvent above a solution. The freezing point of a liquid at 1 atm usually is close to the temperature at which both the liquid and the solid have the same vapor pressure. As Figure 12.8 shows, this is the temperature at which the vapor pressure curves of the liquid and the solid cross. Because the vapor pressure of the solution is lower than that of the pure solid at any given temperature, the vapor pressure curve of the solution crosses that of the solid at a lower temperature than does the curve of the pure liquid, as Figure 12.8 shows. We speak of a freezing point *depression.* It is defined by an approximate relationship almost identical to that for the boiling point elevation:

$$\Delta T = K_f m \qquad (12.16)$$

which differs from Equation 12.15 only in the proportionality constant $K_f$. This proportionality constant is called the *molal freezing point constant,* and it is characteristic of the solvent. Values of $K_f$ are found in Table 12.2. It should be noted that $\Delta T$ is the absolute value of the difference between the freezing point of pure liquid and the freezing point of liquid in a solution.

Generally, the freezing point of a solvent is more sensitive than the boiling point to the addition of solute. For this reason, and because freezing points usually can be measured more accurately than boiling points, the freezing point is often used in studies of solutions.

**Example 12.12**

The Rast method, one of the older techniques for finding the molecular weight of an unknown, measures the freezing point depression of solutions in camphor. Camphor is used because its freezing point is very sensitive to added solute. A solution of 2.342 g of an unknown substance in 49.88 g of camphor freezes at 441.2 K. Using the data from Table 12.2, calculate the molecular weight of the unknown.

**Solution**

We can find the molality of the solution from the freezing point of the solution, the freezing point of pure camphor, and the value of $K_f$ given in Table 12.2 by using the relationship expressed in Equation 12.16:

$$453.0 \text{ K} - 441.2 \text{ K} = (40.0)m$$
$$m = 0.295$$

From the molality, we can find the amount of solute in any mass of solvent. In this solution, the mass of solvent is 49.88 g. Therefore:

$$0.295m = \frac{n_{solute}}{49.88 \text{ g camphor}} \times 1000 \frac{\text{g}}{\text{kg}}$$
$$n_{solute} = 0.0147 \text{ mol solute}$$

Since 2.342 g of solute is 0.0147 mol, the molar mass of the solute is

$$\frac{2.342 \text{ g}}{0.0147 \text{ mol}} = 159 \frac{\text{g}}{\text{mol}}$$

The molecular weight is 159.

One mole of solute does not necessarily form one mole of particles in solution. The solute may dissociate partially or completely to form more than one mole of particles; or some molecules of solute may come together, or associate, partially or completely to form less than one mole of particles. Information about the behavior of a solute sometimes can be obtained from the colligative properties of the solution. Often, this sort of information about solute behavior cannot be obtained easily by any other method.

The simplest way to determine the number of solute particles in a solution is to measure the molecular weight of the solute when it is in solution. This "apparent" molecular weight often differs from the molecular weight calculated from the chemical formula. It is found from appropriate measurements of a colligative property of a solution of known composition. If there is neither association nor dissociation, the apparent molecular weight of the solute in solution will be very close to the molecular weight that is calculated from the formula of the solute. If there is dissociation, the apparent molecular weight in solution will be lower than the molecular weight calculated from the formula. If there is association, the apparent molecular weight will be higher than the one calculated from the formula.

**Example 12.13**   A solution of 1.43 g of acetic acid, $CH_3COOH$, in 12.3 g of benzene freezes at 273.9 K. Use the data in Table 12.2 to determine the behavior of acetic acid dissolved in benzene.

**Solution**   We can use the method of Example 12.12 to calculate the apparent molecular weight of acetic acid in benzene solution:

$$(278.7 - 273.9) \text{ K} = (4.90)m$$
$$m = 0.980$$

$$0.980 = \frac{n_{solute}}{12.3 \text{ g benzene}} \times 1000 \frac{g}{kg}$$

$$n_{solute} = 0.0121 \text{ mol solute}$$

$$\frac{1.43 \text{ g acetic acid}}{0.0121 \text{ mol acetic acid}} = 118 \frac{g}{mol}$$

The apparent molecular weight of the acetic acid in benzene solution is 118. The molecular weight calculated from the formula of acetic acid, $CH_3COOH$, is 60. The discrepancy is caused by the strong tendency of acetic acid molecules in solution to associate in pairs, called dimers, by hydrogen bonding. Each pair of molecules then acts as a single solute particle. The structure of the hydrogen-bonded dimer is

## 12.6   OSMOTIC PRESSURE

Another important colligative property is the **osmotic pressure** of a solution, which can be observed under special circumstances. The observation requires a *semipermeable membrane,* either natural or synthetic, which allows some substances but not others to pass through it.

The mechanisms by which semipermeable membranes work are not always well understood, but the practical uses of such membranes are of great interest. Consider a semipermeable membrane that separates an

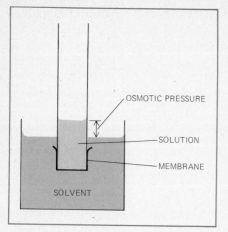

**Figure 12.9**
Osmosis. The membrane allows the passage of solvent but not solute. The level of the solution in the tube rises until the pressure on the membrane from within the tube is great enough to cause water to flow out of the tube at the same rate water flows into the tube. The height of the column of the solution in the tube is a measure of osmotic pressure.

aqueous solution from pure water (Figure 12.9). If the membrane were not there, the solution and the water would mix. Pure water would pass into the tube containing the solution and solution would pass out of the tube, until the concentration of the solute would be the same throughout the liquid phase. We would have another example of the natural tendency of substances to mix, which we shall discuss in detail in Chapter 13.

The semipermeable membrane allows water to flow through it in either direction, but it does not allow solute to pass through it. Pure water flows into the tube containing the solution, solute does not pass out, and the level of the water in the tube rises. It might seem that the water in the tube would never stop rising, since the solution inside the tube can never be as dilute as the pure water outside. But as the level of water rises in the tube, water begins to flow more rapidly the other way, out of the tube. Eventually, the column of solution is high enough so that its pressure causes the rate of flow of water out of the tube to equal the rate of flow into the tube. The level of solution no longer rises. The pressure at which the rates of flow into and out of the solution are equal is called the **osmotic pressure** of the solution.

There are several ways to measure the osmotic pressure of a solution. We can measure the height of the column of liquid when it no longer rises. As we saw in Section 4.2, such a measurement is readily converted to a pressure. Alternatively, we can apply an external pressure just great enough to stop the net flow of water through a semipermeable membrane into the tube. Measuring this applied pressure gives the osmotic pressure of the solution.

The osmotic pressure of dilute solutions is a colligative property. That is, osmotic pressure depends on the amount of solute in the solution but not on the nature of the solute. The relationship between the osmotic pressure of a solution that is dilute enough to follow ideal behavior and the composition of the solution is given by the formula:

$$\pi V = nRT \tag{12.17}$$

where $\pi$ is the osmotic pressure in atmospheres, $V$ is the volume of the solution in liters, $n$ is the amount (in moles) of solute particles in this volume of solution, $T$ is the absolute temperature, and $R$ is 0.0821 L atm mol$^{-1}$ K$^{-1}$, the universal gas constant. The similarity between this relationship and the ideal gas equation is striking. Since the quantity $n/V$ is an expression of concentration, Equation 12.17 can also be written:

$$\pi = cRT \tag{12.18}$$

where $$c = n/V$$

Osmotic pressure is a colligative property of great interest because of the role it plays in the basic processes of life. Osmosis takes place in all living organisms. The wall of a living cell is a semipermeable membrane;

# THE OSMOSIS INDUSTRY

Over the past two decades, industrial and biomedical use of semipermeable membranes has built up an industry whose estimated annual sales exceed $500 million. Both sales and the number of uses are continuing to grow, as designers of semipermeable membranes develop new practical applications for the phenomenon of osmosis.

In the medical field, more than 50 000 Americans with kidney failure now are kept alive by "artificial kidneys," which use semipermeable membranes to purify the blood of waste products that would otherwise cause death. In a process called dialysis, blood is run through a tube made of a semipermeable membrane that is immersed in a solution whose composition is close to that of normal body fluid. Because the concentrations of electrolytes are the same in the blood and in the dialysis solution, they do not cross the membrane. Impurities do because they are in higher concentration in the blood than in the dialysis solution.

Another major use of semipermeable membranes is in water desalination plants. Such plants use reverse osmosis, in which a stream of saline water is fed into a semipermeable membrane that does not allow salts to pass. Fresh water is forced through the membrane by high pressures, while the salts remain behind. The largest reverse-osmosis, membrane-based water purification plant supplies 120 million liters a day of fresh water for Saudi Arabia. A reverse-osmosis plant that would desalinate nearly 400 million liters of water from the Colorado River is being planned.

Semipermeable membranes also are being used for gas separation. Membranes that separate one or more gases from gaseous solutions are more complex than those used for fluids, and the mechanism by which they work is not entirely understood. It is believed that gas molecules dissolve in the membrane, and that different gases work their way through the membrane at variable rates. Typically, a membrane for gas separation consists of a thick, porous layer that provides mechanical strength and a thinner, more selective layer through which somes gases pass more readily than others. One such system is used as part of a process for increasing natural gas production by pumping carbon dioxide into wells to force the gas to the surface. The resulting gas stream is put through a semipermeable membrane that separates the carbon dioxide from the natural gas.

Semipermeable membranes are also expected to play a major role in future applications of biotechnology. For example, the use of bacteria that have been genetically engineered to produce specific substances such as hormones is expected to become a major growth industry. One proposal is to house these bacteria within semipermeable membranes that can be used to control the flow of nutrients and to keep out contaminants.

Osmosis has even been proposed as a way of meeting future energy needs. The proposal is to use giant semipermeable membranes to raise columns of water in a suitably enclosed portion of the ocean. The elevated water would fall through turbines, generating electricity. However, major questions remain to be answered before osmotic power can be generated on a large scale.

one of its roles is to keep some substances out of the cell while allowing others to enter.

In the laboratory, the measurement of osmotic pressure is often used to study the properties of substances with very high molecular weights, such as proteins and nucleic acids. Because of their high molecular weights, it is difficult to get substantial molar quantities of these substances into solution. But a solution of a relatively small quantity of such a dissolved solute develops a large osmotic pressure. For example, a solution that contains 0.1 mol of solute in 1 L of solution (0.1$M$) has an osmotic pressure of 2.46 atm at 300 K. This osmotic pressure is equivalent to the pressure of a column of water more than 20 m high. Measurements of osmotic pressures of solutions as dilute as $10^{-4}M$ can be made with great accuracy.

**Example 12.14**    A solution of 4.68 g of hemoglobin, the oxygen-carrying protein in red blood cells, in 0.125 L of water is found to have an osmotic pressure of 0.0135 atm at 300 K. Calculate the molar mass of hemoglobin.

**Solution**    All the data needed to calculate the amount (in moles) of hemoglobin in the solution are given. Using the relationship expressed in Equation 12.17:

$$(0.0135 \text{ atm})(0.125 \text{ L}) = n \left( 0.0821 \frac{\text{L atm}}{\text{mol K}} \right) (300 \text{ K})$$

$$n = 0.0000685 \text{ mol hemoglobin}$$

Since this amount of hemoglobin has a mass of 4.68 g, the molar mass of hemoglobin is

$$\frac{4.68 \text{ g hemoglobin}}{0.0000685 \text{ mol hemoglobin}} = 68\,300 \frac{\text{g}}{\text{mol}}$$

## 12.7  SOLUTIONS OF ELECTROLYTES

So far, our discussion of colligative properties has focused on solutions of substances that do not ionize. In general, the number of solute particles in such solutions is the same as the number of particles of solute that are dissolved. In some compounds, such as organic acids, association can occur, reducing the number of solute particles.

But the most important solutions with which we shall deal are aqueous solutions of substances that dissociate partially or completely into ions. The colligative properties of these solutions are affected by the increase in the number of solute particles that results from dissociation. These solutions deviate from ideal behavior at much lower concentrations than solutions that do not produce ions. In addition, the electrical properties of water are changed substantially by ionic solutes.

Solutes can be classified by the electrical conductivity of their aqueous solutions. Figure 12.10 shows a simple apparatus for measuring the electrical conductivity of a solution. Using such an apparatus, we find that pure water is a very poor conductor of electricity. Aqueous solutions of substances that do not dissociate into ions are also poor conductors of electricity. These substances, which include most organic compounds and gases such as $O_2$ and $N_2$, are called **nonelectrolytes.**

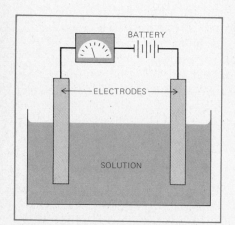

**Figure 12.10**

An apparatus for demonstrating that a solution conducts electricity. The meter measures the current that passes between the electrodes through the solution.

Substances that dissociate into ions in aqueous solution are classified as **electrolytes.** The electrical conductivity of an aqueous solution of an electrolyte usually is much greater than that of water, because charged particles are present.

Electrolytes are divided into two groups: *strong electrolytes* and *weak electrolytes.* Strong electrolytes are almost completely dissociated in aqueous solution at concentrations around $1M$. Substances classified as strong electrolytes include all salts of group IA and group IIA metals, many salts of other metals, strong acids, and strong bases. Strong electrolytes usually are substantially ionic even when they are not in solution, but there are exceptions to this rule. For example, HCl is a polar gas when pure, but a strong electrolyte when in solution.

**Example 12.15**

Using the data in Table 12.2 and assuming ideal solution behavior, calculate the boiling point elevation of a solution prepared from 0.571 g of LiF, a strong electrolyte, and 111 g of $H_2O$.

**Solution**

The boiling point elevation of the solution is determined by the molality of solute particles in the solution. Since LiF is a strong electrolyte, dissociating into ions in aqueous solution, the molality of solute particles is greater than the molality based on the original amount of undissolved solute. Each mole of LiF produces two moles of dissolved solute particles:

$$LiF(s) \longrightarrow Li^+(aq) + F^-(aq)$$

Therefore, to calculate the molality of solute particles in a solution of LiF, we must multiply by a factor of 2 in the usual molality calculations:

$$m = 2 \times \frac{0.571 \text{ g LiF}}{111 \text{ g } H_2O} \times \frac{1 \text{ mol LiF}}{25.9 \text{ g LiF}} \times 1000 \frac{g}{kg}$$

$$= 0.397m$$

which is the total molality of solute particles.

We can now find the boiling point elevation by assuming ideal behavior and using Equation 12.15:

$$\Delta T = (0.512)(0.397) = 0.203 \text{ K}$$

Weak electrolytes are substances that dissociate only partially into ions at concentrations around $1M$. Weak electrolytes increase the conductivity of water, but not as much as do strong electrolytes. The weak electrolytes include some salts of heavy metals, such as mercury and cadmium; weak acids, such as acetic acid and hydrogen cyanide; and weak bases, such as ammonia.

The extent to which an electrolyte dissociates in solution can be determined by measurement of any of the colligative properties of the solution, since these properties depend on the number of solute particles in solution. A careful study of the colligative properties and electrical conductivities of solutions enabled the Swedish chemist Svante Arrhenius (1859–1927) to propose in 1887 that electrolytes dissociate into ions in solution.

**Example 12.16**     A solution of 26.3 g of $CdSO_4$ in 1.00 kg of $H_2O$ has a freezing point 0.285 K lower than pure water. Calculate the percentage of $CdSO_4$ that dissociates into ions in this solution.

**Solution**     Using the value of $K_f$ in Table 12.2 and Equation 12.16, we can find the molality of solute particles from the measured freezing point depression:

$$0.285 \text{ K} = (1.86)m$$
$$m = 0.153$$

The molality of the solution can also be calculated in the usual way from the mass and molecular weight of $CdSO_4$:

$$\frac{26.3 \text{ g CdSO}_4}{1.00 \text{ kg H}_2\text{O}} \times \frac{1 \text{ mol CdSO}_4}{208.5 \text{ g CdSO}_4} = 0.126m$$

The molality found from the observed freezing point depression is larger than the molality calculated from the mass and molecular weight of solute. The larger molality results from the dissociation of the solute in solution.

Let $x$ equal the molality of $CdSO_4$ that undergoes dissociation. Both the molality of $Cd^{2+}$ that is formed and the molality of $SO_4^{2-}$ that is formed will be $x$. The molality of $CdSO_4$ remaining after the reduction of the molality by $x$ is $0.126 - x$. The molalities present before and after dissociation can be placed beneath the appropriate formulas in the chemical reaction:

|  | $CdSO_4$ | $\longrightarrow Cd^{2+}$ | $+ SO_4^{2-}$ |
|---|---|---|---|
| molality before dissociation | 0.126 | 0 | 0 |
| molality after dissociation | $0.126 - x$ | $x$ | $x$ |

The total molality of all particles in solution after dissociation is $0.153m$.

$$(0.126m - x) + x + x = 0.126m + x = 0.153m$$

and
$$x = 0.027m$$

The percentage of $CdSO_4$ that dissociates in solution, assuming ideal behavior, thus is

$$\frac{0.027m}{0.126m} \times 100\% = 21.4\%$$

One factor that influences the extent to which a weak electrolyte dissociates is the concentration of the electrolyte in the solution. As the concentration of the electrolyte decreases, the *percentage* of dissociation increases. Thus, if more water is added to the solution of Example 12.16, the concentrations of all species in solution decrease, but the percentage of $CdSO_4$ that dissociates increases.

### The van't Hoff Factor

As we mentioned, the ideal solution approximation does not work nearly as well for solutions of electrolytes as it does for solutions of nonelectrolytes. The deviation from ideality often is expressed by the use of a quantity $i$, the *van't Hoff factor,* which relates the observed value of a colligative property to the value calculated by one of the ideal relationships. For example, the relationship between observed and ideal value can be expressed for the freezing point depression by:

$$\Delta T = iK_f m \qquad (12.19)$$

where $\Delta T$ is the observed freezing point depression and $m$ is the value of the molality calculated on the basis of no dissociation of the solute. Assuming ideal behavior, the value of $i$ for a strong electrolyte is equal to the number of moles of ions formed from complete dissociation of each mole of undissolved substance. For NaCl, $i$ should have a value of 2, since each mole of NaCl produces two moles of ions or solute particles in solution. For $MgCl_2$, $i$ should be 3, since each mole of $MgCl_2$ produces one mole of $Mg^{2+}$ ions and two moles of $Cl^-$ ions in solution.

Even in very dilute solutions, the value of $i$ deviates from the ideal value. The deviation increases as the concentration of the solution increases. For example, the values of $i$ for solutions of NaCl and $MgSO_4$, each of which dissociates into two ions, are:

|  | Value of $i$ | |
|---|---|---|
|  | **NaCl** | **MgSO$_4$** |
| ideal | 2 | 2 |
| $0.001m$ concentration | 1.97 | 1.82 |
| $0.1m$ concentration | 1.87 | 1.21 |

A number of theories have been developed to explain both the deviation of the value of $i$ from ideality and some other characteristics of solutions of electrolytes. The problem is extremely complex, and it is far from solved for solutions of even moderate concentration, about $0.1m$. Very dilute solutions are handled well by the Debye-Hückel theory, proposed in 1923 by Peter Debye (1884–1966) and Erich Hückel (1896–    ). Their theory is based on the idea that deviations from ideality are caused by electrical interactions between ions of opposite charge in solution. A positive ion tends to have negative ions in its vicinity. The resulting coulombic attractions have the same effect as incomplete dissociation. These attractions lower the value of $i$ below the theoretical value. As expected, the effect is greater in solutions of ions with larger charges. The deviation from ideality is greater in a solution of $0.001m$ $MgSO_4$ than in a solution of $0.001m$ NaCl because the coulombic attraction between the ions of charge $+2$ and $-2$ that are found in the $MgSO_4$ solution is four times greater than the attraction between the ions of charge $+1$ and $-1$ found in the NaCl solution.

**Example 12.17**    A $0.0103M$ solution of $K_2SO_4$ is found to have an osmotic pressure of 0.680 atm at 299 K. Calculate the value of $i$ for the solution.

**Solution**    Equation 12.18 can be rewritten to include $i$:

$$\pi = icRT$$

For this solution:

$$0.680 \text{ atm} = i \left( 0.0103 \, \frac{\text{mol}}{\text{L}} \right) \left( 0.0821 \, \frac{\text{L atm}}{\text{mol K}} \right) (299 \text{ K})$$

$$i = 2.69$$

The calculated value of $i$ is substantially lower than the ideal value $i = 3$, even in this rather dilute solution.

Many of the solutions that are used in the laboratory and many that we shall discuss later are of relatively high concentration and so are described poorly by the ideal solution approximation. More elaborate quantitative treatments of concentrated solutions have been developed, but they are beyond our scope. We shall use the ideal solution approximation throughout.

## 12.8    COLLOIDS

A solution and a suspension are two extreme ways in which two substances can mix. In a solution, the solute particles are so small that they cannot be seen. In a suspension of, say, a solid in a liquid, the particles of solid are large enough to be visible, through an ordinary microscope if not by the unaided eye. If a suspension is allowed to stand, the solid particles settle out after a while because of gravity.

Particle size makes the difference between a solution and a suspension. In a solution, the solute particles are smaller than about 1 nm in diameter. In a suspension, the solid particles are larger than 1000 nm in diameter. There are mixtures in which the size of particles is between the upper limit

**TABLE 12.3** Names of Colloids

| Name | Dispersed Phase | Dispersion Medium | Example |
|------|-----------------|-------------------|---------|
| emulsion | liquid | liquid | milk (fat in water) |
| foam | gas | liquid | suds (carbon dioxide in beer) |
| aerosol | solid | gas | smoke (particles in air) |
|  | liquid | gas | fog (water in air) |

for solutions and the lower limit for suspensions. These mixtures often have properties of both suspensions and solutions. They are called **colloids.**

The particles of a colloid do not settle out because they are too small. They are not directly visible, but particles of colloidal size scatter light. A light beam that passes through a colloid is visible to an observer at right angles to the beam. An everyday example of this scattering effect can be seen when light beams pass through tobacco smoke, a colloid consisting of solid particles in air.

Colloids are classified by the state of their components when they are pure. As we define a solute and a solvent for solutions, for colloids we define a *dispersed phase* (which corresponds to the solute) and a *dispersion medium* (which corresponds to the solvent).

The dispersion medium in many colloids is a liquid. If the dispersed phase is a solid and the system appears to be liquid and flows, it is called a *sol.* If it has a solidlike structure that prevents it from flowing, it is called a *gel.* Milk of magnesia is a sol. Jell-O is a gel when it is cool but a sol when it is warm enough to flow. The names given to some colloids with various dispersed phases and dispersion media are listed in Table 12.3.

Particles of colloidal size generally can be prepared either by condensation of smaller particles or by dispersion of larger particles. A variety of methods have been devised to carry out both of these processes.

**Summary**

$\mathbf{W}$e began this chapter by defining a **solution** as a homogeneous mixture of two or more substances, and noted that a solution can be a solid, a liquid, or a gas. We then defined some important expressions of concentration: the **mass fraction,** which is the mass of solute in a mass of solution; the **mole fraction,** which is the amount (in moles) of solute in an amount of solution; **molality,** which is the amount of solute in one kilogram of solvent; **molarity,** the amount of solute in one liter of solution; and **volume percent,** the volume of solute divided by the sum of the volumes of the components of the solution. We said that the best rule for predicting solubility is that **like dissolves like,** meaning that polar solutes tend to dissolve in polar solvents and nonpolar solutes in nonpolar solvents. We then gave some rules for predicting the **solubility** of ionic compounds in water. We noted that the solubility of salts in water tends to increase but that the solubility of gases decreases as temperature rises. We then introduced the concept of an **ideal solution,** in which all interactions between all components are assumed to be equal. We used that concept to discuss vapor pressure and **Raoult's law,** which says that the vapor pressure of a substance above a solution is equal to its vapor pressure when pure multiplied by its mole fraction in the solution. We then showed that an increased concentration of solute causes a **boiling point elevation** and a **freezing point depression** of a solution. We went on to define **osmotic pressure,** the pressure at which the rates of flow into and out of a solution are equal, and to discuss the importance of osmosis in living organisms. We then classified substances as **electrolytes,** which dissociate into ions in aqueous solution, and **nonelectrolytes,** which do not dissociate, and noted that strong electrolytes dissociate almost completely while weak electrolytes dissociate only partially. Finally, we described **colloids,** whose properties are midway between those of solutions and suspensions.

## Exercises

**12.1** Find the mass fraction and mass percent of glucose in a solution prepared from 71.1 g of $H_2O$ and 14.3 g of glucose.

**12.2** Find the mass of bromine in 5.22 g of a solution of bromine in carbon tetrachloride that is 0.223 bromine by mass.

**12.3** Calculate the mass of pure gold required to prepare 17.8 g of 14-carat gold, which is $\frac{14}{24}$ parts gold by mass.

**12.4** A 4.47-kg sample of water contains 12.9 mg of radium. Express the concentration of radium in parts per million.

**12.5[1]** The density of ethyl alcohol is 0.789 g/cm$^3$ at 293 K. The density of carbon tetrachloride is 1.59 g/cm$^3$ at 293 K. Find the mass percent of ethanol in a solution prepared from equal volumes of these two liquids at that temperature.

**12.6** Find the mole fraction of iodine in a solution prepared from 1.62 g of $I_2$ and 9.87 g of $CCl_4$.

**12.7** Find the mole fraction of tin in

---

[1] The answers to exercises whose numbers are in color can be found in Appendix VII. The star indicates an exercise that is more challenging than average.

an alloy prepared from equal masses of tin and copper.

**12.8** Calculate the mass of glucose (MW = 180) that must be added to 50.0 g of water to prepare a solution in which the mole fraction of glucose is 0.0110.

**12.9** Calculate the mole fraction of ethylene glycol, $C_2H_6O_2$, in a solution that is 63.9% ethylene glycol by mass.

**12.10** The density of mercury is 13.6 g/cm$^3$ at 273 K. A solution of gold in mercury, called an amalgam, is prepared from 20.0 cm$^3$ of mercury and 1.28 g of gold. Calculate the mole fraction of gold in this solution.

**12.11★** Find the mass of urea (MW = 60.1) needed to prepare 50.0 g of a solution in water in which the mole fraction of urea is 0.0770.

**12.12** Find the molality of ascorbic acid (MW = 176) in a solution prepared from 1.94 g of ascorbic acid and 50.1 g of $H_2O$.

**12.13** Find the molality of methanol in a solution prepared from 1.00 mol of methanol ($CH_3OH$) and 2.00 mol of water.

**12.14** Find the molality of water in a solution that is 63.9% ethylene glycol, $C_2H_6O_2$, by mass.

**12.15** Calculate the mass of water and of vanillin (MW = 152) needed to prepare 75.0 g of a solution that is $0.0250m$ in vanillin.

**12.16** Find the molarity of nicotine (MW = 162) in a solution that contains 0.711 g of nicotine in 56.8 cm³ of solution.

**12.17** Calculate the mass of iodine needed to prepare 125 cm³ of a $0.22M$ solution of iodine in carbon tetrachloride.

**12.18** The concentration of fluoride ion in a supply of drinking water is $2.1 \times 10^{-5}M$. Find the volume of drinking water that contains 1.0 g of $F^-$.

**12.19** A solution is prepared by the addition of 5.34 g of NaCl and 2.01 g of $CaCl_2$ to enough water to make 0.750 L of solution. Find the molarity of $Cl^-$ in the solution.

**12.20** A solution is prepared by the addition of 0.33 mol of LiCl and 0.25 mol $Ca_3(PO_4)_2$ to enough water to make 3.0 L of solution. Assuming complete dissociation of each salt into ions, find the total molarity of ions in the solution.

**12.21** Find the volume of a $0.22M$

solution of $H_2SO_4$ that is needed to react completely with 5.7 g of KOH.

**12.22** When Mg metal is added to a solution of acid, the reaction $2H^+(aq) + Mg(s) \rightarrow H_2(g) + Mg^{2+}(aq)$ takes place. Find the volume of $H_2$ liberated at STP when excess Mg is added to 51.6 cm³ of a $0.131M$ solution of HCl.

**12.23** How would you prepare 100 cm³ of a $0.001M$ solution of sucrose in water from a 100-cm³ sample of a $0.1M$ solution of sucrose in water, using only a 100-cm³ graduated cylinder?

**12.24** Equal volumes of a $0.10M$ NaCl solution and a $0.25M$ $Na_2SO_4$ solution are mixed. Find the molarity of $Na^+$ in the resulting solution, assuming that its volume is the sum of the volumes of the two solutions.

**12.25** A 75.0-cm³ volume of a $0.161M$ solution of phenol in water is mixed with a 10.0-cm³ volume of a $0.994M$ solution of phenol in water. Calculate the molarity of phenol in the resulting solution, assuming that its volume is the sum of the volumes of the two solutions.

**12.26** The density of a 14.2% by mass solution of potassium chloride in water is 1.12 g/cm³. Find the molarity of the solution.

**12.27** The density of a $0.640M$ solution of lead nitrate in water is 1.18 g/cm³. Find the mass fraction of lead nitrate in the solution.

**12.28** At a certain temperature, the density of water is 0.991 g/cm³ and the solubility of lithium chloride is 64.1 g/100 g of water. Find the molarity of a saturated solution of lithium chloride in water, assuming no volume change when the LiCl is dissolved in water.

**12.29** A $0.781M$ solution of sodium iodate, $NaIO_3$, has a density of 1.099 g/cm³. Find the molality of the solution.

**12.30** The density of a $0.309m$ solution of urea (MW = 60.06) in water is 1.101 g/cm³. Find the mass of urea in 75.1 cm³ of solution.

**12.31★** A solution of 49.0% $H_2SO_4$ by mass has a density of 1.39 g/cm³ at 293 K. A 25.0-cm³ sample of this solution is mixed with enough water to increase the volume of the solution to 99.8 cm³. Find the molarity of sulfuric acid in this solution.

**12.32** A solution is prepared from 0.250 L of water and 0.330 L of methanol. Find the volume percent of methanol in this solution.

**12.33** The density of water at 293 K is 0.998 g/cm³. The density of acetone at 293 K is 0.792 g/cm³. Find the mole fraction of acetone in a solution that is 50.0% acetone by volume.

**12.34** At 293 K, the density of ethanol is 0.791 g/cm³ and the density of water is 0.998 g/cm³. Find the molality and the mole fraction of ethanol in warm beer that is 6.40 proof.

**12.35** At 293 K, the density of $D_2O$ (heavy water) is 1.105 g/cm³ and the density of $H_2O$ is 0.998 g/cm³. A solution of the two liquids has a density of 1.020 g/cm³. Find the volume percent of $D_2O$ in the solution, assuming that its volume is the sum of the volumes of the two starting liquids.

**12.36★** A solution is prepared by mixing 316 cm³ of $CH_3OH$, whose density is 0.792 g/cm³, with 501 cm³ of water. The molarity of $CH_3OH$ in the resulting solution is $9.39M$. Find the volume of the solution.

**12.37** A saturated solution of lithium sulfate in water is $2.37M$ in lithium sulfate. Find the solubility of lithium sulfate in grams per 100 cm³ of water, assuming no volume change when the solution is formed.

**12.38** Formulate a picture of the solvent-solute interactions in the following solutions: (a) sodium fluoride in water, (b) lithium fluoride in acetone, (c) propane ($C_3H_8$) in hexane ($C_6H_{14}$), (d) water in ethanol ($C_2H_5OH$).

**12.39**  Predict which of the following salts are soluble, which are insoluble, and which are slightly soluble: (a) sodium carbonate, (b) silver bromide, (c) tin nitrate, (d) barium acetate, (e) calcium sulfate, (f) copper sulfide, (g) lead iodide.

**12.40**  The partial pressure of $O_2$ in the atmosphere at sea level is 0.12 atm. Find the mole fraction of $O_2$ in a saturated aqueous solution at 273 K.

**12.41**  Find the pressure of $N_2$ needed to prepare a solution of $N_2$ in water with a mole fraction of $N_2$ equal to 0.10 at 298 K.

**12.42**  Suggest a reason for the use of helium in the breathing mixtures used by deep-sea divers.

**12.43**  Calculate the volume of $CO_2$ released when a bottle containing 500 $cm^3$ of soda water bottled at a pressure of 1.1 atm of $CO_2$ is opened at 298 K and 1 atm of external pressure.

**12.44**  A solution is prepared from 14.13 g of sucrose (MW = 342.30) and 111.8 g of water. Find the vapor pressure of water above this solution at 323.0 K. The vapor pressure of pure water at this temperature is 0.1217 atm.

**12.45**  The vapor pressure of water above a solution of barbituric acid is 0.0248 atm at 296 K. The vapor pressure of pure water at this temperature is 0.0278 atm. Find the mole fraction of barbituric acid in the solution.

**12.46**  The vapor pressure of pure benzene (MW = 78.12) is 0.132 atm at 299 K. Find the mass of bromobenzene (MW = 157.02) that must be dissolved in 50.0 g of benzene to lower its vapor pressure to 0.100 atm.

**12.47**  A solution is prepared from 122 g of acetone, $C_3H_6O$, and 16.9 g of

an unknown compound. The vapor pressure of acetone above the solution is 0.484 atm. The vapor pressure of pure acetone at the same temperature is 0.526 atm. Find the molecular weight of the unknown compound.

**12.48**  At 316 K, the vapor pressure of chloroform is 0.526 atm and the vapor pressure of carbon tetrachloride is 0.354 atm. Find the total vapor pressure above a solution prepared from equimolar amounts of the two.

**12.49**  At 333 K, the vapor pressure of benzene, $C_6H_6$, is 0.521 atm and the vapor pressure of toluene, $C_7H_8$, is 0.184 atm. Find the mole fraction of toluene in the vapor above a solution prepared from equal masses of these two components.

**12.50**  At 337 K, the vapor pressure of ethanol is 0.526 atm and the vapor pressure of water is 0.236 atm. A solution is prepared from equimolar amounts of water and ethanol at this temperature. The vapor above the solution is removed and condensed. The new solution is heated to 337 K and the vapor above the solution is removed and recondensed. The process is repeated once more. Find the composition of the resulting solution.

**12.51***  Isopropyl alcohol and propyl alcohol both have the same formula, $C_3H_8O$. A solution of the two that is 25% isopropyl alcohol by mass has a total vapor pressure of 0.090 atm at a given temperature. A solution of the two that is 75% isopropyl alcohol by mass has a total vapor pressure of 0.123 atm at the same temperature. Find the vapor pressures of the pure alcohols at this temperature.

**12.52**  A solution is prepared from 47.2 g of glucose, $C_6H_{12}O_6$, and 186 g of water. Using the data in Table 12.2, find the boiling point of this solution.

**12.53**  Find the mass of naphthalene, $C_{10}H_8$, that must be dissolved in 200.0 g of benzene to raise the boiling

point of benzene by 1.0 K.

**12.54**  A 3.94-g sample of an unknown substance is dissolved in 74.9 g of $CCl_4$. The boiling point of the resulting solution is 351.9 K. Use the data in Table 12.2 to find the molecular weight of the unknown substance.

**12.55***  A solution of a nonvolatile solute in water boils at 374.87 K. Find the vapor pressure of water above this solution at 348 K. (*Hint:* Table 4.1)

**12.56**  The solubility of two substances, A and B, in water at 275 K is about the same. The solubility of A increases as the temperature decreases, while the solubility of B decreases as the temperature decreases. Which substance is likely to form a solution with the lower eutectic temperature? Explain.

**12.57**  Find the freezing point of a solution that is 40.0% ethanol, $C_2H_6O$, by mass in water.

**12.58**  Pure nitrobenzene freezes at 278.9 K. A solution of 134 mg of urea (MW = 60.06) in 16.8 g of nitrobenzene freezes at 278.0 K. Find $K_f$ for nitrobenzene.

**12.59**  The chief ingredient of antifreeze is ethylene glycol, $C_2H_6O_2$. Find the mass of ethylene glycol that must be dissolved in 4.0 L of water to lower the freezing point of water to 253 K.

**12.60**  A solution of 0.367 g of an unknown substance in 6.64 g of camphor freezes at 447.6 K. Find the molecular weight of the unknown substance.

**12.61**  The freezing point of a 0.01$m$ solution of $CH_3NH_2$ in benzene is 0.048 K lower than the freezing point of pure benzene. The freezing point of a 1.00$m$ solution of $CH_3NH_2$ in benzene is 3.5 K lower than the freezing point of pure benzene. Explain these observations.

**12.62**  Explain the observation that hot water pipes freeze before cold water pipes in the winter.

**12.63** Calculate the osmotic pressure of a solution of 1.00 g of sucrose, $C_{12}H_{22}O_{11}$, in 1.00 L of water at 298 K.

**12.64** A solution is prepared from 134 mg of insulin and enough water to make a total volume of 25.0 cm³. At 298 K, the solution has an osmotic pressure of 0.0230 atm. Find the molecular weight of insulin.

**12.65** The density of a solution of aniline, $C_6H_5NH_2$, in acetone, $C_3H_6O$, that is 2.1% aniline by mass is 0.81 g/cm³ at 288 K. Find the osmotic pressure of the solution at this temperature.

**12.66** Calculate the vapor pressure of water above a solution prepared from 2.78 g of calcium nitrate and 26.0 g of water at 295 K. Assume ideal behavior. The vapor pressure of pure water at this temperature is 0.0261 atm.

**12.67** Find the mass of $Na_2SO_4$ that must be dissolved in 100.0 g of water to lower the freezing point of water by 1.00 K. Assume ideal behavior.

**12.68** The eutectic temperature of sodium chloride solutions in water is 252 K. Find the molality of sodium chloride at this temperature. Assume ideal behavior.

**12.69** A 0.841-g sample of a salt of a dipositive cation and a mononegative anion is dissolved in 50.0 g of water. The solution boils at 373.68 K. Find the formula weight of the salt. Assume ideal behavior.

**12.70** The density of a 0.330$M$ solution of $MgBr_2$ in water at 373 K is 1.055 g/cm³. Calculate the vapor pressure of water above this solution. Assume ideal behavior.

**12.71** Calculate the mass of ammonium chloride that must be dissolved in 1.00 L of aqueous solution for the solution to have an osmotic pressure of 5.00 atm at 298 K. Assume ideal behavior.

**12.72** Seawater is 3.50% salt by mass and has a density of 1.04 g/cm³ at 293 K. Calculate the osmotic pressure, assuming ideal behavior and that all the dissolved salt is sodium chloride.

**12.73** A solution of 5.25 g of a solute in 20 g of benzene freezes at 252.8 K. A solution of 5.25 g of this solute in 20 g of water freezes at 270.7 K. Assuming no dissociation or association in benzene and ideal behavior, calculate the number of particles formed in water solution from each original solute particle.

**12.74★** A metal M, atomic weight 96.0, forms a fluoride salt that can be represented as $MF_x$. To determine $x$ and therefore the formula of the salt, a freezing point experiment is performed. A 7.64-g sample of the salt is dissolved in 100 g of water, and the freezing point of the solution is found to be 268.69 K. Find the formula of the salt, assuming ideal behavior.

**12.75** A solution of a salt AB undergoes 50.00% dissociation in a solution

prepared from 1.000 mol of the salt and 1000 g of water. Find the boiling point of the solution.

**12.76** A solution contains 1.58 g of $HNO_2$ in 47.0 g of water. The freezing point of the solution is 271.80 K. Calculate the fraction of $HNO_2$ that undergoes dissociation to $H^+$ and $NO_2^-$.

**12.77** In water, sulfuric acid dissociates completely into $H^+$ and $HSO_4^-$ ions. The $HSO_4^-$ ion dissociates to a limited extent into $H^+$ and $SO_4^{2-}$. The freezing point of a 0.1000$m$ solution of sulfuric acid in water is 272.76 K. Calculate the molality of $SO_4^{2-}$ in the solution, assuming ideal behavior.

**12.78** Use the appropriate values of $i$ to find the difference in osmotic pressure between 0.1$M$ solutions of NaCl and $MgSO_4$ that have the same molality and molarity at 298 K.

**12.79** An aqueous solution of sodium sulfate, $Na_2SO_4$, is prepared in which its mole fraction is 0.100. The vapor pressure of water above this solution at 323.0 K is 0.0986 atm. The vapor pressure of pure water at this temperature is 0.1217 atm. Find the value of $i$ for sodium sulfate in this solution.

**12.80** Emulsions of oil and water often separate rapidly into two liquid layers. Such emulsions can be stabilized by the addition of a detergent, which lowers the surface tension of water. Suggest an explanation for this action of detergents on emulsions.

# 13

# Chemical Thermodynamics

**Preview** | The energy crisis has made thermodynamics, the subject of this chapter, an important topic for all of us. We begin by defining the basic terms of thermodynamics, such as heat, work, and energy. We then explain the first law of thermodynamics, the law of conservation of energy, and describe how calorimetry—the measurement of heat—is used to quantify observations related to the first law. Next we discuss the second law of thermodynamics, which deals with the spontaneous direction of processes, introducing the concept of entropy and describing its use. We then define free energy, a function that is used to determine whether a chemical change occurs spontaneously, and describe its relationship to the equilibrium constant. Finally, we show how the concepts of thermodynamics are central to the problems facing our technological society.

Thermodynamics can be described as the study of the relationship between heat and other forms of energy. It is one of the major fields of the physical sciences, but its concepts have been used with striking success in many other areas. The ideas of thermodynamics have been applied to subjects as diverse as the fate of the universe, the nature of change in social systems, the design of communications networks, and the use of fossil fuels in a modern society.

In the physical sciences, thermodynamics originated from experimental observations. It is used to study relationships among those properties of matter that are directly measurable, such as volume, temperature, and pressure. Thermodynamics concerns itself not with the atomic structure of matter, but with large collections of atoms or molecules.

For the chemist, thermodynamics provides the answers to several important questions:

1. Can a physical or chemical change be expected in one or more substances under a given set of conditions?
2. How far can such a change proceed?
3. If a chemical or physical change occurs, what are the accompanying energy changes in the system and its surroundings?

In Chapter 8, we saw how we could obtain answers to such questions by studying the heat changes that accompany chemical processes. In this chapter, we shall use the same approach in a more general way, to study both physical and chemical changes.

One basic limitation should be kept in mind from the outset: *Thermodynamics does not concern itself with time.* On the practical level, this limitation can be most important. For example, thermodynamics tells us that if 2 mol of $H_2(g)$ and 1 mol of $O_2(g)$ are mixed at room temperature, the two gases will react to form 2 mol of $H_2O$, with the evolution of a considerable quantity of heat. But thermodynamics does not tell us that if the two gases are left alone, the reaction will not take place in a normal human lifetime. Nor does thermodynamics tell us that the presence of a spark makes this reaction proceed so quickly that there is an explosion. Thermodynamics ignores the time aspect because the spark does not change the extent to which the reaction ultimately proceeds or the quantity of energy that is liberated; it just allows the process to go faster. The question of time in chemical reactions is the subject of the discipline called kinetics, which we shall discuss in Chapter 17.

The rate at which reactions occur is a subject of practical importance for the chemist. Many substances or mixtures of substances can be regarded as thermodynamically unstable. If left alone, they should change. But thermodynamics tells us only whether something *should* happen. Kinetics determines whether that something actually *does* happen at an appreciable rate. For example, dynamite is thermodynamically unstable. But it can be transported and stored because it is undergoing chemical change at an immeasurably slow rate. Only when dynamite is detonated does its instability become evident. Similarly, most of the gaseous oxides of nitrogen are thermodynamically unstable at room temperature. But most of them can be stored for prolonged periods. The gaseous oxides of nitrogen are decomposing to $N_2(g)$ and $O_2(g)$, but the rate of decomposition usually is immeasurably slow and of no practical importance.

Throughout our discussion, you should remember that thermodynamics tells us *whether something should happen, but it does not tell us the length of time that it takes to happen.*

## 13.1    DEFINITIONS

We have mentioned that thermodynamics deals entirely with the bulk properties of matter — volume, temperature, pressure — and not with the internal structure of matter. In thermodynamics, we can ignore the behavior and structure of individual atoms and molecules. We are concerned with large collections of atoms or molecules.

Thermodynamics starts with the definition of a **system,** a specific amount of one or more substances. For the sake of simplicity, the system is separated from the rest of the universe by a *boundary,* which can be a real or imaginary surface. The boundary confines the physical system to a specific location. The part of the universe outside the boundary is called the *surroundings* of the system.

Usually, a boundary prevents mass from entering or leaving the system, so we have a *closed* system. If the boundary also prevents energy from entering or leaving, so that there is no interaction of the system with the surroundings, we are dealing with an *isolated* system.

The physical characteristics of a system that can be perceived or measured, such as its volume, its mass, and its temperature, are the *properties* of the system. When these properties are defined, a **state** of the system is defined. An **equilibrium state** is a state in which the properties of a system do not change unless the system is influenced by the surroundings.

Figure 13.1 shows a sealed container that holds one or more chemical substances. The system consists of the substances and the vapor space inside the container. The boundary is the walls of the container. Everything outside the container is the surroundings. The properties of the system include the quantities of the substances in the container, their temperature, their pressure, and their energy. All the properties, taken together, specify the state of the system. If the values of the properties do not change with time, the system is in an equilibrium state.

We can describe a system in an equilibrium state by specifying the values of certain of its properties called **state functions.** The value of a state function is fixed when the system is in a given state. It is not influenced by the path that the system followed to reach that state. By convention, the symbol for a state function is a capital letter. The equilibrium state of a system can be described by specification of the values of a small number of state functions, two or three in most cases. We are already familiar with

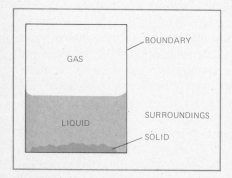

**Figure 13.1**
A system and its surroundings. The system is everything inside the container. The surroundings are everything outside the container. The walls of the container form the boundary between the system and the surroundings.

three state functions, pressure ($P$), temperature ($T$), and volume ($V$). In our discussion of thermodynamics, we shall be interested in four other state functions: internal energy ($E$), enthalpy ($H$), entropy ($S$), and free energy ($G$). If a system is not in an equilibrium state, it is usually necessary to specify the values of more than two or three state functions to describe the state.

Thermodynamics deals with changes in state, that is, with processes in which a system in an initial state goes through a change that brings it to some final state. A convenient way to describe such a change in state is to give the values for the changes that occur in the state functions. Changes in a function are represented by the symbol $\Delta$, the Greek letter delta, placed before the symbol of the function. The symbol $\Delta$ denotes a difference that we find by subtracting the value of the state function in the final state from its value in the initial state. Thus, if temperature, $T$, is the state function of interest:

$$\Delta T = T_{\text{final}} - T_{\text{initial}}$$

We said that the value of a state function depends only on the state of the system and not on how this state is reached. Therefore, the value of the change in a state function does not depend on the path followed by the system between its initial and final states. Hess's law and many of the thermochemical calculations we carried out in Section 8.7 are based on this property of state functions.

Many of the quantities associated with a change in state do depend on the path that the system follows. As an analogy, consider a resident of New York who moves to Los Angeles. The changes in latitude and longitude are both changes in state functions, since neither depends on the path of the journey. But several other quantities called **path functions** depend on the path that is followed. For example, the distance traveled, the work performed by the person making the trip, and the time needed to make the trip depend on the route and the mode of transportation. The traveler can fly from New York directly to Los Angeles, can ride a bike via Arizona or can drive via North Dakota. In each case, the path functions are different. In chemical thermodynamics, by convention, the symbols for path functions are lowercase letters.

### Internal Energy

One important consideration in thermodynamics is the magnitude of the energy changes in a system and its surroundings that are associated with a change in state. We have discussed a number of different kinds of energy. In thermodynamics, we simply define the **internal energy,** $E$, as the total of all the possible kinds of energy of a system. The internal energy of a system is a state function. It is usually not possible to measure the value of $E$ of a system in a given state. But it is often possible to measure or find the value of $\Delta E$ for a change in state of a system.

The SI unit for all forms of energy, including internal energy, is the joule (J). It is defined as $1 \text{ J} = 1 \text{ kg m}^2 \text{s}^{-2}$. Another unit of energy, which is not an SI unit, is called the calorie. It was originally defined as the quantity of heat needed to raise the temperature of 1 g of water from 14.5°C to 15.5°C. It is now defined as $1 \text{ cal} = 4.184 \text{ J}$.

In thermodynamics, we approach $\Delta E$, the change in the internal energy of a system that results from a change in state, by considering the ways in which the energy of a system can change. The energy of a system changes because energy is transferred into or out of a system across its boundary. The two methods of transferring energy across the boundary of a system are **heat** and **work.** These are familiar terms in everyday life, but we shall give them more formal definitions.

## Work

We shall use the term *work* to refer to mechanical work, which is succinctly defined as the product of a force by a displacement. There is a relationship between the internal energy of a system and work. The greater the internal energy, the more work the system can do.

The potential energy of a weight that is suspended 2 m (meters) above the floor of a room is greater than its potential energy when the weight is 1 m above the floor. When the weight falls, it can do more work when it falls 2 m than when it falls 1 m. A mixture of carbon and hydrogen in the form of gasoline and of $O_2$ is of higher potential energy than a mixture of the same quantities of these three elements in the form of carbon dioxide and water. When it takes place in the cylinder of an automobile engine, the change in state from gasoline and $O_2$ to $CO_2$ and $H_2O$, called combustion, is used to do such work as moving the automobile (Figure 13.2).

The units of work are units of force (newtons in SI), multiplied by units of distance (meters in SI). Since $1 \text{ J} = 1\text{N} \times 1 \text{ m}$, the unit of work is the joule, the same as the unit of energy. As we shall see, the quantity of work that accompanies a change in state depends on how the change in state is brought about. Work is a path function and is represented by the lower-case letter $w$.

In dealing with work, we designate not only the quantity of work in joules but also the direction of the work. A system can perform work on the surroundings. Energy is transferred from the system to the surroundings in the form of work, and the sign of $w$ is negative. A system can absorb energy when the surroundings perform work on the system. For this type of energy transfer, the sign of $w$ is positive. More simply, work done by the system on the surroundings has a negative sign ($-w$), while work done on the system by the surroundings has a positive sign ($+w$).

We shall limit ourselves to **expansion work,** which is work caused by changes in volume. For the sake of simplicity, we shall ignore all the other kinds of work, such as electrical work. Since we limit ourselves to considering changes in volume, there are just three possibilities to discuss. If the volume of a system increases as the result of a change in state ($\Delta V > 0$),

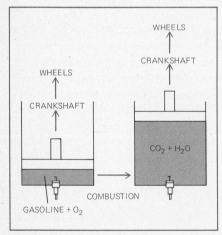

**Figure 13.2**
Both before and after combustion, this system contains hydrogen, carbon, and oxygen. Before combustion, the cylinder contains a mixture of gasoline and $O_2$. In the combustion reaction, the gasoline combines with the $O_2$ to produce $CO_2$ and $H_2O$ and release energy. The piston moves and thus does work. The quantity of work that can be done by a system is related to its internal energy.

the system does work on the surroundings and $w$ has a negative sign $(-w)$. If the volume of a system decreases because of a change in state $(\Delta V < 0)$, the surroundings do work on the system and $w$ has a positive sign $(+w)$. If there is no change in volume as a result of a change in state $(\Delta V = 0)$, then $w = 0$ and no work is done.

The assumption that only expansion work is possible simplifies the thermodynamic description of chemical processes considerably. For example, if a system has no mechanical link with the surroundings (as in Figure 13.1), no change of volume is possible and $w = 0$ for any change of state. Furthermore, changes in state between liquids and solids are not accompanied by significant changes in volume. Thus, we can assume $w = 0$ for all systems that contain only solids or liquids under ordinary conditions.

The simplest type of expansion work occurs when the volume of a system changes at constant pressure. If the pressure on the system is $P_{ext}$, we can show that the work caused by a volume change $\Delta V$ is given by:

$$w = -P_{ext} \Delta V \tag{13.1}$$

In SI, the units of pressure, force per area, can be given as N/m², and the units of volume are m³. The units of $P_{ext} \Delta V$ are therefore N × m = J. Since we are using atmospheres and liters for $P$ and $V$, we use the relationship 1 L atm = 101.3 J to express expansion work in units of joules.

**Example 13.1**

Calculate the work associated with the vaporization of 1.0 mol of water at 373 K and 1.0 atm (Figure 13.4). Assume ideal gas behavior.

**Solution**

The value of $\Delta V$ is the difference in volume between the initial and final states of the system. The final volume is the volume of one mole of water vapor at the specified conditions. We find it by using the ideal gas equation:

$$V = \frac{nRT}{P} = \frac{(1.0 \text{ mol})(0.082 \text{ L atm mol}^{-1} \text{ K}^{-1})(373 \text{ K})}{1.0 \text{ atm}}$$
$$= 31 \text{ L}$$

The initial volume of the liquid water is less than 0.02 L. In general, the volumes of liquid or solid can be neglected in the calculation of $\Delta V$ because they are so much smaller than the volume of the same amount of gas. Thus, $\Delta V = V_{final} - V_{initial} = 31$ L.

$$w = -P_{ext} \Delta V = -(1.0 \text{ atm})(31 \text{ L})\left(101 \frac{\text{J}}{\text{L atm}}\right) = -3100 \text{ J}$$

The negative sign of $w$ indicates that the system does work on the surroundings, because of its increase in volume. The steam engine operates on this principle.

## Heat

Heat, the second method by which energy is transferred across a boundary, is harder to define. Work is the result of measurable mechanical displacements of macroscopic objects. Heat is the result of microscopic effects. The transfer of energy by heat occurs as the result of processes such as thermal conduction or thermal radiation. These processes are best explained by reference to atoms and molecules, whose existence is ignored by classical thermodynamics.

The symbol for heat is $q$, a lowercase letter, because heat is a path function. It is convenient to picture heat as flowing in a direction. If heat flows from the surroundings into the system to raise the energy of the system, its sign is positive, $+q$. If heat flows from the system into the surroundings, lowering the energy of the system, its sign is negative, $-q$. For many years, the calorie was the unit used to express the magnitude of heat. The SI unit of heat is the joule, which is also the unit of work and the unit of energy. We often use the kilojoule, kJ, for large quantities of work, heat, or energy; $1 kJ = 1000$ J.

We can summarize the sign conventions for $q$ and $w$ by noting that they are positive quantities when the energy of a system increases and negative quantities when the energy of the system decreases:

> Heat flows into the system, $q$ is positive.
> Work is done by the surroundings on the system, $w$ is positive.
> Heat flows out of the system, $q$ is negative.
> Work is done by the system on the surroundings, $w$ is negative.

## 13.2    THE FIRST LAW OF THERMODYNAMICS

There are many ways to state an important result derived from countless experiments and observations of the energy changes that accompany changes in state. These different but equivalent statements are known as the **first law of thermodynamics.**

The first law of thermodynamics is sometimes called the *law of conservation of energy.* Most generally, it states that the energy of the universe is constant. More specifically, it states that energy cannot be destroyed or created; it can only be transferred from place to place or transformed from one form to another. The first law gives us a way to keep a balance sheet on the energy change of any system. Since there are only two ways to transfer energy, heat and work, the first law may be stated:

$$\Delta E = q + w \qquad (13.2)$$

Equation 13.2 states that the change in the energy of a system is equal to the sum of the heat that flows into the system and the work done on the system. The equation gives a complete accounting of the energy.

## PERPETUAL MOTION: THE DURABLE FALLACY

It might not seem possible that anyone in today's sophisticated society would invest money in a perpetual motion machine. But the world still holds many inventors who, knowingly or otherwise, are trying to sell a modern version of perpetual motion. The key to these machines is their apparent complexity, and such machines still are capable of causing an occasional stir in Wall Street, where experts are more at home with price-earnings ratios and dividend rates than they are with the laws of thermodynamics. In the early 1970s, for example, the price of one stock went up sharply for a time (before the Securities and Exchange Commission intervened) because the company owned rights to a machine that was said to produce hydrogen from water in a self-sustaining reaction.

The machine contained two steel tanks, each holding granules of an unidentified metal. Supposedly, the metal reacted with steam to bind oxygen and release hydrogen, which could be used as fuel. Several short demonstrations were held, at which the machine did indeed produce hydrogen.

However, this machine cannot pass the ultimate test by being truly self-sustaining. Sooner or later (probably sooner), the reactant metal must be recycled by heating, driving off the oxygen, and regenerating the metal so that it reacts with steam again. The hydrogen-producing reaction in one steel tank must give off enough heat to drive the recycling reaction in the other tank. We can represent the two reactions as:

$$M + H_2O \longrightarrow H_2 + MO + heat$$

$$heat + MO \longrightarrow M + \tfrac{1}{2}O_2$$

The overall process is

$$heat + H_2O \longrightarrow H_2 + \tfrac{1}{2}O_2 + heat$$

Since even the simplest measurement demonstrates that the decomposition of water into $H_2$ and $O_2$ is endothermic, the quantity of heat on the left side of the reaction is greater than the quantity on the right side. Heat is required for the overall process. Since energy cannot be created inside the machine, it must be supplied from an external source. Therefore, the machine cannot be self-sustaining, according to the first law.

Even if the inventor of the machine somehow thought of a way to circumvent the first law, the recycling process would require the conversion of a quantity of heat from the hydrogen-generating reaction into an equal amount of work to regenerate the metal. As we shall see, the second law of thermodynamics states that heat cannot be converted into work without some loss. Again, the machine requires some energy from the surroundings.

The marvelous hydrogen-making device was no more than a perpetual motion machine. The machine might be a useful source of hydrogen, if the energy input required to recycle the reactant is sufficiently small. But no machine could live up to the claims made for this one. A perpetual motion machine by any name will never work.

Another way to state the first law is to say that it is impossible to construct a machine that operates in a cycle and that yields more energy output than it receives as input. In other words, there cannot be a machine that does a greater quantity of work on the surroundings than the quantity of heat it receives from the surroundings. Nevertheless, untold effort has been expended for centuries in attempts to construct such a machine, which is called a perpetual motion machine of the first kind.

Perpetual motion of this kind is impossible because a machine that operates in a cycle must return to its initial state at regular intervals. The value of $\Delta E$ for the cycle must be 0, because $E$ is a state function and the initial and final states of a cycle are identical. If the value of work done is greater than the value of heat absorbed, $\Delta E < 0$. A perpetual motion machine cannot run continuously unless energy is somehow created. Since energy cannot be created, such a machine cannot be built.

There are still occasional claims that a perpetual motion machine of the first kind has been constructed. The first law of thermodynamics rests upon such a firm foundation that these claims are not taken seriously.

Stated loosely, the first law says, "You can't win," or, "You can't get something for nothing." In the form of Equation 13.2, the first law is a convenient method for reasoning both qualitatively and quantitatively about the changes in work, heat, and energy associated with a change in state.

As an example of this type of reasoning, let us consider an *adiabatic* process, one for which $q = 0$. For an adiabatic process, Equation 13.2 becomes:

$$\Delta E = w$$

Suppose we want to know what happens in an adiabatic volume change of an ideal gas. If the gas expands, the system does work on the surroundings and the sign of work is negative. Therefore, $\Delta E$ is negative; the energy of the system decreases. Since the energy of an ideal gas is directly proportional to its temperature (Section 4.6), the temperature falls. The same reasoning allows us to state that an adiabatic compression of an ideal gas raises its temperature.

**Example 13.2**

The volume of a sample of an ideal gas contracts from 8.4 L to 4.2 L as the result of an applied pressure of 1.5 atm. The system evolves 830 J of heat during this contraction (heat flows from system to surroundings). Find $\Delta E$ for this change in state.

**Solution**

To find $\Delta E$, we must know $q$ and $w$. The value of $q$ is given as $-830$ J. The value of $w$ can be calculated:

$$\begin{aligned} w &= -P_{ext}\,\Delta V \\ &= -(1.5 \text{ atm})(4.2 \text{ L} - 8.4 \text{ L})(101 \text{ J L}^{-1} \text{ atm}^{-1}) \\ &= 640 \text{ J} \end{aligned}$$

The sign for $w$ is positive because the surroundings do work on the system.

$$\begin{aligned} \Delta E &= q + w \\ &= -830 \text{ J} + 640 \text{ J} \\ &= -190 \text{ J} \end{aligned}$$

Since the temperature of an ideal gas is proportional to its internal energy, the temperature of the gas also decreases.

The energy change that accompanies a real chemical or physical transformation is of interest to chemists. The relative stability of products and reactants is important both theoretically and practically. Measuring the magnitude of $\Delta E$ is generally a way to obtain this information. The first law of thermodynamics tells us how this measurement can be made. Almost every process is accompanied by the flow of heat, and many processes are accompanied by work. The magnitude of $\Delta E$ can be found if the magnitudes of both $q$ and $w$ are found. The measurement can be simplified if we carry out the transformation at constant volume. If $\Delta V = 0$, $w = 0$, since only expansion work is possible. Equation 13.2 then becomes:

$$\Delta E = q_v \qquad (13.3)$$

where the subscript $v$ indicates constant volume. In other words, the $\Delta E$ of a process can be measured if the process is carried out in a sealed container and the flow of heat into or out of the container is measured.

At first glance, Equation 13.3 might seem to contain an inconsistency. We said that $E$ is a state function, and that the value of $\Delta E$ depends only on the initial and final states of the system, not the path between those states. But $q$ is a path function whose value for a given change in state depends on the path that is followed. Thus, Equation 13.3 defines a state function, $E$, in terms of a path function, $q$. Such a definition is often encountered in thermodynamics. There is no inconsistency because the path associated with the path function is also specified. By specifying the exact path, we give the path function an exact value. In Equation 13.3, the subscript $v$, which says that the constant-volume path is followed, describes the path completely enough to fix the value of $q$.

## Enthalpy

In practice, it is often difficult to carry out a change in state at constant volume. In processes where the amount of gas present changes, the resulting change in pressure may be great enough to require the use of thick-walled, sealed containers called bombs. Most chemical reactions are carried out in containers that are open to the atmosphere. Therefore, most changes occur along constant-pressure paths, at the pressure of the atmosphere.

We must therefore define another thermodynamic state function, enthalpy, $H$, which we discussed briefly in Section 8.7. Enthalpy is related to internal energy by the definition:

$$H = E + PV \qquad (13.4)$$

The change in enthalpy $\Delta H$ is given by:

$$\begin{aligned} \Delta H &= \Delta E + \Delta(PV) \\ &= \Delta E + P\,\Delta V + V\,\Delta P \end{aligned} \qquad (13.5)$$

For a change in state that occurs at constant pressure ($\Delta P = 0$):

$$\Delta H = \Delta E + P \Delta V \tag{13.6}$$

Since $\Delta E = q + w$ and $P \Delta V = -w$, we can write Equation 13.6 as:

$$\Delta H = (q + w) + (-w)$$

or $\Delta H = q$ when the change in state occurs at constant pressure. This relationship usually is written as:

$$\Delta H = q_p \tag{13.7}$$

where the subscript $p$ means constant pressure. Once again a state function is defined in terms of a path function. This time, the path is specified as the constant-pressure path. The $\Delta H$ can be measured by measuring the heat of a process occurring at constant pressure.

The difference between $\Delta H$ and $\Delta E$ is a $PV$ term related to expansion work. For processes that include only liquids and solids, changes in state do not cause significant volume changes; $\Delta V \cong 0$ and $w \cong 0$. For such processes, $\Delta H$ and $\Delta E$ are approximately the same.

The relationship between $\Delta H$ and $\Delta E$ is of interest for reactions in which there are changes in the amounts of gases. Assuming ideal behavior:

$$PV = nRT \quad \text{and} \quad \Delta(PV) = R\,\Delta(nT)$$

If the number of moles of an ideal gas changes in a process at constant $T$, then:

$$\Delta(PV) = RT\,\Delta n$$

and substitution into Equation 13.5 gives:

$$\Delta H = \Delta E + RT\,\Delta n \tag{13.8}$$

If both the temperature and the amount of ideal gas remain constant, $RT\,\Delta n = 0$ and $\Delta E = \Delta H$. Since the energy of a given amount of an ideal gas depends only on the temperature, $\Delta E$ and $\Delta H$ are 0.

---

**Example 13.3**    The value of $\Delta H$ for the reaction:

$$2N_2(g) + O_2(g) \longrightarrow 2N_2O(g)$$

at 298 K is 164 kJ. Calculate $\Delta E$.

**Solution**    In this process, 3 mol of gas changes to 2 mol of gas at constant temperature. Assuming ideal gas behavior, we can use Equation 13.8:

$$164 \text{ kJ} = \Delta E + R(298 \text{ K})(2 \text{ mol} - 3 \text{ mol})$$

To obtain a value for $\Delta E$, we express $R$ in units of $J \text{ mol}^{-1} \text{ K}^{-1}$:

$$R = 8.314 \text{ J mol}^{-1} \text{ K}^{-1}$$

Converting the value of $\Delta H$ from 164 kJ to 164 000 J gives:

$$164\,000 \text{ J} = \Delta E + (8.31 \text{ J mol}^{-1} \text{ K}^{-1})(298 \text{ K})(-1 \text{ mol})$$
$$\Delta E = 166\,000 \text{ J} = 166 \text{ kJ}$$

The difference between $\Delta E$ and $\Delta H$ is only 2000 J, a small fraction of either.

## 13.3  THE MEASUREMENT OF HEAT; CALORIMETRY

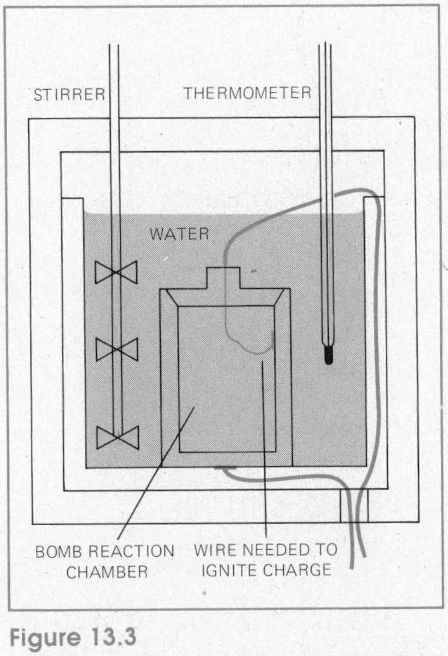

**Figure 13.3**

A bomb calorimeter. The heat produced by a change of state of the system in the bomb causes a change in the temperature of the water. The quantity of heat released by the process is found by measurement of the temperature change of the water.

The branch of science concerned with the measurement of heat is **calorimetry.** A device for carrying out such a measurement is called a **calorimeter.** There are several ways to measure the quantity of heat that is evolved or absorbed during a process. All of them are based on the measurement of some property that changes in a known way in response to heat. One of the first calorimeters was constructed in 1780 by Lavoisier and Laplace, who measured the heat liberated in various processes by measuring the quantity of ice melted by the heat.

Figure 13.3 shows a device called a *bomb calorimeter,* which is often used to measure the heat evolved by exothermic reactions such as combustion. The heat liberated by a process is calculated by measurement of the temperature change in the water surrounding the bomb.

We must know the quantity of heat that causes a given temperature change, $\Delta T$, in the calorimeter so that we can use the value of $\Delta T$ to find the quantity of heat released by a change in state such as a chemical reaction. The relationship between the heat released and the temperature change in the calorimeter is the **heat capacity** of the calorimeter. We usually measure it by finding the temperature change caused by a fixed quantity of electrical energy. If we know the heat capacity of the calorimeter, we can convert a measured temperature rise into a value for the quantity of heat liberated. The heat capacity of the calorimeter can be defined as:

$$\text{heat capacity} = \frac{\text{heat absorbed by the calorimeter}}{\Delta T} \qquad (13.9)$$

**Example 13.4**

The reaction bomb of the calorimeter in Figure 13.3 is charged with 2.456 g of *n*-decane, $C_{10}H_{22}$, and excess $O_2$. It is then sealed and placed in the calorimeter. The temperature of the water in the calorimeter is 296.32 K. The decane is ignited by the resistance wire. After combustion is complete, the temperature of the water is 303.51 K. In a separate experiment, the heat capacity of the calorimeter is found

CHAPTER 13    Chemical Thermodynamics

to be 16.24 kJ/K. Calculate $\Delta E$ and $\Delta H$ for the combustion of one mole of *n*-decane.

**Solution**   A bomb calorimeter operates at constant volume. Therefore, the heat evolved is $q_v = \Delta E$. The combustion reaction is exothermic, since the temperature of the water in the calorimeter rises after the reaction occurs. Thus, $\Delta E$ and $q$ are negative. We can use Equation 13.9, which defines the heat capacity, to find the heat absorbed by the water of the calorimeter:

$$16.24 \ \frac{kJ}{K} = \frac{\text{heat absorbed}}{(303.51 \ K - 296.32 \ K)}$$

$$\text{heat absorbed} = 117 \ kJ$$

For the system of decane and oxygen, $\Delta E = q_v = -117$ kJ. The negative sign indicates that the process is exothermic.

The heat absorbed by the calorimeter is the heat evolved by the combustion of 2.456 g of decane (MW = 142.3), which is

$$2.456 \ \text{g decane} \times \frac{1 \ \text{mol decane}}{142.3 \ \text{g decane}} = 0.0173 \ \text{mol decane}$$

Therefore,

$$\Delta E = \frac{-117 \ kJ}{0.0173 \ \text{mol decane}} = -6760 \ kJ/mol$$

The $\Delta H$ of combustion at 296.3 K can be calculated from these data by assuming ideal gas behavior of the gases in the system:

$$\Delta H = \Delta E + RT \, \Delta n$$

The equation for the combustion of one mole of *n*-decane is

$$C_{10}H_{22}(l) + \frac{31}{2} O_2(g) \longrightarrow 10CO_2(g) + 11H_2O(l)$$

and

$$\Delta n = 10 \ \text{mol} - \frac{31}{2} \ \text{mol} = -5.5 \ \text{mol}$$

$$\Delta H = -6760 \ \frac{kJ}{mol} \times 1000 \ \frac{J}{kJ} + 8.31 \ \frac{J}{\text{mol K}} \times (296.3 \ K)(-5.5 \ \text{mol})$$

$$= -6770 \ \frac{kJ}{mol}$$

## Heat Capacity

The relationship between heat and temperature change is used extensively in thermodynamics. The molar heat capacity of a substance is

13.4 Paths for Changes in State

493

defined as the heat (in joules) that changes the temperature of one mole of a substance by 1 K. Molar heat capacity can be regarded as the proportionality constant that relates heat and temperature change per mole of substance. It is defined by:

$$q = nC\,\Delta T \qquad (13.10)$$

where $C$ is the molar heat capacity and $n$ is the amount (in moles) of the substance whose temperature changes. A quantity that is less frequently used is the *specific heat capacity* of a substance, which is defined as the heat that changes the temperature of one gram of a substance by 1 K.

The value of the heat capacity depends on the path taken by the system in changing from one temperature to another. We shall assume a constant-pressure path in which case we can use the symbol $C_P$ for the heat capacity. Table 13.1 lists the heat capacities for some substances.

**TABLE 13.1** Molar Heat Capacities of Some Substances at Room Temperature

| Substance | Molar Heat Capacity, $C_P$ (J mol$^{-1}$K$^{-1}$) |
|---|---|
| C$_{(graphite)}$ | 8.54 |
| He | 20.8 |
| Al | 24.3 |
| Fe | 24.8 |
| Hg | 27.9 |
| H$_2$ | 28.8 |
| NaCl | 45.3 |
| H$_2$O | 75.2 |
| ethanol | 113 |

**Example 13.5**

Calculate the heat required to raise the temperature of 1.0 kg of water from 298 K to 308 K. Calculate the heat required to bring about the same temperature change in the same mass of mercury.

**Solution**

Equation 13.10 gives the relationship between the heat and the three quantities given: the quantity of material, the temperature change, and the heat capacity. The $C_P$ of H$_2$O is given in Table 13.1 as 75.2 J/mol K. Thus:

$$q = 1000\text{ g} \times \frac{1\text{ mol H}_2\text{O}}{18.0\text{ g H}_2\text{O}} \times 75.2\,\frac{\text{J}}{\text{mol K}} \times (308\text{ K} - 298\text{ K})$$
$$= 42\,000\text{ J}$$

We can do the same calculation for Hg, using its heat capacity and molar mass:

$$q = 1000\text{ g} \times \frac{1\text{ mol Hg}}{200.6\text{ g Hg}} \times 27.9\,\frac{\text{J}}{\text{mol K}} \times (308\text{ K} - 298\text{ K})$$
$$= 1400\text{ J}$$

The difference in the two quantities of heat reflects the high specific heat capacity of water. A relatively large quantity of heat is required to raise the temperature of a given mass of water.

## 13.4 PATHS FOR CHANGES IN STATE

Constant-volume and constant-pressure paths for changes in state are of special interest to chemists, who use them to help measure $\Delta E$ and $\Delta H$ directly from $q$. Many real processes occur by pathways that closely approximate these ideal paths.

It will be useful to define one more idealized path for carrying out a

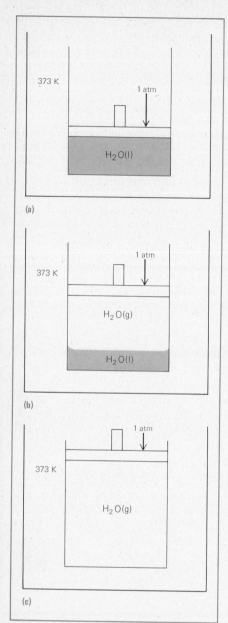

**Figure 13.4**
The evaporation of water against a pressure of 1 atm at 373 K is a real process that closely resembles an ideal process. The liquid water can absorb small amounts of heat, forming water vapor, which raises the piston. All the water can evaporate. The system is in an equilibrium state during the entire process.

change in state. It is called the **reversible path.** A system that undergoes a change in state by a reversible path is always in an equilibrium state (Section 4.2). Every state of the system — the initial state, the final state, and all states in between — is an equilibrium state. A change of state by a reversible path is called a **reversible process.** A change in state in which there are one or more intermediate states that are not equilibrium states is an **irreversible process.**

A truly reversible process exists only in theory. But there are real processes that are good approximations of reversible processes. One such process occurs when ice melts at 273 K. Mixtures consisting of an increasing quantity of liquid and a decreasing quantity of solid are formed until all of the ice melts. At 273 K, any mixture of ice and liquid water is an equilibrium state, provided that there are no localized hot or cold regions in the system. The term *reversible* can be used to describe this process because the absorption of a small quantity of heat causes some ice to melt, while the withdrawal of the same quantity of heat reverses the process, causing the same amount of water to freeze.

Another real process that approximates an ideal reversible process is the evaporation of liquid water at 373 K against an opposing pressure of 1 atm (Figure 13.4). As the liquid water absorbs small quantities of heat, it forms vapor at a pressure of 1 atm (Figure 13.4b). As the amount of vapor increases, the piston rises. Eventually, all the water evaporates; only vapor at 1 atm and 373 K is present. At any stage along the way, the equilibrium state — liquid water and water vapor at a pressure of 1 atm — exists. At any stage along the way, the withdrawal of a small quantity of heat causes the condensation of a small amount of water vapor, with a consequent fall of the piston. (Note that this reversible process involves not only heat but also work.)

The work associated with a reversible path is of special interest. If a system does work on the surroundings by expanding, it does the maximum possible work for the given change in state if the expansion occurs at constant temperature by a reversible path. The opposite is also true. If the surroundings do work on the system, causing it to contract, the minimum work is required if the change in state occurs at constant temperature by a reversible path.

## 13.5 THE SECOND LAW OF THERMODYNAMICS

The first law of thermodynamics tells us that a process in which energy is created or destroyed cannot occur. Another rule, equally basic, is illustrated when two metal blocks, one hot and one cold, are placed in contact. Heat flows from the hot block to the cold block until the two blocks are at the same temperature. We never see heat flow in the other direction, so that the cold block gets colder and the hot block gets hotter. Similarly, if we have two blocks at the same temperature, we never see heat flow so that one becomes hot and the other becomes cold. We can say that the two

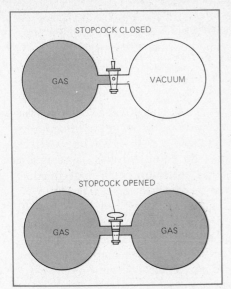

**Figure 13.5**

Gas flows from a filled bulb to an evacuated bulb until the pressures in both bulbs are equal. But when two bulbs containing gas at equal pressure are connected, a flow of gas that produces unequal pressures will not occur.

blocks at the same temperature are in thermal equilibrium.

Now consider the system shown in Figure 13.5. When the stopcock is opened, gas flows into the evacuated bulb until the pressure is the same throughout the system. If the walls of the system are made of a material that does not allow the flow of heat, the process occurs adiabatically; $q = 0$. Since there is no mechanical link with the surroundings, $w = 0$, and therefore the first law tells us that $\Delta E = 0$.

We know from experience that the reverse process does not occur. If the system is left undisturbed, there will never be a net flow of gas to create unequal pressures in the connected bulbs, even though the reverse flow also obeys the first law; $q = 0$, $w = 0$, and $\Delta E = 0$.

Both the processes we have described have a preferred direction. Heat flows from a hot body to a cold body and not the other way. Gas flows from a region of high pressure to a region of low pressure and not the other way. Chemical changes also have the same kind of preferred direction. A mixture of gasoline and oxygen in the presence of a spark forms $H_2O$ and $CO_2$, but a mixture of $CO_2$ and $H_2O$ does not form gasoline and oxygen.

These examples and countless others that illustrate a preferred direction are all *irreversible processes in which a system that is not in an equilibrium state changes spontaneously until it reaches an equilibrium state.* A system in an equilibrium state undergoes no further detectable change unless it is disturbed by a change in the surroundings.

The meaning of the word *spontaneous* in thermodynamics is somewhat different from the ordinary connotation. In thermodynamics, time is not a factor. A system is said to change spontaneously if the change occurs without the intervention of an external agent. But since this definition does not include time, a spontaneous change may occur very slowly in practice.

If the system is not in an equilibrium state, a spontaneous change is not merely possible; it is inevitable. The system eventually will undergo the change in state that is necessary to reach equilibrium. In some cases, we may have to wait millions of years, but if we have patience (and longevity) we will see the system reach equilibrium. A system in an equilibrium state does not undergo any spontaneous change in state if it is left undisturbed. To take the system away from equilibrium, it must be changed in some way by the surroundings. For example, work must be done on the system by the surroundings.

## Entropy

The spontaneous direction of processes is the subject of the second law of thermodynamics. Like the first law, the second law is the result of countless observations and can be stated in many ways. In one statement of the second law, we use a new state function called **entropy**, represented by the symbol $S$. We shall defer briefly a discussion of the definition and significance of $S$ and simply use it here to state the second law.

A change in state is accompanied by a change in entropy, $\Delta S$. In its most general sense, the second law refers to the change in the entropy of the entire universe that results from a change in state of a system. We take the term *universe* to mean the system and its surroundings:

$$\Delta S_{univ} = \Delta S_{sys} + \Delta S_{surr}$$

When an irreversible spontaneous process occurs, the entropy of the universe increases, $\Delta S_{univ} > 0$. When a reversible process occurs, the entropy of the universe remains constant, $\Delta S_{univ} = 0$. At no time does the entropy of the universe decrease. Since the entire universe is undergoing spontaneous change, the second law can be most generally and concisely stated as: *The entropy of the universe is constantly increasing.*

We know what the other state functions, such as $P$, $V$, $T$, $E$, and $H$, measure. But what does $S$ measure? We shall take up this point in detail later. For the moment, it can be said that classical thermodynamics does not require a physical explanation of the concept of entropy. All we need is an operational definition so that we can calculate the entropy change of the system and surroundings that accompanies a process.

A basic definition of the entropy change associated with a change in state of a system at constant temperature is

$$\Delta S_{sys} = \frac{q_{rev}}{T} \tag{13.11}$$

where $q_{rev}$ is the heat that accompanies the change in state if it takes place by a reversible path.

As we shall see, this definition can be applied to both the system and the surroundings. As is true for all state functions, the value of $\Delta S_{sys}$ is independent of the path followed for a change in state. Just as we did previously for $\Delta H$ and $\Delta E$, we can define the change in a state function with a path function if the exact path is specified. For $\Delta S$ we specify the reversible path. Equation 13.11 says that we can determine the entropy change accompanying a change in state at constant temperature by finding the value $q$ would have if the change occurred reversibly, and dividing this value of $q$ by the temperature in kelvins.

### Phase Changes and Entropy

For a phase change occurring at the normal transition temperature and constant pressure of 1 atm, $q_{rev}$ is the $\Delta H$ of the phase change. For example, $q_{rev}$ for the melting of ice or the freezing of water at 273 K is the $\Delta H$ of fusion of ice, 6.01 kJ/mol, or the $\Delta H$ of the freezing of water, $-6.01$ kJ/mol. Thus, we find $\Delta S_{sys}$ for 1 mol of $H_2O(s) \rightarrow 1$ mol of $H_2O(l)$ at 273 K and 1 atm by dividing $\Delta H$ of fusion, which is $q_{rev}$, by the melting point, which is $T$:

$$\Delta S_{sys} = \frac{q_{rev}}{T} = \frac{6010 \text{ J/mol}}{273 \text{ K}} = 22.0 \text{ J mol}^{-1} \text{ K}^{-1}$$

When a solid melts or a liquid boils, $\Delta H(q_{rev})$ is positive. From Equation 13.11, the entropy of the system increases. For exothermic phase changes, such as the condensation of a gas or the freezing of liquid, the entropy of the system decreases. Phase changes provide simple examples of the relationship between entropy changes and spontaneity. Although $\Delta S_{sys}$ is positive for the melting of a solid, a solid does not always melt spontaneously. Below its melting point it remains solid. We must calculate the sum of $\Delta S_{sys}$ and $\Delta S_{surr}$ at the temperature of interest to find whether a process is spontaneous.

The calculation of $\Delta S_{surr}$ is simplified if we assume that the surroundings are much larger than the system, which means that the flow of heat into or out of the system does not change the temperature of the surroundings. It also means that the surroundings remain in an equilibrium state during a change in state of the system. Therefore, the heat that flows into or out of the surroundings because of the change of state is $q_{rev}$ with respect to the surroundings. Thus, we must know two things to calculate $\Delta S_{surr}$: the temperature of the surroundings and the value of $q$ for the change in state of the system. Then:

$$\Delta S_{surr} = \frac{-q}{T} \qquad (13.12)$$

where $T$ is the temperature of the surroundings.

Consider the melting of ice at 272 K, slightly below the melting point. Although in general $\Delta H$ is temperature dependent, it can be shown that $\Delta H$ of fusion and $\Delta S_{sys}$ of fusion change only very slightly with temperature. Therefore, at temperatures close to 273 K, $\Delta S_{sys}$ is still very close to 6010 J/273 K, and $\Delta H$, and therefore $q$ is still very close to 6010 J. Thus at 272 K:

$$\Delta S_{sys} + \Delta S_{surr} = \frac{6010 \text{ J}}{273 \text{ K}} + \frac{(-6010 \text{ J})}{272 \text{ K}} < 0$$

Since the total entropy change is negative, the process does not take place.

Above the melting point of ice, say at 274 K:

$$\Delta S_{sys} + \Delta S_{surr} = \frac{6010 \text{ J}}{273 \text{ K}} + \frac{(-6010 \text{ J})}{274 \text{ K}} > 0$$

The total entropy change is positive and the process is spontaneous.

At the melting point of ice we can write:

$$\Delta S_{sys} + \Delta S_{surr} = \frac{6010 \text{ J}}{273 \text{ K}} + \frac{(-6010 \text{ J})}{273 \text{ K}} = 0$$

The total entropy change is 0 and the process is reversible.

There are many different but equivalent statements of the second law

## WILL THE UNIVERSE DIE?

The German physicist Rudolf Clausius (1822–1888) formulated the laws of thermodynamics in two sentences:*

1. The energy of the universe is constant.
2. The entropy of the universe tends toward a maximum.

The philosophers of nineteenth-century Europe dramatized this formulation into a picture of the "heat death" of the universe. Looking at the stars, they saw a continuing energy transfer process that would end only when all temperature differences were wiped out, the state of maximum entropy. Without a temperature difference, heat cannot be converted to mechanical work. Thus, all chemical, physical, and biological processes must eventually cease.

Recent cosmology has attempted to find an escape from this bleak forecast. One source of hope is the discovery that the universe is expanding. This expansion has given rise to at least two theories in which the universe will not die but will go on eternally.

One theory, formulated in the 1950s, sees a "steady-state" universe. According to this theory, the expansion of the universe is accompanied by the creation of new matter. Only one proton need be created per year in a billion liters of space to keep the mass of the universe constant, it has been calculated.

The alternative to the steady-state theory is the "big bang theory," which says that all the matter in the universe was gathered into one "superatom" some 15 to 20 billion years ago. The explosion of the superatom is believed to have started the expansion of the universe that is still going on. The big bang theory today is accepted by almost all cosmologists, because observations of the universe are consistent with almost all its predictions. In particular, background radiation that is believed to be the "ashes" of the primeval big bang has been detected.

The big bang theory opens another escape route from the heat death of the universe. It is possible that the expansion of the universe will not go on forever. Given enough mass, gravitational attraction will first slow, then reverse the expansion. Over a period of perhaps 50 billion years, all the matter in the universe will once again contract into a new superatom, and a new big bang will start the cycle again. If this picture is true, the universe will go on expanding, contracting, and exploding in an eternal cycle. However, if there is not enough matter in the universe to provide the needed gravitational attraction, expansion will be perpetual and the heat death is inevitable.

Observations made in the 1970s and 1980s appear to support the latter pessimistic picture. Astronomers at a number of observatories, using a variety of techniques, have found that the mass of the matter in and between stars apparently falls far short of what is needed to reverse the expansion of the universe. However, both these observations and our knowledge of the universe are admittedly imperfect. There is still the possibility that the universe will not die the slow death predicted by our current understanding of the laws of thermodynamics.

of thermodynamics. Some of these statements concern what are called perpetual motion machines of the second kind. One way to state the second law is to say that no machine operating in a cycle can convert a quantity of heat from the surroundings into an equal quantity of work on the surroundings. The first law does not forbid the existence of such a machine. The second law does.

Earlier, we gave "You can't win" as a simplified version of the first law. A comparable version of the second law is, "You can't even break even." You can never get back as work all the heat energy that goes into a process.

## 13.6   THE MEANING OF ENTROPY

You need not know the meaning of entropy to predict the spontaneous direction of a process. However, you will understand the behavior of systems better if you know what entropy actually measures. As an approach to the meaning of entropy, think of the microscopic composition of a macroscopic system.

Any macroscopic system consists of a very large number of small units: atoms, ions, or molecules. The thermodynamic properties of the system, such as pressure, temperature, and energy, are determined by the state of its microscopic units. There are many possible microscopic states, specific arrangements of all the units of the system, corresponding to any macroscopic state.

If we were studying temperature, we could in theory describe the state of an ideal gas by specifying the velocity of each molecule of the gas. In practice, there are so many molecules in even a small sample that we cannot make the necessary measurements.

There is another fact to consider. On the microscopic scale, every system changes constantly. Even a system in an equilibrium state, which is macroscopically static, changes continually on the atomic level (Section 4.2). Molecules are in constant motion, and their individual positions, velocities, and energies change from moment to moment. Each change of each molecule creates a new microscopic state of the system. In other words, any given macroscopic state of a system can have many different microscopic states.

*The entropy of a system in a given state is a measure of the number of different microscopic states that correspond to a given macroscopic state.* Entropy increases as the number of microscopic states increases.

This definition can help us understand why the direction of spontaneous change is always toward the equilibrium state, and why entropy is often described as *a measure of increasing disorder, or a measure of probability.*

We can illustrate the point with an analogy, shown in Figure 13.6. Suppose we have five coins. We can define two macroscopic states for the coins. They can be in an orderly arrangement, a matched state, in which either five heads or five tails are showing. Or they can be in a disorderly, unmatched state, in which some of the coins show heads and some show tails. Only two microscopic states are possible for the matched state: all heads or all tails. A much larger number of microscopic states is possible for the unmatched state; we may have one head and four tails, two heads and three tails, and so on. With only five coins, 30 different microscopic states are possible for the unmatched macroscopic state.

Suppose we put the five coins in a box, shake the box, and then look at the coins. Since there are only two microscopic states that correspond to the matched macroscopic state and 30 microscopic states that correspond to the unmatched state, the odds are 15 to 1 that the coins are in the unmatched state. Whether we start with the matched state or the un-

**Figure 13.6**

Only two matched states are possible for a system of five coins, all heads or all tails. Thirty unmatched states, various combinations of heads and tails, are possible. If we shake a box containing the coins, the odds are that an unmatched state will result. As the number of coins is increased, the odds in favor of an unmatched, disorganized state increase sharply.

OR

MATCHED STATES

↓SHAKE

SOME UNMATCHED STATES

matched state, it is equally unlikely that the matched state will occur after we shake the box.

Even with five coins, there is a direction associated with the process of shaking the box. The process "flows" toward the unmatched state. If we increase the number of coins, the direction of the process becomes more obvious. With 100 coins in the box, there are still only two microscopic states that produce a matched macroscopic state, but there are $1.3 \times 10^{30}$ microscopic states that produce the unmatched state. For practical purposes, there is almost no chance of finding a matched state after we shake a box containing 100 coins.

To use the thermodynamic term, the unmatched state in this system is the equilibrium state. We may change the microscopic state when we shake the box, but we almost certainly will not move the system to the matched state. A human hand can put the system in the matched state by doing work, picking up and turning over the necessary number of coins. But again, a shake of the box causes the system to proceed spontaneously to the unmatched equilibrium state.

Now consider a system whose units of structure are molecules, not coins. This page is such a system. The molecules of this page are in constant, random motion at equilibrium. There is always some small probability that the molecules making up the page will move spontaneously in an ordered way, so that the page will turn by itself. You are not advised to wait for this event to happen. It is extremely improbable that the page will turn of itself, even in the 12 billion or so years before the sun burns out. The probability of this nonequilibrium state forming is small enough to ignore.

## Predicting $\Delta S_{sys}$

Once we define entropy in this way, we can reason qualitatively and predict the sign of $\Delta S_{sys}$ for various changes in state. (It should be remembered that the $\Delta S$ of system plus surroundings is always either positive or zero.)

When a substance undergoes an expansion at constant temperature, the entropy of the system increases, $\Delta S_{sys} > 0$. When a given quantity of material occupies a greater volume, it becomes more disordered. More microscopic states are possible because more positions are available for the molecules, and therefore the value of $\Delta S$ is greater. Conversely, $\Delta S_{sys} < 0$ when a substance contracts at constant temperature.

When the temperature of a substance increases at constant volume, $\Delta S_{sys} > 0$. There is a greater distribution of molecular velocities at higher temperatures, and thus a greater number of microscopic states is possible.

Any process in which different substances are mixed without chemical change has $\Delta S_{sys} > 0$, because mixing means increased disorder. Gaseous solutions form by diffusion because the entropy of the system increases as a result of mixing. As we mentioned in Chapter 12, the solubility of one substance in another is often caused largely by the natural tendency for

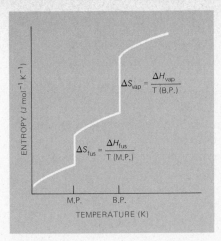

**Figure 13.7**
The entropy of a substance increases with rising temperature. There are sharp increases at the melting point (M.P.) and the boiling point (B.P.).

In the figure:
$$\Delta S_{vap} = \frac{\Delta H_{vap}}{T \text{ (B.P.)}}$$

$$\Delta S_{fus} = \frac{\Delta H_{fus}}{T \text{ (M.P.)}}$$

ENTROPY (J mol$^{-1}$ K$^{-1}$)

TEMPERATURE (K)

two materials to mix. We can now say that solubility is due in part to an increase in the entropy of the system due to mixing.

Any process that increases the number of particles in the system has $\Delta S_{sys} > 0$. For a reaction like $H_2(g) \rightarrow 2H(g)$, which forms two particles from one, the entropy of the system increases. In the reverse process, $2H(g) \rightarrow H_2(g)$, in which the number of particles is reduced, the entropy of the system decreases, $\Delta S_{sys} < 0$.

The change in the entropy of the system associated with a phase change is related to the difference in order in the phases. A solid is more ordered than a liquid, which is more ordered than a gas. Therefore, an increase in the entropy of the system is associated with melting (the change from solid to liquid), boiling (the change from liquid to gas), and sublimation (the change from solid to gas). A decrease in the entropy of the system results from condensation (a change from gas to liquid or solid) and freezing (a change from liquid to solid) (Figure 13.7).

Remember, the entropy of a *system* can decrease only if the entropy of the surroundings increases at least as much.

**Example 13.6**

Predict the sign of $\Delta S_{sys}$ for each of the following processes: (a) $I_2(g) \rightarrow I_2(s)$, (b) $CH_3OH(l) \rightarrow CH_3OH(aq)$, (c) $N_2(g) + 3H_2(g) \rightarrow 2NH_3(g)$, (d) $CH_4(g) + 2O_2(g) \rightarrow CO_2(g) + 2H_2O(l)$.

**Solution**

a. In the change from gas to solid, $\Delta S_{sys}$ is negative because a gas is more disorganized than a solid.

b. When a solution of methanol and water is formed from pure methanol and pure water, $\Delta S_{sys}$ is positive because of the disorder that results from mixing. In the formation of some solutions, factors other than mixing may affect the sign of $\Delta S_{sys}$.

c. In this reaction, there are four molecules of reactants and two molecules of product. This decrease in the number of molecules results in a decrease in the entropy of the system, so $\Delta S_{sys}$ is negative.

d. This reaction does not change the number of molecules, but two molecules of gas are changed to molecules of liquid. The system thus becomes more organized, so the $\Delta S_{sys}$ is negative.

## The Third Law of Thermodynamics

We stated that it is impossible to find the absolute value of $E$ or $H$ of a given state. But we can find the $\Delta E$ or $\Delta H$ associated with a change in state. Therefore, we can choose an arbitrary standard state or zero point for these state functions. Elements in their standard states are assigned zero enthalpy of formation for the sake of convenience.

The choice of a state for the zero entropy point is not arbitrary. It is based on the fact that entropy measures the number of microscopic states that correspond to a given macroscopic state. When there is only one microscopic state, $S = 0$. The designation of this state is often called the

**501**

third law of thermodynamics. The third law is stated: *The entropy of a perfect crystal of a pure substance at the absolute zero of temperature is zero.*

A perfect crystal is a completely ordered arrangement in which each unit of structure has a fixed position. At 0 K, each atom has only one possible energy state, the minimum state. Since the substance is perfectly pure, there is no disorder resulting from mixing. Therefore, a perfect crystal of a pure substance at the absolute zero of temperature is the most ordered system imaginable.

When $S = 0$, there can be only one microscopic state. The fact that log $1 = 0$ suggests that there is a logarithmic relationship between the number of microscopic states and $S$. The relationship can be expressed as:

$$S = k \ln \Omega \qquad (13.13)$$

where $k$ is called *Boltzmann's constant* and is defined as $R/N$, the gas constant divided by Avogadro's number, and $\Omega$ is the number of microscopic states.

The third law makes it possible to find an actual value for the entropy of a substance at a given temperature. This value is called the absolute entropy, $S°$. Some absolute entropies are listed in Appendix II.

The value of $S°$ indicates the relationship between the molecular structure and the entropy of a substance in a given state. Other things being equal, the absolute entropy generally increases as molecular size and complexity increase. For a substance of given molecular complexity, the gas has higher $S°$ than the liquid, and the liquid has higher $S°$ than the solid. The entropy of diamond, in which each carbon atom is bonded to four others in a rigid structure, is lower than that of graphite, where each carbon atom is bonded to only three others in a less rigid structure. Hard substances such as diamond generally have rigid, symmetrical crystal structures with a minimum of disorder. Therefore, hard substances tend to have low $S°$.

The values of $S°$ can be used to find $\Delta S°$, the absolute entropy change of a system undergoing a chemical reaction. This calculation is performed in essentially the same way as the calculation of $\Delta H°$ in Section 8.7.

$$\Delta S° = S°(\text{products}) - S°(\text{reactants}) \qquad (13.14)$$

**Example 13.7**    Use the data in Appendix II to calculate $\Delta S°$ for the reaction:

$$N_2(g) + 3H_2(g) \longrightarrow 2NH_3(g)$$

at 298 K and 1 atm.

**Solution**    The relevant values of $S°$ (in J K$^{-1}$ mol$^{-1}$) are: NH$_3$(g), 193; N$_2$(g), 192; H$_2$(g), 131. Using the relationship expressed in Equation 13.14:

$$\Delta S° = [(2 \text{ mol})(193 \text{ J K}^{-1} \text{ mol}^{-1})] - [(192 \text{ J K}^{-1} \text{ mol}^{-1})$$
$$+ (3 \text{ mol})(131 \text{ J K}^{-1} \text{ mol}^{-1})]$$
$$= -199 \text{ J/K}$$

In the reaction, the amount of gas decreases from four moles to two moles. Since the number of particles decreases, there is less disorder, and entropy decreases.

## 13.7   FREE ENERGY

The use of the total entropy change ($\Delta S_{\text{univ}}$) as a criterion for the spontaneity of reactions is somewhat awkward. It requires two calculations, one for the $\Delta S$ of the system, the other for the $\Delta S$ of the surroundings. It would be convenient to have a state function that is a property of the system alone and that could be used as a criterion for spontaneity.

We shall define a state function that can be used to predict spontaneity for changes in state carried out at *constant temperature and pressure,* the conditions of most chemical reactions. This state function is the **Gibbs free energy,** *G*, which is defined in terms of other state functions of the system as:

$$G = H - TS \tag{13.15}$$

For a change in state at constant temperature,

$$\Delta G = \Delta H - T\Delta S \tag{13.16}$$

Each change in state function is for the *system.*

Equation 13.16 is one of the most important relationships in chemistry. The relationship between $\Delta G$ and the total entropy change ($\Delta S_{\text{univ}}$) can be derived from it. For a process carried out at constant pressure, $\Delta H = q$. Substitution into Equation 13.12 gives a simple relationship between $\Delta H$ and $\Delta S_{\text{surr}}$:

$$\Delta H = -T\Delta S_{\text{surr}}$$

The $\Delta S$ in Equation 13.16 is the $\Delta S_{\text{sys}}$. Therefore, Equation 13.16 can be rewritten as:

$$\Delta G = -T\Delta S_{\text{surr}} - T\Delta S_{\text{sys}} = -T\Delta S_{\text{univ}} \tag{13.17}$$

Therefore, $\Delta G = 0$ *for a reversible change at constant pressure and temperature.* Because of the negative sign in Equation 13.17, $\Delta G$ *is negative for a spontaneous change at constant pressure and temperature.* A change in state with a positive $\Delta G$ can occur only if work is done on the system; it is nonspontaneous.

It is important not to confuse the signs of $\Delta S_{univ}$ and $\Delta G$ for a change in state. For a reversible change, both these quantities are zero. But for a spontaneous change, $\Delta S_{univ}$ is positive while $\Delta G$ is negative. Remember, $\Delta G$ refers to the system.

## Free Energy of Phase Changes

Equation 13.16 tells us a great deal about changes of state. Consider

$$H_2O(l) \longrightarrow H_2O(g)(P = 1 \text{ atm})$$

Since a gas is more disordered than a liquid, $\Delta S$ of the system for this change is positive. The value of $\Delta S_{vap}$ of $H_2O$ has been found to be 108.8 J $K^{-1}$ $mol^{-1}$ at 373 K. The sign of $\Delta H$ for this change is also positive, since heat from the surroundings is needed to make the liquid vaporize. The value of $\Delta H$ has been found to be 40 626 J/mol at 373 K.

Although $\Delta H$ and $\Delta S$ vary with temperature, we can assume that they do not vary appreciably at temperatures close to the boiling point. We can determine the boiling point of water from the values of $\Delta H$ and $\Delta S$, using Equation 13.16. The boiling point can be defined as the temperature at which any quantities of $H_2O(g)$ and $H_2O(l)$ are at equilibrium at 1 atm. Therefore, the changes from $H_2O(l)$ to $H_2O(g)$ and from $H_2O(g)$ to $H_2O(l)$ at 1 atm are reversible changes, for which $\Delta G = 0$. Setting $\Delta G$ in Equation 13.16 equal to zero gives:

$$\Delta H = T \Delta S$$

or
$$T = \frac{\Delta H}{\Delta S} \qquad\qquad (13.18)$$

Substituting the values of $\Delta H_{vap}$ and $\Delta S_{vap}$ for $H_2O$ in Equation 13.18 gives the boiling point:

$$T = \frac{40\ 626 \text{ J/mol}}{108.8 \text{ J mol}^{-1} \text{ K}^{-1}} = 373 \text{ K}$$

Another way to express the same idea is to say that at the boiling point of water, the favorable negative term, which is the product of temperature and entropy, just balances the unfavorable positive enthalpy term. At a temperature below 373 K, the product of $T$ and $\Delta S$ is not great enough to cancel $\Delta H$; that is, $\Delta H > T \Delta S$, and therefore $\Delta G > 0$. The positive value of $\Delta G$ indicates that water does not boil spontaneously below 373 K at 1 atm. We cannot have $H_2O(g)$ with a pressure of 1 atm below 373 K. When the temperature is greater than 373 K, then $T \Delta S > \Delta H$, and $\Delta G < 0$. The negative value of $\Delta G$ indicates that water boils spontaneously above 373 K at 1 atm.

For the reverse process:

$$H_2O(g)(P = 1 \text{ atm}) \longrightarrow H_2O(l)$$

both $\Delta S$ and $\Delta H$ are negative. When $T < 373$ K, the term $\Delta H$ is more important than the term $T \Delta S$. Since $\Delta H < 0$, then $\Delta G < 0$. This process, 1 atm of water vapor condensing to liquid at a temperature below 373 K, is spontaneous. At temperatures above 373 K, the term $T \Delta S$ is more important than the term $\Delta H$. And since $\Delta S < 0$, then $-T \Delta S > 0$ and $\Delta G > 0$. The condensation of 1 atm of water vapor at temperatures above the boiling point of water is not a spontaneous process. But at exactly 373 K, $\Delta H = T \Delta S$, $\Delta G = 0$, and again the process is reversible.

Since $\Delta G = 0$ for the process

$$H_2O(l) \rightleftharpoons H_2O(g)(P = 1 \text{ atm}, T = 373 \text{ K})$$

in either direction, we can say that the free energy of $H_2O(l)$ is equal to the free energy of $H_2O(g)$ at 373 K and 1 atm. Thus, a system containing these two components is in an equilibrium state.

## Enthalpy, Entropy, and Spontaneity

This sort of analysis can be performed on many processes. The first step is to judge whether $\Delta H$ and $\Delta S$ are positive or negative for a given process. By considering the bonds made and broken (Section 8.7) and the phase changes that occur, we often can determine whether $\Delta H$ is positive or negative. A similar judgment for $\Delta S$ is often possible from simple considerations of order and disorder, such as changes in the number of molecules. A process that is exothermic ($\Delta H < 0$) and causes increased disorder ($\Delta S > 0$) must result in a negative $\Delta G$. Such a process is spontaneous. A process that is endothermic ($\Delta H > 0$) and causes increased order ($\Delta S < 0$) must result in a positive $\Delta G$. Such a process is nonspontaneous. We need more information to make the analysis only when $\Delta H$ and $\Delta S$ have the same sign. Table 13.2 tabulates these possibilities.

## Free Energy and Temperature

Although both $\Delta H$ and $\Delta S$ do not change appreciably as the temperature changes, $\Delta G$ can vary considerably. According to Equation 13.16, the value of $\Delta G$ is determined by $\Delta H$ and by $T \Delta S$. Therefore, the contribution of $\Delta S$ to the value of $\Delta G$ becomes relatively more important as the temperature increases. An endothermic process that results in increasing disorder is more likely to proceed spontaneously as the temperature increases. At high enough temperatures, $\Delta S$ becomes the controlling factor. Conversely, an exothermic process that results in increasing order is more likely to proceed spontaneously as the temperature decreases. When $T$ is small, $T \Delta S$ can be small compared to $\Delta H$.

**TABLE 13.2  Temperature and Spontaneity**

| $\Delta H$ | $\Delta S$ | $\Delta G$ |
|---|---|---|
| − | + | − reaction is favored at all temperatures |
| + | − | + reaction is never favored |
| − | − | ± decreasing temperature favors reaction |
| + | + | ± increasing temperature favors reaction |

A good rule of thumb (which has numerous exceptions) is that $\Delta H$ is the controlling factor at temperatures close to those normally found on the surface of the earth. Exothermic processes generally are spontaneous. Endothermic processes generally are not. A simple change that illustrates these points is

$$\tfrac{1}{2}H_2(g) \longrightarrow H(g)$$

The reaction does not proceed to any measurable extent at 300 K. It proceeds to an appreciable extent above 4000 K. Since the reaction breaks one bond and does not make any, it is endothermic, $\Delta H > 0$. Since it produces two particles from one, it increases disorder, $\Delta S > 0$. When $T$ is not large, $T \Delta S$ is not large; $\Delta H$ is the controlling factor and $\Delta G$ has a positive sign. When $T$ is large, $T \Delta S$ is large and $\Delta G$ has a negative value. Table 13.3 lists the values of $\Delta G$, $\Delta H$, and $\Delta S$ at different temperatures at 1 atm for this reaction.

This reaction is just one demonstration of the fact that all compounds break down into their constituent atoms at very high temperatures. We can look to the sun for confirmation. Temperatures inside the sun are so high that no amount of exothermic bond-making can overcome the bond-breaking associated with a large $T \Delta S$ term. There are no molecules in the sun. In fact, the sun is so hot that it consists of plasma, a gas in which all the atoms have been completely ionized.

## Standard Free Energy

In Section 8.7 (page 314), which you should review at this point, we used tabulated values of $\Delta H$ to calculate the enthalpy changes associated with physical and chemical changes. We can use an analogous approach for calculations with $\Delta G$.

Once again, we start by defining a standard state. In this case, we define the standard state of a substance as its most stable state at 1 atm pressure. The symbol for the free energy of a substance in the standard state is $G^\circ$. To obtain the standard free energy change, $\Delta G^\circ$, we subtract the standard free energy of the reactants from the standard free energy of the products at the temperature of the process.

$$\Delta G^\circ = G^\circ(\text{products}) - G^\circ(\text{reactants}) \tag{13.19}$$

**TABLE 13.3**  Thermodynamic Functions for the Reaction $\frac{1}{2}H_2(g) \rightarrow H(g)$

| Temperature (K) | $\Delta H^\circ$ (kJ) | $\Delta S^\circ$ (kJ/K) | $\Delta G^\circ$ (kJ) |
|---|---|---|---|
| 298 | 218 | 0.0494 | 203 |
| 500 | 219 | 0.0527 | 193 |
| 1000 | 222 | 0.0567 | 166 |
| 2000 | 227 | 0.0600 | 107 |
| 3000 | 230 | 0.0612 | 46 |
| 4000 | 231 | 0.0617 | −15 |
| 5000 | 232 | 0.0618 | −80 |

It is helpful to define a $\Delta G^\circ$ for the formation of one mole of a substance in its standard state from its constituent elements in their standard states. The $\Delta G^\circ$ for this specific process is called the *standard free energy of formation,* and its symbol is $\Delta G_f^\circ$. By definition, *the $\Delta G_f^\circ$ of an element in its standard state, is zero.* For example, the $\Delta G^\circ$ of the reaction:

$$H_2(g) + \tfrac{1}{2}O_2(g) \longrightarrow H_2O(l)$$

is $\Delta G_f^\circ$, the standard free energy of formation of water, when the pressures of all the substances in the reaction are 1 atm.

The relationship between the $\Delta G^\circ$ of a change in state and the $\Delta G_f^\circ$ of the substance involved in the change is:

$$\Delta G^\circ = \Sigma\, \Delta G_f^\circ(\text{products}) - \Sigma\, \Delta G_f^\circ(\text{reactants}) \qquad (13.20)$$

Note that this expression is essentially the same relationship as Equation 8.2, which we used earlier for $\Delta H^\circ$.

There are extensive tables listing $\Delta G_f^\circ$ values for many substances. Most of these lists give $\Delta G_f^\circ$ values for a substance at 298 K. One such table is in Appendix II. You can use this information to calculate $\Delta G^\circ$ for a change in state at 298 K in the same way that you would use the data for the standard enthalpy of formation.

**Example 13.8**  Using the data in Appendix II, calculate $\Delta G^\circ$ at 298 K for the combustion of acetylene:

$$C_2H_2(g) + \tfrac{5}{2}O_2(g) \longrightarrow 2CO_2(g) + H_2O(l)$$

**Solution**  The values of $\Delta G_f^\circ$ from Appendix II are $C_2H_2(g)$, 209 kJ/mol; $O_2(g)$, 0 kJ/mol (an element in its standard state); $CO_2(g)$, −394 kJ/mol; $H_2O(l)$, −237 kJ/mol. We can calculate the value of $\Delta G^\circ$ by using the relationship given in Equation 13.20:

$$\Delta G^\circ = \left[(2\text{ mol})\left(-394\,\frac{kJ}{mol}\right) + \left(-237\,\frac{kJ}{mol}\right)\right] - \left(209\,\frac{kJ}{mol} + 0\right)$$

$$= -1230\text{ kJ}$$

The standard free energy of formation of many compounds is negative at 298 K. These compounds form spontaneously from their constituent elements in their standard states. Compounds that have positive values of $\Delta G_f^\circ$ usually are not formed directly from their constituent elements. Indirect methods generally must be used to form them.

We can illustrate the meaning of $\Delta G^\circ$ by using a general reaction:

$$A(g) \rightleftharpoons B(g)$$

Assume that both A and B are ideal gases in a bulb, each at a pressure of 1 atm; both A and B are in their standard states. If the $G^\circ$ of A is the same as the $G^\circ$ of B, then

$$\Delta G^\circ = G_B^\circ - G_A^\circ = 0$$

and the system is in an equilibrium state. No detectable change in composition occurs. But if the $G^\circ$ of B is less than the $G^\circ$ of A, then $\Delta G^\circ < 0$. Some A will convert spontaneously to B. The free energy of A decreases as the amount of A decreases, and the free energy of B increases as the amount of B increases. Eventually, the two free energies are equal, $G_B = G_A$ and $\Delta G = 0$; the system is at equilibrium. The equilibrium state contains both A and B, but it contains more B, the substance with the lower $G^\circ$, than A.

Now assume that the $G^\circ$ of A is less than the $G^\circ$ of B. Now $\Delta G^\circ$ for the reaction is positive, $\Delta G^\circ > 0$. The reaction proceeds "backward," from right to left, spontaneously producing A from B until an equilibrium state is reached. In this equilibrium state, there is more A, the substance with the lower $G^\circ$, than B.

## Free Energy and Equilibrium

The relationship between $\Delta G^\circ$ and the composition of equilibrium states is one of the most important in chemistry. We shall begin the discussion of this relationship by writing a general chemical reaction as:

$$a A(g) + b B(g) + c C(l) \rightleftharpoons d D(g) + e E(g) + f F(s)$$

The lowercase letters are the coefficients of the chemical equation, and the uppercase letters represent different substances whose states are indicated by state symbols. We assume ideal gas behavior. When each component of this reaction is in its standard state — that is, at a pressure of 1 atm — the free energy change of this change in state, $\Delta G = \Delta G^\circ$. At equilibrium, however, the pressure of each component generally is not 1 atm.

To find the relationship between $\Delta G^\circ$ and the composition of the equilibrium state, we must first consider how the free energy of a substance changes with pressure. The free energy of a solid or liquid does *not* change significantly with changes in pressure. The free energy of a gas does change significantly with a change in pressure. We can show (using calculus) that the free energy $G$ of an ideal gas with a pressure or partial pressure $P$ is[1]

$$G = G^\circ + nRT \ln \frac{P}{P^\circ} \tag{13.21}$$

Where $P^\circ = 1$ atm, the standard state. We can use this equation to find the relationship between $\Delta G^\circ$ and $\Delta G$ for the general reaction. It is

$$\Delta G = \Delta G^\circ + RT \ln \frac{P_D{}^d P_E{}^e}{P_A{}^a P_B{}^b} \tag{13.22}$$

Equation 13.22 tells us that the difference between the free energy change for a process in which all the components are at a pressure of 1 atm ($\Delta G^\circ$) and the free energy change under any other set of pressure conditions ($\Delta G$) is a logarithmic pressure term. The form of this term depends on the form of the chemical reaction. The numerator of the fraction includes all the gases on the right side of the general reaction. The denominator includes all the gases on the left side of the reaction. The coefficients of the gases in the reaction appear as exponents of the terms for the pressures of the gases.[2] None of the liquids or solids in the reaction appear in the pressure term, because the free energy of a liquid or solid does not change in any significant way with a change in pressure.

## The Equilibrium Constant

When the pressures of all the components of the system are such that it is in an equilibrium state, we know that $\Delta G = 0$. Therefore, when the system is at equilibrium, Equation 13.22 becomes:

$$\Delta G^\circ = -RT \ln \frac{P_D{}^d P_E{}^e}{P_A{}^a P_B{}^b} \tag{13.23}$$

Here we have the desired relationship between $\Delta G^\circ$ and the composition of the equilibrium state of the system.

Once we write a chemical equation that relates the components of a system, we fix the value of $\Delta G^\circ$ at any given temperature. We can calculate the value of $\Delta G^\circ$ when the temperature is 298 K by using the data in Appendix II.

---

[1] This equation, like many others in thermodynamics, includes a natural logarithm (ln) rather than a logarithm to the base 10 (log). We shall write equations in terms of ln, but remember its relationship to log: $\ln x = 2.303 \log x$.

[2] In this equation and others derived from it, each $P$ is actually the ratio of $P$ to $P^\circ$. Since $P^\circ$ is 1 atm, we can omit it. But because the pressure of each gas is a ratio, the logarithmic pressure term has no units, provided that $P$ is expressed in atmospheres.

Now look at each of the terms in Equation 13.23. At a given temperature, $R$, $\Delta G°$, and $T$ are all constants. Therefore, the pressure term must also be a constant, which is a most interesting conclusion. It means that for any chemical reaction, we can write an expression that includes the pressures of the gases in the system and that has a constant value at a given temperature when the system is at equilibrium. This useful expression is called the **equilibrium constant**. It is usually represented by the symbol $K$. Equation 13.23 usually is written in a simpler form that uses the symbol for the equilibrium constant:

$$\Delta G° = -RT \ln K \qquad (13.24)$$

The form of the equilibrium constant results from the way in which the free energy depends on changes in pressure.[3] The free energy of pure liquids and solids does not change with pressure. Therefore, no information about liquids and solids appears in the expression for $K$. The relationship between the free energy change of a gas and a change in its pressure is logarithmic. Therefore, the coefficients of the gases appear in the expression for $K$ as exponents.

**Example 13.9** Calculate the value of $K$ at 298 K for the reaction:

$$N_2O_4(g) \rightleftharpoons 2NO_2(g)$$

using the data in Appendix II.

**Solution** From Equation 13.20,

$$\Delta G° = \left[(2 \text{ mol})\left(51 \frac{kJ}{mol}\right)\right] - \left[(1 \text{ mol})\left(98 \frac{kJ}{mol}\right)\right]$$
$$= 4 \text{ kJ}$$

From Equation 13.24,

$$4000 \text{ J} = -(8.31 \text{ J mol}^{-1} \text{ K}^{-1})(298 \text{ K}) \ln K$$
$$\ln K = -1.62$$
$$K = 2.0 \times 10^{-1}$$

We shall discuss the use and significance of the equilibrium constant in detail in Chapter 14. For the moment, let us review some of the features of $K$ and $\Delta G$:

**1.** At constant temperature and pressure, *any spontaneous change in state of a system is accompanied by a negative $\Delta G$*, that is, a decrease in the

[3] Remember that $K$, like the logarithmic pressure term in Equation 13.22, has no units if pressure is expressed in atmospheres.

free energy of the system. The equilibrium state is said to be the most stable state of the system; it is *the state of minimum free energy.* There is a clear relationship between the free energy of a system and its stability. The lower the free energy, the greater the stability. A substance with a negative free energy of formation is stable with respect to decomposition to its elements. A substance with a positive free energy of formation is unstable with respect to its elements.

2. For the general chemical reaction on page 508, an equilibrium state can exist for many different quantities of the gases A, B, D, and E. But when the pressure of each gas at equilibrium is substituted into the expression for $K$, the value of $K$ must be the specific value defined by the $\Delta G°$ and $T$ of the system for the equilibrium state. The pressure of each gas depends on the amount of gas in the system, but at equilibrium these pressures must be consistent with the relationship among $K$, $\Delta G°$, and $T$.

3. The amounts of liquid and solid components present at equilibrium do not influence the pressures of the gaseous components at equilibrium. But the nature of the solid and liquid components do determine the value of $\Delta G°$ and therefore of $K$.

    Consider the reaction:

$$CaCO_3(s) \rightleftharpoons CaO(s) + CO_2(g)$$

If a quantity of $CaCO_3(s)$ is heated to 1000 K, the pressure of $CO_2(g)$ at equilibrium is about $4 \times 10^{-2}$ atm. The equilibrium pressure of $CO_2(g)$ is related to the difference between the standard free energy of the products, $CO_2(g)$ and $CaO(s)$, and the reactant, $CaCO_3(s)$, at this temperature. But adding or removing either solid, $CaCO_3(s)$ or $CaO(s)$, does not change the equilibrium pressure.

4. There is a simple relationship between the sign of $\Delta G°$ and the magnitude of $K$. We can put the relationship in tabular form:

| $\Delta G°$ | $< 0$ | $> 1$ | 0 |
|---|---|---|---|
| $K$ | $> 1$ | $< 1$ | 1 |

As an example, consider the process:

$$H_2O(l) \rightleftharpoons H_2O(g)$$

which has $K = P_{H_2O}$. We know that at 373 K, $\Delta G° = 0$. At that temperature, $K = 1$; or, $P_{H_2O} = 1$ atm. At temperatures lower than 373 K, liquid water at 1 atm is more stable than gaseous water at 1 atm; $\Delta G° > 0$ for the process, so $K < 1$. That is, at equilibrium at temperatures below 373 K, the vapor pressure of water is less than 1 atm. At temperatures above 373 K, $\Delta G° < 0$; gaseous water in its standard state at 1 atm is more stable—has a lower free energy—than liquid water at 1 atm. For $H_2O(l)$ and $H_2O(g)$ to be at equilibrium at temperatures above 373 K, the pressure must be greater than 1 atm.

### Solution Equilibrium

We can broaden the relationship between $\Delta G °$, $K$, and the composition of the system to include dissolved substances. For a discussion of solution equilibrium, we need a standard state for solutes. A concentration of $1M$ is used as the standard state of a dissolved solute. The relationship between the free energy and the molarity of an ideal solute is the same as that between the free energy and the pressure of an ideal gas. We can therefore derive a relationship, similar to Equation 13.23, which includes the molarities of solutes as well as the pressures of gases. The molarity of any substance in solution appears in the equilibrium constant expression. In the expression, the molarity is raised to the power of the coefficient of the solute in the chemical equation.

Thus, the equilibrium expression for the reaction in which gaseous $CO_2$ dissolves in water, $CO_2(g) \rightleftharpoons CO_2(aq)$, is

$$K = \frac{[CO_2]}{P_{CO_2}}$$

where, by convention, we indicate the molarity of each species in solution by placing the formula of the species in square brackets.

For the reaction in which solid lead chloride dissolves in water, $PbCl_2(s) \rightarrow Pb^{2+}(aq) + 2Cl^-(aq)$ the equilibrium expression is

$$K = [Pb^{2+}][Cl^-]^2$$

We shall discuss solution equilibria in greater detail in later chapters.

### Temperature Dependence of K

Often we would like to know how $K$ for a given reaction varies with temperature. Equation 13.24 can give this information, if the value of $\Delta G °$ at the temperature of interest is known. In many cases, data are not available for the calculation of $\Delta G °$ at all temperatures. However, while $\Delta G °$ varies considerably with temperature, $\Delta H °$ and $\Delta S °$ are relatively constant over a range of temperatures. On this basis, a relationship between $K$ and $T$ can be derived from Equations 13.15 and 13.24:

$$\frac{\Delta H °(T_2 - T_1)}{T_1 T_2} = R \ln \frac{K_2}{K_1} \qquad (13.25)$$

where $K_1$ and $K_2$ are the equilibrium constants at $T_1$ and $T_2$ for the reaction whose standard enthalpy change is $\Delta H °$. Equation 13.25 allows us to predict how $K$ varies with $T$, if we know whether the reaction is exothermic or endothermic. Suppose that $T_2$ is greater than $T_1$ in Equation 13.25. Then $T_2 - T_1$ is positive. If the reaction is endothermic, $\Delta H°$ is positive. Therefore, $\ln (K_2/K_1)$ is positive, and $(K_2/K_1) > 1$. *The value of K for an endothermic reaction increases with temperature.*

The reasoning is similar for an exothermic reaction. In this case, $\Delta H° < 0$, and the left side of Equation 13.25 is negative; $\ln (K_2/K_1)$ is also negative, and $(K_2/K_1) < 1$. *The value of K for an exothermic reaction decreases with temperature.*

---

**Example 13.10**    Calculate the value of $K$ at 500 K for the system of Example 13.9, $N_2O_4(g) \rightleftharpoons 2NO_2(g)$, using the data in Appendix II.

**Solution**    First we must calculate $\Delta H°$ for the reaction in the usual way, using $\Delta H_f°$ of $NO_2 = 34$ kJ/mol and $\Delta H_f°$ of $N_2O_4 = 10$ kJ/mol:

$$\Delta H° = \Delta H_f°(\text{products}) - \Delta H_f°(\text{reactants})$$

$$\Delta H° = \left[ (2 \text{ mol})\left( 34 \ \frac{\text{kJ}}{\text{mol}} \right) \right] - \left[ 10 \ \frac{\text{kJ}}{\text{mol}} \right] = 58 \text{ kJ} = 58\,000 \text{ J}$$

From Example 13.9, we know that $K_{298} = 2.0 \times 10^{-1}$ atm. Equation 13.25 gives:

$$\frac{58\,000 \text{ J}(500 \text{ K} - 298 \text{ K})}{(500 \text{ K})(298 \text{ K})} = (8.31 \text{ J mol}^{-1} \text{ K}^{-1}) \ln \frac{K_2}{2.0 \times 10^{-1}}$$

$$\ln K_2 = 7.85$$
$$K_2 = 2.6 \times 10^3 \text{ atm}$$

As expected, when the temperature increases, the $K$ of this endothermic reaction increases.

---

## 13.8  THERMODYNAMICS IN THE REAL WORLD

Chemical thermodynamics is more than a set of principles that are applied to idealized laboratory systems. Living things also obey the laws of thermodynamics. Because living things are complex systems, we must use approximations when we apply the laws of thermodynamics to them. But even this imperfect application teaches us some valuable lessons about the activity of living things, up to and including human beings.

We can start with a basic thermodynamic study of an animal. The food that an animal eats goes through a long series of chemical reactions in the body. In these reactions, heat is released. Some of the heat is thrown off into the surroundings and some is used in the body of the animal.

Every process involving work, heat, and energy in the living organism must obey the first law of thermodynamics. In theory, we can do the same first-law bookkeeping for an animal as for an ideal gas. But the processes

An experimental solar energy plant at Daggett, California. More than 1800 mirrors concentrate sunlight on the central boiler to produce steam that is used to generate electricity. Solar energy is one of the most promising alternatives to the use of solar fuels, whose supply is dwindling. *(Georg Gerster, Photo Researchers, Inc.)*

in a living organism are quite complex, and living systems are never really in an equilibrium state. The net processes of life are spontaneous and irreversible, since living organisms are heading relentlessly toward an equilibrium state that can be reached only after the life process stops.

The net result of the processes of life is an increase in the entropy of the universe. To demonstrate this increase rigorously, we must consider such parameters as the heat thrown off by the organism and the heat given off by the sun, which provides the energy for all life on earth. There are lessons to be learned from qualitative reasoning about the thermodynamics of living organisms. We know from the first law that no living thing can exist without food, since it cannot create energy. We know from the second law that the efficiency with which food is used to produce work is less than 100%. We know that all living things eventually die.

The same laws can be applied to our society. A technological society needs energy to run machines. We have used energy for our machines so prodigally that we have a crisis. Thermodynamic reasoning helps us to appreciate the nature and magnitude of the problem and to evaluate possible solutions.

The first law tells us that we cannot create energy. We are limited to a fairly small number of energy sources: fossil fuels, direct use of solar energy, nuclear fission, nuclear fusion.

The second law tells us that we cannot achieve 100% conversion of heat to work. We must regard an energy source as something that releases waste heat. Even in our most efficient available processes, a substantial fraction of the available heat is wasted. A coal- or oil-burning plant turns only 40% of the heat from its fuel into work. A nuclear fission plant turns only 33% of its heat into work. An automobile is even less efficient.

As our society uses more energy, it releases more waste heat. The result of this heat release is a change in the surroundings, the environment. We know that the "heat island" of a large city changes weather patterns, and that the "thermal discharges" of generating plants have some effect on living organisms in a river. While the long-term effects of such changes are difficult to establish, we have learned from hard experience in recent years to be cautious about any large-scale, artificial change in the environment.

In recent years, human progress has been equated with an increase in the available supply of energy. More energy means better living standards for more people. But waste heat is a limiting factor that must be considered when we discuss the utopia of virtually unlimited energy from such sources as nuclear fusion. The generation of virtually unlimited energy from nuclear fusion could release waste heat in quantities that could first melt the polar ice caps and eventually make this planet unlivable.

Solar energy appears to offer a way out. The earth receives energy from the sun and radiates almost the same quantity of energy back into space. It would seem that capturing solar energy for human use is virtually harmless. Even solar energy has some potentially harmful side effects, however. The large-scale use of solar cells, windmills, and other solar energy systems requires enormous consumption of raw materials and could cause damaging changes on a local scale. Only with careful planning will solar energy be able to offer us the limitless, low-pollution energy source that we are seeking.

## Summary

This chapter introduced **thermodynamics,** the study of the relationship between **heat** and other forms of **energy.** We first noted that thermodynamics does not deal with time. It tells us whether a reaction will occur but does not tell us how much time is needed for the reaction to occur. We then defined a **system,** a quantity of material that is separated from the rest of the universe by a **boundary;** an **equilibrium state,** in which the properties of a system do not change unless it is disturbed from the surroundings; and **state functions,** properties that can define the state of a system. We noted that energy can be transferred into or out of a system by **work,** which we defined as the product of a force by a displacement, or **heat,** the movement of microscopic entities. We then discussed the **first law of thermodynamics,** which says that the energy of the universe remains constant. We described **calorimetry,** the discipline concerned with the measurement of heat. Then we distinguished between a **reversible** path, in which a system undergoing change is always in an equilibrium state, and an **irreversible** path, in which a system leaves the equilibrium state. We then described the **second law of thermodynamics.** It uses the state function **entropy,** which can be defined as a measure of **disorder.** The second law says that the entropy of the universe never decreases, so systems tend toward greater disorder. We went on to the **third law of thermodynamics,** which says that the entropy of a perfect crystal at the absolute zero of temperature is zero. We then showed how the state function called **free energy** can be used to predict the spontaneity of changes in state. We introduced the **equilibrium constant,** an expression that gives the relationship between free energy and the composition of a system at equilibrium, and showed how it varies with temperature. Finally, we described how the concepts of thermodynamics can be applied to living organisms and to our technological society.

## Exercises

**13.1**   Indicate which of the following are state functions and which are path functions: (a) your mass, (b) the fuel consumption of a car that travels 6.0 km, (c) your elevation above sea level, (d) the quantity of sweat on your brow from studying thermodynamics.

**13.2**   One mole of an ideal gas at a temperature of 250 K and a pressure of 1.1 atm undergoes a change in state to STP. Find $\Delta T$, $\Delta P$, and $\Delta V$ for this process.

**13.3**   Indicate whether the work accompanying each of the following processes is positive, negative, or zero: (a) opening your chemistry book, (b) the formation of water from $H_2$ and $O_2$ in a sealed container, (c) the formation of $NH_3(g)$ and $HCl(g)$ from $NH_4Cl(s)$ in a container with a mechanical link, (d) the decomposition of HCl to $H_2$ and $Cl_2$ at 500 K in a container with a mechanical link.

**13.4**   One mole of an ideal gas is at STP. Find the work associated with its expansion from 22.4 L to 44.8 L against a constant opposing pressure of 1.0 atm.

**13.5**[4]   Find the work associated with the contraction of an ideal gas from 22.4 L to 11.2 L as the result of a pressure of 1.0 atm.

**13.6**   Find the work associated with the conversion of 450 g of liquid water to steam against a pressure of 2.4 atm.

_____

[4] The answers to exercises whose numbers are in color can be found in Appendix VII. The star indicates an exercise that is more challenging than average.

**13.7**   Indicate whether each of the following processes taking place in a sealed container is accompanied by positive, negative, or zero heat flow: (a) $CH_4(g) + 2O_2(g) \rightarrow CO_2(g) + 2H_2O(l)$, (b) $Br_2(l) \rightarrow 2Br(g)$, (c) $I_2(g) \rightarrow I_2(s)$.

**13.8**   Indicate whether the heat flow accompanying each of the following processes is positive, negative, or zero: (a) an ideal gas contracts as the result of a constant opposing pressure and the temperature remains constant, (b) an ideal gas expands into a vacuum and its temperature remains constant, (c) an ideal gas expands against a constant opposing pressure and its temperature increases. (*Hint:* The internal energy of an ideal gas is proportional to the temperature.)

**13.9**   Predict the temperature change of the system when each of the following processes takes place in an adiabatic container:   (a) $C(s) + O_2(g) \rightarrow CO_2(g)$, (b) $2KI(s) \rightarrow 2K(s) + I_2(s)$, (c) expansion of an ideal gas against a constant opposing pressure.

**13.10**   A sample of an ideal gas expands from 1.8 L to 1.9 L against a constant pressure of 1.2 atm and evolves 480 J of heat. Find $\Delta E$.

**13.11**   A sample of an ideal gas contracts from 8.6 L to 7.1 L as the result of an applied constant pressure of 2.2 atm. The $\Delta E$ of the process is 620 J. Find the value of $q$.

**13.12**   An ideal gas absorbs 780 J of heat, but its temperature does not change. Find $w$. (*Hint:* The internal en-

ergy of an ideal gas is directly proportional to the temperature.)

**13.13**   The $\Delta H_f^\circ$ of $Al_2S_3$ is $-509$ kJ/mol. Find $\Delta E$ for the formation of one mole of $Al_2S_3(s)$ from its elements in their standard states at 298 K and 1 atm.

**13.14**   The $\Delta H$ for the evaporation of water at 373 K and 1 atm is 40.6 kJ/mol. Find $\Delta E$ for the evaporation of 1.0 mol of water under these conditions. Assume ideal gas behavior.

**13.15**   Indicate whether the quantity $\Delta H - \Delta E$ is positive, negative, or essentially zero for each of the following processes:   (a) the sublimation of a solid, (b) the condensation of a gas to a liquid, (c) the fusion of a solid, (d) the dissociation of $H_2$ into H atoms, (e) the combustion of octane at 298 K to carbon dioxide and liquid water.

**13.16**   The value of $\Delta H$ for the reaction   $2SO_3(g) \rightarrow 2SO_2(g) + O_2(g)$   at 303 K is 198 kJ. Find $\Delta E$.

**13.17**   Derive a relationship between $\Delta H$ and $\Delta E$ for a process in which the temperature of a fixed amount of an ideal gas changes. (*Hint:* Follow a procedure similar to the one used to derive Equation 13.8.)

**13.18**   Find the heat capacity of a calorimeter whose temperature is increased 2.09 K by 63.9 kJ of heat.

**13.19**   The heat capacity of a calorimeter of the type shown in Figure 13.3 is 78.90 kJ/K. When a 5.347-g sample of $CH_4$ is burned in the calorimeter   [$CH_4(g) + 2O_2(g) \rightarrow$

$CO_2(g) + 2H_2O(l)]$, the temperature of the calorimeter increases from 296.14 K to 299.88 K. Find $\Delta E$ for the combustion of methane.

**13.20** The $\Delta E$ of the combustion of ethanol to liquid water is 1370 kJ/mol. When 1.765 g of ethanol is combusted in a calorimeter of the type shown in Figure 13.3, the temperature of the calorimeter increases from 294.33 K to 295.84 K. Find the heat capacity of the calorimeter.

**13.21** One teaspoon of peanut butter has a mass of 5.3 g. It is combusted in a calorimeter whose heat capacity is 120 kJ/K. The temperature of the calorimeter rises from 295.2 K to 296.3 K. Find the caloric content of peanut butter. (*Hint:* One food calorie is 4.184 kJ.)

**13.22** The $\Delta H$ of the reaction $2Na(s) + I_2(s) \rightarrow 2NaI(s)$ is $-575.8$ kJ. A mixture of sodium and iodine in a constant-volume calorimeter forms 2.691 g of NaI. The initial temperature of the calorimeter is 288.1 K and its heat capacity is 13.45 kJ. Find the temperature of the calorimeter after the reaction.

**13.23** A 32.55-g sample of ice is placed in a calorimeter whose heat capacity is 2.348 kJ/K. The temperature of the calorimeter drops from 279.25 K to 274.62 K as all the ice melts. Find the heat of fusion of ice.

**13.24** Calculate the heat required to raise the temperature of 25.8 g of mercury from 301.1 K to 302.2 K.[5]

**13.25** Calculate the heat evolved when the temperature of 1.00 mol of liquid ethanol decreases 50.0 K.[5]

**13.26** The heat of combustion of methane is 890.4 kJ/mol. A tank of water is heated by combustion of methane. Find the mass of methane required to raise the temperature of 31.9

kg of water by 43.8 K.[5]

**13.27** The heat of combustion of octane is 1303 kJ/mol. A 7.79-g sample of octane is burned and the heat evolved is used to raise the temperature of a 1.0-kg block of iron. Assume that there is no heat loss and calculate the temperature increase of the iron.[5]

**13.28★** An 8.6-g ice cube is added to a cup of coffee that contains 110 g of liquid at 367 K. The heat of fusion of ice is 6.0 kJ/mol. Find the temperature of the coffee after the ice melts, assuming no other heat losses. Assume that the heat capacity of coffee is the same as that of water.[5]

**13.29** A 1.00-mol amount of helium at STP absorbs 450 J of heat at constant pressure. Assume ideal gas behavior.
(a) Calculate the temperature of the helium.
(b) Calculate its volume.
(c) Calculate the work associated with the process.
(d) Calculate $\Delta E$ for the process.[5]

**13.30** The temperature of 4.35 g of $H_2$ drops from 315 K to 305 K at a constant pressure of 1.00 atm. Calculate $q$, $w$, $\Delta E$, and $\Delta H$ for this process. Assume ideal gas behavior.[5]

**13.31** Predict the direction of spontaneous change for the following systems: (a) one mole of ideal gas at STP in a volume of 20 L, (b) a mixture of $H_2$ and $Cl_2$ at 298 K, (c) an acorn in the soil, (d) one mole of ideal gas at STP in a volume of 22.4 L.

**13.32** The heat of vaporization of $NH_3$ at its boiling point, 239.8 K, is 23.35 kJ/mol. Find $\Delta S_{sys}$ for the vaporization of $NH_3$ at its boiling point.

**13.33** The boiling point of $NH_3$ is 239.8 K and its heat of vaporization is 23.35 kJ/mol. Use Equations 13.11 and 13.12 to find $\Delta S$ of the system and surroundings at (a) 241 K and (b) 239 K for the evaporation of $NH_3$.

Show that these results are consistent with the second law.

**13.34** The $\Delta S_{sys}$ for the evaporation of HF(l) is 0.0257 kJ/mol K and the $\Delta H$ is 7.53 kJ/mol at the boiling point. Find the boiling point of HF.

**13.35** The $\Delta S_{sys}$ for the fusion of platinum is 9.62 J/mol K and the $\Delta S_{sys} + \Delta S_{surr} = 0$ at 2043 K. Find the heat of fusion of platinum.

**13.36** Indicate the sign of $\Delta S_{sys}$ for each of the following phase changes and explain your answers: (a) condensation, (b) freezing, (c) evaporation, (d) fusion, (e) sublimation.

**13.37** For which of the following phase changes must the entropy change of the *surroundings* be positive: (a) $H_2O(g) \rightarrow H_2O(l)$ at 373 K and 1 atm, (b) $C_6H_6(s) \rightarrow C_6H_6(l)$ above the melting point, (c) $CO_2(s) \rightarrow CO_2(g)$ at 298 K and 1 atm.

**13.38** Entropy is sometimes called the arrow of time. Suggest an explanation for this phrase.

**13.39** Indicate the sign of $\Delta S_{sys}$ and $\Delta S_{univ}$ for each of the following processes: (a) the decomposition of water to $H_2$ and $O_2$ by the application of electricity, (b) the formation of $H_2O$ from $H_2$ and $O_2$ and a spark, (c) the reproduction of a virus.

**13.40** Indicate the sign of $\Delta S_{sys}$ for the following changes in state: (a) expansion of a gas at constant $T$, (b) increase in $T$ of a gas at constant $V$, (c) decrease in the pressure of a gas at constant $T$, (d) the mixing of two gases. Explain your answers.

**13.41** Indicate the sign of $\Delta S_{sys}$ for each of the following chemical changes:
(a) $2Cl(g) \rightarrow Cl_2(g)$
(b) $CO_2(g) + C(s) \rightarrow 2CO(g)$
(c) $C_6H_{12}O_6(s) \rightarrow C_6H_{12}O_6(aq)$
(d) $CO_2(g) + MgO(s) \rightarrow MgCO_3(s)$
Explain your answers.

---

[5] The data for this exercise can be found in Table 13.1.

**13.42** Calculate $\Delta S°$ of each of the following reactions at 298 K:
(a) $2HCl(g) \rightarrow H_2(g) + Cl_2(g)$
(b) $2HI(g) \rightleftharpoons H_2(g) + I_2(g)$
(c) $2HI(g) \rightarrow H_2(g) + I_2(s)$
Discuss your results.[6]

**13.43** Calculate the value of $\Delta S_f°$ at 298 K for each of the following gases: (a) NO, (b) $N_2O_3$, (c) $O_3$, (d) $CO_2$.[6]

**13.44** Calculate the $\Delta S_f°$ at 298 K for each of the following solids: (a) KCl, (b) $KClO_4$, (c) diamond, (d) white phosphorus.[6]

**13.45** The boiling point of $CHCl_3$ is 344.4 K. Find the $\Delta H°$ of vaporization at the boiling point from the values of $S°$ given in Appendix II.

**13.46** Find the $\Delta S°$ of combustion of $C_2H_2$ to $CO_2$ and $H_2O(g)$.[6]

**13.47** Use the data in Appendix II to find the boiling point of Hg.

**13.48** Use the data in Appendix II to find the temperature at which the vapor pressure of $I_2(s)$ is 1.0 atm.

**13.49** Discuss the effect of increasing temperature on the extent to which the following reactions proceed:
(a) $N_2O_4(g) \rightarrow 2NO_2(g)$
(b) $CaCO_3(s) \rightarrow CaO(s) + CO_2(g)$
(c) $2CO(g) + O_2(g) \rightarrow 2CO_2(g)$ (exothermic)
(d) $N_2(g) + O_2(g) \rightarrow 2NO(g)$ (endothermic)

**13.50** Discuss the effect of temperature on the formation of the following substances from their elements in their standard states: (a) $O_3(g)$, (b) $N_2O(g)$, (c) NaCl(s).[7]

**13.51** The $\Delta H_f°$ of BaO(s) is $-558$ kJ/mol and the $\Delta S_f°$ is $-97$ J/mol K. Find $\Delta G_f°$.

**13.52** Use the following data to find $\Delta G°$ at 700 K of the reaction $CO(g) + \frac{1}{2}O_2(g) \rightarrow CO_2$: CO(g), $\Delta H° = -110.5$ kJ/mol, $S° = 223$ J/mol K; $CO_2(g)$, $\Delta H° = -394.0$ kJ/mol, $S° = 251$ J/mol K; $O_2(g)$, $S° = 231$ J/mol K, all at 700 K. Calculate the value of $\Delta G°$ at 800 K by assuming that $\Delta H°$ and $\Delta S°$ do not change significantly over this temperature range.

**13.53** Calculate $\Delta G°$ for the following reactions at 298 K:
(a) $NH_3(g) + HCl(g) \rightarrow NH_4Cl(s)$
(b) $BaCO_3(s) \rightarrow BaO(s) + CO_2(g)$
(c) $CH_4(g) + 4Cl_2(g) \rightarrow CCl_4(g) + 4HCl(g)$[7]

**13.54** Each of the following reactions is important in an automobile's catalytic converter. Find $\Delta G°$ for each at 298 K. Predict the effect of increasing temperature on each $\Delta G°$:
(a) $2CO(g) + 2NO(g) \rightleftharpoons N_2(g) + 2CO_2(g)$
(b) $5H_2(g) + 2NO(g) \rightleftharpoons 2NH_3(g) + 2H_2O(g)$
(c) $2H_2(g) + 2NO(g) \rightleftharpoons N_2(g) + 2H_2O(g)$
(d) $2NH_3(g) + 2O_2(g) \rightleftharpoons N_2O(g) + 3H_2O(g)$[7]

**13.55** Calculate the $\Delta G_f$ of CO(g) at 298 K from its elements in their standard states at a pressure of (a) 4.0 atm, (b) 0.01 atm.[7]

**13.56** Calculate $\Delta G$ for the formation of NO(g) at a pressure of 1.0 atm from $N_2$ at 2.0 atm and $O_2$ at 3.0 atm at 298 K.[7]

**13.57** If the standard state is redefined as $P = 2$ atm, what is the $\Delta G_f°$ for each of the following substances: (a) HBr(g), (b) $NO_2(g)$, (c) Cl(g).[7]

**13.58** Calculate $K$ for each reaction of Exercise 13.53 at 298 K.

**13.59** Calculate the $K$ for each reaction of Exercise 13.54 at 298 K. Predict the effect of an increase in temperature on the magnitude of each $K$.

**13.60** The salt $NH_4NO_3$ can decompose by three pathways: (a) to $HNO_3(g) + NH_3(g)$, (b) to $N_2O(g) + H_2O(g)$, (c) to $N_2(g) + O_2(g) + H_2O(g)$. Find the value of $K$ for each mode of decomposition at 298 K.[7]

**13.61** Find the value of $K$ for the decomposition of $H_2(g)$ at each of the temperatures listed in Table 13.3. Explain the trend with increasing temperature.

**13.62** Find the value of $K$ for the reaction $CO(g) + \frac{1}{2}O_2(g) \rightleftharpoons CO_2(g)$ at 298 K. Use the data in Exercise 13.52 to find $K$ at 700 K. Explain the difference.[7]

**13.63** Find $\Delta H°$ of a reaction whose equilibrium constant increases by a factor of 3.0 when the temperature increases from 300 K to 400 K.

**13.64** Use the data in Appendix II and Equation 13.25 to find $K$ for the reaction $CO(g) + \frac{1}{2}O_2(g) \rightleftharpoons CO_2(g)$ at 700 K. Compare this result to the value of $K$ calculated in Exercise 13.62 at 700 K and explain the difference.

**13.65** The $\Delta H°$ of the reaction $N_2(g) + 3H_2(g) \rightleftharpoons 2NH_3(g)$ is $-92$ kJ, and $K$ for the reaction at 300 K is $3.8 \times 10^5$. Find $K$ at 600 K and at 800 K.

**13.66** Use the data in Appendix II to find $K$ for the reaction $H_2(g) \rightleftharpoons 2H(g)$ at 3000 K. Compare your result to the value calculated in Exercise 13.61 and explain the difference.

---

[6] The values of $S°$ that are necessary for this exercise can be found in Appendix II.
[7] The data needed to solve this exercise can be found in Appendix II.

# 14

# Chemical Equilibrium

**Preview**    **E**quilibrium is a concept that is easy to understand intuitively but must be defined precisely in the study of chemistry. In this chapter, we introduce the equilibrium state and explain how a chemical system that is at equilibrium can be both static and constantly changing. We show how the equilibrium constant, *K*, can be used to obtain information about the composition of chemical systems. Then we describe the principle of le Chatelier, which predicts how a system returns to equilibrium after it is stressed. A central part of the chapter deals with calculations that use gas phase equilibrium constants. We then describe and use the solubility product, an equilibrium constant for saturated solutions of sparingly soluble ionic solids, and conclude by outlining the time-tested method of analyzing the composition of solutions by the method of selective precipitation.

The equilibrium state is the ultimate goal of every chemical system. A system that is in the equilibrium state does not change spontaneously. Only an external influence can cause a system that is at equilibrium to undergo a change. If a system is not at equilibrium, it undergoes spontaneous change and attempts to attain the equilibrium state.

All of us have some intuitive knowledge about spontaneous reactions and the equilibrium state. We know that sugar dissolves in water, but only up to a point. If we evaporate some water from the saturated solution of sugar, or if we cool it, sugar crystallizes from the solution. We know that silver reacts with sulfur compounds in the air to form tarnish. We may not know that a small amount of the tarnish can decompose back to silver and sulfur. In this chapter and the next, we shall discuss the criteria that enable us to tell whether a reaction is spontaneous and the extent to which it can proceed. We shall also discuss the qualitative and quantitative features of the equilibrium state.

## 14.1 THE EQUILIBRIUM STATE AND THE EQUILIBRIUM CONSTANT

### The Equilibrium State

In principle, we can easily define an equilibrium state: A system is in an equilibrium state if its macroscopic properties do not change spontaneously, even over a prolonged period of time. In practice, it is often difficult to know whether a system is in an equilibrium state. Many systems reach equilibrium rapidly. Others proceed so slowly to equilibrium that they appear not to be changing at all. For example, a sample of laughing gas, $N_2O$, decomposes spontaneously at room temperature until almost all of it is converted to $N_2$ and $O_2$. However, no detectable amount of $N_2$ or $O_2$ appears in a purified sample of $N_2O$, because the rate of decomposition is immeasurably slow. A mixture of $N_2$ and $O_2$ should spontaneously form a very small quantity of $N_2O$. This process also is not observed, because the rate is too slow. We shall not be concerned with the time needed for a system to reach equilibrium, but only with the state of the system when it reaches equilibrium.

A system at equilibrium appears static. This appearance is deceptive. On the atomic and molecular level, an equilibrium system is dynamic; an enormous number of changes occur continuously. The net result of all these changes on the molecular level is no change on the macroscopic level (Figure 14.1).

*At equilibrium any change and the exact reverse of that change take place at the same rate.* We have already discussed (Section 4.2) equilibrium in a system consisting of a liquid and its vapor in a closed container. At any given temperature, the equilibrium state is characterized by a certain vapor pressure above the liquid. As long as the system is left alone, it does not change macroscopically. However, it *is* changing constantly on

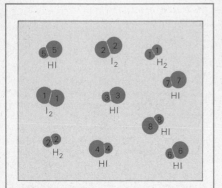

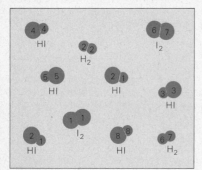

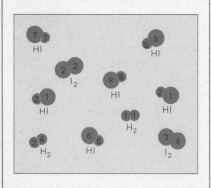

**Figure 14.1**

A system at equilibrium does not change on the macroscopic level. In this system of $H_2$, $I_2$, and HI, the relative amount of each substance does not change. But if we could observe individual molecules, we would see that the system is changing constantly on the microscopic level.

the molecular level. Molecules continually evaporate from the liquid and condense from the vapor. There is no overall change because the rates of evaporation and condensation are equal. If there is an external stress, such as a change in temperature, the concentration of substance in the vapor phase increases or decreases. The system goes to a new equilibrium state, characterized by a different vapor pressure.

## Chemical Change and Equilibrium

Now we shall consider equilibrium states of systems containing substances that are related by simple chemical changes. We shall examine the equilibrium states of these systems, and we shall see how they behave in response to external stress.

Consider the system described by the reaction:

$$H_2(g) + I_2(g) \rightleftharpoons 2HI(g)$$

Equilibrium is established rapidly at 700 K. The equilibrium pressure of each gas depends not only on the temperature, but also on the amount of gaseous material in the system. Putting 10 mol of HI into a container results in a greater equilibrium pressure of HI, $H_2$, and $I_2$ than putting 1 mol of HI into the same container. By the same token, putting 10 mol each of $H_2$ and $I_2$ into a container results in a greater equilibrium pressure of HI, $H_2$, and $I_2$ than putting 1 mol of each in the same container.

The composition of equilibrium states as a function of the quantities of the substances in a system has been studied for more than a century. The results of a typical study are shown in Table 14.1. The equilibrium pressures of the components of the system can be seen to vary considerably. However, the value of the fraction $P_{HI}^2/P_{H_2}P_{I_2}$, which we shall call $K$, is relatively constant.

The data in Table 14.1 reveal two crucial aspects of chemical equilibrium. The value of the expression that we designate as $K$ does not change significantly with changes in the amounts of the substances present in the system. The exact composition of the equilibrium state is determined by

A juggler's act provides a rough analogy for the dynamic equilibrium of a chemical system. The chemical system remains unchanged on the macroscopic level despite a large number of changes on the molecular level. The position of the individual objects that are juggled changes constantly, but the overall system appears to be static. *(Elizabeth Hibbs, Monkmeyer)*

**TABLE 14.1   Equilibrium in the System $H_2(g) + I_2(g) \rightleftharpoons 2HI(g)$**

| Initial Composition (atm) | | | Equilibrium Composition (atm) | | | $K = \dfrac{P_{HI}^2}{P_{H_2}P_{I_2}}$ |
|---|---|---|---|---|---|---|
| $P_{H_2}$ | $P_{I_2}$ | $P_{HI}$ | $P_{H_2}$ | $P_{I_2}$ | $P_{HI}$ | |
| 0.638 | 0.570 | 0 | 0.165 | 0.0978 | 0.945 | 55.3 |
| 0.610 | 0.686 | 0 | 0.103 | 0.179 | 1.013 | 55.7 |
| 0 | 0 | 0.612 | 0.0644 | 0.0654 | 0.482 | 55.2 |
| 0 | 0 | 0.256 | 0.0270 | 0.0274 | 0.202 | 55.2 |

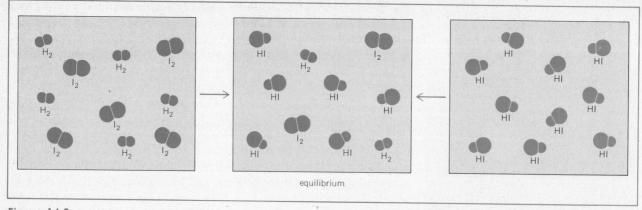

equilibrium

**Figure 14.2**

The equilibrium state can be approached from either direction. A mixture of five $H_2$ molecules and five $I_2$ molecules reaches the same equilibrium state as 10 HI molecules.

the amounts of the substances in the system, but *not* by the direction from which equilibrium is approached (Figure 14.2).

### Determining the Expression for *K*

We shall deal with chemical systems at equilibrium by using the **equilibrium constant, *K***, whose form is based on the equation that describes the chemical change. For a given system at a given temperature, the value of *K* is fixed. If we want to determine the value of *K* for a given system, the first and most important step is to *write a balanced chemical reaction that relates the components of the system*. We then obtain the expression for *K* by following these rules:

1. The form of the expression for *K* depends on how we write the reaction. Usually, *K* is a fraction whose numerator includes the products of the reaction and whose denominator includes the reactants.
2. The expression for *K* includes terms for gases and substances in solution. The terms for gases usually express the partial pressure of the gases in atmospheres (atm), although they sometimes give concentrations of the gases in moles per liter (mol/L). The terms for solutions usually are expressed in units of concentration, such as moles per liter (molarity) of the solutes. By convention, we symbolize molarity by placing the formula of a substance in square brackets, [ ], so $[Br_2]$ means the molarity of bromine in solution and $[Na^+]$ means the molarity of $Na^+$ ion in solution. Unless otherwise indicated, *solution* means an aqueous solution.[1]
3. Terms for pure liquids and solids are not included in the expression for

---

[1] Throughout our discussion we assume ideal gas behavior and ideal solution behavior. The theory of equilibrium can be developed for real gases and solutions by the use of alternative expressions for pressure and concentration that accommodate nonideality. The fugacity of a real gas or the activity of a real solute can be interpreted as the "effective pressure or concentration," and can be used in equilibrium descriptions.

$K$. The concentration of a pure liquid or a pure solid at a given temperature is essentially a constant — that is, the number of moles of solid per liter of solid or of liquid per liter of liquid is the same no matter what the quantity of solid or liquid. Furthermore, the free energy of solids and liquids does not change significantly with ordinary changes in pressure (Section 13.7).

4. The numerator of the expression for $K$ is the product of all the terms written for substances on the right side of the chemical equation. The denominator is the product of all the terms written for substances on the left side of the equation.

5. The coefficients of the chemical equation also affect the value of $K$. The pressure of a gas or the concentration of a solute in the expression for $K$ is raised to the power of its coefficient in the equation. If a gas has the coefficient 2 in an equation, the pressure of the gas is squared in the calculation of the value of $K$. If a substance in solution has the coefficient 3, its molarity is cubed in the calculation of $K$.

**Example 14.1**   Write the expressions for $K$ for the following chemical equations, expressing gases as pressures:

    a. $C(s) + CO_2(g) \rightleftharpoons 2CO(g)$
    b. $NH_3(g) + HCN(l) \rightleftharpoons NH_4^+(aq) + CN^-(aq)$
    c. $5Fe^{2+}(aq) + MnO_4^-(aq) + 8H^+(aq) \rightleftharpoons 5Fe^{3+}(aq) + Mn^{2+}(aq) + 4H_2O$

**Solution**   a. Solids do not appear in the expression for $K$, and the coefficient of the CO appears as an exponent. The expression for $K$ thus is

$$K = \frac{P_{CO}^2}{P_{CO2}}$$

b. Liquids do not appear in the expression for $K$. The expression thus is

$$K = \frac{[NH_4^+][CN^-]}{P_{NH_3}}$$

c. The concentration of water, the solvent, is not included in the expression, which is

$$K = \frac{[Mn^{2+}][Fe^{3+}]^5}{[MnO_4^-][Fe^{2+}]^5[H^+]^8}$$

## Writing Chemical Equations

You can see that it is necessary to indicate the state of each substance when writing a chemical equation. Otherwise, we cannot write the correct expression for the equilibrium constant, $K$. In Section 2.7, we discussed the need to indicate the states of substances when we write equations from word descriptions of chemical changes. *Review that section carefully.* To help you review, we shall repeat the rules to follow when you write a chemical equation.

1. The formula for each substance is followed by a state symbol (Section 2.6) in the equation: (g) for gas, (l) for liquid, (s) for solid, (aq) for substances in aqueous solution. The state of any substance at atmospheric pressure depends directly on temperature. If the word description of a chemical reaction does not mention temperature, assume that the reaction occurs at room temperature, about 300 K, and that each substance is in its appropriate state at that temperature.

2. Substances in aqueous solution can be subdivided into three categories. A substance can be a strong electrolyte that exists almost entirely as dissociated ions in aqueous solution. It can be a weak electrolyte that is partially dissociated into ions in aqueous solutions. Or it can be a nonelectrolyte that does not form ions in aqueous solution.

If a substance is a strong electrolyte, its formula should be written to indicate complete dissociation into ions in aqueous solution. *The formula of an undissociated strong electrolyte should not appear in a chemical reaction with the state symbol (aq). Only the dissociated ions exist in solution.*

There are three main types of strong electrolytes: most salts, strong acids, and strong bases. A salt usually is a compound of a metal ion with either a nonmetal or a polyatomic ion. Table 2.7 lists some important polyatomic ions. All salts are solids at room temperature. Therefore, in reactions occurring at room temperature, a salt that is not in solution gets the state symbol (s). A salt that is in solution is represented as dissociated ions with the state symbol (aq).

Some common strong acids, which also are represented as dissociated ions when in solution, are listed in Table 2.9. As for strong bases, many of them are compounds of a metal and a hydroxide ion, $OH^-$, and are solids. Strong bases in solution are represented as dissociated ions. We shall discuss acids and bases in detail in Chapter 15.

The formulas for weak electrolytes and nonelectrolytes in aqueous solution are written to show them undissociated.

3. Any component of a system that does not change in any way is not included in the chemical reaction. Do not write a chemical reaction that includes the same substance, unchanged, on both sides of the reaction.

For example, when a solution of lead nitrate, which contains $Pb^{2+}$ cations and $NO_3^-$ anions, is mixed with a solution of sodium chloride, which contains $Na^+$ cations and $Cl^-$ anions, a precipitate of lead chloride forms. No other chemical change accompanies this precipitation; the $Na^+$ ions and the $NO_3^-$ ions remain in solution. The chemical equation for this reaction is thus:

$$Pb^{2+}(aq) + 2Cl^-(aq) \rightleftharpoons PbCl_2(s)$$

There is nothing in this equation to show that sodium chloride is the source of the $Cl^-$ ion. The ion could come from any soluble chloride, and the reaction would be the same. Therefore, it is appropriate to omit the $Na^+$ from the equation. For our equilibrium and thermodynamic calculations, we want to know the amount of chloride in the solution, not the exact composition of the starting solution.

---

**Example 14.2**   Write balanced chemical equations and the corresponding expressions for $K$ for the following processes:
  a. the sublimation of iodine
  b. the dissolution of silver sulfate in water
  c. the decomposition of calcium carbonate to calcium oxide and carbon dioxide
  d. the formation of a precipitate of silver chloride after a solution of hydrochloric acid and a solution of silver nitrate are mixed
  e. the conversion of nitrogen dioxide to dinitrogen tetroxide

**Solution**   a. Sublimation is the conversion of a solid to a gas. The equation for the sublimation of iodine is

$$I_2(s) \rightleftharpoons I_2(g) \qquad K = P_{I_2(g)}$$

The expression for the equilibrium constant shows that the pressure of $I_2(g)$ at equilibrium does not depend on the amount of solid present.
  b. Silver sulfate is a solid at room temperature, and it dissociates into ions when it is dissolved in water. The equation is thus:

$$Ag_2SO_4(s) \rightleftharpoons 2Ag^+(aq) + SO_4^{2-}(aq) \qquad K = [Ag^+]^2[SO_4^{2-}]$$

Once again, the amount of solid does not determine the concentration of ions in the solution, so long as there is enough solid to form a saturated solution.
  c. The equation for the reaction is

$$CaCO_3(s) \rightleftharpoons CaO(s) + CO_2(g) \qquad K = P_{CO_2(g)}$$

Neither solid appears in the expression for $K$.
  d. Hydrochloric acid is a strong acid that is a solution of $H^+$ and $Cl^-$ ions. A solution of silver nitrate, a salt that dissociates into ions, contains $Ag^+$ ion and $NO_3^-$ ion. Silver chloride forms from silver ion and chloride ion. The net equation is thus:

$$Ag^+(aq) + Cl^-(aq) \rightleftharpoons AgCl(s)$$

The $H^+$ and $NO_3^-$ ions are properly omitted because neither undergoes a change. The equilibrium constant for this equation is $K = 1/[Ag^+][Cl^-]$. The numerator of the expression for $K$ is 1 when no other terms appear.
  e. The balanced equation for this reaction can be written as:

$$2NO_2(g) \rightleftharpoons N_2O_4(g)$$

For this equation,

$$K = \frac{P_{N_2O_4}}{P_{NO_2}^2}$$

If the equation is written as

$$NO_2(g) \rightleftharpoons \tfrac{1}{2}N_2O_4(g)$$

then

$$K = \frac{P_{N_2O_4}^{1/2}}{P_{NO_2}}$$

and it has a different numerical value. Remember that the numerical value of $K$ can be changed by changing the form of a chemical equation. We normally write equations with the smallest whole-number coefficients, for uniformity.

The value of $K$ conveys a great deal of information about the behavior of chemical systems. It defines the equilibrium state and enables us to find the direction and extent of a reaction.

### The Reaction Quotient Q

Let us consider a specific system, the conversion of nitrogen dioxide to dinitrogen tetroxide. The equation is

$$N_2O_4(g) \rightleftharpoons 2NO_2(g)$$

and

$$K = \frac{P_{NO_2}^2}{P_{N_2O_4}}$$

If the temperature is 373 K, the value of $K$ for this equation is about 80 atm.[2] This brief statement has several implications.

Suppose we measure the pressures of $NO_2$ and $N_2O_4$ in a system at 373 K and substitute the values we obtain into the expression for $K$. If the value obtained for $K$ is 80 atm, we know that the system is in an equilibrium state. If we get a different value, the system is not at equilibrium. It is only when the system is at equilibrium that the expression $P_{NO_2}^2/P_{N_2O_4}$ is $K$. When the system is not at equilibrium, this expression is designated by $Q$ and is called the *reaction quotient*. The expression for $Q$ has the same form as the expression for $K$. It has a different numerical value, which changes as the reaction proceeds and becomes equal to $K$ at equilibrium. Therefore, the value of $Q$ tells us something important about the system.

Suppose we measure the pressures of $NO_2$ and $N_2O_4$ in a system and

---

[2] As we have mentioned (Section 13.7) $K$ does not have units, but we shall assign it the units it seems to have in order to simplify dimensional analysis.

obtain a numerical value for $Q$ that is larger than 80 atm. Now we know that the system is not in an equilibrium state. But we know something more: The system is not in an equilibrium state because there is relatively too much $NO_2$ present. In the expression for $Q$, the pressure of $NO_2$, which is on the right side of the chemical equation, is in the numerator. An increase in the value of the numerator means an increase in the value of $Q$. We can make a general rule: Whenever the value of $Q$ in a given system is larger than the value of $K$, there is an excess of the substances on the right side of the chemical equation.

Suppose that the numerical value of $Q$ is smaller than the value of $K$. Now we know that the system has an excess of the substances on the left side of the equation, in the present case, $N_2O_4(g)$. The pressure of $N_2O_4(g)$ appears in the denominator of the expression for $Q$, and an increase in the value of the denominator means a decrease in the value of $Q$.

By comparing the values of $Q$ and $K$, we can predict the direction of spontaneous change for a chemical system that is not at equilibrium. If $Q$ is larger than $K$, there is a relative excess of the materials on the right side of the chemical equation and the reaction proceeds spontaneously from right to left toward equilibrium. If $Q$ is smaller than $K$, there is a relative excess of the materials on the left side of the equation and the reaction proceeds spontaneously from left to right toward equilibrium.

---

**Example 14.3**   Predict the behavior of each of the following systems prepared from the given pressures of the two gases in atmospheres: (a) $P_{NO_2} = 0.4$, $P_{N_2O_4} = 0.002$; (b) $P_{NO_2} = 0.2$, $P_{N_2O_4} = 0.004$; (c) $P_{NO_2} = 1.0$, $P_{N_2O_4} = 0.002$; (d) $P_{NO_2} = 0.001$, $P_{N_2O_4} = 0.0000$, at 373 K. The reaction is $N_2O_4(g) \rightleftharpoons 2NO_2(g)$, $K = 80$ atm.

**Solution**   We use the same method in each case: Substitute the data into the expression for $Q$ and then compare the result with the value of $K$.

a.
$$\frac{P^2_{NO_2}}{P_{N_2O_4}} = \frac{(0.4 \text{ atm})^2}{0.002 \text{ atm}} = 80 \text{ atm}$$

The system is at equilibrium, since $Q = K$.

b.
$$\frac{(0.2 \text{ atm})^2}{0.004 \text{ atm}} = 10 \text{ atm}$$

Since $Q < K$, the system is not at equilibrium. There is relatively too little of the product, $NO_2(g)$. The reaction will proceed spontaneously from left to right to produce $NO_2(g)$ and use up $N_2O_4(g)$ until $Q = K = 80$ atm.

c.
$$\frac{(1.0 \text{ atm})^2}{0.002 \text{ atm}} = 500 \text{ atm}$$

Since $Q > K$, the system is not at equilibrium. There is an excess of $NO_2(g)$. The reaction will proceed from right to left until equilibrium is reached and $Q = K = 80$ atm.

d. Whenever one of the components that is present at equilibrium does not appear, the reaction will proceed in the direction which produces that substance, in this case from right to left to produce $N_2O_4(g)$.

## The Composition of the Equilibrium State

One reason $K$ is so useful is that its value is independent of the quantities of materials in the system. *Once the composition of one equilibrium state of a system is known, the composition of any other equilibrium state of the same system containing different concentrations or pressures of the same substances at the same temperature can be found.*

**Example 14.4**    An equilibrium state of a system containing $NO_2(g)$ and $N_2O_4(g)$ at an unknown temperature (not 373 K) is found to have $P_{NO_2} = 0.86$ atm and $P_{N_2O_4} = 0.12$ atm. Another system at the same temperature is also at equilibrium, and is found to have $P_{NO_2} = 0.98$ atm. Calculate the pressure of $N_2O_4$ in the second system. Use the reaction $N_2O_4(g) \rightleftharpoons 2NO_2(g)$.

**Solution**    We can find the value of the equilibrium constant for the first system by using the given data:

$$K = \frac{P^2_{NO_2}}{P_{N_2O_4}} = \frac{(0.86 \text{ atm})^2}{(0.12 \text{ atm})} = 6.2 \text{ atm}$$

This value is $K$ for any equilibrium state of these components at this temperature. For the second system, therefore:

$$6.2 \text{ atm} = \frac{(0.98 \text{ atm})^2}{P_{N_2O_4}}$$

$$P_{N_2O_4} = 0.15 \text{ atm}$$

There are dangers in comparing different systems that have different forms for $K$. But we can allow ourselves one useful generalization derived from observation. Suppose we begin with the pressures of all gases equal to 1 atm and the concentrations of all solutes equal to $1M$ so that $Q = 1$. In a system where the value of $K$ is large ($K > 1$), the reaction proceeds from left to right, or toward completion. We can say that such a reaction is favorable as written. Conversely, for a system in which the value of $K$ is small ($K < 1$), the reaction proceeds from right to left, and we can say that such a reaction is unfavorable as written.

It is also important to understand that the composition of the equilibrium state does not change if we vary the amounts of liquid or solid in the system. The expression for $K$ contains no terms for liquids and solids. The presence of any amount of solid or liquid at equilibrium is enough to keep the system at equilibrium.

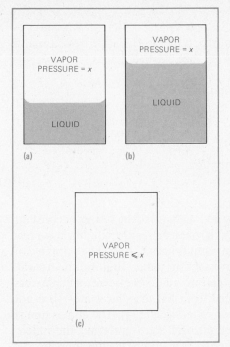

**Figure 14.3**

At a given temperature, the vapor pressure of a system at equilibrium is the same no matter how much or how little liquid is in the system. If the vapor pressure of the system in (a) is said to be $x$, then the vapor pressure of the system in (b), which contains more liquid, is also $x$. Only when the system contains no liquid, as shown in (c), can the pressure of the gas fall below the equilibrium vapor pressure.

For example, the composition of the equilibrium state of a system containing a liquid and its vapor is described by the vapor pressure alone. We can change the value of the equilibrium constant by changing the temperature, but not by adding or removing liquid. As long as some liquid is present, the pressure of vapor is the equilibrium pressure. The pressure of gas can fall below the equilibrium pressure only if the liquid is removed completely (Figure 14.3).

Now look at a saturated solution, the equilibrium state of a system containing dissolved and undissolved material. If the solute is a gas, the composition of the solution varies as the pressure of the undissolved solute varies. But if the solute is a pure liquid or a pure solid, the quantity of undissolved solute does not affect the composition of the system. Suppose we have a system composed of solid glucose, $C_6H_{12}O_6$, dissolved glucose, and water. The equilibrium state of the system can be defined by the appropriate value of $K$ for the reaction: $C_6H_{12}O_6(s) \rightleftharpoons C_6H_{12}O_6(aq)$; $K = [C_6H_{12}O_6]$. The expression for $K$ tells us that the quantity of solid glucose that is present at equilibrium does not influence the quantity of dissolved glucose.

The same is true for a reaction that does not occur in aqueous solution, such as the decomposition of magnesium carbonate. At 800 K, the value of the equilibrium constant for this process is about 0.95 atm. The reaction is $MgCO_3(s) \rightleftharpoons MgO(s) + CO_2(g)$ and $K = P_{CO_2} = 0.95$ atm. The equilibrium state of this system at 800 K is any state that has carbon dioxide at a pressure of 0.95 atm and that also contains any amount of magnesium carbonate and magnesium oxide. Adding or removing even large amounts of solid does not affect the pressure of carbon dioxide at equilibrium. The $P_{CO_2}$ will be below 0.95 atm at 800 K only if insufficient $MgCO_3$ was added to reach equilibrium.

## 14.2   THE PRINCIPLE OF LE CHATELIER

We have emphasized that a system in an equilibrium state will not change unless something is done to stress the system. We shall now examine the effects of external stresses that cause a system to change from the equilibrium state — stresses that change the concentrations or pressures of the components of the system, or the temperature of the system. We shall describe how to predict the way in which the system will change in reaching the new equilibrium state.

This problem was considered by Henri Le Chatelier (1850–1936), who in 1888 enunciated a remarkably simple rule that has come to be called the **principle of Le Chatelier:**

*When a system in an equilibrium state is subjected to an external stress, the system will establish a new equilibrium state, when possible, so as to minimize the external stress.*

Let us look at this principle in action by studying the equilibrium states of some systems.

## The Synthesis of Ammonia

The reaction for the fixation of nitrogen by the Haber process, a major source of nitrogen fertilizers, is

$$3H_2(g) + N_2(g) \rightleftharpoons 2NH_3(g)$$

The expression

$$\frac{P^2_{NH_3}}{P_{N_2} P^3_{H_2}}$$

defines $K$ and $Q$ for this reaction.

Suppose we stress the system by adding $N_2(g)$ to the system at constant volume. Now the pressure of $N_2(g)$ is greater than the equilibrium pressure and the system is no longer at equilibrium; $Q < K$. The principle of Le Chatelier tells us that the system establishes a new equilibrium in a way that minimizes the stress, in this case by removing some of the excess nitrogen. The reaction will proceed from left to right; some—but not all—of the added $N_2(g)$ will be converted to $NH_3(g)$. The conversion is incomplete because if all the added nitrogen were converted, the system would have an excess of ammonia and $Q$ would be greater than $K$.

Now suppose that a quantity of $NH_3(g)$ is added to the system at equilibrium. This time, the stress is the addition of a substance on the right side of the chemical equation. The reaction proceeds from right to left to reach the new equilibrium state.

If we decrease the volume of the system, the reaction proceeds in the direction that produces the smaller number of moles of gas. We can see why by reasoning from the equilibrium constant. Decreasing the volume increases the pressure (or concentration) of each gaseous component. For this reaction, the numerator of the expression for $Q$ includes a pressure term to the second power, while the denominator includes pressure terms to the fourth power. Since the pressure of each gas increases by a constant factor, the denominator of $Q$ increases more than the numerator. The reaction must proceed from left to right to reduce the value of the denominator of $Q$ as the system moves to a new equilibrium state (Figure 14.4). If the volume of the system increases, the reaction proceeds in the direction that increases the number of moles of gas. In this system, the reaction proceeds from right to left.

A change in volume does not always cause a system to move out of an equilibrium state. The system represented by the reaction $2HCl(g) \rightleftharpoons H_2(g) + Cl_2(g)$ is not affected by changes in volume, because the number of moles of gas on each side of the chemical equation is the same.

When the temperature of a system at equilibrium is changed, the system usually is no longer at equilibrium. We can predict how the system adjusts to a temperature change if we know how its equilibrium constant changes with temperature. The variation of $K$ with temperature is related to the sign of $\Delta H°$ (Equation 13.25). For an endothermic process ($\Delta H° > 0$), the value of $K$ increases as the temperature increases. For an exothermic process ($\Delta H° < 0$), the value of $K$ decreases as the temperature increases. The synthesis of ammonia by the Haber process is exothermic, so the value of $K$ decreases as the temperature goes up. The relative amount of $NH_3(g)$ at equilibrium decreases and the reaction proceeds from right to left until the new equilibrium state is reached.

We can reach the same conclusion by using the principle of Le Chatelier. Let us write heat as part of the chemical reaction, as if it were a reagent. An exothermic reaction gives off heat, so the equation can be written as

$$3H_2(g) + N_2(g) \rightleftharpoons 2NH_3(g) + \text{heat}$$

Heat is on the right side of the equation. (Remember, however, that $\Delta H$ is negative for an exothermic reaction.) Assume that we add heat to raise the temperature. The effect is the same as adding $NH_3(g)$, the reagent on the right. The reaction proceeds from right to left to reach equilibrium at the new temperature. Conversely, if we lower the temperature, we remove heat. The reaction proceeds from left to right until equilibrium is restored. We can make the same arguments in reverse for an endothermic system by placing heat on the left of the reaction. Table 14.2 summarizes the changes we have discussed for the Haber process.

### A Saturated Solution of an Ionic Solid

Consider the equilibrium system composed of a saturated solution of lead chloride and some solid lead chloride. The reaction is

$$PbCl_2(s) \rightleftharpoons Pb^{2+}(aq) + 2Cl^-(aq)$$

and

$$K = Q = [Pb^{2+}][Cl^-]^2$$

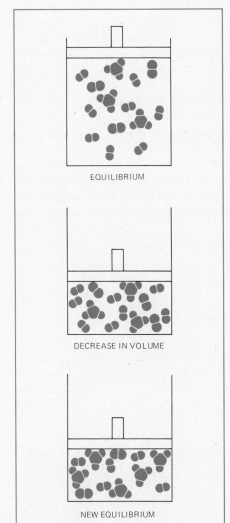

**Figure 14.4**

If a system containing $H_2(g)$, $N_2(g)$, and $NH_3(g)$ is at equilibrium (top), and its volume is reduced (center), the system is no longer at equilibrium. The principle of Le Chatelier tells us that the system will move to a new equilibrium state by spontaneously undergoing reaction to decrease the number of molecules of gas (bottom).

**TABLE 14.2** The Principle of Le Chatelier and the Haber Process

| Change in System | Subsequent Changes in Return to Equilibrium[a] | | |
| --- | --- | --- | --- |
| | $P_{NH_3}$ | $P_{N_2}$ | $P_{H_2}$ |
| add $NH_3(g)$ | − | + | + |
| add $N_2(g)$ or $H_2(g)$ | + | − | − |
| increase total pressure[b] | + | − | − |
| increase total volume[b] | − | + | + |
| increase temperature[b] | − | + | + |

[a] Decrease is −, increase is +.
[b] The effect of a decrease is the opposite.

Suppose the following things happen (Figure 14.5):

1. More solid lead chloride is added to the system. *Result:* The system is still at equilibrium, which is defined only by the concentrations of the dissolved ions. These concentrations are independent of the amount of solid in the system.

2. More water is added. *Result:* The concentration of all the dissolved components is lowered, $Q < K$, and the system is no longer at equilibrium. More $PbCl_2(s)$ dissolves until the concentrations of the dissolved ions are back to the equilibrium value. That is, the reaction proceeds from left to right.

3. The concentration of one or both of the dissolved ions is increased. *Result:* The system is no longer at equilibrium; $Q > K$. The reaction proceeds from right to left until the system returns to equilibrium; some $PbCl_2(s)$ precipitates.

The concentration of dissolved ions can be raised in several ways. The concentration of chloride ion can be raised by addition of any salt that contains chloride ion, such as sodium chloride. The concentration of $Pb^{2+}$ ion can also be increased by addition of any salt that contains

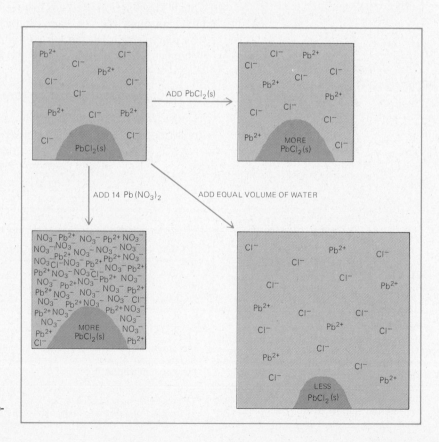

**Figure 14.5**
The principle of Le Chatelier in a saturated solution of an ionic solid.

## THE GREENHOUSE AND THE OCEANS

Over the next decades and centuries, the earth's climate faces an unprecedented change caused by human activities. Primarily because of the burning of fossil fuels, the concentration of carbon dioxide in the atmosphere is increasing steadily. Because carbon dioxide absorbs heat that otherwise would be radiated into space, an increase in atmospheric $CO_2$ concentration enhances the "greenhouse effect," which raises surface temperatures.

A number of expert committees have attempted to predict the extent to which atmospheric $CO_2$ levels will increase in the next century. Their predictions cover a wide range of possibilities, not only because of uncertainty about the amount of fossil fuels that will be burned in the next hundred years, but also about the amount of carbon dioxide that will be absorbed by the oceans. In making the latter estimate, the expert committees must deal with a problem of equilibrium chemistry and Le Chatelier's principle on a global scale. Their studies indicate that carbon dioxide may ultimately have as great an effect on the oceans as on the atmosphere.

As the concentration of carbon dioxide in the atmosphere increases, the surface water of the oceans absorbs more $CO_2$. The principal effect of the added carbon dioxide is to consume carbonate ion:

$$CO_2(aq) + CO_3^{2-}(aq) + H_2O \longrightarrow 2HCO_3^-(aq)$$

As carbonate ion is consumed, more is produced by the dissolution of calcium carbonate, the material that forms the shells of marine organisms such as coral, oysters, and clams:

$$CaCO_3(s) \longrightarrow Ca^{2+}(aq) + CO_3^{2-}(aq)$$

As the calcium carbonate dissolves, more $CO_2$ can be absorbed by the ocean — which, in turn, causes more calcium carbonate to dissolve.

Over the very long run, this phenomenon could have a major effect on $CO_2$ concentrations in the atmosphere. A study made by a National Research Council committee in 1983 said, "If we were to dissolve an average depth of 3 cm of pure calcium carbonate from the ocean floor, then in 1500 years the atmospheric $CO_2$ level would be some 30% lower than with no change in alkalinity." But it can also have a devastating effect on marine organisms and structures that contain calcium carbonate. If 40 cm of calcium carbonate from shallow-water sediments dissolves, the committee estimated, then the peak atmospheric level of $CO_2$ would be reduced by 40%. But this prospect is more a problem in its own right than a solution for the problem of the greenhouse effect. "The disappearance of such massive amounts of carbonate shells and corals and sediments from our shores and shallow seas would present a crisis for man arousing far greater concern than any incremental effects on $CO_2$ levels," the committee wrote. The amount of $CO_2$ that will be absorbed by the oceans and the amount of calcium carbonate that dissolves in response can only be determined by measurements in the coming years, the committee concluded.

lead ion, such as lead nitrate. The source of the lead and chloride ions need not (and should not) be mentioned in the equation, since the accompanying sodium ions or nitrate ions do not take part in the reaction.

The composition of the new equilibrium state is affected by the addition of either chloride ions or lead ions. If chloride ions are added, the new equilibrium state has a higher concentration of chloride ion and a lower concentration of lead ion than the original equilibrium state. The reverse is true if lead ions are added. But in either case, the value of the expression $[Pb^{2+}][Cl^-]^2$ is the same in the new equilibrium state as in the original equilibrium state.

4. The temperature of the system is raised. To predict the result, we must know the sign of $\Delta H°$. Lead chloride is observed to absorb heat when it dissolves in water. The reaction is endothermic and heat can be included on the left side of the equation. When the temperature is raised, the reaction proceeds from left to right. More solid dissolves, and $[Pb^{2+}]$, $[Cl^-]$, and $K$ increase.

The principle of Le Chatelier permits qualitative predictions about the behavior of equilibrium systems. Now we shall see that we can also treat equilibrium systems quantitatively.

## 14.3 CALCULATION OF GAS PHASE EQUILIBRIUM CONSTANTS

The quantitative treatment of equilibrium systems usually requires one of two calculations. Either we must calculate the value of $K$ at a given temperature from measurements of pressure or concentration, or we must calculate the composition of an equilibrium state, reached in a designated way, when the value of $K$ is given. For some problems, a combination of these two kinds of calculations is necessary.

We shall start with gas phase systems for the sake of simplicity, and we shall assume ideal gas behavior throughout. We shall begin with the most basic type of equilibrium problem, in which we are given the pressure of each gas at equilibrium and are asked to calculate the value of $K$.

---

**Example 14.5**    Automobile catalytic converters convert pollutants in exhaust gases to carbon dioxide and water. But these devices can also convert sulfur dioxide in the air to sulfur trioxide, a more worrisome pollutant. A sealed bulb at 1000 K contains sulfur dioxide, $P = 0.15$ atm; sulfur trioxide, $P = 0.23$ atm; and oxygen, $P = 0.73$ atm. The system is at equilibrium. Calculate the value of $K$ for this system.

**Solution**    Before we begin any calculations, we must consider the first, and often the most important, part of a problem in chemical equilibrium: writing a relevant, balanced chemical reaction that relates the components of the system and describes the chemical change. The way in which the equation for the reaction is written determines the form and value of $K$. Once the equation is written, we must write the expression for $K$ that corresponds to the equation. The third step is to substitute the data on the composition of the system at equilibrium into the expression for $K$ and to calculate the numerical value of $K$.

An equation for the relevant chemical change in this example is:

$$2SO_2(g) + O_2(g) \rightleftharpoons 2SO_3(g)$$

The form of the equilibrium constant for this equation is

$$K = \frac{P_{SO_3}^2}{P_{SO_2}^2 P_{O_2}}$$

Since the pressure of each component is given, we can calculate the numerical value of $K$:

$$K = \frac{(0.23 \text{ atm})^2}{(0.15 \text{ atm})^2 (0.73 \text{ atm})} = 3.2 \text{ atm}^{-1}$$

If we had written the equation in reverse, as $2SO_3(g) \rightleftharpoons 2SO_2(g) + O_2(g)$, the answer would have been $K = (1/3.2)$ atm, the reciprocal of the answer given above, since reversing an equation inverts $K$. If we had written the equation with different coefficients, the value of $K$ would be different.

Although $K$ itself actually has no units (Section 13.7), the value of $K$ depends on how we express the quantities of the components of the system. If units of pressure other than atmospheres are used, difficulties can be avoided by conversion of these units to atmospheres before they are substituted into the expression for $K$. We shall follow this practice. By agreement, the standard state and $\Delta G^\circ$ refer to 1 atm. Therefore, the equilibrium constant from $\Delta G^\circ$ must have pressures of gases in atmospheres (Section 13.7).

In some applications, units of concentration are more convenient than units of pressure for the quantities of gas in an equilibrium system. The units of concentration for gases are moles per liter (mol/L), and the standard state is 1 mol/L. We can write an equilibrium constant, designated $K_c$, using units of concentration. The equilibrium constant in which quantities of gases are expressed as pressures can be designated $K_p$ for clarity. There is no direct relationship between $K_c$ and $\Delta G^\circ$ because $K_c$ does not use the standard state of 1 atm.

Using the ideal gas equation, we can derive a relationship between $K_c$ and $K_p$:

$$P = \frac{(n)RT}{(V)}$$

where $n/V$ is concentration. If we substitute for each $P$ in the equilibrium expression, we can show that

$$K_c = K_p \left(\frac{1}{RT}\right)^{\Delta n}$$

where $\Delta n$ is the change in the number of moles of gas when the reaction goes from left to right. For reactions in which the number of moles of gas does not change, $K_c = K_p$.

**Example 14.6**    An equilibrium system contains NO(g), $P = 0.115$ atm; $Cl_2(g)$, $P = 0.181$ atm; and NOCl(g), $P = 0.258$ atm at 500 K. Find $K_c$.

**Solution**    The essential first step is to write an equation that describes the system. One possibility is

$$2NO(g) + Cl_2(g) \rightleftharpoons 2NOCl(g)$$

First we can find $K_p$ by substituting the given pressure data into the equilibrium expression:

$$K = \frac{P_{NOCl}^2}{P_{Cl_2} P_{NO}^2} = \frac{(0.258 \text{ atm})^2}{(0.181 \text{ atm})(0.115 \text{ atm})^2} = 27.8 \text{ atm}^{-1}$$

To convert to $K_c$, we find $\Delta n$, the change in the number of moles of gas as the reaction proceeds from left to right. As the reaction is written, there are 3 mol of gas on the left and 2 mol of gas on the right, so $\Delta n = 2 - 3 = -1$.

$$K_c = K_p \left(\frac{1}{RT}\right)^{-1} = 27.8 \text{ atm}^{-1} \left[ \left\{ 1 \Big/ \left( 0.0821 \frac{\text{L atm}}{\text{mol K}} \right) (500 \text{ K}) \right\} \right]^{-1}$$

$$= 1140 \text{ L/mol}$$

Generally, we do not need to know the pressure of every gas in the equilibrium to find the value of $K$. In the next example, we calculate the value of $K$ from the pressure of one gas and a number of other measurements of the system.

**Example 14.7**    A 7.24-g sample of IBr is placed in a container whose volume is 0.225 L and is heated to 500 K. Some of the IBr decomposes to $I_2$ and $Br_2$. All three substances are in the gas phase. The system is at equilibrium and the measured pressure of $Br_2(g)$ in the system is 3.01 atm. Calculate the value of $K$.

**Solution**    As always, we start by writing a balanced equation from the word description. One possibility is:

$$2IBr(g) \rightleftharpoons I_2(g) + Br_2(g)$$

We can calculate the value of $K$ by finding the pressure of each gas present at equilibrium. Since the partial pressure of a gas is directly proportional to the amount of the gas, we can use the molar relationships expressed in the balanced equation. It states that bromine and iodine are formed in equal amounts. Therefore, $P_{Br_2} = P_{I_2} = 3.01$ atm.

The equation also states that 2 atm of product forms when 2 atm of IBr reacts. Therefore, the total pressure of $I_2$ and $Br_2$ that forms, $P_{I_2} + P_{Br_2}$, equals $P_{IBr}$ that reacts. The arithmetic is simple: 3.01 atm + 3.01 atm = 6.02 atm, which is the pressure of IBr undergoing reaction.

To calculate the pressure of IBr present at equilibium, we need the original pressure of IBr. We can find it by using the ideal gas equation and the data given:

$$P = \frac{(7.24 \text{ g IBr})(0.0821 \text{ L atm mol}^{-1} \text{ K}^{-1})(500 \text{ K})}{(206.8 \text{ g IBr/mol IBr})(0.225 \text{ L})}$$

$$= 6.39 \text{ atm}$$

The pressure of IBr at equilibium is 6.39 atm − 6.02 atm = 0.37 atm.

This type of reasoning is quite common. If we know the quantity of a substance originally in the system and the quantity that undergoes reaction, we know the quantity of the substance that remains.

Now that we have the value of the pressure of each component at equilibrium, we can calculate the value of $K$:

$$K = \frac{P_{I_2} P_{Br_2}}{P_{IBr}^2} = \frac{(3.01 \text{ atm})(3.01 \text{ atm})}{(0.37 \text{ atm})^2} = 66$$

In Example 14.7, we found the value of $K$ starting with the pressures of the components of the system. The same kind of calculation can be done from measurements of total pressure. It is common to find the value of $K$ in this way, since measurements of total pressure usually are the easiest that can be done on a system containing gases.

**Example 14.8**   A container at 1000 K holds carbon dioxide, $P = 0.464$ atm. Graphite is added to the container, and some of the carbon dioxide is converted to carbon monoxide. At equilibrium, the total pressure in the container is 0.746 atm. Calculate the value of $K$.

**Solution**   The equation of interest can be written as:

$$CO_2(g) + C(s) \rightleftharpoons 2CO(g) \qquad K = \frac{P_{CO}^2}{P_{CO_2}}$$

To calculate a value of $K$, we must know the pressures of both $CO_2$ and CO at equilibrium.

As this reaction proceeds from left to right, 2 mol of CO(g) is produced for each mole of reactant $CO_2$(g). Therefore, the total pressure increases. If we let $x$ equal the pressure of $CO_2$ in atmospheres that must react for the system to reach equilibrium, we can say from the stoichiometry of the reaction that $2x$ will be the pressure of CO at equilibrium. Thus, when the system proceeds to equilibrium, we can see that there is a decrease of $x$ atm in the starting pressure of $CO_2$ and an increase of $2x$ atm in the pressure of CO, which started as zero.

|  | $CO_2(g)$ + C(s) $\rightleftharpoons$ | $2CO(g)$ |
|---|---|---|
| **start** | 0.464 atm | 0 |
| **equil** | 0.464 atm − $x$ | $2x$ |

The initial conditions are expressed on the line labeled "start." The equilibrium conditions are given on the line labeled "equil." No entry is needed for the solid, since the solid does not appear in the expression for $K$.

We can find the value for $K$ by solving for $x$, using the additional information that the total pressure is 0.746 atm. Total pressure is the sum of the partial pressures of the gases: $P_T = P_{CO_2} + P_{CO}$. If we express the values of these pressures in terms of $x$, we get:

$$0.746 \text{ atm} = (0.464 \text{ atm} - x) + 2x$$
$$x = 0.282 \text{ atm}$$

This value of $x$ is used to find the pressures of the gases at equilibrium: $P_{CO_2} = 0.464 \text{ atm} - 0.282 \text{ atm} = 0.182 \text{ atm}$, and $P_{CO} = 2(0.282 \text{ atm}) = 0.564$ atm.

We can calculate the value of $K$ using these values for $P$:

$$K = \frac{(0.564 \text{ atm})^2}{(0.182 \text{ atm})} = 1.75 \text{ atm}$$

Many variations of this type of problem are possible, depending on the measurements made and the stoichiometry of the reaction. In addition, many methods can be used in the laboratory to find the data needed to calculate an equilibrium constant. One method is to calculate $K$ by finding how much of a reactant is consumed.

**Example 14.9**    When sulfur in the form of $S_8$ is heated to 900 K, at equilibrium the pressure of $S_8$ falls by 29% from 1.00 atm because some of the $S_8$ is converted to $S_2$. Find the value of $K$ for this reaction.

**Solution**    The chemical equation and the starting and equilibrium pressures are

|  | $S_8(g)$ | $\rightleftharpoons$ $4S_2(g)$ |
|---|---|---|
| **start** | 1.00 atm | 0 |
| **equil** | 1.00 atm $-$ 0.29 atm | 4(0.29 atm) |

The equilibrium pressures are found by this line of reasoning: The pressure of $S_8$ at equilibrium is the starting pressure minus 29% of the original 1.00 atm of $S_8$, that is, 1.00 atm $-$ 0.29 atm. For each mole or atmosphere of $S_8$ that reacts, 4 mole or 4 atm of $S_2$ is formed. If 0.29 atm of $S_8$ reacts, 4(0.29 atm) of $S_2$ is formed.

Using these data, we can calculate the value for $K$:

$$K = \frac{P_{S_2}^4}{P_{S_8}} = \frac{[4(0.29 \text{ atm})]^4}{1.00 \text{ atm} - 0.29 \text{ atm}} = 2.6 \text{ atm}^3$$

## 14.4 CALCULATIONS FROM GAS PHASE EQUILIBRIUM CONSTANTS

In this section we shall see how the equilibrium constant for one state of a system can be used to find the composition of other equilibrium states of the system at the same temperature. Most equilibrium calculations in general chemistry require you to find the composition of an equilibrium state from a set of starting conditions and a given value of an equilibrium constant. Here are some examples.

**Example 14.10**   At 400 K, solid ammonium chloride decomposes to gaseous ammonia and hydrogen chloride. For this process, $K = 6.0 \times 10^{-9}$ atm². Calculate the equilibrium pressures for the two gases at this temperature.

**Solution**   At first glance, it might seem that we cannot solve this problem because no starting quantity of ammonium chloride is given. But ammonium chloride is a solid. The quantity of ammonium chloride is therefore unimportant, as long as there is enough present for the system to reach equilibrium. In problems where the exact quantity of starting material is unimportant, we shall designate this quantity by the letter $a$. Thus, for this example, the equation and starting conditions are:

$$NH_4Cl(s) \rightleftharpoons NH_3(g) + HCl(g)$$

$$\textbf{start} \qquad a \qquad\qquad 0 \qquad\quad 0$$

For the system to reach equilibrium, some ammonium chloride must decompose and some ammonia must form. Let us designate the pressure (in atmospheres) of ammonia that forms by the letter $x$. From the stoichiometry of the reaction, the pressure of hydrogen chloride that forms is also $x$. Now we can describe the equilibrium conditions symbolically while ignoring the quantity of ammonium chloride:

$$\textbf{equil} \qquad\qquad\qquad x \qquad\quad x$$

We know that the expression for $K$ includes only the equilibrium pressures of the two gases:

$$K = P_{NH_3} P_{HCl} = 6.0 \times 10^{-9}\ \text{atm}^2$$

Expressing the pressures in terms of $x$ gives:

$$(x)(x) = 6.0 \times 10^{-9}\ \text{atm}^2$$
$$x = 7.7 \times 10^{-5}\ \text{atm} = P_{NH_3} = P_{HCl}$$

Since the value of $K$ is small, the equilibrium pressures of the two gases are also small.

The mathematics of the problem in Example 14.10 is relatively simple, because $K$ has a relatively simple form. But you may encounter some more involved algebraic calculations in equilibrium problems. A knowledge of chemistry can enable you to simplify the calculations.

**Example 14.11**    The value of $K$ for the reaction in which 2 mol of nitrogen dioxide forms 1 mol of dinitrogen tetroxide at 451 K is $5.33 \times 10^{-3}$ atm$^{-1}$. A 0.460-g sample of nitrogen dioxide is heated to 451 K in a reaction vessel whose volume is 0.500 L. Calculate the pressure of dinitrogen tetroxide at equilibrium.

**Solution**    We must first write the chemical equation and tabulate the starting quantities. But there is a problem. The starting quantity of $NO_2$ is given in units of mass, while $K$ has units of pressure. We must therefore express the given mass of $NO_2$ as a pressure of $NO_2$ at 451 K. We substitute the data into the ideal gas equation:

$$P = \frac{(0.460 \text{ g } NO_2)(0.0821 \text{ L atm mol}^{-1} \text{ K}^{-1})(451 \text{ K})}{(46.0 \text{ g } NO_2/\text{mol } NO_2)(0.500 \text{ L})}$$

$$= 0.741 \text{ atm}$$

Now we can write the equation:

$$2NO_2(g) \rightleftharpoons N_2O_4(g)$$

**start**    0.741 atm             0

To avoid fractions, let $2x$ be the pressure in atmospheres of $NO_2$ that is consumed to reach equilibrium. Then $x$ is the pressure in atmospheres of $N_2O_4$ formed. The equilibrium pressures are:

**equil**    0.741 atm $- 2x$         $x$

When we substitute these equilibrium pressures into the expression for $K$, the result is a quadratic equation:

$$K = \frac{P_{N_2O_4}}{P_{NO_2}^2} = \frac{x}{(0.741 \text{ atm} - 2x)^2} = 5.33 \times 10^{-3} \text{ atm}^{-1}$$

This expression can be rearranged to:

$$(2.13 \times 10^{-2})x^2 - 1.0158x + 2.93 \times 10^{-3} = 0$$

and can be solved using the quadratic formula

$$x = \frac{-b \pm \sqrt{b^2 - 4ac}}{2a}$$

The value of $x$ found in his way is $2.88 \times 10^{-3}$ atm. In using the quadratic formula, we obtain two values for $x$. Only one makes chemical sense. In this case, the other value for $x$ is 47.7 atm, much larger than the initial pressure of $NO_2$. In other problems, one of the values for $x$ may be negative, a chemical impossibility.

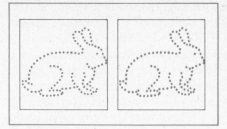

**Figure 14.6**
When a relatively small number is subtracted from or added to a much larger one, the result can be approximated by the large number. Can you tell the difference between the two pictures? The second one has three dots fewer than the first one.

The computation with the quadratic formula is often unnecessary. We can find a simpler way by applying some chemical reasoning to the problem.

We start with the fact that the value for $K$ is small. We learned earlier that a small value for $K$ indicates that only a small amount of the starting material undergoes reaction. The unknown ($x$) usually is assigned to a quantity of material that undergoes reaction. Therefore, the value of $x$ is small relative to the quantity of starting material whenever the numerical value of $K$ is small. In this example, for instance, $2x$, the pressure of $NO_2$ that reacts, is only 2(0.00288 atm) compared to 0.741 atm of $NO_2$.

A very large number is not changed appreciably by the subtraction or addition of a very small number. In such a case, we can go on using the larger number as if no subtraction or addition took place (Figure 14.6). If you are in a traffic jam and estimate idly that you are surrounded by 800 cars, you will not change your estimate if three cars pull off the road. Rather than revising your estimate to 797 cars, you assume that 800 is still a good approximation. That approximation also holds if five cars pull onto the highway. We can state this numerically as $800 \approx 800 - 3 \approx 800 + 5$. We can apply similar reasoning to examples in chemical equilibrium.

Let us apply it to this system. We start with 0.741 atm of $NO_2$. Since $K$ is small, we know that a very small amount of $NO_2$, designated as $2x$, undergoes reaction. Since $2x$ is much smaller than 0.741, we can say that $0.741 \approx 0.741 - 2x$. This approximation helps us greatly, since substitution into the expression for $K$ gives a much simpler expression:

$$\frac{x}{(0.741 \text{ atm})^2} = 5.33 \times 10^{-3} \text{ atm}^{-1}$$

The previous expression was a quadratic equation. This one is not. It can be solved directly, giving $x = 2.93 \times 10^{-3}$ atm. We check the approximation and find it to be a valid one since this value is quite close to the value obtained with the quadratic formula.

There are many problems in which the method of approximation used in Example 14.11 cannot be used. There are guidelines to help you decide when it is appropriate to use approximations.

To begin with, *approximations can be used only for sums and differences, never for products or quotients.* Any time a multiplication or division must be performed, approximation is ruled out. Thus, we can say $2 + x \approx 2$ or $2 - x \approx 2$ if $x$ is suitably small, but we cannot say $2x \approx 2$ or $2/x \approx 2$.

How small must $x$ be? In this text, we shall say arbitrarily that the approximation is valid when the error that is introduced is less than 5%. This level of approximation is useful for us and is consistent with the overall approximate nature of the ideal gas and solution approximations. In other contexts a different level of approximation might be chosen. We check on the approximation by using the calculated value of the unknown. To give a specific instance, in Example 14.11 we used the approximation that $0.741 - 2x = 0.741$ and found that $x = 2.93 \times 10^{-3}$, or

0.00293. If this value is used to verify the approximation, we find that we have said that: $0.741 - 2(0.00293) = 0.735 \approx 0.741$. The error—that is, the difference between 0.735 and 0.741—is less than 1%.

The value of $K$ often helps us to decide whether an approximation is reasonable. If the value of $K$ for a system is either large enough ($> 10^2$) or small enough ($< 10^{-2}$), it *may* be possible to make approximations.

*In any problem where an approximation has been made, it is crucial to check the approximation using the original data and the value of the unknown that has been calculated on the basis of the approximation.* Only by checking in this way can you be sure that it was appropriate to make an approximation.

**Example 14.12**

When phosgene, $COCl_2$, is heated to 600 K, it partially decomposes to form carbon monoxide and chlorine. The value of $K$ for this process is $4.10 \times 10^{-3}$ atm. Calculate the equilibrium pressures of the components of the system produced by a starting pressure of phosgene of 0.124 atm.

**Solution**

The balanced chemical equation and starting conditions are

$$COCl_2(g) \rightleftharpoons CO(g) + Cl_2(g)$$
$$\textbf{start} \qquad 0.124 \text{ atm} \qquad 0 \qquad 0$$

Let $x$ be the pressure (in atmospheres) of $COCl_2$ that must react for the system to reach equilibrium. The equilibrium line is:

$$\textbf{equil} \qquad 0.124 \text{ atm} - x \qquad x \qquad x$$

Substituting these pressures into the expression for $K$ gives:

$$K = \frac{P_{CO}P_{Cl_2}}{P_{COCl_2}} = \frac{(x)(x)}{(0.124 \text{ atm} - x)} = 4.10 \times 10^{-3} \text{ atm}$$

To avoid the use of the quadratic formula, we turn to the approximation $0.124$ atm $- x \approx 0.124$ atm. The equation becomes:

$$\frac{x^2}{0.124 \text{ atm}} = 4.10 \times 10^{-3} \text{ atm}$$
$$x = 2.25 \times 10^{-2} \text{ atm}$$

Checking the approximation, we find that we have said that $0.124 - 0.0225 = 0.102 \approx 0.124$. This error of 20% is unacceptable. But we can use this admittedly erroneous value of $x$ to improve the approximation. The correct value of $0.124 - x$ is closer to $0.124 - 0.0225$ than it is to 0.124 alone, since $x$ is at least in the neighborhood of 0.0225. We now write the equation as:

$$\frac{x^2}{(0.124 \text{ atm} - 0.0225 \text{ atm})} = 4.10 \times 10^{-3} \text{ atm}$$
$$x = 2.04 \times 10^{-2} \text{ atm}$$

This value of $x$ is closer to the correct one. We can repeat the approximation a second time, using this value of $x$.

$$\frac{x^2}{(0.124 \text{ atm} - 0.0204 \text{ atm})} = 4.10 \times 10^{-3} \text{ atm}$$

$$x = 2.06 \times 10^{-2} \text{ atm}$$

This method of solution is called the method of successive approximations. Solution of the equation:

$$\frac{(x)(x)}{(0.124 \text{ atm} - x)} = 4.10 \times 10^{-3} \text{ atm}$$

by the quadratic formula gives $x = 2.06 \times 10^{-2}$ atm.

Let us now consider chemical reactions with large equilibrium constants. If the equilibrium constant is large enough, the methods of approximation can also be used to simplify calculations.

When a reaction with a large equilibrium constant proceeds to equilibrium from left to right, most of the reactants are consumed. If we let the unknown be the quantity of reactant that is consumed, the value of the unknown would be too large to allow the use of approximations. But if a reaction has a large equilibrium constant, the reverse reaction has a small one. If the reaction with the large $K$ could be made to proceed from right to left to reach equilibrium, the value of the unknown would be relatively small.

Thus we can make approximations if we approach the equilibrium state of the system from the direction that corresponds to the small $K$, or the side of the equation that is closer to the final equilibrium state. When $K$ is large, equilibrium should be approached from right to left. The final answer will be the same no matter how we approach equilibrium, since the equilibrium state is independent of the direction of approach.

To make use of this rule, we introduce an extra step into calculations for systems where $K$ is large (usually larger than $10^2$) and the starting conditions are such that the system reaches equilibrium from left to right. In the extra step, we take the reaction past the equilibrium state to completion, and then we approach the equilibrium state from right to left. In this way, the unknown that we assign will be relatively small. The completion step in which the system passes equilibrium bears no relation to the actual behavior of the system. Its only purpose is to simplify algebra for us. The completion step is simply a stoichiometric calculation, sometimes of the limiting reagent type that we discussed in Section 3.3.

Example 14.13   At 700 K, sulfur dioxide is converted almost completely to sulfur trioxide by oxygen. For a reaction of 2 mol of sulfur dioxide, $K = 8.24 \times 10^4$ atm$^{-1}$. Calculate the pressures at equilibrium when 0.490 atm of $SO_2$ and 0.245 atm of $O_2$ are brought together at 700 K.

Solution     We write the relevant balanced equation and tabulate the starting quantity:

$$2SO_2(g) \quad + O_2(g) \quad \rightleftharpoons 2SO_3(g)$$

|  | | |
|---|---|---|
| **start** | 0.490 atm | 0.245 atm | 0 |

Since $K$ is large, almost all these starting quantities of $SO_2$ and $O_2$ are consumed when the system reaches equilibrium. Let us take the reaction past equilibrium to completion. We now need to tabulate the pressures of all the substances in the system at completion. The reaction indicates that 2 mol (or 2 atm) of $SO_2$ reacts with 1 mol (or 1 atm) of $O_2$ to produce 2 mol (or 2 atm) of $SO_3$. The starting pressure of $SO_2$ in this system is exactly twice that of $O_2$. Therefore:

| **complete** | 0 | 0 | 0.490 atm |
|---|---|---|---|

Now we allow the system to approach equilibrium from right to left. If $2x$ is the pressure (in atmospheres) of $SO_3$ that reacts when the system proceeds to equilibrium, the equilibrium line is:

| **equil** | $2x$ | $x$ | 0.490 atm $- 2x$ |
|---|---|---|---|

Since $x$ is relatively small, 0.490 atm $- 2x \approx 0.490$ atm. When we use this approximation, substitution into the expression for $K$ gives a manageable equation:

$$K = \frac{P^2_{SO_3}}{P^2_{SO_2}P_{O_2}} = \frac{(0.490 \text{ atm})^2}{(2x)^2(x)} = 8.24 \times 10^4 \text{ atm}^{-1}$$

$$4x^3 = \frac{(0.490 \text{ atm})^2}{8.24 \times 10^4 \text{ atm}^{-1}}$$

$$x = 0.00900 \text{ atm}$$

We check the validity of the approximation by using the calculated value of $x$. We find that 0.490 atm $- 2(0.00900$ atm$) = 0.472$ atm $\approx 0.490$ atm, about a 4% error. Using the value of $x$ gives: $P_{SO_2} = 0.018$ atm, $P_{O_2} = 0.0090$ atm.

If you try to solve this problem by approaching equilibrium from left to right, you obtain a cubic equation that is virtually unsolvable and cannot be simplified by chemically valid approximations. By introducing a purely imaginary completion step, we have greatly simplified our calculations. But it must again be emphasized that the completion step is entirely imaginary. It tells us nothing about the way in which the system proceeds to equilibrium.

We started with a system whose components were present in stoichiometric amounts. Now it is time for a problem in which a reaction with a large equilibrium constant starts with nonstoichiometric quantities of reactants.

Example 14.14     For the reaction in which 2 mol of hydrogen bromide forms from hydrogen and bromine at 700 K, $K = 5.5 \times 10^8$. A mixture of 0.34 mol of hydrogen and 0.22 mol of bromine is heated to 700 K. Calculate the composition of the system at equilibrium.

**Solution**     We write the balanced chemical reaction and tabulate the starting conditions:

$$H_2(g) \qquad + Br_2(g) \quad \rightleftharpoons 2HBr(g)$$

| | | | |
|---|---|---|---|
| **start** | 0.34 mol | 0.22 mol | 0 |

Since there is no change in the number of moles of gas in this reaction, $K_p = K_c$ and we can use moles of gas directly rather than converting to units of pressure. And since $K$ is large for the reaction as written, we introduce the extra completion step.

The reaction shows that 1 mol of hydrogen reacts with 1 mol of bromine to produce 2 mol of hydrogen bromide. Less bromine than hydrogen is present, so bromine is the limiting reagent. Therefore, 0.22 mol of $Br_2$ and 0.22 mol of $H_2$ are consumed and 2(0.22) mol of HBr is formed, giving the completion line:

**complete**     0.34 mol − 0.22 mol    0                    0.44 mol

We allow the system to reach equilibrium from right to left. The amount of HBr (in moles) that reacts as the system approaches equilibrium from the right is designated as $2x$. The equilibrium line is

**equil**     0.12 mol + $x$          $x$              0.44 mol − $2x$

Since $x$ is relatively small, we can make the approximations 0.12 mol + $x \approx$ 0.12 mol and 0.44 mol − $2x \approx$ 0.44 mol. The expression for $K$ then gives:

$$K = \frac{(0.44 \text{ mol})^2}{(0.12 \text{ mol})(x)} = 5.5 \times 10^8$$

$$x = 2.9 \times 10^{-9} \text{ mol}$$

Because $x$ is exceedingly small compared to the other quantities in the system, the approximations are valid. The value of $x$ is now used to calculate the composition of the equilibrium state: $n_{H_2} = 0.12$ mol; $n_{Br_2} = 2.9 \times 10^{-9}$ mol; $n_{HBr} = 0.44$ mol.

---

The previous examples dealt with systems that start in a nonequilibrium state and move to equilibrium. The same sort of calculations can be used to deal with systems that are in equilibrium, are perturbed, and then return to a new equilibrium state.

---

**Example 14.15**     At 700 K, carbon dioxide and hydrogen react to form carbon monoxide and water. For this process, $K = 0.11$. A mixture of 0.45 mol of $CO_2$ and 0.45 mol of $H_2$ is heated to 700 K. (a) Find the amount of each gas at equilibrium. (b) After the system reaches equilibrium, another 0.34 mol of $CO_2$ and 0.34 mol of $H_2$ are added to the system. Find the composition of the new equilibrium state.

**Solution**     a. The relevant chemical equation with the data tabulated is

$$CO_2(g) \qquad + H_2(g) \qquad \rightleftharpoons CO(g) + H_2O(g)$$

| | | | |
|---|---|---|---|
| **start** | 0.45 mol | 0.45 mol | 0        0 |

As in Example 14.14, we need not convert moles to units of pressure. Let $x$ be the amount of $CO_2$ that undergoes reaction when the system reaches equilibrium.

**equil**    0.45 mol $- x$    0.45 mol $- x$    $x$    $x$

Substitution into the expression for $K$ gives:

$$K = \frac{n_{CO}n_{H_2O}}{n_{CO_2}n_{H_2}} = \frac{(x)(x)}{(0.45\ \text{mol} - x)(0.45\ \text{mol} - x)} = 0.11$$

If the equation is rewritten as

$$\frac{x^2}{(0.45\ \text{mol} - x)^2} = 0.11$$

an easy method of solution is available. Take the square root of both sides:

$$\frac{x}{(0.45\ \text{mol} - x)} = 0.33$$

and $x = 0.11$ mol $= n_{CO} = n_{H_2O}$; 0.45 mol $-$ 0.11 mol $=$ 0.34 mol $= n_{CO_2} = n_{H_2}$.

b. When more $CO_2$ and $H_2$ are added to the system, we have a new set of starting conditions, which can be tabulated in the usual way:

|  | $CO_2(g)$ | $+ H_2(g)$ | $\rightleftharpoons CO(g)$ | $+ H_2O(g)$ |
|---|---|---|---|---|
| **start** | 0.34 mol | 0.34 mol | 0.11 mol | 0.11 mol |
|  | +0.34 mol | +0.34 mol |  |  |

The principle of Le Chatelier tells us that the reaction proceeds from left to right. Once again we let $x$ be the amount of $CO_2$ that must undergo reaction for the system to reach equilibrium.

**equil**    0.68 mol $- x$    0.68 mol $- x$    0.11 mol $+ x$    0.11 mol $+ x$

We again use the expression for $K$ to find $x$:

$$\frac{(0.11\ \text{mol} + x)^2}{(0.68\ \text{mol} - x)^2} = 0.11$$

Taking the square root of both sides of the equation gives $x = 0.09$ mol; 0.11 mol $+$ 0.09 mol $=$ 0.20 mol $= n_{CO} = n_{H_2O}$, 0.68 mol $-$ 0.09 mol $=$ 0.59 mol $= n_{CO_2} = n_{H_2}$. As a check, we substitute the calculated amounts into the expression for $K$ to see if we obtain the correct value of $K$:

$$\frac{(0.20\ \text{mol})(0.20\ \text{mol})}{(0.59\ \text{mol})(0.59\ \text{mol})} = 0.11$$

**Example 14.16**    An equilibrium mixture contains $N_2O_4$, $P = 0.28$ atm, and $NO_2$, $P = 1.1$ atm, at 350 K. The volume of the container is doubled. Calculate the equilibrium pressures of the two gases when the system reaches a new equilibrium.

**Solution**   The reaction of interest and the conditions are

$$N_2O_4(g) \rightleftharpoons 2NO_2(g)$$

**equil**     0.28 atm        1.1 atm

We use the data to calculate the value of $K$:

$$K = \frac{P_{NO_2}^2}{P_{N_2O_4}} = \frac{(1.1 \text{ atm})^2}{(0.28 \text{ atm})} = 4.3 \text{ atm}$$

According to Boyle's law, doubling the volume of the container decreases the pressure of each gas by a factor of two. The system is no longer at equilibrium. We find the pressure of each gas by dividing the equilibrium pressure by two, and we tabulate the data as starting conditions:

**start**     $\dfrac{0.28 \text{ atm}}{2}$        $\dfrac{1.1 \text{ atm}}{2}$

We can easily verify that these are no longer equilibrium pressures by substitution into the expression for $K$.

The principle of Le Chatelier tells us that the reaction in a system whose pressure is lowered proceeds in the direction that produces more moles of gas. For this reaction, the direction is from left to right: $N_2O_4$ is consumed. We let $x$ be the pressure of $N_2O_4$ consumed and write a new equilibrium line:

$$N_2O_4(g) \rightleftharpoons 2NO_2(g)$$

**equil**     $\dfrac{0.28 \text{ atm}}{2} - x$        $\dfrac{1.1 \text{ atm}}{2} + 2x$

We use the expression for $K$ to solve for $x$:

$$K = \frac{(0.55 \text{ atm} + 2x)^2}{(0.14 \text{ atm} - x)} = 4.3 \text{ atm}$$

We must use the quadratic formula to find $x$; $x = 0.045$ atm. Thus, the new equilibrium pressures are $P_{NO_2} = 0.55 \text{ atm} + 2x = 0.55 \text{ atm} + 2(0.045 \text{ atm}) = 0.64$ atm and $P_{N_2O_4} = 0.14 \text{ atm} - x = 0.14 \text{ atm} - 0.045 \text{ atm} = 0.095$ atm.

We can check these values by substituting into the expression for $K$:

$$K = \frac{(0.64 \text{ atm})^2}{0.095} = 4.3 \text{ atm}$$

You now have been exposed to many problems in gas phase equilibrium. Most of these problems have features in common. A review of these features provides a guide to some of the best methods of solving problems in chemical equilibrium.

1. *Write a balanced chemical equation that expresses the chemical changes that occur in the system.*

2. To keep the calculation orderly, tabulate the initial conditions. The statement of the problem may give the quantities of some or all of the substances of the system. After making any necessary unit conversions, list each value under the appropriate formula for the substance.

3. The statement of the problem will say whether the system starts in an equilibrium state. If it does, see whether a numerical value can be listed under every substance that appears in the expression for $K$, remembering that only gases and dissolved substances appear in the expression. If there are no unknowns on this line, you can substitute the data into the expression for $K$ and determine the value of $K$.

4. If the initial conditions are not equilibrium conditions, determine the direction in which the reaction proceeds to reach equilibrium. If one of the components is initially absent, the reaction must proceed in the direction to produce that component. In other cases, the relative values of $Q$ and $K$ indicate the direction.

5. Assign an unknown. The unknown usually is related to the quantity of one of the substances that is consumed when the system moves from the initial nonequilibrium state to equilibrium. It is expressed in units appropriate to the problem.

6. Once you have tabulated the initial conditions of a nonequilibrium state on a "start" line and assigned an unknown, tabulate the composition of the equilibrium state algebraically on an "equil" line. Below each substance in the chemical equation that appears in the expression for $K$, write a term for its pressure or concentration at equilibrium. The number of unknowns on the equilibrium line can provide clues to the solution of the problem.

7. To find the values of the unknowns on the equilibrium line, we must be given an equal number of bits of data about the composition of the equilibrium state. For example, when the equilibrium line has one unknown, we need one bit of data, usually the value of $K$, to find the value of the unknown.

In the next example, we develop an "equil" line with two unknowns, and we are given two bits of data about the composition of the equilibrium state. We use both of them to find the values of the two unknowns.

---

**Example 14.17**

For the decomposition of 1 mol of carbon tetrachloride vapor to graphite and chlorine at 700 K, $K = 0.76$ atm. Calculate the starting pressure of carbon tetrachloride that will produce a total pressure of 1.0 atm at equilibrium.

**Solution**

The only information given about the initial conditions is that we start with an unknown quantity of $CCl_4$. Let us represent this starting quantity by an unknown, $y$, the required starting pressure of $CCl_4$ in atmospheres. The relevant reaction and the starting conditions are

$$CCl_4(g) \rightleftharpoons 2Cl_2(g) + C(s)$$

$$\text{start} \quad\quad y \quad\quad\quad 0$$

The system now proceeds to equilibrium as an unknown quantity of $CCl_4$ reacts. Let $x$ be this quantity in atmospheres. The equilibrium line is

**equil**      $y - x$          $2x$

The equilibrium line has two unknowns, so we must look for two bits of data that are given about the composition of the equilibrium state. They are the value of $K$ and the total pressure at equilibrium. First we eliminate one of the unknowns from the "equil" line by using the total pressure:

$$P_T = P_{CCl_4} + P_{Cl_2} = (y - x) + 2x = 1.0 \text{ atm}$$

which can be rearranged to $y = 1.0 \text{ atm} - x$. Substituting the expression for $y$ in terms of $x$ on the "equil" line gives:

$$CCl_4 \rightleftharpoons 2Cl_2(g) + C(s)$$
**equil**   $1.0 \text{ atm} - 2x$   $2x$

Now that the equilibrium line has only one unknown, we can substitute pressure terms into the expression for $K$ to solve for $x$:

$$K = \frac{(2x)^2}{(1.0 \text{ atm} - 2x)} = 0.76 \text{ atm}$$

Since $K$ is neither particularly large nor particularly small, an approximation cannot be used. Applying the quadratic formula gives $x = 0.29 \text{ atm}$. Since $y = 1.0 \text{ atm} - x$, then $y = 1.0 \text{ atm} - 0.29 \text{ atm} = 0.71 \text{ atm}$, the starting pressure of carbon tetrachloride needed to produce a total pressure of 1.0 atm at equilibrium.

## 14.5  SOLUTIONS OF SPARINGLY SOLUBLE SUBSTANCES: THE SOLUBILITY PRODUCT

The reasoning that was used in Sections 14.3 and 14.4 for gas phase equilibrium also can be used for equilibrium systems consisting of water and a sparingly soluble ionic solid.

We limit ourself to these systems because our discussion of chemical equilibrium is based on the ideal solution approximation. This approach gives poor results for solutions of appreciably soluble ionic substances, in which the concentration of ions is high. Such systems can be studied quantitatively only by methods that are beyond our scope.

As you might expect, the first step in treating equilibrium systems of solutions is to write the chemical equation describing the appropriate reaction. The second step is to find the equilibrium constant for the reaction. Sometimes it must be calculated from the given data; in other cases, it can be found in standard reference works. Once it is known, the equilibrium constant is used to find the composition of equilibrium states that are reached from different starting conditions.

The chemical equations for aqueous solutions of sparingly soluble ionic solids all have the same form: The undissolved solid is on the left, the dissociated ions on the right. Some examples are

silver chloride:    $AgCl(s) \rightleftharpoons Ag^+(aq) + Cl^-(aq);$
$K_{sp} = [Ag^+][Cl^-]$

barium fluoride:    $BaF_2(s) \rightleftharpoons Ba^{2+}(aq) + 2F^-(aq);$
$K_{sp} = [Ba^{2+}][F^-]^2$

ferric hydroxide:    $Fe(OH)_3(s) \rightleftharpoons Fe^{3+}(aq) + 3OH^-(aq);$
$K_{sp} = [Fe^{3+}][OH^-]^3$

calcium phosphate:    $Ca_3(PO_4)_2(s) \rightleftharpoons 3Ca^{2+}(aq) + 2PO_4^{3-}(aq);$
$K_{sp} = [Ca^{2+}]^3[PO_4^{3-}]^2$

None of the expressions for the equilibrium constants of these reactions has a denominator, since the left sides of the equations have only solids, which are not included in the expression for $K$. The expression for $K$ includes the product of the concentrations of at least two ions, because the right side of each equation must contain at least two ions. Each concentration has an exponent equal to the coefficient of the ion in the equilibrium.

The equilibrium constant for the dissolution of a sparingly soluble ionic solid in water is called the **solubility product.** It is abbreviated $K_{sp}$. The expression for the $K_{sp}$ for each solid is shown next to the equilibrium above. Values have been determined for the solubility products of many substances. Table 14.3 lists some of them.

To write correct chemical equations, you must know which ions are formed when a given ionic substance dissolves in water. It is advisable at this point to review the information on polyatomic ions in Table 2.7 and the procedure for writing net ionic equations in Section 14.1. We shall use net equations throughout.

Once the correct chemical equation is written, you must be able to find the value of the equilibrium constant. Some explanation about the way in which we shall use the solubility product, $K_{sp}$, is necessary.

In addition to using the ideal solution approximation, we also make some simplifying assumptions about the chemistry of solutions of ionic substances. We assume that the only process of concern is the one in which the sparingly soluble material dissolves. Other chemical reactions can occur; an ion can react with water or with other ions. For the moment, we shall ignore these complications.

It is important not to confuse the term *solubility product* with the term *solubility*. The solubility product is the equilibrium constant of a specific reaction. The solubility is the quantity of a substance that dissolves in a given quantity of water. These two terms have different numerical values for a given reaction, but they are related. The relationship can be calculated.

**TABLE 14.3**   Solubility Products of Some Sparingly Soluble Ionic Solids at Room Temperature

| Substance | Solubility Product[a] | Substance | Solubility Product[a] |
|---|---|---|---|
| AgBr | $7.7 \times 10^{-13}$ | CuCN | $1.0 \times 10^{-11}$ |
| AgCN | $1.2 \times 10^{-16}$ | CuCl | $1.0 \times 10^{-6}$ |
| $Ag_2CO_3$ | $6.2 \times 10^{-12}$ | CuI | $5.1 \times 10^{-12}$ |
| AgCl | $1.6 \times 10^{-10}$ | $Cu_2S$ | $2 \times 10^{-47}$ |
| $Ag_2CrO_4$ | $1.1 \times 10^{-12}$ | CuS | $6 \times 10^{-36}$ |
| AgI | $8.5 \times 10^{-17}$ | $Fe(OH)_3$ | $4 \times 10^{-38}$ |
| $AgIO_3$ | $3.0 \times 10^{-8}$ | $Hg_2Cl_2$ | $1.3 \times 10^{-18}$ |
| $AgNO_2$ | $1.6 \times 10^{-4}$ | $Hg_2I_2$ | $2.5 \times 10^{-26}$ |
| $Ag_3PO_4$ | $1.6 \times 10^{-18}$ | $HgI_2$ | $8.8 \times 10^{-12}$ |
| $Ag_2S$ | $1.6 \times 10^{-49}$ | HgS | $4 \times 10^{-53}$ |
| $Ag_2SO_3$ | $1.5 \times 10^{-14}$ | $MgCO_3$ | $2.6 \times 10^{-5}$ |
| $Ag_2SO_4$ | $1.6 \times 10^{-5}$ | $Mg(OH)_2$ | $1.2 \times 10^{-11}$ |
| $Al(OH)_3$ | $2 \times 10^{-32}$ | $Mg_3(PO_4)_2$ | $1.0 \times 10^{-13}$ |
| $AlPO_4$ | $5.8 \times 10^{-19}$ | $Mn(OH)_2$ | $1.9 \times 10^{-13}$ |
| $Au(OH)_3$ | $5.5 \times 10^{-46}$ | $Ni(OH)_2$ | $6.5 \times 10^{-18}$ |
| $BaCO_3$ | $8.1 \times 10^{-9}$ | NiS | $3 \times 10^{-19}$ |
| $BaCrO_4$ | $2.4 \times 10^{-10}$ | $PbBr_2$ | $4.6 \times 10^{-6}$ |
| $BaF_2$ | $1.7 \times 10^{-6}$ | $PbCO_3$ | $3.3 \times 10^{-14}$ |
| $BaSO_4$ | $1.1 \times 10^{-10}$ | $PbCl_2$ | $1.6 \times 10^{-5}$ |
| $CaCO_3$ | $8.7 \times 10^{-9}$ | $PbI_2$ | $7.1 \times 10^{-9}$ |
| $CaF_2$ | $3.4 \times 10^{-11}$ | $Pb(IO_3)_2$ | $1.2 \times 10^{-13}$ |
| $Ca(IO_3)_2$ | $1.2 \times 10^{-10}$ | SnS | $1 \times 10^{-25}$ |
| $Ca(OH)_2$ | $5.5 \times 10^{-6}$ | $PbO_2$ | $3.2 \times 10^{-66}$ |
| $Ca_3(PO_4)_2$ | $2.0 \times 10^{-29}$ | $PbSO_4$ | $1.2 \times 10^{-8}$ |
| $Cd(OH)_2$ | $5.9 \times 10^{-15}$ | $SrCO_3$ | $7.0 \times 10^{-10}$ |
| CdS | $3.6 \times 10^{-29}$ | $SrF_2$ | $7.9 \times 10^{-10}$ |
| $Co(OH)_2$ | $2 \times 10^{-16}$ | $SrSO_4$ | $3.8 \times 10^{-7}$ |
| CoS | $3 \times 10^{-26}$ | $Zn(CN)_2$ | $2.6 \times 10^{-13}$ |
| $Cr(OH)_2$ | $1.0 \times 10^{-17}$ | $ZnCO_3$ | $1.4 \times 10^{-11}$ |
| $Cr(OH)_3$ | $6 \times 10^{-31}$ | $Zn(OH)_2$ | $1.8 \times 10^{-14}$ |
| CuBr | $4.2 \times 10^{-8}$ | ZnS | $1.2 \times 10^{-23}$ |

[a] All ion concentrations are in moles per liter.

**Example 14.18**   We prepare a saturated solution of lead sulfate by shaking an excess of the solid with water until no more dissolves. By measuring the starting quantity of lead sulfate and the amount remaining undissolved, we find that $1.1 \times 10^{-4}$ mol of $PbSO_4$ dissolves in 1 L of water at 300 K to produce a saturated solution. Calculate the value of $K_{sp}$ for lead sulfate.

**Solution**   Simpler methods are available, but we shall use the same method that we have utilized for previous equilibrium calculations.

The relevant chemical equation is

$$PbSO_4(s) \rightleftharpoons Pb^{2+}(aq) + SO_4^{2-}(aq)$$

| | | | |
|---|---|---|---|
| **start** | $a$ | 0 | 0 |

We use the symbol $a$ for the starting amount of $PbSO_4(s)$. The exact value is not needed, since it is not included in the expression for $K_{sp}$.

Let $x$ be the amount (in moles) of lead sulfate that dissolves in 1 L of solution when the system comes to equilibrium, so that the concentrations of dissolved species will have units of moles per liter ($M$). We shall calculate and express all equilibrium constants for aqueous solutions using the molarities of the dissolved species.

$$PbSO_4(s) \rightleftharpoons Pb^{2+}(aq) + SO_4^{2-}(aq)$$

**equil**                          $x$          $x$

We need not list the amount of solid lead sulfate remaining after the system reaches equilibrium, but we can see that it is $a - x$.

In this problem, we are given that $x = 1.1 \times 10^{-4}$ mol/L ($M$). We can determine the value for $K_{sp}$ by substituting the given value of $x$ into the expression for $K_{sp}$:

$$K_{sp} = [Pb^{2+}][SO_4^{2-}] = (x)(x) = \left(1.1 \times 10^{-4} \frac{mol}{L}\right)^2$$

$$= 1.2 \times 10^{-8} \frac{mol^2}{L^2}$$

We shall assign units to $K_{sp}$ to simplify dimensional analysis, although strictly speaking, it has no units.

You should note that the solution of the problem took for granted the correct interpretation of the stoichiometry of the reaction. By this time, you should know that 1 mol of lead sulfate dissolves to produce 1 mol of $Pb^{2+}$ ion and 1 mol of $SO_4^{2-}$ ion. You must always pay close attention to stoichiometry in these calculations.

**Example 14.19**

We prepare a saturated solution of calcium fluoride by shaking an excess of the solid with water until no more dissolves. By measuring the starting quantity of calcium fluoride and the amount remaining undissolved, we find that $4.1 \times 10^{-4}$ mol of $CaF_2$ dissolves in 2 L of water at 300 K to produce a saturated solution. Calculate the value of $K_{sp}$ for calcium fluoride.

**Solution**

The relevant chemical reaction and starting conditions are

$$CaF_2(s) \rightleftharpoons Ca^{2+}(aq) + 2F^-(aq)$$

**start**      $a$             0             0

To write the equilibrium line, we must remember that the expression for $K_{sp}$ includes the molarities of the substances in the equation. By letting $x$, the unknown, stand for the solubility of $CaF_2$—that is, the number of moles of calcium fluoride that dissolve in 1 L of water—we express the molarities of the dissolved species in terms of $x$. The chemical equation tells us that 2 mol of fluoride ion and 1 mol of calcium ion are formed when 1 mol of calcium fluoride dissolves. The equilibrium line reads:

$$CaF_2(s) \rightleftharpoons Ca^{2+}(aq) + 2F^-(aq)$$

**equil** $\qquad\qquad\quad x \qquad\quad 2x$

The value of $x$ can be found from the information given in the problem. If $4.1 \times 10^{-4}$ mol of $CaF_2$ dissolves in 2 L of water, then half that amount, $x$, dissolves in 1 L. Thus, $x = (4.1 \times 10^{-4}/2)$ mol/L $= 2.05 \times 10^{-4}$ mol/L. Using this value of $x$, we can find the value of $K_{sp}$:

$$K_{sp} = [Ca^{2+}][F^-]^2 = (x)(2x)^2 = 4\left(2.05 \times 10^{-4}\, \frac{mol}{L}\right)^3$$

$$= 3.5 \times 10^{-11}\, \frac{mol^3}{L^3}$$

Although this example is almost the same as Example 14.18, it is easy to go wrong unless careful attention is paid to details. The chemical equation says that 2 mol of fluoride ion is formed when 1 mol of solid $CaF_2$ dissolves. If $x$ mol/L of solid dissolves, the concentration of $F^-$ ion is $2x$ mol/L. The expression for $K_{sp}$ states that the concentration of $F^-$ ion must be raised to the second power in the calculations; $2x$ to the second power is $(2x)^2$, or $4x^2$. Carelessness can give an erroneous value.

---

**Example 14.20**

The solubility product of lead iodide is $7.1 \times 10^{-9}$ mol³/L³ at 298 K. Calculate the solubility of this salt in moles per liter and find the molarity of the ions in a saturated solution of $PbI_2$.

**Solution**

The relevant equation and the starting conditions are

$$PbI_2(s) \rightleftharpoons Pb^{2+}(aq) + 2I^-(aq)$$

**start** $\qquad\quad a \qquad\quad\;\; 0 \qquad\quad\; 0$

You can think of these starting conditions as describing an experimental procedure for finding the solubility of lead iodide. Some lead iodide is placed in water and the quantity that dissolves when the system reaches equilibrium is measured. Let $x$ be the amount (in moles) of solid lead iodide that dissolves in 1 L of water. The equilibrium concentrations (in moles per liter) of the ions then are

$$PbI_2(s) \rightleftharpoons Pb^{2+}(aq) + 2I^-(aq)$$

**equil** $\qquad\qquad\qquad x \qquad\quad 2x$

The value of the unknown can be found by substitution of the concentrations into the expression for $K_{sp}$:

$$K_{sp} = [Pb^{2+}][I^-]^2 = (x)(2x)^2 = 7.1 \times 10^{-9}\, \frac{mol^3}{L^3}$$

$$x = 1.2 \times 10^{-3}\, \frac{mol}{L}$$

The value of $x$ is the solubility of lead iodide, the amount in moles of solid $PbI_2$ that dissolves in 1 L of water to form a saturated solution. The value of $x$ can be used to find the molarity of each ion in the solution: $[Pb^{2+}] = x = 1.2 \times 10^{-3}$ mol/L and $[I^-] = 2x = 2.4 \times 10^{-3}$ mol/L.

Now suppose that the concentration of $I^-$ ion in the equilibrium system of Example 14.20 is raised by the addition of $I^-$ ions from a different source than $PbI_2$; for example, from a highly soluble iodide salt. The principle of Le Chatelier tells us that some solid lead iodide will precipitate from the saturated solution, removing part of the added $I^-$ ion. Stated another way, the solubility of lead iodide is lower in a solution that contains some iodide ion from another source than it is in pure water. We get the same result if $Pb^{2+}$ ions from another source are in the solution; the solubility of lead iodide is lowered. The $Pb^{2+}$ ion and the $I^-$ from other sources are called *common ions* of lead iodide. The *common ion effect* is a general rule that can be applied to sparingly soluble ionic solids: *The solubility of an ionic solid in a solution containing one or more of its common ions is lower than its solubility in pure water.* We shall assume that the presence of unrelated ions in water does not affect the solubility of an ionic substance.

If the value of the solubility product is known, it is possible to find how the presence of common ions in a solution affects the solubility of a sparingly soluble ionic solid.

**Example 14.21**

Use the data given in Example 14.20 to calculate the solubility of lead iodide in a $0.10M$ solution of sodium iodide.

**Solution**

In a $0.10M$ solution of sodium iodide, $[Na^+] = 0.10M$ and $[I^-] = 0.10M$. If lead iodide is added, the common ion is $I^-$. The relevant reaction and starting conditions are

$$PbI_2(s) \rightleftharpoons Pb^{2+}(aq) + 2I^-(aq)$$
**start**     $a$        $0$        $0.10M$

The reaction must proceed from left to right for the system to reach equilibrium. If we let $x$ be the amount (in moles) of solid $PbI_2$ that dissolves in 1 L of the sodium iodide solution, the molarities of the ions in solution at equilibrium are

$$PbI_2(s) \rightleftharpoons Pb^{2+}(aq) + 2I^-(aq)$$
**equil**            $x$        $0.10M + 2x$

If we substitute the concentrations into the expression for $K_{sp}$, we get:

$$K_{sp} = (x)(0.10M + 2x)^2 = 7.1 \times 10^{-9}M^3$$

We can simplify the calculations by making an approximation. Since $K_{sp}$ is small, the amount of lead iodide that dissolves is small, especially because of the common ion effect. We can make the approximation that $0.10M + 2x \approx 0.10M$. Now the equation becomes:

$$(x)(0.10M)^2 = 7.1 \times 10^{-9}M^3$$

$$x = 7.1 \times 10^{-7}\ \frac{\text{mol}}{\text{L}}$$

If you compare this value with the solubility of lead iodide in pure water that was calculated in Example 14.20, you can see that lead iodide is less soluble in a solution containing iodide ion.

You will note that $Na^+$ was omitted from the calculations, because it is not part of the net ionic equation. It is not a common ion in this solution.

We can also calculate the effect of the common ion $Pb^{2+}$ in solution on the solubility of lead iodide.

**Example 14.22**

Calculate the solubility of lead iodide in a $0.10M$ solution of lead nitrate.

**Solution**

A solution of lead nitrate contains $Pb^{2+}$ ions and $NO_3^-$ ions. The common ion is $Pb^{2+}$, whose initial concentration is $0.10M$. Following the same procedure as in Examples 14.20 and 14.21, we write:

|  | $PbI_2(s) \rightleftharpoons$ | $Pb^{2+}(aq)$ | $+ 2I^-(aq)$ |
|---|---|---|---|
| **start** | $a$ | $0.10M$ | $0$ |
| **equil** |  | $0.10M + x$ | $2x$ |

Substitution into the expression for $K_{sp}$ gives:

$$(0.10M + x)(2x)^2 = 7.1 \times 10^{-9}M^3$$

Making the approximation that $0.10M + x \approx 0.10M$, the equation becomes $(0.10M)(2x)^2 = 7.1 \times 10^{-9}M^3$ and $x = 1.3 \times 10^{-4}$ mol/L.

Comparing this result to those of the two previous examples, we see that the presence of either $Pb^{2+}$ ion or $I^-$ ion reduces the solubility of lead iodide. But a given concentration of $Pb^{2+}$ reduces the solubility less than the same concentration of $I^-$ ion. Since 2 mol of $I^-$ ions and only 1 mol of $Pb^{2+}$ ions are formed when 1 mol of $PbI_2$ dissolves, there is a smaller effect from the $Pb^{2+}$ ion.

## Precipitation Reactions

Precipitation reactions are processes in which a sparingly soluble material comes out of solution. Both in the laboratory and in industrial processes, it is often necessary to know whether and to what extent a precipitation will occur. The solubility product can help us to obtain that information.

For example, the $K_{sp}$ of AgCl is $1.6 \times 10^{-10}M^2$ at 298 K. Thus, a solution that contains $Ag^+$ ion and $Cl^-$ ion, from whatever source, is at

## SOLUTION, PRECIPITATION, AND CAVES

Vast, majestic underground cave systems, such as Carlsbad Caverns and Mammoth Cave, with all their incredibly beautiful and varied decorative formations, are created by a process of solution and precipitation that takes place over many centuries.

Caves are created because rocks are soluble in natural waters. Limestone, dolomite, gypsum, and anhydrite, in particular, dissolve rather readily under the proper conditions. Limestone, for example, is made up primarily of calcium carbonate, $CaCO_3$. Pure, running water always has some carbon dioxide from the atmosphere dissolved in it. When water comes in contact with limestone, it can dissolve the calcium carbonate to form a solution of calcium bicarbonate:

$$CaCO_3(s) + CO_2 + H_2O \rightleftharpoons Ca^{2+} + 2HCO_3^-$$

The extent to which this reaction takes place is extremely sensitive to the partial pressure of $CO_2$. Caves can form rapidly because the partial pressure of $CO_2$ is much greater in air trapped in soil than in the atmosphere. As water trickles through the soil, it is exposed to the higher partial pressure of $CO_2$, and its ability to dissolve calcium carbonate is increased manyfold. The less saturated the water and the faster it flows, the more rapidly solution proceeds. Thus, as calcium carbonate goes into solution, more water flows into the resulting space, and the cave grows more rapidly.

As more calcium carbonate is dissolved, the natural solution comes closer to saturation. If the water should emerge into an air-filled chamber, where the partial pressure of $CO_2$ is lower than in the soil, precipitation can occur to form stalactites, stalagmites, and other familiar features of caves. The air in the cave may also be warmer than ambient air above. The higher temperature reduces the solubility of $CO_2$ in water. As $CO_2$ comes out of solution, the aqueous solution of $Ca(HCO_3)_2$ becomes saturated, and $CaCO_3$ precipitates.

If the water drips from a cave ceiling, a precipitate of $CaCO_3$ can form both on the ceiling and on the floor below. The cylindrical deposit that forms on the ceiling as water falls drop by drop is called a stalactite. The bulbous deposit that forms on the floor where the splashing drops fall is called a stalagmite. Given enough time, the stalactite and stalagmite can meet and form a column. Changing conditions over long periods of time produce the fantastic variety of formations found in many caves. Standing pools of water can become saturated with $CaCO_3$ as $CO_2$ is released, precipitating limestone in shapes resembling lily pads. "Cave coral" consists of similar precipitates formed where moisture adheres to cave floors or walls. Water running down cave walls can lead to the growth of delicate mineral traceries called draperies or flowstone. In some caves, precipitation can occur on such a scale that large passages that were formed by dissolution of rock are closed again by the deposition of minerals from solution.

equilibrium only when $[Ag^+][Cl^-] = 1.6 \times 10^{-10}M^2$. In a solution in which $[Ag^+][Cl^-] < 1.6 \times 10^{-10}M^2$, the solution is not saturated and the system is not at equilibrium. If solid silver chloride is added, some of it dissolves until the product of the concentrations reaches the value of $K_{sp}$.

If the product of the two ion concentrations exceeds the value of $K_{sp}$, $[Ag^+][Cl^-] > 1.6 \times 10^{-10}M^2$, the system is not at equilibrium. It proceeds spontaneously to equilibrium by lowering the concentration of the dissolved ions. A precipitate of silver chloride forms, spontaneously and rapidly. Precipitation stops when the concentrations of the ions are reduced and the product of the concentrations equals $1.6 \times 10^{-10}M^2$.

In many precipitation reactions, the information of interest is qualitative. Typically, we want to know whether a given precipitate forms when certain solutions of ions are prepared. If more than one precipitate is possible, we want to know which precipitate forms. As before, the information conveyed by solubility product data can simplify our calculations.

**Example 14.23**   A solution is prepared from 0.00042 mol of $Hg_2(NO_3)_2$, 0.0023 mol of $CaCl_2$, and 2.0 L of water. Does a precipitate form from this solution?

**Solution**   The mercury(I) nitrate is a source of two ions, $Hg_2^{2+}$ and $NO_3^-$. The calcium chloride is a source of two ions, $Ca^{2+}$ and $Cl^-$. We first determine whether any sparingly soluble salts form from these ions. All nitrates are soluble, so sodium nitrate will not precipitate. But $Hg_2Cl_2$ is sparingly soluble. Table 14.3 gives its $K_{sp}$ as $1.3 \times 10^{-18}$.

To find if $Hg_2Cl_2$ precipitates, we find the concentrations of the two ions of interest and substitute them into the $K_{sp}$ expression. If the calculated value of $K_{sp}$ is greater than $1.3 \times 10^{-18}$, precipitation occurs.

$$[Hg_2^{2+}] = \frac{0.0023 \text{ mol}}{2.0 \text{ L}} = 0.0012 M$$

$$[Cl^-] = \frac{2(0.0023 \text{ mol})}{2.0 \text{ L}} = 0.0023 M$$

$$[Hg_2^{2+}][Cl^-]^2 = (0.0012 M)(0.0023 M)^2$$
$$= 6.4 \times 10^{-9}$$

This value is greater than the $K_{sp}$ of $Hg_2Cl_2$, so a precipitate of $Hg_2Cl_2$ does form from this solution.

A common method of preparing a solution in which the solubility product of a sparingly soluble ionic solid is exceeded is to mix two solutions. One of these solutions contains a relatively soluble salt of the desired cation. The other contains a relatively soluble salt of the desired anion. For example, when solutions of appropriate concentrations of sodium chloride and silver nitrate are mixed, the value of the $K_{sp}$ of silver chloride, $[Ag^+][Cl^-]$, is exceeded and a precipitate of AgCl(s) forms almost immediately. The chemical equation we write to describe precipitation includes neither the nitrate ion nor the sodium ion, since neither undergoes change.

**Example 14.24**   A 15-cm³ volume of a $0.050M$ solution of magnesium bromide is mixed with a 25-cm³ volume of a $0.050M$ solution of lead nitrate at 298 K. Will a precipitate form?

**Solution**   The magnesium bromide solution contains two ions, $Mg^{2+}$ and $Br^-$. The lead nitrate solution contains two ions, $Pb^{2+}$ and $NO_3^-$. Two precipitates are possible by an exchange of ions: lead bromide, $PbBr_2$, and magnesium nitrate, $Mg(NO_3)_2$. Lead bromide is sparingly soluble. At 298 K, $K_{sp}$ for $PbBr_2$ is $4.6 \times 10^{-6} M^3$. We must calculate the concentrations of the two ions of interest when the solutions are mixed and substitute those concentrations into the expression for $K_{sp}$. When two solutions are mixed, the volume of the resulting solution is approximately the sum of the original volumes. That is, $V_T = V_1 + V_2$, where $V_T$ is the volume of the resulting solution and $V_1$ and $V_2$ are the volumes of the original solutions. The increase in volume decreases the concentration of ions in the resulting solution. We can find the new molarity of an ion by using a dilution factor. For ions in solution 1, the dilution factor is $V_1/V_T$. For ions in solution 2, it is $V_2/V_T$. We multiply the original concentration by the dilution factor to find the concentration in the new solution. The new molarity of an ion from solution 1, $M'_1$, is given by:

$$M'_1 = M_1 \frac{V_1}{V_T}$$

Similarly, the new molarity of an ion from solution 2, $M'_2$, is given by:

$$M'_2 = M_2 \frac{V_2}{V_T}$$

Now we can find the concentrations of the ions of interest in the solution after the mixing. The original concentration of $Br^-$ ion in a $0.050M$ solution of $MgBr_2$ is $2(0.050M) = 0.10M$. Therefore:

$$M'_1 = [Br^-] = (0.10M) \left( \frac{15 \text{ cm}^3}{15 \text{ cm}^3 + 25 \text{ cm}^3} \right) = 3.8 \times 10^{-2} M$$

The original concentration of $Pb^{2+}$ ion in a $0.050M$ solution of $Pb(NO_3)_2$ is $0.050M$. The concentration after mixing is

$$M'_2 = [Pb^{2+}] = (0.050M) \left( \frac{25 \text{ cm}^3}{15 \text{ cm}^3 + 25 \text{ cm}^3} \right) = 3.1 \times 10^{-2} M$$

Substituting these values into the expression for the solubility product of lead bromide gives:

$$[Pb^{2+}][Br^-]^2 = (3.1 \times 10^{-2} M)(3.8 \times 10^{-2} M)^2 = 4.5 \times 10^{-5} M^3$$

This value is larger than $4.6 \times 10^{-6} M^3$, the value of $K_{sp}$. Therefore a precipitate of $PbBr_2$ forms.

It is possible to have a solution yield first one solid and then another by careful addition of solutions of selected ions. This procedure is called *selective precipitation*. It is important in the synthesis of inorganic substances and in both quantitative and qualitative analysis of mixtures of ions in solutions. We can carry out selective precipitation calculations by using solubility product data.

**Example 14.25**     A solution contains a mixture of silver ion, $[Ag^+] = 0.10M$, and mercury(I) ion, $[Hg_2^{2+}] = 0.10M$. You are asked to separate the two ions. How would you go about it?

**Solution**     Selective precipitation can be used to separate the two ions. The values of the solubility products (Table 14.3) indicate that both these cations form iodide salts that are sparingly soluble. For AgI, $K_{sp} = 8.5 \times 10^{-17}M^2$ at 298 K. For $Hg_2I_2$, $K_{sp} = 2.5 \times 10^{-26}M^3$ at 298 K. (Note that the mercury(I) cation in solution is unusual; it exists as a polyatomic ion $[Hg—Hg]^{2+}$.) To use selective precipitation successfully, we must meet several conditions. We must first add iodide ion to the solution to cause precipitation of either silver iodide or mercury(I) iodide. If the precipitation of one of these iodides is virtually complete before the other begins to precipitate, the two ions can be separated.

To see whether selective precipitation is feasible, we must determine which iodide precipitates first. Then we determine how much of this cation remains in solution when the second iodide begins to precipitate.

We can use appropriate $K_{sp}$ expressions to calculate the concentration of iodide ion that must be reached for each solid to precipitate. For simplicity, we shall assume that we add a solution of salt, such as sodium iodide, that is concentrated enough so that its addition causes a negligible change in the volume of the solution.

For silver iodide, $[Ag^+][I^-] = 8.5 \times 10^{-17}M^2$ at equilibrium. Since the $[Ag^+] = 0.10M$, we can find the concentration of $I^-$ ion at equilibrium: $[I^-] = 8.5 \times 10^{-17}M^2/0.10M = 8.5 \times 10^{-16}M$.

As iodide ion is added gradually, its concentration increases until it reaches this value. The system is then saturated with respect to silver iodide. If more iodide is added, silver iodide will precipitate.

For mercury(I) iodide, a similar calculation tells us:

$$K_{sp} = [Hg_2^{2+}][I^-]^2 = (0.10M)[I^-]^2 = 2.5 \times 10^{-26}M^3$$
$$[I^-] = 5.0 \times 10^{-13}M$$

Precipitation of silver iodide begins at a lower concentration of iodide ion than does precipitation of mercury(I) iodide, so silver iodide precipitates first. Selective precipitation may be possible; but it is practical only if most of the silver iodide has already precipitated when the concentration of iodide ion becomes great enough for the precipitation of mercury(I) ion to begin. We have determined that mercury(I) iodide begins precipitating when the concentration of $I^-$ exceeds $5.0 \times 10^{-13}M$. Now we must calculate the concentration of silver ion remaining in solution when the iodide ion concentration reaches this value. We find out by substituting this value of $[I^-]$ into the expression for $K_{sp}$ of silver iodide:

$$K_{sp} = [Ag^+][I^-] = [Ag^+](5.0 \times 10^{-13}M) = 8.5 \times 10^{-17}M^2$$
$$[Ag^+] = 1.7 \times 10^{-4}M$$

We now know that when there is enough iodide ion in the solution to cause the precipitation of mercury(I) iodide, almost all the $Ag^+$ has precipitated as AgI. Only $(1.7 \times 10^{-4}M/0.10M) \times 100\% = 0.17\%$ of the original $Ag^+$ remains in solution. In other words, 99.83% of the silver ion is removed from solution. If the silver iodide is separated just at this point, we are left with a solution of mercury(I) ion that is only slightly contaminated by a trace impurity of silver ion.

In this case, selective precipitation works well.

As you saw in Example 14.25, it is sometimes necessary to think about more than one solubility product at a time. When a system contains several species, it is at equilibrium only when the concentration of the dissolved ions satisfies all the relevant solubility product expressions. Consider an aqueous solution that contains ions from two sparingly soluble ionic solids. If the two substances have an ion in common, the interdependence of a solubility products can be used to calculate the composition of the system at equilibrium.

## 14.6  QUALITATIVE ANALYSIS BY SELECTIVE PRECIPITATION

It has been traditional in introductory chemistry laboratory courses to have students perform a qualitative analysis of an unknown mixture of cations in solution, using methods based on solubility differences. Instruments now can do such analyses almost automatically. However, a review of the traditional method of qualitative analysis is still quite instructive.

In the ideal method of separating and identifying two or more cations in solution, we first find an anion that forms a sparingly soluble solid with only one of the cations. A source of this anion is added to the solution, and the precipitate that forms is removed from the solution. Then a second anion, which forms an insoluble solid with only one of the remaining cations, is added, and the second precipitate is removed. This procedure is repeated until all the ions have been separated and identified.

Suppose we have a solution that may contain any or all of the cations $Ag^+$, $Cd^{2+}$, and $Ba^{2+}$ (Figure 14.7). The first step is to consult a table of solubility product data, such as Table 14.3. A complete table of $K_{sp}$ values will show that only silver forms a sparingly soluble chloride, since no $K_{sp}$ values will be listed for either cadmium chloride or barium chloride.

When a solution containing chloride ion is added to the solution to be tested, a precipitate of silver chloride forms if $Ag^+$ ion is present. We remove the precipitate from the solution. The small $K_{sp}$ indicates that almost all the silver ion will precipitate if a reasonable amount of $Cl^-$ ion is added.

Again, we consult a complete table of $K_{sp}$ values. It shows that cadmium sulfide is sparingly soluble. There is no listing for barium sulfide; it is soluble. We add a solution containing sulfide ion to the solution that may contain $Cd^{2+}$ and $Ba^{2+}$. If $Cd^{2+}$ is present, cadmium sulfide precipitates. The solid is removed and we are left with a solution that may contain only the barium ion. Another consultation with the $K_{sp}$ table suggests the addition of a solution containing carbonate anion to precipitate barium carbonate to test for $Ba^{2+}$ in solution.

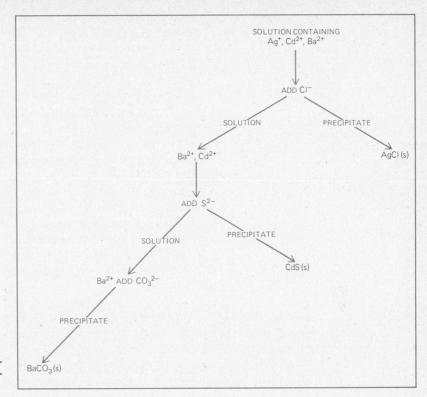

**Figure 14.7**
A flowchart for the separation and identification of $Ag^+$, $Cd^{2+}$, and $Ba^{2+}$ in aqueous solution.

Careful selection of the sequence in which the solutions are added is vital for the success of this procedure. For example, silver sulfide is insoluble. If the concentration of silver ion is not greatly reduced by the chloride precipitation of the first step, a precipitate will form when sulfide ion is added to test for cadmium, even if cadmium is absent. That precipitate will be silver sulfide.

### Groups of Cations

You might have to test a solution containing more than three ions. It is usually not possible to find an anion that forms a sparingly soluble solid with only one of the cations. In such a situation, you will find it convenient to group the cations by the order in which they are precipitated.

Group I includes cations that are precipitated by chloride ion. It includes $Ag^+$, $Hg_2^{2+}$, and $Pb^{2+}$. When a source of chloride ion is added to a solution that may contain one or more of these cations, a precipitate that may include $AgCl$, $Hg_2Cl_2$, and $PbCl_2$ forms. This precipitate is separated from the solution and appropriate procedures are used to separate and identify the mixture of precipitates.

Group II includes cations that form sulfides of extremely low solubility. Suitable procedures are available for keeping $[S^{2-}]$ very low and for precipitating only those cations that form very insoluble sulfides. Some cations in group II are $Cu^{2+}$, $Cd^{2+}$, $Hg^{2+}$, and $Sn^{4+}$.

Group III includes cations that form less insoluble sulfides than group II. The concentration of $S^{2-}$ ion is increased if the solution is made alkaline, so this group of cations also includes those that form insoluble hydroxides. Some important cations in group III are $Zn^{2+}$, $Ni^{2+}$, and $Co^{2+}$, which precipitate as sulfides, and $Al^{3+}$, $Fe^{3+}$, and $Cr^{3+}$, which precipitate as hydroxides.

Group IV includes cations that form insoluble carbonates. These cations derive from alkaline-earth metals. They include $Mg^{2+}$, $Ca^{2+}$, $Sr^{2+}$, and $Ba^{2+}$.

Group V includes all the cations that remain after all the others have precipitated. The important members of group V are $NH_4^+$, $Na^+$, and $K^+$.

There are procedures that allow separation and identification of all the cations in each of the groups. These procedures are based in part on solubility differences. They also are based on other types of aqueous equilibria, which we shall discuss in Chapters 15 and 18.

**Summary**

In this chapter, we continued our discussion of **chemical equilibrium.** We noted that a system in the equilibrium state does not change spontaneously on the macroscopic level, although it constantly changes on the molecular level. We showed how to determine the expression for **K,** the **equilibrium constant,** by first writing a balanced chemical reaction and then using the terms for the pressure of gases and concentrations of solutes, and we noted that K excludes terms for liquids and solids. We mentioned the **reaction quotient** Q, the equivalent of K for a nonequilibrium system. We then described the **principle of Le Chatelier,** which says that a system in the equilibrium state responds to a stress in a way that minimizes the stress. Next we showed how to calculate gas phase equilibrium constants, and how such a constant for one state of a system can be used to find the composition of other equilibrium states. We introduced the **solubility product,** the equilibrium constant for a solution of a sparingly soluble ionic solid in water, and showed how it can be used in calculations. We then described **precipitation** reactions, in which the addition of selected ionic solutes to a solution can cause desired solutes to precipitate out of the solution. We concluded by describing the classic problem of performing a **qualitative analysis** of an unknown mixture of cations, using methods based on differences in the solubility of cations.

# Exercises

**14.1** Classify each of the following processes as spontaneous or nonspontaneous:
(a) $2N_2(g) + O_2(g) \rightarrow 2N_2O(g)$ at 298 K.
(b) A ball rolls down a hill.
(c) Water freezes at 270 K.
(d) A tree comes into leaf in the spring.

**14.2** Consider the general reaction $aA(g) + bB(g) \rightleftharpoons cC(g)$. Explain why the composition of the equilibrium state will be the same whether we begin with $a$ moles of A and $b$ moles of B or with $c$ moles of C, for any values of $a$, $b$, and $c$.

**14.3** Write the expressions for the equilibrium constants of each of the following processes:
(a) $2NO(g) + O_2(g) \rightleftharpoons 2NO_2(g)$
(b) $P_4(s) + 6Cl_2(g) \rightleftharpoons 4PCl_3(l)$
(c) $Al_2(CO_3)_3(s) \rightleftharpoons 2Al^{3+}(aq) + 3CO_3{}^{2-}(aq)$
(d) $Ag^+(aq) + Br^-(aq) \rightleftharpoons AgBr(s)$
(e) $HCO_3{}^-(aq) + HNO_2(aq) \rightleftharpoons H_2O + NO_2{}^-(aq) + CO_2(g)$

**14.4**[3] Write reactions that correspond to each of the following equilibrium expressions:
(a) $P_{SO_3}^2/P_{SO_2}^2 P_{O_2}$
(b) $P_{IF}^2/P_{F_2}$
(c) $[H^+][NO_3^-]P_{NO}^2/[HNO_2]^3$
(d) $1/[Ca^{2+}][F^-]^2$

**14.5** Write balanced chemical equations for each of the following processes: (a) the condensation of carbon dioxide at 298 K and 1 atm, (b) the dissolution of barium fluoride in water, (c) the conversion of dinitrogen trioxide to nitric oxide and nitrogen dioxide in the gas phase, (d) the decomposition of magnesium carbonate to form magnesium oxide and carbon dioxide.

**14.6** Write balanced net equations for each of the following processes taking place in aqueous solution:
(a) A solution of sodium sulfide is mixed with a solution of copper(II) nitrate and a precipitate of copper(II) sulfide forms.
(b) A solution of hydrogen sulfide is mixed with a solution of zinc iodide and a precipitate of zinc sulfide forms.
(c) A solution of hydrochloric acid is mixed with a solution of sodium bicarbonate and carbon dioxide is evolved.

**14.7** Write the expression for $K$ for each of the following phase changes: (a) the condensation of water vapor at 298 K, (b) the condensation of water vapor at 198 K, (c) the melting of sodium chloride.

**14.8** The following processes differ only in the states of the reactants and products. Write the expression for the $K$ of each: (a) pure ammonia and pure hydrogen bromide form ammonium bromide, (b) pure ammonia and hydrobromic acid form ammonium bromide, (c) a solution of ammonia and pure hydrogen bromide forms ammonium bromide, (d) ammonia in solution and hydrobromic acid form ammonium bromide.

**14.9** Write five different reactions whose equilibrium constant is $K = P_{Cl_2}$.

**14.10** At 1000 K the reaction $2SO_2(g) + O_2(g) \rightleftharpoons 2SO_3(g)$ has $K = 3.3$ atm$^{-1}$. Predict the direction of reaction of each of the following systems that contain the given pressures of the three gases in atmospheres: (a) $P_{SO_2} = P_{SO_3} = P_{O_2} = 0.10$ atm; (b) $P_{SO_2} = P_{O_2} = 0.10$ atm, $P_{SO_3} = 0.33$ atm; (c) $P_{SO_2} = 0.30$ atm, $P_{SO_3} = 0.10$ atm, $P_{O_2} = 0.034$ atm; (d) $P_{SO_2} = 0.14$ atm, $P_{SO_3} = 0.17$ atm, $P_{O_2} = 0.00$ atm.

**14.11** Predict the direction of chemical change in the system $Cl_2(g) \rightleftharpoons 2Cl(g)$, originally at equilibrium, when we stress it in each of the following ways: (a) add $Cl(g)$, (b) add $Cl_2(g)$, (c) decrease the temperature, (d) increase the volume.

**14.12** Propose an explanation for Henry's law (Equation 12.9) using the principle of Le Chatelier.

**14.13** The reaction between CO and $H_2O$ in the gas phase to form $H_2$ and $CO_2$ is endothermic. Predict the changes that take place when this system is stressed in each of the following ways: (a) CO is removed, (b) $CO_2$ is removed, (c) the temperature is increased, (d) the volume is decreased, (e) the pressure is decreased at constant temperature.

**14.14** In Section 9.2 we discussed the fact that diamonds are denser than graphite and high pressures are used in the preparation of artificial diamonds. Use the principle of Le Chatelier to account for these facts.

**14.15** Ideally, the combustion of gasoline in an automobile engine should give only $CO_2$ and $H_2O$. Unless the engine operates at very high temperatures, however, some CO, an undesirable air pollutant, also forms. Nevertheless, there are some serious drawbacks to the operation of automobile engines at very high temperatures. Chief among them is the formation of $NO(g)$, a very harmful pollutant. Explain the relationship between the temperature and the formation of $NO(g)$ during the operation of an automobile engine. (*Hint:* Appendix II)

**14.16** When 2 atm of $NO(g)$ and 0.5 atm of $O_2(g)$ are introduced into a bulb at a certain temperature, the following

---

[3] The answers to exercises whose numbers are in color can be found in Appendix VII. The star indicates an exercise that is more challenging than average.

equilibria are rapidly established:

$$NO(g) + \tfrac{1}{2}O_2(g) \rightleftharpoons NO_2(g)$$
$$K = 1 \times 10^6$$
$$2NO_2(g) \rightleftharpoons N_2O_4(g)$$
$$K = 1 \times 10^2$$

Calculate the pressures of all gases in the system at equilibrium with pressures of at least 0.1 atm.

**14.17** The temperature of the system in Exercise 14.16 is raised by a small amount. Indicate whether the relative pressure of each component of the system increases, decreases, or stays the same.

**14.18** The volume of the system of Exercise 14.16 is reduced at constant temperature. Indicate whether the relative pressure of each component of the system increases, decreases, or stays the same.

**14.19** Indicate whether each of the following events will increase, decrease, or not affect the solubility of magnesium fluoride, which evolves heat when it dissolves in water: (a) sodium fluoride is added to the solution, (b) silver nitrate is added to the solution, (c) the temperature is raised.

**14.20** The equilibrium constant for the reaction $F_2(g) \rightleftharpoons 2F(g)$ is $6.3 \times 10^{-2}$ at 1100 K. The pressure of $F_2(g)$ in a system at equilibrium is 0.87 atm. Find the pressure of $F(g)$ in the system.

**14.21** The reaction $2NO(g) + Cl_2(g) \rightleftharpoons 2NOCl(g)$ has an equilibrium constant of 0.26 atm$^{-1}$ at 700 K. In a certain system at equilibrium the pressure of NO is exactly twice the pressure of NOCl. Find the pressure of $Cl_2(g)$ in this system.

**14.22** The reaction $H_2(g) + CO_2(g) \rightleftharpoons H_2O(g) + CO(g)$ has an equilibrium constant of 1.5 at 1300 K. At equilibrium in this system the ratio of $CO_2$ to $H_2O$ is 2.0 : 3.0 and the pressure of CO is 0.44 atm. Find the pressure of $H_2$.

**14.23** At 670 K the system $CO(g) + Cl_2(g) \rightleftharpoons COCl_2(g)$ is found to have $P_{Cl_2} = 0.15$ atm, $P_{CO} = 0.30$ atm, and $P_{COCl_2} = 1.05$ atm. Find K.

**14.24** At 1100 K the system $2CO(g) \rightleftharpoons CO_2(g) + C(s)$ is found to contain 0.340 atm of CO, 0.106 atm of $CO_2$ and 0.250 mol of C. Find K.

**14.25** A system at equilibrium at 1200 K contains HCN(g) at a pressure of 0.0017 atm and $N_2(g)$ and $H_2(g)$ each at a pressure of 5.2 atm. Write a suitable reaction for this system and find its K.

**14.26** At 1100 K a system at equilibrium contains $ClF_3(g)$, $P = 0.46$ atm; $F_2(g)$, $P = 0.33$ atm; and $Cl_2(g)$, $P = 1.3$ atm. Write a suitable reaction for the system and find its K.

**14.27** The reaction $2NO(g) + Cl_2(g) \rightleftharpoons 2NOCl(g)$ has an equilibrium constant of 0.26 atm$^{-1}$ at 700 K. Find $K_c$.

**14.28** At 1300 K a system at equilibrium contains $BrF_5(g)$, $P = 5.7$ atm; $Br_2(g)$, $P = 0.21$ atm; and $F_2(g)$, $P = 0.31$ atm. Write a suitable reaction for this system and find its $K_c$.

**14.29** A 500-cm$^3$ bulb contains 0.044 mol of Br(g) and 0.55 mol of $Br_2(g)$. Write a suitable reaction to describe this system and find the $K_c$.

**14.30★** Derive the relationship between $K_p$ and $K_c$. (*Hint: $PV = nRT$.*)

**14.31** A 0.126-mol amount of $NO_2(g)$ is placed in a sealed 1.00-L container and heated to 700 K. At equilibrium there remains 0.101 mol of $NO_2(g)$. Find $K_p$ for the reaction $NO_2(g) \rightleftharpoons NO(g) + \tfrac{1}{2}O_2(g)$.

**14.32** The equilibrium constant for the reaction $2NO_2(g) + Cl_2(g) \rightleftharpoons 2NO_2Cl(g)$ can be measured by placing 2.50 atm of $NO_2Cl(g)$ at 410 K in a container and allowing the system to come to equilibrium. The pressure of $Cl_2(g)$ is then found to be 1.20 atm.

Find the value of K for the reaction.

**14.33** A 0.221-g sample of $ClF_3$ is introduced into a 500-cm$^3$ bulb and heated to 500 K. It vaporizes completely and undergoes partial decomposition to $Cl_2(g)$ and $F_2(g)$. The pressure of $Cl_2(g)$ is found to be 0.086 atm. Find K for the decomposition of $ClF_3(g)$ at 500 K.

**14.34** A pressure of 1.068 atm of $N_2O_3(g)$ is placed in a container at 475 K. After the reaction $N_2O_3(g) \rightleftharpoons NO(g) + NO_2(g)$ reaches equilibrium the total pressure is 1.673 atm. Find K.

**14.35** A mixture of 0.455 atm of NO and 0.144 atm of $O_2$ is prepared at 650 K. After the reaction $2NO(g) + O_2(g) \rightleftharpoons 2NO_2(g)$ comes to equilibrium the total pressure is found to be 0.517 atm. Find K for the reaction.

**14.36** At 650 K 73.0% of a sample of $PH_3(g)$ initially at a pressure of 0.967 atm decomposes to $P_2(g)$ and $H_2(g)$. Write a reaction for the system and find its K.

**14.37** For each of the following systems indicate the relationship, if any, between the fraction of the reactant that decomposes and its initial pressure:
(a) $2HI(g) \rightleftharpoons H_2(g) + I_2(g)$
(b) $N_2O_4(g) \rightleftharpoons 2NO_2(g)$
(c) $S_8(g) \rightleftharpoons 4S_2(g)$

**14.38** The equilibrium constant for the reaction $S_2(g) + C(s) \rightleftharpoons CS_2(g)$ at 950 K is 8.74. Calculate the pressure of the two gases at equilibrium when 1.28 atm of $S_2(g)$ and excess C(s) come to equilibrium. Repeat the calculation for the equilibrium state reached when 1.28 atm of $CS_2(g)$ is maintained at this temperature.

**14.39** The equilibrium constant for the reaction $NH_4CO_2NH_2(s) \rightleftharpoons 2NH_3(g) + CO_2(g)$ is $8.8 \times 10^{-4}$ at 400 K. Find the equilibrium pressure of the two gases at this temperature.

**14.40** The equilibrium constant for

the dissociation of $I_2(g)$ to $2I(g)$ at 800 K is $2.95 \times 10^{-5}$ atm. Find the pressure of $I(g)$ at equilibrium when 1.00 atm of $I_2(g)$ is placed in the bulb and allowed to dissociate. Calculate the pressure of $I(g)$ when the pressure of $I_2(g)$ at equilibrium is 1.00 atm.

**14.41** A sample of $MgCO_3(s)$ is introduced into a sealed container of volume 0.221 L and heated to 650 K until equilibrium is reached. The $K$ for the reaction $MgCO_3(s) \rightleftharpoons MgO(s) + CO_2(g)$ is $2.6 \times 10^{-2}$ atm at this temperature. Find the mass of MgO present in this system at equilibrium.

**14.42** The equilibrium constant for the reaction $2PH_3(g) \rightleftharpoons P_2(g) + 3H_2(g)$ is $1.04 \times 10^{-6}$ atm$^2$ at 500 K. (a) Find the composition of the equilibrium state that results from a starting pressure of $PH_3(g)$ of 1.00 atm. (b) Find the fraction of $PH_3$ that decomposes.

**14.43** The equilibrium constant for the reaction $2NO(g) + I_2(g) \rightleftharpoons 2NOI(g)$ at 500 K is $7.48 \times 10^{-5}$ atm$^{-1}$. Find the pressure of NOI at equilibrium from initial pressures of 0.333 atm of NO and $I_2$.

**14.44** A mixture of $PH_3(g)$, $P = 0.40$ atm, and $H_2(g)$, $P = 0.50$ atm, is in a reaction bulb at 500 K. Use the data given in Exercise 14.42 to find the pressure of $P_2(g)$ at equilibrium.

**14.45** At 700 K the $K$ for the reaction $SO_3(g) \rightleftharpoons SO_2(g) + \frac{1}{2}O_2(g)$ is $3.48 \times 10^{-3}$ atm$^{1/2}$. Find the composition of the equilibrium state when 0.139 atm of $SO_3(g)$ decomposes at this temperature.

**14.46** The equilibrium constant for the reaction $CO(g) + H_2O(g) \rightleftharpoons CO_2(g) + H_2(g)$ is 0.629 at 1260 K. Determine the composition of the equilibrium state when 0.608 atm of $H_2O$ and 0.608 atm of CO are introduced into a reaction bulb at this temperature.

**14.47** At 1000 K the $K$ for the reaction $2CO(g) \rightleftharpoons C(s) + CO_2(g)$ is 0.575

atm$^{-1}$. Find the pressure of $CO_2$ at equilibrium from an initial pressure of CO of 1.00 atm.

**14.48** At 600 K the equilibrium constant for the formation of $F_2(g)$ from $2F(g)$ is $4.17 \times 10^7$ atm$^{-1}$. Find the equilibrium pressure of $F(g)$ remaining from an initial pressure of 0.00880 atm of $F(g)$.

**14.49** The $K$ for the reaction $2NO(g) + Cl_2(g) \rightleftharpoons 2NOCl(g)$ is $1.40 \times 10^5$ at a certain temperature. Determine the composition of the equilibrium state that forms from initial pressures of NO = 0.160 atm and $Cl_2 = 0.080$ atm.

**14.50** Use the data given in Exercise 14.49 to find the composition of the equilibrium state that forms from initial pressures of NO and $Cl_2$ of 1.00 atm each.

**14.51** A system at equilibrium at 700 K contains 0.18 atm of $H_2$, 0.29 atm of $CH_4$, and some graphite. An additional 0.13 atm of $H_2$ is introduced. Calculate the composition of the system when it returns to equilibrium.

**14.52** A system at equilibrium contains 0.34 atm of $Br(g)$ and 0.48 atm of $Br_2(g)$ at 1600 K. The volume of the container is doubled. Find the pressure of each gas when the system returns to equilibrium.

**14.53** The equilibrium constant for the reaction $SO_2(g) + NO_2(g) \rightleftharpoons SO_3(g) + NO(g)$ is 3.0 at a certain temperature. Find the amount of NO that must be added to 1.2 mol of $SO_3$ to form 0.4 mol of $SO_2$.

**14.54** The equilibrium constant for the reaction $Cl_2(g) \rightleftharpoons 2Cl(g)$ at 2000 K is 0.55 atm. A sample of $Cl_2$ is placed in a reaction bulb of volume 50.0 cm$^3$, which is then heated to 2000 K. The total pressure is found to be 1.36 atm. Find the mass of $Cl_2$ originally placed in the bulb.

**14.55** Write solubility product ex-

pressions for each of the following solids: (a) $MgCO_3$, (b) $BaF_2$, (c) $Pb(IO_3)_2$, (d) $Cr(OH)_3$.

**14.56** Find the solubility (in moles per liter) in water of each of the following solids: (a) AgBr, (b) $CaF_2$, (c) $Ag_2SO_3$.[4]

**14.57** Find the concentration of each ion in a saturated solution of: (a) $Hg_2I_2$, (b) $Ca_3(PO_4)_2$.[4]

**14.58** The $[CrO_4^{2-}]$ in a saturated solution of $CuCrO_4$ is $2.21 \times 10^{-3} M$. Find the solubility product of copper(II) chromate.

**14.59** The $[Bi^{3+}]$ in a saturated solution of $BiI_3$ is $1.32 \times 10^{-5} M$. Find the solubility product of bismuth iodide.

**14.60** The solubility of $Ba(IO_3)_2$ in pure water is $7.28 \times 10^{-4}$ mol/L. Find the $K_{sp}$ of barium iodate.

**14.61** The solubility of $Ba_3(AsO_4)_2$ in pure water is $3.7 \times 10^{-11}$ mol/L. Find the $K_{sp}$ of barium arsenate.

**14.62** A system at equilibrium contains solid zinc cyanide and $Zn^{2+}$ in solution. The $[Zn^{2+}] = 0.051 M$. Find the $[CN^-]$.[4]

**14.63** Find the solubility of $PbSO_4$ in (a) pure water, (b) $0.10 M$ lead nitrate solution, (c) $0.10 M$ sodium sulfate solution.[4]

**14.64** Find the solubility of $SrF_2$ in (a) pure water, (b) $0.20 M$ strontium nitrate solution, (c) $0.20 M$ sodium fluoride solution, (d) $0.20 M$ sodium nitrate solution.[4]

**14.65** Find the solubility of $Ag_3PO_4$ in (a) pure water, (b) $0.05 M$ silver nitrate solution, (c) $0.15 M$ sodium phosphate solution.[4]

**14.66** Find the $[Cu^+]$ necessary to just begin precipitation of CuCl from a solution in which the $[Cl^-] = 4.5 \times 10^{-2}$.[4]

---

[4] The data necessary for this exercise can be found in Table 14.3.

**14.67**   Find the $[SO_4^{2-}]$ necessary to just begin the precipitation of $Ag_2SO_4$ from a solution in which $[Ag^+] = 0.21M$.[4]

**14.68**   A 20-cm³ volume of $0.0104M$ $Cd(NO_3)_2$ solution is mixed with a 20-cm³ volume of $0.000641M$ NaOH solution. Describe the changes that take place.[4]

**14.69**   A volume of 125 cm³ of $0.10M$ $Hg(NO_3)_2$ solution is mixed with a volume of 125 cm³ of $0.40M$ NaI solution. Find the amount (in moles) of precipitate that forms and the concentrations of the ions in solution at equilibrium.[4]

**14.70**★   You have 1.00 L of a $0.100M$ solution of calcium nitrate from which you wish to precipitate 99.0% of the $Ca^{2+}$ ion. Calculate the volume of a $0.100M$ $Na_2SO_4$ solution required for this purpose.[4]

**14.71**   A solution contains a mixture of chloride ion, $[Cl^-] = 0.10M$, and iodide ion, $[I^-] = 0.010M$. Evaluate the feasibility of a separation of the ions from this solution by selective precipitation carried out by the slow addition of concentrated $Hg_2(NO_3)_2$ solution.[4]

**14.72**   A solution contains a mixture of lead(II) ion, $[Pb^{2+}] = 0.10M$, and silver ion, $[Ag^+] = 0.010M$. Evaluate the feasibility of a separation of these ions from this solution by selective precipitation carried out by the slow addition of concentrated NaBr solution.[4]

**14.73**   A solution is saturated with respect to both $SrCO_3$ and $SrF_2$. The $[CO_3^{2-}] = 1.2 \times 10^{-3}M$. Calculate $[F^-]$.[4]

**14.74**   Write net ionic equations for the precipitation reactions of the group III cations (*not* group IIIA of the periodic table).

**14.75**   Using only the data in Table 14.3, propose a procedure for the separation of the following cations: $Na^+$, $Ag^+$, $Pb^{2+}$, $Ba^{2+}$, $Zn^{2+}$, and $Co^{2+}$.

**14.76**   Using only the data in Table 14.3, propose a procedure for the separation of the following anions: $NO_3^-$, $NO_2^-$, $F^-$, $Cl^-$, $CN^-$, and $SO_4^{2-}$.

# 15

# Acids and Bases

Preview

**A**cids and bases are substances we encounter constantly in everyday life. In this chapter, we provide a precise chemical description of acids and bases, starting with two ways of defining them: the Brønsted theory, based on proton transfer, and the Lewis theory, based on electron pair transfer. We then show how acids and bases are classified as strong or weak and describe how their strengths are related to their structures. Next we discuss a variety of calculations dealing with acids and bases in aqueous solution, considering the composition of acid-base systems at equilibrium, neutralization reactions, titrations, buffer solutions, and finally polyprotic acids.

Whether or not you have studied chemistry before, you almost certainly have picked up some knowledge about acids and bases. Most of us know that an acid is something that is put into automobile batteries, that causes heartburn, and that is neutralized by products advertised on television. Folk wisdom adds that a "strong" acid will burn the skin. Bases are not as well defined in everyday life, but you may have heard that a base feels soapy, that it can also burn the skin, that it can be used to unclog drains, and that it has something to do with washing. The words *alkali, alkaline,* and *caustic* also are associated with bases in some way. This chapter will expand your knowledge of acids and bases as they are defined and used in chemistry.

## 15.1    DEFINITIONS OF ACIDS AND BASES

When we mentioned acids and bases in Section 2.3, we used a temporary working definition: An acid is a species that increases the concentration of hydrogen ions ($H^+$) in aqueous solution, and a base is a species that increases the concentration of hydroxide ions ($OH^-$) in aqueous solution. This definition was first proposed in 1890 by Svante Arrhenius (1859–1927). It was generally accepted for 30 years and was used to develop many important quantitative relationships in the chemistry of acids and bases.

If you assume that an acid gives rise to $H^+$ ions and a base gives rise to $OH^-$ ions, the reaction of one with the other is explained by the equation

$$H^+(aq) + OH^-(aq) \rightleftharpoons H_2O$$

The Arrhenius definition was a major advance that helped to correlate a great deal of experimental data, in a field not previously noted for precise definitions. Originally, *acid* was used to refer to substances with a sour taste; the word itself, in English and other languages, is derived from words meaning "sour." A base, or alkali, was any substance that destroyed an acid, forming a salt and water. Given this background, the Arrhenius definition was an important step forward.

As time passed, however, the Arrhenius definition began to seem less satisfactory. For one thing, there seem to be two kinds of bases. Metal hydroxides such as sodium hydroxide produce hydroxide ion in water by ionic dissociation. Bases such as ammonia, $NH_3$, produce hydroxide ions in aqueous solution by undergoing a reaction with water:

$$NH_3(aq) + H_2O \rightleftharpoons NH_4^+(aq) + OH^-(aq)$$

But more important, the Arrhenius definition is narrow. A great many chemical processes take place in solvents other than water. The Arrhenius definition applies only to acids and bases in aqueous solution.

### The Brønsted Theory

A more general definition of acids and bases was proposed in 1923, primarily by J. N. Brønsted (1879–1947). According to Brønsted, **an acid is a species that can donate a proton, and a base is a species that can accept a proton.** This definition can be represented by the general reaction:

$$X \rightleftharpoons Y + H^+$$

which does not attempt to show mass or electrical charge balance. In this equation, X is the acid, Y is the base, and $H^+$, a hydrogen atom without an electron, is a proton. Together, X and Y are called a **conjugate acid-base pair.** There is a close relationship between X and Y. Remove a proton from X and you have Y. Add a proton to Y and you have X. We can call Y the conjugate base of X, and X the conjugate acid of Y.

In this general representation, we do not show any electrical charges that X and Y may bear. For example, if X is a neutral molecule such as HCl, then Y will be $Cl^-$, an anion of charge $-1$. If X is a cation of charge $+1$ such as $NH_4^+$, then Y will be $NH_3$, a neutral molcule. In a reaction of specific species, we must always include the necessary charges.

The Brønsted definition is more general than the Arrhenius definition. It works in liquid ammonia, alcohol, benzene, or any other solvent. Furthermore, it does not require any differentiation between the classes of substances defined as either acids or bases. Among the acids are electrically neutral substances such as HCl, HCN, and $H_2SO_4$; anions such as $HSO_4^-$, $HCO_3^-$, $H_2PO_4^-$, and $HPO_4^{2-}$; and cations such as $NH_4^+$ and $CH_3NH_3^+$. Similarly, the bases include electrically neutral substances such as $NH_3$ and $CH_3NH_2$; anions such as $OH^-$, $HCO_3^-$, and $HPO_4^{2-}$; and cations such as $Mg(OH)^+$ and $Al(OH)^{2+}$.

There is a problem in applying the Brønsted definition to acid-base systems in solution. The problem arises because free protons, $H^+$, cannot exist in solution to any great extent. The protons very often interact with the solvent, which acts as a base by accepting the protons. The same reasoning can be applied to a base in solution. Since there are no free protons in solution, the base must obtain a proton from a proton source. Very often, the proton source is the solvent.

We must therefore regard all acid-base reactions in solution as proton transfer reactions. As a consequence, an acid can act as an acid only if there is a base present. A base can act as a base only if there is an acid present. Any proton transfer reaction includes two acid-base conjugate pairs—the original acid-base pair, plus another pair to accept the proton from the acid or donate the proton to the base. Often, this second acid-base pair is derived from the solvent. Table 15.1 lists some conjugate acid-base pairs.

Each substance in the acid column is converted to its partner in the base column by removal of a proton. Each substance in the base column is

TABLE 15.1    Conjugate Acid-Base Pairs

| Name of Acid | Acid | Base | Name of Base |
|---|---|---|---|
| hydrochloric acid | HCl | $Cl^-$ | chloride |
| hydrocyanic acid | HCN | $CN^-$ | cyanide |
| hypochlorous acid | HClO | $ClO^-$ | hypochlorite |
| sulfuric acid | $H_2SO_4$ | $HSO_4^-$ | hydrogen sulfate |
| hydrogen sulfate | $HSO_4^-$ | $SO_4^{2-}$ | sulfate |
| ammonium | $NH_4^+$ | $NH_3$ | ammonia |
| ammonia | $NH_3$ | $NH_2^-$ | amide |
| water | $H_2O$ | $OH^-$ | hydroxide |
| oxonium or hydronium | $H_3O^+$ | $H_2O$ | water |
| carbonic acid | $H_2CO_3$ | $HCO_3^-$ | bicarbonate |
| bicarbonate | $HCO_3^-$ | $CO_3^{2-}$ | carbonate |
| phosphoric acid | $H_3PO_4$ | $H_2PO_4^-$ | dihydrogen phosphate |
| dihydrogen phosphate | $H_2PO_4^-$ | $HPO_4^{2-}$ | hydrogen phosphate |
| hydrogen phosphate | $HPO_4^{2-}$ | $PO_4^{3-}$ | phosphate |
| methyl ammonium | $CH_3NH_3^+$ | $CH_3NH_2$ | methyl amine |

converted to its partner in the acid column by the addition of a proton. Any two conjugate acid-base pairs can be combined in an acid-base reaction. For example, in the equation

$$\underset{\text{Acid}}{HCl(aq)} + \underset{\text{Base}}{CN^-(aq)} \rightleftharpoons \underset{\text{Base}}{Cl^-(aq)} + \underset{\text{Acid}}{HCN(aq)}$$

HCl and $Cl^-$ are one conjugate acid-base pair and $CN^-$ and HCN are the other pair. The proton donated by the acid HCl is accepted by the base $CN^-$ in the forward reaction and the proton accepted by the base $Cl^-$ is donated by the acid HCN in the reverse reaction.

## Amphoteric Substances

Some substances appear in both columns of Table 15.1. These substances are both acids and bases. Each can donate or accept a proton. Such substances are said to be *amphoteric*. Bicarbonate ion is amphoteric. In the reaction

$$HCO_3^-(aq) + NO_2^-(aq) \rightleftharpoons CO_3^{2-}(aq) + HNO_2(aq)$$

bicarbonate ion acts as an acid, donating a proton to the nitrite ion, the base. In the reaction

$$HCO_3^-(aq) + HNO_2(aq) \rightleftharpoons H_2CO_3(aq) + NO_2^-(aq)$$

bicarbonate ion acts as a base, accepting a proton from nitrous acid, the acid. Bicarbonate ion can even react with itself:

$$\overset{\text{Base} \longleftarrow \hspace{3.5cm} \longrightarrow \text{Acid}}{\underset{\text{Acid} \longleftarrow \hspace{2.5cm} \longrightarrow \text{Base}}{HCO_3^-(aq) + HCO_3^-(aq) \rightleftharpoons CO_3^{2-}(aq) + H_2CO_3(aq)}}$$

The transfer of a proton from one bicarbonate to the other forms both $CO_3^{2-}$, the conjugate base, and $H_2CO_3$, the conjugate acid.

It is of special importance that water is amphoteric. Water appears in both columns of Table 15.1 Water can act as an acid, donating a proton to a base, as in

$$CN^-(aq) + H_2O \rightleftharpoons HCN(aq) + OH^-(aq)$$
or
$$NH_3(aq) + H_2O \rightleftharpoons NH_4^+(aq) + OH^-(aq)$$

Water can also act as a base, accepting a proton, as in

$$H_2O + HCN(aq) \rightleftharpoons H_3O^+(aq) + CN^-(aq)$$

Water can react with itself, by a process that can be represented as

$$2H_2O \rightleftharpoons H_3O^+(aq) + OH^-(aq)$$

This process, in which one water molecule gains a proton from the other, is called the *autoprotolysis* of water. Since water is by far the most important solvent for inorganic materials, aqueous solutions will dominate our discussion of acid-base reactions.

If you look back over the last few pages, you will see that the Brønsted definition of acids and bases is more general and more useful than the Arrhenius definition. The Brønsted definition can be summarized in two sentences: An acid is a proton donor, and a base is a proton acceptor. Acid-base reactions in solution require two conjugate acid-base pairs, because free protons do not exist in solution.

### The Lewis Theory

The Brønsted definition has its shortcomings. Since it defines an acid as a proton donor, it excludes from the category of acids substances that have no protons to donate. This limitation was overcome by a more general definition of acids and bases, based on electronic structure, which was proposed by G. N. Lewis.

By the Brønsted definition, a species must accept a proton to be classified as a base. To accept a proton, the base must form a chemical bond, which requires that the base donate the two electrons needed to form the bond. In the Lewis definition, therefore, a base is a species that has a

$$: NH_3(aq) + H_2O \rightleftharpoons NH_4^+(aq) + OH^-(aq)$$

Arrhenius:   $NH_3$ is a base because it increases the concentration of $OH^-$.
Brønsted:   $NH_3$ is a base because it accepts $H^+$ (from $H_2O$).
Lewis:   $NH_3$ is a base because it donates its electron pair to an electron pair acceptor ($H^+$ of $H_2O$).

**Figure 15.1**
Ammonia as a base in the Arrhenius, Brønsted, and Lewis systems.

nonbonding valence electron pair that can be used to form a chemical bond. More simply, *a Lewis base is an electron pair donor* (Figure 15.1).

The Lewis definition significantly broadens the category of things that can be classified as acids. A base donates electrons to form the bond with the proton. Therefore, the proton accepts the electrons. *A Lewis acid is a species that is an electron pair acceptor.* A proton is the simplest Lewis acid. Many other species fit the definition, including a number of cations.

A metal cation usually can accept electrons to form bonds. In almost any reaction in which a metal cation forms a bond, we can say that the cation acts as a Lewis acid, accepting electrons from a Lewis base. In the reaction

$$Ag^+(aq) + 2NH_3(aq) \rightleftharpoons Ag(NH_3)_2^+(aq)$$

the silver cation is a Lewis acid and $NH_3$ is a Lewis base. In

$$Zn^{2+}(aq) + 2OH^-(aq) \rightleftharpoons Zn(OH)_2(s)$$

the zinc cation is a Lewis acid and hydroxide ion is the Lewis base.

Many reactions can be understood as the combination of a Lewis acid and a Lewis base. In fact, a common method of classifying chemical reagents uses the Lewis acid and Lewis base definition. A Lewis acid, which seeks electrons, is called an *electrophile,* from the Greek for "lover of electrons." A Lewis base, which seeks substances that are electron deficient, is classified as a *nucleophile,* from the Greek for "lover of nuclei." In the reaction between $Zn^{2+}$ and $OH^-$, the cation $Zn^{2+}$ is an electrophile and the anion $OH^-$ is a nucleophile. When an electrophile and a nucleophile find each other, a reaction can occur.

Many Lewis acids are electrically neutral. They can accept electrons and form additional bonds either because they have incomplete octets or because octet expansion is possible. Many compounds of the elements in column IIIA of the periodic table are Lewis acids. The best known of them are the halides of boron and aluminum, such as $BF_3$, $BCl_3$, $AlCl_3$, and $AlBr_3$. The central atom of such a compound has an incomplete octet and can form another bond. We classify these compounds as electrophiles because they seek more electrons to achieve complete octets. They react with Lewis bases, which are nucleophiles:

$$BF_3 + :NH_3 \rightleftharpoons F_3B—NH_3$$
$$AlCl_3 + Cl^- \rightleftharpoons AlCl_4^-$$

Many compounds of the transition metals, other heavy metals, and some nonmetals are Lewis acids because of octet expansion. Examples are

$$PtCl_4 + 2Cl^- \rightleftharpoons PtCl_6{}^{2-}$$
$$PF_5 + :NH_3 \rightleftharpoons F_5P—NH_3$$

We shall discuss some of these reactions in greater detail in Chapter 18.

The Lewis definition allows us to classify many substances as acids or bases and helps us to understand many chemical reactions. But the Brønsted definition is quite adequate for the chemistry of acids and bases in aqueous solution. Since the rest of this chapter deals primarily with aqueous equilibria, we shall use the Brønsted definition.

## 15.2  STRENGTHS OF ACIDS AND BASES

We listed some weak and strong acids in Section 2.3. The words *weak* and *strong* are commonplace in talk about acids and bases, but the exact meaning of these words is often blurred in everyday use. The Brønsted definition lets us formulate a simple yet precise description of the strength of acids and bases.

The description starts with the equilibrium $X \rightleftharpoons Y + H^+$. We define the relative strength of the acid X and the base Y by the relative amount of each substance present when the system is at equilibrium. The stronger the acid X, the further to the right is the position of equilibrium and the larger the value of the equilibrium constant $K$. The stronger the base Y, the further to the left is the position of equilibrium. *The strength of an acid and its conjugate base are connected. The stronger the acid, the weaker its conjugate base, and vice versa.*

This definition of acid and base strengths is fundamental, but it is not practical, because free protons do not exist in solution. A practical definition of acid and base strength requires an equilibrium that we can measure. Such an equilibrium must include another acid-base conjugate pair to accept the proton from the acid and to donate the proton to the base.

### Acid Strength in Water

Since most of our discussion will deal with aqueous solutions, it is convenient to have water as one component of the other conjugate acid-base pair when we define acid and base strength. The equilibrium reaction that we shall use to define the relative strengths of acids is thus:

$$HA(aq) + H_2O \rightleftharpoons A^-(aq) + H_3O^+(aq)$$

The formula HA represents an electrically neutral acid and A$^-$ represents its conjugate base. Note that the mass and electrical charges are balanced in this equation.

The value of the equilibrium constant $K$ for the equation is a measure of the strength of the acid. The larger the value of $K$, the stronger the acid. For example, we can measure the value of $K$ for the reaction:

$$HClO(aq) + H_2O \rightleftharpoons ClO^-(aq) + H_3O^+(aq)$$

$$K = \frac{[ClO^-][H_3O^+]}{[HClO]} = 3.2 \times 10^{-8}$$

and find that it is larger than the value of $K$ for the reaction:

$$HCN(aq) + H_2O \rightleftharpoons CN^-(aq) + H_3O^+(aq)$$

$$K = \frac{[CN^-][H_3O^+]}{[HCN]} = 4.93 \times 10^{-10}$$

Therefore, hypochlorous acid is a stronger acid than hydrocyanic acid.

The value of the equilibrium constant for this type of reaction also gives us information about the relative strengths of the two acids in the equilibrium, HA and $H_3O^+$. The larger the value of $K$, the stronger is acid HA relative to $H_3O^+$, the conjugate acid of water.

## The Acid Dissociation Constant

As always, the expression for the equilibrium constant includes terms that designate the molarities of the dissolved substances. But since water is the solvent, its concentration is very large $(55M)$[1] and relatively constant. For this reason, it is customary to include the term $[H_2O]$ in $K$, especially in dilute solutions. The general expression for $K$ is

$$K = \frac{[A^-][H_3O^+]}{[HA]}$$

which is the equilibrium constant for the general reaction

$$HA(aq) + H_2O \rightleftharpoons A^-(aq) + H_3O^+(aq)$$

This expression is often called the **acid dissociation constant** of the acid HA, and usually is represented by the symbol $K_A$. We shall use it whenever possible to express the strength of acids in water. However, this "constant" is an actual constant only in dilute solutions, under conditions in which the ideal solution approximation can be used. We shall assume

---

[1] The density of $H_2O$ is 1 g/cm$^3$. The mass of 1 L of $H_2O$ is 1000 g. There is, therefore, 1000 g $H_2O \times 1$ mol $H_2O/18$ g $H_2O = 55$ mol $H_2O$ in 1 L.

| TABLE 15.2 | Relative Strengths of Acids and Their Conjugate Bases | |
|---|---|---|
| | **Acid** | **Conjugate Base** |
| strongest | $HIO_3$ | $IO_3^-$ | weakest |
| | $HClO_2$ | $ClO_2^-$ | |
| | $HNO_2$ | $NO_2^-$ | |
| | $HF$ | $F^-$ | |
| | $HCOOH$ | $HCOO^-$ | |
| | $HCN$ | $CN^-$ | |
| weakest | $H_2O$ | $OH^-$ | strongest |

that the aqueous solutions are dilute enough so that $K_A$ is a constant.

The value of the acid dissociation constant $K_A$ also indicates the relative base strength of the conjugate base $A^-$. As the value of the $K_A$ of HA increases, the base strength of $A^-$ decreases. Table 15.2 lists some acids and their conjugate bases in order of their strengths.

In practice, $K_A$ is used to define the strength only of those acids that are weaker than $H_3O^+$. Stronger acids than $H_3O^+$ exist almost completely as their conjugate bases when dissolved in water. An example is nitric acid. In water, the equilibrium

$$HNO_3(aq) + H_2O \rightleftharpoons H_3O^+(aq) + NO_3^-(aq)$$

is essentially completely to the right. There is no measurable undissociated $HNO_3$, and $K_A$ cannot be found. We can say only that in water, nitric acid is immeasurably strong, and that its conjugate base, $NO_3^-$, is immeasurably weak. For this reason, strong acids (Table 2.9) in solution are always written as dissociated ions (Section 2.6).

### The Leveling Effect

If the equilibrium between any strong acid and water lies completely to the right, we cannot evaluate relative strengths of strong acids by comparing them to $H_3O^+$. In aqueous solutions, all acids stronger than $H_3O^+$ appear equally strong. This phenomenon is called the **leveling effect,** because water is said to level the strengths of all these acids, making them appear identical.

We can assess the relative strengths of acids that are strong in water by using a solvent that is a weaker base than water. One such solvent is methanol, which can also accept a proton:

$$CH_3OH + H^+ \rightleftharpoons CH_3OH_2^+$$

This cation is a stronger acid than the hydronium ion. When nitric acid, a strong acid in water, is dissolved in methanol, the transfer of a proton from nitric acid to methanol is not complete, since nitric acid is a weaker acid than $CH_3OH_2^+$. It is possible to measure the equilibrium constant of the reaction:

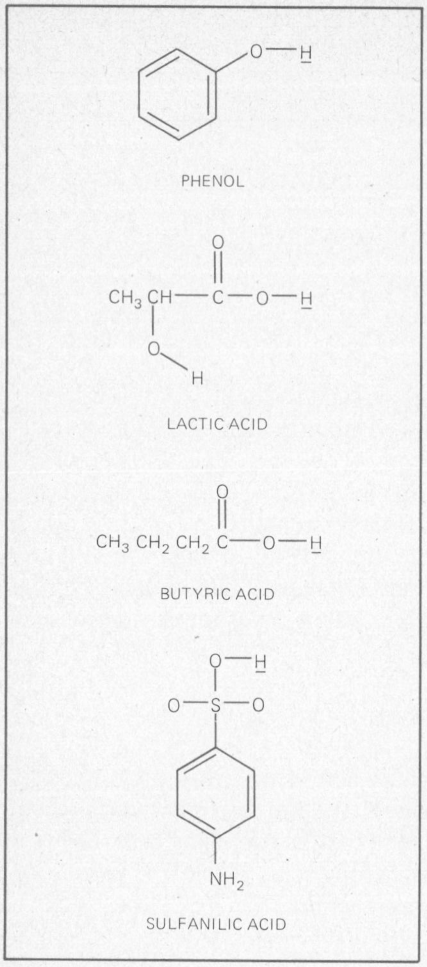

$$HNO_3 + CH_3OH \rightleftharpoons NO_3^- + CH_3OH_2^+$$

Perchloric acid, $HClO_4$, is another strong acid. The equilibrium in water:

$$HClO_4(aq) + H_2O \rightleftharpoons H_3O^+(aq) + ClO_4^-(aq)$$

lies completely to the right. We cannot tell whether perchloric acid or nitric acid is stronger in water, but we can make the evaluation by studying the equilibrium between perchloric acid and methanol:

$$HClO_4 + CH_3OH \rightleftharpoons ClO_4^- + CH_3OH_2^+$$

This equilibrium lies completely to the right. The corresponding equilibrium for nitric acid does not. Methanol differentiates between the two acids; perchloric acid is the stronger of the two. This evaluation is another reminder that the terms *strong* and *weak* are relative terms. In water, both of these acids are strong. In methanol, perchloric acid is strong and nitric acid is relatively weak.

Only a few common acids are stronger than $H_3O^+$. Many of them can be found in Table 2.9. As a working rule, you should assume that any acid that is not specifically described as being strong is weak. Since the value of $K_A$ for strong acids cannot be measured by any practical method, we shall list values of $K_A$ for weak acids only. Any acid that is listed in a table of acid dissociation constants is a weak acid. Table 15.3 lists the acid dissociation constants of some common weak acids. (See also Figure 15.2).

**Figure 15.2**
The structures of some organic acids. The acidic hydrogen atom is underlined.

## Base Strength in Water

The approach that we used for acids in water solution also can be used for bases. To define the relative strengths of bases in water, we shall use the equilibrium reaction

$$B(aq) + H_2O \rightleftharpoons BH^+(aq) + OH^-(aq)$$

We assume that the base B is an electrically neutral compound and write $BH^+$ to represent its conjugate acid. The electrical charges in this equation are thus balanced.

If we include the $[H_2O]$ in $K$ the equilibrium expression is

$$K = \frac{[BH^+][OH^-]}{[B]}$$

**TABLE 15.3** Dissociation Constants of Weak Acids at Room Temperature

| Electrically Neutral Inorganic Acids | $K_A$ $(M)^a$ |
|---|---|
| hydrogen peroxide, $H_2O_2$ | $2.4 \times 10^{-12}$ |
| hypoiodous acid, HIO | $2.3 \times 10^{-11}$ |
| hydrocyanic acid, HCN | $4.93 \times 10^{-10}$ |
| hypobromous acid, HBrO | $2.06 \times 10^{-9}$ |
| hypochlorous acid, HClO | $3.2 \times 10^{-8}$ |
| hydrofluoric acid, HF | $3.53 \times 10^{-4}$ |
| nitrous acid, $HNO_2$ | $4.5 \times 10^{-4}$ |
| chlorous acid, $HClO_2$ | $1.1 \times 10^{-2}$ |
| periodic acid, $H_5IO_6$ | $2.3 \times 10^{-2}$ |
| iodic acid, $HIO_3$ | $1.69 \times 10^{-1}$ |

| Electrically Neutral Organic Acids | |
|---|---|
| saccharin, $C_7H_5NO_3S$ | $2.1 \times 10^{-12}$ |
| phenol, $C_6H_5OH$ | $1.28 \times 10^{-10}$ |
| butyric acid, $C_3H_7COOH$ | $1.54 \times 10^{-5}$ |
| acetic acid, $CH_3COOH$ | $1.76 \times 10^{-5}$ |
| acrylic acid, $C_2H_3COOH$ | $5.6 \times 10^{-5}$ |
| uric acid, $C_5H_4N_4O_3$ | $1.3 \times 10^{-4}$ |
| lactic acid, $C_3H_6O_3$ | $1.37 \times 10^{-4}$ |
| formic acid, HCOOH | $1.77 \times 10^{-4}$ |
| sulfanilic acid, $C_6H_7NO_3S$ | $5.9 \times 10^{-4}$ |
| chloroacetic acid, $ClCH_2COOH$ | $1.40 \times 10^{-3}$ |
| dichloroacetic acid, $Cl_2CHCOOH$ | $3.32 \times 10^{-2}$ |
| trichloroacetic acid, $Cl_3CCOOH$ | $2 \times 10^{-1}$ |
| picric acid, $C_6H_3N_3O_7$ | $4.2 \times 10^{-1}$ |

$^a$ Although equilibrium constants do not carry units, we shall write them with units in this chapter to help you with dimensional analysis.

**TABLE 15.4** Relative Strengths of Bases and Their Conjugate Acids

| | Base | Conjugate Acid | |
|---|---|---|---|
| strongest | $CH_3NH_2$ | $CH_3NH_3^+$ | weakest |
| ↓ | $NH_3$ | $NH_4^+$ | ↓ |
| | $N_2H_4$ | $N_2H_5^+$ | |
| | $NH_2OH$ | $NH_2OH_2^+$ | |
| weakest | $H_2O$ | $H_3O^+$ | strongest |

This equilibrium expression is often called the **base dissociation constant** of the base B. It usually is represented by the symbol $K_B$.

The value of the base dissociation constant $K_B$ also indicates the relative acid strength of the conjugate acid $BH^+$. As the value of the $K_B$ increases, the acid strength of $BH^+$ decreases. Table 15.4 lists some bases and their conjugate acids in order of decreasing strength.

In practice, $K_B$ is used to define the strength of only those bases that are weaker than $OH^-$. Bases that are stronger than $OH^-$ exist almost com-

| TABLE 15.5 Dissociation Constants of Weak Bases at Room Temperature | |
|---|---|
| **Electrically Neutral Inorganic Bases** | $K_B$ (M) |
| hydroxylamine, $NH_2OH$ | $1.1 \times 10^{-8}$ |
| hydrazine, $N_2H_4$ | $1.7 \times 10^{-6}$ |
| ammonia, $NH_3$ | $1.79 \times 10^{-5}$ |
| **Electrically Neutral Organic Bases** | |
| aniline, $C_6H_5NH_2$ | $4.27 \times 10^{-10}$ |
| imidazole, $C_3H_4N_2$ | $9.01 \times 10^{-8}$ |
| nicotine, $C_{10}H_{14}N_2$ | $1.05 \times 10^{-6}$ |
| morphine, $C_{17}H_{19}NO_3$ | $1.62 \times 10^{-6}$ |
| codeine, $C_{18}H_{21}NO_3$ | $1.63 \times 10^{-6}$ |
| ephedrine, $C_{10}H_{15}NO$ | $1.38 \times 10^{-4}$ |
| methyl amine, $CH_3NH_2$ | $3.70 \times 10^{-4}$ |

pletely as their conjugate acids when dissolved in water because the $K_B$ equilibrium lies almost completely to the right. Their conjugate acids are immeasurably weak in water. Table 15.5 lists the base dissociation constants of some common weak bases. (See also Figure 15.3)

Many metal hydroxides are ionic solids and liberate $OH^-$ ions when they dissolve in water, because they undergo ionic dissociation. If the dissociation is essentially complete, the metal hydroxide can be called a strong base. But we can obtain a strongly alkaline solution only if the hydroxide is appreciably soluble in water. For example, calcium hydroxide, $Ca(OH)_2$, dissociates completely in solution and is a strong base. However, it cannot be used to prepare strongly alkaline solutions because it is relatively insoluble. Compounds such as sodium hydroxide (NaOH), potassium hydroxide (KOH), and barium hydroxide ($Ba(OH)_2$) dissociate completely and are appreciably soluble. They can be used to prepare a strongly alkaline solution.

Many metal oxides, especially those of metals in columns IA and IIA of the periodic table, form strongly alkaline solutions. A typical reaction of such a *basic oxide* is

$$Na_2O(s) + H_2O \longrightarrow 2Na^+(aq) + 2OH^-(aq)$$

These basic oxides behave as if they were hydroxides with the water removed. They are called *anhydrides* of hydroxides, from *anhydrous* ("without water").

## Formulas of Acids and Bases

In this chapter, we shall divide acids into two groups, strong and weak, based on their reaction with water. The formula for any strong acid in

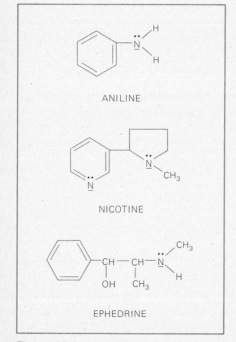

**Figure 15.3**

The structures of some organic bases. The basic nitrogen atom is underlined.

solution will be written to show complete dissociation. The conjugate bases of strong acids are so weak that they do not participate in acid-base reactions in aqueous solution. An acid dissociation constant $K_A$ will be used to show the extent of dissociation of a weak acid in solution. If we have a value for $K_A$, the acid is weak. We shall assume that weak acids in solution are essentially undissociated.

We shall also divide bases into two groups, weak and strong. The formulas of weak bases in aqueous solution are written to show no dissociation. If we have a value for $K_B$, the base is weak. The formulas of metal hydroxides, especially those of group IA and IIA metals, are written to show dissociation in aqueous solution. They are strong bases.

Thus, the major species in an aqueous solution of $NH_3$ is $NH_3(aq)$. But the major species in an aqueous solution of NaOH are $Na^+(aq)$ and $OH^-(aq)$. The metal cations in solutions of metal hydroxides generally do not participate in acid-base reactions.

The acid dissociation constants in Table 15.3 and the base dissociation constants in Table 15.5 are for electrically neutral compounds only. Many acids or bases are ions. Although the acid or base equilibrium constants for these ions are not listed, there are ways of finding them.

If an anion is the conjugate base of a weak acid, we shall find its base strength by using the $K_A$ of the acid. If a cation is the conjugate acid of a weak base, we shall find its acid strength by using the $K_B$ of the base (Section 15.5). The conjugate bases of acids stronger than $H_3O^+$ and the conjugate acids of bases stronger than $OH^-$ are so weak that they do not participate in acid-base reactions in aqueous solutions.

An important first step in acid-base chemistry is to write chemical equations showing the correct form of the participating acids and bases. You should review the rules for writing net ionic equations that are given in Section 2.7.

**Example 15.1**  Write the chemical equations for the reactions that occur when the following substances are mixed: (a) a solution of hydrochloric acid and a solution of sodium hydroxide; (b) a solution of hydrochloric acid and a solution of ammonia; (c) a solution of hydrogen cyanide and a solution of sodium hydroxide; (d) a solution of hydrogen cyanide and a solution of ammonia; (e) hydrogen chloride and ammonia, not in solution.

**Solution**  We include in the chemical equation only those substances that change.

a. Since hydrochloric acid is a strong acid, the equation is written to show its complete dissociation into ions, $H_3O^+$ and $Cl^-$, in aqueous solution. Sodium hydroxide is a metal hydroxide, so in solution it is written as dissociated ions, $Na^+$ and $OH^-$. In the reaction that occurs, $H_3O^+$ transfers a proton to $OH^-$:

$$H_3O^+(aq) + OH^-(aq) \rightleftharpoons 2H_2O$$

The $Cl^-$ ion, the conjugate base of a strong acid, is too weak to take part in any reaction that could compete with this process. The metal cation, $Na^+$, does not participate in acid-base reactions.

b. The strongest acid in this mixture is $H_3O^+$ from the hydrochloric acid in solution. The base is $NH_3$, which is weak. The reaction is

$$H_3O^+(aq) + NH_3(aq) \rightleftharpoons H_2O + NH_4^+(aq)$$

in which a proton is transferred from the hydronium ion to ammonia, forming water, the conjugate base of hydronium ion, and ammonium ion, the conjugate acid of ammonia.

c. Hydrocyanic acid is a weak acid, and so does not dissociate appreciably in solution. Sodium hydroxide dissociates into sodium ions and hydroxide ions. The reaction is

$$HCN(aq) + OH^-(aq) \rightleftharpoons CN^-(aq) + H_2O$$

A proton is transferred from the HCN to the $OH^-$, forming cyanide ion, the conjugate base of HCN, and water, the conjugate acid of $OH^-$. As in part (a), the sodium cation does not participate in acid-base reactions in water.

d. Here both the acid and the base are weak. Neither is written as dissociated ions. The reaction is

$$HCN(aq) + NH_3(aq) \rightleftharpoons CN^-(aq) + NH_4^+(aq)$$

e. When not in solution, both hydrogen chloride and ammonia are gases. An acid-base reaction, the transfer of a proton, does occur. The product is ammonium chloride, a salt that is a solid. The reaction is

$$HCl(g) + NH_3(g) \rightleftharpoons NH_4Cl(s)$$

in which a proton is transferred from the hydrogen chloride to the ammonia, forming chloride ion, the conjugate base of HCl, and ammonium ion, the conjugate acid of $NH_3$. These are the component ions of the salt ammonium chloride.

## 15.3  ACID-BASE STRENGTH AND CHEMICAL STRUCTURE

The strength of an acid depends on the relative stability of the acid and its conjugate base in solution. Their stability is related to their structure. In many cases there are so many structural factors to consider that it is difficult for us to make accurate predictions. We must measure the strength of the acid and then try to explain it.

### Binary Hydrides

There are two clear trends in the acidity of the binary hydrides of the nonmetals. Their acidity increases going down a column of the periodic table and increases from left to right across a row of the periodic table.

Consider the series of hydrides of the halogens: HF, HCl, HBr, and HI.

Experimentally, we find that acid strength increases as we go down the column; HF is the weakest acid and HI is the strongest. We explain this trend by the decrease in the strength of the hydrogen-halogen bond. The H—F bond is the strongest and the H—I bond is the weakest.

Now consider the hydrides of the nonmetals in the second row of the periodic table: $CH_4$, $NH_3$, $H_2O$, and HF. Experimentally, we find that acid strength increases from left to right. While $CH_4$ is an extremely weak acid, HF is an acid of measurable strength in water. This trend is the reverse of what we would predict on the basis of bond strength. We explain the experimental results by the increase in electronegativity as we go from carbon to fluorine. As electronegativity increases, the electron pair of the H—X bond is held more closely by X (where X is an atom of any element). Dissociation into $H^+$ and $X^-$ thus becomes increasingly favored over the covalent H—X bond.

### Oxyacids

Many inorganic acids are called *oxyacids* because they are composed of hydrogen, oxygen, and a third element, usually a nonmetal or a semimetal. The general formula for an oxyacid can be written as $H_mXO_n$, where X is the third element. The subscript $m$ indicates the number of acidic hydrogen atoms in a molecule of the oxyacid, and the subscript $n$ indicates the number of oxygen atoms. The acidic hydrogen atoms in an oxyacid are always bonded to an oxygen atom. Some structures of oxyacids are

$$\text{Cl—O—H} \qquad \text{H—O—N=O} \qquad \text{H—O—}\overset{\displaystyle \overset{O}{\|}}{\underset{\displaystyle \underset{O}{\|}}{\text{S}}}\text{—O—H}$$

The strength of an oxyacid can be related to the polarity of the O—H bond that undergoes cleavage. If the central X atom is highly electronegative, electrons are held less tightly by the H atom and the oxyacid is stronger. In the series HClO, HBrO, and HIO, there is a decrease in acid strength (Table 15.3) that parallels the decrease in electronegativity from Cl to I. Many of the very strong acids, such as $HClO_4$ and $HNO_3$, have strongly electronegative central atoms. Many of the weaker oxyacids, such as $H_3PO_4$, $H_2CO_3$, and $H_3BO_3$, have less electronegative central atoms.

The polarity of the O—H bond, and thus the acid strength of an oxyacid, also is influenced by the number of oxygen atoms in the molecule or, more generally, the oxidation number of the central atom. If the central atom has a high oxidation number, it withdraws electrons more effectively from the OH group, making the bond more polar. A good example of this effect is the increase in acidity in the series HClO, $HClO_2$, $HClO_3$, and $HClO_4$.

A rule that incorporates these effects allows us to predict the relative strengths of many oxyacids. The value of $n - m$, where $n$ and $m$ are the subscripts in the general formula for oxyacids, gives the strength of an oxyacid. Generally, a large value of $n - m$ indicates that an oxyacid is strong. For $HClO_4$, $n - m = 3$, and $HClO_4$ is a very strong acid. For $H_2SO_4$, $n - m = 2$, and $H_2SO_4$ is a strong acid. For $H_3PO_4$, $n - m = 1$, and $H_3PO_4$ is a weak acid ($K_A \approx 10^{-2}$). For HClO, $n - m = 0$, and HClO is a very weak acid ($K_A \approx 10^{-8}$).

The vast majority of organic acids are oxyacids and are weak. We often can correlate the acid strength of organic acids with their structures by studying the polarity and the strength of the bond to the acidic proton. Their bond usually is an O—H bond, but there are many exceptions to this rule.

## Bases

To accept a proton, a base must make available a nonbonding electron pair that can form a bond with the proton. Therefore, we would expect that the strength of a base is related to the availability of an electron pair. For example, $:NH_3$ is a base of measurable strength in water, but $:NF_3$ has no detectable basic properties. The lone pair in $:NH_3$ is available; but the electronegative F atoms in $:NF_3$ attract electrons so strongly that the lone pair is not available for bonding to a proton.

The availability of electron pairs does not readily explain why $:NH_3$ is a stronger base than $:PH_3$. In this case, a better explanation refers to the conjugate acids. The N—H bond in $NH_4^+$ is much stronger than the P—H bond in $PH_4^+$. Therefore, the position of the equilibrium

$$PH_3 + H^+ \rightleftharpoons PH_4^+$$

lies more to the left than does the corresponding equilibrium for $NH_3$. Since relatively little $PH_4^+$ is present at equilibrium, $PH_3$ is a weak base.

These kinds of explanations apply to many organic bases, including those listed in Table 15.5. These bases often have complex structures, but we can regard them as derivatives of $:NH_3$. In most cases, the nonbonding electron pair of a neutral weak base of measurable strength in aqueous solution lies on an N atom.

For any trend in acid strength, we find the opposite trend for the conjugate bases. We have seen that acidity increases as we go down column VIIA from F to I. There is a corresponding decrease in base strength in the ions $F^-(aq)$, $Cl^-(aq)$, $Br^-(aq)$, and $I^-(aq)$ as we go down the column. In the same way, we see decreasing base strength in the series $NH_3$, $N_2H_4(H_2N—NH_2)$, $NH_2OH$ (Table 15.5) because of the increasing electronegativity of the group substituted for the H atom of $NH_3$. The strength of the conjugate acids therefore decreases in the reverse order: $NH_3OH_2^+$, $N_2H_5^+$, $NH_4^+$.

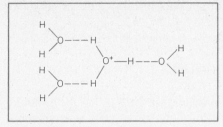

Figure 15.4
A representation of the $H_9O_4^+$ ion.

## 15.4 ACID-BASE REACTIONS IN AQUEOUS SOLUTIONS

Some simplifications are possible in expression of acid and base equilibria. We already have made one. Since the $[H_2O]$ does not change appreciably during a reaction, it can be included in the value of $K$ and omitted from the equilibrium expression. We can make another simplification by writing $H^+(aq)$ instead of $H_3O^+(aq)$.

Like most cations, the $H_3O^+$ ion is not free in aqueous solution; it is hydrated (Section 11.3). There is evidence that one important form of the hydronium ion has the composition $H_9O_4^+$. In this species, the $H_3O^+$ ion is associated with three water molecules (Figure 15.4). Since $H_3O^+$ is itself not an accurate representation for the hydronium ion in aqueous solution, we can choose the simplest possible symbol and write $H^+(aq)$ to represent all hydrated forms of the proton. It must be understood, however, that free $H^+$ does not actually exist in solution.

Using these simplifications, let us write equations for the reactions that correspond to some of the dissociation constants given in Tables 15.3 and 15.5. The reaction that defines $K_A$ for acetic acid[2] is represented as

$$HOAc(aq) \rightleftharpoons H^+(aq) + OAc^-(aq) \qquad K = 1.76 \times 10^{-5}$$

For nitrous acid it is

$$HNO_2(aq) \rightleftharpoons H^+(aq) + NO_2^-(aq) \qquad K = 4.5 \times 10^{-4}$$

For methyl amine, the reaction that defines $K_B$ is

$$CH_3NH_2(aq) + H_2O \rightleftharpoons CH_3NH_3^+(aq) + OH^-(aq)$$
$$K = 3.70 \times 10^{-4}$$

Water does not appear in the acid dissociation reactions, but you should remember that water is the base that accepts the proton in these reactions. All hydrated forms of the proton are represented by $H^+(aq)$.

### The Water Equilibrium

The equilibrium for the autoprotolysis of water is written $H_2O \rightleftharpoons H^+(aq) + OH^-(aq)$. This equation shows only one water molecule, which acts as an acid. A second water molecule, which acts as a base, is omitted from the equation. The equilibrium constant for this reaction is

$$K = \frac{[H^+][OH^-]}{[H_2O]}$$

[2] The abbreviation $OAc^-$ is used for acetate ion, $CH_3CO_2^-$.

In a dilute solution, we can define an equilibrium constant equal to $K[H_2O]$. The resulting equilibrium expression, called the *ion product of water* and represented by the symbol $K_W$, is extremely important:

$$K_W = [H^+][OH^-]$$

At room temperature (298 K), its value is $1.0 \times 10^{-14}M^2$. You should memorize this value. It gives useful information about any aqueous solution at equilibrium. No matter what else occurs in the solution at equilibrium, the concentration of hydrated protons multiplied by the concentration of hydroxide ion is $1.0 \times 10^{-14}M^2$ at 298 K:

$$[H^+][OH^-] = 1.0 \times 10^{-14}M^2$$

In water that contains no added bases or acids, $[H^+] = [OH^-] = x$, so at room temperature:

$$x^2 = 1.0 \times 10^{-14}M^2 \quad \text{and} \quad x = 1.0 \times 10^{-7}M$$

A solution in which both the $[H^+]$ and the $[OH^-]$ are equal is called a neutral solution. At 298 K the $[H^+]$ and $[OH^-]$ in a neutral solution are each $1.0 \times 10^{-7}M$. In practice, a neutral solution is almost impossible to obtain. When water is exposed to the atmosphere, some carbon dioxide dissolves in it. Since carbon dioxide reacts with $H_2O$ to form carbonic acid, $H_2CO_3$, the solution is no longer neutral. If water contains a dissolved acid, the $[H^+]$ has a value greater than $1.0 \times 10^{-7}M$ and the $[OH^-]$ therefore has a value less than $1.0 \times 10^{-7}M$. Such a solution is said to be acidic. If water contains a dissolved base, the value of $[OH^-]$ is greater than $1.0 \times 10^{-7}M$ and the $[H^+]$ is less than $1.0 \times 10^{-7}M$. Such a solution is said to be basic or alkaline.

**Example 15.2**    Calculate the $[H^+]$ and the $[OH^-]$ in a solution prepared from 0.030 mol of HI and enough water to form 0.50 L of solution.

**Solution**    Hydriodic acid, HI, is a strong acid that dissociates completely in solution. The solution has 0.030 mol HI/0.50 L = 0.060$M$ in HI. Since 1 mol of HI forms 1 mol of $H^+$ in solution, the $[H^+] = 0.060M$. From the expression for $K_W$:

$$[H^+][OH^-] = (0.060M)[OH^-] = 1.0 \times 10^{-14}M^2$$

$$[OH^-] = \frac{1.0 \times 10^{-14}M^2}{0.060M} = 1.7 \times 10^{-3}M$$

**Example 15.3**    Calculate the $[H^+]$ and the $[OH^-]$ in a solution that is 1.0$M$ in potassium hydroxide.

Solution   Although the concentration of potassium hydroxide is high, we shall use the ideal solution approximation. Potassium hydroxide is a strong base that dissociates completely in solution, and thus $[OH^-] = 1.0M$. From the expression for the ion product of water:

$$[H^+][OH^-] = [H^+](1.0M) = 1.0 \times 10^{-14}M^2$$
$$[H^+] = 1.0 \times 10^{-14}M$$

We see that the concentration of hydroxide ion is low when the solution is acidic, and that the concentration of protons is low when the solution is basic. As the acidity of a solution increases, the $[H^+]$ increases and the $[OH^-]$ decreases. As the basicity of a solution increases, the $[OH^-]$ increases and the $[H^+]$ decreases; they are inversely proportional.

### The Symbol p

Equilibrium constants, such as $K_A$, $K_B$, and $K_W$ itself, are small numbers. The $[H^+]$ and $[OH^-]$ are also small numbers, except in relatively concentrated solutions of strong acids or strong bases. Any small number can be written as the product of two numbers (Appendix IV). One is an exponential, 10 raised to a negative power. The other is an ordinary number, usually with one integer to the left of the decimal point.

But such numbers are not easy to work with, so a more convenient way of expressing them is used: the negative logarithm. To express the numerical value of a given quantity as its negative logarithm, we use the symbol p. When the symbol p precedes the symbol for a quantity, it means "take the negative logarithm of the quantity."

Thus, pH means the negative logarithm of the $[H^+]$, pOH means the negative logarithm of the $[OH^-]$, $pK_A$ means the negative logarithm of $K_A$, the acid dissociation constant, and so on.

Example 15.4   Find the negative logarithms of the following numbers: (a) $1.0 \times 10^{-7}$, (b) $3.64 \times 10^{-12}$, (c) 0.25, (d) 1.3.

Solution   In each case, we find the negative logarithms of two terms and add them.
   a. $-\log 1.0 = 0.00$; $-\log 10^{-7} = 7$; $-\log(1.0 \times 10^{-7}) = 7.00$.
   b. $-\log 3.64 = -0.561$; $-\log 10^{-12} = 12$; $-\log(3.64 \times 10^{-12}) = 12 - 0.561 = 11.439$.
   c. First convert 0.25 to an exponential form, $2.5 \times 10^{-1}$. Then $-\log 2.5 = -0.40$; $-\log 10^{-1} = 1$; $-\log(2.5 \times 10^{-1}) = -0.40 + 1 = 0.60$.
   d. 1.3 can be written $1.3 \times 10^0$. $-\log 1.3 = -0.11 + 0 = -0.11$.

Some adjustment in thinking is needed when we use negative logarithms. The smaller a negative logarithm, the larger the quantity it represents. A pH of 7 represents a larger $[H^+]$ than a pH of 9. A pH of 2 represents a $[H^+]$ of $1.0 \times 10^{-2}$, and a pH of $-1$ represents a $[H^+]$ of $1.0 \times$

$10^1$, a larger concentration. The p$K_A$ of acetic acid is 4.74 and the p$K_A$ of HCN is 9.40. Acetic acid has the smaller negative logarithm, and therefore has the larger $K_A$ and is the stronger acid.

Remember that logarithms represent an exponential scale. Each integer represents a factor of 10. The [H$^+$] of a solution whose pH is 7 is 10 times greater than the [H$^+$] of a solution of pH 8 and 1000 ($10^3$) times greater than the [H$^+$] of a solution of pH 10.

The symbol p is especially convenient for expressing the acidity or basicity of aqueous solutions. Since multiplying numbers is the same as adding their logarithms, we can write

$$[H^+][OH^-] = K_W = 1.0 \times 10^{-14}$$
$$pH + pOH = pK_W = 14$$

The sum of the pH and the pOH in any aqueous solution at room temperature is 14. Given one, the other is easily found by subtraction from 14. Acidic solutions can be described as having a low pH. Basic solutions have high pH. In a neutral solution, pH = pOH = 7. These relationships are shown in Table 15.6. The pH of some common materials are shown in Table 15.7.

---

**Example 15.5**    Calculate the pH and the pOH of a solution prepared from 0.033 mol of perchloric acid and enough water to form 250 cm$^3$ of solution.

**Solution**    Since HClO$_4$ is a strong acid, it dissociates completely in solution. Therefore, there is 0.33 mol of H$^+$ in the 250 cm$^3$ of solution.

$$[H^+] = \frac{0.033 \text{ mol H}^+}{250 \text{ cm}^3} \times \frac{1000 \text{ cm}^3}{1 \text{ L}} = 0.132M$$

The pH is the negative logarithm of this number.

$$pH = -\log 1.32 \times 10^{-1} = 1 - 0.12 = 0.88$$
$$pOH = 14 - pH = 14 - 0.88 = 13.12$$

---

**Example 15.6**    Calculate the pH and the pOH of a solution prepared from 0.00135 mol of barium hydroxide and enough water to form 100.0 cm$^3$ of solution.

**Solution**    Since each mole of Ba(OH)$_2$ in solution forms 2 mol of OH$^-$,

$$[OH^-] = \frac{(2)(0.00135 \text{ mol Ba(OH)}_2)}{100 \text{ cm}^3} \times \frac{1000 \text{ cm}^3}{1 \text{ L}} = 2.70 \times 10^{-2}M$$
$$pOH = -\log 2.70 \times 10^{-2} = 2 - 0.431 = 1.569$$
$$pH = 14 - pOH = 14 - 1.569 = 12.431$$

TABLE 15.6   The pH Scale

| pH | $[H^+](M)$ | $[OH^-](M)$ | pOH | |
|---|---|---|---|---|
| -1 | 10 | $1 \times 10^{-15}$ | 15 | acid |
| 0 | 1 | $1 \times 10^{-14}$ | 14 | |
| 1 | $1 \times 10^{-1}$ | $1 \times 10^{-13}$ | 13 | |
| 2 | $1 \times 10^{-2}$ | $1 \times 10^{-12}$ | 12 | |
| 3 | $1 \times 10^{-3}$ | $1 \times 10^{-11}$ | 11 | |
| 4 | $1 \times 10^{-4}$ | $1 \times 10^{-10}$ | 10 | |
| 5 | $1 \times 10^{-5}$ | $1 \times 10^{-9}$ | 9 | |
| 6 | $1 \times 10^{-6}$ | $1 \times 10^{-8}$ | 8 | |
| 7 | $1 \times 10^{-7}$ | $1 \times 10^{-7}$ | 7 | neutral |
| 8 | $1 \times 10^{-8}$ | $1 \times 10^{-6}$ | 6 | |
| 9 | $1 \times 10^{-9}$ | $1 \times 10^{-5}$ | 5 | |
| 10 | $1 \times 10^{-10}$ | $1 \times 10^{-4}$ | 4 | |
| 11 | $1 \times 10^{-11}$ | $1 \times 10^{-3}$ | 3 | |
| 12 | $1 \times 10^{-12}$ | $1 \times 10^{-2}$ | 2 | |
| 13 | $1 \times 10^{-13}$ | $1 \times 10^{-1}$ | 1 | |
| 14 | $1 \times 10^{-14}$ | 1 | 0 | |
| 15 | $1 \times 10^{-15}$ | 10 | -1 | alkaline |

TABLE 15.7   The pH of Some Common Materials

| Material | pH | Material | pH |
|---|---|---|---|
| gastric juice | 1.4 | rainwater | 6.5 |
| lemon juice | 2.1 | pure water | 7.0 |
| soft drinks | 3.0 | saliva | 7.0 |
| sauerkraut | 3.5 | blood | 7.4 |
| beer | 4.5 | egg | 7.8 |
| black coffee | 5 | household ammonia | 11.9 |
| cow's milk | 6.5 | | |

An instrument found in almost any chemistry laboratory is the pH meter, which measures the pH of a solution electrochemically. When you use a pH meter, you will sometimes have to convert a given value of pH into a $[H^+]$.

---

**Example 15.7**    The pH of a solution is found to be 8.78. Calculate the $[H^+]$ and the $[OH^-]$.

**Solution**    This problem asks us to find the number whose negative logarithm is 8.78. We first write 8.78 as the difference between the next highest integer and a decimal: $9 - (1 - 0.78) = 9 - 0.22$. In the conversion to a negative logarithm, the integer becomes the negative exponent of 10. Here, the integer is 9, so we write $10^{-9}$. We find the number by which this exponential term is multiplied by finding the antilog of 0.22, which is 1.66. Thus:

$$\text{antilog}(-8.78) = 1.66 \times 10^{-9}$$
$$[H^+] = 1.66 \times 10^{-9} M$$

$$[OH^-] = \frac{1.00 \times 10^{-14}M^2}{[H^+]} = \frac{1.00 \times 10^{-14}M^2}{1.66 \times 10^{-9}M} = 6.0 \times 10^{-6}M$$

## Common Types of Acid-Base Reactions

Chemists often must write balanced equations for reactions that occur in aqueous solutions of acids and bases. In writing these equations, remember that reactions between acids and bases are proton transfers from acids to bases.

We usually start with a word description of an experiment or a problem. One key to writing the equation we need is to go from the word description to a set of reactants. The relevant equilibrium usually is the reaction between the strongest acid and 'the strongest base present in *appreciable* concentration. While more than one equilibrium reaction takes place in any aqueous solution, we generally are interested in the reaction whose equilibrium constant $K$ has the largest value.

Let us start with a simple word description: "a solution of acetic acid." This statement defines the reactants present in *appreciable* concentration in the system: acetic acid, HOAc, and water, $H_2O$. The system has two acids, HOAc and $H_2O$. The acetic acid is the stronger, and it is included in the equilibrium of interest. The system has one base, $H_2O$. Therefore, the acid-base reaction is the transfer of a proton from the strongest acid in the system, HOAc, to the only base, $H_2O$:

$$HOAc(aq) + H_2O \rightleftharpoons OAc^-(aq) + H_3O^+(aq)$$

or, more simply,

$$HOAc(aq) \rightleftharpoons H^+(aq) + OAc^-(aq)$$

Another equilibrium that occurs in this system is the ionization of water,

$$H_2O \rightleftharpoons H^+(aq) + OH^-(aq)$$

The $K$ for this reaction is much smaller than the $K_A$ for the dissociation of acetic acid, because acetic acid is a stronger acid than water (Table. 15.4).

We do not write reactions that include the other species in the solution, such as $OH^-$ and $OAc^-$, as reactants, even if the equilibrium constants of the reactions are larger than the $K_A$ of acetic acid. *No species may appear as a reactant in an equation unless it is present in appreciable concentration when the system is initially constituted.* Generally, these species are mentioned in the word description of the problem.

Remember the rules for the correct representation of solutions of ionic substances. If a strong acid is included in the set of reactants, the acid is written as dissociated ions. The equation for a reaction of a solution of hydrobromic acid may include $H^+(aq)$ and $Br^-(aq)$ as reactants. But

## ACID RAIN AND OUR ENVIRONMENT

An unintentional experiment on a global scale is providing dramatic evidence of the sensitivity of living organisms and ecosystems to even slight changes in pH.

For the past two decades, the acidity of rain and snow falling on large areas of the United States and Europe has been increasing, apparently because more and more pollutants such as sulfur dioxide and nitrogen oxides are being produced from the combustion of fossil fuels. The acid rain has caused a marked decline in the numbers of animals of some sensitive species, such as salmon and trout, and severe damage to plants.

Some acidity is normal in precipitation, since gaseous carbon dioxide in the atmosphere dissolves in water to produce a slightly acid solution. The minimum pH value expected for water in equilibrium with atmospheric $CO_2$ is about 5.6. But the pH of rain and snow falling on much of northern Europe and the eastern United States in recent years has often been as low as 5 and has sometimes been as low as 3. The increased acidity of the rain results from the presence of sulfuric acid and nitric acid.

The effects of acid precipitation have been documented particularly well in Scandinavia. Pollutants from England and other industralized nations drift north and east with prevailing winds and are deposited on the Scandinavian nations. About 5000 lakes in Sweden are estimated to have pH values of 5.0 or below, and fish populations have been seriously affected. Studies have shown that the mean acidity of precipitation falling on southern Norway has fallen to a pH value of 4.6, and that the number of lakes without salmon and trout has increased alarmingly.

The same effect has been observed in the lakes of New York's Adirondack mountains. A study in the 1930s found that only 4% of 217 mountain lakes had a pH of 5.0 or under and had no fish populations. In the early 1970s, a survey found that half of the 217 lakes had pH values below 5.0, and that 90% of the lakes with low pH contained no fish.

Recent studies of forests in the northeastern United States and Scandinavia have found widespread damage to trees that appears to be attributable to acid precipitation, but the relationship has been disputed.

One major reason for controversy about the damaging effects of acid rain is that the subject has become a major economic and political issue that crosses both state and national boundaries. In the United States, electric utilities and industries in the Midwest have disputed scientific studies that attributed the damage done by acid precipitation in the Northeast to emissions from their smokestacks. In Europe, the Scandinavian countries' contention that emissions from other nations are responsible for the Scandinavian acid rain problem is also being disputed.

No quick solution is in sight. Even if governments and industries agree on the source of the acid precipitation, it will take years and many billions of dollars to install the equipment needed to bring about a drastic reduction in air pollutants. While discussions about the steps needed to reduce emissions of nitrogen oxides and sulfur oxides continue, the problem of acid rain remains with us.

HBr(aq) cannot be a reactant, because it does not exist in any appreciable concentration in solution. If the problem mentions a strong base such as sodium hydroxide in solution, the equation may include $Na^+$(aq) or $OH^-$(aq), as reactants. But NaOH(aq) undissociated in solution may not be a reactant. For salts in solution, for example $NH_4Cl$, the salt is written as dissociated ions. The equation for a reaction of a solution of ammonium chloride may include $NH_4^+$(aq) or $Cl^-$(aq), but not $NH_4Cl$(aq) as a reactant.

Thus, the acid-base system described by the phrase "a solution of potassium acetate" contains $K^+$ ion, which does not enter into acid-base equilibria; $OAc^-$ ion, which is a base that is stronger than water; and water itself, which is both an acid and a base. The system does *not* initially contain appreciable concentrations of acetic acid, $H^+$, $OH^-$, or undissociated KOAc. None of these materials may appear as reactants. The relevant reaction that we shall use to find the composition of the equilibrium state of this system is the reaction between $H_2O$, the strongest acid present in appreciable concentration, and $OAc^-$, the strongest base.

You might wonder how we know that $OAc^-$ is a stronger base than water. We know that acetic acid is a weaker acid than $H_3O^+$ because its $K_A$ is listed in Table 15.3. Therefore, the conjugate base of acetic acid, $OAc^-$, is stronger than $H_2O$, which is the conjugate base of $H_3O^+$.

As usual, a proton is transferred from the acid to the base:

$$H_2O + OAc^-(aq) \rightleftharpoons OH^-(aq) + HOAc(aq)$$

## Categories of Acid-Base Systems

With these considerations in mind, we can define some of the important categories of acid-base systems and then write the relevant equilibrium reaction for each category. Once we do this, we can write the relevant equilibrium for any system that is identified as belonging to one of these categories. The categories are

1. A solution of a neutral weak acid. The relevant equilibrium for problem solving in this system is the reaction between water, acting as a base, and the weak acid. The equilibrium constant for this reaction is $K_A$. One such system is a solution of chlorous acid:

$$HClO_2(aq) \rightleftharpoons ClO_2^-(aq) + H^+(aq)$$

2. A solution of a neutral weak base. The relevant equilibrium is the reaction between water, acting as an acid, and the weak base. The equilibrium constant for this reaction is $K_B$. One such system is a solution of aniline:

$$C_6H_5NH_2(aq) + H_2O \rightleftharpoons C_6H_5NH_3^+(aq) + OH^-(aq)$$

3. A solution of a salt containing an anion that is a conjugate base of a weak acid. The relevant equilibrium is the reaction between water, acting as an acid, and the anion, acting as a base. We must be careful to represent the solution correctly. The sources of basic anions such as $CN^-$, $OAc^-$, and $ClO^-$ are usually salts, and a solution of such an anion is obtained by dissolving the appropriate salt in water. Solutions of salts are written to show dissociated ions, but the cations of these salts usually do not take part in the relevant reaction. In a solution of NaClO, the relevant reaction is

$$ClO^-(aq) + H_2O \rightleftharpoons HClO(aq) + OH^-(aq)$$

In a solution of potassium fluoride, the relevant reaction is:

$$F^-(aq) + H_2O \rightleftharpoons HF(aq) + OH^-(aq)$$

This kind of reaction sometimes is called a hydrolysis reaction. Its equilibrium constant can be designated as $K_h$. We shall not use this notation.

4. A solution of a salt containing a cation that is the conjugate acid of a weak base. The relevant equilibrium is the reaction between water, acting as a base, and the cation, acting as an acid. In a solution of the salt $NH_4Cl$, the relevant equation is

$$NH_4^+(aq) \rightleftharpoons NH_3(aq) + H^+(aq)$$

As usual, water does not appear in the equation when it acts as a base.

5. A solution of an acid mixed with a solution of a base. The reaction that takes place is called a **neutralization reaction** when water is not a reactant in the relevant equilibrium reaction. We shall discuss neutralization reactions in Section 15.6.

6. A solution of a weak acid mixed with a solution of its conjugate base. These components form what is called a **buffer solution,** which we shall discuss in Section 15.7. Buffer systems are conveniently handled by use of the reaction that defines the $K_A$ of the weak acid. One buffer system is the buffer solution formed when a solution of acetic acid is mixed with a solution of sodium acetate. The salt is a source of acetate ion, the conjugate base of acetic acid. The reaction between the strongest acid (HOAc) and the strongest base ($OAc^-$) results in no change. The relevant reaction is between the strongest acid and the next-strongest base ($H_2O$):

$$HOAc(aq) \rightleftharpoons H^+(aq) + OAc^-(aq) \qquad K_A = 1.76 \times 10^{-5}$$

Now that we have described some of the most common acid-base systems, we can solve some problems relating to the composition of the equilibrium states that we obtain when we prepare these systems.

## 15.5  WEAK ACIDS AND WEAK BASES

To successfully use experimental data obtained for aqueous solutions of acids and bases, you must be able to carry out the appropriate calculations. The best way to approach these calculations is by example.

**Example 15.8**  While hydrofluoric acid etches glass and causes severe burns, it is correctly described as a weak acid. You will find its $K_A$ in Table 15.3. Calculate the pH of a solution prepared from 0.245 mol of HF and enough water to form 1.00 L of solution.

**Solution**    Acid-base equilibrium problems of this type are attacked in the same way as problems in other types of equilibrium systems described in Chapter 14. *We first write the relevant balanced chemical equation,* the dissociation of the acid:

$$HF(aq) \rightleftharpoons H^+(aq) + F^-(aq)$$

We get the equilibrium constant $K_A$ for this reaction from Table 15.3: $K_A = 3.53 \times 10^{-4}M$.

As we did in Chapter 14, we then write a starting line that lists the initial concentrations of all the materials appearing in the equilibrium reaction. Except for special cases that we shall specify, the concentration of $H^+$ (or of $OH^-$) from the autoprotolysis of water is not included on the start line. For this system, the start line is

$$\begin{array}{cccc} & HF(aq) & \rightleftharpoons H^+(aq) + & F^-(aq) \\ \textbf{start} & 0.245M & 0 & 0 \end{array}$$

The reaction proceeds to equilibrium by the transfer of protons from HF to water, the so-called dissociation of HF. Let $x$ represent the [HF] that dissociates. The equilibrium line is

$$\begin{array}{cccc} & HF(aq) & \rightleftharpoons H^+(aq) + & F^-(aq) \\ \textbf{equil} & 0.245M - x & x & x \end{array}$$

The expression for the equilibrium constant of the reaction gives:

$$K_A = \frac{[H^+][F^-]}{[HF]} = \frac{(x)(x)}{(0.245M - x)} = 3.53 \times 10^{-4}M$$

Since $K_A$ is small (as it usually is in these problems) we can use the approximation $0.245 - x \approx 0.245$ to solve for $x$. We find $x = 9.30 \times 10^{-3}M$. We always check our approximation. In this case, 0.009 is almost 4% of 0.245. Since pH is generally measured to two significant figures the approximation is a good one. The value of $x$ is both the $[H^+]$ and the $[F^-]$. The pH is the negative logarithm of the $[H^+]$.

$$pH = -\log 9.30 \times 10^{-3} = 3 - \log 9.30 = 3 - 0.97 = 2.03$$

This problem could have been phrased in a different way. You might have been asked to calculate the fraction of HF that dissociates in a solution that is originally $0.245M$ in HF. The fraction that dissociates, $f$, is the amount of HF that dissociates divided by the original amount:

$$f = \frac{x}{0.245M} = \frac{(9.30 \times 10^{-3}M)}{0.245M} = 0.038$$

Only a small fraction of HF dissociates, because HF is a weak acid.

**Example 15.9**   Solutions of ammonia, a weak base, suitable for cleaning windows can be bought in any supermarket. One such solution is 9.9% ammonia by mass and has a density of 0.99 g/cm³. Calculate the pH of the solution.

**Solution**   We must first convert the concentration from units of mass percent to units of molarity, since all equilibrium constant data are given in units of molarity.

$$\frac{1000 \text{ cm}^3}{1 \text{ L}} \times \frac{0.99 \text{ g}}{1 \text{ cm}^3} = \frac{990 \text{ g solution}}{1 \text{ L solution}}$$

$$\frac{990 \text{ g solution}}{1 \text{ L solution}} \times \frac{9.9 \text{ g NH}_3}{100 \text{ g solution}} \times \frac{1 \text{ mol NH}_3}{17 \text{ g NH}_3} = \frac{5.8 \text{ mol NH}_3}{1 \text{ L solution}}$$

$$= 5.8M$$

The relevant reaction is between ammonia, the base, and water, the acid:

$$NH_3(aq) + H_2O \rightleftharpoons NH_4^+(aq) + OH^-(aq) \qquad K_B = 1.79 \times 10^{-5}M$$

The rest of the calculation follows the pattern outlined in Example 15.8.

$$NH_3(aq) + H_2O \rightleftharpoons NH_4^+(aq) + OH^-(aq)$$

|  | | | |
|---|---|---|---|
| **start** | 5.8M | 0 | 0 |
| **equil** | 5.8M − x | x | x |

Substituting into the expression for the equilibrium constant gives

$$K = \frac{(x)(x)}{(5.8M - x)} = 1.79 \times 10^{-5}M$$

$$x = 1.02 \times 10^{-2}M$$

We find the value of x by making the approximation that $5.8M - x \approx 5.8M$. Since $x = [OH^-] = 1.02 \times 10^{-2}M$, the approximation is valid and pOH = 2 − log 1.02 = 2 − 0.009 = 1.99. The pH = 14 − 1.99 = 12.01.

Since this solution has a high concentration of $NH_3$, the answer is only approximate. The ideal solution approximation does not serve very well in such a concentrated solution.

The same sort of calculation carried out for neutral acids and bases in the previous two examples can be done for anionic bases or cationic acids. We must introduce an extra step in which we find the K of the relevant equilibrium from the appropriate $K_A$ or $K_B$. This extra step is based on the principle (Section 13.7) that when equations are added their equilibrium constants are multiplied to produce the equilibrium constant of the new equation.

**Example 15.10**   In a suspected case of cyanide poisoning, the medical examiner looks for burns in the mouth of the victim, caused by the alkaline cyanide ion. Find the equilibrium constant for the reaction of $CN^-(aq)$ acting as a base.

**Solution**    The relevant reaction is between cyanide ion, the base, and water, the acid:

$$CN^-(aq) + H_2O \rightleftharpoons HCN(aq) + OH^-(aq)$$

We need the value of the equilibrium constant of this reaction, which is $K_B$ of $CN^-$. Table 15.3 lists the $K_A$ of its neutral conjugate acid, HCN. We can find the relationship between the $K_B$ of $CN^-$ and the $K_A$ of HCN by manipulating reactions with known equilibrium constants to obtain the reaction between $CN^-(aq)$ and $H_2O$.

Since the relevant reaction has $CN^-(aq)$ on the left, we start by writing in reverse the reaction that defines $K_A$ of HCN:

$$CN^-(aq) + H^+(aq) \rightleftharpoons HCN(aq)$$

The equilibrium constant of this reaction is the reciprocal of $K_A$:

$$\frac{1}{K_A} = \frac{1}{4.93 \times 10^{-10}M}$$

We obtain the reaction of interest by combining this equation with the equation for the water equilibrium:

$$\begin{array}{r} CN^-(aq) + \cancel{H^+(aq)} \rightleftharpoons HCN(aq) \\ H_2O \rightleftharpoons \cancel{H^+(aq)} + OH^-(aq) \\ \hline CN^-(aq) + H_2O \rightleftharpoons HCN(aq) + OH^-(aq) \end{array}$$

Since we add these two equations, their equilibrium constants are multiplied to produce the equilibrium constant of the new equation. The equilibrium constant $K_B$ of this reaction is therefore

$$K_B = K_W \times \frac{1}{K_A} = \frac{1.00 \times 10^{-14}M^2}{4.93 \times 10^{-10}M} = 2.03 \times 10^{-5}M$$

The expression for $K_B$ that we obtained in Example 15.10 is general for the reaction of a conjugate base of a weak acid with water. It is

$$K_B = \frac{K_W}{K_A}$$

where $K_A$ is the acid dissociation constant of the weak acid related to the conjugate base.

We can use a similar method to find the equilibrium constant for the reaction of a conjugate acid of a weak base with water. For example, $NH_4^+$, the conjugate acid of $NH_3$, dissociates in water:

$$NH_4^+(aq) \rightleftharpoons NH_3(aq) + H^+(aq)$$

The $K$ of this reaction is $K_A$ of $NH_4^+$. The expression for this $K_A$ and others of its type is

$$K_A = \frac{K_W}{K_B}$$

where $K_B$ is the base dissociation constant of the weak base related to the acid.

You should note an instructive feature of these two equilibrium expressions. They can both be written as

$$K_A K_B = K_W$$

The product of $K_A$, a measure of the strength of an acid, with $K_B$, a measure of the base strength of the conjugate base of the acid, is a constant, $K_W$. Thus, as $K_A$ increases and the acid gets stronger, $K_B$ decreases and its conjugate base gets weaker.

### Measurement of $K_A$

We can often determine the dissociation constant of a weak acid by using a pH meter to measure the pH of a solution of a known concentration of the acid. Once this value is known, the pH is converted into $K_A$, the acid dissociation constant, by a calculation that is a reverse of the kind done in Examples 15.8 and 15.9.

---

**Example 15.11**

The parent compound of the barbiturates, drugs that are widely used as central nervous system depressants, is barbituric acid, an organic acid. Aside from its practical use, barbituric acid is interesting because its acidic proton is bonded to a carbon atom, rather than an oxygen atom as is usually the case. We shall symbolize barbituric acid as HBar.

A solution is prepared in which the initial concentration of HBar is $0.25M$. The pH of this solution is found to be 2.31 when equilibrium is reached. Calculate $K_A$ of HBar.

**Solution**

The relevant equilibrium and the tabulated concentrations are

$$\text{HBar(aq)} \rightleftharpoons \text{H}^+\text{(aq)} + \text{Bar}^-\text{(aq)}$$

|        | HBar(aq)     | H⁺(aq) | Bar⁻(aq) |
|--------|--------------|--------|----------|
| start  | $0.25M$      | 0      | 0        |
| equil  | $0.25M - x$  | $x$    | $x$      |

We can find the value of $x$, which is the $[\text{H}^+]$, by converting the pH to a $[\text{H}^+]$ by the method used in Example 15.7.

$$2.31 = 3 - (1 - 0.31) = 3 - 0.69$$
$$[\text{H}^+] = \text{antilog}\,[-(3 - 0.6)] = 4.9 \times 10^{-3}M = x = [\text{Bar}^-]$$

We can find the equilibrium concentrations by using this value of $x$. We can then substitute the equilibrium concentrations into the expression for $K_A$.

$$K_A = \frac{[\text{H}^+][\text{Bar}^-]}{[\text{HBar}]} = \frac{(4.9 \times 10^{-3}M)^2}{(0.25M - 4.9 \times 10^{-3}M)} = 9.8 \times 10^{-5}M$$

The strength of an anionic weak base can be measured by the same kind of experiment. However, the result is expressed as the $K_A$ of the conjugate acid of the weak base.

**Example 15.12**    Lactic acid gets its name from milk, in which it is found. Salts of this organic acid are called lactates. One salt, calcium lactate, is often used as a source of calcium for rapidly growing animals.

Calcium lactate can be represented as $Ca(Lac)_2$. A saturated solution of $Ca(Lac)_2$ contains 0.26 mol of this salt in 1.0 L of solution. The pOH of such a solution is found to be 5.60 at 373 K. Assuming the ideal solution approximation and complete dissociation of the salt, calculate $K_A$ of lactic acid.

**Solution**    The relevant equilibrium in a solution of lactate ion and the tabulated concentrations are

$$Lac^-(aq) + H_2O \rightleftharpoons HLac(aq) + OH^-(aq)$$

|  | | | |
|---|---|---|---|
| **start** | $0.52M$ | $0$ | $0$ |
| **equil** | $0.52M - x$ | $x$ | $x$ |

Since each mole of calcium lactate forms 2 mol of lactate ion, the starting concentration of lactate is $2 \times 0.26M = 0.52M$. The concentration of hydroxide ion and of lactic acid at equilibrium can be found from the pOH:

$$[OH^-] = \text{antilog}\,(-5.60) = 2.5 \times 10^{-6}M = x = [HLac]$$

We can use the approximation that $0.52M - x \approx 0.52M$ and substitute the equilibrium concentrations into the expression for $K_B$.

$$K_B = \frac{[HLac][OH^-]}{[Lac^-]} = \frac{(2.5 \times 10^{-6}M)^2}{0.52M} = 1.2 \times 10^{-11}M$$

In Example 15.10, we found the relationship

$$K_B = \frac{K_W}{K_A}$$

for an anionic conjugate base of a weak acid in water.

Thus, $K_A$ for lactic acid can be found from the value of $K_B$ calculated above:

$$1.2 \times 10^{-11}M = \frac{1.0 \times 10^{-14}M^2}{K_A}$$

$$K_A = 8.3 \times 10^{-4}M$$

The common ion effect was mentioned in our discussion of solubility products. It can also be important in acid-base systems. The dissociation of a weak acid can be repressed by the presence of a different acid in the solution, and the reaction of a base with water can be repressed by the presence of a different base.

**Example 15.13**   The artificial sweetener saccharin is a weak organic acid that can be represented as HSac. The concentration of Sac⁻ ions at equilibrium in an acidic solution is lower than in a neutral solution. Since lemon juice is acidic, less ionization of HSac occurs in tea with lemon than in plain tea.

A $2.8 \times 10^{-4}$ mol amount of saccharin is added to a glass of tea and lemon whose volume is 150 cm³ and pH is 2.00. Calculate the [Sac⁻] at equilibrium.

**Solution**   This problem differs from those presented until now because there is a significant starting concentration of H⁺. A pH of 2.0 corresponds to $[H^+] = 0.01M$. The starting $[HSac] = 2.8 \times 10^{-4}$ mol/0.150 L $= 0.0019M$.

The relevant equation is the dissociation of the weak acid:

|  | HSac(aq) | $\rightleftharpoons$ H⁺(aq) | + Sac⁻(aq) |
|---|---|---|---|
| **start** | $0.0019M$ | $0.01M$ | 0 |
| **equil** | $0.0019M - x$ | $0.01M + x$ | $x$ |

The value of $K_A$ is listed in Table 15.3 as $2.1 \times 10^{-12}M$. Since this value is small, and since the ionization of saccharin is repressed by the common ion, we can expect $x$ to be very small. Thus, we approximate that $0.01M + x \approx 0.01M$, and that $0.0019M - x \approx 0.0019M$. Using these approximations and substituting the equilibrium concentrations into the expression for $K_A$ gives

$$K_A = \frac{(0.01M)(x)}{0.0019M} = 2.1 \times 10^{-12}M$$

and $\qquad\qquad x = 4.0 \times 10^{-13}M = [Sac^-]$ at equilibrium

This value is substantially lower than [Sac⁻] in a neutral solution. The [Sac⁻] in a solution of $0.0019M$ saccharin can be calculated to be $6.3 \times 10^{-8}M$. Note that $x$ is very small compared to 0.0019, and the approximations are valid.

The pH of a solution can affect the position of an acid-base equilibrium. We can take advantage of this phenomenon by adjusting experimental conditions in advance to obtain a given composition at equilibrium.

**Example 15.14**   Aniline, $C_6H_5NH_2$, is a weak organic base. Its conjugate acid is $C_6H_5NH_3^+$, which is called the anilinium ion. The acidic proton is on the nitrogen atom in the anilinium ion. Aniline is a major intermediate in many chemical processes, especially the manufacture of synthetic dyes. The success of a process often depends on careful control of the relative concentrations of the base and its conjugate acid.

In one such process, the concentration of anilinium ion must be no higher than $1.0 \times 10^{-9}M$ in a solution that is $0.10M$ in aniline. Find the necessary concentration of sodium hydroxide to keep the $[C_6H_5NH_3^+]$ below $1.0 \times 10^{-9}M$.

**Solution**   Let the desired $[OH^-] = x$. The relevant equilibrium for a solution of a weak base is

$$C_6H_5NH_2(aq) + H_2O \rightleftharpoons C_6H_5NH_3{}^+(aq) + OH^-(aq)$$

| | | | |
|---|---|---|---|
| **start** | $0.10M$ | $0$ | $x$ |
| **equil** | $0.10M - y$ | $y$ | $x + y$ |

Here $y$ is the concentration of aniline that reacts when the system reaches equilibrium. Since the problem states that the equilibrium concentration of anilinium ion is $1.0 \times 10^{-9}M$, this is the value of $y$. Since $y$ is so small, $0.10M - y \approx 0.10M$ and $x + y \approx x$. Table 15.5 gives the value of the $K_B$ of the aniline as $4.27 \times 10^{-10}M$.

Substituting the concentrations on the "equal" line into the expression for $K_B$ gives

$$K_B = \frac{(1.0 \times 10^{-9}M)(x)}{0.10} = 4.27 \times 10^{-10}$$

and $x = 0.043M$, the concentration of hydroxide ion needed to keep the concentration of anilinium ion at the desired level.

## 15.6    NEUTRALIZATION AND TITRATIONS

By the Brønsted definitions, the term *neutralization* can be used to describe the reaction of any acid with any base. But we shall use it in a more precise sense and say that a neutralization reaction occurs when an aqueous solution of an acid stronger than water is mixed with an aqueous solution of a base stronger than water. Neutralization procedures are generally carried out with stoichiometric quantities of acid and base. In this section, we shall discuss some quantitative aspects of neutralization procedures.

Since both acids and bases can be classified as either strong or weak, there are four classes of neutralization reactions: (1) a strong acid with a strong base, (2) a strong acid with a weak base, (3) a weak acid with a strong base, and (4) a weak acid with a weak base.

1. *Strong Acid-Strong Base.* A solution of hydrochloric acid is mixed with a solution of sodium hydroxide. The strongest base in the system is $OH^-$ and the strongest acid is $H^+$. The reaction occurs between the strongest acid and the strongest base:

$$H^+(aq) + OH^-(aq) \rightleftharpoons H_2O$$

This is the net reaction that occurs when a solution of any strong acid is mixed with a solution of any strong base. It is the equation for the autoprotolysis of water, written in reverse, so its equilibrium constant is the reciprocal of $K_W$.

$$K = \frac{1}{K_W} = \frac{1}{1.0 \times 10^{-14}M^2} = 1.0 \times 10^{14}M^{-2}$$

The large value of the equilibrium constant is typical of most neutralization reactions.

2. *Weak Acid-Strong Base.* A solution of sodium hydroxide is mixed with a solution of formic acid. The strongest base in the system is $OH^-$ and the strongest acid is HCOOH, formic acid. The neutralization reaction is

$$OH^-(aq) + HCOOH(aq) \rightleftharpoons H_2O + HCOO^-(aq)$$

The equilibrium constant for this reaction is related to the $K_A$ of the weak acid and to $K_W$. If we combine the equation that defines $K_A$ with the reverse of the equation that defines $K_W$ we get

$$HCOOH(aq) \rightleftharpoons \cancel{H^+(aq)} + HCOO^-(aq) \quad K_A$$

$$\cancel{H^+(aq)} + OH^-(aq) \rightleftharpoons H_2O \qquad\qquad \frac{1}{K_W}$$

$$\overline{HCOOH(aq) + OH^-(aq) \rightleftharpoons H_2O + HCOO^-(aq) \quad K_A\left(\frac{1}{K_W}\right)}$$

The sum of the two equations is the neutralization reaction. Therefore, the product of their equilibrium constants is the $K$ of the neutralization:

$$K = \frac{K_A}{K_W}$$

The expression $K_A/K_W$ defines the equilibrium constant for the neutralization of weak acid with strong base in solution.

3. *Strong Acid-Weak Base.* A solution of nitric acid is mixed with a solution of ammonia. The strongest acid in the system is $H^+(aq)$ and the strongest base is $NH_3(aq)$. The reaction for the neutralization is

$$H^+(aq) + NH_3(aq) \rightleftharpoons NH_4^+(aq)$$

If we add the equation that defines $K_B$ to the reverse of the equation for $K_W$, we get the neutralization equation. Therefore, the equilibrium constant for this type of neutralization is $K = K_B/K_W$, where $K_B$ is the dissociation constant of the weak base.

4. *Weak Acid-Weak Base.* A solution of methyl amine is mixed with a solution of HClO. The strongest acid in the system is HClO(aq) and the strongest base is $CH_3NH_2(aq)$. The equation for the neutralization is

$$HClO(aq) + CH_3NH_3(aq) \rightleftharpoons ClO^-(aq) + CH_3NH_3^+(aq)$$

This equation is the sum of three equations whose equilibrium constants are known:

$$HClO(aq) \rightleftharpoons H^+(aq) + ClO^-(aq) \qquad\qquad K_A$$
$$CH_3NH_2(aq) + H_2O \rightleftharpoons CH_3NH_3^+(aq) + OH^-(aq) \qquad K_B$$

$$H^+(aq) + OH^-(aq) \rightleftharpoons H_2O \qquad\qquad\qquad \frac{1}{K_W}$$

The equilibrium constant for the neutralization reaction is the product of these three constants:

$$K = \frac{K_A K_B}{K_W}$$

Such an equilibrium constant often is not very large. For this reason, this type of neutralization is the least common.

This four-part classification will be useful to you in calculations relating to neutralization reactions. You will find it helpful to decide which of the four types of neutralization is occurring, and therefore which dissociation constants are needed to calculate the relevant equilibrium constant.

## Titration

Neutralization reactions are common in chemistry laboratories. Because the equilibrium constants of most neutralization reactions are large, these reactions can be treated as if they proceed to completion. Neutralization reactions often are used for quantitative analysis of acidic or basic substances.

Most often, a neutralization is carried out by a technique called *titration,* in which a measured volume of one solution is added gradually to a measured quantity of another solution. Figure 15.5 shows a typical titration apparatus. The addition of the solution is usually regulated by a buret, which is a graduated tube with a stopcock at the bottom. The stopcock is used to control the flow of the solution out of the buret, while the graduations on the buret indicate the volume of solution that has been added at any point during the titration.

For a successful neutralization, we need a way to know when the desired stoichiometric quantity of the solution in the buret is added to the other solution. There are a number of methods, all based on measuring a property of the system that changes as one solution is added to another. Either instruments or colored substances called *indicators* can be used to make these measurements. We shall discuss indicators later in this chapter.

We shall consider two quantitative factors in a neutralization: the stoichiometry of the reaction and the composition of the system when the neutralization is complete.

## Stoichiometry of Neutralization

We approach the stoichiometry of a neutralization reaction in the same way as we approach the stoichiometry of any other chemical reaction. We start with a balanced chemical equation that gives the molar ratios of the substances in the reaction. We then consider the composition of the solution after the acids and bases have been mixed and equilibrium is reached.

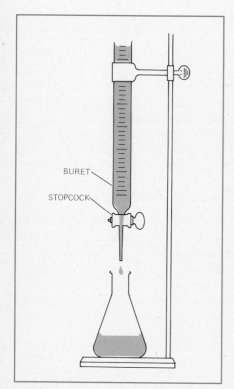

**Figure 15.5**

Titration, as it is usually done in the laboratory. The flask contains a measured volume of solution. A measured volume of another solution is added gradually. The rate of flow is controlled by the stopcock in the buret.

**Example 15.15**   A volume of 24.4 cm³ of 0.117$M$ nitric acid solution is needed to neutralize 50.1 cm³ of a sodium hydroxide solution. What is the concentration of the solution of sodium hydroxide?

**Solution**   The equation for the neutralization is $H^+(aq) + OH^-(aq) \rightleftharpoons H_2O$. One mole of protons neutralizes one mole of hydroxide ions. Therefore, the amount of protons in 24.4 cm³ of nitric acid solution is equal to the amount of hydroxide ion in 50.1 cm³ of sodium hydroxide solution. From the definition of molarity we can write

$$\text{amount (moles)} = \text{molarity} \left( \frac{\text{moles}}{\text{liter}} \right) \times \text{volume (liters)}$$

where the volume is measured in liters.
Using this relationship and the given data, we get:

$$n_{H^+} = \left( 0.117 \, \frac{\text{mol HNO}_3}{\text{L}} \right) \times \frac{1 \text{ mol H}^+}{1 \text{ mol HNO}_3} \times 24.4 \text{ cm}^3 \times \frac{1 \text{ L}}{1000 \text{ cm}^3}$$
$$= 2.85 \times 10^{-3} \text{ mol H}^+$$

From the stoichiometry, the amount of $OH^-$ in the 50.1 cm³ of sodium hydroxide solution must also be $2.85 \times 10^{-3}$ mol $OH^-$. Therefore, the concentration of the sodium hydroxide solution is

$$M = \frac{n_{OH^-}}{V} = \frac{(2.85 \times 10^{-3} \text{ mol OH}^-)}{50.1 \text{ cm}^3} \times \frac{1 \text{ mol NaOH}}{1 \text{ mol OH}^-} \times \frac{1000 \text{ cm}^3}{\text{L}}$$
$$= 0.057M$$

which is the concentration of NaOH.

Example 15.15 demonstrates a useful generalization about the neutralization of any strong acid with any strong base: The number of moles of protons must equal the number of moles of hydroxide ions. There are several ways in which this generalization can be expressed, including

$$n_{H^+} = n_{OH^-} \tag{15.1}$$
or
$$(M_{H^+})(V_{H^+}) = (M_{OH^-})(V_{OH^-}) \tag{15.2}$$

**Example 15.16**   What volume of a solution of 0.129$M$ hydrochloric acid is required to neutralize 0.237 g of barium hydroxide?

**Solution**   Since the mass of the barium hydroxide is given, we can find the amount of barium hydroxide by using the table of atomic weights:

$$n_{Ba(OH)_2} = 0.237 \text{ g Ba(OH)}_2 \times \frac{1 \text{ mol Ba(OH)}_2}{171.4 \text{ g Ba(OH)}_2}$$
$$= 1.38 \times 10^{-3} \text{ mol Ba(OH)}_2$$

Since each mole of $Ba(OH)_2$ is a source of 2 mol of hydroxide ion,

$$n_{OH^-} = (2 \text{ mol } OH^-/\text{mol } BaOH)_2)(1.38 \times 10^{-3} \text{ mol } Ba(OH)_2)$$
$$= 2.76 \times 10^{-3} \text{ mol } OH^-$$

We can combine Equations 15.1 and 15.2 to derive a relationship suitable for this example:

$$(M_{H^+})(V_{H^+}) = n_{OH^-}$$

Since HCl has only one proton to donate, the molarity of $H^+$ is the same as the molarity of the hydrochloric acid. Using the data and the relationship above gives:

$$\left(0.129 \frac{\text{mol}}{\text{L}}\right)(V_{H^+}) = 2.76 \times 10^{-3} \text{ mol}$$

and
$$V_{H^+} = 0.0214 \text{ L} = 21.4 \text{ cm}^3$$

Equations 15.1 and 15.2 can be generalized even further. Neutralizations of even weak acids or weak bases proceed essentially to completion. Therefore, the symbol $n_{H^+}$ can also be taken to mean the number of moles of available protons from the acid in the neutralization.

For example, in solution, acetic acid, a weak acid, is largely undissociated. But if a stoichiometric quantity of a strong base such as hydroxide ion is present, essentially all of the protons of the acetic acid will be transferred to the hydroxide ion. For neutralization calculations, we assume that a solution of $0.1M$ acetic acid has 0.1 mol of protons available in each liter of solution.

We can use the same reasoning for solutions of acids that have more than one available proton per molecule. In a neutralization of sulfuric acid, each mole of $H_2SO_4$ can be a source of 2 mol of protons. Similarly, 1 mol of phosphoric acid, $H_3PO_4$, can be a source of 3 mol of protons. Such polyprotic acids, as they are called, will be discussed in more detail in Section 15.8.

We can also generalize the meaning of the symbol $n_{OH^-}$. We can take it to mean either the total number of moles of available hydroxide ion in the solution or the total number of moles of protons that can be accepted by the base in the solution. Some bases can accept more than one proton per molecule. Each mole of the anionic base $CO_3^{2-}$, carbonate ion, can accept 2 mol of protons and form carbonic acid, $H_2CO_3$. Each phosphate anion, $PO_4^{3-}$, can accept up to three protons to form $H_3PO_4$.

This generalized definition of $n_{H^+}$ and $n_{OH^-}$ simplifies stoichiometric calculations of neutralization reactions between polyprotic acids and bases that can accept more than one proton.

Neutralization procedures have many applications other than the analysis of the composition of mixtures. A classic type of experiment is illustrated in the next example.

**Example 15.17**   You are asked to calculate the molecular weight of a solid weak acid of unknown composition. The acid donates one proton per molecule. When a 1.02-g sample of the acid is dissolved in water, the resulting solution requires 48.0 cm³ of a 0.241$M$ solution of sodium hydroxide for neutralization.

**Solution**   The number of moles of acid is equal to the number of moles of hydroxide ion:

$$n_{OH^-} = (M_{OH^-})(V_{OH^-}) = \left(0.241\ \frac{mol\ OH^-}{L}\right)(48.0\ cm^3)\left(\frac{1\ L}{1000\ cm^3}\right)$$

$$= 0.0116\ mol\ OH^- = n_{H^+}$$

We now know the mass and the amount (in moles) of the acid. The molar mass and molecular weight of the acid can thus be calculated:

$$\mathcal{M} = \frac{1.02\ g\ acid}{0.0116\ mol\ acid} = 88.2\ \frac{g}{mol} \qquad MW = 88.2$$

In Example 15.17, you were given the information that the unknown acid has only one proton to donate. This kind of information about an unknown acid usually is not available. In such cases, we often measure what is called the equivalent weight of the acid, the molecular weight divided by the number of protons that one molecule of the acid donates.

$$equivalent\ weight = \frac{molecular\ weight}{available\ protons\ per\ molecule}$$

In the case of a monoprotic acid, which has only one proton per molecule to donate, the molecular weight and the equivalent weight are identical. For a diprotic acid, which has two protons per molecule to donate, the equivalent weight is half the molecular weight. For an acid that has three protons per molecule to donate, the equivalent weight is one-third of the molecular weight.

We can define the equivalent weight as the "weight" of one equivalent. One *equivalent* is the quantity of acid that furnishes 1 mol of protons to a base, or the quantity of a base that can accept 1 mol of protons. There is even a concentration unit based on the equivalent. *Normality* ($N$) is analogous to molarity. The normality of a solution is the number of equivalents in one liter of solution. An experiment of the kind described in Example 15.17 can give only the equivalent weight of an acid if the composition of the acid is unknown. Only if the number of acidic protons per molecule is known does the experiment give us the molecular weight.

### The Equivalence Point

In a titration, the point at which the exact volume of the solution in the buret required for neutralization has been added to the other solution is called the *equivalence point*. The object of many neutralization calculations is to determine the equilibrium composition of the system at the equivalence point. Equivalence-point calculations use the same procedure as other equilibrium calculations. Most methods of monitoring a neutralization measure the pH at the equivalence point.

The pH at the equivalence point in a neutralization is determined by the nature of the acid and the base. In the neutralization of any strong acid with any strong base, the pH at the equivalence point is 7.00 and the solution is neutral. In the neutralization of a weak acid by a strong base, the pH at the equivalence point is greater than 7 and the solution is alkaline; the exact pH depends on the concentrations of the solutions and on the nature of the weak acid. In the neutralization of a strong acid by a weak base, the pH of the solution at the equivalence point is less than 7 and the solution is acidic.

**Example 15.18**

Calculate the pH at the equivalence point when a solution of $0.10M$ HF is titrated with a solution of $0.10M$ NaOH. Calculate the pH at the equivalence point when a $0.10M$ solution of NaF is titrated with a $0.10M$ solution of HCl.

**Solution**

The two parts of the problem are related. In each part, the first step is to write the relevant reaction, the starting conditions, and the equilibrium line. In both parts, we must be careful to correct the concentrations of the original solutions to account for the increased volume of the solution at the equivalence point.

The solutions of the acids and of the bases have the same initial concentrations. Therefore, equal volumes of each solution are required for neutralization. As a result, the volume of the solution at the equivalence point is twice the original volume of either solution, and the concentrations are half those of the original solutions.

$$HF(aq) + OH^-(aq) \rightleftharpoons F^-(aq) \quad + H_2O$$

| | | | |
|---|---|---|---|
| **start** | $\dfrac{0.10M}{2}$ | $\dfrac{0.10M}{2}$ | $0$ |

The value of $K$ for this reaction is

$$K = \frac{K_A}{K_W} = \frac{3.53 \times 10^{-4}M}{1.0 \times 10^{-14}M^2} = 3.53 \times 10^{10}M^{-1}$$

which we find by using the data in Table 15.3. The value of $K$ is large for this reaction, so a completion step is helpful:

$$HF(aq) + OH^-(aq) \rightleftharpoons F^-(aq) \quad + H_2O$$

| | | | |
|---|---|---|---|
| **complete** | $0$ | $0$ | $0.050M$ |
| **equil** | $x$ | $x$ | $0.050M - x$ |

As usual, the concentration of water is not included.

From the concentration terms on the equilibrium line and the expression for $K$, we obtain:

$$K = \frac{(0.050M - x)}{x^2} = 3.53 \times 10^{10}M^{-1}$$

Making the approximation that $0.050M - x \approx 0.050M$, we find that

$$x = 1.19 \times 10^{-6}M = [OH^-]$$

and the approximation is valid.

$$pOH = 6 - \log 1.19 = 5.92$$
$$pH = 14 - pOH = 14 - 5.92 = 8.08$$

The second part of the problem is handled in the same way. The relevant equation and tabulated data are

$$F^-(aq) + H^+(aq) \rightleftharpoons HF(aq)$$

|        | | |
|--------|--------|--------|
| **start** | $\dfrac{0.10M}{2}$ | $\dfrac{0.10M}{2}$ | 0 |

The value of $K$ for this reaction, which is the reverse of acid dissociation, is

$$K = \frac{1}{K_A} = \frac{1}{3.53 \times 10^{-4}M} = 2.83 \times 10^3 M^{-1}$$

$$F^-(aq) + H^+(aq) \rightleftharpoons HF(aq)$$

|             |     |     |            |
|-------------|-----|-----|------------|
| **complete** | 0   | 0   | $0.050M$   |
| **equil**    | $x$ | $x$ | $0.050M - x$ |

$$K = \frac{(0.050M - x)}{x^2} = 2.83 \times 10^3 M^{-1}$$

Since $x$ is small, we try the approximation that $0.050M - x \approx 0.050M$. We find that $x = 4.2 \times 10^{-3}M$, almost 10% of 0.050. The approximation is not a good one. We can get the exact solution using the quadratic formula:

$$x = 4.03 \times 10^{-3}M = [H^+]$$
$$pH = 3 - \log 4.03 = 2.39$$

## Indicators

Clearly, in a neutralization procedure we need some way to detect the equivalence point. The most direct method is to use a pH meter to measure the pH. But a simpler and cheaper method is to use a substance called an **indicator.** An indicator usually is a highly colored organic molecule of complex structure, whose color in aqueous solution changes as the pH of the solution changes.

Tea is an everyday example of an indicator. Its color changes as an acid,

**TABLE 15.8**  Acid-Base Indicators

| Common Name | $pK_{In}$ | pH Interval[a] | Color Acid | Alkaline |
|---|---|---|---|---|
| cresol red[b] |  | 0.2–1.8 | red | yellow |
| thymol blue[b] | 1.6 | 1.2–2.8 | red | yellow |
| tropeoline OO |  | 1.3–3.0 | red | yellow |
| methyl yellow | 3.3 | 2.8–4.0 | red | yellow |
| bromphenol blue | 3.8 | 3.0–4.6 | yellow | purple |
| methyl orange | 3.5 | 3.1–4.4 | red | yellow |
| bromcresol green | 4.7 | 3.8–5.4 | yellow | blue |
| methyl red | 5.0 | 4.2–6.2 | red | yellow |
| chlorphenol red | 6.0 | 4.8–6.4 | yellow | red |
| bromcresol purple | 6.1 | 5.2–6.8 | yellow | purple |
| bromthymol blue | 7.1 | 6.0–7.6 | yellow | blue |
| phenol red | 7.8 | 6.4–8.0 | yellow | red |
| neutral red | 6.8 | 6.8–8.0 | red | yellow-brown |
| cresol red[c] | 8.1 | 7.2–8.8 | yellow | red |
| cresol purple[c] | 8.3 | 7.4–9.0 | yellow | purple |
| thymol blue[c] | 8.9 | 8.0–9.6 | yellow | blue |
| phenolphthalein | 9.3 | 8.0–9.8 | colorless | red-violet |
| thymolphthalein |  | 9.3–10.5 | colorless | blue |
| alizarin yellow |  | 10.1–12.0 | yellow | violet |

[a] Will vary with the observer.
[b] Acid range; the indicator has two color change intervals.
[c] Alkaline range; the indicator has two color change intervals.
Source: Gilbert H. Ayres, *Quantitative Chemical Analysis,* Second Edition, New York: Harper & Row, 1968.

lemon juice, is added. Many indicators are used in the laboratory. We can represent any one of them as a weak acid of the general formula HIn, where In represents indicator. Like any weak acid, an indicator dissociates in aqueous solution:

$$HIn(aq) \rightleftharpoons H^+(aq) + In^-(aq)$$

If HIn and its conjugate base, In$^-$, have different colors, the relative quantities of acid and conjugate base determine the color of a solution. Since these relative quantities depend on the [H$^+$], the color of the solution changes as the pH changes.

Perhaps the best-known laboratory indicator is litmus, whose acid form, HIn, is pink, and whose base form, In$^-$, is blue. According to the principle of Le Chatelier, when the [H$^+$] is high, the dissociation of HIn is repressed. The relatively high [HIn] causes the solution to appear pink. When the [H$^+$] is low, the dissociation of HIn is enhanced, and the relatively high [In$^-$] causes the solution to appear blue.

Suppose we are titrating a solution of an acid with a solution of a base, using litmus as an indicator. The pH increases, until at some point we see the color of the indicator change from pink to blue. We call this point the *end point* of the titration. We can define the exact point of the color

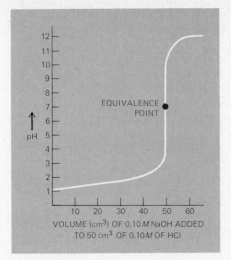

**Figure 15.6**

A titration curve for the neutralization of a strong acid with a strong base. A small volume of titrant near the equivalence point causes a large change in pH.

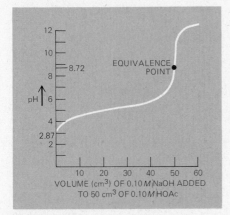

**Figure 15.7**

A titration curve for the neutralization of a weak acid with a strong base. The slope of the curve is not as steep at the equivalence point as the slope of the curve for the neutralization of a strong base with a strong acid.

change as the pH at which $[HIn] = [In^-]$. (However, our eyes are not sensitive enough to detect very subtle color changes. We actually see a color change over a pH interval that depends on our ability to detect one color in the presence of another.) The pH at the point where $[HIn] = [In^-]$ depends on the value of the $K_A$ for the indicator equilibrium. This equilibrium constant is sometimes symbolized by $K_{In}$.

$$K_{In} = \frac{[H^+][In^-]}{[HIn]}$$

At the color change, $[In^-] = [HIn]$, and this expression becomes

$$K_{In} = [H^+] \quad \text{or} \quad pK_{In} = pH$$

In other words, if the $pK_{In}$ of an indicator is equal to the pH at the equivalence point, the indicator changes color at the equivalence point. Table 15.8 lists some indicators and their $pK_{In}$ values.

In Example 15.18, a suitable indicator for the titration of HF by NaOH is one with $pK_{In} = 8.08$, the pH of the solution at the equivalence point. Similarly, a suitable indicator for the titration of NaF with HCl is one with $pK_{In} = 2.38$.

Some problems are possible in the use of indicators. It may not be possible to find an indicator with a $pK_{In}$ and an end point that coincide exactly with the pH at the equivalence point. The experimenter's color perception may not be perfect, so the color change is not detected exactly at the equivalence point. Given these sources of imprecision, will there be a large difference between the measured volume at the end point and the actual volume at the equivalence point?

We can answer this question by studying how the pH of a solution changes as the titration proceeds. We can calculate the pH during a titration and present the results graphically. Such a plot of pH against volume of added solution is called a *titration curve*. The curve shown in Figure 15.6 describes the titration of a strong acid such as HCl with a strong base such as NaOH. Note the steepness of the slope around the equivalence point. The fact that the curve is so steep means that very large changes in pH are caused by small volumes of added solution. Therefore, a large error in monitoring the pH will cause only a small error in measuring the volume of added solution.

Titration curves for the neutralization of a weak acid with a strong base (Figure 15.7) or of a weak base with a strong acid are not as steep around the equivalence point. In these titrations, therefore, the use of an indicator whose $pK_{In}$ is relatively close to the pH at the equivalence point is desirable.

**Example 15.19**

Ephedrine, $C_{10}H_{15}NO$, is a weak organic base. It is a central nervous system stimulant that is often used in nasal sprays because it is also a powerful decongestant. Table 15.5 lists the value of the $K_B$ of ephedrine as $1.38 \times 10^{-4}M$. To monitor the composition of nasal sprays, solutions of ephedrine in water whose concentrations are around $0.2M$ can be titrated with hydrochloric acid solutions of about the same concentration. Which of the indicators listed in Table 15.8 is most suitable for use in this titration?

**Solution**    The most suitable indicator for the titration is one whose $pK_{In} = pH$ at the equivalence point. We can calculate the pH at the equivalence point, using Eph to represent ephedrine and $EphH^+$ to represent its conjugate acid.

$$Eph(aq) + H^+(aq) \rightleftharpoons EphH^+(aq)$$

| | | | |
|---|---|---|---|
| **start** | $\dfrac{0.2M}{2}$ | $\dfrac{0.2M}{2}$ | 0 |

$$K = \frac{K_B}{K_W} = \frac{1.4 \times 10^{-4}M}{1.0 \times 10^{-14}M^2} = 1.4 \times 10^{10}M^{-1}$$

| | | | |
|---|---|---|---|
| **complete** | 0 | 0 | $0.1M$ |
| **equil** | $x$ | $x$ | $0.1M - x$ |

Since $K = 1.4 \times 10^{10}M^{-1}$, $x$ is small, and we approximate: $0.1M - x \approx 0.1M$. Therefore

$$\frac{0.1M}{x^2} = 1.4 \times 10^{10}M^{-1}$$

and                                $x = 2.7 \times 10^{-6}M = [H^+]$

and the approximation is valid.

$$pH = 6 - \log 2.7 = 5.6$$

The indicator whose $pK_{In}$ is closest to this pH is chlorphenol red, with $pK_{In} = 6.0$.

## 15.7   BUFFERS

While pure water has a pH of 7, it is almost impossible to obtain a sample whose pH is even close to 7. Very small traces of dissolved acidic or basic impurities change the pH of water substantially, and such impurities are present in virtually every sample of water.

When water is exposed to air, some of the carbon dioxide in the atmosphere dissolves. When such a solution is saturated, $[CO_2]$ is about $1 \times 10^{-5}M$. As we shall see in Section 15.8, carbon dioxide acts as a weak acid when it is dissolved in water. The pH of water that is saturated with respect to the carbon dioxide in the air is about 5.7.

Changes in pH reflect exponential changes in $[H^+]$. Many chemical processes, including a large number of reactions in aqueous solutions, are sensitive to variations in $[H^+]$. Many reactions proceed rapidly at a given

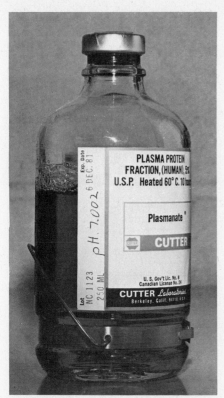

The importance of pH in physiological solutions is demonstrated by the prominent display of the pH of a unit of blood plasma destined for transfusion. *(Lester Bergman & Assoc.)*

pH but much more slowly if the pH is several units higher or lower. In some reactions, variations in the pH of a solution cause changes in the products.

Many reactions in living organisms proceed normally only in a narrow pH range. A change in the pH of the aqueous solutions in a living organism can be fatal. Living organisms thus need a mechanism to keep the pH of their physiological solutions relatively constant.

The method used by a chemist to prevent large variations in pH even when relatively large amounts of acid or base are added to a system is essentially the same as the method used in biological systems. In a living organism, as in a laboratory, the way to minimize the decrease in the pH of a solution when acid is added to a system is to have some base already present in the system to neutralize the acid. Similarly, the way to minimize the increase in the pH when base is added is to have some acid in the system. In other words, a system will keep a relatively constant pH if it already contains a quantity of acid and a quantity of base.

*A solution that is relatively resistant to changes in pH is called a buffer solution.* We can also say that the solution is buffered. The simplest way to achieve substantial concentrations of both acid and base in the same solution is to use a conjugate acid-base pair. The most common form of buffer solution contains a weak acid and its conjugate base or a weak base and its conjugate acid. *The chemical reaction that is conveniently used to calculate the equilibrium composition or chemical behavior of such buffer solutions is either the one that defines $K_A$, the dissociation constant of the weak acid, or the one that defines $K_B$, the dissociation constant of the weak base.*

For example, a solution can be prepared from formic acid, a weak acid, and sodium formate, a salt of its conjugate base. Transfer of a proton from formic acid to formate ion does not result in a change in equilibrium concentrations.

$$HCOOH(aq) + HCOO^-(aq) \rightleftharpoons HCOO^-(aq) + HCOOH(aq)$$

Therefore the composition of an aqueous solution of formic acid and formate ion at equilibrium — or, more generally, of any aqueous solution of a weak acid and its conjugate base — depends primarily on the quantities of each that are originally added to the solution.

The buffer solution consisting of formic acid and sodium formate is most conveniently described by the reaction that defines $K_A$ of formic acid:

$$HCOOH(aq) \rightleftharpoons HCOO^-(aq) + H^+(aq)$$

The pH of a buffer solution is determined mainly by two factors: the value of $K_A$ of the weak acid (or $K_B$ of the weak base) and the ratio of the concentrations of the weak acid and its conjugate base.

The action of a buffer can be understood qualitatively if we apply the principle of Le Chatelier. If acid ($H^+$) is added to the formic acid-formate buffer, the pH of the solution does not change appreciably, since the acid dissociation equilibrium reaction of formic acid proceeds from right to left, consuming most of the added $H^+$. If the amount of added $H^+$ is not too large, the ratio of formic acid to formate does not change very much. If base is added to the buffer, some of the $H^+$ in the solution is consumed by reaction with the base. But the acid dissociation reaction proceeds from left to right to restore most of the lost $H^+$ and return the system to equilibrium. Again, if the amount of added base is not too large, the ratio of formic acid to formate does not change appreciably.

**Example 15.20**

Calculate the pH of a buffer solution prepared by dissolving 40.2 g of $NH_4Cl$ and 20.3 g of $NH_3$ in enough water to form 1.00 L of solution.

**Solution**

Since $NH_4Cl$ is a salt that is a source of $NH_4^+$ cations, the relevant equilibrium is

$$NH_3(aq) + H_2O \rightleftharpoons NH_4^+(aq) + OH^-(aq)$$

From the data in Table 15.5,

$$K_B = \frac{[NH_4^+][OH^-]}{[NH_3]} = 1.79 \times 10^{-5}M$$

From data given in the problem and the appropriate molar masses:

$$[NH_4^+] = 40.2 \text{ g } NH_4Cl \times \frac{1 \text{ mol } NH_4Cl}{53.5 \text{ g } NH_4Cl} \times \frac{1}{1 \text{ L}} = 0.751M$$

$$[NH_3] = 20.3 \text{ g } NH_3 \times \frac{1 \text{ mol } NH_3}{17.0 \text{ g } NH_3} \times \frac{1}{1 \text{ L}} = 1.19M$$

The $[OH^-]$ can be calculated by substitution of this concentration data into the equilibrium expression for $K_B$ of $NH_3$:

$$\frac{(0.751M)[OH^-]}{(1.19M)} = 1.79 \times 10^{-5}M$$

$$[OH^-] = 2.84 \times 10^{-5}M$$
$$pOH = 4.55, pH = 9.45$$

A full-fledged equilibrium calculation is not necessary in a problem of this type. If we did such a calculation:

| | $NH_3(aq) + H_2O \rightleftharpoons$ | $NH_4^+(aq)$ | $+ OH^-(aq)$ |
|---|---|---|---|
| **start** | $1.19M$ | $0.751M$ | $0$ |
| **equil** | $1.19M - x$ | $0.751M + x$ | $x$ |

we would approximate that $1.19 - x \approx 1.19$ and $0.751M + x \approx 0.751$. These approximations would lead to the same expression for the $[OH^-]$ that we had by direct substitution.

Example 15.20 illustrates how the pH of a buffer solution depends on the *ratio* of the concentrations of the acid and the base, not on their absolute quantities. In Example 15.20 the ratio of the concentration of ammonium to ammonia is $0.751:1.19$, and the pH of the buffer is 9.45. Any solution containing these two substances in this ratio will have a pH = 9.45, assuming the ideal solution approximation is valid.

## The Henderson-Hasselbalch Equation

The relationship between the pH of the solution, the $pK_A$ of the weak acid used to prepare it, and the ratio of the concentrations of the weak acid, HA, and its conjugate base, $A^-$, can be expressed by conversion of the equilibrium expression that defines $K_A$ into negative logarithmic form:

$$K_A = \frac{[H^+][A^-]}{[HA]}$$

or, rearranging terms:

$$[H^+] = \frac{K_A[HA]}{[A^-]}$$

Taking the negative logarithms of both sides of this equation gives

$$-\log [H^+] = -\log \left(\frac{K_A[HA]}{[A^-]}\right) = -\log K_A + \log \frac{[A^-]}{[HA]}$$

or
$$pH = pK_A + \log \left(\frac{[A^-]}{[HA]}\right) \tag{15.3}$$

Since a buffer solution must maintain a relatively constant pH when either acid or base is added, it is usually desirable for the buffer to have roughly comparable concentrations of the weak acid and its conjugate base. In the best case, when $[HA] = [A^-]$, $\log [A^-]/[HA] = 0$, and pH = $pK_A$.

Equation 15.3 is sometimes called the Henderson-Hasselbalch equation. It is a chemically valid approximation when the ratio of weak acid to its conjugate base is neither very large nor very small, that is, when the pH of the solution does not differ greatly from the $pK_A$ of the acid and the ratio of $[A^-]:[HA]$ is between 0.1 and 10.

The Henderson-Hasselbalch equation is important to biologists because their experiments often require solutions that are buffered at or close to a given pH. For a buffer to contain roughly comparable concentrations of the weak acid and its conjugate base, a weak acid with a $pK_A$ close to the pH desired for the buffer solution is needed.

**Example 15.21** A buffer solution of pH = 3.00 is needed. Using the data in Table 15.3, suggest an appropriate weak acid-conjugate base pair, and determine the ratio of the concentrations of the two substances in the solution at this pH.

**Solution**

A pH of 3.00 corresponds to $[H^+] = 1.0 \times 10^{-3}M$. We should select a weak acid with a $K_A$ as close to $1.0 \times 10^{-3}M$ as possible. Chloroacetic acid, $ClCH_2COOH$, has $K_A = 1.40 \times 10^{-3}$, $pK_A = 2.85$, and is therefore suitable. Substituting into Equation 15.3:

$$3.00 = 2.85 + \log \frac{[ClCH_2COO^-]}{[ClCH_2COOH]}$$

$$\frac{[ClCH_2COO^-]}{[ClCH_2COOH]} = 1.4$$

If we select a weak acid whose $pK_A$ differs by more than 1.0 from the pH desired for the buffer solution, one of the two components used to prepare the buffer will have to be present in at least tenfold excess. Working with buffers solutions that have disparate concentrations of the two conponents usually is quite inconvenient.

## Buffer Capacity

The *capacity* of a buffer refers to the amount of acid or base that can be added to the buffer solution without changing its pH appreciably. The capacity of the buffer solution is related to the quantities of the weak acid and its conjugate base that are used to prepare the solution. A solution of $0.1M$ HCOOH and $0.1M$ HCOO$^-$ has roughly one-tenth the capacity of a buffer solution of $1M$ HCOOH and $1M$ HCOO$^-$, even though both solutions have the same pH. The pH of the buffer solution depends on the *ratio* of the two components, their relative amounts. The capacity of the buffer solution depends on the *concentration* of the two components, their absolute amounts.

We must often calculate the change in the pH of a buffer solution caused by the addition of a given amount of an acid or a base.

**Example 15.22**

Proteins play an important role in buffering the blood, lymph, and other physiological fluids of animals. Proteins are large molecules, portions of which act as weak acids or weak bases. A weak organic nitrogen-containing base called imidazole is found in protein molecules and often plays a role in their buffering action. Imidazole can accept a proton to form the imidazolium ion, in the same way that ammonia can accept a proton to form the ammonium ion. The base dissociation equilibrium can be symbolized as

$$ImidN(aq) + H_2O \rightleftharpoons ImidNH^+(aq) + OH^-(aq)$$

where ImidNH$^+$ is the imidazolium ion and ImidN is imidazole. Table 15.5 gives $K_B$ as $9.01 \times 10^{-8}M$.

A study of the complex protein buffer system could start with a study of the relatively simple imidazole system. A buffer prepared from the imidazolium ion–imidazole acid-base system ideally has equal concentrations of both substances. A simple way to equalize the concentrations is to start with a solution of imidazole in water and to neutralize half of the dissolved imidazole with a strong acid. If half of a quantity of imidazole is converted to the imidazolium ion by the strong acid, one obtains the ideal buffer solution, containing equal concentrations of acid and base.

A solution of 1.00 mol of imidazole and 0.50 mol of $HNO_3$ in a 1.00-L volume of solution is prepared. Calculate the pH of this solution. A sample of 4.00 g of NaOH is dissolved in the solution. Calculate the pH of this new solution.

**Solution**  Since the amount of strong acid, $HNO_3$, is half the amount of imidazole, half of the imidazole is converted to the imidazolium ion. At equilibrium the concentration of imidazole is equal to the concentration of imidazolium ion: [ImidN] = [ImidNH$^+$] = 0.50$M$.

$$K_B = \frac{(0.50M)[OH^-]}{(0.50M)} = 9.01 \times 10^{-8}M$$

Therefore the [OH$^-$] is equal to $K_B$.

$$pOH = 7.05, \quad pH = 6.95$$

When NaOH is added to the solution, the reaction between the imidazolium ion, the acid of the buffer, and the hydroxide ion neutralizes essentially all of the added hydroxide ion. As a result, the ratio of imidazole to imidazolium ion changes. The amount of hydroxide added is

$$4.00 \text{ g NaOH} \times \frac{1 \text{ mol NaOH}}{40.0 \text{ g NaOH}} = 0.100 \text{ mol NaOH}$$

Since 0.10 mol of OH$^-$ consumes 0.10 mol of imidazolium ion to form 0.10 mol of imidazole, the concentration of imidazolium ion after it reacts with the hydroxide ion is $0.50M - 0.10M = 0.40M$, and the concentration of imidazole is $0.50M + 0.10M = 0.60M$.

$$\frac{(0.40M)[OH^-]}{(0.60M)} = 9.01 \times 10^{-8}M$$

$$[OH^-] = 1.4 \times 10^{-7}M \quad pOH = 6.87 \quad pH = 7.13$$

Because the solution is buffered, the addition of a substantial amount of a strong base increases the pH by only 0.18. The same quantity of sodium hydroxide would change the pH of a liter of pure water from 7 to 13.

Two other buffer systems are also found in the fluids of living organisms. The more important of these systems is based on dissolved carbon dioxide and bicarbonate ion, $HCO_3^-$. Carbon dioxide might not appear to be an acid, since it has no proton to donate. But when $CO_2$ is dissolved in water, a small portion of it forms carbonic acid, $H_2CO_3$, by the reaction

$$CO_2(aq) + H_2O \rightleftharpoons H_2CO_3(aq)$$

whose equilibrium constant, $K$, is estimated to be $3.7 \times 10^{-3}$.

The conjugate base of carbonic acid is the bicarbonate ion, $HCO_3^-$:

$$H_2CO_3(aq) \rightleftharpoons HCO_3^-(aq) + H^+(aq)$$

The overall reaction relating carbon dioxide and bicarbonate ion is the sum of these two reactions:

$$CO_2(aq) + H_2O \rightleftharpoons HCO_3^-(aq) + H^+(aq)$$

This equilibrium is used for calculations in this buffer system. The value of the equilibrium constant, $K$, for this reaction is $4.30 \times 10^{-7} M$; $pK = 6.37$. The $pK$ of the equilibrium for this buffer system as well as the $pK$ for the imidazole equilibrium are both close to 7.

A second class of physiological buffers is based on the dihydrogen phosphate anion, $H_2PO_4^-$, which is a weak acid, and $HPO_4^{2-}$, its conjugate base. The acid dissociation equilibrium of the dihydrogen phosphate anion is

$$H_2PO_4^-(aq) \rightleftharpoons HPO_4^{2-}(aq) + H^+(aq)$$

$K = 6.23 \times 10^{-8}$ and $pK = 7.21$. Again, the $pK$ is close to physiological pH.

Both these buffer systems are complex, because their components are part of acid-base systems that can transfer more than one proton. Such systems are the subject of the next section.

## 15.8   POLYPROTIC ACIDS

*Polyprotic acids* have more than one proton per molecule to donate. They include sulfuric acid, $H_2SO_4$, a diprotic acid; carbonic acid, $H_2CO_3$, also a diprotic acid; and phosphoric acid, $H_3PO_4$, a triprotic acid.

One way to study equilibria of polyprotic acids is to assume that the acid molecule donates one proton at a time to the base. The acid is assumed to transfer its first proton completely before any transfer of the second proton begins. An example is $H_2S$, hydrogen sulfide, which is a diprotic acid in water. Two equilibria, each for the transfer of a single proton, are written to describe its behavior:

$$H_2S(aq) \rightleftharpoons HS^-(aq) + H^+(aq)$$
$$HS^-(aq) \rightleftharpoons S^{2-}(aq) + H^+(aq)$$

The equilibrium constant for the first equilibrium, the transfer of the first proton from $H_2S$, is designated $K_1$. The equilibrium constant for the second reaction is designated $K_2$. Thus, $K_2$ of $H_2S$ is the equilibrium constant of a reaction that does not include $H_2S$. The reactant is $HS^-$, the

## BREATH, BLOOD, AND pH

"Take a deep breath" is a common bit of advice for someone in trouble. But there are certain circumstances in which deep breathing can cause trouble. Hyperventilation—breathing too deeply for too long—can cause symptoms that mimic heart disease and that can bring sudden death to divers.

When a swimmer takes a number of deep breaths before diving into the water, several things happen inside the body. More air flows into the lungs, and the composition of the gases in the lungs changes. The partial pressure of $O_2$ goes up and the partial pressure of $CO_2$ goes down. Breathing is controlled by the nerve cells in the carotid body, in the carotid artery of the head, which monitors blood $O_2$, $CO_2$, and pH. As the blood content of $O_2$ goes up and the $CO_2$ concentration goes down, the pH rises, since a loss of $CO_2$ decreases the carbonic acid content of the blood plasma.

The effect of this alkalosis, as it is called, can be a constriction of the blood vessels, including those in the brain. At the same time, the reduced $CO_2$ blood level and increased $O_2$ level affect the carotid body so that the swimmer can wait much longer before breathing. Too long, in fact, because the $O_2$ pressure may fall below the level needed to sustain consciousness. There have been several cases recorded of divers who have suddenly lost consciousness in this way. Unless immediate help is available, the swimmer may simply sink to the bottom and drown.

Even for persons on dry land, a small change in blood pH can cause problems. The normal pH of arterial blood is about 7.40. An increase to only 7.45 can be produced by the wrong kind of breathing habits. If an individual breathes too often and too deeply, $CO_2$ is lost from the body and arterial pH rises. The resulting symptoms can include light-headedness, agitation, a burning or prickling sensation in the limbs, fainting, and severe chest pains.

In some cases, the pains are so severe that the individual is diagnosed as having heart disease. The usual diagnosis is angina pectoris, a condition that causes excruciating chest pains. In true angina, the pain results from partial blockage of the coronary arteries, which reduces the flow of blood to the heart muscle. In false angina, the pain can result from hyperventilation.

If hyperventilation is known to be the problem, the cure may be as simple as having the patient breathe into a paper bag. Exhaled breath has a higher pressure of $CO_2$ than ordinary air, so exhaled air helps bring the blood $CO_2$ level back to normal. But for individuals who hyperventilate by force of habit, weeks or even months of training may be needed to restore a normal breathing pattern that will keep blood pH within the normal range.

conjugate base of $H_2S$. The hydrogen sulfide anion, $HS^-$, is *amphoteric*. It can accept a proton and return to $H_2S$, so it can be a base. It can donate a proton and form $S^{2-}$, so it can be an acid. All anions formed from the transfer of some of the acidic protons of a polyprotic acid are amphoteric.

Phosphoric acid is a triprotic acid. Thus there are three equilibria and three equilibrium constants to consider in a description of its behavior and the behavior of anions derived from it:

$$H_3PO_4(aq) \rightleftharpoons H_2PO_4^-(aq) + H^+(aq) \qquad K_1$$
$$H_2PO_4^-(aq) \rightleftharpoons HPO_4^{2-}(aq) + H^+(aq) \qquad K_2$$
$$HPO_4^{2-}(aq) \rightleftharpoons PO_4^{3-}(aq) + H^+(aq) \qquad K_3$$

Both $H_2PO_4^-$ and $HPO_4^{2-}$ are amphoteric.

Sulfuric acid is unusual among polyprotic acids because it is a strong acid with respect to the loss of its first proton. A solution of sulfuric acid starts as $H^+$ and $HSO_4^-$. But the $HSO_4^-$ anion is a relatively weak acid:

$$HSO_4^-(aq) \rightleftharpoons SO_4^{2-}(aq) + H^+(aq) \qquad K_2 = 1.20 \times 10^{-2}M$$

Table 15.9 lists acid dissociation constants for some polyprotic acids. Although the symbols for these constants have numerical subscripts, each one is actually an acid dissociation constant and thus can be regarded as a $K_A$. For all polyprotic acids, $K_1$ is greater than $K_2$. For a triprotic acid, $K_2$ is greater than $K_3$. The loss of a proton by the acid creates a negative charge. The loss of a second proton adds more negative charge. It is more unfavorable to add negative charge to an entity that is already negatively charged than to a neutral entity. Thus, each successive proton is substantially more difficult to remove. It is for this reason that polyprotic acids seem to transfer one proton at a time.

Equilibrium calculations for systems including polyprotic acids generally are more complicated than those for monoprotic acids. The calculations are manageable if suitable approximations are made, as we show in the next example.

**TABLE 15.9**  Dissociation Constants of Polyprotic Acids at Room Temperature

| Inorganic Acids | | $K\ (M)$ |
|---|---|---|
| carbon dioxide, $CO_2$ or | $K_1$ | $4.30 \times 10^{-7}$ |
| carbonic acid, $H_2CO_3$ | $K_2$ | $5.61 \times 10^{-11}$ |
| chromic acid, $H_2CrO_4$ | $K_1$ | $1.8\ \times 10^{-1}$ |
| | $K_2$ | $3.20 \times 10^{-7}$ |
| hydrogen sulfide, $H_2S$ | $K_1$ | $1.1\ \times 10^{-7}$ |
| | $K_2$ | $1.0\ \times 10^{-12}$ |
| phosphoric acid, $H_3PO_4$ | $K_1$ | $7.52 \times 10^{-3}$ |
| | $K_2$ | $6.23 \times 10^{-8}$ |
| | $K_3$ | $4.7\ \times 10^{-13}$ |
| phosphorous acid, $H_3PO_3$ | $K_1$ | $1.0\ \times 10^{-2}$ |
| | $K_2$ | $2.6\ \times 10^{-7}$ |
| selenic acid, $H_2SeO_4$ | $K_1$ | large |
| | $K_2$ | $1\ \ \times 10^{-2}$ |
| sulfuric acid, $H_2SO_4$ | $K_1$ | large |
| | $K_2$ | $1.20 \times 10^{-2}$ |
| sulfurous acid, $H_2SO_3$ | $K_1$ | $1.54 \times 10^{-2}$ |
| | $K_2$ | $1.02 \times 10^{-7}$ |

| Organic Acids | | |
|---|---|---|
| adipic acid | $K_1$ | $3.80 \times 10^{-5}$ |
| | $K_2$ | $3.89 \times 10^{-6}$ |
| citric acid | $K_1$ | $8.4\ \times 10^{-4}$ |
| | $K_2$ | $1.8\ \times 10^{-5}$ |
| | $K_3$ | $5.0\ \times 10^{-7}$ |
| oxalic acid | $K_1$ | $5.36 \times 10^{-2}$ |
| | $K_2$ | $5.42 \times 10^{-5}$ |

**Example 15.23**

A 0.20-mol amount of $H_2S(g)$ is dissolved in water to form a solution of volume 2.00 L. Calculate the concentrations of $H_2S$ and all the ions in the solution at equilibrium, using the data in Table 15.9.

**Solution**

At equilibrium, the solution, like any aqueous solution, contains $H^+$ and $OH^-$ ions in addition to dissolved $H_2S$. Since $H_2S$ is an acid, we can predict that the $[H^+]$ is greater than $1.0 \times 10^{-7}M$ and that the $[OH^-]$ is less than $1.0 \times 10^{-7}M$. The concentrations of these ions at equilibrium are related to the acid strength of $H_2S$—that is, to the magnitudes of $K_1$ and $K_2$. In addition, the transfer of protons from the $H_2S$ molecules to water molecules results in the formation of $HS^-$ and $S^{2-}$ anions. The concentrations of these anions are also related to the magnitudes of $K_1$ and $K_2$.

As always, the first step in the equilibrium calculation is to write the relevant equation. We could use the equilibrium that defines $K_1$, the equilibrium that defines $K_2$, or an equilibrium that combines them, such as:

$$H_2S(aq) \rightleftharpoons S^{2-}(aq) + 2H^+(aq) \qquad K = K_1K_2$$

Unfortunately, none of these equilibrium reactions includes all the species that interest us. We must either use a number of equilibrium reactions simultaneously or work with the reaction that is most important, leaving the others for later consideration. Whenever possible, we shall use the second approach.

The equilibrium that seems most important is the one that has the largest equilibrium constant and has as reactants species present in appreciable concentration. For $H_2S$ in solution, it is the equilibrium that defines $K_1$. The relevant reaction is

| | $H_2S(aq)$ | $\rightleftharpoons$ | $HS^-(aq)$ | $+ H^+(aq)$ |
|---|---|---|---|---|
| **start** | $\dfrac{0.20 \text{ mol}}{2.00 \text{ L}}$ | | 0 | 0 |
| **equil** | $0.10M - x$ | | $x$ | $x$ |

The expression for $K_1$ gives

$$K_1 = \frac{[H^+][HS^-]}{[H_2S]} = \frac{(x)(x)}{(0.10 - x)} = 1.1 \times 10^{-7}M$$

Making the usual approximation for a weak acid that $0.10M - x \approx 0.10M$, and solving, $x = 1.0 \times 10^{-4}M$, and the approximation is valid. $[H^+] = [HS^-] = 1.0 \times 10^{-4}M$; $[H_2S] = 0.10M - 1.0 \times 10^{-4}M = 0.10M$.

We now have the equilibrium concentrations of all the species appearing in the $K_1$ equilibrium. To calculate the concentrations of the other ions in the solution, we can use the other equilibria that are established in the system. For example, to find the $[OH^-]$, we substitute the calculated $[H^+]$ into the expression for $K_W$. Since the product of these two concentrations in any aqueous solution at equilibrium is $1.0 \times 10^{-14}M^2$, $[OH^-]$ is fixed once $[H^+]$ is found:

$$[OH^-](1.0 \times 10^{-4}M) = 1.0 \times 10^{-14}M^2$$
$$[OH^-] = 1.0 \times 10^{-10}M$$

The concentration of sulfide ion, $[S^{2-}]$, is found by an analogous procedure. A reaction that includes sulfide ion and has a known equilibrium constant is needed. The equation for $K_2$ is most convenient:

$$HS^-(aq) \rightleftharpoons S^{2-}(aq) + H^+(aq)$$

$$K_2 = \frac{[S^{2-}][H^+]}{[HS^-]} = 1.0 \times 10^{-12}M$$

Substitution of the calculated concentrations into this expression gives:

$$\frac{[S^{2-}](1.0 \times 10^{-4}M)}{(1.0 \times 10^{-4}M)} = 1.0 \times 10^{-12}M$$

and

$$[S^{2-}] = 1.0 \times 10^{-12}M$$

The method described in Example 15.23 can be used for any solution prepared from a polyprotic acid, in which $K_2 \ll K_1$. *The relevant reaction for calculations of the equilibrium composition of solutions of polyprotic acids is the reaction that defines $K_1$.* The first step in such calculations is to find the concentrations of all the substances that appear in the equation for this reaction, using the customary procedure for equilibrium calculations. The next step is to find the concentrations of the substances that are present in the system at equilibrium but do not appear in this equation. To find these concentrations, we substitute the concentrations found in the first step into the equilibrium constant expression for other equilibria, called secondary equilibria, that occur in the solution. An approximation is implicit in this procedure. It is assumed that the concentrations determined in the first step are not changed by the less important equilibria.

This approximation was a good one for the system in Example 15.23. By using $K_1$, we found that the $[HS^-]$ in a solution of $0.10M\ H_2S$ is $1.0 \times 10^{-4}M$. The transfer of the second proton does not really change this value. The value of $K_2$ is so small that only an inconsequential fraction of $HS^-$ undergoes reaction in the $K_2$ equilibrium. Similarly, the $H^+$ that is formed by the second ionization of $H_2S$ is inconsequential compared to the $H^+$ formed by the first ionization.

The lesser importance of the second equilibria helps to simplify the calculations. The full-fledged equilibrium calculation must be done to find the concentration of species that appear in the most important relevant equilibrium. To calculate the concentration of species that appear only in the secondary equilibria, we simply substitute concentrations directly into the expressions for $K$. Figure 15.8 helps us see why we can neglect the concentrations of these species. At most pH values, only one or two species are present in significant amounts.

**Example 15.24**   The importance of the carbon dioxide-bicarbonate buffer in physiological fluids was mentioned earlier. Blood is one such fluid. A sample of blood plasma is found to have $[HCO_3^-] = 2.4 \times 10^{-2}M$ and $[CO_2] = 1.2 \times 10^{-3}M$. The value of $K_1$ for carbon dioxide under physiological conditions is $7.9 \times 10^{-7} M$ and the value of $K_2$ is $1.0 \times 10^{-10} M$. Calculate the pH of the blood plasma sample and the $[CO_3^{2-}]$ in the solution.

**Solution**   The relevant equilibrium is:

$$CO_2(aq) \quad + H_2O \rightleftharpoons HCO_3^-(aq) \quad + H^+(aq)$$

| | | |
|---|---|---|
| **start** | $1.2 \times 10^{-3} M$ | $2.4 \times 10^{-2} M$ | $0$ |
| **equil** | $1.2 \times 10^{-3} M - x$ | $2.4 \times 10^{-2} M + x$ | $x$ |

Making the approximation that $1.2 \times 10^{-3} M - x \approx 1.2 \times 10^{-3} M$ and that $2.4 \times 10^{-2} M + x \approx 2.4 \times 10^{-2} M$, the expression for $K_1$ gives:

$$\frac{(2.4 \times 10^{-2} M)(x)}{(1.2 \times 10^{-3} M)} = 7.9 \times 10^{-7} M$$

and $x = 4.0 \times 10^{-8} M$, the approximation is valid. $[H^+] = 4.0 \times 10^{-8} M$, and pH = 7.40.

To find the $[CO_3^{2-}]$, we use a secondary equilibrium in which it is included, such as $K_2$. Using the concentrations that we have just found, the $K_2$ expression gives:

$$K_2 = \frac{[CO_3^{2-}][H^+]}{[HCO_3^-]} = \frac{[CO_3^{2-}](4.0 \times 10^{-8} M)}{(2.4 \times 10^{-2} M)} = 1.0 \times 10^{-10} M$$

and                  $$[CO_3^{2-}] = 6.0 \times 10^{-5} M$$

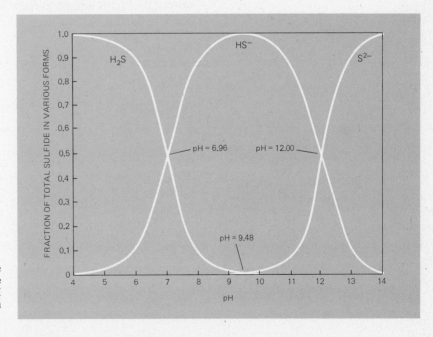

**Figure 15.8**
The fraction of solute species found in sulfide systems at different pH values. You can see the species present in significant amount at any pH from a graph such as this one. This information enables you to make helpful approximations.

Like the carbon dioxide-bicarbonate buffer that is the subject of Example 15.24, many other physiological buffers are composed of substances related to polyprotic acids. For these buffers, *the relevant equilibrium is the reaction that most simply relates the two species of which the buffer is composed.* We can apply this rule to another important system, one composed of substances related to phosphoric acid, a triprotic acid. Consider the buffer system composed of the dihydrogen phosphate anion, $H_2PO_4^-$, an acid, and its conjugate base, the hydrogen phosphate dianion, $HPO_4^{2-}$. The relevant equilibrium for calculations on this system is the simplest equilibrium that relates these two anions:

$$H_2PO_4^- \rightleftharpoons HPO_4^{2-} + H^+$$

This is the equation that defines $K_2$ for phosphoric acid. Under physiological conditions, the value of $K_2$ is $1.6 \times 10^{-7} M$ and $pK_2 = 6.8$. As you might expect, this value is close to the range of physiological pH.

### Simultaneous Equilibria

The complete description of solutions that contain species related to polyprotic acids often requires us to consider a number of different equilibrium reactions. A quantitative approach to such systems can be quite complex, and a number of methods have been developed to carry out the calculations. Special methods of calculation may also be needed for other equilibrium systems, such as those that include relatively insoluble materials as well as ions in solution. In Appendix VI, we outline method for solving problems related to the composition of systems in which a number of simultaneous reactions take place.

## Summary

Summary We began this chapter on acids and bases with the **Arrhenius definition,** which describes an acid as a species that increases the concentration of hydrogen ions in aqueous solution and a base as a species that increases the concentration of hydroxide ions. We then gave the more general **Brønsted definition,** which says that an acid is a **proton donor** and a base is a **proton acceptor.** We noted that the Brønsted definition is best understood by regarding all acid-base reactions in solution as **proton transfers** involving two **acid-base conjugate pairs.** We added that some substances can either accept or donate protons; they are said to be **amphoteric.** Water is amphoteric, and **autoprotolysis,** in which a proton is transferred from one water molecule to another, is an important reaction. We then introduced the **Lewis definition,** which says that a base is an electron pair donor and an acid is an electron pair acceptor. We went on to the concept of acid and base **strengths,** noting that the strength of acids and bases is a measure of the extent to which they donate or accept protons, and we described the **acid dissociation constant,** which is a measure of the strength of acids in water, and the **base dissociation constant,** which serves the same purpose for bases. We stressed that the strength of an acid and its conjugate base are connected: The stronger the acid, the weaker its conjugate base, and vice versa. We divided acids into two groups, strong and weak, and showed how the strengths of different acids are related to their chemical structures. In a discussion of acid-base reactions in water, we introduced the concepts of the **water equilibrium** and of **pH,** a valuable measure of acidity. We defined some important categories of acid-base systems and showed how a relevant equilibrium reaction can be written for each system. We demonstrated some essential calculations for aqueous solutions of **weak acids and bases.** Then we described **neutralization,** the reaction that occurs when aqueous solutions of an acid and a base are mixed, and added that most neutralizations are carried out by **titration,** the addition of a measured amount of one solution to another, and that the **equivalence point** is the point at which the exact volume of solution needed for neutralization has been added. We described **indicators,** substances whose color changes are used to detect the equivalence point. We defined a **buffer** solution as one that is relatively resistant to changes in pH. Finally, we dealt with **polyprotic acids,** which can donate more than one proton per molecule.

## Exercises

**15.1** Classify each of the following species as a Brønsted acid, a Brønsted base, or both: (a) HBr, (b) $CH_3^-$, (c) $PH_4^+$, (d) $CO_3^{2-}$, (e) $HS^-$, (f) $H_4IO_6^-$.

**15.2** Indicate which of the following substances are unlikely to act as either Brønsted acids or bases: (a) HNO, (b) $H_3PO_2$, (c) $CH_4$, (d) $N_2O$, (e) Pb.

**15.3** Write equations for the reactions that take place in aqueous solution between the following species: (a) hypochlorous acid and methyl amine, (b) acetic acid and ammonia, (c) fluoride ion and ammonium ion, (d) bicarbonate ion and cyanide ion.

**15.4** Write equations for each of the following substances reacting with itself in an acid-base reaction: (a) $HS^-$, (b) $H_2O_2$.

**15.5**[3] Indicate which of the following

---

[3] The answers to exercises whose numbers are in color can be found in Appendix VII. The star indicates an exercise that is more challenging than average.

species are Lewis acids: (a) $PH_3$, (b) $NH_4^+$, (c) $BI_3$, (d) $Cr^{3+}$, (e) $PtCl_4$.

**15.6**   Write equations for the acid-base reactions that take place between the following species: (a) ammonia and aluminum fluoride, (b) $Zn(OH)_2$ and $2OH^-(aq)$, (c) ethylene ($C_2H_4$) and HBr.

**15.7**   All Brønsted bases are Lewis bases but not all Brønsted acids are Lewis acids. Give two examples of Brønsted acids that are Lewis acids and two that are not. Explain.

**15.8**   Classify each of the following species as a weak or strong acid or base: (a) $HNO_3$, (b) $HNO_2$, (c) $NO_3^-$, (d) $NO_2^-$, (e) $NH_2^-$, (f) $PO_4^{3-}$, (g) $NH_2NH_3^+$.

**15.9**   Designate the stronger acid in each of the following pairs: (a) $HBrO_3$ and $HClO_2$, (b) $CH_3NH_3^+$ and $CH_3NH_2$, (c) $CH_3OH_2^+$ and $H_3O^+$.

**15.10**   Designate the stronger base in each of the following pairs: (a) $BrO_3^-$ and $ClO_2^-$, (b) $CH_3NH_2$ and $CH_3NH^-$, (c) $CH_3OH$ and $H_2O$.

**15.11\***   When $HNO_3$ is dissolved in concentrated sulfuric acid it can act as a base and accept a proton from the sulfuric acid. Draw the Lewis structure of the conjugate acid of $HNO_3$ and write an equation for the acid-base reaction.

**15.12**   Four solutions are prepared with 1 mol of $NH_3$ in 0.1 L of methyl amine, sulfuric acid, acetic acid, and water, respectively. List the solutions in order of increasing concentration of $NH_3$ at equilibrium.

**15.13**   Write net ionic equations for the reactions that occur when the following solutions are mixed: (a) a solution of hydriodic acid with a solution of potassium hydroxide, (b) a solution of nitric acid with a solution of barium

hydroxide, (c) a solution of potassium hydroxide and a solution of methyl amine, (d) a solution of ammonia and a solution of hydrobromic acid, (e) a solution of ammonia and a solution of nitrous acid.

**15.14**   Write net ionic equations for the reactions that occur when the following solutions are mixed: (a) a solution of hypochlorous acid and a solution of sodium hydroxide, (b) a solution of nitrous acid and a solution of methyl amine, (c) a solution of hydrazine and a solution of hydrochloric acid, (d) a solution of hydrocyanic acid and a solution of calcium hydroxide.

**15.15**   Write net ionic equations for the reactions that occur when the following solutions are mixed: (a) a solution of sodium acetate and a solution of hydrochloric acid, (b) a solution of ammonium chloride and a solution of calcium hydroxide, (c) a solution of ammonium bromide and a solution of sodium nitrite, (d) a solution of sodium cyanide and a solution of potassium hydroxide, (e) a solution of methylammonium bromide and a solution of potassium cyanide.

**15.16**   Write the net ionic equation that defines the $K_A$ of each of the following species: (a) phenol, (b) iodic acid, (c) anilinium ion, (d) hydroxyl-ammonium ion.

**15.17**   Write the net ionic equation that defines the $K_B$ of each of the following species: (a) hydrazine, (b) ephedrine, (c) lactate ion, (d) fluoride ion.

**15.18**   Use the data in Table 15.3 to list the following bases in order of increasing base strength: $CN^-$, $HCOO^-$, $IO_3^-$, $IO^-$, $C_6H_5O^-$.

**15.19**   Arrange the following hydrides in order of increasing acidity: $NH_3$, $H_2O$, $H_2S$, $H_2Se$, $H_2Te$.

**15.20**   Arrange the following ions in order of increasing basicity: $AsH_2^-$, $Br^-$, $GeH_3^-$, $I^-$, $HSe^-$.

**15.21**   Suggest an explanation for the order of acidities of the three chloro derivatives of acetic acid listed in Table 15.3.

**15.22**   All three simple oxyacids of phosphorus have acid dissociation constants ($K_A$) of comparable magnitude although the relative number of H and O atoms seems different in each. Use the information in Figure 9.13 to account for this observation.

**15.23**   Predict the relative acidities and basicities of the following compounds: $NH_2F$, $NH_2Cl$, $NH_2Br$, $NH_2I$, $NH_3$.

**15.24\***   Acidity in the gas phase can be defined by the equilibrium constant for a dissociation such as: $CH_3NH_3^+(g) \rightleftharpoons CH_3NH_2(g) + H^+(g)$. In the gas phase $CH_3NH_3^+$ is more acidic than $(CH_3)_3NH^+$ while in solution the reverse is true. Account for this observation.

**15.25**   Arrange the following oxyacids in order of increasing strength: $HClO_4$, $H_4SiO_4$, $H_2SO_3$, $H_2SO_4$.

**15.26**   Arrange the following bases in order of increasing strength: $MnO_4^-$, $H_2BO_3^-$, $HSeO_4^-$, $H_2PO_4^-$.

**15.27**   Calculate the $[H^+]$ and the $[OH^-]$ in each of the following solutions: (a) $2.4M$ KOH, (b) $0.35M$ $HClO_4$, (c) $2.4 \times 10^{-3}M$ $Ba(OH)_2$, (d) $5.5 \times 10^{-5}M$ HBr.

**15.28**   Calculate the pH and the pOH of each of the following solutions: (a) $2M$ NaOH, (b) $0.5M$ $HNO_3$, (c) $0.0003M$ $Ba(OH)_2$, (d) $2.4 \times 10^{-5}M$ HCl.

**15.29**   Calculate the $[H^+]$ in solutions of the following pH: (a) 6.45, (b) 7.85, (c) $-0.48$.

**15.30**   Calculate the $[OH^-]$ in solutions of the following pH: (a) 7.30, (b) 6.70, (c) 13.75, (d) 14.87.

**15.31**   Calculate the pH of a solution

prepared from 0.22 mol of acetic acid and enough water to make 1.0 L of solution.[4]

**15.32** Calculate the pH and pOH of a solution prepared from 0.16 mol of hydrazine and enough water to make 1.0 L of solution.[5]

**15.33** Which solution has a lower pH: 0.10$M$ acetic acid or 0.20$M$ HClO? Find the difference in pH between the two solutions.[4]

**15.34** Write the equation and calculate the value of $K_B$ for the reaction in which the conjugate base of each of the following acids accepts a proton from water: (a) $HIO_3$, (b) HBrO, (c) butyric acid (Figure 15.2), (d) anilinium ion.[4,5]

**15.35** Write the equation and calculate the value of $K_A$ for the reaction in which the conjugate acid of each of the following bases donates a proton to water: (a) hydrazine, (b) morphine, (c) imidazole, (d) phenoxide ion $(C_6H_5O^-)$.[4,5]

**15.36** Calculate the $[H^+]$ and $[OH^-]$ in a solution prepared from 0.68 mol of $NaNO_2$ and enough water to make 0.30 L of solution.[4]

**15.37** Calculate the pH of a solution prepared from 0.11 mol of methylammonium chloride and enough water to make 0.51 L of solution.[5]

**15.38** Calculate the [HCN] in a solution prepared from 0.37 mol of NaCN and enough water to make 0.98 L of solution.[4]

**15.39** A solution prepared from 0.35 mol of an organic acid and enough water to form 1.0 L of solution is found to have a pH of 4.79. Find $K_A$ of the acid.

**15.40** A solution prepared from 0.22 mol of an organic base in enough water to form 0.24 L of solution is found to have a pH of 8.88. Find $K_B$ of the base.

**15.41** A solution of 0.23 mol of the chloride salt of protonated quinine, a weak organic base used in the treatment of malaria, in enough water to form 1.0 L of solution has a pH of 4.58. Find $K_B$ of quinine.

**15.42** A solution of the sodium salt of ascorbic acid (AcbOH), commonly known as vitamin C, is prepared from 0.063 mol of the salt in enough water to make 0.25 L of solution. The $[OH^-]$ is found to be $5.7 \times 10^{-6}$. Find the $K_A$ of ascorbic acid.

**15.43** Calculate the amount of potassium formate that must be dissolved in 0.25 L of water to produce a solution of pH = 8.60.[4]

**15.44** Calculate the amount of ammonium chloride that must be dissolved in 0.75 L of water to produce a solution of pH = 5.25.[5]

**15.45** Calculate the fraction of $HClO_2$ that dissociates in a solution prepared from 0.10 mol of the acid and enough water to make (a) 1.0 L, (b) 0.10 L of solution.[4]

**15.46** Calculate the $[F^-]$ in a solution prepared from 0.010 mol of HF, 0.020 mol of HBr, and enough water to make 1.0 L of solution.[4]

**15.47** Calculate the fraction of nicotine that becomes protonated in a solution that has an initial concentration of nicotine of 0.011$M$ at the following pHs: (a) 10, (b) 14.[5]

**15.48** Calculate the amount of $IO^-$ in 0.50 L of solution prepared from 1.0 mol of HIO and $1.0 \times 10^{-5}$ mol of $HNO_3$.[4]

**15.49** Calculate the amount of hydrochloric acid that must be added to 1.0 L of a solution prepared from 1.0 mol of iodic acid in order to keep the $[IO_3^-]$ below 0.2$M$.[4]

**15.50** Calculate the amount of HF that must be added to 1.0 L of a solution prepared from 0.17 mol of KF to produce a solution with pH = 3.57.

**15.51** Write the relevant equilibrium and find the value of its $K$ for the reaction that takes place when each of the following pairs of solutions are mixed: (a) barium hydroxide and hydrochloric acid, (b) potassium hydroxide and hydrogen peroxide, (c) imidazole and hydrochloric acid, (d) hydroxylamine and acetic acid.[4,5]

**15.52** Write the relevant equilibrium and find the value of its $K$ for the reaction that takes place when each of the following pairs of solutions are mixed: (a) sodium cyanide and nitric acid, (b) ammonium chloride and calcium hydroxide, (c) hydrazinium chloride and sodium nitrite, (d) sodium chloride and ammonia.[4,5]

**15.53** A volume of 23.4 cm³ of 0.309$M$ nitric acid is required to neutralize 24.9 cm³ of an NaOH solution. Find the concentration of the NaOH solution.

**15.54** A volume of 18.5 cm³ of $H_2SO_4$ solution is required to neutralize 3.92 g of KOH. Find the concentration of the $H_2SO_4$ solution.

**15.55** A sample of orange juice requires 14.2 cm³ of 0.107$M$ KOH solution for neutralization. Find the amount (in moles) of available protons in the sample of juice.

**15.56** Calculate the volume of 0.281$M$ hydrochloric acid solution necessary to neutralize 10.1 cm³ of a solution prepared from 7.44 g of potassium nitrite and enough water to make 1.0 L of solution.

**15.57** Find the concentration of a solution of $(NH_4)_2SO_4$ that requires 48.2 cm³ of a 0.646$M$ sodium hydroxide solution for neutralization of a 27.2-cm³ volume.

**15.58★** A 25.0-cm³ volume of barium

---

[4] The necessary data for this exercise are in Table 15.3.
[5] The necessary data for this exercise are in Table 15.5.

hydroxide solution requires 39.2 cm³ of a 0.189$M$ nitric acid solution for neutralization. A 15.0-cm³ volume of a phosphoric acid solution requires 26.2 cm³ of the barium hydroxide solution for complete neutralization. Calculate the concentration of the phosphoric acid solution.

**15.59**   A 0.567-g sample of an unknown acid requires 42.1 cm³ of a 0.119$M$ sodium hydroxide solution for neutralization. Calculate the equivalent weight of the acid. Assuming that the acid donates three protons per mole, calculate the molecular weight of the acid.

**15.60**   A 0.811-g sample of ephedrine, a base that accepts only one proton per mole, requires 14.2 cm³ of a 0.347$M$ solution of hydrochloric acid for neutralization. Find the molecular weight of ephedrine.

**15.61**★   Morphine, the well-known narcotic painkiller, accepts one proton per molecule. Its composition is $C_{17}H_{19}NO_3$. The major source of morphine is opium. A 0.446-g sample of opium is found to require 9.74 cm³ of a 0.0202$M$ solution of hydrochloric acid for neutralization. Assuming that morphine is the only acid or base present in opium, calculate the percentage of morphine by weight in the sample of opium.

**15.62**   Calculate the pH at the equivalence point in the titration of 0.20$M$ formic acid solution with 0.20$M$ potassium hydroxide solution.[4]

**15.63**   Calculate the pH at the equivalence point in the titration of 0.20$M$ $NH_3$ solution with 0.20$M$ HBr solution.[5]

**15.64**   Calculate the pH at the equivalence point in the titration of solid $NaClO_2$ with 0.234$M$ HCl solution.[4]

**15.65**   Calculate the pH at the equivalence point in the titration of solid $C_6H_5NH_3Cl$ by 0.110$M$ NaOH solu-

tion.[5]

**15.66**   Calculate the pH at the equivalence point in the titration of 50.0 cm³ of a 0.441$M$ solution of methyl amine by a 0.132$M$ solution of formic acid.[4,5]

**15.67**★   A 100-cm³ sample of a 0.100$M$ hydrochloric acid solution is titrated with a 0.100$M$ solution of sodium hydroxide. Calculate the pH after the addition of the following volumes of the solution of base to the acid: (a) 50.0 cm³, (b) 90.0 cm³, (c) 99.0 cm³, (d) 99.9 cm³, (e) 100.1 cm³.

**15.68**★   Repeat the calculation of Exercise 15.67 for the titration of a 0.100$M$ solution of $NH_4Cl$ with a 0.100$M$ solution of NaOH.[5]

**15.69**   Select a suitable indicator from those listed in Table 15.8 for each of the following titrations: (a) nitric acid with calcium hydroxide, (b) 0.10$M$ chlorous acid with 0.10$M$ KOH, (c) 0.10$M$ $NH_3$ with 0.10$M$ HI.[4,5]

**15.70**   Select a suitable indicator from those listed in Table 15.8 for each of the following titrations: (a) solid sodium acetate with 0.10$M$ HCl, (b) 0.10$M$ iodic acid with 0.10$M$ $CH_3NH_2$, (c) 0.10$M$ ammonium bromide with 0.10$M$ sodium hydroxide.[4,5]

**15.71**   Use the principle of Le Chatelier to explain how an acetic acid-acetate buffer solution acts to keep the pH relatively constant upon addition of small amounts of HCl or NaOH.

**15.72**   Calculate the pH of a buffer solution prepared from 0.23 mol of acetic acid and 0.37 mol of sodium acetate in enough water to make 1.0 L of solution.[4]

**15.73**   Calculate the pH of a buffer solution prepared from 0.19 mol of ammonia and 0.13 mol of ammonium sulfate in enough water to make 1.0 L of solution.[5]

**15.74**   Calculate the mass of sodium formate that must be dissolved in 1.0 L

of a 1.1$M$ formic acid solution to prepare a buffer of pH = 3.65. Assume no volume changes.[4]

**15.75**   Calculate the relative masses of lactic acid and calcium lactate necessary to prepare a buffer solution of pH = 4.00.

**15.76**   Select a suitable conjugate acid-base pair from the listing in Table 15.3 to prepare a buffer of pH = 9.50. Find the molar ratio of the acid and base needed for this pH.

**15.77**   Calculate the pH of a buffer solution prepared by adding 50.0 cm³ of 0.102$M$ HCl solution to 25.0 cm³ of 0.389$M$ $NaNO_2$ solution.[4]

**15.78**   Calculate the pH of a buffer solution prepared from 1.0 mol of $NH_3$ and 1.0 mol of $NH_4^+$ in enough water to make 1.0 L of solution. Calculate the pH of the buffer after the addition of 0.10 mol of HCl. Repeat the calculation for a buffer prepared from 0.10 mol of $NH_3$ and 0.10 mol of $NH_4^+$.[5]

**15.79**   Calculate the pH of a buffer solution prepared from 1.0 mol of formic acid and 1.0 mol of sodium formate in enough water to make 1.0 L of solution. Calculate the pH of the buffer after the addition of 0.10 mol of NaOH. Repeat the calculation for a buffer prepared from 0.10 mol of formic acid and 0.10 mol of sodium formate.[4]

**15.80**   A student chooses phenol and sodium phenoxide as the conjugate acid-base pair to prepare a buffer of pH = 8.00. What is wrong with this selection?

**15.81**   Explain the shape of the titration curve in Figure 15.7 on the basis of the buffer solutions that form during the titration.

**15.82**   Buffer solutions are often prepared with salts of anions formed from dissociation of phosphoric acid. Calculate the pH of the buffer solution prepared from equimolar quantities of each of the following salts: (a) $NaH_2PO_4$ and $Na_2HPO_4$, (b) $Na_2HPO_4$ and $Na_3PO_4$.[6]

**15.83** Calculate the concentration of all species at equilibrium in a solution prepared from 1.00 mol of $H_2SO_4$ and enough water to make 0.500 L of solution.[6]

**15.84** The formula of oxalic acid is $H_2C_2O_4$ and both protons are acidic. Calculate the concentration of all species at equilibrium in a solution prepared from 0.69 mol of oxalic acid in enough water to make 1.0 L of solution.[6]

**15.85** A solution is $0.30M$ in HCl and a 0.22-mol amount of $H_2S(g)$ dissolves in 1.0 L of the solution. Find the $[S^{2-}]$ at equilibrium.[6]

**15.86** Use the data in Example 15.24

---

[6] The necessary data for this exercise are in Table 15.9.

to find the $[HCO_3^-]$ and $[CO_2]$ when the pH of the blood drops to 7.25. Assume the total of the two concentrations does not change.

**15.87** When strong acid is added to a solution of sodium carbonate the evolution of a gas is observed. Identify the gas and explain its formation using the principle of Le Chatelier.

**15.88** Using only water or $HCrO_4^-$ or both as reactants, list all the equilibria that are established in a solution of $NaHCrO_4$ and evaluate the equilibrium constant of each.[6]

**15.89** A solution of $0.35M$ sodium carbonate is alkaline. Find the pH of the solution and the $[HCO_3^-]$ and $[H_2CO_3]$ at equilibrium.[6]

**15.90★** In a solution of sodium bicarbonate the most important equilibrium is the one in which the amphoteric bicarbonate ion reacts with itself. Find the $K$ of this reaction and the pH

of a $0.20M$ sodium bicarbonate solution.

**15.91★** Calculate the solubility of $Cd(OH)_2$ in $1.0M$ $HNO_3$ solution. The $K_{sp}$ is $5.9 \times 10^{-15}$. (*Hint:* The relevant equilibrium includes the solid and $H^+$ as reactants.)

**15.92★** Calculate the solubility of $Fe(OH)_3$ in pure water. The $K_{sp}$ is $4 \times 10^{-38}$. (*Hint:* pOH = 7, consider the common ion effect.)

**15.93★** Calculate the solubility of $Au(OH)_3$ when pOH = 7. Calculate its solubility when pOH = 15. The $K_{sp}$ is $5.5 \times 10^{-46}$.

**15.94★** The $K_{sp}$ of $BaCO_3$ is $5.1 \times 10^{-9}$. Find the $K$ for the reaction $BaCO_3(s) + CO_2(aq) + H_2O \leftrightarrows Ba^{2+}(aq) + 2HCO_3^-(aq)$. The $[CO_2]$ in a saturated solution is $0.042M$. Find the solubility of barium carbonate in a saturated solution of $CO_2$ in water.[6]

**16**

# Oxidation-
# Reduction

**Preview**    **W**e put oxidation-reduction reactions to use every time we turn on an electronic game, a pocket calculator, or any other device that uses a battery. In this chapter, we first visualize an oxidation-reduction reaction as a chemical change that results from electron transfer between the reactants, and we then show how such a reaction can be described as the sum of two half-reactions. We discuss electrolysis, the conversion of electrical energy into chemical energy, and show how Faraday's laws relate the extent of chemical change to the quantity of electricity. Next we introduce galvanic cells, which convert chemical energy to electrical energy, and we describe the different kinds of galvanic cells and how they operate. A major part of the chapter deals with electric potential—how it is measured in cells and its relation to free energy, to the equilibrium constant, and to chemical behavior. We show how reduction potentials are tabulated in the emf series and describe the relationship between potential and concentration which is given by the Nernst equation. Finally, we describe some important electron transfer processes that take place in corroding metals and in living organisms.

**A**ll the aqueous equilibrium processes that we have discussed so far have one feature in common: There is no change in the number of electrons associated with each participant in the reaction. For example, when $H^+$ is transferred from an acid to a base, there is no change in the number of electrons of either the acid or the base, since $H^+$ has no electrons.

Now we shall discuss reactions that cause changes in the number of electrons of some or all of the participants. These changes can be visualized as the result of actual or formal electron transfer between the substances that undergo reaction. **A chemical change that occurs as the result of an electron transfer is called an oxidation-reduction, or redox, process.**

Oxidation-reduction reactions occur all around us—and, in fact, within us. The energy that enables you to read these words comes from biological oxidation-reduction reactions. Most of the energy that powers our technological society comes from oxidation-reduction reactions. Let us examine them in detail.

## 16.1 OXIDATION-REDUCTION REACTIONS

In Section 7.6 we discussed the concept of an oxidation state or oxidation number. We described how every atom in a molecule or an ion can be assigned an oxidation state, a positive or negative integer related to the electronic configuration of the atom in the molecule or ion. If you do not have a good understanding of these ideas, a review of Section 7.6 is strongly recommended.

An oxidation-reduction reaction can be defined as a chemical change in which there are changes in oxidation states. The oxidation state of one or more atoms increases and the oxidation state of one or more atoms decreases. If the oxidation state of an atom increases, that atom is said to be oxidized. If the oxidation state of an atom decreases, that atom is said to be reduced.

A change in oxidation state can be regarded as equivalent to a gain or loss of electrons. The oxidation state of an atom increases when it loses electrons and decreases when it gains electrons. We can see the relationship most clearly by studying changes in a single atom and its ions. In $Mn^{2+}$, the oxidation state of manganese is $+2$. If the $Mn^{2+}$ cation *loses* an electron, its charge increases by $+1$, and it becomes $Mn^{3+}$. The oxidation state of $Mn^{3+}$ is $+3$, and the $Mn^{2+}$ cation is said to have been oxidized to $Mn^{3+}$. If $Mn^{2+}$ *gains* two electrons, its charge is changed by $-2$; $Mn^{2+}$ becomes Mn. We call this a reduction, because when $Mn^{2+}$ is reduced to Mn, its oxidation state is reduced from $+2$ to 0.

*Oxidation is the loss of electrons; reduction is the gain of electrons. Under ordinary chemical conditions, one of these processes cannot occur unless the other process also occurs.*

Oxidation-reduction processes are visualized as electron transfer processes. Electrons are transferred from the substance that is oxidized to the substance that is reduced. The substance that is oxidized is called a **reducing agent,** because it brings about the reduction of another substance. Similarly, a substance that is reduced by gaining electrons causes the oxidation of the substance that loses the electrons. The substance that is reduced is therefore called an **oxidizing agent.** The basic idea is summed up simply: A substance that is oxidized is a reducing agent; a substance that is reduced is an oxidizing agent.

For example, the reaction:

$$Cl_2(aq) + 2Br^-(aq) \longrightarrow Br_2(aq) + 2Cl^-(aq)$$

is an oxidation-reduction reaction. The oxidation state of chlorine decreases from 0 to $-1$, while the oxidation state of bromine increases from $-1$ to 0. The $Cl_2$ is reduced and the $Br^-$ is oxidized. The $Cl_2$ is an oxidizing agent that causes the oxidation of $Br^-$. The $Br^-$ is a reducing agent that causes the reduction of $Cl_2$.

## The Half-Reaction Concept

An oxidation cannot occur without an accompanying reduction. But it is often convenient to divide oxidation-reduction processes in half artificially to help us analyze them. This artificial division gives us two half-reactions. You can distinguish a half-reaction from a normal reaction easily. The half-reaction includes one or more electrons, whose symbol is $e^-$.

There are two kinds of half-reactions, one for oxidations and one for reductions. In an oxidation half-reaction, the electrons appear as products on the right side of the chemical equation. Electrons are lost in an oxidation. The half-reaction for the oxidation of manganese from the $+2$ to the $+3$ state is

$$Mn^{2+}(aq) \longrightarrow Mn^{3+}(aq) + e^-$$

In a reduction half-reaction, the electrons appear as reactants on the left side of the equation. Electrons are gained in a reduction. The half-reaction for the reduction of manganese from the $+2$ to the 0 oxidation state is

$$Mn^{2+}(aq) + 2e^- \longrightarrow Mn(s)$$

Every half-reaction must be balanced in two ways. We must balance not only the mass of all the elements but also the total charge. To balance the charge in a half-reaction, we take the charge of an electron to be $-1$. The two half-reactions that we used as examples are balanced in both respects.

An oxidation half-reaction cannot occur unless a reduction half-reaction occurs simultaneously. A complete oxidation-reduction reaction consists of a suitable combination of an oxidation half-reaction and a reduction half-reaction. *The equation for an oxidation-reduction reaction is written correctly only if it is balanced with respect to both mass and charge and if it contains no electrons.*

Half-reactions have several uses. They are used to help balance oxidation-reduction reactions and to compare the strength of oxidizing agents or reducing agents (Section 16.5). They also help clarify the relationship between chemical energy and electrical energy.

## Balancing Oxidation-Reduction Reactions

The methods for writing balanced chemical equations that were presented in Section 2.4 should be applied whenever possible to the writing of equations for oxidation-reduction reactions. Inspection alone will suffice to balance the charges as well as the masses in some oxidation-reduction equations. One example is the reaction in which silver cation in solution reacts with zinc metal to form zinc cation in solution and silver metal:

$$Ag^+(aq) + Zn(s) \longrightarrow Ag(s) + Zn^{2+}(aq)$$

The mass of the elements in this expression is balanced. The charge is not, since the left side of the expression has a net charge of $+1$ while the right side has a net charge of $+2$. We can balance the charge by giving the $Ag^+$ ion the coefficient 2. Now the mass is no longer balanced. The mass balance can be restored if we give the $Ag(s)$ the coefficient 2. The balanced equation is

$$2Ag^+(aq) + Zn(s) \longrightarrow 2Ag(s) + Zn^{2+}(aq)$$

It is not possible to write balanced equations for many oxidation-reduction reactions by inspection alone. There are several formal procedures for writing balanced oxidation-reduction equations when the reactants and products are known. We shall discuss the two methods that are most commonly used. In this discussion, we shall follow the usual conventions for writing chemical equations in net ionic form. The formulas of strong electrolytes such as salts, strong acids, and strong bases will be written to show that they dissociate into ions in solution. The formulas for weak electrolytes and nonelectrolytes will be written to show that they do not dissociate appreciably in solution. Only substances that undergo change will appear in the chemical equation. Coefficients will be in their simplest form.

## The Method of Half-Reactions

The **method of half-reactions** divides the oxidation-reduction process into two half-reactions. Each is first balanced with respect to mass and charge. They are then combined in a way that eliminates the electrons, giving the overall balanced equation for the oxidation-reduction process.

To begin, we must identify the reactant that is oxidized and the product that it forms as well as the reactant that is reduced and the product that it forms. The description of the oxidation-reduction process usually gives enough information to make the identification possible. We shall then have two pairs of substances that provide the skeletons for the two half-reactions.

Suppose we are told that an acidic solution of sodium dichromate, $Na_2Cr_2O_7$, is mixed with a solution of potassium bromide, KBr; that a reaction occurs; and that the products are identified as bromine, $Br_2$, and chromium(III) cation, $Cr^{3+}$, in solution. The two starting materials are salts in solution, so we write their formulas to show their dissociation into ions: $Na^+ + Cr_2O_7^{2-}$ and $K^+ + Br^-$. We can see that the bromide ion is oxidized, since its oxidation state increases from $-1$ in the ion to 0 in $Br_2$. The dichromate ion is reduced, since the oxidation state of Cr in $Cr_2O_7^{2-}$ is $+6$ and in the $Cr^{3+}$ cation, the product, it is $+3$. We can now write the two skeleton half-reactions:

$$Br^-(aq) \longrightarrow Br_2(aq)$$
$$Cr_2O_7^{2-}(aq) \longrightarrow Cr^{3+}(aq)$$

We do not include the sodium and potassium ions, which are unchanged in the process.

Each skeleton half-reaction must now be converted into a half-reaction balanced with respect to both mass and charge. We do so by following a fixed sequence of steps.

1. Balance the number of atoms of all elements except H and O, using coefficients in the usual way. In this case, we write

$$2Br^-(aq) \longrightarrow Br_2(aq)$$
$$Cr_2O_7^{2-}(aq) \longrightarrow 2Cr^{3+}(aq)$$

2. Balance the oxygen by adding the necessary number of oxygen atoms, in the form of $H_2O$, to the oxygen-deficient side of the expression. In this case, only the second expression must be balanced for oxygen:

$$Cr_2O_7^{2-}(aq) \longrightarrow 2Cr^{3+}(aq) + 7H_2O$$

3. Balance the hydrogen by adding the appropriate number of hydrogen atoms, in the form of $H^+$, to the hydrogen-deficient side of the expression. Again in this case, only the second expression must be balanced for hydrogen:

$$14H^+(aq) + Cr_2O_7^{2-}(aq) \longrightarrow 2Cr^{3+}(aq) + 7H_2O$$

4. Balance each half-reaction for charge by adding the appropriate number of electrons to the more positive side of the expression. The number

of electrons added should equal the difference in charge between the two sides of the expression. In this case, the left side of the first expression has a charge of $-2$ and the right side has a charge of 0. We add two electrons to the right side:

$$2Br^-(aq) \longrightarrow Br_2(aq) + 2e^-$$

The left side of the second expression has a charge of $+12$ and the right side has a charge of $+6$. We add six electrons to the left side:

$$6e^- + 14H^+(aq) + Cr_2O_7{}^{2-}(aq) \longrightarrow 2Cr^{3+}(aq) + 7H_2O$$

This gives us two balanced half-reactions, one for the reduction and one for the oxidation. We must add the two to get the overall equation for the process. But the overall equation cannot include electrons. Therefore, *we must make sure that both half-reactions have the same number of electrons.* If they do, the electrons will be eliminated from the overall equation. When, as in this case, the half-reactions do not have the same number of electrons, we multiply one or both of the half-reactions by the necessary factor. Inspection shows that both half-reactions will include six electrons if the oxidation half-reaction is multiplied by three. The half-reactions can then be combined:

$$6Br^-(aq) \longrightarrow 3Br_2(aq) + \cancel{6e}$$
$$\underline{6e + 14H^+(aq) + Cr_2O_7{}^{2-}(aq) \longrightarrow 2Cr^{3+}(aq) + 7H_2O}$$
$$6Br^-(aq) + 14H^+(aq) + Cr_2O_7{}^{2-}(aq) \longrightarrow$$
$$3Br_2(aq) + 2Cr^{3+}(aq) + 7H_2O$$

There is a procedure for finding the factors by which the half-reactions are multiplied to give them both the same number of electrons. First, find the smallest number that is divisible by the number of electrons in each half-reaction. Then divide the number of electrons in each half-reaction into this number. The division gives the factor by which each half-reaction must be multiplied.

For example, suppose we wish to add two half-reactions that have, respectively, $2e^-$ and $5e^-$. The lowest number divisible by both is 10. The factor for the first half-reaction is $10/2 = 5$, and the factor for the second half-reaction is $10/5 = 2$. After multiplication by these factors, each half-reaction has $10e^-$. When the half-reactions are combined, the electrons will cancel.

Three steps remain after the half-reactions are combined. The coefficients of the equation must be reduced to the lowest terms. The reaction must also be checked for substances that appear unchanged on both sides of the equation. Such substances should be eliminated from the equation. Finally, *check* to be sure that the equation is indeed balanced with respect to both mass and charge.

An additional step is needed for systems that are designated as alkaline. When an oxidation-reduction reaction occurs in alkaline solution, there is no appreciable concentration of $H^+$. We need an extra step to replace $H^+$ by $OH^-$ in the equation. We add the equation $H^+ + OH^- \rightarrow H_2O$ (or the reverse, $H_2O \rightarrow H^+ + OH^-$) multiplied by the coefficient of the $H^+$ that we are replacing. This procedure makes both sides of the equation equal in $H^+$, which thus cancels. After this step, the equation usually has $H_2O$ on both sides, so it must be made net by appropriate subtraction of $H_2O$. This procedure has no chemical or physical significance, but it gives the right answer.

**Example 16.1**    When an alkaline solution of potassium permanganate is mixed with an alkaline solution of sodium sulfide, a yellow precipitate of sulfur and a brown precipitate of manganese dioxide form. Write a balanced equation for this oxidation-reduction reaction.

**Solution**    The two skeleton half-reactions are

$$S^{2-}(aq) \longrightarrow S(s)$$
$$MnO_4^-(aq) \longrightarrow MnO_2(s)$$

First we balance these skeleton half-reactions. The first one, the oxidation, already is balanced with respect to mass, but electrons must be added:

$$S^{2-}(aq) \longrightarrow S(s) + 2e^-$$

For the second half-reaction, the reduction, $H_2O$ and $H^+$ must be added to balance the mass:

$$MnO_4^-(aq) + 4H^+(aq) \longrightarrow MnO_2(s) + 2H_2O$$

and then electrons must be added to balance the charge:

$$MnO_4^-(aq) + 4H^+(aq) + 3e^- \longrightarrow MnO_2(s) + 2H_2O$$

The oxidation half-reaction has $2e^-$ and the reduction half-reaction has $3e^-$. The smallest number that is divisible by both 2 and 3 is 6. Therefore, the oxidation half-reaction is multiplied by $6/2 = 3$ and the reduction half-reaction is multiplied by $6/3 = 2$. The two half-reactions are then combined:

$$3S^{2-}(aq) \longrightarrow 3S(s) + 6e^-$$
$$\underline{2MnO_4^-(aq) + 8H^+(aq) + 6e^- \longrightarrow 2MnO_2(s) + 4H_2O}$$
$$3S^{2-}(aq) + 2MnO_4^-(aq) + 8H^+(aq) \longrightarrow 3S(s) + 2MnO_2(s) + 4H_2O$$

Since this is an alkaline solution and $8H^+$ appears in our equation, we add $8H_2O \rightarrow 8H^+ + 8OH^-$. The $8H^+$ on the left side of the equation is canceled by the $8H^+$ on the right side. The equation is now:

$$3S^{2-}(aq) + 2MnO_4^-(aq) + 8H_2O \longrightarrow 3S(s) + 2MnO_2(s) + 4H_2O + 8OH^-(aq)$$

This equation is not net, because $H_2O$ appears on both sides. We get the final net balanced equation by subtracting $4H_2O$ from both sides:

$$3S^{2-}(aq) + 2MnO_4^-(aq) + 4H_2O \longrightarrow 3S(s) + 2MnO_2(s) + 8OH^-(aq)$$

### The Method of Change in Oxidation Numbers

In an oxidation-reduction process, one or more elements will have an increase in oxidation number and one or more elements will have a decrease in oxidation number. *In any balanced equation for an oxidation-reduction process, the total increase in oxidation numbers must equal the total decrease in oxidation numbers.* In the **method of change in oxidation numbers,** we use this principle to write correct equations by balancing the changes in oxidation numbers.

We start by using the description of the process to write a single skeleton expression that includes the elements that undergo a change in oxidation number.

Let us take the process in which an acidic solution of potassium permanganate, $KMnO_4$, is added to a solution of sodium chloride. The products are chlorine and manganese(II) cation. The two starting materials are salts in solution, so their formulas are written to show them as dissociated ions, $K^+ + MnO_4^-$ and $Na^+ + Cl^-$. Since neither $Na^+$ nor $K^+$ undergoes change in the process, they are left out of the equation. The skeleton expression includes the two ions that undergo reaction and the given products:

$$MnO_4^-(aq) + Cl^-(aq) \longrightarrow Mn^{2+}(aq) + Cl_2(aq)$$

1. First identify those elements whose oxidation numbers increase and those whose oxidation numbers decrease. We see that the oxidation number of Mn decreases from $+7$ to $+2$, a change of five units for each Mn atom. The oxidation number of $Cl^-$ increases from $-1$ to $0$, a change of one unit for each Cl atom.
2. Now make the total increase in oxidation number equal the total decrease in oxidation number. To do this, we find the element that undergoes the smaller total change in oxidation number. We then find the number of moles of this element that balance the change undergone by one mole of the other element. In this process, one mole of Mn undergoes a change of five units and one mole of Cl undergoes a change of one unit. Therefore, the increase and decrease in oxidation numbers will be balanced if there are five moles of Cl for each mole of Mn. We add the appropriate coefficient to Cl, while at the same time we balance all the elements in the expression except H and O:

$$MnO_4^-(aq) + 5Cl^-(aq) \longrightarrow Mn^{2+}(aq) + \tfrac{5}{2}Cl_2(aq)$$

3. The next step is to balance for charge. Find the difference in charge between the two sides of the expression and add this number of $H^+$ cations to the more negative side of the equation. In the above expression, the charge on the left is $-6$ and the charge on the right is $+2$. The difference is $+8$, so we add $8H^+$ to the left side, the more negative side of the expression:

$$MnO_4^-(aq) + 5Cl^-(aq) + 8H^+(aq) \longrightarrow Mn^{2+}(aq) + \tfrac{5}{2}Cl_2(aq)$$

4. We now balance for oxygen by adding $H_2O$ to the oxygen-deficient side of the expression:

$$MnO_4^-(aq) + 5Cl^-(aq) + 8H^+(aq) \longrightarrow$$
$$Mn^{2+}(aq) + \tfrac{5}{2}Cl_2(aq) + 4H_2O$$

The equation should now be balanced. Check to be sure that it is net and that the coefficients are in lowest terms. If there is a fractional coefficient, as there is in this case, you may multiply to eliminate the fraction. Here, we multiply by two:

$$2MnO_4^-(aq) + 10Cl^-(aq) + 16H^+(aq) \longrightarrow$$
$$2Mn^{2+}(aq) + 5Cl_2(aq) + 8H_2O$$

If the solution is described as alkaline, we need an extra step to replace $H^+$ by $OH^-$ in the equation. We add the equation $H^+ + OH^- \rightarrow H_2O$ (or $H_2O \rightarrow H^+ + OH^-$) multiplied by the coefficient of the $H^+$ that we are replacing. This procedure makes both sides of the equation equal in $H^+$, which thus cancels. The equation now usually has $H_2O$ on both sides, so it must be made net by appropriate subtraction of $H_2O$. This procedure has no chemical or physical significance, but it gives the right answer.

For example, in alkaline solution the products of the reaction of $KMnO_4$ and $NaCl$ are $Cl_2$ and $MnO_2(s)$. The balanced overall equation is

$$6Cl^-(aq) + 8H^+(aq) + 2MnO_4^-(aq) \longrightarrow$$
$$3Cl_2(aq) + 2MnO_2(s) + 4H_2O$$

Now we add $8H_2O \longrightarrow 8H^+(aq) + 8OH^-(aq)$ and cancel the $8H^+(aq)$ from each side to get

$$6Cl^-(aq) + 2MnO_4^-(aq) + 8H_2O \longrightarrow$$
$$3Cl_2(aq) + 2MnO_2(s) + 8OH^-(aq)$$

We get the final net equation by subtracting $4H_2O$ from each side:

$$6Cl^-(aq) + 2MnO_4^-(aq) + 4H_2O \longrightarrow$$
$$3Cl_2(aq) + 2MnO_2(s) + 8OH^-(aq)$$

## 16.2  OXIDATION-REDUCTION PROCESSES IN AQUEOUS SOLUTION

Many oxidation-reduction processes proceed differently in aqueous solution than the aqueous equilibria that we discussed in Chapters 14 and 15. Proton transfer reactions and many precipitation reactions occur very quickly, virtually at the instant that the solutions are mixed. By contrast, many oxidation-reduction processes proceed slowly in aqueous solution. We cannot predict what will happen on the basis of equilibrium constants alone. A given system may take so long to reach its equilibrium state that equilibrium is never achieved. In a multistep process, one step may occur so slowly that the system stops at this intermediate state. A large value for $K$ does *not* guarantee the formation of products under a given set of conditions. To know the outcome of an oxidation-reduction process, we need experimental observations.

Some oxidation-reduction processes in aqueous solution proceed smoothly and rapidly. The equilibrium constants of these processes are often much larger than those of proton transfer or precipitation reactions. For example, the value of $K$ at 298 K for the reaction

$$10I^-(aq) + 2MnO_4^-(aq) + 16H^+(aq) \rightleftharpoons$$
$$2Mn^{2+}(aq) + 8H_2O + 5I_2(aq)$$

is about $3 \times 10^{164}$.

### Oxidation-Reduction Titrations

Oxidation-reduction processes in aqueous solution are extremely useful in analytical chemistry. If an element exists in more than one oxidation state, a procedure based on an oxidation-reduction process can be used to measure the amount of that element in a sample. One widely used method is oxidation-reduction titration, which is similar to the acid-base titrations discussed in Section 15.6.

In an oxidation-reduction titration, the solution in the buret, called the titrant, is a solution of an oxidizing agent or a reducing agent. The equivalence point is reached when the stoichiometric quantity of the reagent in the buret has been added to the solution of the sample, that is, when the oxidation-reduction reaction is complete. The equivalence point can be determined by several methods. Some are based on the electrical properties of the solution and some on a change in color of a substance in the solution.

The choice of a titrant depends on the oxidation state of the element that is being analyzed. If the element is in a low oxidation state, a solution of an oxidizing agent is the titrant. If it is in a high oxidation state, a solution of a reducing agent is the titrant. The titrant must be powerful enough to oxidize or reduce the sample completely, and the reaction must

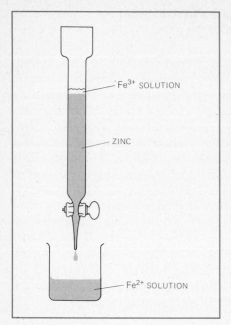

**Figure 16.1**
A Jones reductor. Iron in the $+3$ oxidation state is reduced to the $+2$ oxidation state when it is passed through a column of specially treated zinc.

occur rapidly. Preliminary treatment often is needed to ensure that all of the element that is being analyzed is in the same oxidation state at the start. It is also important to ensure that the titrant reacts only with the element of interest.

In practice, the titrant usually is an oxidizing agent. Some common oxidizing agents are solutions of potassium permanganate, $MnO_4^-$; potassium dichromate, $Cr_2O_7^{2-}$; cerium(IV) sulfate, $Ce^{4+}$; potassium bromate, $BrO_3^-$; and iodine, $I_2$. Many of these substances are colored and so can be used as self-indicators. Some common reducing agents are solutions of iron(II) cation, $Fe^{2+}$; sodium arsenite, $AsO_3^{3-}$; sodium oxalate, $C_2O_4^{2-}$; oxalic acid, $H_2C_2O_4$; and sodium thiosulfate, $S_2O_3^{2-}$.

One common oxidation-reduction titration is the determination of the quantity of iron in a sample of iron ore. The first step is to dissolve the iron ore in a mixture of perchloric acid and phosphoric acid. In this solution, the iron ore exists as a complex ion:

$$Fe_2O_3(s) + 2H^+(aq) + 2H_3PO_4(aq) \rightleftharpoons 2Fe(HPO_4)^+(aq) + 3H_2O$$

If we are to carry out an oxidation-reduction titration using an oxidizing agent, the iron must first be reduced from its $+3$ oxidation state in this ore to the $+2$ oxidation state. The solution is passed through a column filled with specially treated zinc, a so-called Jones reductor (Figure 16.1). The reaction can be represented as:

$$2Fe^{3+}(aq) + Zn(s) \rightleftharpoons 2Fe^{2+}(aq) + Zn^{2+}(aq)$$

The titration is carried out by oxidation of the $Fe^{2+}$ cation. The oxidizing agent is a solution of potassium permanganate, and the reaction between the two substances is rapid and essentially complete:

$$MnO_4^-(aq) + 5Fe^{2+}(aq) + 8H^+(aq) \rightleftharpoons Mn^{2+}(aq) + 5Fe^{3+}(aq) + 4H_2O$$

According to the stoichiometry of this process, 0.2 mol of permanganate ion is needed to oxidize 1 mol of $Fe^{2+}$ ion. The equivalence point in the titration is reached when this relative quantity of $MnO_4^-$ is added from the buret. The oxidizing agent is the indicator for the titration. A solution of $KMnO_4$ is purple. Its reduction product, the $Mn^{2+}$ ion, is pale pink. As the titrant is added to the sample, the color of the titrant changes. When all the $Fe^{2+}$ has been oxidized to $Fe^{3+}$, the permanganate no longer changes color. The addition of the first drop of titrant solution past the equivalence point causes an easily detected purple coloration. The amount of iron in the sample is determined by measurement of the amount of $MnO_4^-$ added at that point.

Oxidation-reduction calculations may seem more difficult than neutralizations, because we encounter more complex equations and more complicated stoichiometry. But if we start with a balanced chemical equation and use the molar relationships that arise from the equation, we can do any calculation without introducing extra concepts.

**Example 16.2**   A sample of a substance whose only oxidizable material is tin in the $+2$ state is titrated with a dichromate solution that is prepared by dissolving $1.226$ g of $K_2Cr_2O_7$ in enough water to give a total volume of $0.250$ L. A $0.821$-g sample of the substance requires a volume of $23.9$ cm$^3$ of the titrant to reach the equivalence point. The product of the oxidation of the $Sn^{2+}$ is the $Sn^{4+}$ cation. The reduction product of the dichromate is the $Cr^{3+}$ cation. Calculate the percent of tin (by mass) in the substance.

**Solution**   The half-reactions for this oxidation-reduction are

$$Sn^{2+}(aq) \longrightarrow Sn^{4+}(aq) + 2e^-$$
$$Cr_2O_7{}^{2-}(aq) + 14H^+(aq) + 6e^- \longrightarrow 2Cr^{3+}(aq) + 7H_2O$$

We combine the half-reactions to get the overall reaction for the process:

$$3Sn^{2+}(aq) + Cr_2O_7{}^{2-}(aq) + 14H^+(aq) \longrightarrow 3Sn^{4+}(aq) + 2Cr^{3+}(aq) + 7H_2O$$

This equation tells us that 1 mol of dichromate oxidizes 3 mol of tin.
We find the concentration of the solution of dichromate:

$$[Cr_2O_7{}^{2-}] = 1.226 \text{ g} \times \frac{1 \text{ mol K}_2Cr_2O_7}{294.2 \text{ g K}_2Cr_2O_7} \times \frac{1}{0.250 \text{ L}} = 0.01667M$$

The amount of dichromate required to reach the equivalence point was

$$\text{amount Cr}_2O_7{}^{2-} = 0.01667 \frac{\text{mol}}{\text{L}} \times 23.9 \text{ cm}^3 \times \frac{1 \text{ L}}{1000 \text{ cm}^3}$$
$$= 3.98 \times 10^{-4} \text{ mol Cr}_2O_7{}^{2-}$$

From the stoichiometry of the reaction, the amount of tin in the sample is

$$\text{amount Sn}^{2+} = \frac{3 \text{ mol Sn}^{2+}}{1 \text{ mol Cr}_2O_7{}^{2-}} \times 3.98 \times 10^{-4} \text{ mol Cr}_2O_7{}^{2-}$$
$$= 1.20 \times 10^{-3} \text{ mol Sn}^{2+}$$

The mass of the tin in the sample is

$$\text{mass Sn} = 1.20 \times 10^{-3} \text{ mol Sn} \times \frac{118.7 \text{ g Sn}}{1 \text{ mol Sn}} = 1.42 \times 10^{-1} \text{ g Sn}$$

and the percentage of tin in the sample is

$$\text{percentage Sn} = \frac{1.42 \times 10^{-1} \text{ g Sn}}{0.821 \text{ g sample}} \times 100\% = 17.3\%$$

If you examine this problem, you will see that once we write the balanced equation for the oxidation-reduction reaction on which the titration is based, all the remaining steps are familiar from neutralization problems.

## 16.3  ELECTROLYSIS AND FARADAY'S LAWS

Electrolysis is a chemical change that results from the interaction between matter and an electric current. Electrolysis is carried out in an apparatus called an electrolytic cell. Figure 16.2 shows a simplified electrolytic cell. The reacting chemical system, which usually is a liquid or a liquid solution, is connected to the source of electricity through two electrodes. The electrodes usually are solid rods of inert conducting material, such as graphite or platinum.

Generally, the chemical change that takes place in an electrolysis is an oxidation-reduction process. *The reduction takes place at the cathode* and *the oxidation takes place at the anode*—not only in electrolytic cells but, as we shall see, in *all* electrochemical cells.

An electric current is a flow of electrons. In an electrolytic cell, the electrons enter the system at the cathode, where the reduction—a gain of electrons—takes place. When electrons enter the system, there must be a way for electrons to leave the system. Otherwise, there will be a buildup of negative charge that will cause the flow of electrons to stop. Electrons leave the system at the anode, where the oxidation—a loss of electrons—takes place. In this way, the circuit is completed.

The chemical changes of an electrolysis are the result of the way in which ionic substances interact with electricity. In a metal, an electric current is simply the flow of the electrons of the metal, with no other changes. In an ionic substance, the application of an electric current causes electron transfer from the circuit to ions at the cathode and from

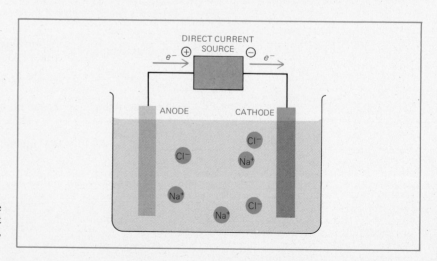

**Figure 16.2**
Electrolysis. Electrons enter the system at the cathode. They are transferred to the system at the cathode and leave the system at the anode, completing the circuit.

ions to the circuit at the anode. These two electron transfers result in an oxidation-reduction process.

The electrolysis of molten sodium chloride (Figure 16.3) is an example. Electrons are accepted by the sodium cations at the surface of the cathode. The process, which can be represented by a half-reaction:

$$Na^+ + e^- \longrightarrow Na$$

is the reduction of sodium from the $+1$ oxidation state to the 0 oxidation state. As the process proceeds, $Na^+$ cations migrate to the cathode to accept the electrons that enter the system. The electrical circuit is completed by the release of electrons from $Cl^-$ anions to the anode, a process in which chlorine is oxidized from the $-1$ oxidation state to the 0 state:

$$Cl^- \longrightarrow \tfrac{1}{2}Cl_2 + e^-$$

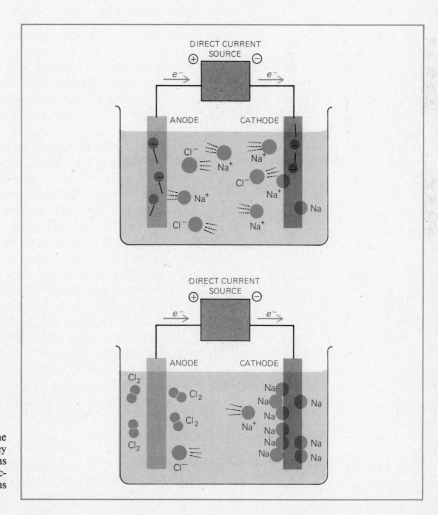

**Figure 16.3**

Electrolysis of molten sodium chloride. The $Na^+$ cations migrate to the cathode, where they gain electrons and are reduced. The $Cl^-$ anions migrate to the anode, where they release electrons and are oxidized. This flow of electrons maintains the current.

As this process proceeds, $Cl^-$ anions migrate to the anode to release electrons that leave the system, preventing the buildup of negative charge. Thus, we see that the current in the molten salt is maintained by the movement of ions, not electrons. Cations move to the cathode to be reduced, anions move to the anode to be oxidized, and the charge passes through the medium. The direct current source is the driving force for these reactions.

In an electrolysis, *the half-reaction that occurs at the cathode is a reduction and the half-reaction that occurs at the anode is an oxidation.* As in any oxidation-reduction process, we obtain the overall chemical change by adding the two half-reactions, eliminating the electrons in the addition. In this case the overall reaction is

$$Na^+ + Cl^- \longrightarrow Na + \tfrac{1}{2}Cl_2$$

## Electrical Energy and Chemical Energy

We can think of an electrolysis as the conversion of electrical energy to chemical energy. In the electrolysis of sodium chloride, for example, a chemical system in an equilibrium state is changed to a nonequilibrium state by an external stress, the electric current. In the system

$$Na^+ + Cl^- \rightleftharpoons Na + \tfrac{1}{2}Cl_2$$

the equilibrium lies far to the left. The sodium and chlorine exist virtually completely as ions. The electric current supplies the energy that forces the system to the right. If the current is turned off, the system does not return to equilibrium immediately, because the sodium metal and $Cl_2$ that form are kept apart by the distance between the electrodes of the electrolytic cell. But when the Na and $Cl_2$ are brought together, they react violently, forming NaCl and releasing the large quantity of energy that it took to form Na and $Cl_2$ from the ions by the original electrolysis. We can imagine that the electrical energy that went into the electrolysis was stored in the nonequilibrium system of Na and $Cl_2$ as chemical energy and was then released as thermal energy.

Electrolysis has many practical uses because it is a good way to force chemical systems into nonequilibrium states. Electrolysis is often used to prepare materials that would not otherwise be available. For example, aluminum, the world's most widely used nonferrous metal, is produced in vast quantities by a process in which $Al_2O_3$, from bauxite, is electrolyzed at high temperatures in a solution of molten cryolite, $Na_3AlF_6$. Before Charles Hall developed this process in 1886, aluminum was a rare and costly metal.

Electrolysis can occur not only in molten salts but also in aqueous solutions, even those with relatively low concentrations of ions. The processes at the electrodes in aqueous solutions often differ chemically from those in molten salts. Water molecules can accept electrons from the cathode, so the reaction that occurs at the cathode depends on the dissolved cations. If a cation accepts electrons readily, the cathode process probably will be its reduction. If a cation does not accept electrons readily, the reduction of water may be the cathode process.

In the electrolysis of a *solution* of sodium chloride, for example (Figure 16.4), the reduction of water is the cathode process:

$$2H_2O + 2e^- \longrightarrow H_2(g) + 2OH^-(aq)$$

The $Na^+$ cation is more difficult to reduce than water itself. But in a solution containing $Ag^+$ cations, which readily accept electrons, the reduction process is

$$Ag^+(aq) + e^- \longrightarrow Ag(s)$$

and silver metal plates out at the cathode.

In acidic solutions with relatively high concentrations of $H^+$, the cathode process for the reduction of water can be written in a simpler way as:

$$H^+(aq) + e^- \longrightarrow \tfrac{1}{2}H_2(g)$$

The oxidation that occurs at the anode of aqueous solutions depends on the anions. The anode process in a solution of sodium chloride generally is

$$Cl^-(aq) \longrightarrow \tfrac{1}{2}Cl_2(g) + e^-$$

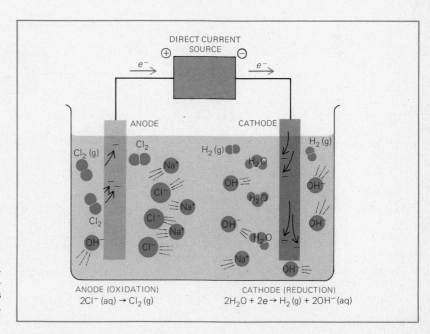

**Figure 16.4**

Electrolysis of a solution of sodium chloride. The $Cl^-$ anions migrate to the anode and release electrons, but the presence of water prevents the $Na^+$ cations from accepting electrons at the cathode. The cathode process is, instead, the reduction of $H_2O$.

Chlorine gas usually is manufactured by the electrolysis of brine solutions, concentrated aqueous solutions of sodium chloride.

If the anion is sulfate, $SO_4^{2-}$, or nitrate, $NO_3^-$, which are extremely difficult to oxidize, the anode reaction usually is the oxidation of water:

$$2H_2O \longrightarrow O_2(g) + 4H^+(aq) + 4e^-$$

In an alkaline solution, the oxidation can be written as

$$4OH^-(aq) \longrightarrow O_2(g) + 2H_2O + 4e^-$$

If neither the anion nor the cation in a solution participates in the electrode processes (Figure 16.5), only the water molecules react. The electrolysis of water is much less expensive when electrolytes such as $NaNO_3$ are present. The ions $Na^+$ and $NO_3^-$ do not undergo chemical change, but they greatly facilitate the electrolysis of water because they carry the current. In such a solution, the electrolysis of water can be represented by the sum of the two half-reactions for the oxidation and reduction of $H_2O$:

$$6H_2O \longrightarrow 2H_2(g) + O_2(g) + 4H^+(aq) + 4OH^-(aq)$$

In this reaction, $H^+$ is produced at the anode and $OH^-$ at the cathode. These ions diffuse through the solution, neutralizing each other. Since the ions in the reaction form four $H_2O$ molecules, the overall reaction for the electrolysis of water can be written as

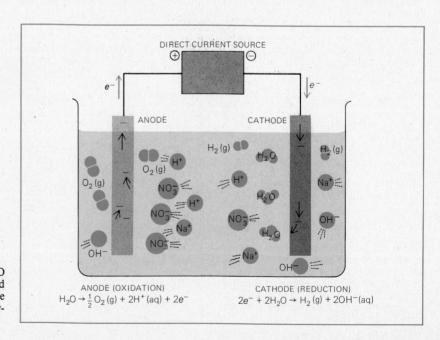

**Figure 16.5**
Electrolysis of water. At the anode, the $H_2O$ molecule loses electrons to form oxygen gas and protons. At the cathode, the $H_2O$ molecule gains electrons to form hydrogen gas and hydroxide ion.

ANODE (OXIDATION)
$$H_2O \rightarrow \tfrac{1}{2}O_2(g) + 2H^+(aq) + 2e^-$$

CATHODE (REDUCTION)
$$2e^- + 2H_2O \rightarrow H_2(g) + 2OH^-(aq)$$

$$2H_2O \longrightarrow 2H_2(g) + O_2(g)$$

Again, an electric current has caused a chemical system to go from equilibrium to nonequilibrium. In this reaction, the energy from the electrical current is stored in the mixture of hydrogen and oxygen. It can be released when a spark causes the two gases to combine explosively, forming water.

### Faraday's Laws

In the early 1830s, Michael Faraday (1791–1867) systematically investigated the newly discovered process of electrolysis, discovering relationships that today are called Faraday's laws.

One of Faraday's achievements was to demonstrate that the chemical changes that occur when an electric current is passed through a given liquid are related to the quantity of electricity and the nature of the substance. Faraday described the relationship between a quantity of electricity and the mass of a substance that undergoes chemical change. In modern terms, Faraday's laws are

1. The mass of a substance formed or consumed in an electrolysis is proportional to the amount of charge passing through it.
2. The mass of a substance formed or consumed in an electrolysis is also proportional to its atomic or molecular weight.
3. The mass of a substance formed or liberated in an electrolysis is inversely proportional to the number of electrons per mole needed to cause the indicated change in oxidation state.

To use these relationships, we need units of quantity for electricity. An electric current is a flow of electrons. Each electron has the same negative charge. The quantity of electricity is most simply designated as the total charge of the electrons in the current. The unit of current in the SI is the ampere, whose symbol is A. We define a current of 1 A as the flow of 1 coulomb, C, of charge, ($6.24 \times 10^{18}$ electrons), for 1 second, s, past a given point. This relationship between charge, current, and time usually is written as

$$i = \frac{q}{t} \tag{16.1}$$

where $i$ is current, $t$ is time and $q$ is charge, measured respectively in coulombs, amperes, and seconds.

Once we know that a quantity of electricity (or charge) is a specific number of electrons, we can express the relationships of Faraday's laws in simple chemical terms. Thus, the half-reaction for the cathode process in the electrolysis of molten sodium chloride,

$$Na^+ + e^- \longrightarrow Na$$

can be described as the combination of 1 mol of sodium cations and 1 mol of electrons to form 1 mol of sodium metal. The half-reaction for the anode process:

$$Cl^- \longrightarrow \tfrac{1}{2}Cl_2 + e^-$$

can be described as the formation of 0.5 mol of chlorine and 1 mol of electrons from 1 mol of chloride ion.

Here we have an explanation for Faraday's first law. It is evident why a given quantity of charge produces a given quantity of sodium metal or of chlorine. The transfer of 1 mol of electrons through molten sodium chloride causes the formation of 1 mol of sodium metal and 0.5 mol of chlorine. By treating the symbol $e^-$ in a half-reaction as if it were the symbol for an ordinary chemical reagent, we can understand electrolysis without introducing any new concepts.

We can also understand why the quantity of material formed or consumed in electrolysis is inversely proportional to the number of electrons per mole required to bring about the indicated change in oxidation state. One mole of electrons will cause the formation of 1 mol of sodium metal in the electrolysis of molten sodium chloride. But in the electrolysis of a molten magnesium salt,

$$Mg^{2+} + 2e^- \longrightarrow Mg$$

one mole of electrons forms only 0.5 mol of magnesium metal. Twice as many electrons per mole are needed to bring about the indicated change in oxidation state, because the charge on the magnesium ion is twice the charge on the sodium ion.

We can see this effect even more clearly in the electrolysis of metal salts that display more than one oxidation state. The reduction of tin in the $+4$ oxidation state to the metal requires 4 mol of electrons per mole of tin:

$$Sn^{4+} + 4e^- \longrightarrow Sn$$

Since the atomic weight of tin is 118.7, when 1 mol of electrons is passed through a system containing $Sn^{4+}$, a mass of $(118.7 \text{ g Sn/mol Sn})/(4 \text{ mol } e^-/\text{mol Sn}) = 29.68$ g Sn forms. But in the reduction of $Sn^{2+}$ to the metal, only 2 mol of electrons are required per mole of tin:

$$Sn^{2+} + 2e^- \longrightarrow Sn$$

A mass of $(118.7 \text{ g Sn/mol Sn})/(2 \text{ mol } e^-/\text{mol Sn}) = 59.35$ g Sn metal forms from 1 mol of electrons.

A unit other than the mole is more convenient when we are dealing with electrons. When we talk about "a mole," we usually indicate the mass of the substance: 1 mol of Na is 22.990 g, and of $Cl_2$ is 70.906 g. It is more convenient to express a mole of electrons as charge in coulombs, rather than as mass in grams, because the mass of electrons is so small. We usually express 1 mol of electrons as 96 500 C of charge (a better value is $9.6487 \times 10^4$ C), a quantity that is called the **Faraday constant.** The charge on a single electron is $1.6022 \times 10^{-19}$ C. If we multiply this charge by Avogadro's number, we get

$$\left(1.6022 \times 10^{-19} \frac{C}{e^-}\right)\left(6.0221 \times 10^{23} \frac{e^-}{mol\ e^-}\right) = 9.6486 \times 10^4 \frac{C}{mol\ e^-}$$

which agrees closely with the value of Faraday's constant measured from mass-charge relationships in electrolysis.

The quantity of material formed by 96 500 C of charge is sometimes called an equivalent. We shall use moles, not equivalents, for our calculations.

---

**Example 16.3**

When a solution of potassium iodide is electrolyzed, $I_2$ is produced at the anode and $H_2$ at the cathode. What mass of each substance is formed when a current of 5.20 A is applied for 46 min?

**Solution**

The quantity of charge that flows through the solution determines the quantity of products in an electrolysis. The first step, therefore, is to convert the data we are given on current and time to a quantity of charge, using Equation 16.1:

$$q = it = (5.20\ A)(46\ min)\left(60\ \frac{s}{min}\right) = 14\ 400\ C$$

Note that the number of minutes is multiplied by 60 to express time in seconds, the units of time for this relationship.

Now the balanced half-reactions for the two electrode processes must be written. These half-reactions will give us the number of moles of electrons needed to bring about a change in oxidation state per mole of substance.

$$\text{anode:} \qquad I^-(aq) \longrightarrow \tfrac{1}{2}I_2(aq) + e^-$$
$$\text{cathode:} \quad H_2O + e^- \longrightarrow \tfrac{1}{2}H_2(g) + OH^-(aq)$$

In each process, 96 500 C will form a half-mole of the product of interest. Using this information, we can find the mass of each product that is formed by 14 400 C:

$$\text{mass } I_2 = 14\ 400\ C \times \frac{0.5\ mol\ I_2}{96\ 500\ C} \times \frac{254\ g\ I_2}{1\ mol\ I_2} = 19.0\ g\ I_2$$

We can use the same sort of calculation to find the mass of $H_2(g)$ that is formed, using the molar mass of $H_2$, 2.02 g/mol:

$$\text{mass } H_2 = 0.150\ g\ H_2$$

---

In Example 16.3, we determined the quantity of material formed in an electrolysis by measuring the total quantity of charge. We can reverse the reasoning to get information about current or time by measuring the mass of products formed in an electrolysis.

**Example 16.4**    The production of aluminum metal from bauxite ore, $Al_2O_3$, by electrolysis is the cornerstone of the aluminum industry. One step in this complex process is the reduction of aluminum from the $+3$ to the $0$ oxidation state. Calculate the time needed for a current of 11.2 A to deposit 454 g of metallic aluminum.

**Solution**    The half-reaction for the reduction of aluminum:

$$Al^{3+} + 3e^- \longrightarrow Al$$

shows that 3 mol of electrons, or 3(96 500) C of charge, are needed to form 1 mol of Al. The charge required to form the desired quantity thus is

$$454 \text{ g Al} \times \frac{1 \text{ mol Al}}{27.0 \text{ g Al}} \times \frac{3(96\ 500) \text{ C}}{1 \text{ mol Al}} = 4.87 \times 10^6 \text{ C}$$

To calculate the time needed to obtain this quantity of charge from a current of 11.2 A, we use Equation 16.1:

$$4.87 \times 10^6 \text{ C} = (11.2 \text{ A})(t)$$

and
$$t = 4.35 \times 10^5 \text{ s}    \text{or}    121 \text{ hours}$$

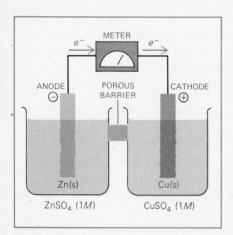

**Figure 16.6**
A galvanic cell. An oxidation process occurs in the half-cell on the left and a reduction process occurs in the half-cell on the right. The electrons travel through the wire at the top, creating an electric current. The meter measures the quantity of current. The porous barrier allows movement of ions to preserve neutrality but hinders the solutions from mixing.

## 16.4  GALVANIC CELLS

Oxidation-reduction processes can occur even when the reactants are separated from one another. This remarkable phenomenon occurs because an electron transfer from one reactant to another can take place through an electrical conductor, usually a metal wire.

An apparatus for carrying out an oxidation-reduction process in this way is called a **galvanic** or **voltaic cell** (Figure 16.6). You will notice a resemblance to the electrolytic cells mentioned in Section 16.3. The major difference is that an electrolytic cell has an external source of current but the galvanic cell does not. The current is created by the electron transfer of an oxidation-reduction process within the cell. The galvanic cell shown in Figure 16.6 has a barrier, usually not found in an electrolytic cell. It hinders the solutions in the anode and cathode compartments from mixing but allows ion flow between the two solutions.

We can best regard the galvanic cell as a combination of two half-cells, each consisting of an electrode in contact with a solution. A half-reaction occurs in each half-cell. Such a half-reaction is often called an *electrode process,* which gives a different meaning to the term electrode. Instead of referring only to the bar of solid attached to an external electrical connection, *electrode* in a galvanic cell often refers to an entire half-cell.

## Operation of a Galvanic Cell

If we connect the electrodes in the galvanic cell of Figure 16.6, we observe distinct chemical changes. The zinc electrode shrinks and the copper electrode grows. At the same time, the $[Zn^{2+}]$ becomes greater than $1M$ and the $[Cu^{2+}]$ becomes less than $1M$. These changes can be described by a pair of half-reactions:

$$Zn(s) \longrightarrow Zn^{2+}(aq) + 2e^-$$
$$Cu^{2+}(aq) + 2e^- \longrightarrow Cu(s)$$

The first half-reaction is an oxidation. *Oxidation always takes place at the anode* of a galvanic cell. The second half-reaction is a reduction. *Reduction always takes place at the cathode* of a galvanic cell. In an electrolytic cell, the electrons come from an external source. In a galvanic cell, the electrons originate in the reacting chemical system. Electrons flow in the external circuit from the anode to the cathode in the galvanic cell. Therefore, the anode must be negatively charged and the cathode must be positively charged. The electrons liberated by the anodic oxidation travel through the wire to the cathode, where they are accepted by the species that is reduced.

It seems that as this cell operates and more $Zn^{2+}$ is formed, the positive charge of the anode compartment should increase. At the same time, the positive charge of the cathode compartment should decrease as more $Cu^{2+}$ ion is removed from solution. But electrical neutrality must be maintained if the cell is to continue to function. Negative charge must enter the anode compartment and leave the cathode compartment. This transfer of charge does, indeed, occur. The $SO_4^{2-}$ anion passes from the cathode compartment to the anode compartment, through the porous barrier. The $SO_4^{2-}$ anions do not undergo any chemical change; their migration serves only to preserve electrical neutrality and complete the electric circuit.

We can describe the overall operation of this galvanic cell by writing an equation that is the sum of the two half-reactions for the electrode processes of the two half-cells:

$$Zn(s) + Cu^{2+}(aq) \rightleftharpoons Zn^{2+}(aq) + Cu(s)$$

The description given earlier indicates that the reaction will proceed from left to right when each cation has a concentration of $1M$. Since the reaction proceeds spontaneously from left to right, in the equilibrium state there is a preponderance of the substances on the right side of the equation. The value of the equilibrium constant, $K$, of the system is large, so the $[Zn^{2+}]$ will be much greater than the $[Cu^{2+}]$ at equilibrium.

The reaction that takes place in the galvanic cell is precisely the spontaneous reaction that takes place when the components of the cell are mixed in a beaker. The difference is that in a beaker the electron transfer does not take place through a wire.

While proceeding spontaneously to an equilibrium state, a galvanic cell produces an electric current that can be used to perform work. In other words, a galvanic cell is a device that converts chemical energy into electrical energy. This capability is put to practical use on an enormous scale. The batteries that power our automobile electrical systems, our flashlights, tape recorders, portable radios, golf carts—even our spacecraft—are galvanic cells. This wide range of applications helps explain why so much effort has gone into inventing different kinds of galvanic cells. All these cells are similar in essence. Every galvanic cell has a set of components that are not at equilibrium and that proceed spontaneously to equilibrium by an electron transfer. The electron transfer occurs through an electrical connection between the electrodes, rather than directly. Once the system reaches equilibrium, the cell stops producing current.

### Notation for Galvanic Cells

A conventional notation has been developed to describe galvanic cells. In this notation, the cell pictured in Figure 16.6 is represented as

$$Zn(s)|Zn^{2+}(aq,1M)\|Cu^{2+}(aq),1M)|Cu(s)$$

The single vertical lines indicate boundaries between phases. The double vertical line indicates a barrier that allows the movement of ions but not the mixing of solutions. By convention, the anode half-cell, where oxidation takes place, is on the left; the cathode, where reduction takes place, is on the right. The notation for this system does not include the $SO_4^{2-}$ anions, since they serve only to maintain neutrality.

Many different kinds of electrodes and barriers can be used in galvanic cells. Figure 16.7 shows a galvanic cell that can be represented as

$$Pt(s)|Sn^{2+}(aq,1M), Sn^{4+}(aq,1M)\|Fe^{2+}(aq,1M), Fe^{3+}(aq,1M)|Pt(s)$$

In this system, the oxidation half-reaction that occurs in the anode half-cell is

$$Sn^{2+}(aq) \longrightarrow Sn^{4+}(aq) + 2e^-$$

and the reduction half-reaction that occurs in the cathode is

$$Fe^{3+}(aq) + e^- \longrightarrow Fe^{2+}(aq)$$

The overall cell reaction is obtained by adding the half-reactions so as to cancel the electrons from the equation:

$$Sn^{2+}(aq) + 2Fe^{3+}(aq) \longrightarrow Sn^{4+}(aq) + 2Fe^{2+}(aq)$$

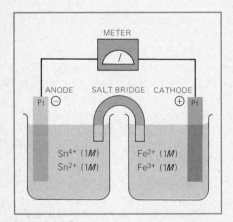

**Figure 16.7**

A tin-iron galvanic cell with inert electrodes. Oxidation occurs in the tin electrode at the left, reduction in the iron electrode at the right. The salt bridge hinders the solutions from mixing but allows the flow of anions into the anode compartment and cations into the cathode compartment to maintain electrical neutrality. These ions are usually from the salt of the salt bridge. Common salts are KCl and KNO$_3$.

## THE STALLED ELECTRIC AUTOMOBILE

In the 1970s, both government and industrial research laboratories were mounting major efforts to develop "superbatteries" that were regarded as one important answer to an energy crisis. The search for a superbattery was based on two beliefs. One was that the price of oil would go on rising steadily, so electric automobiles would become competitive with gasoline-powered vehicles. The other was the knowledge that lead-acid batteries, which dominated the market for rechargeable electric sources, were not good enough for the new generation of electric automobiles.

Ambitious plans were made. The government said that it planned to have a fleet of more than 5000 electric automobiles operating on superbatteries by the mid-1980s. General Motors announced plans to market an electric town car, suitable for shopping or commuting, by the middle of the decade.

The government's electric-powered fleet never materialized, and General Motors has never marketed its electric town car. The primary reason was that the price of oil leveled off in the 1980s, as conservation and a global recession reduced petroleum demand. Without a continuing increase in oil prices, electric vehicles could not compete with the current generation of high-mileage gasoline-powered automobiles. Some progress toward a superbattery was made, but not enough to justify a major expenditure on the project.

Today's lead-acid batteries have a power density of 30 watt hours per kilogram (watt hr/kg) and a life of about 700 charge-discharge cycles. A typical 1700-kg electric car today needs 500 kg of lead-acid batteries to achieve a range of 30 km at a steady speed of 11 km an hour, a performance that is markedly inferior to that of gasoline-powered vehicles. The goal of industrial and government research efforts was a superbattery that would give a major increase in driving range with a reduction of 50% or more in battery weight.

One candidate was a zinc-nickel oxide battery. In 1979, General Motors said that it had a research model of such a battery with a power density of more than 60 watt hr/kg. Another proposed superbattery uses molten sodium for the anode, a mixture of molten sulfur and sodium polysulfide as the cathode, and a solid ceramic material that is an excellent conductor of sodium ions, as the barrier between electrodes. One sodium-sulfur battery, operating at about 600 K, achieved a power density of more than 120 watt hr/kg.

A major difficulty of those systems was their high operating temperatures. That problem was overcome by a zinc-chlorine system with an aqueous electrolyte. One unusual feature is the method used to store chlorine when the battery is charged. The chlorine gas is chilled in the presence of the water to form solid chlorine hydrate. To discharge the battery, the frozen chlorine hydrate is heated. The chlorine evaporates and can then take part in the reaction.

It is still possible that quiet, nonpolluting electric vehicles powered by future versions of these superbatteries may fill city streets and highways. At the moment, however, economic realities make large-scale production of electric automobiles seem a distant prospect.

In this cell, ions move between the two solutions through a barrier called a *salt bridge,* a tube filled with a concentrated solution of a salt, such as potassium chloride, that does not participate in the cell reaction. A salt bridge allows ions to pass as needed to maintain electrical neutrality but prevents the solutions from mixing.

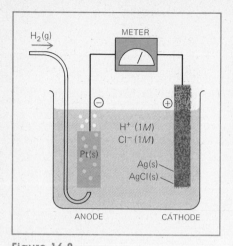

**Figure 16.8**
In this galvanic cell, the anode at the left has a gas electrode, consisting of a stream of hydrogen bubbles in contact with both the solution of $H^+$ and the platinum strip. The cathode at the right has an electrode consisting of silver and one of its insoluble salts, silver chloride.

## Types of Galvanic Cells

Galvanic cells can have many kinds of electrodes. A galvanic cell can have a gas electrode, consisting of a strip of an inert conductor in contact with both the liquid solution and a stream of gas whose pressure is kept constant. Other cells can have a metal-insoluble salt electrode, in which the metal of the electrode is coated with one of the sparingly soluble salts of the metal. The salt is in contact with the solution containing the anion of the metal salt.

Figure 16.8 shows a galvanic cell whose anode is a gas electrode and whose cathode is a metal-insoluble salt electrode. The cell can be represented as

$$Pt(s)|H_2(g)(1\ atm)|H^+(aq, 1M),\ Cl^-(aq, 1M)|AgCl(s)|Ag(s)$$

The anode half-reaction is

$$H_2(g) \longrightarrow 2H^+(aq) + 2e^-$$

This reaction takes place because a stream of hydrogen bubbles, at a pressure of 1 atm, is in contact with both a $1M$ solution of $H^+$ and a platinum strip. This system is called the *standard hydrogen electrode*. We shall discuss it in greater detail later in this chapter.

The metal of the cathode is silver, and the salt is therefore a silver salt. Since the anion of the silver salt is chloride, the solution contains chloride ion. The half-reaction is

$$AgCl(s) + e^- \longrightarrow Ag(s) + Cl^-(aq)$$

This cell does not contain a porous barrier or salt bridge because the reactants are insoluble and do not mix. The overall reaction of the cell is

$$H_2(g) + 2AgCl(s) \longrightarrow 2H^+(aq) + 2Ag(s) + 2Cl^-(aq)$$

Galvanic cells such as this one that have the same solution at both electrodes are quite common.

---

**Example 16.5**    Write the two half-cell processes and the overall reaction that takes place in the cell:

$$Zn(s)|NH_3(aq, 1M),\ Zn(NH_3)_4^{2+}(aq, 1M)\|Br^-(aq, 1M)|Br_2(l)|Pt(s)$$

**Solution**    By convention, the half-cell on the left is the anode, in which the oxidation takes place. The material undergoing a change in oxidation state is zinc, which goes from the 0 oxidation state in the metal to the +2 state in the complex ion. The balanced half-reaction is

$$Zn(s) + 4NH_3(aq) \longrightarrow Zn(NH_3)_4{}^{2+}(aq) + 2e^-$$

The cathode is on the right, separated from the anode by the porous barrier that is represented by the double vertical line. In the cathode, bromine is reduced from the 0 oxidation state in $Br_2(l)$ to the $-1$ state in the bromide ion. The half-reaction is

$$Br_2(l) + 2e^- \longrightarrow 2Br^-(aq)$$

We obtain the overall cell reaction by adding the half-reactions so as to cancel the electrons from the equation:

$$Zn(s) + 4NH_3(aq) + Br_2(l) \longrightarrow Zn(NH_3)_4{}^{2+}(aq) + 2Br^-(aq)$$

## Batteries

A battery is one or more galvanic cells. Batteries have several unique advantages. They convert chemical energy to electrical energy and then to work with much greater efficiency than does a steam engine or a diesel engine. The energy of a galvanic cell can be stored until it is needed if the electrodes simply are kept unconnected. The most familiar galvanic cell is the dry cell (Figure 16.9), which is available in several forms, including the flashlight battery. The anode, the source of electrons, is the zinc of the cell wall. The oxidation process at the anode is

$$Zn(s) \longrightarrow Zn^{2+} + 2e^-$$

The cathode is a rod of graphite surrounded by a paste of $MnO_2$ and carbon. The cathode reaction is complex but can be represented as

$$2NH_4{}^+ + 2MnO_2(s) + 2e^- \longrightarrow 2MnO(OH) + 2NH_3$$

Both electrodes are in contact with a wet paste of ammonium chloride, zinc chloride, water, and an inert filler, so the "dry cell" is not really dry.

The lead storage battery used in automobiles is really a number of galvanic cells connected in series. The galvanic cells are designed so that the cell reactions that produce current can be reversed by application of an external current. The battery is rechargeable and thus long lived.

The chemistry of the automobile battery is based on changes in the oxidation state of lead. In one half-cell, the oxidation state of lead is changed from 0 to $+2$. In the other half-cell, the oxidation state is changed from $+4$ to $+2$. The lead in the 0 oxidation state is in the form of a spongy alloy. The lead in the $+2$ oxidation state is in the form of $PbSO_4$, a white insoluble salt that can sometimes be seen as a coating on dead batteries. Lead in the $+4$ oxidation state is in the form of $PbO_2$, lead dioxide, which is also insoluble.

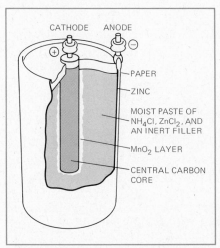

**Figure 16.9**
A dry cell. The anode is the zinc that makes up the cell wall, while the cathode is the graphite rod and the moist paste that surrounds it.

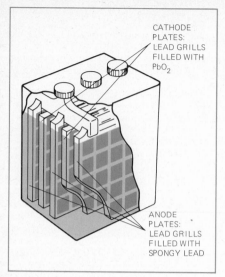

**Figure 16.10**
A lead storage battery from an automobile. The anodes are the plates filled with spongy lead, while the cathodes are plates filled with lead dioxide. The plates are immersed in a solution of sulfuric acid.

The anodes of a lead storage battery (Figure 16.10) are filled with the spongy lead alloy. The cathodes are filled with lead dioxide. Both electrodes are in contact with a solution of $H_2SO_4$, sulfuric acid, in water. The solution is about 38% sulfuric acid by mass, and its density is 1.30 g/cm³. The sulfuric acid participates in the cell reaction. Its role is to ensure that lead in the +2 oxidation state is produced in the form of the insoluble $PbSO_4$.

The chemical system of a charged lead storage battery is not at equilibrium, and an electric current flows when the plates of lead and of lead dioxide are connected. The oxidation process that occurs in the anode is

$$Pb(s) + SO_4^{2-} \longrightarrow PbSO_4(s) + 2e^-$$

The reduction that occurs when the lead dioxide plate, the cathode, receives the electrons released by the oxidation of lead is

$$PbO_2(s) + SO_4^{2-} + 4H^+ + 2e^- \longrightarrow PbSO_4(s) + 2H_2O$$

You will note that both these processes produce $PbSO_4$, in which lead is in the +2 oxidation state. The overall reaction for the cell is

$$Pb(s) + PbO_2(s) + 2SO_4^{2-} + 4H^+ \underset{\text{Charge}}{\overset{\text{Discharge}}{\rightleftharpoons}} 2PbSO_4(s) + 2H_2O$$

As the cell discharges electricity and proceeds toward equilibrium, insoluble lead sulfate is deposited on both plates and sulfuric acid is consumed. The consumption of $H_2SO_4$ reduces the density of the solution in the battery. The extent to which a battery has been discharged can be determined by measurement of the density of the solution with a hydrometer.

An automobile battery will last for several years. Its long life is based on the ability of the battery to function not only as a galvanic cell but also as an electrolytic cell. As a galvanic cell, the lead storage battery converts chemical energy into electrical energy. As an electrolytic cell, the battery converts electrical energy into chemical energy. The cycle can be repeated many times. When an automobile engine runs, it drives an electrical generator or alternator, a source of electric current. This current flows into the battery in a direction that reverses the two half-cell reactions and thus the overall cell reaction.

The direct conversion of chemical energy to electrical energy that takes place in galvanic cells is highly efficient, but it is also expensive. Fossil-fuel plants are only about 40% efficient, which means that most of their energy is lost to the surroundings. But even in these days of high energy prices, coal, oil, and natural gas are much cheaper than the reactants used in galvanic cells — so much cheaper that batteries are economically competitive only in specialized applications.

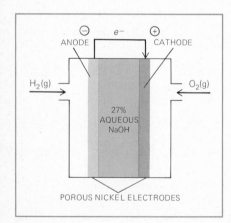

**Figure 16.11**
A fuel cell. Hydrogen gas is oxidized at the anode and oxygen gas is reduced at the cathode, causing a flow of electrons and an electric current. The only product is pure water.

## HOW LIBERTY RUSTED

After standing for nearly a century in New York harbor, the Statue of Liberty needed a major restoration project to restore it to sound condition for its centennial celebration. The major cause of the damage to Liberty was an unintended oxidation-reduction reaction that resulted from the combination of a damp harbor atmosphere and the combination of materials from which the statue was made.

The skin of the statue consists of about 300 sections of copper plate, each about 25 mm thick, held together by about 300 000 rivets. These copper plates are held in place by an ingenious structure that was designed by Gustave Eiffel. The supporting structure for the Statue of Liberty was Eiffel's largest work at the time. It was designed only a few years before the building of the Eiffel Tower.

Eiffel built a three-part structure. The central core is a pylon with four main columns braced by crossbeams. A secondary frame extends from the core, transmitting the strains exerted by the skin to the central pylon. The 300 copper plates, weighing a total of 80 tons, are held in place by the tertiary structure, called the armature. It consists of 1350 iron ribs, each shaped to fit the surface of the statue, attached to the skin by copper saddles.

When the statue was built, the iron ribs were insulated from the copper skin by a pitchlike material. Over the decades, however, the insulation wore away, so that the iron came into contact with the copper. As a result, the corrosion of the iron began to proceed rapidly. The electrons donated by the iron as it is oxidized are easily transferred to the oxygen and water vapor of the damp harbor atmosphere through the copper skin, a very good conductor of electrons. The copper-iron contact points provide an excellent pathway for the transfer of electrons.

The resulting anodic reaction caused the iron ribs to rust at an accelerated rate. When engineers inspected the statue in 1982, they found that many of the iron bars in the armature had rusted away almost completely. As rust built up, many of the copper saddles and rivets were pulled out of place, leaving holes in the copper skin.

The damage to the torch was so extensive that the statue's upraised arm had to be removed so that it could be rebuilt. To prevent the kind of oxidation-reduction reaction that had caused the accelerated rusting, engineers planned to use either a nonferrous metal or plastic-coated metal replacements for the damaged ribs. Once the repairs were made, they said, Liberty would be structurally sound for another century and more.

### Fuel Cells

In recent years, a great deal of research has gone into development of the fuel cell, which potentially combines the high efficiency of the galvanic cell and the low cost of fossil-fuel plants. The fuel cell uses the combustion reaction as the oxidation-reduction process of a galvanic cell. In a typical fuel cell, hydrogen combines with oxygen, producing electricity with high efficiency and low pollution. Unfortunately, such practical problems as the extremely high cost of electrodes have limited the use of fuel cells to applications where cost is not an important factor, notably in spacecraft such as the space shuttle.

The Bacon hydrogen-oxygen fuel cell, diagramed in Figure 16.11, has been used on space flights. The two half-reactions are

$$2H_2(g) \longrightarrow 4H^+ + 4e^-$$
$$O_2(g) + 4H^+ + 4e^- \longrightarrow 2H_2O$$

and the overall reaction is the familiar formation of water:

$$2H_2(g) + O_2(g) \longrightarrow 2H_2O(l)$$

Porous electrodes and other techniques are used to make the cell processes occur rapidly.

## 16.5   CELL POTENTIAL

It is the difference in potential energy between the two electrodes of a galvanic cell that causes charge to flow. This energy difference is most conveniently described as **potential,** which is the *potential energy per unit charge.* The unit of charge is the coulomb, C; the unit of potential energy is the joule, J; and the unit of potential is the volt, V:

$$1\ V = 1\ \frac{J}{C}$$

Charge flows only if there is a difference in potential between two points. In a galvanic cell, electrons flow from the anode to the cathode because the potential at the anode is more negative than at the cathode. The potential at any one point cannot be measured, but the difference in potential between two points can be measured with a voltmeter. As Figure 16.6 shows, a voltmeter can be placed in the line between the electrodes of a galvanic cell to measure the potential difference between the half-cells. The reading obtained in this way is called the potential of the cell; it is represented by the symbol $\mathscr{E}$.[1]

Voltmeter measurements show that the cell potential is affected by a number of factors: the nature of the substances that make up each half-cell, the concentrations of dissolved ions and molecules, the pressures of gases, and the temperature. These are the same factors that determine both the free energy difference between substances and the value of the equilibrium constant, $K$, for a chemical system.

A direct and useful relationship between free energy and potential can be derived. Free energy (Section 13.7) gives us a great deal of information about the equilibrium state of a chemical system and about the amount of work that can be obtained from a chemical system that is proceeding to an equilibrium state. We cannot measure the free energy of a system easily,

---

[1] The value of the potential measured in this way varies with the current of the cell. Ideally, we want the potential of the cell when the current is 0. This potential can be measured with an instrument called a potentiometer. The cell potential when the current is 0 is called the electromotive force (emf) of the cell. We shall use *potential* synonymously with *emf.*

Improvements in battery technology are needed to make the electric automobile a viable alternative to the gasoline-fueled auto. This experimental vehicle is powered by 900 lb of advanced zinc-nickel oxide batteries which can do the same job as 2000 lb of conventional lead-acid batteries. *(General Motors)*

but we can measure the potential of a galvanic cell. The potential in volts is proportional to the free energy change, so our measurements of the potential of galvanic cells gives us the same information that we would get by measuring the free energy change.

## Standard Potential

To compare potentials of different cells, we must define a standard state for a galvanic cell. This standard state is defined in terms of the concentration of dissolved substances and the pressure of gases, which together determine the value of the cell potential. The standard state for cell potential is the same as the standard state for free energy. In each, the concentration of all dissolved species is $1 M$,[2] and the pressure of all gases is 1 atm. When these conditions of concentration and pressure are met, the cell displays what is called **standard potential,** whose symbol is $\mathscr{E}°$. Since the $\mathscr{E}°$ of a galvanic cell is temperature dependent, we shall assume that all measurements are made at 298 K, unless otherwise stated. The cells shown in Figures 16.6, 16.7, and 16.8 are all in the standard state, and their measured potential is the standard potential.

[2] The standard state is actually unit activity, which we are approximating as unit molarity.

## Potential and Free Energy

The relationship between $\mathcal{E}°$ for a galvanic cell and $\Delta G°$, the standard free energy change, for the chemical reaction of the cell is

$$\Delta G° = -nF\mathcal{E}° \qquad (16.2)$$

where $\Delta G°$ is measured in joules, $\mathcal{E}°$ is measured in volts, $F$ is the Faraday, the charge on 1 mol of electrons, $9.6487 \times 10^4$ C, and $n$ is the number of moles of electrons transferred by the balanced cell reaction.

Calculating $n$ for a chemical reaction is not always easy. In the reaction for the cell of Figure 16.6,

$$Zn(s) + Cu^{2+}(aq) \longrightarrow Zn^{2+}(aq) + Cu(s)$$

we can see that the reaction occurs by the transfer of 2 mol of electrons. But the reaction of Example 16.1:

$$3S^{2-}(aq) + 2MnO_4^-(aq) + 4H_2O \longrightarrow 3S(s) + 2MnO_2(s) + 8OH^-(aq)$$

requires more analysis. The best approach is to break down the overall reaction into its two constituent half-reactions, in effect reversing the procedure by which the balanced overall reaction is obtained. The reaction of Example 16.1 is the sum of these half-reactions:

$$3S^{2-}(aq) \longrightarrow 3S(s) + 6e^-$$
$$2MnO_4^-(aq) + 4H_2O + 6e^- \longrightarrow 2MnO_2(s) + 8OH^-(aq)$$

You can see that six electrons are transferred by each half-reaction, so $n = 6$.

---

**Example 16.6**    We find that $\mathcal{E}°$ for the cell

$$Cu(s)|Cu^{2+}(aq, 1M)\|Ag^+(aq, 1M)|Ag(s)$$

is 0.4594 V. Determine the cell reaction and calculate $\Delta G°$.

**Solution**    The best approach is to write the two half-cell reactions. Since the anode half-cell is by convention the left side of the expression, the oxidation half-reaction is

$$Cu(s) \longrightarrow Cu^{2+}(aq) + 2e^-$$

and the reduction is

$$Ag^+(aq) + e^- \longrightarrow Ag(s)$$

Each half-reaction must have the same number of electrons. In this case, if we multiply the reduction half-reaction by two, each half-reaction then has two electrons. When the half-reactions are combined, the electrons cancel and the overall reaction is

$$Cu(s) + 2Ag^+(aq) \longrightarrow Cu^{2+}(aq) + 2Ag(s)$$

The number of electrons in each half-reaction is the number of moles of electrons transferred in the overall reaction. That is, $n = 2$. We now calculate $\Delta G°$, using the relationship expressed in Equation 16.2:

$$\Delta G° = -2 \text{ mol } e^- \left( 96\ 490\ \frac{C}{mol} \right)(0.4594 \text{ V}) = -88\ 660 \text{ J}$$

Note that the unit of free energy is the joule, since 1 V = 1 J/C.

You should note that while $\mathscr{E}°$ is an intensive property (Section 1.2), both $n$ and $\Delta G°$ are extensive properties. For example, in the system of Example 16.6, the value of $n$ for the reaction

$$\tfrac{1}{2}Cu(s) + Ag^+(aq) \longrightarrow \tfrac{1}{2}Cu^{2+}(aq) + Ag(s)$$

is only 1, and $\Delta G°$ for this reaction is $-88\ 660/2$ J. But $\mathscr{E}°$ for the reaction is still 0.4594 V.

In Chapter 13 we discussed the meaning of the standard free energy change, $\Delta G°$, for a chemical system. The sign of the free energy change gives the direction of spontaneous change, while the magnitude of $\Delta G°$ gives the extent to which the system must change from the standard state to reach equilibrium. The value of $\Delta G°$ also gives the amount of work that can be obtained from a reacting chemical system. In fact (Section 13.7), the magnitude of the equilibrium constant $K$ is directly related to the magnitude of the standard free energy change of the system:

$$\Delta G° = -RT \ln K \qquad\qquad (16.3)$$

We can get the same kind of information from the standard potential by using Equation 16.2. We know that the operation of a galvanic cell corresponds to a chemical reaction. The measured standard potential, $\mathscr{E}°$, of the galvanic cell gives us information about this reaction. To start with, note that Equation 16.2 includes a negative sign, which affects our interpretation of $\mathscr{E}°$. *The direction of spontaneous change gives a negative value for $\Delta G°$ but a positive value for $\mathscr{E}°$.*

## The Meaning of $\mathscr{E}°$

In the expression for a galvanic cell, the anode half-cell is written first, on the left by convention. If the measured $\mathscr{E}°$ of the cell is positive, we know that the system changes spontaneously from the standard state to equilibrium when the oxidation occurs in the half-cell that is written first. If the measured $\mathscr{E}°$ of the cell is negative, we know that we have written the expression for the cell incorrectly. Two examples will demonstrate. For the cell

$$Zn(s)|Zn^{2+}(aq,1M)\|Cu^{2+}(aq,1M)|Cu(s)$$

$\mathscr{E}° = 1.1030$ V. The positive value indicates that the system proceeds to equilibrium by an oxidation in the zinc half-cell and a reduction in the copper half-cell. The overall reaction

$$Zn(s) + Cu^{2+}(aq) \rightleftharpoons Zn^{2+}(aq) + Cu(s)$$

proceeds from left to right to reach the equilibrium state when all the components of the system are in their standard states.

But when we construct the cell

$$Zn(s)|Zn^{2+}(aq,1M)\|Mg^{2+}(aq,1M)|Mg(s)$$

we find that the standard potential of the cell is $-1.612$ V. The oxidation occurs in the magnesium half-cell and the reduction in the zinc half-cell. The negative value of the potential tells us that it is best to rewrite the expression for the cell as

$$Mg(s)|Mg^{2+}(aq,1M)\|Zn^{2+}(aq,1M)|Zn(s)$$

to indicate that the oxidation takes place in the anode, which is the half-cell written first. In its rewritten form, the cell has $\mathscr{E}° = 1.612$ V. The overall reaction is

$$Mg(s) + Zn^{2+}(aq) \rightleftharpoons Mg^{2+}(aq) + Zn(s)$$

The reaction proceeds from left to right to reach equilibrium when all the components of the system start in their standard states.

There are some rare cases where a galvanic cell has $\mathscr{E}° = 0$. In such a case, the system is at equilibrium when all its components are in their standard states. The equilibrium constant, $K$, for such a cell reaction has a numerical value of 1. In this system there is no potential difference between the electrodes when the concentration of each dissolved species is $1M$ and the pressure of each gas is 1 atm, so there is no net transfer of charge.

We can say that the sign of $\mathscr{E}°$ indicates the direction of spontaneous change and the magnitude of $\mathscr{E}°$ indicates the extent to which the system must change from the standard state to reach equilibrium. The magnitude of $\mathscr{E}°$ can also be used to find the value of $K$ for the cell reaction. We can derive this relationship by combining the relationship between $\Delta G°$ and $K$ (Equation 16.3) with the relationship between $\Delta G°$ and $\mathscr{E}°$ (Equation 16.2):

$$-nF \mathcal{E}^\circ = -RT \ln K$$

or $\quad\quad\quad\quad\quad \mathcal{E}^\circ = \dfrac{RT}{nF} \ln K \quad\quad\quad\quad\quad$ (16.4)

Equation 16.4 is used to calculate the equilibrium constant at a given temperature of any reaction for which $\mathcal{E}^\circ$ can be measured. Generally, we assume that the temperature is 298 K. We can obtain a simpler version of Equation 16.4 by using 298 K for $T$ and calculating the numerical value of the term $RT/F$:

$$\mathcal{E}^\circ = \dfrac{0.02568}{n} \ln K \quad\quad\quad\quad (16.5)$$

The units of $RT/F$ are joules/coulombs, or volts.

---

**Example 16.7**   The cell

$$Pt(s)|Fe^{2+}(aq,1M), Fe^{3+}(aq,1M)\|MnO_4^-(aq,1M),Mn^{2+}(aq,1M),$$
$$H^+(aq,1M)|Pt(s)$$

is found to have a potential of 0.721 V at 298 K. Calculate $K$ for the cell reaction.

**Solution**   Since all the components of the system are in their standard states, the measured potential is $\mathcal{E}^\circ$. Therefore, we can use Equation 16.5 to calculate the value of $K$.
   To be sure of the cell reaction and to help find the value of $n$, it is best to write the half-reactions.
   At the anode:

$$Fe^{2+}(aq) \longrightarrow Fe^{3+}(aq) + e^-$$

At the cathode:

$$MnO_4^-(aq) + 8H^+(aq) + 5e^- \longrightarrow Mn^{2+}(aq) + 4H_2O$$

We multiply the oxidation half-reaction by 5 to equalize the number of electrons. The electrons then cancel when the half-reactions are combined to obtain the overall reaction:

$$5Fe^{2+}(aq) + MnO_4^-(aq) + 8H^+(aq) \longrightarrow 5Fe^{3+}(aq) + Mn^{2+}(aq) + 4H_2O$$

Since each half-reaction has $5e^-$, when we combine them $n = 5$. Substituting the data into Equation 16.5 gives

$$0.721 \text{ V} = \dfrac{0.0257 \text{ V}}{5} \ln K$$

$$\ln K = 140.27$$
$$K = 8.3 \times 10^{60}$$

Electrochemical measurements usually are needed to evaluate equilibrium constants of such large magnitude. Potential measurements often are used for this purpose.

## 16.6  ELECTRODE POTENTIALS

No one has yet devised a way to measure the potential of a single electrode, and it seems probable that no one ever will. If such a method were available, it would be an easy matter to find the potentials of cells that have never been constructed and the equilibrium constants of reactions that have never been run. We would simply compare the potentials of different half-cells.

But the same thing can be done indirectly. Cell potential is the *difference* between two electrode potentials. If we know the relative potentials of the two electrodes, we can find the difference in potential between them: the cell potential. The problem is comparable to finding the distance between the fifth floor of a building and the eighth floor. If we know that the fifth floor is 18 m above the ground floor and the eighth floor is 27 m above the ground floor, we know that the two floors are 9 m apart. The absolute height of either floor above sea level can be ignored.

We can use the same approach to assign potentials to electrodes and to calculate the potential difference between electrodes in a galvanic cell. By convention, the "ground floor" is the *standard hydrogen electrode* whose electrode potential is defined as *exactly 0 V*. This electrode consists of $H_2$ gas at a pressure of 1 atm, in contact with both a platinum strip and a solution where $[H^+] = 1M$. The half-reaction for this electrode process is

$$2H^+(aq) + 2e^- \longrightarrow H_2(g)$$

and the half-cell is

$$Pt(s)|H_2(g)(1 \text{ atm})|H^+(aq, 1M)$$

Suppose we want to find the potential of the $Zn(s)|Zn^{2+}(aq, 1M)$ half-cell relative to the standard hydrogen half-cell. We must assign a numerical value to the potential of the half-reaction:

$$Zn^{2+}(aq) + 2e^- \longrightarrow Zn(s)$$

We can construct a galvanic cell of these two half-cells and measure the magnitude and sign of its potential.

To start with, note that both half-reactions are written as reductions. It is convenient to write all half-reactions in the same way, and there is an international agreement to write all electrode processes as reductions when expressing their relative potentials. The sign conventions for electrode potentials are also based on this agreement. In a galvanic cell, of course, only one of the half-cells will proceed as a reduction. The half-reaction at the other electrode will be an oxidation, the reverse of what is written.

The galvanic cell made of these two half-cells is

$$Zn(s)|Zn^{2+}(aq,1M)\|H^+(aq,1M)|H_2(g)(1\ atm)|Pt(s)$$

The magnitude of the standard potential of the cell is found to be 0.7628 V with the standard hydrogen electrode positive (the cathode). Therefore, by convention the standard reduction potential of the zinc half-cell is $-0.7628$ V. Suppose we wish to find the potential of the $Ag(s)|Ag^+(aq,1M)$ half-cell. We construct a galvanic cell of this half-cell and the standard hydrogen half-cell. The cell is

$$Pt(s)|H_2(g)(1\ atm)|H^+(aq,1M)\|Ag^+(aq,1M)|Ag(s)$$

The magnitude of the standard potential of this cell is found to be 0.7996 V with the standard hydrogen electrode negative (the anode). Therefore, the standard reduction potential of the silver half-cell is $+0.7996$ V.

Since the measured potential of a galvanic cell is the difference in potential between its two half-cells, we can give a relationship between the two half-cell potentials and the potential of the cell. By the convention we are using, *the potential of the anode half-reaction must be subtracted from the potential of the cathode half-reaction, when both are written as reductions.* That is

$$\mathscr{E}_{cell} = \mathscr{E}_{cathode} - \mathscr{E}_{anode} \tag{16.6}$$

The anode potential is subtracted from the cathode potential in Equation 16.6 because we write both half-reactions as reductions. The anode half-reaction is an oxidation, not a reduction. The reduction half-reaction that we write to define the potential must be reversed to give the correct anode half-reaction. If a half-reaction is reversed, he sign of its $\mathscr{E}°$ is also reversed.

For the zinc cell we have discussed, the actual half-reaction that takes place at the anode and that is combined to get the cell reaction is

$$Zn(s) \longrightarrow Zn^{2+}(aq) + 2e^- \qquad \mathscr{E}° = +0.7628\ V$$

To summarize briefly: An electrode process corresponds to a half-cell or a half-reaction that we write as a reduction. When the substance being reduced in this half-reaction is more difficult to reduce than $H^+$, the potential of the half-reaction is negative. The negative value of the potential increases as the difficulty of reducing the substance in the half-cell increases.

When a substance is easier to reduce than $H^+$, the potential of the half-reaction will be positive. The positive value of the potential increases with the ease of reducing the substance.

## Tabulating Standard Potentials

By making a series of such comparisons, we can build up a tabulation of standard potentials. For example, the cell

$$Cr(s)|Cr^{3+}(aq,1M)\|H^+(aq,1M)|H_2(g)(1\ atm)|Pt(s)$$

is found to have $\mathcal{E}° = +0.74$ V. The cell reaction is

$$2Cr(s) + 6H^+(aq) \longrightarrow 2Cr^{3+}(aq) + 3H_2(g)$$

The chromium half-cell is the anode. Therefore, the potential of the reduction half-reaction

$$Cr^{3+}(aq) + 3e^- \longrightarrow Cr(s)$$

is negative: $\mathcal{E}° = -0.74$ V. This value is not as negative as that for the zinc electrode, so we can see that $Cr^{3+}$ is more easily reduced than $Zn^{2+}$. We now have standard potentials for three half-reactions, which we can list in numerical order:

$$\begin{array}{ll} Zn^{2+}(aq) + 2e^- \longrightarrow Zn(s) & \mathcal{E}° = -0.7628\ V \\ Cr^{3+}(aq) + 3e^- \longrightarrow Cr(s) & \mathcal{E}° = -0.74\ V \\ 2H^+(aq) + 2e^- \longrightarrow H_2(g) & \mathcal{E}° = 0.0000\ V \end{array}$$

Again, the cell

$$Pt(s)|H_2(g)(1\ atm)|H^+(aq,1M)\|Ag^+(aq,1M)|Ag(s)$$

is found to have $\mathcal{E}° = +0.7996$ V with the standard hydrogen electrode as the anode. The cell reaction is

$$2Ag^+(aq) + H_2(g) \longrightarrow 2Ag(s) + 2H^+(aq)$$

Therefore, the half-reaction $Ag^+(aq) + e^- \rightarrow Ag(s)$ has $\mathcal{E}° = +0.7996$ V, the same as the cell, and $Ag^+$ is more easily reduced than $H^+$. We get this result by substitution into Equation 16.6:

$$0.7996\ V = \mathcal{E}_{cathode} - 0$$
$$\mathcal{E}_{cathode} = +0.7996\ V$$

Another example is the cell

$$Pt(s)|H_2(g)(1\ atm)|H^+(aq,1M)\|Fe^{3+}(aq,1M),\ Fe^{2+}(aq,1M)|Pt(s)$$

for which $\mathscr{E}° = +0.770$ V with the hydrogen electrode as the anode. The cell reaction is

$$H_2(g) + 2Fe^{3+}(aq) \longrightarrow 2H^+(aq) + 2Fe^{2+}(aq)$$

Again, $Fe^{3+}$, the substance in the cathode, is more easily reduced than $H^+$. The potential of the reduction half-reaction

$$Fe^{3+}(aq) + e^- \longrightarrow Fe^{2+}(aq)$$

is positive; it has the value of the cell potential, $\mathscr{E}° = +0.770$ V.

## The emf Series

We can continue our list of half-reactions and their standard potentials with these two electrode processes:

$$\begin{array}{ll} Zn^{2+}(aq) + 2e^- \rightleftharpoons Zn(s) & \mathscr{E}° = -0.7628 \text{ V} \\ Cr^{3+}(aq) + 3e^- \rightleftharpoons Cr(s) & \mathscr{E}° = -0.74 \text{ V} \\ 2H^+(aq) + 2e^- \rightleftharpoons H_2(g) & \mathscr{E}° = 0.0000 \text{ V} \\ Fe^{3+}(aq) + e^- \rightleftharpoons Fe^{2+}(aq) & \mathscr{E}° = 0.770 \text{ V} \\ Ag^+(aq) + e^- \rightleftharpoons Ag(s) & \mathscr{E}° = 0.7996 \text{ V} \end{array}$$

A tabulation of this kind, listing reduction half-reactions in order of increasing standard potential, is sometimes called an **emf** (electromotive force) **series** or an **electrochemical series**. It can be used to predict and correlate a large body of data on chemical behavior. Table 16.1 is a more comprehensive list of this kind. Let us summarize some of the conventions we have used or implied about these half-reactions and their standard potentials.

1. The potentials in this series are standard potentials, since all gases in the systems are at a pressure of 1 atm and all dissolved species are at a concentration of $1M$.
2. By definition, the standard hydrogen electrode has a standard potential of zero. All other electrode potentials are measured relative to this electrode, directly or indirectly.
3. All the half-reactions in the series are written as reductions. But it is understood that when two half-reactions are coupled in a cell or in a chemical system, one will reverse and proceed as an oxidation.
4. The magnitude of the standard electrode potential is a measure of the extent to which the half-reaction proceeds from left to right. The more positive the value of $\mathscr{E}°$, the greater the tendency of the reaction to proceed from left to right. The more negative the value of $\mathscr{E}°$, the smaller the tendency of the reaction to proceed from left to right.
5. The potential of a half-reaction is not related to the coefficients of the equation, which means that the value of $\mathscr{E}°$ does not change if both

## TABLE 16.1 Standard Reduction Potentials at 298 K

| Half-Reaction | $\mathscr{E}°$ (V) | Half-Reaction | $\mathscr{E}°$ (V) |
|---|---|---|---|
| $Li^+(aq) + e^- \rightleftharpoons Li(s)$ | −3.045 | $I_2(s) + 2e^- \rightleftharpoons 2I^-(aq)$ | 0.535 |
| $Rb^+(aq) + e^- \rightleftharpoons Rb(s)$ | −2.925 | $MnO_4^-(aq) + e^- \rightleftharpoons MnO_4^{2-}(aq)$ | 0.564 |
| $K^+(aq) + e^- \rightleftharpoons K(s)$ | −2.924 | $O_2(g) + 2H^+(aq) + 2e^- \rightleftharpoons H_2O_2(aq)$ | 0.682 |
| $Cs^+(aq) + e^- \rightleftharpoons Cs(s)$ | −2.923 | $Fe^{3+}(aq) + e^- \rightleftharpoons Fe^{2+}(aq)$ | 0.770 |
| $Ba^{2+}(aq) + 2e^- \rightleftharpoons Ba(s)$ | −2.90 | $Hg_2^{2+}(aq) + 2e^- \rightleftharpoons 2Hg(l)$ | 0.7961 |
| $Sr^{2+}(aq) + 2e^- \rightleftharpoons Sr(s)$ | −2.89 | $Ag^+(aq) + e^- \rightleftharpoons Ag(s)$ | 0.7996 |
| $Ca^{2+}(aq) + 2e^- \rightleftharpoons Ca(s)$ | −2.76 | $2NO_3^-(aq) + 4H^+(aq) + 2e^- \rightleftharpoons N_2O_4(g) + 2H_2O$ | 0.81 |
| $Na^+(aq) + e^- \rightleftharpoons Na(s)$ | −2.7109 | $2Hg^{2+}(aq) + 2e^- \rightleftharpoons Hg_2^{2+}(aq)$ | 0.905 |
| $Mg^{2+}(aq) + 2e^- \rightleftharpoons Mg(s)$ | −2.375 | $NO_3^-(aq) + 3H^+(aq) + 2e^- \rightleftharpoons HNO_2(aq) + H_2O$ | 0.94 |
| $Al^{3+}(aq) + 3e^- \rightleftharpoons Al(s)$ | −1.706 | $NO_3^-(aq) + 4H^+(aq) + 3e^- \rightleftharpoons NO(g) + 2H_2O$ | 0.96 |
| $Be^{2+}(aq) + 2e^- \rightleftharpoons Be(s)$ | −1.70 | $HNO_2(aq) + H^+(aq) + e^- \rightleftharpoons NO(g) + H_2O$ | 0.99 |
| $Zn(NH_3)_4^{2+}(aq) + 2e^- \rightleftharpoons Zn(s) + 4NH_3(aq)$ | −1.04 | $AuCl_4^-(aq) + 3e^- \rightleftharpoons Au(s) + 4Cl^-(aq)$ | 0.994 |
| $Mn^{2+}(aq) + 2e^- \rightleftharpoons Mn(s)$ | −1.029 | $Br_2(l) + 2e^- \rightleftharpoons 2Br^-(aq)$ | 1.065 |
| $Cr^{2+}(aq) + 2e^- \rightleftharpoons Cr(s)$ | −0.91 | $Cu^{2+}(aq) + 2CN^-(aq) + e^- \rightleftharpoons Cu(CN)_2^-(aq)$ | 1.12 |
| $V^{3+}(aq) + 3e^- \rightleftharpoons V(s)$ | −0.89 | $MnO_2(s) + 4H^+(aq) + 2e^- \rightleftharpoons Mn^{2+}(aq) + 2H_2O$ | 1.208 |
| $Zn^{2+}(aq) + 2e^- \rightleftharpoons Zn(s)$ | −0.7628 | $O_2(g) + 4H^+(aq) + 4e^- \rightleftharpoons 2H_2O$ | 1.229 |
| $Cr^{3+}(aq) + 3e^- \rightleftharpoons Cr(s)$ | −0.74 | $Au^{3+}(aq) + 2e^- \rightleftharpoons Au^+(aq)$ | 1.29 |
| $Ag_2S(s) + 2e^- \rightleftharpoons 2Ag(s) + S^{2-}(aq)$ | −0.7051 | $Cr_2O_7^{2-}(aq) + 14H^+(aq) + 6e^- \rightleftharpoons 2Cr^{3+}(aq) + 7H_2O$ | 1.33 |
| $S(s) + 2e^- \rightleftharpoons S^{2-}(aq)$ | −0.508 | $Cl_2(g) + 2e^- \rightleftharpoons 2Cl^-(aq)$ | 1.3583 |
| $Ni^{2+}(aq) + 2e^- \rightleftharpoons Ni(s)$ | −0.23 | $Au^{3+}(aq) + 3e^- \rightleftharpoons Au(s)$ | 1.42 |
| $Sn^{2+}(aq) + 2e^- \rightleftharpoons Sn(s)$ | −0.1364 | $MnO_4^-(aq) + 8H^+(aq) + 5e^- \rightleftharpoons Mn^{2+}(aq) + 4H_2O$ | 1.491 |
| $Pb^{2+}(aq) + 2e^- \rightleftharpoons Pb(s)$ | −0.1263 | $Mn^{3+}(aq) + e^- \rightleftharpoons Mn^{2+}(aq)$ | 1.51 |
| $Fe^{3+}(aq) + 3e^- \rightleftharpoons Fe(s)$ | −0.036 | $HClO(aq) + H^+(aq) + e^- \rightleftharpoons \frac{1}{2}Cl_2(g) + H_2O$ | 1.63 |
| $2H^+(aq) + 2e^- \rightleftharpoons H_2(g)$ | 0 | $HClO_2(aq) + 2H^+(aq) + 2e^- \rightleftharpoons HClO(aq) + H_2O$ | 1.64 |
| $AgBr(s) + e^- \rightleftharpoons Ag(s) + Br^-(aq)$ | 0.0713 | $MnO_4^-(aq) + 4H^+(aq) + 3e^- \rightleftharpoons MnO_2(s) + 2H_2O$ | 1.679 |
| $Sn^{4+}(aq) + 2e^- \rightleftharpoons Sn^{2+}(aq)$ | 0.15 | $N_2O(g) + 2H^+(aq) + 2e^- \rightleftharpoons N_2(g) + H_2O$ | 1.77 |
| $Cu^{2+}(aq) + e^- \rightleftharpoons Cu^+(aq)$ | 0.158 | $H_2O_2(aq) + 2H^+(aq) + 2e^- \rightleftharpoons 2H_2O$ | 1.776 |
| $Hg_2Cl_2(s) + 2e^- \rightleftharpoons 2Hg(l) + 2Cl^-(aq)$ | 0.2676 | $O_3(g) + 2H^+(aq) + 2e^- \rightleftharpoons O_2(g) + H_2O$ | 2.07 |
| $Cu^{2+}(aq) + 2e^- \rightleftharpoons Cu(s)$ | 0.3402 | $F_2(g) + 2e^- \rightleftharpoons 2F^-(aq)$ | 2.87 |

### Alkaline Solution

| Half-Reaction | $\mathscr{E}°$ (V) | Half-Reaction | $\mathscr{E}°$ (V) |
|---|---|---|---|
| $Al(OH)_4^-(aq) + 3e^- \rightleftharpoons Al(s) + 4OH^-(aq)$ | −2.35 | $ClO_4^-(aq) + H_2O + 2e^- \rightleftharpoons ClO_3^-(aq) + 2OH^-(aq)$ | 0.17 |
| $Mn(OH)_2(s) + 2e^- \rightleftharpoons Mn(s) + 2OH^-(aq)$ | −1.47 | $2ClO^-(aq) + 2H_2O + 2e^- \rightleftharpoons Cl_2(g) + 4OH^-(aq)$ | 0.40 |
| $Cr(OH)_3(s) + 3e^- \rightleftharpoons Cr(s) + 3OH^-(aq)$ | −1.3 | $MnO_4^-(aq) + 2H_2O + 3e^- \rightleftharpoons MnO_2(s) + 4OH^-(aq)$ | 0.58 |
| $Zn(OH)_4^{2-}(aq) + 2e^- \rightleftharpoons Zn(s) + 4OH^-(aq)$ | −1.216 | $ClO_2(aq) + H_2O + 2e^- \rightleftharpoons ClO^-(aq) + 2OH^-(aq)$ | 0.59 |
| $Zn(OH)_2(s) + 2e^- \rightleftharpoons Zn(s) + 2OH^-(aq)$ | −1.17 | $ClO_3^-(aq) + 3H_2O + 6e^- \rightleftharpoons Cl^-(aq) + 6OH^-(aq)$ | 0.62 |
| $2H_2O + 2e^- \rightleftharpoons H_2(g) + 2OH^-(aq)$ | −0.8277 | $O_3(g) + H_2O + 2e^- \rightleftharpoons O_2(g) + 2OH^-(aq)$ | 1.24 |
| $O_2(g) + 2H_2O + 2e^- \rightleftharpoons H_2O_2(aq) + 2OH^-(aq)$ | −0.146 | | |
| $NO_3^-(aq) + H_2O + 2e^- \rightleftharpoons NO_2^-(aq) + 2OH^-(aq)$ | 0.01 | | |

sides of a half-reaction are multiplied by a factor. It also means that the number of electrons in the half-reaction does not affect the value of $\mathscr{E}°$; potential is potential energy *per charge*. The half-reactions

$$Zn^{2+}(aq) + 2e^- \longrightarrow Zn(s)$$
$$\tfrac{1}{2}Zn^{2+}(aq) + e^- \longrightarrow \tfrac{1}{2}Zn(s)$$
$$2Zn^{2+}(aq) + 4e^- \longrightarrow 2Zn(s)$$

all have $\mathscr{E}° = -0.7628$ V.

6. When the direction of a half-reaction is reversed, the sign of its potential is reversed. Thus, an unfavorable reduction half-reaction with a

large negative potential is a favorable oxidation half-reaction with a large positive potential. The more negative $\mathscr{E}°$ is, the greater the tendency is for the reaction to reverse.

## Cell Potentials from Electrode Potentials

Any two half-reactions from the series on page 663 or from Table 16.1 can be selected to be the half-cells of a galvanic cell. We can find the standard potential of this cell simply by using the listed values of the standard potentials of the two half-reactions. We must first decide which half-reaction takes place as an oxidation and is the anode half-reaction in the galvanic cell. The positions of the half-reactions are the key to making this decision. Consistent with point 4 above, *the higher a half-reaction is in the series, the more readily it becomes the anode of the galvanic cell.*

Once the anode is chosen, we can find the standard potential of the cell by using Equation 16.6. For example, the galvanic cell made of the half-cells at the top and bottom of the series is

$$Zn(s)|Zn^{2+}(aq,1M)\|Ag^{+}(aq,1M)|Ag(s)$$

Its standard potential is given by

$$\mathscr{E}°_{cell} = (0.7996 \text{ V}) - (-0.7628 \text{ V}) = 1.5624 \text{ V}$$

Since these two half-reactions are the farthest apart in the series, the cell they form has the greatest standard potential of any galvanic cell we could construct using the half-cells in this series.

## Chemical Behavior from Electrical Potential

There is a cell reaction associated with every galvanic cell. The reaction can occur either in the cell or in an ordinary chemical system, where electron transfer occurs directly, rather than through an external connection. The spontaneous reaction that occurs when substances are mixed is the same as the spontaneous reaction that occurs in the galvanic cell. The data in an emf series can be used to obtain a great deal of information, qualitative and quantitative, about the behavior of chemical systems.

An overall chemical reaction is a combination of two half-reactions. We can use the relative positions of the two half-reactions in the emf series to predict the spontaneous direction of any reaction that starts with all the components in their standard states.

In any emf series, we put the reduction half-reaction with the most negative $\mathscr{E}°$ at the top. The half-reactions at the top of the emf series have the greatest tendency to reverse and proceed from right to left as oxidations. The closer a reduction half-reaction is to the top of the series, the less favorable it is when written as a reduction and the more favorable it is in reverse as an oxidation.

Consider a solution with $[Zn^{2+}] = 1M$, $[H^+] = 1M$, and $P_{H_2} = 1$ atm, in contact with zinc metal. The spontaneous direction of the reaction

$$Zn(s) + 2H^+(aq) \rightleftharpoons Zn^{2+}(aq) + H_2(g)$$

is from left to right. The zinc half-reaction is higher in the emf series, so it reverses and proceeds as an oxidation. When zinc metal is in contact with an acidic solution, the reaction will proceed substantially to the right. The zinc metal will dissolve until almost all the $H^+$ is consumed.

The same reasoning can be applied to the combination of the next half-reaction in our series and the standard hydrogen half-reaction. The spontaneous direction of the reaction

$$2Cr(s) + 6H^+(aq) \rightleftharpoons 2Cr^{3+}(aq) + 3H_2(g)$$

is from left to right. Chromium metal dissolves in acid solution.

But the combination of the silver half-reaction and the hydrogen half-reaction gives

$$2Ag^+(aq) + H_2(g) \rightleftharpoons 2Ag(s) + 2H^+(aq)$$

whose spontaneous direction is from left to right. Silver will not dissolve appreciably in acid solution.

We can generalize from these examples. Any metal that appears in a half-reaction above the standard hydrogen half-reaction in the emf series will dissolve in acid solution.[3] Any metal that appears below the standard hydrogen half-reaction in the emf series will not dissolve appreciably in acid.

## Oxidizing and Reducing Agents

Once we know the spontaneous direction of a reaction, we know the relative strengths of the oxidizing and reducing agents in the system. For example, the spontaneous direction of the reaction

$$Zn(s) + 2Ag^+(aq) \rightleftharpoons Zn^{2+}(aq) + 2Ag(s)$$

is from left to right when the components are in their standard states.

This observation tells us that of the two reducing agents in the system, $Zn(s)$ and $Ag(s)$, $Zn(s)$ is the stronger. It also tells us that of the two oxidizing agents in the system, $Ag^+$ and $Zn^{2+}$, $Ag^+$ is the stronger.

The same kind of information can be obtained directly from an emf series. Each half-reaction in the series gives a relationship between two oxidation states of an element. The substance on the left of the reduction

---

[3] Some metals above hydrogen in the emf series dissolve very slowly or not at all because of surface effects (Section 16.8).

**TABLE 16.2**   Relative Strengths of Oxidizing Agents and Reducing Agents

| | Oxidizing Agent | | Reducing Agent | | Standard Potential (V) |
|---|---|---|---|---|---|
| increasing strength ↓ | $Zn^{2+}(aq) + 2e^-$ | $\rightleftharpoons$ | $Zn(s)$ | increasing strength ↑ | $-0.7628$ |
| | $Cr^{3+}(aq) + 3e^-$ | $\rightleftharpoons$ | $Cr(s)$ | | $-0.74$ |
| | $2H^+(aq) + 2e^-$ | $\rightleftharpoons$ | $H_2(g)$ | | $0.0000$ |
| | $Fe^{3+}(aq) + e^-$ | $\rightleftharpoons$ | $Fe^{2+}(aq)$ | | $0.770$ |
| | $Ag^+(aq) + e^-$ | $\rightleftharpoons$ | $Ag(s)$ | | $0.7996$ |

half-reaction is the element as an oxidizing agent; the substance on the right is the element as a reducing agent. The standard potential of the half-reaction is a measure of the relative strengths of the oxidizing agent and the reducing agent. If the value of $\mathscr{E}°$ is large and positive, we have a strong oxidizing agent and a weak reducing agent.

A substance on the right side of a half-reaction toward the top of the emf series is easily oxidized and therefore is a strong reducing agent. A substance on the left side of a half-reaction toward the top of the emf series is difficult to reduce and therefore is a weak oxidizing agent. In our small series, $Zn(s)$ is the strongest reducing agent and $Zn^{2+}$ is the weakest oxidizing agent. As we go down the list, we encounter progressively weaker reducing agents on the right of the half-reactions and progressively stronger oxidizing agents on the left of the half-reactions. Thus $Ag^+$, at the bottom, is the most powerful oxidizing agent of the group, while $Ag(s)$ is the weakest reducing agent. Table 16.2 shows these relationships.

A number of chemical correlations can be made from the relative positions of substances in the emf series. In general, we can say that an oxidizing agent will oxidize a reducing agent that is above it in the emf series. Similarly, an oxidizing agent will be reduced by any reducing agent that appears above it in the emf series (Figure 16.12). This generalization does not tell us anything about the kinetics of these processes. A process that is predicted to be favorable or spontaneous may occur very slowly.

**Figure 16.12**
The reaction between an oxidizing agent and a reducing agent above it in the emf series is favorable. The reaction of an oxidizing agent with a reducing agent below it in the emf series is unfavorable.

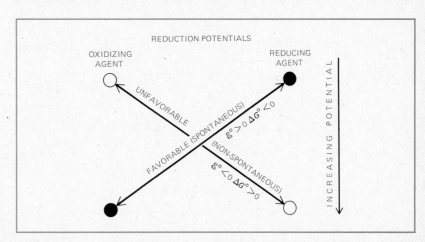

## Reaction Potential from Electrode Potential

These qualitative arguments can be made quantitatively. We can calculate a numerical value for the standard potential of any reaction that is a combination of two half-reactions of known standard potential. And if we know the standard potential, we can calculate both the standard free energy change, $\Delta G°$, and the equilibrium constant, $K$, for the reaction. We can also use the sign of $\mathcal{E}°$ to determine the spontaneous direction of the reaction when all the substances in it are in their standard states. The reaction is spontaneous from left to right when $\mathcal{E}°$ is positive and from right to left when $\mathcal{E}°$ is negative.

We find $\mathcal{E}°$ for a reaction by the same procedure used to find $\mathcal{E}°$ for a cell. The first step is to select the reduction half-reaction that reverses and proceeds as an oxidation. Once again, the half-reaction that is closer to the top in the emf series is the one to reverse. The difference between the two potentials, taken directly from the series, gives the value of $\mathcal{E}°$. No other arithmetic operations are necessary. It is important to remember that the value of the standard potential is independent of the number of electrons in the half-reactions. When calculating $\mathcal{E}°$ for a reaction, we can ignore the number of electrons in each half-reaction.

**Example 16.8**    Calculate the standard potential, the standard free energy, and the equilibrium constant for the reaction

$$2Cr^{3+}(aq) + 3Zn(s) \rightleftharpoons 2Cr(s) + 3Zn^{2+}(aq)$$

**Solution**    We must first determine the two half-reactions that are combined to make this overall process. The reduction is

$$Cr^{3+}(aq) + 3e^- \rightleftharpoons Cr(s)$$

and the oxidation is

$$Zn(s) \rightleftharpoons Zn^{2+}(aq) + 2e^-$$

Both these half-reactions are listed in the series. The standard potential of the overall reaction is the listed potential of the half-reaction that occurs as a reduction minus the listed potential of the half-reaction that occurs as the oxidation. (We subtract, rather than add, the potentials because the series lists only reduction half-reactions. The reduction that is higher in the series is reversed to obtain the overall reaction.)

$$\mathcal{E}°_{reaction} = \mathcal{E}°_{reduction} - \mathcal{E}°_{oxidation}$$

which is equivalent to Equation 16.6.

$$\mathcal{E}° = (-0.74 \text{ V}) - (-0.76 \text{ V}) = 0.02 \text{ V}$$

Equation 16.2 is used to find the value of $\Delta G°$. The value of $n$ is found by inspection, which indicates that when the two half-reactions are combined in the overall reaction, the total number of electrons in each is six. Thus,

$$\Delta G° = -(6 \text{ mol } e^-)\left(96\ 500\ \frac{\text{C}}{\text{mol } e^-}\right)(0.02 \text{ V}) = -12\ 000 \text{ J}$$

The value of $K$ can be found from Equation 16.5:

$$0.02 \text{ V} = \frac{0.026 \text{ V}}{6} \ln K$$

$$K = 1 \times 10^2$$

**Example 16.9**

Using the data in Table 16.1, predict whether $Cl_2$ disproportionates in alkaline solution.

**Solution**

Chlorine can exist in many oxidation states. Chlorine in the 0 oxidation state, as in $Cl_2$, can act as an oxidizing agent and be reduced. Chlorine can also be oxidized to one of several positive oxidation states. So $Cl_2$ can react with itself in an oxidation-reduction reaction, which means that it is possible for chlorine to disproportionate.

For the half-reaction

$$Cl_2(g) + 2e^- \rightleftharpoons 2Cl^-(aq)$$

Table 16.1 lists $\mathscr{E}° = +1.3583$ V. This large positive reduction potential shows that $Cl_2$ is a relatively powerful oxidizing agent. The table also lists the half-reaction

$$2ClO^-(aq) + 2H_2O + 2e^- \rightleftharpoons Cl_2(g) + 4OH^-(aq)$$

with $\mathscr{E}°$ of 0.40 V. The reverse of this half-reaction is an oxidation of $Cl_2$. When the two half-reactions are combined, the overall reaction is

$$2Cl_2(g) + 4OH^-(aq) \rightleftharpoons 2Cl^-(aq) + 2ClO^-(aq) + 2H_2O$$

or, more simply

$$Cl_2(g) + 2OH^-(aq) \rightleftharpoons Cl^-(aq) + ClO^-(aq) + H_2O$$

The standard potential of this reaction is positive: 1.36 V − 0.40 V = 0.96 V. Therefore, the reaction proceeds spontaneously from left to right. In other words, $Cl_2$ is unstable in alkaline solution; it is converted to chloride anion and hypochlorite anion. Chlorine does disproportionate, in a reaction that is put to practical use to keep swimming pools clear of algae. The reaction can usually proceed further, forming $ClO_3^-$, if we wait long enough.

We can even use reduction potentials to calculate the equilibrium constants of reactions that are not oxidation-reductions, if they are the combination of two half-reactions listed in the emf series. Some examples of such reactions are the dissociation of complex ions into the metal ion and its ligands, and the dissolution of a sparingly soluble solid in water. For example, Table 16.1 lists a standard potential for the half-reaction

$$Zn(NH_3)_4^{2+}(aq) + 2e^- \rightleftharpoons Zn(s) + 4NH_3(aq)$$

which is the reduction of zinc from the $+2$ oxidation state in a complex ion to zinc metal in the 0 oxidation state. Table 16.1 also lists the half-reaction

$$Zn^{2+}(aq) + 2e^- \rightleftharpoons Zn(s)$$

in which zinc again is reduced from the $+2$ oxidation state to zinc metal in the 0 oxidation state. When the first half-reaction is reversed and these two half-reactions are combined to give an overall reaction, we find that not only the two electrons but also the Zn(s) cancel out:

$$Zn^{2+}(aq) + 4NH_3(aq) \rightleftharpoons Zn(NH_3)_4^{2+}(aq)$$

The overall reaction thus is a change from one form of zinc in the $+2$ oxidation state to another form of zinc in the $+2$ oxidation state. This process is not an oxidation-reduction reaction in the usual sense, although the reaction does occur as an oxidation-reduction if the cell

$$Zn(s)|NH_3(aq, 1M), Zn(NH_3)_4^{2+}(aq, 1M)\|Zn^{2+}(aq, 1M)|Zn(s)$$

is constructed. A transfer of two electrons can be visualized, since two electrons in each half-reaction cancel when the half-reactions are combined. We have the standard potential for each half-reaction, so we can find the standard potential of the overall reaction:

$$\mathcal{E}^\circ = (-0.76 \text{ V}) - (-1.04 \text{ V}) = 0.28 \text{ V}$$

Since we have $n = 2$, we can also find $\Delta G^\circ$ and $K$.

---

**Example 16.10**

Using the data in Table 16.1, calculate the solubility product of silver bromide.

**Solution**

The reaction of interest that defines the $K_{sp}$ is

$$AgBr(s) \rightleftharpoons Ag^+(aq) + Br^-(aq)$$

One of the half-reactions in Table 16.1 includes solid silver bromide:

$$AgBr(s) + e^- \rightleftharpoons Ag(s) + Br^-(aq) \qquad \mathcal{E}^\circ = 0.0713 \text{ V}$$

This half-reaction is the reduction of silver from the $+1$ to the 0 oxidation state. To obtain an overall reaction, we must combine it with another half-reaction that includes the same change in oxidation state. We find in Table 16.1 the half-reaction

$$Ag^+(aq) + e^- \rightleftharpoons Ag(s) \qquad \mathcal{E}° = 0.7996 \text{ V}$$

The overall reaction that we are trying to form must have $Ag^+$ on the right. This half-reaction has $Ag^+$ on the left, so it is the one that we must reverse. Therefore, we change the sign of the $\mathcal{E}°$ of this half-reaction. In tabular form:

$$
\begin{array}{ll}
AgBr(s) + e^- \rightleftharpoons \cancel{Ag(s)} + Br^-(aq) & \mathcal{E}° = \phantom{-}0.0713 \text{ V} \\
\underline{\cancel{Ag(s)} \rightleftharpoons Ag^+(aq) + e^-} & \underline{\mathcal{E}° = -0.7996 \text{ V}} \\
AgBr(s) \rightleftharpoons Ag^+(aq) + Br^-(aq) & \mathcal{E}° = -0.7283 \text{ V}
\end{array}
$$

The number of moles of electrons transferred, $n$, is 1, the number of electrons that cancel when the half-reactions are combined. We calculate $K$ by using Equation 16.5:

$$-0.7283 \text{ V} = \frac{0.02568 \text{ V}}{1} \ln K$$

$$K = 4.82 \times 10^{-13}$$

The small equilibrium constant is consistent with the negative $\mathcal{E}°$.

## Half-Reaction Potentials from Electrode Potentials

Many elements exist in a number of different oxidation states. The emf series does not list the values of all the standard potentials of every possible half-reaction for these elements. But we can obtain these values, as long as each oxidation state of interest appears in a half-reaction in the series. For example, the emf series in Table 16.1 lists the half-reaction for the reduction of tin from the $+4$ to the $+2$ oxidation state. It also lists the half-reaction for the reduction of tin from the $+2$ to the 0 oxidation state. The series thus gives us the information needed to calculate the standard potential of the half-reaction in which tin is reduced directly from the $+4$ to the 0 oxidation state.

To obtain this kind of information, we must combine half-reactions in the series to make a new half-reaction that is not in the series. The standard potential of the new half-reaction cannot be found simply by a combination of the potentials of the original half-reactions. We could convert the $\mathcal{E}°$ of each half-reaction to the corresponding $\Delta G°$, combine the $\Delta G°$ values to find the $\Delta G°$ of the new half-reaction, and then convert this $\Delta G°$ back to $\mathcal{E}°$. But we shall use a different method. The $\mathcal{E}°$ of each half-reaction is multiplied by the number of electrons in that half-reaction. The resulting values are combined. Finally, we divide the resulting potential by the number of electrons in the new half-reaction to get its potential. The procedure is symbolized by

$$\mathscr{E}_T^\circ = \frac{n_1 \mathscr{E}_1^\circ + n_2 \mathscr{E}_2^\circ + \cdot \cdot \cdot}{n_T} \qquad (16.7)$$

where $n_1$ is the number of electrons in the first half-reaction and $\mathscr{E}_1^\circ$ is its standard potential, $n_2$ is the number of electrons in the second half-reaction and $\mathscr{E}_2^\circ$ is its standard potential (and so on for as many half-reactions as are being combined), and $n_T$ is the number of electrons that appear in the resulting half-reaction. Equation 16.7 is derived from Equation 16.2. In using this relationship, be careful to make the sign of each $\mathscr{E}°$ consistent with the way in which its half-reaction is combined to make the new half-reaction.

**Example 16.11**    Using the data in Table 16.1, find the standard potential of the half-reaction for the reduction of iron from the $+2$ to the 0 oxidation state.

**Solution**    We begin by writing the desired half-reaction:

$$Fe^{2+}(aq) + 2e^- \rightleftharpoons Fe(s)$$

We then inspect the table for half-reactions that include these oxidation states of iron. Listed are

$$Fe^{3+}(aq) + 3e^- \rightleftharpoons Fe(s) \qquad \mathscr{E}° = -0.036 \text{ V}$$
$$Fe^{3+}(aq) + e^- \rightleftharpoons Fe^{2+}(aq) \qquad \mathscr{E}° = \phantom{-}0.770 \text{ V}$$

Applying the reasoning used earlier to combine reactions, we see that the second half-reaction (and the sign of its $\mathscr{E}°$) should be reversed and added to the first to produce the desired half-reaction. The $\mathscr{E}°$ of the new half-reaction can be calculated by Equation 16.7:

$$\mathscr{E}_T^\circ = \frac{(3)(-0.036 \text{ V}) + (1)(-0.770 \text{ V})}{2}$$
$$= -0.439 \text{ V}$$

## 16.7  POTENTIAL AND CONCENTRATION. THE NERNST EQUATION

By defining the standard state, we can make meaningful comparisons of many different galvanic cells and chemical systems. But in practice, we often encounter galvanic cells or chemical systems whose components are not in the standard state. We need to know how variations from the standard state of a system affect its potential. We can then convert the standard potential of any given system to the potential under nonstandard conditions, obtaining valuable information about the behavior of chemical systems in nonstandard states.

The relationship of the potential, $\mathscr{E}$, of a nonstandard system to its standard potential, $\mathscr{E}°$, and its concentration and pressure terms was described by Walter Nernst (1864–1941) and is usually called the Nernst equation:

$$\mathscr{E} = \mathscr{E}° - \frac{RT}{nF} \ln Q \qquad (16.8)$$

At 298 K, Equation 16.8 can be simplified to

$$\mathscr{E} = \mathscr{E}° - \frac{0.02568}{n} \ln Q \qquad (16.9)$$

where $n$ has its usual meaning as the number of moles of electrons transferred in the process and $Q$ is called the reaction quotient (Section 14.1). The expression for $Q$ has exactly the same form as the expression for the equilibrium constant, $K$. We use a different symbol because the concentrations and pressures in the expression for $Q$ are not those of the equilibrium state, as they are in the expression for $K$.

The Nernst equation can be derived rather easily from relationships between the free energy and the composition of a system and the expression $\Delta G = -nF\mathscr{E}$.

The Nernst equation clarifies the relationship between $\mathscr{E}°$ and $K$. A system in which electron transfer occurs is at equilibrium only if there is no net transfer of electrons. Net electron transfer occurs because of a potential difference. Therefore, when net electron transfer stops and the system is at equilibrium, there is no potential difference; $\mathscr{E} = 0$. When this happens, $Q = K$, and the Nernst equation becomes

$$0 = \mathscr{E}° - \frac{RT}{nF} \ln K$$

which is the same as Equation 16.4.

The Nernst equation can be used to find the potential of a half-cell, a galvanic cell, or any chemical system if we know (1) the standard potential of the system and (2) concentrations and pressures of the relevant constituents of the system. The following procedure can be used in many calculations of this kind:

1. Write the reaction that occurs in the half-cell, the cell, or the chemical system.
2. Use the half-cell half-reaction or break the overall reaction into two half-reactions.
3. If possible, find the $\mathscr{E}°$ for these half-reactions in the emf series, and for a chemical system or a cell, calculate the $\mathscr{E}°$ for the overall reaction.

4. Find $n$, the number of moles of electrons transferred in the process. This number is also the number of electrons that are canceled when the half-reactions are combined to give the overall reaction.
5. Write the expression for $Q$ for the half-reaction or the overall reaction, keeping in mind the fact that $Q$ and $K$ have the same form.
6. Substitute all the data, including that on the composition of the system, into the Nernst equation.

**Example 16.12**    Find the reduction potential of the half-cell

$$Pt(s)|Cu^{2+}(aq,0.22M), Cu^{+}(aq,0.043M)$$

**Solution**    The half-cell half-reaction is

$$Cu^{2+}(aq) + e^{-} \rightleftharpoons Cu^{+}(aq)$$

and its $\mathcal{E}°$ is given in Table 16.1 as 0.158 V.

The value of $n$, the number of electrons transferred, is clearly 1 for this half-reaction.

The expression for $Q$ is $[Cu^{+}]/[Cu^{2+}]$.

We can substitute these data into the Nernst equation

$$\mathcal{E} = 0.158 \text{ V} - \frac{0.0257}{1} \ln \frac{(0.043)}{(0.22)}$$

$$= 0.20 \text{ V}$$

This answer is consistent with the chemistry of this system. The positive value of $\mathcal{E}°$ tells us that $Cu^{+}(aq)$ is stable relative to $Cu^{2+}(aq)$ and will predominate at equilibrium. In this half-cell, the system is even further from equilibrium than in the standard state because the $[Cu^{2+}]$ is substantially larger than the $[Cu^{+}]$. Therefore, $\mathcal{E}$ is greater than $\mathcal{E}°$.

To use the Nernst equation for a complete cell, we must perform some additional analysis of the cell reaction, as we see in the next example.

**Example 16.13**    Calculate the potential of the cell

$$V(s)|V^{3+}(aq,0.0011M)\|Ni^{2+}(aq,0.24M)|Ni(s)$$

**Solution**    The cell reaction is

$$2V(s) + 3Ni^{2+}(aq) \rightleftharpoons 2V^{3+}(aq) + 3Ni(s)$$

The two half-reactions can be found in Table 16.1:

$$V^{3+}(aq) + 3e^{-} \rightleftharpoons V(s) \qquad \mathcal{E}° = -0.89 \text{ V}$$
$$Ni^{2+}(aq) + 2e^{-} \rightleftharpoons Ni(s) \qquad \mathcal{E}° = -0.23 \text{ V}$$

The first half-reaction reverses and is the anode, so $\mathscr{E}°$ for the overall reaction is:

$$\mathscr{E}°_T = (-0.23\text{ V}) - (-0.089\text{ V}) = 0.66\text{ V}$$

To obtain the overall reaction, we reverse the vanadium half-reaction. We multiply it by two and the nickel half-reaction by three. The number of electrons that cancel when the two half-reactions are combined is six, so $n = 6$.
The expression for $Q$ for the overall reaction is

$$Q = \frac{[V^{3+}]^2}{[Ni^{2+}]^3}$$

Using the calculated value of $\mathscr{E}°$ and the given data on concentration, the potential of the cell can be calculated with the help of the Nernst equation:

$$\mathscr{E} = 0.66\text{ V} - \frac{0.0257\text{ V}}{6}\ln\frac{(0.0011)^2}{(0.24)^3}$$

$$\mathscr{E} = 0.70\text{ V}$$

The values of $\mathscr{E}°$ and of $\mathscr{E}$ tell us a great deal about the behavior of the chemical system that corresponds to the cell reaction. The large positive value of $\mathscr{E}°$ means that the reaction has a large equilibrium constant and proceeds substantially to completion as written. Since $\mathscr{E}$ is even larger than $\mathscr{E}°$ the system is further from equilibrium than it would be at standard concentrations.

The Nernst equation can also be used to convert measured potentials to standard potentials.

**Example 16.14**   In investigating the properties of rare and expensive metals such as rhenium, element 75, it is both impractical and uneconomical to prepare cells with standard concentrations. To find the standard potential of the $Re^{3+}(aq)|Re(s)$ electrode, the following cell is constructed:

$$Pt(s)|Re(s)|Re^{3+}(aq,0.0018M)\|Ag^+(aq,0.010M)|Ag(s)$$

The potential of this cell is found to be 0.42 V with the Re electrode as the anode. Calculate the standard potential of the half-reaction $Re^{3+}(aq) + 3e^- \rightleftharpoons Re(s)$.

**Solution**   Table 16.1 lists the $\mathscr{E}°$ for the $Ag^+(aq)|Ag(s)$ electrode as 0.80 V. We can find the $\mathscr{E}°$ of the other electrode by finding the $\mathscr{E}°$ of the cell.
The cell reaction is

$$Re(s) + 3Ag^+(aq) \rightleftharpoons Re^{3+}(aq) + 3Ag(s)$$

The two half-reactions are

$$Re(s) \rightleftharpoons Re^{3+}(aq) + 3e^-$$
$$Ag^+(aq) + e^- \rightleftharpoons Ag(s)$$

We obtain the overall reaction by multiplying the silver half-reaction by three. Thus, $n = 3$. The expression for $Q$ is

$$Q = \frac{[Re^{3+}]}{[Ag^+]^3}$$

We now can solve for the standard potential of the cell:

$$0.42 \text{ V} = \mathcal{E}° - \frac{0.0267 \text{ V}}{3} \ln \frac{(0.0018)}{(0.010)^3}$$

$$\mathcal{E}° = 0.48 \text{ V}$$

Since

$$\mathcal{E}°_T = \mathcal{E}°_{cathode} - \mathcal{E}°_{anode}$$
$$0.48 \text{ V} = 0.80 \text{ V} - \mathcal{E}°_{anode}$$

and

$$\mathcal{E}°_{anode} = 0.32 \text{ V}$$

which is the standard reduction potential of the $Re^{3+}(aq)|Re(s)$ half-reaction.

Measurements of potential are very convenient ways of finding concentrations, as we see in the next example.

**Example 16.15**    The reduction potential of a $Pt(s)|Cl_2(g)|Cl^-(aq)$ electrode is found to be 1.421 V when the pressure of $Cl_2$ is 0.247 atm. Find the $[Cl^-]$ in this half-cell.

**Solution**    The half-reaction is $Cl_2(g) + 2e^- \rightleftharpoons 2Cl^-(aq)$ and its $\mathcal{E}°$ is given in Table 16.1 as 1.3583 V. The value of $n$, the number of electrons transferred, is 2. The expression for $Q$ is $[Cl^-]^2/P_{Cl_2}$.

We can substitute these data into the Nernst equation

$$1.421 \text{ V} = 1.3583 \text{ V} - \frac{0.02568}{2} \ln \frac{[Cl^-]^2}{0.247}$$

$$[Cl^-] = 0.0432 M$$

## 16.8  SOME IMPORTANT ELECTRON TRANSFER PROCESSES

Electron transfer processes play major roles in biological systems and in technology. While the electron transfer processes of living organisms and technology are considerably more complex than those of the simple chemical systems we have discussed in this chapter, the general principles are the same.

### Corrosion

Corrosion, the deterioration of metals caused by chemical reactions on their surfaces, costs our society billions of dollars each year. The most

## ELECTROPLATING

The silver-plated spoon and the chromium-plated automobile bumper are familiar objects that are made by the well-known process of electroplating, in which a thin layer of metal is deposited on an electrically conducting surface in an electrolytic cell. Electroplating is more than a century old, but some of its most inventive and useful applications have been developed relatively recently.

For example, most of us know that the "tin can" is actually a steel can that has been plated with a protective layer of tin. But an increasing percentage of tin cans now are made of corrosion-resistant "tin-free steel," which is actually plated with chromium. The tinless tin can became possible in the 1960s, with the development of a method of putting an extremely thin coating of chromium on steel very quickly. Ordinary chromium plating is about $2 \times 10^{-4}$ mm thick and takes several minutes of plating time to produce. The chromium plating on a tinless tin can is about $1.5 \times 10^{-6}$ mm thick and is applied in a plating process that takes one-third of a second.

Metal-plated plastics are also a development of the 1960s, which saw the introduction of acrylonitrile-butadiene-styrene (ABS) plastic, which is readily plated. The plastic is etched chemically, usually by being dipped into an acid solution. After further treatment, it is coated with copper or nickel and then is placed in an electrolytic cell for the final plating. The process produces lightweight parts that can bear moderate stress and are useful in many applications where weight reduction is desired.

A specialized branch of electroplating called electrotyping is commonly used to print books and magazines. To make a book plate, an impression of the type is made by using wax, a soft plastic, or a lead sheet. This mold is coated with graphite to make it electrically conducting and is then put into an electroplating bath of copper sulfate or a similar salt. A thin layer of copper, nickel, or another metal is deposited on the surface of the mold, which is then removed. The thin metal shell is backed with a thicker layer to give it strength. A million or more pages can be printed from a single such plate.

Continual improvements are being made in electroplating technology, and there is an ongoing effort to develop new plating methods that will serve specialized needs. For example, the successful replacement of worn or damaged joints in the human body by artificial hips, knees, and the like depends on the existence of materials that are strong, light, and capable of withstanding the strong corrosive effects of saline body fluids. Several alloys for artificial joints have been developed, but the attempt to produce better materials goes on. One material that has shown promise in experiments is made by plating a lightweight alloy with a thin coat of tantalum, a metal that is unusually resistant to corrosion by body fluids and that is completely nonirritating.

widely used metals — iron, aluminum, copper, nickel — all undergo corrosion when in contact with the air, unless they are protected in some way. Only a few so-called noble metals, such as gold and platinum, do not undergo corrosion.

Corrosion processes are electron transfers, oxidation-reduction processes that occur when the surface of a metal is in contact with the atmosphere. A potential difference exists on the surface of a metal because of small differences in the composition of the metal, which arise from lattice defects, impurities, or even partial oxidation. This potential difference makes it possible for a corrosion process resembling the operation of a galvanic cell to take place on the metal surface. The site of the oxidation and the site of the reduction can be separated in space. Electrons flow between these sites through the metal in the way that electrons flow through the external electrical connection of a galvanic cell. Water vapor that condenses on the metal surface provides the solution through which the ions flow. Corrosion usually occurs faster near the sea, because droplets of water in the air contain dissolved salts such as sodium chloride, forming an ionic solution that conducts electricity better than pure water does.

The anode reaction in corrosion is the oxidation of the metal. This process forms metal ions that dissolve in the ion-conducting medium, the atmospheric moisture in contact with the metal surface. In the corrosion of iron, for example, the oxidation is

$$Fe(s) \longrightarrow Fe^{2+} + 2e^-$$

We can predict that *active metals,* which are high in the emf series, should be most susceptible to corrosion. Cesium and rubidium, which are toward the top of the emf series, corrode very quickly in moist air. Other metals near the top of the emf series usually are stored away from air, to prevent the swift corrosion that occurs when they come in contact with the atmosphere for any period of time. There are exceptions to this rule, however; some active metals corrode quite slowly because they form protective oxide coatings.

Given an oxidation, there must be a reduction to consume the electrons released in the anode process. Some of the important cathode, or reduction, half-reactions in corrosion are

$$2H^+ + 2e^- \longrightarrow H_2(g) \qquad \text{(acid solution)}$$
$$O_2 + 4H^+ + 4e^- \longrightarrow 2H_2O \qquad \text{(acid solution)}$$
$$O_2 + 2H_2O + 4e^- \longrightarrow 4OH^- \qquad \text{(neutral or alkaline solution)}$$

Figure 16.13 shows schematically how iron corrodes, with the last half-reaction as the reduction process.

**Figure 16.13**
The corrosion of iron in moist air. The reduction half-reaction occurs in the oxygen-rich area in the center. The oxidation process occurs in the oxygen-poor areas at the right and the left. The net result is a combination of iron with oxygen to form rust.

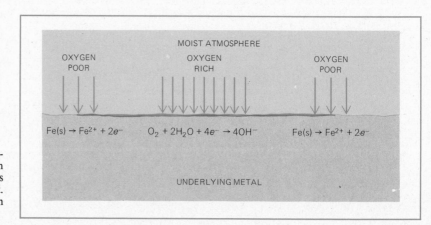

The standard potentials of the half-reactions in which dissolved $O_2$ is reduced are positive, so these reductions are more favorable than those involving $H^+$ or $H_2O$ alone. Metals corrode faster at higher partial pressures of oxygen, because more oxygen then dissolves into the layer of moisture that is in contact with the metal. There is an interesting aspect of corrosion called the principle of differential aeration: If a part of a metal surface is exposed to a relatively high concentration of $O_2$, corrosion occurs in another region of the metal. The potential difference that leads to corrosion requires the physical separation of the oxidation and reduction processes, by analogy with a galvanic cell. The reduction process in corrosion, which is the reduction of $O_2$, occurs in the region where the concentration of $O_2$ is highest. Therefore the oxidation half-reaction, which does the real damage, takes place elsewhere.

In the case of iron, the $Fe^{2+}$ cations that form migrate toward the cathodic regions of the metal surface. There they react with water or $OH^-$ to form $Fe(OH)_2$, which undergoes further oxidation to form $Fe(OH)_3$, the familiar reddish material called rust. Meanwhile, the anodic region of the metal surface undergoes the real damage: Holes appear, the surface is eaten away, and the metal's structural strength is weakened. The worst damage seems to occur when the reduction process liberates $H_2(g)$, apparently because this gas penetrates below the surface and further weakens the metal.

The corrosion of an automobile is a familiar demonstration of the principle of differential aeration. When some of the paint that protects the metal of the automobile chips off, corrosion does not occur at the site of the chipping. Rust does form at this spot, because it is the place where the reduction half-reaction occurs. The real damage is done at the anode region, which is a site near the exposed area (Figure 16.14).

We can also demonstrate corrosion by differential aeration by embedding metal rods in sand, leaving the upper part of the rods in water. As Figure 16.15 shows, it is the segment of the rod in the sand that corrodes, not the segment in the water. The sand prevents $O_2$ from reaching the metal. The reduction occurs in the oxygen-rich water and the damaging oxidation process occurs in the sand.

A metal can corrode if the potential of the half-reaction for its oxidation is relatively more positive than any other oxidation that can occur when it is exposed to moisture in the atmosphere. This concept is the basis of several methods for preventing corrosion.

The most widely used method is to use paint or a similar coating to prevent $O_2$ and $H_2O$ from coming in contact with the metal surface. Another group of methods uses what are called corrosion inhibitors, substances that form surface films that interfere with the flow of charge needed for corrosion to occur. There are anodic inhibitors, which act at the anode and interfere with the metal dissolution reaction. These include inorganic salts such as chromates, phosphates, and carbonates, as well as compounds containing organic nitrogen or sulfur. There are also cathodic inhibitors, which interfere with the reduction process that occurs at the

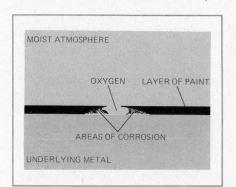

**Figure 16.14**
How an auto body corrodes. While the driving force for corrosion is the reduction process that occurs at the spot where the auto's paint has chipped and metal is in contact with moist air, the structural damage is done at the sites where the oxidation process occurs.

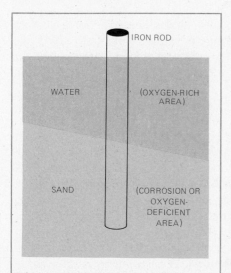

**Figure 16.15**
In the corrosion of a metal bar half-buried in sand, the reduction process occurs in the segment of the bar exposed to oxygen-carrying water. The segment of the bar that corrodes is under the sand, where the oxidation process occurs.

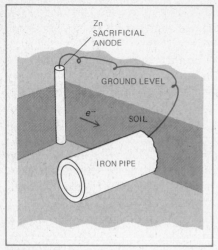

**Figure 16.16**
Cathodic protection. The iron pipe is connected to a "sacrificial anode" made of zinc, a metal that is higher in the emf series than iron. The zinc corrodes and the iron does not.

cathodic region of the metal surface. These include salts of magnesium, zinc, or nickel.

Another corrosion-control method is cathodic protection, which prevents a metal from corroding by connecting it to another metal that is more active. The more active metal, which is higher in the emf series, is the one that corrodes. An iron pipe can be protected from corrosion if it is connected to a bar of zinc, as shown in Figure 16.16. Since the potential for the oxidation of zinc is more positive than the potential for the oxidation of iron, it is the zinc bar that corrodes. The zinc is called the sacrificial anode. It prevents corrosion by pumping electrons into the iron, thus preventing the electron-releasing oxidation of iron from occurring.

Surprisingly, corrosion can also be inhibited if electrons are pumped out of the metal. This method is based on a rather complex phenomenon called *passivation.* The basic idea is to force the corrosion process to occur so rapidly that an oxide film with special properties forms on the metal surface and prevents further corrosion.

Passivation occurs naturally. Some metals such as Al or Cr that are high in the emf series are relatively corrosion resistant because of passivation. Iron corrodes completely in dilute nitric acid, but it does not corrode in concentrated nitric acid because of passivation. The rapid reaction that takes place in the concentrated acid causes formation of an oxide film that protects the iron from further attack.

## Biological Oxidations

Living things eat food to get energy that enables them to perform work on the surroundings and to maintain the nonequilibrium state called life. The major source of this energy is the oxidation of food, a process that, for convenience, can be represented by the overall reaction for the oxidation of glucose:

$$C_6H_{12}O_6 + 6O_2 \longrightarrow 6CO_2 + 6H_2O + \text{energy}$$

The $\Delta G°$ for this process is very large. Its value is $-2870$ kJ/mol, and 24 mol of electrons is transferred for each mole of glucose that is oxidized. If such a large quantity of free energy were made available suddenly as heat—as it is when we set fire to a lump of sugar—most of it would be wasted by the organism, and it even could cause damage. In living cells, the oxidation of glucose and other molecules that are biological energy sources does not occur in one step. Biological oxidations take place in a complex, multistep pathway. Each step releases only a small quantity of the total free energy change of the process. This free energy is not released primarily as heat. Rather, it is used to drive a chemical system away from equilibrium.

We can compare this biological process with the process by which electrical energy is converted to chemical energy in an electrolytic cell. In electrolysis, an applied current drives a chemical system away from equilibrium. When the system is allowed to return to equilibrium, the stored energy becomes available. In a living organism, energy is also used to drive a system away from equilibrium. When the system is allowed to return to equilibrium, the energy is released at a convenient moment and in a convenient quantity.

The free energy from each of many steps in the oxidation of glucose is used to synthesize an unstable, energy-rich molecule called adenosine triphosphate (usually abbreviated ATP) from constituents in the cell. The formation of ATP is a process proceeding away from equilibrium. Under physiological conditions, $\Delta G$ for the synthesis of ATP is about $+40$ kJ/mol. Ideally, each step in the oxidation of glucose should release roughly this quantity of energy.

Most details of the oxidation of glucose in the living cell have been worked out by biochemists. Each intermediate, partially oxidized form of glucose has been identified, and we can write the entire sequence of steps, starting with glucose and ending with $CO_2$ and $H_2O$, for a number of biological oxidation pathways. These pathways are quite complex and have many intermediates. We shall focus on one typical step, in which malic acid, $C_4H_6O_5$, which is formed after many steps from the partial oxidation of glucose, is further oxidized to another organic acid, oxalo-acetic acid, $C_4H_4O_5$:

$$C_4H_6O_5 + \tfrac{1}{2}O_2 \longrightarrow C_4H_4O_5 + H_2O$$

This overall reaction can be broken into two half-reactions. We can write both these half-reactions as reductions, giving their potentials at physiological conditions:

$$C_4H_4O_5 + 2H^+ + 2e^- \longrightarrow C_4H_6O_5 \qquad \mathcal{E} = -0.17 \text{ V}$$
$$\tfrac{1}{2}O_2 + 2H^+ + 2e^- \longrightarrow H_2O \qquad \mathcal{E} = \phantom{-}0.82 \text{ V}$$

The number of electrons transferred in this process is $n = 2$. Generally, two electrons are transferred in each step of the oxidation of glucose. We can calculate $\mathcal{E}$ for the overall reaction in the usual way, by noting that the half-reaction with the negative potential reverses. Thus,

$$\mathcal{E} = 0.82 \text{ V} + (+0.17 \text{ V}) = 0.99 \text{ V}$$

Since we know the value of the potential and the number of electrons transferred, we can calculate the free energy change for the reaction:

$$\Delta G = -nF\mathcal{E} = -(2 \text{ mol } e^-)\left(\frac{96\,500 \text{ C}}{\text{mol } e^-}\right)(0.99 \text{ V})$$
$$= -190\,000 \text{ J}$$

### The Electron Transfer Chain

Thus, the quantity of energy that is released when two electrons are transferred from malic acid to $O_2$ is 190 kJ/mol. But this quantity is more energy than a biological system can use efficiently. Such electron transfers from intermediates in the oxidation of glucose do not occur directly. It has been shown that there is an electron transfer chain of at least seven steps, in which the two electrons are transferred from one substance to another as they travel from malic acid to $O_2$. Many of these transfer steps release a quantity of energy that is small enough to be used efficiently to synthesize ATP.

We could say that combining the malic acid oxidation half-reaction directly with the $O_2$ reduction half-reaction is like jumping from the roof of a building. Sending the electrons along the electron transfer chain, in this analogy, is like walking down a flight of stairs. The same quantity of energy is released; in the one case impractically, in the other case in usable form. To "walk" this energy downstairs via the electron transport chain, the malic acid oxidation half-reaction is combined with a reduction half-reaction of much lower potential than that of the $O_2$ reduction. This half-reaction can be represented as

$$\text{oxid}_1 + 2e^- \longrightarrow \text{red}_1$$

where $\text{oxid}_1$ represents the oxidized form of the first substance in the electron transport chain and $\text{red}_1$ represents the reduced form. The reaction is

$$\text{malic acid} + \text{oxid}_1 \longrightarrow \text{oxaloacetic acid} + \text{red}_1$$

Because the potential of this reduction half-reaction is lower than that of the $O_2$ reduction half-reaction, the potential of this reaction is lower than that of the overall reaction with $O_2$, and the free energy release is lower.

At this point, malic acid and oxaloacetic acid have fulfilled their function, and the pair of electrons passes on to the next step, which can be represented as

$$\text{oxid}_2 + 2e^- \longrightarrow \text{red}_2$$

where $\text{oxid}_2$ and $\text{red}_2$ are, respectively, the oxidized and reduced forms of the next substances in the chain. The potential of this reduction half-reaction must be more positive than that of the first reduction half-reaction, so the overall reaction for the second step:

$$\text{oxid}_2 + \text{red}_1 \longrightarrow \text{red}_2 + \text{oxid}_1$$

can occur spontaneously. This process reverses the first reduction by transferring the two electrons from $\text{red}_1$, the reduced form of the substance in the first step, to $\text{oxid}_2$. The effect of this reversal is the regeneration of $\text{oxid}_1$, which allows another electron transfer from malic acid.

Each subsequent step of the electron transfer chain is roughly similar to this one. In each step, a half-reaction has a more positive potential than the preceding half-reaction. The preceding half-reaction reverses to transfer two electrons, and a small, manageable quantity of free energy is released. At the end of the chain, the electrons are transferred to $O_2$ in a process that can be represented as

$$\tfrac{1}{2}O_2 + 2H^+ + 2\ red_6 \longrightarrow H_2O + 2\ oxid_6$$

The potential of the half-reaction

$$2\ oxid_6 + 2e^- \longrightarrow 2\ red_6$$

has been found to be 0.5 V, so the potential of the last overall reaction is

$$\mathscr{E} = 0.82 - 0.5 = 0.3\ V$$

and the corresponding free energy change is

$$\Delta G = -(2\ mol\ e^-)\left(\frac{96\ 500\ C}{mol\ e^-}\right)(0.3\ V) = -60\ kJ$$

Thus, the electron transfer chain utilizes a series of substances that can be oxidized and reduced reversibly. The sequence in which the substances are utilized is important, since each reduction half-reaction must have a more positive potential than its predecessor in the chain.

Not all the details of these processes have been worked out. The substances in the electron transport chain have complex molecular structures that require a great deal of study. But it is known that the electron transfers appear to occur primarily in cell structures called mitochondria. It appears that one function of the mitochondria is to hold the substances rather rigidly in line so that electron transfer will occur in the most effective way.

**Summary**

This chapter dealt with **oxidation-reduction** reactions, which occur as the result of **electron transfers.** Oxidation is the **loss** of electrons and reduction is the **gain** of electrons, and one cannot normally occur without the other. The substance that is oxidized is called a **reducing agent,** and the substance that is reduced is called an **oxidizing agent.** We noted that it is convenient to divide an oxidation-reduction process into two **half-reactions,** one for the oxidation and one for the reduction, and we described how to balance the overall reaction for such a process by the method of half-reactions or the method of change in oxidation numbers.

We introduced **electrolysis,** an oxidation-reduction process that is brought about by the application of an electric current. The **cathode** is where reduction takes place and the **anode** is where oxidation takes place. Electrolysis can be regarded as the conversion of electrical energy to chemical energy. We discussed **Faraday's laws,** which quantify the changes that occur in electrolysis. Then we described the **galvanic cell,** a device for converting chemical energy into electrical energy, and said that it is best regarded as two half-cells, one for oxidation and one for reduction. **Batteries** are galvanic cells. We then defined **cell potential** as the energy difference between the two electrodes of a galvanic cell, and noted that the potential of different cells can be compared in terms of a **standard state** of the components of a cell. We said that a **standard hydrogen electrode,** whose electrode potential is defined as exactly 0 V, is used to find the potential difference between electrodes and showed how a tabulation of potentials can be made in what is called an **emf series.** We gave the **Nernst equation,** and showed how it can be used to find the potential of a galvanic cell. Finally we described **corrosion,** an electron transfer process that is important in technology, and showed how living organisms use electron transfers in the metabolic processes that provide the energy for life.

## Exercises

**16.1** Indicate for each of the following descriptions of chemical change whether an oxidation-reduction reaction takes place:
(a) Hydrogen peroxide decomposes to water and oxygen.
(b) Sodium nitrate and sulfuric acid form sodium sulfate and nitric acid.
(c) Calcium fluoride forms when solutions of calcium nitrate and sodium fluoride are mixed.
(d) Sodium iodide and bromine form sodium bromide and iodine.
(e) Sodium chromate and sulfuric acid form sodium dichromate.

**16.2** Classify each of the following changes as an oxidation or a reduction and write a balanced half-reaction for the change: (a) $Cl_2(g)$ to $Cl^-(aq)$, (b) $HClO_2(aq)$ to $ClO^-(aq)$, (c) $MnO_4^-(aq)$ to $MnO_4^{2-}(aq)$, (d) $Cr^{3+}(aq)$ to $Cr_2O_7^{2-}(aq)$, (e) $Sn(s)$ and $Cl^-(aq)$ to $SnCl_6^{2-}(aq)$.

**16.3** Balance each of the following skeleton expressions of chemical change taking place in acid solution: (a) $Fe^{2+}(aq) + Sn^{4+}(aq) \rightarrow Fe^{3+}(aq) + Sn^{2+}(aq)$, (b) $H_2S(aq) + Hg^{2+}(aq) \rightarrow S(s) + Hg(l)$, (c) $Cr_2O_7^{2-}(aq) + NO_2(g) \rightarrow Cr^{3+}(aq) + NO_3^-(aq)$, (d) $MnO_4^-(aq) + Br^-(aq) \rightarrow Mn^{2+}(aq) + BrO^-(aq)$.

**16.4** Balance each of the following skeleton expressions of chemical change taking place in acid solution: (a) $H_2O_2(aq) + NO(aq) \rightarrow NO_3^-(aq)$, (b) $O_3(g) + Au(s) + Cl^-(aq) \rightarrow O_2(g) + AuCl_4^-(aq)$, (c) $Ag(s) + Br^-(aq) + Hg_2^{2+}(aq) \rightarrow AgBr(s) + Hg(l)$.

**16.5** Balance each of the following expressions of chemical change taking place in alkaline solution: (a) $CN^-(aq) + ClO_3^-(aq) \rightarrow CNO^-(aq) + Cl^-(aq)$, (b) $H_2(g) + Cu^{2+}(aq) \rightarrow Cu(s)$, (c) $MnO_4^-(aq) + Cl_2(g) \rightarrow MnO_2(s) + ClO^-(aq)$.

**16.6** Balance each of the following skeleton expressions of chemical change taking place in alkaline solution: (a) $Zn(s) + I_2(aq) \rightarrow Zn(OH)_4^{2-} + I^-(aq)$, (b) $PH_3(g) + CrO_4^{2-}(aq) \rightarrow Cr(OH)_4^-(aq) + P_4(s)$, (c) $F_2(g) + O_2(aq) \rightarrow F^-(aq) + O_3(g)$.

**16.7**[4] Write balanced reactions for the following disproportionations (reaction of a substance with itself): (a) $HClO(aq) \rightarrow Cl^- + HClO_2(aq)$, (b) $MnO_2(s) \rightarrow Mn^{2+}(aq) + MnO_4^-(aq)$ alkaline, (c) $N_2O(aq) \rightarrow N_2(g) + NO_2^-(aq)$, (d) $Sn^{2+}(aq) \rightarrow Sn(s) + Sn(OH)_6^{2-}$ alkaline.

**16.8** Find the amount of $Cu^+$ in a sample that requires 32.2 cm³ of 0.129$M$ $KMnO_4$ solution to reach the equivalence point. The products are $Cu^{2+}$ and $Mn^{2+}$.

**16.9** A 25.1-cm³ sample of a solution of NaBr requires 24.9 cm³ of a 0.129$M$ $KMnO_4$ solution to reach the equivalence point for its oxidation to $BrO^-$ in alkaline solution. The $KMnO_4$ is reduced to $MnO_2(s)$. Find the concentration of the NaBr solution.

**16.10** A 0.535-g sample of impure tin is dissolved in strong acid to give $Sn^{2+}$. The $Sn^{2+}$ is then titrated with a 0.0448$M$ solution of $I_2$. A volume of 34.4 cm³ of $I_2$ solution is required to reach the equivalence point. Find the percent by mass of tin in the sample, assuming that it contains no other reducing agents.

**16.11** Sodium oxalate, $Na_2C_2O_4$, in solution is oxidized to $CO_2(g)$ by $MnO_4^-$, which is reduced to $Mn^{2+}$. A 25.1-cm³ volume of a solution of $MnO_4^-$ is required to titrate a 0.153-g sample of sodium oxalate. Find the concentration of the $MnO_4^-$ solution.

**16.12**★ A 0.286-g sample of $Na_2C_2O_4$

---

[4] The answers to exercises whose numbers are in color can be found in Appendix VII. The star indicates an exercise that is more challenging than average.

---

requires 25.8 cm³ of a $KMnO_4$ solution for titration. The $MnO_4^-$ is reduced to $Mn^{2+}$ and the $C_2O_4^{2-}$ is oxidized to $CO_2$. This $KMnO_4$ solution is used to titrate a sample of $As_2O_3$, which is converted to $H_3AsO_4$ in acid solution while the $KMnO_4$ again forms $Mn^{2+}$. A sample of $As_2O_3$ requires 44.8 cm³ of $KMnO_4$ solution for titration. Find the mass of $As_2O_3$.

**16.13** Write equations for the anode and cathode process and the overall reaction in the electrolysis of each of the following: (a) molten calcium iodide, (b) aqueous silver bromide, (c) aqueous potassium nitrate, (d) molten zinc oxide.

**16.14** The electrolysis of bauxite in the Hall process is carried out with graphite electrodes. In addition to aluminum the products are oxygen, carbon monoxide, and carbon dioxide. Write half-reactions for the processes by which these products form.

**16.15** Calculate the quantity of charge required for the following conversions: (a) 0.076 mol of $Sn^{4+}$ to $Sn^{2+}$, (b) 0.076 mol of $Sn^{4+}$ to $Sn(s)$, (c) 0.076 mol of $Sn(s)$ to $Sn^{2+}$.

**16.16** Calculate the quantity of charge necessary to produce 1.6 L of $O_2$ at STP from the electrolysis of water.

**16.17** The electrolysis of molten rubidium chloride is carried out with a current of 9.78 A. Find the time needed to form 17.5 g of rubidium metal.

**16.18** Find the mass of chlorine produced from a sodium chloride solution upon the application of a current of 8.55 A for 17.6 min.

**16.19** The atomic weight of a metal is 91.2 and it forms a bromide of unknown composition. Electrolysis of the molten bromide produces 3.5 g of the metal and 15.3 g of bromine. Find the empirical formula of the bromide.

**16.20** A metal forms the chloride

$MCl_4$. Electrolysis of the molten chloride by a current of 1.81 A for 25.6 min deposits 1.09 g of the metal. Find the atomic weight of the metal.

**16.21** A current of 1.00 A is applied to 1.0 L of 1.0$M$ HCl solution for 24 hr. Find the pH of the solution.

**16.22**★ In electroplating, a thin layer of a metal is applied to a solid surface by an electrolytic process. An object whose surface area is 78.6 cm² is to be plated with an even layer of gold $8.00 \times 10^{-4}$ cm thick. The density of gold is 19.3 g/cm³. The object is placed in a solution of $Au(NO_3)_3$ and a current of 2.75 A is applied. Find the time required for the electroplating to be completed, assuming that the layer of gold builds up evenly.

**16.23** Find the number of electrons required to form 10.0 g of calcium metal from the electrolysis of molten calcium chloride.

**16.24** A galvanic cell is constructed of two half-cells. The anode is made of a piece of lead in contact with a solution of $Pb(NO_3)_2$. The cathode is made of a piece of copper in contact with a solution of $Cu(NO_3)_2$. Write the two half-cell reactions and the overall reaction for the cell. Describe the flow of electrons and the movement of ions as the cell operates, assuming that the two half-cells are connected by a porous barrier.

**16.25** Use cell notation to describe the following cells:
(a) The anode is the standard hydrogen electrode and the cathode is the standard Au(s)|Au³⁺ half-cell.
(b) The anode is the nickel metal, solid nickel(II) sulfide half-cell and the cathode is the nickel, nickel(II) nitrate (1$M$) half-cell.
(c) The anode is $Zn|Zn(NH_3)_4^{2+}$ and the cathode is solid $I_2$ on platinum; all concentrations are 1$M$.

**16.26** Write a cell corresponding to the following reactions: (a) $3Mg(s) +$

$2Fe^{3+}(aq) \rightarrow 3Mg^{2+}(aq) + 2Fe(s)$, (b) $H_2(g) + Cl_2(g) \rightarrow 2H^+(aq) + 2Cl^-(aq)$, (c) $Ni(s) + Hg_2Cl_2(s) \rightarrow Ni^{2+}(aq) + 2Hg(l) + 2Cl^-(aq)$.

**16.27** Write a cell corresponding to the reaction $Zn^{2+}(aq) + 4NH_3(aq) \rightarrow Zn(NH_3)_4^{2+}(aq)$.

**16.28** Find the quantity of charge delivered and the current produced when 0.638 g of zinc in a dry cell is consumed in 45 min.

**16.29** Calculate the mass of $PbSO_4(s)$ formed in a lead storage battery when it delivers a current of 13.0 A for exactly one hour.

**16.30** The $\mathcal{E}°$ of the cell $Cr(s)|Cr^{3+}\|Sn^{2+}|Sn(s)$ is 0.604 V. Find $\Delta G°$ for the cell reaction.

**16.31** The $\Delta G°$ of the reaction $Cu(s) + Br_2(l) \rightarrow Cu^{2+}(aq) + 2Br^-(aq)$ is $-140$ kJ. (a) Find $\mathcal{E}°$ of the corresponding cell. (b) Find $\Delta G°$ for the reaction $2Cu(s) + 2Br_2(l) \rightarrow 2Cu^{2+}(aq) + 4Br^-(aq)$ and $\mathcal{E}°$ for the corresponding cell.

**16.32★** The cell $Pt(s)|Cu^+(1M)$, $Cu^{2+}(1M)\|Cu^+(1M)|Cu(s)$ has $\mathcal{E}° = 0.364$ V. The cell $Cu(s)|Cu^{2+}(1M)\|Cu^+(1M)|Cu(s)$ has $\mathcal{E}° = 0.182$ V. Write the cell reaction for each cell. Calculate $\Delta G°$ for each reaction from these data.

**16.33** The $\mathcal{E}°$ for the cell $V(s)|V^{3+}(1M)\|Ag^+(1M)Ag(s)$ is 1.69 V. Calculate $K$ for the cell reaction at 298 K.

**16.34** The $\mathcal{E}°$ for the reaction of $F_2(g)$ and $Br^-(aq)$ to form $F^-(aq)$ and $Br_2(l)$ is 1.81 V. Calculate $K$ for the reaction at 298 K.

**16.35** The $K$ for the reaction $2Cr(s) + 6H^+(aq) \rightleftharpoons 2Cr^{3+}(aq) + 3H_2(g)$ is $2.18 \times 10^{92}$. Calculate $\mathcal{E}°$ for the cell $Cr(s)|Cr^{3+}(1M)\|H^+(1M)|H_2(g)|Pt(s)$.

**16.36★** The $K_{sp}$ of $PbBr_2$ is $4.6 \times 10^{-6}$. Find the $\mathcal{E}°$ of the cell $Pb(s)|PbBr_2(s)|Br^-(1M)\|Pb^{2+}(1M)|Pb(s)$.

**16.37** Two cells are constructed in which the standard hydrogen electrode is one of the half-cells. When the other half-cell is the $Cr(s)|Cr^{2+}$ electrode the $\mathcal{E}°$ is found to be 0.91 V with the hydrogen electrode as the cathode. When the other half-cell is the $Ag(s)|Ag^+$ electrode the $\mathcal{E}°$ is found to be 0.80 V with the hydrogen electrode as the anode. Find $\mathcal{E}°$ for the cell made of the two metal half-cells and designate the anode and the cathode.

**16.38** Two cells are constructed in which one of the half-cells is the $Zn(s)|Zn^{2+}$ electrode. When the other half-cell is the $Pb(s)|Pb^{2+}$ electrode the $\mathcal{E}°$ is found to be 0.64 V with the zinc electrode as the anode. When the other half-cell is the $V(s)|V^{3+}$ electrode $\mathcal{E}°$ is found to be 0.13 V with the zinc electrode as the cathode. Find the $\mathcal{E}°$ of the cell made from the lead and vanadium electrodes and designate the anode and the cathode.

**16.39** List the following species in order of increasing strength as reducing agents: $Fe(s)$, $Cr(s)$, $Au^+$, $H_2O_2$, $Sn^{2+}$, $I^-$, $Br^-$.[5]

**16.40** List the following species in order of increasing strength as oxidizing agents: $H_2O_2$, $Sn^{2+}$, $Br_2$, $Cl_2$, $O_2$, $Cr^{3+}$, $Zn^{2+}$, $Sn^{4+}$.[5]

**16.41** Calculate the standard potential of the following cells:[5] (a) $Mg(s)|Mg^{2+}(1M)\|Cr^{3+}(1M)|Cr(s)$, (b) $Cu(s)|Cu^{2+}(1M)\|Hg_2^{2+}(1M)|Hg(l)|Pt(s)$, (c) $Ni(s)|Ni^{2+}(1M)\|Mn^{2+}(1M)$, $Mn^{3+}(1M)|Pt(s)$.

**16.42** Calculate the standard potential of the following alkaline cells:[5] (a) $Zn(s)|Zn(OH)_4^{2-}(1M)$, $OH^-(1M)\|H_2(g)$ (1 atm)$|Pt(s)$, (b) $Cr(s)|Cr(OH)_3(s)|OH^-(1M)\|ClO_3^-(1M)$, $Cl^-(1M)|Pt(s)$.

**16.43** Predict whether there should be substantial reaction when the fol-

lowing substances are brought together in their standard states: (a) Zn and Mg, (b) $F_2$ and Li, (c) $H_2O_2$ and $Cl_2$, (d) $Cr^{3+}$ and Pb.[5]

**16.44** Predict whether there should be substantial reaction when the following substances are brought together in their standard states in a solution of pH $= 14$: (a) $ClO_3^-$ and $NO_2^-$, (b) $ClO_3^-$ and $ClO_4^-$, (c) Cr and $Al(OH)_4^-$.[5]

**16.45** Indicate which of the following metals should dissolve to an appreciable extent in a solution of pH $= 0$: Hg, Ba, Al, Ag, Cu.[5]

**16.46** Indicate which of the following metals should dissolve to appreciable extent in a solution of pH $= 14$: Mn, K, Be, Ni.[5]

**16.47** We can predict from the emf series that Al should dissolve in $1M$ acid solution, yet we generally do not observe such a reaction. Suggest an explanation for this behavior of aluminum.

**16.48** Predict which of the following metal cations should disproportionate appreciably in $1M$ aqueous solution: $Cr^{3+}$, $Mn^{2+}$, $Sn^{2+}$, $Hg_2^{2+}$.[5]

**16.49** Predict which of the following species should disproportionate in $1M$ aqueous solution: HClO, $ClO_3^-$(alkaline), $H_2O_2$.[5]

**16.50** Use the emf series of Table 16.1 to find the $\mathcal{E}°$ of the following cell reactions: (a) $Ni(s) + 2Ag^+(aq) \rightleftharpoons Ni^{2+}(aq) + 2Ag(s)$, (b) $2Hg(l) + 2Cl^-(aq) + Zn^{2+}(aq) \rightleftharpoons Hg_2Cl_2(s) + Zn(s)$, (c) $2Br^-(aq) + F_2(g) \rightleftharpoons Br_2(l) + 2F^-(aq)$.

**16.51** Calculate $\Delta G°$ from $\mathcal{E}°$ for each of the reactions in Exercise 16.50.

**16.52** Calculate $K$ from $\mathcal{E}°$ for each of the reactions in Exercise 16.50.

**16.53** Use the data in Table 16.1 to calculate $\Delta G°$ for the following reactions: (a) $H_2O_2(aq)$ forms from

---

[5] The data necessary for this exercise are found in Table 16.1.

$O_2(g) + H_2O$, (b) $MnO_4^-(aq)$ and $I^-(aq)$ form $I_2(s)$ and $Mn^{2+}(aq)$ in aqueous solution, (c) $O_3(g)$ and $NO_2^-(aq)$ form $O_2(g)$ and $NO_3^-(aq)$ in alkaline solution.

**16.54** The reaction $Zn(NH_3)_4^{2+}(aq) + 4OH^-(aq) \rightleftharpoons Zn(OH)_4^{2-}(aq) + 4NH_3(aq)$ is not an oxidation-reduction. Nevertheless use the data in Table 16.1 to find $\mathcal{E}^\circ$, $\Delta G^\circ$, and $K$ for this reaction.

**16.55** Calculate $K_{sp}$ for $Ag_2S(s)$ using the data in Table 16.1.

**16.56*** The $K_{sp}$ for $AgCl(s)$ is $1.6 \times 10^{-10}$. Find the $\mathcal{E}^\circ$ of the half-reaction $AgCl(s) + e^- \rightarrow Ag(s) + Cl^-(aq)$.[5]

**16.57*** Find the solubility of $Zn(OH)_2(s)$ in $1M$ $OH^-$ solution.[5]

**16.58** Calculate $\Delta G^\circ$ and $K$ for the following disproportionations: (a) $Cr^{3+}$ to $Cr(s)$ and $Cr_2O_7^{2-}$ in acidic solution, (b) $HNO_2$ to $NO(g)$ and $NO_3^-$ in acidic solution.

**16.59** Find $\mathcal{E}^\circ$ for the following half-reactions that are not listed explicitly in Table 16.1 using the half-reactions that are listed in the table: (a) $Cr^{3+}(aq) + e^- \rightarrow Cr^{2+}(aq)$, (b) $Hg^{2+}(aq) + 2e^- \rightarrow Hg(l)$, (c) $Au^+(aq) + e^- \rightarrow Au(s)$.

**16.60** Find $\mathcal{E}^\circ$ of the following half-reactions: (a) $MnO_4^{2-}(aq)$ to $MnO_2(s)$, (b) $N_2O_4(g)$ to $NO(g)$ in aqueous solution, (c) $ClO_4^-$ to $Cl^-(aq)$.[5]

**16.61*** Find $\mathcal{E}^\circ$, $\Delta G^\circ$, and $K$ for the reaction in which $MnO_4^{2-}$ and $HNO_2$ react to form $N_2O_4(g)$ and $Mn^{2+}$ in aqueous solution.[5]

**16.62** Calculate $\mathcal{E}$ for the cell $Ni(s)|Ni^{2+}(0.11M)\|Cu^{2+}(0.24M)|Cu(s)$.[5]

**16.63** Calculate $\mathcal{E}$ for the cell $Pt(s)|Br_2(l)|Br^-(0.95M)\|Cl^-(0.082M)|Cl_2(g)(0.35 \text{ atm})|Pt(s)$.[5]

**16.64** Calculate $\mathcal{E}$ for the cell

$Pb(s)|Pb^{2+}(1.0M)\|H^+(aq)|H_2(g)(1.0 \text{ atm})|Pt(s)$ at: (a) pH = 0, (b) pH = 3, (c) pH = 7.[5]

**16.65** The potential of the cell $In(s)|In^{2+}(aq, 0.019M)\|H^+(1.0M)|H_2 (g, 1.0 \text{ atm})|Pt(s)$ is 0.27 V. Calculate $\mathcal{E}^\circ$ of the cell. Calculate $\mathcal{E}^\circ$ of the $In(s)|In^{2+}$ electrode.

**16.66** The potential of the cell $U(s)|U^{3+}(aq, 0.0046M)\|Zn^{2+}(aq, 0.31M)|Zn(s)$ is 1.06 V. Find $\mathcal{E}^\circ$ of the $U(s)|U^{3+}$ electrode.[5]

**16.67** Find the ratio of $[Cu^+]$ to $[Cu^{2+}]$ when the potential of the $Cu^+|Cu^{2+}$ electrode is 0.00 V.[5]

**16.68** Calculate the reduction potential of the half-cell $Pt(s)|H_2(g)(1 \text{ atm})|H^+(aq)$ at pH = 1, pH = 3, and pH = 7.

**16.69** The potential of the cell $I_2(s)|I^-(aq, 0.0961M)\|Br^-(aq, 0.121M)|Br_2(l)$ is 0.524 V. Find $K$ of the reaction $Br_2(l) + 2I^-(aq) \rightleftharpoons I_2(s) + 2Br^-(aq)$.

**16.70** The $\mathcal{E}$ of the cell $Pt(s)|Sn^{2+} (0.46M)$, $Sn^{4+}(0.0012M)\|H^+(1.0M)$, $MnO_4^-(0.58M)$, $Mn^{2+}(0.0014M)|Pt(s)$ is 1.45 V. Find $K$ of the reaction $2MnO_4^-(aq) + 16H^+(aq) + 5Sn^{2+}(aq) \rightleftharpoons 2Mn^{2+}(aq) + 5Sn^{4+}(aq) + 4H_2O$.

**16.71*** The $\mathcal{E}$ of the cell $Mg(s)|MgCO_3(s)|CO_3^{2-}(0.23M)\|Mg^{2+}(0.14M)|Mg(s)$ is 0.127 V. Find the $K_{sp}$ of $MgCO_3$.

**16.72** A concentration cell is a galvanic cell in which both half-cells are composed of the same substances, but in different concentrations. Since $\mathcal{E}$ is determined by concentrations, a potential difference and a current are established. Consider a cell in which both half-cells contain manganese metal electrodes and $Mn^{2+}$ in solution. The $[Mn^{2+}]$ is $1.0M$ in one half-cell and $0.10M$ in the other. Identify the anode

and the cathode and calculate $\mathcal{E}$ of the cell.

**16.73** Determine $\mathcal{E}$ of a concentration cell with $Zn(s)$ electrodes and $[Zn^{2+}] = 0.45M$ in one half-cell and $0.0025M$ in the other.

**16.74** Find the $[Sn^{2+}]$ in a half-cell of a concentration cell with $Sn(s)$ electrodes and the other half-cell having $[Sn^{2+}] = 1.0M$, when $\mathcal{E} = 0.010$ V.

**16.75** Write the anode and cathode processes and the overall reaction for the corrosion of vanadium under alkaline conditions.

**16.76** Write the anode and cathode processes and the overall reaction for the corrosion of iron in dilute nitric acid. Write the anode and cathode processes and the overall reaction for the passivation of iron in concentrated nitric acid. Assume that $NO_2(g)$ is one of the products.

**16.77** Show the anode and cathode processes and the overall reaction that takes place when the corrosion of iron is prevented by cathodic protection with aluminum. The conditions are acidic and oxygen rich.

**16.78** Find the potential that corresponds to a $\Delta G$ of $-40$ kJ/mol for a one-electron transfer and for a two-electron transfer.

**16.79*** An inorganic electron transport chain can be constructed by using the couples $Sn^{2+}$, $Sn^{4+}$; $Fe^{2+}$, $Fe^{3+}$; $Cu^{2+}$, $Cu^+$; and $Au^{3+}$, $Au^+$ suitably arranged. Suppose we wish to use this chain to transfer 2 mol of electrons from $H_2(g)$ to $O_2(g)$ in acid solution. List the couples in the order in which they should be used. Write the overall reaction for each step and the overall reaction for the entire process. Calculate $\mathcal{E}^\circ$ and $\Delta G^\circ$ for each step of the process.[5]

# 17

# Chemical Kinetics

**Preview**    **N**ow we turn to the topic of time in chemistry, focusing on the rate of chemical reactions. We first describe how reaction rates are measured, explaining the difference between the average rate and the instantaneous rate. Next we show how a rate law gives the relationship between the concentrations of substances in a reaction and the rate of the reaction. We then describe two rate laws that depend on the concentration of a single substance, and their corresponding rate equations: the first-order rate law, which introduces the concept of half-life, and the one-term second-order rate law. Next we deal with reaction mechanisms, detailed molecular descriptions of chemical changes, and their relationship to rate laws. We then discuss the activation energy of a reaction and its relationship to the effect of temperature on rate. Finally, we discuss the acceleration of reaction rates by catalysts in industrial processes and in living organisms.

O ur discussion of chemical reactions in previous chapters has left out one crucial factor: time. With the knowledge we now have, and given the appropriate data, we can predict whether a chemical system *should* change, the nature of the change, and the composition of the system at the equilibrium state. But the omission of time means that we cannot say whether any chemical system *will* change in the way we predict.

Until now, we have studied thermodynamics, which tells us whether or not a chemical change is possible. We shall now study chemical kinetics, which tells us how much time is needed for a chemical change to occur. We cannot get complete information about a chemical change without applying both chemical thermodynamics and chemical kinetics.

Thermodynamics tells us that $H_2O$ does not spontaneously dissociate to form $H_2$ and $O_2$ under ordinary conditions. The position of equilibrium in the reaction

$$H_2O(l) \rightleftharpoons H_2(g) + \tfrac{1}{2}O_2(g)$$

lies far to the left ($K = 3 \times 10^{-42}$). If we want this change to occur, we must force it to happen — for example, by passing an electric current through the water. Thermodynamics also tells us that the reverse process, in which $H_2$ and $O_2$ form water, is spontaneous; all we have to do is wait, and water should eventually be formed. Thermodynamics does not tell us that it will be a very long wait, because the reaction occurs very slowly. Kinetics gives us this information. Kinetics also tells us that $H_2$ and $O_2$ will react very quickly if a spark is applied to the mixture of the two gases — so quickly that an explosion may occur.

There are many such examples in everyday life. Thermodynamics tells us that the contents of an egg will change from liquid to solid spontaneously. Kinetics tells us that the best way to make this change occur is to put the egg in boiling water. Thermodynamics tells us that nitrous oxide, the propellant in a can of whipped cream, is unstable and forms $N_2$ and $O_2$ spontaneously. Kinetics tells us that the nitrous oxide dissociates so slowly that we need not rush to use a can of whipped cream.

We can sum up by saying that chemical kinetics deals primarily with the answers to three questions:

**1.** At what rate does a chemical system undergo change under a given set of conditions?

**2.** How will a change in conditions affect the rate at which a chemical change occurs?

**3.** Given the answers to the first two questions, what information is available about the details of the chemical change?

## 17.1  RATE OF REACTION

Expressions that define the rate at which events occur are commonplace in everyday life. An individual drives a car at 90 km/hr, reads 50 pages of a book in an hour, loses 1 kg a week on a diet, drinks half a glass of milk a minute. All these expressions have one feature in common. Each of them describes a change that occurs in a given interval of time. The speed of an automobile, for example, expresses a change in position, measured in kilometers, in an interval of one hour. If an individual drives at 90 km/hr, the position of the automobile in 1 hr changes by 90 km. Similarly, if someone reads a book for 1 hr, the number of pages that have been read increases by 50.

The rate of a chemical reaction can also be expressed as a given change in a given interval of time. The rate usually is expressed in terms of the change in the concentration or the pressure of one component. It may be expressed as a decrease in the concentration of a reactant or an increase in the concentration of a product. The rate of change of a chemical reaction is always expressed as a positive quantity. The time interval usually is 1 sec (although other time intervals are sometimes more convenient).

For reactions in solutions, the rate is the change in concentration (usually the molarity) of a component per second. For reactions in the gas phase, the rate is the change in the pressure of a component per second. The symbol $\Delta$ is used to express change, in the way in which we have used it previously: $\Delta$ means the final value minus the initial value.

For example, the rate at which the reaction

$$Cl_2 + 2I^- \longrightarrow 2Cl^- + I_2$$

occurs in solution can be expressed as the change in concentration of any of the four components of the system. If we write $\Delta[Cl_2]$, we mean the concentration of $Cl_2$ at the end of a given time interval minus the concentration of $Cl_2$ at the beginning of the time interval:

$$\Delta[Cl_2] = [Cl_2]_{final} - [Cl_2]_{initial}$$

The time interval can also be represented as $\Delta t = t_{final} - t_{initial}$. Thus, the rate at which $Cl_2$ changes in this reaction is $-\Delta[Cl_2]/\Delta t$. The negative sign is necessary because $Cl_2$ is disappearing, so the final concentration is lower than the initial concentration. By writing the negative sign, we fulfill the requirement that the rate must be a positive quantity. The units of the rate at which $Cl_2$ disappears in solution are moles per liter per second (mol $L^{-1}$ $sec^{-1}$). (We shall use "sec" as the abbreviation for second in this chapter, although the SI abbreviation is "s.")

We could also express the rate of this reaction as the rate of disappearance of $I^-$, which is $-\Delta[I^-]/\Delta t$. The rate of disappearance of $I^-$ is not the same as the rate of disappearance of $Cl_2$ in this reaction. The stoichiometry of the reaction shows that 2 mol of $I^-$ is consumed for each mole of $Cl_2$

that is consumed. Therefore, the rate of disappearance of $I^-$ is twice that of $Cl_2$. That is,

$$\frac{-\Delta[I^-]}{\Delta t} = 2\left(\frac{-\Delta[Cl_2]}{\Delta t}\right)$$

We could also express the rate of this process as the rate of appearance of one of the products. The form of the expression is the same, but a negative sign is not necessary, since the final concentration of the product is greater than the initial concentration. Again, the stoichiometry of the reaction indicates that $Cl^-$ will appear twice as fast as $I_2$. Thus

$$\frac{\Delta[Cl^-]}{\Delta t} = 2\left(\frac{\Delta[I_2]}{\Delta t}\right) = 2\left(\frac{-\Delta[Cl_2]}{\Delta t}\right) = \frac{-\Delta[I^-]}{\Delta t}$$

In the case of a reaction of gases,

$$2O_3(g) \longrightarrow 3O_2(g)$$

the changes in the quantities of the components can also be expressed as changes in pressure, and the rate can be given in units of atmospheres per second (atm/sec). For this reaction, the rate expressions are

$$3\left(\frac{-\Delta P_{O_3}}{\Delta t}\right) = 2\left(\frac{\Delta P_{O_2}}{\Delta t}\right) \quad \text{or} \quad \frac{1}{2}\left(\frac{-\Delta P_{O_3}}{\Delta t}\right) = \frac{1}{3}\left(\frac{\Delta P_{O_2}}{\Delta t}\right)$$

The pressure of $O_2$ increases faster than the pressure of $O_3$ decreases, so the total pressure of the system increases as the reaction proceeds.

For a general reaction

$$aA + bB \longrightarrow cC + dD$$

the rates of appearance and disappearance of the components in the reaction are given by

$$\frac{1}{a}\left(\frac{-\Delta A}{\Delta t}\right) = \frac{1}{b}\left(\frac{-\Delta B}{\Delta t}\right) = \frac{1}{c}\left(\frac{\Delta C}{\Delta t}\right) = \frac{1}{d}\left(\frac{\Delta D}{\Delta t}\right)$$

where the lowercase letters are the coefficients of the reaction.

**Example 17.1**   When ammonia is treated with $O_2$ at elevated temperatures, the rate of disappearance of ammonia is found to be $3.5 \times 10^{-2}$ mol $L^{-1}$ sec$^{-1}$ during a measured time interval. Calculate the rate of appearance of nitric oxide and water.

**Solution**   Since the relative rates of appearance of the products are governed by the coefficients of the chemical equation, the first step is to write the equation for the reaction:

$$4NH_3(g) + 5O_2(g) \longrightarrow 4NO(g) + 6H_2O(g)$$

Since NO and $NH_3$ have the same coefficient, the rate of appearance of nitric oxide is the same as the rate of disappearance of ammonia:

$$\frac{\Delta[NO]}{\Delta t} = 3.5 \times 10^{-2} \text{ mol L}^{-1} \text{ sec}^{-1}$$

The rate of appearance of $H_2O$ is 6/4 or 1.5 times as fast:

$$\frac{\Delta[H_2O]}{\Delta t} = 1.5(3.5 \times 10^{-2} \text{ mol L}^{-1} \text{ sec}^{-1}) = 5.3 \times 10^{-2} \text{ mol L}^{-1} \text{ sec}^{-1}$$

To know the numerical value of a rate expression, we must know the concentration or the pressure of a substance at two different times during the course of a reaction. This information can be obtained experimentally, if we have (1) a method of measuring time, (2) a method of measuring concentration or pressure, and (3) a method of keeping the conditions —especially the temperature—constant.

Table 17.1 lists some kinetic data for the reaction

$$CH_3Cl(aq) + I^-(aq) \longrightarrow CH_3I(aq) + Cl^-(aq)$$

taking place in aqueous solution at 298 K.

The data in Table 17.1 can be used to calculate the rate of the reaction for the time intervals that are given. For example, the rate of disappearance of $I^-$ from the start of the reaction to 180 min later is given by

$$\frac{-\Delta[I^-]}{\Delta t} = \frac{-(0.45 \text{ mol/L} - 0.50 \text{ mol/L})}{(180 \text{ min} - 0)(60 \text{ sec/min})} = 4.6 \times 10^{-6} \text{ mol L}^{-1} \text{ sec}^{-1}$$

with the factor 60 sec/min included in the denominator to convert the time, given in minutes, to seconds. This calculation can be done for any time interval. For example, the rate of disappearance of $I^-$ in the interval from 360 min after the start of the experiment to 1440 min after the start is given by

$$\frac{-\Delta[I^-]}{\Delta t} = \frac{-(0.27 \text{ mol/L} - 0.41 \text{ mol/L})}{(1440 \text{ min} - 360 \text{ min})(60 \text{ sec/min})}$$
$$= 2.2 \times 10^{-6} \text{ mol L}^{-1} \text{ sec}^{-1}$$

In many kinetics experiments, we do not measure concentration directly. Instead, we measure a property that can be related to the concentration of one substance in the system. In a gas phase system, we may measure the total pressure of the gases. A total-pressure measurement usually is a good way to determine the rate of a process that causes a change in the number of moles of gas in a system.

**TABLE 17.1    Disappearance of $I^-$**

| $[I^-]$ $(M)$ | $t$ (min) |
|---|---|
| 0.50 | 0 |
| 0.45 | 180 |
| 0.41 | 360 |
| 0.35 | 720 |
| 0.27 | 1440 |

**Example 17.2**   At high temperatures, $N_2O$ decomposes to $N_2$ and $O_2$. If the change in total pressure with time is measured, the following data are obtained from an initial pressure of $N_2O$ of 0.29 atm at 970 K:

| $P_T$ (atm) | 0.29 | 0.33 | 0.36 | 0.39 | 0.41 |
|---|---|---|---|---|---|
| $t$ (sec) | 0 | 300 | 900 | 2000 | 4000 |

Find the rate of disappearance of $N_2O$ and the rate of appearance of $O_2$ for the first 300 sec and the last 2000 sec of this process.

**Solution**   We can find the rate of disappearance of nitrous oxide by using the total pressure to find the pressure of nitrous oxide at any given time. The relationship between $P_T$ and $P_{N_2O}$ is derived from the equation for the reaction:

$$2N_2O(g) \longrightarrow \quad 2N_2(g) + O_2(g)$$
$$\textbf{start} \quad 0.29 \text{ atm} \qquad 0 \qquad 0$$

Let $2x$ be the pressure in atmospheres of $N_2O$ that reacts in the time interval of interest. The pressure of each gas at the end of this interval can be tabulated:

$$\textbf{time} \quad 0.29 \text{ atm} - 2x \qquad 2x \qquad x$$

The measured total pressure is the sum of these pressures:

$$P_T = P_{N_2O} + P_{N_2} + P_{O_2} = 0.29 \text{ atm} - 2x + 2x + x$$
$$= 0.29 \text{ atm} + x$$

The total pressure after 300 sec is given as 0.33 atm. Therefore

$$0.33 \text{ atm} = 0.29 \text{ atm} + x$$

and $x = 0.04$ atm. At time = 300 sec,

$$P_{N_2O} = 0.29 \text{ atm} - 2x = 0.29 \text{ atm} - 2(0.04 \text{ atm}) = 0.21 \text{ atm}$$

This value can be used to find the rate of disappearance of $N_2O$:

$$\frac{-\Delta P_{N_2O}}{\Delta t} = \frac{-(0.21 \text{ atm} - 0.29 \text{ atm})}{300 \text{ sec}} = 2.7 \times 10^{-4} \frac{\text{atm}}{\text{sec}}$$

There are a number of ways to find the rate of appearance of $O_2$ in this time interval. The stoichiometry of the reaction shows that the rate of appearance of $O_2$ is half the rate of disappearance of $N_2O$. Thus

$$\frac{\Delta P_{O_2}}{\Delta t} = \frac{1}{2} \left( 2.7 \times 10^{-4} \frac{\text{atm}}{\text{sec}} \right) = 1.4 \times 10^{-4} \frac{\text{atm}}{\text{sec}}$$

Another method finds the rate of appearance of $O_2$ directly from the $P_{O_2}$ at the start and end of the given time interval. At the start, no $O_2$ is present. After 300 sec, the $P_{O_2} = 0.04$ atm:

$$\frac{\Delta P_{O_2}}{\Delta t} = \frac{0.04 \text{ atm} - 0}{300 \text{ sec} - 0} = 1.3 \times 10^{-4} \frac{\text{atm}}{\text{sec}}$$

Similarly, we find the rate for the last 2000 sec by finding the pressures at the start and finish of this interval.

At 2000 sec,

$$P_T = 0.39 \text{ atm} = 0.29 \text{ atm} + x \quad \text{and} \quad x = 0.10 \text{ atm} = P_{O_2}$$

At 4000 sec,

$$P_T = 0.41 \text{ atm} = 0.29 \text{ atm} + x \quad \text{and} \quad x = 0.12 \text{ atm} = P_{O_2}$$

The rate of appearance of $O_2$ can be calculated directly:

$$\frac{\Delta P_{O_2}}{\Delta t} = \frac{0.12 \text{ atm} - 0.10 \text{ atm}}{4000 \text{ sec} - 2000 \text{ sec}} = 1.0 \times 10^{-5} \frac{\text{atm}}{\text{sec}}$$

The rate of disappearance of $N_2O$ is twice as fast:

$$\frac{-\Delta P_{N_2O}}{\Delta t} = 2.0 \times 10^{-5} \frac{\text{atm}}{\text{sec}}$$

## Average Rates and Instantaneous Rates

In both Example 17.1 and 17.2, the rates of the processes become slower with time. Most chemical reactions become slower as they proceed. *The rate of most chemical reactions depends in some way on the concentration of one or more of the reactants.* Since the reactants are consumed as the reaction proceeds, their concentrations decrease and the rate of the reaction decreases.

Since the quantities of the reactants in a chemical reaction decrease continuously as the reaction proceeds, the rate of the reaction also decreases continuously. *The rate measured for a time interval is only the average rate for that interval.* Therefore, the rate during a measured time interval is called the *average rate.* The rate at any one instant during the interval is called the *instantaneous rate.* The average rate and the instantaneous rate are equal for only one instant in any time interval. The instantaneous rate changes continuously. At first, the instantaneous rate is higher than the average rate. At the end of the interval, the instantaneous rate is lower than the average rate.

We can illustrate the relationship between average rate and instantaneous rate by using data for a reaction that takes place in polluted air. This process is the reaction between ozone, $O_3$, and the hydrocarbon ethylene, $C_2H_4$, in the gas phase, which can be represented as

$$O_3(g) + C_2H_4(g) \longrightarrow O_2(g) + C_2H_4O(g)$$

The rate of disappearance of $O_3$ can be followed easily at 303 K. Table 17.2 lists the data.

Suppose we want to know the instantaneous rate of this reaction 30 sec after the start; or, to put it another way, the instantaneous rate when $[O_3] = 1.63 \times 10^{-5}M$. We can approximate this instantaneous rate by calculating an average rate from the data. For example, the average rate of disappearance of $O_3$ for the entire time interval is

$$\frac{-\Delta[O_3]}{\Delta t} = \frac{-(1.10 \times 10^{-5}\ mol/L - 3.20 \times 10^{-5}\ mol/L)}{60.0\ sec - 0\ sec}$$

$$= 3.50 \times 10^{-7}\ mol\ L^{-1}\ sec^{-1}$$

This value is an approximation of the instantaneous rate. But it is not a very good approximation. A better value is the average rate between 10 sec and 50 sec:

$$\frac{-\Delta[O_3]}{\Delta t} = \frac{-(1.23 \times 10^{-5}\ mol/L - 2.42 \times 10^{-5}\ mol/L)}{50.0\ sec - 10.0\ sec}$$

$$= 2.98 \times 10^{-7}\ mol\ L^{-1}\ sec^{-1}$$

And a still better approximation is the average rate for the interval between 20 sec and 40 sec:

$$\frac{-\Delta[O_3]}{\Delta t} = \frac{-(1.40 \times 10^{-5}\ mol/L - 1.95 \times 10^{-5}\ mol/L)}{40.0\ sec - 20.0\ sec}$$

$$= 2.75 \times 10^{-7}\ mol\ L^{-1}\ sec^{-1}$$

As the time interval becomes smaller, the average rate becomes a better and better approximation of the instantaneous rate. The average rate will be the same as the instantaneous rate when the time interval is zero. While we cannot measure two concentrations during a time interval of zero seconds, we can say that the average rate gets closer to the instantaneous rate as the time interval gets closer to zero:

$$\text{instantaneous rate} = \lim_{\Delta t \to 0} \frac{-\Delta[A]}{\Delta t}$$

The instantaneous rate is the limit of the average rate as $\Delta t$ approaches 0. Using the notation of the calculus, we write the expression for this limit as $-d[A]/dt$.

The relationship between instantaneous rate and average rate can be shown graphically. Figure 17.1 shows a plot of $[O_3]$ against time. The solid curve shows the continuous decrease in $[O_3]$ as the reaction proceeds. The

| **TABLE 17.2** Disappearance of $O_3$ | |
|---|---|
| $[O_3]$ (mol/L) | $t$ (sec) |
| $3.20 \times 10^{-5}$ | 0 |
| $2.42 \times 10^{-5}$ | 10.0 |
| $1.95 \times 10^{-5}$ | 20.0 |
| $1.63 \times 10^{-5}$ | 30.0 |
| $1.40 \times 10^{-5}$ | 40.0 |
| $1.23 \times 10^{-5}$ | 50.0 |
| $1.10 \times 10^{-5}$ | 60.0 |

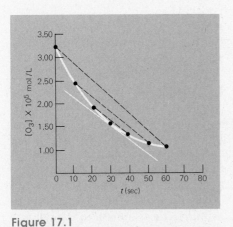

**Figure 17.1**
The solid curve shows the rate of disappearance of ozone with time in the reaction with $C_2H_4$. The instantaneous rate of the reaction at any given time is found from the tangent to the curve at that time. The slope of each dashed line, $-\Delta[O_3]/\Delta t$, shows the average rate of the reaction for a given interval. As the intervals are shortened, the slopes of the dashed lines come closer to the slope of the tangent to the curve.

slopes of the dashed lines show the average rates for the time points they connect. The slope of a dashed line is $-\Delta[O_3]/\Delta t$. The instantaneous rate at any time is the slope of the tangent to the curve at the point corresponding to that time. As the time interval represented by an average rate line grows smaller, the slope of the line becomes more nearly equal to the slope of the tangent to the curve at the 30-sec mark.

The curve for the disappearance of $O_3$ shows that the rate of the process slows with time. At the start, the curve falls sharply. As the reaction proceeds, it levels off. Tangents to the beginning segment of the curve have a steeper slope than the tangents to the end segments of the curve.

## 17.2    REACTION RATE AND CONCENTRATION

We have seen that the rate of a reaction is related to the concentration of one or more of the substances in a system. One important goal of a kinetic investigation is to establish the exact relationship between concentration and the rate of reaction for a given system. If we know how the concentration affects the rate of reaction for a specific system, we can get practical information about the best way to conduct the reaction and valuable insights into the way that the chemical change occurs.

*The relationship between rate and concentration for a given reaction can be established only by experiment.* Fortunately, a few relatively simple relationships describe most reactions. The equation that describes such a relationship is called a **rate law.** For example, the rate of the reaction

$$2H_2O_2(aq) \longrightarrow 2H_2O + O_2(g)$$

has been found to be directly proportional to the concentration of hydrogen peroxide. We can write the rate law for the reaction as:

$$\text{rate} = k[H_2O_2]$$

where "rate" is the instantaneous rate of disappearance or appearance of a specified substance and $k$ is the proportionality constant between the rate and the concentration and is called the **specific rate constant.** The units of the specific rate constant are not the same as the units of rate, and you should be careful not to confuse the specific rate constant with the rate. In this rate law, for example, the units of rate are mol $L^{-1}$ sec$^{-1}$, and the units of concentration are mol $L^{-1}$. Substituting these units into the rate law gives

$$\text{mol } L^{-1} \text{ sec}^{-1} = k \text{ mol } L^{-1}$$

The units of mol $L^{-1}$ can be canceled from both sides of the equation, so that the units of $k$ are sec$^{-1}$ for this rate law.

## CHLOROFLUOROCARBONS, THE OZONE LAYER, AND KINETICS

The debate about the extent to which the earth's ozone layer is damaged by chlorofluorocarbons that are used as propellants in some aerosol spray cans and as the working liquid in refrigerators is essentially a debate about chemical kinetics. Decisions on banning the chlorofluorocarbons are based on knowledge about the rates at which a number of chemical reactions occur in the atmosphere.

The chlorofluorocarbons are inert gases often sold under the trade name Freon. The two most widely used are Freon 11, $CFCl_3$, and Freon 12, $CF_2Cl_2$. Early in the 1970s, it was found that these gases drift slowly to the upper atmosphere, where ultraviolet radiation from the sun initiates their dissociation. The chlorine atoms produced by this reaction can bring about the conversion of $O_3$ to $O_2$:

$$Cl + O_3 \longrightarrow ClO + O_2$$
$$ClO + O \longrightarrow Cl + O_2$$

A single chlorine atom can cause the destruction of tens of thousands of $O_3$ molecules.

The extent to which the ozone layer is destroyed depends on the rate at which chlorofluorocarbons reach the upper atmosphere, the region of the ozone layer. If the chlorofluorocarbon molecules participate in other reactions in the lower atmosphere, only negligible amounts of chlorine atoms may be freed in the ozone layer. The chlorofluorocarbons can participate in more than seven different reactions in the lower atmosphere. A major research effort has been mounted to determine how many of these reactions actually occur and the rates at which they take place.

In 1979, the National Academy of Sciences issued a report, based in large part on the direct determination of several key rate constants, which concluded that continued release of chlorofluorocarbons at the then current rate would lead to an eventual ozone depletion rate of between 15% and 18%. A later report, issued in 1982, lowered that estimate to a long-term reduction of 5% to 9%, given a continuation of current production levels. The 1982 report said the lowered estimate of ozone destruction was due to improved measurements of chemical reaction rates. In particular, laboratory studies gave improved measurements of rate constants for several reactions affecting the concentration of hydroxyl radical in the upper atmosphere.

The 1982 report also noted that other gases in the atmosphere could affect the ozone layer. Increases in emissions of oxides of nitrogen from jet-engine exhausts, from combustion, and from agriculture could increase the rate of ozone depletion, the report said, while a doubling of nitrous oxide concentrations could produce ozone reductions of 10% to 15%. The global increase in carbon dioxide levels, however, would slow the rate of ozone depletion by lowering stratospheric temperatures, reducing reaction rates for atmospheric compounds, and altering atmospheric concentrations of water vapor.

No matter what the outcome of the chlorofluorocarbon story, the ozone layer will still get close attention. Even if the total amount of ozone in the atmosphere does not change, a redistribution of ozone at different levels of the atmosphere could have a significant effect on global weather.

The rate law for the gas phase decomposition of nitrosyl chloride

$$2NOCl(g) \longrightarrow 2NO(g) + Cl_2(g)$$

is found to be

$$\text{rate} = k[NOCl]^2$$

The units of the rate constant $k$ in this rate law can be found by substitution:

$$\text{mol L}^{-1} \text{ sec}^{-1} = k(\text{mol L}^{-1})^2$$

$$k = \frac{\text{mol L}^{-1} \text{ sec}^{-1}}{\text{mol}^2 \text{ L}^{-2}} = \text{L mol}^{-1} \text{ sec}^{-1}$$

## Rate Order

The only difference between the two rate laws above is that they have different exponents on the concentration terms. These exponents define a central characteristic, the **order** of a rate law or a reaction. In the first case, the exponent on the term for the concentration of hydrogen peroxide is 1. We say that this is a *first-order reaction.* In the second rate law, the exponent on the term for the concentration of nitrosyl chloride is 2. We say that this is a *second-order reaction.*

The exponents of the concentration terms also are used to define the order of more complex reactions, those whose rate laws include the products of two or more concentration terms. The gas phase reaction

$$NO_2(g) + CO(g) \longrightarrow NO(g) + CO_2(g)$$

follows the rate law

$$\text{rate} = k[NO_2][CO]$$

Each of the two concentration terms has the exponent 1. The sum of the exponents on the concentration terms is 2, and the reaction is said to be second order. We can be more precise and say that the reaction is first order in nitrogen dioxide, first order in carbon monoxide, and second order overall. In other words, the overall order of a reaction is the sum of all the exponents on the concentration terms in the rate law. We note that the units of the specific rate constant in this second-order rate law are liters per mole per second ($\text{L mol}^{-1} \text{ sec}^{-1}$), the same units as in the second-order rate law for the NOCl reaction above.

The gas phase reaction

$$2NO(g) + O_2(g) \longrightarrow 2NO_2(g)$$

follows the rate law

$$\text{rate} = k[NO]^2[O_2]$$

The sum of the exponents is 3, so this is a third-order rate law; it is second order in NO and first order in $O_2$. The units of $k$ are $\text{L}^2 \text{ mol}^{-2} \text{ sec}^{-1}$.

Fractional exponents appear in some rate laws. Under certain conditions, the rate law for the reaction

$$H_2(g) + Br_2(g) \longrightarrow 2HBr(g)$$

is                                            rate $= k[H_2][Br_2]^{1/2}$

This can be called a three-halves-order reaction. It is first order in hydrogen and half-order in bromine.

*The rate law of a reaction cannot be found by inspection of the equation. The rate law can be found only by experiment.* In simple cases, we can find the dependence of the rate of a reaction on concentration by measuring rates at different concentrations. One method is based on the measurement of initial rates. An *initial rate* is the average rate during a relatively short time interval beginning at $t = 0$. Often we can find the order of the reaction by obtaining values of the initial rate at different concentrations. This method is especially useful for reactions that are complicated by secondary reactions involving the products. Table 17.3, which lists data for the reaction between nitric oxide and ozone at 340 K, an important reaction related to the formation of smog and the destruction of the ozone layer, can be used to demonstrate this method.

In the first three entries in the table, the [NO] is held constant and the $[O_3]$ is varied. The rate can be seen to vary as the $[O_3]$ changes. Specifically, the rate changes by the same factor as the $[O_3]$. When the $[O_3]$ is doubled, the rate doubles. When the $[O_3]$ increases by a factor of 1.5, as it does from the second entry to the third, the rate increases by the same factor of 1.5. The rate of this reaction, therefore, is proportional to the first power of the $[O_3]$.

In the last three entries in the table, the $[O_3]$ is held constant while the [NO] is varied. Again, the rate varies. When [NO] doubles from $2.1 \times 10^{-6}$ mol/L to $4.2 \times 10^{-6}$ mol/L, the rate also doubles, from $4.8 \times 10^{-5}$ mol $L^{-1}$ sec$^{-1}$ to $9.6 \times 10^{-5}$ mol $L^{-1}$ sec$^{-1}$. Therefore, the rate of this reaction is also proportional to the first power of the [NO]. The rate law for this reaction is: rate $= k[O_3][NO]$.

When each concentration changes by a factor, the rate changes by the product of the factors. For example, each concentration changes by a factor of three between the first and last entries in Table 17.3. The rate changes by a factor of nine, from $1.6 \times 10^{-5}$ mol $L^{-1}$ sec$^{-1}$ to $14.4 \times 10^{-5}$ mol $L^{-1}$ sec$^{-1}$.

**TABLE 17.3**   Initial Rates of the Reaction $O_3(g) + NO(g) \rightarrow O_2(g) + NO_2(g)$

| $[O_3]$ (mol $L^{-1}$) | [NO] (mol $L^{-1}$) | Initial Rate (mol $L^{-1}$ sec$^{-1}$) |
|---|---|---|
| $2.1 \times 10^{-6}$ | $2.1 \times 10^{-6}$ | $1.6 \times 10^{-5}$ |
| $4.2 \times 10^{-6}$ | $2.1 \times 10^{-6}$ | $3.2 \times 10^{-5}$ |
| $6.3 \times 10^{-6}$ | $2.1 \times 10^{-6}$ | $4.8 \times 10^{-5}$ |
| $6.3 \times 10^{-6}$ | $4.2 \times 10^{-6}$ | $9.6 \times 10^{-5}$ |
| $6.3 \times 10^{-6}$ | $6.3 \times 10^{-6}$ | $14.4 \times 10^{-5}$ |

**Example 17.3**   The oxidation of manganate ion to permanganate ion by periodate ion in alkaline solution can be represented by the equation

$$2MnO_4^{2-}(aq) + H_3IO_6^{2-}(aq) \longrightarrow 2MnO_4^-(aq) + IO_3^-(aq) + 3OH^-(aq)$$

The data for the initial rate of the reaction measured as the rate of appearance of $MnO_4^-$ as a function of concentration at 310 K are

| $[MnO_4^{2-}]$ (mol L$^{-1}$) | $[H_3IO_6^{2-}]$ (mol L$^{-1}$) | Initial Rate (mol L$^{-1}$ sec$^{-1}$) |
|---|---|---|
| $1.6 \times 10^{-4}$ | $3.1 \times 10^{-4}$ | $2.6 \times 10^{-6}$ |
| $3.2 \times 10^{-4}$ | $3.1 \times 10^{-4}$ | $1.0 \times 10^{-5}$ |
| $6.4 \times 10^{-4}$ | $3.1 \times 10^{-4}$ | $4.1 \times 10^{-5}$ |
| $1.6 \times 10^{-4}$ | $4.7 \times 10^{-4}$ | $2.6 \times 10^{-6}$ |
| $1.6 \times 10^{-4}$ | $6.2 \times 10^{-4}$ | $2.6 \times 10^{-6}$ |

Find the rate law for this reaction.

**Solution**    In the first three entries in the table, only the $[MnO_4^{2-}]$ changes. When the $[MnO_4^{2-}]$ changes by the factor $3.2/1.6 = 2$, the initial rate changes by the factor $10/2.6 \approx 4$. When the $[MnO_4^{2-}]$ changes by the factor $6.4/3.2 = 2$, the initial rate changes by the factor $4.1/1.0 \approx 4$. We see that when the concentration of $MnO_4^{2-}$ doubles (a factor of 2), the rate quadruples (a factor of $2^2$). When the concentration is increased by a factor of $6.4/1.6 = 4$, the rate therefore increases by the factor $4^2 = 16$. The rate is proportional to the $[MnO_4^{2-}]$ raised to a power. If we let $x$ be the power, then $2^x = 4$ and $x = 2$. The rate is proportional to $[MnO_4^{2-}]^2$.

Is the rate also proportional to the $[H_3IO_6^{2-}]$? In the last two entries of the table, the $[MnO_4^{2-}]$ is constant and the $[H_3IO_6^{2-}]$ varies. The rate does not vary; therefore, the rate does not depend on the $[H_3IO_6^{2-}]$. We can say that the reaction is zero order in periodate ion. The rate law of the reaction is

$$\text{rate} = k[MnO_4^{2-}]^2$$

It is not unusual to find that the rate law does not reflect the overall stoichiometry of an equation and that one or more of the reactants may not appear in the rate law.

We cannot always vary the concentrations of all the reactants. For example, the reaction

$$CH_3I(aq) + H_2O \longrightarrow CH_3OH(aq) + H^+(aq) + I^-(aq)$$

follows the rate law

$$\text{rate} = k[CH_3I][H_2O]$$

This reaction is carried out in aqueous solution. The concentration of $H_2O$ does not drop appreciably as the reaction proceeds. Under these conditions, we can regard the concentration of $H_2O$ as a constant, and we can define a new specific rate constant, $k'$, which is equal to $k[H_2O]$. The rate law then becomes

$$\text{rate} = k[\text{CH}_3\text{I}]$$

A rate law of this kind is called a pseudo-first-order rate law. While it appears to be first order, it actually is second order. We encounter such rate laws when a large excess of one of the reactants leads to an apparent simplification of the rate law.

## Rate Equations

The rate law can be used to calculate the course of a reaction with time. For such calculations, it is convenient to convert the rate law into a form that gives the relationship between the initial concentration of a reactant (that is, the concentration at $t = 0$) and its concentration at any time during the course of the reaction. For the simpler rate laws, this conversion is readily accomplished by the methods of the integral calculus. For this reason, the forms of the rate laws that give the relationships between concentration and time are often called *integrated rate equations*. We shall use the two simplest integrated rate equations, the equation for the first-order rate and the equation for the one-term second-order rate.

## First Order

Consider a general first-order reaction, which can be represented as

$$A \longrightarrow B$$

It follows the rate law

$$\text{rate} = k[\text{A}]$$

Assume the reaction begins at time $t = 0$, with a concentration of A that can be represented as $c_0$. Then the relationship between the concentration of A at any other time, the specific rate constant $k$, and the initial concentration $c_0$ is given by the integrated first-order equation

$$\ln \frac{c}{c_0} = -kt \tag{17.1}$$

where $c$ is the concentration of A at time $t$.[1]

The integrated first-order rate equation defines the relationship between the specific rate constant $k$, the time $t$, and the *ratio* of the concen-

---

[1] Those familiar with the calculus will be able to follow the derivation of this equation. The first-order rate law is actually a differential equation: $-dc/dt = kc$. This equation can be rearranged to $dc/c = -k \, dt$. Both sides of the equation can be integrated from $t = 0$ to $t$:

$$\int_{c_0}^{c} \frac{dc}{c} = -k \int_{0}^{t} dt \quad \text{or} \quad \ln \frac{c}{c_0} = -kt$$

tration $c$ at this time to the initial concentration, $c/c_0$. A ratio is dimensionally independent — that is, it has no units — and so the values of $k$ and $t$ do not depend on the units of concentration. Therefore, Equation 17.1 can be used with any measurable property of a system that is proportional to concentration. Data on the decrease in pressure of a reactant, on the decrease in color intensity of a reactant, or even, as we shall see, on the decrease in radioactivity of a reactant, can be used directly in Equation 17.1. Thus, $c$ in the integrated first-order rate equation refers to any property that is proportional to concentration.

---

Example 17.4

The decomposition of $Cl_2O_7$ at 400 K in the gas phase to $Cl_2$ and $O_2$ follows first-order kinetics.
   a. After 55 sec at 400 K, the pressure of $Cl_2O_7$ falls from 0.062 atm to 0.044 atm. Calculate the specific rate constant.[2]
   b. Calculate the pressure of $Cl_2O_7$ after 100 sec of decomposition at this temperature.
   c. Calculate the time required for the pressure of $Cl_2O_7$ to fall to one-tenth of its original value.

Solution

a. The overall reaction is

$$2Cl_2O_7(g) \longrightarrow 2Cl_2(g) + 7O_2(g)$$

We are given that $c_0 = 0.062$ atm and that at $t = 55$ sec, $c = 0.044$ atm. Since we have values for three of the four terms in the first-order rate equation, we can find the value of the fourth:

$$\ln \frac{0.044 \text{ atm}}{0.062 \text{ atm}} = -k(55 \text{ sec})$$

$$k = 6.2 \times 10^{-3} \text{ sec}^{-1}$$

b. Now that the value of the specific rate constant is known, it can be used to find the pressure of reactant that remains at any given time. Thus:

$$\ln \frac{c}{0.062 \text{ atm}} = -(6.2 \times 10^{-3} \text{ sec}^{-1})(100 \text{ sec})$$

This equation can be solved more easily if we use the relationship $\ln (a/b) = \ln a - \ln b$ to rewrite it as

$$\ln c - \ln 0.062 = -(6.2 \times 10^{-3} \text{ sec}^{-1})(100 \text{ sec})$$
$$c = 0.033 \text{ atm}$$

c. When the pressure falls to one-tenth of its original value, it will be 0.0062

---

[2] In practice, an experiment designed to find the value of the specific rate constant would not rely on only one data point, but would use a number of concentrations at different times. Often the value of $k$ is then found graphically. A plot of $\ln c/c_0$ against $t$ is a straight line whose slope is $-k$ for a first-order reaction.

atm. We could substitute this value, along with the values of $c_0$ and $k$, into the rate equation to find $t$. However, if we recognize that the fraction $c/c_0$ has the value 0.1 when the pressure falls to one-tenth its original value, the problem is more easily solved. We can write

$$\ln 0.1 = -(6.2 \times 10^{-3} \text{ sec}^{-1})t$$

and
$$t = 370 \text{ sec}$$

In Example 17.4(c), we calculated the time in which a given fraction of the starting quantity of a reactant is consumed. For first-order reactions only, *the time in which a given fraction of a reactant is consumed is independent of the starting quantity of reactant.* We can show why this is so.

Suppose we want to know the time in which 20% of the starting quantity of $Cl_2O_7$ in Example 17.4 will decompose. When 20% decomposes, there remains $c_0 - 0.2c_0 = 0.8c_0$. Substitution of this value into the rate equation gives:

$$\ln \frac{0.8c_0}{c_0} = -kt$$

or
$$\ln 0.8 = -kt$$

Since $c_0$ does not appear in the equation, the time $t$ is independent of $c_0$. The value of $t$ is determined only by $k$.

**Half-Life**

Very often, we want to find the time $t$ required for half the starting quantity of a reactant to be consumed, the *half-life* of the reaction. The expression for finding the half-life is based on the simple observation that when half the starting quantity of material is gone, the other half remains. Using the first-order rate equation, we find that

$$\ln 0.5 = -kt_{1/2}$$

where $t_{1/2}$ is the half-life. This is usually written as

$$t_{1/2} = \frac{0.693}{k} \tag{17.2}$$

Given this simple relationship between the half-life and the specific rate constant, we can use the half-life as a convenient expression for the rate of a first-order process. Equation 17.2 can also be used to calculate $k$ from the half-life.

In many cases, the time elapsed from the start of a process is given in terms of the half-life. We can easily calculate that the half-life of the process in Example 17.4 is

$$t_{1/2} = \frac{0.693}{6.2 \times 10^{-3} \text{ sec}^{-1}} = 110 \text{ sec}$$

We can use the half-life to describe the extent and duration of a first-order reaction. For example, if we say that the reaction has proceeded for three half-lives, we mean that $t = 3(110) = 330$ sec. After the first half-life (110 sec), $\frac{1}{2}$ the original quantity remains. At the end of the second half-life (220 sec), $\frac{1}{2}$ of $\frac{1}{2}$, or $\frac{1}{4}$ of the original quantity of material remains. At the end of the third half-life, at 330 sec, $\frac{1}{2}$ of $\frac{1}{4}$, or $\frac{1}{8}$ of the original quantity of material remains. Thus, after three half-lives, the pressure of $Cl_2O_7$ in Example 17.4 is $\frac{1}{8} \times (0.062 \text{ atm}) = 0.0078$ atm. For the general case, after $n$ half-lives, $(\frac{1}{2})^n$ of the original quantity of material remains.

The half-life concept is commonly used to describe the rate at which radioactive decay processes occur (Section 20.2). Radioactive decay is a process in which an element changes either to a different element or to a different isotope of the same element. It is a first-order process. The easiest way to measure the rate of decay is to count the number of particles (usually alpha or beta particles) emitted by the sample during a given time interval, using instruments such as the Geiger counter or the scintillation counter. The value of the count is directly proportional to the amount of radioactive substance present, and it can be used directly in the first-order rate equation for $c_0$ and $c$.

### Radiocarbon Dating

A radioactive isotope of carbon that is of great value in several fields of science is $^{14}C$, which decays with a half-life of 5760 years. Carbon-14 is produced continually in the atmosphere by the bombardment of nitrogen by cosmic rays. The rate at which $^{14}C$ decays is balanced by the rate at which it forms, giving a fairly constant concentration of $^{14}C$ in the atmosphere for relatively long periods of time. It was once believed that the atmospheric concentration of $^{14}C$ never changes. Recent evidence indicates that the concentration does vary over very long time periods, apparently because of changes in solar activity. The $^{14}C$ in the atmosphere forms $^{14}CO_2$, which is incorporated in plants and is transferred to animals that eat the plants. When an organism dies, the incorporation of $^{14}C$ stops. The ratio of $^{14}C$ to $^{12}C$, the stable isotope of carbon, decreases steadily after death. The length of time since the death of an organism can be determined by a measurement of the ratio of $^{14}C$ to $^{12}C$. Radiocarbon dating, the measurement of the ratio of $^{14}C$ to $^{12}C$ in organic matter, is widely used to measure the age of any carbon-containing matter — wood, bone, and so on — that has been dead for periods ranging from several hundred years to about 50,000 years.

The discovery that the atmospheric concentration of $^{14}C$ apparently was higher several thousand years ago has required some revision of radiocarbon dates. The discovery was made because of discrepancies between ages determined by radiocarbon dating and those arrived at by dendrochronology, a method in which ages can be measured by analysis of tree ring patterns. It has now been established that the tree ring chronology is accurate and the radiocarbon dates are in error.

One practical use of radiocarbon dating is the authentication of works of art. Example 17.5 shows how radiocarbon dating can be used by art historians.

---

**Example 17.5**

A museum curator, doubtful about the authenticity of an Egyptian papyrus painting that purports to be from the twenty-first century B.C., asks for a radiocarbon test. A small piece of the papyrus is burned to $CO_2$, which is collected. A Geiger counter measures 14.7 counts per minute (c.p.m.) per gram of carbon, compared to 15.3 c.p.m. from $^{14}C$ in a living organism. It is assumed that the $^{14}C$ content of the rushes from which the papyrus was made was 15.3 c.p.m. when they were alive. Is the painting genuine?

**Solution**

The specific rate constant for the first-order radioactive decay of $^{14}C$ can be found from the half-life:

$$k = \frac{0.693}{5760 \text{ yr}} = 1.20 \times 10^{-4} \text{ yr}^{-1}$$

Since $k$, $c_0$, and $c$ are known, the time elapsed since the death of the rushes can be found by the first-order rate equation:

$$\ln \frac{14.7 \text{ c.p.m.}}{15.3 \text{ c.p.m.}} = -(1.20 \times 10^{-4} \text{ yr}^{-1})t$$

$$t = 333 \text{ yr}$$

This measurement shows clearly that the painting is a forgery. Assuming a steady-state concentration of $^{14}C$ of 15.3 c.p.m., the radioactive count of a 3000-year-old papyrus should be

$$\ln \frac{c}{15.3 \text{ c.p.m.}} = -(1.20 \times 10^{-4} \text{ yr}^{-1})(3000 \text{ yr})$$

$$c = 10.7 \text{ c.p.m. per gram of carbon}$$

---

**Second Order**

Now let us consider reactions that follow the one-term second-order rate law. The general reaction can be represented as $A \rightarrow B$. The rate law that it follows is

$$\text{rate} = k[A]^2$$

The integrated rate equation for this rate law is

$$\frac{1}{c} - \frac{1}{c_0} = kt \tag{17.3}$$

where the initial concentration is $c_0$, the concentration at time $t$ is $c$, and $k$ is the specific rate constant. Usually, $c$ and $c_0$ will be expressed in units of mol/L, so $k$, as we have seen, has the units L mol$^{-1}$ sec$^{-1}$.

In the second-order equation, the units—and therefore the numerical value of $k$—depend on the units in which $c$ and $c_0$ are expressed. The value of the specific rate constant for an equation that follows a second-order rate law can be found by measurement of the concentration of a reactant with time. Once the specific rate constant is known, it can be used either to find the concentration of a reactant with time or to find the time needed to reach a given concentration of a reactant.

**Example 17.6**

The decomposition of nitric oxide to $N_2$ and $O_2$ in the gas phase at elevated temperatures has been studied extensively, because it plays a role in atmospheric chemistry. It has been found that the reaction is second order in nitric oxide at 1370 K.

a. Over a period of 2000 sec, the concentration of NO falls from an initial value of $2.8 \times 10^{-3}$ mol/L to $2.0 \times 10^{-3}$ mol/L. Find the value of the specific rate constant.

b. Find the concentration of NO after the reaction proceeds for another 2000 sec.

The reaction is

$$2NO(g) \longrightarrow N_2(g) + O_2(g)$$

**Solution**

a. The specific rate constant is found from the second-order rate equation:

$$\frac{1}{2.0 \times 10^{-3} \text{ mol/L}} - \frac{1}{2.8 \times 10^{-3} \text{ mol/L}} = k(2000 \text{ sec})$$

$$k = 7.1 \times 10^{-2} \text{ L mol}^{-1} \text{ sec}^{-1}$$

Again, in practice more than one measurement of concentration would be made. If the reaction follows this rate law, a plot of $1/c$ against $t$ is a straight line whose slope is $k$.

b. We can use the value of the specific rate constant found in part a to determine the concentration at any given time, using the rate equation:

$$\frac{1}{c} - \frac{1}{2.8 \times 10^{-3} \text{ mol/L}} = (7.1 \times 10^{-2} \text{ L mol}^{-1} \text{ sec}^{-1})(2000 \text{ sec} + 2000 \text{ sec})$$

$$c = 1.6 \times 10^{-3} \text{ mol/L}$$

The half-life of a first-order reaction is not affected by the starting concentrations. In a second-order process, the half-life depends on both the starting concentrations and the specific rate constant and is therefore not nearly as useful. The rate equation shows this relationship.

The concentration at $t_{1/2}$ is $c_0/2$. Thus

$$\frac{1}{c_0/2} - \frac{1}{c_0} = kt_{1/2}$$

Solving for $t_{1/2}$ gives

$$t_{1/2} = \frac{1}{kc_0} \qquad (17.4)$$

In Example 17.6, the half-life of the NO when it has an initial concentration of $2.8 \times 10^{-3}$ mol/L is:

$$t_{1/2} = \frac{1}{(7.1 \times 10^{-2} \text{ L mol}^{-1} \text{ sec}^{-1})(2.8 \times 10^{-3} \text{ mol L}^{-1})} = 5000 \text{ sec}$$

Thus, after 5000 sec the concentration of NO is $(2.8 \times 10^{-3}$ mol/L$)/2 = 1.4 \times 10^{-3}$ mol/L. But this concentration of NO will not be reduced by half after another 5000 sec. By Equation 17.4, the time required to reduce this concentration by half is

$$t_{1/2} = \frac{1}{(7.1 \times 10^{-2})(1.4 \times 10^{-3})} = 10\,000 \text{ sec}$$

The second-order rate slows down faster than the first-order rate. Therefore, the half-life of a reactant that follows a second-order rate law increases as the concentration decreases.

## 17.3   RATE LAWS AND REACTION MECHANISMS

A reaction mechanism is a description on a molecular level of all the changes that reactants undergo in a chemical reaction. At the most sophisticated level, a reaction mechanism describes the movement of the electrons and nuclei of the reactants, giving a continuous description of the changes in chemical bonding. But we must have a more elementary description before we can develop such a detailed reaction mechanism.

### Elementary Reactions

Many chemical reactions proceed in a number of simple steps called *elementary reactions*. For example, kinetic studies suggest that the conversion of ozone to oxygen:

$$2O_3(g) \longrightarrow 3O_2(g)$$

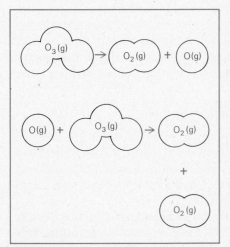

**Figure 17.2**
The two-step mechanism by which ozone is converted to molecular oxygen. In the first step, an $O_3$ molecule decomposes to an $O_2$ molecule and an O atom. In the second step, another $O_3$ molecule reacts with the O atom to form two $O_2$ molecules.

proceeds in two steps:

1.    $O_3 \rightleftharpoons O_2 + O$
2.    $O_3 + O \longrightarrow 2O_2$

as shown in Figure 17.2. Each step is an elementary reaction. One product of the first elementary reaction is atomic oxygen, $O(g)$, which is an *intermediate* and is not included among the final products of the overall reaction. Since the overall reaction takes place in more than one step, we can describe it as a complex reaction.

It is important to understand the distinction between the overall reaction, which gives the stoichiometry of the process, and the elementary reactions, which together provide a two-step description of a possible reaction mechanism for the conversion of $O_3$ to $O_2$. While the overall reaction shows two $O_3$ molecules as reactants, the reaction does not occur by a collision between the molecules. The overall reaction does not occur exactly as written. But the two *elementary reactions occur exactly as written.* In the first step, one $O_3$ molecule does break into two parts, an $O_2$ molecule and an O atom. In the second step, an $O_3$ molecule and an O atom come together to form two $O_2$ molecules.

## Molecularity

Molecularity is a term that applies only to elementary reactions. *The molecularity of an elementary reaction is generally the number of reactant molecules (or atoms or ions) that come together to form the products. It is the sum of the coefficients of the reactants in the elementary reaction.* The first elementary reaction in the conversion of ozone to oxygen is said to be *unimolecular,* since the reactant is one $O_3$ molecule. The second elementary reaction is said to be *bimolecular,* since the reactants are one $O_3$ molecule and one O atom. Most elementary reactions are unimolecular or bimolecular. Termolecular elementary reactions, in which three species are reactants, are rare. Elementary reactions with molecularity of four or higher are unknown. The reason is simple. An elementary reaction describes an actual collision. The simultaneous collision of three or more molecules (or atoms or ions) is extremely improbable. In thermodynamic terms, $\Delta S$ for a termolecular (or higher) collision is quite negative.

## Rate Laws of Elementary Reactions

We said earlier that the rate law and reaction order cannot be predicted from the stoichiometry of a complex reaction. Now you can understand why. The rate law is determined by the elementary reactions that make up the complex reaction. The stoichiometry of a complex reaction gives no information about its elementary reactions.

However, we can write the rate law for an elementary reaction directly from its molecularity. A unimolecular reaction follows a first-order rate law, and a bimolecular reaction follows a second-order rate law. Thus, the rate law for the first elementary reaction in the conversion of $O_3$ to $O_2$ is

$$\text{rate} = k_1[O_3]$$

and the rate law for the second elementary reaction is

$$\text{rate} = k_2[O_3][O]$$

*The order and the molecularity of an elementary reaction are equal.* The rate law for any elementary reaction is written as the product of the reactants raised to the power of their coefficients. For example, the rate law for the elementary reaction

$$2CH_3(g) \longrightarrow C_2H_6(g)$$

is

$$\text{rate} = k[CH_3]^2$$

Some ordinary chemical processes occur essentially in a single step. One of them is the reaction

$$CH_3Cl + I^- \longrightarrow CH_3I + Cl^-$$

which occurs in solution. The process can be pictured as an elementary reaction in which an iodide ion collides directly with the methyl chloride molecule, forming methyl iodide and displacing a chloride ion, as shown in Figure 17.3. Since this reaction is a bimolecular process, it should follow a second-order rate law if it does indeed proceed as a single elementary step. The rate law for the reaction has been found by observation to be

$$\text{rate} = k[CH_3Cl][I^-]$$

This second-order rate law is consistent with the picture in which the reaction occurs in a single elementary step. But it does not prove that the reaction occurs in this way. If any experiment found that this reaction is not second order, we would have to say that it occurs in more than one step.

In the case of another reaction:

$$2NO(g) + Cl_2(g) \longrightarrow 2NOCl$$

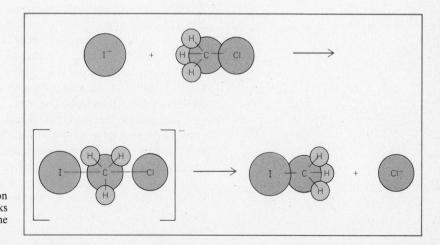

**Figure 17.3**

The one-step reaction in which the iodide ion reacts with methyl chloride. The $I^-$ ion attacks the $CH_3Cl$ molecule from one side, while the chloride ion leaves from the other side.

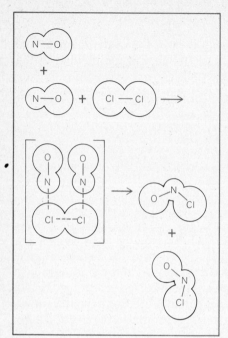

**Figure 17.4**
The reaction of nitric oxide and chlorine in the gas phase is believed to occur in a single step. Two NO molecules collide with one $Cl_2$ molecule to form two NOCl molecules.

which takes place in the gas phase, the reaction is believed to occur in a single step: The chlorine molecule collides with the two nitric oxide molecules to form the products directly, as shown in Figure 17.4. Experiments have found that this reaction occurs by a third-order rate law:

$$rate = k[NO]^2[Cl_2]$$

which is consistent with the single-step mechanism.

The rate law of a complex reaction is related to the rate laws of its elementary steps. Therefore, we can use an experimentally determined rate law for an overall reaction to propose elementary reactions by which the chemical change comes about. A proposed mechanism must be consistent with the observed kinetics. Kinetics is one of many experimental approaches to the investigation of reaction mechanisms. It is usually not possible, however, to "prove" that a proposed mechanism is correct. The best we can do is to prove that a proposed mechanism is incorrect.

In the reactions between $CH_3Cl$ and $I^-$ or between $Cl_2$ and NO, for example, the observed rate laws are consistent with single-step mechanisms. But the rate laws do not prove that these are the actual mechanisms. While the reaction:

$$H_2(g) + I_2(g) \longrightarrow 2HI(g)$$

follows a second-order rate law:

$$rate = k[H_2][I_2]$$

which is consistent with a single-step mechanism, there is other experimental evidence that proves that the single-step mechanism is incorrect.

## Rate Laws from Reaction Mechanisms

It is usually quite difficult to formulate a possible mechanism from a rate law. It can also be difficult to derive the rate law required by a proposed mechanism. But the required rate law can be found easily for a mechanism that includes what is called a **rate-determining step.**

When one elementary reaction in a complex mechanism proceeds at a much slower rate than any of the others, the slow elementary reaction is the rate-determining step. The rate of this step determines the overall rate of the complex reaction, no matter whether the rate-determining step comes first, last, or in the middle of the series of elementary reactions. We can think of a mountain-climbing team composed of three experts and a beginner, all roped together. The beginner must climb at a slower rate than any of the experts. Therefore, the speed of the party is determined by the speed of the beginner, no matter in what order the climbers are roped together.

We find a simple example of a rate-determining step in the gas phase reaction

$$NO_2(g) + CO(g) \longrightarrow NO(g) + CO_2(g)$$

which is believed to occur by a two-step mechanism below 500 K:

1. $$NO_2 + NO_2 \xrightarrow{k_1} NO_3 + NO \qquad \text{slow}$$
2. $$NO_3 + CO \xrightarrow{k_2} NO_2 + CO_2 \qquad \text{fast}$$

The designations "slow" and "fast" indicate the relative rates of the steps. We follow custom by writing the specific rate constant of each elementary step above the arrow, with a subscript indicating the number of the step.

The first step is rate determining, which means that the rate of the overall reaction is the rate of step 1. The rate law for step 1 is

$$\text{rate} = k_1[NO_2]^2$$

which is also the rate law for the overall reaction. The observed rate law for the overall reaction is second order in $NO_2$ and is therefore consistent with the proposed mechanism. Because the unusual species $NO_3$ is a reactive intermediate, we expect that step 2 will be faster than step 1.

Obtaining the overall rate law from the mechanism of a reaction is simplified if we realize that *the rate law for the overall reaction is not influenced by steps that occur after the rate-determining step.* In the simplest case, the rate-determining step comes first, as in the reaction

$$H_2O_2(aq) + 3I^-(aq) + 2H^+(aq) \longrightarrow 2H_2O + I_3^-(aq)$$

which is believed to occur by the mechanism:

1. $$H_2O_2 + I^- \longrightarrow H_2O + IO^- \qquad \text{slow}$$
2. $$IO^- + H^+ \rightleftharpoons HOI \qquad \text{fast}$$
3. $$HOI + H^+ + I^- \rightleftharpoons H_2O + I_2 \qquad \text{fast}$$
4. $$I^- + I_2 \rightleftharpoons I_3^- \qquad \text{fast}$$

The sum of all the steps gives the overall stoichiometry of the reaction. But steps 2, 3, and 4 do not influence the overall rate because they occur after the rate-determining slow step. The overall rate law is the rate law of step 1:

$$\text{rate} = k[H_2O_2][I^-]$$

which can be verified experimentally.

It is not as easy to derive the overall rate law for a mechanism in which the first step is not the rate-determining step. Such a mechanism often begins with what is called a fast preequilibrium, the rapid and reversible formation of an intermediate. The intermediate is then consumed in the rate-determining step. We find such a mechanism in the decomposition of nitric oxide, the least stable of all the oxides of nitrogen. While NO is stable at ordinary pressures at room temperature, it decomposes at pressures in excess of 100 atm by the reaction

$$3NO(g) \longrightarrow N_2O(g) + NO_2(g)$$

The proposed mechanism for this reaction is

| | | |
|---|---|---|
| 1. | $2NO \rightleftharpoons (NO)_2$ | fast |
| 2. | $(NO)_2 + NO \xrightarrow{k_2} N_2O + NO_2$ | slow |

The rate law for the overall reaction is the rate law of step 2, the rate-determining step:

$$\text{rate} = k_2[(NO)_2][NO]$$

The usefulness of this rate law is limited because the concentration of $(NO)_2$ cannot be measured. It is not possible to make a direct measurement of the concentration of NO dimer, as $(NO)_2$ is called, because it is an intermediate with a very short lifetime. Therefore, we cannot readily carry out an experimental test of the rate law. But there is a way out of this impasse. The value of $[(NO)_2]$ is related to the value of the starting concentration of NO. Since step 1 of the mechanism occurs relatively quickly, equilibrium is established between NO and $(NO)_2$. As usual, the composition of the equilibrium state is expressed by an equilibrium constant:

$$K_1 = \frac{[(NO)_2]}{[NO]^2}$$

Solving for $[(NO)_2]$:

$$[(NO)_2] = K_1[NO]^2$$

We now have expressed the concentration of the NO dimer in terms of a concentration that can be measured. Substituting this expression into the rate law for the rate-determining step gives:

$$\text{rate} = k_2[NO]K_1[NO]^2$$

Since both $k_2$ and $K_1$ are constants, we can let $k' = k_2K_1$, which gives a simple one-term third-order rate law for the overall reaction:

$$\text{rate} = k'[NO]^3$$

While a single-step mechanism is also consistent with this rate law, other experimental evidence led to the proposal of this two-step mechanism.

**Example 17.7**   The oxidation of iodide ion by hypochlorite ion:

$$ClO^-(aq) + I^-(aq) \longrightarrow Cl^-(aq) + IO^-(aq)$$

has been postulated to occur by the three-step mechanism:

| | | |
|---|---|---|
| 1. | $ClO^- + H_2O \rightleftharpoons HClO + OH^-$ | fast |
| 2. | $I^- + HClO \xrightarrow{k_2} HIO + Cl^-$ | slow |
| 3. | $OH^- + HIO \rightleftharpoons H_2O + IO^-$ | fast |

What rate law is required by this mechanism?

**Solution**   The rate-determining step is step 2. Its rate law is

$$\text{rate} = k_2[I^-][HClO]$$

This rate law includes the concentration of HClO, which is not one of the original reactants. To convert the rate law to a more useful form, we must find the relationship between the concentration of HClO and the concentration of substances originally present in the solution. Step 1 controls the [HClO], and step 1 is described by the equilibrium constant

$$K_1 = \frac{[HClO][OH^-]}{[ClO^-]}$$

Solving for [HClO] gives

$$[HClO] = \frac{K_1[ClO^-]}{[OH^-]}$$

Substituting this expression into the rate law gives

$$\text{rate} = \frac{k_2 K_1[I^-][ClO^-]}{[OH^-]}$$

This is the overall rate law for the reaction. The product $k_2 K_1$ is a constant. The reaction is first order in iodide ion, first order in hypochlorite ion, and minus first order in hydroxide ion — that is, as the concentration of $OH^-$ increases, the rate of the reaction decreases. This decrease in rate can be attributed to the decrease in the concentration of HClO as the solution becomes more alkaline. The rate-determining step proceeds more slowly as the concentration of HClO decreases. This sort of information can help us to determine the best conditions for carrying out this reaction.

One of the most interesting examples of the relationship between rate law and mechanism is provided by the reaction

$$H_2(g) + I_2(g) \longrightarrow 2HI(g)$$

We mentioned earlier that this reaction is first order in each reactant and follows a second-order rate law:

$$\text{rate} = k[H_2][I_2]$$

This rate law presents something of a problem: Is this reaction elementary or complex? While the rate law for the reaction between hydrogen and iodine is consistent with an elementary reaction, we may still find that the reaction is complex.

The reaction between $H_2$ and $I_2$ was the subject of one of the first kinetic investigations ever performed. Because second-order kinetics were observed, it was proposed that the reaction proceeds by a bimolecular single-step mechanism. For decades, most chemistry textbooks included this reaction as the standard example of a simple bimolecular process. However, this mechanism was disproven in the 1960s. The mechanism that is now accepted as being consistent with recent experimental work is

1.                     $I_2 \rightleftharpoons 2I$         fast
2.               $I + H_2 \rightleftharpoons H_2I$       fast
3.             $H_2I + I \longrightarrow 2HI$        slow

with step 3 as the rate-determining step.

## 17.4  REACTION RATE AND TEMPERATURE

Chemical reactions require time to occur. A study of the reasons can give us valuable information about the way in which chemical changes come about.

Consider a simple bimolecular elementary reaction that occurs in the gas phase at 273 K:

$$NO(g) + O_3(g) \longrightarrow NO_2(g) + O_2(g)$$

This reaction is highly favored thermodynamically. Why doesn't it occur instantaneously? One answer is that the reaction takes place only when two molecules collide, and that time is needed for the molecules to "find" each other. Thus, the rate of formation of products is limited by the rate at which collisions occur between molecules of NO and molecules of $O_3$.

We can calculate the exact rate of collision by using the kinetic theory

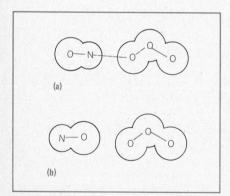

**Figure 17.5**

(a) The bimolecular elementary reaction of nitric oxide and ozone cannot occur unless the nitrogen atom of the NO molecule collides with the $O_3$ molecule. (b) If the oxygen end of the NO molecule collides with the $O_3$ molecule, the products do not form.

of gases that was discussed in Chapter 4. This rate depends on the size of the molecules, their concentration, and their velocity. Their velocity, in turn, depends on the mass of the molecules and the temperature of the system. In this system, $10^{31}$ collisions between molecules occur per liter per second at standard conditions.

The rate at which the reaction actually occurs tells us that only a small fraction of the collisions result in the formation of products. One reason is that the reaction will not take place unless the two reactant molecules collide in a way that allows the chemical change to occur, as Figure 17.5 shows in a simplified way. The products form only if a bond forms between the N atom of the NO molecule and an O atom of the $O_3$ molecule. Therefore, the $O_3$ molecule must collide with the nitrogen end of the NO molecule. The fraction of the collisions in which the molecules have the proper orientation is called the *steric factor,* and it varies from reaction to reaction.

There is an even more important reason why most collisions do not result in a chemical change. When two molecules approach closely, there is a natural electrostatic repulsion between them. For a reaction to occur, this repulsion must be overcome. The colliding molecules usually must achieve relatively close contact for a change in bonding to occur. If this change is to happen, the molecules must approach each other with relatively high kinetic energies. Even more energy is needed so that the electronic changes that lead to the formation of products can take place. When a collision takes place that is energetic enough to allow the formation of products, the system is in a state called the **activated complex** or **transition state.**

The difference between the mean energy of all the collisions of the reactants and the average energy of collisions in which reaction takes place is called the **activation energy,** usually represented by $E_a$. The energy relationship of reactants, products, and the activated complex for a spontaneous process is shown schematically in Figure 17.6. The activation energy gives the highest potential energy state that the system must pass through to form products. For most reactions, the activation energy can be thought of as a barrier to the occurrence of the reaction. The greater the activation energy, the slower the rate of the reaction.

This interpretation is often given in graphical form, as in the energy diagram for the reaction of ozone and nitric oxide that is shown in Figure 17.7. Such a diagram shows how the potential energy changes as reactants proceed to products; the "reaction coordinate" can be interpreted as the extent to which the reaction occurs.

An energy diagram gives only a crude picture of a chemical reaction, but it does give useful graphical information about some important characteristics of a reaction. Figure 17.7 shows that the reaction of ozone and nitric oxide is highly exothermic, and that the activation energy of the reaction is relatively low.

Energy diagrams can be drawn for complex reactions with any number

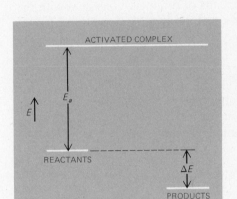

**Figure 17.6**
The relationship between the energy of reactants, products, and the activated complex in a spontaneous reaction. Even though the products have a lower energy than the reactants, the reaction occurs only when a collision is energetic enough to produce an activated complex.

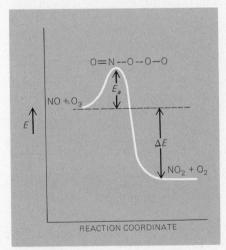

**Figure 17.7**
An energy diagram for the reaction of $O_3$ and NO. The peak, or maximum, is the activated complex or transition state. The minima are the reactants and the products.

of steps. Figure 17.8 shows some examples. Each elementary reaction has an energy barrier. The rate-determining step has the energy barrier that is highest above the energy of the original reactants. The energy diagram for a complex reaction has one maximum for each elementary reaction. Between each two maxima is a minimum that corresponds to the existence of an intermediate. In the energy diagram, these minima are shown with higher energies than the reactants or products, consistent with the great reactivity and short lifetimes that are typical of reactive intermediates.

## The Arrhenius Law

A relationship between the activation energy and the specific rate constant was proposed by Arrhenius in 1889. It has since been confirmed by many experiments. The relationship is the **Arrhenius law,** which can be written as:

$$k = Ae^{-E_a/RT} \quad \text{or} \quad \ln k = \ln A - \frac{E_a}{RT} \tag{17.5}$$

where $k$ is the specific rate constant, $T$ is the temperature, and $A$ is a constant called the *frequency factor,* which is related to the steric factor

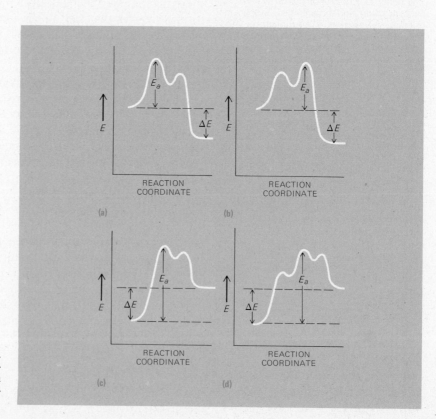

**Figure 17.8**

Energy diagrams for four complex reactions. Each maximum represents the activated complex of an elementary reaction; each minimum between maxima corresponds to an intermediate.

and the collisional frequency. The frequency factor is characteristic of the reaction and has the same units as the specific rate constant. The Arrhenius law makes the approximation that $E_a$ and $A$ do not change with temperature. This approximation is good if the temperature range is not too large.

If we look at the Arrhenius law term by term, we can see that it expresses much of what we have learned about reaction rates. The specific rate constant $k$ is a convenient expression of relative rate; a reaction with a fast rate has a specific rate constant $k$ whose value is high. Equation 17.5 shows that the value of $k$ of a reaction is related directly to the value of $A$, the frequency factor. The value of $A$ increases if the frequency of collisions increases or if the steric factor increases.

The second term in the Arrhenius law, $E_a/RT$, gives us a good deal of information about the factors that influence reaction rate. For example, the value of this term depends on the value of $E_a$. The term usually has a positive value, because $E_a$ is almost always positive. Since this term is subtracted from log $A$ in Equation 17.5, you can see that the value of $k$ decreases as the value of $E_a/RT$ increases. In other words, an increase in the value of $E_a$ decreases the rate of the reaction.

Conversely, an increase in temperature $T$, which is in the denominator, reduces the value of the term and thus increases the value of $k$. This increase is consistent with the observation that the rates of almost all reactions increase as the temperature rises.

It can be shown that the sensitivity of the rate of a reaction to changes in temperature depends on the value of $E_a$. If the value of $E_a$, the activation energy, is high, the rate of a reaction is more sensitive to temperature changes. The usual estimate is that the rate of a reaction in solution approximately doubles for every increase of 10 K in temperature. This generalization is roughly accurate around room temperature, although there are many exceptions to the rule. The generalization is not very useful for gas phase reactions, where we find a much wider range of values of $E_a$.

The easiest way to measure $E_a$ is to measure the value of $k$ at two different temperatures and substitute the values into the Arrhenius law. Better values of $E_a$ can be obtained if we make a number of such measurements. A plot of ln $k$ against $1/T$ gives a straight line whose slope is $-E_a/R$ and whose $y$ intercept is ln $A$. Figure 17.9 shows such a plot for the reaction between ozone and nitric oxide. The value of $A$, the frequency factor in the Arrhenius law, can be found if we know the value of $E_a$ and of $k$ at any temperature.

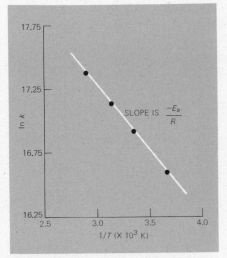

**Figure 17.9**

We can determine the activation energy of a reaction by measuring the specific rate constant $k$ at several different temperatures, substituting the values of $k$ into the Arrhenius law, and then making a plot of ln $k$ against $1/T$. The slope of the line gives the activation energy. Here we show such a plot for the reaction between $O_3$ and NO.

---

**Example 17.8**    One of the earliest reactions to be investigated kinetically was the hydrolysis of sucrose in acidic solutions. This reaction is known as the inversion of cane sugar and is

$$C_{12}H_{22}O_{11}(aq) + H_2O \longrightarrow C_6H_{12}O_6(aq) + C_6H_{12}O_6(aq)$$
$$\text{Sucrose} \qquad\qquad\qquad \text{Glucose} \qquad \text{Fructose}$$

The reaction follows a second-order rate law:

$$\text{rate} = k[C_{12}H_{22}O_{11}][H^+]$$

At 300 K, $k = 2.12 \times 10^{-4}$ L mol$^{-1}$ sec$^{-1}$, and at 310 K, $k = 8.46 \times 10^{-4}$ L mol$^{-1}$ sec$^{-1}$.

a. Calculate the value of $E_a$ and $A$.

b. Calculate the value of the specific rate constant at 320 K.

**Solution**    a. At 300 K, the Arrhenius law for this reaction is

$$\ln 2.12 \times 10^{-4} = \ln A - \frac{E_a}{(8.31 \text{ J mol}^{-1} \text{ K}^{-1})(300 \text{ K})}$$

At 310 K, it is:

$$\ln 8.46 \times 10^{-4} = \ln A - \frac{E_a}{(8.31 \text{ J mol}^{-1} \text{ K}^{-1})(310 \text{ K})}$$

We can eliminate one of the two unknowns by subtracting the first equation from the second to obtain:

$$\ln \frac{8.46 \times 10^{-4}}{2.12 \times 10^{-4}}$$

$$= \frac{E_a}{(8.31 \text{ J mol}^{-1} \text{ K}^{-1})(300 \text{ K})} - \frac{E_a}{(8.31 \text{ J mol}^{-1} \text{ K}^{-1})(310 \text{ K})}$$

$$= \frac{E_a}{(8.31 \text{ J mol}^{-1} \text{ K}^{-1})} \left[ \frac{1}{300 \text{ K}} - \frac{1}{310 \text{ K}} \right]$$

$$= \frac{E_a}{(8.31 \text{ J mol}^{-1} \text{ K}^{-1})} \left[ \frac{310 \text{ K} - 300 \text{ K}}{(300 \text{ K})(310 \text{ K})} \right]$$

$$E_a = 107\,000 \frac{\text{J}}{\text{mol}}$$

We can then substitute this value into either of the equations to find the value of $A$:

$$\ln 2.12 \times 10^{-4} = \ln A - \frac{107\,000 \text{ J mol}^{-1}}{(8.31 \text{ J mol}^{-1} \text{ K}^{-1})(300 \text{ K})}$$

$$\ln A = 34.509$$
$$A = 9.71 \times 10^{14} \text{ L mol}^{-1} \text{ sec}^{-1}$$

b. Once the frequency factor $A$ and the activation energy $E_a$ are known, the Arrhenius law can be used to find the specific rate constant at any other temperature. At 320 K:

$$\ln k = \ln(9.71 \times 10^{14}) - \frac{107\,000 \text{ J mol}^{-1}}{(8.31 \text{ J mol}^{-1} \text{ K}^{-1})(320 \text{ K})}$$

$$= -5.773$$
$$k = 3.11 \times 10^{-3} \text{ L mol}^{-1} \text{ sec}^{-1}$$

The algebraic manipulation carried out in part a of Example 17.8 can be used to derive a relationship between the activation energy $E_a$ and the specific rate constants that are measured at any two temperatures. If the temperatures are $T_1$ and $T_2$ and the specific rate constants are $k_1$ and $k_2$, the relationship is

$$\ln \frac{k_2}{k_1} = \frac{E_a}{R}\left[\frac{T_2 - T_1}{T_1 T_2}\right] \tag{17.6}$$

or, solving for $E_a$:

$$E_a = R\left[\frac{T_1 T_2}{T_2 - T_1}\right]\left[\ln \frac{k_2}{k_1}\right] \tag{17.7}$$

These equations resemble Equation 13.25, which gives the temperature dependence of the equilibrium constant. The existence of an activation energy helps us to understand the way in which temperature affects the rate of a reaction. As we have mentioned, the rate of many reactions increases substantially, often by about a factor of two or more, for every temperature increase of 10 K. This increase in rate does not result from an increase in the frequency of collisions. For example, when the temperature increases from 300 K to 310 K, the frequency of collisions increases by a factor of only 1.015. Neither does the rate increase result from an increase in the mean energy of the system, which goes up by a factor of only 1.03 for a 10-K temperature increase.

To understand the reason for the increase in rate, we must consider the distribution of molecular speeds that we discussed in Section 4.10. A distribution of molecular speeds also means a distribution of molecular energies. In any system, the energies of most molecules are close to the mean molecular energy. But there are a few molecules whose energies are much lower or much higher than the mean.

The activation energy of a reaction is usually much higher than the mean energy of the reacting system. Only a small fraction of the molecules in the system have energies sufficiently above the mean to react. When the temperature of the system increases by 10 K, the mean energy does not increase very much. But the number of molecules that are energetic enough to react is increased substantially, as Figure 17.10 shows.

We are faced with an apparent difficulty. If high-energy reactant molecules are converted to product, the system logically will soon have only reactant molecules whose energies are close to the mean, and the reaction will stop. But in practice, many reactions do not stop, even if thermal energy is not provided by the surroundings. The energy released when the activated complex forms product is distributed among the other molecules in the system by collisions. As a result, other reactant molecules become sufficiently energetic to undergo reaction.

Until now, we have discussed bimolecular processes. The same reasoning can be applied with little change to the relatively small number of

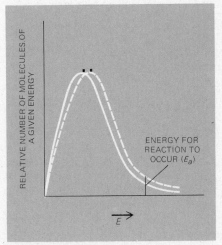

**Figure 17.10**
When temperature increases, the curve that gives the distribution of molecular energies shifts to the right. While the mean energy of all the molecules in a system goes up only slightly when the temperature rises, the fraction of the molecules that are energetic enough to react increases substantially.

termolecular processes. In a termolecular process, the activated complex contains three species rather than two. In these processes, collisions are much less frequent; we find that termolecular processes must have low activation energies to occur at reasonable rates.

A different line of reasoning is not necessary for unimolecular processes, in which a single reactant molecule goes to the product. The Arrhenius law also applies to unimolecular processes, and activation energies can be measured for them. In a unimolecular process, the activated complex has only one molecule. When a molecule attains the energy of the activated complex, the products can form. The molecule attains this energy by collisions with other molecules in the system, as in bimolecular and termolecular processes. A unimolecular process must be preceded by collisions between the molecule that is to react and other molecules in the system, in order to raise the energy of the reacting molecule to the level of the activated complex.

This discussion helps us give a more precise definition of molecularity. We can say that the molecularity of an elementary reaction is the number of molecules in the activated complex.

## 17.5   REACTION RATE AND EQUILIBRIUM

At first glance, it would seem that there is no need to discuss the kinetics of a system that is at equilibrium, since the overall composition of the system does not change. But a system at equilibrium is undergoing many changes at the molecular level, so its kinetics are of interest to us.

There is an important relationship between reaction rate expressions and equilibrium constants. It is *not* a simple, direct relationship, in which any reaction with a large equilibrium constant proceeds at a fast rate. We have already encountered a contrary example, the reaction in which water is formed from $H_2$ and $O_2$. The equilibrium constant of this reaction is large, but the reaction can be exceedingly slow.

To find the relationship between reaction rate and equilibrium, let us consider the process.

$$2NOCl(g) \underset{k_{-1}}{\overset{k_1}{\rightleftharpoons}} 2NO(g) + Cl_2(g)$$

At equilibrium, the concentrations of the three gases do not change with time. However, the forward reaction in which NOCl is consumed occurs constantly, as does the reverse reaction in which NOCl is formed. Since the concentration of NOCl does not change, its rate of disappearance by the forward reaction must equal its rate of appearance by the reverse reaction. Both reactions have been found to be elementary reactions, and we can therefore write their rate laws.

The rate law for the forward reaction is

$$\text{rate} = k_1[NOCl]^2$$

and the rate law for the reverse reaction is

$$\text{rate} = k_{-1}[NO]^2[Cl_2]$$

(The symbol $k_{-1}$ is commonly used for the specific rate constant of a reaction that is the reverse of a reaction whose specific rate constant is $k_1$.)
At equilibrium, the two rates must be equal. That is,

$$k_1[NOCl]^2 = k_{-1}[NO]^2[Cl_2]$$

This equation can be rearranged to give:

$$\frac{k_1}{k_{-1}} = \frac{[NO]^2[Cl_2]}{[NOCl]^2}$$

The concentration expression on the right side of this equation is identical to the expression for $K$. Therefore, for an elementary reaction:

$$K = \frac{k_1}{k_{-1}} \tag{17.8}$$

Equation 17.8 defines the relationship between the equilibrium constant and the specific rate constants of a system at equilibrium. If a reaction is elementary and the specific rate constants of the forward and reverse reactions are known, Equation 17.8 can be used to calculate the equilibrium constant. If the equilibrium constant and one specific rate constant are known, Equation 17.8 can be used to calculate the other specific rate constant.

**Example 17.9**   The specific rate constant for the bimolecular decomposition of NOCl at 473 K is found to be $7.8 \times 10^{-2}$ L mol$^{-1}$ sec$^{-1}$. The specific rate constant for the third-order reaction of nitric oxide with chlorine at this temperature is found to be $4.7 \times 10^2$ L$^2$ mol$^{-2}$ sec$^{-1}$. Calculate the equilibrium constant for the reaction: $2NOCl(g) \rightleftharpoons 2NO(g) + Cl_2(g)$.

**Solution**   Since we are given the specific rate constants for the forward and reverse reactions, Equation 17.8 can be used to find the value of $K$:

$$K = \frac{7.8 \times 10^{-2} \text{ L mol}^{-1} \text{ sec}^{-1}}{4.7 \times 10^2 \text{ L}^2 \text{ mol}^{-2} \text{ sec}^{-1}} = 1.7 \times 10^{-4} \frac{\text{mol}}{\text{L}}$$

The relationship between the equilibrium constants and the specific rate constants of complex reactions differs somewhat from the relationship for elementary reactions. It is based on a general principle that is called *the principle of detailed balancing* when it is applied to large-scale systems and *the principle of microscopic reversibility* when it is applied on the molecular level. It can be stated: *When equilibrium is reached in a reaction system, any elementary process and the reverse of that process must occur, on the average, at the same rate.*

We can show how this principle is used to find the relationship between the equilibrium constant and the relevant specific rate constants by analyzing the reaction system:

$$2NO_2(g) + F_2(g) \rightleftharpoons 2NO_2F(g)$$

It has been suggested that the mechanism for the formation of $NO_2F$ includes two elementary reactions:

1.  $\qquad NO_2 + F_2 \xrightarrow{k_1} NO_2F + F \qquad$ slow

2.  $\qquad NO_2 + F \xrightarrow{k_2} NO_2F \qquad$ fast

The initial rate law for the reaction is:

$$\text{rate} = k_1[NO_2][F_2]$$

Figure 17.8(a) shows the energy diagram for these two elementary reactions. Until now, we have read such energy diagrams only from left to right. But at equilibrium, the steps in this mechanism proceed both forward and backward, since the $NO_2F$ decomposes back to $NO_2$ and $F_2$. On the energy diagram, the system can be visualized as proceeding in both directions along the same energy pathway. The principle of detailed balancing tells us that the forward rate equals the reverse rate in each step at equilibrium. Since these are elementary processes, we can write the rate laws for each step:

$$\text{step 1 forward rate} = k_1[NO_2][F_2]$$
$$\text{step 1 reverse rate} = k_{-1}[NO_2F][F]$$

and since these rates are equal:

a.  $\qquad k_1[NO_2][F_2] = k_{-1}[NO_2F][F]$

$$\text{step 2 forward rate} = k_2[NO_2][F]$$
$$\text{step 2 reverse rate} = k_{-2}[NO_2F]$$

and since these rates are equal:

b.  $\qquad k_2[NO_2][F] = k_{-2}[NO_2F]$

To find the relationship of the equilibrium constant $K$ to these specific rate constants, equations a and b can be combined. After the equations are multiplied and rearranged, we have:

$$\frac{k_1 k_2}{k_{-1} k_{-2}} = \frac{[NO_2F][F][NO_2F]}{[NO_2][F_2][NO_2][F]} = \frac{[NO_2F]^2}{[NO_2]^2[F_2]} = K$$

For this mechanism, the equilibrium constant is the product of the specific rate constants of the forward reactions divided by the product of the specific rate constants of the reverse reactions.

### Kinetic and Thermodynamic Control

We encounter a different aspect of the relationship between kinetics and equilibrium in systems where more than one product can form from a given set of reactants. This situation can be represented by the general scheme:

a. $\qquad\qquad\qquad\qquad\qquad$ $A + B \rightleftharpoons C$
b. $\qquad\qquad\qquad\qquad\qquad$ $A + B \rightleftharpoons D$

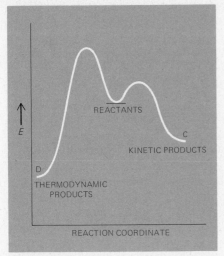

**Figure 17.11**
An energy diagram for a reaction in which one product, C, forms faster and another product, D, is more stable. Product C can be formed if the reaction is stopped after a short time. Product D can be formed if the reaction is allowed to proceed to equilibrium.

   In such reaction systems, the product that forms fastest is not necessarily the most stable. Let us assume that product C forms fastest and product D is more stable, a situation represented by the energy diagram in Figure 17.11, and that reactions a and b are reversible. It is then possible to form selectively either C or D by controlling the way in which the reaction is carried out.

   The product that forms faster, C, is called the *product of kinetic control*. We can obtain it by allowing the reaction to proceed for only a relatively short time, or by removing C as it forms. The more stable product, D, is called the *product of thermodynamic control*. It is obtained if we allow the system to proceed to equilibrium. Since D is more stable, it will predominate at equilibrium.

   We can understand the buildup of product D by examining the relative rates of the reverse reactions. While C forms faster, it also decomposes faster when conditions permit the reaction to reverse. When D finally forms, it decomposes to the reactants more slowly than C. Thus, when D forms, it tends to remain. The energy diagram in Figure 17.11 shows that the energy barriers for the reverse reactions are consistent with this explanation.

## 17.6  CATALYSIS

*A catalyst is a substance that speeds up the rate of a reaction without being consumed.* A catalyst enters into a reaction but emerges from it unchanged. It may, for example, form an intermediate with one or more of the reactants. The intermediate then decomposes to form the products and to regenerate the catalyst.

   Catalysts are important in many reactions, organic and inorganic. We know the general mode of action of catalysts. The height of the energy

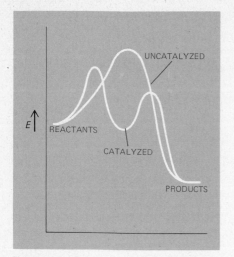

**Figure 17.12**
A catalyst makes a reaction go faster by providing a pathway with a lower energy barrier (bottom line) between reactants and products than the pathway that is normally available (top line).

barrier along the pathway from reactants to products controls the rate of a reaction. As Figure 17.12 shows, a catalyst quickens the rate of a reaction by providing a different pathway with a lower energy barrier between reactants and products.

From Figure 17.12, we see that a catalyst does not change the relative energy of the reactants and the products, which means that *a catalyst has no effect on the position of equilibrium*. We also see that the system can go either way on the lower-energy pathway opened up by the catalyst. Therefore, *a catalyst speeds up the rate of a reaction in both the forward and reverse directions*. By so doing, a catalyst causes a system to reach equilibrium faster.

We can divide catalyzed reactions into two main categories: *homogeneous reactions*, which occur entirely in one phase, usually in the gas phase or in solution, and *heterogeneous reactions*, which occur at the interface between two phases — for example, on the surface of a solid that is in contact with a solution or a gas.

## Homogeneous Catalysis

The mass of substances released into the atmosphere by humans is small compared to the total mass of the atmosphere. But some of these substances can cause relatively large changes because they act as catalysts.

One harmful process is the formation of sulfur trioxide by the oxidation of the sulfur dioxide that is released when fossil fuels are burned. This process leads to the formation of acidic sulfates, which are believed to be harmful to human health. The oxidation reaction is

$$2SO_2(g) + O_2(g) \longrightarrow 2SO_3(g)$$

Normally, the rate of this reaction is slow. A termolecular collision is needed for the reaction to occur directly in a single step, and termolecular collisions generally are not favored. But oxides of nitrogen, which also are produced when fossil fuels are burned, have been found to catalyze the oxidation by the following steps:

$$2NO(g) + O_2(g) \longrightarrow 2NO_2(g)$$
$$NO_2(g) + SO_2(g) \longrightarrow NO(g) + SO_3(g)$$

The sulfur dioxide is oxidized by the $NO_2$, a faster process than its direct oxidation by $O_2$. The NO is not consumed in the overall process. The $SO_3$ then combines with $H_2O$ in the atmosphere to produce corrosive sulfuric acid, $H_2SO_4$, and other sulfates.

Homogeneous catalysis in solution is also quite common. One important kind is acid-base catalysis, which often occurs when a substance reacts through its conjugate acid or conjugate base as an intermediate. The hydrolysis of nitramide, $NH_2NO_2$, for example, is believed to occur by the mechanism:

$$NH_2NO_2(aq) + OH^-(aq) \rightleftharpoons H_2O + NHNO_2^-(aq)$$
$$NHNO_2^-(aq) \longrightarrow N_2O(g) + OH^-(aq)$$

The rate of reaction increases as the concentration of hydroxide ion increases. You can see that the hydroxide ion qualifies as a catalyst because it is not consumed in the overall reaction.

Another reaction for which catalysis is important is the decomposition of hydrogen peroxide in aqueous solution:

$$2H_2O_2(aq) \longrightarrow 2H_2O + O_2(g)$$

Even though the position of equilibrium lies far to the right for this system, the reaction is slow if the $H_2O_2$ is very pure. But it can be catalyzed in a number of different ways, some of which are based on the ability of $H_2O_2$ to be both an oxidizing agent and a reducing agent.

A number of oxidation-reduction couples catalyze the decomposition of hydrogen peroxide. One group of catalysts that has been studied extensively is ions of transition metals that display more than one oxidation state. The catalysis of hydrogen peroxide decomposition by these ions is an important biological process.

In particular, the catalysis of the decomposition of $H_2O_2$ by cations of iron is of great interest. The overall changes are

$$H_2O_2(aq) + 2H^+(aq) + 2Fe^{2+}(aq) \longrightarrow 2H_2O + 2Fe^{3+}(aq)$$
$$H_2O_2(aq) + 2Fe^{3+}(aq) \longrightarrow O_2(g) + 2H^+(aq) + 2Fe^{2+}(aq)$$

The sum of these reactions is simply

$$2H_2O_2(aq) \longrightarrow 2H_2O + O_2(g)$$

The details of these steps are rather complex and have not yet been worked out completely.

## Enzyme Catalysis

The phenomenon we call life would be impossible without catalysis. Most of the reactions that are essential to life take place very slowly outside the living organism. But inside living cells, biological catalysts called *enzymes* speed the rates of these reactions, often by enormous factors.

Enzymes are large proteins — indeed, they are the most important and numerous class of proteins. More than a thousand enzymes, grouped into six main classifications, have been identified. The name of an enzyme group, or of an individual enzyme, consists of a stem indicating the reaction that is catalyzed and the suffix *-ase*. For example, the major groups include the oxidoreductases, which catalyze oxidation-reduction reactions; the lyases, which catalyze addition reactions; and the hydrolases, which catalyze hydrolysis reactions.

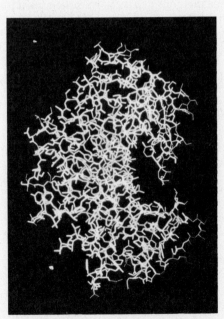

A computer-generated image of the molecular structure of the enzyme pepsin. The active site is only a small portion of the enzyme, but the folding of the entire molecule determines the conformation at the active site. *(Blundell, Photo Researchers, Inc.)*

726    CHAPTER 17   Chemical Kinetics

Generally, enzymes can function as catalysts only within a narrow temperature range, so there is a need for temperature control in living organisms. In addition, enzymes can function only in a narrow pH range, so there is a need for buffering and close control of pH in most living organisms.

An enzyme often is associated with a smaller nonprotein molecule, called a *coenzyme* or prosthetic group, which is needed for the enzyme to perform its catalytic function. Undoubtedly the best-known group of coenzymes are the vitamins, which take part in a number of essential processes in the human body.

Only a small region of the protein molecule, the *active site,* participates in the enzyme's catalytic action. In a sense, the structure of the entire molecule is built around the active site. The enzyme normally maintains a fixed three-dimensional structure that keeps the appropriate portions of the molecule at the active site, so that the enzyme can act as a catalyst.

The reactant in an enzyme-catalyzed reaction is called the substrate. There is a high degree of specificity in both the types of reactions that enzymes catalyze and the substrates whose reactions they catalyze. Often, an enzyme will catalyze a reaction for one substrate but not for another that is chemically quite similar. In many such cases, it is possible to note three-dimensional structural differences between the substrates.

The first step in an enzyme-catalyzed reaction is the rapid formation of an enzyme-substrate complex. Since this complex is easily broken up, it is assumed that the substrate is held to the active site of the enzyme by weak interactions, such as hydrogen bonding or ion-dipole attractions. The reaction can be represented as

$$E + S \rightleftharpoons ES$$

where E is the enzyme, S is the substrate, and ES is the complex. The equilibrium constant for this reaction usually is large, between $10^3$ and $10^6$. After the enzyme-substrate complex is formed, there are reactions that form the product and regenerate the enzyme. We can represent such reactions schematically as

$$ES \longrightarrow P + E$$

where P is the product.

The formation of the ES complex is often described as the insertion of a "key," the substrate, into a "lock," the active site of the enzyme, as shown in Figure 17.13. The point of this lock-and-key analogy is that only a substrate with precisely the right shape will "unlock" the enzyme's catalytic capabilities. The precise fit allows several catalytic groups that are part of the enzyme to work on the substrate simultaneously.

The structure of the active site and the nature of the enzyme-substrate complex have been the subject of much recent research. Investigations of this kind are unusually demanding, because of the complexity of enzyme

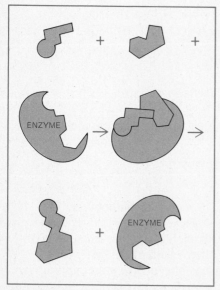

**Figure 17.13**
The "lock-and-key" theory of enzyme action, shown schematically. The substrates serve as the "key" fitting into the "lock," the enzyme's active site, where they are held in an orientation that allows the reaction to occur swiftly. In the first step, the enzyme and the substrates form a complex, which decomposes in the second step.

structures. One of the most remarkable achievements of the past few decades has been the determination of the complete structure of several enzymes by X-ray crystallography, a feat that makes possible a reasonably good understanding of the way in which an enzyme works. Research on enzyme catalysis is one of the most active areas in chemistry today.

Research thus far has pointed up the remarkable efficiency of many enzymatic processes. For example, one of the most efficient enzymes known is catalase, which catalyzes the familiar reaction.

$$2H_2O_2 \longrightarrow 2H_2O + O_2$$

One molecule of catalase will convert more than one million molecules of hydrogen peroxide to water and oxygen in one minute at room temperature. Since hydrogen peroxide is a necessary by-product of many biological oxidation steps and is also quite poisonous, its rapid removal is essential for the well-being of living organisms.

The mode of action of catalase has not been worked out in detail. But it is known that iron in the $+3$ oxidation state is part of the catalase molecule and takes part in the catalytic process, presumably in a way similar to the inorganic catalysis we have discussed.

There are measurements that give us a rough idea of the effectiveness of this biological catalysis. The uncatalyzed decomposition of $H_2O_2$ is reported to have $k = 1 \times 10^{-7}$ sec$^{-1}$ and $E_a = 75$ kJ/mol at room temperature. For the decomposition brought about with catalase, $k = 4 \times 10^7$ sec$^{-1}$ and $E_a = 8$ kJ/mol. Thus, for a given set of conditions, the presence of catalase increases the rate of decomposition of $H_2O_2$ by a factor of more than $10^{14}$.

## Heterogeneous Catalysis

Solids that increase the rate of chemical reactions because of their surface properties are called heterogeneous catalysts. There is a wide variety of heterogeneous catalysts, including metals, metal oxides, metal sulfides, and salts. They are widely used to catalyze reactions of gases or liquids on the surfaces of solids, including many processes that are essential to modern chemical industry. For example, most of the reactions by which petroleum is refined are catalyzed by solids such as alumina, $Al_2O_3$; silica, $SiO_2$; and metal salts. The manufacture of ammonia by the Haber process:

$$N_2(g) + 3H_2(g) \longrightarrow 2NH_3(g)$$

is catalyzed by iron, molybdenum, or other metals. There is a whole group of reactions, called Fischer-Tropsch reactions, in which organic products are produced from $H_2$ and CO. One of these reactions, catalyzed by zinc oxide, is used for the production of methanol:

$$CO(g) + 2H_2(g) \longrightarrow CH_3OH(g)$$

## THE CATALYSTS IN YOUR CAR

The catalytic converters that help reduce the emissions of carbon monoxide, hydrocarbons, and nitrogen oxides from many of today's automobiles are unusual because they actually are two converters in one: a reducing converter and an oxidizing converter, working in coordination to change pollutants into harmless gases.

The catalysts in the converters are primarily platinum and palladium, deposited in a thin layer on a porous material that has a very large surface area. Both converters use the same catalysts, but a wholly different set of reactions occurs in each.

The control system really begins at the carburetor, which is set for a rich fuel-to-air mixture that has two effects: It lowers the combustion temperature, so that less NO is produced, and it produces an abundance of hydrocarbons (HC), CO, and $H_2$. These gases, all reducing agents, are needed as reactants in the reducing converter, the first on line. Some of the reactions that occur in the reducing converter are

$$2CO + 2NO \longrightarrow N_2 + 2CO_2$$
$$5H_2 + 2NO \longrightarrow 2NH_3 + 2H_2O$$
$$2H_2 + 2NO \longrightarrow N_2 + 2H_2O$$
$$CO + H_2O \longrightarrow CO_2 + H_2$$

For the oxidizing catalyst to do its job, the oxygen-poor output of the reducing converter must be made oxygen-rich. A stream of air is pumped into the exhaust as it enters the oxidizing converter, where these reactions take place:

$$2CO + O_2 \longrightarrow 2CO_2$$
$$4HC + 5O_2 \longrightarrow 4CO_2 + 2H_2O$$
$$2H_2 + O_2 \longrightarrow 2H_2O$$

Ideally, the exhaust gases that emerge from the oxidizing converter contain only $CO_2$, $N_2$, $H_2O$, and $O_2$. In reality, the undesirable gases are not removed completely. One of the products of the reducing converter is ammonia, $NH_3$. Not all of this ammonia is converted in the reducing converter. In the oxidizing converter, ammonia is converted to nitrogen oxides by the reactions:

$$4NH_3 + 5O_2 \longrightarrow 4NO + 6H_2O$$
$$2NH_3 + 2O_2 \longrightarrow N_2O + 3H_2O$$

There are further complications. When the engine starts, it takes several minutes for the oxidizing converter to reach its operating temperature of 650 K. The system thus must be engineered so that the reducing converter can bring about the oxidizing reactions temporarily. And since the tetraethyl lead used in gasoline can coat the catalyst and "poison" it, automobiles with catalytic converters can use only lead-free gasoline.

The reaction of $H_2$ with CO can also be catalyzed with nickel to form methane, $CH_4$, or with cobalt to form more complex hydrocarbons. Heterogeneous catalysis is also used in the oxidation of ammonia to nitric oxide, the first step in the Ostwald process (Section 9.3) for the manufacture of nitric acid:

$$4NH_3(g) + 5O_2(g) \longrightarrow 4NO(g) + 6H_2O(g)$$

which is catalyzed by an alloy of platinum and rhodium. The oxidation of $SO_2$ to $SO_3$, a step in the contact process for the manufacture of sulfuric acid, is catalyzed by $V_2O_5$. A list of industrial processes that use heterogeneous catalysis could run for many pages.

## Adsorption

We can visualize the mechanism of reactions that take place on the surfaces of solids as a sequence of steps:

1.   The reactants find their way to the surface of the catalyst.
2.   The reactants are adsorbed on the surface.
3.   The adsorbed substances react.
4.   The products of the reaction are desorbed from the surface.
5.   The products go back into the gas or liquid phase.

The first and last steps usually occur quickly and are not related to the reaction mechanism; we shall not consider them. The second step, adsorption, is the process by which a substance becomes attached to the surface of a solid. There are two ways in which a molecule can be adsorbed on the surface of a solid catalyst. In *physical adsorption,* relatively weak forces, typical of nonbonding interactions, hold the molecule on the surface. Physical adsorption is not a major factor in catalytic action. The second kind of adsorption, *chemisorption,* has most of the characteristics of a chemical reaction between the reactant molecule and the solid. The strength of the forces holding the reactant molecule to the surface is as great as that of chemical bonds, and chemisorption often changes the reactant molecule in a way that makes it easier for the desired reaction to occur. An appreciable energy of activation is associated with chemisorption in many cases, so reactions on solid surfaces must often be carried out at high temperatures.

The overall rate of a surface-catalyzed reaction usually is controlled by the rate at which reactions occur between the adsorbed molecules on the surface of the catalyst. But, depending on the reaction conditions, the overall rate can also be controlled by the rate of adsorption of the reactants or the rate of desorption of the products. We can see the controlling effect of adsorption and desorption in the reversible reaction for the synthesis of ammonia, which takes place on a metal surface:

$$N_2(g) + 3H_2(g) \rightleftharpoons 2NH_3(g)$$

At relatively high pressures and temperatures between 700 K and 900 K, the rate of the forward reaction is controlled by the rate of chemisorption of $N_2$, while the rate of the reverse reaction is controlled by the rate of desorption of $N_2$.

The rates of reaction in this system in the absence of catalyst are extremely slow. It is not until the temperature rises above 1600 K that the homogeneous decomposition of $NH_3$ occurs, and then only at a moderate rate. The uncatalyzed thermal reaction of $N_2$ and $H_2$ has never been observed. The reaction is exothermic, so the equilibrium position moves further to the left as the temperature goes up. In fact, the uncatalyzed reaction between $N_2$ and $H_2$ can occur at a measurable rate only at a temperature so high that no detectable $NH_3$ is found at equilibrium.

The major action of the catalyst in this reaction appears to be the activation of $N_2$. We have said that the $N\equiv N$ triple bond is one of the strongest bonds known, and that reactions with $N_2$ often are extremely slow because so much energy is needed to break, or even to weaken, the bond. A suitable catalyst weakens or even breaks the bond but does not adsorb the $N_2$ molecule too tightly for other reactions to occur. The beginning of the catalyzed synthesis of ammonia can be represented by the sequence:

$$N_2(g) \rightleftharpoons N_2 \text{ (adsorbed)}$$
$$N_2 \text{ (adsorbed)} \rightleftharpoons 2N \text{ (adsorbed)}$$
$$H_2(g) \rightleftharpoons H_2 \text{ (adsorbed)}$$
$$H_2 \text{ (adsorbed)} \rightleftharpoons 2H \text{ (adsorbed)}$$

When these chemisorption steps have taken place, the surface of the metal is covered with the equivalent of atomic nitrogen and atomic hydrogen, which can react with each other readily. An overall change in which three atoms of H and one atom of N combine can occur in steps. The $NH_3$ that is formed on the surface desorbs and goes into the gas phase, allowing more $NH_3$ to form:

$$N \text{ (adsorbed)} + 3H \text{ (adsorbed)} \rightleftharpoons NH_3 \text{ (adsorbed)}$$
$$NH_3 \text{ (adsorbed)} \rightleftharpoons NH_3(g)$$

Some surface-catalyzed reactions are among the few known examples of reactions that can be zero order overall. One of them is the decomposition of nitrous oxide gas on a hot platinum surface:

$$2N_2O(g) \longrightarrow 2N_2(g) + O_2(g)$$

When the concentration of $N_2O$ is high enough, the surface of the metal is covered completely with adsorbed gas. An increase in the pressure of the gas above the solid does not affect the rate, since the surface is already saturated with $N_2O$ and the reaction cannot proceed any faster. Since an increase in concentration of the reactant has no effect on the rate, the reaction is zero order.

## Summary

$\mathsf{T}$he subject of this chapter was **kinetics,** which deals with the **rate** at which chemical reactions occur. We observed that the rate of most chemical reactions depends in some way on the concentration of one or more reactants, and we distinguished between the **average rate** of a reaction, the rate measured for a total time span, and the **instantaneous rate,** the limit of the average rate as the time span approaches zero.

We said that the equation that describes the relationship between rate and concentration is called a **rate law.** A key part of a rate law is the **specific rate constant,** the proportionality constant between rate and concentration. We noted that a rate law includes the exponents on the concentration terms and that the **order** of a rate law is described in terms of those exponents. A rate law is **first-order** if the exponent is 1 and **second-order** if the sum of the exponents is 2.

We defined **half-life,** the time required for half the starting quantity of a reactant to be consumed. We noted that the half-life of a first-order reaction is the only one that remains constant as the concentration decreases. We said that many chemical reactions proceed in a number of simple steps called **elementary reactions.** To help explain the importance of elementary reactions, we noted that **molecularity** is the number of reactant molecules in such a reaction, and that the molecularity and the order of an elementary reaction are equal. We said that the overall mechanism by which a reaction occurs can often be understood by finding the **rate-determining step,** the slowest elementary reaction. We pointed out that most chemical reactions go faster as temperature increases, and explained that the rate increases because more reactants can exceed the **activation energy,** the energy barrier that must be overcome for a reaction to occur. We described the **Arrhenius law,** the relationship between activation energy and rate constant. We showed that in a system at equilibrium, the equilibrium constant is related to the specific rate constant of the forward and back reactions. Finally, we discussed **catalysts,** substances that speed up the rates of reactions without being consumed.

## Exercises

**17.1** Propose units that can conveniently be used to express the rates of the following processes: (a) swimming the English Channel, (b) talking, (c) eating peanuts, (d) breathing.

**17.2** Write expressions that can be used to express the rate of the following reaction and show the relationship between these expressions: $N_2(g) + 3H_2(g) \rightarrow 2NH_3(g)$.

**17.3** Write expressions that can be used to express the rate of the following reaction and show the relationship between these expressions: $Cr_2O_7{}^{2-}(aq) + 6H^+(aq) + 2NO(aq) \rightarrow 2Cr^{3+}(aq) + 3H_2O + 2NO_3{}^-(aq)$.

**17.4** The rate of appearance of $SO_3(g)$ from the reaction of $SO_2(g)$ and $O_2(g)$ is $3.6 \times 10^{-5}$ atm/sec under certain conditions. Find the rate of disappearance of $O_2$ and $SO_2$.

**17.5**[4] The rate of decomposition of $BrF_5(g)$ to $Br_2(g)$ and $F_2(g)$ under certain conditions is $8.9 \times 10^{-4}$ atm/sec. Find the rate of appearance of $F_2$ and $Br_2$.

**17.6** When $I^-$ is oxidized by $MnO_4{}^-$ in aqueous acid the products are $Mn^{2+}$, $I_2$, and $H_2O$. The rate of appearance of $I_2$ is found to be $7.8 \times 10^{-3}$ mol $L^{-1}$ $sec^{-1}$. Find the rate of disappearance of $MnO_4{}^-$ and $I^-$.

---

[4] The answers to exercises whose numbers are in color can be found in Appendix VII. The star indicates an exercise that is more challenging than average.

**17.7**  Write three possible rate expressions that you could use to express the rate of decomposition of $CaCO_3(s)$ to $CaO(s)$ and $CO_2(g)$.

**17.8**  The following data are obtained in a study of the rate of disappearance of $CH_3OH(aq)$ in the reaction $CH_3OH(aq) + H^+(aq) + Br^-(aq) \rightarrow CH_3Br(aq) + H_2O$:

| $[CH_3OH]$ $(M)$ | $t$ (min) |
|---|---|
| 1.20 | 0 |
| 0.992 | 75.0 |
| 0.874 | 125 |
| 0.678 | 225 |
| 0.317 | 525 |

Find the average rate of disappearance of $CH_3OH$ for the time interval between each measurement and for the total time interval.

**17.9**  The following data are obtained in a study of the rate of appearance of $NOCl(g)$ in the reaction $2NO(g) + Cl_2(g) \rightarrow 2NOCl(g)$:

| $P_{NOCl}$ (mmHg) | $t$ (sec) |
|---|---|
| 0 | 0 |
| 83 | 40 |
| 137 | 81 |
| 175 | 120 |
| 242 | 250 |

Find the average rate of appearance of $NOCl(g)$ for the time interval between each measurement and for the total time.

**17.10**  The following total pressure measurements are obtained in a study of the rate of the reaction $2CO(g) \rightarrow CO_2(g) + C(s)$:

| $P_T$ (atm) | $t$ (min) |
|---|---|
| 0.329 | 0 |
| 0.313 | 6.68 |
| 0.295 | 16.6 |
| 0.276 | 29.9 |

Find the average rate of disappearance of CO for the time interval between each measurement and for the total time interval. Find the average rate of appearance of $CO_2$ for the total time interval.

**17.11**  The gas phase decomposition of $N_2O_5$ to $NO_2$ and $O_2$ is monitored by measurements of total pressure. The following data are obtained:

| $P_T$ (atm) | $t$ (sec) |
|---|---|
| 0.154 | 0 |
| 0.215 | 54.0 |
| 0.260 | 105 |
| 0.315 | 203 |
| 0.346 | 311 |

Find the average rate of appearance of $NO_2$ for the time interval between each measurement and for the total time interval.

**17.12**  Use the data given in Exercise 17.8 to estimate the instantaneous rate of disappearance of $CH_3OH$ at 100 minutes.

**17.13**  Before beginning the kinetic study of a reaction, a researcher might make some guesses about the rate law of a reaction. For example, for the reaction $2NO(g) + O_2(g) \rightarrow 2NO_2(g)$, some simple rate laws suggest themselves. Write the rate law that corresponds to each of the following word descriptions: (a) first order in nitric oxide, (b) second order in nitric oxide, (c) first order in nitric oxide, first order in oxygen, (d) second order in nitric oxide, first order in oxygen. Indicate the overall order of each rate law.

**17.14**  The rate law for the reaction $2Cl^-(aq) + 2H^+(aq) + H_2O_2(aq) \rightarrow Cl_2(aq) + 2H_2O$ is third order overall and first order in each reactant. Write the equation for the rate law.

**17.15**  Suggest units that can be used for the specific rate constant of a gas phase reaction that is (a) first order in each of three reactants, (b) second order in one reactant and overall.

**17.16**  The compound $CH_3CH_2Cl$ decomposes to $C_2H_4$ and HCl at elevated temperatures. Find the rate law for this reaction from the following rate data:

| $P_{CH_3CH_2Cl}$ (atm) | initial rate (atm/sec) |
|---|---|
| $1.8 \times 10^{-2}$ | $7.5 \times 10^{-4}$ |
| $4.3 \times 10^{-2}$ | $1.8 \times 10^{-3}$ |
| $7.8 \times 10^{-2}$ | $3.3 \times 10^{-3}$ |

**17.17**  At 400 K, oxalic acid decomposes according to the following reaction: $H_2C_2O_4(g) \rightarrow CO_2(g) + HCOOH(g)$. The rate of this reaction can be studied by measurements of the total pressure. Find the rate law of the reaction from the following measurements that give the total pressure reached after 20 000 sec from the indicated starting pressures of oxalic acid.

| $P_{H_2C_2O_4}$ (atm) | $P_T$ (atm) |
|---|---|
| $6.58 \times 10^{-3}$ | $9.46 \times 10^{-3}$ |
| $9.21 \times 10^{-3}$ | $1.32 \times 10^{-2}$ |
| $1.11 \times 10^{-2}$ | $1.60 \times 10^{-2}$ |

**17.18**  Acetaldehyde, $CH_3CHO$, decomposes to CO and $CH_4$ in the gas phase. Find the rate law of the reaction from the following data on initial rates:

| $P_{CH_3CHO}$ (atm) | initial rate (atm/sec) |
|---|---|
| 0.213 | $1.34 \times 10^{-3}$ |
| 0.416 | $3.79 \times 10^{-3}$ |
| 0.107 | $4.77 \times 10^{-4}$ |

**17.19**  The rate of the reaction $H_2PO_2^-(aq) + OH^-(aq) \rightarrow HPO_3^{2-}(aq) + H_2(g)$ can be measured in several ways. Suggest two relatively simple techniques that could be used for this purpose.

**17.20**  The following measurements of initial rate are made for varying concentrations of $OH^-$ ion and $H_2PO_2^-$ ion:

| [OH⁻] $(M)$ | [H₂PO₂⁻] $(M)$ | initial rate (mol L⁻¹ sec⁻¹) |
|---|---|---|
| 0.21 | 0.35 | $5.2 \times 10^{-4}$ |
| 0.28 | 0.35 | $9.3 \times 10^{-4}$ |
| 0.35 | 0.35 | $1.5 \times 10^{-3}$ |
| 0.21 | 0.52 | $7.8 \times 10^{-4}$ |
| 0.21 | 0.69 | $1.0 \times 10^{-3}$ |

Find the rate law for the reaction.

**17.21** The following data are obtained in a study of the rate of the reaction $SO_2Cl_2(g) \rightarrow SO_2(g) + Cl_2(g)$:

| $P_{SO_2Cl_2}$(atm) | $t$ (sec) |
|---|---|
| 0.500 | 0 |
| 0.457 | 500 |
| 0.418 | 1000 |
| 0.349 | 2000 |
| 0.244 | 4000 |

The reaction follows a first-order rate law. Find the value of the specific rate constant.

**17.22** The specific rate constant for the first-order decomposition of hydrogen peroxide to water and $O_2(g)$ is found to be $1.20 \times 10^{-5}$ sec⁻¹ at a given temperature. A solution is prepared in which the $[H_2O_2] = 0.884M$. (a) Find the time required for the $[H_2O_2]$ to fall to $0.553M$. (b) Find the time required for the $[H_2O_2]$ to fall to $0.0884M$. (c) Find the $[H_2O_2]$ after 36 hr.

**17.23** Under certain conditions the rate of the reaction $I_2(aq) + H_2O \rightarrow I^-(aq) + HIO(aq) + H^+(aq)$ is first order in $I_2(aq)$. The $[I_2]$ can be measured indirectly by measurement of the optical density of the brown solution of $I_2$ in water. The value of the optical density is proportional to the $[I_2]$. The optical density of a solution of $I_2$ is found to be 0.48 when the solution is first prepared. After 75 sec the optical density is 0.26 and after 125 sec it is 0.18. (a) Find the specific rate constant. (b) Find the time required for the optical density to drop to 0.39.

**17.24★** The rate of decomposition of $N_2O_3(g)$ to $NO_2(g)$ and $NO(g)$ is followed by measuring the $[NO_2]$. The following data are obtained:

| [NO₂] (mol/L) | $t$ (sec) |
|---|---|
| 0 | 0 |
| 0.193 | 884 |
| 0.316 | 1610 |
| 0.427 | 2460 |
| 0.784 | 50 000 |

Show graphically that the reaction follows a first-order rate law. Calculate the specific rate constant.

**17.25★** The specific rate constant for the first-order decomposition of $N_2O_5(g)$ to $NO_2(g)$ and $O_2(g)$ is $6.37 \times 10^{-3}$ sec⁻¹, at a given temperature. (a) Find the time required for the total pressure in a system containing $N_2O_5$ at an initial pressure of 0.100 atm to rise to 0.134 atm; (b) to 0.200 atm. (c) Find the total pressure after 100 sec of reaction.

**17.26** Derive a general expression for the time required for the concentration of a reactant in a first-order reaction to fall by one-fourth.

**17.27** Find the fraction of the starting concentration of a reactant that remains when a first-order reaction proceeds for eight half-lives.

**17.28** Find the half-life of the reaction of Exercise 17.22.

**17.29** The half-life for radioactive decay of ⁶⁰Co, the radioisotope of cobalt used extensively in the treatment of cancer, is 5.26 yrs. (a) Find the specific rate constant for the decay. (b) Find the time required for 99% of a sample of ⁶⁰Co to decay.

**17.30** A bit of vegetable dye scraped from a cave painting is converted to $CO_2$. The radioactivity of the $CO_2$ is measured as 0.947 c.p.m. Find the age of the cave painting, using the data in Example 17.5.

**17.31** At 900 K the decomposition of $NO_2(g)$ to $NO(g)$ and $O_2(g)$ is second order in $NO_2$. The following data are obtained from a kinetic study of this reaction:

| $P_{NO_2}$ (atm) | $t$ (min) |
|---|---|
| 0.074 | 0 |
| 0.051 | 317 |
| 0.038 | 667 |
| 0.029 | 1090 |
| 0.011 | 4030 |

(a) Find the specific rate constant of the reaction. (b) Find the $P_{NO_2}$ after 10 000 min. (c) Find the time required to build up a $P_{NO}$ of 0.050 atm.

**17.32** The reaction $NO(g) + O_3(g) \rightarrow NO_2(g) + O_2(g)$ is second order overall and first order in each reactant. To simplify the treatment of the data, a kinetic study is carried out beginning with a gaseous mixture in which $P_{NO} = P_{O_3} = 0.0094$ atm. (a) Why is the treatment of the data simplified by starting with equal pressures of each reactant? (b) Use the following data to find the specific rate constant:

| $P_{NO}$ (atm) | $t$ (sec) |
|---|---|
| 0.0094 | 0 |
| 0.0077 | 287 |
| 0.0052 | 1052 |
| 0.0039 | 1840 |

(c) Find the $P_{O_3}$ after 2500 sec.

**17.33** Derive a general expression for the time required for the concentration of a reactant in a one-term second-order reaction to fall by one-fourth.

**17.34** The following data are obtained for the reaction $A(g) \rightarrow B(g)$:

| [A] (mol/L) | $t$ (sec) |
|---|---|
| 0.470 | 0 |
| 0.405 | 250 |
| 0.356 | 500 |
| 0.319 | 750 |
| 0.286 | 1000 |

Determine whether the reaction is first or second order and calculate $k$.

**17.35** Consider the reaction $A(g) \rightarrow B(g)$. The starting pressure of A is 0.20 atm. After 60 sec, the pressure of A is found to be 0.10 atm. Calculate the pressure of A after another 60 sec if the reaction is (a) first order in A, (b) second order in A, (c) zero order in A.

**17.36** Write rate laws for the following gas phase elementary processes: (a) $COCl + Cl \rightarrow CO + Cl_2$, (b) $2ClO \rightarrow Cl_2 + O_2$, (c) $2NO + I_2 \rightarrow 2NOI$, (d) $C_4H_8 \rightarrow 2C_2H_4$.

**17.37** The reaction $C_4H_9Br + OH^- \rightarrow C_4H_8 + Br^- + H_2O$ is believed to take place in polar solvents by the mechanism:

$$C_4H_9Br \xrightarrow{k_1} C_4H_9^+ + Br^- \quad \text{slow}$$
$$C_4H_9^+ + OH^- \xrightarrow{k_2} C_4H_8 + H_2O \quad \text{fast}$$

Write the rate law that is consistent with this mechanism.

**17.38** The reaction in the gas phase in which $NO_2Br$ decomposes to form $Br_2$ and $NO_2$ is believed to take place by the following mechanism:

1. $NO_2Br \xrightarrow{k_1} NO_2 + Br \quad$ slow
2. $NO_2Br + Br \xrightarrow{k_2} NO_2 + Br_2 \quad$ fast

Write the overall reaction and the rate law that are consistent with this mechanism.

**17.39** One proposed, although probably incorrect, mechanism for the oxidation of nitric oxide is

1. $NO(g) + O_2(g) \rightleftharpoons NO_3(g) \quad$ fast
2. $NO_3(g) + NO(g) \xrightarrow{k_2} 2NO_2(g) \quad$ slow

Write the overall reaction and the rate law that are consistent with this mechanism. An alternate mechanism is a one-step process. Write the rate law consistent with this mechanism.

**17.40** The proposed two-step mechanism for the decomposition of $O_3(g)$ is

1. $O_3(g) \rightleftharpoons O_2(g) + O(g)$
2. $O(g) + O_3(g) \xrightarrow{k_2} 2O_2(g)$

(a) Write the rate law consistent with the first step being rate determining. (b) Write the rate law consistent with the second step being rate determining. (c) Suggest an experiment to distinguish between these rate laws.

**17.41** The proposed mechanism for the formation of hydrogen bromide can be written in a simplified form as:

1. $Br_2(g) \rightleftharpoons 2Br(g) \quad$ fast
2. $Br(g) + H_2(g) \xrightarrow{k_2} HBr(g) + H(g) \quad$ slow
3. $H(g) + Br_2(g) \xrightarrow{k_3} HBr(g) + Br(g) \quad$ fast

This mechanism is consistent with a fractional-order rate law. Write the rate law.

**17.42** Find the rate law consistent with the three-step mechanism given in Section 17.3 for the reaction of $H_2$ and $I_2$.

**17.43** The gas phase reaction $COCl + Cl \rightarrow CO + Cl_2$ is an elementary reaction. It is found that only a very small fraction of the collisions between reactant molecules lead to the formation of products. Discuss in terms of molecular structure some of the factors that may result in so many nonreactive collisions.

**17.44** The activation energy of the gas phase reaction $2NOCl \rightarrow 2NO + Cl_2$ is 103.0 kJ and the frequency factor

is $1.0 \times 10^{10}$ L $mol^{-1}$ $sec^{-1}$. Find the specific rate constant at 700 K.

**17.45** The specific rate constant $k$ for the reaction $C_4F_8(g) \rightarrow 2C_2F_4(g)$ at 750 K is $2.25 \times 10^{-6}$ $sec^{-1}$ and the activation energy is 310 kJ/mol. Find the frequency factor.

**17.46** The hydrolysis of sucrose, $C_{12}H_{22}O_{11} + H_2O \rightarrow 2C_6H_{12}O_6$, has a specific rate constant of $5.4 \times 10^{-5}$ $sec^{-1}$ at 280 K and a specific rate constant of $2.4 \times 10^{-4}$ $sec^{-1}$ at 300 K. Find the activation energy of the reaction from these data.

**17.47** The activation energy of the gas phase reaction $NO + Cl_2 \rightarrow NOCl + Cl$ is 84.9 kJ/mol and the specific rate constant at 500 K is 5.07 L $mol^{-1}$ $sec^{-1}$. Find the specific rate constant at 400 K.

**17.48** The specific rate constant of a reaction doubles when the temperature is increased from 400 K to 420 K. Find the activation energy of the reaction.

**17.49** Dutch paintings of the seventeenth century have pigments that are insoluble in virtually all solvents because they have become 99% polymerized after 300 yr at an average room temperature of 298 K. The twentieth-century art forger Meegeren achieved the same effect by baking his paintings at 373 K for 2 weeks. Assuming the polymerization process is first order, calculate the activation energy of this process.

**17.50★** The rate of the reaction $2NO(g) + O_2(g) \rightarrow 2NO_2(g)$ *decreases* with increasing temperature. Assuming that this reaction is an elementary process, what can you say about the activation energy of the reaction? Propose an explanation in terms of electronic spins for the unusual activation energy.

**17.51** Draw energy diagrams for the following types of reactions: (a) a one-step spontaneous process, (b) a two-step nonspontaneous process in which the second step is rate determining, (c) a three-step spontaneous process in which each step is faster than the preceding one.

**17.52** At 500 K the $K$ for the reaction $NO(g) + Cl_2(g) \rightleftharpoons NOCl(g) + Cl(g)$ is $1.2 \times 10^{-9}$. The specific rate constant for the reaction of NO and $Cl_2$ is $5.1 \times 10^{-3}$ L/mol sec. Calculate the specific rate constant for the reaction between NOCl and Cl.

**17.53** The formation of phosgene, $COCl_2$, from chlorine and carbon monoxide is believed to proceed by the mechanism:

1.    $Cl(g) + CO(g) \underset{k_{-1}}{\overset{k_1}{\rightleftharpoons}} COCl(g)$

2. $COCl(g) + Cl_2(g) \underset{k_{-2}}{\overset{k_2}{\rightleftharpoons}} COCl_2(g) + Cl(g)$

Show that this mechanism is consistent with the overall reaction $Cl_2(g) + CO(g) \rightleftharpoons COCl_2(g)$. Derive an expression for the $K$ of this reaction in terms of the specific rate constants.

**17.54\*** Special techniques have been developed for the measurement of specific rate constants of very fast reactions. With these techniques, the $k$ at 298 K of the reaction $H_3O^+(aq) + OH^-(aq) \rightarrow 2H_2O$ has been found to be $1.4 \times 10^{11}$ L mol$^{-1}$ sec$^{-1}$. Calculate the $k$ for the reverse reaction. (*Hint:* Use $[H_2O] = 55.5M$ to correct $K_W$.)

**17.55** A material that is structurally similar to a substrate in a reaction catalyzed by an enzyme may act as an enzyme inhibitor, interfering markedly with the functioning of the enzyme. Explain this observation in terms of the lock-and-key analogy.

**17.56** Suggest a reason why powdered charcoal rather than graphite is used in water and air purifiers.

**17.57** Propose a mechanism for the decomposition of nitrous oxide to $N_2$ and $O_2$ catalyzed by platinum by analogy with the mechanism for the formation of $NH_3$. Predict the extent to which nitrous oxide can form from $N_2$ and $O_2$ at elevated temperatures in the presence of a hot platinum surface.

**17.58** The chemisorption of $H_2(g)$ on a metal surface is found to liberate 50 kJ/mol. Estimate the strength of the metal-to-H bond.

# 18

# Coordination Chemistry

**Preview**  This chapter deals with the fascinating and important substances called coordination compounds. We begin by describing the structure of a coordination compound: a core, called a complex ion, consisting of a metal atom and a group of tightly held ligands; and other ions, loosely associated with the complex ion to preserve electrical neutrality. We discuss the different kinds of isomerism found in coordination compounds and then introduce the valence bond theory and the crystal field theory, which are used to describe the bonding in coordination compounds. Next, we define the distinction between lability of coordination compounds, whih   refers to their rate of reaction, and stability, which refers to their relative energy. After discussing equilibria involving complex ions in aqueous solution, we conclude by describing the role that coordination compounds play in industrial catalysis and in living organisms.

The study of coordination compounds is an active and exciting area of current chemical research. Coordination compounds have a broad range of interesting theoretical and practical features. In industry, they are used as catalysts in a number of processes. In biology, they play central roles in many life processes.

## 18.1   STRUCTURE OF COORDINATION COMPOUNDS

A coordination compound includes a metal atom surrounded by a set of other atoms or groups of atoms called **ligands.** To a large extent, the metal atom and its surrounding ligands act as a single chemical entity.

The single entity represented by the metal atom and its coordinated ligands is called a *coordination complex.* It may be either neutral or electrically charged. We usually can determine the charge or oxidation number of the metal atom by studying the nature of the ligands and the net charge on the complex.

For example, we can see that in the complex ion $[Co(NH_3)_6]^{3+}$, cobalt is in the $+3$ oxidation state. (By convention, the complex is placed within brackets to emphasize that it acts as a single chemical unit.) The ligands are $NH_3$ molecules, which are electrically neutral. Therefore, the charge of the complex gives the oxidation state of the metal atom. In $[Fe(CN)_6]^{4-}$, each of the six CN ligands has a charge of $-1$ and the complex has a net charge of $-4$. Therefore, the iron atom is in the $+2$ oxidation state. In the complex $[Fe(CN)_6]^{3-}$, the net charge of $-3$ shows that the oxidation state of the iron atom is $+3$.

One puzzling feature of coordination compounds to early investigators was the association of the metal atom with more ligands than the number corresponding to what they called the "valence" of the metal. For example, it was believed that iron in the $+2$ oxidation state should have a valence of only two, as it does in a compound such as $FeCl_2$ or $Fe(OH)_2$. Therefore, the structure of a species such as $[Fe(CN)_6]^{4-}$ was a mystery.

The great historical figure in coordination chemistry is Alfred Werner (1866–1919), who was awarded the Nobel Prize in 1913. Werner provided much of the framework for understanding the structures and behavior of coordination compounds by proposing that the metal displays two kinds of valences.

We can understand in modern terms the difference between his two kinds of valence by considering a coordination compound that includes the complex ion $[Co(NH_3)_6]^{3+}$. To form a neutral compound, this cation must be combined with anions of total charge $-3$, possibly three $Cl^-$ ions. The formula of such a compound is written $[Co(NH_3)_6]Cl_3$. The structure of this compound is represented in Figure 18.1.

The six $NH_3$ ligands are closely associated with the metal and are said to be in the *coordination sphere* of the metal. The three $Cl^-$ anions are outside the coordination sphere and can be regarded as less important. The "coordination valence" of the metal is 6. This valence is satisfied by

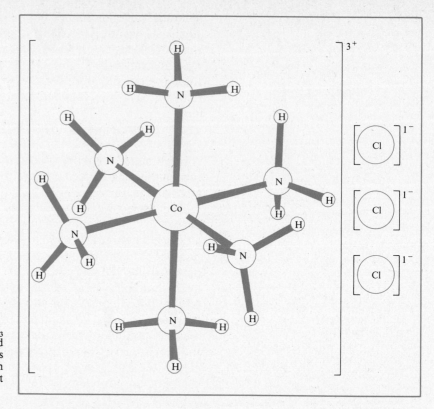

**Figure 18.1**

The coordination compound $[Co(NH_3)_6]Cl_3$ consists of the complex ion $[Co(NH_3)_6]^{3+}$ and three $Cl^-$ anions. The three $Cl^-$ ions are not as closely associated with the central cobalt atom as are the six $NH_3$ ligands. The figure does not show the actual arrangement of the $Cl^-$ ions.

the six closely associated $NH_3$ ligands. In modern terms, the coordination valence of the metal is its *coordination number*. The *"ionic valence"* of the cobalt in this compound is its oxidation state; it is $+3$. This valence is satisfied by the three $Cl^-$ anions that are necessary for electrical neutrality.

The identity of the anions that satisfy the ionic valence of the metal and bring about electrical neutrality is usually not important. Only their charge matters. These anions are called *counter ions* when they are not in the coordination sphere of the metal.

The difference between the ligands and the counter ions often becomes apparent when the coordination compound is dissolved in water. The ligands tend to remain closely associated with the metal, while the counter ions dissociate. Thus, when $[Co(NH_3)_6]Cl_3$ is dissolved in water and a solution of silver nitrate is added, all the chloride is precipitated as AgCl. But when $[Co(NH_3)_5Cl]Cl_2$ is dissolved in water, only two-thirds of the chloride is precipitated as AgCl. When $[Co(NH_3)_4Cl_2]Cl$ is dissolved in water only one-third of the chloride is precipitated as AgCl. The chlorides that reside in the coordination sphere of the metal do not dissociate in water.

In many complexes, some or all of the ligands are anions. These anionic ligands also satisfy some or all of the ionic valence of the metal. Thus the complex ion $[Co(NH_3)_5Cl]^{2+}$ of cobalt in the $+3$ oxidation state requires only two $Cl^-$ counter ions for neutrality. The complex $[Co(NH_3)_3Cl_3]$ of cobalt in the $+3$ oxidation state is already neutral and requires no counter ions. Sometimes cationic counter ions neutralize a complex ion that has a net negative charge from its anionic ligands. An example is $K_2[PtCl_6]$, in which the platinum is in the $+4$ oxidation state. The complex ion bears a charge of $-2$, and two $K^+$ cations are the counter ions.

Our discussion of coordination chemistry will focus on the metal and the ligands in its coordination sphere. We shall be especially interested in the coordination number of the metal and how it affects the structure and behavior of the complex. We shall see that most metals tend to have one, or at most a few, characteristic coordination numbers.

All metals form coordination complexes. The most interesting—and often the most stable—complexes are those of the transition metals, which are found in groups IB and IIIB through VIIIB of the periodic table. A transition metal is defined as one that either has an incomplete $d$ electronic subshell or readily forms a cation with an incomplete $d$ electronic subshell. The incompletely filled $d$ orbitals can participate in the bonding of the ligands, allowing the formation of complexes that would not otherwise form. Transition metals can also achieve high oxidation states. The electrostatic attraction between highly charged metal cations and anionic ligands favors the formation of complexes of transition metals.

## Ligands

Many different molecules and ions can serve as ligands in coordination complexes. All are electron pair donors and therefore Lewis bases, while the metal cations are Lewis acids. Most ligands are anions or substances that contain nonbonding valence electron pairs. The common anionic ligands include the halide ions, $F^-$, $Cl^-$, $Br^-$, and $I^-$; conjugate bases of oxyacids, such as $SO_4^{2-}$, $NO_2^-$, $NO_3^-$, and $OH^-$; and a variety of other anions, such as $S^{2-}$ and $CN^-$. The common neutral ligands include $H_2O$ and other nonmetallic oxides in which oxygen, with its nonbonding valence electrons, is the electron donor. Nitrogen is an electron donor atom in many ligands; the simplest such ligand is $NH_3$. Some neutral ligands of biological importance are $O_2$, $N_2$, and CO.

Ligands that have more than one donor atom in a single molecule can form complexes of special interest. One such molecule, which has two donor atoms, is ethylenediamine, $NH_2CH_2CH_2NH_2$. Each of the two nitrogen atoms has a nonbonding valence electron pair, so each can coordinate with a metal. When both of these nitrogen atoms coordinate with the same metal atom, as shown in Figure 18.2, a structure called a *chelate* ring (from the Greek word for "claw") is formed. A ligand of this kind is called a *bidentate* (two-toothed) *ligand* or a *chelating agent*. In chemical formulas and equations, the abbreviation "en" is used to represent the ethylenediamine ligand. Thus, the complex ion with the structure

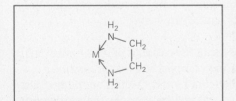

**Figure 18.2**

A chelate ring consisting of a metal atom and the bidentate ligand ethylenediamine. Each of the two nitrogen atoms in the ligand coordinates with the metal atom, forming a ring structure.

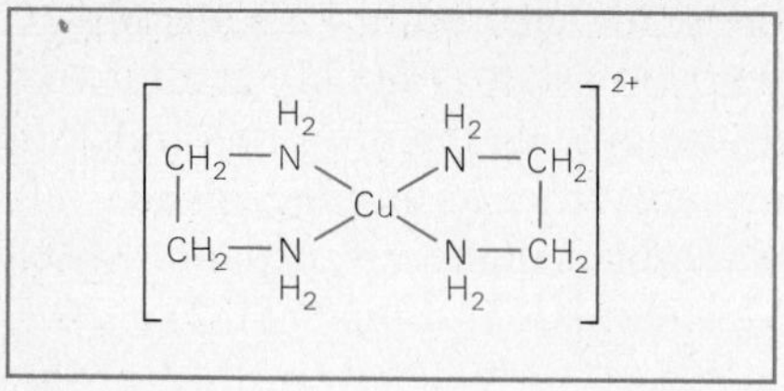

**Figure 18.3**
The complex ion $[Cu(en)_2]^{2+}$, consisting of a central copper(II) ion and two ethylenediamine ligands that form two chelate rings.

shown in Figure 18.3 is written as $[Cu(en)_2]^{2+}$.

Complex ions with chelate rings often are unusually stable, as we can see from the large equilibrium constant for the reaction

$$[Ni(NH_3)_6]^{2+} + 3en \rightleftharpoons [Ni(en)_3]^{2+} + 6NH_3$$

The complex ion on the right side of the equation has three chelate rings, while the complex ion on the left side has none. The equilibrium constant for the reaction is about $10^{10}$, meaning that the equilibrium lies far on the side of the complex with the chelate rings. Both complex ions have six nitrogen atoms coordinated on a nickel atom, but the complex ion with the chelate rings is much more stable than the one with the $NH_3$ ligands. This enhanced stability is known as the *chelate effect*. It can be explained on the basis of entropy. The $\Delta H$ for the equilibrium is essentially 0, but $\Delta S$ is positive since seven particles form from four particles.

There are ligands that contain three to six donor atoms. Some quadridentate ligands—that is, ligands with four donor atoms in a single molecule—play essential roles in the chemistry of living organisms. Each of these ligands includes a large ring of atoms. A molecule containing such a ring is called a macrocycle. Some organic macrocycles called porphyrins have rings that include four nitrogen atoms, which can all act as donors toward a single metal atom. Complexes with these macrocyclic ligands tend to be even more stable than complexes with simple chelate rings. Many important proteins include complexes between porphyrins and iron. One such coordination complex is heme. Heme-containing proteins include hemoglobin and myoglobin, which participate in biological oxygen transfer; cytochromes, which participate in biological energy transport; and enzymes such as catalase. We shall discuss biological complexes further in Section 18.6.

## Nomenclature

The name of a coordination compound must specify the metal, the ligands, and the counter ions in the compound. But that is not all. The name must also tell us the oxidation state of the metal, which are the ligands, and which are the counter ions. Chemists name coordination compounds by the following set of agreed-upon rules:

1. If the coordination compound is a salt, the cation is named first, even if it is the counter ion rather than the complex ion (example a).
2. Ligands are named first, in alphabetical order. Metals are named last (examples b and c). The suffix *-ate* is used for complex anions (examples a and d). No suffix is used for a complex cation or a neutral compound (examples b and c). The suffix *-ate* sometimes is added to a prefix derived from the Latin name (Table 2.2) of the metal (example e).

3. When an anion is a ligand, the final *e* in its name is replaced by *o*, as in *nitrato, acetato,* and *sulfido*. A number of common anions do not follow this rule. They include the ligands fluoro ($F^-$), chloro ($Cl^-$), bromo ($Br^-$), iodo ($I^-$), and cyano ($CN^-$), in which the final *ide* is replaced by *o*. Neutral ligands usually have the same name as the molecule. Some exceptions to this rule are aquo ($H_2O$), ammine ($NH_3$), carbonyl (CO), and nitrosyl (NO).

4. A Greek prefix (Section 2.3) is used to indicate the number of a specific ligand present in a compound. If the name of the ligand is complicated or if it already has a multiplying prefix, it is enclosed in parentheses in the name of the compound (example f). The prefixes bis, tris, tetrakis, and so on are used in this case.

5. The oxidation state of the metal is indicated by a Roman numeral in parentheses after the name of the metal. Some examples that illustrate these rules are

**a.** $Na_2[PtCl_4]$       sodium tetrachloroplatinate(II)
**b.** $[Pt(NH_3)_2Cl_2]$       diamminedichloroplatinum(II)
**c.** $[Pt(NH_3)_3Cl]Cl$       triamminechloroplatinum(II) chloride
**d.** $(NH_4)_2[PtCl_6]$       ammonium hexachloroplatinate(IV)
**e.** $K_3[Fe(CN)_6]$       potassium hexacyanoferrate(III)
**f.** $[Ni(en)_3]Br_2$       tris(ethylenediamine)nickel(II) bromide
**g.** $[Cr(H_2O)_4Cl_2]Cl$       tetraaquodichlorochromium(III) chloride

There are other rules that enable us to name more complicated coordination compounds and to give information about their three-dimensional structure. These rules are beyond our scope.

### Geometry

The ligands in the coordination sphere around a metal atom are held rather rigidly in place. Because of this fixed orientation, we can describe the geometry of coordination complexes in the same way as we described the geometry of ordinary covalent compounds of nonmetals (Chapter 8). We specify points that correspond to the centers of the donor atoms that are complexed with the metal atom.

The geometry of metal complexes is related to the coordination number of the metal. There are ideal geometries associated with specific coordination numbers, and many complexes have geometries that are reasonably close to these ideals. However, the exact geometry is influenced by such factors as the nature of the metal and the ligand, and we find more than one geometry for some coordination numbers.

Complexes with metals whose coordination number is 2 are relatively uncommon. They are formed by the metal ions $Cu^+$, $Ag^+$, $Au^+$, and $Hg^{2+}$, each of which has a complete *d* electronic shell. The valence electronic configuration for $Cu^+$ is $3d^{10}$; for $Ag^+$, $4d^{10}$; for $Au^+$, $5d^{10}$; and for $Hg^{2+}$,

**Figure 18.4**
The complex ion $[Ag(NH_3)_2]^+$. The coordination number of the silver ion is 2, and the complex is linear, with the center of the metal ion and the centers of the two donor atoms of the ligands lying on a straight line. Linear geometry is typical of complexes of metals whose coordination number is 2.

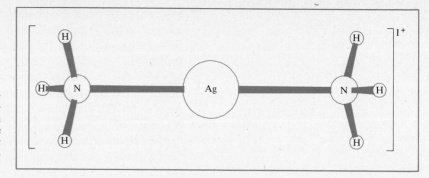

$5d^{10}$. The complexes formed by these ions typically are linear, as shown in Figure 18.4. Some typical complexes are $[CuCl_2]^-$, $[Ag(NH_3)_2]^+$, $[Au(CN)_2]^-$, and $HgCl_2$.

Complexes with a metal whose coordination number is 4 are more common. Two ideal geometries are possible when four ligands lie around a metal atom. One is the familiar tetrahedral geometry, encountered in covalent compounds of the nonmetals (Figure 8.6). The points at the centers of the donor atoms of the four ligands lie at the corners of a regular tetrahedron. The center of the metal atom is at the center of the tetrahedron. Figure 18.5 shows the tetrahedral geometry of $[Zn(NH_3)_4]^{2+}$.

Some complexes of metals with coordination number 4 have square planar geometry. The points at the centers of the four donor atoms lie in a plane at the corners of a square. The center of the metal atom is in the same plane and at the center of the square. Figure 18.6 shows $[Pt(NH_3)_4]^{2+}$, which has square planar geometry.

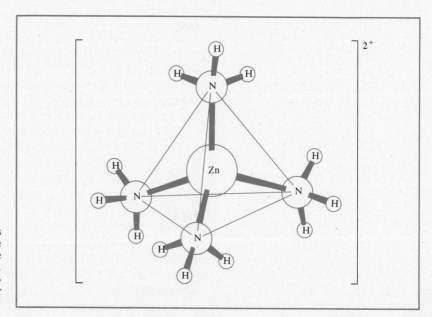

**Figure 18.5**
The coordination complex $[Zn(NH_3)_4]^{2+}$ has tetrahedral geometry. The four $NH_3$ ligands lie at the corners of a regular tetrahedron, with the metal atom at the center of the tetrahedron. Tetrahedral geometry is found in many complex ions of metals whose coordination number is 4.

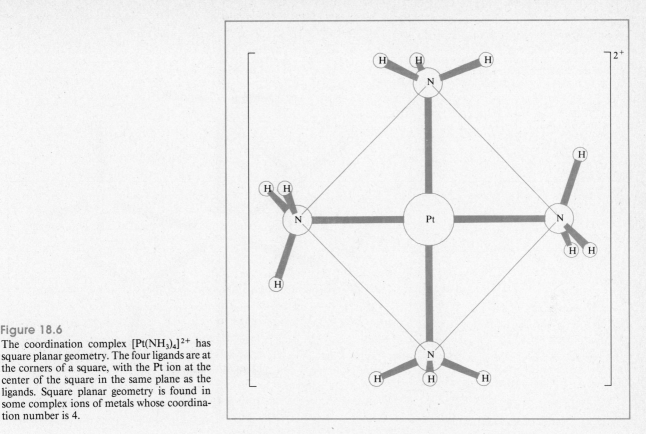

**Figure 18.6**
The coordination complex $[Pt(NH_3)_4]^{2+}$ has square planar geometry. The four ligands are at the corners of a square, with the Pt ion at the center of the square in the same plane as the ligands. Square planar geometry is found in some complex ions of metals whose coordination number is 4.

Four-coordinate complexes of nontransition metals, such as $[AlCl_4]^-$, $[Zn(CN)_4]^{2-}$, and $SnBr_4$, almost always display tetrahedral geometry, sometimes in a distorted form. Tetrahedral geometry minimizes repulsions between ligands, so this geometry is especially favored when the ligands are relatively large and the metal is relatively small, as in $[MnO_4]^-$, $VCl_4$, $[FeCl_4]^-$, and $[NiBr_4]^{2-}$. But in some four-coordinate complexes of transition metals with relatively small ligands, square planar geometry is favored. The size of the ligands and the metal as well as the valence electronic configuration influence the geometry. Square planar geometry is most common for metals whose valence electronic configuration is $d^8$. Thus, complexes of $Ni^{2+}$, whose valence electronic configuration is $3d^8$, display square planar geometry, but only when the ligands are small. Square planar geometry is common for complexes of the larger metals, such as $Pd^{2+}$ (valence electronic configuration $4d^8$), $Pt^{2+}$ ($5d^8$), and $Au^{3+}$ ($5d^8$).

The coordination number 6 is by far the most common for transition metal complexes. Some transition metal ions appear to form only six-coordinate complexes. These ions include $Cr^{3+}$, $Co^{3+}$, and $Pt^{4+}$. With rare exceptions, six-coordinate complexes have octahedral geometry. The centers of the six donor atoms lie at the corners of an octahedron, either

**Figure 18.7**
The complex ion $[Co(NH_3)_6]^{3+}$ has octahedral geometry. The six $NH_3$ ligands lie at the corners of a regular octahedron, with the $Co^{3+}$ ion at the center of the octahedron. Octahedral geometry is found in complex ions of metals whose coordination number is 6.

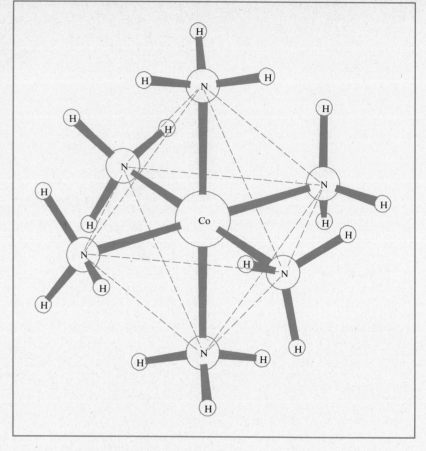

regular or asymmetrical (Figure 8.8). The center of the metal ion is at the center of the octahedron. Figure 18.7 shows the structure of $[Co(NH_3)_6]^{3+}$, which has regular octahedral geometry.

The six corners of an octahedron lie at the ends of three mutually perpendicular lines of equal length, as Figure 18.8 shows. The six ligands are at the six corners of the octahedron. All six positions are equivalent. The three mutually perpendicular lines of Figure 18.8 are familiar to you as the $x$, $y$, and $z$ coordinate axes in three dimensions.

There are coordination complexes with other coordination numbers, including 3, 5, and 7 through 12. These complexes are less often encountered, so we shall not discuss them.

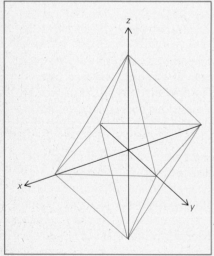

**Figure 18.8**
In a coordination complex with octahedral geometry, the six ligands lie at the ends of three lines that are of equal length and mutually perpendicular. The center of the metal ion is the point where the lines intersect.

## 18.2  ISOMERISM IN COORDINATION COMPOUNDS

Two chemical compounds that have the same molecular formula but different structures are called **isomers.** We find several kinds of isomerism in coordination compounds. For example, isomerism occurs when one of the ligands can be a counter ion and one of the counter ions can be a ligand. The molecular formula of the coordination compound does not change, but its structure does, in this *ionization isomerism*. The coordination compounds $[Co(NH_3)_5Cl]SO_4$ and $[Co(NH_3)_5SO_4]Cl$ are ionization isomers. Both of these compounds are complexes of cobalt in the $+3$ oxidation state. In the first one, the $Cl^-$ is the ligand and the $SO_4^{2-}$ is the counter ion. In the second, the $SO_4^{2-}$ is the ligand and the $Cl^-$ is the counter ion.

Closely related to ionization isomerism is *hydrate isomerism,* which involves the water of hydration that is often included in ionic crystals (Section 11.3). Hydrate isomers have the same total number of water molecules, but these molecules are bonded differently in the isomers. Three hydrate isomers are $[Cr(H_2O)_6]Cl_3$, which is gray-blue; $[Cr(H_2O)_5Cl]Cl_2 \cdot H_2O$, which is light green; and $[Cr(H_2O)_4Cl_2]Cl \cdot 2H_2O$, which is dark green.

We find a different kind of isomerism in *ambidentate ligands*. An ambidentate ligand can complex with a metal atom in two or more ways. One well-known ambidentate ligand is $NO_2^-$. It can form nitrito complexes, in which one of the oxygen atoms is the donor, or nitro complexes, in which the nitrogen atom is the donor. Two compounds that differ only in the way that the $NO_2^-$ ligand is attached are $[(NH_3)_5CoONO]Cl_2$, which is the red nitrito complex, and $[(NH_3)_5CoNO_2]Cl_2$, which is the yellow nitro complex. These compounds, and others that have the same kind of isomerism, are called *linkage isomers.*

Cyanide, $C{\equiv}N^-$, is another ambidentate ligand that forms linkage isomers. The carbon atom usually coordinates with the metal to form cyano complexes, but there are isocyano complexes in which the nitrogen atom of the cyanide coordinates with the metal. The thiocyanate anion, $SCN^-$, also gives rise to linkage isomers. Either the sulfur atom or the nitrogen atom can be the donor atom.

### Stereoisomerism

An especially interesting kind of isomerism, found in many coordination compounds, is called **stereoisomerism.** Two compounds are said to be stereoisomers when they differ only in the arrangement of their atoms in space. Two stereoisomers have the same molecular formula and the same bonding and yet are different.

The most important type of stereoisomerism found in coordination complexes is called *geometric isomerism*. Geometric isomers have the same overall geometry, composition, and bonding. They differ only in the spatial arrangement of the ligands around the metal ion.

Geometric isomerism can occur in four-coordinate complexes with square planar geometry. The complex $[Pt(NH_3)_2Cl_2]$ exists as two geometric isomers. Each isomer has a central platinum ion surrounded by the same four ligands, which lie at the corners of a square. Yet the isomers are two different compounds with different properties. To see how this differ-

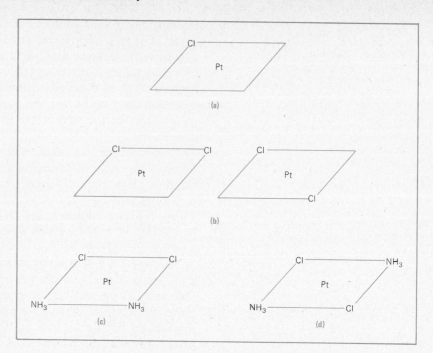

**Figure 18.9**

The geometric isomers of the complex [Pt(NH$_3$)$_2$Cl$_2$], which has square planar geometry. With a chloride ligand in one corner of the square, the second chloride ligand can occupy either the opposite corner or an adjacent corner. Because the two positions are not equivalent, the complex has two isomers whose formulas are the same but whose structures are different.

ence can come about, let us consider how the two pairs of ligands can be distributed on the corners of a square.

Suppose we put a chloride ligand at one corner of an empty square, as shown in Figure 18.9(a). The second chloride can go either on the corner on the same edge of the square or on the diagonally opposite corner (Figure 18.9(b)). Because these two positions are not equivalent, two arrangements are possible (Figures 18.9(c) and 18.9(d)). Each arrangement corresponds to a different geometric isomer. The isomer with the two like ligands on the same edge of the square (Figure 18.9(c)) is called the *cis* isomer; the isomer with the two like ligands on opposite corners of the square (Figure 18.9(d)) is called the *trans* isomer. These isomers are different substances with different properties.

Geometric isomerism is possible only when no more than two positions of the square are occupied by identical ligands. If there are three identical ligands, as in the complex [Pt(NH$_3$)$_3$Cl]$^+$, all the arrangements are equivalent and geometric isomerism is impossible. Geometric isomerism cannot occur in a four-coordinate complex with tetrahedral geometry. After the first ligand is put at a corner of the tetrahedron, all three remaining positions are equivalent.

Geometric isomerism can occur in six-coordinate octahedral geometry. Consider the ion [Pt(NH$_3$)$_4$Cl$_2$]$^{2+}$, a complex of platinum in the +4 state. We first put a chloride ligand at one corner of the octahedron (Figure 18.10(a)). The second chloride ligand can be placed on the same axis as the first chloride ligand, which puts it at the opposite corner of the

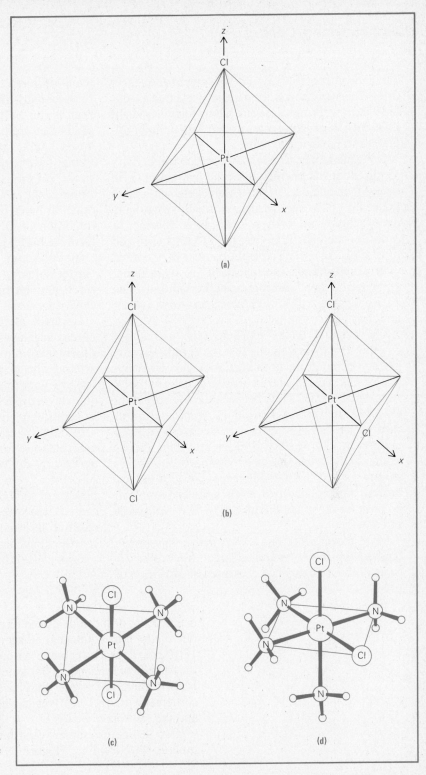

**Figure 18.10**
The geometric isomers of the complex $[Pt(NH_3)_4Cl_2]^{2+}$, which has octahedral geometry. With a chloride ligand at one corner of the octahedron, the second chloride ligand can occupy either the opposite end of the same axis, creating a *trans* isomer, or one of the other two axes, creating a *cis* isomer. The remaining positions are then occupied by $NH_3$ ligands to give the two isomers: *trans* (c) and *cis* (d).

## Platinum and Cancer

The *cis* isomer of $[Pt(NH_3)_2Cl_2]$ is a potent anticancer drug. The *trans* isomer is not. Many research groups have been working to explain why a relatively slight difference in structure is associated with a major difference in biological activity.

It is generally believed that the anticancer activity results from the interaction of the compound with deoxyribonucleic acid (DNA), the genetic material of the cell. Cancer cells divide continually, eventually overwhelming the body by their sheer volume. In order for the cells to divide, the DNA of each cell must replicate itself. The DNA consists of two long, intertwined strands. In replication, the strands first separate and then form complementary strands. The *cis* isomer stops DNA synthesis; the *trans* isomer does not.

The original theory explained the activity of *cis*-$[Pt(NH_3)_2Cl_2]$ by saying that the complex is carried in the bloodstream to the tumor site, where hydrolysis takes place. The theory said that the hydrolysis product formed a link between the two strands of the DNA, preventing replication. According to this theory, the *trans* isomer was ineffective because it underwent hydrolysis at a faster rate than the *cis* isomer and so never reached the cancer. But further studies showed that the differences between the hydrolysis rates of the two isomers are too small to account for the major differences in their anticancer activity.

More recent work has shown that both isomers bind to the DNA, but in different ways. The *cis* isomer binds in a way that causes severe distortion of the DNA molecule while the *trans* isomer does not. Both isomers bind to DNA because their chloride ligands can be replaced by nitrogen-containing ligands that are part of the DNA molecule. It is believed that the inhibition of DNA synthesis by the *cis* isomer results from replacement of both chloride ligands by ligands that are close to each other in one strand of the DNA molecule. For unknown reasons, only this mode of binding seems to result in marked inhibition of DNA synthesis. The *cis* isomer can form such bonds because the distance between its chloride ligands is about the same as the distance between binding sites on the DNA strand. The chloride ligands on the *trans* isomer are further apart, so it cannot form such a bifunctional linkage.

Cisplatin, as the drug is named, has proved to be effective against several forms of cancer and now is widely used in treatment despite several potentially limiting side effects. The drug can cause severe nausea and vomiting in some patients, but the use of antiemetic drugs has eased that problem somewhat. In addition, platinum compounds tend to cause severe kidney damage because they are reduced to metallic platinum in the kidney. It has been found that the damage can be reduced and the dosage increased substantially by giving a diuretic which speeds the passage of the compound through the kidney. Because of the success of this drug, a search for other effective anticancer compounds is going on among coordination complexes of other group VIIIB metals such as palladium, ruthenium, and rhodium.

octahedron; or it can be placed on one of the other axes, which puts it at one of the four equivalent adjacent corners of the octahedron (Figure 18.10(b)). Thus, two geometric isomers are possible. The isomer that has the two chloride ligands at opposite corners (Figure 18.10(c)) is called the *trans* isomer. The isomer that has the two chloride ligands at adjacent corners (Figure 18.10(d)) is called the *cis* isomer. The *trans* isomer is green and the *cis* isomer is pink.

Two geometric isomers exist for the octahedral complex whose formula is $[Pt(NH_3)_3Cl_3]^+$ (Figure 18.11). One isomer has one chloride li-

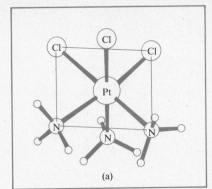
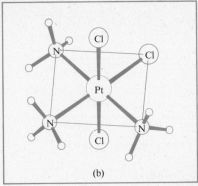

**Figure 18.11**
The geometric isomers of the complex $[Pt(NH_3)_3Cl_3]^+$, which has octahedral geometry. The isomer in (a) has one chloride ligand on each of its three axes. The isomer in (b) has two chloride ligands on one axis, with the third on another axis.

gand on each of its three axes. The other isomer has two chloride ligands on the same axis, with the third chloride ligand on a different axis. More than two geometric isomers can exist when a complex has more than two different ligands. However, geometric isomerism in a six-coordinate octahedral complex is possible only when there are no more than four identical ligands.

### Optical Isomerism and Chirality

*Optical isomerism* is another kind of stereoisomerism that is also of great interest in organic chemistry and biochemistry.

Two optical isomers differ only in a property called *chirality* or *handedness.* There are many chiral objects around us. One of them is the human hand. You can see that your hands are not identical any time you try to put your right hand into a left-hand glove. If you hold your right hand before a mirror, the reflection you see appears to be a left hand. This is true of any chiral object: the object itself and its mirror image are different. One is left-handed, the other is right-handed, and they cannot be superimposed. Wood screws are also chiral objects. A screw may have a thread that goes to the left or to the right. The screwdriver is turned clockwise or counterclockwise, depending on the chirality of the screw.

Chirality occurs on the molecular level. We can have two isomers that are mirror images of each other, a left-handed molecule and a right-handed molecule. These are called optical isomers. Most simple molecules do not display optical isomerism, but many complex molecules exist as optical isomers.

While two optical isomers generally have virtually identical physical and chemical properties, differences manifest themselves when the isomers interact with something that is also chiral. The usual method for distinguishing between two optical isomers uses plane-polarized light, which is rotated differently by two optical isomers.

The most direct way to determine whether a specific coordination complex is chiral is to draw a picture of the complex and of its mirror

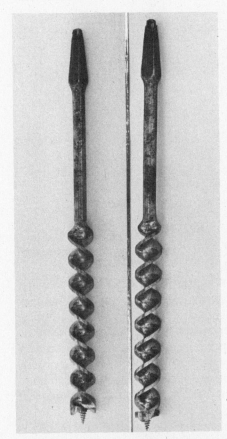
A drilling bit and its mirror image. The bit is a chiral object, as we can see by comparing it with the mirror image. The bit has a right-hand spiral, while the image has a left-hand spiral, and the two cannot be superimposed. *(Beckwith Studios)*

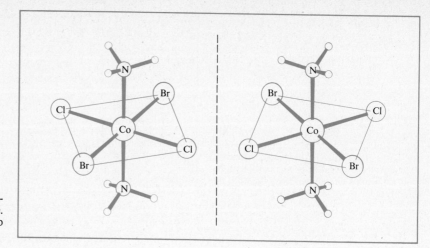

**Figure 18.12**
One geometric isomer of the hypothetical complex $[Co(NH_3)_2Cl_2Br_2]^-$ and its mirror image. The two complexes can be superimposed, so they are identical.

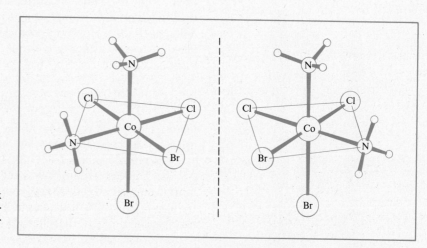

**Figure 18.13**
Another geometric isomer of the complex $[Co(NH_3)_2Cl_2Br_2]^-$ and its mirror image. Because the two complexes cannot be superimposed, they are optical isomers

reflection and compare the two. In Figure 18.12, we see such drawings for one of the geometrical isomers of a hypothetical complex $[Co(NH_3)_2Cl_2Br_2]^-$. The complex and its mirror image are superimposable, so this complex does not display optical isomerism. Figure 18.13 shows a drawing of another geometric isomer of this complex and its mirror image. The two complexes are not superimposable. Therefore, they are optical isomers. Any time we recognize that a coordination complex has chirality, we know that there are two optical isomers of that complex.

A second way to determine whether a specific coordination complex is chiral is based on the relationship between chirality and symmetry. Most symmetrical objects do not display optical isomerism. Asymmetrical objects can display optical isomerism. Often, it is easier to determine whether a molecule is symmetrical than to draw its mirror image and test for superimposability. Some simple rules can be stated about the symmetry of complex ions of monodentate ligands. Linear complexes and square planar complexes never display optical activity. A tetrahedral complex displays optical isomerism only if it has four different ligands. Octahedral complexes of monodentate ligands are capable of optical isomerism only when they have at least three different ligands, and where ligands of the same type are mutually *cis*.

## 18.3 BONDING IN COORDINATION COMPOUNDS

The description of the bonding of nonmetal compounds in Section 7.4 does not allow us to understand the geometry, the colors, and the magnetic properties of coordination compounds. Several different methods of describing the bonding in coordination compounds have been developed.

We start with the observation that both the electrons in the metal-ligand bond come from the ligand. This kind of bond was discussed in Section 7.4. It is called a *dative bond* or a *coordinate covalent bond*. The latter name comes from the existence of such bonds in coordination compounds.

In the traditional Lewis formulation, a coordinate covalent bond implies the existence of formal charges. In each of these bonds, there is a formal charge of $+1$ on the donor atom and a formal charge of $-1$ on the acceptor atom. In a complex, the metal ion is the acceptor. The complex $[Co(NH_3)_6]^{3+}$, for example, has six ligand-metal bonds. There are six formal positive charges, one on each ligand, and therefore six formal negative charges on the metal ion. Since the metal has an ionic charge of $+3$, the formal charge of $-6$ results in a net charge of $-3$ on the metal. However, such a net negative charge is contrary to chemical common sense, since metals rarely bear negative charges.

The metal atom in this complex does not bear a negative charge because the coordinate covalent bond has ionic character. The donor atom of the ligand usually is highly electronegative, so the bonding electron pair is more closely associated with the ligand. These bonds are partially negative on the donor atom and partially positive on the metal atom. The polarity of these bonds is opposite to the polarity of the formal charges. The ionic bonding thus helps offset the formal charges.

### Valence Bond Theory

A more detailed description of the bonding in coordination compounds, called the *valence bond* (VB) theory, has been developed by Linus Pauling (1901 –    ). While it has several shortcomings, the VB theory does give valuable insights into the bonding in coordination compounds.

The theory starts by formulating hybrid orbitals for the metal atom. It assumes that the number of hybrid orbitals formed by the metal atom is the same as the coordination number of the metal. These hybrid orbitals are used to form $\sigma$ bonds with the ligands. Each hybrid orbital overlaps with an orbital of a ligand and holds the electron pair donated by the ligand, forming a coordinate covalent bond. The valence electrons of the metal atom are distributed among the remaining unhybridized orbitals of the metal in a way consistent with the Aufbau principle.

The hybrid orbitals used to describe the bonding in a specific complex are selected primarily on the basis of the observed geometry and magnetic properties of the complex. Each of the ideal geometries observed in complexes corresponds to a type of hybridization. A six-coordinate octahedral geometry requires six symmetrically placed hybrid orbitals that can be formed from $s$, $p$, and $d$ orbitals. Octahedral geometry can be accommodated by a hybridization of $d^2sp^3$. The six component atomic orbitals — one $s$ orbital, three $p$ orbitals, and two $d$ orbitals — overlap to form six equivalent hybrid orbitals.

Tetrahedral geometry is consistent with four $sp^3$ hybrid orbitals, as we mentioned in our discussion of nonmetal compounds. The hybridization for square planar geometry calls for four $dsp^2$ orbitals, formed from one $s$ orbital, two $p$ orbitals, and one $d$ orbital. The two kinds of four-coordinate geometry correspond to different types of hybridization. Therefore, we usually cannot formulate a valence bond picture of a four-coordinate transition metal complex without knowing its geometry. This difficulty is a weakness of the valence bond theory.

Valence bond theory has other serious weaknesses. For example, the colors and many of the chemical properties of complexes are difficult or impossible to explain with valence bond theory. While the valence bond theory was the best that was available for many years, it has recently been replaced by several new theories. In order of increasing complexity, these are the crystal field theory, the ligand field theory, and the molecular orbital theory. We shall discuss only the crystal field theory.

## Crystal Field Theory

Crystal field theory differs from valence bond theory in its treatment of the ligand-metal bond. Valence bond theory emphasizes the covalent character of the bond. It acknowledges that the bond is partially ionic but does not deal directly with the ionic character. The crystal field theory takes just the opposite approach. It treats the bond as completely ionic, omitting consideration of the bond's covalent character.

Crystal field theory assumes that the only interaction between the metal atom and the ligand is electrostatic. It treats each ligand as a point of negative charge. The arrangement of the ligands that minimizes the repulsions between these negative point charges can be calculated by the equations of classical electrostatics. This concept is essentially the same approach as VSEPR, which we used for compounds of nonmetals in Section 8.2. Octahedral geometry is expected for a six-coordinate complex, tetrahedral geometry for a four-coordinate complex, and linear geometry for a two-coordinate complex.

Crystal field theory also considers the effect of the ligands on the relative energy of the $d$ orbitals of the central metal atom. Since the six-coordinate octahedral geometry is the most common for transition metal complexes, we shall begin by discussing crystal field theory for this geometry.

## Octahedral Complexes

When an atom of a transition metal is put into a spherical negative electrical field, the total energy of the five $d$ orbitals of the atom increases. The increase is caused by electrostatic repulsions between the electrons in the $d$ orbitals and the negative field.

We get a slightly different result when we study the $d$ orbitals of a transition metal atom in a complex ion. Once again, the atom is in a negative electrical field, this one created by the negatively charged ligands. But in a complex with octahedral geometry, the individual $d$ orbitals are affected differently by the electric field of the ligands, as we can see in Figure 18.14. The six ligands are on the $x$, $y$, and $z$ axes. The $d$ orbitals can be oriented with respect to these axes. You can see that the $d_{z^2}$ and $d_{x^2-y^2}$ orbitals (Figure 18.14(a)) point directly at the ligands, while the $d_{xy}$, $d_{xz}$, and $d_{yz}$ orbitals (Figure 18.14(b)) have their lobes of maximum electron density between the ligands.

The repulsion between the ligands and the two $d$ orbitals that lie on the coordinate axes is greater than the repulsion between the ligands and the other three $d$ orbitals. Thus, the energy of the $d_{z^2}$ and $d_{x^2-y^2}$ orbitals is higher in the octahedral field than in a spherical electric field. But it can be shown that the total energy of the $d$ orbitals is the same in the octahedral field as in the spherical field. Therefore, the energy of the $d_{xy}$, $d_{xz}$, and $d_{yz}$ orbitals must be lower in the octahedral field than in the spherical field, as Figure 18.15 shows.

In an octahedral complex, therefore, the $d$ orbitals of the metal atom are divided into two groups. There are two orbitals with relatively high energy and three orbitals with relatively low energy. Because the octahe-

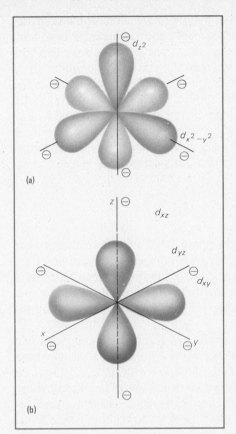

**Figure 18.14**
The effect of the octahedral field of a complex ion on the energies of the orbitals of the metal ion. (a) Two of the $d$ orbitals, the $d_{z^2}$ and the $d_{x^2-y^2}$, are on the $x$-, $y$-, and $z$-axes. The repulsion between these two orbitals and the ligands is relatively large, so the energy of these two orbitals is relatively high. (b) The $d_{xy}$ orbital is one of the three $d$ orbitals that lie between the axes. The repulsion between these three orbitals and the ligands is relatively small, so the energy of these three orbitals is relatively low.

**Figure 18.15**
When an isolated atom (a) is placed in a spherical electrical field, the energy of its orbitals increases equally (b). When the same atom is placed in the octahedral field of a complex ion, the energy increase is the same, but is distributed unequally among its orbitals (c). Three of the orbitals have lower energies than do the other two.

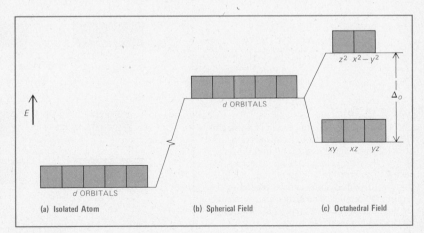

dron is symmetrical, the three low-energy orbitals all have the same energy, and the two high-energy orbitals have the same energy. The total increase in the energy of the $d_{z^2}$ and $d_{x^2-y^2}$ orbitals equals the total decrease in the energy of the $d_{xy}$, $d_{xz}$, and $d_{yz}$ orbitals. The difference in energy between the high-level and the low-level orbitals is called the crystal field splitting energy, whose symbol is $\Delta_o$.

The energy of each of the two high-energy orbitals is increased by $\frac{3}{5}\Delta_o$ above their energy in a spherical field, while the energy of each of the three low-energy orbitals is decreased by $\frac{2}{5}\Delta_o$ below their energy in a spherical field. The total increase is equal to the total decrease, so

$$(2)\left(\frac{3}{5}\Delta_o\right) = (3)\left(\frac{2}{5}\Delta_o\right)$$

The splitting of the $d$ orbitals predicted by the crystal field theory not only helps explain a number of properties of complexes but also simplifies the description of the electronic structure of many complexes. All that is necessary is to assign the valence electrons of the metal cation to the $d$ orbitals of the metal; the electron pairs of the ligands need not be considered.

The Aufbau principle can be applied to formulate some electronic configurations, as Figure 18.16 shows. For a complex of a cation whose valence electronic configuration is $d^1$, such as $Ti^{3+}$ ($3d^1$), $V^{4+}$ ($3d^1$), or $Mo^{5+}$ ($4d^1$), the single $d$ electron is placed in any one of the three low-energy $d$ orbitals. For a complex of a cation with valence electronic configuration $d^2$, such as $V^{3+}$ ($3d^2$), $Nb^{3+}$ ($4d^2$), or $W^{4+}$ ($5d^2$), the two $d$ electrons are placed in two of the low-energy orbitals; in accordance with Hund's

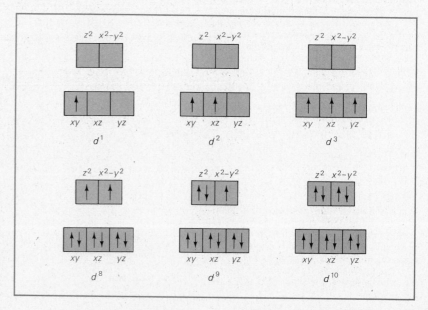

**Figure 18.16**
The Aufbau principle can be used in a straightforward way to determine electronic configurations for some metal ions in coordination complexes. In these complexes, $d$ electrons are placed in vacant low-energy orbitals. Six electrons with paired spins can be placed in the low-energy orbitals. The higher-energy orbitals are then filled. The boxes represent orbitals and the arrows represent electrons.

rules, their spins are parallel. Similarly, the three $d$ electrons in $d^3$ cations such as $V^{2+}$ $(3d^3)$ or $Cr^{3+}$ $(3d^3)$ are in the three low-energy orbitals and have parallel spins.

The Aufbau principle can also be applied to formulate the electronic structure of complexes of cations with electronic configuration $d^8$, such as $Ni^{2+}$ $(3d^8)$ or $Au^{3+}$ $(5d^8)$. As Figure 18.16 shows, the three low-energy orbitals are filled by three pairs of electrons. The remaining two electrons are in the two high-energy orbitals and have parallel spin. Figure 18.16 also shows the electronic configurations of complexes of $d^9$ cations such as $Cu^{2+}$ and $d^{10}$ cations such as $Zn^{2+}$. With the exception of atoms whose configuration is $d^{10}$, octahedral complexes of cations with any of the other electronic configurations shown in Figure 18.16 have one or more unpaired electrons and thus are expected to be paramagnetic. This prediction is confirmed by experimental observation.

The electronic structure of complexes of metal cations whose electronic configuration is $d^4$, $d^5$, $d^6$, or $d^7$ cannot be formulated solely by application of the Aufbau principle. More information is needed because $\Delta_o$, the energy difference between the high-energy and low-energy groups of orbitals, is relatively very small. Therefore, we often have a choice of orbitals in which an electron can be placed.

## High-Spin States and Low-Spin States

Let us examine the electronic configuration of a $d^4$ cation in an octahedral complex. The first three $d$ electrons are placed in the three low-energy orbitals, with parallel spins. According to the Aufbau principle, we should place the fourth electron in the unfilled orbital of lowest energy, in this case one of the three low-energy orbitals. This placement would create the electronic configuration, shown in Figure 18.17, which has only two

**Figure 18.17**

In the low-spin state of a metal ion, the low-energy orbitals must be filled before any electrons are placed in the high-energy orbitals. In the high-spin state, all the orbitals are given one electron each before additional electrons are placed in the low-energy orbitals.

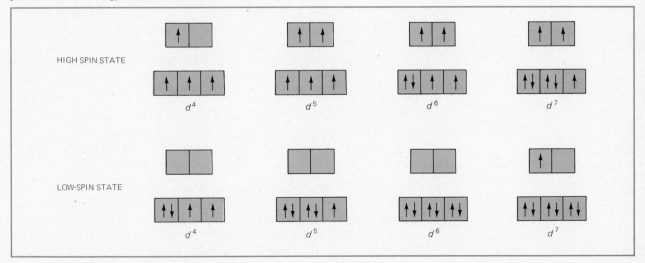

unpaired electrons and is called a *low-spin state.* But it is also possible to put the fourth electron in one of the two high-energy orbitals, as is also shown in Figure 18.17, keeping its spin parallel to that of the other three electrons. This arrangement, called the *high-spin state,* has four unpaired electrons.

As Figure 18.17 shows, we have the same choice in all the electronic configurations from $d^4$ to $d^7$. There is a low-spin state, in which the three low-energy orbitals are filled before any electrons are placed in the high-energy orbitals. And there is a high-spin state, in which the dominant effect is to have the maximum number of electrons with parallel spins. In the high-spin state, electrons are placed in high-energy orbitals as soon as the low-energy orbitals are half filled. The other low-energy orbitals are not filled until both the high-energy orbitals are half filled.

The electronic configuration of any given complex is the one that is most stable. For each spin state, relative stability is a balance of two opposing effects. The low-spin state has many of its electrons in low-energy orbitals, which is a stabilizing effect. But many of its electrons have paired spins, which is a destabilizing effect. The high-energy spin state is stabilized because more of its electrons have unpaired spins but is destabilized because one or two of its electrons are in high-energy orbitals. The relative stability of high-spin and low-spin states is determined primarily by the magnitude of $\Delta_o$. If $\Delta_o$ is relatively large, the low-spin state is more stable because the difference in energy between the high-energy and low-energy orbitals is large. If $\Delta_o$ is small, the energy difference between high-energy and low-energy orbitals is small, and the high-spin state is more stable.

For any given metal cation, the magnitude of $\Delta_o$ depends on the nature of the ligands. The crystal field theory is not very helpful in determining the nature of the ligands, since it treats them only as point charges. But the effect of different ligands on the magnitude of $\Delta_o$ can be determined experimentally. Ligands that cause a large $\Delta_o$ are called *strong field ligands,* while those that cause a small $\Delta_o$ are called *weak field ligands.* Some common ligands can be listed in descending order of $\Delta_o$. The following list is called a *spectrochemical series,* because it is based primarily on the study of the absorption of light by complexes.

strong field                                                              weak field
ligands; low spin                                              ligands; high spin

$$CO > CN^- > NO_2^- > en > NH_3 > H_2O > OH^- > F^-$$
$$> Cl^- > Br^- > I^-$$

Ligands at the left of the series are always strong field ligands, while those at the right are always weak field ligands. Those in the middle can be either strong field or weak field ligands, depending on the metal with which they are complexed. The order of ligands in the spectrochemical series cannot be predicted by electrostatic arguments alone. The listing is based on experiment and can be understood only if we also consider the details of the covalent bonding between the ligands and the metal.

### Magnetism

The behavior of a substance in an external magnetic field can tell us a great deal about its electronic configuration (Section 6.2). A substance is *diamagnetic* (very weakly repelled by an external magnetic field) when all its electrons have paired spins. A substance is *paramagnetic* (more strongly attracted by an external magnetic field) when it has one or more electrons with unpaired spins. All substances with an odd number of electrons are paramagnetic, but not all substances with an even number of electrons are diamagnetic.

Using the spectrochemical series, it is possible to understand, and in many cases even to predict, whether a coordination compound with an even number of electrons is diamagnetic or paramagnetic. Using the principle that strong field ligands cause low-spin states to be more stable and that weak field ligands cause high-spin states to be more stable, we can determine whether all the electronic spins are paired. For example, the low-spin state of $d^6$ ions such as $Co^{3+}$, $Fe^{2+}$, and $Pt^{4+}$ is diamagnetic. In each of these ions, all six $d$ electrons are paired in the three low-energy orbitals. The high-spin states of these same ions have four unpaired electrons and thus are paramagnetic. We expect the complexes of these ions with strong field ligands to be diamagnetic and complexes with weak field ligands to be paramagnetic. The experimental procedure shown in Figure 6.9 verifies this prediction.

The complexes of $Co^{3+}$ are cases in point. The complex ion $[Co(NH_3)_6]^{3+}$ is diamagnetic, and the complex ion $[CoF_6]^{3-}$ is paramagnetic. The spectrochemical series shows that $NH_3$ is a relatively strong field ligand which causes a large $\Delta_o$ and favors low-spin states. The $F^-$ ligand is shown to be a weak field ligand which favors the high-spin state. You can see that the low-spin state, in which all the electron spins are paired, is favored in the complex of $Co^{3+}$ with $NH_3$, while the high-spin state, in which four electrons have parallel spins, is favored in the complex of $Co^{3+}$ with $F^-$.

Similarly, the complex $[Fe(CN)_6]^{4-}$ is diamagnetic, because the strong field ligand $CN^-$ favors the low-spin state of the $Fe^{2+}$ ion, whose valence electronic configuration is $3d^6$. But the complex $[Fe(NH_3)_6]^{2+}$ is paramagnetic. In complexes of $Co^{3+}$, $NH_3$ is a strong field ligand. But in complexes of $Fe^{2+}$, the $NH_3$ ligand behaves as a weak field ligand and favors the high-spin state. Some ligands can be either strong field or weak field, depending on the metal cation with which they complex.

### Colors

The splitting of $d$ orbitals described in crystal field theory also helps account for some of the optical properties of coordination compounds, such as their color. Most pure chemical substances are colorless, which means that they do not absorb visible light. They usually absorb radiation in the ultraviolet region, from 100 nm to 400 nm. Most complexes of transition metals are exceptional. They are colored, which means that they absorb some wavelengths of electromagnetic radiation in the visible region, from 400 nm to 750 nm.

We discussed the way in which atoms or molecules absorb radiation in Section 5.7. Absorbed radiation excites an electron from its ground state orbital to a higher-energy orbital that is either vacant or half filled. The wavelength $\lambda$ of the absorbed radiation is inversely proportional to $\Delta E$, the difference in energy between the ground state orbital and the excited state orbital:

$$\lambda = \frac{hc}{\Delta E}$$

where $h$ is Planck's constant and $c$ is the speed of light.

Visible light has relatively long wavelengths, and therefore $\Delta E$ is small when visible light is absorbed. Thus, a molecule cannot absorb visible light unless it has a vacant or half-filled orbital whose energy is relatively close to that of the occupied orbitals. Most substances do not have such low-lying orbitals. But two low-lying orbitals are created in a complex when the ligands split the $d$ orbitals of the metal. The existence of these orbitals (the $d_{z^2}$ orbital and the $d_{x^2-y^2}$ orbital) allows the absorption of visible light. Therefore, the compound is colored.

Figure 18.18(a) shows the ground state electronic configuration of a $d^2$ ion. The absorption of light causes an electron to be promoted into a vacant higher-energy $d$ orbital, so that an excited state is formed (Figure 18.18(b)).

The energy difference between the high-energy and low-energy $d$ orbitals in complexes is $\Delta_o$, which can be substituted for $\Delta E$ in the relationship above to give:

$$\lambda = \frac{hc}{\Delta_o}$$

The value of $\Delta_o$ is such that the wavelength of absorbed radiation usually is in the visible range. The value of $\Delta_o$ can be found by measurement of the wavelength of the absorbed radiation. This is the method used to construct the spectrochemical series.

We can apply the crystal field theory to complexes with other coordination numbers and geometries, using the same basic approach as in the octahedral case. Valuable as it is, the crystal field theory has a weakness. It does not take into account the covalent interactions between the metal ion and the ligands. Both the ligand field theory and the molecular orbital theory include these interactions, as well as the differences in energy between the orbitals of the metal atom. These theories thus are both more satisfactory and considerably more complex than the crystal field theory.

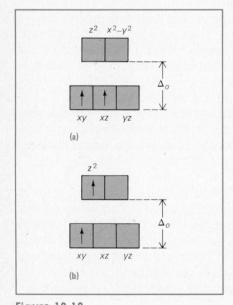

**Figure 18.18**
When a molecule absorbs visible light, one of its electrons moves from (a) a low-energy ground state orbital into (b) a higher-energy orbital. The difference between the energies of the two orbitals is $\Delta E$.

## 18.4   LABILITY AND STABILITY OF COORDINATION COMPLEXES

One of the most fundamental reactions of a coordination complex is the replacement of one ligand by another. A complex can be classified as either *labile* or *inert* on the basis of the rate at which its ligands can be replaced. If ligand substitution is essentially complete in one minute at room temperature when the concentration of reactants is about $0.1M$, a complex is classified as labile. If ligand substitution takes longer under these conditions, the complex is classified as inert.

Lability is a kinetic property. It refers to the rate at which a reaction takes place. Lability is not related directly to stability, which is a thermodynamic characteristic that refers to the relative energy of the complex.

A classic example of this distinction between lability and stability is the behavior of the complex $[Co(NH_3)_6]^{3+}$. When this complex ion is dissolved in aqueous acid, the equilibrium

$$[Co(NH_3)_6]^{3+}(aq) + 6H^+(aq) + 6H_2O$$
$$\rightleftharpoons [Co(H_2O)_6]^{3+}(aq) + 6NH_4^+(aq)$$

lies far to the right, because the six ammonia ligands are neutralized by the six protons. The value of the equilibrium constant $K$ is about $10^{25}$. But this reaction may take weeks or even months to occur when a solution of the complex ion with the $NH_3$ ligands is acidified. The $[Co(NH_3)_6]^{3+}$ ion is relatively unstable in acid solution. But it is also inert, so the reaction occurs very slowly. Complexes of $Cr^{3+}$, of $Co^{3+}$, and of $Pt^{2+}$ in low-spin states are also unstable but inert. Chemists were able to investigate the properties of these complexes in the early days of coordination chemistry because the complexes react slowly.

The three cyanide complexes $[Ni(CN)_4]^{2-}$, $[Mn(CN)_6]^{3-}$, and $[Cr(CN)_6]^{3-}$, all of which are extremely stable, also demonstrate the distinction between lability and stability. We cannot easily convert stable complexes to other complexes to measure their lability. Although a reaction such as

$$[Ni(CN)_4]^{2-}(aq) + 4H_2O \rightleftharpoons [Ni(H_2O)_4]^{2+}(aq) + 4CN^-(aq)$$

may have a fast rate, the position of equilibrium lies so far to the left that we cannot measure the rate, since no detectable amount of $[Ni(H_2O)_4]^{2+}$ forms. We need a method that enables us to measure ligand exchange without trying to convert a more stable complex to a less stable one.

One such method measures the rate at which $CN^-$ ligands that are complexed to the metal ion are replaced by other $CN^-$ ligands. Radioactive cyanide is used to make this measurement. Radioactive $^{14}CN^-$ is added to an aqueous solution of $[Ni(CN)_4]^{2-}$. The reaction is

$$[Ni(CN)_4]^{2-}(aq) + 4^{14}CN^-(aq) \rightleftharpoons [Ni(^{14}CN)_4]^{2-}(aq) + 4CN^-(aq)$$

The equilibrium is established in less than a minute; the complex, while stable, is also quite labile. The same reaction with $[Mn(CN)_6]^{3-}$ takes several hours to occur. For the $[Cr(CN)_6]^{3-}$ ion, which is both stable and extremely inert, the reaction is not completed for several months. We thus have three complexes, all of which are quite stable but which differ in lability.

To some extent, the lability of complexes can be correlated with size and electronic characteristics:

1. A small metal atom tends to form relatively inert complexes, because it holds the ligands rather tightly.
2. Complexes that have electrons in high-energy $d$ orbitals tend to hold ligands relatively weakly, and so are labile.
3. A complex with an empty low-energy $d$ orbital tends to be labile, because an incoming ligand can approach the metal ion along this empty orbital without encountering much electrostatic repulsion from the electrons of the metal.
4. Complexes that do not have unfilled low-energy $d$ orbitals or electrons in high-energy $d$ orbitals are inert. Thus, octahedral complexes of metal cations of electronic configuration $d^3$ and low-spin states of $d^4$, $d^5$, and $d^6$ ions are inert.

## 18.5  COMPLEX IONS IN AQUEOUS EQUILIBRIA

When we deal with aqueous systems that contain dissolved metal cations, especially those of the transition metals and the heavier metals in the center of the periodic table, we must consider equilibria in which complex ions in solution are formed. The most common neutral ligand in complex ions is the water molecule. You should assume that all metal cations are hydrated in aqueous solution. We shall not consider these water ligands in our discussion of aqueous equilibria.

Many complex ions form readily in systems containing a metal cation and appropriate ligands. Some measure of the relative stability of complex ions in water is needed to evaluate their role in aqueous equilibrium reactions. One approach is to assume that complex ions are in equilibrium with the free metal cation and the ligands, and to write reactions such as

$$Cu(NH_3)_4^{2+}(aq) \rightleftharpoons Cu^{2+}(aq) + 4NH_3(aq)$$
$$Al(OH)_6^{3-}(aq) \rightleftharpoons Al^{3+}(aq) + 6OH^-(aq)$$

We shall use this approach for the moment, but it must be noted that these reactions are not exactly correct on two counts: They do not include the water of hydration and they assume that all the ligands are separated from the metal cation simultaneously. In fact, most complex ions lose their ligands one at a time, forming other complex ions that also are stable.

**TABLE 18.1** Instability Constants of Complex Ions at Room Temperature

| Metal Ion | Ligand | Complex Ion | $K_i$ |
|---|---|---|---|
| $Ag^+$ | $NH_3$ | $[Ag(NH_3)_2]^+$ | $6.3 \times 10^{-8}$ |
| | $Br^-$ | $[AgBr_4]^{3-}$ | $5.0 \times 10^{-10}$ |
| | $Cl^-$ | $[AgCl_4]^{3-}$ | $1.0 \times 10^{-6}$ |
| | $CN^-$ | $[Ag(CN)_2]^-$ | $7.9 \times 10^{-22}$ |
| | en | $[Ag(en)_2]^+$ | $1.0 \times 10^{-8}$ |
| | $I^-$ | $[AgI_3]^{2-}$ | $1.3 \times 10^{-14}$ |
| | $S_2O_3^{2-}$ | $[Ag(S_2O_3)_2]^{3-}$ | $1.0 \times 10^{-13}$ |
| $Al^{3+}$ | $F^-$ | $[AlF_6]^{3-}$ | $2.0 \times 10^{-20}$ |
| | $OH^-$ | $[AlOH]^{2+}$ | $1.3 \times 10^{-9}$ |
| $Cd^{2+}$ | $NH_3$ | $[Cd(NH_3)_4]^{2+}$ | $1.0 \times 10^{-7}$ |
| | $Br^-$ | $[CdBr_4]^{2-}$ | $1.6 \times 10^{-4}$ |
| | $Cl^-$ | $[CdCl_3]^-$ | $5.0 \times 10^{-3}$ |
| | $CN^-$ | $[Cd(CN)_4]^{2-}$ | $1.6 \times 10^{-19}$ |
| | en | $[Cd(en)_3]^{2+}$ | $6.3 \times 10^{-13}$ |
| | $I^-$ | $[CdI_4]^{2-}$ | $7.9 \times 10^{-7}$ |
| $Co^{2+}$ | $NH_3$ | $[Co(NH_3)_6]^{2+}$ | $1.0 \times 10^{-5}$ |
| | en | $[Co(en)_3]^{2+}$ | $1.3 \times 10^{-14}$ |
| $Co^{3+}$ | $NH_3$ | $[Co(NH_3)_6]^{3+}$ | $1.0 \times 10^{-34}$ |
| $Cu^+$ | $NH_3$ | $[Cu(NH_3)_2]^+$ | $1.6 \times 10^{-11}$ |
| | $Cl^-$ | $[CuCl_2]^-$ | $2.0 \times 10^{-5}$ |
| | $CN^-$ | $[Cu(CN)_2]^-$ | $1.0 \times 10^{-24}$ |
| | en | $[Cu(en)_2]^+$ | $1.6 \times 10^{-11}$ |
| | $I^-$ | $[CuI_2]^-$ | $1.6 \times 10^{-9}$ |
| $Cu^{2+}$ | $NH_3$ | $[Cu(NH_3)_4]^{2+}$ | $5.0 \times 10^{-14}$ |
| | en | $[Cu(en)_2]^{2+}$ | $1.0 \times 10^{-20}$ |
| $Fe^{2+}$ | $CN^-$ | $[Fe(CN)_6]^{4-}$ | $1.0 \times 10^{-37}$ |
| | en | $[Fe(en)_3]^{2+}$ | $2.5 \times 10^{-10}$ |
| $Fe^{3+}$ | $Cl^-$ | $[FeCl_4]^-$ | $1.0 \times 10^{-2}$ |
| | $CN^-$ | $[Fe(CN)_6]^{3-}$ | $1.0 \times 10^{-44}$ |
| | $SCN^-$ | $[Fe(SCN)_6]^{3-}$ | $8.0 \times 10^{-10}$ |
| $Hg^{2+}$ | $NH_3$ | $[Hg(NH_3)_4]^{2+}$ | $5.0 \times 10^{-20}$ |
| | $Br^-$ | $[HgBr_4]^{2-}$ | $1.0 \times 10^{-20}$ |
| | $Cl^-$ | $[HgCl_4]^{2-}$ | $1.3 \times 10^{-15}$ |
| | $CN^-$ | $[Hg(CN)_4]^{2-}$ | $3.2 \times 10^{-42}$ |
| | $I^-$ | $[HgI_4]^{2-}$ | $1.6 \times 10^{-30}$ |
| | $SO_4^{2-}$ | $[Hg(SO_4)_2]^{2-}$ | $4.0 \times 10^{-3}$ |
| $Ni^{2+}$ | $NH_3$ | $[Ni(NH_3)_6]^{2+}$ | $2.0 \times 10^{-9}$ |
| | $CN^-$ | $[Ni(CN)_4]^{2-}$ | $1.0 \times 10^{-31}$ |
| | en | $[Ni(en)_3]^{2+}$ | $7.9 \times 10^{-20}$ |
| $Pb^{2+}$ | $Br^-$ | $[PbBr_4]^{2-}$ | $1.0 \times 10^{-3}$ |
| | $Cl^-$ | $[PbCl_3]^-$ | $1.6 \times 10^{-2}$ |
| | $OH^-$ | $[PbOH]^+$ | $6.3 \times 10^{-7}$ |
| | $I^-$ | $[PbI_4]^{2-}$ | $1.3 \times 10^{-4}$ |
| $Sn^{2+}$ | $Br^-$ | $[SnBr_3]^-$ | $5.0 \times 10^{-2}$ |
| | $OH^-$ | $[Sn(OH)_3]^-$ | $4.0 \times 10^{-26}$ |
| $Sn^{4+}$ | $Cl^-$ | $[SnCl_6]^{2-}$ | $3.2 \times 10^{-2}$ |
| | $F^-$ | $[SnF_6]^{2-}$ | $1.0 \times 10^{-18}$ |
| $Zn^{2+}$ | $NH_3$ | $[Zn(NH_3)_4]^{2+}$ | $4.0 \times 10^{-10}$ |
| | $CN^-$ | $[Zn(CN)_4]^{2-}$ | $6.3 \times 10^{-21}$ |
| | $OH^-$ | $[Zn(OH)_4]^{2-}$ | $2.5 \times 10^{-16}$ |

To be more accurate, we should write only stepwise reactions, such as

$$Al(OH)_6^{3-}(aq) \rightleftharpoons Al(OH)_5^{2-}(aq) + OH^-(aq)$$
$$Al(OH)_5^{2-}(aq) \rightleftharpoons Al(OH)_4^-(aq) + OH^-(aq)$$

We shall disregard these complications.

The equilibrium constant for the reaction in which all the ligands leave the metal cation is called $K_i$, where $i$ stands for instability. The magnitude of $K_i$ is a convenient measure of the stability of a complex ion. When $K_i$ is small, there is very little tendency for the complex to dissociate, so it is relatively stable. If $K_i$ is large, the complex is relatively unstable. Table 18.1 lists values of $K_i$ for some complex ions.

Given the necessary $K_i$ data, equilibrium calculations in which complex ion formation plays a role can be done by the same procedures that we used previously.

**Example 18.1**    A 0.29-mol amount of $NH_3$ is dissolved in 0.45 L of a 0.36$M$ silver nitrate solution. Calculate the equilibrium concentrations of all species.

**Solution**    The species present in appreciable concentration when the system is prepared are $H_2O$, $NH_3$, $Ag^+$, and $NO_3^-$. To select the equilibrium that we shall use in our calculations on this system, we must consider that $Ag^+$ forms a complex ion with $NH_3$ ligands. The equilibria of interest are

a. $\qquad\qquad\quad H_2O \rightleftharpoons H^+(aq) + OH^-(aq) \qquad K_W = 1.0 \times 10^{-14}M^2$
b. $\quad NH_3(aq) + H_2O \rightleftharpoons NH_4^+(aq) + OH^-(aq) \qquad K_B = 1.8 \times 10^{-5}M$
c. $Ag^+(aq) + 2NH_3(aq) \rightleftharpoons Ag(NH_3)_2^+(aq)$

Equation c is the reverse of the equation that defines $K_i$; it is the formation of the complex ion from the metal cation and the ligands. Therefore:

$$K = \frac{1}{K_i} = \frac{1}{6.3 \times 10^{-8}M^2} = 1.6 \times 10^7 M^{-2}$$

Since its equilibrium constant is much larger than the others, reaction c is the equilibrium we shall use in our calculations. The starting concentration of $NH_3$ is 0.29 mol/0.45 L = 0.64$M$.

|  | $Ag^+(aq)$ | $+ 2NH_3(aq)$ | $\rightleftharpoons Ag(NH_3)_2^+(aq)$ |
|---|---|---|---|
| **start** | 0.36$M$ | 0.64$M$ | 0 |
| **complete** | 0.36$M - 0.32M$ | 0.64$M - 0.64M$ | 0.32$M$ |
| **equil** | 0.04$M + x$ | 2$x$ | 0.32$M - x$ |

Assuming that $x$ is very small relative to 0.04, and substituting into the equilibrium expression for the reaction:

$$K = \frac{(0.32M)}{(2x)^2(0.04M)} = 1.6 \times 10^7 M^{-2}$$

and $\qquad\qquad x = 3.5 \times 10^{-4}M$

The approximation is valid.

The concentrations of the species in the equilibrium are

$$[NH_3] = 2x = 7.0 \times 10^{-4} M$$
$$[Ag^+] = 0.04 M$$
$$[Ag(NH_3)_2^+] = 0.32 M$$

The concentrations of other ions in the solution can be found from the appropriate expressions for the secondary equilibria. For reaction b

$$K_B = \frac{[NH_4^+][OH^-]}{[NH_3]} = 1.8 \times 10^{-5} M$$

Since the concentrations of $OH^-$ and $NH_4^+$ must be identical (provided that the pH of the solution is far from 7), according to the stoichiometry of reaction b and since the $[NH_3]$ is known from the first part of the calculation, we can write

$$\frac{(x)(x)}{(7.0 \times 10^{-4} M)} = 1.8 \times 10^{-5} M$$

where $x = [NH_4^+] = [OH^-]$ and $x = 1.1 \times 10^{-4} M$.

The remaining ion in solution is $H^+$. We can find its concentration by using the $[OH^-]$ in the water equilibrium expression:

$$[H^+] = \frac{1.0 \times 10^{-14} M^2}{1.1 \times 10^{-4} M} = 8.9 \times 10^{-11} M$$

The major ion in this solution is the silver ammonia complex ion.

---

Solid hydroxides of metals that form stable complex ions with hydroxide as the ligand often are soluble in solutions of hydroxide, although they may be insoluble in water. These compounds are called amphoteric hydroxides, since they are also soluble in solutions of strong acids.

---

**Example 18.2**

Calculate the solubility of $Zn(OH)_2$ in $1.0 M$ NaOH solution. The $K_{sp}$ of zinc hydroxide is $1.8 \times 10^{-14} M^3$.

**Solution**

Table 18.1 shows that $Zn^{2+}$ forms a stable complex ion, $Zn(OH)_4^{2-}$, with hydroxide as the ligand. Therefore, when $Zn(OH)_2(s)$ is added to a solution containing an appreciable concentration of hydroxide ion, the relevant equilibrium is the one that leads to formation of a stable complex ion:

$$Zn(OH)_2(s) + 2OH^-(aq) \rightleftharpoons Zn(OH)_4^{2-}(aq)$$

This reaction is the sum of two reactions whose equilibrium constants are known:

$$Zn(OH)_2(s) \rightleftharpoons Zn^{2+}(aq) + 2OH^-(aq) \qquad K_{sp} = 1.8 \times 10^{-14} M^3$$

$$Zn^{2+}(aq) + 4OH^-(aq) \rightleftharpoons Zn(OH)_4^{2-}(aq) \qquad \frac{1}{K_i} = \frac{1}{3.3 \times 10^{-16} M^4}$$

The equilibrium constant for the reaction is the product of these two equilibrium constants:

$$K = \frac{K_{sp}}{K_i} = \frac{1.8 \times 10^{-14} M^3}{3.3 \times 10^{-16} M^4} = 5.5 \times 10^1 M^{-1}$$

The $[Zn(OH)_4{}^{2-}]$ is a convenient measure of the solubility of $Zn(OH)_2(s)$, since one mole of the complex ion is formed for every mole of the solid that dissolves. We determine $[Zn(OH)_4{}^{2-}]$ in the usual way:

$$Zn(OH)_2(s) + 2OH^-(aq) \rightleftharpoons Zn(OH)_4{}^{2-}(aq)$$

|        |            |      |
|--------|------------|------|
| **start** | $1.0M$ | $0$ |
| **equil** | $1.0M - 2x$ | $x$ |

Substitution into the expression for $K$ gives

$$K = \frac{x}{(1.0M - 2x)^2} = 5.5 \times 10^1 M^{-1}$$

Since $K$ is neither large nor small, we use the quadratic formula to solve this equation, obtaining $x = 0.46M$, which is the $[Zn(OH)_4{}^{2-}]$. Thus, the solubility of zinc hydroxide in $1.0M$ hydroxide solution is 0.46 mol/L, substantially larger than the solubility of zinc hydroxide in water, which is $1.7 \times 10^{-5}$ mol/L.

## 18.6  SOME APPLICATIONS OF COORDINATION CHEMISTRY

Interest in the chemistry of coordination compounds has increased greatly in recent years. The vital role they play in many biological processes has been recognized, and new uses for coordination compounds in industry are being developed.

### Catalysis

Coordination compounds are used as catalysts for many industrial processes. They have the major advantage of being homogeneous catalysts, which means that the entire reaction system, reactants and catalyst alike, is in the same phase. A suitable homogeneous catalyst is better than a heterogeneous catalyst in several ways. Homogeneous catalysts require simpler laboratory apparatus, generally are more efficient than heterogeneous catalysts, and allow the reaction to be carried out at lower temperatures.

The details of catalysis by coordination compounds tend to be complicated, as we can see by studying the catalysis of hydrogenation reactions by coordination compounds. In the hydrogenation of compounds with multiple bonds (Section 9.1), two hydrogen atoms are added to a double bond, converting it to a single bond. The reaction proceeds readily in the presence of a heterogeneous catalyst, such as finely divided platinum. The

$H_2$ is chemisorbed (Section 17.6) on the surface of the metal. We can generalize such a reaction as:

$$H—H + \overset{\diagdown}{\underset{\diagup}{}}C=C\overset{\diagup}{\underset{\diagdown}{}} \longrightarrow H—\overset{\diagdown}{\underset{\diagup}{}}C—C\overset{\diagup}{\underset{\diagdown}{}}—H$$

Many transition metal complexes can act as homogeneous catalysts for hydrogenation. The catalysis can occur in several ways. For example, a cyanide complex of Co(II) reacts with $H_2$ to form a complex in which the cobalt is oxidized to Co(III) and an $H^-$, hydrido, ligand is added:

$$2[Co(CN)_5]^{3-} + H_2 \longrightarrow 2[Co(CN)_5H]^{3-}$$

While the reactivity of $H_2$ is relatively low, the reactivity of the product complex is quite high, so the complex can donate the equivalent of $H^-$ to a double bond. The formation of the hydrido complex thus provides a pathway for hydrogenation.

Some hydrogenation reactions are catalyzed by a chloride complex of rhodium, which can be represented as $RhClL_3$, where L represents a ligand of complicated structure with a phosphorus donor atom. This homogeneous catalysis takes place in a different way. Under the conditions of the reaction, $H_2$ adds to the complex oxidatively to form $RhClH_2L_3$, which has two hydrido ligands. The next step is the formation of a complex in which one of the ligands is replaced by the compound that is to be hydrogenated. The loosely held $\pi$ electrons of a double bond can act as donor electrons, so compounds with $\pi$ electrons are often found as ligands in coordination compounds:

$$RhClH_2L_3 + \overset{\diagdown}{\underset{\diagup}{}}C=C\overset{\diagup}{\underset{\diagdown}{}} \longrightarrow \overset{C}{\underset{C}{\|}}{\rightarrow}RhClH_2L_2 + L$$

This complex not only has the H atom in a reactive form but also brings the two reactants close together.

A major research effort is being made to achieve nitrogen fixation through catalysis by coordination compounds. We have already described the great stability of the $N_2$ molecule, the drastic conditions required to make it react, and the importance of nitrogen fixation in the world economy. A method that makes nitrogen fixation possible under mild conditions will save a great deal of energy. Homogeneous catalysis through coordination complexes in which $N_2$ is a ligand seems to offer the most promise.

Complexes of the general form $M(N_2)_2(L)_4$, where M is either tungsten or molybdenum, and the other ligands, represented by L, are phosphorus compounds, have been found to react under mild conditions with acid to give ammonia:

$$M(N_2)_2(L)_4 + H^+ \longrightarrow NH_3 + N_2 + \text{other products}$$

Another reaction of current interest is

$$H_2O(l) \longrightarrow H_2(g) + \tfrac{1}{2} O_2(g)$$

This reaction is not favored thermodynamically, and it can occur only if energy is supplied to the system. If energy in the form of light from the sun could be used to bring about the reaction, we would have a convenient way of producing hydrogen gas, a useful fuel, from solar energy. Since oceans and lakes do not evolve $H_2$ and $O_2$ when they are exposed to sunlight, we know that this reaction does not occur under ordinary conditions. Some coordination complexes, especially those of ruthenium, have been found to catalyze the reaction. If research can develop a catalysis process that is economically viable, ordinary water could become an important energy source.

## Biological Complexes

We said at the beginning of this chapter that a number of biologically important substances contain a coordination complex of iron and porphyrin called heme. Its structure is shown in Figure 18.19. The major features of this complex structure are the central iron atom and the quadridentate macrocyclic ligand that complexes the iron. All the carbon atoms of the ring system, the coordinating nitrogen atoms, and the central iron atom lie in the same plane. The planar structure of heme influences the structure and activity of heme-containing proteins.

One heme-containing protein is cytochrome $c$, which participates in the biological electron transport chain (Section 16.8). The function of cytochrome $c$ is to donate or accept one electron. The iron atom of heme takes part in this electron transport, since it can exist in both the $+2$ and $+3$ oxidation states. (The ability of metal ions to exist in several different oxidation states, accepting or donating electrons as required, is one reason for their importance in biological processes.) But why should the iron, which performs what seems to be a simple function, be a part of a molecule as complex as cytochrome $c$ (molecular weight 12 500)? There is no complete answer to this question as yet, but we can get a clue from the fact that the $K_{sp}$ of iron(III) hydroxide is extremely small. Uncomplexed iron(III) ion will not dissolve in water at physiological pH. Enzyme-catalyzed reactions are homogeneous reactions. The reactants as well as the catalyst must be in solution. It appears that many metals that are important in biological processes cannot dissolve and serve as homogeneous catalysts unless they are part of a complex.

But we must also explain why the metal complex is part of a large, complicated molecule. We can get a partial explanation from the study of hemoglobin, the oxygen-carrying substance of the blood. Hemoglobin contains four heme groups embedded in a protein whose molecular

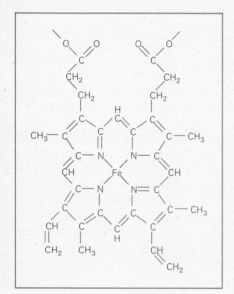

**Figure 18.19**

The structure of heme, a biologically important coordination complex of iron and porphyrin. The iron atom, the nitrogen atoms, and the carbon atoms of the rings all lie in the same plane.

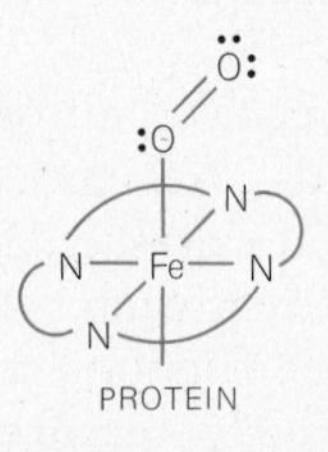

**Figure 18.20**

A schematic structure for the heme group of a hemoglobin molecule. Four of the six coordination sites of the iron are occupied by the quadridentate porphyrin ligand. The fifth site is occupied by a ligand belonging to the protein. The sixth site is available for the transport of $O_2$.

## Essential Elements: The List Grows

How many elements are essential to life? After more than a century of painstaking investigation, the question still cannot be answered with certainty. To demonstrate that an element is essential, all traces of the element must be kept from the organism. Many elements are essential only in exceedingly small quantities, so it is almost impossible to exclude them totally. In recent years, new techniques for keeping experimental animals in a completely isolated sterile environment have lengthened the list of essential elements.

Although only 10 elements account for more than 99% of the matter in living organisms, experiments have suggested that at least 29 other elements, including 21 metals, are essential to life. These elements are listed below in the table of essential elements.

To determine trace-element requirements, rats or other small animals are kept in enclosures that are made of plastic to eliminate contaminants from metal, glass, and rubber. Air entering the enclosures is filtered to remove trace substances that might be present in dust. The animals are fed ultrapure amino acids, and their diet is carefully checked for metal contaminants. Controlled amounts of known essential elements are included in the diet. If a deficiency disease develops, small amounts of other elements can be added to determine which element corrects the deficiency.

For example, test animals on a diet that has no trace of vanadium suffer about a 30% growth retardation. The addition of one-tenth of a part per million of vanadium to the diet restores normal growth. One-half of a part per million of fluorine in the diet has been found to be essential for normal growth. Thirty parts per million of silicon and two parts per million of tin are needed for normal development.

The precise role of these trace elements in the body's metabolic processes is being investigated. It is presumed that in many cases coordination complexes of the trace metals are part of enzymes, but the nature of many of these metalloenzymes is unknown. It could be that the presence of one trace mineral in the body is necessary for the utilization of another. For example, it is known that copper is essential for the metabolism of iron. Ceruloplasmin, a copper-containing protein in the blood, promotes the release of iron from the liver. The iron complexes with transferrin, a protein in the blood serum, which enables the iron to be incorporated into the hemoglobin molecule. Similarly complex relationships may exist between the newly discovered essential trace elements and other elements in the body.

| H | | | | | | | | | | | | | | | | | He |
|---|---|---|---|---|---|---|---|---|---|---|---|---|---|---|---|---|---|
| Li | Be | | THE ESSENTIAL ELEMENTS | | | | | | | | | B | C | N | O | F | Ne |
| Na | Mg | | Make up more than 99% of living matter | Confirmed essential trace elements | Suspected essential trace elements | | | | | | | Al | Si | P | S | Cl | Ar |
| K | Ca | Sc | Ti | V | Cr | Mn | Fe | Co | Ni | Cu | Zn | Ga | Ge | As | Se | Br | Kr |
| Rb | Sr | Y | Zr | Nb | Mo | Tc | Ru | Rh | Pd | Ag | Cd | In | Sn | Sb | Te | I | Xe |
| Cs | Ba | La | Hf | Ta | W | Re | Os | Ir | Pt | Au | Hg | Tl | Pb | Bi | Po | At | Rn |

weight is 64 000. The iron of each heme group is in the $+2$ oxidation state. Four of the six coordination sites of each iron are occupied by the quadridentate porphyrin ligand. A fifth site attaches the heme to the protein, and the sixth site is vacant. It is the vacant site to which $O_2$ molecules are attached as they are carried through the blood by hemoglobin; Figure 18.20 shows a possible schematic structure for this complex. Heme alone forms a complex with $O_2$, but this complex is not stable. When the heme that forms an oxygen complex is embedded in a large protein molecule, the bulk of the protein protects the complex from further attack. We can say that the iron cation is complexed to keep it in solution and that the complex is embedded in a protein to protect the relatively unstable oxygen complex.

We find another reason why metals are complexed in biological molecules by studying vitamin $B_{12}$, a cobalt-containing substance that is a prosthetic group (Section 17.6) for a number of enzymes. The cobalt ion is complexed by a quadridentate ligand that is quite similar to the porphyrin ligand of hemoglobin. It appears that vitamin $B_{12}$ is active only if cobalt is in the $+1$ oxidation state, a state that is normally not displayed by free cobalt. Apparently, cobalt can be in the $+1$ oxidation state only when it is complexed. Thus, another reason for complexing metals is to achieve oxidation states that are not readily possible in the uncomplexed metals.

Many other metals that are essential to biological processes usually are in coordination complexes. Chlorophyll, the green plant pigment that plays a fundamental role in photosynthesis, is a complex of magnesium. Enzymes that participate in nitrogen fixation in bacteria contain both iron and molybdenum. Digestive enzymes contain zinc, and enzymes that catalyze electron transport processes contain copper. Other transition metals — vanadium, chromium, manganese, nickel — usually in the form of coordination complexes, are biologically important. The field of bioinorganic chemistry, the biological chemistry of the metals and some of the heavier nonmetals, is one of the most active in chemical research today.

**Summary**

**W**e began this chapter on coordination chemistry by saying that a **coordination compound** includes a metal atom surrounded by a set of atoms or groups of atoms called **ligands.** We noted that the metal displays two kinds of valences, the **coordination valence** or coordination number, for ligands that are in the coordination sphere, and the **ionic valence,** for counter ions, which are outside the coordination sphere. We listed the rules for naming coordination complexes. Then we described the **geometries** associated with different coordination numbers: **linear** for metals with coordination number 2, **tetrahedral** or **square planar** for coordination number 4, **octahedral** for coordination number 6. We then said that several kinds of **isomerism** — different structures for compounds with the same molecular formula — can occur in coordination compounds. There are **stereoisomers,** which differ only in the spatial arrangement of their atoms, and **optical isomers,** in which molecules differ in their handedness, or **chirality.** We then noted that there are several methods for describing the bonding in coordination compounds: the **valence bond theory,** which emphasizes the covalent nature of the metal-ligand bond, the **crystal field theory,** which assumes only electrostatic interactions between the metal and the ligand. We noted the difference between the **lability** of a coordination compound, the rate at which its ligands can be replaced, and its **stability,** the relative energy of the compound. We concluded by discussing some applications of coordination compounds, which are used as catalysts and play important roles in living organisms.

# Exercises

**18.1** Find the oxidation state of the metal in each of the following complex ions: (a) $[Au(OH)_4]^-$, (b) $[CrF_4O]^-$, (c) $[Co(NH_3)_5Br]^+$, (d) $[Cr(CN)_6]^{3-}$.

**18.2** Find the oxidation states of the metals in the following compounds: (a) $Na[AgF_4]$, (b) $K_2[Ni(NH_3)_2Cl_4]$, (c) $[Ni(NH_3)_6]Br_2$, (d) $[Cr(ONO)(NH_3)_5]Cl_2$.

**18.3** Find the coordination number of the metal in each of the following ions: (a) $[PtCl_3CN]^{2-}$, (b) $[Zn(en)_2]^{2+}$, (c) $[Cr(NH_3)_3(H_2O)_3]^{3+}$.

**18.4**[1] Find the coordination number of the transition metal in each of the following coordination compounds: (a) $[Co(NH_3)_5Cl]Cl_2$ (b) $[Co(NH_3)_4Cl_2]Cl$ (c) $Na_2[Co(NO)Cl_5]$

**18.5** The reaction $Co^{2+}(aq) + 6NH_3(aq) \rightleftharpoons [Co(NH_3)_6]^{2+}$ has a $K$ of $1.0 \times 10^5$. The $K$ for the reaction $Co^{2+}(aq) + 3en(aq) \rightleftharpoons [Co(en)_3]^{2+}$ is $7.7 \times 10^{13}$. Find $K$ and $\Delta G°$ for the reaction $[Co(NH_3)_6]^{2+} + 3en(aq) \rightleftharpoons [Co(en)_3]^{2+} + 6NH_3(aq)$.

**18.6** Find the charge on a complex ion formed by iron(II) and the following ligands: (a) four $H_2O$ and two $CN^-$, (b) five $CN^-$ and one $Cl^-$, (c) three $OH^-$ and three $NH_3$.

**18.7** The coordination number of Co(II) is 6. Write the formulas of four different neutral coordination compounds of Co(II) with $NH_3$ and $Cl^-$ ligands.

**18.8** Name the following coordination compounds: (a) $[Cr(NH_3)_6]Br_2$, (b) $K_2[PtCl_6]$, (c) $K[Au(OH)_4]$.

**18.9** Name the following coordina-

[1] The answers to exercises whose numbers are in color can be found in Appendix VII. The star indicates an exercise that is more challenging than average.

tion compounds that have more than one type of ligand: (a) $[Cr(NH_3)_3(H_2O)_3]Cl_3$ (b) $K[CrF_4O]$ (c) $Na_2[SnBrCl_5]$

**18.10** Write the formulas of the following coordination compounds: (a) sodium hexachloroplatinate(IV), (b) potassium tetracyanonickelate(II), (c) tetramminecopper(II) bromide.

**18.11** Write the formulas of the following coordination compounds that have more than one type of ligand: (a) sodium dibromodichloroplumbate(II), (b) diamminedichloroplatinum(II), (c) sodium diaquotetrahydroxyaluminate.

**18.12** Predict the most likely geometry for each of the following complexes: (a) $[Ag(CN)_2]^-$, (b) $[AgI_3]^{2-}$, (c) $[AgCl_4]^{3-}$, (d) $[Ag(en)_3]^+$.

**18.13** Predict the most likely geometry for the following complexes: (a) $[PbI_4]^{2-}$, (b) $[Pt(NH_3)_4]^{2+}$, (c) $[NiF_4]^{2-}$, (d) $[Au(NH_3)_2Cl_2]^+$.

**18.14** Propose two possible geometries for the compound $WF_5$.

**18.15** Write the formulas of two ionization isomers of a complex with the composition $Cr(NH_3)_5BrCl$.

**18.16** Write the formulas of all the ionization isomers of a complex with the composition $PbBr_2Cl_2(NH_3)_2$ and a coordination number of 4.

**18.17** Write the formulas of the hydrate isomers of a complex with the composition $Ru(H_2O)_6Cl_4$.

**18.18** Draw the Lewis structures of the two linkage isomers of the complex $[AgCl_2CN]^{2-}$.

**18.19** Draw three linkage isomers of a complex ion with the composition $[M(NO_2)_2]^-$ (where M is a metal).

**18.20** Draw the structures of the geometric isomers of the complex $[Pd(CN)_2(NH_3)_2]$, which has square planar geometry. Use the method of representation of Figure 18.9.

**18.21** Draw the structures of the geometric isomers of the complex ion $[NiBrClF_2]^{2-}$, which has square planar geometry.

**18.22** Indicate the number of geometric isomers that exist for each of the following complexes: (a) $[Pd(NH_3)Cl_3]$ (square planar), (b) $[AlCl_2F_2]^-$ (tetrahedral), (d) $[SnBrClF]^-$ (planar), (d) $[PtBrClFI]^{2-}$ (square planar).

**18.23** Draw the structures of the geometric isomers of the complex $[Ru(NH_3)_3Cl_3]$ using the method of representation of Figure 18.10.

**18.24** Draw the structures of the geometric isomers of the complex $[Cr(CN)_4(H_2O)_2]^-$ using the method of representation of Figure 18.10.

**18.25** Draw the structures of the geometric isomers of the complex $[Fe(H_2O)_2Cl_2I_2]^-$ using the method of representation of Figure 18.10.

**18.26** Indicate which of the following objects are chiral: (a) a propeller, (b) a football, (c) a nail, (d) a scissors, (e) a screwdriver, (f) a triangle with three unequal angles.

**18.27** Indicate which of the following hypothetical complexes display optical isomerism: (a) $[SnClBrIF]$ (tetrahedral), (b) $[PdClBrIF]$ (square planar), (c) $[AuClBr]^-$, (d) $Co(NH_3)_3Cl_3]$, (e) $[Co(NH_3)_4Cl_2]$.

**18.28** One isomer of the hypothetical complex ion $[Ru(NH_3)_2(CN)_2Cl_2]^-$ is chiral. Draw the two mirror images of this isomer.

**18.29** Indicate the hybridization predicted by the valence bond theory for the metal in the following complexes: (a) $[Ag(CN)_2]^-$, (b) $[SnBr_3]^-$, (c) $[Zn(OH)_4]^{2-}$, (d) $[PtCl_4]^{2-}$, (e) $[CoF_5]^{3-}$, (f) $[Fe(CN)_6]^{2-}$.

**18.30** Indicate the number of electrons in the partially filled $d$ subshells of the following ions: (a) $Au^{3+}$, (b) $Rh^{4+}$, (c) $Ru^{4+}$, (d) $Fe^{2+}$, (e) $Co^{2+}$, (f) $Ni^{2+}$, (g) $Cu^{2+}$.

**18.31** Indicate the ground state valence electronic configurations of the following isolated ions, showing electron spins: (a) $Os^{6+}$, (b) $Re^{4+}$, (c) $Ru^{2+}$, (d) $Tc^{2+}$.

**18.32** Indicate the valence electronic configurations of the following ions in their low-spin state: (a) $Cr^{2+}$, (b) $Tc^{2+}$, (c) $Fe^{2+}$, (d) $Ir^{2+}$.

**18.33** Indicate the valence electronic configurations of the following ions in their high-spin states: (a) $Re^{3+}$, (b) $Os^{3+}$, (c) $Rh^{3+}$, (d) $Pt^{3+}$.

**18.34** Predict the magnetic properties of each of the following complexes and justify your predictions:
(a) $[Cr(CN)_6]^{4-}$
(b) $[CrF_6]^{4-}$
(c) $[Cu(CN)_4]^{2-}$
(d) $[CdCl_4]^{2-}$

**18.35** Predict the magnetic properties of each of the following complexes. Justify your predictions: (a) $[AuCl_4]^-$, (b) $[FeCl_4(H_2O)_2]^-$, (c) $[RuF_6]^{2-}$, (d) $[Ir(NH_3)_6]^{2+}$.

**18.36** Use the spectrochemical series to predict the magnetic properties of each of the following complexes. Justify your predictions: (a) $[Co(en)_3]^{3+}$, (b) $[Ru(CN)_6]^{4-}$, (c) $[Ru(H_2O)_6]^{2+}$, (d) $[Mn(CN)_6]^{5-}$.

**18.37** Given that the value of $\Delta_o$ for $[Co(H_2O)_6]^{2+}$ is 221 kJ/mol, predict the wavelength of the radiation that is absorbed by this complex.

**18.38** The $[Ni(H_2O)_6]^{2+}$ ion absorbs light of wavelength 580 nm. Find the value of $\Delta_o$ that corresponds to this wavelength.

**18.39** Suggest an explanation for the observation that $[Fe(CN)_6]^{3-}$ is poisonous while $[Fe(CN)_6]^{4-}$ is not.

**18.40** The $[Ni(H_2O)_6]^{2+}$ ion is bright green and the $[Ni(NH_3)_6]^{2+}$ ion is violet. When a solution of the latter ion in water is acidified, the color changes from violet to green. Write reactions for the changes that take place as a result of the addition of acid. Discuss the lability and stability of the two ions.

**18.41★** Halide complexes of metal M of the form $[MX_6]^{3-}$ are found to be stable in aqueous solution. Suggest an experiment to measure their lability that does not employ radioactive labels.

**18.42** Suggest an explanation for the observation that $d^6$ metal cations in low-spin states are relatively inert.

**18.43** A solution is prepared from 1.00 mol of $Hg(NO_3)_2$, 1.00 mol of NaI, and enough water to make 1.00 L of solution. Find the concentration of all species at equilibrium.[2]

**18.44** A solution is prepared from 0.10 mol of $CdI_2$, 0.20 mol of NaI, and enough water to make 1.0 L of solution. Find the concentration of all species at equilibrium.[2]

**18.45** The $K_{sp}$ of $Zn(OH)_2$ is $1.8 \times 10^{-14}$. Find the solubility of $Zn(OH)_2$ in a $2.0M$ solution of NaOH.[2]

**18.46** The $K_{sp}$ of AgBr is $5.2 \times 10^{-13}$. Find the solubility of AgBr in a $1.0M$ solution of $NH_3$.[2]

**18.47** The $K_{sp}$ of $PbBr_2$ is $4.6 \times 10^{-6}$. Find the solubility of $PbBr_2$ in a $1.0M$ solution of NaBr.[2]

**18.48** Indicate changes in oxidation states taking place in the reaction:
$$2[Co(CN)_5]^{3-} + H_2 \rightarrow$$
$$2[Co(CN)_5H]^{3-}.$$

**18.49** Coordination compounds of copper have been found in some important biological systems. Suggest some possible functions for the copper in such compounds.

---

[2] Some of the data necessary for this exercise can be found in Table 18.1.

# 19

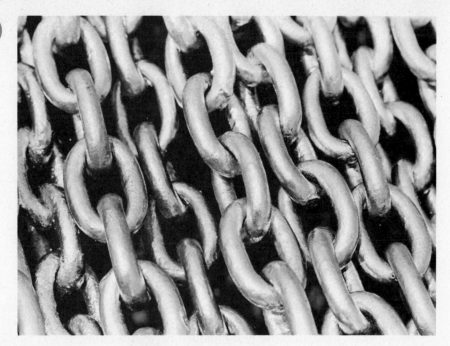

# Metals

**Preview**

This chapter will help you to understand why metals play such a central role in technology. We first list the characteristic properties of metals, such as electrical and thermal conductivity, ductility, and malleability, and then show how the kind of chemical bonding found in metals can explain these properties. We then introduce two pictures of metallic bonding, the electron sea model and the band theory, and we show how we can also use the band theory to understand the behavior of semiconductors. We next discuss the metallurgical processes that begin with metal ore and end with metals in industrially useful forms. Finally, we give detailed descriptions of two families of metals that have been crucial in human history: the group VIIIB metals, which include iron and nickel, and the group IB metals, which include gold and silver.

We have encountered metals frequently in our discussions of such subjects as the chemistry of the nonmetals, the chemistry of aqueous solutions, and the chemistry of coordination compounds. In this chapter, we shall discuss the metallic state and the general characteristics of metals, with a closer look at the chemistry of the metals in groups IB and VIIIB of the periodic table.

## 19.1    THE METALLIC STATE

We began by dividing the chemical elements into three groups: the non-metals, which are found toward the right and top of the periodic table; the semimetals, which run in a diagonal band from boron at the upper left of the table to tellurium at the lower right; and the metals, which lie toward the left and the bottom of the periodic table (Figure 2.4). We can identify the metals by a number of chemical and physical properties that they have in common.

1. *Metals are good conductors of electricity.* Some metals conduct electricity better than others — the metals of group IB, copper, silver, and gold, are the best conductors, with aluminum and beryllium next — but every metal is a conductor. In metals, moreover, conduction of electricity takes place without the transfer of material or the chemical changes that occur in electrolysis or ionic conduction. The conductivity of metals *decreases* as the temperature increases.
2. *Metals are good conductors of heat.* This property is evident when we compare metals to nonmetals, which tend to be thermal insulators. But there is a wide range of thermal conductivity among metals. The metals of group IB are the best thermal conductors. Aluminum is the best electrical and thermal conductor of the other metals.
3. *Metals have luster.* Their surfaces, when clean, are shiny. The metallic luster results from the reflection of all wavelengths of visible light from the surface of most metals. (Gold and copper are exceptions because they absorb some frequencies of visible radiation.) When metals appear gray or black, it is because a chemical reaction has taken place on their surfaces.
4. *Metals are deformable and plastic.* A piece of metal can be pounded into a thin sheet, so metals are said to be *malleable.* A piece of metal can be drawn into a fine wire, so metals are said to be *ductile.* Most nonmetallic solids do not have these properties. For example, a piece of sodium chloride is brittle; if it is hit with a hammer, it shatters. Some nonmetallic solids can be deformed: Rubber stretches, for instance. But generally, these nonmetallic solids are elastic; when the force that causes the deformation is removed, the solid returns to its original shape. Metals can be deformed well past the elastic range.
5. *Metals display the photoelectric effect.* When metals are exposed to radiation of short wavelength, they emit electrons (Section 5.6). A

related phenomenon, the thermionic effect, was observed in 1883 by Thomas Edison, who found that a metal wire emitted electrons when heated. The discovery of the thermionic effect led to the invention of electronic tubes and the birth of the electronics industry.

6. *All metals but mercury are crystalline solids at room temperature.* Metals generally have relatively high densities and high melting and boiling points. One reason metals are so dense is that the atoms of their crystals are in close-packed arrangements (Section 10.5). Cubic closest packing or hexagonal closest packing, in which each atom is surrounded by 12 identical and equidistant atoms, is common. Body-centered cubic packing is found in about 20 metals, most of them in the top left part of the periodic table.

7. *When a metal combines chemically with another element, the other element usually is a nonmetal and the metal generally changes to a positive oxidation state.* The bonding between a metal and a nonmetal usually is ionic, with a substantial electron transfer from the metal to the nonmetal. Metals have relatively low electron affinities, so when they form covalent bonds, those bonds generally are not very strong. While nonmetals frequently exist as relatively small homonuclear molecules that are gases at room temperature, metals do not have any great tendency to form such molecules.

## 19.2  THE METALLIC BOND

Ordinary covalent bonding is not important between metal atoms, but there are strong forces between the atoms in a metallic crystal. Data on the heat of atomization, the quantity of heat needed to form 1 mol of separated gaseous atoms, demonstrate the existence of these forces, as we can see by examining the data for sodium. Like many other metals, sodium in the gas phase can exist as a diatomic molecule with relatively weak covalent bonds. The strength of the sodium-sodium bond in the $Na_2$ molecule is 72.4 kJ/mol. The heat of atomization is the heat required to form 1 mol of atoms. For $Na_2$, it is thus 72.4/2 = 36.2 kJ/mol, since 1 mol of $Na_2$ forms 2 mol of Na.

The heat of atomization for solid sodium is much higher, 108 kJ/mol. The forces holding sodium atoms together in the solid are stronger than the covalent bonds in the $Na_2$ molecule. The same is true of all other metals, as you can see in Table 19.1, which lists the heats of atomization of diatomic molecules and of solid metals at 298 K.

The force that holds metal atoms together in the crystal is known as the *metallic bond.* Any theory that attempts to describe the metallic bond must account for the strength of this force and the unique properties of metals.

Compared to nonmetals, metals have a small number of $s$ and $p$ electrons in their valence shells. For example, the valence shell electronic configurations of the metals in the second row of the periodic table are Na,

TABLE 19.1    Heats of Atomization (kJ/mol) at 298 K

| Metal | Symbol | Atomic Number | Diatomic[a] | Solid |
|---|---|---|---|---|
| lithium | Li | 3 | 52 | 162 |
| potassium | K | 19 | 24.7 | 90.0 |
| copper | Cu | 29 | 98 | 339 |
| zinc | Zn | 30 | 13 | 131 |
| rubidium | Rb | 37 | 22.6 | 81.6 |
| silver | Ag | 47 | 82 | 285 |
| cadmium | Cd | 48 | 4.4 | 112 |
| tin | Sn | 50 | 96 | 301 |
| cesium | Cs | 55 | 21.8 | 78.2 |
| gold | Au | 79 | 109 | 368 |
| lead | Pb | 82 | 48 | 197 |

[a] The heat of atomization of a diatomic molecule is half the bond energy.

$3s^1$; Mg, $3s^2$; Al, $3s^23p^1$. Those of the nonmetals are P, $3s^23p^3$; S, $3s^23p^4$; Cl, $3s^23p^5$. The nonmetals can form only a limited number of covalent bonds, because of the relatively small number of electrons required to fill their valence shells. Thus, chlorine exists as $Cl_2$, with one covalent bond; sulfur as $S_8$, with two covalent bonds for each S atom; and phosphorus as $P_4$, with three covalent bonds for each P atom. In each case, the number of atoms that are bonded to a given atom is determined by the number of unpaired valence electrons.

In metal atoms we find an entirely different situation, which can be described as *delocalized bonding.* There are more vacant orbitals than electrons in the valence shell of a metal atom. These vacant orbitals allow the formation of metal-metal bonds that are not localized between two atoms. Instead, the bonding electrons are shared among the vacant orbitals in a number of neighboring atoms.

For example, a sodium atom has one valence electron in a $3s$ orbital, but it has four valence orbitals, the $3s$ and three $3p$ orbitals. The sodium atom can form only one bond because it has only one valence electron. That bond is not localized between two sodium atoms in the solid. It is spread out between the sodium atom and all its neighboring sodium atoms. Since solid sodium has a body-centered cubic crystal structure, we can say that the bond formed by a sodium atom is spread between this atom, its eight nearest neighbors, its six next-nearest neighbors and even, to some extent, more remote sodium atoms.

We get a similar picture for the other metals. Magnesium has a valence electron configuration of $3s^2$, which means that it has no unpaired valence electrons. But an excited state of magnesium whose electronic configuration is $3s^13p^1$ can be used to form bonds. Thus, each excited magnesium atom has two unpaired electrons and four valence orbitals available. Solid magnesium has a hexagonal closest-packed crystal structure, which gives each atom 12 nearest neighbors. The two bonding electrons of each atom are delocalized between these 12 nearest neighbors and, to a lesser extent, between more remote magnesium atoms. This delocalization of electron density lowers the total energy, since it increases the distance between electrons. It more than compensates for the energy required to produce the excited state of the magnesium atom.

Many of the observed properties of metals can be explained by electron delocalization. The bonding electrons, which are not closely associated with one or two atoms, can move freely through the crystal. The high electrical conductivity of metals, the photoelectric effect, and the thermionic effect are all consistent with a picture in which the electrons are relatively free to move in the metallic crystal.

## Metallic Valence

The strength of the metallic bond depends on a number of factors, including the geometry of the crystal, the ionization energy of the metal, and the electron affinity of the metal. But the most important factor by far is *the number of electrons that a metal atom has available for bonding, called the* **metallic valence.** The greater the metallic valence, the stronger the forces holding the atoms of the crystal together. The metallic valence of sodium is 1, of magnesium is 2, and of aluminum is 3. Therefore, atoms are held more tightly in solid aluminum than in magnesium, and more tightly in magnesium than in sodium.

The heat of atomization would seem to be the most direct measurement of the strength of the metallic bond. For these three metals, the heats of atomization are as expected: greatest in aluminum (326 kJ/mol), less in magnesium (146 kJ/mol), lowest in sodium (108 kJ/mol). However, the picture is not so simple. Each metal crystal has a different geometry. Differences in geometry affect the heats of atomization. In addition, energy must be added for both magnesium and aluminum to form the excited state that permits the maximum number of electrons to participate in bonding. Thus, the measured heats of atomization of magnesium and aluminum understate the magnitude of the forces holding nearest neighbors together, as the thermochemical diagram in Figure 19.1 shows.

Many properties of metals reflect the strength of the metallic bond. One such property is hardness, which is a measure of resistance to a change in shape. For a metal to change shape, some atoms must move away from one another. The hardness of a metal is thus related to the strength of the metallic bond. From their electronic structures, we would predict that aluminum is harder than magnesium and that magnesium is harder than sodium. The prediction is verified by observation.

The melting point of a metal is also related to the strength of the metallic bond: the stronger the bond, the higher the melting point. Thus, we find that aluminum has the highest melting point of the three metals, 933 K, with magnesium next at 923 K, and sodium last at 371 K. (The trend is not completely regular because of the influence of crystal structure and other factors.)

We find the same correlation between metallic valence and metallic

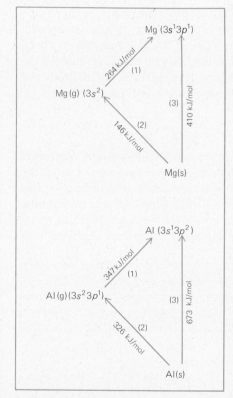

**Figure 19.1**

Thermochemical diagrams of the atomization of Mg and Al. Step (1) is the energy required to promote a $3s$ electron into a $3p$ orbital. Step (2) is the measured heat of atomization. Step (3) is the sum of steps (1) and (2) and is a better measure of the forces holding the atoms together in the metallic crystal than step (2) alone.

properties in the transition metals. If we look at the fourth row of the periodic table, we find that potassium, with the smallest metallic valence, is a soft, low-density metal with a low melting point. The next metal is calcium, which is harder and denser than potassium and has a higher melting point. The trend of increasing hardness, density, and melting point continues with the first transition metal, scandium, through titanium, vanadium, and chromium. Table 19.2 summarizes the data for these six metals and other metals in the fourth row.

Each element in this row has one more valence electron than does its predecessor. The metallic valences go up in a regular way; 1 for K, 2 for Ca, 3 for Sc, 4 for Ti, 5 for V, and 6 for Cr. However, the metallic valence does not increase beyond six, which is the maximum number of orbitals that can be used effectively for bonding. Thus, the next few metals all have metallic valences of 6, and all have similar hardness, density, and melting points. Transition metals such as chromium, manganese, iron, cobalt, and nickel are used on a large scale because of such properties as hardness and density, which are related to high metallic valences.

We find a decrease in hardness, density, and melting point starting with copper and continuing with zinc and gallium. The decrease reflects a drop in metallic valences that is caused by the filling of the $d$ orbitals and the pairing of electrons. There are irregularities in all these trends, since the properties of metals are affected by such factors as crystal geometry.

## The Electron Sea Model

Our picture of the metallic bond thus far has been rather crude. We can get a better picture by using what is called the *electron sea model,* in which the metal atom and all the electrons that do not participate in bonding are considered separately from the electrons that do participate in bonding. The electron sea model pictures the crystal lattice as a network of metal atoms, which bear positive charge because some of their electrons have

**TABLE 19.2** Properties of Fourth-Row Metals

| Metal | Symbol | Atomic Number | Melting Point (K) | Density (g/cm³) | Heat of Atomization (kJ/mol) |
|---|---|---|---|---|---|
| potassium | K | 19 | 337 | 0.86 | 90.0 |
| calcium | Ca | 20 | 1110 | 1.55 | 178 |
| scandium | Sc | 21 | 1812 | 3.0 | 375 |
| titanium | Ti | 22 | 1941 | 4.51 | 469 |
| vanadium | V | 23 | 2173 | 6.1 | 515 |
| chromium | Cr | 24 | 2148 | 7.19 | 397 |
| manganese | Mn | 25 | 1518 | 7.43 | 285 |
| iron | Fe | 26 | 1809 | 7.86 | 416 |
| cobalt | Co | 27 | 1768 | 8.9 | 428 |
| nickel | Ni | 28 | 1726 | 8.9 | 430 |
| copper | Cu | 29 | 1356 | 8.96 | 339 |
| zinc | Zn | 30 | 693 | 7.14 | 131 |
| gallium | Ga | 31 | 302 | 5.91 | 276 |

been separated off for bonding. The lattice of positive metal ions is pictured as being immersed in a "sea" of negative electricity formed by the bonding electrons, as Figure 19.2 shows. While the atoms maintain fixed positions, the bonding electrons lose all connection with their source atoms and move freely throughout the entire metal crystal, forming the electron sea.

We can explain the photoelectric effect and the thermionic effect by saying that the ejected electrons jump off the surface of the electron sea because of an input of energy. The conductivity of electricity can be explained if we say that an electron enters the electron sea and creates a disturbance, which is transmitted through the sea until an electron leaves at the "opposite shore." The overall neutrality of the metal is maintained, and the net result is the entrance of one electron at one end of the metal and the departure of another electron at the other end, as shown in Figure 19.3. This picture is consistent with the observed decrease in the electrical

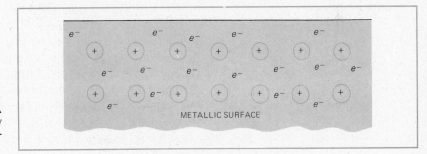

**Figure 19.2**
The electron sea model of the metallic bond. A metal crystal is pictured as a lattice of positively charged ions surrounded by a "sea" of completely delocalized bonding electrons.

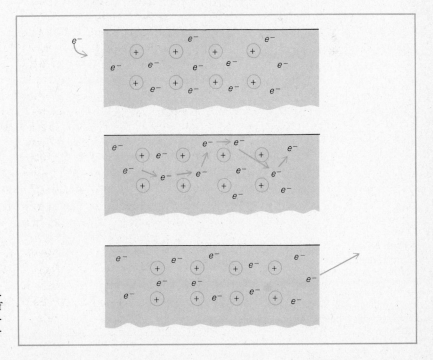

**Figure 19.3**
Electrical conductivity as pictured by the electron sea model. An electron enters one end of the "sea" formed by delocalized bonding electrons and causes a "ripple." Eventually an electron leaves at the other end.

conductivity of metals with an increase in temperature. As the temperature rises, the thermal motion of ions in the lattice increases, interfering with the movement of a "wave" through the electron sea and reducing electrical conductivity.

The electron sea model is also consistent with the plastic deformability of metals. Figure 19.4(a) shows an idealized picture of deformation: One plane of atoms slips with respect to another. Since the bonds between atoms are pictured as being completely delocalized, this slippage does not create any bonding problems. There are no bonds to be broken, and since all the units of the crystal lattice have identical charges, any unit is an acceptable nearest neighbor for any other. Deformation is more difficult in an ionic crystal. When the units in an ionic crystal move with respect to eath other, substantial electrical repulsions are created, as shown in Figure 19.4(b). Deformation is even more difficult in a crystal such as diamond, whose units of structure are covalently bonded. Strong covalent bonds must be broken if those atoms are to be moved with respect to each other.

The electron sea picture is oversimplified because it overemphasizes delocalization of bonding electrons. If we followed the model to its logical conclusion, we would assume that metals have no mechanical strength because they have no resistance at all to applied stresses. But we know that metals are hard and resist deformation. When we study the properties of metals, we must also consider the picture in which there are attractions between nearest neighbors in the metal crystal lattice.

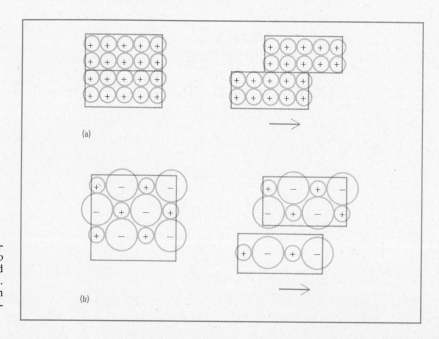

**Figure 19.4**
(a) Metal deformation is pictured as the slippage of one layer of metal atoms with respect to the adjoining layer. Metals are easily deformed because the bonding electrons are delocalized. (b) In ionic bonding, by contrast, deformation of ionic crystals creates substantial electrical repulsions.

## 19.3   THE BAND THEORY

For another description of bonding in metals, we must focus on molecular orbitals formed by overlap of the atomic orbitals that participate in the metal bond. Two atomic orbitals overlap to form two molecular orbitals: a bonding orbital, of lower energy than either of the two original non-bonding atomic orbitals; and an antibonding orbital, of higher energy than either of the two original orbitals (Section 8.3). The two electrons of the covalent bond are found in the bonding orbital. The decrease in energy of the bonding orbital relative to the atomic orbital equals the increase in energy of the antibonding orbital.

Let us apply this description to the bonding of two lithium atoms. When the two atoms come together, the 2s orbitals of the atoms overlap, forming two molecular orbitals that extend over both lithium atoms. The 2s electron of each atom is in the lower-energy bonding molecular orbital. If more than two atoms come together, more than two atomic orbitals overlap to form molecular orbitals. As more atoms come together to form a lithium crystal, more 2s orbitals overlap and form molecular orbitals that extend over more and more lithium atoms. As the number of atoms increases, the difference in energy between the molecular orbitals that form becomes steadily smaller, as Figure 19.5 shows schematically. When there is a very large number of atoms, as in a crystal, there are many molecular orbitals in a relatively narrow energy range. We can say that there is a *band of orbitals.*

The number of levels in an orbital band is the same as the number of atoms. A band can hold two electrons for each of its levels. For lithium, the 2s band is only half full. Since each orbital of the band extends over all the atoms, the orbital band theory allows us to picture the electrons moving freely through the crystal, as in the electron sea model.

The orbital band concept can help us understand some properties of lithium, such as its luster. In general, metals are shiny because they easily absorb and emit all wavelengths of visible light. The half-filled orbital band in lithium includes a number of closely spaced levels, both occupied and unoccupied. All wavelengths of visible light can be absorbed and emitted because of electronic transitions between these levels. The band model also accounts for the electrical conductivity of lithium. When an electric current is applied, electrons enter vacant levels in the band. Since each level extends over the entire crystal, the vacant levels provide a route by which electric current may pass easily through the metal.

We encounter a difficulty when we apply the band theory to beryllium, the next metal in the periodic table. Beryllium has twice as many valence electrons as lithium. Therefore, the 2s orbital band will be filled. It can be shown that a filled orbital band does not allow the conduction of electricity. Yet the electrical conductivity of beryllium is greater than that of lithium.

Conduction occurs because the 2s orbital is not the only atomic orbital that overlaps to form bands. Empty bands can be formed by the overlap of

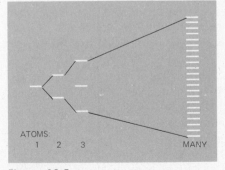

**Figure 19.5**

The band theory of metal bonding. When metal atoms come together, their atomic orbitals overlap. When there are very many metal atoms, there are very many molecular orbitals in a narrow band of energy levels.

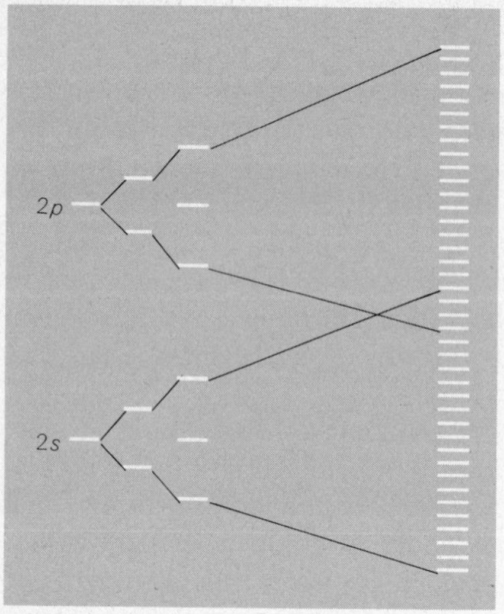

**Figure 19.6**

The overlap of a conduction band and a valence band. In alkaline-earth metals, the $2s$ and $2p$ orbitals form bands whose energy levels overlap. The metals conduct electricity because of the overlap of the $2p$ conduction band with the $2s$ valence band.

the $2p$ orbitals of the beryllium. We can thus differentiate between two types of orbital bands. The valence band holds the valence electrons. If it is unfilled, as in lithium, it can be used to conduct electricity. If the valence band is filled, as in beryllium, it cannot be used to conduct electricity. We then turn our attention to the other orbital bands that are formed by the overlap of valence atomic orbitals. These orbital bands, which are higher in energy than the valence band, are called the *conduction bands*. They can be used to conduct electricity.

While the $2s$ and $2p$ atomic orbitals have different energies, the spread in energy that results from band formation makes it possible for parts of the bands to fall in the same energy range. In beryllium and the other alkaline-earth metals, the $2s$ and $2p$ bands have overlapping regions of energy, as Figure 19.6 shows. *If the valence band and the conduction band in a solid overlap, the solid is an electrical conductor* because the conduction band contains some electrons. All metals have either an unfilled valence band or overlap between the valence band and another band that can be the conduction band. In general, substances are electrical conductors when there are more valence orbitals than valence electrons. The only element that has more valence orbitals than valence electrons but does not have metallic properties is boron.

**The Band Energy Gap**

Band theory also gives an explanation of the properties of insulators, materials with very low electrical conductivity, and semiconductors, materials whose electrical conductivity is substantially larger than that of insulators but substantially smaller than that of metals. In both an insulator and a semiconductor, the energies of the conduction band and of the valence band are so different that there is no overlap. There is an energy gap, sometimes called a forbidden zone, between the bands. The difference between an insulator and a semiconductor can be traced to the magnitude of the energy gap, $\Delta E_0$, which can be taken as the difference in energy between the top of the valence band and the bottom of the conduction band. Figure 19.7 is a schematic representation of the relative energies of the valence and conduction bands in insulators (a), semiconductors (b), and metals (c) and (d).

A material with a large $\Delta E_0$ is an insulator because not enough energy is available under normal conditions to promote electrons from the valence band to the conduction band. There are ways in which insulators can be made to carry current. An insulator can be placed in an intense electrical field, which may provide enough energy to raise some electrons into the conduction band. Some insulators are photoconductors. They conduct electricity when they are exposed to certain frequencies of ultraviolet radiation. The frequency of the radiation must be high enough to make $h\nu$ equal to or greater than $\Delta E_0$, so that electrons are promoted into the conduction band when the radiation is absorbed.

In semiconductors, $\Delta E_0$ is smaller than in insulators. The thermal

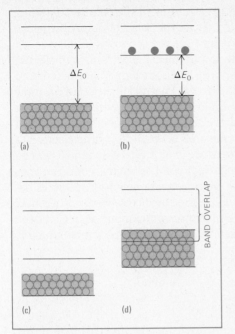

**Figure 19.7**
According to the band theory, the energy gap $\Delta E_0$ between the energy of the valence band and that of the conduction band determines the electrical properties of a substance. In an insulator (a), the energy gap is so wide that no electrons can move into the conduction band. In a semiconductor (b), the gap is narrower and a few electrons can move into the conduction band. In a metal, the valence band is not filled (c) or the valence band and conduction band overlap (d), so there is no energy gap and the metal is a good conductor.

energy available under ordinary conditions can promote electrons out of the valence band into the conduction band. The semiconductor then conducts electricity for two reasons: There are electrons in the conduction band, and the valence band is no longer completely filled. The orbital band model predicts that the electrical conductivity of a semiconductor will increase as the temperature rises. Indeed, the increase in conductivity with an increase in temperature is one feature that distinguishes a semiconductor from a metal.

In general, the magnitude of $\Delta E_0$ decreases as we go down any given column of the periodic table. Table 19.3 gives the trend for group IVA. As we move down the column from a nonmetallic insulator to semimetallic conductors to metals, we find a decrease in the value of $\Delta E_0$.

## Semiconductors

There are two kinds of semiconductors. Materials that are semiconducting because of a relatively small value of $\Delta E_0$ are called intrinsic semiconductors. Materials that are semiconductors because they contain trace impurities are called extrinsic semiconductors. Silicon is a well-known extrinsic semiconductor. It is listed as an insulator in Table 19.3, but silicon must be of extreme purity to be an insulator. Even the tiniest traces of impurities increase conductivity enough to make silicon a semiconductor. For example, 10 atoms of boron in 1 000 000 atoms of silicon increase its conductivity by a factor of 1000.

By introducing impurities into a crystal, we create additional energy levels close to the existing bands. We do not create new bands, but the impurities add electronic levels, which are called impurity levels.

One way to increase conductivity is to add an element that has more electrons than the element in the pure crystal. We can add a small quantity of arsenic or phosphorus to silicon, a process called doping. Both phosphorus ($3s^2 3p^3$) and arsenic ($4s^2 4p^3$) have five valence electrons. Only four valence electrons bond a P or As atom into the lattice. As Figure 19.8(a) shows, the fifth electron resides in an impurity level that is fairly close in energy to the conduction band of silicon. Thermal energy can promote these electrons into the conduction band, so silicon doped with either arsenic or phosphorus is a semiconductor. Since the conductivity

| TABLE 19.3   Energy Gap of Elements in Group IVA | | |
|---|---|---|
| **Substance** | **Electrical Conductivity** | **$\Delta E_0$ (kJ/mol)** |
| diamond | insulator | 502 |
| silicon | insulator | 105 |
| germanium | semiconductor | 58 |
| tin (gray) | semiconductor | 8 |
| tin (white) | metal | 0 |
| lead | metal | 0 |

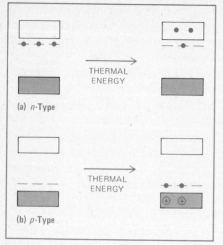

**Figure 19.8**

(a) In an *n*-type semiconductor, such as silicon doped with arsenic, one electron of each added atom is not used for bonding. The electron is in an impurity level and can be moved into the conduction band by thermal energy. The flow of these electrons results in electrical conduction. (b) In a *p*-type semiconductor, such as silicon doped with boron, electrons are promoted by thermal energy from the filled valence band into vacant impurity levels. Electricity is conducted by the movement of electrons into positive holes in the silicon lattice.

results from the presence of negative electrons in the conduction band, silicon doped in this way is called an *n*-type semiconductor, where *n* stands for "negative."

We can also add elements with fewer valence electrons than silicon, such as boron. In this case, the impurity levels are empty and are close to the valence band, as Figure 19.8(b) shows. Thermal energy can promote electrons from the valence band to the impurity levels. When electrons are promoted out of the valence band, the band is no longer completely filled and the conduction of electricity is possible. Since the loss of an electron to the impurity level causes the silicon lattice to be positively charged, we can picture the lattice as containing positive holes. Electricity is conducted as electrons move into these positive holes, leaving new positive holes behind them. In a sense, we can visualize the current as being carried by a reverse flow of the positive holes. An extrinsic semiconductor in which the impurity is relatively electron deficient is called a *p*-type semiconductor, where *p* stands for "positive."

## 19.4   METALLURGY

Almost all the metals we use come from ores that are mined from the crust of the earth. We define an *ore* as a mineral or other naturally occurring material that is a usable source of a metal. The technology that produces metal from ore is *metallurgy*. Ore usually undergoes some preliminary treatment, called *dressing,* which is followed by the actual extraction of the metal, called *winning,* and purification of the metal, called *refining.*

### Dressing; Preliminary Treatment

Ore usually is mined in the form of large chunks of metal-containing rock, mixed with unwanted materials such as sand or clay, collectively called gangue. The first step in dressing the ore is to crush the chunks into smaller particles. Next, as much as possible of the gangue is separated from the ore, most often by a process called flotation, in which air is blown through a mixture of the ore and water to form a foam that floats on top of the water. The particles of metal-containing material adhere to the foam and are skimmed off, while the gangue sinks to the bottom.

Other dressing methods are available. One technique that is familiar to moviegoers is panning, in which flowing water is used to wash away sand and clay from gold-containing particles. Some metals combine readily with mercury to form solutions called *amalgams.* After the amalgam forms, it is poured off and subjected to further purification. Ores of gold and silver are commonly dressed with mercury.

Some ores undergo further dressing, either to remove impurities or to convert the metal-containing substances into compounds from which the metal is more easily won. Such a step is necessary in the production of aluminum. Iron oxide is a common contaminant of bauxite, the major

## THE ELECTRONICS EXPLOSION

In the 1960s, you could buy a hand-held calculator that would add, subtract, multiply, and divide for anywhere between $500 and $1000. In the early 1970s, as prices came down, there was speculation that a four-function çalculator might someday be available for less than $100. Today, a basic four-function calculator can be had for $10 or less, and $50 will buy a programmable calculator whose capabilities approach those of a small computer.

The price revolution is the result of a technological revolution. The calculator of the 1960s used integrated electronic circuits that contained about a dozen transistors or similar components on a single chip. Today, mass-produced chips only a few millimeters square contain several thousand such components. This very large scale integration (VLSI) has been achieved largely with metal-oxide-semiconductor (MOS) technology, which gets its name from the materials that are used.

A single MOS semiconductor consists of two semiconductor regions, called the source and the drain, which are made of silicon heavily doped with impurities. The silicon semiconductor is coated with silicon oxide, which acts as an electrical conductor. A metal is used as a "gate" electrode, which is placed between the source and the drain. When the potential is high enough, a thin region of silicon under the gate becomes a conduction path for electrons, so the gate potential controls the amount of current in the transistor.

Very large scale integration is possible because the method used to produce one transistor can be used to make many thousands of other components simultaneously on a single chip. In fact, hundreds of chips can be produced at once from a wafer about 5 cm square by a multistep chemical fabrication process.

In the first few steps of the process, a layer of silicon oxide is grown on the wafer and a layer of a photoresistant material, which hardens on exposure to ultraviolet radiation, is applied. Ultraviolet irradiation followed by acid etching creates the pattern of a circuit on the wafer. Phosphorus atoms then are diffused into exposure areas of silicon to form the source and drain areas. Perhaps a dozen more such steps are needed to produce a finished circuit.

Such a circuit is so small that the wavelength of the visible light used for some steps has become the limiting factor. In current technology, the smallest features are about 5 microns ($\mu$) wide. Visible light cannot be used for features smaller than about 2 $\mu$ wide, so research has begun on the possible use of electron beams, which have produced features only 1 $\mu$ wide experimentally. Even further miniaturization might be possible with the use of X rays, which have been used to fabricate features as small as 0.1 $\mu$ wide.

Continual advances in VLSI have led to circuits of microscopic size and advanced capability. The past few years have seen the introduction of the microprocessor, a single chip that has the capability of a computer that sold for $50,000 or more in the 1950s. Microprocessors today are being used in automobiles, appliances, electronic games, and home computers which cost only a few hundred dollars. As costs continue to drop, the electronic revolution continues.

ore of aluminum. The iron oxide is removed before the metal is won from the ore. Bauxite, which is actually aluminum oxide, is soluble in sodium hydroxide solution, because aluminum forms a complex ion with hydroxide ligands. Iron does not, and iron oxides are insoluble in basic solution. To eliminate the iron oxides, the aluminum oxide is dissolved away from the rest of the ore by the process:

$$Al_2O_3(s) + 2OH^-(aq) + 3H_2O \longrightarrow 2Al(OH)_4^-(aq)$$

The solution is separated and then neutralized. The resulting precipitate of aluminum hydroxide is heated to drive off water. This leaves $Al_2O_3$, from which the metallic aluminum is produced.

For the winning of a metal, it is often desirable to have the metal as its oxide. In the case of carbonate ores, the oxides can be readily formed by heating:

$$ZnCO_3(s) \longrightarrow ZnO(s) + CO_2(g)$$

Sulfide ores can be converted to oxides by roasting, heating the ore in air:

$$2PbS(s) + 3O_2(g) \longrightarrow 2PbO(s) + 2SO_2(g)$$

## Extraction

To some extent, both the process by which a given metal is won from its ore and the nature of the compounds of the metal in the ore can be correlated with the reduction potential of the metal (Section 16.6 and Table 16.1). Table 19.4 summarizes this relationship.

Metals with large negative reduction potentials do not occur in uncombined form in nature. Compounds of such metals are stable with respect to the uncombined metal. The alkali metals and most alkaline-earth metals, which have large negative reduction potentials, react readily with oxygen in the air to form oxides. These metals are never found uncombined in nature. Interestingly, they are not found in the form of simple oxides, either. These simple oxides are so basic that they react readily with water and other acidic oxides, such as carbon dioxide. Solutions of NaOH are difficult to store because they react with atmospheric $CO_2$. For this reason, NaOH solutions must be standardized just before they are used for titrations. If free sodium is exposed to the atmosphere, we can represent the reactions that take place as:

$$4Na(s) + O_2(g) \longrightarrow 2Na_2O(s)$$
$$Na_2O(s) + CO_2(g) \longrightarrow Na_2CO_3(s)$$

We do not find sodium oxide in the earth's crust, but we do find large deposits of sodium carbonate. The alkali metals and alkaline-earth metals are also found as chlorides, sulfates, and complex oxides.

Because these metals form relatively stable compounds, they are not easily extracted from their ores by conventional chemical techniques. Direct thermal decomposition is rarely practical. The usual method is electrolysis. Since the metals react with water, the electrolysis generally is performed with the molten salts.

Metals such as magnesium and aluminum, which form less basic oxides than the alkali metals and other alkaline-earth metals, are found in nature as simple oxides and silicates, in which they are combined with both silicon and oxygen. We mentioned that the most common ore of

**TABLE 19.4** Reduction Potential[a] and Metallurgy

| Metal | Symbol | $\mathscr{E}°$ (V) | Occurrence | Extraction |
|-------|--------|------|------------|------------|
| lithium | Li | −3.045 | chlorides, | electrolytic |
| rubidium | Rb | −2.925 | carbonates | |
| potassium | K | −2.924 | sulfates, and | |
| barium | Ba | −2.90 | complex oxides | |
| strontium | Sr | −2.89 | | |
| calcium | Ca | −2.76 | | |
| sodium | Na | −2.711 | | |
| magnesium | Mg | −2.375 | complex oxides, | electrolytic or |
| aluminum | Al | −1.706 | silicates, some | chemical |
| beryllium | Be | −1.70 | simple oxides and carbonates | |
| manganese | Mn | −1.029 | simple oxides and | chemical reduction |
| vanadium | V | −0.89 | sulfides, some | with carbon |
| zinc | Zn | −0.7628 | complex oxides | |
| chromium | Cr | −0.74 | | |
| iron | Fe | −0.44 | | |
| cobalt | Co | −0.27 | | |
| nickel | Ni | −0.23 | | |
| tin | Sn | −0.1364 | | |
| lead | Pb | −0.1263 | | |
| bismuth | Bi | 0.32 | simple oxides and | chemical or |
| copper | Cu | 0.3402 | sulfides, free | physical |
| mercury | Hg | 0.7961 | metal (native) | |
| silver | Ag | 0.7996 | | |
| platinum | Pt | 1.23 | | |
| gold | Au | 1.42 | | |

[a] From the most common oxidation state.

aluminum is bauxite, which is hydrated aluminum oxide. Magnesium is often found as the silicate called asbestos, whose formula is $Mg_6Si_4O_{11}(OH)_6 \cdot H_2O$, or as talc, whose formula is $Mg_3Si_4O_{10}(OH)_2$.

Electrolysis is the standard method for preparing aluminum from bauxite, although some nonelectrolytic processes have been developed recently. Magnesium is produced either by the electrolysis of molten magnesium chloride or by the ferrosilicon process, which is based on a chemical reduction of magnesium oxide. A plentiful mineral called dolomite, a magnesium calcium carbonate whose formula can be written $MgCO_3 \cdot CaCO_3$, is converted to the oxides by heating. The oxides are reduced by ferrosilicon, an alloy of iron and silicon, in a reaction that can be written as:

$$2MgO \cdot CaO + Si(Fe) \longrightarrow 2Mg + Ca_2SiO_4 + Fe$$

and is carried out at about 900 K.

Metals with less negative reduction potentials, such as manganese and all the metals below it in Table 19.4, usually are found as simple oxides or as sulfides. Compounds of these metals generally are less stable than compounds of metals with more negative reduction potentials. These metals thus are more easily won from the ores. Chemical reduction is used most often.

This method can be employed to produce zinc. The most common ore of zinc is sphalerite, which is zinc sulfide, ZnS. The preliminary treatment of the ore is roasting:

$$2ZnS(s) + 3O_2(g) \longrightarrow 2ZnO(s) + 2SO_2(g)$$

In a method of chemical extraction called the distillation process, the zinc oxide is reduced at high temperature with carbon:

$$ZnO(s) + C(s) \longrightarrow Zn(g) + CO(g)$$

At elevated temperatures, the zinc metal is a gas that is collected in relatively pure form by condensation. This kind of reaction is used to win metals with reduction potentials close to that of zinc. Carbon is the reducing agent of choice. It is inexpensive and it forms gaseous oxidation products that are easily removed.

Metals that have less negative reduction potentials than zinc are easier to extract from their ores because their compounds are even less stable. Tin and lead are two such metals. Both are relatively inexpensive because they are easily prepared from their ores.

Metals with positive reduction potentials can be found native — uncombined — in the earth's crust. The most familiar metals, those that were known earliest in history, are those with positive or only slightly negative reduction potentials. Metals such as gold and copper are not abundant in the earth's crust, but the ease with which they can be obtained from their ores made them available to ancient civilizations. Abundance is generally no indication of a metal's role in history. Almost all the common metals, those most frequently encountered in our lives — tin, gold, lead, copper, silver, mercury — are not abundant, but they are easily prepared. The major exception is iron, which is both abundant and easily won, and which has played a prominent role in history.

## Refining

Many methods are used to refine metals. Some are physical methods, based on a difference in phase transition temperature or in solubility between a metal and its impurities. Zinc, which has a relatively low boiling point (1180 K), can be separated from iron (B.P. 3272 K) and lead (B.P. 2010 K), both common impurities, by distillation, a method that is useful for all metals with low boiling points.

A method used to prepare extremely pure samples of metals and other solids that do not decompose when they melt is zone refining, which is shown schematically in Figure 19.9.

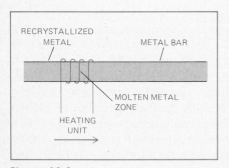

**Figure 19.9**
Zone refining. A small section of a metal bar is melted by the heating unit. As the heating unit moves along the bar, impurities move with it, since they are more soluble in liquid metal than in solid metal. The impurities are thus concentrated in one zone, which can be removed when the heating unit reaches the end of the bar. Successive passes can give metal of extremely high purity.

Metals can also be refined by electrolysis, a process that is especially important in copper production. The smelting of copper ore produces impure copper that is formed into large anodes, weighing about 300 kg each, which then are suspended in a solution of copper(II) sulfate and sulfuric acid. The cathodes are thin sheets of pure copper. When current is passed through the cell, the half-reactions are

$$\text{anode:} \qquad Cu(s) \longrightarrow Cu^{2+} + 2e^-$$
$$\text{cathode:} \quad Cu^{2+} + 2e^- \longrightarrow Cu(s)$$

The pure copper that collects on the cathode is removed and recast.

Another technique used to prepare very pure metals is the conversion of the metal to a volatile compound that is purified by distillation. This technique is used for the purification of nickel, with nickel carbonyl as the intermediate:

$$Ni \text{ (impure)} + 4CO(g) \longrightarrow Ni(CO)_4(g) \xrightarrow[\Delta]{} Ni \text{ (pure)} + 4CO(g)$$

The cyanide process can be used to refine some precious metals that form relatively stable cyanide complexes. When impure gold is treated with cyanide solution through which air is bubbled, a complex ion forms:

$$4Au(s) + 8CN^-(aq) + O_2(g) + 2H_2O \longrightarrow 4Au(CN)_2^-(aq) + 4OH^-(aq)$$

The gold is recovered from the complex ion by treatment with a metal such as zinc:

$$2Au(CN)_2^-(aq) + Zn(s) \longrightarrow 2Au(s) + Zn(CN)_4^{2-}(aq)$$

## 19.5   THE GROUP VIIIB METALS

The metals of group VIIIB can be divided into two subgroups. One subgroup consists of iron, cobalt, and nickel, which are in the fourth row of the periodic table, the first row of the transition elements. Many properties of these three elements differ from the properties of the six so-called platinum metals in the other subgroup, which are in the second and third rows of the transition metals. The six group VIIIB metals in these rows are ruthenium (Ru), rhodium (Rh), palladium (Pd), osmium (Os), iridium (Ir), and platinum (Pt).

With the exception of platinum and palladium, the group VIIIB metals have ground state valence electronic configurations in which the outermost $s$ subshell is filled and the outermost $d$ subshell is partly filled. The $+2$ oxidation state plays the dominant role in the chemistry of these metals, but the $+3$ state is also common, especially in iron and cobalt. The higher oxidation states are more important in the chemistry of the platinum metals. For example, ruthenium is found in all the oxidation states

The high temperatures required to obtain iron from its ores in a blast furnace are evident in this scene. (*Inland Steel*)

from $+1$ to $+8$. The $+3$, $+4$, $+6$, and $+8$ states are the most important. For osmium, the principal oxidation state is $+8$.

All the elements of group VIIIB have a metallic valence of 6. The resulting strong metallic bonding is reflected in the strength and hardness that makes these metals so valuable. The six platinum metals are much more inert than the first three group VIIIB metals — so much so that they are often called noble metals.

The group VIIIB metal that has had the widest range of uses is iron, the fourth most abundant element in the earth's crust (after oxygen, silicon, and aluminum). Modern civilization is built on a foundation of iron, particularly the alloys and the intermetallic compounds we call steel.

## Metallurgy of Iron

Extracting iron from its ores and producing iron alloys is one of the central activities of any industrial society. The major iron ores are the oxides hematite, $Fe_2O_3$, and magnetite, $Fe_3O_4$, and the carbonate siderite, $FeCO_3$. Once the ores are mined, they are roasted. The iron is won from the ores in a blast furnace (Figure 19.10). If the ores are oxides, the iron is liberated by reduction with carbon, in the form of coke, at high temperatures. The reactions are

$$2C(s) + O_2(g) \longrightarrow 2CO(g)$$
$$3CO(g) + Fe_2O_3(s) \longrightarrow 2Fe(l) + 3CO_2(g)$$

The gases formed in the reactions escape through the top of the blast furnace. The liquid iron percolates to the bottom of the furnace, where it is poured off.

Substantial quantities of impurities in iron ore are removed by the introduction of a material called *flux*, which causes the impurities to fuse. The fused impurities combine with the flux to form a glassy substance called *slag*. The slag, which is less dense than liquid iron, forms a layer atop the iron that can be removed easily. If the ore contains sand or clay, limestone, $CaCO_3$, is used as the flux. If the ore contains limestone as an impurity, sand or clay — actually silica, $SiO_2$ — is used as the flux. In either case, the slag is primarily a calcium silicate whose formation can be represented by the reactions:

$$CaCO_3(s) \longrightarrow CaO(s) + CO_2(g)$$
$$CaO(s) + SiO_2(s) \longrightarrow CaSiO_3(l)$$

The molten iron that collects at the bottom of a blast furnace contains about 3% or 4% carbon and a number of other impurities, including small amounts of silicon and sulfur. This iron is cast into bars that are called pigs. While pig iron is the cheapest form of iron, its usefulness is limited because it is brittle.

Wrought iron is essentially pig iron that has been further purified. It has

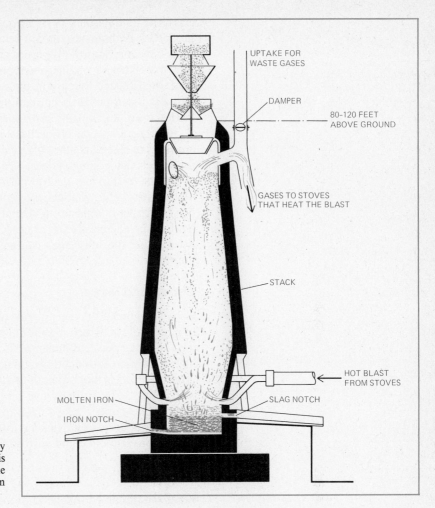

**Figure 19.10**
A blast furnance. Iron is won from its ore by reduction with carbon. The molten iron is poured off from the bottom. Impurities in the iron combine with flux to form a glassy calcium silicate called slag, which is separated out.

more desirable mechanical properties than pig iron. Most of the impurities are removed when the pig iron is heated in a furnace. A bed of iron oxide causes the oxidation of impurities and is itself reduced to relatively pure iron. Carbon becomes carbon monoxide, and sulfur becomes sulfur dioxide. Gases such as these escape from the iron, while impurities such as silicon and phosphorus form a slag that is poured off. The end product, wrought iron, is 99.5% pure.

## Steel

Steel is manufactured by further purification of the iron. A stream of heated air is blown into the iron to oxidize the impurities. The furnace is lined with an oxide such as lime (CaO) or silica to absorb other oxidation products. Once the iron is purified, carefully measured amounts of carbon can be added to form carbon steel. There are several kinds of carbon steel, classified by the quantity of carbon they contain and their mechanical properties. Mild steels contain relatively low concentrations of carbon. They have mechanical properties similar to those of wrought iron and are used to make wire, chains, pipes, and sheet iron. Medium steels, containing from 0.2% to 0.6% carbon, are used for buildings, bridges, and railroad rails. High-carbon steels, containing up to 1.5% carbon, are used for razor blades, tools, and cutting instruments.

Steels with special properties can be produced by addition of small amounts of elements other than carbon. Stainless steels, for example, are mild steels that contain at least 12% chromium and 7% to 9% nickel. The resistance of stainless steel to corrosion makes it useful for cutlery and other instruments. Manganese steel is wear resistant and is used for such things as safes and grinding machinery. Chromium-vanadium steels are both strong and elastic and are used for, among other things, automobile axles. Other specialty steels contain tungsten, molybdenum, cobalt, and titanium.

## Cobalt and Nickel

Cobalt and nickel are constituents of many alloys. Together, they are used in alnico alloys, which also contain aluminum and iron and which are used to make powerful magnets. Cobalt is added to steels designed to be highly resistant to oxidation and corrosion. Nickel is a constituent of a number of steels that are both tough and ductile. Nickel is often alloyed with copper. The ordinary five-cent piece, the "nickel," is such an alloy; other nickel-copper alloys have many uses based on their exceptional resistance to corrosion. Nickel is often used to plate other metals, making them more attractive and corrosion resistant.

## Ferromagnetism

When we discussed magnetic properties in Section 6.2 we divided substances into two groups; those that are paramagnetic and are attracted by an external magnetic field, and those that are diamagnetic and are repelled by an external magnetic field. A relatively small group of paramagnetic substances display an unexpectedly large attraction when they are exposed to a relatively weak external magnetic field. These substances have another unusual characteristic — they remain magnetized when the external field is removed. Such substances are said to be *ferromagnetic.* Iron, cobalt, and nickel are ferromagnetic at room temperature. Only one other pure element, gadolinium, is ferromagnetic at room temperature, although some alloys and compounds are ferromagnetic.

Only solids are ferromagnetic. The phenomenon occurs when all the units of structure in a small region of the crystal, called a *domain,* are lined up so that their magnetic moments are parallel. A typical domain is about 0.01 mm across and contains about $10^6$ units. Ordinarily, the moments of domains are randomly oriented throughout a crystal. When even a weak external magnetic field is applied, the direction of magnetization of many

# METALLIC GLASSES AND THEIR USES

If a metal alloy is cooled extremely rapidly, it does not solidify with the regular crystal structure that is typical of metals. Instead, an amorphous solid is formed in which the atoms are distributed randomly, much as they are in glass. Such metallic glasses were first discovered about two decades ago. Now, because of their unique properties, they are being put to widespread use in a number of applications.

Some metals are much stronger in the glassy form than when they have the regular metal structure. Others have superior resistance to corrosion or fatigue. The group of metallic glasses that is getting the most commercial attention is iron-based alloys that are magnetized very easily. They are being tested in transformers, where they promise a great reduction in energy losses.

About 20 million distribution transformers are operating in the United States today. Typically seen mounted on utility poles or concrete pads, these transformers reduce voltage from the level at which it is distributed on power lines, generally, 14 200 volts, to the 110-V or 220-V level necessary for home use. A significant amount of energy now is lost in the form of heat in the process of magnetizing the core of these transformers so that the lower voltage can be in-ducted in the secondary coils. Because metallic glasses are so easily magnetized, transformer cores made of these alloys would generate much less waste heat. It has been estimated that complete replacement of all existing transformer cores by metallic glass cores would save 12 billion killowatt-hours a year, the amount of electricity needed for a city of 4 million people.

One problem is that these metallic glasses tend to be brittle and unusually vulnerable to stress. The iron-based materials are produced by spraying molten metal through a slit onto a fast-turning copper wheel. When the molten metal hits the wheel, it cools at a rate of about $1 \times 10^6$ K/s. A thin ribbon of metallic glass can be formed at rates of up to 2 km/min.

Other metallic glasses are potentially valuable because they have a high resistance to wear. One experimental technique uses a laser to melt a thin layer of an alloy on the surface of a metal part, to serve as a wear-resistant coating. Still another technique uses metals in the gaseous state. The metal atoms condense on a cold surface. This "sputtering" technique is being tested as a way of putting coatings on metal parts that must function in highly corrosive environments.

domains becomes aligned with the field, so the substance has a strong magnetic polarization. When the external magnetic field is removed, the domains usually retain their aligned direction. To make the solid lose its strong magnetization, the domains must be returned to their random orientation. The solid can be melted, or heated, or even pounded with a hammer to destroy the ordered configuration and make it lose its magnetization.

## Compounds of Iron, Cobalt, and Nickel

We mentioned that the $+2$ oxidation state is frequently encountered in compounds of iron, cobalt, and nickel. The $+3$ oxidation state is also found in many compounds of iron. Cobalt generally can be in the $+3$ oxidation state only when it is stabilized by surrounding ligands, while only a few compounds of nickel in the $+3$ oxidation state are known. There are many compounds of these three metals, with a broad range of practical applications.

Iron forms three important oxides, FeO, $Fe_2O_3$, and $Fe_3O_4$. The last, named magnetite, is a black solid that was the first ferromagnetic substance to be discovered. It is called a mixed oxide and is best thought of as a combination of FeO and $Fe_2O_3$. The iron(II) oxide is unstable; it disproportionates to iron and magnetite:

$$4FeO(s) \longrightarrow Fe(s) + Fe_3O_4(s)$$

The ease with which Fe(II) is oxidized to Fe(III) by air creates practical difficulties for anyone who wants to work with iron in the $+2$ oxidation state. If a piece of iron(II) hydroxide, $Fe(OH)_2$, a white solid, is left standing on a table, it rapidly changes color, first to dark green and then to black, as $O_2$ in the air oxidizes the Fe(II) to Fe(III). The dark colors seen in this mixture and in magnetite are typical of many solids that contain two oxidation states of the same element. If the hydroxide is left standing long enough, its continued reaction with $O_2$ in the air eventually causes the color to change to red-brown, as the Fe(II) is converted completely to Fe(III). The formula $Fe(OH)_3$ is often written for the $+3$ hydroxide of iron, although no such compound has ever been isolated. The reddish brown substance formed in this way, or by precipitation from alkaline solutions of $Fe^{3+}$, actually has a composition close to FeO(OH).

Cobalt also forms three oxides whose composition and behavior resemble those of the oxides of iron. The colors of these oxides make them useful as pigments in ceramics and pottery. If a solution of $Co^{2+}$ is treated with a base, it precipitates $Co(OH)_2$, which can be either pink or blue, depending on the conditions. Solid $Co(OH)_2$ also is air-oxidized, although not as readily as the comparable hydroxide of iron. The brown solid that forms as Co(II) is oxidized to Co(III). It has a composition close to CoO(OH).

The nickel(II) oxide, NiO, and the hydroxide, $Ni(OH)_2$, are easily prepared. They do not oxidize in air, since higher oxidation states of nickel are much more difficult to form than those of iron and cobalt. Both NiO and $Ni(OH)_2$ are green.

Iron, cobalt, and nickel all react directly with carbon monoxide to form compounds called carbonyls, in which the metal is formally in the 0 oxidation state. Carbonyls are interesting chemically and have a number of practical uses. Iron pentacarbonyl, $Fe(CO)_5$, is a yellow liquid with a trigonal bipyramidal structure, with the Fe atom at the center. It is soluble in organic solvents and is used widely as a homogeneous catalyst in industry. The most important carbonyl of cobalt, $Co_2(CO)_8$, has a bond between the two cobalt atoms. It is used extensively to catalyze organic reactions. Nickel tetracarbonyl, $Ni(CO)_4$, is an extremely toxic substance that is used to prepare high-purity nickel powders and coatings. It is an intermediate in the refining of nickel.

Other compounds of iron, cobalt, and nickel—halides, sulfates, cyanides, sulfides, and so on—are well known and have many applications. Among them are $FeSO_4$, which is used in water purification and ink

| TABLE 19.5 | Cost$^a$ of Pure Metals (dollars/g) in 1984 |
|---|---|
| Metal | Cost |
| ruthenium | 3 |
| rhodium | 18 |
| palladium | 6 |
| osmium | 20 |
| iridium | 24 |
| platinum | 13 |
| copper | 0.0016 |
| silver | 0.32 |
| gold | 12.5 |

$^a$ At the time of this writing, the prices of some precious metals have been fluctuating. Prices may be quite different by publication time.

manufacturing; $FeCl_3 \cdot 6H_2O$, which is used medically to treat anemia and to stop bleeding; $CoCl_2 \cdot 6H_2O$, which can be used as an invisible ink, since a small amount of heat causes it to lose some water of hydration and to change from colorless to blue; and $NiSO_4$, which is used in nickel-plating baths.

### The Platinum Metals

Ruthenium, rhodium, palladium, osmium, iridium, and platinum are grouped together as the platinum metals because their properties are so similar. All six metals are dense, have high melting points, and are highly resistant to chemical attack, particularly oxidation by $O_2$ in the air. The combination of chemical inertness, desirable metallic properties, low abundance, and difficulty of refining makes these "noble metals" extremely expensive. Table 19.5 lists the prevailing 1984 prices of the platinum metals and some other metals in pure form.

The platinum metals occur in native form, so there is no great difficulty in winning them from their ores. But their abundance is so low that there are major difficulties in concentrating the ores. An even greater difficulty arises from the fact that all six metals occur together. Since they have similar properties and are unreactive chemically, they are not easily separated.

The platinum metals have many applications. Platinum and palladium are used extensively as catalysts in the petroleum industry and in many industrial processes. Both metals are used for switches and relays in telecommunications systems. Alloys of Pt and Pd are used in dentistry, while an alloy of gold and platinum, "white gold," and an alloy of platinum and iridium are used in jewelry.

The four less common platinum metals have found a variety of specialized uses. Any of them can be alloyed with platinum. The resulting alloys are quite wear resistant and are used for such parts as electrical contacts, fountain pen tips, and phonograph needles, in which the high cost of the alloy is offset by the small quantity that is needed. All four metals are used as catalysts in chemical manufacturing. Jewelry makers plate silver objects with rhodium; medical supply companies use iridium for hypodermic needles and other medical accessories. A compound of osmium and oxygen, osmium tetroxide, $OsO_4$, is widely used as a mild oxidizing agent and as a stain for tissue samples in biology and medicine. Osmium tetroxide is highly toxic and must be handled with great care.

## 19.6   THE GROUP IB METALS

Group IB contains the three coinage metals: copper, silver, and gold. Even though the neutral atoms of these three elements have completed $d$ subshells, all three are classified as transition metals because they can form cations with incomplete $d$ subshells. The electronic configuration of these metals resembles that of the group IA metals, since both have a half-filled $s$ shell. The configuration of copper is $4s^13d^{10}$, that of silver is $5s^14d^{10}$, and that of gold is $6s^15d^{10}$. In the group IB metals, however, the electron in the $s$ shell lies outside a completed $d$ subshell, while in the IA metals it lies outside a complete noble gas electronic shell. A group IA metal that loses one electron becomes a monocation with a noble gas electronic configuration. A group IB metal that loses an electron does not have a noble gas electronic configuration, so the physical and chemical properties of the two groups of metals are quite different.

## The Metals and Alloys

Copper, silver, and gold have been familiar from the earliest times — not because they are abundant metals, but because they were easily found and removed from the earth's crust. Gold may have been the first metal discovered and used. Copper objects more than 10 000 years old have been found by archeologists. As early as 5000 B.C., copper ores were being smelted on a fairly wide scale in the Middle East. Silver, unlike copper and gold, is rarely found in native form and so was not used as widely in early civilizations. In fact, silver was so rare that some ancient cultures prized it more than gold.

Aside from the cultural and sacramental value placed on gold and silver, the group IB metals have properties that make them valuable for many practical uses. Pure copper is an excellent electrical and thermal conductor and is widely used in electronics and in cookware. If pure copper is first heated and then cooled, a process called annealing, it is soft enough to be drawn into wire or hammered into shape. Conveniently, the metal hardens as it is worked and can be softened again by reheating.

Alloys in which copper is the chief constituent are used more widely than pure copper, because the alloys usually cost less and have more desirable properties. One major technological advance of early civilizations was the discovery of bronze, an alloy of copper and tin that is easier to cast, harder, and less malleable than pure copper. Bronze dominated early technology before iron came into use. Brass is an alloy of copper and zinc with multitudinous uses. Alloys of copper, nickel, and zinc, called nickel silver or German silver, have high resistance to corrosion and wear and are used as bases for silver plating and in costume jewelry.

Silver, until recently, was used primarily in coins, in the form of coinage silver, or sterling silver, an alloy of 90% silver and 10% copper. In recent years, industrial demand for silver has raised its price to the point where its use for coinage is impractical. Silver has a number of desirable properties. In pure form, silver has the highest electrical and thermal conductivity of any metal, and it is more ductile and malleable than all metals but gold and palladium. Silver is even valuable as a germicide. A solution of silver that is harmless to higher organisms can kill microbes on contact.

Gold today remains important primarily as an international monetary standard, a role it has held since the late eighteenth century. Although governments and economists have worked diligently to change the situation, gold still is regarded by many as the ultimate medium of exchange —an opinion that gains strength in times when paper monies deteriorate in value. But gold has properties that have led to its increasing use in electronics and in spacecraft. Gold is the most malleable and ductile of all the metals; it can be beaten into a foil, gold leaf, only $1 \times 10^{-6}$ cm thick. The electrical and thermal conductivity of gold is exceeded only by that of silver and copper. Thin films of gold reflect a high percentage of infrared radiation and thus are excellent heat shields; you may have seen pictures of spacecraft wrapped in gold leaf. Many heat-sensitive electronic devices are electroplated with gold. For other electronic uses, gold can be drawn into extremely fine wire. One gram of gold can form a wire about 3 km long.

Pure gold is too soft to be usable in jewelry or coinage, so it usually is alloyed with copper, silver, nickel, zinc, palladium, or other metals. The composition of a gold alloy is usually expressed as the number of parts of gold in 24 parts of alloy. Each part is called a karat. Thus, pure gold is 24 karat. An alloy of 75% gold and 25% copper is 18-karat gold.

While copper, silver, and gold all have the same valence electronic configuration and are similar in some physical properties, there are many differences in their chemical behavior that are not all explained simply.

All three metals are "noble"; their reduction potentials are positive, they resist corrosion, and they are not readily attacked by air. As we mentioned earlier, silver tarnishes when exposed to air that contains compounds of sulfur. When copper is exposed to moist air that contains sulfur compounds, it slowly forms a green patina that is a combination of $CuCO_3$, $Cu(OH)_2$, and some $CuSO_4$. The Statue of Liberty, which is made of copper, is covered by such a patina.

## Compounds

Each of the three metals has a half-filled $s$ electronic subshell, so we would not ordinarily expect them to be unreactive. Their relatively low reactivity results from their relatively high ionization energies, which can be explained by a comparison with the group IA metals. In the group IA metals, the single $s$ electron is shielded from the nucleus by a completed electronic shell. In the group IB metals, the single electron is shielded from the nucleus by a completed $d$ subshell, which does a less effective job. The negatively charged electron in a group IB metal is more tightly held by the nucleus.

There is also a contrast in second ionization energies of metals in the two groups. The second ionization energy of a group IA metal is very high, since an electron must be removed from a completed electronic shell. The second and third ionization energies of group IB metals are much lower, since the electrons are removed from the $d$ subshell. Thus, higher oxidation states of the group IA metals are unknown, while $+2$ and $+3$ oxidation states of the group IB metals are common.

The most common oxidation states are $+2$ for copper, $+1$ for silver, and $+3$ for gold. However, the $+1$ state is well known for copper, as are the $+2$ state for silver and the $+1$ state for gold. An interplay of factors determines the relative stability of the oxidation states of each metal.

The electronic configuration associated with the $+1$ state would seem to be best for all three metals, since it has a closed electronic subshell: $3d^{10}$ for $Cu^+$, $4d^{10}$ for $Ag^+$, and $5d^{10}$ for $Au^+$. But these monocations may not be as stable as dications or trications where there is electrostatic stabilization caused by solvation, or where they have anionic neighbors in a crystal.

In the case of copper, the $Cu^+$ cation is more stable than the $Cu^{2+}$ cation in the gas phase, but the opposite is true in aqueous solutions or in crystalline salts. For silver, the $Ag^+$ cation is the most stable under most conditions, probably because the second ionization energy of silver is relatively high, at least in comparison with copper. Gold has an unusually low third ionization energy that is consistent with the stability of the $+3$ ionization state. Table 19.6 lists the ionization energies of the three metals, with those of potassium, a group IA metal, shown for comparison.

There are many compounds of copper in both the $+1$ and the $+2$ oxidation states. In compounds of the $+1$ oxidation state, the copper is usually covalently bonded or complexed. Thus, the copper halides, CuCl, CuBr, and CuI, can be regarded as covalent compounds. Copper(I) is present in biological systems, stabilized by complexation with ligands in protein molecules. Compounds of copper in the $+2$ oxidation state are ionic solids or complex ions with coordination numbers of 4, 5, or at most, 6. In aqueous solution, the $Cu^+$ ion is unstable with respect to the $Cu^{2+}$ ion. The equilibrium

$$2Cu^+(aq) \rightleftharpoons Cu^{2+}(aq) + Cu(s)$$

lies far to the right.

Some compounds of copper with important practical uses are CuCl, a catalyst and desulfurizing agent in the petroleum industry; $CuCl_2$, a textile dye and a crop fungicide; $Cu_2O$, which is used in marine paints and electronic components; and CuO, an insoluble black solid which is used as an oxidizing agent. The most widely used copper compound is blue

**TABLE 19.6**  Ionization Energies of Group IB Metals (kJ/mol)

| Metal | Electronic Configuration | First | Second | Third |
|-------|--------------------------|-------|--------|-------|
| copper | $[Ar]3d4s^{10}$ | 745 | 1958 | 3554 |
| silver | $[Kr]4d5s^{10}$ | 731 | 2074 | 3361 |
| gold | $[Xe]4f^{14}5d6s^{10}$ | 890 | 1978 | 2940 |
| potassium | $[Ar]4s$ | 419 | 3050 | 4410 |

vitriol, hydrated copper(II) sulfate, $CuSO_4 \cdot 5H_2O$, which is a soil additive, a fungicide, a wood preservative, a component of electric cells, and a starting material for the manufacture of other copper compounds.

Many simple compounds and complexes of silver(I), are known. There are only two known simple compounds of silver(II), $AgF_2$ and $AgO$, although a number of complexes of silver(II) have been identified. One useful compound of silver(I) is silver nitrate, $AgNO_3$, which is unusual among simple silver salts because it is soluble in water; almost all the others are not. Silver nitrate is a starting material for the production of other silver salts. Another silver(I) compound, silver oxide, $Ag_2O$, is used as a mild oxidizing agent.

Gold compounds are not nearly as useful as those of copper and silver. Pure gold is needed for most applications. However, the gold cyanides, such as $KAu(CN)_2$, are used for electroplating, and some gold compounds are used medically to treat arthritis and tuberculosis—uses that are limited by the toxicity of gold. Most gold compounds are unstable thermally, decomposing to gold metal if they are heated. The only oxide of gold that is well characterized is $Au_2O_3$, which must be prepared indirectly because gold is the only metal that does not react with $O_2$, even at high temperatures.

## Photography

Unquestionably the most important silver compounds are the silver halides, $AgCl$, $AgBr$, and $AgI$, which are the light-sensing materials in photographic film. More than 30% of the industrial silver in the United States goes into photography.

When the silver halides are exposed to light, the silver is reduced from the $+1$ to the 0 oxidation state, as in the reaction

$$AgBr(s) \longrightarrow Ag(s) + \tfrac{1}{2} Br_2(l)$$

This process is called a photoreduction, that is, a reduction brought about by light. In photography, the photoreduction occurs in a photographic emulsion, which consists of small grains of a silver halide (usually silver bromide) suspended in gelatin and coated on a strip of cellulose acetate. Dyes in the emulsion make the silver bromide sensitive to a greater range of wavelengths of light.

When film is exposed, a small number of the molecules on the surface of the grain of silver halide undergo photoreduction. Grains in which this occurs are said to be sensitized. They can be reduced much more easily than grains of silver halide that are not struck by the light. Photosensitization is a highly effective process. Reduction of only five atoms of silver to the 0 oxidation state on the surface of a grain of silver halide with $10^{10}$ atoms is enough to sensitize the grain. When the exposed film is treated with a mild reducing agent, all the sensitized grains of the halide are reduced to silver, while the grains that were not struck by light remain unchanged. This exposure to a reducing agent is commonly called developing. Organic reducing agents such as hydroquinone are the most widely used developers.

The next step is to remove the undeveloped grains of silver halide by formation of a complex ion:

$$AgBr(s) + 2S_2O_3{}^{2-}(aq) \longrightarrow Ag(S_2O_3)_2{}^{3-}(aq) + Br^-(aq)$$

The process is called fixing. The most common fixer is a solution of sodium thiosulfate, $Na_2S_2SO_3 \cdot 5H_2O$, customarily called hypo. When the silver(I) complex is washed away, all that remains on the film are grains of metallic silver, which are most numerous in areas where the incident light was strongest. The film is then called a negative. A positive print is made when light is passed through the negative to strike another photographic emulsion on print paper.

The sensitivity of silver halides to light is not the reason they are uniquely suited for photography. Many other substances are equally light sensitive. The silver halides are unique because a very small amount of light sensitizes a large quantity of the silver compound to further reduction. In developing, the silver metal image formed by the initial exposure to light is intensified by as much as a factor of $10^{11}$.

**Summary**

We began this chapter by describing the chemical and physical properties of **metals.** They are elements that conduct heat and electricity well, are deformable and plastic, display the photoelectric effect and—aside from mercury—are solids at room temperature. We noted that when a metal combines chemically with another element, the other metal usually is a nonmetal. We described the **delocalized bonding** in metals, in which a number of atoms share bonding electrons, and said that the most important factor in the strength of the **metallic bond** is the **metallic valence,** the number of electrons that a metal atom has available for bonding. We then described the **electron sea** model, in which a lattice of metal ions is pictured as being immersed in a "sea" of relatively free electrons. For a different perspective on bonding in metals, we outlined the **band theory,** in which energy bands are formed by the overlap of bonding orbitals. We showed how the band theory can explain the electrical conductivity of metals and of **semiconductors.** Moving to the practical uses of metals, we described how metals are obtained from ores through metallurgy. We concluded by discussing the properties and uses of a number of metals in groups VIIIB and IB of the periodic table.

## Exercises

**19.1** List six characteristic properties of a material that allow us to classify it as a metal.

**19.2** Use the data in Table 19.1 to find the energy released when $K_2(g)$ forms $K(s)$.

**19.3**[1] The heat of atomization of $Bi(s)$ is 207 kJ/mol, while the heat absorbed when $Bi(s)$ is converted to $Bi_2(g)$ is 125 kJ/(mol of $Bi(s)$). Calculate the bond energy of the $Bi-Bi$ bond in $Bi_2(g)$.

**19.4** Consider a five-atom portion of a crystal of sodium, one sodium atom and four nearest neighbors. Draw ordinary contributing structures with localized bonds, that in sum describe the bonding in this portion of the metallic crystal.

**19.5** The $\Delta H_f^\circ$ of metal atoms in the gas phase is often used as an estimate of metallic bond strength. Indicate for

each of the following metals whether the estimate is a good one or is substantially in error: (a) lithium, (b) silver, (c) zinc, (d) vanadium.

**19.6** Propose an explanation for the fact that a greater energy is required for aluminum to reach an excited state for metallic bonding than is required for magnesium.

**19.7** Compare the metallic valence of the main group metals to their most common oxidation states.

**19.8** The ideal density of a metal is defined as the density it would have if its atomic weight were 50. Thus the ideal density is 50 divided by the volume of 1 mol of the metal. Use the data in Table 19.2 to make a plot of ideal density on the $y$-axis against number of valence electrons on the $x$-axis.

**19.9** Predict the trend in the melting point of the metals from rubidium to palladium in the periodic table.

**19.10** Indicate which of the following metals have an incompletely filled va-

lence band: (a) K, (b) Ag, (c) Cu, (d) Zn, (e) Ba.

**19.11** Propose an explanation for the thermal conductivity of metals based on the band theory.

**19.12** Graphite, although a nonmetal, is a good conductor of electricity. Propose an explanation for this observation.

**19.13**★ Tin exists in two allotropic forms. Gray tin has a diamond structure and white tin has a close-packed structure. Predict which allotrope is (a) denser, (b) a conductor of electricity. Predict the valence and the electronic configuration of tin in each allotrope.

**19.14** Calculate the minimum frequency of radiation required to promote an electron from the valence band of diamond into a conduction band.

**19.15** The conductivity of a $p$-type semiconductor increases with temperature in a fairly narrow temperature range above room temperature and

---

[1] The answers to exercises whose numbers are in color can be found in Appendix VII. The star indicates an exercise that is more challenging than average.

799

then levels off. Explain this observation.

**19.16**   Write balanced reactions for the following processes: (a) the roasting of molybdenum(IV) sulfide, (b) formation of the oxide from aluminum carbonate, (c) reduction of tin(IV) oxide with carbon, (d) reduction of beryllium oxide with ferrosilicon.

**19.17**   Cobaltite, CoAsS, is a mineral that is a source of cobalt. Propose a chemical procedure to win cobalt from this mineral. Discuss some of the hazards of the procedure.

**19.18**   The major ore of uranium is pitchblende, $U_3O_8$. The uranium can be won from the ore by reduction with carbon, calcium, or aluminum. Write equations for these three processes.

**19.19★**   The $K_i$ for $[Ag(CN)_2]^-$ is $7.9 \times 10^{-22}$ and that of $[Cu(CN)_2]^-$ is $1.0 \times 10^{-24}$. The standard reduction potential, $\mathscr{E}°$, is 0.80 V for $Ag^+$ and 0.52 V for $Cu^+$. Calculate $K$ at 298 K for the reaction $[Ag(CN)_2]^-(aq) + Cu(s) \rightleftharpoons [Cu(CN)_2]^-(aq) + Ag(s)$.

**19.20**   The first ionization energies of iron, cobalt, nickel, and the first three platinum metals are all about the same; the ionization energies of osmium, iridium, and platinum are substantially higher. Suggest an explanation for this observation.

**19.21**   Discuss the formation of slag in the metallurgy of iron in terms of the Lewis theory of acids and bases.

**19.22**   Write chemical equations for the following processes: (a) the air oxidation of Fe(II) hydroxide, (b) the mixing of a solution of cobalt(II) nitrate with a solution of sodium hydroxide, (c) the formation of the carbonyl of cobalt from the metal and carbon monoxide.

**19.23**   Find the formal oxidation state of the group VIIIB metal in the following compounds, all of which have been prepared:   (a)   $K_4[Ni_2(CN)_6]$,   (b) $Na_2[Ni(CO)_6]$,   (c)   $BaFeO_4$,   (d) $Na_2[Fe(CO)_4]$.

**19.24**   One of the few substances that attacks platinum metal is aqua regia, a mixture of nitric acid and hydrochloric acid. The platinum goes into solution as the $[PtCl_6]^{2-}$ ion while a brown gas is evolved. Write the reaction for this process.

**19.25**   Predict the geometry of $OsO_4$ and $RuO_4$ and discuss the hybridization and bonding in these oxides.

**19.26**   Although both the group IA and IB metals have a half-filled $s$ subshell, the IB metals have markedly higher first ionization energies, densities, and melting points and markedly lower second and third ionization energies than the corresponding IA metals. Explain these observations.

**19.27**   The standard reduction potential, $\mathscr{E}°$, for the $Cu^{2+}|Cu(s)$ couple is 0.34 V. For the $Cu^{2+}|Cu^+$ couple it is 0.16 V. Find the value of $K$ for the reaction $2Cu^+(aq) \rightleftharpoons Cu^{2+}(aq) + Cu(s)$.

**19.28★**   The $K_i$ of $[Cu(NH_3)_2]^+$ is $1.6 \times 10^{-11}$ and the $K_i$ of $[Cu(NH_3)_4]^{2+}$ is $5.0 \times 10^{-14}$. Use these data and the result of Exercise 19.27 to find the $K$ for the reaction $2[Cu(NH_3)_2]^+(aq) \rightleftharpoons [Cu(NH_3)_4]^{2+}(aq) + Cu(s)$.

# 20

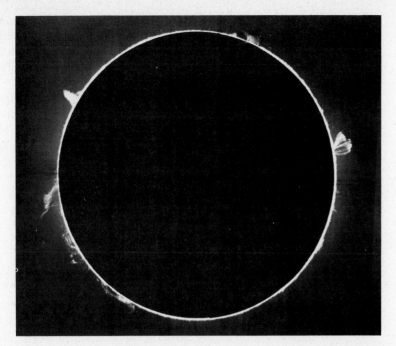

# Nuclear Chemistry

**Preview**    In this chapter, we deal with a subject that is rarely out of the headlines: nuclear reactions. After describing the atomic nucleus and introducing the notation used to specify the number of protons and neutrons found in different nuclei, we discuss the various processes of radioactive decay, in which nuclei spontaneously emit particles and radiation. We next discuss other nuclear reactions and how they can be used to produce new elements in the laboratory, and then show how the naturally occurring elements were created by a series of nuclear reactions that took place over billions of years in the interiors of stars. We conclude by describing how nuclear reactions are put to practical use in weapons, in nuclear generating plants, and in the effort to use fusion reactions as a source of energy.

When the twentieth century began, the atom was regarded as the basic and unchangeable unit of matter. The idea that one element could be transformed into another was regarded with the same scorn as the efforts of medieval alchemists to prepare gold from base metals. That belief was to change rapidly.

The discovery of the electron (Section 5.2) demonstrated that atoms are made up of smaller particles. The nuclear model of the atom (Section 5.4) provided a detailed view of atomic structure. The discovery of natural radioactivity demonstrated that at least some atoms could change spontaneously. A whole new branch of chemistry, the field of nuclear chemistry, was opened.

Nuclear chemistry is concerned with changes in which the nucleus of the atom participates in a major way. Ordinary chemical changes are the result of electronic changes. They generally are sensitive to external conditions, such as temperature and pressure, while most of the reactions in nuclear chemistry are not.

In nuclear chemistry, the transmutation of elements is a normal event. Such chemical changes are called *nuclear reactions*. As early as 1903, Ernest Rutherford and others concluded that transmutation occurs in radioactive elements. By 1919, Rutherford had achieved the first artificial transmutation of an element, bombarding nitrogen atoms with alpha particles to produce oxygen atoms and protons. Since then, the study of nuclear reactions has become a major area of research. The nuclear technology that has grown out of this research has had an incalculable impact on human beings.

## 20.1  THE ATOMIC NUCLEUS

The atomic nucleus contains protons and neutrons, fundamental particles that are called *nucleons* (Section 5.4). Every atomic nucleus contains protons, which are positively charged particles. The positive charge of the proton is equal in magnitude to the negative charge of the electron; its value is $1.60219 \times 10^{-19}$ C. The mass of the proton is $1.67251 \times 10^{-24}$ g, which is about 1800 times the mass of the electron ($9.1083 \times 10^{-28}$ g). Relative to an atom of $^{12}C$, whose mass is defined as exactly 12, the proton has a mass of 1.007280. The atomic number of an element and the positive charge of its nucleus are both equal to the number of protons in the nucleus.

All nuclei except $^1H$ also contain one or more neutrons. The neutron is an electrically neutral particle whose mass relative to $^{12}C$ is 1.008665, slightly greater than that of the proton. In the notation that has been developed to describe atoms, the number of neutrons in the nucleus is represented by the symbol $N$ and the number of protons is represented by the symbol $Z$, which is the atomic number. The symbol $A$ represents the mass number, whose value is close to that of the relative isotopic mass.

The mass number is defined as the sum of protons and neutrons in the nucleus:

$$A = Z + N \qquad (20.1)$$

We write the mass number as a superscript and the atomic number as a subscript to the left of the elemental symbol X:

$$_Z^A X$$

Two different atomic species with the same number of protons but different numbers of neutrons are said to be **isotopes** of the same chemical element. Isotopes can be stable or unstable. A stable isotope does not undergo radioactive decay. An unstable isotope undergoes radioactive decay and eventually is transformed into a stable isotope, usually of another element.

Most elements of atomic number 83 and under have more than one stable isotope. The average is three stable isotopes per element. Tin has the largest number of stable isotopes, ten.

The word isotope generally is used when two or more different nuclear species are discussed. When we are talking about a single nuclear species with a given value of $A$ and $Z$, the term **nuclide** is preferred. Thus, we can say either that $_8^{16}O$ and $_8^{18}O$ are isotopes of oxygen or that $_8^{16}O$ is a nuclide and $_8^{18}O$ is a nuclide.

We can define isotopes as nuclides that have the same $Z$ but different $A$. **Isobars** are nuclides with the same $A$ but different $Z$; $_{92}^{235}U$, $_{93}^{235}Np$, and $_{94}^{235}Pu$ are isobars. **Isotones** are nuclides with the same $N$ but different $Z$. For example, $_6^{14}C$, $_7^{15}N$, and $_8^{16}O$ are isotones, since each nucleus contains eight neutrons.

Almost all the mass of the atom is concentrated in the nucleus, which is quite small compared to the whole atom. Atoms can be described as spheres whose radii are in the range of $10^{-8}$ cm. The radius of the nucleus, which can be regarded as approximately spherical, is a little less than $10^{-12}$ cm. The size of the nuclear radius $R$ is directly proportional to the cube root of $A$, the mass number of the nucleus:

$$R = r_0 A^{1/3} \qquad (20.2)$$

where $r_0$ is a proportionality constant whose value is about $1.3 \times 10^{-13}$ cm.

All nuclei have virtually the same density. If we assume that the nucleus is spherical, we can calculate a value for the nuclear density of about $1.8 \times 10^{14}$ g/cm$^3$. This value is very large compared to the density of everyday materials. For example, osmium, the densest element, has a density of only 22.6 g/cm$^3$. We can get a better idea of nuclear density by saying that the mass of a 1-cm cube of nuclear matter would be greater than the combined mass of all the automobiles in the United States. In recent years, astronomers have discovered stars, called neutron stars, whose density is about the same as the density of the atomic nucleus. A neutron star with twice the mass of the sun would have a radius of about 10 km, less than the size of Manhattan Island.

## Nuclear Binding Energy

A nucleus consists of positively charged protons and neutral neutrons, packed closely together. The repulsive coulombic forces between the protons would make the nucleons fly apart, but there are nuclear forces of attraction stronger than the repulsive forces. The forces of attraction between nucleons are very strong at small distances but fall off rapidly as distance increases.

We can find the magnitude of the force that holds the nucleus together by applying what is perhaps the best-known equation in science, the equivalence between mass and energy given by Einstein's theory of special relativity:

$$E = mc^2 \qquad (20.3)$$

where $E$ is the energy, $m$ is mass, and $c$ is the velocity of electromagnetic radiation. A convenient unit of energy to use on the nuclear scale is the electronvolt, eV, whose magnitude is approximately $1.6022 \times 10^{-19}$ J. Common multiples of the electronvolt are:

$$\text{kiloelectronvolt (keV)} = 10^3 \text{ eV}$$
$$\text{megaelectronvolt (MeV)} = 10^6 \text{ eV}$$
$$\text{gigaelectronvolt (GeV)} = 10^9 \text{ eV}$$

A convenient unit of mass on the nuclear scale is the unified atomic mass unit (u), which is exactly one-twelfth the mass of an atom of $^{12}$C, or $1.6605655 \times 10^{-27}$ kg.

We can find the energy equivalent of this mass if we use $2.998 \times 10^8$ m/sec as the velocity of electromagnetic radiation. By Equation 20.3, the energy equivalent of one unified atomic mass unit is 931 MeV; that is, 1 u = 931 MeV.

Careful measurements of relative atomic masses show that their values are slightly less than the sum of the masses of their nucleons. This difference can be regarded as the amount of energy liberated when nucleons are brought together in the nucleus, the *nuclear binding energy*. The existence of nuclear binding energy is of enormous importance in the universe.

Stars consist mostly of hydrogen nuclei, or protons, with some alpha particles, or helium nuclei. An alpha particle, $^4_2\text{H}^{2+}$, has a mass of 4.00151 u. If we add up the relative masses of the two protons and two neutrons in the alpha particle, we get:

$$\text{protons: } 2 \times 1.007277 \text{ u} = 2.014554 \text{ u}$$
$$\text{neutrons: } 2 \times 1.008665 \text{ u} = \underline{2.017330 \text{ u}}$$
$$\text{total} = 4.031884 \text{ u}$$

The difference between the calculated mass and the measured mass is

$$4.03189 \text{ u} - 4.00151 \text{ u} = 0.03038 \text{ u}$$

The energy equivalent of this mass is

$$E = 0.03038 \text{ u} \times \frac{931 \text{ MeV}}{1 \text{ u}} = 28.3 \text{ MeV}$$

The binding energy in a single nucleus of $_2^4\text{He}$ thus is 28 MeV. This binding energy is the amount of energy needed to break the helium nucleus into its constituent nucleons. It is also the energy that is liberated when the four nucleons come together in a reaction to form the helium nucleus:

$$2_1^1\text{H} + 2_0^1 n \longrightarrow {}_2^4\text{He}$$

The symbol $n$ represents the neutron, whose $Z = 0$ and $A = 1$.

The magnitude of the nuclear binding energy is enormous compared with chemical bond energy. We are accustomed to measuring the bond energy of chemical bonds in molar amounts. To convert from the binding energy of a single nucleus to molar quantities of nuclei, we must use Avogadro's number, $6.022 \times 10^{23}$. Thus, 1 mol of helium nuclei has a binding energy of (28.3 MeV/nucleus) $\times$ ($6.022 \times 10^{23}$ nuclei/mol) = $1.70 \times 10^{25}$ MeV/mol. In more familiar energy units, the binding energy is $2.72 \times 10^9$ kJ/mol, roughly the bond energy of $10^7$ mol of a fairly strong chemical bond. In other words, the binding energy of the nucleus is 10 million times greater than the bond energy of a chemical bond.

As the mass number of the nucleus increases, *total* binding energy increases. However, the average binding energy — the binding energy *per nucleon* — does not vary over a large range. To determine the value of the average binding energy, we divide total binding energy by the mass number. For helium, average binding energy is 28.3/4 = 7.1 MeV. With the exception of the lightest nuclei, the average binding energy per nucleon of a nucleus is in the range from 7.4 to 8.8 MeV.

The binding energy per nucleon reaches a peak for nuclei whose mass numbers are near 60, nuclei of iron and nickel. These nuclei tend to be the most stable, which helps to account for the relatively high abundance of iron and nickel in the core of the earth and in meteorites. As Figure 20.1 shows, the average binding energy per nucleon falls off rapidly from this maximum.

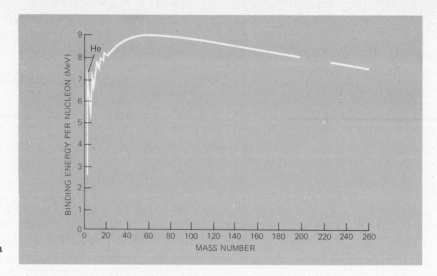

**Figure 20.1**

The average binding energy per nucleon as a function of mass number.

## The Stability of Nuclei

A number of factors are responsible for the reduction in average binding energy per nucleon with increasing $Z$. A major factor is the coulombic repulsion between protons. Since these repulsions act over a much greater distance than the attractive forces between nucleons, coulombic repulsions become more important as more protons are added to the nucleus.

The destabilizing effect of coulombic repulsions can be offset to some extent by the presence of neutrons. The additional binding energy of the neutrons, which is not accompanied by any increase in coulombic repulsions, can confer stability on nuclei with many protons. All stable nuclei with $Z > 20$ contain more neutrons than protons.

Another factor that affects average binding energy is the relative number of protons and neutrons in a nucleus. Nuclei in which the number of protons is about the same as the number of neutrons have greater average binding energy than do nuclei in which there is a large difference between the number of protons and the number of neutrons. An empirical correction can be made for the difference between the number of protons and the number of neutrons, $N - Z$.

Another factor has a major effect on the binding energy. The most stable nuclei are those with even numbers of both protons and neutrons. Even-even nuclei, as these are called, tend to be more stable than nuclei that have an even number of protons and an odd number of neutrons and that are called even-odd nuclei. These even-odd nuclei, in turn, are more stable than odd-even nuclei, which have an odd number of protons and an even number of neutrons. Least stable of all are odd-odd nuclei, which have odd numbers of both protons and neutrons. Of the known stable nuclides, 164 are even-even, 55 are even-odd, 50 are odd-even and only 4 are odd-odd. The four stable odd-odd nuclei are all of very light elements: $^{2}_{1}H$, $^{6}_{3}Li$, $^{10}_{5}B$, and $^{14}_{7}N$.

An analogy can be made between the pairing of electrons and the pairing of nucleons. Just as increased stability is associated with pairs of electrons of opposite spin, increased stability is associated with pairs of like nucleons. Nuclei with certain "magic numbers" of protons or neutrons seem to be especially stable. Stability is associated with 2, 8, 20, 28, 50, 82, and 126 of either protons or neutrons. The unusual stability of some of the light nuclei first called attention to this phenomenon. For example, the $^4_2$He nucleus, with two protons and two neutrons, is exceptionally stable. So are the $^{16}_8$O nucleus (eight protons, eight neutrons), and the $^{40}_{20}$Ca nucleus (20 protons, 20 neutrons).

The stable isotopes of the heavier elements also provide evidence for the existence of magic numbers of nucleons. Elements for which $Z$ is a magic number tend to have a relatively large number of stable isotopes. For example, we find the largest number of stable isotopes for tin, $Z = 50$. The natural radioactive decay of all heavy elements eventually ends at lead, $Z = 82$. As for neutrons, there are an unusually large number of isotones in which $N = 50$ and $N = 80$. Also, $N = 126$ for the two heaviest stable nuclides known, $^{208}_{82}$Pb and $^{209}_{83}$Bi. Many nuclear reactions can be described in terms of the formation of nuclides with magic numbers of protons and neutrons. (You should note that while there are magic numbers for $Z$ and $N$, there are no magic numbers for $A$.)

The nuclear shell model was developed in 1949 to explain the existence of magic numbers and other observations about nuclei. In many ways, the description of the nucleus that is given by the nuclear shell model is similar to our description of the electronic structure of the atom.

The nuclear shell model places nucleons into energy levels that are grouped into shells. Special stability is associated with closed shells that have "magic numbers" of nucleons. The interaction between individual nucleons are not considered. The nuclear shell model does not agree very well with some of the measured magnetic and electric properties of nuclei, but it is quite helpful in rationalizing many aspects of nuclear reactions.

## 20.2 RADIOACTIVITY

A radioactive decay process is the transformation of a relatively unstable nuclide to another nuclide with the accompanying emission of particles or electromagnetic radiation. In such a process, the unstable or *parent* nuclide decays into the *daughter* nuclide, which may itself be either radioactive or stable.

Radioactive decay reactions, unlike ordinary chemical reactions, are not affected by ordinary changes in temperature and pressure. With minor exceptions, their rates do not depend on the chemical or physical state of the reacting substance. The rate of a nuclear decay reaction depends only on the nature of the nucleus. Thus, as we mentioned in Section 17.2, a radioactive decay reaction is an ideal example of a first-order rate process. The half-life concept, which applies to first-order reactions, is

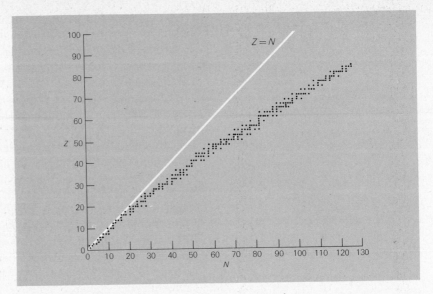

**Figure 20.2**

A plot of $Z$ against $N$ for all known stable nuclei. The solid line corresponds to $Z = N$. The ratio of neutrons to protons in stable nuclides increases with atomic number.

commonly used to express the rate of a radioactive decay process and the relative stability of any given nuclide.

A nucleus decays because it is unstable. A major source of nuclear instability is coulombic repulsion between protons. As we have seen, this destabilizing force is counterbalanced by the presence of neutrons.

The relationship between nuclear stability and nuclear composition can be conveniently represented by a graph such as that of Figure 20.2, a plot of $Z$ against $N$ for all the stable nuclides. When $Z$ is small, all the stable nuclides lie close to the $Z = N$ line. As $Z$ increases, the stable nuclides fall in a zone below the $Z = N$ line, reflecting the increase in the ratio of neutrons to protons that is necessary to offset the coulombic repulsions of the protons. The stable nuclides end at $Z = 83$. In a plot of this sort, the region in which the stable nuclides are found can be regarded as a "zone of stability." The relative stability of a nucleus and its mode of decay often can be predicted from its location relative to the zone of stability.

As we have noted, stability reaches a maximum in nuclei whose mass numbers are about 60. The most stable nucleus of all is believed to be $^{56}_{26}Fe$, which can be regarded as the thermodynamic equilibrium state for all nuclear matter. We can even say that all other nuclei are being transformed spontaneously into $^{56}_{26}Fe$. However, the rate at which these transformations occurs is so slow that these processes are of no practical importance for most nuclei.

The phenomenon of radioactive decay was discovered through the detection of the radiation emitted during decay. It was found that three kinds of radiation can be emitted. They are named $\alpha$ rays, $\beta$ rays, and $\gamma$ rays.

### α Decay

An α particle is $_2^4\text{He}^{2+}$. The loss of an α particle from a nucleus therefore lowers the mass number $A$ of that nucleus by four and its atomic number $Z$ by two, and increases the ratio of neutrons to protons. Alpha decay is observed particularly often in nuclei with large mass numbers.

A typical α decay can be represented by an equation such as

$$_{92}^{238}\text{U} \longrightarrow {}_{90}^{234}\text{Th} + {}_2^4\text{He}$$

The parent nuclide, uranium, element 92, decays to the daughter nuclide, thorium, element 90. You should note that equations for nuclear reactions must be balanced. The sum of the mass numbers and the sum of the nuclear charges on both sides of the equation are equal. Thus, both $A$ and $Z$ are conserved in a nuclear reaction.

A heavy nucleus decays by emitting a four-nucleon fragment, rather than a single nucleon, because of the unusually high binding energy of the $_2^4\text{He}$ nucleus compared to other light nuclides.

We can calculate the energy changes in α-decay reactions from the atomic masses of the parent and daughter nuclides. There is a loss of mass, $\Delta M$, in any decay process. The loss in mass has an energy equivalent of $1\text{ u} = 931\text{ MeV}$. Atomic masses are used in these energy calculations because the mass and binding energy of the electrons must be taken into account. Table 20.1 lists the atomic masses of some important nuclides.

The energy change in the decay of $_{92}^{238}\text{U}$ into $_{90}^{234}\text{Th}$ is

$$\begin{aligned}\Delta M &= (\text{mass } {}_{92}^{238}\text{U}) - (\text{mass } {}_{90}^{234}\text{Th} + \text{mass } {}_2^4\text{He})\\ &= 238.0508\text{ u} - (234.0436\text{ u} + 4.0026\text{ u})\\ &= 0.0046\text{ u}\end{aligned}$$

$$E = 0.0046\text{ u} \times 931\,\frac{\text{MeV}}{\text{u}} = 4.3\text{ MeV}$$

Most of this energy appears as the kinetic energy of the emitted α particle. A small amount, about 0.1 MeV in this reaction, appears as the recoil energy of the nucleus.

### γ Emission

In many radioactive decay processes, γ rays, electromagnetic radiation of higher energy and shorter wavelength than X rays, are emitted. The emission of γ rays results from the existence of excited nuclear states that are analogous to excited electronic states. A parent nuclide may decay to an excited nuclear state of the daughter nuclide. After the excited daughter nuclide forms, it decays to its ground state with the emission of γ rays. Figure 20.3 diagrams these energy relationships for the decay of $_{88}^{226}\text{Ra}$ to $_{86}^{222}\text{Rn}$.

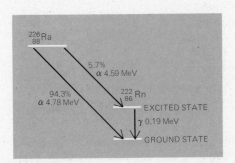

**Figure 20.3**

The decay of $_{88}^{226}\text{Ra}$ to $_{86}^{222}\text{Rn}$ may occur in two steps. The parent nuclide first emits an α particle and decays to an excited state of the daughter nuclide. The excited daughter nuclide then decays to the ground state with the emission of γ rays. As the diagram shows, most of the energy release in the process takes place in the first step. The decay may also occur directly in one step.

**TABLE 20.1**    Relative Atomic Masses of Particles and Nuclides

| Nuclide | Percent Abundance | Mass[a] | Nuclide | Percent Abundance | Mass[a] | Species | Mass[a] |
|---|---|---|---|---|---|---|---|
| $^1_1H$ | 99.985 | 1.0078252 | $^{35}_{17}Cl$ | 75.53 | 34.968854 | $e^-$ | 0.000548580 |
| $^2_1H$ | 0.015 | 2.0141022 | $^{37}_{17}Cl$ | 24.47 | 36.965896 | $^1_1p^+$ | 1.007276470 |
| $^3_1H$ | | 3.0160494 | $^{40}_{18}Ar$ | 99.600 | 39.962384 | $^1_0n$ | 1.008665012 |
| $^3_2He$ | trace | 3.0160299 | $^{39}_{19}K$ | 93.10 | 38.963714 | | |
| $^4_2He$ | 100 | 4.0026036 | $^{40}_{19}K$ | 0.0118 | 39.964008 | | |
| $^6_3Li$ | 7.42 | 6.015126 | $^{40}_{20}Ca$ | 96.97 | 39.962589 | | |
| $^7_3Li$ | 92.58 | 7.016005 | $^{55}_{25}Mn$ | 100 | 54.938054 | | |
| $^{10}_5B$ | 19.61 | 10.012939 | $^{56}_{26}Fe$ | 91.66 | 55.93493 | | |
| $^{11}_5B$ | 80.39 | 11.0093051 | $^{59}_{27}Co$ | 100 | 58.933189 | | |
| $^{12}_6C$ | 98.893 | 12 (exactly) | $^{58}_{28}Ni$ | 67.88 | 57.93534 | | |
| $^{13}_6C$ | 1.107 | 13.003354 | $^{60}_{28}Ni$ | 26.23 | 59.93078 | | |
| $^{14}_6C$ | | 14.0032419 | $^{87}_{37}Rb$ | 27.85 | 86.90918 | | |
| $^{14}_7N$ | 99.634 | 14.0030744 | $^{87}_{38}Sr$ | 7.02 | 86.9089 | | |
| $^{15}_7N$ | 0.366 | 15.000108 | $^{127}_{53}I$ | 100 | 126.90447 | | |
| $^{16}_8O$ | 99.759 | 15.9949149 | $^{129}_{53}I$ | | 128.90498 | | |
| $^{17}_8O$ | 0.0374 | 16.999133 | $^{206}_{82}Pb$ | 23.6 | 205.97446 | | |
| $^{18}_8O$ | 0.2039 | 17.9991598 | $^{207}_{82}Pb$ | 22.6 | 206.97590 | | |
| $^{18}_9F$ | | 18.000950 | $^{208}_{82}Pb$ | 52.3 | 207.97664 | | |
| $^{19}_9F$ | 100 | 18.9984032 | $^{209}_{83}Bi$ | 100 | 208.98042 | | |
| $^{20}_{10}Ne$ | 90.92 | 19.9924404 | $^{232}_{90}Th$ | 100 | 232.03821 | | |
| $^{23}_{11}Na$ | 100 | 22.989773 | $^{234}_{90}Th$ | | 234.0436 | | |
| $^{24}_{12}Mg$ | 78.70 | 23.985045 | $^{234}_{92}U$ | 0.0056 | 234.0409 | | |
| $^{27}_{13}Al$ | 100 | 26.981535 | $^{235}_{92}U$ | 0.7205 | 235.04393 | | |
| $^{28}_{14}Si$ | 92.21 | 27.976927 | $^{238}_{92}U$ | 99.274 | 238.0508 | | |
| $^{31}_{15}P$ | 100 | 30.973763 | $^{237}_{93}Np$ | | 237.04803 | | |
| $^{32}_{16}S$ | 95.0 | 31.972074 | $^{239}_{94}Pu$ | | 239.05216 | | |

[a] Relative to $^{12}C$, which has a mass of exactly 12.

Just as we can find the electronic energy level structure of an atom by studying its emission spectrum (Section 5.5), we can find the energy level structure of a nucleus by studying its $\gamma$-ray spectrum.

The emission of $\gamma$ rays accompanies not only $\alpha$ decay but also most other nuclear transformations. The emission of $\gamma$ rays represents an energy change within the nucleus, a decay from a high-energy state to a lower-energy state. By itself, the emission of $\gamma$ rays does not result in any change in either $A$ or $Z$ of the nucleus. We find $\gamma$-ray emission only in nuclear transformations in which ground state daughter nuclides are not formed directly.

## β Decay

There are many radioactive decay processes in which there is no change in the value of the mass number $A$ but there is a change in the value of $Z$, and therefore in the value of $N$. Such a process is called $\beta$ decay. Beta decay can occur in several different ways, but only one type of $\beta$ decay is observed in naturally occurring radioactive nuclides. It is called $\beta^-$ or negatron decay, and it generally occurs in nuclides that are unstable because the ratio of neutrons to protons is too high. Nuclides that lie to the right of the zone of stability in Figure 20.2 undergo $\beta^-$ decay.

The most efficient way to correct an excess of neutrons over protons in a nucleus is to convert a neutron to a proton. This conversion is $\beta^-$ decay. It can be represented as:

$$n \longrightarrow p^+ + e^-$$

where $n$ represents the neutron and $p^+$ the proton. Electrons cannot exist in a nucleus. The electron is created at the instant it is emitted, just as a photon that is emitted as the result of an electronic change in an atom is created at the instant of emission.

Thus, a negatron decay is the emission of an electron with an increase of 1 in the value of $Z$ and a decrease of 1 in the value of $N$ of the nucleus. You will note that $\beta^-$ decay causes an *increase* in atomic number; the daughter nuclide is one element higher in atomic number than the parent nuclide. A typical reaction is

$$^{227}_{89}\text{Ac} \longrightarrow \, ^{227}_{90}\text{Th} + \, ^{0}_{-1}\beta$$

By giving the $\beta$ particle (the electron) a subscript of $-1$, we balance the equation with respect to both $A$ and $Z$. A $\beta$-decay process changes the ratio of neutrons to protons in the nucleus. Beta decay often changes an odd-odd nucleus to an even-even nucleus.

## The Neutrino

In addition to the electron, another fundamental particle called the neutrino is emitted during $\beta^-$ decay. A neutrino has no electrical charge, no measurable mass when at rest, negligible magnetic properties, and minimal interactions with other particles. Experimental evidence for the existence of the neutrino who was not obtained until 1955.

A complete energy accounting of a $\beta$-decay process must include the energy carried away by the neutrino. Thus, the transformation of a neutron to a proton can be represented by:

$$^{1}_{0}n \longrightarrow \, ^{1}_{1}p^+ + \, ^{0}_{-1}\beta + \nu$$

where $\nu$ represents the neutrino. The $\beta$ decay of the unstable nuclide $^{14}_{6}\text{C}$, which is formed in the atmosphere by cosmic rays and is used in radiocarbon dating (Section 17.2), is represented by:

$$^{14}_{6}C \longrightarrow ^{14}_{7}N + ^{0}_{-1}\beta + \nu$$

## The Positron

Artificially produced unstable nuclides can undergo another type of $\beta$ decay, called $\beta^+$ decay or positron decay. Nuclides in which the ratio of neutrons to protons is too low — that is, nuclides that lie to the left of the zone of stability of Figure 20.2 — are converted to more stable daughter nuclides by such a process. In $\beta^+$ decay, a proton is converted to a neutron, with the formation of two other particles: a positron, which is identical to an electron except that its charge is positive; and a neutrino, similar to but not identical to the neutrino that is formed in $\beta^-$ processes. The process is

$$^{1}_{1}p^+ \longrightarrow ^{1}_{0}n + ^{0}_{1}\beta + \nu$$

In positron decay, there is no change in mass number. There is an increase in $N$, and a decrease in $Z$. An example of $\beta^+$ decay is

$$^{17}_{9}F \longrightarrow ^{17}_{8}O + ^{0}_{1}\beta + \nu$$

The positron is called the antiparticle of the electron. When the two collide, their masses are converted to energy in a process called annihilation. We can represent the process as:

$$e^- + e^+ \longrightarrow 2\gamma$$

## Electron Capture

Another mode of $\beta$ decay is called electron capture. In electron capture, a proton is converted to a neutron when the nucleus captures one of the electrons of the atom:

$$p^+ + e^- \longrightarrow n + \nu$$

The only radiation emitted from the nucleus in electron capture is the elusive neutrino. Electron capture is a mode of decay available to many nuclides in which the ratio of neutrons to protons is too low, and it competes with $\beta^+$ decay. An example of electron capture is

$$^{57}_{27}Co + ^{0}_{-1}e^- \longrightarrow ^{57}_{26}Fe + \nu$$

The electron that is captured by the nucleus usually is in an orbital of low principal quantum number, since these orbitals lie closest to the nucleus. The loss of the electron creates a vacancy that is filled by an

**TABLE 20.2** Radioactive Decay Processes

| Process | Nuclear Condition | Emitted Radiation | Change in $A$ | Change in $Z$ | Change in $N$ |
|---|---|---|---|---|---|
| $\alpha$ decay | excess mass | $^{4}_{2}\text{He}^{2+}$ (alpha particle) | $-4$ | $-2$ | $-2$ |
| $\beta^{-}$ decay | neutron-to-proton ratio too high | $^{0}_{-1}e^{-}$ (electron) | 0 | $+1$ | $-1$ |
| $\beta^{+}$ decay | neutron-to-proton ratio too low | $^{0}_{1}e^{+}$ (positron) | 0 | $-1$ | $+1$ |
| Electron capture | neutron-to-proton ratio too low | none (except neutrino) | 0 | $-1$ | $+1$ |
| $\gamma$ emission | excited nuclear state | $\gamma$ rays | 0 | 0 | 0 |

electron from an orbital of higher energy. When the electron drops from the higher-energy orbital, the atom emits electromagnetic radiation in the X-ray part of the spectrum. This emission occurs after the nuclear transformation.

The energy changes associated with $\beta$-decay processes can be calculated from the atomic masses of parent and daughter nuclides, by procedures similar to those used for $\alpha$ decay. The use of atomic masses, which include the masses and binding energies of the electrons, eliminates the problem of keeping track of the electronic changes that accompany the nuclear change except in $\beta^{+}$ decay. Table 20.2 summarizes all the radioactive decay processes that we have discussed.

**Example 20.1**    The unstable nuclide $^{40}_{19}\text{K}$, which occurs naturally and has been used in radioisotope dating, can undergo all three types of $\beta$ decay that we have discussed. The relative atomic masses of the three isobars of $A = 40$ that are related by $\beta$ decay are $^{40}_{19}\text{K} = 39.964008$, $^{40}_{18}\text{Ar} = 39.962384$, and $^{40}_{20}\text{Ca} = 39.962589$. Calculate the energy change for each of the three types of decay.

**Solution**    The equation for $\beta^{-}$ decay is

$$^{40}_{19}\text{K} \longrightarrow {}^{40}_{20}\text{Ca} + {}^{0}_{-1}\beta$$

$$\Delta M = (\text{mass } {}^{40}_{19}\text{K}) - (\text{mass } {}^{40}_{20}\text{Ca})$$
$$= 39.964008 \text{ u} - 39.962589 \text{ u}$$
$$= 0.001419 \text{ u}$$

$$E = 0.001419 \text{ u} \times 931 \frac{\text{MeV}}{\text{u}}$$

$$= 1.32 \text{ MeV}$$

The equation for electron capture is

$$^{40}_{19}\text{K} + {}^{0}_{-1}e^{-} \longrightarrow {}^{40}_{18}\text{Ar}$$

$$\Delta M = (\text{mass } {}^{40}_{19}\text{K}) - (\text{mass } {}^{40}_{18}\text{Ar})$$
$$= 39.964008 \text{ u} - 39.962384 \text{ u}$$
$$= 0.001624 \text{ u}$$

$$E = 0.001624 \text{ u} \times 931 \frac{\text{MeV}}{\text{u}}$$

$$= 1.51 \text{ MeV}$$

CHAPTER 20 Nuclear Chemistry

The equation for $\beta^+$ decay is

$$^{40}_{19}\text{K} \longrightarrow {}^{40}_{18}\text{Ar} + {}^{0}_{1}\beta$$

In $\beta^+$ decay, because of the use of atomic masses, the mass of two more electrons than are included in the mass of the daughter nuclide must be included in our calculation for mass balance.

$$\Delta M = (\text{mass } {}^{40}_{19}\text{K}) - [\text{mass } {}^{40}_{18}\text{Ar} + (2) \text{ mass } e^-]$$
$$= 39.964008 \text{ u} - [39.962384 \text{ u} + (2)(0.0005486 \text{ u})]$$
$$= 0.000527 \text{ u}$$

$$E = 0.000527 \text{ u} \times 931 \frac{\text{MeV}}{\text{u}}$$
$$= 0.490 \text{ MeV}$$

## Natural Radioactivity

The radioactive nuclides that occur naturally on earth are divided into three categories. One small group, of which $^{14}_{6}\text{C}$ is an important member, consists of nuclides that are formed by the action of cosmic rays on stable nuclides. A second group consists of nuclides with half-lives on the same order of magnitude as the life of the earth. Among these nuclides are $^{40}_{19}\text{K}$ (half-life $1.3 \times 10^9$ years), $^{87}_{37}\text{Rb}$ (half-life $5 \times 10^{11}$ years), $^{232}_{90}\text{Th}$ (half-life $1.39 \times 10^{10}$ years), $^{235}_{92}\text{U}$ (half-life $7.13 \times 10^8$ years), and $^{238}_{92}\text{U}$ (half-life $4.51 \times 10^9$ years).

The other naturally occurring radioactive nuclides have shorter half-lives. They are formed by radioactive decay processes that begin with one of the long-lived nuclides. Such a decay process results in what is called a *radioactive family*, or *series*. Three such series are found in nature. Each series starts with a parent nuclide that decays through a series of daughter radioactive nuclides to a stable end product. Table 20.3 summarizes the starting and ending points of these three series. We shall discuss only the thorium series.

From Table 20.3, we see that the net change as a result of the entire decay process in the thorium series is a reduction of $232 - 208 = 24$ in the mass number $A$. The only mechanism for reduction of mass number is

**TABLE 20.3** Natural Radioactive Decay Series

| Parent | Half-Life (years) | End Product | Name |
|---|---|---|---|
| $^{238}_{92}\text{U}$ | $4.51 \times 10^9$ | $^{206}_{82}\text{Pb}$ | uranium series |
| $^{235}_{92}\text{U}$ | $7.13 \times 10^8$ | $^{207}_{82}\text{Pb}$ | actinium series |
| $^{232}_{90}\text{Th}$ | $1.39 \times 10^{10}$ | $^{208}_{82}\text{Pb}$ | thorium series |

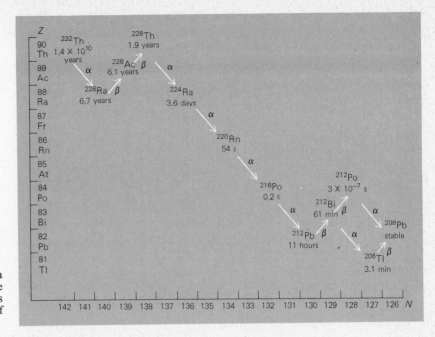

**Figure 20.4**
The thorium series, in which $^{232}_{90}$Th decays by a multistep path to the stable nuclide $^{208}_{82}$Pb. The vertical axis represents the number of protons in a nucleus, the horizontal axis the number of neutrons.

the emission of $\alpha$ particles, with $A = 4$. Therefore, the total reduction in $A$ must be divisible by four and the reduction in mass must occur in steps of four mass units each. In the thorium series, there is a total of six such steps, each an $\alpha$ decay.

Other processes also occur. Six $\alpha$ decays should reduce the value of $Z$ by 12. In fact, $Z$ is only reduced by eight in going from Th to Pb. Four $\beta^-$-decay processes must also occur to account for the observed reduction in the value of $Z$.

There are branches in the series when a parent nuclide can follow different paths to form a daughter product. In each branch, the intermediate nuclide is different. For example, $^{212}_{83}$Bi decays to $^{206}_{82}$Pb. It may do so by first forming $^{208}_{81}$Tl by $\alpha$ decay, which then forms $^{208}_{82}$Pb by $\beta^-$ decay. Or $^{212}_{83}$Bi may first form $^{212}_{84}$Po by $\beta^-$ decay, which then forms $^{208}_{82}$Pb by $\alpha$ decay. Figure 20.4 shows a convenient way of diagraming the thorium series.

In addition to the naturally occurring radioactive nuclides, more than a thousand others have been created artificially. At least one unstable isotope of every element is known.

## 20.3  NUCLEAR REACTIONS

In addition to the radioactive decay processes, there are many other nuclear reactions in which a nucleus is converted to one or more different nuclei as the result of an interaction with another nucleus or with an elementary particle. A nuclear reaction can be represented by an equation similar to those used for ordinary chemical reactions.

The first nuclear reaction carried out in the laboratory, Rutherford's formation of oxygen from nitrogen by bombardment with $\alpha$ particles, is represented by:

$$^{14}_{7}\text{N} + ^{4}_{2}\text{He} \longrightarrow ^{17}_{8}\text{O} + ^{1}_{1}\text{H}$$

Another nuclear reaction is the fusion reaction of deuterium:

$$^{2}_{1}\text{H} + ^{2}_{1}\text{H} \longrightarrow ^{3}_{2}\text{He} + ^{1}_{0}n$$

Both mass number $A$ and atomic number $Z$ are conserved in nuclear reactions.

The energy changes that accompany nuclear reactions are of great interest. Just as ordinary chemical reactions are described as exothermic or endothermic, so nuclear reactions are described as *exoergic,* energy-releasing, or *endoergic,* energy-absorbing. The transmutation of nitrogen to oxygen, for example, is an endoergic reaction; the energy of the products is higher than of the reactants. For this reaction to occur, energy must be supplied in the form of the kinetic energy of the bombarding $\alpha$ particles. However, the reaction will not occur unless the kinetic energies of the $\alpha$ particles are substantially greater than the energy gained in this reaction. There is a coulombic barrier to the reaction that results from the repulsion between the positively charged nucleus and a positively charged particle that approaches it.

For the short-range nuclear forces to be effective, the bombarding $\alpha$ particle must come very close to the nucleus. To do so, it must have enough energy to surmount the coulombic barrier. The magnitude of the barrier increases with both the positive charge of the bombarding particle and the charge of the nucleus. For the heaviest elements, the barrier is about 25 MeV for the dipositive $\alpha$ particle and about 12 MeV for the monopositive proton or the deuteron, an isotope of hydrogen that has one proton and one neutron.

Even exoergic reactions in which a positively charged particle must interact with the nucleus have substantial barrier energies. The reaction for the fusion of deuterium is exoergic, but it has a coulombic barrier of about 0.45 MeV. This barrier is relatively small because the two colliding particles have a charge of only $+1$ each. However, even this small barrier has caused major practical difficulties in the effort to use fusion reactions to generate electricity.

By the standard of ordinary chemical reactions, the kinetic energy needed for a positively charged particle to penetrate the coulombic barrier of a nucleus is huge. An average kinetic energy of 0.01 MeV corresponds to a temperature of about 100 000 000 K. The coulombic barrier can be overcome with machines called accelerators, which use electromagnetic fields to accelerate charged particles.

There is no coulombic barrier for the reaction of nuclei with neutrons, because neutrons have no electric charge. "Thermal" neutrons, so called because their kinetic energies are roughly the same as those of gas molecules at ordinary temperatures, react readily with nuclei. These slow-moving thermal neutrons take part in the reactions that occur in the core of a nuclear reactor.

## Varieties of Nuclear Reactions

Nuclear reactions have been initiated in many ways: by $\gamma$ rays, X rays, electrons, and neutrons, and by a series of positively charged particles with low masses, such as protons, deuterons, and $\alpha$ particles. The development of high-energy accelerators has also made it possible to use heavy nuclei with positive charges, such as $^{10}_{5}B$, $^{12}_{6}C$, $^{16}_{8}O$, or even heavier nuclides as bombarding particles. The emissions in nuclear reactions include X rays, neutrons, and protons. Many nuclear reactions result in the emission of more than one kind of particle. In one class of nuclear reactions called *fission reactions,* the reactant nuclide forms two or more product nuclides of smaller mass number.

There is a simple notation for representing nuclear reactions. The symbols of the bombarding particle and of the emitted particle are placed in parentheses between the symbols of the reactant nucleus and the product nucleus. The expression for a given reaction lists, from left to right, the reactant nucleus, the bombarding particle, the emitted particle, and finally, the product nucleus. The symbols used in this system include $n$ for the neutron, $p$ for the proton, $d$ for the deuteron, $\alpha$ for the alpha particle, $e$ for the electron, and $\gamma$ for the gamma ray. Thus, the reaction in which nitrogen is transmuted into oxygen by bombardment with $\alpha$ particles is represented as:

$$^{14}_{7}N(\alpha,p)^{17}_{8}O$$

## Preparing Radioactive Isotopes

A number of radioactive isotopes are prepared by bombardment of stable nuclei with slow neutrons. The most common kind of slow-neutron reaction is the $(n,\gamma)$ reaction, also called *neutron capture.* The product is an isotope of the parent nuclide that contains one neutron more than the target nuclide. The product nuclide thus is often unstable, because the ratio of neutrons to protons is too high, so the nuclide usually undergoes $\beta^{-}$ decay. A radioisotope of almost any element can be prepared by bombardment of a stable isotope of the element with neutrons to cause a $(n,\gamma)$ reaction. One such reaction is $^{59}_{27}Co(n,\gamma)^{60}_{27}Co$, which also can be represented as:

$$^{59}_{27}Co + ^{1}_{0}n \longrightarrow ^{60}_{27}Co^{*} \longrightarrow ^{60}_{27}Co + \gamma$$

An intermediate excited state of the $^{60}_{27}Co$ nucleus decays to the ground state by emitting energy in the form of $\gamma$ rays. The product nuclide is an artificial radioisotope that undergoes $\beta^-$ decay with a half-life of 5.26 years:

$$^{60}_{27}\text{Co} \longrightarrow {}^{60}_{28}\text{Ni} + {}^{0}_{-1}\beta + \gamma$$

Radioactive cobalt is often used in radiation therapy against cancer.

The $(n,\gamma)$ nuclear reaction is also one basis for a method of chemical analysis of small samples. In the method, called neutron activation analysis, the sample to be analyzed is exposed to thermal neutrons. The radiation emitted by the products of the resulting nuclear reactions provides extremely accurate information on the constituent elements of the starting material. Neutron activation analysis is widely used in such fields as archeology and space exploration. A quantity of material as small as $10^{-12}$ g can be detected in a sample by neutron activation analysis.

The $(n,\gamma)$ reaction is the most important one caused by bombardment with slow neutrons. However, nuclides of low atomic number can also undergo $(n,p)$ and $(n,\alpha)$ reactions when bombarded with slow neutrons. One important $(n,p)$ reaction is responsible for the formation of $^{14}_{6}\text{C}$ from $^{14}_{7}\text{N}$ in the atmosphere. The reaction is $^{14}_{7}\text{N}(n,p)^{14}_{6}\text{C}$ or:

$$^{14}_{7}\text{N} + {}^{1}_{0}n \longrightarrow {}^{15}_{7}\text{N}^{*} \longrightarrow {}^{14}_{6}\text{C} + {}^{1}_{1}\text{H}$$

The $^{14}_{6}\text{C}$ nuclide is unstable and reverts to the starting nuclide by $\beta^-$ decay, with a half-life of 5760 years:

$$^{14}_{6}\text{C} \longrightarrow {}^{14}_{7}\text{N} + {}^{0}_{-1}\beta$$

Another group of nuclear reactions are produced by bombardment with medium-energy particles including faster neutrons or positively charged particles such as protons, deuterons, $\alpha$ particles, or nuclei of higher mass number. The higher energy of the bombarding particles leads to a much wider range of reactions.

A medium-energy $(\alpha,n)$ reaction led to the discovery of the neutron in 1932. The reaction is $^{9}_{4}\text{Be}(\alpha,n)^{12}_{6}\text{C}$, which can be expressed as:

$$^{9}_{4}\text{Be} + {}^{4}_{2}\text{He} \longrightarrow {}^{12}_{6}\text{C} + {}^{1}_{0}n$$

This reaction still is used as a neutron source for nuclear experiments.

Simple alpha-capture $(\alpha,\gamma)$ reactions have also been observed for light nuclides. A large number of different reactions are possible when the proton is the bombarding particle. Examples of $(p,\alpha)$, $(p,n)$, $(p,\gamma)$, and even $(p,d)$ reactions are known. There is a similarly large variety of nuclear particles emitted in nuclear reactions in which the deuteron, $^{2}_{1}\text{H}$, is the bombarding particle. The exoergic reactions in which deuterium or tritium is bombarded with deuterons are of special interest in the effort to use thermonuclear power as an energy source.

## EXPLORING THE BRAIN WITH ISOTOPES

For decades, neurologists have wanted a method by which they could study brain activity in detail without the need for surgery. One such method that is rapidly moving from research into medical practice is called PET, for positron emission tomography, which depends on the availability of special radioactively tagged compounds.

PET uses compounds that have been tagged with nuclides that emit positrons, such as carbon-11. The compound is administered to an individual by injection or inhalation. Inside the body, a positron emitted by the compound quickly is annihilated by an encounter with an electron, creating two high-energy photons that shoot off in opposite directions. Detectors that are located 180° apart on a rotating assembly register only pairs of photons that arrive simultaneously. Data from the detectors are processed by a computer to give a picture of the distribution of the radioactive compound in the brain.

One widely used compound is glucose, which supplies all the energy used by brain cells. By measuring the rate of glucose metabolism in different parts of the brain, researchers can study the differences between normal and abnormal brain functioning. For example, glucose metabolism has been found to be unusually low in the frontal regions of patients with schizophrenia. Depressed rates of glucose metabolism have also been found in senile dementia, while manic disorders are accompanied by more rapid glucose metabolism.

One problem with using glucose is that it is metabolized so quickly that measurements must be made within 5 minutes after it is administered. Many researchers have begun using analogues of glucose that are metabolized much more slowly and so accumulate in brain tissue.

PET already is being used in medical practice. For example, some forms of epilepsy require surgery because they cannot be treated with any available drugs. Neurosurgeons at several medical centers in the United States are using PET to locate the site of the brain abnormality that causes epilepsy, so that healthy tissue is not removed during an operation. PET scans also are being used to determine whether brain tumors are malignant. Ordinary X-ray examinations often cannot distinguish between benign brain tumors and the malignant, life-threatening cancers called gliomas. Studies have found that the rate of glucose metabolism is much greater in gliomas than in benign tumors.

Perhaps the most exciting prospect is that PET scans can allow scientists to study how different parts of the brain take part in thought and emotion. While the technique needs considerable refinement before neuroscientists will be able to give a detailed anatomy of logical thought processes and emotions such as fear, progress is being made toward that goal.

### Synthetic Elements

In the past few decades, nuclear reactions have been used to prepare elements that are not found on earth. One such element is technetium, element 43, whose most stable isotope, $_{43}^{97}\text{Tc}$, has a half-life of only $2.6 \times 10^6$ years, much less than the age of the earth. Technetium was first made artificially as the product of a $(d,n)$ nuclear reaction.

$$_{42}^{96}\text{Mo} + _{1}^{2}\text{H} \longrightarrow _{43}^{97}\text{Tc} + _{0}^{1}n$$

Technetium is also obtained as a by-product of nuclear reactors.

Element 61, promethium, also is not found on earth. All of its isotopes are unstable. The least unstable is $^{145}_{61}Pm$, whose half-life is 17.7 years. Promethium is prepared from a stable isotope of samarium in two steps. The first step produces an unstable isotope of samarium:

$$^{144}_{62}Sm + {}^{1}_{0}n \longrightarrow {}^{145}_{62}Sm + \gamma$$

In the second step, this unstable isotope decays by electron capture to form $^{145}_{61}Pm$.

The only other member of the first 92 elements that is not found naturally occurring on earth is astatine. It was first prepared artificially by bombardment of bismuth with $\alpha$ particles in the reaction:

$$^{209}_{83}Bi + {}^{4}_{2}He \longrightarrow {}^{211}_{85}At + 2{}^{1}_{0}n$$

This isotope of astatine is one of the longer-lived ones. Its half-life is about 7.2 hours.

All of the known elements whose atomic number is greater than 92, the transuranium elements, have been produced artificially by nuclear reactions. As yet, there is no definite evidence that any of these elements exists on earth at present, although they are known to exist in stars.

The first transuranium element to be produced artificially was neptunium, Np, whose most stable isotope, $^{237}_{93}Np$, has a half-life of about $2.2 \times 10^6$ years. Neptunium is formed when $^{238}_{92}U$ is bombarded by fast neutrons; the product of this reaction forms $^{237}_{93}Np$ by $\beta^-$ decay:

$$^{238}_{92}U + {}^{1}_{0}n \longrightarrow {}^{237}_{92}U + 2{}^{1}_{0}n$$
$$^{237}_{92}U \longrightarrow {}^{237}_{93}Np + {}^{0}_{-1}\beta$$

This isotope of Np is the parent of a radioactive decay series similar to the three decay series of naturally occurring radioisotopes. This series is not found in the earth's crust because the half-lives of the parent nuclide and other nuclides in the series are much shorter than the age of the earth. However, the series has been well established in the laboratory, and it probably existed on earth during the early years of the planet.

The transuranium element of greatest practical importance is plutonium, which is produced in large quantities in nuclear reactors. The most important isotope of plutonium is $^{239}_{94}Pu$, which can be used as fuel for a nuclear reactor or as explosive material. The widespread use of plutonium as nuclear reactor fuel has been the subject of intense debate in recent years. It is feared that nuclear weapons could become available to any nation, or even that plutonium could be stolen and fashioned into a bomb by terrorist groups if reprocessing of spent fuel rods to extract plutonium becomes routine.

Plutonium is prepared by the reaction of $^{238}_{92}U$ with a neutron, followed by two $\beta^-$-decay reactions:

$$^{238}_{92}U + ^1_0n \longrightarrow ^{239}_{92}U + \gamma$$
$$^{239}_{92}U \longrightarrow ^{239}_{93}Np + ^0_{-1}\beta$$
$$^{239}_{93}Np \longrightarrow ^{239}_{94}Pu + ^0_{-1}\beta$$

This isotope of plutonium decays by $\alpha$ emission with a half-life of 24 360 years. Another isotope, $^{244}_{94}Pu$, which has a half-life of $7.6 \times 10^7$ years, is believed to be the real parent nuclide of the thorium series. Since the half-life of $^{244}_{94}Pu$ is relatively short compared to the life of the earth, all of it that may once have existed on earth is believed to have decayed to the presently observed parent, $^{232}_{90}Th$, by a series of $\alpha$-emission and $\beta$-emission reactions.

Still heavier elements can be made by bombardment of suitable targets with charged particles. Nuclei of light atoms sometimes are used for these reactions. For example, curium, element 96, can be prepared by the reaction:

$$^{232}_{90}Th + ^{12}_6C \longrightarrow ^{240}_{96}Cm + 4^1_0n$$

Fermium, element 100, can be prepared by the reaction:

$$^{238}_{92}U + ^{16}_8O \longrightarrow ^{249}_{100}Fm + 5^1_0n$$

The number of transuranium elements that have been made in the laboratory continues to increase. In 1974, scientists from both the Soviet Union and the United States claimed to have synthesized element 106, which has recently been given the systematic name unnilhexium. The United States report said that the element was formed by the reaction $^{249}_{98}Cf(^{18}_8O,4n)^{263}_{106}X$. The Soviet report said the element had been created by several reactions, including some in which several isotopes of lead were bombarded with nuclei of $^{54}_{24}Cr$. In 1976, the Soviet Union claimed to have formed element 107, which would be called unnilheptium, by bombarding $^{209}_{83}Bi$ with nuclei of $^{54}_{24}Cr$. Attempts to create superheavy elements continue today.

## 20.4   THE CHEMISTRY OF THE UNIVERSE

One of the most challenging areas of scientific research is the investigation of the origin and evolution of the universe. A central part of this research deals with the chemical composition of the stars and the origin of the elements. Only limited data are available. Theories are tentative and complex. Nevertheless, there seems to be a broad understanding of the nuclear reactions that occur on the cosmic scale and that have created the chemical elements.

It is currently accepted that the universe as we know it originated in a "big bang" some 10 to 18 billion years ago. According to this theory, all the matter in the universe was packed into one mass, which somehow exploded to send matter streaming out. The expansion of the universe which began with the big bang still continues.

It is believed that during this expansion, clouds of gas condensed to form stars that were composed primarily of hydrogen. Such a star is called a *first-generation star.*

The prevailing picture of stellar evolution begins with the contraction of a cloud of gas, whose molecules are pulled together by gravitational attraction to form a star. As the gas cloud contracts, its temperature and density increase.

When the temperature in a first-generation star rises to about $10^7$ K, hydrogen nuclei can begin reacting with each other. The reaction is

$$^1_1H + ^1_1H \longrightarrow {}^2_1H + {}^0_1\beta + \nu$$

This is a relatively slow process that requires surprisingly high temperatures. In a much faster nuclear reaction, the deuteron that is formed in the proton-proton reaction combines with another proton:

$$^2_1H + ^1_1H \longrightarrow {}^3_2He + \nu$$

After enough $^3_2He$ accumulates, this nuclide can react with itself:

$$^3_2He + ^3_2He \longrightarrow {}^4_2He + 2^1_1H$$

The overall process, therefore, is the fusion of four protons into a single $\alpha$ particle. This process is called *hydrogen burning.* As all the hydrogen is converted to helium, the energy output of the star decreases and gravitational contraction begins again. When the temperature at the core of the star rises to $10^8$ K, the helium nuclei that have become the dominant part of the core can begin to undergo fusion reactions. The basic helium reaction is

$$3^4_2He \longrightarrow {}^{12}_6C$$

Some of the $^{12}_6C$ formed by this helium-burning process can react with $\alpha$ particles to form oxygen nuclei:

$$^{12}_6C + ^4_2He \longrightarrow {}^{16}_8O + \gamma$$

When helium burning is complete, the core of the star contains carbon and oxygen nuclei. Shrinkage begins again, until the core temperature reaches $6 \times 10^8$ K. At this point, carbon-burning reactions begin. Two of these reactions are

$$^{12}_6C + ^{12}_6C \longrightarrow {}^{23}_{11}Na + {}^1_1H$$
$$^{12}_6C + ^{12}_6C \longrightarrow {}^{20}_{10}Ne + {}^4_2He$$

The products of these reactions undergo further reactions very quickly at the elevated temperatures in the cores of the stars. When carbon burning is finished, the core of the star consists of $^{16}_{8}O$, $^{20}_{10}Ne$, $^{24}_{12}Mg$, and other nuclides with mass numbers in this range.

Gravitational collapse then continues and the temperature of the star increases once more. The nuclear reactions that occur lead to the formation of many different nuclei, the heaviest being $^{32}_{16}S$. However, there is a limit to this progression. Even though the temperature of the core of the star reaches $4 \times 10^9$ K, reactions between heavy nuclei become difficult. Alternate multistep reaction paths become important in the formation of still heavier nuclei.

The interior of the star proceeds toward a condition called nuclear statistical equilibrium, which has a superficial resemblance to ordinary chemical equilibrium. Reactions occur quickly, and the species that comes to be predominant in the interior of the star is the nucleus with the greatest binding energy per nucleon, $^{56}_{26}Fe$. The interior of the star can no longer release energy by nuclear reactions, and nuclei with mass numbers greater than 56 cannot form.

The star itself becomes unstable. It may undergo what is called a *supernova explosion,* expelling much of its material in a high-temperature shock wave that is accompanied by a number of nuclear reactions. At this point, the regions toward the surface of the star are of interest. Even though the core of the star may consist entirely of $^{56}_{26}Fe$, there are cooler outer regions that contain the lighter elements, even hydrogen. When the supernova explosion occurs, all these elements are expelled. Much of the evidence used to develop this model of stellar evolution comes from observations of the distribution of the elements in stars, interstellar clouds, and planets.

The most abundant nuclide is $^{1}_{1}H$. The second most abundant is $^{4}_{2}He$. Other light nuclides, such as deuterium, $^{3}_{2}He$, and isotopes of lithium, beryllium, and boron, which we believe are only transient intermediates in the evolution of a first-generation star, are not abundant. There is a relatively great abundance of the nuclides $^{12}_{6}C$ and $^{16}_{8}O$, which we believe form from hydrogen-burning reactions, and of $^{56}_{26}Fe$, the most stable nuclide.

## Second-Generation Stars

The formation of nuclides heavier than $^{56}_{26}Fe$ takes place in second-generation stars, those that have formed from interstellar gas that includes the products of first-generation stars.

Because second-generation stars contain these elements in reasonable abundance, a whole new range of nuclear reactions in which these elements are energy sources is possible. These reactions are believed to play an important part in the energy-producing budget of stars like the sun.

One crucial sequence of reactions is called the *carbon-nitrogen cycle.* In this sequence, four protons are converted to an $\alpha$ particle, with an energy release of about 26 MeV. The presence of carbon and nitrogen makes this

**TABLE 20.4  Stellar Evolution**

1. Contraction of interstellar hydrogen to form a first-generation star.
2. Nuclear reactions to form nuclides up to $^{56}_{26}$Fe.
3. Supernova explosion; the nuclides in the star are expelled and become interstellar material.
4. Contraction of interstellar material to form a second-generation star.
5. Nuclear reactions to form nuclides up to $^{56}_{26}$Fe; neutrons are released by some of these reactions.
6. Formation of many more nuclides as heavy as lead and bismuth as the result of slow reactions with slow neutrons.
7. Supernova explosion or other catastrophic event in which nuclides as heavy as thorium and uranium are formed by rapid multiple neutron-capture reactions. The nuclides are expelled from the star and become interstellar material.
8. Contraction of interstellar material to form a third-generation star.
9. Repeated cycles of nuclear reactions and eventual supernova explosions to produce succeeding generations of stars.

fusion process faster than the direct proton-proton reactions in first-generation stars. The important steps are

$$^{12}_{6}C + ^{1}_{1}H \longrightarrow ^{13}_{7}N + \gamma$$
$$^{13}_{7}N \longrightarrow ^{13}_{6}C + ^{0}_{1}\beta + \nu \ (\beta^+ \text{ decay})$$
$$^{13}_{6}C + ^{1}_{1}H \longrightarrow ^{14}_{7}N + \gamma$$
$$^{14}_{7}N + ^{1}_{1}H \longrightarrow ^{15}_{8}O + \gamma$$
$$^{15}_{8}O \longrightarrow ^{15}_{7}N + ^{0}_{1}\beta + \nu \ (\beta^+ \text{ decay})$$
$$^{15}_{7}N + ^{1}_{1}H \longrightarrow ^{12}_{6}C + ^{4}_{2}He$$

In a sense, the $^{12}_{6}C$ acts as a catalyst. It opens up a faster pathway for the reaction and it is regenerated at the end of the sequence. Some further branching steps in the sequence have the effect of converting $^{12}_{6}C$ to $^{14}_{7}N$. Thus, when hydrogen burning by this sequence is completed, the core of the star contains a considerable quantity of $^{14}_{7}N$, as well as helium.

Helium burning in second-generation stars produces a fairly large number of neutrons. The formation of many nuclides with mass number greater than 56 is believed to occur by neutron capture on a relatively slow time scale. Starting with $^{56}_{26}$Fe, heavier nuclides are believed to form step by step through neutron capture and $\beta^-$ decay. The sequence is believed to end with the formation of lead and bismuth, which cannot react to form nuclides of higher mass number.

This model is consistent with many observations of the relative abundance of heavier nuclides. However, several heavier elements, most notably uranium and thorium, cannot form by such reactions. These nuclides are believed to be formed in supernova explosions of stars that are second generation and later. Nuclei can react with neutrons very rapidly during such an explosion to form nuclides of mass number up to 270. Support for this theory comes from the observation of rapid multiple neutron-capture reactions in the explosion of thermonuclear bombs, where $^{254}_{98}$Cf is formed through neutron capture by $^{238}_{92}$U.

The matter expelled by the supernova explosion of a second-generation star can serve as the raw material for the formation of a third-generation star. Table 20.4 outlines the major steps in stellar evolution. Uranium and thorium exist in our solar system, so our sun is believed to be at least a third-generation star. The known half-lives of the isotopes of uranium and thorium that are found in our solar system enable us to say that the supernova explosion that provided the raw material for the formation of the sun and the planets could not have occurred much more than 5 billion years ago.

## 20.5  ENERGY FROM NUCLEAR REACTIONS; FISSION AND FUSION

Because nuclear reactions release much larger quantities of energy than do ordinary chemical reactions, they have enormous potential for both military and peaceful applications. However, there are several barriers in the way. The probability that a bombarding particle will interact with a target nucleus is rather low because the nucleus is a small target. Furthermore, it is often difficult to produce enough bombarding particles of the desired energy. It is not always easy to obtain enough of the target nuclide to produce a large amount of energy. Finally, it is not always easy to create the conditions in which a desired nuclear reaction will occur.

Many of these difficulties have been overcome in the past few decades. The major impetus for advances in this field has been the military demand for more powerful weapons of destruction. Weapons research has been followed by the effort to develop peaceful uses of nuclear reactions, primarily for the generation of electricity.

The most important nuclear reaction for commercial energy production is the fission reaction, in which a nucleus splits into two nuclei of more or less equal mass with the accompanying release of energy. The existence of fission reactions was first recognized by German physicists in the late 1930s. The discovery led to the Manhattan Project of World War II, in which usable fission weapons — atomic bombs — were developed in less than 4 years by an all-out wartime effort.

In nature, only the fission of the very heaviest elements occurs rapidly enough to be detectable. Compared to other modes of decay, fission is a relatively improbable process for most elements. The half-life of $^{232}_{90}\text{Th}$ is $1.39 \times 10^{10}$ years with respect to $\alpha$ decay and $10^{21}$ years with respect to fission. The half-life of $^{238}_{92}\text{U}$ is $7.13 \times 10^8$ years with respect to $\alpha$ decay and $6 \times 10^{15}$ years with respect to fission. The half-life with respect to fission does decrease rapidly with increasing atomic number. Thus, the half-life of $^{244}_{96}\text{Cm}$ is $1.4 \times 10^7$ years with respect to fission. The half-life with respect to fission of $^{254}_{100}\text{Fm}$ is only 100 days. For element 104, it is believed to be 0.3 s, although this result is still open to question.

The data on average binding energy per nucleon shown in Figure 20.1 indicate that nuclei with mass numbers greater than about 60 are unstable with respect to two smaller nuclei. These heavier nuclei should thus undergo spontaneous fission. They generally do not, because there is an energy barrier to fission that is analogous to a large energy of activation in ordinary chemical reactions. This energy barrier can be overcome if nuclei with a suitable amount of energy are supplied, usually by bombardment of the nuclei with particles. In some heavier elements, the bombarding particles do not have to be of very high energy. The most important fission reactions are those produced in heavy elements by thermal neutrons, which are not high-energy particles. The nuclides that undergo fission when they absorb thermal neutrons include the $^{235}_{92}\text{U}$ used in the atomic bomb dropped on Hiroshima and the $^{239}_{94}\text{Pu}$ used in the atomic bomb dropped on Nagasaki.

When such a heavy nucleus absorbs a thermal neutron, it undergoes fission because it becomes sufficiently excited to overcome the energy barrier to fission. The average binding energy per nucleon in the two fission product nuclei increases substantially compared with the average energy per nucleon in the original nucleus. As a result, fission is accompanied by a large release of energy, in the range of 200 MeV per fission.

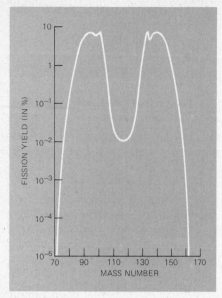

**Figure 20.5**
The relative yields of different fission products for the slow-neutron fission of $^{235}$U, plotted as a function of mass number.

The fission of $^{235}_{92}$U demonstrates this release of energy. When the nucleus absorbs a thermal neutron, it forms the compound nucleus $^{236}_{92}$U. The average binding energy per nucleon of a nucleus of this mass number is about 7.6 MeV. Assume that the fission products are two nuclei of mass number 118 each. In a nucleus of this mass number, the average binding energy per nucleon is about 8.5 MeV. The energy released by the fission thus is

$$E = 2(118)(8.5 \text{ MeV}) - (236)(7.6 \text{ MeV}) = 210 \text{ MeV}$$

Most of this energy appears as kinetic energy of the fission products, but some energy is released in a different form in subsequent processes. The fission products generally have an excess of neutrons to protons and are therefore unstable. Some of the energy of fission is released when the fission products undergo $\beta^-$ decay because of their instability.

Energy calculations are complicated by the fact that fission reactions usually do not result in two product nuclides of equal size. For example, the fission of $^{235}_{92}$U forms products ranging from zinc ($Z = 30$) to terbium ($Z = 65$), with mass numbers from 72 to 161. Figure 20.5 shows the percentages of fission product nuclei of different mass number produced when $^{235}_{92}$U fissions by thermal neutron absorption.

Neutrons and $\gamma$ rays also are emitted when a heavy excited nucleus undergoes fission. In the thermal neutron fission of $^{235}_{92}$U, an average of 2.5 neutrons per nucleus is emitted. In the fission of $^{239}_{94}$Pu, an average of 2.9 neutrons per nucleus is emitted. Most of these neutrons are available to cause fission of other nuclei, which is one of the most important characteristics of these fission reactions.

## Chain Reactions

If a nucleus absorbs one neutron and releases several neutrons, a chain reaction can occur. That is, the fission of a single nucleus releases neutrons that cause the fission of several nuclei. Each of these nuclei, in turn, produces several neutrons that cause the fission of more nuclei. Such a chain reaction is basic to both military and peaceful uses of fission energy. In an atomic bomb, the chain reaction is allowed to cascade. A large number of nuclei undergo fission almost simultaneously, releasing destructively large amounts of energy. In a nuclear reactor, the chain reaction is controlled, so that a manageable amount of energy is released slowly over a long period of time.

Chain reactions are rare in nature. For a chain reaction to occur, a single fission must release one or more neutrons that go on to initiate other fissions. More than one neutron must be released per fission in naturally occurring samples, because some of the neutrons are absorbed by nuclei that cannot undergo fission. For example, naturally occurring uranium contains two isotopes, $^{238}$U and $^{235}$U, in a ratio of about 140:1. When a nucleus of $^{235}$U absorbs a neutron, it undergoes fission. A nucleus of $^{238}$U does not; instead, it undergoes $\beta^-$ decay. A chain reaction is not

## THE MISSING NEUTRINOS

The most unusual observatory in the history of astronomy is a 400 000-L tank of dry-cleaning fluid 1500 m underground in a mine at Lead, South Dakota. This observatory has given astronomers their best view of the processes occurring deep in the interior of the sun. The results obtained from the observatory indicate that there could be basic flaws in the accepted theory of the nature of the solar interior.

The observatory was built to detect neutrinos produced in certain of the solar reactions, such as steps in the carbon-nitrogen cycle. The neutrinos emitted in a solar fusion reaction such as:

$$\,^2_1H + \,^1_1H \longrightarrow \,^3_2He + \nu$$

pass readily through the sun. Most of them also pass through the earth without reacting with other particles. Even though solar neutrinos are emitted in enormous numbers, only a few interactions occur. These interactions can easily be hidden by interaction caused by cosmic rays. The observatory was placed underground to screen out all but solar neutrino interactions.

The tank in the mine contains perchloroethylene, $C_2Cl_4$. Every so often, a neutrino will react with a chlorine atom to form a radioactive argon atom:

$$\nu + \,^{37}_{17}Cl \longrightarrow \,^{37}_{18}Ar + \,^{\ 0}_{-1}\beta^-$$

Once every hundred days, inert gas is bubbled through the tank, and the number of neutrino interactions is measured by a count of the number of $^{37}Ar$ atoms that are swept out by the gas. The chemical technique has been so perfected that even a few radioactive argon atoms can be isolated and counted.

If the existing model of the solar interior is correct, about one $^{37}Ar$ atom a day should be produced in the tank. The results obtained over several years are well below that level. Apparently, one $^{37}Ar$ atom is produced only every two and a half or three days.

This result has led to many speculative theories about the sun and the neutrino. It has been suggested that the sun is a variable star, whose changing energy output is responsible for the discrepancy between the predicted number of neutrinos and the observed number. It has also been proposed that neutrinos behave differently over large distances from the way they do in laboratory observations, or that the neutrino, which is assumed to be massless, may indeed have a small mass. Each of these theories could explain the mystery of the missing neutrinos, but none of them has been widely accepted.

As the 1980s began, several other solar neutrino experiments were planned. One of them, already underway, will utilize the capture of neutrinos by $^{71}Ga$ to form radioactive $^{71}Ge$. About 18 000 kg of gallium, roughly 2 years' total production of this expensive metal, eventually will be used in the experiment. Another experiment would use 200 000 L of nearly saturated aqueous lithium chloride solution. The $^7Li$ nuclei capture an occasional neutrino to form $^7Be$ nuclei. Until such experiments are completed, doubts about our understanding of the sun will remain.

possible in most deposits of naturally occurring uranium because most of the neutrons emitted when $^{235}U$ nuclei fission are absorbed by the $^{238}U$ nuclei or other substances. However, evidence that a natural chain reaction occurred two billion years ago in a uranium deposit in what is now the Gabon Republic in West Africa has been discovered. An unusual sequence of events apparently allowed the concentration of $^{235}U$ to increase, so that the chain reaction could occur. There are two major pieces of evidence for the existence of this natural fission reactor: The concentration of $^{235}U$ is lower than it should be, and the deposit contains elements, in particular xenon, that are characteristic products of fission reactions. As far as is known, this chain reaction was a unique event.

Even when the $^{235}U$ concentration is relatively high, a chain reaction may not occur if the sample of uranium is so small that many neutrons can escape without striking another nucleus. The minimum quantity of material required to sustain a chain reaction is called the *critical mass.*

One other requirement must be met for a chain reaction to occur. The fast neutrons emitted in fission reactions must be slowed down. The probability that a nucleus will capture a bombarding particle and undergo a reaction is called the nuclear cross section. Nuclear cross sections are greater for the capture of slow-moving thermal neutrons than for fast neutrons. For example, the cross section of $^{235}U$ is 500 times greater for thermal neutrons than for fast neutrons. However, fission reactions release fast neutrons. Therefore, fissionable material usually is mixed with a material called a *moderator,* which slows fast neutrons to thermal speeds. Light substances that do not undergo reactions with neutrons generally are used as moderators. The first nuclear reactor, or "pile," as it was called, used graphite as a moderator. Most nuclear generating plants in the United States use ordinary water as the moderator.

The uncontrolled chain reaction that takes place in an atomic bomb must occur with the fast neutrons released by fission, since there is no time to slow the neutrons with a moderator. An atomic bomb is set off when pieces of fissionable material are brought together very rapidly to form a critical mass. Care must be taken that the beginning of the chain reaction does not blow the pieces apart again, causing a small premature explosion called a fizzle. A great deal of ingenuity has been devoted to producing nuclear explosives that use the minimum amount of fissionable material and have the maximum amount of destructive power.

In a nuclear reactor, the chain reaction must be controlled within fairly strict limits. The primary method of control is the use of rods made of a material that absorbs neutrons readily. Control rods often are made of $^{10}_{5}B$, whose neutron-absorbing capabilities are excellent.

A typical nuclear power plant (Figure 20.6) in the United States has a

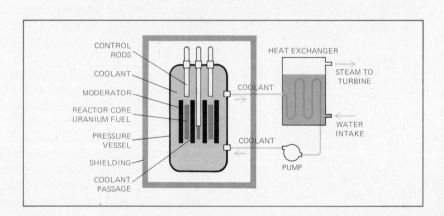

**Figure 20.6**
Essential components of a nuclear reactor.

# THE NUCLEAR WASTE PROBLEM

When coal is burned to generate electricity, its waste products include sulfur oxides and ash. When nuclear fuel is "burned" to generate electricity, its waste products include a host of radioactive elements, some of them potentially useful and many of them quite dangerous. An intense debate about the most desirable way to handle nuclear wastes has been going on for several years.

One proposal is that the nuclear fuel be processed to remove fissionable material, which could then be used in new fuel rods. A typical fuel rod contains about 3.3% $^{235}$U and 97.7% $^{238}$U. As the fissionable $^{235}$U is consumed, some of the $^{238}$U is transformed by neutron absorption and $\beta^-$ decay into transuranium elements, including $^{239}$Pu and $^{240}$Pu, which can be used as fissionable fuel. To obtain the plutonium, the fuel rod is dissolved in acid, and the resulting solution is processed chemically to isolate the fissionable fuels. Plutonium and $^{235}$U are then fashioned into a so-called mixed oxide fuel for reactors.

The opposition to such fuel recycling is based partly on the fear that the plutonium obtained in this process could be used to make nuclear weapons, either by governments bent on aggression or by terrorist groups. In addition, fuel reprocessing leaves behind some liquid wastes of extremely high radioactivity that will be dangerous for many thousands of years.

For example, the fission products in the liquid wastes include $^{90}$Sr, whose half-life is 29 years, and $^{137}$Cs, whose half-life is 30 years. It will take 400 years for the radioactivity from these fission products to decrease to a reasonably safe level. But even after 400 years, there will be substantial danger from other radioactive elements such as $^{241}$Am and $^{229}$Th, whose half-lives are measured in thousands or even millions of years.

The most commonly discussed plan is to transform these liquid wastes into glassy ceramic rods, each about 3 m long and 30 cm in diameter, which would be put into salt beds, granite, or other geological deposits, where they presumably would remain untouched for the many thousands of years needed for the radioactive elements to decay. The wastes from a single 1000-MW nuclear reactor could be contained in 10 such rods.

Several studies have concluded that this method of waste disposal is feasible, and that safe geological deposits can be found. Critics, however, have pointed out that these wastes must be stored safely for a much longer period of time than any human society has ever existed, and that safety cannot be guaranteed for such periods.

While the debate goes on, spent fuel rods from U.S. reactors are being held in storage. The issue may not be resolved until the end of the century. In 1980, the government outlined a plan for identifying safe storage sites by 1985. Under this plan, actual use of the sites would not begin until 1995 at the earliest. Thus, the ultimate fate of nuclear wastes still is far from settled.

core consisting of about 40 000 fuel rods, weighing a total of 120 000 kg, made of uranium whose $^{235}$U content has been enriched to 3%. The core is immersed in water, which acts as the moderator as well as a coolant (temperatures in the core can reach 600 K). Control rods are interspersed among the fuel rods. The core is enclosed in a steel vessel perhaps 12 m high and 5 m in diameter, with steel walls 30 cm thick. A number of safety features designed to prevent overheating of the core and escape of radioactive material are essential parts of a nuclear power plant. Proponents of nuclear energy say that these safety devices make a serious accident highly improbable. But the accident that occurred at the Three Mile Island nuclear generating plant in Pennsylvania in 1979 indicated that there are flaws in the safety systems. The Three Mile Island incident not only led to a temporary moratorium in the licensing of new nuclear power plants but also raised serious doubts about the long-term future of nuclear energy.

## Fusion Weapons

A major effort now is being made to obtain usable energy from fusion, the process that occurs in stars. Just as the maximum energy release in fission occurs in the heaviest elements, the largest energy release in fusion processes occurs in the lightest elements. The coulombic repulsion barrier also is smallest for the lighter elements, whose nuclei have fewer protons.

Attention is focused on the fusion of heavier isotopes of hydrogen, such as deuterons, as sources of energy. A fusion reaction between two deuterons can take two different courses:

$$^2_1H + ^2_1H \longrightarrow ^3_2He + ^1_0n + 3.3 \text{ MeV}$$
$$^2_1H + ^2_1H \longrightarrow ^3_1H + ^1_1H + 4.0 \text{ MeV}$$

Such fusion reactions take place only at extremely high temperatures. They are called *thermonuclear reactions.* To start a thermonuclear reaction, the kinetic energy of the reacting nuclei must be raised to the equivalent of a temperature of 100 000 000 K. However, the average temperature can be lower. Most of the nuclei will have energies too low for fusion to occur, but a few will have the energy necessary for fusion. Once some appreciable fraction undergoes a fusion reaction, enough energy is liberated to raise the kinetic energy of other nuclei to the level needed for fusion, so the fusion reaction becomes self-sustaining.

The temperature necessary to begin the thermonuclear reaction in a hydrogen bomb is achieved by explosion of an atomic bomb. This fission bomb is surrounded by deuterium, tritium, and $^6_3Li$. The explosion of the fission bomb raises the temperature of these nuclides to $10^7$ K. Fusion reactions occur, releasing energy that causes a rapid temperature increase and more fusion reactions. The reactions that occur in a hydrogen bomb include:

$$^2_1H + ^3_1H \longrightarrow ^4_2He + ^1_0n + 17.6 \text{ MeV}$$
$$^6_3Li + ^1_0n \longrightarrow ^4_2He + ^3_1H + 4.8 \text{ MeV}$$
$$^3_2He + ^2_1H \longrightarrow ^4_2He + ^1_1H + 18.3 \text{ MeV}$$

The lithium reaction is important because it replenishes the supply of tritium, which is consumed in the fusion process.

To add explosive power, the material undergoing fusion is encased in a shell of ordinary uranium. The fusion reactions emit fast neutrons that can cause a fission chain reaction in the $^{238}U$ in the shell. This fission reaction cannot take place with slow neutrons. Thus, a thermonuclear bomb of the most modern design has three stages: nuclear fission of a few kilograms of $^{239}_{94}$Pu, nuclear fusion of about 150 kg of lithium, deuterium, and tritium, and nuclear fission of about 500 kg of $^{238}_{92}$U. The total mass of such a bomb is about 1500 kg. The energy released is equivalent to the explosion of 20 000 000 000 kg of TNT. There is no practical limit to the explosive yield of a modern thermonuclear weapon.

## Fusion Reactors

In principle, thermonuclear reactions are a source of unlimited energy for our needs. Large amounts of deuterium, the principal fuel proposed for a fusion generating plant, can be obtained at low cost from ordinary seawater. However, the task of creating the conditions under which a controlled thermonuclear reaction can occur are formidable. The control of thermonuclear reactions has been a high-priority research goal for more than two decades, but widespread use of peaceful thermonuclear energy still is believed to be decades away.

The principal problem is the difficulty of producing and maintaining matter in a physical state quite different from that which is encountered at ordinary temperatures. At very high temperatures, there is virtually complete ionization of all atoms, forming a state of matter called *plasma*. Plasma is a homogeneous mixture of atomic nuclei and electrons, moving rapidly and randomly. At the temperatures necessary for fusion reactions, a plasma loses energy very quickly by emitting electromagnetic radiation. A fusion reaction cannot be self-sustaining unless the rate at which energy is produced by fusion exceeds the rate at which energy is lost from the plasma as radiation. A thermonuclear reactor will produce surplus energy only when the temperature of the plasma is at least $10^8$ K.

An added complication is the loss of energy by the plasma through

The Tokomak Fusion Test Reactor at Princeton University, one of the major U.S. fusion energy research facilities. Inside the reactor is a plasma, which is prevented from touching the steel walls by intense magnetic fields. (*Princeton University Plasma Physics Laboratory*)

thermal conduction. Because of their free electrons, plasmas are extremely good conductors of heat. A plasma conducts heat more than $10^6$ times better than a metal at room temperature. Therefore, the plasma must not be allowed to come in contact with the walls of a container, or it will lose energy so rapidly that no significant number of fusion reactions can occur. In most thermonuclear experiments, the plasma is kept from the walls of the container by magnetic fields, in what is called magnetic confinement. We can picture the plasma, a gas consisting of charged particles, as being enclosed in a cage whose bars are the lines of force of a magnetic field. Several different types of magnetic confinement are being tried. One is a toroidal (doughnut-shaped) device called the tokomak, originated by the Soviet Union and adapted by laboratories in the United States. Another is a so-called magnetic mirror, in which the plasma can be pictured as washing back and forth between two "walls" of magnetic force.

A successful fusion reactor must satisfy what are called the Lawson criteria, concerning the product of $n$, the plasma density in particles per cubic centimeter, and $\tau$, the confinement time of the plasma in seconds. For deuteron-deuteron reactions, $n\tau$ must exceed $10^{16}$. For deuteron-tritium reactions, $n\tau$ must exceed $10^{14}$. In 1984, leaders of the United States fusion research effort were predicting that the Lawson criteria would be met in a fusion reactor that achieved "scientific break-even," in which the amount of energy released by fusion reactions is equal to the energy required to cause fusion, during this decade.

Magnetic confinement was the only approach to controlled thermonuclear fusion until the 1960s. In that decade, research began on a method called inertial confinement, in which light from lasers or beams of ions would be used to compress tiny pellets of nuclear fuel, to produce fusion reactions. In essence, the inertial-confinement approach would produce a series of miniature thermonuclear explosions whose energy could be captured by the walls of the container in which the explosions occurred. Recent work indicates that the inertial-confinement approach does not seem practical.

If the effort to harness thermonuclear fusion succeeds, we shall have a source of energy essentially without limit. However, as we mentioned in Chapter 13, such an energy source could create problems as severe as any it solves. The challenge posed by the technology that has arisen from advances in nuclear science is one of the gravest that the human race has ever faced.

**Summary**

In this chapter, we first noted that the atomic **nucleus** contains two kinds of particles, positively charged **protons** and electrically neutral **neutrons.** We defined the **atomic number,** the number of protons in a nucleus, and the **mass number,** the number of protons and neutrons in a nucleus. We noted that **isotopes** are species with the same number of protons but different numbers of neutrons. Then we described the **binding energy,** the powerful force that holds protons and neutrons together in the nucleus, as being 10 million times greater than the energy of a chemical bond. We mentioned that the stability of nuclei is lessened by coulombic repulsions between protons and is strengthened by the presence of neutrons, adding that the stablest neuclei have even numbers of both protons and neutrons, while the least stable have odd numbers of both. We then described **radioactivity,** the process by which some nuclei decay, and discussed **alpha** decay, in which a helium ion is emitted from a nucleus; **gamma** decay, in which gamma rays are emitted; and **beta** decay, which can occur in several different ways and in which the mass number of a nucleus remains the same but the atomic number changes. We discussed the natural decay processes that occur on earth and the elements that take part in them. Then we described how both atomic number and mass number can be changed by **nuclear reactions,** and how radioactive isotopes for research, industry, and medicine can be prepared by such reactions. We went on to show how a series of reactions inside stars produce heavier elements, starting with hydrogen and helium. Finally, we discussed efforts to obtain explosive energy from **fission** in atomic bombs and **fusion** in thermonuclear weapons, and programs to obtain useful energy from fission and fusion.

**Exercises**

**20.1** Indicate the number of neutrons and protons in each of the following nuclides: (a) lithium-7, (b) neon-22, (c) plutonium-239, (d) platinum-194.

**20.2** Use the conventional notation to list five isotopes of (a) carbon, (b) tin.

**20.3** Use the conventional notation to list five isobars with (a) $A = 40$, (b) $A = 234$.

**20.4**[1] Use the conventional notation to list five isotones with (a) $N = 14$, (b) $N = 143$.

**20.5** Find the radius of (a) a nucleus of $^4$He, (b) a nucleus of $^{247}$Bk, (c) the ratio of the two radii.

**20.6** Find the mass of 1 mol of nuclei of $^2$H.

**20.7** Calculate the density of a nucleus of $^{12}$C.

**20.8** Calculate the mass of a 1-cm cube of $^{12}$C nuclei.

**20.9** Find the energy equivalent of the mass of (a) an electron, (b) a proton.

**20.10** Find the energy equivalent of 1 mol of $^{12}$C.

**20.11** The heat of combustion of gasoline is about $3 \times 10^4$ kJ/L. Calculate the mass loss on combustion of 1 L of gasoline.

**20.12** The mass of $^7$Li$^{3+}$ = 7.014359.

Use the relative masses of the proton and the neutron given in Table 20.1 to find the binding energy of this nuclide.

**20.13★** Find the binding energy in $^3$He.[2]

**20.14** Classify each of the following nuclides as even-even, even-odd, odd-even, or odd-odd: (a) $^{56}$Fe, (b) $^{57}$Fe, (c) $^{58}$Co, (d) $^{59}$Co.

**20.15** Explain the observation that no stable nuclide of tin is known in which both $Z$ and $N$ are magic numbers.

**20.16** Write reactions for the $\alpha$ decay of the following nuclides: (a) $^8$Li, (b) $^{190}$Pt, (c) $^{192}$Pt, (d) $^{239}$Pu.

[1] The answers to exercises whose numbers are in color can be found in Appendix VII. The star indicates an exercise that is more challenging than average.

[2] The necessary data for this exercise are listed in Table 20.1.

**20.17**   Calculate the mass loss and energy release in the $\alpha$ decay of $^{239}$Pu.[2]

**20.18**   The relative mass of $^{242}$Cm is 242.0588. Calculate the mass loss and energy release when this nuclide undergoes $\alpha$ decay to $^{234}$U.

**20.19**   Find the energy change that corresponds to the emission of $\gamma$ rays of wavelength $1.0 \times 10^{-12}$ m.

**20.20**   Write reactions for the $\beta^-$ decay of the following nuclides: (a) $^{28}$Al, (b) $^{56}$Mn, (c) $^{147}$Pm, (d) $^{223}$Fr.

**20.21**   Write reactions for the $\beta^+$ decay of the following nuclides: (a) $^{15}$O, (b) $^{56}$Co, (c) $^{94}$Tc, (d) $^{206}$Bi.

**20.22**   Write reactions for electron capture by the following nuclides: (a) $^{26}$Al, (b) $^{60}$Cu, (c) $^{209}$Po.

**20.23**   Predict the most likely type of $\beta$ decay for the following nuclides: (a) $^{17}$F, (b) $^{21}$F, (c) $^{197}$Hg, (d) $^{206}$Hg.

**20.24**   Calculate the mass loss and energy release for the $\beta^-$ decay of $^{87}$Rb.[2]

**20.25**   Calculate the mass loss and energy release for the $\beta^-$ decay of $^{14}$C.[2]

**20.26**   Find the mass loss and energy release when $^{10}$C (relative mass = 10.016830) undergoes $\beta^+$ decay.[2]

**20.27**   Find the mass loss and energy release when $^{15}$O (relative mass = 15.003072) undergoes $\beta^+$ decay.[2]

**20.28**   Find the mass loss and energy release when $^{209}$Po (relative mass = 208.9829) decays by electron capture.[2]

**20.29**   The nuclide $^{18}$F decays by both electron capture and $\beta^+$ decay. Find the difference in the energy released by these two processes.[2]

**20.30**   Find the energy released when 1.0 mol of $^3$H undergoes $\beta^-$ decay.[2]

**20.31**   Write equations for each of the first three $\alpha$-decay steps in the thorium series.

**20.32**   Write equations for three of the $\beta^-$ decay steps in the thorium series.

**20.33**   Suggest an explanation for the absence of any isotopes of platinum or gold as intermediates in the thorium series.

**20.34\***   Suggest an explanation for the observation that $^{236}$Np (relative mass = 236.0466) undergoes electron capture but not $\beta^+$ decay to form $^{236}$U (relative mass = 236.0457).

**20.35**   Calculate $\Delta M$ and $E$ for the reaction $^{59}$Co$(n,v)^{60}$Co, given that the relative mass of $^{60}$Co is 59.93355.[2]

**20.36**   Calculate $\Delta M$ and $E$ for the two-step process in which $^{59}$Co is converted to $^{60}$Ni by neutrons.[2]

**20.37**   Find $\Delta M$ and $E$ for the formation of $^{14}$C in the atmosphere.[2]

**20.38**   Write the reaction and calculate $\Delta M$ and $E$ for the process $^{10}$B$(n,\alpha)^7$Li.[2]

**20.39**   Write the reactions, including the compound nuclei, for the following processes: (a) $^6$Li$(n,\alpha)^3$H, (b) $^7$Li$(n,v)^8$Li, (c) $^{45}$Sc$(n,p)^{45}$Ca.

**20.40**   Write the reactions, including the compound nuclei, for the following processes: (a) $^{31}$P$(d,p)^{32}$P, (b) $^{10}$B$(p,v)^{11}$C, (c) $^9$Be$(d,2p)^9$Li.

**20.41**   Write the reactions, including the compound nuclei, for the following processes: (a) $^{106}$Pd$(\alpha,p)^{109}$Ag, (b) $^6$Li$(^3$He$,n)^8$B, (c) $^{141}$Pm$(^{12}$C,$4n)^{149}$Tb.

**20.42**   Suggest two possible pathways, using any bombarding particles you choose, for the conversion of $^{197}$Au to $^{208}$Pb.

**20.43**   The nuclide $^{241}$Am has been made from the reaction between $^{238}$U

and $^4$He. Suggest a pathway for the formation of this nuclide.

**20.44**   The nuclide $^{247}$Es can be made by bombardment of $^{238}$U in a reaction that emits five neutrons. Identify the bombarding particle.

**20.45\***   Find $E$ for the overall hydrogen burning process in a star.[2]

**20.46**   Write a series of steps by which $^{64}$Zn could form from $^{56}$Fe in a second-generation star.

**20.47**   Write a series of steps by which $^{238}$U might form from $^{208}$Pb in a supernova explosion.

**20.48**   Explain why a fission chain reaction does not ordinarily take place in natural uranium ores.

**20.49**   Explain the function of each of the following in a nuclear reactor: (a) critical mass, (b) moderator, (c) control rod, (d) coolant.

**20.50**   Predict the effect of the removal of each of the things mentioned in Exercise 20.49 on the operation of a nuclear reactor.

**20.51**   Account for the formation of plasma at high temperatures in terms of the second law of thermodynamics.

**20.52**   Given that the energy released in the fusion of two deuterons to a $^3$He and a neutron is 3.3 MeV and in the fusion to tritium and a proton it is 4.0 MeV, calculate the energy change in the process $^3$He$(n,p)^3$H. Suggest an explanation for the fact that this process occurs at much lower temperatures than either of the first two.

**20.53**   Discuss the possible effects of virtually unlimited energy from controlled fusion in terms of the conversion of heat to work and the second law of thermodynamics.

# 21

# Organic Chemistry

**Preview**

This chapter deals with the chemistry of a single element: carbon, which is the basis not only for life on earth but also for much of the modern chemical industry. Beginning with the simplest organic compounds, the alkanes, we go on to describe increasingly more complicated compounds: the alkenes, the alkynes, and the cyclic hydrocarbons, which are built around rings of carbon atoms. We then show how much of the chemical behavior of organic compounds can be described in terms of the behavior of their functional groups, and we introduce a number of functional groups that contain oxygen, nitrogen, and sulfur. Finally, we describe the basic and applied research done by organic chemists.

**O**rganic chemistry is the study of the compounds of carbon. Until the middle of the nineteenth century, organic chemistry was defined as the chemistry of living organisms and the substances isolated from them. It was believed that anything alive possessed a "vital force" that was absent in nonliving things. All organic compounds were believed to possess this vital force. The downfall of the vital force theory was the result of many experiments that demonstrated that a sample of a substance from a living thing and a sample of the same substance from a nonliving source are indistinguishable. Today, organic chemistry deals with many substances that have nothing to do with living organisms.

Almost without exception, organic compounds contain hydrogen in addition to carbon. A very few contain one of the halogens instead of hydrogen. Many also contain one or more other elements, such as oxygen, nitrogen, sulfur, and phosphorus. A special class of compounds called organometallics contain a metal as well.

More than 3 million organic compounds are known, ten times more than all the compounds of the other elements. Aside from sheer numbers, organic compounds are important because much of the chemistry of life is organic chemistry. In addition, a substantial part of the modern chemical industry is devoted to production of synthetic organic compounds: plastics, pharmaceuticals, paints, detergents, fibers, and all the other products that are an essential part of our daily lives.

The existence of such a great number of organic compounds results in large part from the ability of a carbon atom to form strong bonds with other carbon atoms, building chains or rings of carbon atoms. This property, called *catenation,* is not unique to carbon. Other elements near carbon in the periodic table can form rings or chains. But those formed by carbon are unusually stable and unreactive, in part because of the relative strength of the carbon-carbon bond. Not only is the typical carbon-carbon bond energy substantial, almost 350 kJ/mol, but it is also close to or larger than the bond energies of carbon with other elements.

## 21.1  THE ALKANES; SATURATED HYDROCARBONS

Compounds that contain only carbon and hydrogen are called *hydrocarbons.* A hydrocarbon whose structure includes no double bonds and no rings is called an *alkane.* an organic compound with no multiple bonds is said to be *saturated,* and a compound with one or more multiple bonds is said to be *unsaturated.* All alkanes thus are saturated. The alkanes are the simplest group of organic compounds in terms of chemical behavior.

The simplest alkane, with one carbon atom, is methane. The alkane with two carbon atoms is ethane, and the alkane with three carbon atoms is propane. Their structures are

$$
\begin{array}{ccc}
& \quad \text{H} & \quad \text{H} \quad \text{H} & \quad \text{H} \quad \text{H} \quad \text{H} \\
& | & | \quad\; | & | \quad\; | \quad\; | \\
\text{H}-\text{C}-\text{H} & \quad \text{H}-\text{C}-\text{C}-\text{H} & \quad \text{H}-\text{C}-\text{C}-\text{C}-\text{H} \\
& | & | \quad\; | & | \quad\; | \quad\; | \\
& \text{H} & \text{H} \quad \text{H} & \text{H} \quad \text{H} \quad \text{H}
\end{array}
$$

|  Methane  |  Ethane  |  Propane  |

In these compounds, as in almost every other organic compound known, each carbon atom has a total of four covalent bonds. We can use this fact to simplify the represention of organic molecules.

A first simplification is to eliminate the lines that indicate bonds between carbon and hydrogen atoms. Instead, all the hydrogen atoms that are bonded to a given carbon atom in a molecule are grouped together. The lines showing bonds between carbon atoms are retained. Thus, methane is represented as $CH_4$, ethane is represented as $CH_3-CH_3$, and propane is represented as $CH_3-CH_2-CH_3$. The Lewis structures can be written from this simplified representation if we distribute the hydrogen atoms around the carbon atoms so that each carbon atom has a total of four bonds.

We can make a further simplification and eliminate the lines showing the carbon-carbon bonds. The structure of ethane then is written $CH_3CH_3$, and the structure of propane becomes $CH_3CH_2CH_3$. Since we know that each carbon atom has four bonds and each hydrogen atom has only one bond, an unambiguous Lewis structure can be written for each molecule from the simplified representation.

The relationship between the compositions of different alkanes can be seen even in the first three compounds in the group. Starting with methane, we can *imagine* that a $CH_2$ group is inserted into one of the $C-H$ bonds of $CH_4$ to form ethane. We can also imagine a $CH_2$ group being inserted into one of the $C-H$ bonds of ethane to form propane. In this way, we can construct an entire series of alkanes, each with one more $CH_2$ group than its predecessor. The general formula for any alkane is $C_nH_{2n+2}$, where $n$ is the number of carbon atoms.

But we quickly encounter a situation in which more than one structure is possible. Propane has two different kinds of hydrogens. Six hydrogen atoms are bonded to the two carbon atoms at the ends of the three-carbon chain (the $CH_3$ groups). Two are bonded to the carbon atom in the middle of the chain (the $CH_2$ group). We can imagine the next alkane in the series being formed by insertion of another $CH_2$ group into a $C-H$ bond of propane. Since there are two types of $C-H$ bonds, two alkanes are possible. They both have the composition $C_4H_{10}$ and they are both called butane. Their structures are

$$
CH_3CH_2CH_2CH_3 \qquad\qquad CH_3CHCH_3 \\
\qquad\qquad\qquad\qquad\qquad\qquad\quad | \\
\qquad\qquad\qquad\qquad\qquad\qquad\quad CH_3
$$

|   *n*-Butane   |   Isobutane   |

## Chain Isomerism

As we mentioned earlier (Section 18.2), two compounds with the same composition are called **isomers.** Isomerism is common in organic chemistry. The type of isomerism observed in butane is frequently encountered. It is called *chain isomerism,* because the difference between the isomers of two butanes is in the chain of carbon atoms. The butane that has a straight chain of four carbon atoms is called *n*-butane, where *n* stands for "normal." The butane that has a chain of three carbon atoms with a one-carbon branch is called isobutane.

The number of isomers of an alkane increases rapidly as the number of carbon atoms increases. There are three isomers, called pentanes, whose composition is $C_5H_{12}$. They are

$$CH_3CH_2CH_2CH_2CH_3 \qquad CH_3\underset{\underset{CH_3}{|}}{CH}CH_2CH_3 \qquad CH_3\overset{\overset{CH_3}{|}}{\underset{\underset{CH_3}{|}}{C}}CH_3$$

| *n*-Pentane | Isopentane | Neopentane |

The names of the simple alkanes illustrate some basic principles in naming organic compounds. A clear and unambiguous system of names is necessary because of the great number and complexity of organic compounds.

Both suffixes and prefixes in the names of organic compounds convey specific information about their structures. All the compounds we have discussed so far have names that end in the suffix *-ane.* This suffix conveys the information that the compound has no multiple bonds. The prefix *alk-* is used to name groups of compounds that contain only carbon and hydrogen atoms. However, *alk-* is not used for individual compounds in the group. Instead, we use prefixes that indicate the number of carbon atoms in the compound. Table 21.1 lists some of these prefixes.

If necessary, additional prefixes and suffixes can be added to indicate additional details about the molecular structure of a compound. As the number of carbon atoms increases, it becomes a formidable task to write and name all the isomers of a given composition. For example, there are 336 319 isomers with the composition $C_{20}H_{42}$. Rather than studying the

**TABLE 21.1    Prefixes for Organic Compounds**

| Prefix | Number of C Atoms | Prefix | Number of C Atoms |
|--------|-------------------|--------|-------------------|
| meth   | 1                 | pent   | 5                 |
| eth    | 2                 | hex    | 6                 |
| prop   | 3                 | hept   | 7                 |
| but    | 4                 | oct    | 8                 |

alkanes one by one, it is better to describe some properties that they have in common. Then we can try to find correlations between properties and composition within the group.

Every alkane contains only two kinds of bonds, C—C bonds and C—H bonds. Both of these bonds are quite strong, so the alkanes tend to be unreactive. An alkane generally undergoes reaction only at elevated temperatures or in other extreme conditions. The most important reaction for the alkanes is combustion, the rapid, high-temperature reaction with $O_2$.

We can group alkanes and other types of organic compounds into *homologous series* on the basis of their structures. A homologous series is a group of organic compounds that differ only in the number of $CH_2$ units in the longest chain. The simplest homologous series is that of the straight-chain alkanes. The first five members of the series are methane, ethane, propane, *n*-butane, and *n*-pentane. Each succeeding member of the series — *n*-hexane, *n*-heptane, and so on — has a chain that is longer by one carbon atom.

Another homologous series consists of alkanes that each have a one-carbon branch located on the next-to-last carbon atom of the chain. We can call this series the *iso* series. We already have mentioned isobutane and isopentane, the first two members of the series.

The members of a homologous series not only have similar chemical properties; even their physical properties change in a regular way as more $CH_2$ groups are added. One striking example of this regularity is the increase of 29 K in the boiling point of the alkanes with each added $CH_2$ group.

## 21.2   UNSATURATED HYDROCARBONS

We mentioned that an organic compound with one or more multiple bonds is said to be unsaturated. The simplest class of unsaturated compounds is the *alkenes*.

The alkenes contain only carbon and hydrogen, as the prefix *alk-* indicates. An alkene has only one carbon-carbon double bond and does not contain any rings of carbon atoms. The suffix *-ene* means "double bond."

The two simplest alkenes are ethylene, $C_2H_4$, and propylene, $C_3H_6$. The location of the double bond is shown when the simplified structures of alkenes are written:

$$CH_2{=}CH_2 \qquad CH_2{=}CHCH_3$$
$$\text{Ethylene} \qquad\quad \text{Propylene}$$

Since each carbon atom can have only four bonds, the presence of a double bond between two carbon atoms means that the compound has two fewer hydrogen atoms than the corresponding alkane. The general formula for an alkane is $C_nH_{2n+2}$, so the general formula for an alkene is $C_nH_{2n}$.

## Positional Isomerism

The introduction of a double bond into a carbon chain creates more isomeric possibilities. Chain isomerism is found in the butenes, just as in the butanes. For example, two butenes with different chains are

$$CH_2=CHCH_2CH_3 \qquad CH_2=C\begin{smallmatrix}CH_3\\CH_3\end{smallmatrix}$$

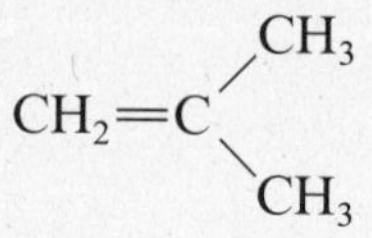

A double bond can be introduced into a four-carbon chain in one of two positions, giving rise to *positional isomers.* One positional isomer of the straight-chain butene is shown above. The other one is

$$CH_3CH=CHCH_3$$

By numbering the carbon atoms of the chain in the indicated way, we can differentiate between the two positional isomers. Thus, one isomer is named 1-butene, because the double bond starts at the carbon atom numbered 1. The other is named 2-butene, because the double bond starts at the carbon atom numbered 2. The chain isomer is called isobutene, to differentiate it from the two straight-chain positional isomers.

## Geometric Isomerism

There are actually four isomeric alkenes with the composition $C_4H_8$. Two isomers of 2-butene differ in the spatial arrangement of the substituents on the double bond. In one isomer, both $CH_3$ groups are on the same side of the double bond. This compound is called *cis*-2-butene. In the other isomer, the two $CH_3$ groups are on opposite sides of the double bond. This compound is called *trans*-2-butene. Their structures are

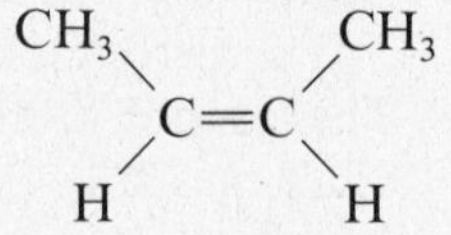

cis-2-Butene 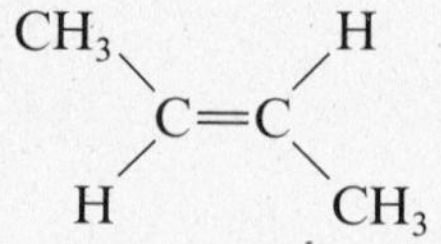          trans-2-Butene

These two isomers of 2-butene are called *geometric isomers* (Section 18.2). They differ in the spatial arrangement of their atoms but not in the way that the atoms are attached to each other.

Geometric isomerism occurs in alkenes whenever the two carbon atoms connected by the double bond bear two different substituents, because there is no rotation around the double bond at room temperature. In 2-butene, for example, each carbon atom of the double bond bears one hydrogen atom and one $CH_3$ group, so geometric isomerism occurs. In 1-butene, however, the carbon atom that is numbered 1 bears two identical substituents, two hydrogen atoms, so there is no geometric isomerism in 1-butene.

The geometric isomer that has similar substituents on the same side of the double bond is called the *cis* isomer. The one in which the similar substituents are on opposite sides of the double bond is called the *trans* isomer.

## Polyenes

Compounds with more than one double bond are called *polyenes*. The polyenes can be classified and named by the number of double bonds they contain. For example, the compounds that consist of only carbon and hydrogen, have no rings of carbon atoms, and have two double bonds are called *alkadienes*. You will note that this name includes both the prefix *alk-,* indicating that the compounds contain only carbon and hydrogen, and the prefix *di-*, which indicates the presence of two double bonds. Since an alkadiene has two double bonds, it has two fewer hydrogen atoms than the alkene with the same number of hydrogen atoms. The general formula for an alkadiene is $C_nH_{2n-2}$. We indicate the location of the double bonds by numbering the carbon atoms. For example, the compound whose name is 1,3-butadiene has the structure:

$$\underset{1}{CH_2}=\underset{2}{CH}-\underset{3}{CH}=\underset{4}{CH_2}$$

## Chemistry of Alkenes

The alkenes are much more reactive than the alkanes. Even under mild conditions, they undergo a wide variety of reactions, most of which result in changes at the double bond. There are a number of reagents that add to double bonds but do not cleave the carbon chain. Some examples of this type of reaction, which is called an *addition reaction,* are shown in Figure 21.1(a). Other reactions result in the cleavage of the carbon chain at the site of the double bond. Figure 21.1(b) shows a reaction of this type. Still other reactions result in changes at carbon atoms adjacent to the double bond or changes in the location of the double bond in the chain.

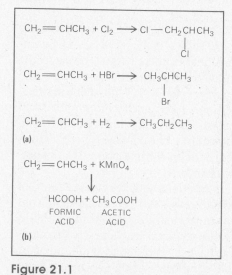

**Figure 21.1**

Reactions of alkenes. In the first three reactions, the carbon skeleton remains intact. The addition is only to the $\pi$ bond of the double bond. In the fourth reaction, the alkene is cleaved, and two products are formed.

## Functional Groups

The double bond in an alkene is one of the simplest examples of a **functional group.** A functional group can be defined as a small group of atoms in a molecule that differ in some way from an alkane structure. The presence of any atom other than carbon or hydrogen or of any multiple bond indicates the presence of a functional group. The most important generalization in organic chemistry is that *the chemical behavior of compounds is determined primarily by the behavior of their functional groups.*

This generalization is a great simplifying principle. Rather than trying to learn the chemistry of a huge number of individual compounds, we can learn the chemistry of a relatively small number of functional groups. For example, the chemistry of all the alkenes is similar because they all have the same functional group, a double bond.

However, the overall structure of a compound does influence its chemical behavior. There are differences in the chemical behavior of alkenes that result from differences in their overall structures. One important part of organic chemistry is the study of the way such structural differences lead to observed differences in chemical behavior.

## Alkynes

The suffix *-yne* means "triple bond." An alkyne, therefore, is a hydrocarbon with a triple bond. The simplest alkyne is acetylene $C_2H_2$; the next is propyne, $C_3H_4$:

$$HC\equiv CH \qquad CH_3C\equiv CH$$
Acetylene         Propyne

An alkyne has four fewer hydrogen atoms than the corresponding alkane. The general formula of the alkynes thus is $C_nH_{2n-2}$, which is identical to that of the alkadienes.

There are four compounds with a straight chain of four carbon atoms and the composition $C_4H_6$. They are shown in Figure 21.2. The compounds numbered **1** and **2** are positional isomers named, respectively, 1-butyne and 2-butyne. The compounds numbered **3** and **4** are also positional isomers which are named, respectively, 1,3-butadiene and 1,2-butadiene. But if we compare compounds **1** and **2** with compounds **3** and **4**, we find a different kind of isomerism. These compounds are *functional isomers.* They have identical compositions but contain different functional groups. Compounds **1** and **2** contain a triple bond, while compounds **3** and **4** contain two double bonds.

Triple bonds undergo the addition reactions and cleavage reactions of the sort that double bonds undergo. However, there are some significant differences in the behavior of the two functional groups. The most important difference appears in what are called terminal acetylenes, compounds such as propyne and 1-butyne, which have the triple bond at the end of the chain of carbon atoms. Such a triple bond bears a single hydrogen atom that is relatively acidic. These compounds are weak acids. The other hydrogen atoms in these hydrocarbons, as well as most hydrogen atoms bonded to carbon, are virtually nonacidic and can be removed only by the strongest bases.

$$CH_3CH_2C\equiv CH$$
**1**

$$CH_3C\equiv CCH_3$$
**2**

$$CH_2=CHCH=CH_2$$
**3**

$$CH_2=C=CHCH_3$$
**4**

**Figure 21.2**
Some isomers of $C_4H_6$.

## 21.3  CYCLIC HYDROCARBONS

Carbon atoms can form rings as well as chains. The name of a compound that contains one or more rings of carbon atoms usually includes the prefix *cyclo-*. Compounds that contain only carbon and hydrogen and have at least one ring of carbon atoms are called *cyclic hydrocarbons*. The class of compounds called the *cycloalkanes* are those that have a ring of carbon atoms *(cyclo-)*, contain only carbon and hydrogen atoms *(-alk-)*, and have no multiple bonds *(-ane)*. The cycloalkenes and the cycloalkadienes are just two of the many other classes of cyclic hydrocarbons.

We can imagine that a cycloalkane is formed by removal of a hydrogen atom from each end of an alkane chain and then joining of the two ends of the chain. The general formula of the cycloalkanes is $C_nH_{2n}$, the same as that of the alkenes. Thus, there are two compounds with the composition $C_3H_6$. One is propylene. The other, which has a ring of three carbon atoms, is called cyclopropane.

Figure 21.3 shows several different representations of the simplest cycloalkane, cyclopropane. The first structure (Figure 21.3(a)), which shows all the bonds, is rarely used. In the second structure (Figure 21.3(b)), the practice of omitting the C—H bonds but including the C—C bonds is followed. The representation that is most commonly used is that of Figure 21.3(c), in which the cyclic structure is represented by a polygon. This greatly simplified representation can be used because it is understood that each carbon atom has four bonds.

The sides of the polygon in Figure 21.3(c) represent carbon-carbon single bonds. Each corner represents a carbon atom. It is assumed that the molecule has just enough hydrogen atoms to allow each carbon atom to form four bonds. Substituent atoms or groups are shown only if they are not single hydrogen atoms. For example, the compound whose composition is $C_4H_8$ and whose name is cyclobutane is represented as:

□

The representation indicates that cyclobutane has a ring of four carbon atoms, each with two singly bonded hydrogen atoms attached to it.

The name of an unsubstituted cycloalkane is derived from the prefix *cyclo-* and the name of the alkane with the same number of carbon atoms. For example, the cycloalkane whose composition is $C_6H_{12}$ is called cyclohexane. It has a six-carbon ring and its structure is represented as a regular hexagon.

We can learn the principles that are followed in naming compounds whose rings bear substituents other than hydrogen atoms by studying specific compounds. A compound whose composition is $C_4H_8$ is

▽—$CH_3$

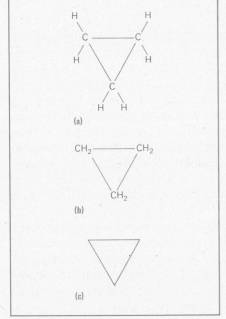

**Figure 21.3**
Three ways of representing the simplest cycloalkane, cyclopropane.

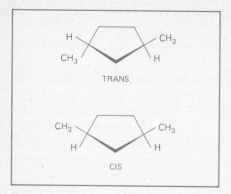

**Figure 21.4**
The geometric isomers of 1,3-dimethylcyclopentane.

This compound is named methyl cyclopropane. Cyclopropane denotes the three-carbon-atom ring. The relationship between methane, $CH_4$, and the $CH_3$ substituent on the ring is reflected in the name *methyl*. The suffix *-yl* designates a polyatomic group formed by the loss of one hydrogen atom from the molecule named in the stem. The loss of the hydrogen atom allows the attachment of the substituent, sometimes called a *radical*, to another atom, such as the carbon atom of a chain or ring. Many organic compounds are named by designating the type and location of radicals on a chain or ring of carbon atoms.

Compounds with ring structures that bear more than one substituent are quite common. Geometric isomerism, analogous to the geometric isomerism found in alkenes, can exist in compounds of this kind. If there are two substituents on the same side of the ring, we have the *cis* isomer. If the substituents are on opposite sides of the ring, we have the *trans* isomer. Figure 21.4 shows the *cis* and *trans* isomers of 1,3-dimethylcyclopentane.

Like the alkanes, the cycloalkanes tend to be unreactive. Some carbon atoms of a ring may be joined by double bonds. Even triple bonds are found in some rings, although they are rare. A cyclic hydrocarbon with one double bond is called a *cycloalkene.* To represent the structure of a cycloalkene, we indicate the location of the double bond in the polygon that represents the ring structure. For example, the structure of the simplest cycloalkene, cyclopropene, is represented as:

In part because of the reactivity of the double bond, cyclopropene is much more reactive than cyclopropane, the corresponding cycloalkane.

## Polycyclic Compounds

Polycyclic compounds have more than one ring as part of their structures. One major class of polycyclic compounds is the *steroids,* which are derivatives of the system of four rings shown in Figure 21.5. This type of system, in which the rings share some carbon atoms, is called a fused ring system. Although the representation in Figure 21.5 appears to be much more complex than that of a simple cycloalkane, it is interpreted in the same way.

All steroids are physiologically active. A number of steroids occur naturally in plants and animals and many others have been synthesized. Perhaps the best-known and certainly the most abundant steroid in the human body is cholesterol, which makes up about one-sixth of the dry weight of the body and has been implicated in diseases of the cardiovascular system.

Many other steroids play major roles in the human body. For example, the steroid sex hormones regulate sexual functions in both men and women. Female sex hormones are called estrogens. Male hormones are

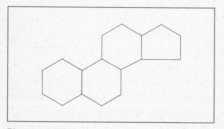

**Figure 21.5**
The system of four rings on which the steroids are based.

## STEROIDS FROM PLANTS

One day in 1943, a chemist named Russell E. Marker walked into the office of a small Mexican pharmaceutical company named Laboratorios Hormona carrying two pickle jars wrapped in newspaper. In the jars were more than two kilograms of a steroid hormone called progesterone, worth about $160,000 at the going price and representing a substantial percentage of the existing annual world production of progesterone. Marker had produced the hormone on his own, by a method of his own devising, and that day in Mexico City was a significant milestone in modern pharmaceutical history.

At that time, almost all steroid hormones used in medicine were synthesized by a small number of European companies, which used cholesterol from animal sources as a starting material. The European "hormone cartel" kept supplies limited and prices high.

Marker had been working on the synthesis of steroid hormones while he was a professor of steroid chemistry at Pennsylvania State University. He concentrated on steroid substances from plants, called sapogenins, whose structure resembles that of cholesterol. In 1939, he had worked out a way of approaching the structure of the steroid hormones by chemical treatment of the sapogenins. The following year, he found a way of synthesizing progesterone from one kind of sapogenin called diosgenin, taken from a plant of the genus *Dioscorea.* When American pharmaceutical companies refused to provide financial support for his research, Marker resigned his position and went to Mexico. He had found that a high yield of diosgenin could be obtained from the root of a yam called *cabeza de negro,* which grows in the jungles of Mexico.

Marker rented a rundown shack in Mexico City and went to work, helped only by unskilled laborers. Legend has it that he was so suspicious that he usually slept with a pistol next to his research notebook. Marker's efforts to keep his work secret were successful. When he appeared with his two kilograms of progesterone, it was a stunning surprise.

The owners of Laboratorios Hormona quickly joined with Marker to found a new company to produce progesterone. The company, Syntex, today is one of the world's major pharmaceutical manufacturers. At that time, its main asset was Marker's knowledge. He quickly produced enough progesterone to break the cartel's hold on the world market. When Marker first appeared on the scene, progesterone was selling at $80 a gram. By 1945, the price was down to $18 a gram. In 1952, Syntex was able to fill a contract for more than 9000 kg of progesterone at 48 cents a gram.

Marker, ever the individualist, stalked out of Syntex in 1945 after an argument with his partners, taking his knowledge with him. He did not reappear for two decades. It took several months of research for others to work out the details of Marker's process, enabling production to be resumed.

Today more than half the world's supply of steroid hormones comes from the kind of plant compound that caught the attention of Russell Marker in the 1930s.

called androgens. A third type of steroid sex hormone, called progesterone, is active not only in regulating female sex function but also in regulating the many physiological changes that occur during pregnancy.

Oral contraceptives, the most widely used birth control method in the United States, include two synthetic steroids. One is a progestational hormone, the other is an estrogen. Another group of synthetic steroids which are frequently used in medicine are anti-inflammatory agents derived from the corticosteroids, hormones secreted by the adrenal gland.

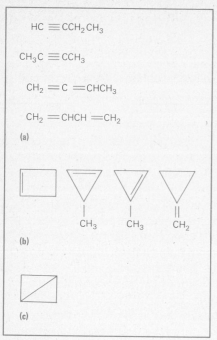

$HC \equiv CCH_2CH_3$

$CH_3C \equiv CCH_3$

$CH_2 = C = CHCH_3$

$CH_2 = CHCH = CH_2$

(a)

(b)

(c)

**Figure 21.6**
All the isomers of $C_4H_6$.

## Composition and Structure

From the information we already have, we can work out the relationship between the composition of a hydrocarbon and the kind of structure or structures that are possible for the compound. We can start with the alkanes, whose general formula is $C_nH_{2n+2}$. Any compound with this composition is a hydrocarbon with no rings and no multiple bonds. If the number of hydrogen atoms is reduced by two, the general formula becomes $C_nH_{2n}$. A compound with this composition can have either one ring or one double bond. If we reduce the number of hydrogen atoms by two again, the general formula becomes $C_nH_{2n-2}$. A compound with this composition can have two double bonds, or two rings, or one double bond and one ring, or one triple bond. Table 21.2 summarizes these relationships. You can see that the number of possible structures increases rapidly as the number of hydrogen atoms in the general formula decreases.

**TABLE 21.2** Possible Structures of Hydrocarbons

| Formula | Structure |
|---|---|
| $C_nH_{2n}$ | one double bond or one ring |
| $C_nH_{2n-2}$ | two double bonds; or one triple bond; or two rings; or one double bond and one ring |
| $C_nH_{2n-4}$ | three double bonds; or one double bond and one triple bond; or three rings; or two rings and one double bond; or one ring and two double bonds; or one ring and one triple bond |

**Example 21.1**    Draw the structures of all the isomers of $C_4H_6$.

**Solution**    We note from the formula that this compound has the general formula $C_nH_{2n-2}$. Table 21.2 gives the possible combinations of rings and multiple bonds.

We must also consider the possible arrangements of the four carbon atoms. They may be arranged in a straight chain of four atoms. They may form a straight chain of three atoms with a one-atom branch. They may form a four-membered ring. Or they may form a three-membered ring with a one-atom substituent.

The straight chain of carbon atoms can accommodate either one triple bond or two double bonds. Figure 21.6(a) shows the possible positional isomers of this arrangement. The branched-chain structure cannot accommodate two double bonds or one triple bond. There are no branched-chain isomers of $C_4H_6$.

Either a four-carbon ring or a three-carbon ring can accommodate a double bond. There is positional isomerism for the three-membered ring structure, since there are different ways of placing the double bond. Figure 21.6(b) shows the possible ring structures. Finally, there is one possible structure with two rings, shown in Figure 21.6(c).

## 21.4   AROMATIC HYDROCARBONS

Benzene is a most unusual hydrocarbon. It has the composition $C_6H_6$, and the six carbon atoms form a six-membered ring. The rules summarized in Table 21.2 indicate that this ring should contain three additional $\pi$ bonds and that a reasonable structure to write for benzene is

However, this structure is inconsistent with the experimental evidence. The six carbon atoms of the $C_6H_6$ molecule have been found to lie in the same plane and to have identical bonds whose length is midway between those of single and double bonds. All the carbon atoms of benzene are identical. So are all the hydrogen atoms and all the C—H bonds.

Benzene cannot be described by an ordinary structural formula. Rather benzene is a resonance hybrid (Section 7.5) and is best described as a blend of two equivalent contributing structures:

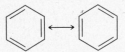

Neither of these structures exists. We know that they are imaginary and write them to help us represent the structure of benzene conveniently.

The difference between these two contributing structures is in the distribution of the six $\pi$ electrons in benzene. The structure of benzene sometimes is written with a circle to represent the six $\pi$ electrons:

As Figure 21.7(a) shows, each carbon atom of the benzene ring is $sp^2$ hybridized. The $sp^2$ hybrid orbitals form $\sigma$ bonds with the three attached atoms. Each carbon atom has an unhybridized $p$ orbital that is perpendicular to the plane of the carbon and hydrogen atoms. These six $p$ orbitals are parallel to each other. They can overlap to form six molecular orbitals that extend over all of the six carbon atoms of the ring. The six $\pi$ electrons of the benzene rings are in three of these orbitals and can be regarded as circulating around the ring. The $p$ orbitals have two lobes, so there is electron density both above and below the plane of the ring, as Figure 21.7(b) shows.

Benzene has been found to be even more stable than can be explained

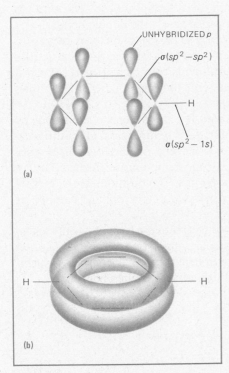

**Figure 21.7**

The benzene molecule. Each carbon atom in the ring is $sp^2$ hybridized (a). The $p$ orbitals have two lobes, so there is electron density above and below the plane of the ring (b).

by simple resonance energy. Experimental evidence indicates that unusually great stability is associated with the presence of six $\pi$ electrons in a cyclic structure like that of benzene. This unusual stability is called *aromaticity*. The six-membered benzene ring is called an aromatic ring, and benzene can be regarded as the parent of all compounds containing such rings. A hydrocarbon with one or more benzene rings is called an *aromatic hydrocarbon* and is unusually stable.

The chemical properties of aromatic compounds reflect their stability. Even though benzene appears to have double bonds, its chemistry is quite different from that of the alkenes. While alkenes readily add a variety of substances under mild conditions, benzene requires more extreme conditions for reaction. When reaction does occur, it is not addition but substitution. A typical substitution reaction of benzene takes place when it is treated with $Cl_2$. The products are HCl and chlorobenzene, $C_6H_5Cl$:

This reaction contrasts sharply with the reaction of propylene with $Cl_2$ (Figure 21.1), where the $Cl_2$ simply adds across the double bond.

Many compounds consist of a benzene ring with one or more substituents. These structures are represented in the same way as the structures of substituted cycloalkanes: The substituents are attached to the appropriate corners of the polygon that represents the ring. Figure 21.8(a) shows toluene, a simple substituted benzene.

Disubstituted benzenes—those with two substituents—are named by using a special system developed for these compounds. Only three arrangements for two substituents on a benzene ring are possible. Figure 21.8(b) shows these arrangements for the compounds called xylenes, in which both substituents are methyl groups. The positions of the methyl groups are indicated by the prefixes ortho *(o)*, meta *(m)*, and para *(p)*, as shown in Figure 21.8.

Many relatively simple substituted benzenes are physiologically active and are used as drugs or for other applications. Most of these substituted benzenes contain oxygen and/or nitrogen. Aspirin, acetylsalicylic acid, probably the most familiar of all drugs, is a substituted benzene.

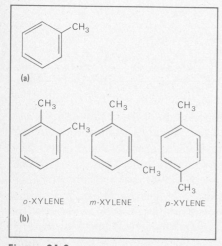

**Figure 21.8**
Toluene, a simple substituted benzene (a), and the three xylenes (b).

Many other simple substituted benzenes are used as painkillers. Benzocaine is a local anesthetic. Phenacetin is an analgesic. Oil of wintergreen (methyl salicylate) is applied externally for muscle aches. The ampheta-mines are a group of simple benzene derivatives that are central nervous system stimulants. A related compound is ephedrine, which is used as a nasal decongestant. A number of benzene derivatives are central nervous system depressants; the best known of them is phenobarbital. Mescaline, the hallucinogen found in peyote cactus, is a substituted benzene of fairly simple structure. So is adrenaline, a central nervous system stimulant found in the human body. Eugenol, or oil of cloves, is a toothache remedy; vanillin is a familiar flavoring agent. The list of substituted benzenes could go on for pages.

### Polynuclear Aromatic Hydrocarbons

Compounds with more than one benzene ring are called polynuclear aromatic hydrocarbons. In most of these compounds, atoms are shared between rings. The simplest such compound is naphthalene, the familiar moth repellent, which is a resonance hybrid of three contributing structures:

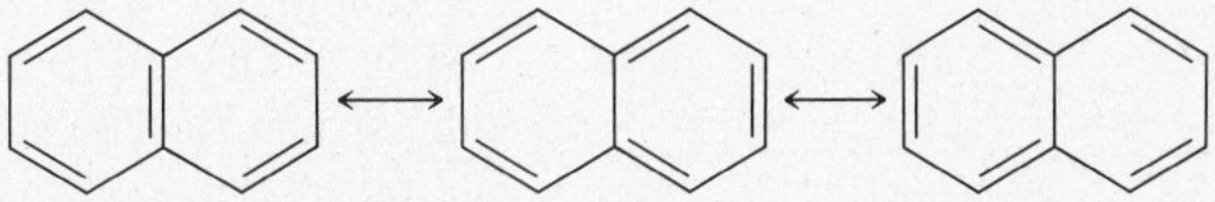

Coal tar contains a number of polynuclear aromatic hydrocarbons. Such compounds are also found in other materials, most notably in the "tars" of cigarette smoke. Many polynuclear aromatic hydrocarbons are carcinogens; that is, they cause cancer. There seems to be a correlation between the $\pi$ electronic structure of a polynuclear aromatic hydrocarbon and its carcinogenicity.

### Coal and Oil

Coal and petroleum consist primarily of hydrocarbons. When we burn coal or oil, we get heat from the combustion reactions of these hydrocarbons. By distillation of crude petroleum, it is possible to obtain hydrocarbons that burn at different temperatures and that can be used for different purposes. The gas fraction of petroleum, which boils below 310 K, consists of $C_1$ through $C_5$ hydrocarbons. The fraction that boils in the range from 310 K to 450 K is gasoline, which consists of more than a hundred $C_6$ through $C_{10}$ hydrocarbons. Kerosene, which consists of $C_{11}$ and $C_{12}$ hydrocarbons, boils from 450 K to 500 K. Fractions of petroleum that boil at higher temperatures are called oils and lubricants. Vaseline is the last fraction to distill. It consists of $C_{26}$ through $C_{38}$ hydrocarbons that boil in the range from 680 K to 790 K. The residue that remains after all these fractions are distilled from petroleum is the thick, black substance called asphalt.

## 21.5   OXYGEN FUNCTIONAL GROUPS

We have said that much of organic chemistry can be systematized by identification of the reactive portions of molecules, especially functional groups. By describing the chemistry associated with a specific functional group, we can in large measure describe the chemistry of compounds that have this functional group.

Many common functional groups include one or more oxygen atoms. It is convenient to classify oxygen-containing functional groups on the basis of their oxidation states. Formal oxidation numbers usually are not assigned in organic chemistry. Rather, the oxidation state is based on the number of oxygen atoms and hydrogen atoms in the functional group.

### Alcohols and Ethers

There are two related functional groups in the lowest oxidation state associated with oxygen functional groups. One is the functional group —OH, which is found in *alcohols.* The other is the functional group —O—, which is found in *ethers.* Both these classes of compounds have the general formula $C_nH_{2n+2}O$.

The —OH group in an alcohol replaces a hydrogen atom on an alkane structure. The simplest alcohol, for example, has the composition $CH_4O$, which usually is written $CH_3OH$. This compound is called methyl alcohol, methanol, or wood alcohol. It is a poison that causes blindness. The next alcohol in the series, the one that is most familiar to us, can be represented as $CH_3CH_2OH$ and is called ethyl alcohol, ethanol, or grain alcohol. It is the alcohol in beverages.

Alcohols, particularly those of higher molecular weight, tend to have similar properties. We can understand many aspects of their chemical behavior by studying the relationship between the alcohols and water. An alcohol can be regarded as an organic derivative of water in which one of the hydrogen atoms of the water molecule is replaced by an organic radical. An alcohol is both a weak acid and a weak base whose acid strength and base strength are somewhat less than those of water. The conjugate bases of the simple alcohols are the *alkoxides.* We can prepare salts of alkoxide anions and metal cations, such as $CH_3CH_2ONa$, which is called sodium ethoxide and is analogous to sodium hydroxide. Since its conjugate acid, ethanol, is a slightly weaker acid than water, sodium ethoxide is a slightly stronger base than sodium hydroxide.

### Reactions of Alcohols

Alcohols undergo many different types of reactions. Substitution reactions, in which the —OH group is replaced by a different group, are easily carried out in many alcohols. A typical example is the reaction of an alcohol with hydrochloric acid:

$$\underset{\text{OH}}{\text{CH}_3\overset{|}{\text{C}}\text{HCH}_3} + \text{HCl} \longrightarrow \underset{\text{Cl}}{\text{CH}_3\overset{|}{\text{C}}\text{HCH}_3} + \text{HOH}$$

in which the —Cl group replaces the —OH group and water forms as the other product.

Dehydrations, in which an alcohol loses water, can be carried out with a dehydrating agent. When ethanol is treated with $H_2SO_4$, it loses $H_2O$ to form ethylene:

$$\text{CH}_3\text{CH}_2\text{OH} \overset{H^+}{\longrightarrow} \text{CH}_2{=}\text{CH}_2 + \text{H}_2\text{O}$$

The $H^+$ can be written above the yield sign in this reaction to indicate that it is required but not consumed in the reaction.

Oxidation reactions can be carried out to convert an alcohol to a compound with an oxygen-containing functional group of higher oxidation state. Alcohols can also be converted to alkanes. This conversion is regarded as a reduction, because the alcohol loses an oxygen atom.

The group of alcohols in which the —OH group is attached directly to a benzene ring is called the *phenols*. The simplest compound of the group is called phenol and has the structure:

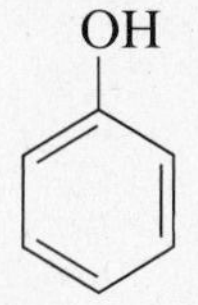

Phenol was the first substance to be used as an antiseptic. Although phenol has been replaced by more effective substances, antiseptics still are rated by their "phenol number."

The chemical properties of the phenols are somewhat different from those of the alcohols. The most striking difference is in acidity. Phenols are much stronger acids than alcohols. The $K_A$ of a typical phenol is $10^{-10}$. The $K_A$ of a typical alcohol is only about $10^{-18}$. This increase in acidity results primarily from resonance stabilization of the phenoxide ion, the conjugate base of phenol:

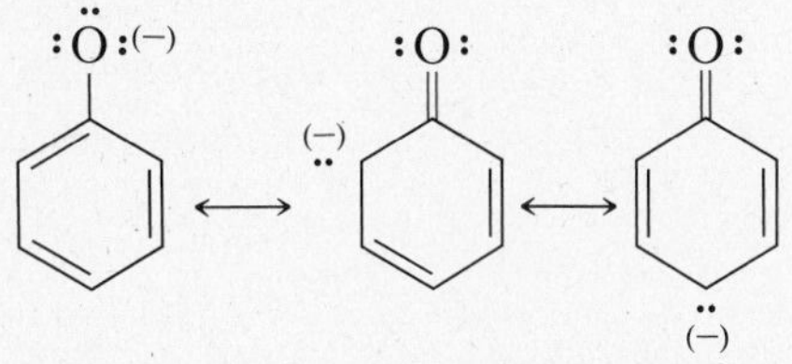

The ethers are quite unreactive. They undergo chemical change only when treated with powerful reagents under vigorous conditions. In an ether, the —O— functional group is attached to two organic groups. The simplest ether has the structure $CH_3$—O—$CH_3$. It is called dimethyl ether. Dimethyl ether and ethyl alcohol have the same composition, $C_2H_6O$, and they are functional isomers.

## Aldehydes and Ketones

The oxygen-containing functional groups in the next-higher oxidation state are the aldehydes and ketones. Both these classes of compounds have the *carbonyl* functional group, a carbon atom connected by a double bond to an oxygen atom:

$$\underset{-C-}{\overset{O}{\parallel}}$$

If the carbon atom of the carbonyl bears a hydrogen atom, the class of compounds is called the *aldehydes,* with the general formula:

$$\underset{O}{\overset{\phantom{O}}{R-C-H}}$$
$$\parallel$$

The simplest aldehyde, $CH_2$=O, is called formaldehyde. The next-simplest one is

$$\overset{O}{\underset{CH_3C-H}{\parallel}}$$

called acetaldehyde.

A class of compounds in which the carbonyl functional group bears two carbon substituents and no hydrogen atom is the *ketones.* The general formula of the ketones is

$$\underset{O}{\overset{\phantom{O}}{R-C-R}}$$
$$\parallel$$

The simplest ketone is called acetone. Its structure is

$$\overset{O}{\underset{CH_3-C-CH_3}{\parallel}}$$

Functional isomerism exists between aldehydes and ketones. For example, a functional isomer of acetone is the three-carbon aldehyde called propionaldehyde,

$$CH_3CH_2\underset{\displaystyle \underset{O}{\|}}{C}{-}H$$

Both the aldehydes and the ketones have the general formula $C_nH_{2n}O$. By comparing this general formula to that of the alcohols and ethers, we can see that the aldehydes and the ketones are in a higher oxidation state. An aldehyde or ketone has two fewer hydrogen atoms than does the corresponding alcohol or ether. Therefore, to convert an alcohol into a ketone or aldehyde, an oxidizing agent should be used. To convert a ketone or an aldehyde into an alcohol, a reducing agent should be used.

There is a great deal of evidence indicating that ketones and aldehydes are in equilibrium with other compounds called *enols*. The equilibrium for acetone and its corresponding enol is:

$$CH_3{-}\underset{\displaystyle \underset{O}{\|}}{C}{-}CH_3 \rightleftharpoons CH_3{-}\underset{\displaystyle \underset{OH}{|}}{C}{=}CH_2$$

The difference between these compounds is the shift of a hydrogen atom from a $CH_3$ group in acetone to the O atom in the enol, with the necessary accompanying change in the position of the double bond. These two compounds are functional isomers. The equilibrium is established rapidly at room temperature. A rapidly established equilibrium between two isomers is called a *tautomerism*. This one is called *keto-enol tautomerism*. Generally, the equilibrium lies far on the side of the ketone.

Just as an alcohol loses $H_2O$ to form an alkene, an aldehyde or a ketone can lose $H_2O$ to form an alkyne. We can visualize this change more easily by including the appropriate enol as an intermediate. Thus, the conversion of acetone to propyne through the dehydration of the enol is

$$CH_3\underset{\displaystyle \underset{O}{\|}}{C}CH_3 \rightleftharpoons CH_3\underset{\displaystyle \underset{OH\ H}{|\ \ |}}{C}{=}CH \longrightarrow CH_3C{\equiv}CH + H_2O$$

The reverse reaction can occur under appropriate conditions. Thus, aldehydes and ketones can be prepared from alkynes by the addition of $H_2O$, just as alcohols can be prepared from alkenes. The addition or loss of $H_2O$ is not an oxidation-reduction reaction. Therefore, we can say that alcohols and alkenes are in the same oxidation state, and that aldehydes, ketones, and alkynes are in the same oxidation state.

## Carboxylic Acids and Derivatives

The *carboxylic acid* functional group is the next-higher oxidation state of oxygen-containing functional groups. The carboxylic acid functional group appears to be a combination of the alcohol and the carbonyl groups. Its structure is

$$-\overset{\displaystyle \|}{\underset{\displaystyle O}{C}}-OH$$

The two simplest carboxylic acids are formic acid and acetic acid:

$$H-\overset{\displaystyle \overset{O}{\|}}{C}-OH \qquad CH_3-\overset{\displaystyle \overset{O}{\|}}{C}-OH$$

Formic acid        Acetic acid

The hydrogen atom on the —OH portion of the carboxylic acid is an acidic proton. The conjugate base of a carboxylic acid is called a *carboxylate anion*. These anions are stabilized by resonance:

$$R-\overset{\displaystyle \underset{:O:}{\|}}{C}-\ddot{O}:^- \longleftrightarrow R-\overset{\displaystyle \underset{:O:^-}{|}}{C}=\ddot{O}$$

The derivatives of the carboxylic acids are classes of compounds in which the —OH group of the acid is replaced by an atom or a group other than carbon or hydrogen. The *acid halides* are compounds in which the —OH group is replaced by a halogen. A simple acid halide derived from acetic acid is acetyl chloride, whose structure is

$$CH_3-\overset{\displaystyle \underset{O}{\|}}{C}-Cl$$

The *amides* are compounds in which the —OH group is replaced by $NH_2$. A simple amide, derived from acetic acid, is acetamide, whose structure is

$$CH_3-\overset{\displaystyle \underset{O}{\|}}{C}-NH_2$$

The proteins, one of the most important classes of compounds in living organisms, are complex amides.

The most commonly encountered derivatives of carboxylic acids are the *esters*. An ester is formed when the H atom of the —OH part of the

| **TABLE 21.3**  Oxidation State of Carbon Compounds | | | | |
|---|---|---|---|---|
| **Increasing Oxidation State $\longrightarrow$** | | | | |
| alkanes | alkenes | alkynes | carboxylic acids and derivatives | carbon dioxide and derivatives |
| $C_nH_{2n+2}$ | $C_nH_{2n}$ | $C_nH_{2n-2}$ | $C_nH_{2n}O_2$ | $CO_2$ |
| | alcohols and ethers | aldehydes and ketones | | |
| | $C_nH_{2n+2}O$ | $C_nH_{2n}O$ | | |

functional group is replaced by an organic radical. The general formula of an ester can be represented as:

$$R-\underset{\underset{O}{\|}}{C}-O-R'$$

where R and R′ are two radicals. Two simple esters of acetic acid are

$$CH_3-\underset{\underset{O}{\|}}{C}-O-CH_3 \qquad CH_3-\underset{\underset{O}{\|}}{C}-O-CH_2CH_3$$

Methyl acetate                Ethyl acetate

The formation of an ester from an alcohol and a carboxylic acid is called *esterification*. A typical esterification is the formation of ethyl acetate from acetic acid and ethanol:

$$CH_3-\underset{\underset{O}{\|}}{C}-OH + CH_3CH_2-OH \rightleftharpoons CH_3-\underset{\underset{O}{\|}}{C}-O-CH_2CH_3 + H_2O$$

The other product is water. Virtually complete formation of ethyl acetate is assured by removal of the water as it forms, which drives the equilibrium to the right. This method is based on the principle of Le Chatelier.

The general formula of the carboxylic acids and the esters is $C_nH_{2n}O_2$. The presence of two oxygen atoms accounts for the increase in oxidation state relative to the ketones.

A still higher oxidation state than the carboxylic acid is found in $CO_2$ and its derivatives. Table 21.3 summarizes the oxidation states that we have discussed.

## 21.6  FUNCTIONAL GROUPS WITHOUT OXYGEN

Many nonmetallic elements other than oxygen are found in organic compounds. They include the halogens, nitrogen, sulfur, and phosphorus.

## THE BRAIN'S OWN OPIATES

Why should human beings be so sensitive to opiates, substances that come from plants? The answer to this question has come from the discovery that there are natural opiates in the brain, a discovery that has provided a deeper understanding of how the brain works and of physiological drug addiction. It may also open the way to synthesis of new drugs to alleviate pain without the danger of addiction.

The first step toward the discovery was made in the early 1970s, when opiate receptors were isolated from the membranes of brain cells. Opiates such as morphine or heroin produce their effects by binding to these sites and altering the operations of the brain cells. The existence of these receptors suggested that the brain normally contains some molecule that is similar in shape and function to the opiates isolated from plants. Two different kinds of such molecules have been found in the brain.

One kind of molecule has been named enkephalin, from the Greek for "in the head." The enkephalins that have been isolated to date are peptides, compounds related to proteins. They are quite different in overall composition from the alkaloids, but quite similar to morphine and other opiates in shape. It is believed that the enkephalins bind to receptors at the ends of nerve cells in the brain and inhibit the activity of the cells, thus deadening the sensation of pain.

The discovery of the enkephalins has led to a new theory about how two major features of drug addiction, tolerance and physical dependence, come about. Tolerance refers to the need for increasing amounts of an addictive drug to elicit the same response. Physical dependence refers to the severe withdrawal symptoms that develop when the drug is not administered. According to the theory, opiate receptors in the brain usually are exposed to a constant level of enkephalin. When heroin or another addictive drug is administered, these molecules bind to unoccupied opiate receptors. If the drug is taken habitually, enkephalin production decreases, so that larger amounts of the addictive drug are needed to produce the same effects. When administration of the drug is stopped, the withdrawal symptoms occur during the period in which the body has not yet resumed its normal production of enkephalins.

A second kind of natural opiates is the endorphins, a name that is an abbreviation for "endogenous morphines." The endorphins are produced by the pituitary, a gland at the base of the brain.

Enkephalins have been found to have the same pain-killing potency as morphine. Attempts are being made to make analogues of the enkephalins that would ease pain but would not be addictive. The existence of these natural opiates has raised hopes that this long-sought goal may now be within reach.

The discovery has also helped explain some kinds of human behavior. For example, it has been found that running raises the levels of endorphins in the brain. This increase has been linked to a phenomenon called "runner's high," an unusual feeling of euphoria experienced by some individuals who run long distances daily. Indeed, some researchers say that the increase in endorphins explains why some runners become extremely upset if they are deprived of their daily exercise. These runners are believed to be addicted to the brain's own opiates.

### Organic Halides

A halogen atom that is attached to a carbon atom can be regarded as a functional group. Compounds with halogens as functional groups are called *organic halides*. The properties of some simple chlorides are shown in Table 21.4.

There are many differences in the properties of halides in which the halogen is attached to a benzene ring and those in which the halogen is on

| **TABLE 21.4** Simple Organic Chlorides | | | |
|---|---|---|---|
| **Name** | **Formula** | **Melting Point (K)** | **Boiling Point (K)** |
| carbon tetrachloride | $CCl_4$ | 250 | 350 |
| chloroform | $CHCl_3$ | 210 | 335 |
| methylene chloride | $CH_2Cl_2$ | 178 | 313 |
| methyl chloride | $CH_3Cl$ | 175 | 249 |
| ethyl chloride | $CH_3CH_2Cl$ | 137 | 285 |
| vinyl chloride | $CH_2{=}CHCl$ | 119 | 260 |
| chlorobenzene | $C_6H_5Cl$ | 228 | 405 |

a nonaromatic carbon atom. Similarly, a halogen on the $sp^2$ carbon of a double bond has properties quite different from those of one on an $sp^3$ carbon. The properties of the halides can be influenced even by the number of hydrogen atoms bonded to the $sp^2$ carbon, by whether the carbon is part of a ring, and by the size of such a ring. Polyhalogenated hydrocarbons are used as refrigerants and as dry-cleaning solvents.

## Nitrogen-Containing Functional Groups

Many functional groups contain one or more nitrogen atoms. The *amides* are nitrogen-containing carboxylic acid derivatives. The *amines* are a large class of compounds that are organic derivatives of ammonia, just as alcohols and ethers are organic derivatives of water. We can divide the amines into three groups on the basis of the number of hydrogen atoms in ammonia that are replaced by organic radicals.

In primary amines, which can be represented as $RNH_2$, one hydrogen atom is replaced by an organic radical. In secondary amines, which can be represented as

$$RNR'$$
$$|$$
$$H$$

two hydrogen atoms are replaced. In tertiary amines, which can be represented as

$$RNR'$$
$$|$$
$$R''$$

all three hydrogen atoms are replaced by organic groups. Compounds in which the nitrogen atom of the amine is attached directly to a benzene ring are called aromatic amines or *anilines.*

The amines generally are weak bases. We discussed some aspects of their acid-base chemistry in Chapter 15.

A group of amines, often of complex structure, that are found in many plants are the *alkaloids.* The alkaloids are physiologically active. Often, they are psychoactive—that is, they can influence the way in which the brain functions. Alkaloids often play important roles in medicine and in religion, in both primitive and advanced societies.

The opium alkaloids, for example, are a group of more than 20 amines found in a resin that is exuded by the opium poppy, which is grown primarily in Turkey and Southeast Asia. The most abundant of the opium alkaloids is morphine, which is named for Morpheus, the Greek god of sleep. Morphine and many of its numerous derivatives, such as codeine, are extremely effective painkillers. Unfortunately, they are also physiologically addictive. For many decades, chemists have attempted to develop a synthetic morphine derivative that kills pain but is not addictive. All their efforts have failed. Heroin is one product of the effort to find a synthetic nonaddictive painkiller. It was found to be addictive only after it was used medically.

The ergot alkaloids are found in ergot, a fungus that grows on rye and other cereals. The ergot alkaloids have been used for centuries in obstetrics, for such purposes as the induction of labor. It is only in recent years that the ergot alkaloids have been generally replaced by other drugs, although they are still used in the treatment of migraine headaches.

The ergot alkaloids are deadly poisons. Epidemic poisonings called St. Anthony's fire, caused by the presence of ergot in rye bread, killed tens of thousands of persons in Europe in medieval times. The ergot alkaloids can cause severe physical and mental damage; the symptoms of ergotism include insanity and gangrene of the arms or legs.

In the past few years, one derivative of the ergot alkaloids that was prepared by an organic chemist seeking a nonaddictive painkiller has become quite well known. This synthetic compound is an amide derivative of lysergic acid, which is commonly called LSD, from the initials of its German name. LSD is the most powerful hallucinogen known.

The reserpine alkaloids were first isolated from the root of a plant that grows in India. Scientific interest in the plant was aroused by a tradition, perhaps thousands of years old, of using the roots to treat insanity. The alkaloids isolated from the roots of this Indian plant include some of the most powerful tranquilizers known.

The list of plant alkaloids is long. The quinine alkaloids include quinine itself, the drug of choice for treatment of several kinds of malaria. Drugs such as belladonna, nicotine, cocaine, and scopolamine are all alkaloids. In general, these drugs were discovered in the same way as the reserpine alkaloids, by modern scientists who followed the lead given by folk medicine. Today, the search for new drugs continues in relatively unexplored regions of the world, carried out by scientists who are called ethnobotanists or ethnopharmacologists.

## Sulfur-Containing Functional Groups

A number of sulfur-containing functional groups are of interest to both organic chemists and biologists. We can classify them according to the oxidation states of the sulfur atom in the functional group. They range from organic derivatives of $H_2S$, in which the sulfur is in the $-2$ oxidation state, to organic derivatives of sulfuric acid, in which the sulfur is in the $+6$ oxidation state.

Organic derivatives of $H_2S$ can be regarded as the sulfur analogues of the alcohols and ethers. The mercaptans, whose general formula is RSH, correspond to the alcohols; the sulfides, whose general formula is RSR$'$, correspond to the ethers. The disulfides, whose general formula is RSSR$'$, are a class of organosulfur compounds in which the sulfur is in the $-1$ oxidation state.

Sulfur-containing functional groups are found in a number of physiologically active molecules. The mercaptans, which are also called thiols, are distinguishable by their smell. Butanethiol, $CH_3CH_2CH_2CH_2SH$, is the active principle of skunk odor, for example. Other organosulfur compounds are responsible for many of the unpleasant odors associated with industrial pollution.

The disulfide functional group is found in many proteins. The S—S group often acts as a bridge linking distant parts of the protein chain and thus plays a major role in maintaining the characteristic three-dimensional structures of proteins.

## 21.7   WHAT ORGANIC CHEMISTS DO

A large fraction of practicing chemists consider themselves to be organic chemists. The research efforts of many organic chemists fall into a few well-defined categories.

### Structure Determination

New substances with complex structures are continually being isolated from natural sources, primarily from plants. These compounds may help chemists learn more about biological processes; many of them are studied for possible use in medicine. The first question that is asked about a newly discovered compound is usually: What is its molecular structure? More and more, organic chemists have come to rely on instruments to help them work out these structures (Section 8.1).

Infrared spectroscopy is used primarily to identify functional groups and, to a limited extent, to identify some features of the compound's carbon skeleton. Ultraviolet spectroscopy is helpful in establishing some structural details of compounds with aromatic rings or multiple bonds, such as the location of double bonds relative to each other. A technique called nuclear magnetic resonance spectroscopy is quite valuable to the organic chemist. It gives information about functional groups and about the carbon skeleton of a molecule by identifying the chemically different types of hydrogen atoms in a molecule.

One of the most useful techniques in determining structures is mass spectrometry. A small sample of the unknown substance is bombarded with relatively high-energy electrons. The molecule fragments into pieces whose masses can be measured very accurately by the mass spectrometer. The correct structure of the unknown substance can be established by interpretation of the fragmentation pattern of the molecule. Often, a computer is used to help in the analysis.

The most direct instrumental method of structural determination is X-ray crystallography (Section 10.6). It is not yet used routinely by organic chemists, because analysis of X-ray crystallography diffraction patterns may take many months.

Chemical methods are also used to determine structure. The conversion of an unknown substance to a known substance by a well-defined reaction pathway is a good way to establish the structure of the unknown. The chemical behavior of an unknown substance can provide useful information about the nature of its functional groups or of its carbon skeleton. The chemical fragmentation of a complex molecule into smaller molecules, which are more easily identified, is another useful technique.

Despite the continued improvement in instrumentation, determining the structure of a natural product still can be a formidable task. It is therefore impressive to learn that the determination of exceedingly complex structures, such as that of strychnine, an alkaloid that is a well-known poison, was carried out without the use of modern instrumentation. Figure 21.9 shows the structure of strychnine. The molecule is even more complex than this two-dimensional structure indicates. In addition, there are many isomers, both geometric and optical, which are represented by this two-dimensional structure. It was a major achievement to single out strychnine from all the rest of the isomers.

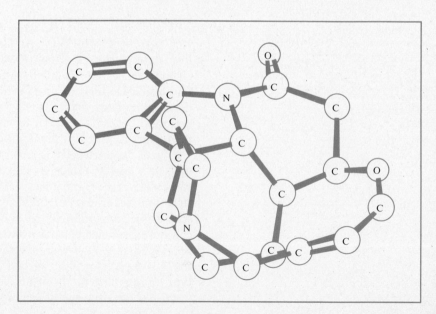

**Figure 21.9**
The structure of strychnine. The hydrogen atoms are omitted.

# PLASTICS THAT CONDUCT ELECTRICITY

Plastics traditionally have been electrical insulators, not conductors. In recent years, however, chemists and physicists at a number of laboratories have developed families of polymers, the long-chain molecules that make up plastics, that conduct electricity just as well as some metals do. Although there still are some major obstacles to be overcome, these new conductors have great commercial possibilites.

One group of conducting plastics consists of organic polymers. The first such polymer to be discovered was polyacetylene, the simplest conjugated organic polymer—conjugated meaning that it has alternating double and single bonds between the carbons. Polymers are insulators because their electrons are tightly bound. A carbon atom in polyacetylene forms single bonds with one adjacent carbon atom and one hydrogen atom, and one double bond with a carbon atom. One electron in the double bond is in a nonbonding orbital, and thus is potentially free to travel along the polymer chain. The addition of positive or negative ions to polyacetylene in a process called doping allows the electrons to travel, producing the flow of electrons that constitutes an electric current.

A second group of conducting organic materials consists of compounds called charge-transfer salts. Their components are planar, or flat, molecules that can be stacked like poker chips. In one such material, called (TTF) (TCNQ), stacks of two different planar molecules are close together at angles to each other. Their orientation causes an overlap of orbitals that allows the transfer of charge between the two parts of the compound. Each TTF molecule contributes roughly half an electron to each TCNQ molecule, producing a flow of electrons up or down the stack.

One promising set of materials derived from charge-transfer salts is made from phthalocyanines, inexpensive compounds used in pigments or dyes. A phthalocyanine consists of several interlinked rings with a metal atom in the center. The molecules can be connected by a chain of alternating oxygen and silicon or germanium atoms to form a polymer. The polymer is then doped to make it a conductor. The properties of these polymers can be altered by the use of various phthalocyanine molecules or dopants.

Different problems must be overcome to make each material suitable for practical applications. Conducting organic polymers are easily oxidized, so when they are exposed to air they rapidly lose their conductivity. Conductors made from charge-transfer salts are much more stable than those derived from organic polymers, but have only about one-tenth of their electrical conductivity. One goal is to develop batteries made of plastics that not only are far lighter than conventional materials but also are strong enough to use as structural units, such as automobile bumpers, fenders, or doors.

## Synthesis of Compounds

The synthesis of organic compounds of complex structure from simple and available starting materials is a major area of activity in organic chemistry. Some compounds prepared by synthetic organic chemists are only of theoretical interest. But many of these compounds are prepared with practical aims in mind, particularly in industrial research.

Most new drugs are substances made in the laboratory. Chemists sometimes design a molecule for a specific purpose, basing the design on existing molecules. More often, the synthesis of a new drug is a matter of trial and error, in which a large number of compounds are prepared and tested to determine if any one of them is useful. In an effort to reduce the waste of the trial-and-error approach, a great deal of effort in organic chemistry is devoted to the study of the relationship between molecular structure and physiological activity.

The plastics industry also relies heavily on synthetic organic compounds. The list of commercial synthetic products is almost endless: solvents, paints, gasoline additives, insecticides, herbicides, fibers, and many more.

Several steps lie between the synthesis of a compound and its commercial use. Economically feasible methods of synthesis must be found. A considerable effort often is expended to maximize the yield of a compound. A difference of a few percent in the yield of the material may be the difference between marketing a product and abandoning it.

## New Reactions

Organic chemists are constantly developing new kinds of reactions for bringing about chemical changes. The reactions of synthetic organic chemistry can be divided into two broad categories. One category consists of functional-group interconversion reactions, in which there is a change in the functional group but not in the carbon skeleton. The second category includes reactions in which a carbon skeleton is built up or changed.

Functional-group interconversion reactions usually are fairly easy to manage. We have described some of them — the conversion of an alkene to a dichloride, a bromide, or an alkane (Figure 21.1); the conversion of a carboxylic acid to an ester; the oxidation of an alcohol to an aldehyde; the oxidation of an aldehyde to a carboxylic acid; and the reduction of a ketone to an alcohol.

There are a number of reasons for the ongoing search for new kinds of reactions that give high yields of the desired product and are accompanied by a minimum of side reactions. In many reactions, the organic chemist tries to achieve a change in just one functional group of a molecule that has several functional groups. A number of such specific reactions have been developed.

The second type of reaction, in which a carbon skeleton is built up, is generally much more difficult to conceive and develop. The synthetic chemist must construct a complex carbon skeleton in a way that makes it possible to introduce functional groups at specific sites.

The existence of geometric and optical isomers creates problems in the formation of carbon chains and rings. One of the most challenging areas in organic synthesis is to carry out stereospecific synthesis, that is, to make only the desired isomer. It is difficult to build up a carbon skeleton in a way that leads not only to the correct location of functional groups but also to their correct orientation in space.

An especially active area in organic synthesis is the preparation of natural products of complex structure from simple starting materials. Such syntheses may sometimes seem unnecessary to those outside the field, because they require an enormous expenditure of time and effort for preparation of a small amount of a compound that may be easily available from a natural source. However, these syntheses are invaluable in adding to chemical knowledge. The synthesis of a compound is sometimes the last and most elegant step in a structure determination. To prove that a structure proposed for a substance is correct, one must synthesize the compound with the proposed structure. The identity of the synthetic compound and the natural compound provides the final proof that the proposed structure is correct.

A major motive for synthesizing a natural product of complex structure is pure intellectual curiosity. For example, morphine, the major alkaloid of opium, is available in large quantity from the natural source. Nevertheless, years of work by many excellent chemists were devoted to synthesizing morphine, whose structure is shown in Figure 21.10. Although a great deal of chemistry was learned along the way, the synthesis of morphine is primarily an impressive intellectual achievement. More than 20 steps, beginning with simple starting materials, were required for the synthesis of morphine.

## Theory and Practice

Research in organic chemistry is also aimed at developing and refining the basic theoretical picture that allows chemists to rationalize, systematize, and predict the behavior of organic molecules. One approach is to apply the methods of quantum mechanics, with suitable simplifications, to formulation of the structure of organic molecules. A detailed picture of the electrons and the orbitals in which they are found makes it possible to understand, and in some cases to predict, chemical behavior. This approach has been especially successful for molecules with $\pi$ electrons.

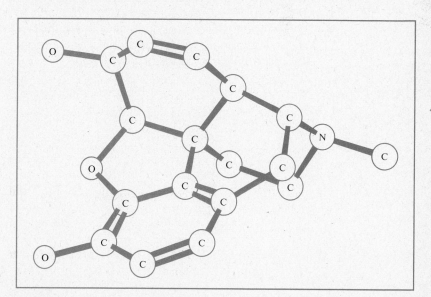

**Figure 21.10**
The structure of morphine. The hydrogen atoms are omitted.

Organic chemists also devote considerable attention to the study of reaction mechanisms. In organic chemistry, a reaction that appears to be relatively simple often proceeds by an involved pathway. The reaction mechanism is a detailed listing of all the changes that the system undergoes along the pathway from reactants to products. Kinetic methods, which we discussed in Chapter 17, are only some of the techniques used to elucidate reaction mechanisms. Great progress has been made in working out details of many reaction mechanisms. However, much remains to be done.

**Summary**

**W**e began this chapter by defining organic chemistry as the study of the compounds of **carbon.** Most organic compounds also contain hydrogen, and many contain oxygen, nitrogen, sulfur, or other elements. We went on to discuss **hydrocarbons,** compounds that contain only hydrogen and carbon. We discussed **alkanes,** hydrocarbons with no double bonds and no rings, and noted that **isomers** — two compounds with the same composition but different structures — are common in alkanes and almost all other organic compounds. We described **chain isomerism,** in which the main chains of carbon atoms differ. We then discussed **alkenes,** hydrocarbons with one or more double bonds, and their isomers: **positional isomers,** in which the multiple bonds have different locations; and **geometrical isomers,** in which the spatial arrangement of groups is different. We then defined a **functional group** as a small group of atoms that differ in some way from an alkane structure, and then discussed **alkynes,** hydrocarbons with triple bonds. We described **cyclic hydrocarbons,** which contain rings of carbon atoms, and **aromatic hydrocarbons,** which have six-carbon benzene rings. We mentioned that the presence of six $\pi$ electrons in a benzene ring makes aromatic hydrocarbons unusually stable. We then discussed compounds that contain oxygen functional groups, such as **alcohols** and **ethers,** and other functional groups, including those that contain halides, nitrogen, and sulfur. We concluded by describing how organic chemists determine the structure of compounds and synthesize them.

# Exercises

**21.1**[1]   Indicate which of the following compounds are alkanes: (a) $C_6H_{12}$, (b) $C_6H_{14}$, (c) $C_{100}H_{202}$, (d) $C_{20}H_{38}$.

**21.2**   Draw structures for: (a) a straight-chain alkane of seven carbons, (b) an alkane with a straight chain of five carbons and a two-carbon branch.

**21.3**   Draw structures for (a) $n$-octane, (b) isononane.

**21.4**   Draw structures for (a) $n$-hexane, (b) isohexane, (c) neohexane.

**21.5**   Draw the structures of two chain isomers of $C_6H_{14}$ in addition to the ones mentioned in Exercise 21.4.

**21.6**   Draw the structures of five isomers of $C_7H_{16}$ that have a straight chain of five carbon atoms.

**21.7**   Draw the structures of the first three members of a homologous series of alkanes in which there is a $CH_3$ substituent two atoms from the carbon at the end of the chain.

**21.8**   The boiling point of $n$-hexane is 342 K. Predict the boiling point of $n$-octane.

**21.9**   Suggest an explanation for the observation that the boiling point of neopentane is 25 K lower than the boiling point of $n$-hexane.

**21.10**   Name and draw the structures of the positional isomers of $n$-octene, disregarding geometric isomerism.

**21.11**   Show that the name neohexene can reasonably correspond to only one alkene.

**21.12**   Find the number of compounds that are positional isomers of isooctene and draw their structures.

**21.13**   Indicate which of the following alkenes exist as geometric isomers: (a) propylene, (b) 1-heptene, (c) 2-hexene, (d) 1,3-butadiene.

**21.14**   Draw the structures of the two geometric isomers of 2-pentene.

**21.15**   $CH_3CH{=}CHCH{=}CHCH_3$ is a compound with three geometric isomers. Draw their structures and suggest names to differentiate between them.

**21.16**   $CH_3CH_2CH{=}CHCH{=}CHCH_3$ is a compound with four geometric isomers. Compare it to the compound in Exercise 21.15 and account for the different number of isomers of these two compounds.

**21.17**   A compound has a straight-chain structure, contains no triple bonds, and has the composition $C_5H_8$. Draw all the isomers with this composition and indicate the relationships between them.

**21.18**   Name the following compounds: (a) $CH_2{=}C{=}CHCH_3$, (b) $CH_2{=}CHCH{=}CHCH{=}CH_2$.

**21.19**   Write the products of the following reactions: (a) ethylene and $Br_2$, (b) ethylene and HCl, (c) isobutene and $Cl_2$, (d) $cis$-3-octene and $H_2$.

**21.20**   An alkene of unknown structure is found to yield only acetic acid on treatment with $KMnO_4$. Propose a structure for the alkene.

**21.21**   The addition of HCl to 1-butene can form two products while the addition of HCl to $cis$-2-butene can form only one product. Explain this observation using structural formulas.

**21.22**   Draw the structures of all the isomers of $C_5H_8$ that have a straight chain of carbon atoms. Name each isomer.

**21.23**   An alkyne such as 1-butyne reacts with $Br_2$ in excess to form a product that does not contain any multiple bonds. Write the product of the reaction of 1-butyne with $Br_2$.

**21.24**   Somewhat surprisingly, it is found that alkynes are less reactive toward reagents such as $Cl_2$ than are alkenes. Predict the composition of the products that are isolated after treatment of 1 mol of 2-pentyne with 1 mol of $Cl_2$.

**21.25**   Draw and name all the cyclic isomers of $C_5H_{10}$, disregarding chirality.

**21.26**   Indicate which of the isomers that you found in Exercise 21.25 are chiral.

**21.27**   The $cis$ isomer of 1,2-dimethylcyclopropane is found to be less stable than the $trans$ isomer. Suggest an explanation for this observation.

**21.28**⋆   Draw all the isomers of $C_5H_8$.

**21.29**   Find the composition of a hydrocarbon that has the steroid ring system, but no other substituents.

**21.30**   Find the number of geometric isomers that are possible for the unsubstituted steroid ring system. (*Hint:* Two rings may be fused in two ways.)

**21.31**   Draw the structures of the three isomers of dichlorobenzene ($C_6H_4Cl_2$) and name them.

**21.32**⋆   Cyclooctatetraene, $C_8H_8$, is a compound with an eight-membered ring and four alternating double bonds, so that superficially its structure resembles benzene. It displays no special stability, and is not regarded as a resonance hybrid. (a) Suggest a possible geometry for this compound. (b) Suggest a reason for its dissimilarity to benzene.

[1] The answers to exercises whose numbers are in color can be found in Appendix VII. The star indicates an exercise that is more challenging than average.

**21.33**★    The removal of a proton from cyclopentadiene forms the anion $C_5H_5^-$, which is surprisingly stable for such a species. Suggest an explanation for this stability.

**21.34**    There are two compounds that have three aromatic rings and the formula $C_{14}H_{10}$. Draw their structures.

**21.35**    Draw all the straight-chain isomers of $C_4H_{10}O$, indicating which are functional isomers and which are positional isomers.

**21.36**    Write the chemical reaction that takes place when solutions of ammonia and phenol are mixed. (a) Name the product that forms. (b) Use the data in Table 15.3 to calculate the $K$ for this reaction.

**21.37**    Suggest a sequence of reactions that can be used to convert ethanol to $Br—CH_2CH_2—Br$.

**21.38**    Suggest a sequence of reactions that can be used to convert $n$-propyl alcohol, $CH_3CH_2CH_2OH$, to acetic acid.

**21.39**    Draw all the isomers with the composition $C_4H_4O$ and a straight chain of four carbon atoms. Indicate which are functional isomers.

**21.40**    Suggest an explanation for the observation that aldehydes are quite easily oxidized, while drastic conditions are required to oxidize a ketone. (*Hint:* Consider the bonds that must be broken.)

**21.41**    The ketone that has a straight chain of four carbon atoms is called 2-butanone. Draw the structures of the enols that can be formed from this ketone.

**21.42**    The formation of an enol from a ketone in basic solution proceeds through the intermediacy of an enolate anion which is formed by the loss of a proton from the ketone. Draw the two contributing structures of this anion and use them to account for the reversibility of enol formation.

**21.43**    Draw the structures of all the monofunctional isomers with the composition $C_3H_6O_2$.

**21.44**    The simple carboxylic acid $CH_3CH_2CH_2COOH$ is called butyric acid and is responsible for the odor of rancid butter. Draw the structures of its derivatives.

**21.45**    An acid derivative called an anhydride is formed by the elimination of a water molecule, as two acid molecules join through an oxygen atom. The anhydride of acetic acid is called acetic anhydride. Draw its structure.

**21.46**★    Draw the structures of all the amines with the composition $C_4H_{11}N$. (*Hint:* There are eight of them.)

**21.47**    Either from your general knowledge or on the basis of their names, suggest possible physiological effects of the following alkaloids: (a) nicotine, (b) caffeine, (c) emetine, (d) mescaline.

**21.48**    Suggest a general approach that can be used to construct a six-membered ring from smaller noncyclic compounds.

**21.49**    Find the composition of morphine from the structural formula given in Figure 21.10.

**21.50**    Alkanes react with $Cl_2$ when exposed to light. A Cl atom replaces a hydrogen on an alkane chain. Draw the structures of all the monochloride products that will form from the reaction of isopentane with $Cl_2$.

# Appendix I
# Physical and Chemical Constants and Conversion Factors

## Fundamental Constants

| Quantity | Symbol | Value[a,b] |
|---|---|---|
| unified atomic mass unit | $u$ (or amu) | $1.660\ 565\ 5 \times 10^{-27}$ kg |
| Avogadro constant[c] | $N$ or $N_A$ | $6.022\ 097\ 8 \times 10^{23}$ mol$^{-1}$ |
| electronic charge | $e$ | $1.602\ 189\ 2 \times 10^{-19}$ C |
| Faraday constant | $F$ | $9.648\ 456 \times 10^4$ C mol$^{-1}$ |
| gas constant | $R$ | $8.314\ 41$ J mol$^{-1}$ K$^{-1}$ |
| | | $0.082\ 056\ 8$ L atm mol$^{-1}$ K$^{-1}$ |
| mass of electron | $m_e$ | $9.109\ 534 \times 10^{-31}$ kg |
| mass of neutron | $m_n$ | $1.674\ 954\ 3 \times 10^{-27}$ kg |
| mass of proton | $m_p$ | $1.672\ 648\ 5 \times 10^{-27}$ kg |
| Planck constant | $h$ | $6.626\ 176 \times 10^{-34}$ J s |
| Rydberg constant | $R_H$ | $1.096\ 775\ 78 \times 10^7$ m$^{-1}$ |
| speed of light in a vacuum | $c$ | $2.997\ 924\ 58 \times 10^2$ m s$^{-1}$ |

[a] Values are from the *CODATA Bulletin*, December 1973.
[b] A table of these constants to four significant figures can be found inside the back cover.
[c] The value of the Avogadro constant is from R. B. Deslattes, *Ann. Rev. Phys. Chem.* **31**:435 (1980).

## Conversion Factors

| Quantity | Conversion Factor |
|---|---|
| energy | 1 cal = 4.184 J |
| | 1 L atm = 101.3 J |
| | 1 eV = $1.602 \times 10^{-19}$ J |
| | 1 eV = $9.648 \times 10^4$ J/mol |
| | 1 cm$^{-1}$ = 11.96 J/mol |
| pressure | 1 atm = $1.013 \times 10^5$ Pa |
| | 1 atm = 760 mmHg |
| | 1 atm = 760 torr |
| temperature | 0°C = 273.15 K |

# Appendix II
# Values of Thermodynamic Properties
# of Substances at 298 K and 1 atm

| Substance[a,b] | $\Delta H_f^\circ$ (kJ/mol) | $\Delta G_f^\circ$ (kJ/mol) | $S^\circ$ (J mol$^{-1}$ K$^{-1}$) |
|---|---|---|---|
| Ag(s) | 0 | 0 | 43 |
| Ag(g) | 284 | 246 | 173 |
| AgBr(s) | −100 | −97 | 107 |
| AgCl(s) | −127 | −110 | 96 |
| AgF(s) | −205 | −187 | 84 |
| AgI(s) | −62 | −66 | 115 |
| Ag$_2$O(s) | −31 | −11 | 122 |
| Al(s) | 0 | 0 | 28 |
| Al(g) | 330 | 286 | 164 |
| AlCl$_3$(s) | −704 | −629 | 111 |
| Al$_2$O$_3$(s) | −1670 | −1576 | 51 |
| Ar(g) | 0 | 0 | 155 |
| As(s) gray | 0 | 0 | 35 |
| As(g) | 302 | 261 | 174 |
| As$_4$(g) | 144 | 92 | 314 |
| AsCl$_3$(g) | −262 | −249 | 327 |
| As$_2$O$_3$(s) | −657 | −578 | 107 |
| As$_2$O$_5$(s) | −925 | −782 | 105 |
| Au(s) | 0 | 0 | 48 |
| Au(g) | 366 | 326 | 180 |
| Au(OH)$_3$(s) | −425 | −317 | 190 |
| B(s) | 0 | 0 | 6.5 |
| B(g) | 573 | 529 | 153 |
| BCl$_3$(g) | −395 | −380 | 290 |
| BF$_3$(g) | −1137 | −1120 | 254 |
| B$_2$H$_6$(g) | 36 | 87 | 232 |
| B$_2$O$_3$(s) | −1273 | −1193 | 54 |
| Ba(s) | 0 | 0 | 67 |
| Ba(g) | 180 | 146 | 170 |
| BaCl$_2$(s) | −860 | −811 | 126 |
| BaCO$_3$(s) | −1216 | −1138 | 112 |
| BaO(s) | −558 | −529 | 70.3 |
| Be(s) | 0 | 0 | 9 |
| Be(g) | 324 | 287 | 136 |
| BeCl$_2$(s) | −490 | −445 | 83 |
| BeO(s) | −610 | −580 | 14 |
| Bi(s) | 0 | 0 | 57 |
| Bi(g) | 208 | 168 | 187 |
| BiCl$_3$(s) | −379 | −317 | 184 |
| Bi$_2$O$_3$(s) | −574 | −494 | 151 |

[a] Substances are listed alphabetically by formula, except the standard state of an element is listed first. Formulas are written in the usual way.
[b] Substances with the state symbol (aq) are in aqueous solution at 1$M$ concentration.

| Substance[a,b] | $\Delta H_f^\circ$ (kJ/mol) | $\Delta G_f^\circ$ (kJ/mol) | $S^\circ$ (J mol$^{-1}$ K$^{-1}$) |
|---|---|---|---|
| $Br_2(l)$ | 0 | 0 | 152 |
| $Br_2(g)$ | 31 | 3 | 245 |
| $Br(g)$ | 112 | 82 | 175 |
| $Br^-(aq)$ | −121 | −104 | 83 |
| $BrF(g)$ | −59 | −74 | 229 |
| $BrF_3(g)$ | −256 | −229 | 292 |
| $BrF_5(g)$ | −429 | −350 | 320 |
| $BrO^-(aq)$ | −94 | −33 | 42 |
| $BrO_3^-(aq)$ | −67 | 2 | 163 |
| $BrO_4^-(aq)$ | 13 | | |
| $C(s)$ graphite | 0 | 0 | 5.69 |
| $C(s)$ diamond | 1.9 | 2.9 | 2.43 |
| $C(g)$ | 718 | 671 | 158 |
| $C_2(g)$ | 808 | 780 | 200 |
| $CBr_4(g)$ | 50 | 36 | 358 |
| $CCl_4(g)$ | −107 | −58 | 310 |
| $CH_4(g)$ | −75 | −51 | 186 |
| $C_2H_2(g)$ | 227 | 209 | 201 |
| $C_2H_4(g)$ | 52 | 68 | 220 |
| $C_2H_6(g)$ | −85 | −33 | 230 |
| $CHCl_3(g)$ | −100 | −69 | 296 |
| $CHCl_3(g)^c$ | | | 303 |
| $CHCl_3(l)$ | −135 | −74 | 203 |
| $CHCl_3(l)^c$ | | | 215 |
| $CH_2O(g)$ | −116 | −110 | 219 |
| $CO(g)$ | −110.5 | −137.3 | 197.6 |
| $CO_2(g)$ | −393.5 | −394.4 | 213.7 |
| $CS_2(g)$ | 115 | 67 | 238 |
| $Ca(s)$ | 0 | 0 | 42 |
| $Ca(g)$ | 178 | 144 | 155 |
| $CaCl_2(s)$ | −795 | −750 | 114 |
| $CaCO_3(s)$ | −1207 | −1129 | 93 |
| $CaF_2(s)$ | −1220 | −1167 | 69 |
| $CaO(s)$ | −636 | −604 | 40 |
| $Cd(s)$ | 0 | 0 | 51 |
| $Cd(g)$ | 112 | 77 | 168 |
| $CdCl_2(s)$ | −389 | −343 | 251 |
| $CdO(s)$ | −255 | −225 | 55 |
| $Cl_2(g)$ | 0 | 0 | 223 |
| $Cl(g)$ | 121 | 105 | 165 |
| $Cl^-(aq)$ | −167 | −131 | 57 |
| $ClF(g)$ | −51 | −56 | 218 |
| $ClF_3(g)$ | −159 | −123 | 281 |
| $ClF_5(g)$ | −240 | | |
| $ClO(g)$ | 109 | 98 | 227 |
| $ClO_2(g)$ | 103 | 120 | 257 |
| $Cl_2O(g)$ | 80 | 98 | 266 |
| $Cl_2O_7(l)$ | 272 | | |
| $ClO^-(aq)$ | −107 | −37 | 42 |
| $ClO_2^-(aq)$ | −67 | 17 | 101 |
| $ClO_3^-(aq)$ | −104 | −3 | 162 |
| $ClO_4^-(aq)$ | −128 | −9 | 182 |
| $Co(s)$ | 0 | 0 | 30 |

$^c$ At the boiling point, 334 K.

| Substance[a,b] | $\Delta H_f^\circ$ (kJ/mol) | $\Delta G_f^\circ$ (kJ/mol) | $S^\circ$ (J mol$^{-1}$ K$^{-1}$) |
|---|---|---|---|
| Co(g) | 425 | 380 | 179 |
| CoCl$_2$(s) | $-318$ | $-274$ | 103 |
| CoO(s) | $-238$ | $-214$ | 53 |
| Cr(s) | 0 | 0 | 24 |
| Cr(g) | 397 | 352 | 173 |
| CrCl$_2$(s) | $-395$ | $-356$ | 115 |
| CrCl$_3$(s) | $-552$ | $-481$ | 123 |
| Cr$_2$O$_3$(s) | $-1140$ | $-1058$ | 81 |
| CrO$_3$(s) | $-590$ | $-513$ | 72 |
| Cs(s) | 0 | 0 | 85 |
| Cs(g) | 79 | 51 | 175 |
| CsCl(s) | $-447$ | $-419$ | 100 |
| Cs$_2$O(s) | $-318$ | $-290$ | 124 |
| Cu(s) | 0 | 0 | 33 |
| Cu(g) | 338 | 299 | 166 |
| CuCl(s) | $-137$ | $-120$ | 86 |
| CuCl$_2$(s) | $-220$ | $-176$ | 108 |
| CuO(s) | $-155$ | $-127$ | 44 |
| Cu$_2$O(s) | $-167$ | $-146$ | 101 |
| F$_2$(g) | 0 | 0 | 203 |
| F$^-$(aq) | $-333$ | $-279$ | $-14$ |
| FSO$_3$H(l) | $-800$ | | |
| Fe(s) | 0 | 0 | 27 |
| Fe(g) | 418 | 372 | 180 |
| FeCl$_2$(s) | $-342$ | $-302$ | 118 |
| FeCl$_3$(s) | $-399$ | $-334$ | 142 |
| FeO(s) | $-272$ | $-251$ | 61 |
| Fe$_2$O$_3$(s) | $-824$ | $-742$ | 87 |
| Fe$_3$O$_4$(s) | $-1118$ | $-1015$ | 146 |
| Ge(s) | 0 | 0 | 3 |
| Ge(g) | 377 | 336 | 168 |
| GeCl$_4$(g) | $-496$ | $-457$ | 348 |
| GeH$_4$(g) | 91 | 113 | 217 |
| GeO$_2$(s) | $-551$ | $-497$ | 55 |
| H$_2$(g) | 0 | 0 | 131 |
| H(g) | 218 | 203 | 115 |
| H$^+$(g) | 1536 | 1517 | 109 |
| H$^-$(g) | 149 | 133 | 109 |
| HBr(g) | $-36$ | $-53$ | 198 |
| HBrO(aq) | $-113$ | $-82$ | 142 |
| HBrO$_3$(aq) | $-40$ | 2 | 163 |
| HCl(g) | $-92$ | $-95$ | 187 |
| HClO(aq) | $-121$ | $-80$ | 142 |
| HClO$_2$(aq) | $-52$ | 6 | 188 |
| HClO$_3$(aq) | $-98$ | 3 | 162 |
| HClO$_4$(l) | $-41$ | 84 | 188 |
| HClO$_4$(aq) | $-129$ | 8.6 | 182 |
| HF(g) | $-273$ | $-273$ | 174 |
| HI(g) | 26 | 2 | 206 |
| HIO(aq) | $-138$ | $-99$ | 95 |
| HIO$_3$(aq) | $-230$ | $-133$ | 167 |
| H$_5$IO$_6$(s) | $-834$ | | |
| HNO$_2$(aq) | $-119$ | $-56$ | 153 |
| HNO$_3$(g) | $-135$ | $-75$ | 266 |

| Substance[a,b] | $\Delta H_f^\circ$ (kJ/mol) | $\Delta G_f^\circ$ (kJ/mol) | $S^\circ$ (J mol$^{-1}$ K$^{-1}$) |
|---|---|---|---|
| $HNO_3(aq)$ | $-207$ | $-111$ | 146 |
| $H_2O(g)$ | $-242$ | $-229$ | 189 |
| $H_2O(l)$ | $-286$ | $-237$ | 70 |
| $HH_2PO_2(aq)$ | $-592$ | | |
| $H_2HPO_3(aq)$ | $-955$ | | |
| $H_3PO_4(s)$ | $-1260$ | $-1126$ | 110 |
| $H_2S(g)$ | $-20$ | $-34$ | 206 |
| $H_2SO_3(aq)$ | $-633$ | $-538$ | 232 |
| $H_2SO_4(l)$ | $-814$ | $-690$ | 157 |
| $H_2SO_4(aq)$ | $-908$ | $-745$ | 20 |
| $He(g)$ | 0 | 0 | 126 |
| $Hg(l)$ | 0 | 0 | 76 |
| $Hg(g)$ | 61 | 32 | 175 |
| $HgCl_2(s)$ | $-224$ | $-179$ | 146 |
| $Hg_2Cl_2(s)$ | $-265$ | $-211$ | 192 |
| $HgO(s)$ red | $-91$ | $-59$ | 70 |
| $I_2(s)$ | 0 | 0 | 116 |
| $I_2(g)$ | 62 | 19 | 261 |
| $I(g)$ | 107 | 70 | 181 |
| $I^-(aq)$ | $-57$ | $-52$ | 107 |
| $IF(g)$ | $-95$ | $-118$ | 236 |
| $IF_3(g)$ | $-485$ | | |
| $IF_5(g)$ | $-840$ | $-768$ | 328 |
| $IF_7(g)$ | $-961$ | $-828$ | 346 |
| $I_2O_5(s)$ | $-158$ | | |
| $IO^-(aq)$ | $-108$ | $-38$ | $-5$ |
| $IO_3^-(aq)$ | $-220$ | $-128$ | 118 |
| $IO_4^-(aq)$ | $-145$ | | |
| $K(s)$ | 0 | 0 | 65 |
| $K(g)$ | 90 | 61 | 160 |
| $KBr(s)$ | $-394$ | $-379$ | 97 |
| $HBrO_4(s)$ | $-287$ | | |
| $KCl(s)$ | $-437$ | $-408$ | 83 |
| $KClO_4(s)$ | $-432$ | $-300$ | 151 |
| $KF(s)$ | $-529$ | $-533$ | 67 |
| $KI(s)$ | $-328$ | $-322$ | 104 |
| $KIO_4(s)$ | $-461$ | $-395$ | 159 |
| $Kr(g)$ | 0 | 0 | 164 |
| $Li(s)$ | 0 | 0 | 29 |
| $Li(g)$ | 161 | 128 | 138 |
| $LiCl(s)$ | $-402$ | $-377$ | 59 |
| $Li_2O(s)$ | $-596$ | $-560$ | 38 |
| $Mg(s)$ | 0 | 0 | 33 |
| $Mg(g)$ | 148 | 113 | 149 |
| $MgCl_2(s)$ | $-642$ | $-592$ | 90 |
| $MgCO_3(s)$ | $-1096$ | $-1012$ | 66 |
| $MgO(s)$ | $-601$ | $-570$ | 27 |
| $Mn(s)$ | 0 | 0 | 32 |
| $Mn(g)$ | 281 | 238 | 174 |
| $MnCl_2(s)$ | $-481$ | $-441$ | 118 |
| $Mn_2O_7(s)$ | $-728$ | | |
| $N_2(g)$ | 0 | 0 | 192 |
| $N(g)$ | 473 | 456 | 153 |
| $NF_3(g)$ | $-114$ | $-83$ | 261 |

| Substance[a,b] | $\Delta H_f^\circ$ (kJ/mol) | $\Delta G_f^\circ$ (kJ/mol) | $S^\circ$ (J mol$^{-1}$ K$^{-1}$) |
|---|---|---|---|
| $N_2F_4(g)$ | −7 | 81 | 301 |
| $NH_3(g)$ | −46 | −16 | 193 |
| $NH_3(aq)$ | −80 | −27 | 111 |
| $N_2H_4(l)$ | 50 | 149 | 121 |
| $NH_4Br(s)$ | −270 | −175 | 110 |
| $NH_4Cl(s)$ | −314 | −203 | 95 |
| $NH_4NO_3(s)$ | −365 | −184 | 151 |
| $NH_2OH(s)$ | −107 | | |
| $NO(g)$ | 90 | 87 | 211 |
| $NO_2(g)$ | 34 | 51 | 240 |
| $N_2O(g)$ | 82 | 104 | 220 |
| $N_2O_3(g)$ | 84 | 139 | 312 |
| $N_2O_4(g)$ | 10 | 98 | 304 |
| $N_2O_5(s)$ | −42 | 114 | 178 |
| $Na(s)$ | 0 | 0 | 51 |
| $Na(g)$ | 109 | 78 | 154 |
| $NaBr(s)$ | −360 | −347 | 84 |
| $NaCl(s)$ | −411 | −384 | 72 |
| $NaF(s)$ | −574 | −544 | 51 |
| $NaI(s)$ | −288 | −282 | 91 |
| $Na_2O(s)$ | −416 | −377 | 73 |
| $Ne(g)$ | 0 | 0 | 146 |
| $O_2(g)$ | 0 | 0 | 205 |
| $O(g)$ | 249 | 232 | 161 |
| $O^+(g)$ | 1567 | 1547 | 155 |
| $O^-(g)$ | 110 | 89 | 158 |
| $O_2(aq)$ | −12 | 16 | 111 |
| $O_2^+(g)$ | 1184 | 1166 | 205 |
| $O_3(g)$ | 142 | 163 | 239 |
| $OH(g)$ | 42 | 34 | 184 |
| $OH^-(g)$ | −133 | | |
| $OH^-(aq)$ | −230 | −157 | −11 |
| $P(s)$ red | 0 | 0 | 23 |
| $P(s)$ white | 18 | 12 | 41 |
| $P(s)$ black | −17 | −19 | 23 |
| $P(g)$ | 334 | 290 | 163 |
| $P_2(g)$ | 179 | 116 | 218 |
| $P_4(g)$ | 77 | 36 | 280 |
| $PBr_3(l)$ | −167 | −164 | 240 |
| $PBr_5(s)$ | −293 | | |
| $PCl_3(l)$ | −300 | −260 | 217 |
| $PCl_5(s)$ | −400 | | |
| $PH_3(g)$ | 23 | 25 | 210 |
| $P_4O_6(s)$ | −1593 | | |
| $P_4O_{10}(s)$ | −2940 | −2676 | 228 |
| $POCl_3(l)$ | −578 | −509 | 222 |
| $Pb(s)$ | 0 | 0 | 65 |
| $Pb(g)$ | 194 | 163 | 175 |
| $PbCl_2(s)$ | −359 | −314 | 136 |
| $PbO(s)$ | −217 | −188 | 69 |
| $PbO_2(s)$ | −277 | −217 | 69 |
| $Rb(s)$ | 0 | 0 | 77 |
| $Rb(g)$ | 86 | 64 | 170 |
| $RbCl(s)$ | −433 | −404 | 92 |

| Substance[a,b] | $\Delta H_f^\circ$ (kJ/mol) | $\Delta G_f^\circ$ (kJ/mol) | $S^\circ$ (J mol$^{-1}$ K$^{-1}$) |
|---|---|---|---|
| $Rb_2O(s)$ | $-330$ | $-297$ | 110 |
| $S_8(s)$ rhombic | 0 | 0 | 32 |
| $S(g)$ | 277 | 238 | 168 |
| $S_2(g)$ | 128 | 79 | 228 |
| $S_8(g)$ | 102 | 50 | 431 |
| $S_2Cl_2(l)$ | $-60$ | | |
| $SF_4(g)$ | $-780$ | $-731$ | 292 |
| $SF_6(g)$ | $-1220$ | $-1105$ | 292 |
| $SOCl_2(g)$ | $-210$ | $-198$ | 310 |
| $SO_2Cl_2(g)$ | $-364$ | $-320$ | 312 |
| $SO(g)$ | 6.9 | $-20$ | 222 |
| $SO_2(g)$ | $-297$ | $-300$ | 248 |
| $SO_3(g)$ | $-396$ | $-371$ | 257 |
| $Sb(s)$ | 0 | 0 | 46 |
| $Sb(g)$ | 262 | 222 | 180 |
| $SbCl_3(s)$ | $-382$ | $-324$ | 184 |
| $SbH_3(g)$ | 155 | 148 | 233 |
| $Sb_2O_5(s)$ | $-972$ | $-829$ | 125 |
| $Si(s)$ | 0 | 0 | 19 |
| $Si(g)$ | 450 | 395 | 168 |
| $SiCl_4(g)$ | $-657$ | $-617$ | 331 |
| $SiH_4(g)$ | 34 | 57 | 204 |
| $SiO_2(s)$ quartz | $-911$ | $-856$ | 41 |
| $Sn(s)$ white | 0 | 0 | 52 |
| $Sn(s)$ gray | $-2$ | 0.1 | 44 |
| $Sn(g)$ | 302 | 267 | 168 |
| $SnCl_2(s)$ | $-325$ | | |
| $SnH_4(g)$ | 163 | 188 | 228 |
| $SnO(s)$ | $-285$ | $-257$ | 56 |
| $SnO_2(s)$ | $-581$ | $-520$ | 52 |
| $Sr(s)$ | 0 | 0 | 52 |
| $Sr(g)$ | 164 | 131 | 165 |
| $SrCl_2(s)$ | $-829$ | $-781$ | 115 |
| $SrCO_3(s)$ | $-1220$ | $-1140$ | 97 |
| $SrO(s)$ | $-592$ | $-562$ | 54 |
| $Xe(g)$ | 0 | 0 | 170 |
| $XeF_2(g)$ | $-130$ | $-96$ | 260 |
| $XeF_4(g)$ | $-215$ | $-138$ | 316 |
| $XeF_6(g)$ | $-294$ | | |
| $XeO_3(g)$ | 502 | 561 | 287 |
| $Zn(s)$ | 0 | 0 | 42 |
| $Zn(g)$ | 131 | 95 | 161 |
| $ZnCl_2(s)$ | $-415$ | $-369$ | 111 |
| $ZnO(s)$ | $-348$ | $-318$ | 44 |

# Appendix III
## Average Bond Energies at 298 K

### Single Bonds[a]

| Bond | Bond Energy (kJ/mol) | Bond | Bond Energy (kJ/mol) |
|------|------|------|------|
| H—H | 436 | O—Si | 452 |
| H—C | 413 | O—P | 335 |
| H—N | 391 | O—Cl | 218 |
| H—O | 463 | O—Br | 201 |
| H—F | 563 | O—I | 201 |
| H—Si | 318 | F—F | 158 |
| H—P | 322 | F—Si | 586 |
| H—S | 368 | F—P | 503 |
| H—Cl | 432 | F—S | 327 |
| H—Br | 366 | F—Cl | 253 |
| H—I | 299 | F—Br | 249 |
| C—C | 346 | F—I | 280 |
| C—N | 305 | Si—Si | 176 |
| C—O | 358 | Si—Cl | 396 |
| C—F | 489 | P—P | 201 |
| C—P | 264 | P—Cl | 322 |
| C—S | 272 | P—Br | 264 |
| C—Cl | 328 | P—I | 184 |
| C—Br | 285 | S—S | 251 |
| C—I | 218 | S—Cl | 271 |
| N—N | 163 | Cl—Cl | 243 |
| N—O | 222 | Cl—Br | 219 |
| N—F | 275 | Cl—I | 211 |
| N—Cl | 192 | Br—Br | 193 |
| O—O | 146 | Br—I | 178 |
| O—F | 193 | I—I | 151 |

[a] Elements are listed in order of increasing atomic number.

### Double Bonds[a]

| Bond | Bond Energy (kJ/mol) | Bond | Bond Energy (kJ/mol) |
|------|------|------|------|
| C=C | 615 | N=N | 418 |
| C=N | 615 | N=O | 607 |
| C=O | 749 | O=O | 498 |
| C=O | 803[b] | O=P | 504 |
| C=S | 536 | O=S | 498[c] |

[b] In carbon dioxide.
[c] In sulfur dioxide.

### Triple Bonds[a]

| Bond | Bond Energy (kJ/mol) |
|------|------|
| C≡C | 812 |
| C≡N | 890 |
| N≡N | 946 |
| P≡P | 490 |

# Appendix IV
# Mathematical Procedures

### Exponential Notation

The numbers we encounter in chemistry often are very large or very small. A convenient way to express these numbers uses exponential notation, in which a number is expressed as the product of a simple number and a power of 10:

$$N \times 10^x$$

where $N$ is the simple number and $x$ is the exponent, or power to which 10 is raised.

In general, we shall find it convenient for the value of $N$ to be between 1 and 10, so that $N$ has only one digit to the left of the decimal point. For numbers larger than 10, the value of $x$ is positive; for numbers smaller than 1, it is negative.

To convert an ordinary number into exponential notation, the decimal point must be moved so that there is only one digit to the left of the decimal point. We move the decimal point to the left for large numbers and to the right for small numbers. The value of $x$ in the exponential is equal to the number of places that the decimal point is moved. When the decimal point is moved to the left, the exponent is positive; when it is moved to the right, the exponent is negative. Some examples are

| Number | $N$ | Number of Places Decimal Point Is Moved | $x$ | Exponential Notation |
|--------|-----|------------------------------------------|-----|----------------------|
| 4567 | 4.567 | 3, left | 3 | $4.567 \times 10^3$ |
| 0.004567 | 4.567 | 3, right | $-3$ | $4.567 \times 10^{-3}$ |
| 100 000 | 1 | 5, left | 5 | $1 \times 10^5$ |
| 0.000001 | 1 | 6, right | $-6$ | $1 \times 10^{-6}$ |

The same procedure can be used to convert from exponential notation back to an ordinary number or to move the decimal point in $N$. When a decimal point is moved one place to the left, the value of $x$ is increased by one. Thus:

$$45.67 \times 10^2 = 4.567 \times 10^3$$
$$456.7 \times 10^{-5} = 4.567 \times 10^{-3}$$

When a decimal point is moved one place to the right, the value of $x$ is decreased by one. Thus:

$$0.1 \times 10^{-5} = 1 \times 10^{-6}$$
$$0.4567 \times 10^3 = 4.567 \times 10^2$$

To convert from exponential notation to an ordinary number, the decimal point must be moved until the exponent $x = 0$, since $10^0 = 1$. For example, to convert $2.35 \times 10^5$ to an ordinary number, the exponent must be reduced from 5 to 0. Therefore, the decimal point must be moved five places to the right:

$$2.35 \times 10^5 = 235\ 000$$

Arithmetic operations with exponentials are carried out by the following rules:

1. When exponentials are multiplied, the exponents are added and the $N$s are multiplied. Thus:

$$(2.3 \times 10^3)(4.8 \times 10^2) = 11 \times 10^5 = 1.1 \times 10^6$$
$$(4.5 \times 10^{-5})(4.5 \times 10^{-3}) = 20 \times 10^{-8} = 2.0 \times 10^{-7}$$

2. When exponentials are divided, the exponent in the denominator is subtracted from the exponent in the numerator, and the $N$ in the numerator is divided by the $N$ in the denominator. Thus:

$$\frac{(4.5 \times 10^{-5})}{(6.5 \times 10^{-3})} = 0.69 \times 10^{-2} = 6.9 \times 10^{-3}$$

$$\frac{(2.3 \times 10^3)}{(4.8 \times 10^2)} = 0.48 \times 10^1 = 4.8$$

$$\frac{(4.5 \times 10^{-5})}{(6.5 \times 10^3)} = 0.69 \times 10^{-8} = 6.9 \times 10^{-9}$$

3. When an exponential is raised to a power, $N$ is raised to the power and $x$ is multiplied by the power. Thus:

$$(3.3 \times 10^7)^2 = (3.3)^2 \times 10^{14} = 1.1 \times 10^{15}$$
$$(6.8 \times 10^{-8})^3 = (6.8)^3 \times 10^{-24} = 310 \times 10^{-24} = 3.1 \times 10^{-22}$$

4. When a root of an exponential is extracted, the root of $N$ is extracted in the usual way and the exponent $x$ is multiplied by the fraction that

corresponds to the root: one-half for the square root, one-third for the cube root, etc. To ensure that the exponent in the result is a whole number, the exponential must sometimes be rewritten so that it is divisible by the root. Thus:

$$\sqrt[2]{3.4 \times 10^{-5}} = \sqrt[2]{34 \times 10^{-6}} = \sqrt[2]{34} \times 10^{-3} = 5.8 \times 10^{-3}$$
$$(4.7 \times 10^{6})^{1/4} = (470 \times 10^{4})^{1/4} = (470)^{1/4} \times 10^{1} = 4.7 \times 10^{1}$$

5. When exponentials are added or subtracted, their exponents must have the same value. The exponent thus is unchanged, while one $N$ is subtracted from or added to the other. Thus:

$$1.3 \times 10^{-2} + 4.2 \times 10^{-3} = 1.3 \times 10^{-2} + 0.42 \times 10^{-2} = 1.7 \times 10^{-2}$$
$$1.30 \times 10^{5} - 4.0 \times 10^{3} = 1.30 \times 10^{5} - 0.04 \times 10^{5} = 1.26 \times 10^{5}$$

### Logarithms

A logarithm is an exponent. A logarithm to the base 10 of a number is the exponent to which 10 must be raised to obtain the number. That is,

$$a = 10^{q} \quad \text{and} \quad \log a = q$$

where $a$ is the number and $\log a$ is the logarithm to the base 10. We shall use logarithms to the base 10 as much as possible.

Another important system of logarithms uses the number $e$, whose value is 2.71828 . . . as its base. Logarithms to the base $e$ are called natural logarithms. The natural logarithm of a number $a$ is written as $\ln a$. The factor 2.303 converts between the two systems of logarithms:

$$\ln a = 2.303 \log a$$

A logarithm has two parts: the *characteristic,* which is the part of the logarithm to the left of the decimal point, and the *mantissa,* the part of the logarithm to the right of the decimal point. In the logarithm 2.456, 2 is the characteristic and 0.456 is the mantissa. In the logarithm 102.3, 102 is the characteristic and 0.3 is the mantissa.

One advantage of the form of exponential notation that we use is that there is a direct correspondence between an exponential and a logarithm to the base 10. The characteristic corresponds to the $x$, the exponent of 10. The mantissa is the logarithm of any simple number $N$ whose value is between 1 and 10. This correspondence can be illustrated by some examples.

**1.** Taking the logarithm of a number:

$$\log (2.3 \times 10^3) = 3 + \log 2.3 = 3 + 0.36 = 3.36$$

The value of log 2.3 can be obtained from a table of logarithms or from a calculator with a "log" key.

$$\log 0.00047 = \log (4.7 \times 10^{-4}) = -4 + \log 4.7$$
$$= -4 + 0.67 = -3.33$$

**2.** Finding the antilogarithm, the number whose logarithm is given:

$$\text{antilog} (5.42) = \text{antilog} (0.42) \times 10^5 = 2.6 \times 10^5$$

Again, the value of the antilog can be obtained from a table of logarithms or from a calculator with an antilog, or "$10^x$" key.

$$\text{antilog} (-6.78) = \text{antilog} (-7 + 0.22)$$
$$= \text{antilog} (0.22) \times 10^{-7} = 1.7 \times 10^{-7}$$

Since a table of logarithms gives a mantissa as a positive number, the negative logarithm must be rewritten as the sum of a negative integer and a positive decimal (Examples 15.4 – 15.7).

Significant figures and units in the correct expression of numerical quantities are extremely important in chemistry. The value of a logarithm usually is determined by the units in which its antilogarithm is given, but the logarithm itself has no units. The number of significant figures in a logarithm can be taken as the number of figures in the mantissa. Thus, the logarithm 2.34 has two significant figures and the logarithm 102.3 has one significant figure.

Arithmetic operations using logarithms are carried out in the same way as the operations for any exponents. The relevant operations can be summarized by few relationships:

When two numbers are multiplied, their logarithms are added:

$$\log (ab) = \log a + \log b$$

When two numbers are divided, the logarithm of the denominator is subtracted from the logarithm of the numerator:

$$\log \frac{a}{b} = \log a - \log b$$

When a number is raised to a power, its logarithm is multiplied by the power:

$$\log a^n = n \log a$$

When a root is extracted from a number, its logarithm is multiplied by the fractional exponent corresponding to the root:

$$\log a^{1/n} \times \left(\frac{1}{n}\right) \log a$$

### Graphs

Experimental data obtained in the laboratory are often presented by means of a graph. When two variables are measured, it is convenient to show one of them along the vertical axis (the $y$-axis) and the other along the horizontal axis (the $x$-axis). Each pair of measurements is a point on the graph; the points are connected by a smooth curve.

The shape of the smooth curve sometimes reveals more clearly than the numbers alone the relationship between the two variables. For example, Figure 4.9 shows the relationship between the pressure $P$ and the volume $V$ of an ideal gas. The shape of the curve is consistent with an inverse proportionality; that is, the product of the two variables is a constant $k$: $PV = k$. We see similar examples in Section 17.2, where we discuss graphical methods for finding rate laws.

Straight-line plots are of special interest. A straight-line plot is observed when the two variables being studied are directly proportional. Figures 4.11 and 4.12 show the straight lines that are obtained when volume ($V$) and temperature ($T$) are the variables.

The general equation for a straight line usually is represented as:

$$y = mx + b$$

where $x$ and $y$ are the two variables, $m$ is the slope of the line, and $b$ is its intercept with the $y$-axis. When the intercept with the $y$-axis is at the origin, the equation becomes $y = mx$, which is the simple relationship of a direct proportionality. Such an equation is often encountered in chemical studies. For example, volume is directly proportional to absolute temperature: $V = kT$. The slope of the line gives the value of $k$.

To find the slope of a straight line from a graph, we pick two points on the line and find the change in each variable between the two points. The slope is then:

$$\frac{\Delta y}{\Delta x}$$

Figure IV.1 is a plot of the volume of 1 mol of an ideal gas at a pressure of 1 atm against the Celsius temperature. The value of $b$, the $y$ intercept, can be seen to be 22.4 L. The slope can be found from the graph, as shown in the figure.

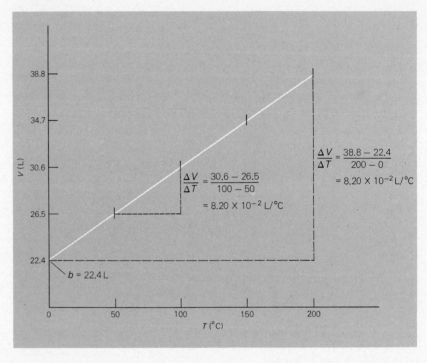

Figure IV.1

## Exercises[1]

**IV.1** Express the following numbers in exponential notation: (a) 9876, (b) 0.009 876, (c) 1 000 001, (d) 0.000 040 40, (e) 1.0004.

**IV.2** Convert the following exponential expressions into ordinary numbers: (a) $2.345 \times 10^5$, (b) $2.345 \times 10^{-5}$, (c) $1.0 \times 10^6$, (d) $1.00 \times 10^{-1}$.

**IV.3** Perform the following calculations and express your result in the

[1] Do not use a calculator for Exercises 1–4, since they will then serve no purpose. Use a calculator to check your answers if you wish.

usual exponential form: (a) $(4.5 \times 10^5)(6.5 \times 10^3)$, (b) $(4.5 \times 10^{-5})(6.5 \times 10^{-3})$, (c) $(4.5 \times 10^5)/(6.5 \times 10^3)$, (d) $(6.5 \times 10^{-3})(4.5 \times 10^{-5})$, (e) $1/(4.5 \times 10^5)$, (f) $1/(6.5 \times 10^{-3})$.

**IV.4** Find the square and the square root, the cube and the cube root of the following exponentials: (a) $4.5 \times 10^{-5}$, (b) $6.5 \times 10^3$.

**IV.5** Find the logarithms of the following numbers: (a) 2.33, (b) 101, (c) 0.0233, (d) 0.001 01.

**IV.6** Find the antilogarithms of the

following numbers: (a) 0.289, (b) 3.289, (c) −2.67, (d) 1.000, (e) 101.1.

**IV.7** Use logarithms to perform the following calculations: (a) (4.567)(7.654), (b) 8.765/5.678, (c) $\sqrt[3]{87.9}$, (d) $(0.00452)^4$.

**IV.8** Find the slope and the $y$ intercept of the straight line obtained by plotting the following data:

| $P$ (atm) | 0.804 | 0.856 | 0.910 | 0.965 |
| $T$ (°C) | 20 | 40 | 60 | 80 |

# Appendix V
# Equilibrium Constants and Reduction Potentials for Aqueous Solutions

**TABLE 1** Solubility Products of Some Sparingly Soluble Ionic Solids at Room Temperature

| Substance | Solubility Product[a] | Substance | Solubility Product[a] |
|---|---|---|---|
| AgBr | $5.2 \times 10^{-13}$ | CuCN | $1.0 \times 10^{-11}$ |
| AgCN | $1.2 \times 10^{-16}$ | CuCl | $3.2 \times 10^{-7}$ |
| $Ag_2CO_3$ | $8.2 \times 10^{-12}$ | CuI | $1.1 \times 10^{-12}$ |
| AgCl | $1.6 \times 10^{-10}$ | $Cu_2S$ | $3 \times 10^{-48}$ |
| $Ag_2CrO_4$ | $1.1 \times 10^{-12}$ | CuS | $6 \times 10^{-36}$ |
| AgI | $8.5 \times 10^{-17}$ | $Fe(OH)_3$ | $4 \times 10^{-38}$ |
| $AgIO_3$ | $3.0 \times 10^{-8}$ | $Hg_2Cl_2$ | $1.3 \times 10^{-18}$ |
| $AgNO_2$ | $1.6 \times 10^{-4}$ | $Hg_2I_2$ | $2.5 \times 10^{-26}$ |
| $Ag_3PO_4$ | $1.6 \times 10^{-18}$ | $HgI_2$ | $8.8 \times 10^{-12}$ |
| $Ag_2S$ | $6 \times 10^{-50}$ | HgS | $4 \times 10^{-53}$ |
| $Ag_2SO_3$ | $1.5 \times 10^{-14}$ | $MgCO_3$ | $1.6 \times 10^{-6}$ |
| $Ag_2SO_4$ | $1.6 \times 10^{-5}$ | $Mg(OH)_2$ | $8.9 \times 10^{-12}$ |
| $Al(OH)_3$ | $2 \times 10^{-32}$ | $Mg_3(PO_4)_2$ | $1.0 \times 10^{-13}$ |
| $AlPO_4$ | $5.8 \times 10^{-19}$ | $Mn(OH)_2$ | $1.9 \times 10^{-13}$ |
| $Au(OH)_3$ | $5.5 \times 10^{-46}$ | $Ni(OH)_2$ | $6.5 \times 10^{-18}$ |
| $BaCO_3$ | $5.1 \times 10^{-9}$ | NiS | $3 \times 10^{-19}$ |
| $BaCrO_4$ | $1.2 \times 10^{-10}$ | $PbBr_2$ | $4.6 \times 10^{-6}$ |
| $BaF_2$ | $1.0 \times 10^{-6}$ | $PbCO_3$ | $3.3 \times 10^{-14}$ |
| $BaSO_4$ | $1.3 \times 10^{-10}$ | $PbCl_2$ | $1.6 \times 10^{-5}$ |
| $CaCO_3$ | $4.8 \times 10^{-9}$ | $PbI_2$ | $7.1 \times 10^{-9}$ |
| $CaF_2$ | $1.7 \times 10^{-10}$ | $Pb(IO_3)_2$ | $1.2 \times 10^{-13}$ |
| $Ca(OH)_2$ | $5.5 \times 10^{-6}$ | $PbO_2$ | $3.2 \times 10^{-66}$ |
| $Ca_3(PO_4)_2$ | $2.0 \times 10^{-29}$ | $PbSO_4$ | $1.2 \times 10^{-8}$ |
| $CaSO_4$ | $1.2 \times 10^{-6}$ | SnS | $1 \times 10^{-25}$ |
| $Cd(OH)_2$ | $5.9 \times 10^{-15}$ | $SrCO_3$ | $7.0 \times 10^{-10}$ |
| CdS | $2 \times 10^{-28}$ | $SrF_2$ | $7.9 \times 10^{-10}$ |
| $Co(OH)_2$ | $2 \times 10^{-16}$ | $SrSO_4$ | $3.2 \times 10^{-7}$ |
| CoS | $4 \times 10^{-21}$ | $Zn(CN)_2$ | $2.6 \times 10^{-13}$ |
| $Cr(OH)_2$ | $1.0 \times 10^{-17}$ | $ZnCO_3$ | $1.4 \times 10^{-11}$ |
| $Cr(OH)_3$ | $6 \times 10^{-31}$ | $Zn(OH)_2$ | $1.8 \times 10^{-14}$ |
| CuBr | $5.2 \times 10^{-9}$ | ZnS | $2 \times 10^{-24}$ |

[a] All ion concentrations are in moles per liter.

**TABLE 2** Dissociation Constants of Weak Acids at Room Temperature

| Electrically Neutral Inorganic Acids | $K_A$ (M) |
|---|---|
| hydrogen peroxide, $H_2O_2$ | $2.4 \times 10^{-12}$ |
| hypoiodous acid, HIO | $2.3 \times 10^{-11}$ |
| hydrocyanic acid, HCN | $4.93 \times 10^{-10}$ |
| hypobromous acid, HBrO | $2.06 \times 10^{-9}$ |
| hypochlorous acid, HClO | $3.2 \times 10^{-8}$ |
| hydrofluoric acid, HF | $3.53 \times 10^{-4}$ |
| nitrous acid, $HNO_2$ | $4.5 \times 10^{-4}$ |
| chlorous acid, $HClO_2$ | $1.1 \times 10^{-2}$ |
| periodic acid, $H_5IO_6$ | $2.3 \times 10^{-2}$ |
| iodic acid, $HIO_3$ | $1.69 \times 10^{-1}$ |

| Electrically Neutral Organic Acids | |
|---|---|
| saccharin | $2.1 \times 10^{-12}$ |
| phenol | $1.28 \times 10^{-10}$ |
| butyric acid | $1.54 \times 10^{-5}$ |
| acetic acid | $1.76 \times 10^{-5}$ |
| acrylic acid | $5.6 \times 10^{-5}$ |
| uric acid | $1.3 \times 10^{-4}$ |
| lactic acid | $1.37 \times 10^{-4}$ |
| formic acid | $1.77 \times 10^{-4}$ |
| sulfanilic acid | $5.9 \times 10^{-4}$ |
| chloroacetic acid | $1.40 \times 10^{-3}$ |
| dichloroacetic acid | $3.32 \times 10^{-2}$ |
| trichloroacetic acid | $2 \times 10^{-1}$ |

**TABLE 3** Dissociation Constants of Weak Bases at Room Temperature

| Electrically Neutral Inorganic Bases | $K_B$ (M) |
|---|---|
| hydroxylamine, $NH_2OH$ | $1.1 \times 10^{-8}$ |
| hydrazine, $N_2H_4$ | $1.7 \times 10^{-5}$ |
| ammonia, $NH_3$ | $1.79 \times 10^{-5}$ |

| Electrically Neutral Organic Bases | |
|---|---|
| aniline, $C_6H_5NH_2$ | $4.27 \times 10^{-10}$ |
| imidazole | $9.01 \times 10^{-8}$ |
| nicotine | $1.05 \times 10^{-6}$ |
| morphine | $1.62 \times 10^{-6}$ |
| codeine | $1.63 \times 10^{-6}$ |
| ephedrine | $1.38 \times 10^{-4}$ |
| methyl amine, $CH_3NH_2$ | $3.70 \times 10^{-4}$ |

| TABLE 4    Dissociation Constants of Polyprotic Acids at Room Temperature | |
|---|---|
| **Inorganic Acids** | $K\,(M)$ |
| $\begin{cases}\text{carbon dioxide, } CO_2 \text{ or} \\ \text{carbonic acid, } H_2CO_3\end{cases}$ | $K_1$  $4.30 \times 10^{-7}$ <br> $K_2$  $5.61 \times 10^{-11}$ |
| chromic acid, $H_2CrO_4$ | $K_1$  $1.8\ \times 10^{-1}$ <br> $K_2$  $3.20 \times 10^{-7}$ |
| hydrogen sulfide, $H_2S$ | $K_1$  $1.1\ \times 10^{-7}$ <br> $K_2$  $1.0\ \times 10^{-12}$ |
| phosphoric acid, $H_3PO_4$ | $K_1$  $7.52 \times 10^{-3}$ <br> $K_2$  $6.23 \times 10^{-8}$ <br> $K_3$  $4.7\ \times 10^{-13}$ |
| phosphorous acid, $H_3PO_3$ | $K_1$  $1.0\ \times 10^{-2}$ <br> $K_2$  $2.6\ \times 10^{-7}$ |
| selenic acid, $H_2SeO_4$ | $K_1$  large <br> $K_2$  $1\ \ \ \times 10^{-2}$ |
| sulfuric acid, $H_2SO_4$ | $K_1$  large <br> $K_2$  $1.20 \times 10^{-2}$ |
| sulfurous acid, $H_2SO_3$ | $K_1$  $1.54 \times 10^{-2}$ <br> $K_2$  $1.02 \times 10^{-7}$ |
| **Organic Acids** | |
| adipic acid | $K_1$  $3.80 \times 10^{-5}$ <br> $K_2$  $3.89 \times 10^{-6}$ |
| citric acid | $K_1$  $8.4\ \times 10^{-4}$ <br> $K_2$  $1.8\ \times 10^{-5}$ <br> $K_3$  $5.0\ \times 10^{-7}$ |
| oxalic acid | $K_1$  $5.36 \times 10^{-2}$ <br> $K_2$  $5.42 \times 10^{-5}$ |

**TABLE 5** Instability Constants of Complex Ions at Room Temperature

| Metal Ion | Ligand | Complex Ion | $K_i^a$ |
|-----------|--------|-------------|---------|
| $Ag^+$ | $NH_3$ | $[Ag(NH_3)_2]^+$ | $6.3 \times 10^{-8}$ |
| | $Br^-$ | $[AgBr_4]^{3-}$ | $5.0 \times 10^{-10}$ |
| | $Cl^-$ | $[AgCl_4]^{3-}$ | $1.0 \times 10^{-6}$ |
| | $CN^-$ | $[Ag(CN)_2]^-$ | $7.9 \times 10^{-22}$ |
| | en | $[Ag(en)_2]^+$ | $1.0 \times 10^{-8}$ |
| | $I^-$ | $[AgI_3]^{2-}$ | $1.3 \times 10^{-14}$ |
| | $S_2O_3^{2-}$ | $[Ag(S_2O_3)_2]^{3-}$ | $1.0 \times 10^{-13}$ |
| $Al^{3+}$ | $F^-$ | $[AlF_6]^{3-}$ | $2.0 \times 10^{-20}$ |
| | $OH^-$ | $[AlOH]^{2+}$ | $1.3 \times 10^{-9}$ |
| $Cd^{2+}$ | $NH_3$ | $[Cd(NH_3)_4]^{2+}$ | $1.0 \times 10^{-7}$ |
| | $Br^-$ | $[CdBr_4]^{2-}$ | $1.6 \times 10^{-4}$ |
| | $Cl^-$ | $[CdCl_3]^-$ | $5.0 \times 10^{-3}$ |
| | $CN^-$ | $[Cd(CN)_4]^{2-}$ | $1.6 \times 10^{-19}$ |
| | en | $[Cd(en)_3]^{2+}$ | $6.3 \times 10^{-13}$ |
| | $I^-$ | $[CdI_4]^{2-}$ | $7.9 \times 10^{-7}$ |
| $Co^{2+}$ | $NH_3$ | $[Co(NH_3)_6]^{2+}$ | $1.0 \times 10^{-5}$ |
| | en | $[Co(en)_3]^{2+}$ | $1.3 \times 10^{-14}$ |
| $Co^{3+}$ | $NH_3$ | $[Co(NH_3)_6]^{3+}$ | $1.0 \times 10^{-34}$ |
| $Cu^+$ | $NH_3$ | $[Cu(NH_3)_2]^+$ | $1.6 \times 10^{-11}$ |
| | $Cl^-$ | $[CuCl_2]^-$ | $2.0 \times 10^{-5}$ |
| | $CN^-$ | $[Cu(CN)_2]^-$ | $1.0 \times 10^{-24}$ |
| | en | $[Cu(en)_2]^+$ | $1.6 \times 10^{-11}$ |
| | $I^-$ | $[CuI_2]^-$ | $1.6 \times 10^{-9}$ |
| $Cu^{2+}$ | $NH_3$ | $[Cu(NH_3)_4]^{2+}$ | $5.0 \times 10^{-14}$ |
| | en | $[Cu(en)_2]^{2+}$ | $1.0 \times 10^{-20}$ |
| $Fe^{2+}$ | $CN^-$ | $[Fe(CN)_6]^{4-}$ | $1.0 \times 10^{-37}$ |
| | en | $[Fe(en)_3]^{2+}$ | $2.5 \times 10^{-10}$ |
| $Fe^{3+}$ | $Cl^-$ | $[FeCl_4]^-$ | $1.0 \times 10^{-2}$ |
| | $CN^-$ | $[Fe(CN)_6]^{3-}$ | $1.0 \times 10^{-44}$ |
| | $SCN^-$ | $[Fe(SCN)_6]^{3-}$ | $8.0 \times 10^{-10}$ |
| $Hg^{2+}$ | $NH_3$ | $[Hg(NH_3)_4]^{2+}$ | $5.0 \times 10^{-20}$ |
| | $Br^-$ | $[HgBr_4]^{2-}$ | $1.0 \times 10^{-20}$ |
| | $Cl^-$ | $[HgCl_4]^{2-}$ | $1.3 \times 10^{-15}$ |
| | $CN^-$ | $[Hg(CN)_4]^{2-}$ | $3.2 \times 10^{-42}$ |
| | $I^-$ | $[HgI_4]^{2-}$ | $1.6 \times 10^{-30}$ |
| | $SO_4^{2-}$ | $[Hg(SO_4)_2]^{2-}$ | $4.0 \times 10^{-3}$ |
| $Ni^{2+}$ | $NH_3$ | $[Ni(NH_3)_6]^{2+}$ | $2.0 \times 10^{-9}$ |
| | $CN^-$ | $[Ni(CN)_4]^{2-}$ | $1.0 \times 10^{-31}$ |
| | en | $[Ni(en)_3]^{2+}$ | $7.9 \times 10^{-20}$ |
| $Pb^{2+}$ | $Br^-$ | $[PbBr_4]^{2-}$ | $1.0 \times 10^{-3}$ |
| | $Cl^-$ | $[PbCl_3]^-$ | $1.6 \times 10^{-2}$ |
| | $OH^-$ | $[PbOH]^+$ | $6.3 \times 10^{-7}$ |
| | $I^-$ | $[PbI_4]^{2-}$ | $1.3 \times 10^{-4}$ |
| $Sn^{2+}$ | $Br^-$ | $[SnBr_3]^-$ | $5.0 \times 10^{-2}$ |
| | $OH^-$ | $[Sn(OH)_3]^-$ | $4.0 \times 10^{-26}$ |
| $Sn^{4+}$ | $Cl^-$ | $[SnCl_6]^{2-}$ | $3.2 \times 10^{-2}$ |
| | $F^-$ | $[SnF_6]^{2-}$ | $1.0 \times 10^{-18}$ |
| $Zn^{2+}$ | $NH_3$ | $[Zn(NH_3)_4]^{2+}$ | $4.0 \times 10^{-10}$ |
| | $CN^-$ | $[Zn(CN)_4]^{2-}$ | $6.3 \times 10^{-21}$ |
| | $OH^-$ | $[Zn(OH)_4]^{2-}$ | $2.5 \times 10^{-16}$ |

$a$ All ion concentrations are in moles per liter.

**TABLE 6** Standard Reduction Potentials at 298 K

| Half-Reaction | $\mathcal{E}°$ (V) | Half-Reaction | $\mathcal{E}°$ (V) |
|---|---|---|---|
| $Li^+(aq)^a + e^- \rightleftharpoons Li(s)$ | $-3.045$ | $I_2(s) + 2e^- \rightleftharpoons 2I^-(aq)$ | 0.535 |
| $Rb^+(aq) + e^- \rightleftharpoons Rb(s)$ | $-2.925$ | $MnO_4^-(aq) + e^- \rightleftharpoons MnO_4^{2-}(aq)$ | 0.564 |
| $K^+(aq) + e^- \rightleftharpoons K(s)$ | $-2.924$ | $O_2(g) + 2H^+(aq) + 2e^- \rightleftharpoons H_2O_2(aq)$ | 0.682 |
| $Cs^+(aq) + e^- \rightleftharpoons Cs(s)$ | $-2.923$ | $Fe^{3+}(aq) + e^- \rightleftharpoons Fe^{2+}(aq)$ | 0.770 |
| $Ba^{2+}(aq) + 2e^- \rightleftharpoons Ba(s)$ | $-2.90$ | $Hg_2^{2+}(aq) + 2e^- \rightleftharpoons 2Hg(l)$ | 0.7961 |
| $Sr^{2+}(aq) + 2e^- \rightleftharpoons Sr(s)$ | $-2.89$ | $Ag^+(aq) + e^- \rightleftharpoons Ag(s)$ | 0.7996 |
| $Ca^{2+}(aq) + 2e^- \rightleftharpoons Ca(s)$ | $-2.76$ | $2NO_3^-(aq) + 4H^+(aq) + 2e^- \rightleftharpoons N_2O_4(g) + 2H_2O$ | 0.81 |
| $Na^+(aq) + e^- \rightleftharpoons Na(s)$ | $-2.7109$ | $2Hg^{2+}(aq) + 2e^- \rightleftharpoons Hg_2^{2+}(aq)$ | 0.905 |
| $Mg^{2+}(aq) + 2e^- \rightleftharpoons Mg(s)$ | $-2.375$ | $NO_3^-(aq) + 3H^+(aq) + 2e^- \rightleftharpoons HNO_2(aq) + H_2O$ | 0.94 |
| $Al^{3+}(aq) + 3e^- \rightleftharpoons Al(s)$ | $-1.706$ | $NO_3^-(aq) + 4H^+(aq) + 3e^- \rightleftharpoons NO(g) + 2H_2O$ | 0.96 |
| $Be^{2+}(aq) + 2e^- \rightleftharpoons Be(s)$ | $-1.70$ | $HNO_2(aq) + H^+(aq) + e^- \rightleftharpoons NO(g) + H_2O$ | 0.99 |
| $Zn(NH_3)_4^{2+}(aq) + 2e^- \rightleftharpoons Zn(s) + 4NH_3(aq)$ | $-1.04$ | $AuCl_4^-(aq) + 3e^- \rightleftharpoons Au(s) + 4Cl^-(aq)$ | 0.994 |
| $Mn^{2+}(aq) + 2e^- \rightleftharpoons Mn(s)$ | $-1.029$ | $Br_2(l) + 2e^- \rightleftharpoons 2Br^-(aq)$ | 1.065 |
| $Cr^{2+}(aq) + 2e^- \rightleftharpoons Cr(s)$ | $-0.91$ | $Cu^{2+}(aq) + 2CN^-(aq) + e^- \rightleftharpoons Cu(CN)_2^-(aq)$ | 1.12 |
| $V^{3+}(aq) + 3e^- \rightleftharpoons V(s)$ | $-0.89$ | $MnO_2(s) + 4H^+(aq) + 2e^- \rightleftharpoons Mn^{2+}(aq) + 2H_2O$ | 1.208 |
| $Zn^{2+}(aq) + 2e^- \rightleftharpoons Zn(s)$ | $-0.7628$ | $O_2(g) + 4H^+(aq) + 4e^- \rightleftharpoons 2H_2O$ | 1.229 |
| $Cr^{3+}(aq) + 3e^- \rightleftharpoons Cr(s)$ | $-0.74$ | $Au^{3+}(aq) + 2e^- \rightleftharpoons Au^+(aq)$ | 1.29 |
| $Ag_2S(s) + 2e^- \rightleftharpoons 2Ag(s) + S^{2-}(aq)$ | $-0.7051$ | $Cr_2O_7^{2-}(aq) + 14H^+(aq) + 6e^- \rightleftharpoons 2Cr^{3+}(aq) + 7H_2O$ | 1.33 |
| $S(s) + 2e^- \rightleftharpoons S^{2-}(aq)$ | $-0.508$ | $Cl_2(g) + 2e^- \rightleftharpoons 2Cl^-(aq)$ | 1.3583 |
| $Ni^{2+}(aq) + 2e^- \rightleftharpoons Ni(s)$ | $-0.23$ | $Au^{3+}(aq) + 3e^- \rightleftharpoons Au(s)$ | 1.42 |
| $Sn^{2+}(aq) + 2e^- \rightleftharpoons Sn(s)$ | $-0.1364$ | $MnO_4^-(aq) + 8H^+(aq) + 5e^- \rightleftharpoons Mn^{2+}(aq) + 4H_2O$ | 1.491 |
| $Pb^{2+}(aq) + 2e^- \rightleftharpoons Pb(s)$ | $-0.1263$ | $Mn^{3+}(aq) + e^- \rightleftharpoons Mn^{2+}(aq)$ | 1.51 |
| $Fe^{3+}(aq) + 3e^- \rightleftharpoons Fe(s)$ | $-0.036$ | $HClO(aq) + H^+(aq) + e^- \rightleftharpoons \frac{1}{2}Cl_2(g) + H_2O$ | 1.63 |
| $2H^+(aq) + 2e^- \rightleftharpoons H_2(g)$ | 0 | $HClO_2(aq) + 2H^+(aq) + 2e^- \rightleftharpoons HClO(aq) + H_2O$ | 1.64 |
| $AgBr(s) + e^- \rightleftharpoons Ag(s) + Br^-(aq)$ | 0.0713 | $MnO_4^-(aq) + 4H^+(aq) + 3e^- \rightleftharpoons MnO_2(s) + 2H_2O$ | 1.679 |
| $Sn^{4+}(aq) + 2e^- \rightleftharpoons Sn^{2+}(aq)$ | 0.15 | $N_2O(g) + 2H^+(aq) + 2e^- \rightleftharpoons N_2(g) + H_2O$ | 1.77 |
| $Cu^{2+}(aq) + e^- \rightleftharpoons Cu^+(aq)$ | 0.158 | $H_2O_2(aq) + 2H^+(aq) + 2e^- \rightleftharpoons 2H_2O$ | 1.776 |
| $Hg_2Cl_2(s) + 2e^- \rightleftharpoons 2Hg(l) + 2Cl^-(aq)$ | 0.2676 | $O_3(g) + 2H^+(aq) + 2e^- \rightleftharpoons O_2(g) + H_2O$ | 2.07 |
| $Cu^{2+}(aq) + 2e^- \rightleftharpoons Cu(s)$ | 0.3402 | $F_2(g) + 2e^- \rightleftharpoons 2F^-(aq)$ | 2.87 |

| **Alkaline Solution** | | | |
|---|---|---|---|
| $Al(OH)_4^-(aq) + 3e^- \rightleftharpoons Al(s) + 4OH^-(aq)$ | $-2.35$ | $NO_3^-(aq) + H_2O + 2e^- \rightleftharpoons NO_2^-(aq) + 2OH^-(aq)$ | 0.01 |
| $Mn(OH)_2(s) + 2e^- \rightleftharpoons Mn(s) + 2OH^-(aq)$ | $-1.47$ | $ClO_4^-(aq) + H_2O + 2e^- \rightleftharpoons ClO_3^-(aq) + 2OH^-(aq)$ | 0.17 |
| $Cr(OH)_3(s) + 3e^- \rightleftharpoons Cr(s) + 3OH^-(aq)$ | $-1.3$ | $2ClO^-(aq) + 2H_2O + 2e^- \rightleftharpoons Cl_2(g) + 4OH^-(aq)$ | 0.40 |
| $Zn(OH)_4^{2-}(aq) + 2e^- \rightleftharpoons Zn(s) + 4OH^-(aq)$ | $-1.216$ | $MnO_4^-(aq) + 2H_2O + 3e^- \rightleftharpoons MnO_2(s) + 4OH^-(aq)$ | 0.58 |
| $Zn(OH)_2(s) + 2e^- \rightleftharpoons Zn(s) + 2OH^-(aq)$ | $-1.17$ | $ClO_2^-(aq) + H_2O + 2e^- \rightleftharpoons ClO^-(aq) + 2OH^-(aq)$ | 0.59 |
| $2H_2O + 2e^- \rightleftharpoons H_2(g) + 2OH^-(aq)$ | $-0.8277$ | $ClO_3^-(aq) + 3H_2O + 6e^- \rightleftharpoons Cl^-(aq) + 6OH^-(aq)$ | 0.62 |
| $O_2(g) + 2H_2O + 2e^- \rightleftharpoons H_2O_2(aq) + 2OH^-(aq)$ | $-0.146$ | $O_3(g) + H_2O + 2e^- \rightleftharpoons O_2(g) + 2OH^-(aq)$ | 1.24 |

$^a$ Substances followed by (aq) are $1M$ in aqueous solution.

# Appendix VI
## The Main Reaction Approximation for the Solution of Problems in Simultaneous Equilibria

A number of different equilibrium reactions occur even in the simplest aqueous solutions. We have seen that there are several equilibrium reactions in solutions of weak acids or weak bases. In such solutions, the equilibrium reaction between the strongest acid and the strongest base is chosen as the relevant one for calculations. For example, in a solution of HCN, the relevant equilibrium is the one that defines $K_A$. The water equilibrium occurs simultaneously, but we ignore it until we wish to calculate $[OH^-]$, which appears in $K_W$ but not in $K_A$.

Many different equilibrium reactions also occur in an aqueous solution of a polyprotic acid. The procedure used for solutions of a polyprotic acid can be applied to any aqueous systems in which a large number of different equilibrium reactions take place simultaneously. We first choose a *main equilibrium reaction,* in polyprotic acids, the reaction that defines $K_1$, and we find the equilibrium concentrations of all the substances in this reaction. We then determine the concentrations of all the other species in solution by choosing secondary equilibria that include these substances and substituting the concentrations found from the primary equilibrium into the secondary equilibria.

Without such a procedure, calculations on the composition of these systems can require the solution of many simultaneous equations in many unknowns. A simple method is desirable, but we are faced with a problem. Out of all the simultaneous equilibria that occur in a system, how do we identify the one that qualifies as a main equilibrium? Some guidelines help us make the choice:

1. *To qualify as the main equilibrium, a reaction must have an equilibrium constant at least 100 times larger than the equilibrium constant of any other reaction that can be considered as a possible main equilibrium.* In a solution of phosphoric acid, for example, the ionization of the first proton, $H_3PO_4 \rightleftharpoons H_2PO_4^- + H^+$, is the main equilibrium because $K_1$ is more than $10^6$ larger than $K_2$, more than $10^{10}$ larger than $K_3$, and more than $10^{11}$ larger than $K_W$.
2. *To qualify as the main equilibrium, a reaction must include as reactants only those substances that are present in appreciable concentrations*

*when the solution is prepared.* In practice, this usually means that *substances that are not specifically mentioned in the description of the system of interest do not appear on the left side of any reaction that can be considered as a candidate for the main equilibrium.*

For example, many equilibria occur in a solution of hydrocyanic acid, HCN. Many of these equilibria have large equilibrium constants. The equilibrium constant for one of these reactions,

$$H_2O + CN^-(aq) \rightleftharpoons HCN(aq) + OH^-(aq)$$

is larger than $K_A$, the equilibrium constant for the main equilibrium. But this reaction is not a candidate for the main equilibrium because there is no appreciable concentration of $CN^-$ when a solution of HCN is first prepared. By rule 2, cyanide ion cannot appear on the left side of any reaction that can be used as the main equilibrium for this system. Only $H_2O$ and/or HCN can appear on the left side of such a reaction. You will encounter many instances in which the application of this rule limits the number of reactions to consider in selecting a main equilibrium.

The procedure for selecting the main equilibrium is

1. List all the substances that are present in appreciable concentrations when the solution is prepared. A careful reading of the description of the solution is highly desirable in this step.
2. List all the reactions that are possible if these substances are the only reactants. Each reactant is classified by its behavior in aqueous systems. Among the possible reactions are the transfer of a proton from an acid to a base, the dissolution of a sparingly soluble material, the precipitation of a sparingly soluble material, the formation of a complex ion, or even an oxidation-reduction. Each reactant can therefore be identified as a proton donor, an acid; a proton acceptor, a base; both a donor and an acceptor, an amphoteric reactant; or neither an acid nor a base. In addition, you must say whether each ion in solution is a possible component of a relatively insoluble material or of a complex ion. The solubility rules given in Chapter 12 and the information in Chapter 18 will help you make this identification. Use the information in Table 16.1 to decide if species that are oxidizing or reducing agents in water are present. Having made this classification, list all the reactions that are possible between the allowed reactants.
3. Use all the available data on equilibrium constants to obtain the equilibrium constant for each reaction that is listed as a possible main equilibrium.
4. Compare all these numerical values. If one equilibrium constant is at

least $10^2$ larger than any of the others, it identifies the desired main equilibrium reaction. The method described here can be used only if one equilibrium constant is at least $10^2$ larger than any of the others.

If the method does identify a main reaction, the concentrations of all the substances present in the solution at equilibrium can be found by the methods described in the following examples.

**Example VI.1**   Calculate the concentrations of all species present at equilibrium in an aqueous solution that has a total volume of 1.0 L and contains 0.10 mol of sodium bicarbonate.

**Solution**   1. Following the rules outlined above, we list all the substances present in appreciable concentration. In addition to $H_2O$, the solution of sodium bicarbonate, a salt, also contains the dissociated ions $Na^+$ and $HCO_3^-$.

2. We now classify each of these substances according to its behavior in aqueous equilibria. All sodium salts are soluble, and $Na^+$ ion does not enter into acid-base reactions in water, it does not form complexes, and it is not an oxidizing agent in water. Therefore, the $Na^+$ ion can be disregarded. We can disregard all group IA metal cations in water; no possible main equilibrium will contain them as reactants. The $HCO_3^-$ ion is amphoteric: It is an acid because it has a proton to donate and it is a base because it can accept a proton. Water is also amphoteric. We need consider only acid-base equilibria, that is, proton transfer reactions, and we can list all the possibilities:

a. $H_2O \rightleftharpoons H^+(aq) + OH^-(aq)$; water is both the acid and the base.
b. $HCO_3^-(aq) + H_2O \rightleftharpoons H_2CO_3(aq) + OH^-(aq)$; water is the acid and bicarbonate is the base.
c. $HCO_3^-(aq) \rightleftharpoons CO_3^{2-}(aq) + H^+(aq)$; water is the base and bicarbonate is the acid. (As usual, water is not included in the equation when it acts as a base.)
d. $HCO_3^-(aq) + HCO_3^-(aq) \rightleftharpoons H_2CO_3(aq) + CO_3^{2-}(aq)$; bicarbonate is both the acid and the base.

3. Table 15.8 lists the $K_1$ and $K_2$ for carbon dioxide, which we have represented as $H_2CO_3$ for the sake of simplicity. The value of $K_W$ is also known, so we have sufficient data to evaluate the equilibrium constant for each of the equilibria.

a. $K_W = 1.0 \times 10^{-14}$.
b. $K = K_W/K_1 = (1.0 \times 10^{-14})/(4.3 \times 10^{-7}) = 2.3 \times 10^{-8}$; note that $K_1$ is used here because the reaction includes carbonic acid and bicarbonate ion, as $K_1$ does.
c. This reaction defines $K_2$, whose value is $5.6 \times 10^{-11}$.
d. To find the equilibrium constant of this disproportionation, we must find a combination of reactions with known equilibrium constants that will

give the reaction. We see that $HCO_3^-$ forms $CO_3^{2-}$, just as it does in the $K_2$ reaction. We also see that $HCO_3^-$ forms $H_2CO_3$, which is equivalent to $CO_2$, in the reverse of the $K_1$ reaction. The combination of these reactions is

$$HCO_3^-(aq) \rightleftharpoons CO_3^{2-}(aq) + H^+(aq) \qquad K_2$$

$$HCO_3^-(aq) + H^+(aq) \rightleftharpoons H_2CO_3(aq) \qquad \frac{1}{K_1}$$

$$\overline{\qquad\qquad\qquad\qquad\qquad\qquad\qquad\qquad\qquad\qquad}$$

$$2HCO_3^-(aq) \rightleftharpoons H_2CO_3(aq) + CO_3^{2-}(aq) \qquad K = \frac{K_2}{K_1}$$

$$K = \frac{5.6 \times 10^{-11}}{4.3 \times 10^{-7}} = 1.3 \times 10^{-4}$$

4. Comparing the values obtained in step 3, we see that the equilibrium constant for the disproportionation is almost $10^4$ larger than the second largest one, that for reaction **b**. Therefore, reaction **d** is the main reaction. It can be used to carry out the usual equilibrium calculation.

$$2HCO_3^-(aq) \rightleftharpoons H_2CO_3(aq) + CO_3^{2-}(aq)$$

|        | $2HCO_3^-(aq)$ | $H_2CO_3(aq)$ | $CO_3^{2-}(aq)$ |
|--------|-----------------|---------------|-----------------|
| **start** | $0.10M$ | $0$ | $0$ |
| **equil** | $0.10M - 2x$ | $x$ | $x$ |

Substituting into the expression for $K$:

$$K = \frac{x^2}{(0.10M - 2x)^2} = 1.3 \times 10^{-4}$$

Taking the square root of both sides of this equation, we find that $x = 1.1 \times 10^{-3}M$.

Therefore,

$$[H_2CO_3] = [CO_3^{2-}] = 1.1 \times 10^{-3}M$$
$$[HCO_3^-] = 0.10M - (2)(0.0011M) = 0.10M$$

Both $H^+$ and $OH^-$ are also present in the solution. To find the concentration of these ions, we use the secondary equilibria in which they appear, for instance

$$K_2 = \frac{[H^+][CO_3^{2-}]}{[HCO_3^-]} = 5.6 \times 10^{-11}M$$

The $[H^+]$ is found by substituting the concentration of bicarbonate ion and carbonate ion, calculated from the main equilibrium, directly into this expression:

$$\frac{[H^+](1.1 \times 10^{-3}M)}{(0.10M)} = 5.6 \times 10^{-11}M$$

and

$$[H^+] = 5.1 \times 10^{-9}M$$

The $[OH^-]$ is found in the usual way, by substituting into the expression for $K_W$:

$$[OH^-] = \frac{1.0 \times 10^{-14}M^2}{5.1 \times 10^{-9}M} = 2.0 \times 10^{-6}M$$

This approach can also be used for more complicated systems that include undissolved substances.

**Example VI.2**    Some magnesium hydroxide, $Mg(OH)_2(s)$, is added to a $1.0M$ solution of nitric acid. The $K_{sp}$ of the magnesium hydroxide is $8.9 \times 10^{-12}$. Find the main equilibrium reaction for this system and evaluate its equilibrium constant.

**Solution**    1. List all the species present in appreciable quantity when the solution is prepared: $Mg(OH)_2(s)$, $H_2O$, $H^+$, and $NO_3^-$. No $HNO_3$ is present, since nitric acid, being a strong acid, is ionized completely in solution.

2. Each substance is classified by its behavior in aqueous equilibria. The $H_2O$ is amphoteric, and the $H^+$ is a strong acid — in fact, the strongest in the system. The $NO_3^-$ is not part of any equilibrium of interest; it is too weak a base to react with water. Nitrate salts — including magnesium nitrate, which could form in this system — are all soluble. The $Mg(OH)_2(s)$ can dissolve in water.

We thus have a system that starts with an undissolved substance and ions in solution. In such a system, we must also consider the reactions between the ions ($Mg^{2+}$ and $OH^-$) that form when the solid dissolves and the ions that are already in solution.

The simple equilibria are:

a. $\quad\quad H_2O \rightleftharpoons H^+(aq) + OH^-(aq) \quad\quad K_W = 1.0 \times 10^{-14}$

b. $\quad Mg(OH)_2(s) \rightleftharpoons Mg^{2+}(aq) + 2OH^-(aq) \quad\quad K_{sp} = 8.9 \times 10^{-12}$

We can assume that $Mg^{2+}$ does not react further, but we must consider the reaction between the $OH^-$, produced when $Mg(OH)_2(s)$ dissolves, and $H^+$, the strongest acid in the system. The product of the reaction is $H_2O$. The reactants are $H^+$, which is present in the original system, and the added $Mg(OH)_2(s)$, which is the source of $OH^-$. The equation is:

c. $\quad\quad Mg(OH)_2(s) + 2H^+(aq) \rightleftharpoons Mg^{2+}(aq) + H_2O$

To evaluate the equilibrium constant for this reaction, we need a suitable combination of reactions with known equilibrium constants. Reaction **b**, which defines $K_{sp}$, and reaction **c** both have solid magnesium hydroxide as a reactant and the magnesium cation as a product. Reaction **c** has water as a product and $H^+$ as a reactant, the reverse of reaction **a**. But the coefficient is 2 in reaction **a** and 1 in reaction **c**. To obtain the suitable equilibrium constant, we must therefore reverse reaction **a**, multiply the resulting equation by 2, and raise its equilibrium constant to the second power. The combination is thus:

$$Mg(OH)_2(s) \rightleftharpoons Mg^{2+}(aq) + 2OH^-(aq)$$
$$\underline{2H^+(aq) + 2OH^-(aq) \rightleftharpoons 2H_2O}$$
$$Mg(OH)_2(s) + 2H^+(aq) \rightleftharpoons Mg^{2+}(aq) + 2H_2O$$
$$K_{sp} = 8.9 \times 10^{-12}$$
$$\frac{1}{K_W^2} = \frac{1}{1.0 \times 10^{-28}}$$
$$K = \frac{K_{sp}}{K_W^2}$$
$$= 8.9 \times 10^{16}$$

Reaction c has a very large equilibrium constant and is the main equilibrium. This calculation tells us that magnesium hydroxide, which has a small $K_{sp}$, is relatively insoluble in water but is quite soluble in solutions of strong acids. If excess magnesium hydroxide is added to a solution of strong acid, it will dissolve until the strong acid is neutralized and the solution is slightly alkaline. The large value of $K$ for the neutralization of a strong acid by a strong base accounts for the dissolution of magnesium hydroxide in the acid solution. The numerical value of the solubility of magnesium hydroxide in this solution can be found by use of the main equilibrium to determine the $[Mg^{2+}]$ at equilibrium.

A more complex system is treated in the next example.

**Example VI.3**   Solid CuI is added to a $0.4M$ solution of HCN. The $K_{sp}$ of CuI(s) = $1.1 \times 10^{-12}$, the $K_{sp}$ of CuCN(s) = $1.0 \times 10^{-11}$. The $K_A$ of HCN = $4.9 \times 10^{-10}$ and the $K_i$ of $Cu(CN)_2^- = 1.0 \times 10^{-24}$. Find the main reaction and evaluate its equilibrium constant.

**Solution**   1. List all the substances present in appreciable quantities when the system is prepared: CuI(s), HCN, and $H_2O$.

2. Classify each substance according to its chemical behavior. The CuI(s) is a sparingly soluble ionic solid that can dissolve in water, the HCN is a weak acid, and the $H_2O$ is an acid and a base. Also consider the $Cu^+$ ions formed when the solid dissolves; $Cu^+$ can form the sparingly soluble CuCN(s) or the complex ion $Cu(CN)_2^-$, if a source of $CN^-$ ion (such as HCN) is available.

3. We list possible reactions, including not only the simple reactions but also those between the ions that form. The simple possibilities are:

a.   $CuI(s) \rightleftharpoons Cu^+(aq) + I^-(aq)$   $K_{sp} = 1.1 \times 10^{-12}$
b.   $H_2O \rightleftharpoons H^+(aq) + OH^-(aq)$   $K_W = 1.0 \times 10^{-14}$
c.   $HCN(aq) \rightleftharpoons H^+(aq) + CN^-(aq)$   $K_A = 4.9 \times 10^{-10}$

The more complex reactions are those between the $Cu^+$ from the CuI and the $CN^-$ from the HCN.

d.    $CuI(s) + HCN(aq) \rightleftharpoons CuCN(s) + H^+(aq) + I^-(aq)$

$$K = \frac{K_{sp}(CuI)K_A}{K_{sp}(CuCN)} = \frac{(1.1 \times 10^{-12})(4.9 \times 10^{-10})}{(1.0 \times 10^{-11})} = 5.4 \times 10^{-11}$$

e.    $CuI(s) + 2HCN(aq) \rightleftharpoons Cu(CN)_2^-(aq) + 2H^+(aq) + I^-(aq)$

$$K = \frac{K_{sp}(CuI)K_A^2}{K_i} = \frac{(1.1 \times 10^{-12})(4.9 \times 10^{-10})^2}{1.0 \times 10^{-24}} = 2.6 \times 10^{-7}$$

Since the equilibrium constant for this reaction is more than $10^2$ larger than the next largest one, it can be used as the main reaction for the purpose of calculation.

In some equilibrium systems the main reaction may change when the conditions change.

**Example VI.4**   Excess $Al(OH)_3(s)$ is added to solutions at (a) pH = 1.00, (b) pH = 7.00, (c) pH = 13.00. Find the amount of $Al(OH)_3(s)$ that dissolves in 1 L of each solution. Find the pH of each solution at equilibrium. The $K_{sp}$ of $Al(OH)_3(s)$ is $2.0 \times 10^{-32}$ and $K_i$ of the complex ion $[Al(OH)_4]^-(aq)$ is $5.0 \times 10^{-35}$. Assume that no other important complex ion formation takes place.

**Solution**   a. The species present in appreciable amounts are $Al(OH)_3(s)$, $H_2O$, and $H^+$. Possible reactions are:

$Al(OH)_3(s) \rightleftharpoons Al^{3+}(aq) + 3OH^-(aq)$

$K_{sp} = 2.0 \times 10^{-32}$

$Al(OH)_3(s) + 3H^+(aq) \rightleftharpoons Al^{3+}(aq) + 3H_2O$

$$K = \frac{K_{sp}}{K_W^3} = \frac{2.0 \times 10^{-32}}{(1.0 \times 10^{-14})^3} = 2.0 \times 10^{10}$$

The second one is the main reaction.

$$Al(OH)_3(s) + 3H^+(aq) \rightleftharpoons Al^{3+}(aq) + 3H_2O$$

**start**                                      $0.10M$

Since $K$ is large, we introduce a completion step.

**complete**                                      $\dfrac{0.10M}{3}$

Let $x = [Al^{3+}]$ that reacts to reach equilibrium.

**equil**                        $3x$                    $0.033M - x$

$$K = \frac{(0.33M - x)}{(3x)^3} = 2.0 \times 10^{10}$$

We can use the approximation $0.033 - x \approx 0.033$ to find $x = 3.9 \times 10^{-5}M$, and the approximation is a valid one. We can now say that 0.033 mol of $Al(OH)_3(s)$ dissolves in 1 L of the solution.

$$[H^+] = 3x = 3(3.9 \times 10^{-5}M) = 1.2 \times 10^{-4}M$$
$$pH = 3.92$$

b. The species present in appreciable amounts are $Al(OH)_3(s)$ and $H_2O$. Possible reactions are

$$Al(OH)_3(s) \rightleftharpoons Al^{3+}(aq) + 3OH^-(aq) \qquad K_{sp} = 2.0 \times 10^{-32}$$
$$H_2O \rightleftharpoons H^+(aq) + OH^-(aq) \qquad K_W = 1.0 \times 10^{-14}$$

The water equilibrium is the main reaction in this case. In solutions where the pH is close to 7, the water equilibrium may be the main reaction when no other process has an equilibrium constant of comparable magnitude. A solution of a very insoluble metal hydroxide is one common example of such a system.

The main reaction tells us that at equilibrium $[OH^-] = 1.0 \times 10^{-7}M$. Substituting this value into the secondary equilibrium allows us to find the $[Al^{3+}]$ at equilibrium.

$$K_{sp} = 2.0 \times 10^{-32} = [Al^{3+}][OH^-]^3 = [Al^{3+}](10^{-7}M)^3$$
$$[Al^{3+}] = 2.0 \times 10^{-11}M$$

From this concentration we can say the $2.0 \times 10^{-11}$ mol of $Al(OH)_3(s)$ dissolves in 1 L of solution. This small amount of dissolved solid does not change the pH of the solution in any significant way from its original value of 7.00. You should note that the solubility of $Al(OH)_3$ calculated in this way differs from the solubility that would be found by the usual solubility product calculation (Section 14.5).

c. The species present in appreciable amounts are $Al(OH)_3(s)$, $H_2O$, and $OH^-$. Possible reactions are

$$Al(OH)_3(s) \rightleftharpoons Al^{3+}(aq) + 3OH^-(aq)$$
$$Al(OH)_3(s) + OH^-(aq) \rightleftharpoons [Al(OH)_4]^-(aq)$$

$$K_{sp} = 2.0 \times 10^{-32}$$
$$K = \frac{K_{sp}}{K_i} = \frac{2.0 \times 10^{-32}}{5.0 \times 10^{-35}} = 4.0 \times 10^2$$

The second one is the main reaction.

$$Al(OH)_3(s) + OH^-(aq) \rightleftharpoons [Al(OH)_4]^-(aq)$$
start                           $0.10M$

Let $x = [OH^-]$ that reacts to reach equilibrium.

$$\text{equil} \qquad 0.10M - x \quad x$$

$$K = \frac{x}{0.10M - x} = 4.0 \times 10^2$$

$$x = \frac{40}{401} = 0.10M$$

We can say the 0.10 mol of $Al(OH)_3(s)$ dissolves in 1 L of solution.

$$[OH^-] = 0.10 - \frac{40}{401} = 2.5 \times 10^{-4}M$$

$$pH = 10.40$$

## Exercises

**VI.1**[1]   Calculate the pH of a solution that is prepared from 0.10 mol of $Na_2HPO_4$ and enough water to make 1 L of solution.[2]

**VI.2**   Calculate the pH of a solution prepared from 0.10 mol of $NaH_2PO_4$ and enough water for 1 L of solution.[2]

**VI.3**   Calculate the solubility of $Al(OH)_3$ in 1.0$M$ nitric acid solution. The $K_{sp}$ is $2 \times 10^{-32}$.

**VI.4**   The $K_{sp}$ of $BaF_2$ is $1.0 \times 10^{-6}$. Find its solubility in 0.10$M$ hydrochloric acid solution. ($K_A$ of HF = $3.53 \times 10^{-4}$.)

**VI.5**   The $K_i$ of $[Zn(CN)_4]^{2-}$ is $6.3 \times 10^{-21}$. Find the solubility of $Zn(OH)_2$, whose $K_{sp}$ is $1.8 \times 10^{-14}$, in 0.20$M$ NaCN solution.

**VI.6**   Find the main reaction and evaluate its equilibrium constant in a system prepared from solid AgCl, $K_{sp} = 1.6 \times 10^{-10}$, and a solution that is 1.0$M$ in $NH_3$ and 1.0$M$ in $NH_4Br$. The $K_i$ of $[Ag(NH_3)_2]^+$ is $6.3 \times 10^{-8}$ and the $K_{sp}$ of AgBr is $5.2 \times 10^{-13}$.

**VI.7**   Use appropriate data from Tables 14.3 and 18.1 to find the main reaction and to evaluate its equilibrium constant in a system containing a 1.0$M$ solution of mercury(II) nitrate and excess solid silver iodide.

**VI.8**   Find the concentrations of all species at equilibrium in a solution prepared from 1.0 mol of sodium carbonate and enough water to make 1 L of solution.[2]

**VI.9**   Find the solubility of $Mg(OH)_2(s)$ in a solution that is 1.0$M$ in $NH_4Cl$.[2]

**VI.10**   Find the concentration of $CN^-$ in a solution prepared from 1.0 L of 0.40$M$ NaCN and excess $Zn(OH)_2(s)$.[2]

**VI.11**   Find the pH of a solution prepared from 1.0 L of 0.10$M$ $Ba(OH)_2$ and excess $Zn(OH)_2(s)$.

**VI.12**   Find the concentrations of all species at equilibrium in a solution prepared by mixing of equal volumes of a 0.10$M$ solution of $Na_2SO_4$ and a 0.10$M$ solution of $AgNO_3$.[2]

**VI.13**   Calculate the concentration of all species at equilibrium in a solution prepared by addition of excess AgCN(s) to 0.10$M$ HCN solution.[2]

[1] The answers to exercises whose numbers are in color can be found in Appendix VII.
[2] The necessary data for this exercise can be found in Appendix V.

# Appendix VII
## Answers to Selected Exercises

### Chapter 1

**1.3** (a) observation; (b) hypothesis; (c) theory; (d) law. **1.8** (a) area; (b) frequency of a wave; (c) velocity; (d) heat; (e) pressure. **1.15** $10^4$ m$^3$. **1.23** 21.9 mi/hr. **1.25** 30.4 atm. **1.28** 8.26 lb/gal. **1.32** 574. **1.37** (a) 2; (b) 3; (c) 1; (d) 4; (e) 2.

### Chapter 2

**2.5** 1.75 g C. **2.10** 0.83 g O. **2.16** $S_2O$. **2.20** (a) RbF; (b) SrF$_2$; (c) BF$_3$; (d) SiF$_4$; (e) NF$_3$; (f) OF$_2$; (g) ClF. **2.31** (a) $-1$; (b) $-2$; (c) $-2$. **2.40** (a) platinum(IV) cyanide; (b) mercury(I) sulfate; (c) tin(II) nitrite; (d) copper(II) chlorate. **2.47** (a) dinitrogen pentoxide; (b) tetraphosphorus hexoxide; (c) chlorine trifluoride; (d) tetrasulfur tetranitride. **2.52** (a) $2N_2O_5 \rightarrow 2N_2O_4 + O_2$; (b) $4AsCl_5 \rightarrow As_4 + 10Cl_2$; (c) $2NI_3 \rightarrow N_2 + 3I_2$; (d) $2HBrO_3 \rightarrow Br_2O_5 + H_2O$. **2.54** (a) $2NH_4NO_3 \rightarrow 2N_2 + O_2 + 4H_2O$; (b) $2CH_4 + 3O_2 \rightarrow 2CO + 4H_2O$; (c) $4Al + 3MnO_2 \rightarrow 3Mn + 2Al_2O_3$; (d) $4NH_3 + 5O_2 \rightarrow 4NO + 6H_2O$. **2.56** (a) NaHSO$_4$; (b) I$_2O_5$; (c) HClO$_2$; (d) PH$_3$. **2.60** (a) K$^+$(aq), Cl$^-$(aq); (b) Ca$^{2+}$(aq), NO$_3^-$(aq); (c) Na$^+$(aq), H$_2$PO$_4^-$(aq); (d) Cu$^{2+}$(aq), Br$^-$(aq). **2.64** (a) $2H_2O(l) \rightarrow 2H_2(g) + O_2(g)$; (b) $CaCO_3(s) \rightarrow CaO(s) + CO_2(g)$; (c) $2CO(g) + O_2(g) \rightarrow 2CO_2(g)$; (d) $2H_2O_2(aq) \rightarrow 2H_2O + O_2(g)$. **2.68** (a) $2H^+(aq) + CO_3^{2-}(aq) \rightarrow CO_2(g) + H_2O$; (b) $Ba^{2+}(aq) + CO_2(g) + H_2O \rightarrow BaCO_3(s) + 2H^+(aq)$. **2.71** $4.53 \times 10^5$ J.

### Chapter 3

**3.6** $2.5 \times 10^{24}$ g. **3.16** 3.97152. **3.19** 82.6. **3.23** $^{63}$Cu 69.169%, $^{65}$Cu 30.831%. **3.28** (a) 187 g; (b) 1016 g. **3.34** 31.0. **3.40** 0.020 g. **3.48** 8.60 cents/mol N. **3.55** H$_3$PO$_3$. **3.60** QZ$_2$, Q$_2$Z$_5$, 7:8. **3.65** 73. **3.69** C$_3$H$_4$O$_2$. **3.74** 0.745 mol NO$_2$. **3.80** 0.817 g UF$_6$. **3.87** 0.333 g PH$_3$. **3.91** 1.01 g HI. **3.94** Ca$_3$(PO$_4$)$_2$. **3.100** 59.5. **3.105** 20.4 g KCl. **3.108** 0.0666 mol AgBr. **3.116** 1.28 g/cm$^3$. **3.123** 74.2 g PCl$_3$. **3.124** 11.8 g AgI. **3.125** 0.4 g C$_3$H$_8$.

### Chapter 4

**4.3** (a) increases; (b) decreases; (c) increases; (d) remains the same. **4.9** 550 cm$^3$. **4.16** 3160 mmHg. **4.22** 1:3 by amount; 1:4.9 by mass. **4.23** 1.5 atm. **4.27** 280 K. **4.34** 0.46 g/L. **4.37** 429 L O$_2$. **4.41** 0.390 g Ar. **4.46** 0.72 atm H. **4.51** 3700 J. **4.55** 0.91. **4.59** 1.48 g/L. **4.63** $3.6 \times 10^{12}$ K. **4.67** $1.5 \times 10^{-4}$ L at STP; $3.0 \times 10^{-2}$ L at 200 atm.

### Chapter 5

**5.6** 50.537. **5.13** $9.7 \times 10^{-12}$ m. **5.19** $3.3 \times 10^{15}$ s$^{-1}$. **5.24** $1.7 \times 10^{24}$. **5.30** $7.03 \times 10^5$ m/s. **5.36** $1.8 \times 10^{-38}$ m. **5.41** If the energy is known very accurately, the uncertainty in time is great. Thus, we do not know when the electron has the given energy.

### Chapter 6

**6.3** 10. **6.8** 1$s$; 2$s$; 2$p$, 3$s$; 3$p$, 3$d$, 4$s$; 4$p$, 4$d$, 5$s$; 5$p$, 4$f$, 5$d$, 6$s$; 6$p$, 5$f$, 6$d$, 7$s$. **6.14** 5$g$, 6$f$, 7$d$, 8$s$, 8$p$; 50 electrons; 168. **6.19** (a) antimony; (b) germanium and selenium; (c) molybdenum; (d) cesium, platinum, gold. **6.22** (a) [Kr]; (b) [Ar]3$d^{10}$4$s^2$4$p^4$; (c) [Ar]3$d^{10}$4$s^2$ (d) [Ne]; (e) [He]. **6.26** 120. **6.33** [Rn]5$f^{14}$6$d^4$7$s^2$ or [Rn]5$f^{14}$6$d^5$7$s^1$. **6.34** (a) phosphorus; (b) gallium; (c) cobalt; (d) argon. **6.40** The increase in nuclear charge from barium to radium is relatively large, 32. The extra electrons are $f$ electrons, which do not screen the nucleus very well from the outer electrons. **6.44** (a) N$^{3-}$, O$^{2-}$, F$^-$, Ne, Na$^+$, Mg$^{2+}$; (b) same; (c) Mg$^{2+}$, Na$^+$, Ne, F$^-$, O$^{2-}$, N$^{3-}$. **6.50** 295 K, 950 K, 85 cm$^3$/mol; (a) H$_2$(g) + Fr$^+$; (b) FrCl; (c) Fr$_2$O. All reactions will be violent. **6.54** (a) $K_2O(s) + H_2O \rightarrow 2K^+(aq) + 2OH^-(aq)$; (b) $BaO(s) + H_2O \rightarrow Ba^{2+}(aq) + 2OH^-(aq)$; (c) $CO_2(g) + H_2O \rightarrow H_2CO_3(aq)$; (d) $SO_3(g) + H_2O \rightarrow H^+(aq) + HSO_4^-(aq)$. **6.57** (a) One if by land; (b) Atoms are fun.

### Chapter 7

**7.5** (a) Rb$^+$ and Cl$^-$; (b) Na$^+$ and Cl$^-$; (c) Ca$^{2+}$ and O$^-$. **7.10** $5.8 \times 10^{-9}$ N. **7.15** (a) (ii); (b) (i); (c) (iii). **7.18** The extra lattice energy in O$^{2-}$ solids that results from the increased charge more than offsets the extra energy needed to add an electron to O$^-$. The addition of an electron to O$^{2-}$, however, to form O$^{3-}$ is exceedingly unfavorable because the extra electron has to go into the next

# Appendix VII  Answers to Selected Exercises (continued)

noble gas shell. **7.22** (a) ionic; (b) polar; (c) nonpolar; (d) nonpolar; (e) polar.

**7.28** (a) H—Ö—Cl:;

(b) H—S̈—S̈—H; (d) H—N̈=Ö.

**7.31** (a) [:Ö—Ö:]⁻; (b) [:C≡N:]⁻;

(d) H—C—Ö:⁻.
          ‖
         :O:

**7.35** (structure) C=C—Ö—H with H's

(structure) H—C—C=Ö with H's

(structure) C—C with H's and :O:

**7.38** (a) and (c). **7.42** (b) and (c).

**7.46** Br—C≡N: and :C̄=N⁺—Br. The first one is more likely to be correct.

**7.51** (structure) C=N⁺=N̄:

(structure) C̄—N⁺≡N:

**7.53** (a) +3; (b) +6; (c) +6; (d) +2.

**7.58** [:S̈—S̈—Ö:]²⁻ with O above and below

## Chapter 8

**8.5** (a) ideal; (b) other; (c) other; (d) close; (e) other. **8.8** (b) and (c).
**8.12** (a) octahedral; (b) octahedral; (c) trigonal bipyramidal; (d) trigonal

bipyramidal.

**8.14** (structures F—S—F with I's and F)

The first is likely to be more stable because the large I atoms are further apart. **8.19** The electron pairs of the bonds are very close to the F atoms. Therefore the interactions with the nonbonding electron pairs are less serious and the $OF_2$, which has only two F atoms, can spread out to keep the F atoms apart. **8.25** C—H bonds 1s of H with $2sp^3$ of C; C—Cl $2sp^3$ of C with $3p$ of Cl. **8.36** (a) $sp^3$; (b) $sp^3$; (c) $sp^3$; (d) $sp^2$.

**8.40** (structure) C=N⁺=N̄⁻ with H's

**8.44** (a) trigonal bipyramidal; (b) the four F atoms are at the two axial and two of the three equatorial positions of a trigonal bipyramid; (c) T-shaped; (d) linear. **8.49** There is no simple or obvious explanation we can think of. It has been suggested that the inner electrons of the barium may influence the geometry.
**8.53** $CHF_3$, $CHCl_3$, $CHBr_3$, $CHI_3$.
**8.58** (a) same as $C_2$, diamagnetic; (b) same as $O_2$, paramagnetic; (c) same as $Ne_2$, diamagnetic; (d) same as $O_2$, paramagnetic. **8.63** 45 000 L.
**8.67** −1132 kJ. **8.72** 56.1 kJ.
**8.76** 2244 kJ. **8.80** 536 kJ.
**8.85** 2162 kJ. **8.86** 126 kJ/mol.

## Chapter 9

**9.4** (a) $CH_4(g) + H_2O(g) \rightarrow 3H_2(g) + CO(g)$; (b) $2H^+(aq) + Mg(s) \rightarrow H_2(g) + Mg^{2+}(aq)$; (c) $2Na(s) + 2H_2O \rightarrow 2Na^+(aq) + 2OH^-(aq) + H_2(g)$; (d) $LiH(s) +$

$H_2O \rightarrow H_2(g) + Li^+(aq) + OH^-(aq)$.

**9.8** (structure of diborane-like B-H bridged molecule)

**9.11** 692 kJ/mol. **9.16** 86.2 L of $CO_2$. **9.20** $NH_2OH$ is more stable because $HN_3$ readily evolves nitrogen. **9.25** (a) $HNO_3$; (b) $HNO_2$; (c) HNO. **9.32** (a) $P_4O_6(s) + 6H_2O \rightarrow 4H_3PO_3(aq)$; (b) $P_4(s) + 3O_2(g) \rightarrow P_4O_6(s)$; (c) $PCl_3(l) + 3H_2O \rightarrow 3H^+(aq) + 3Cl^-(aq) + H_3PO_3(aq)$; (d) $PBr_5(s) + 4H_2O \rightarrow 5H^+(aq) + 5Br^-(aq) + H_3PO_4(aq)$.
**9.36** $P_4(s) + 6Br_2(l) \rightarrow 4PBr_3(l)$, $2PBr_3(l) + O_2(g) \rightarrow 2POBr_3(l)$.
**9.41** (a) $2Na(s) + H_2O_2(l) \rightarrow Na_2O_2(s) + H_2(g)$; (b) $K(s) + H_2O_2(l) \rightarrow KO_2H(s) + \frac{1}{2}H_2(g)$; (c) $4KO_2(s) + 2CO_2(g) \rightarrow 2K_2CO_3(s) + 3O_2(g)$. **9.49** S̈=S̈=Ö, :S≡S̈—Ö:, :S̈—S≡O:. **9.53** $H_2SO_4$ and $H_2SO_3$. **9.61** (a) Linear; (b) square planar; (c) distorted octahedron.
**9.66** The increase in boiling point with increase in atomic number is due to increasing polarizability and London forces with increase in the number of electrons.

## Chapter 10

**10.6** 33 300 J. **10.12** (a) From solid to vapor at just below 195 K; (b) from solid to vapor at 195 K; (c) from solid to triple point at 216 K to vapor above 216 K; (d) from solid to liquid just above 216 K to vapor at about 260 K; (e) from solid to liquid at around 220 K, at 304 K the critical point the liquid and vapor become identical. **10.19** 39.0 cm³/mol.

**10.25**   0.1387 nm. **10.33**   184.2.
**10.38**   0.0535 nm.

## Chapter 11

**11.5**   (a); (b); (e).

**11.9**

$$H-\overset{\displaystyle H}{\underset{\displaystyle H}{N^+}}-H \quad [F\text{---}H\text{---}F]^-$$

**11.14**   $H-O-\overset{\displaystyle \|}{C}-\overset{\displaystyle \|}{C}=O$

with $O$ and $O$ bridged by $H$ below.

**11.20**   Lithium fluoride is less soluble in water than lithium chloride because it has a very large lattice energy. Cesium fluoride is more soluble than cesium chloride because the fluoride anion has more solvation energy than the chloride anion. In cesium salts where the cation is relatively large, the lattice energy does not vary as much with changes in the anion. **11.25**   $2.9 \times 10^7$ g of Ra.
**11.30**   0.49 L of $O_2$.

## Chapter 12

**12.5**   33.2%. **12.11**   10.9 g of urea.
**12.18**   2500 L of water.
**12.24**   $0.30M$. **12.31**   $1.74M$.
**12.36**   0.831 L of solution.
**12.43**   0.46 L of $CO_2$. **12.51**   0.140 atm for isopropyl and 0.0735 atm for propyl. **12.55**   0.359 atm.
**12.61**   The $1.00m$ solution should show a freezing point depression 100 times that of the $0.01m$ solution. It shows less because of the association of the $CH_3NH_2$ at high concentrations due to hydrogen bonding.
**12.68**   $5.6m$ in NaCl. **12.74**   $MF_5$.

## Chapter 13

**13.5**   1130 J. **13.12**   $-780$ J.

**13.19**   55.2 kJ/g $CH_4$. **13.28**   354 K.
**13.34**   293 K. **13.40**   (a) +; (b) +;
(c) +; (d) +. **13.47**   616 K.
**13.51**   $-529$ kJ/mol. **13.60**   (a) $5.0 \times 10^{-17}$; (b) $6.3 \times 10^{29}$; (c) $1.1 \times 10^{48}$. **13.65**   $3.7 \times 10^{-3}$ at 600 K; $3.7 \times 10^{-5}$ at 800 K.

## Chapter 14

**14.4**   (a) $2SO_2(g) + O_2(g) \rightleftharpoons 2SO_3(g)$;
(b) $F_2(g) + I_2(s) \rightleftharpoons 2IF(g)$; (c) $3HNO_2(aq) \rightleftharpoons H^+(aq) + NO_3^-(aq) + 2NO(g) + H_2O$; (d) $Ca^{2+}(aq) + 2F^-(aq) \rightleftharpoons CaF_2(s)$. **14.11**   (a) right to left; (b) left to right; (c) right to left; (d) left to right. **14.19**   (a) decrease; (b) increase; (c) decrease.
**14.26**   $2ClF_3(g) \rightleftharpoons Cl_2(g) + 3F_2(g)$;
0.22. **14.32**   $1.45 \times 10^{-3}$.
**14.38**   $P_{CS_2} = 1.15$ atm, $P_{S_2} = 0.13$ atm; same. **14.43**   $1.67 \times 10^{-3}$ atm.
**14.51**   $P_{CH_4} = 0.35$ atm, $P_{H_2} = 0.20$.
**14.56**   (a) $7.2 \times 10^{-7}$ mol/L; (b) $3.5 \times 10^{-4}$ mol/L; (c) $1.6 \times 10^{-5}$ mol/L. **14.62**   $[CN^-] = 2.3 \times 10^{-6}M$.
**14.70**   1.04 L. **14.73**   $3.7 \times 10^{-2}M$.

## Chapter 15

**15.5**   (c), (d), and (e). **15.11**   $HNO_3 + H_2SO_4 \rightleftharpoons HSO_4^- + H_2NO_3^+$;

$$H-\overset{..}{\underset{..}{O}}-\overset{+}{N}-\overset{..}{\underset{..}{O}}-H.$$
$$\overset{\displaystyle \|}{:O:}$$

$$H-O-N-O-H.$$

**15.17**   $NH_2NH_2(aq) + H_2O \rightleftharpoons NH_2NH_3^+(aq) + OH^-(aq)$; (b) $C_{10}H_{15}NO(aq) + H_2O \rightleftharpoons C_{10}H_{16}NO^+(aq) + OH^-(aq)$; (c) $C_3H_5O_3^-(aq) + H_2O \rightleftharpoons C_3H_6O_3(aq) + OH^-(aq)$; (d) $F^-(aq) + H_2O \rightleftharpoons HF(aq) + OH^-(aq)$.
**15.24**   Trimethylammonium ion is poorly solvated due to the bulky methyl groups around the positively charged N. It is therefore relatively unstable and thus a stronger acid than methylammonium ion in

solution. In the gas phase the trimethylammonium ion is more stable because of electron release by the methyl groups which helps delocalize the positive charge.
**15.30**   (a) $2.0 \times 10^{-7}M$; (b) $5.0 \times 10^{-8}M$; (c) $0.56M$; (d) $7.4M$.
**15.36**   $[OH^-] = 7.1 \times 10^{-6}M$
**15.43**   0.071 mol. **15.50**   0.13 mol of HF. **15.55**   $1.52 \times 10^{-3}$ mol.
**15.58**   $0.173M$. **15.61**   12.6%
**15.67**   (a) 1.48; (b) 2.28; (c) 3.30; (d) 4.30; (e) 9.70. **15.68**   (a) 9.25; (b) 10.21; (c) 10.86; (d) 10.98; (e) 10.99.
**15.75**   0.83 g of lactic acid for 1 g of calcium lactate. **15.82**   (a) 7.21; (b) 12.33. **15.86**   $[CO_2] = 1.7 \times 10^{-3}M$; $[HCO_3^-] = 2.3 \times 10^{-2}M$.
**15.90**   $K = 1.30 \times 10^{-4}$, pH = 8.33.
**15.91**   0.50 mol/L. **15.92**   $4 \times 10^{-17}$ mol/L. **15.93**   $5.5 \times 10^{-25}$ at pOH = 7; 0.55 mol/L at pOH = 15.
**15.94**   $K = 3.9 \times 10^{-5}$; solubility = $7.0 \times 10^{-3}$ mol/L.

## Chapter 16

**16.7**   (a) $2HClO(aq) \rightarrow Cl^-(aq) + HClO_2(aq) + H^+(aq)$; (b) $5MnO_2(s) + 4H^+(aq) \rightarrow 3Mn^{2+}(aq) + 2H_2O + 2MnO_4^-(aq)$. **16.12**   0.367 g of $As_2O_3$. **16.17**   2020 s. **16.22**   649 s.
**16.32**   $2Cu^+(aq) \rightleftharpoons Cu(s) + Cu^{2+}(aq)$, $-35.1$ kJ. **16.36**   0.16 V. **16.42**   (a) 0.388 V; (b) 1.9 V. **16.49**   $ClO_3^-$(alkaline), $H_2O_2$. **16.56**   0.22 V.
**16.57**   0.45 mol/L. **16.61**   $\mathscr{E}° = 0.65$ V, $\Delta G° -251$ kJ, $K = 9.3 \times 10^{43}$.
**16.67**   471. **16.71**   $1.6 \times 10^{-6}$.
**16.79**   (1) $Sn^{4+} + H_2(g) \rightarrow Sn^{2+} + 2H^+$, $\mathscr{E}° = 0.15$ V, $\Delta G° = -30$ kJ; (2) $2Cu^{2+} + Sn^{2+} \rightarrow 2Cu^+ + Sn^{4+}$, $\mathscr{E}° = 0.01$ V, $\Delta G° = -2$ kJ; (3) $2Cu^+ + 2Fe^{3+} \rightarrow 2Cu^{2+} + 2Fe^{2+}$, $\mathscr{E}° = 0.61$ V, $\Delta G° = -120$ kJ; (4) $2Fe^{2+} + Au^{3+} \rightarrow 2Fe^{3+} + Au^+$, $\mathscr{E}° = 0.52$ V, $\Delta G° = -100$ kJ; (5) $Au^+ + \frac{1}{2}O_2(g) + 2H^+ \rightarrow Au^{3+} + H_2O$,

$\mathcal{E}° = -0.06$ V, $\Delta G° = 10$ kJ. Overall, $H_2(g) + \frac{1}{2}O_2(g) \rightarrow H_2O$.

## Chapter 17

**17.5**  $F_2$, $2.2 \times 10^{-3}$ atm/sec; $Br_2$, $4.5 \times 10^{-4}$ atm/sec. **17.11**  $7.41 \times 10^{-4}$ atm/sec; $5.88 \times 10^{-4}$ atm/sec; total, $4.12 \times 10^{-4}$ atm/sec. **17.17**  Rate $= kP_{H_2C_2O_4}$. **17.24**  $3.20 \times 10^{-4}$ sec$^{-1}$. **17.25**  (a) 41 sec; (b) 172 sec; (c) 0.171 atm. **17.30**  23 200 yr. **17.36**  (a) $kP_{COCl}P_{Cl}$; (b) $kP_{ClO}^2$; (c) $kP_{NO}^2 P_{I_2}$; (d) $kP_{C_4H_8}$. **17.41**  $k(K^{1/2})P_{Br_2}^{1/2}P_{H_2}$. **17.46**  52 kJ/mol. **17.50**  The activation energy is negative. In the transition state, two unpaired electrons, one from each NO, start pairing in bonds. **17.54**  $4.55 \times 10^{-7}$ L mol$^{-1}$ sec$^{-1}$.

## Chapter 18

**18.4**  (a) 6; (b) 6; (c) 6. **18.10**  (a) $Na_2[PtCl_6]$; (b) $K_2[Ni(CN)_4]$; (c) $[Cu(NH_3)_4]Br_2$. **18.17**  $[Ru(H_2O)_6]Cl_4$; $[Ru(H_2O)_5Cl]Cl_3H_2O$; $[Ru(H_2O)_4Cl_2]Cl_2(H_2O)_2$; $[Ru(H_2O)_3Cl_3]Cl(H_2O)_3$; $[Ru(H_2O)_2Cl_4](H_2O)_4$. **18.22**  (a) one; (b) one; (c) one; (d) three. **18.27**  (a). **18.33**  (a) $[Xe]4f^{14}5d_{xy}5d_{xz}5d_{yz}5d_{z^2}$; (b) $[Xe]4f^{14}5d_{xy}5d_{xz}5d_{yz}5d_{z^2}5d_{x^2-y^2}$; (c) $[Kr]4d_{xy}^2 4d_{xz}4d_{yz}4d_{z^2}4d_{x^2-y^2}$; (d) $[Xe]4f^{14}5d_{xy}^2 5d_{xz}^2 5d_{yz}5d_{z^2}5d_{x^2-y^2}$. **18.38**  $3.43 \times 10^{-19}$ J. **18.41**  Mix two complexes such as $MCl_6^{3-}$ and $MBr_6^{3-}$ in solution and measure the rate of ligand exchange. **18.45**  0.94 mol/L.

## Chapter 19

**19.3**  164 kJ. **19.6**  The nuclear charge of aluminum ($Z = 13$) is greater than that of magnesium ($Z =$

12), so the electron has more to gain in Al by being closer to the nucleus in the $s$ orbital. Also, since in Al there already is an electron in the $p$ subshell, the exciting of a second electron into the subshell generates electron-electron repulsions. **19.13**  (a) White tin, it is close-packed, the diamond structure is more open; (b) white tin; the covalently bonded diamond structure has a large $\Delta E_0$ between bands and is a poor conductor; $5sp^3$ hybrids for diamond in gray tin, unhybridized $5s^25p^2$ in close-packed white tin. **19.19**  $4.3 \times 10^7$. **19.23**  (a) +1; (b) −2; (c) +6; (d) −2. **19.28**  $6.3 \times 10^{-3}$.

## Chapter 20

**20.4**  (a) $^{26}_{12}Mg$, $^{27}_{13}Al$, $^{28}_{14}Si$, $^{29}_{15}P$, $^{30}_{16}S$. **20.10**  $1.0785062 \times 10^6$ J. **20.13**  7.71 MeV. **20.19**  $2.0 \times 10^{-13}$ J. **20.25**  0.0001675 u; 0.156 MeV. **20.30**  $1.75 \times 10^6$ J/mol. **20.34**  The mass difference between the two nuclides is less than the total mass lost in the annihilation of an electron and a positron. **20.42**  $^{197}_{79}Au + ^7_3Li \rightarrow ^{204}_{82}Pb$, $^{204}_{82}Pb + 4^1_0n \rightarrow ^{208}_{82}Pb$; $^{197}_{79}Au + 2^4_2He \rightarrow ^{205}_{83}Bi \rightarrow ^{205}_{82}Pb + ^0_1\beta$; $^{205}_{82}Pb + 3^1_0n \rightarrow ^{208}_{82}Pb$. **20.45**  −24.7 MeV. **20.52**  0.7 MeV; no coulombic barrier for collision with a neutron.

## Chapter 21

**21.1**  (b); (c). **21.8**  400 K. **21.13**  (c). **21.19**  (a) $BrCH_2CH_2Br$; (b) $CH_3CH_2Cl$; (c) Cl
|
$\overset{|}{C}CH_3$
|
$CH_3$

(d) $CH_3(CH_2)_6CH_3$. **21.24**  0.5 mol of 2,2,3,3-tetrachloropentane and 0.5

mol of 2-pentyne remain unreacted. **21.28**  $HC\equiv C-\overset{\displaystyle CH_3}{\underset{\displaystyle |}{C}}HCH_3$

$CH_2=C=C\overset{\displaystyle CH_3}{\underset{\displaystyle CH_3}{<}}$

$CH_2=CH-C\overset{\displaystyle CH_2}{\underset{\displaystyle CH_3}{<}}$

plus isomers of Exercise 21.22. **21.32**  (a) Nonplanar, a tublike structure:

(b) It has eight $\pi$ electrons rather than six. **21.33**  It has six $\pi$ electrons in a cyclic conjugated system. **21.38**  $CH_3CH_2CH_2OH + H^+ \rightarrow CH_3CH=CH_2 + H_2O$, $CH_3CH=CH_2 + KMnO_4 \rightarrow$

$CH_3COOH + HCOOH.$

**21.42** $H_2C{=}C{-}CH_3 \leftrightarrow$
$\quad\quad\quad\quad \underset{O^-}{|}$

$H_2C^-{-}\underset{\underset{O}{\|}}{C}{-}CH_3$ Protonation of the

structure on the left gives the enol, of the structure on the right gives the ketone.

**21.46** $CH_3{-}\underset{\underset{CH_2CH_3}{|}}{N}{-}CH_3$

$CH_3CH_2\underset{\underset{H}{|}}{N}CH_2CH_3$

$CH_3CH_2CH_2\underset{\underset{H}{|}}{N}CH_3$

$\underset{CH_3}{\overset{CH_3}{\diagdown}}CHNCH_3$

$CH_3CH_2CH_2CH_2NH_2$

$CH_3CH_2\underset{\underset{NH_2}{|}}{C}HCH_3$

$CH_3\underset{\underset{CH_3}{|}}{C}HCH_2NH_2$

$CH_3\underset{\underset{CH_3}{|}}{\overset{\overset{NH_2}{|}}{C}}CH_3$

**21.49** $C_{17}H_{19}NO_3.$

**Appendix VI**

**VI.1** 9.78. **VI.4** 0.046 mol/L.
**VI.6** $AgCl(s) + Br^-(aq) \rightleftharpoons$
$AgBr(s) + Cl^-(aq)$, $K = 3.1 \times 10^2.$
**VI.10** $6.0 \times 10^{-3}M.$
**VI.13** $[Ag(CN)_2^-] = [H^+] = 2.7 \times 10^{-3}M$, $[HCN] = 0.10M$, $[OH^-] = 3.7 \times 10^{-12}M$, $[CN^-] = 1.8 \times 10^{-8}M$, $[Ag^+] = 6.6 \times 10^{-9}M.$

# Glossary

**Absolute entropy** *(13.6)*  The value of the entropy of a substance at a given temperature. See also Entropy.

**Absolute zero** *(4.3)*  The zero point on the Kelvin temperature scale. The temperature at which the volume of an ideal gas is zero and at which all thermal motion ceases.

**Acid, Arrhenius** *(15.1)*  A substance that releases hydrogen ions ($H^+$) in aqueous solution.

**Acid, Brønsted** *(15.1)*  A substance that can donate a proton.

**Acid, Lewis** *(15.1)*  A substance that can accept an electron pair to form a chemical bond.

**Acid-base indicator** *(15.6)*  A substance whose color in solution changes as the pH of the solution changes. Usually an organic molecule of complex structure.

**Acid dissociation constant** *(15.2)*  An expression of the strength of acids in water, represented by the symbol $K_A$.

**Activated complex** *(17.4)*  The highest energy state between the reactants and products in an elementary reaction.

**Activation energy** *(17.4)*  The difference between the mean energy of all the collisions of reactant molecules and the average energy of collisions in which reaction takes place.

**Adiabatic process** *(13.2)*  A thermodynamic process in which no heat flows into or out of a system.

**Adsorption** *(17.6)*  The process by which a substance sticks to a solid surface.

**Alcohol** *(21.5)*  An organic compound that contains the —OH functional group.

**Aldehyde** *(21.5)*  An organic compound that has the carbonyl functional group whose carbon atom bears a hydrogen atom (C—H).
$$\overset{\|}{\underset{O}{}}$$

**Alkali metal** *(6.6)*  An element in column IA of the periodic table. The alkali metals are Li, Na, K, Rb, and Cs.

**Alkaline-earth metal** *(6.6)*  An element in column IIA of the periodic table. The alkaline-earth metals are Be, Mg, Ca, Sr, and Ba.

**Alkane** *(21.1)*  A hydrocarbon whose structure has no double or triple bonds.

**Alkene** *(21.2)*  A hydrocarbon with a carbon-carbon double bond.

**Alkyne** *(21.2)*  A hydrocarbon with a triple bond.

**Allotropes** *(9.2)*  Different forms of the same element, especially in the same phase. For example, diamond and graphite are allotropes of carbon.

**Alloy** *(19.5)*  A solid solution of two or more metals.

**Alpha decay** *(20.2)*  The emission of an alpha particle ($^4_2He^{2+}$) from an atomic nucleus.

**Amalgam** *(19.4)*  A solution of a metal in mercury.

**Ambidentate ligand** *(18.2)*  A ligand that can complex with a metal atom in two or more ways.

**Amorphous solid** *(10.3)*  A material that appears to have the mechanical properties of a solid but does not have the regular crystalline structure of a solid. Also called a glass.

**Ampere** *(16.3)*  The unit of electrical current. One ampere is the flow of one coulomb of charge for one second past a given point.

**Amphoteric compound** *(15.1)*  A substance that can either donate or accept a proton and thus can act either as an acid or a base.

**Angular momentum quantum number ($\ell$)** *(6.1)*  The quantum number that determines the shape of an orbital and the angular momentum of an electron occupying that orbital. It can have integral values from 0 to $n - 1$, where $n$ is the principal quantum number.

**Anion** *(2.1)*  An atom or group of atoms that carries a negative electric charge.

**Anode** *(16.3)*  The electrode at which oxidation takes place.

**Antibonding molecular orbital** *(8.3)*  One of the orbitals that is formed when two atomic orbitals overlap in a covalent bond. It is of higher energy than either of the atomic orbitals. See also Bonding molecular orbital.

**Aromatic hydrocarbon** *(21.4)*  A hydrocarbon whose structure has one or more benzene rings.

**Arrhenius equation** *(17.4)*  An empirical expression of the relationship between activation energy, the temperature, and the specific rate constant of a reaction.

**Atmospheric pressure** *(4.2)*  The pressure exerted by the air. At sea level, atmospheric pressure is 1 atm or

101 325 Pa, enough to support a column of mercury 0.76 m high.

**Atom** *(2.1)*   The basic unit of an element. It has a small but massive nucleus, which is surrounded by electrons.

**Atomic mass unit (u)** *(3.1)*   A unit used to express the relative mass of atoms. It is equal to one-twelfth the mass of a single atom of $^{12}C$.

**Atomic number** *(5.4)*   The number of protons in the nucleus of an atom.

**Atomic weight** *(3.1)*   The mass of an atom relative to $^{12}C$, expressed as a weighted average that reflects the natural abundances of the isotopes of each element.

**Aufbau principle** *(6.2)*   A method of working out the electronic configurations of atoms. Each succeeding electron is placed in the lowest-energy hydrogenlike orbital that is available.

**Autoprotolysis** *(15.4)*   A process in which a proton is transferred from one molecule of an amphoteric substance to another.

**Average bond energy** *(8.7)*   The quantity of energy needed to break a bond between atoms in a substance that is in the gas phase. It can be obtained only from complete dissociation of the gaseous substance to gaseous atoms. See also Bond dissociation energy.

**Avogadro's law** *(4.3)*   A relationship that states that equal volumes of different gases under the same conditions of temperature and pressure contain equal numbers of molecules.

**Avogadro's number $(N_A)$** *(3.1)*   The number of entities in 1 mol of a substance. It is $6.0221 \times 10^{23}$.

**Base, Arrhenius** *(15.1)*   A substance that produces hydroxide ions ($OH^-$) in aqueous solution.

**Base, Brønsted** *(15.1)*   A substance that can accept a proton.

**Base, Lewis** *(15.1)*   A substance that has a nonbonding valence electron pair that can be used to form a chemical bond. More simply, an electron pair donor.

**Base dissociation constant** *(15.3)*   An expression of the strengths of bases in aqueous solution, represented by the symbol $K_B$.

**Beta decay** *(20.2)*   A radioactive decay process in which there is a change in the atomic number of an atom resulting from emission of an electron or a positron from the nucleus.

**Biochemical oxygen demand (BOD)** *(11.5)*   The rate at which the microbial population of a body of water consumes oxygen.

**Blackbody radiation** *(5.5)*   The spectrum of radiation emitted at elevated temperatures by a blackbody, which reflects none of the light falling on it.

**BOD** *(11.5)*   See Biochemical oxygen demand.

**Body-centered lattice** *(10.4)*   A crystal structure in which

each unit cell has a lattice point in the center, in addition to the eight lattice points at the corners of the cell.

**Boiling curve** *(10.2)*   A curve representing values of pressure and temperature at which the vapor and liquid of a substance are in equilibrium.

**Boiling point** *(4.2, 10.2)*   The temperature at which the vapor pressure of a liquid is equal to the external pressure on the liquid.

**Bond angle** *(8.1)*   The angle formed by two lines connecting a central atom with two atoms that are bonded to it.

**Bond dissociation energy** *(8.7)*   The quantity of energy required to break one bond in a molecule. See also Average bond energy.

**Bond length** *(8.1)*   The distance between the centers of two bonded atoms.

**Bond order** *(8.5)*   The number of electrons shared between two bonded atoms.

**Bonding molecular orbital** *(8.3)*   One of the orbitals that is formed when two atomic orbitals overlap in a covalent bond. It is of lower energy than either of the two atomic orbitals. See also Antibonding molecular orbital.

**Born-Haber cycle** *(8.7)*   A method of analyzing the formation of an ionic solid as a series of thermochemical steps.

**Boyle's law** *(4.3)*   An observation, published by Robert Boyle in 1662, that states that at constant temperature the volume of a sample of gas is inversely proportional to its pressure.

**Buffer capacity** *(15.7)*   The amount of an acid or a base that can be added to a buffer solution without changing its pH appreciably.

**Buffer solution** *(15.7)*   A solution that is relatively resistant to changes in pH.

**Calorie** *(1.5)*   A non-SI unit of energy, equal to 4.184 J.

**Calorimeter** *(13.3)*   A device for measuring the heat changes that accompany physical and chemical changes.

**Capillary action** *(10.1)*   The force that operates to raise the level of a liquid inside a tube of small diameter.

**Carboxylic acid** *(21.5)*   An oxygen-containing functional group in organic compounds. Its structure is $-\overset{\displaystyle \|}{\underset{\displaystyle O}{C}}-OH$.

**Catalyst** *(17.6)*   A substance that increases the rate of a reaction without being consumed by the reaction.

**Catenation** *(9.2, 21.2)*   The tendency of atoms to form long chains or rings with other atoms of the same element.

**Cathode** *(16.3)*   The electrode at which reduction takes place.

**Cation** *(2.1)*   An atom or group of atoms that carries a positive electric charge.

**Chain isomers** *(21.1)*   Organic compounds that are identical in chemical composition but differ in the structure of their chains of carbon atoms.

**Chain reaction** *(9.1)*   A self-sustaining reaction whose products contribute to the continuation of the reaction.

**Charles and Gay-Lussac, law of** *(4.3)*   The observation that the volume of a sample of an ideal gas is directly proportional to its absolute temperature at constant pressure. The volume of an ideal gas is reduced by 1/273 for every reduction of 1 K in temperature.

**Chelate effect** *(18.1)*   The increased stability that occurs when polydentate ligands form a ring structure with a metal in a coordination complex.

**Chirality** *(18.2)*   A form of isomerism in which two molecules differ only in their symmetry. Chiral molecules have nonidentical mirror images, called optical isomers.

**Cis isomer** *(21.2)*   An isomer that has two identical substituents on the same side of a double bond. See also Trans isomer and Geometrical isomers.

**Clathrate** *(11.3)*   A compound in which one component is enclosed in a cage structure formed by the molecules of the other component. Many substances form clathrate hydrates in which a molecule of the substance is enclosed in a cage made of water molecules.

**Colligative property** *(12.5)*   Any property of a solution that depends only on the number of solute particles, rather than on the nature of the particles.

**Colloid** *(12.8)*   A mixture of substances whose particles are larger than those of a solute but smaller than those of a suspension.

**Common ion effect** *(14.5)*   The decrease in the solubility of an ionic solid that is caused by the presence of one or more of its ions in the solution.

**Compound** *(1.2)*   A pure substance, composed of two or more elements in definite proportions, that can be decomposed by simple chemical change.

**Concentration** *(3.4, 12.2)*   The relative quantities of solute and solvent in a solution.

**Condensation** *(4.2)*   The process by which a vapor becomes a liquid or a solid.

**Condensed phase** *(4.1)*   Matter in the liquid or the solid state.

**Conjugate acid-base pair** *(15.1)*   An acid and the base that is derived by the removal of a proton from the acid.

**Conservation of energy** *(13.2)*   The law that states that energy cannot be destroyed or created but can only be transferred from place to place or transformed from one form to another. Also called the first law of thermodynamics.

**Conservation of mass** *(2.4)*   The law that states that mass cannot be created or destroyed. The law applies to ordinary chemical reactions but not to nuclear reactions, in which mass may be transformed to energy or vice versa.

**Constant composition, law of** *(2.2)*   The law that all samples of a pure substance contain the same elements in the same proportions.

**Coordinate-covalent bond** *(7.3)*   A covalent bond in which one atom appears to contribute both of the bonding electrons. Also called a dative bond.

**Coordination complex** *(18.1)*   A compound consisting of a metal atom and one or more ligands.

**Coordination number** *(11.3, 18.1)*   In a lattice, the number of nearest neighbors of each ion. In a coordination compound, the number of ligands around the metal atom.

**Coordination sphere** *(18.1)*   The close association of a metal atom and its ligands in a coordination compound. See also Counter ion.

**Corrosion** *(16.8)*   The deterioration of metals caused by chemical reactions on their surfaces.

**Coulomb's law** *(5.2, 7.1)*   The observation that the force of attraction or repulsion between two electrical charges is inversely proportional to the square of the distance between them and directly proportional to the magnitude of the charges.

**Counter ion** *(18.1)*   An ion that is outside the coordination sphere of a metal in a coordination complex. It simply maintains electrical neutrality.

**Covalent bond** *(7.2)*   A chemical bond that is characterized by the sharing of the bonding electrons between the bonded atoms.

**Covalent network solid** *(10.3)*   A crystal structure in which the forces of attraction are chemical bonds between the atoms in the crystal.

**Critical mass** *(20.4)*   The minimum amount of material needed to sustain a nuclear chain reaction.

**Critical pressure** *(4.12)*   The pressure that must be applied to liquefy a gas at its critical temperature.

**Critical temperature** *(4.12)*   The temperature above which a gas cannot be liquefied by an increase in pressure.

**Crystalline state** *(10.3)*   The arrangement of the units of structure of a solid in a regularly repeating three-dimensional array. Most true solids occur in the crystalline state.

**Cubic closest packing** *(10.5)*   An arrangement in which each atom of a solid is surrounded by 12 identical and equidistant atoms, called nearest neighbors.

**Dalton's law of partial pressures** *(4.5)*   The pressure exerted by each gas in a mixture of gases is the same as if the gas were present alone. An alternate statement is that the total pressure of the mixture is the sum of the partial pressures of all the gases.

**Dative bond** *(7.4)*   See Coordinate-covalent bond.

**de Broglie wavelength** *(5.8)*   The wavelength that is associated with any moving particle.

**Degenerate orbitals** *(8.6)*   Orbitals of equal energy.

**Dehydrating agent** *(9.4)*   A substance used to remove water from other substances.

**Density** *(3.4)*   The mass of a substance divided by its volume.

**Diamagnetism** *(6.2, 18.3)*   The phenomenon in which a substance is slightly repelled by a magnetic field. See also Paramagnetism.

**Dielectric constant** *(11.3)*   A measure of the ability of a substance to dissipate an applied electric field. It is related to the polarity of the substance.

**Diffraction pattern** *(10.6)*   The wave interference patterns that are formed when electromagnetic radiation passes through a crystalline structure.

**Diffusion** *(4.7)*   The process by which a gas or any other substance spreads through another substance. See also Effusion.

**Dipole, bond** *(7.3, 8.5)*   The pair of opposite charges that is formed in a covalent bond between two atoms of different elements. See also Polar bond.

**Dipole moment** *(8.5)*   A measure of the magnitude of the charges that are separated and the distance separating them in a molecule.

**Distillation** *(12.4)*   A method of separating and purifying liquids by heating them to the boiling point and condensing the vapor.

**Effusion** *(4.9)*   The process by which a gas flows through a small hole in its container.

**Electrolysis** *(6.8, 16.3)*   The production of chemical changes by the passage of an electric current through a substance.

**Electrolyte** *(2.3, 12.7)*   A substance that exists as ions in solution.

**Electromotive force series (emf series)** *(16.6)*   A tabulation of reduction half-reactions in order of increasing standard potential. It can be used to calculate the standard potential of galvanic cells and to predict chemical behavior.

**Electron** *(5.2)*   An elementary particle with a negative electric charge equal to the positive charge of the proton and with a mass 1/1837 that of a proton.

**Electron affinity** *(6.5)*   The energy required to remove an electron that has been added to an atom in the gas phase. Alternately, the energy released when an electron is captured by a neutral atom.

**Electronegativity** *(7.3)*   A property that is used to describe the ability of an atom to attract electrons when it is bonded to other atoms.

**Element** *(1.2)*   A substance that cannot be decomposed into simpler substances by chemical means.

**Elementary reaction** *(17.3)*   A single-step reaction that takes place on the molecular level as written.

**Empirical formula** *(3.2)*   A chemical formula that shows only the relative numbers of different atoms in a substance, not the actual number of atoms in a molecule of the substance.

**Endothermic process** *(8.7)*   A process that absorbs heat.

**Energy** *(2.8)*   The ability to do work.

**Enthalpy *(H)*** *(8.7, 13.2)*   A state function in thermodynamics that describes the heat content of a substance. The actual heat content of a substance cannot be determined, but relative enthalpies can be defined and enthalpy changes can be measured.

**Entropy *(S)*** *(13.5, 13.6)*   A state function in thermodynamics that can be described as a measure of increasing disorder. It can be defined as the number of microscopic states of a system that correspond to a given macroscopic state of the system.

**Equilibrium constant** *(14.1)*   An expression that is related to the pressures of gases and concentrations of solutes in a system that is at chemical equilibrium. Its value is a constant for a given system at a given temperature.

**Equilibrium state** *(4.2)*   A state in which no net change occurs in a system with time.

**Equivalence point** *(15.6)*   The point in a neutralization reaction when stoichiometric quantities of acid and base have been mixed.

**Equivalent weight** *(15.6)*   The quantity of an acid that donates 1 mol of protons or of a base that accepts 1 mol of protons. For an acid that donates one proton per molecule, the molecular weight and the equivalent weight are identical. For an acid that donates two protons per molecule, the equivalent weight is half the molecular weight.

**Ester** *(21.5)*   An organic compound that is formed when the H atom of the —OH group of a carboxylic acid is replaced by an organic radical.

**Ether** *(21.5)*   An organic molecule with a single oxygen atom that is attached to two organic groups.

**Eutectic temperature** *(12.5)*   The lowest temperature at which a liquid solution of a given solute can exist.

**Evaporation** *(4.2)*   The process by which a substance changes from the liquid phase to the gas phase.

**Excited state** *(5.7)*   A state in which the energy of an atom or molecule is higher than in the ground state.

**Exothermic process** *(8.7)*   A process that releases heat.

**Expanded octet** *(7.4)*   The association of an atom in a molecule with more than eight valence electrons in covalent bonding.

**Extensive property** *(1.2)*   A property of a sample of a substance such as shape, size, or length.

**Extrinsic semiconductor** *(19.3)*   A material that requires the addition of small amounts of another substance to be a semiconductor. See also Intrinsic semiconductor.

**Face-centered lattice** *(10.4)*   A crystal structure in which each unit cell has one or more lattice points centered in the faces, in addition to the eight lattice points at the corners of the cell.

**Faraday's laws** *(16.3)*   The laws, described by Michael Faraday, giving the relationship between the electric current and the mass of a substance that undergoes chemical change in an electrolysis.

**Ferromagnetism** *(19.5)*   The ability of certain materials to become and remain magnetized when exposed to a magnetic field.

**First law of thermodynamics** *(13.2)*   See Conservation of energy.

**Fission reaction** *(20.4)*   A nuclear reaction in which a nuclide forms two or more product nuclides of smaller mass number.

**Flocculation** *(11.6)*   A water purification treatment in which a compound such as aluminum sulfate is added to water. The gelatinous precipitate that forms traps the impurities.

**Formal charge** *(7.4)*   A device for describing the source of the electrons in a coordinate-covalent bond by writing pluses or minuses next to the symbols in Lewis structures. As the name implies, a formal charge does not indicate actual distribution of charge within a molecule; it is a bookkeeping device.

**Free energy** *(13.7)*   See Gibbs free energy.

**Freezing point** *(10.2)*   The temperature at which the solid and liquid phases of a substance are in equilibrium.

**Frequency factor** *(A)* *(17.4)*   In the Arrhenius equation, a term related to the steric factor and the collisional frequency.

**Fuel cell** *(16.4)*   A galvanic cell in which oxygen reacts with another element, usually hydrogen, at ordinary temperatures to generate electricity.

**Functional group** *(21.2)*   In an organic compound, an atom or group of atoms that differs from the basic alkane structure. The chemical behavior of an organic compound is determined primarily by its functional groups.

**Functional isomers** *(21.2)*   Compounds that are identical in chemical composition but have different functional groups.

**Fusion reaction** *(20.4)*   A nuclear reaction in which nuclei of lighter elements unite to form heavier nuclei, with the release of energy.

**Galvanic cell** *(16.4)*   An apparatus in which the reactants in an oxidation-reduction reaction are separated into two half-cells. The electron transfer takes place through a wire connecting the cells.

**Gamma emission** *(20.2)*   The emission of high-energy photons, called gamma rays, during a radioactive decay process.

**Gas** *(4.1)*   A substance in the vapor phase whose atoms or molecules are kept apart by thermal motion.

**Gas constant** *(4.4)*   In the ideal gas equation, a constant relating pressure, volume, amount, and temperature. Its symbol is $R$ and its value is 0.082056 L atm mol$^{-1}$ K$^{-1}$.

**Gel** *(12.8)*   A colloid in the solid state.

**Geometrical isomers** *(18.2, 21.2)*   Compounds that are identical in chemical composition but differ in the spatial arrangement of their atoms.

**Gibbs free energy** *(G)* *(13.7)*   In thermodynamics, a state function that can be used to describe the tendency of a chemical reaction to proceed spontaneously.

**Glass** *(10.2)*   See Amorphous solid.

**Ground state** *(5.7)*   The lowest energy state of an atom or molecule.

**Haber process** *(9.3)*   A catalytic process for nitrogen fixation that combines $N_2$ and $H_2$ under high pressure and temperature to produce ammonia.

**Half-cell** *(16.4)*   A container holding an electrolyte solution in which an electrode is immersed. A galvanic cell consists of two half-cells, with an oxidation process occurring in one and a reduction process in the other.

**Half-life** *($t_{1/2}$)* *(17.2)*   The time required for half of the starting quantity of a reactant to be consumed.

**Half-reaction** *(16.1)*   A reaction in which electrons are a reactant or a product. A half-reaction is an artificial division of an oxidation-reduction reaction into two parts.

**Halide** *(6.9)*   A binary compound of a halogen and another element.

**Halogen** *(9.7)*   An element in column VIIA of the periodic table. The halogens are F, Cl, Br, I, and At.

**Heat** *(13.1)*   In thermodynamics, a method by which energy is transferred across a boundary by microscopic processes such as thermal conduction or thermal radiation.

**Heat capacity** *(11.1, 13.3)*   The quantity of heat required to change the temperature of 1 mol of a substance by 1 K.

**Heat of condensation** *(8.7)*   The enthalpy change associated with the process by which a gas becomes a liquid.

**Heat of fusion** *(8.7)*   The enthalpy change associated with the process by which a solid becomes a liquid.

**Heat of sublimation** *(8.7)*   The enthalpy change associated with the process by which a solid becomes a gas.

**Heat of vaporization** *(8.7)*   The enthalpy change associated with the process by which a liquid becomes a gas.

**Heisenberg uncertainty principle** *(5.8)*   The statement that
it is impossible to determine simultaneously both the
exact position and the exact momentum of a particle.

**Henderson-Hasselbalch equation** *(15.7)*   An equation for
calculation of the pH of a solution that contains known
amounts of an acid and its conjugate base. It is

$$pH = pK_A + \log \frac{[A^-]}{[HA]}$$

**Henry's law** *(12.3)*   An expression for the relationship
between the partial pressure of a gas above a liquid and
the solubility of the gas in the liquid. For a gas $A$ it is
$X_A = kP_A$, where $P_A$ is the partial pressure of the gas, $X_A$
is the mole fraction of the gas in solution, and $k$ is a
proportionality constant.

**Hess's law** *(8.7)*   The statement that the heat change
accompanying a chemical change is the same whether
the change occurs in one step or many steps.

**Hexagonal closest packing** *(10.5)*   See Cubic closest
packing.

**Homologous series** *(21.1)*   A group of organic compounds
that differ only in the number of $CH_2$ units in the
longest carbon chain.

**Hund's rule** *(6.2)*   A rule used in assigning ground state
electronic configurations. It states that electrons go into
unoccupied orbitals of a subshell rather than pairing
with electrons in partially occupied orbitals of the
subshell. When two or more electrons are in half-filled
orbitals, the spins of the electrons are parallel.

**Hybrid orbital** *(8.4)*   An orbital of intermediate type that
is formed by the mixing of atomic orbitals. Hybrid
orbitals help us to explain the formation of bonds and
the geometry of molecules.

**Hydrated species** *(11.3)*   A species that holds one or more
water molecules.

**Hydride** *(6.9)*   A binary compound of hydrogen and
another element.

**Hydrocarbon** *(9.1, 21.1)*   A compound that contains only
hydrogen and carbon.

**Hydrogen bond** *(11.2)*   A bond produced by the coulombic
attraction between a hydrogen atom bonded to a very
electronegative element and another atom of great
electronegativity. It is weaker than a covalent or ionic
bond but much stronger than other intermolecular forces.

**Hydrogenation** *(9.1)*   The addition of hydrogen atoms to a
molecule. One important kind of hydrogenation is the
addition of a hydrogen atom to a double bond, converting it to a single bond.

**Hydrolysis** *(2.5)*   The reaction of an atom, molecule, or
ion with water.

**Ideal gas** *(4.3)*   An imaginary gas whose behavior can be
explained completely by ideal gas laws, such as Boyle's
law and the law of Charles and Gay-Lussac.

**Ideal gas equation** *(4.4)*   An equation expressing the
relationship between pressure, volume, amount, and
temperature of an ideal gas. It is $PV = nRT$, where $P$ is
pressure, $V$ is volume, $n$ is the amount (in moles) of the
gas, $R$ is the ideal gas constant, and $T$ is temperature (in K).

**Ideal solution** *(12.4)*   A hypothetical solution in which the
interactions among all the components of the solution
are identical.

**Instantaneous reaction rate** *(17.1)*   The rate of a reaction
at any one instant.

**Intensive property** *(1.2)*   A property of any sample of a
substance such as density, melting point, and boiling point.

**Interhalogen compound** *(9.7)*   Binary compounds of one
halogen with another.

**Intermolecular forces** *(8.8)*   Attractive forces between
molecules that are weaker than covalent or ionic bonds.
They include hydrogen bonds, dipole-dipole interactions,
and London forces.

**Internal energy** *(13.1)*   In thermodynamics, a state
function, whose symbol is $E$, that describes the total of
all the possible kinds of energy of a system.

**International System of Units (SI)** *(1.5)*   A system of
measurement, based on the metric system, that was
completed in 1969 and has been adopted by many international scientific bodies.

**Intrinsic semiconductor** *(19.3)*   A material that is a
semiconductor without the addition of any other substance. See also Extrinsic semiconductor.

**Ion** *(2.1)*   An atom or group of atoms that carries an
electric charge.

**Ionic bond** *(7.1)*   A chemical bond formed by electrostatic
attraction between two atoms. In an ideal ionic bond,
one atom loses one or more electrons to become a cation
and the other atom gains one or more electrons to
become an anion.

**Ionization energy** *(6.4)*   The energy needed to remove the
electron with the highest energy from the ground state of
an atom in the gas phase.

**Isoelectronic species** *(6.4)*   Species that have the same
electronic configuration. For example, a singly ionized
oxygen atom and a neutral nitrogen atom are isoelectronic.

**Isolated system** *(13.1)*   In thermodynamics, a system in
which neither mass nor energy can enter or leave, so that
there is no interaction with the surroundings.

**Isomers** *(21.2)*   Two chemical compounds that have the
same molecular formula but different molecular
structures.

**Isotopes** *(3.1, 5.4, 20.1)*   Atoms of an element that differ in atomic mass. The nuclei of isotopes have the same number of protons but different numbers of neutrons.

**Kelvin temperature scale** *(4.3)*   A scale in which the zero point is absolute zero and the degree is 1/273.16 of the temperature of the triple point of water.

**Ketone** *(21.5)*   An organic compound that has the carbonyl functional group bearing two carbon substituents and no hydrogen atoms. The general formula is

$$R-\overset{\underset{\parallel}{O}}{C}-R$$

**Kinetic energy** *(2.8)*   Energy of motion. The work that a moving object can perform as it is brought to rest.

**Lability** *(18.4)*   A kinetic classification that refers to the rate at which ligands of a coordination compound can be replaced.

**Lanthanoids** *(6.6)*   The elements from atomic numbers 57 through 71. Also called the rare-earth metals.

**Lattice energy** *(7.1)*   The energy released by the formation of an ionic crystal lattice from isolated ions in the gas phase.

**Law of conservation of mass.**   See Conservation of mass.

**Law of constant composition.**   See Constant composition.

**Law of multiple proportions.**   See Multiple proportions.

**Le Chatelier, principle of** *(14.2)*   The statement that when a system in an equilibrium state undergoes a change because of an external stress, the system will reach a new equilibrium state in a way that minimizes the external stress.

**Leveling effect** *(15.2)*   The phenomenon that all acids stronger than the conjugate acid of the solvent and all bases stronger than the conjugate base of the solvent appear to have the same strength.

**Lewis structure** *(7.4)*   A way of describing the structures of covalent molecules by drawing diagrams in which bonds are represented by lines and pairs of nonbonding valence electrons are represented by dots.

**Ligancy** *(11.3)*   See Coordination number.

**Limiting reagent** *(3.3)*   The reactant that is exhausted first when the reactants for a process are not present in stoichiometric amounts. The amount of the limiting reagent determines the amounts of products.

**Liquid** *(4.1)*   A substance in the condensed phase that has a definite volume but not a definite shape.

**Liquid crystal** *(10.7)*   A phase of matter distinct from the solid or the liquid. A liquid crystal can undergo changes in state comparable to the melting or freezing of solids and liquids.

**London forces** *(8.8)*   Intermolecular forces that arise from induced dipoles or momentary irregularities in the distribution of electron densities in molecules.

**Magnetic quantum number ($m_\ell$)** *(6.1)*   A quantum number whose values correspond to allowed orientations of orbitals. Its values are restricted by the value of the angular momentum quantum number $\ell$. The magnetic quantum number can have integral values ranging from $-\ell$ to $+\ell$, including 0.

**Main group element** *(6.6)*   See Representative element.

**Mass** *(3.1)*   A measure of the resistance of an object to acceleration. It is defined by $F = ma$, where $F$ is force, $a$ is acceleration, and $m$ is mass.

**Mass number** *(5.4, 20.1)*   The sum of protons and neutrons in an atomic nucleus, represented by the symbol $A$.

**Mass percent** *(3.4, 12.2)*   An expression that describes the mass of solute in a given mass of solution. Mass percent is the mass of solute divided by the mass of solution multiplied by 100%.

**Matter** *(1.2)*   Anything possessing mass and occupying space.

**Melting curve** *(10.2)*   All the points corresponding to values of pressure and temperature at which the solid and liquid phases of a substance are in equilibrium.

**Meniscus** *(10.1)*   The curved surface of a liquid in a vessel. A liquid that wets the surface has a concave meniscus and a liquid that does not wet the surface has a convex meniscus.

**Metallic bond** *(19.2)*   A kind of bonding in which there are more vacant orbitals than electrons, so bonding electrons are shared among the vacant orbitals in a number of neighboring atoms. The electrons in a metallic bond are highly delocalized.

**Metallic valence** *(19.2)*   The number of electrons that a metal atom has available for bonding.

**Metallurgy** *(19.4)*   The technology that applies the principles of physical science to produce metals in usable form.

**Microscopic reversibility, principle of** *(17.5)*   The statement that when equilibrium is reached in a reaction system, any elementary process and the reverse of that process must occur, on the average, at the same rate.

**Miscibility** *(12.1)*   The mutual solubility of two or more substances in one another in all proportions.

**Molal boiling constant** *(12.5)*   A measure of the way in which the boiling point of a solvent changes in response to the addition of various amounts of a solute. Its symbol is $K_b$.

**Molal freezing point constant** *(12.5)*   A measure of the way in which the freezing point of a solvent changes in response to the addition of various amounts of a solute. Its symbol is $K_f$.

**Molality (m)** *(12.2)*    An expression of concentration: the amount (in moles) of solute in 1 kg of solvent.

**Molarity (M)** *(12.2)*    The most commonly used expression of concentration of a solution. It is the amount (in moles) of solute in 1 L of solution.

**Mole** *(3.1)*    The SI unit of amount. It is the amount of a substance that contains as many entities as there are atoms in exactly 0.012 kg of $^{12}C$. There are $6.0221 \times 10^{23}$ entities in a mole. See also Avogadro's number.

**Molecular orbital (MO) method** *(8.6)*    A method of describing chemical bonding. It assumes that there is a set of allowed orbitals for a molecule. To describe bonding in a molecule, the set of allowed orbitals is determined and electrons are assigned to the orbitals in order of increasing energy.

**Molecularity** *(17.3)*    A term that applies only to elementary reactions. The molecularity of an elementary reaction is the number of reactant species that come together to form the products.

**Mole fraction (X)** *(12.1)*    An expression of concentration of a solution, defined as the amount (in moles) of the solute divided by the total amount (in moles) of the solute and the solvent.

**Multiple bond** *(7.4)*    A bond in which atoms share two or three pairs of electrons.

**Multiple proportions, law of** *(2.3)*    The law that states that if two elements can form more than one compound, the masses of one element that combine with a fixed mass of the other element are in the ratio of small whole numbers.

**Nematic state** *(10.7)*    The state in which the molecules of a substance all have the same orientation but are not in equispaced planes.

**Nernst equation** *(16.7)*    The relationship between the potential of any galvanic cell or chemical system, its standard potential, its temperature, and a concentration and pressure term.

**Neutralization** *(15.6)*    Usually, the reaction that occurs when stoichiometric amounts of an aqueous solution of an acid and an aqueous solution of a base are mixed. More generally, the reaction of any acid with any base.

**Neutron** *(5.4)*    A subatomic particle without electrical charge, whose mass is slightly greater than that of a proton. It is found in all atomic nuclei except those of $^{1}H$.

**Nitrogen fixation** *(9.3)*    The conversion of $N_2$ in the atmosphere to nitrogen compounds that are usable biologically. It is carried out naturally by microorganisms and industrially by the Haber process.

**Nonbonding electrons** *(7.4)*    Valence electrons that are not included in the covalent bonds between atoms. Also called lone pairs.

**Nonelectrolytes** *(2.3)*    Compounds that do not dissociate into ions when they dissolve in water.

**Normal boiling point** *(4.2, 10.2)*    The temperature at which the vapor pressure of a liquid is 1 atm.

**Nucleon** *(20.3)*    A fundamental particle found in the atomic nucleus. Protons and neutrons are nucleons.

**Nucleus** *(5.4)*    The small, massive, positively charged core of an atom, containing protons and neutrons. Almost all of the mass of an atom is in the nucleus.

**Nuclide** *(20.1)*    A single nuclear species with a specific atomic number and mass number.

**Octet rule** *(7.4)*    The rule that each atom in a molecule achieves a noble gas electronic configuration by covalent bonding.

**Optical isomerism** *(18.2)*    See Chirality.

**Order** *(17.2)*    In kinetics, a description of a rate law or a reaction that is based on the exponents of the concentration terms in the rate law. The exponent 1 corresponds to first order; the exponent 2 to second order.

**Osmotic pressure** *(12.6)*    The pressure that must be exerted on a solution separated from a pure solvent by a semipermeable membrane to maintain equilibrium.

**Oxidation** *(7.6, 16.1)*    A chemical reaction in which a species loses electrons, so its oxidation state increases. An oxidation is always accompanied by a reduction. See Oxidation-reduction process.

**Oxidation number** *(7.6)*    A positive or negative integer that is assigned to an atom in a molecule. Oxidation numbers are related to electronic distribution and are used to convey information about the ability of a species to combine with other species.

**Oxidation-reduction process** *(16.1)*    A chemical change that occurs as the result of an electron transfer. The species that gains electrons is said to be reduced and the species that loses electrons is said to be oxidized.

**Oxide** *(6.9)*    A binary compound of any element with oxygen.

**Oxidizing agent** *(16.1)*    A substance that gains electrons and causes the oxidation of another substance. In an oxidation-reduction process, the substance that is reduced is the oxidizing agent.

**Paracrystalline state** *(10.7)*    See Liquid crystal.

**Paramagnetism** *(6.2, 9.5, 18.3)*    The phenomenon in which a substance is strongly attracted by a magnetic field. See also Diamagnetism.

**Partial pressure** *(4.5)*    The pressure exerted by each gas in a mixture of gases. The total pressure of the mixture is the sum of the partial pressures.

**Path function** *(13.1)*   In thermodynamics, a quantity whose value depends on the path followed by a system in a change of state.

**Pauli exclusion principle** *(6.2)*   The rule that no two electrons in a given atom can have the same four quantum numbers.

**Periodic table** *(6.6)*   A table of the elements in order of increasing atomic number, arranged according to a regular periodic occurrence of chemical and physical properties.

**Phase diagram** *(10.2)*   A graphic display that summarizes all the pressure and temperature data needed to determine the relationships among the solid, liquid, and gaseous states of a substance. A phase diagram displays the various combinations of pressure and temperature at which different phases are at equilibrium.

**Photoelectric effect** *(5.6)*   The emission of electrons from the surface of a metal that is struck by light or other electromagnetic radiation.

**Photon** *(5.6)*   A quantum of electromagnetic radiation.

**Physical properties** *(1.2)*   The characteristics by which we describe a substance, including color, density, hardness, and so on.

**Pi orbital** *(8.3)*   A bonding orbital formed by the overlap of two parallel $p$ orbitals. In a pi orbital, the most probable region for the electrons of the bond is outside the region of the sigma bond. See also Sigma orbital.

**Planck's constant** *(5.5)*   The constant that relates the energy of a quantum of energy to its frequency. Its symbol is $h$ and its value is $6.6262 \times 10^{-34}$ J s.

**Polar bond** *(7.3)*   A covalent bond between two different atoms, in which one atom has a relative excess of negative charge and the other has a relative excess of positive charge, producing an electrical polarity.

**Polarizability** *(8.8)*   The sensitivity of the electron distribution in a molecule to change by an outside influence.

**Polyprotic acid** *(15.8)*   An acid that has more than one proton per molecule to donate.

**Polyprotic base** *(15.8)*   A base that can accept more than one proton per molecule.

**Potential energy** *(2.8)*   The capability of doing work that matter has by virtue of its position in a field of force.

**Pressure** *(4.2)*   The force exerted on a given unit of area.

**Principal quantum number (n)** *(5.7)*   The quantum number that is most important in designating the energy of an electron in an atom. It may have integer values 1, 2, 3, 4. . . .

**Proton** *(5.4)*   A nucleon that has a positive electrical charge equal to the negative charge on the electron and has a mass 1837 times greater than that of the electron.

**Quantum** *(5.5)*   A discrete packet of radiant energy. The value of a quantum of radiant energy is $h\nu$ where $h$ is Planck's constant and $\nu$ is the frequency of the radiation. Radiant energy is always emitted in integral multiples of a quantum.

**Radial probability function** *(6.3)*   A representation that describes the probability of finding an electron at a given distance from the nucleus of an atom.

**Radiant energy** *(2.8)*   Energy emitted in the form of electromagnetic waves, which consist of a combined electric field and magnetic field propagated through space.

**Radioactivity** *(20.2)*   The emission of radiant energy or particles by atomic nuclei.

**Rare-earth element** *(6.6)*   See Lanthanoids.

**Rate constant** *(17.1)*   The proportionality constant between the rate of a reaction and the concentration of one or more of the substances in the system. Its symbol is $k$. See also Rate law.

**Rate-determining step** *(17.3)*   The slowest elementary reaction in a complex mechanism, which determines the overall rate of the complex reaction.

**Rate law** *(17.2)*   An equation that describes the relationship between the rate of a reaction and the concentration of one or more of the substances in the system.

**Reaction mechanism** *(17.3)*   A description on the molecular level of all the changes that reactants undergo during their transformation to products in a chemical reaction.

**Reaction quotient** *(14.1, 16.7)*   An expression, whose symbol is $Q$, that describes the pressures or concentrations of materials in a chemical system that is not at equilibrium.

**Reaction rate** *(17.1)*   The rate at which a chemical reaction occurs. It usually is expressed in terms of the change in the pressure or concentration of one component of the system per unit time.

**Redox reaction** *(16.1)*   See Oxidation-reduction process.

**Reducing agent** *(16.1)*   A substance that loses electrons and causes the reduction of another substance. In an oxidation-reduction process, the substance that is oxidized is the reducing agent.

**Reduction** *(16.1, 17.6)*   A chemical reaction in which a species gains electrons, so its oxidation state decreases. See also Oxidation-reduction process.

**Refining** *(19.4)*   The purification of a metal after it has been extracted from its ore.

**Representative element** *(6.6)*   An element in the $s$ or $p$ blocks of the periodic table. The representative elements occupy the A columns on the right and left of the periodic table. Also called main group elements.

**Resonance** *(7.5)*   A phenomenon that occurs when more than one location is possible for a multiple bond in a molecule or ion. A molecule or ion exhibits resonance when its actual structure is a blend of the possible Lewis structures that may be drawn. See also Resonance hybrid.

**Resonance energy** *(8.7)*   The extra stability observed in a resonance hybrid.

**Resonance hybrid** *(7.5)*   A molecule or ion that displays resonance. A resonance hybrid is a blend of the Lewis structures that can be drawn for the molecule.

**Reversible path** *(13.4)*   In thermodynamics, a change in state in which every intermediate state of the system is an equilibrium state.

**Salinity** *(11.4)*   The total mass in grams of dissolved salts in 1 kg of water.

**Salt** *(2.3)*   A compound of a metal and a nonmetal, or of a metal and a negative polyatomic group. Any compound of anions and cations.

**Salt bridge** *(16.4)*   In a galvanic cell, a barrier that allows ions to pass between the half-cells but does not allow their solutions to mix.

**Saturated compound** *(21.1)*   In organic chemistry, any compound with no multiple bonds.

**Saturated solution** *(12.1)*   A solution at equilibrium that contains the maximum possible amount of solute.

**Schrödinger wave equation** *(5.9)*   An equation that gives the relationship between the total energy of an atom and the wave function that describes the probability of an electron being at a given point in the atom.

**Second law of thermodynamics** *(13.5)*   The law that states that the entropy of the universe is constantly increasing. An alternative statement of the second law is that an engine operating in a cycle cannot convert a quantity of heat from the surroundings into an equal quantity of work on the surroundings.

**Selective precipitation** *(14.5)*   A method of obtaining specific solids from a solution of different ions by addition of appropriate ions.

**Semiconductors** *(10.8, 19.3)*   Materials whose electrical conductivity is intermediate between that of metals and that of nonmetals and increases with increasing temperature. Also called semimetals.

**Semimetals** *(2.3)*   See Semiconductors.

**Semipermeable membrane** *(12.6)*   A membrane that selectively allows some species to pass through.

**Sigma orbital** *(8.3)*   A molecular orbital that has cylindrical symmetry with respect to an imaginary line joining the nuclei of the two bonded atoms.

**Significant figure** *(1.5)*   A way of expressing the uncertainty in a measurement by the number of digits used to write the number. For example, there is one significant figure in 7 cm, two significant figures in 7.0 cm and three significant figures in 7.00 cm.

**Single bond** *(7.4)*   A covalent bond in which the bonded atoms share one pair of electrons.

**Smectic state** *(10.7)*   In a liquid crystal, the state in which all the molecules have the same orientation and are in equispaced planes but are arranged irregularly within the planes.

**Sol** *(12.8)*   A liquid colloid in which the dispersion phase is a liquid and the dispersed phase is a solid.

**Solid** *(4.1)*   A condensed phase in which the basic entities of a substance usually are arranged in an ordered, crystalline array.

**Solubility** *(12.1)*   A measure of the maximum quantity of a substance that can dissolve in a given quantity of solvent under a specified set of conditions.

**Solubility product constant** *(14.5)*   The equilibrium constant for the dissolution of a sparingly soluble ionic solid in water.

**Solute** *(3.4, 12.1)*   A substance that is dissolved in a solvent.

**Solution** *(1.2, 3.4, 12.1)*   A homogeneous mixture of two or more substances. A solution can be composed of liquids, solids, or gases.

**Solvent** *(3.4, 12.1)*   The main component of a solution, in which the solute is dissolved.

**Specific heat** *(13.3)*   The quantity of heat that changes the temperature of 1 g of a substance by 1 K.

**Specific rate constant** *(17.1)*   See Rate constant.

**Spectrochemical series** *(18.3)*   A list of ligands in ascending order of crystal field splitting energy, based primarily on the absorption of light by complexes.

**Spin quantum number ($m_s$)** *(6.1)*   The quantum number that designates the spin of an electron. It has only two possible values, $+\frac{1}{2}$ and $-\frac{1}{2}$.

**Standard free energy change ($\Delta G°$)** *(13.7)*   The difference between the free energy of the reactants in their standard states and the free energy of the products in their standard states.

**Standard heat of formation ($\Delta H_f°$)** *(8.7)*   The enthalpy change for the process in which 1 mol of a substance at 1 atm is formed from its constituent elements in their standard states.

**Standard molar volume of an ideal gas** *(4.4)*   The volume of 1 mol of an ideal gas at standard temperature and pressure, 22.4 L.

**Standard potential** *(16.5)*   The potential of a galvanic cell in which the concentration of all dissolved species is $1M$ and the pressure of all gases is 1 atm.

**Standard state** *(8.7)*   A pressure of 1 atm. Also the state of an element that is most stable at a pressure of 1 atm and a specific temperature.

**State function** *(13.1)* In thermodynamics, a property of a system whose value is not influenced by the path that the system followed to reach the given state.

**Stereochemistry** *(18.1)* The branch of chemistry that is concerned with the three-dimensional structure, bond angles, and bond lengths of molecules.

**Stoichiometric amount** *(3.3)* The amount of a reactant that can combine completely with another reactant, leaving no excess of either.

**Strong electrolyte** *(2.3, 12.7)* An electrolyte that dissociates essentially completely in aqueous solution.

**Sublimation** *(4.2)* The change in state from solid to gas.

**Sublimation curve** *(10.2)* All the points corresponding to values of pressure and temperature at which vapor and solid are in equilibrium.

**Supercooling** *(10.2)* The phenomenon in which a liquid does not freeze when its temperature is below the freezing point.

**Superheating** *(10.2)* The phenomenon in which a liquid does not boil when its temperature is raised above the boiling point.

**Surface tension** *(10.1)* The force that pulls the surface of a liquid inward and resists an increase in surface area.

**Theoretical yield** *(3.3)* The amount of product obtained if reactants combine completely. Most reactions give less than the theoretical yield.

**Thermochemistry** *(8.1)* The branch of chemistry that measures the heat effects associated with chemical and physical processes.

**Third law of thermodynamics** *(13.6)* The law that states that the entropy of a perfect crystal of a pure substance at the absolute zero of temperature is zero.

**Three-center bond** *(9.1)* A covalent bond in which two electrons are shared by three atoms. Three-center bonds are found in the boranes.

**Threshold frequency** *(5.6)* In the photoelectric effect, the frequency of incident radiation at or above which electrons are emitted by the metal.

**Titration** *(15.6)* A neutralization procedure in which a measured volume of a solution of one reactant is added slowly from a buret to a measured amount of the other reactant.

**Trans isomer** *(18.2, 21.2)* An isomer that has two identical substituents on opposite sides of a double bond or ring. See also Cis isomer.

**Transition elements** *(6.6)* The elements in columns IB and IIIB through VIIIB, which have incomplete *d* subshells or can readily give rise to cations with incomplete *d* subshells.

**Transition state** *(17.4)* See Activated complex.

**Triple point** *(10.2)* The point marking the temperature and pressure at which the solid, liquid, and gas phases of a substance are all in equilibrium.

**Triplet electronic state** *(9.5)* An electronic state in which there are two unpaired electrons with parallel spins.

**Unit cell** *(10.4)* The basic three-dimensional pattern that is repeated to form a crystal lattice.

**Unsaturated compound** *(9.1)* An organic compound that has one or more multiple bonds.

**Valence band** *(19.3)* In the band theory of metal bonding, the band that holds the valence electrons.

**Valence bond (VB) method** *(8.3)* A method that describes bonding primarily in terms of localized bonding electrons.

**Valence electrons** *(6.3)* The outermost electrons of an atom, often in an incomplete subshell. The chemical behavior of an element depends primarily on the behavior of its valence electrons.

**Valence shell electron pair repulsion (VSEPR) model** *(8.2)* A model that describes the observed geometry of molecules by a picture in which the repulsions of the valence electrons are minimized.

**Van't Hoff factor** *(12.7)* A quantity, whose symbol is *i*, that relates the observed value of a colligative property of a solution to the value of the property for a nonelectrolyte.

**Vaporization** *(4.1)* The process in which a liquid becomes a gas.

**Vapor pressure** *(4.2)* The pressure of vapor above the liquid or solid phase of a substance at equilibrium.

**Viscosity** *(10.1)* The resistance of liquids to flow.

**Warming curve** *(10.2)* A plot of the increase in the temperature of a substance as heat is added at a constant rate.

**Water of crystallization** *(11.3)* The water molecules that accompany ionic solids when they crystallize from aqueous solution.

**Wave function** *(5.9)* A mathematical expression that describes the wave packet associated with a moving electron.

**Wavelength** *(5.5)* The distance between two adjacent peaks of a wave. Its symbol is $\lambda$.

**Weak electrolyte** *(2.3, 12.7)* An electrolyte that is only partially dissociated into ions.

**Work** *(13.1)* The product of a force by a displacement.

**X rays** *(5.3)* Electromagnetic radiation of relatively short wavelength and high frequency.

# Index

Molar mass, 66
Mole
  calculations with, 70–75
  defined, 63–64
  use by chemists, 67–69
Molecular crystals, 397
Molecular formula, 79–81
Molecular geometry, 282–284
  ideal, 284
  linear, 284
  with lone pairs, 288–290
  in multiple bonding, 286–288
  nonideal, 297–298
  octahedral, 286
  tetrahedral, 285
  trigonal bipyramidal, 285, 290
  trigonal planar, 284
Molecularity, 708
Molecular orbital method, 305–314, 752
Molecular speeds, 138–140
Molecular weight, 68
Molecule, 27–28
  models of, 280, 283
Mole fraction, 447
Mole percent, 447
Momentum, 24
Morphine, 856, 858, 862
Moseley, H. G. J., 160
Multiple bonding, 254
Multiple proportions, law of, 32–34
Myoglobin, 740

Nagasaki, 825
Naphthalene, 453, 849
Nematic state, 413
Neon, 382
  shells, 198
Neptunium, 820
Nernst, Walter, 673
Nernst equation, 672–676
Neutralization, 598–607
Neutralization reaction, 591
Neutrino, 811–812
  missing, 827
Neutron, 28, 29, 160
  mass of, 802
  thermal, 816–817
Neutron activation analysis, 818
Neutron capture, 817
Neutron stars, 804
Newton, 16, 110
Newton, Isaac, 162
Nickel, 787
  compounds of, 792
Nickel silver, 794
Nitrates, solubility of, 454

Nitric acid, 353, 356–357
Nitric oxide, 355–356
  decomposition of, 712
Nitrides, 349
Nitrogen
  compounds of, 352–357
  discovery and occurrence of, 349
  triple bond of, 730
Nitrogen cycle, 349–350
Nitrogen dioxide, 356
Nitrogen fixation, 349–350, 765
Nitrosyl cation, 355
Nitrous acid, Lewis structure, 256–257
Nitrous oxide, 354–355, 563
  decomposition of, 730
Noble gas compounds, 383–384
Noble gas electronic configurations,
    247–248
Noble gases, 222, 381–384
Noble gas shells, 198
Nonelectrolyte, 470, 524
Nonmetals, 35–36, 336–384
Nonpolar compounds, 253
Normal boiling point, 394
Normality, 449, 603
n-type semiconductor, 782
Nuclear binding energy, 804–805
Nuclear charge
  effective, 209
  and electron energy, 207–208
Nuclear chemistry, defined, 802
Nuclear cross section, 828
Nuclear fission, 514, 830
Nuclear fusion, 514, 515, 831
Nuclear reactions, notation for, 817
Nuclear reactor, 828–830
Nuclear statistical equilibrium, 823
Nuclear waste, 829
Nucleon, 160, 802
  magic numbers of, 807
Nucleophile, 572
Nuclide, 803
  parent and daughter, 807

Octahedral geometry, 743–744
Octane, 452
Octet rule, 255
  and Lewis structure, 258–261
Octets
  expanded, 258, 262–264
  incomplete, 258, 261–262
Oil of cloves, 849
Oil of wintergreen, 848
Oleum, 375
Opiates, 856, 858
Opium, 858

Oral contraceptives, 845
Orbitals, 184–185, 188
  antibonding molecular, 292
  bonding, 779
  bonding molecular, 292
  degenerate, 307
  hybrid, 293
  nonbonding, 779
  nonpenetrating, 207
  overlap of, 291–293
  penetrating, 206
  pi, 296
  probability distributions, 206–207
  set of allowed, 191–192
  shape and location, 192–194
  shells, 198
  sigma, 296
Organic chemists, 859–863
Organic compounds
  prefixes, 838
  synthesis of, 861–862
Osmium, 787, 793
Osmosis, 468–469
Osmotic pressure, 467–470
Ostwald process, 353, 357, 728
Oxaloacetic acid, 681
Oxidation, 274–275
Oxidation numbers, 271–275
  method of change in, 633–634
Oxidation-reduction reactions, 274–275,
    627–634
  titrations, 635–638
Oxidation state, 274, 627
Oxides, 364
  amphoteric, 231
  defined, 39
  periodic trends, 229–233
Oxidizing agents, 628
  as titrants, 635–636
Oxyacids, 256, 581–582
Oxyanions, 39
Oxyfluorides, 380
Oxygen, 8, 364–369
Oxygen molecule, MO picture of, 310
Oxygen-oxygen bond, 256
Ozone, 314, 365–366, 367
Ozone layer, 366, 697

*p*, 585–586
Palladium, 787, 793
Paracrystalline state, 413
Paramagnetism, 202, 203, 309, 365, 757
Pascal, 16, 110
Passivation, 680
Path functions, 483, 489
Patina, 795